*STRUCTURE OF MATTER SERIES*

*MARIA GOEPPERT MAYER*

*Advisory Editor*

# COLLISION THEORY

# COLLISION THEORY

Marvin L. Goldberger

*Eugene Higgins Professor of Physics*
*Princeton University*

Kenneth M. Watson

*Professor of Physics*
*University of California, Berkeley*

JOHN WILEY & SONS, Inc.
New York · London · Sydney

SECOND PRINTING, MARCH, 1965

Library of Congress Catalog Card Number: 64-17819

PRINTED IN THE UNITED STATES OF AMERICA

# PREFACE

The analysis of collision phenomena plays a central role in almost all investigations into the structure of matter on the microscopic scale. In fact, most of our detailed information on the interaction between particles is obtained from scattering experiments.

It is not surprising that the theory of scattering is taught in all courses on quantum mechanics and then again in nuclear physics, atomic physics, etc. What is surprising is that there are so few books on collision theory. Until recently there was only the classic monograph of Mott and Massey, *The Theory of Atomic Collisions* (Oxford University Press, London, 1933, 2nd edition, 1948) which educated several generations of physicists. In the last fifteen years, there have been many developments in the subject not to be found in any textbooks.

We attempt in this book to give a systematic and careful description of the basic principles of collision theory and to illustrate them with interesting examples. Most "experts" agree that in the mathematical formulation of the quantum theory of scattering one should construct wave packets for projectile and target, follow the time development of the system, and see what happens. We have done precisely this in our presentation of the theory and have introduced modern, compact formal methods only after assuring ourselves that there are no serious conceptual problems involved.

Since it is our purpose to talk about problems of real physical interest, we make no pretense of strict mathematical rigor. Instead we have substituted mathematical reasonableness and physical rigor. Mathematics is an interesting intellectual sport, but it should not be allowed to stand in the way of obtaining sensible information about physical processes. This attitude may offend some purists, who should feel free to consult the mathematical literature whenever the mood strikes them.

Naturally, in any book of finite size (which this one almost is), it is necessary for the authors to make a decision about what should be included and what omitted. The topics we have included are the ones we feel are the most important and the ones we feel competent to write about. The omission about which we feel most strongly is that concerning many of the modern developments in relativistic quantum theory associated with the names of Dyson, Feynman, and Schwinger. There have, however, been a number of recent books on quantum field theory in which this work is reviewed. We have included a chapter on dispersion theory (in which some of the important concepts of quantum field theory are treated); this chapter hopefully will serve as an introduction to the growing literature on the application of dispersion theoretic methods to dynamical calculations of strongly interacting systems.

Much of what we have presented in this book is widely known. There are, however, a number of things which we believe to be new; we hope there are enough of these that even our erudite friends will be entertained.

We have attempted to make the presentation reasonably self-contained to a reader who has had a good introductory course in quantum mechanics and who knows a little quantum electrodynamics. The most important conceptual problems in scattering theory are treated in Chapter 3 and the remaining formal structure taken up in Chapters 4 and 5. Individual chapters, after Chapter 5, will be fairly understandable to the reader who has not read all the preceding ones.

This book is an outgrowth of various lecture series given by us over the years and of our personal research in the field. Some of the material appears in lecture notes written in collaboration with K. A. Brueckner and M. Gell-Mann. We have also profited from notes prepared by B. S. DeWitt and F. E. Low. We are particularly indebted to Professor Low for a careful reading of the manuscript and for many helpful suggestions. The critical comments and encouragement of Professor E. P. Wigner are gratefully acknowledged.

Valuable editorial advice was given us by Mrs. Mildred Goldberger. The proofreading of the final manuscript by Doctors John Gillespie and Hugh Kelly has been of great help. The patience and unusual care of Mrs. Bonnie Sivers in typing our manuscript contributed significantly to the completion of this book in a reasonable time. Both authors wish to acknowledge partial support from the United States Air Force during the writing of this book. Finally, one of us (KMW) would like to express his appreciation to Princeton University, where he was a Visiting Professor of Physics during the writing of much of this book.

*February 1964*

MARVIN L. GOLDBERGER
KENNETH M. WATSON

# CONTENTS

# 1. INTRODUCTION

In this chapter we shall give a brief review of a number of the general features of quantum mechanics. There are two reasons for doing this. First, we shall establish certain conventions for notation, and second, we shall dispose of some constantly recurring and often troublesome questions of normalization and phase assignment.

## 1.1 THE PRINCIPLES OF QUANTUM MECHANICS

The state of a given quantum mechanical system is represented by a vector $\Psi$, in a linear vector space, which is called a *state vector*. By "state of the system" we mean that the result of a maximum number of *compatible observations* on the system is specified. The mathematical specification of what constitutes compatible observations will be given shortly. The addition of two vectors $\Phi$ and $\Psi$, each of which represent possible states, gives a new state vector $\chi$,

$$\chi = \Psi + \Phi, \tag{1}$$

in the same space. This is an expression of the superposition principle in quantum mechanics.

A metric, or measure of length, is introduced in the linear vector space with the definition of a complex number called the scalar product of two vectors.

$$(\Phi, \Psi) = (\Psi, \Phi)^* \tag{2}$$

Let $\Psi$, $\Phi$, and $\chi$ be any three vectors in the space and $\lambda$ be a complex number, $\lambda^*$ its complex conjugate. The vector $\lambda\Psi$ represents the multiplication of the vector $\Psi$ (that is, each of its components) by $\lambda$. Then we have the relations

$$\begin{aligned}(\Phi, \lambda\Psi) &= \lambda(\Phi, \Psi),\\ (\lambda\Phi, \Psi) &= \lambda^*(\Phi, \Psi),\\ (\chi, \Phi + \Psi) &= (\chi, \Phi) + (\chi, \Psi).\end{aligned} \tag{3}$$

The length of vectors is of no consequence so that the state vector $\lambda\Psi$ represents the same state as $\Psi$.

A set of $n$ vectors $\Phi_1, \Phi_2, \ldots, \Phi_n$ is said to be *linearly independent* if the relation

$$\sum_{i=1}^{n} \lambda_i \Phi_i = 0 \tag{4}$$

can be satisfied only if the $\lambda_i = 0$ ($i = 1, 2, \ldots, n$). A set of linearly independent vectors $\{\Phi_i\}$ is said to be *complete* if any vector $\Psi$ in the space can be expressed in terms of the $\{\Phi_i\}$ according to

$$\Psi = \sum_{i=1}^{n} c_i \Phi_i, \tag{5}$$

where the $c_i$ are complex numbers. We assume that Eqs. (4) and (5) are meaningful for infinite-dimensional vector spaces ($n = \infty$).

A set of vectors $\{\Phi_i\}$ ($i = 1, 2, \ldots$) is said to be *normalized* when

$$(\Phi_i, \Phi_i) = 1 \qquad (i = 1, 2, \ldots). \tag{6}$$

They are called *orthonormal* when

$$(\Phi_i, \Phi_j) = \delta_{ij} \qquad (i, j = 1, 2, \ldots), \tag{7a}$$

where $\delta_{ij}$ is the Kronecker symbol: $\delta_{ij}$ is equal to 1 if $i = j$ and equal to 0 otherwise. All these concepts may be carried over to spaces in which the vectors are labeled by a continuous variable (or set of variables) if Eq. (7*a*) is replaced by

$$(\Phi_i, \Phi_j) = \delta(i - j), \tag{7b}$$

where $\delta(i - j)$ is the Dirac *δ-function*, and all summations are replaced by integrations. The important characteristic of this "function" is expressed by the equation

$$\int d(i) f(i) (\Phi_i, \Phi_j) = f(j) \tag{8}$$

for any function $f(i)$ continuous in the neighborhood of $i = j$. It is easy to see that the vectors belonging to an orthonormal set are linearly independent, for if the $\Phi_i$ constitute such a set, forming the scalar product of Eq. (4) with any $\Phi_j$ leads to $\lambda_j = 0$.

Complete sets of orthonormal vectors are of special importance in quantum mechanics. Let $\{\Phi_i\}$ and $\{\Psi_\lambda\}$ represent two different complete orthonormal sets of vectors spanning the same space. An arbitrary vector $\chi$ may be expanded in terms of either set of basis vectors,

$$\chi = \sum_i c_i \Phi_i, \qquad \chi = \sum_\lambda d_\lambda \Psi_\lambda, \tag{9}$$

where the $c_i$, $d_\lambda$ are complex numbers. A particular choice for $\chi$ might be any one of the basis vectors $\Phi_i$, in which case we have $c_j = \delta_{ji}$. We then write the second of Eqs. (9) as

$$\Phi_i = \sum_\lambda U_{\lambda i}\Psi_\lambda, \tag{10}$$

where the $U_{\lambda i}$ are the particular set of numbers $d_\lambda$ which result in this case. If relation (10) is substituted into Eq. (9) we find

$$\chi = \sum_{i\lambda} c_i U_{\lambda i}\Psi_\lambda = \sum_\lambda d_\lambda \Psi_\lambda,$$

or (11)

$$d_\lambda = \sum_i U_{\lambda i} c_i.$$

The relation between the basis vectors expressed by Eq. (10) has its counterpart in Eq. (11), which gives the relation between the corresponding components (with respect to the two bases) of the given vector. These equations describe a change of basis vectors and are referred to as a "transformation of coordinates" or a "change of representation." This freedom of representation, implied by the superposition principle, permits us to change sets of basis vectors at our convenience in describing state vectors.

A quantity $A$ is said to operate on the vector $\Phi$ if $A\Phi$ is a new vector (in the same space). It is called a *linear operator* if

$$A(\lambda_1\Psi_1 + \lambda_2\Psi_2) = \lambda_1 A\Psi_1 + \lambda_2 A\Psi_2, \tag{12}$$

where $\lambda_1$, $\lambda_2$ are complex numbers and $\Psi_1$, $\Psi_2$ are vectors. For the time being we restrict our attention to such operators. The action of an operator $A$ on an arbitrary vector may be specified in terms of its effect on a complete orthonormal set $\{\Phi_i\}$. We may express the vector $A\Phi_i$ in terms of the $\{\Phi_j\}$ as

$$A\Phi_i = \sum_j \Phi_j A_{ji}, \tag{13}$$

where the $A_{ji}$ are complex numbers, the "coordinates" of the vector $A\Phi_i$ in our basis system. If an arbitrary vector $\chi$ is expressed in terms of the $\{\Phi_i\}$, we have

$$\chi = \sum_i c_i \Phi_i,$$

$$\Gamma \equiv A\chi = A\sum_i c_i\Phi_i = \sum_i c_i A\Phi_i = \sum_{i,j} \Phi_j A_{ji} c_i \equiv \sum d_j \Phi_j.$$

The components $d_j$ of $\Gamma = A\chi$ in the prescribed basis system are therefore related to the components of $\chi$ through

$$d_j = \sum_i A_{ji} c_i. \tag{14a}$$

Evidently this may be regarded as a matrix equation relating the column matrices $d$ and $c$ by means of a square matrix $A$ and written simply as

$$d = Ac. \tag{14b}$$

Note that in any given representation (that is, with a specified set of basis vectors) a vector is completely defined by its components. The realization of the operator $A$ in terms of a specific matrix representation, as in Eq. (14*a*) or (14*b*), provides a unique definition of the operator and does not require a new symbol.

The *adjoint*, or *Hermitian conjugate*, of a matrix is defined by

$$A_{ij}^{\dagger} = (A_{ji})^*. \tag{15}$$

From Eq. (13) we find

$$A_{ji} = (\Phi_j, A\Phi_i),$$

and thus from Eq. (15)

$$A_{ji} = (\Phi_i, A^{\dagger}\Phi_j)^* = (A^{\dagger}\Phi_j, \Phi_i). \tag{16}$$

The transpose of $A$ is denoted by $\tilde{A}$ and is given by

$$\tilde{A}_{ij} = A_{ji}.$$

A matrix or operator is said to be *Hermitian* if it is equal to its Hermitian conjugate.

$$A^{\dagger} = A. \tag{17}$$

It is one of the postulates of quantum mechanics that to every physically observable quantity there is associated a Hermitian operator.

A vector $\Psi_\lambda$ is called an *eigenvector* of the operator $A$ when

$$A\Psi_\lambda = a_\lambda \Psi_\lambda, \tag{18}$$

where $a_\lambda$ is a number which is called an eigenvalue. When $A$ is a Hermitian operator the $a_\lambda$ are real, as may be seen by noting that

$$a_\lambda(\Psi_\lambda, \Psi_\lambda) = (\Psi_\lambda, A\Psi_\lambda)$$

and

$$\begin{aligned} a_\lambda^*(\Psi_\lambda, \Psi_\lambda)^* &= a_\lambda^*(\Psi_\lambda, \Psi_\lambda) = (\Psi_\lambda, A\Psi_\lambda)^* = (A^{\dagger}\Psi_\lambda, \Psi_\lambda)^* \\ &= (A\Psi_\lambda, \Psi_\lambda)^* = (\Psi_\lambda, A\Psi_\lambda). \end{aligned}$$

For an $n$-dimensional vector space there are precisely $n$ eigenvalues $a_\lambda$, not all of which need be distinct. If they are distinct, the eigenvectors $\Psi_\lambda$ are automatically orthogonal, since

$$(a_\lambda - a_{\lambda'})(\Psi_{\lambda'}, \Psi_\lambda) = (\Psi_{\lambda'}, A\Psi_\lambda) - (A\Psi_{\lambda'}, \Psi_\lambda) = 0.$$

Thus if $a_\lambda \neq a_{\lambda'}$ we conclude that $(\Psi_\lambda, \Psi_{\lambda'}) = 0$. When some of the eigenvalues are repeated, so that $a_\lambda = a_{\lambda'}$ for $\lambda \neq \lambda'$, they are said to be *degenerate*. The corresponding eigenvectors, which are not the same, may be combined linearly to form an orthogonal set among themselves, and they are of course automatically orthogonal to the eigenvectors belonging to different eigenvalues. The degeneracy of eigenvalues is inevitably associated with the existence of certain symmetries in the physical system under consideration which result in distinguishable situations yielding the same values for certain measurable quantities.

It will always be assumed that the eigenvectors of a given observable, corresponding to different physical states, have been chosen so as to be mutually orthogonal.

It is a postulate of quantum mechanics that the result of an observation of $A$ can yield one and only one of the eigenvalues $a_\lambda$ of the operator. Furthermore, if the system is described by a state vector $\Psi$, the probability $P$ of finding $a_\lambda$ is

$$P = \sum_{\lambda'} |(\Psi_{\lambda'}, \Psi)|^2 \tag{19a}$$

where the sum runs over all $\lambda'$ for which the eigenvalue $a_{\lambda'} = a_\lambda$. Finally, if $A$ corresponds to an observable (and hence by hypothesis is a Hermitian operator), its eigenfunctions are postulated to form a complete, orthonormal set in the space on which $A$ operates.

It is convenient for many purposes in quantum mechanics to imagine that the basis system spanning the vector space is fixed and to associate the time development of a state with the coefficients of the state vectors in this basis. Thus the state vector itself may be regarded as a function of time, $\Psi(t)$, and it is meaningful to ask for the time rate of change of the probability of finding the system in a particular eigenstate [or degenerate group of states, as in Eq. (19$a$)]. Thus from Eq. (19$a$) we compute such a *transition probability per unit time* $\dot{P}(t)$, as

$$\dot{P}(t) = \frac{d}{dt} \sum_{\lambda'} |(\Psi_{\lambda'}, \Psi(t))|^2. \tag{19b}$$

A dynamical principle is required for specifying the development of the system in time. It is postulated that there exists a Hermitian operator $H$, called the *Hamiltonian* or energy operator and characteristic of the system considered, such that

$$d\Psi(t) = -dt \frac{i}{\hbar} H\Psi(t), \tag{20a}$$

where $d\Psi$ is the change in the state vector occurring during a very small time interval $dt$. Expressed as a differential equation, Eq. (20*a*) is the Schrödinger equation

$$i\hbar \frac{d\Psi(t)}{dt} = H\Psi(t). \tag{20b}$$

Since this is a first-order differential equation in time, $\Psi(t)$ is determined for all $t$ if the state of the system is given at any one time.

A number of properties of operators are frequently encountered, and we shall review them briefly. An operator $A$ is *nonsingular* if it has an inverse $A^{-1}$ such that

$$A^{-1}A = AA^{-1} = I, \tag{21}$$

where $I$ is the identity operator. If the adjoint of an operator $U$ is equal to its inverse, then $U$ is called *unitary*.

$$U^{-1} = U^{\dagger}. \tag{22}$$

The eigenvalues of a unitary matrix are easily seen to have modulus unity.

A Hermitian operator $\Lambda$ is termed a *projection operator* if it has the property

$$\Lambda^2 = \Lambda \cdot \Lambda = \Lambda. \tag{23}$$

In general the operator $\Lambda$ will lead to a null vector, except for a subset $s_\lambda$ of vectors in a space; it therefore projects an arbitrary vector $\Psi$ onto the subspace spanned by the $s_\lambda$. A typical example is

$$\begin{aligned} \Lambda\Psi_\lambda &= \Psi_\lambda \qquad (\lambda \text{ a member of set } s_\lambda), \\ \Lambda\Psi_{\lambda'} &= 0 \qquad (\lambda' \text{ not a member of } s_\lambda). \end{aligned} \tag{24}$$

The *commutator* of two operators $A$ and $B$ is defined as

$$[A, B] \equiv AB - BA. \tag{25a}$$

The *anticommutator* of $A$ and $B$ is

$$\{A, B\} \equiv AB + BA. \tag{25b}$$

When $[A, B] = 0$, it is said that $A$ and $B$ *commute*.

If two observables $A$ and $B$ commute, they may be simultaneously diagonalized. Recall that the matrix of an operator $A$ with respect to the basis defined by the eigenvectors of $A$ itself has a particularly simple form: $A_{\lambda'\lambda} = (\Psi_{\lambda'}, A\Psi_\lambda) = a_\lambda(\Psi_{\lambda'}, \Psi_\lambda) = a_\lambda\delta_{\lambda'\lambda}$, that is, a diagonal matrix. The meaning of the above statement concerning $A$ and $B$ is, then, that they both have the same set of eigenvectors. To see this, let

$$A\Psi_{\lambda 0} = a_{\lambda 0}\Psi_{\lambda 0};$$

then

$$AB\Psi_{\lambda 0} = BA\Psi_{\lambda 0} = a_{\lambda 0}B\Psi_{\lambda 0}.$$

Thus $B\Psi_{\lambda 0}$ is an eigenfunction of $A$ with eigenvalue $a_{\lambda 0}$. By the fundamental expansion theorem,

$$B\Psi_{\lambda 0} = \sum_{\lambda} \Psi_{\lambda} b_{\lambda\lambda 0},$$

where the sum is taken over all $\lambda$ for which $a_\lambda = a_{\lambda 0}$. If this sum is finite, the square matrix $b_{\lambda\lambda 0}$ may be diagonalized to give the eigenvalues of $B$. The new eigenfunctions will be linear combinations of the $\Psi_\lambda$ for which $a_\lambda = a_{\lambda 0}$ and are thus simultaneous eigenfunctions of $A$ and $B$. This argument can evidently be extended to show that any number of commuting observables may be simultaneously diagonalized by a single set of orthonormal vectors.

It is convenient for some purposes to consider two vector spaces simultaneously, such that to every vector $\Phi$ in one there is associated $\Phi^\dagger$, called the *dual* of $\Phi$. It is, in fact, $\Phi^\dagger$ that is involved in the scalar product $(\Phi, \Psi)$ as we have defined it. For our purposes the dual to a vector $\Phi$ may be obtained as follows: In a given basis system $\Phi$ is represented as a column matrix with elements $\{c_i\}$. Take the complex conjugate of each element and regard the resultant set as a row matrix $(c_i^*)$. That is,

$$\Phi = \begin{pmatrix} c_1 \\ c_2 \\ \cdot \\ \cdot \\ \cdot \end{pmatrix}, \qquad \Phi^\dagger = (c_1^* \; c_2^* \cdots). \tag{26}$$

To illustrate this notation, we consider the following examples: Let $\chi = A\Phi$; then $\chi^\dagger = \Phi^\dagger A^\dagger$. A scalar product $(\Phi, \Psi)$ may be written simply as $\Phi^\dagger\Psi$. A projection operator onto a state $\Psi_\lambda$ may be written as

$$\Lambda_\lambda \equiv \Psi_\lambda \Psi_\lambda^\dagger \tag{27}$$

provided we interpret $\Lambda_\lambda\Psi$ as

$$\Lambda_\lambda\Psi = \Psi_\lambda(\Psi_\lambda, \Psi), \qquad \Psi^\dagger\Lambda_\lambda = (\Psi, \Psi_\lambda)\Psi_\lambda^\dagger. \tag{28}$$

This representation is quite consistent with that of operators being realized by matrices which can then multiply or be multiplied by appropriately oriented row or column matrices. Evidently the projection operator onto the entire space spanned by a set $\{\Psi_\lambda\}$ is equivalent to the unit operator, which may be written as

$$\sum_\lambda \Lambda_\lambda = \sum_\lambda \Psi_\lambda \Psi_\lambda^\dagger = 1. \tag{29}$$

The matrix $U$ of Eqs. (10) and (11) is unitary; consequently, a change of representation in quantum mechanics is called a *unitary transformation.* To show that $U^\dagger = U^{-1}$, we recall the definition

$$U_{\lambda i} = (\Psi_\lambda, \Phi_i) = \Psi_\lambda^\dagger \Phi_i.$$

Note that $U$ has only "mixed" matrix elements, one being associated with the complete orthonormal set $\{\Psi_\lambda\}$ and the other with the $\{\Phi_i\}$. We then have

$$(U^\dagger U)_{ij} = \sum_\lambda U_{i\lambda}{}^\dagger U_{\lambda j} = \sum_\lambda (\Psi_\lambda, \Phi_i)^*(\Psi_\lambda, \Phi_j)$$

$$= \sum_\lambda \Phi_i{}^\dagger \Psi_\lambda \Psi_\lambda{}^\dagger \Phi_j = \Phi_i{}^\dagger \Phi_j = \delta_{ij}, \tag{30}$$

where we have used Eq. (29).

We note the convenient notation of Dirac for vectors. Here we write

$$|\lambda\rangle \equiv \Psi_\lambda, \qquad \langle\lambda| \equiv \Psi_\lambda{}^\dagger. \tag{31}$$

The scalar product is written as

$$\langle\lambda' \mid \lambda\rangle \equiv (\Psi_{\lambda'}, \Psi_\lambda), \tag{32}$$

and the matrix element of an operator $A$ becomes

$$\langle\lambda'| A |\lambda\rangle \equiv (\Psi_{\lambda'}, A\Psi_\lambda). \tag{33}$$

We shall frequently use both notations, but we shall maintain a distinction between them. If we wish to refer to some specific eigenvector $\Psi_\lambda$, we shall in general use the notation $|\lambda\rangle$ only if a definite representation (that is, choice of basis vectors) is not implied. If we have a definite representation in mind, we shall use the symbol $\Psi_\lambda$. For example, to represent a particle moving with a momentum $\mathbf{p}$ shall generally write $|\mathbf{p}\rangle$, but if we are using, say, a coordinate representation, we write $\psi_\mathbf{p} = e^{i\mathbf{p}\cdot\mathbf{x}/\hbar}$. Stated more abstractly, using Eq. (29), we would write

$$|\mathbf{p}\rangle = \sum_\mathbf{x} |\mathbf{x}\rangle\langle\mathbf{x} \mid \mathbf{p}\rangle = \sum_\mathbf{x} |\mathbf{x}\rangle\, \psi_\mathbf{p}(\mathbf{x});$$

and whether we use $|\mathbf{p}\rangle$ or $\psi_\mathbf{p}$ to convey the description of a state is purely a matter of convenience.

We frequently encounter functions of an operator $A$. This is usually given meaning in terms of a Taylor series expansion of the function

$$F(A) = c_0 + c_1 A + c_2 A^2 + \cdots \tag{34}$$

More generally, if $A$ has a complete set of eigenvectors $|\lambda\rangle$ such that

$$A|\lambda\rangle = a_\lambda |\lambda\rangle,$$

we may define $F(A)$ with the aid of the projection operators $\Lambda_\lambda$ onto the $|\lambda\rangle$ as

$$F(A) = \sum_\lambda F(a_\lambda)\Lambda_\lambda = \sum_\lambda |\lambda\rangle\, F(a_\lambda)\, \langle\lambda|. \tag{35}$$

## 1.2 SOME SPECIFIC OPERATORS

We now discuss in some detail a number of operators which occur frequently in quantum mechanics. Of particular importance are those that commute with the Hamiltonian of a system as a result of some general invariance property of physical systems.[1]

This is simply illustrated by the operator $\mathscr{P}$ that represents the total momentum of a system of $N$ particles. It is convenient to use a representation in which the coordinate operators of all the particles are diagonal, the so-called Schrödinger representation [see the discussion following Eq. (33)], in which the expansion coefficients of an arbitrary state vector are simply functions of the particle coordinates; they are customarily referred to as wave functions. The operator $\mathscr{P}$ may then be written as

$$\mathscr{P} = \sum_{i=1}^{N} \mathbf{p}_i, \qquad \mathbf{p}_i = -i\hbar\, \nabla_{\mathbf{r}_i}. \tag{36}$$

Consider now the infinitesimal unitary operator

$$U(\epsilon) = 1 - \frac{i\epsilon}{\hbar} \boldsymbol{\lambda} \cdot \mathscr{P},$$

where $\epsilon$ is infinitesimal and $\boldsymbol{\lambda}$ is a constant vector. Observe that

$$\mathbf{r}'_i \equiv U^{-1}\mathbf{r}_i U = \mathbf{r}_i + i\epsilon[\boldsymbol{\lambda}\cdot\mathscr{P}, \mathbf{r}_i] = \mathbf{r}_i + \epsilon\boldsymbol{\lambda}, \tag{37}$$

where $\mathbf{r}_i$ is the position coordinate of the $i$th particle. Then if $H(\mathbf{r}_1, \mathbf{r}_2, \ldots, \mathbf{r}_n)$ is the Hamiltonian (or some other operator) for the system, we have

$$\begin{aligned} H' = U^{-1}HU &= H(\mathbf{r}_1, \ldots, \mathbf{r}_N) + i\frac{\epsilon}{\hbar}[\boldsymbol{\lambda}\cdot\mathscr{P}, \mathbf{H}] \\ &= H(\mathbf{r}_1 + \epsilon\boldsymbol{\lambda}, \ldots, \mathbf{r}_N + \epsilon\boldsymbol{\lambda}). \end{aligned} \tag{38}$$

[1] We use the accepted convention of referring to a "general invariant" as representing an exact integral of a class of Hamiltonians which themselves describe only approximately specific systems. Strictly speaking, no completely general invariants, or even exact Hamiltonians, are known in physics. Truly isolated systems are, however, extremely useful idealizations.

For isolated systems $H$ is expected to depend only on coordinate *differences*, $(\mathbf{r}_i - \mathbf{r}_j)$ $(i, j = 1, \ldots, N)$, so $H'$ should be identical to $H$. We conclude then from (38) that

$$[\mathscr{P}, H] = 0, \tag{39}$$

which, in turn, implies that $\mathscr{P}$ may be diagonalized simultaneously with $H$. For reasons to be discussed later, operators that commute with the Hamiltonian are referred to as *constants of the motion*, or as *conserved quantities*.

An operator that generates *finite* displacements is

$$D(\boldsymbol{\lambda}) \equiv e^{-(i/\hbar)\boldsymbol{\lambda}\cdot\mathscr{P})}. \tag{40}$$

To see this, we remark that

$$\begin{aligned}\mathbf{r}'_i &= D^{-1}(\boldsymbol{\lambda})\mathbf{r}_i D(\boldsymbol{\lambda})\\ &= \mathbf{r}_i + \left[\frac{i}{\hbar}\boldsymbol{\lambda}\cdot\mathscr{P}, \mathbf{r}_i\right] + \frac{1}{2!}\frac{1}{\hbar^2}[i\boldsymbol{\lambda}\cdot\mathscr{P}, [i\boldsymbol{\lambda}\cdot\mathscr{P}, \mathbf{r}_i]] + \cdots\\ &= \mathbf{r}_i + \boldsymbol{\lambda},\end{aligned} \tag{41}$$

because all terms after the first commutator vanish. The original expansion in terms of multiple commutators is a special case of the following theorem. Let $A$ and $B$ be any two operators and $\Gamma$ be the operator $e^{aA}Be^{-aA}$; then

$$\begin{aligned}\Gamma = e^{aA}Be^{-aA} &= B + a[A, B] + \frac{a^2}{2!}[A, [A, B]]\\ &+ \frac{a^3}{3!}[A, [A, [A, B]]] + \cdots\end{aligned} \tag{42}$$

To prove this we imagine making a power series expansion in $a$; we note that

$$\frac{d\Gamma}{da} = e^{aA}[A, B]\,e^{-aA}, \qquad \frac{d^2\Gamma}{da^2} = e^{aA}\,[A, [A, B]]\,e^{-aA},$$

and so on. On substitution into

$$\Gamma = \sum_n \frac{1}{n!}a^n\left(\frac{d^n\Gamma}{da^n}\right)_{a=0}$$

we obtain Eq. (42). Finally, if $F(\mathbf{r}_1, \ldots, \mathbf{r}_N)$ is a function of the $\mathbf{r}_i$ (and may be expanded in a Taylor series) we have from (41)

$$F' = D^{-1}FD = F(\mathbf{r}_1 + \boldsymbol{\lambda}, \ldots, \mathbf{r}_N + \boldsymbol{\lambda}). \tag{43}$$

The operator $\mathscr{P}$ which reflects the rectangular coordinate axes of all particles (in the $3N$-dimensional configuration space used to describe an $N$-particle system) is called the parity operator.

$$\mathbf{r}'_i = \mathscr{P}^{-1}\mathbf{r}_i\mathscr{P} = -\mathbf{r}_i. \tag{44}$$

For the known strong interactions in nature (which are usually involved in scattering experiments), $\mathscr{P}$ is conserved, or

$$[\mathscr{P}, H] = 0, \tag{45}$$

where $H$ is the appropriate Hamiltonian. If the eigenfunctions of $H$ are $\psi_\lambda(\mathbf{r}_1, \ldots, \mathbf{r}_N)$, we may choose them to be also eigenfunctions of $\mathscr{P}$,

$$\mathscr{P}\psi_\lambda = e^{i\omega_\lambda}\psi_\lambda, \tag{46a}$$

where $\omega_\lambda$ is a number. Since the result of two reflections $\mathscr{P}^2$ must leave a system unchanged, and by Eq. (46*a*)

$$\mathscr{P}^2\psi_\lambda = e^{2i\omega_\lambda}\psi_\lambda,$$

we conclude that the $\omega_\lambda$ are real. Consider now an arbitrary state

$$\psi = \sum_\lambda c_\lambda \psi_\lambda.$$

Since $\mathscr{P}^2\psi$ may differ from $\psi$ only by an unobservable phase, say $e^{iu}$ with $u$ real,

$$\mathscr{P}^2\psi = e^{iu}\psi = \sum_\lambda c_\lambda e^{2i\omega_\lambda}\psi_\lambda$$

so that $e^{2i\omega_\lambda} = e^{iu}$ independent of $\lambda$. Since the phase $u$ is unobservable, there is no loss of generality in setting $e^{iu} = 1$, from which we deduce

$$e^{i\omega_\lambda} = \pm 1, \qquad \mathscr{P}\psi_\lambda = (\pm 1)\psi_\lambda. \tag{46b}$$

Those states with eigenvalues $+1$ are said to have *positive parity* and those with $-1$ *negative parity*.

The *angular momentum* of a system is represented by an Hermitian operator $\mathbf{J}$ whose cartesian components satisfy

$$[J_x, J_y] = i\hbar J_z, \qquad [J_y, J_z] = i\hbar J_x, \qquad [J_z, J_x] = i\hbar J_y, \tag{47a}$$

or, using symbolic cross-product notation,

$$\mathbf{J} \times \mathbf{J} = i\hbar\mathbf{J}. \tag{47b}$$

If $\epsilon$ is a positive infinitesimal and $\hat{\mathbf{n}}$ is a fixed unit vector, we construct a unitary rotation operator $R(\epsilon)$ defined by

$$R(\epsilon) = 1 - \frac{i\epsilon}{\hbar}\hat{\mathbf{n}} \cdot \mathbf{J}. \tag{48}$$

The properties of this operator which are needed for the present purpose are

$$\begin{aligned} \mathbf{r}'_i &= R^{-1}\mathbf{r}_i R = \mathbf{r}_i + \epsilon\hat{\mathbf{n}} \times \mathbf{r}_i, \\ \mathbf{p}'_i &= R^{-1}\mathbf{p}_i R = \mathbf{p}_i + \epsilon\hat{\mathbf{n}} \times \mathbf{p}_i. \end{aligned} \tag{49}$$

These relations will be discussed in more detail later (see Section 2.2) but may be directly verified by writing $\mathbf{J} = \Sigma_i \mathbf{r}_i \times \mathbf{p}_i$ and using the commutation relations between $\mathbf{r}$ and $\mathbf{p}$. Evidently $\mathbf{r}'$ is the vector that results from a rotation of $\mathbf{r}$ about the axis defined by $\hat{\mathbf{n}}$ through an angle $\epsilon$, defined as positive when the rotation has the sense of a right-hand screw.

Returning now to the $N$-particle Hamiltonian $H(\mathbf{r}_i, \mathbf{p}_i)$ we have

$$\begin{aligned} H' = R^{-1}HR &= H(\mathbf{r}_i + \epsilon\hat{\mathbf{n}} \times \mathbf{r}_i, \mathbf{p}_i + \epsilon\hat{\mathbf{n}} \times \mathbf{p}_i) \\ &= H + \frac{i\epsilon}{\hbar}[\hat{\mathbf{n}} \cdot \mathbf{J}, H]. \end{aligned} \tag{50}$$

If $H$ is invariant with respect to coordinate rotations, $H' = H$ and we conclude from Eq. (50) (since the direction of $\hat{\mathbf{n}}$ is arbitrary)

$$[\mathbf{J}, H] = 0. \tag{51}$$

The operator that generates a rotation through a finite angle $\omega$ about $\hat{\mathbf{n}}$ is

$$R(\omega) = e^{-i(\omega/\hbar)\hat{\mathbf{n}}.\mathbf{J}}. \tag{52}$$

This follows from the group property of rotations.

$$R(\omega + \epsilon) = R(\omega)R(\epsilon),$$

or

$$dR(\omega) = R(\omega + \epsilon) - R(\omega) = -\frac{i}{\hbar}\epsilon\hat{\mathbf{n}} \cdot \mathbf{J}R(\omega).$$

Thus

$$\frac{dR}{d\omega} = -\frac{i}{\hbar}\hat{\mathbf{n}} \cdot \mathbf{J}R(\omega),$$

and Eq. (52) follows immediately by integration.

The square of the angular momentum operator,

$$J^2 = J_x^2 + J_y^2 + J_z^2, \tag{53}$$

is easily shown to commute with $J_x, J_y, J_z$.[2] This means that $\mathbf{J}^2$ and one component of $\mathbf{J}$, say $J_z$, may be diagonalized simultaneously. The resulting eigenfunctions will be written as

$$|j, m\rangle \equiv \mathscr{Y}_j^m. \tag{54}$$

By purely algebraic manipulations we find

$$\mathbf{J}^2\mathscr{Y}_j^m = \hbar^2 j(j+1)\mathscr{Y}_j^m, \qquad J_z\mathscr{Y}_j^m = \hbar m\mathscr{Y}_j^m, \tag{55}$$

[2] This may be seen by brute force directly from the commutation relations (47*a*), or more elegantly by remarking that just as for $\mathbf{r}_i$ and $\mathbf{p}_i$ [according to Eq. (49)] we have $\mathbf{J}'_i = \mathbf{J}_i + \epsilon\hat{\mathbf{n}} \times \mathbf{J}_i$, from which we find that $\mathbf{J}^2$ is a rotational invariant.

where $-j \leq m \leq j$ and $j$ is a positive integer or half-integer; $m$ is correspondingly integral or half-integral.

If a system consists of $n$ *separate* parts, each having an angular momentum operator $\mathbf{J}_j$ $(i = 1, 2, \ldots, n)$ and satisfying

$$[\mathbf{J}_i, \mathbf{J}_k] = 0 \qquad (i \neq k = 1, 2, \ldots, n),$$

then

$$\mathbf{J} \equiv \sum_{i=1}^{n} \mathbf{J}_i \tag{56}$$

satisfies the commutation rules (47$a$) by virtue of the fact that the individual $\mathbf{J}_i$ do. The eigenvalue spectrum is just that of (55), and the problem arises of relating the eigenvalues and eigenfunctions of the total system to those of the individual parts.

To study this it is sufficient to consider the case

$$\mathbf{J} = \mathbf{J}_1 + \mathbf{J}_2; \tag{57}$$

the general case follows by induction. Since $\mathbf{J}_1$ and $\mathbf{J}_2$ commute, we have the respective eigenvalues $(j_1, m_1)$ and $(j_2, m_2)$ with corresponding eigenfunctions $\overset{1}{\mathscr{Y}}{}^{m_1}_{j_2}$, $\overset{2}{\mathscr{Y}}{}^{m_2}_{j_2}$. The possible eigenvalues of $\mathbf{J}^2$ and $J_z$ are then

$$\begin{aligned} &\mathbf{J}^2 = j(j+1)\hbar^2, \qquad J_z = m\hbar, \\ &j = j_1 + j_2, j_1 + j_2 - 1, \ldots, |j_1 - j_2|, \qquad m = j, j-1, \ldots, -j. \end{aligned} \tag{58}$$

An outline of the argument leading to (58) is as follows. Corresponding to the maximum eigenvalues of $J_{1z}$, $J_{2z}$ there can be only one eigenfunction of the total system $\overset{1}{\mathscr{Y}}{}^{m_1}_{j_2} \overset{2}{\mathscr{Y}}{}^{m_2}_{j_2}$, and for it $J_z = (j_1 + j_2)\hbar$; and thus the maximum $j$ is $j_1 + j_2$ also. The eigenvalue $(j_1 + j_2 - 1)\hbar$ can evidently be reached in two ways: by reducing either $J_{1z}$ or $J_{2z}$ by $\hbar$. One of the eigenvalues corresponds to a member of the set of eigenfunctions for which $j = j_1 + j_2$, but the other must then be associated with a new one for which the maximum $J_z$ is $j_1 + j_2 - 1$, and hence its $j$ is $j_1 + j_2 - 1$. The argument continues from here in an obvious manner.

The eigenfunctions of the total system $\mathscr{Y}_j{}^m$ are related to $\overset{1}{\mathscr{Y}}{}_{j_1}^{m_1}$, $\overset{2}{\mathscr{Y}}{}^{m_2}_{j_2}$ by

$$\begin{aligned} \mathscr{Y}_j{}^m &= \sum_{m_1} \langle j_1, j_2; m_1, m - m_1 \mid j, m\rangle \overset{1}{\mathscr{Y}}{}^{m_1}_{j_1} \overset{2}{\mathscr{Y}}{}^{m_2}_{j_2} \\ &= \sum_{\substack{m_1, m_2 \\ (m_1 + m_2 = m)}} \langle j_1, j_2; m_1, m_2 \mid j, m\rangle \, \overset{1}{\mathscr{Y}}{}^{m_1}_{j_1} \overset{2}{\mathscr{Y}}{}^{m_2}_{j_2}. \end{aligned} \tag{59$a$}$$

The coefficients $\langle j_1, j_2; m_1, m_2 \mid j, m\rangle$ may be chosen to be real and are defined to vanish whenever $m_1 + m_2 \neq m$. They are called *Wigner* coefficients, or *vector addition* coefficients.[3] (Wigner was the first to give

[3] See, for example, E. P. Wigner [4].

their explicit form.) Tables for some special Wigner coefficients are given in Appendix A.

Some useful formal properties of the Wigner coefficients are

$$\langle j_1, j_2; m_1, m_2 \mid j, m\rangle = (-1)^{j-j_1-j_2}\langle j_2, j_1; m_2, m_1 \mid j, m\rangle, \qquad (60a)$$

$$\sum_{m_1} \langle j_1, j_2; m_1, m - m_1 \mid j, m\rangle\langle j_1, j_2; m_1, m' - m_1 \mid j', m'\rangle = \delta_{jj'}\delta_{mm'}, \qquad (60b)$$

$$\sum_{j} \langle j_1, j_2, m_1; m - m_1 \mid j, m\rangle\langle j_1, j_2; m'_1, m - m'_1 \mid j, m\rangle = \delta_{m_1 m'_1}, \qquad (60c)$$

$$\sum_{m_1} \langle j_1, j_2; m_1, m_2 \mid j, m_1 + m_2\rangle\langle j_1, j'_2; m_1, m_2 \mid j, m_1 + m_2\rangle = \frac{2j+1}{2j_2+1}\,\delta_{j_2, j'_2}. \qquad (60d)$$

A few relations of a different character are

$$\begin{aligned}\langle j_1, j_2; m_1, m_2 \mid j, m\rangle &= \langle j_2, j_1; -m_2, -m_1 \mid j, -m\rangle \\ &= (-1)^{j_1+j_2-j}\langle j_1, j_2; -m_1, -m_2 \mid j, -m\rangle \qquad (61) \\ &= (-1)^{j_1-m_1}\left(\frac{2j+1}{2j_2+1}\right)^{1/2}\langle j_1, j; m_1, -m \mid j_2, -m_2\rangle \\ &= (-1)^{j_2+m_2}\left(\frac{2j+1}{2j_1+1}\right)^{1/2}\langle j, j_2; -m, m_2 \mid j_1, -m_1\rangle.\end{aligned}$$

With the aid of (60$c$), for example, we may invert Eq. (59$a$):

$$\overset{1}{\mathscr{Y}}{}^{m_1}_{j_1}\,\overset{2}{\mathscr{Y}}{}^{m-m_1}_{j_2} = \sum_{j} \langle j_1, j_2; m_1, m - m_1 \mid j, m\rangle\,\mathscr{Y}^m_j. \qquad (59b)$$

Equations (60$b$) and (60$c$) state merely the general properties of a unitary transformation between the basis systems. Equations (60$a$), (60$d$), and (61) can be obtained from the general expression of Wigner [4, p. 191], which is

$$\begin{aligned}\langle j_1, j_2; m_1, m_2 \mid j, m\rangle &= \delta_{m, m_1+m_2} \\ &\times \left[\frac{(j+j_1-j_2)!(j-j_1+j_2)!(j_1+j_2-j)!(j+m)!(j-m)!(2j+1)}{(j+j_1+j_2+1)!(j_1-m_1)!(j_2-m_2)!(j_1+m_1)!(j_2+m_2)!}\right]^{1/2} \\ &\times \sum_k \frac{(-1)^{k+j_2+m_2}}{k!(j+m-k)!}\,\frac{(j+j_2+m_1-k)!(j_1-m_1+k)!}{(j-j_1+j_2-k)!(k+j_1-j_2-m)!}. \qquad (62)\end{aligned}$$

Here the sum on $k$ runs over all integral values that give nonnegative arguments of the factorials,

In simple cases the Wigner coefficients may be computed directly with the aid of the *raising* and *lowering* operators.

$$J^{\pm} = J_x \pm iJ_y, \qquad (63a)$$

which have the properties

$$J^{\pm}\mathcal{Y}_j^{\,m} = \hbar[(j \mp m)(j \pm m + 1)]^{1/2}\, \mathcal{Y}_j^{\,m\pm 1}, \tag{63b}$$
$$J^{+}\mathcal{Y}_j^{\,j} = 0, \qquad J^{-}\mathcal{Y}_j^{\,-j} = 0.$$

A practical method for constructing the Wigner coefficients when one of the $j$'s is small is based on the construction of a set of projection operators. This method also provides a useful operatorial representation of the eigenfunctions. Consider the operator $\Sigma \equiv \mathbf{J}_1 \cdot \mathbf{J}_2$; for given $j_1$, $j_2$ the eigenvalues of $\Sigma$ corresponding to any allowable resultant $j$ are given by

$$\Sigma_j = \frac{[j(j+1) - j_1(j_1+1) - j_2(j_2+1)]\hbar^2}{2}, \tag{64a}$$

and they are all different. Now, construct the projection operator

$$\Lambda_j = \frac{\prod\limits_{l \neq j} (\Sigma - \Sigma_l)}{\prod\limits_{l \neq j} (\Sigma_j - \Sigma_l)}. \tag{64b}$$

Clearly this operator gives zero unless the vector on which it operates contains the eigenvalue $j$, and it gives unity for one corresponding to eigenvalue $j$. We obtain then from Eq. (59$b$)

$$\Lambda_j \overset{1}{\mathcal{Y}}{}_{j_1}^{m_1} \overset{2}{\mathcal{Y}}{}_{j_2}^{m-m_1} = \langle j_1, j_2; m_1, m - m_1 \mid j, m\rangle\, \mathcal{Y}_j^{\,m}. \tag{64c}$$

Taking the scalar product of this equation with itself and using the facts that the $\Lambda_j$ are Hermitian and that $\Lambda_j^2 = \Lambda_j$, we find (since the $\mathcal{Y}_j^{\,m}$ are normalized to unity)

$$(\overset{1}{\mathcal{Y}}{}_{j_1}^{m_1} \overset{2}{\mathcal{Y}}{}_{j_2}^{m-m_1}, \Lambda_j \overset{1}{\mathcal{Y}}{}_{j_1}^{m_1} \overset{2}{\mathcal{Y}}{}_{j_2}^{m-m_1}) = \langle j_1, j_2; m_1, m - m_1 \mid j, m\rangle^2. \tag{64d}$$

Thus from the diagonal element of $\Lambda_j$ we obtain the square of the Wigner coefficient. Equation (64$c$) gives an operatorial form of Eq. (59$a$) that is frequently convenient. The evaluation of (64$d$) is not difficult if one of the $j$'s is small; we use $\Sigma = \frac{1}{2}J_1^{+}J_2^{-} + \frac{1}{2}J_1^{-}J_2^{+} + J_{1z}J_{2z}$ and the matrix elements (63$b$). To obtain the relative phases, we write (64$c$) with two different $m_1$ values and take the scalar product of the resulting equations with each other:

$$\left(\overset{1}{\mathcal{Y}}{}_{j_1}^{m'_1} \overset{2}{\mathcal{Y}}{}_{j_2}^{m-m'_1}, \Lambda_j \overset{1}{\mathcal{Y}}{}_{j_1}^{m_1} \overset{2}{\mathcal{Y}}{}_{j_2}^{m-m_1}\right) = \langle j_1, j_2; m'_1, m - m'_1 \mid j, m\rangle$$
$$\times \langle j_1, j_2; m_1, m - m_1 \mid j, m\rangle. \tag{64e}$$

Examples of angular momentum operators are given by the orbital angular momenta for particles,

$$\mathbf{l}_i = \mathbf{r}_i \times \mathbf{p}_i \qquad (i = 1, 2, \ldots, N) \tag{65}$$

and their resultant angular momentum

$$\mathbf{L} = \sum_{i=1} \mathbf{l}_i. \tag{66}$$

If the particles have spin operators $\mathbf{S}_i$, their total spin is

$$\mathbf{S} = \sum_{i=1}^{N} \mathbf{S}_i, \tag{67}$$

and the total angular momentum of the $N$-particles is

$$\mathbf{J} = \mathbf{L} + \mathbf{S}. \tag{68}$$

For a spin one-half particle, we shall usually use the explicit representation [consistent with (63*b*)]

$$\mathbf{S} = \tfrac{1}{2}\,\boldsymbol{\sigma}\,\hbar,$$

$$\sigma_x = \begin{pmatrix} 0 & 1 \\ 1 & 0 \end{pmatrix}, \qquad \sigma_y = \begin{pmatrix} 0 & -i \\ i & 0 \end{pmatrix}, \qquad \sigma_z = \begin{pmatrix} 1 & 0 \\ 0 & -1 \end{pmatrix}. \tag{69a}$$

Eigenfunctions of $\sigma_z$ will be written as $u_{\pm 1/2}$. Thus

$$\sigma_z u_{\pm 1/2} = \pm u_{\pm 1/2}. \tag{69b}$$

## 1.3 EIGENSTATES OF A FREE PARTICLE

We consider a single noninteracting particle in a box of volume $\mathscr{V}$. Let its coordinate be $\mathbf{r}$, its momentum $\mathbf{p}$, and its angular momentum $\mathbf{l}$.[4] If the particle has spin, we suppose this to be uncoupled from its orbital motion so that the spin dynamics separates.

First, we imagine the box $\mathscr{V}$ to be rectangular with the coordinate origin at one corner and the coordinate axes lying along the edges $\mathscr{L}_x$, $\mathscr{L}_y$, $\mathscr{L}_z$ of the box (see Fig. 1.1). If the walls are impenetrable, the eigenfunctions for the particle in the coordinate representation are

$$\chi_{\mathbf{p}}(\mathbf{r}) = \eta \sin \frac{p_1 x}{\hbar} \sin \frac{p_2 y}{\hbar} \sin \frac{p_3 z}{\hbar},$$
$$p_1 = \frac{n_1 \pi}{\mathscr{L}_x}\hbar, \qquad p_2 = \frac{n_2 \pi}{\mathscr{L}_y}\hbar, \qquad p_3 = \frac{n_3 \pi}{\mathscr{L}_z}\hbar. \tag{70}$$

Here the $n_i$ are positive integers and $\eta$ is a normalization constant.

[4] This analysis also applies to the relative motion of two particles. In this case, **r**, **p**, **l** refer respectively to their relative coordinate, their relative momentum, and the angular momentum about their center of mass.

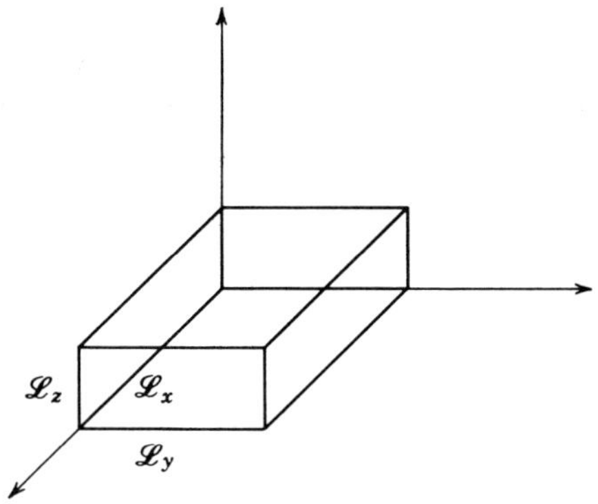


Fig. 1.1 Quantization box for the wave functions of Eq. (70).

If the box has macroscopic dimensions, observations on a system in the box ordinarily involve sums over large numbers of states of the form (70). In such a case it is convenient to approximate the sums by integrals over $p_1, p_2, p_3$. The weight of a small "volume element" in this $p$-space in terms of increments $\delta n_1$, $\delta n_2$, $\delta n_3$ is evidently

$$\delta n_1\, \delta n_2\, \delta n_3 = \frac{\mathscr{L}_x \mathscr{L}_y \mathscr{L}_z}{\pi^3 \hbar^3}\, \delta p_1\, \delta p_2\, \delta p_3.$$

It is convenient to speak of $\delta p_1\, \delta p_2\, \delta p_3$ as a volume element $d^3p$ in *momentum space*. It is also desirable to extend the range of variables to negative values of the $p$'s. Such an extension merely changes the phase of the wave function $\chi_{\mathbf{p}}$ and does not lead to a new state. In extending sums over states to negative $n_i$ we count every state $2^3 = 8$ times which means we should take only one-eighth of the entire momentum space. Doing so we obtain the rule for transforming sums to integrals;

$$\sum_{n_1, n_2, n_3} \leftrightarrow \mathscr{V} \int \frac{d^3p}{(2\pi\hbar)^3} \tag{71}$$

where now all directions and magnitudes of $\mathbf{p}$ are summed over in the integral, and we have set $\mathscr{L}_x \mathscr{L}_y \mathscr{L}_z = \mathscr{V}$.

To a good approximation Eq. (71) is valid for a large box of (almost) arbitrary shape, since the number of states in $d^3p$ is proportional to the volume of the box. An arbitrary box may be subdivided into parallelopipeds of the sort we have considered, and Eq. (71) may be applied to each. The total number of states will be the sum from each sub-box and the sum will again have the form (71).

The actual eigenfunctions in a box of arbitrary shape are generally, of course, quite complicated. Fortunately they are rarely needed, since when the box is "large" we can form *wave packets*[5] having the form

$$\chi_{\mathbf{p}}(\mathbf{r}) = \frac{1}{\mathcal{V}^{1/2}} e^{i\mathbf{p}\cdot(\mathbf{r}/\hbar)} \tag{72}$$

over the region of space where the particle is interacting or being observed. Of course $\chi_{\mathbf{p}}(\mathbf{r})$ as given in (72) will not be of the correct form near the walls of the box where specific boundary conditions must be satisfied. In general this will be unimportant as long as the region of interaction or observation is not near the walls.

The eigenfunctions of a free particle in a spherical box of radius $R$ are simple and of great importance. In this case, if the boundary conditions have spherical symmetry, $\mathbf{l}$ commutes with the Hamiltonian. The eigenfunctions have the form

$$\chi(k, l, m) = aj_l(kr)Y_l^m(\theta, \phi) \tag{73}$$

where $k$ is related to the energy by $2mE = \hbar^2k^2$ (and hence $\hbar\mathbf{k} = \mathbf{p}$, the momentum), $a$ is a normalizing constant, and $\theta$, $\phi$ are polar angles of $\mathbf{r}$. We do not explicitly indicate the $\mathbf{r}$ dependence when we write $\chi(k, l, m)$, but as stated earlier we shall imply this being in a coordinate representation unless otherwise noted. The function $j_l(kr)$ is the spherical Bessel function[6] with the properties

$$\begin{aligned} j_l(kr) &= \sqrt{\pi/2kr}\, J_{l+1/2}(kr). \\ j_l(kr) &\cong \frac{(kr)^l}{(2l+1)(2l-1)(2l-3)\cdots 1} \\ &\equiv \frac{(kr)^l}{(2l+1)!!}, \qquad kr \ll \sqrt{2(2l+3)}, \\ j_l(kr) &\cong \frac{\sin(kr - l\pi/2)}{kr}, \qquad kr \gg l^2. \end{aligned} \tag{74}$$

The spherical harmonic $Y_l^m(\theta, \phi)$ is an eigenfunction of $\mathbf{l}^2$ and $l_z$:

$$\begin{aligned} \mathbf{l}^2 Y_l^m &= \hbar^2 l(l+1) Y_l^m, \\ l_z Y_l^m &= \hbar m Y_l^m. \end{aligned} \tag{75}$$

[5] Wave packets will be discussed in more detail in Chapter 3.
[6] See, for example, Section 6.3 for a review of the properties of Bessel functions.

For reference we list a few properties of the spherical harmonics.

$$P_l(\cos\theta) = \frac{1}{2^l l!}\frac{d^l}{d(\cos\theta)^l}(\cos^2\theta - 1)^l,$$

$$Y_l^m(\theta, \phi) = (-1)^m e^{im\phi}\left[\frac{2l+1}{4\pi}\frac{(l-m)!}{(l+m)!}\right]^{1/2} \times \sin^m\theta \frac{d^m}{d(\cos\theta)^m} P_l(\cos\theta) \qquad (m > 0), \tag{76}$$

$$Y_l^{-m} = (-1)^m [Y_l^m]^* \qquad (m > 0),$$

$$Y_l^0(\theta) = \sqrt{(2l+1)/4\pi}\, P_l(\cos\theta).$$

The spherical harmonics $Y_l^m$, as we have defined them, are consistent with the previously established phase convention for the raising and lowering operators [Eq. (63)]. The orthogonality relations are

$$\int d\Omega Y_l^{m*}(\theta, \phi) Y_{l'}^{m'}(\theta, \phi) = \int_0^{2\pi} d\phi \int_0^{\pi} \sin\theta\, d\theta\, Y_l^{m*} Y_{l'}^{m'} = \delta_{ll'}\delta_{mml'}, \tag{77}$$

$$\int_{-1}^{+1} dx P_l(x) P_{l'}(x) = \frac{2}{2l+1}\delta_{ll'}.$$

($d\Omega = \sin\theta\, d\theta\, d\phi$ is an element of solid angle). If $\mathbf{r}$ and $\mathbf{r}'$ are two vectors with polar angles $(\theta, \phi)$, $(\theta', \phi')$ respectively, and if the angle between $\mathbf{r}$ and $\mathbf{r}'$ is $\Theta$, the addition theorem states

$$\sqrt{(2l+1)/4\pi}\, Y_l^0(\Theta) = \sum_{m=-l}^{+l} Y_l^{m*}(\theta', \phi') Y_l^m(\theta, \phi). \tag{78}$$

Another useful relation is[7]

$$Y_l^m(\theta, \phi) Y_{l'}^{m'}(\theta, \phi) = \sum_{L=|l-l'|}^{l+l'} \left[\frac{(2l+1)(2l'+1)}{4\pi(2L+1)}\right]^{1/2} \langle l, l'; 0, 0 \mid L, 0\rangle \langle l, l'; m, m' \mid L, m+m'\rangle\, Y_L^{m+m'}(\theta, \phi). \tag{79}$$

Let us return now to Eq. (73) and assume that the walls of the box are impenetrable. For simplicity we imagine that $kR \gg l^2$ so that we may use the asymptotic form of $j_l(kr)$ given in Eq. (74). To satisfy the boundary conditions we have

$$k_n R - \frac{l\pi}{2} = n\pi \qquad (n = 0, 1, \ldots). \tag{80}$$

[7] See Blatt and Weisskopf [5, Appendix A].

Here the integer $n$ is called the radial quantum number. The number of states in the range $\delta n$ is

$$\delta n = \delta k_n \frac{R}{\pi}. \tag{81}$$

The sum over states is

$$\sum_{n=0}^{\infty} \sum_{l=0}^{\infty} \sum_{m=-l}^{l} \leftrightarrow \frac{\mathscr{V}}{(2\pi)^3} \int d\Omega_k k^2 \delta k \leftrightarrow \sum_{\mathbf{k}}, \tag{82}$$

where we recall that $k = p/\hbar$ and

$$\delta k = \delta n \frac{\pi}{R}; \tag{83}$$

$d\Omega_{\mathbf{k}}$ is the element of solid angle about the vector $\mathbf{k}$. Finally, the normalization factor $a$ in Eq. (73) is

$$a = \sqrt{2/R}\, k_n. \tag{84}$$

The transformation from the plane wave to the spherical wave description is given by

$$\chi(k_n, l, m) = \sum_{\mathbf{k}'} \chi_{\mathbf{k}'} \langle \mathbf{k}'| T |k_n, l, m\rangle \equiv a j_l(k_n r) Y_l^{\,m},$$

$$\chi_{\mathbf{k}'} = \sum_{n,l,m} \chi(k_n, l, m)\langle k_n, l, m| T^{-1} |\mathbf{k}'\rangle = \frac{1}{\sqrt{\mathscr{V}}} e^{i\mathbf{k}'\cdot\mathbf{r}},$$

$$\langle \mathbf{k}'| T |k_n, l, m\rangle = \delta_{k'k_n} \frac{4\pi}{\mathscr{V}^{1/2}} \left(\frac{1}{k}\sqrt{R/2}\right)(-i)^l Y_l^{m}(\hat{\mathbf{k}}') \tag{85a}$$

$$= \langle k_n l, m| T^{-1} |\mathbf{k}'\rangle^*$$

This relation is just the familiar expansion

$$e^{i\mathbf{k}\cdot\mathbf{r}} = \sum_{l=0}^{\infty} \sqrt{4\pi(2l+1)}\, i^l j_l(kr) Y_l^0(\hat{\mathbf{k}}\cdot\hat{\mathbf{r}}). \tag{85b}$$

We have introduced in Eq. (85*a*) a notation to be used frequently:

$$Y_l^{\,m}(\hat{\mathbf{k}}) \equiv Y_l^{\,m}(\theta, \phi), \tag{86}$$

where $\theta$, $\phi$ are the spherical angles of the unit vector $\hat{\mathbf{k}}$ with respect to some coordinate system.

The wave function of a complex particle with internal degrees of freedom moving with a momentum $\mathbf{p} = \hbar\mathbf{k}$ is written

$$\chi_{\gamma,\mathbf{k}} = \mathscr{V}^{-1/2} e^{i\mathbf{k}\cdot\mathbf{r}} g_\gamma(\xi), \tag{87}$$

where $g_\gamma(\xi)$ is the wave function describing the internal state $\gamma$ which has dynamical variables $\xi$.

It is often more convenient to use instead of normalization in a volume $\mathscr{V}$ (so-called box normalization), normalization in the continuous spectrum. The passage to the limit of infinite volume $\mathscr{V}$ implies that the allowable **k** vectors take on all values from the outset. In this case we replace Eq. (87) by

$$\chi_{\gamma,\mathbf{k}} = (2\pi)^{-3/2} e^{i\mathbf{k}\cdot\mathbf{r}} g_\gamma(\xi). \tag{88}$$

Use of Eq. (88) is equivalent to the replacement of the sum in (71) directly by an integral. To see this consider two state vectors $\Phi_1$, $\Phi_2$ and form, using Eq. (87), the following expression for the scalar product of $\Phi_1$ and $\Phi_2$.

$$\begin{aligned}(\Phi_1, \Phi_2) &= \sum_{\mathbf{k},\gamma} (\Phi_1, \chi_{\gamma,\mathbf{k}})(\chi_{\gamma,\mathbf{k}}, \Phi_2) \\ &= \sum_\gamma \int \frac{\mathscr{V} d^3k}{(2\pi)^3} \left(\Phi_1, \frac{e^{i\mathbf{k}\cdot\mathbf{r}} g_\gamma}{\mathscr{V}^{1/2}}\right)\left(\frac{e^{i\mathbf{k}\cdot\mathbf{r}} g_\gamma, \Phi_2}{\mathscr{V}^{1/2}}\right) \\ &= \sum_\gamma \int d^3k \left(\Phi_1, \frac{e^{i\mathbf{k}\cdot\mathbf{r}} g_\gamma}{(2\pi)^{3/2}}\right)\left(\frac{e^{i\mathbf{k}\cdot\mathbf{r}}}{(2\pi)^{3/2}} g_\gamma, \Phi_2\right).\end{aligned} \tag{89}$$

In the last line we see just the wave function $\chi_{\gamma,\mathbf{k}}$ in the form given in Eq. (88). The normalization condition for these wave functions is

$$\begin{aligned}(\chi_{\gamma,\mathbf{k}}, \chi_{\gamma',\mathbf{k}'}) &= \delta_{\gamma\gamma'} \int \frac{d^3r}{(2\pi)^3} e^{-i(\mathbf{k}-\mathbf{k}')\cdot\mathbf{r}} \\ &= \delta_{\gamma\gamma'} \delta(\mathbf{k} - \mathbf{k}').\end{aligned} \tag{90}$$

## 1.4 RELATIVISTIC QUANTUM MECHANICS

In this section we review in a brief and elementary manner the Lorentz transformation and some of its implications. We treat also the relativistic quantum mechanics of a single particle. No effort is made here to treat relativistic *field theories*.[8]

It will sometimes be convenient to label the rectangular components of a right-handed coordinate system as

$$x_1 = x, \qquad x_2 = y, \qquad x_3 = z, \tag{91}$$

and for the time coordinate $t$ to use

$$x_0 = ct, \qquad x_4 = ict, \tag{92}$$

where $c$ is the velocity of light.

[8] See, however, Chapter 10.

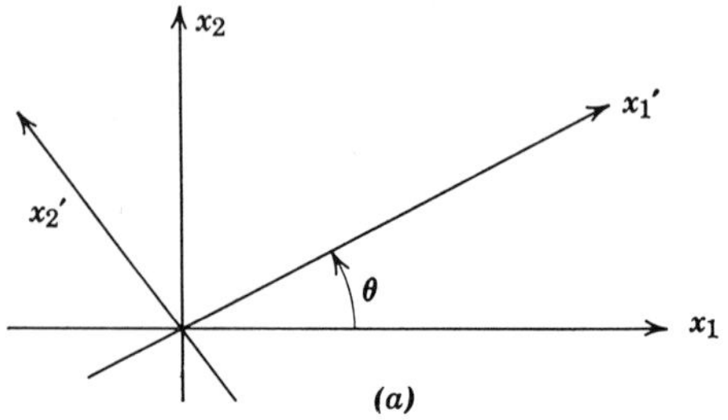


(a)

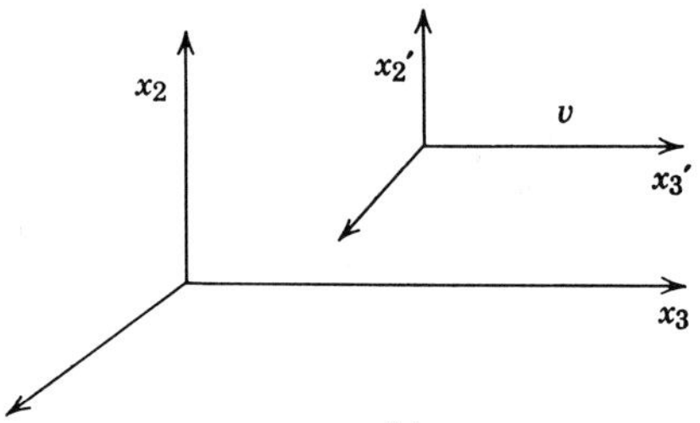


(b)

Fig. 1.2 (a) Coordinate rotation illustrating Eqs. (93). (b) Transformation to moving coordinate system illustrating Eqs. (94).

For a rotation of the *coordinate system* through an angle $\theta$ about the $z$-axis (as in Fig. 1.2$a$), the coordinates of a fixed point $P$ in the new (primed) system in terms of the old (unprimed) are given by

$$\begin{aligned} x'_1 &= \cos\theta\, x_1 + \sin\theta\, x_2, \\ x'_2 &= -\sin\theta\, x_1 + \cos\theta\, x_2, \\ x'_3 &= x_3, \qquad x'_0 = x_0. \end{aligned} \tag{93}$$

The infinitesimal form of the relations given in Eq. (49) corresponds to a rotation of the axes through an angle $\epsilon$. The inverse relation to (93), giving the $x$'s in terms of the $x'$ 's, follows simply by changing $\theta$ to $-\theta$ and interchanging primed and unprimed coordinates.

Next, imagine that the primed system is moving with velocity $v$ in the direction of the $x_3$-axis and that $x'_1$, $x'_2$ are parallel to $x_1$, $x_2$ as illustrated in Fig. 1.2$b$. Then

$$\begin{aligned} x'_0 &= \cosh w x_0 - \sinh w x_3, \\ x'_3 &= -\sinh w x_0 + \cosh w x_3, \\ x'_1 &= x_1, \qquad x'_2 = x_2. \end{aligned} \tag{94}$$

The "angle" $w$ is related to the velocity by

$$\cosh w = \gamma, \qquad \sinh w = \beta\gamma, \qquad \gamma = (1 - \beta^2)^{-1/2}, \qquad \beta = \frac{v}{c}. \tag{95}$$

The inverse relation is obtained by changing $w$ to $-w$ and interchanging primed and unprimed coordinates.

The general proper Lorentz transformation results from combining a number of transformations such as (93) and (94).[9] This may generally be written in the form

$$x'_\mu = a_{\mu\nu} x_\nu$$

and

$$x'^2_\mu = x_\nu^{\,2}, \tag{96a}$$

where $\mu$, $\nu$ run from 1 to 4, $x_4 = ict$ as before, and we use the usual *summation convention* whereby summation is taken over repeated dummy indices [such as $\nu$ in the first of (96*a*) and both $\mu$ and $\nu$ in the second]. We shall further adopt the convention that the components of vectors are labeled with Greek indices if they range from 1 to 4 and with Latin ones if the range is 1 to 3. The transformations discussed earlier, namely space rotation and ordinary Lorentz transformations, may be extended to those involving reflections of space alone, time alone, or both.

$$\begin{aligned} x'_i &= -x'_i, & x'_4 &= x_4, \\ x'_4 &= -x'_4, & x'_i &= x_i, \\ x'_\mu &= -x_\mu. \end{aligned} \tag{96b}$$

From Eq. (96*a*) we obtain

$$x'^2_\mu = a_{\mu\nu} a_{\mu\lambda} x_\nu x_\lambda = x_\lambda^{\,2},$$

which implies that

$$a_{\mu\nu} a_{\mu\lambda} = \delta_{\nu\lambda}. \tag{97a}$$

If we multiply the first of Eqs. (96*a*) by $a_{\mu\lambda}$ and sum over $\mu$, using (97*a*) we find

$$x_\lambda = a_{\mu\lambda} x'_\mu; \tag{98}$$

this is the inverse of Eq. (96*a*), so that regarding $a$ as a matrix we have

$$(a^{-1})_{\mu\nu} = a_{\nu\mu}. \tag{97b}$$

[9] Equations (93) and (94) are each one-parameter transformations, depending on $\theta$ and $w$ respectively. As will be seen from Eq. (99*b*) the general Lorentz transformation depends on six parameters.

It follows incidentally by forming the determinant of the matrix relation (97$a$), since $\det \tilde{A} = \det A$, that $|\det a|^2 = 1$, $\det a = \pm 1$; those transformations $a_{\mu\nu}$ for which $\det a = +1$ are called proper Lorentz transformations. Finally, on forming $x'^2_\mu = x_\nu{}^2$ from Eq. (98), we find

$$a_{\lambda\mu} a_{\nu\mu} = \delta_{\lambda\nu}. \tag{97c}$$

Of particular importance and simplicity are the infinitesimal Lorentz transformations which reduce continuously to the identity operation (and thus have $\det a = +1$):

$$a_{\mu\nu} = \delta_{\mu\nu} + \epsilon_{\mu\nu}, \tag{99a}$$

with the $\epsilon_{\mu\nu}$ being infinitesimal. From (97$a$) and (97$c$) we find

$$a_{\mu\nu} a_{\mu\lambda} = \delta_{\nu\lambda} = \delta_{\nu\lambda} + \epsilon_{\nu\lambda} + \epsilon_{\lambda\nu}, \tag{100}$$

and thus

$$\epsilon_{\lambda\nu} = -\epsilon_{\nu\lambda}. \tag{99b}$$

There are consequently only six independent parameters characterizing the proper Lorentz transformation. The transformation corresponding to (93) has, for small $\theta$, $\epsilon_{12} = \theta = -\epsilon_{21}$ (all others zero), whereas that of (94) has, for small $w$, $\epsilon_{34} = iw = -\epsilon_{43}$ (all others zero).

The energy and momentum of a particle form a four vector and transform the same way as the coordinates $x_\mu$ do; for example, under a simple Lorentz transformation such as Eq. (94) we have

$$\begin{aligned} p'_0 &= \gamma(-\beta p_3 + p_0), \\ p'_3 &= \gamma(p_3 - \beta p_0), \\ p'_2 &= p_2, \qquad p'_1 = p_1, \end{aligned} \tag{101}$$

where $p_0 = c(\mathbf{p}^2 + M^2c^2)^{1/2}$ and $M$ is the rest-mass of the particle.

We now undertake a discussion of the relativistic quantum theory for particles moving in given external force fields. This is necessarily a somewhat restricted topic since *in general* whenever relativistic effects are important particle creation is also important. A proper treatment can be given only (if at all) within the framework of quantum field theory.

The Schrödinger equation for a relativistic spinless particle moving in an external potential $V$ is taken to be

$$[c(M^2c^2 + p^2)^{1/2} + V]\psi = i\hbar \frac{\partial \psi}{\partial t}. \tag{102a}$$

We do not write the more familiar Klein-Gordon equation simply because the latter has no place in a one-particle theory. Actually the difference between the two equations (except for a profound difference in principle) is not so great under many circumstances. Consider Eq. (102*a*) for a stationary state and imagine iterating it as follows (write $\omega = c\sqrt{M^2c^2 + p^2}$):

$$(\omega - V)(\omega + V)\psi = E(\omega - V)\psi = E(E - 2V)\psi,$$
$$\{c^2p^2 + M^2c^4 + [\omega, V]\}\psi = (E - V)^2\psi. \tag{102b}$$

This is exactly the Klein-Gordon equation with scalar (fourth component of a four vector) potential except for the term involving the commutator $[\omega, V]$. Under the conditions that (102*a*) might be expected to be valid physically, that is, that a one-particle theory makes sense, $|V| \ll Mc^2$ and $V$ changes slowly in a deBroglie wavelength, then the solutions of our equation and of the Klein-Gordon equation are essentially the same. The stationary form of Eq. (102*a*) is what we encounter in a rather natural way starting with a quantized field theory of mesons interacting with an external field; it appears to be a useful and practical equation for many applications, such as the scattering of pi-mesons by nuclei.

An alternative description of spinless particles is obtained by using the *Kemmer* equation, but we shall not go into it here. This same theory can be extended to cover spin one particles.

For the calculation of particle scattering we must evaluate the density of particles and particle fluxes at large distances from the region in which the potential $V$ acts on the particles. It is easily seen from Eq. (102*a*) (for $V = 0$) that the respective density and current

$$n(\mathbf{x}) = \tfrac{1}{2}[(\omega^{1/2}\psi^*)(\omega^{-1/2}\psi) + (\omega^{-1/2}\psi^*)(\omega^{1/2}\psi)],$$
$$\mathbf{j}(\mathbf{x}) = \frac{c^2}{2i}[(\omega^{-1/2}\psi^*)(\omega^{-1/2}\nabla\psi) - (\omega^{-1/2}\nabla\psi^*)(\omega^{-1/2}\psi)] \tag{103a}$$

are conserved. For smooth wave packets corresponding to a rather sharply defined particle energy $\epsilon_k = c(M^2c^2 + k^2)^{1/2}$, Eqs. (103*a*) may be represented by the operators

$$n(\mathbf{x}) = \delta(\mathbf{x} - \mathbf{r}),$$
$$\mathbf{j}(\mathbf{x}) = \frac{c^2}{\epsilon_k}\,\mathrm{Re}\left[\frac{1}{i}\,\delta(\mathbf{x} - \mathbf{r})\nabla_r\right]. \tag{103b}$$

Here $\mathbf{r}$ is the position operator for a particle and Re $[\cdot\cdot\cdot]$ means that we must keep the real part of the expectation value of $[\cdot\cdot\cdot]$. We do not use the brackets when the extent of the operand is clear. The expressions

(103*b*) appear as natural generalizations of the corresponding expressions in nonrelativistic quantum mechanics;

$$n(\mathbf{x}) = \delta(\mathbf{x} - \mathbf{r}),$$
$$\mathbf{j}(\mathbf{x}) = \frac{1}{M}\,\mathrm{Re}\left[\frac{1}{i}\,\delta(\mathbf{x} - \mathbf{r})\nabla_r\right]. \tag{103c}$$

For a "smooth" wave packet moving freely with a velocity **v** these expressions may be put into a more convenient form;[10]

$$n(\mathbf{x}) = \delta(\mathbf{x} - \mathbf{r}),$$
$$\mathbf{j}(\mathbf{x}) = \mathbf{v}n(\mathbf{x}), \tag{103d}$$

valid for relativistic or nonrelativistic particles.

The dynamics of relativistic spin one-half particles is described by the Dirac equation. Furthermore, the rather subtle kinematical transformation properties of spin one-half systems are built directly into the equations, so that a more extensive discussion than that given for spin zero is warranted.

If the particle has a charge $e$ and interacts with the electromagnetic field, described by a vector potential $A_\mu = (\mathbf{A}, iA_0)$, the Dirac equation takes the form[11]

$$\left[c\boldsymbol{\alpha}\cdot\left(p - \frac{e}{c}\mathbf{A}\right) + \beta Mc^2 + eA_0\right]\psi = i\hbar\frac{\partial\psi}{\partial t}. \tag{104}$$

Here $\boldsymbol{\alpha}$ and $\beta$ are the familiar Hermitian Dirac matrices which satisfy the anticommutation relations [see Eq. (25*b*)]

$$\{\alpha_i, \alpha_j\} = 2\delta_{ij} \qquad (i, j = 1, 2, 3),$$
$$\{\alpha_i, \beta\} = 0, \tag{105a}$$
$$\alpha_1{}^2 = \alpha_2{}^2 = \alpha_3{}^2 = \beta^2 = 1.$$

An explicit matrix realization of the $\alpha_i$, $\beta$ cannot be effected by matrices of fewer than four rows and columns; consequently $\psi$ has four components.

[10] This will be described in Chapter 3.

[11] If we wish to describe a particle like a proton which does not have a normal magnetic moment $e\hbar/2Mc$ but instead has an additional anomalous moment $\mu_a\, e\hbar/2Mc$, Eq. (104) becomes

$$\left[c\boldsymbol{\alpha}\cdot\left(\boldsymbol{p} - \frac{e}{c}\boldsymbol{A}\right) + \beta Mc^2 + eA_0 - \mu_a\frac{e\hbar}{2Mc}\beta(\boldsymbol{\sigma}\cdot\boldsymbol{B} - i\boldsymbol{\alpha}\cdot\boldsymbol{E})\right]\psi = i\hbar\dot{\psi},$$

where **B** and **E** are the magnetic and electric fields.

Unless otherwise specified we shall assume the representation

$$\alpha_k = \begin{pmatrix} 0 & \sigma_k \\ \sigma_k & 0 \end{pmatrix} \qquad (k = 1, 2, 3),$$

$$\beta = \begin{pmatrix} 1 & 0 \\ 0 & -1 \end{pmatrix} \equiv \begin{pmatrix} 1 & 0 & 0 & 0 \\ 0 & 1 & 0 & 0 \\ 0 & 0 & -1 & 0 \\ 0 & 0 & 0 & -1 \end{pmatrix}. \tag{105b}$$

Here the $\sigma_k$ are the $2 \times 2$ Pauli matrices as given by Eq. (69). In the first expression for $\beta$ we have written 1 for the unit $2 \times 2$ matrix. It is sometimes convenient to introduce an auxiliary set of $2 \times 2$ matrices $\rho_1, \rho_2, \rho_3$ which are numerically identical to the Pauli matrices (69), in terms of which $\alpha_k = \rho_1 \times \sigma_k$ (indicating a direct product). Expand $\rho_3$ into a $4 \times 4$ matrix by forming the direct product of $\rho_3$ with the unit matrix and it then becomes identical with $\beta$.

In regions of space where no forces act on the particles, the respective particle and current densities are

$$\begin{aligned} n(\mathbf{x}) &= \psi^\dagger(\mathbf{x})\psi(\mathbf{x}), \\ \mathbf{j}(\mathbf{x}) &= c\psi^\dagger(\mathbf{x})\boldsymbol{\alpha}\psi(\mathbf{x}). \end{aligned} \tag{106}$$

The conservation of these quantities (that is, $\partial n/\partial t + \nabla \cdot \mathbf{j} = 0$) is easily verified using Eq. (104). The reader may also convince himself that these expressions are equivalent to Eqs. (103*d*) for "smooth" wave packets.

The four components of $\psi$ may be imagined to form a column matrix,

$$\psi = \begin{pmatrix} \psi_1 \\ \psi_2 \\ \psi_3 \\ \psi_4 \end{pmatrix},$$

and from these elements a row matrix may be formed:

$$\psi^\dagger = (\psi_1{}^* \; \psi_2{}^* \; \psi_3{}^* \; \psi_4{}^*). \tag{107}$$

Another row matrix we shall encounter is

$$\bar{\psi} = \psi^\dagger\beta. \tag{108}$$

[Readers who have an aversion to explicit matrix manipulations may feel free to regard $\psi^\dagger$ as a column matrix like $\psi$; after all, there is only one possible way to multiply $\psi^\dagger$ into $\beta$ as in (108): $\bar{\psi}_\sigma = \psi_{\sigma'}{}^*\beta_{\sigma'\sigma}$, which may be equally well written as $\bar{\psi}_\sigma = (\tilde{\beta})_{\sigma\sigma'}\psi_{\sigma'}{}^*$, and whether we form the elements of $\psi$ into a row or column is a matter of choice.]

The Dirac equation may be written more compactly and in a more covariant manner by introducing the notation

$$\begin{aligned} \gamma_k &= i\alpha_k\beta \qquad (k = 1, 2, 3), \\ \gamma_4 &= \beta. \\ \{\gamma_\mu, \gamma_\nu\} &= 2\delta_{\mu\nu}, \qquad \gamma_\mu{}^\dagger = \gamma_\mu. \end{aligned} \tag{109}$$

(It is easily seen that $\gamma_k = \rho_2 \times \sigma_k$.) We obtain

$$\begin{aligned} \left[\gamma_\mu\left(\frac{\partial}{\partial x_\mu} - \frac{ie}{\hbar c} A_\mu\right) + \kappa\right]\psi &= 0, \\ \left[\tilde{\gamma}_\mu\left(\frac{\partial}{\partial x_\mu} + \frac{ie}{\hbar c} A_\mu\right) - \kappa\right]\tilde{\bar{\psi}} &= 0, \end{aligned} \tag{110a}$$

where $\kappa = Mc/\hbar$.

The invariance of these equations (for simplicity, set $A_\mu = 0$) under the Lorentz transformation (96$a$)

$$x_\mu \rightarrow x'_\mu = a_{\mu\nu}x_\nu$$

may be seen as follows: We would like to have, in the primed system, the equations

$$\begin{aligned} \left(\gamma_\mu \frac{\partial}{\partial x'_\mu} + \kappa\right)\psi' &= 0, \\ \left(-\tilde{\gamma}_\mu \frac{\partial}{\partial x'_\mu} + \kappa\right)\tilde{\bar{\psi}}' &= 0. \end{aligned} \tag{110b}$$

Under our transformation $\psi(x) \rightarrow \psi'(x')$ we may expect that since $\psi$ is not a scalar quantity (that is, it has indices) the components become scrambled, so that

$$\psi'(x') = S\psi(x), \tag{111}$$

where $S$ is a $4 \times 4$ matrix.[12] Remarking that

$$\partial/\partial x'_\mu = (\partial x_\nu/\partial x'_\mu)\, \partial/\partial x_\nu = a_{\mu\nu}\, (\partial/\partial x_\nu)$$

[12] Equation (111) is not in disagreement with Eq. (2.17) in the next chapter, since the Dirac $\psi$ has spin and positive-negative energy degrees of freedom, which are implied here rather than written explicitly.

[from (98)], we see that the first equation of (110*b*) becomes identical with the first of (110) (with $A_\mu = 0$) if there exists a matrix $S$ such that

$$S^{-1}\gamma_\mu a_{\mu\nu} S = \gamma_\nu$$

or

$$S^{-1}\gamma_\mu S = a_{\mu\nu}\gamma_\nu \tag{112a}$$

[using (97*c*)]. The second of Eqs. (110*b*) is consistent if

$$\bar{\psi}' = (\widetilde{\tilde{S}^{-1}})\tilde{\bar{\psi}} = \bar{\psi}S^{-1}. \tag{112b}$$

We observe that the relation $\bar{\psi}' = \psi'^\dagger\gamma_4$ imposes a relation between $S$ and $S^{-1}$:

$$\begin{aligned}\bar{\psi}' &= \psi'^\dagger\gamma_4 = (S\psi)^\dagger\gamma_4 = \psi^\dagger S^\dagger\gamma_4 \\ &= \bar{\psi}\gamma_4 S^\dagger\gamma_4 \equiv \bar{\psi}S^{-1}.\end{aligned}$$

Therefore $S^{-1}$ is not quite a unitary matrix; instead

$$S^{-1} = \gamma_4 S^\dagger \gamma_4. \tag{113}$$

A fifth matrix may be formed from the $\gamma_\mu$'s, namely $\gamma_5 = \gamma_1\gamma_2\gamma_3\gamma_4$. With our Hermitian $\gamma_\mu$'s, $\gamma_5$ is also Hermitian ($\gamma_5^\dagger = \gamma_5$) and $\{\gamma_5, \gamma_\mu\} = 0$, $\gamma_5^2 = 1$. In terms of the alternating tensor $\epsilon_{\mu\nu\lambda\sigma}$, which is $+1$ $(-1)$ if $\mu\nu\lambda\sigma$ is an even (odd) permutation of 1234, we may write

$$\gamma_5\epsilon_{\mu\nu\lambda\sigma} = \tfrac{1}{8}[\gamma_\mu, \{\gamma_\nu, [\gamma_\lambda, \gamma_\sigma]\}].$$

From these five matrices and the unit matrix we may form sixteen linearly independent matrices (corresponding to the sixteen elements of an arbitrary $4 \times 4$ matrix). They are $1(1)$, $\gamma_\mu(4)$, $(\gamma_\mu\gamma_\nu - \gamma_\nu\gamma_\mu)/2i(6)$, $\gamma_\mu\gamma_5(4)$, $\gamma_5(1)$, where the numbers in parentheses count the members of each group. Using Eqs. (111) and (112) we easily establish the following transformation properties:

$$\begin{aligned}
\bar{\psi}'\psi' &= \bar{\psi}\psi \qquad \text{(scalar)},\\
\bar{\psi}'\gamma_\mu\psi' &= a_{\mu\nu}\bar{\psi}\gamma_\nu\psi \qquad \text{(vector)},\\
\bar{\psi}'\frac{\gamma_\mu\gamma_\nu - \gamma_\nu\gamma_\mu}{2i}\psi' &= a_{\mu\lambda}a_{\nu\sigma}\bar{\psi}\frac{\gamma_\lambda\gamma_\sigma - \gamma_\sigma\gamma_\lambda}{2i}\psi \qquad \text{(tensor)},\\
\bar{\psi}'\gamma_\mu\gamma_5\psi' &= (\det a)a_{\mu\nu}\bar{\psi}\gamma_\nu\gamma_5\psi \qquad \text{(axial vector)},\\
\bar{\psi}'\gamma_5\psi' &= (\det a)\bar{\psi}\gamma_5\psi \qquad \text{(pseudoscalar)},
\end{aligned} \tag{114}$$

where the names describe the tensorial nature of the quantities formed.

The matrix $S$ corresponding to the infinitesimal transformation Eq. (99*a*) is given explicitly by

$$\begin{aligned}S &= 1 + \epsilon_{\lambda\nu}\gamma_\lambda\gamma_\nu,\\ S^{-1} &= 1 - \epsilon_{\lambda\nu}\gamma_\lambda\gamma_\nu.\end{aligned} \tag{115}$$

Since any proper Lorentz transformation can be generated from the infinitesimal transformations, the matter of the general existence of an $S$ is explicitly demonstrated by Eq. (115).

It is well known that the Dirac equation always has two sets of solutions which may be interpreted as referring to particle and antiparticle and which differ in the sign of their charge (and anomalous magnetic moment, if any). These two types of solutions are interrelated by a unitary transformation called *charge conjugation*. In particular, the charge conjugate wave function $\psi^c$ is defined by

$$\begin{aligned} \psi^c(x) &= C\tilde{\bar{\psi}}(x), \\ \tilde{\bar{\psi}}^c(x) &= C^{-1}\psi(x), \end{aligned} \tag{116}$$

(it is convenient here to regard the $\psi$'s as column symbols) where $C$ satisfies the relations

$$\begin{aligned} C\tilde{\gamma}_\mu C^{-1} &= -\gamma_\mu, \\ C\tilde{\alpha}_k C^{-1} &= -\alpha_k, \\ C\tilde{\beta} C^{-1} &= -\beta. \end{aligned} \tag{117}$$

Let us display explicitly that if $C$ has the properties expressed by (117) the particle-antiparticle interpretation is consistent. Multiply the second of Eqs. (110) by $C$ and use the first of Eqs. (117) and find

$$\left[\gamma_\mu\left(\frac{\partial}{\partial x_\mu} + \frac{ie}{\hbar c} A_\mu\right) + \kappa\right]\psi^c = 0,$$

which is of exactly the same form as the first Eq. (110*a*) if we *change the sign of the charge*. Similarly, the first of Eqs. (110*a*) may be transformed into the second with $\psi \rightarrow \psi^c$ and again with the opposite sign of the charge.

We next show that the unitary matrix $C$ is skew symmetric;

$$C = -\tilde{C}. \tag{118}$$

To see this we note from Eq. (117)

$$\begin{aligned} -\gamma_\mu &= C(\widetilde{-C\tilde{\gamma}_\mu C^{-1}})C^{-1} \\ &= -C\tilde{C}^{-1}\gamma_\mu \tilde{C} C^{-1}, \end{aligned}$$

or
$$[\tilde{C}C^{-1}, \gamma_\mu] = 0.$$

Thus $\tilde{C}C^{-1}$ can differ from the unit matrix only by a numerical factor, which as we now show is independent of any choice of representation. If we change the $\gamma_\mu$ by a unitary transformation $T$,

$$\gamma_\mu \rightarrow \gamma'_\mu = T^{-1}\gamma_\mu T,$$

and insist on the existence of a matrix $C'$ with all the properties attributed to $C$, we have

$$-\tilde{\gamma}'_\mu = C'^{-1}\gamma'_\mu C' = C'^{-1}T^{-1}\gamma_\mu TC' = -\tilde{T}\tilde{\gamma}_\mu\tilde{T}^{-1}$$
$$= \tilde{T}C^{-1}\gamma_\mu C\tilde{T}^{-1}.$$

Therefore

$$C = TC'\tilde{T}$$

(a so-called similarity transformation). Then

$$\tilde{C}C^{-1} = T\tilde{C}'\tilde{T}\tilde{T}^{-1}C'^{-1}T^{-1} = T\tilde{C}'C'^{-1}T^{-1}.$$

Hence if $\tilde{C}C^{-1} = \lambda I$, $\tilde{C}'C'^{-1} = \lambda I$ for the *same* $\lambda$. In a particular representation, $\tilde{C} = -C$ [see Eq. (119)], so this is generally true.

For our standard representation we note $\tilde{\alpha}_1 = \alpha_1$, $\tilde{\alpha}_2 = -\alpha_2$, $\tilde{\alpha}_3 = \alpha_3$, $\tilde{\beta} = \tilde{\gamma}_4 = \beta$; $C$ must anticommute with $\alpha_1$, $\alpha_3$, $\beta$ and commute with $\alpha_2$, which leads instantly to

$$C = \alpha_2 \qquad \text{(standard representation)} \tag{119}$$

A representation sometimes used is the *Majorana* representation in which $\gamma_4^* = -\gamma_4$, $\gamma_k^* = \gamma_k$ ($k = 1, 2, 3$), and here evidently

$$C = C^{-1} = \gamma_4. \tag{120}$$

**REFERENCES**


1. For a very clear description of the mathematical methods of Section 1.1 see, for example, B. Friedman, *Principles and Techniques of Applied Mathematics*, John Wiley and Sons, New York, 1956.
2. For a general review of quantum mechanics, see L. I. Schiff, *Quantum Mechanics*, McGraw-Hill Book Co., New York, 1949; P. A. M. Dirac, *The Principles of Quantum Mechanics*, The Clarendon Press, Oxford, 1947.
3. For the foundations of quantum theory, see J. von Neumann, *Mathematical Foundations of Quantum Mechanics*, Princeton University Press, Princeton, New Jersey, 1955.
4. E. P. Wigner, *Group Theory and its Applications to the Quantum Mechanics of Atomic Spectra*, Academic Press, New York and London, 1959.
5. J. M. Blatt and V. F. Weisskopf, *Theoretical Nuclear Physics*, John Wiley and Sons, New York, 1952.

# 2. SYMMETRY OPERATIONS ON THE SCHRÖDINGER EQUATION

In this chapter we shall expand the discussion of Section 1.2 to consider the effect of symmetry operations on the Schrödinger equation. Such considerations play a very important role in quantum mechanics. They serve both as a guide to the possible forms for equations of motion that describe particles with given transformation properties and as an aid to the solution of specific quantum mechanical problems.

## 2.1 COORDINATE TRANSLATIONS

In order to introduce the ideas as simply as possible, we begin by investigating the behavior of an isolated system under translations, which was mentioned in Section 1.2. Suppose the system is described by a state vector $\Psi$; in the coordinate representation the wave function $\Psi(\mathbf{r})$ is given by $\langle \mathbf{r} \mid \Psi \rangle$. Here $\mathbf{r}$ may represent the set of position coordinates of $N$ particles. If the entire *physical system* is bodily transported a distance $\boldsymbol{\lambda}$, the new situation will be described by a state vector $\Psi'$, with the wave function $\Psi'(\mathbf{r})$ being given now by $\langle \mathbf{r} \mid \Psi' \rangle$. Note that we are moving the system, not the coordinate frame which is used to label the wave functions. This is the so-called *active* point of view, as contrasted to the *passive* one, in which the coordinate system itself is changed and the system left alone. Clearly a displacement of the system by an amount $\boldsymbol{\lambda}$ is equivalent to leaving the system alone and moving the coordinate origin by $-\boldsymbol{\lambda}$.

We shall assume that there exists a unitary operator $D(\boldsymbol{\lambda})$ which relates the two state vectors $\Psi$ and $\Psi'$, and we write

$$\Psi \to \Psi' = D(\boldsymbol{\lambda})\Psi. \tag{1}$$

We have then

$$\Psi'(\mathbf{r}) = \langle \mathbf{r} \mid \Psi' \rangle = \langle \mathbf{r} \mid D(\boldsymbol{\lambda})\Psi \rangle = \int d^3r' \langle \mathbf{r} \mid D(\boldsymbol{\lambda}) \mid \mathbf{r}' \rangle \Psi(\mathbf{r}'), \tag{2}$$

where $d^3r' = d^3r'_1 \cdots d^3r'_N$, if there are $N$ particles in the system. But what is $\Psi'(\mathbf{r})$, after all? It can only be $\Psi$ evaluated at the point $\mathbf{r} - \boldsymbol{\lambda}$.

To see this, imagine that $\Psi$ describes a state localized at a point $\mathbf{b}$; then $\langle \mathbf{r} \mid \Psi \rangle = \delta(\mathbf{r} - \mathbf{b}) = \Psi(\mathbf{r})$. The translated system is one localized at $\mathbf{b} + \boldsymbol{\lambda}$; thus $\langle \mathbf{r} \mid \Psi' \rangle = \delta(\mathbf{r} - \boldsymbol{\lambda} - \mathbf{b}) \equiv \Psi'(\mathbf{r})$. We see, then, that in general

$$\begin{aligned}\Psi'(\mathbf{r}) &= \Psi(\mathbf{r} - \boldsymbol{\lambda})\\ &= e^{-\boldsymbol{\lambda}\cdot\nabla}\Psi(\mathbf{r}) = \exp\left(-\frac{i}{\hbar}\boldsymbol{\lambda}\cdot\frac{\hbar}{i}\nabla\right)\Psi(\mathbf{r}). \end{aligned} \tag{3}$$

But we know quite generally that if $\mathscr{P}$ is the total momentum operator of the system,

$$\langle \mathbf{r} \mid f(\mathscr{P})\Psi \rangle = f(-i\hbar\nabla)\langle \mathbf{r} \mid \Psi \rangle = f(-i\hbar\nabla)\Psi(\mathbf{r}), \tag{4}$$

so we conclude that

$$D(\boldsymbol{\lambda}) = e^{-(i/\hbar)\boldsymbol{\lambda}\cdot\mathscr{P}}. \tag{5}$$

As an example of the use of $D(\boldsymbol{\lambda})$, suppose we wish to extract from an arbitrary state $\Phi$ that portion for which the momentum operator has a particular numerical value, $\mathbf{P}'$, say. Then we translate the state by an amount $\boldsymbol{\lambda}$ and form

$$\int \frac{d^3\lambda}{(2\pi\hbar)^3}\, e^{(i/\hbar)\mathbf{P}'\cdot\boldsymbol{\lambda}}\, D(\boldsymbol{\lambda})\Phi, \tag{6}$$

which is easily seen to yield the desired quantity.

As a trivial illustration of the operation by $D(\boldsymbol{\lambda})$ on a state vector, consider the plane wave of momentum $\mathbf{p}$, $\Psi(\mathbf{r}) = e^{i\mathbf{p}\cdot\mathbf{r}/\hbar}$. Then the expression (5) leads to

$$D(\boldsymbol{\lambda})\Psi(\mathbf{r}) = e^{i\mathbf{p}\cdot(\mathbf{r}-\boldsymbol{\lambda})/\hbar}, \tag{7}$$

which is obviously equivalent to having displaced the coordinate system by a distance $-\boldsymbol{\lambda}$. It is also easy to show that $D(\boldsymbol{\lambda})$ satisfies the group property. Let $\boldsymbol{\lambda}_1$ and $\boldsymbol{\lambda}_2$ be two successive displacements. Then, as we see from Eq. (5),

$$D(\boldsymbol{\lambda}_2)D(\boldsymbol{\lambda}_1) = D(\boldsymbol{\lambda}_1 + \boldsymbol{\lambda}_2). \tag{8}$$

If we had considered a translation in time and dealt only with state vectors and wave functions that obeyed the Schrödinger equation, we would have been led to an operator $D_t(\tau)$, where

$$D_t(\tau) = e^{-\tau(\partial/\partial t)} = \exp\left(\frac{i}{\hbar}\tau i\hbar\frac{\partial}{\partial t}\right) = e^{(i/\hbar)\tau H} \tag{9}$$

and $H$ is the Hamiltonian operator. Thus the operator that generates a space-time translation by an amount $\lambda_\mu$ is

$$D(\lambda_\mu) = e^{-(i/\hbar)\lambda_\mu P_\mu}, \tag{10}$$

where $P_\mu$ is the total energy momentum operator $P_\mu = (\mathscr{P}, iH/c)$ and $\lambda_4 = i\tau$.

We have been speaking as if the operators themselves are not changing, but the state vectors on which they act are being transformed. It is, of course, possible to take the other point of view, but the physical content of the theory must remain unchanged. Clearly we must demand that for the coordinate operator $\mathbf{q}$, for example, which has as eigenvalues the particle coordinates $\mathbf{r}$,

$$(\Psi', \mathbf{q}\Psi') = (\Psi, \mathbf{q}^T\Psi) \tag{11}$$

where the superscript $T$ means the transformed operator. Then for space translations, substituting $\Psi' = D\Psi$, we have

$$\begin{aligned} D^{-1}(\boldsymbol{\lambda})\mathbf{q}D(\boldsymbol{\lambda}) = \mathbf{q}^T &= e^{(i/\hbar)\mathscr{P}\cdot\boldsymbol{\lambda}}\mathbf{q}e^{-(\lambda i/\hbar)\mathscr{P}\cdot\boldsymbol{\lambda}} \\ &= \mathbf{q} + \boldsymbol{\lambda}, \end{aligned} \tag{12}$$

which is just what we expect. Changing the operators is sometimes referred to as a Heisenberg transformation, contrasted with the Schrödinger type in which the states are transformed.

Suppose now we have a relation between state vectors of the form

$$\Phi = A\Psi, \tag{13}$$

and our translated state vectors are, as before, given by $\Psi' = D(\boldsymbol{\lambda})\Psi$, and so on. Then

$$\Phi' = D(\boldsymbol{\lambda})\Phi = D(\boldsymbol{\lambda})A\Psi = D(\boldsymbol{\lambda})AD^{-1}(\boldsymbol{\lambda})D(\boldsymbol{\lambda})\Psi = A'\Psi', \tag{14}$$

where

$$A' \equiv D(\boldsymbol{\lambda})AD^{-1}(\boldsymbol{\lambda}). \tag{15}$$

Under circumstances where $A' = A$ we say that the operator $A$ is invariant under the operation $D(\boldsymbol{\lambda})$. That is, $AD(\boldsymbol{\lambda}) = D(\boldsymbol{\lambda})A$, so that the translation through $\boldsymbol{\lambda}$ and the operation by $A$ on a state may be carried out in any order. Note, incidentally, that if $A$ is the coordinate operator $\mathbf{q}$, $\mathbf{q}' = \mathbf{q} - \boldsymbol{\lambda}$; this is reasonable, since if we "push up" the state by $\boldsymbol{\lambda}$, we must "pull back" the operators by $-\boldsymbol{\lambda}$ to have the same net effect. It is quite reasonable to regard $A'$ in Eq. (15) as a function of $\boldsymbol{\lambda}$ and note the relation

$$i\hbar\nabla_\lambda A'(\boldsymbol{\lambda}) = [\mathscr{P}, A'(\boldsymbol{\lambda})], \tag{16a}$$

or for a four-dimensional displacement

$$i\hbar\frac{\partial}{\partial\lambda_\mu}A'(\lambda_\nu) = [P_\mu, A'(\lambda_\nu)]. \tag{16b}$$

Let us summarize what we have done in a slightly different manner. Under the change of coordinates $\mathbf{r} \to \mathbf{r}'$, we have $\psi(\mathbf{r}) \to \psi'(\mathbf{r}')$ such that

$$\psi'(\mathbf{r}') \equiv \psi[\mathbf{r}(\mathbf{r}')], \tag{17}$$

where, of course, $\mathbf{r}$ on the right must be expressed in terms of $\mathbf{r}'$. For the example described above, $\mathbf{r} \to \mathbf{r}' = \mathbf{r} + \boldsymbol{\lambda}$, and we have, just as before,

$$\psi'(\mathbf{r} + \boldsymbol{\lambda}) = e^{(i/\hbar)\boldsymbol{\lambda}\cdot\mathscr{P}}\psi'(\mathbf{r}) = \psi(\mathbf{r}) \tag{18a}$$

or

$$\psi'(\mathbf{r}') = \psi(\mathbf{r}' - \boldsymbol{\lambda}) = e^{-(i/\hbar)\boldsymbol{\lambda}\cdot\mathscr{P}}\psi(\mathbf{r}'). \tag{18b}$$

These forms are equivalent, since what is at stake is a comparison of *functional forms* of the wave function before and after a transformation. In Eq. (18*b*) $\psi$ is a function of some variable, call it $\mathbf{y}$, and $\psi'$ is a function of a variable called $\mathbf{z}$; we are invited to equate these two functions when $\mathbf{y}$ has the *numerical value* $\mathbf{r}' - \boldsymbol{\lambda}$ while $\mathbf{z}$ has the *numerical value* $\mathbf{r}'$. This is an elementary point, but it is at the root of much confusion in the subject of transformations.

Now we want quite generally to relate $\psi$ and $\psi'$ by an operator, call it $U$, such that

$$U\psi(\mathbf{r}) = \psi'(\mathbf{r}). \tag{19}$$

Again we emphasize what this means. On the right of (19) the new function $\psi'$ is to be evaluated when its *argument* has the *numerical value* $\mathbf{r}$. Since $\psi'$ is defined by $\psi'(\mathbf{r}') \equiv \psi(\mathbf{r})$, if we write $\mathbf{r} \to \mathbf{r}' = T\mathbf{r}$, we have $\psi(\mathbf{r}) = \psi'(T\mathbf{r})$ or $\psi'(\mathbf{r}) = \psi(T^{-1}\mathbf{r})$. Thus we have, for the determination of $U$, the relation

$$U(T)\psi(\mathbf{r}) = \psi(T^{-1}\mathbf{r}). \tag{20}$$

Consider now the effect of two successive transformations $\mathbf{r} \to \mathbf{r}_1 = T_1\mathbf{r}$, $\mathbf{r}_1 \to \mathbf{r}_2 = T_2\mathbf{r}_1$ or $\mathbf{r} \to T_2T_1\mathbf{r}$. Then according to (20), we should have

$$U_{21}\psi(\mathbf{r}) = \psi[(T_2T_1)^{-1}\mathbf{r}], \tag{21a}$$

where $U_{21} = U(T_2)U(T_1)$. To follow this in detail, we have

$$\begin{aligned} U(T_1)\psi(\mathbf{r}) &= \psi(T_1^{-1}\mathbf{r}) \equiv \chi(\mathbf{r}), \\ U(T_2)\chi(\mathbf{r}) &= \chi(T_2^{-1}\mathbf{r}). \end{aligned} \tag{21b}$$

Now the instruction given in the second of Eqs. (21*b*) is to replace $\mathbf{r}$ wherever it may occur in $\chi(\mathbf{r})$ by $T_2^{-1}\mathbf{r}$; therefore

$$\begin{aligned} U(T_2)U(T_1)\psi(\mathbf{r}) &= \psi(T_1^{-1}T_2^{-1}\mathbf{r}) \\ &= \psi[(T_2T_1)^{-1}\mathbf{r}]. \end{aligned} \tag{21c}$$

We have defined the operations here so that with any matrix realization of the operators the normal rules of matrix multiplication hold.

## 2.2 COORDINATE ROTATIONS

We turn next to a description of rotations. Let us imagine rotating the entire system by an infinitesimal amount $\epsilon$ in the positive (that is, right-handed) sense about a unit vector $\hat{\mathbf{n}}$. (Assume for the time being that we are dealing with a system of particles without spin.) Under this rotation, every point $\mathbf{r}$ in the system is moved to $\mathbf{r}' = T\mathbf{r}$, where

$$\begin{aligned} \mathbf{r} \to \mathbf{r}' &= T\mathbf{r} = \mathbf{r} + \epsilon\hat{\mathbf{n}} \times \mathbf{r}, \\ T^{-1}\mathbf{r} &= \mathbf{r} - \epsilon\hat{\mathbf{n}} \times \mathbf{r}. \end{aligned} \tag{22}$$

Let us call the infinitesimal operator that generates this transformation $\mathscr{R}$ and write

$$\mathscr{R} = 1 + \epsilon\hat{\mathbf{n}} \cdot \mathbf{R}. \tag{23}$$

From the fact that $\mathscr{R}$ must be a unitary operator (since the physical content of the systems before and after rotation must be the same), we find immediately that $\mathbf{R}$ must be skew Hermitian: If $\psi \to \psi' = \mathscr{R}\psi$,

$$(\psi', \psi') = (\mathscr{R}\psi, \mathscr{R}\psi) = (\psi, \psi),$$

then

$$\epsilon(\hat{\mathbf{n}} \cdot \mathbf{R}\psi, \psi) + \epsilon(\psi, \hat{\mathbf{n}} \cdot \mathbf{R}\psi) = \epsilon\hat{\mathbf{n}} \cdot (\psi, (\mathbf{R}^\dagger + \mathbf{R})\psi) = 0, \tag{24}$$

so

$$\mathbf{R} = -\mathbf{R}^\dagger.$$

An explicit formula for $\mathscr{R}$ may be derived by considering the relation

$$\begin{aligned} \mathscr{R}\psi(\mathbf{r}) &= \psi'(\mathbf{r}) = \psi(\mathbf{r} - \epsilon\hat{\mathbf{n}} \times \mathbf{r}), \\ (1 + \epsilon\hat{\mathbf{n}} \cdot \mathbf{R})\psi(\mathbf{r}) &= \psi(\mathbf{r}) - \epsilon\hat{\mathbf{n}} \times \mathbf{r} \cdot \nabla\psi(\mathbf{r}) \\ &= \psi(\mathbf{r}) - \epsilon\hat{\mathbf{n}} \cdot \mathbf{r} \times \nabla\psi(\mathbf{r}), \end{aligned} \tag{25}$$

where we retain only first-order terms. We have, then, since the direction of $\hat{\mathbf{n}}$ is arbitrary,

$$\mathbf{R}\psi(\mathbf{r}) = -\mathbf{r} \times \nabla\psi(\mathbf{r}) = -\frac{i}{\hbar}\mathbf{L}\psi(\mathbf{r}), \tag{26}$$

where $\mathbf{L}$ is the orbital angular momentum operator; hence

$$\mathbf{R} = -\frac{i}{\hbar}\mathbf{L}. \tag{27}$$

Using the commutation relations for $\mathbf{L}$ we may show that $[R_x, R_y] = R_z$ (cyclic).

It is instructive to rederive these results in another way: To find the commutation relations that $\mathbf{R}$ satisfies, we invite ourselves to consider two successive rotations about the $x$ and $y$-axes in opposite orders. In general, for an arbitrary rotation, we have, using the first line of Eq. (25) and Eq. (21$c$),

$$\begin{aligned}\mathscr{R}(\epsilon, \hat{\mathbf{n}}_y)\mathscr{R}(\epsilon, \hat{\mathbf{n}}_x)\psi(\mathbf{r}) &= \psi[\mathbf{r} - \epsilon\hat{\mathbf{n}}_y \times \mathbf{r} - \epsilon\hat{\mathbf{n}}_x \times (\mathbf{r} - \epsilon\hat{\mathbf{n}}_y \times \mathbf{r})],\\ \mathscr{R}(\epsilon, \hat{\mathbf{n}}_x)\mathscr{R}(\epsilon, \hat{\mathbf{n}}_y)\psi(\mathbf{r}) &= \psi[\mathbf{r} - \epsilon\hat{\mathbf{n}}_x \times \mathbf{r} - \epsilon\hat{\mathbf{n}}_y \times (\mathbf{r} - \epsilon\hat{\mathbf{n}}_x \times \mathbf{r})].\end{aligned} \tag{28}$$

Now forming the difference between these quantities, we have

$$\begin{aligned}[\mathscr{R}(\epsilon, \hat{\mathbf{n}}_x), \mathscr{R}(\epsilon, \hat{\mathbf{n}}_y)]\psi(\mathbf{r}) &= \epsilon^2[R_x, R_y]\psi(\mathbf{r})\\ &= \psi[\mathbf{r} - \epsilon(\hat{\mathbf{n}}_x + \hat{\mathbf{n}}_y) \times \mathbf{r} + \epsilon^2\hat{\mathbf{n}}_y \times (\hat{\mathbf{n}}_x \times \mathbf{r})]\\ &\quad -\psi[\mathbf{r} - \epsilon(\hat{\mathbf{n}}_y + \hat{\mathbf{n}}_x) \times \mathbf{r} + \epsilon^2\hat{\mathbf{n}}_x \times (\hat{\mathbf{n}}_y \times \mathbf{r})]\\ &= \epsilon^2[\hat{\mathbf{n}}_y \times (\hat{\mathbf{n}}_x \times \mathbf{r}) - \hat{\mathbf{n}}_x \times (\hat{\mathbf{n}}_y \times \mathbf{r})] \cdot \nabla\psi(\mathbf{r})\\ &= \epsilon^2[\hat{\mathbf{n}}_x\hat{\mathbf{n}}_y \cdot \mathbf{r} - \hat{\mathbf{n}}_y\hat{\mathbf{n}}_x \cdot \mathbf{r}] \cdot \nabla\psi(\mathbf{r})\\ &= -\epsilon^2[(\hat{\mathbf{n}}_x \times \hat{\mathbf{n}}_y) \times \mathbf{r}] \cdot \nabla\psi(\mathbf{r})\\ &= -\epsilon^2\hat{\mathbf{n}}_z \times \mathbf{r} \cdot \nabla\psi(\mathbf{r}) = -\epsilon^2\hat{\mathbf{n}}_z \cdot \mathbf{r} \times \nabla\psi(\mathbf{r})\\ &= \epsilon^2 R_z\psi(\mathbf{r}).\end{aligned} \tag{29}$$

[Note that the second-order terms arising from $\epsilon(\hat{\mathbf{n}}_x + \hat{\mathbf{n}}_y) \times \mathbf{r}$ cancel in the commutator.] Thus we conclude that

$$[R_x, R_y] = R_z. \tag{30}$$

Obviously the equations obtained by cyclic permutations of $x$, $y$, $z$ are also valid. These are summarized in the vector equation

$$\mathbf{R} \times \mathbf{R} = \mathbf{R}. \tag{31}$$

If we write the skew Hermitian operator $\mathbf{R}$ as $\mathbf{R} = (i\hbar)^{-1}\mathbf{L}$, where $\mathbf{L}$ is Hermitian, we find

$$\mathbf{L} \times \mathbf{L} = i\hbar\mathbf{L}, \tag{32}$$

which we recognize as the commutation relation for the angular momentum operator.

The generalization of these considerations to more complicated systems involving particles with spin is suggested by a consideration of the behavior of a vector field under rotations. The complication here is, of course, that the components of the vector are also transformed by the rotation. If we perform a rotation on the system such that $\mathbf{r} \rightarrow \mathbf{r}' = T\mathbf{r}$, (by which we mean $r'_i = T_{ik}r_k$) we have for a vector field $\mathbf{A}(\mathbf{r})$.

$$A'_i(\mathbf{r}') = T_{ik}A_k(\mathbf{r}), \tag{33}$$

where we use the notation $A_x = A_1$, and so on. In this equation we imply, of course, that $\mathbf{r}$ is to be expressed as $T^{-1}\mathbf{r}'$ on the right-hand side of (33). Dropping the prime on the coordinate we seek again an operator $\mathscr{R}$ such that

$$A_i(\mathbf{r}) \to A'_i(\mathbf{r}) = \mathscr{R}A_i(\mathbf{r}) = T_{ik}A_k(T^{-1}\mathbf{r}). \tag{34}$$

Now for an infinitesimal rotation through an angle $\epsilon$ about a unit vector $\hat{\mathbf{n}}$ we have

$$T_{ik}A_k = (\delta_{ik} + \epsilon\hat{\mathbf{n}}_j\epsilon_{ijk})A_k, \tag{35}$$

where $\epsilon_{ijk} = \pm 1$ depending on whether $ijk$ is an even or odd permutation of 123. The complete transformation, Eq. (34), may now be written as

$$\mathscr{R}A_i(\mathbf{r}) = (\delta_{ik} + \epsilon\hat{\mathbf{n}}_j\epsilon_{ijk})(1 - \epsilon\hat{\mathbf{n}} \cdot \mathbf{r} \times \nabla)A_k(\mathbf{r}). \tag{36}$$

We introduce a set of three matrices $(S_j)_{ik}$ $(j = 1, 2, 3)$, which act on the *components* of the vectors, by the definition

$$(S_j)_{ik} = i\hbar\epsilon_{ijk}, \tag{37}$$

and (recalling that $\mathbf{L} = -i\hbar\mathbf{r} \times \nabla$) rewrite Eq. (36) as

$$\mathscr{R}\mathbf{A}(\mathbf{r}) = \left(1 - \frac{i}{\hbar}\epsilon\hat{\mathbf{n}} \cdot \mathbf{J}\right)\mathbf{A}(\mathbf{r}), \tag{38}$$

where

$$\mathbf{J} = \mathbf{L} + \mathbf{S} \tag{39}$$

and 1 stands for the unit dyadic with matrix elements $\delta_{ik}$.

It is easy to show that $\mathbf{S}$ satisfies the commutation relation

$$\mathbf{S} \times \mathbf{S} = i\hbar\mathbf{S}, \tag{40}$$

and consequently the quantity $\mathbf{J}$ also satisfies

$$\mathbf{J} \times \mathbf{J} = i\hbar\mathbf{J}. \tag{41}$$

We then find that $(S^2)_{ik} = 2\hbar^2\delta_{ik}$, corresponding to the "spin" angular momentum of unity associated with a vector field. We therefore identify $\mathbf{J}$ with the total angular momentum operator, which evidently generates rotations in both spin and ordinary space. The operator that generates finite rotations of the system by an angle $\theta$ about $\hat{\mathbf{n}}$ is evidently

$$\mathscr{R}(\theta, \hat{\mathbf{n}}) = e^{-(i/\hbar)\theta\,\hat{\mathbf{n}}\cdot\mathbf{J}}. \tag{42}$$

As we have mentioned earlier in connection with translations, for some symmetry operation such that $\mathbf{r} \to \mathbf{r}' = T\mathbf{r}$ there will be associated an operator $U(T)$ on the state vectors. For the transformed operator $\mathscr{O}'$, corresponding to some operator $\mathscr{O}$, we have

$$\mathscr{O}' = U(T)\mathscr{O}U^{-1}(T). \tag{43}$$

A situation of particular importance is that for which $\mathcal{O}' = \mathcal{O}$, in which case we say that $\mathcal{O}$ is invariant under the transformation. Suppose $\mathcal{O}$ is the Hamiltonian of the system and has such an invariance property. Then from the Schrödinger equation we deduce

$$U\left(i\hbar \frac{\partial \Psi}{\partial t}\right) = UH\Psi = HU\Psi = i\hbar \frac{\partial}{\partial t} U\Psi, \tag{44}$$

if $U$ does not depend explicitly on time. Thus $U\Psi$ is also a solution of the Schrödinger equation. In a similar fashion, for the stationary states of $H$ (call them $\Psi_\lambda$ corresponding to energy $E_\lambda$) $U\Psi_\lambda$ is also an eigenstate of $H$ belonging to the same eigenvalue $E_\lambda$. If a complete set of commuting observables are all invariant under $U$, it follows that their eigenfunction $\Psi_\lambda$ is equal to its transform $U\Psi_\lambda$ to within a phase factor:

$$U\Psi_\lambda = e^{i\phi_\lambda}\Psi_\lambda, \tag{45}$$

where $\phi_\lambda$ is real if $U$ is unitary.

Consider next a time-dependent state vector $\Psi(t)$ such that

$$\Psi(t) = \phi_i(t) + \phi_f(t), \tag{46}$$

where we assume that as $t \to \pm\infty$, $\Psi(t)$ approaches the limits

$$\begin{aligned} \Psi(-\infty) &= \phi_i(-\infty), \\ \Psi(+\infty) &= \phi_f(+\infty). \end{aligned} \tag{47}$$

This is evidently a system undergoing a transition from one state to another, for example a scattering process or the decay of an unstable particle. Imagine now that $\Psi$ is an eigenfunction of a unitary operator $U$, so that

$$U\Psi(t) = e^{i\phi}\Psi(t) = e^{i\phi}[\phi_i(t) + \phi_f(t)]. \tag{48}$$

If $U$ is linear, both the initial and final states of the system undergoing the transition are eigenfunctions of $U$ corresponding to the same eigenvalue. Such invariance principles play an important role in classifying the states into which systems can make transitions.

Consider now the expectation value of an operator $\mathcal{O}$ in a state $\Psi$.

$$\langle \mathcal{O} \rangle = \langle \Psi, \mathcal{O}\Psi \rangle. \tag{49}$$

If $U$ is a transformation for which

$$\begin{aligned} \mathcal{O}' &= U\mathcal{O}U^{-1}, \\ \Psi' &= U\Psi, \end{aligned} \tag{50}$$

we have

$$(\Psi', \mathcal{O}'\Psi') = (U\Psi, U\mathcal{O}\Psi) = (\Psi, U^\dagger U\mathcal{O}\Psi); \tag{51}$$

and if $U$ is unitary,

$$(\Psi', \mathscr{O}'\Psi') = (\Psi, \mathscr{O}\Psi). \tag{52}$$

Thus the expectation value of $\mathscr{O}$ may be computed at our convenience either as $\langle\mathscr{O}\rangle$ or $\langle\mathscr{O}'\rangle'$ (in an obvious notation).

We now discuss several important examples of transformations, some of which are less obvious than the simple coordinate transformations we have discussed in detail.

## 2.3 COORDINATE REFLECTIONS

In Section 1.2 the parity operator $\mathscr{P}$ was introduced (Eq. 1.44). It is considered quite likely that all Hamiltonians describing the electromagnetic and nuclear interactions of physical systems commute with $\mathscr{P}$. Such is not the case for the weak interactions, for example, in $\beta$-decay and strange-particle decays for which it is known that $\mathscr{P}$ does *not* commute with the Hamiltonian. For the remainder of this section we shall assume that $[\mathscr{P}, H] = 0$, however. Then according to Eq. (1.46*b*) we may write

$$\mathscr{P}\Psi = \pm\Psi \tag{53}$$

for an eigenstate $\Psi$ of the operator $\mathscr{P}$.

The parity operator for the Dirac equation (1.104) is

$$\mathscr{P} = \eta_p\beta\,\mathscr{P}_0, \tag{54}$$

where $\mathscr{P}_0$ reflects the coordinate $\mathbf{r}$ into $-\mathbf{r}$; $\eta_p$ is of modulus one and may be taken as unity without loss of generality. This may be seen as follows: Starting from the Dirac equation ($\dot{\psi} = \partial\psi/\partial t$).

$$i\hbar\dot{\psi}'(\mathbf{r}', t) = (c\boldsymbol{\alpha}\cdot\mathbf{p}' + \beta mc^2)\psi'(\mathbf{r}', t), \tag{55}$$

we make the transformation $\mathbf{r}' = -\mathbf{r}$ and obtain

$$i\hbar\dot{\psi}'(-\mathbf{r}, t) = (-c\boldsymbol{\alpha}\cdot\mathbf{p} + \beta mc^2)\psi'(-\mathbf{r}, t). \tag{56}$$

To restore the equation to its original form, we multiply by $\beta$ and obtain

$$i\hbar\beta\dot{\psi}'(-\mathbf{r}, t) = (c\boldsymbol{\alpha}\cdot\mathbf{p} + \beta mc^2)\beta\psi'(-\mathbf{r}, t), \tag{57}$$

so that $\beta\psi'(-\mathbf{r}, t)$ is a solution of the ordinary Dirac equation. Stated another way, the form of the equation is preserved under the simultaneous operation of $\mathbf{r}\to-\mathbf{r}$ and multiplication by $\beta$. Thus

$$\psi(\mathbf{r}, t)\to\psi'(\mathbf{r}, t) = \mathscr{P}\psi(\mathbf{r}, t) = \eta_p\beta\psi(-\mathbf{r}, t). \tag{58}$$

If $\psi(\mathbf{r}, t)$ is a solution of the equation, that is, if it represents a possible state, then $\beta\psi(-\mathbf{r}, t)$ is also a possible state. We note that the current operator has the property

$$(\psi'(\mathbf{r}), \boldsymbol{\alpha}\psi'(\mathbf{r})) = -(\psi(-\mathbf{r}), \boldsymbol{\alpha}\psi(-\mathbf{r})). \tag{59}$$

An example of a state of definite parity in the Dirac theory is the ground state of the hydrogen atom, which has the form

$$\psi(\mathbf{r}) = [f(r) + \boldsymbol{\alpha}\cdot\mathbf{r}g(r)]\begin{pmatrix} a \\ b \\ 0 \\ 0 \end{pmatrix},$$

where $a$, $b$ are constants and $f$, $g$ are functions of $r = |\mathbf{r}|$. We note that $\mathscr{P}\psi(\mathbf{r}) = +\psi(\mathbf{r})$, so this is a state of even parity.

## 2.4 CHARGE CONJUGATION

The operation of charge conjugation has been discussed in Section 1.4. The transformation $C$ of Eq. (1.116) changes particles into antiparticles. We now explore this in more detail. Let us consider the explicit free-particle solution of the Dirac equation. It is well known that for a given value of the three-vector momentum $\mathbf{p}$ there are four solutions; two correspond to positive energy, with $z$-component of the spin $\pm\frac{1}{2}$, two to negative energy and spin $\pm\frac{1}{2}$. We have

$$\begin{aligned} Hu_{\mathbf{p},\epsilon,s} \equiv (\boldsymbol{\alpha}\cdot\mathbf{p} + \beta m)u_{\mathbf{p},\epsilon,s} &= \epsilon u_{\mathbf{p},\epsilon,s}, \\ (\boldsymbol{\alpha}\cdot\mathbf{p} + \beta m)u_{\mathbf{p},-\epsilon,s} &= -\epsilon u_{\mathbf{p},-\epsilon,s}. \end{aligned} \tag{60}$$

Here $u_{\mathbf{p},\epsilon,s}$ represents a particle with momentum $\mathbf{p}$, energy $\epsilon$, spin orientation $s$, and $\epsilon = +(\mathbf{p}^2 + m^2)^{\frac{1}{2}}$.[1] These equations are explicitly solved by writing

$$u_{p,\lambda} = N(\epsilon + H\beta)w_\lambda, \tag{61}$$

where $\lambda$ stands for the combinations $\epsilon$, $s$ and takes on four values, $\epsilon$, $+\frac{1}{2}(1)$, $\epsilon$, $-\frac{1}{2}(2)$, $-\epsilon$, $+\frac{1}{2}(-1)$, and $-\epsilon$, $-\frac{1}{2}(-2)$, and $N$ is a normalization constant; the $w_\lambda$ are the four basis spinors

$$w_1 = \begin{pmatrix}1\\0\\0\\0\end{pmatrix},\quad w_2 = \begin{pmatrix}0\\1\\0\\0\end{pmatrix},\quad w_{-1} = \begin{pmatrix}0\\0\\1\\0\end{pmatrix},\quad w_{-2} = \begin{pmatrix}0\\0\\0\\1\end{pmatrix}. \tag{62}$$

[1] It is convenient in this section to choose our units in such a way that $\hbar = c = 1$. This widely used convention can cause no confusion, since the appropriate factors of $\hbar$ and $c$ may be uniquely restored by simple dimensional arguments.

We call $w_1$, $w_{-1}$ "spin up," $w_2$, $w_{-2}$ "spin down." We may, of course, choose linear combinations of $w_1$, $w_2$ to form eigenstates of the spin along an arbitrary direction.

A covariant version of the solutions $u_{\mathbf{p},\lambda}$ is easily given. For this purpose it is more convenient to associate the momentum $-\mathbf{p}$ with the negative-energy states. With this understanding, if $p$ is the four vector $p = (\mathbf{p}, ip_0)$, $p_0 = \epsilon$, and $\gamma \cdot p \equiv \gamma_\mu p_\mu$, we write the Dirac equation in the form

$$[i\gamma \cdot p + ms(\lambda)]u_\lambda(p) = 0, \qquad s(\lambda) = \frac{\lambda}{|\lambda|}, \tag{63}$$

where we have written the spinor as $u_\lambda(p)$ to emphasize that we have made a new convention for the momentum associated with the negative-energy states ($\lambda = -1, -2$). Our explicit solution becomes[2]

$$u_\lambda(p) = N[-i\gamma \cdot ps(\lambda) + m]w_\lambda. \tag{64}$$

The normalization constant $N$ is determined by the requirement that $\bar{u}_{\lambda'}(p)u_\lambda(p) = s(\lambda)\delta_{\lambda'\lambda}$, where $\bar{u}_{\lambda'} = \tilde{u}_{\lambda'}{}^*\gamma_4$ as in Eq. (1.108). We have

$$\bar{u}_{\lambda'}(p) = N\bar{w}_{\lambda'}[-i\gamma \cdot ps(\lambda') + m] \tag{65}$$

and, after an easy calculation making use of the properties of the $w_\lambda$, find

$$N = [2m(m + p_0)]^{-1/2}. \tag{66}$$

It is perhaps worth remarking that the spinor $u_\lambda(p)$, with $\lambda = 1, 2$, is just that obtained by making a Lorentz transformation in the direction $-\mathbf{p}$ on a spinor having spin $\lambda$ in the rest system.

Next we consider the charge conjugate spinors which, in accordance with Eq. (1.116), are defined by

$$v_{-\lambda}(p) = C\tilde{\bar{u}}_\lambda(p). \tag{67}$$

The reason for writing $v_{-\lambda}$ will become clear in a moment. We apply the operator $i\gamma \cdot p$ to $v_{-\lambda}$ and obtain, using $\gamma_\mu C = -C\tilde{\gamma}_\mu$,

$$\begin{aligned} i\gamma \cdot pv_{-\lambda}(p) &= -Ci\tilde{\gamma} \cdot p\tilde{\bar{u}}_\lambda(p) = -C\widetilde{(\bar{u}_\lambda i\gamma \cdot p)} \\ &= ms(\lambda)C\tilde{\bar{u}}_\lambda = ms(\lambda)v_{-\lambda}(p). \end{aligned}$$

Thus we have

$$[i\gamma \cdot p + ms(\lambda)]v_\lambda(p) = 0, \tag{68}$$

[2] It is sometimes convenient to use basis spinors which are eigenfunctions of $\boldsymbol{\sigma} \cdot \hat{\mathbf{n}}$ where $\hat{\mathbf{n}}$ is a unit vector in the direction of $\mathbf{p}$, which has polar angles $\theta$, $\phi$ with respect to some fixed system. These are easily seen to be $\chi_\lambda = (1 + \boldsymbol{\sigma} \cdot \hat{\mathbf{n}}\sigma_z) \times w_\lambda/[2(1 + \cos\theta)]^{\frac{1}{2}}$ and $\boldsymbol{\sigma} \cdot \hat{\mathbf{n}}\chi_1 = \chi_1$, $\boldsymbol{\sigma} \cdot \hat{\mathbf{n}}\chi_2 = -\chi_2$. The complete new spinors, called *helicity spinors*, are $u_\lambda(p) = N[-i\gamma \cdot ps(\lambda) + m]\chi_\lambda$.

so that $v_\lambda$ with $\lambda = 1, 2$, say, may be at most a linear combination of $u_1$ and $u_2$ (since they satisfy the same equation). Note also that a positive-energy charge conjugate (antiparticle) solution is obtained by applying the $C$-operation to a negative-energy particle solution.

To avoid any misunderstanding concerning signs, notation, and so on, we write explicitly the spinors

$$u_\lambda(p) = N[-i\gamma \cdot ps(\lambda) + m]w_\lambda$$

and

$$v_{-\lambda}(p) = C\tilde{\bar{u}}_\lambda(p) = N[-i\gamma \cdot ps(\lambda) + m]C\tilde{\bar{w}}_{-\lambda}:$$

These are

$$u_{1,2} = N\begin{pmatrix}(\epsilon + m)\chi_{1,2} \\ \boldsymbol{\sigma} \cdot \mathbf{p}\chi_{1,2}\end{pmatrix}, \qquad u_{-1,-2} = N\begin{pmatrix}\boldsymbol{\sigma} \cdot \mathbf{p}\chi_{1,2} \\ (\epsilon + m)\chi_{1,2}\end{pmatrix},$$

$$\begin{aligned}\tilde{\bar{u}}_{1,2} &= N(\tilde{\chi}^*_{1,2}(\epsilon + m), -\tilde{\chi}^*_{1,2}\boldsymbol{\sigma} \cdot \mathbf{p}), \\ \tilde{\bar{u}}_{-1,-2} &= N(\tilde{\chi}^*_{1,2}\boldsymbol{\sigma} \cdot \mathbf{p}, -\tilde{\chi}^*_{1,2}(\epsilon + m)),\end{aligned} \tag{69}$$

where

$$\begin{aligned}\chi_1 &= \begin{pmatrix}1 \\ 0\end{pmatrix}, \qquad \text{spin ``up,''} \\ \chi_2 &= \begin{pmatrix}0 \\ 1\end{pmatrix}, \qquad \text{spin ``down.''}\end{aligned} \tag{70}$$

We emphasize,, for example, that $\tilde{\chi}^*\sigma_2 = (\widetilde{\tilde{\sigma}_2\chi^*})$, with $\chi = a\chi_1 + b\chi_2$, is the row matrix

$$(a^*, b^*)\begin{pmatrix}0 & -i \\ i & 0\end{pmatrix} = (ib^*, -ia^*). \tag{71}$$

The $v_\lambda(p)$, constructed according to Eqs. (67) and (1.119) are then

$$\begin{aligned}v_{1,2} &= -N\begin{pmatrix}(\epsilon + m)\sigma_2\chi^*_{1,2} \\ \boldsymbol{\sigma} \cdot \mathbf{p}\sigma_2\chi^*_{1,2}\end{pmatrix}, \\ v_{-1,-2} &= N\begin{pmatrix}\boldsymbol{\sigma} \cdot \mathbf{p}\sigma_2\chi^*_{1,2} \\ (\epsilon + m)\sigma_2\chi^*_{1,2}\end{pmatrix}.\end{aligned} \tag{72}$$

Noting that $\sigma_2\chi_1{}^* = i\chi_2{}^*$, $\sigma_2\chi_2{}^* = -i\chi_1{}^*$ and that the $\chi$'s are real, we see that

$$v_1 = -iu_2, \qquad v_2 = iu_1, \qquad v_{-1} = iu_{-2}, \qquad v_{-2} = -iu_{-1}; \tag{73}$$

the $v$'s and $u$'s satisfy the same equation but the spin components and phases are reshuffled. We see, for example, that $v_1$, which is formed by applying $C$ to $\bar{u}_{-1}$, has spin down (because of the $\sigma_2$-operator) in contrast to $\bar{u}_1$, which has spin up.

We collect for reference the relation between what may be called antiparticle and particle operators. By this we mean the following: If $\psi^C = C\tilde{\bar{\psi}}$ is the charge conjugate wave function, we define the charge conjugate operator $\theta^c$ corresponding to an operator $\theta$, by the requirement that

$$(\psi^C, \theta^c\psi^C) = (\psi, \theta\psi). \tag{74}$$

Using $\psi^C = C\tilde{\bar{\psi}} = C\tilde{\gamma}_4\psi^*$, $\psi^{C*} = \tilde{\gamma}_4 C^{-1}\psi$, we find

$$\theta^c = C(\widetilde{\gamma_4\theta\gamma_4})C^{-1} = \gamma_4 C\tilde{\theta}C^{-1}\gamma_4. \tag{75}$$

For example, if $\theta = \boldsymbol{\sigma}$, then $\theta^c = -\boldsymbol{\sigma}$; if $\theta = \boldsymbol{\alpha}$, then $\theta^c = +\boldsymbol{\alpha}$; if $\theta = \beta\boldsymbol{\sigma}$, then $\theta^c = \beta\boldsymbol{\sigma}$.

We close this section with a final remark about the parity of antiparticle states as compared to particle states. Suppose $\psi(\mathbf{r})$ is a particle state with a definite parity and that we construct from it the charge conjugate state $\psi^C(\mathbf{r})$. Then with $\mathscr{P}$ the parity operator, we have

$$\begin{aligned} \mathscr{P}\psi^C(\mathbf{r}) &= \gamma_4\psi^C(-\mathbf{r}) = \gamma_4 C\tilde{\bar{\psi}}(-\mathbf{r}) \\ &= -C\tilde{\gamma}_4\tilde{\bar{\psi}}(-\mathbf{r}) = -C[\widetilde{\overline{\mathscr{P}\psi(\mathbf{r})}}]. \end{aligned} \tag{76}$$

Thus if $\mathscr{P}\psi = \pm\psi$, $\mathscr{P}\psi^C = \mp\psi^C$; particle and antiparticle in the same orbital state have opposite parity. For example, the $S$-states of positronium have odd parity.

## 2.5 DISPLACEMENT IN TIME

The Schrödinger equation describes the development in time of the state vector $\Psi(t)$. The equation[3]

$$i\dot{\Psi}(t) = H\Psi(t) \tag{77}$$

may be formally integrated to yield

$$\Psi(t) = e^{-i(t-t_0)H}\Psi(t_0), \tag{78}$$

where $\Psi(t_0)$ is the value of the state vector at time $t = t_0$; the transformation of $\Psi$ from time $t_0$ to time $t$ is effected by the unitary operator $e^{-i(t-t_0)H}$.

Next we suppose that the Hamiltonian $H$ may be decomposed into an "unperturbed" part $K$ and a "perturbation" $V$.

$$H = K + V. \tag{79}$$

[3] We shall now set $\hbar = 1$ and restore it only periodically when there appears to be some danger of forgetting entirely about quantum limitations. Thus all energies have the dimension $(\text{time})^{-1}$; momenta, $(\text{length})^{-1}$.

We introduce states $\chi_a$ which are a complete set of eigenstates of $K$.

$$K\chi_a = E_a\chi_a, \tag{80}$$

where we may imagine adjoining a complete set of commuting observables to $K$ to characterize the $\chi_a$ precisely. The time-dependent eigenstates of

$$i\dot{\Phi}_a = K\Phi_a \tag{81}$$

are

$$\Phi_a(t) = e^{-iKt}\chi_a = e^{-iE_at}\chi_a. \tag{82}$$

It is very convenient to "remove" the time dependence of $\Psi(t)$ associated with the $K$ part of $H$, since it is (by assumption) quite trivial. This is achieved by a time-dependent unitary transformation to what is called the *interaction representation.*[4] (We shall discuss representations more generally at the end of this section.) We introduce

$$\Psi_I(t) \equiv e^{iKt}\Psi(t), \tag{83}$$

where the subscript $I$ stands for interaction representation. On differentiating this definition of $\Psi_I$, we obtain

$$\begin{aligned} i\dot{\Psi}_I &= -e^{iKt}K\Psi(t) + e^{iKt}i\dot{\Psi}(t) \\ &= e^{iKt}(-K + H)\Psi(t) \\ &= e^{iKt}V\Psi(t) \\ &= (e^{iKt}V\,e^{-iKt})\Psi_I(t). \end{aligned}$$

On writing

$$V(t) \equiv e^{iKt}V\,e^{-iKt} \tag{84}$$

we have

$$i\dot{\Psi}_I(t) = V(t)\Psi_I(t). \tag{85}$$

Since Eq. (85) for $\Psi_I(t)$ was obtained from the Schrödinger equation (77) by means of a unitary transformation, this equation is physically equivalent to the original one.

Let us suppose that $\Psi_I(t)$ may be obtained from $\Psi_I(t_0)$, an arbitrary state vector at time $t_0$, by a linear operator $U(t, t_0)$:

$$\Psi_I(t) = U(t, t_0)\Psi_I(t_0), \tag{86}$$

where by definition $U(t_0, t_0) = 1$. It is easy to find a formula for $U$. We write

$$\begin{aligned} U(t, t_0)\Psi_I(t_0) &= \Psi_I(t) = e^{iKt}\Psi(t) \\ &= e^{iKt}\,e^{-iH(t-t_0)}\Psi(t_0) \\ &= e^{iKt}\,e^{-iH(t-t_0)}\,e^{-iKt_0}\Psi_I(t_0); \end{aligned}$$

[4] This representation has played an important role in the development of field theory. See, for example, Refs. [1], [2], and [3].

and, since $\Psi_I(t_0)$ is an arbitrary vector, we obtain

$$U(t, t_0) = e^{iKt}\, e^{-iH(t-t_0)}\, e^{-iKt_0}. \tag{87a}$$

Special cases of this representation are

$$\begin{aligned} U(t, 0) &= e^{iKt}\, e^{-iHt}, \\ U(0, t) &= e^{iHt}\, e^{-iKt}. \end{aligned} \tag{87b}$$

It is trivial to see that if we had chosen to define the interaction representation state vector to coincide with the Schrödinger state at some time $\tau$ (rather than at zero, as we did) we would have found

$$\begin{aligned} U_\tau(t, t_0) &= e^{iK(t-\tau)}\, e^{-iH(t-t_0)}\, e^{-iK(t_0-\tau)} \\ &= e^{-iK\tau} U(t, t_0)\, e^{iK\tau} = U(t-\tau, t_0-\tau), \end{aligned} \tag{88}$$

which differs from $U(t, t_0)$ by just a unitary transformation.

If we imagine $U(t, t_0)$ to operate on a function, say one of the $\chi_a$ or a superposition of them, we may give the following interpretation to the result: The operator $e^{-iKt_0}$ describes propagation from time $t = 0$ to $t_0$ according to the "unperturbed" Hamiltonian, $K$; this is followed by propagation from $t_0$ to $t$ with the full Hamiltonian and finally the resulting state is propagated backward in time to the time $t = 0$. It is, so to speak, the configuration obtained after a time $t$ with full interaction, reduced back to $t = 0$ as if the interaction were turned off. We might picture this in terms of a classical trajectory where we start out in a certain direction $\chi_a$ and move uniformly (no force) for a time $t_0$; the trajectory then curves in some manner for a time $t - t_0$ (under the influence of $V$). After reaching $t$ we draw a tangent to the trajectory and move backward in time in uniform motion along this tangent for a time $t$.

We may obtain a differential equation for $U(t, t_0)$, involving only quantities expressed in the interaction representation, by simply substituting $\Psi_I(t) = U(t, t_0)\Psi_I(t_0)$ into the equation satisfied by $\Psi_I$ [Eq. (85)] and using the fact that $\Psi_I(t_0)$ is arbitrary. We find

$$i\frac{\partial}{\partial t} U(t, t_0) = V(t)U(t, t_0). \tag{89}$$

The same result is found by differentiating Eq. (87*a*) with respect to $t$.

$$\begin{aligned} i\frac{\partial}{\partial t} U(t, t_0) &= e^{iKt}(-K + H)\, e^{-iH(t-t_0)}\, e^{-iKt_0} \\ &= e^{iKt} V e^{-iKt} e^{iKt}\, e^{-iH(t-t\ )} e^{-iKt_0} \\ &= V(t)U(t, t_0). \end{aligned}$$

Differentiation with respect to $t_0$ leads to

$$-i\frac{\partial}{\partial t_0}U(t,t_0) = U(t,t_0)V(t_0). \tag{90}$$

This is seen to be the same as the equation satisfied by $U^\dagger(t, t_0)$:

$$-i\frac{\partial}{\partial t}U^\dagger(t,t_0) = U^\dagger(t,t_0)V(t), \tag{91}$$

from which we conclude that

$$U^\dagger(t,t_0) = U(t_0,t). \tag{92}$$

This may be verified directly from Eq. (87*a*). Another general relation is

$$U(t,t_0) = U(t,t')U(t',t_0), \tag{93}$$

which is called the group property of $U$. This also follows directly from Eq. (87*a*), or from

$$\begin{aligned}\Psi_I(t) &= U(t,t')\Psi_I(t')\\ &= U(t,t')U(t',t_0)\Psi_I(t_0)\\ &= U(t,t_0)\Psi_I(t_0).\end{aligned}$$

If we set $t = t_0$ in Eq. (93), we find

$$U(t_0,t_0) = U(t_0,t)U(t,t_0) = 1, \tag{94}$$

or

$$U(t_0,t) = U^{-1}(t,t_0) = U^\dagger(t,t_0). \tag{95}$$

Thus

$$U^\dagger(t,t_0) = U^{-1}(t,t_0), \tag{96}$$

so $U$ is a unitary operator. This may also be shown directly from our differential equations for $U$ and $U^\dagger$.

$$\begin{aligned}i\frac{\partial}{\partial t}U^\dagger(t,t_0)U(t,t_0) &= U^\dagger i\frac{\partial}{\partial t}U(t,t_0) - (-i)\left[\frac{\partial}{\partial t}U^\dagger(t,t_0)\right]U(t,t_0)\\ &= U^\dagger V(t)U - U^\dagger V(t)U = 0,\end{aligned}$$

using Eqs. (89) and (91). Thus we conclude that $U^\dagger(t, t_0)U(t, t_0)$ is independent of time, and since it is obviously unity for $t = t_0$ we have $U^\dagger U = 1$ for all time. By considering the derivative with respect to $t_0$ of $U(t, t_0)U^\dagger(t, t_0)$, we come also to $U(t, t_0)U^\dagger(t, t_0) = 1$, which completes the demonstration of unitarity.

The differential equations for $U(t, t_0)$, which give the dependence on both variables, may be conveniently replaced by integral equations which incorporate the boundary condition $U(t, t) = 1$. We simply integrate the equations (89) and (90) formally to obtain

$$\begin{aligned} U(t, t_0) &= 1 - i\int_{t_0}^{t} dt' V(t')U(t', t_0), \\ U(t, t_0) &= 1 + i\int_{t}^{t_0} dt' U(t, t')V(t'). \end{aligned} \tag{97}$$

When these equations are combined with the explicit form of $U(t, t_0)$ we find the useful expressions

$$\begin{aligned} U(t, 0) &= 1 - i\int_0^t dt'\, e^{iKt'} V\, e^{-iHt'}, \\ U(0, t) &= 1 + i\int_0^t dt'\, e^{iHt'} V e^{-iKt'}. \end{aligned} \tag{98}$$

[Note that the appearance of $V$ rather than $V(t')$ is *not* a misprint!] These equations are important for our subsequent work.

It is easy to generate an expansion of $U$ in powers of $V$ by solving the integral equations by iteration. We then obtain

$$U(t, t_0) = 1 + \sum_{n=1}^{\infty} (-i)^n \int_{t_0}^{t} dt_n V(t_n) \int_{t_0}^{t_n} dt_{n-1} V(t_{n-1}) \cdots \int_{t_0}^{t_2} dt_1 V(t_1). \tag{99}$$

In general the operators $V(t)$ do not commute for different values of the time, so that the indicated order must be maintained. The characteristic structure is such that the operators are arranged in the order of increasing time as we read from right to left. We may, however, formally extend the region of integration of each integral to $(t, t_0)$ provided we always maintain the indicated chronological order and also compensate for counting the true region $n!$ times. We arrive then at the form first given by Dyson [2].

$$U(t, t_0) = 1 + \sum_{n=1}^{\infty} \frac{(-i)^n}{n!} \int_{t_0}^{t} dt_n \int_{t_0}^{t} dt_{n-1} \cdots \int_{t_0}^{t} dt_2 \int_{t_0}^{t} dt_1 T[V(t_n) \cdots V(t_1)] \tag{100}$$

where the time-ordering symbol $T$ is just the instruction given before, namely, that we arrange the $n$ factors $V(t_i)$ in the order of increasing time, reading from right to left. This form suggests an ordered exponential, which we may write symbolically as

$$U(t, t_0) = T\left\{\exp\left[-i\int_{t_0}^{t} dt' V(t')\right]\right\}, \tag{101}$$

the precise definition of which is just Eq. (100). In some respects this behaves much as an ordinary exponential; for example,

$$i\frac{\partial}{\partial t}U(t, t_0) = V(t)T\left\{\exp\left[-i\int_{t_0}^{t} dt'V(t')\right]\right\}.$$

The last topic to be considered in this section is that of the various "representations," mentioned earlier. Until now we have for the most part used what is called the Schrödinger representation, in which the observables are taken to be time-independent Hermitian operators. The time dependence of the expectation values or matrix elements of these operators comes from the explicit time dependence of the state vectors. We have, then, for $\langle A\rangle = (\Psi, A\Psi)$ the relation

$$\begin{aligned} i\frac{d}{dt}\langle A\rangle &= (-i\dot{\Psi}, A\Psi) + (\Psi, Ai\dot{\Psi}) \\ &= (-H\Psi, A\Psi) + (\Psi, AH\Psi) \end{aligned} \tag{102}$$

on using $i\dot{\Psi} = H\Psi$. Therefore

$$i\frac{d}{dt}\langle A\rangle = (\Psi, [A, H]\Psi). \tag{103}$$

We note that if $[A, H] = 0$, the time derivative of any matrix element of $A$ vanishes; this is the reason operators that commute with the Hamiltonian are called constants of the motion.

Now suppose that a second state vector $\Phi(t)$ is related to $\Psi(t)$ by a unitary operator $W$ according to $\Phi = W\Psi$. Then

$$\begin{aligned} \langle A\rangle = (\Psi, A\Psi) &= (W^{\dagger}\Phi, AW^{-1}\Phi) = (\Phi, WAW^{-1}\Phi) \\ &= (\Phi, A'\Phi), \end{aligned} \tag{104}$$

where

$$A' \equiv WAW^{-1}. \tag{105}$$

We assume that $W$ may depend on $t$ and so obtain the equation for the time dependence of $\Phi$.

$$i\frac{\partial\Phi}{\partial t} = i\frac{\partial W}{\partial t}\Psi + Wi\frac{\partial\Psi}{\partial t} = i\frac{\partial W}{\partial t}W^{-1}\Phi + WHW^{-1}\Phi. \tag{106}$$

On writing $H' \equiv WHW^{-1}$ and $L \equiv i\,(\partial W/\partial t)W^{-1}$, we have

$$i\frac{\partial\Phi}{\partial t} = (H' + L)\Phi. \tag{107}$$

Equation (105) may be differentiated to give

$$i \frac{\partial A'}{\partial t} = i \frac{\partial W}{\partial t} A W^{-1} + iWA \frac{\partial W^{-1}}{\partial t} = i \frac{\partial W}{\partial t} W^{-1} A' + iA'W \frac{\partial W^{-1}}{\partial t}. \quad (108)$$

Now

$$\frac{\partial W}{\partial t} W^{-1} = -W \frac{\partial W^{-1}}{\partial t},$$

from differentiating $WW^{-1} = 1$, so Eq. (108) reduces to

$$i \frac{\partial A'}{\partial t} = [L, A']. \quad (109)$$

The various so-called representations correspond to different ways of simplifying Eqs. (107) and (109) for $\Phi$ and $A$, respectively.

First, we recover the *Schrödinger representation* by choosing $W$ independent of time, which means $L = 0$. We have, then,

$$i \frac{\partial \Phi}{\partial t} = H'\Phi, \qquad \frac{\partial A'}{\partial t} = 0.$$

This is just our starting point, except for having made a time-independent unitary transformation.

The next choice is $L = -H'$; this provides the *Heisenberg representation.* We have then

$$i \frac{\partial W}{\partial t} = -H'W = -WH,$$

$$i \frac{\partial \Phi}{\partial t} = 0,$$

and

$$i \frac{\partial A'}{\partial t} = [A', H'].$$

It is clear that $W = e^{itH}$ is a solution of the $W$ equation, so that $H' = H$, and

$$A' = e^{itH} A \, e^{-itH},$$

$$i \frac{\partial}{\partial t} A' = [A', H]. \quad (110)$$

In this representation the state vectors are constant in time, and the time dependence resides in the operators.

Finally, we introduce the *interaction representation* by the choice $L = -K$, where we imagine that $H = K + V$, as before. This leads to $W = e^{iKt}$ and the set of equations

$$H' = K + V',$$
$$V' \equiv V(t) = e^{iKt} V e^{-iKt},$$
$$i \frac{\partial \Phi}{\partial t} = V(t)\Phi,$$

and

$$i \frac{\partial A'}{\partial t} = [A', K]. \tag{111}$$

These are the equations used in the earlier part of this section.

It is sometimes useful to have the relations between operators in the Heisenberg representation and the interaction representation. If we write subscripts $H$ for Heisenberg, $S$ for Schrödinger, and $I$ for interaction, they are

$$A_H(t) = e^{iHt} A_S e^{-iHt},$$
$$A_I(t) = e^{iKt} A_S e^{-iKt},$$

and

$$\begin{aligned} A_H(t) &= e^{iHt} e^{-iKt} A_I(t) e^{iKt} e^{-iHt} \\ &= U(0, t) A_I(t) U(t, 0), \end{aligned} \tag{112}$$

where we have arranged that the operators coincide at $t = 0$. (Any other reference time would do.)

## 2.6 TIME REVERSAL INVARIANCE

All known interactions of systems of particles satisfy the condition of "time reversal invariance."[5] This is presumably true for both strong (that is, electromagnetic and nuclear) and weak ($\beta$-decay, strange-particle decays, and so on) interactions. Loosely speaking, this means that a system will "run backward as well as forward." In the classical mechanics of conservative systems this is easily seen. The Newtonian equations of motion involve only second-order time derivatives, so the transformation $t \to -t$ leaves the equations of motion unchanged. If $\mathbf{r}(t)$ is a possible trajectory for a particle, so is $\mathbf{r}(-t)$.

More generally, in either classical or quantum theory we may describe time reversal as follows: Let the system be in a state $a$ at time $t = 0$,

[5] We assume that we are dealing with an isolated system. If external fields are considered, in general the sources of these fields must be included as part of the system if time reversal invariance is to obtain.

The operation of time reversal in quantum mechanics was introduced by Wigner [4].

specified by giving the values of a complete set of observables. If we reverse the sign of all momenta, angular momenta, and spins of the system, we obtain a new state which we label as $-a$. This is called the "time-reversed state of $a$." In quantum theory this will be induced by an operator (to be called $T$) which leads to the time reversal transformation, $a \to -a$.

Suppose that the system which is in state $a$ at time $t = 0$ is isolated for a time $\tau$. At $t = \tau$ it will be found in a state $b$, determined by the equations of motion. Let it now be transformed from $b$ to $-b$ and allowed to proceed (isolated) for a second interval $\tau$. If at the end of the second interval the system is found to be in the state $-a$, we say that the system is "invariant under time reversal." This is at present considered to be one of the fundamental symmetry properties of physical systems.

Let us consider the transformation $t \to -t$ on the Schrödinger equation. Imagine first that we are dealing with a single spinless particle moving in a potential. We write

$$i\hbar \frac{\partial}{\partial t'} \Psi(\mathbf{r}, t') = \left[ -\frac{\hbar^2}{2m} \nabla^2 + V(\mathbf{r}) \right] \Psi(\mathbf{r}, t') \tag{113}$$

and make the transformation $t' = -t$. We note then that the equation may be restored to its original form by taking the complex conjugate (assume $V$ is real)

$$i\hbar \frac{\partial}{\partial t} \Psi^*(\mathbf{r}, -t) = \left[ -\frac{\hbar^2}{2m} \nabla^2 + V(\mathbf{r}) \right] \Psi^*(\mathbf{r}, -t). \tag{114}$$

Thus, if $\Psi(\mathbf{r}, t)$ is a solution of the Schrödinger equation, so is $\Psi'(\mathbf{r}, t) = \Psi^*(\mathbf{r}, -t)$. This result depends on the fact that we are working in the coordinate representation. All statements about time reversal are really representation dependent, since the operation of complex conjugation must be explained explicitly. We note, for example, that in momentum space we would have

$$\Psi(\mathbf{r}, t) = \int d^3p\, \phi(p, t)\, e^{(i/\hbar)\mathbf{p}.\mathbf{r}},$$

$$\Psi^*(\mathbf{r}, -t) = \int d^3p\, \phi^*(-\mathbf{p}, -t)\, e^{(i/\hbar)\mathbf{p}.\mathbf{r}},$$

$$\Psi'(\mathbf{r}, t) = \int d^3p\, \phi'(\mathbf{p}, t)\, e^{(i/\hbar)\mathbf{p}.\mathbf{r}}.$$

Therefore

$$\phi'(\mathbf{p}, t) = \phi^*(-\mathbf{p}, -t). \tag{115}$$

Thus in momentum space, in addition to taking the complex conjugate of the wave function and changing $t$ to $-t$, we must change $\mathbf{p}$ to $-\mathbf{p}$, which is consistent with the general remarks made earlier.

We turn now to the general case, where we write

$$i\hbar \frac{\partial \Psi}{\partial t'}(t') = H\Psi(t'). \tag{116}$$

If we make the transformation $t' = -t$, followed by operation with an operator $T_0$ which effects complex conjugation, we obtain

$$i\hbar \frac{\partial}{\partial t} \Psi^*(-t) = H^*\Psi^*(-t). \tag{117}$$

By $\Psi^*$ we mean the following. In any representation a state vector $\Phi$ is represented as a vector with components $\{\phi_i\}$. By $\Phi^*$ we mean the vector in the *same* basis system with components $\{\phi_i^*\}$. Unless $H$ is real, Eq. (117) does not yet have the original form of the Schrödinger equation. Let us suppose, however, that there exists a unitary operator $W$ such that

$$WH^*W^{-1} = H. \tag{118}$$

Then operating on Eq. (117) with $W$ leads to

$$\hbar i \frac{\partial}{\partial t} W\Psi^*(-t) = HW\Psi^*(-t), \tag{119}$$

and we conclude that if $\Psi(t)$ is a solution of the Schrödinger equation, so is $\Psi'(t)$ defined by

$$\Psi'(t) = W\Psi^*(-t). \tag{120a}$$

We call this $\Psi'$ the "time-reversed wave function $\Psi(t)$." If a $W$ can be found for which Eq. (118) is valid, the Schrödinger equation is invariant under the operation

$$T = WT_0, \tag{121}$$

where $T_0$ is the operator of complex conjugation (that is, $T_0\Psi = \Psi^*$). Since $T_0^2 = 1$, we have $T_0 = T_0^{-1}$.

To summarize, the necessary and sufficient condition for a physical system characterized by a Hamiltonian $H$ to be invariant under the operation of time reversal is the requirement that

$$THT^{-1} = H, \tag{122}$$

where $T^{-1} = T_0 W^{-1}$.[6] We shall show later that this definition of time reversal invariance is consistent with the physical discussion given at the beginning of this section. Equation (120) may be written in terms of $T$ as

$$\Psi'(t) = T\Psi(-t). \tag{120b}$$

The scalar product of two state vectors $\Phi'(t) = T\Phi(-t)$ and $\Psi'(t) = T\Psi(-t)$ is

$$\begin{aligned}(\Phi'(t), \Psi'(t)) &= (T\Phi(-t), T\Psi(-t)) = (T_0\Phi(-t), W^\dagger W T_0 \Psi(-t)) \\ &= (\Phi(-t), \Psi(-t))^* \\ &= (\Psi(-t), \Phi(-t)),\end{aligned} \tag{123}$$

since $W^\dagger = W^{-1}$.

We define the time-reversed operator $Q'$, corresponding to $Q$, by the requirement

$$\begin{aligned}(\Psi(t), Q\Psi(t)) &= (\Psi'(-t), Q'\Psi'(-t)) \\ &= (T\Psi(t), Q'T\Psi(t)) \\ &= (T^{-1}T\Psi(t), T^{-1}Q'T\Psi(t))^* \\ &= (T^{-1}Q'T\Psi(t), \Psi(t)),\end{aligned} \tag{124}$$

which leads to

$$Q^\dagger = T^{-1}Q'T \quad \text{or} \quad Q' = TQ^\dagger T^{-1}. \tag{125}$$

The appropriate time-reversed operator is usually easy to construct by inspection. The coordinate operator is to be left unchanged, and thus $W$ must commute with $\mathbf{r}$. Since $\mathbf{r}$ is Hermitian, we demand that

$$\mathbf{r}' = T\mathbf{r}T^{-1} = WT_0\mathbf{r}T_0W^{-1} = W\mathbf{r}W^{-1} = \mathbf{r}. \tag{126}$$

From physical considerations, we must require that the linear momentum shall change sign under the $T$-transformation. Since

$$T_0\mathbf{p}T_0^{-1} = T_0(-i\hbar\nabla)T_0^{-1} = i\hbar\nabla = -\mathbf{p},$$

we see that we must have $[\mathbf{p}, W] = 0$, so that

$$\mathbf{p}' = T\mathbf{p}T^{-1} = -\mathbf{p}. \tag{127}$$

The angular momentum $\mathbf{L} = \mathbf{r} \times \mathbf{p}$ must change sign, and our results for the $T$-transform of $\mathbf{r}$ and $\mathbf{p}$ indeed confirm this:

$$\begin{aligned}\mathbf{L}' &= T\mathbf{r} \times \mathbf{p}T^{-1} = \mathbf{r} \times T\mathbf{p}T^{-1} \\ &= -\mathbf{r} \times \mathbf{p} = -\mathbf{L}.\end{aligned} \tag{128}$$

[6] Note that $T_0$ and hence $T$ is not a linear operator since $T_0(a_1\Psi_1 + a_2\Psi_2) = a_1^*T_0\Psi + a_2^*T_0\Psi$. On the other hand $W$ is taken to be unitary. We refer to $T$ as an anti-unitary operator.

Suppose now that a particle has a spin angular momentum **S**. It would be quite insane to have the spin and orbital angular momentum transform differently; hence we must require that

$$\mathbf{S}' = T\mathbf{S}T^{-1} = W\mathbf{S}^*W^{-1} = -\mathbf{S}. \tag{129}$$

For a spin one-half particle it is easy to verify that in our standard representation

$$W = i\sigma_2. \tag{130}$$

This is, incidentally, the correct matrix in the Dirac theory, where (again using our standard representation) it may also be expressed as $-i\gamma_5 C$, with $C$ the charge conjugation matrix. Obviously the phase $i$ in (130) is arbitrary. It is easy to see that $T^2$ is a multiple of the unit operator and that it has eigenvalues $\pm 1$. For spinless particles the appropriate value is $+1$; for spin one-half, it is $-1$.

Let us now consider a Hamiltonian $K$ for which $TKT^{-1} = K$, and which has a complete set of eigenfunctions $\chi_a$.

$$K\chi_a = \epsilon_a\chi_a.$$

Suppose also that there is a complete set of observables $A_\lambda$ ($\lambda = 1, 2, \ldots$), of which $K$ is a member, in terms of which the $\chi_a$ are precisely specified. Then

$$A_\lambda\chi_a = a_\lambda\chi_a; \tag{131}$$

the $A_\lambda$ may describe the angular momentum, spin, and so on, of the states $\chi_a$. For most observables we have

$$A'_\lambda = TA_\lambda{}^\dagger T^{-1} = TA_\lambda T^{-1} = \pm A_\lambda \tag{132}$$

for each $A_\lambda$. Operating on (131) with $T$ we have then

$$A_\lambda\chi'_a = \pm a_\lambda\chi'_a, \tag{133}$$

with $\chi'_a \equiv T\chi_a$. This means that when the states are classified according to observables with the simple time reversal properties expressed by (132), the states $\chi_a$ may be chosen in such a manner that the $\chi'_a$ belong to the *same set*. We may then write

$$\chi'_a = \mu_{-a}\chi_{-a}, \tag{134}$$

where $\chi_{-a}$ is one of the $\chi_a$ and $\mu_{-a}$ is a phase factor; $|\mu_a| = 1$. The state $\chi_{-a}$ is in fact obtained by reversing the direction of momentum, angular momentum, and spin of $\chi_a$, as follows from Eqs. (127), (128), and (129). For example, if $\chi_a = e^{i\mathbf{p}\cdot\mathbf{r}}$, then $\chi'_a = \chi_a = e^{-i\mathbf{p}\cdot\mathbf{r}}$. For a particle of spin one-half, described by a wave function $\chi_\alpha$ and with a definite helicity defined as

$$\boldsymbol{\sigma}\cdot\mathbf{p}\chi_\alpha = -i\boldsymbol{\sigma}\cdot\boldsymbol{\nabla}\chi_\alpha = \epsilon\chi_\alpha, \qquad \epsilon = \pm|\mathbf{p}|, \tag{135}$$

we have [on using Eqs. (127), (130)] the result that the helicity of $\chi'_\alpha = T\chi_\alpha$ is unchanged. Therefore, $\chi'_\alpha$ must be that $\chi_\alpha$ which has its spin *and* momentum reversed with respect to the original one.

We are now ready to return to the original physical definition of time reversal, as described at the beginning of this section. For a given system, invariant under $T$, let the state at $t = 0$ be $\chi_a$:

$$\Psi(0) = \chi_a. \tag{136}$$

After a time $\tau$ the state is

$$\Psi(\tau) = e^{-i\tau H}\chi_a. \tag{137}$$

The time-reversed state of $\Psi(\tau)$ is

$$\begin{aligned}\Psi'(\tau) &= T\Psi(\tau) = e^{i\tau H}T\chi_a \\ &= e^{i\tau H}\mu_{-a}\chi_{-a}.\end{aligned} \tag{138}$$

After a second interval $\tau$ the system is in a state

$$\Psi'(2\tau) = e^{-i\tau H}\Psi'(\tau) = \mu_{-a}\chi_{-a}. \tag{139}$$

This is just the time-reversed initial state, as stated earlier.

We next consider a system in a state $\Psi_J{}^M$ of definite angular momentum $J$ and $z$-component $M$; we assume that all other quantum numbers are not changed by time reversal and suppress them. Let us apply the $T$-operator to

$$J_z\Psi_J{}^M = M\Psi_J{}^M$$

and note that $TJ_zT^{-1} = -J_z$, so

$$J_z\Psi'^M_J = -M\Psi'^M_J.$$

Thus

$$\Psi'^M_J = T\Psi_J{}^M = \mu_J{}^{-M}\Psi_J{}^{-M}, \tag{140}$$

where $|\mu_J{}^{-M}| = 1$. Now on applying $J^\mp = J_x \mp iJ_y$ to $\Psi_J{}^M$ [see Eq. (1.63*a*)], we obtain

$$J^\mp\Psi_J{}^M = [(J \pm M)(J \mp M + 1)]^{1/2}\Psi_J^{M\mp1} \tag{141}$$

However, we note $TJ^\mp T^{-1} = -J^\pm$, so that

$$\begin{aligned}TJ^-\Psi_J{}^M &= -J^+T\Psi_J{}^M \\ &= -\mu_J{}^{-M}J^+\Psi_J{}^{-M} \\ &= -\mu_J{}^{-M}[(J + M)(J - M + 1)]^{1/2}\Psi_J^{-M+1}.\end{aligned} \tag{142}$$

Using (141) and the equation $T\Psi_J^{M-1} = \mu_J^{-M+1}\Psi_J{}^{-M+1}$, we have also

$$TJ^-\Psi_J{}^M = \mu_J^{-M+1}[(J + M)(J - M + 1)]^{1/2}\Psi_J^{-M+1}.$$

We conclude, then, that

$$\mu_J^{-M} = -\mu_J^{-M+1},$$

which in turn implies that

$$\mu_J^{-M} = (-1)^M \mu_J, \qquad |\mu_J| = 1. \tag{143}$$

Thus it is possible to define the phases of the $\Psi_J^M$ so that

$$T\Psi_J^M = (-1)^M \mu_J \Psi_J^{-M}, \tag{144}$$

a very useful result.

As an application of Eq. (144), we consider the application of the $T$-operator to a wave function $\Psi_J^M$ resulting from combining two states $\Phi_{j_1}^{m_1}$ and $\Phi_{j_2}^{m_2}$. According to Eq. (1.59$a$),

$$\Psi_J^M = \sum_{\substack{m_1, m_2 \\ (m_1+m_2=M)}} \langle j_1, j_2; m_1, m_2 \,|\, J, M\rangle \Phi_{j_1}^{m_1} \Phi_{j_2}^{m_2}. \tag{145}$$

Let us choose the phases such that

$$\begin{aligned} T\Phi_{j_1}^{m_1} &= (-1)^{m_1}\Phi_{j_1}^{-m_1}, \\ T\Phi_{j_2}^{m_2} &= (-1)^{m_2}\Phi_{j_2}^{-m_2}; \end{aligned} \tag{146}$$

that is, our choice is such that the $\mu_J$ of Eq. (144) is unity. Since the Wigner coefficient is real we obtain [on using the second relation (1.61)]

$$\begin{aligned} T\Psi_J^M &= (-1)^{j_1+j_2-J} \sum_{\substack{m_1, m_2 \\ m_1+m_2=M}} \langle j_1, j_2; -m_1, -m_2 \,|\, J, -M\rangle \\ &\quad \times (-1)^{m_1+m_2}\Phi_{j_1}^{-m_1}\Phi_{j_2}^{-m_2} \\ &= (-1)^{j_1+j_2-J}(-1)^M \Psi_J^{-M}, \end{aligned} \tag{147}$$

which is consistent with Eq. (144).

It is sometimes useful to redefine the Wigner coefficients in such a manner that the factor $(-1)^{j_1+j_2-J}$ does not appear in Eq. (147). Evidently the appropriate new coefficient is

$$\langle j_1, j_2; m_1, m_2 \,|\, J, M\rangle_T = i^{j_1+j_2-J}\langle j_1, j_2; m_1, m_2 \,|\, J, M\rangle. \tag{148}$$

This new Wigner coefficient, as just defined, has the disadvantage of not being in general real. On the other hand it is often the most convenient form to use when we are concerned with the operation of time reversal.

## REFERENCES


1. S. Tomonaga, *Progr. Theoret. Phys. (Kyoto)* **1,** 27 (1946); J. Schwinger, *Phys. Rev.* **74,** 1439 (1948).
2. F. J. Dyson, *Phys. Rev.* **75,** 486 (1949).
3. J. M. Jauch and R. Rohrlich, *The Theory of Photons and Electrons,* Addison-Wesley Publishing Co., Cambridge, Mass., 1955.
4. E. P. Wigner, *Göttingen Nachr.* **31,** 546 (1932).

# 3. THE DESCRIPTION OF SCATTERING PROCESSES

In this chapter we shall consider the relation of the time-dependent wave function $\Psi(t)$ to the physical measurements to be made on the system described by $\Psi(t)$. We have in mind, for example, two systems that are initially isolated but are later permitted to come into contact and thus to interact with each other. The interaction between these systems, while they are in contact, will in general lead to a change in their states. After a second interval of time the systems will have separated, and measurements carried out on these now isolated systems will indicate the effects of their mutual interaction. Alternatively, we may consider an unstable system, prepared at time $t_0$, which decays into two or more parts at a subsequent time. A measurement of the decay products provides a means for studying the interaction which led to the decay of the initial system.

Scattering experiments are then characterized by three time intervals. During the first interval the systems are prepared and brought into contact. The second interval is that during which they interact. In the third interval the systems produced in the interaction, reaction products, travel to the experimental measuring apparatus. The analysis of the first and third intervals involves only a kinematical study and can be done in a general and straightforward manner. The description of the interaction of the systems involves the most fundamental and difficult problems of physics.

In this chapter we shall address ourselves to the kinematical problem. Although relatively straightforward, this is an essential step in discussing the relation between experimental observations on a system and a theoretical description in terms of a state function $\Psi(t)$.

We shall begin our discussion with a description in Section 3.1 of wave packets. The construction of a wave packet corresponds to the preparation of the system at an initial time $t_0$. The time-dependent Schrödinger equation provides the description of the propagation of the wave packet into and during the period of interaction and finally to the experimental counting apparatus. In Section 3.2 we shall study the interacting systems, and in Section 3.3 we shall obtain the scattering cross section. Finally, in Section 3.4 we shall describe kinematical relations relevant to the scattering.

Our discussion may appear to be very long and circuitous to people familiar only with the customary "hit and run" treatment of scattering theory given in many elementary quantum mechanics texts. Indeed, we shall find ultimately the standard results which are usually qualified by the assurance that they can be justified by the consideration of wave packets. Our aim is to provide such assurance. We are *not* concerned with questions of mathematical rigor; in many cases this can be supplied, but it is largely irrelevant if not downright uninteresting. What we do care about is that our mathematical description of a scattering process should correspond accurately to the real physical world and that there are no lapses in physical rigor.

## 3.1 PREPARATION OF THE INITIAL STATE

### Description of Particles

Let us first consider a single, isolated physical system, such as an "elementary particle," or a molecule, which has been localized at a certain time within a volume $\mathscr{V}_0$. The Hamiltonian for the system is written as $K$. A complete set of eigenfunctions of $K$ are $\chi_{a,\mathbf{q}}$, where

$$K\chi_{a,\mathbf{q}} = \epsilon_{a,\mathbf{q}}\chi_{a,\mathbf{q}} \tag{1}$$

and $\epsilon_{a,\mathbf{q}}$ is the energy of the system when it is in the state $(a, \mathbf{q})$. The notion of an isolated system involves an idealization which can ordinarily be achieved with (almost) arbitrary precision. We emphasize, however, that $\chi_{a,\mathbf{q}}$ is considered to describe a physical system rather than an idealized system such as a "bare particle." There are no difficulties associated with the question of the existence of the Hamiltonian $K$. Indeed, it can be explicitly exhibited in a certain representation.[1]

Among a complete set of commuting observables $A_\lambda$ $(\lambda = 1, 2, \ldots, c)$ for the system is the total momentum $\mathscr{P}$, such that

$$\mathscr{P}\chi_{a,\mathbf{q}} = \mathbf{q}\chi_{a,\mathbf{q}}. \tag{2}$$

Application of the displacement operator $D = e^{-i\mathscr{P}\cdot\boldsymbol{\lambda}}$ [Eq. (1.40)] to $\chi_a$ leads to

$$D^{-1}\chi_{a,\mathbf{q}} = e^{i\mathbf{q}\cdot\boldsymbol{\lambda}}\chi_{a,\mathbf{q}},$$

from Eq. (2). It is therefore possible to write $\chi_{a,\mathbf{q}}$ as

$$\chi_{a,\mathbf{q}} = \frac{e^{i\mathbf{q}\cdot\mathbf{r}}}{(2\pi)^{3/2}}\, g_a, \tag{3}$$

[1] That is, $K = \Sigma_{a,\mathbf{q}}\, \epsilon_{a,\mathbf{q}} n_{a,\mathbf{q}}$ where $n_{a,\mathbf{q}}$ is the "occupation number" of the state $(a, \mathbf{q})$.

where **r** is a coordinate referring to the "position" of the system and $g_a$ is a wave function describing its internal state and satisfying the condition $D^{-1}g_a = g_a$. For example, if the system is an electron, $g_a$ is a Pauli spinor; if it is a molecule, $g_a$ is the internal wave function describing rotational, vibrational, and electronic states. It is often convenient to choose **r** as the center-of-mass or "barycentric" coordinate. The factor $(2\pi)^{-3/2}$ is explicitly written in Eq. (3) in accordance with the continuum normalization convention of Section 1.3. Then $g_a$ is also normalized.

It is perhaps well to describe at this point our use of the term *particle*. Roughly speaking, we shall call a physical system a particle if it is confined to a limited region of space by its own binding forces. More precisely, the system described by a particular $\chi_{a,\mathbf{q}}$ is a *particle* if and only if the internal wave function $g_a$ decreases rapidly to zero outside a finite region of its configuration space. In mathematical terms, the system is a particle if $g_a$ is normalizable to unity; that is, if $(g_a, g_a) = 1$. It is then not a "particle" if a continuum normalization $(g_{a'}, g_a) = \delta(a' - a)$ is required.[2]

The angular momentum of a particle in its own rest-frame is called its *spin*. In this coordinate system $\mathbf{q} = 0$ [in Eq. (3)], so $\chi_{a,0} = (2\pi)^{-3/2}g_a$ and in accordance with the description of angular momentum given in Section 1.2,

$$\mathbf{J}^2 g_a = j(j + 1)g_a,$$

where **J** is the angular momentum operator for the particle. It will be convenient to write **S**, instead of **J**, for the angular momentum operator for the internal state variables. Thus **S**, the spin operator, will be defined from

$$\mathbf{S}^2 g_a = S(S + 1)g_a, \tag{4}$$

where $S$ is called the particle spin. $S$ may take integral or half-integral values.

It is convenient to extend this definition of particle to include *unstable particles*. By an unstable particle we mean that the physical system under consideration can be usefully described by an *approximate* Hamiltonian $K$, as in Eq. (1), such that the resulting $g_a$ are normalizable to unity, as we have described. This pseudo-physical system then satisfies the conditions for being a particle. On the other hand, if the system is an unstable particle, the exact eigenfunctions $g_a^e$ of the complete Hamiltonian $K^e$ are not normalizable to unity. It is often convenient to refer to an unstable particle as a particle when the approximate wave functions $g_a$ provide an adequate description of the system during a portion of the time that an experiment is being performed. (Often the description may be adequate during the entire experiment.)

[2] Whereas the $\chi_{a,\mathbf{q}}$ form a complete set, it is not in general true that only those states $\chi_{a,\mathbf{q}}$, which describe particles as here defined, form a complete set.

We shall use the term particle, then, to describe unstable particles as well, with the understanding that this is a matter of convenience and that the decay rate of the unstable particles is not so great as to render our considerations meaningless.

**The Construction of a Wave Packet**

To localize the particles under consideration within a volume $\mathscr{V}_0$, we prepare a *wave packet* of states $\chi_{a,\mathbf{q}}$:

$$X_{a,\mathbf{p}} \equiv \int a(\mathbf{q} - \mathbf{p}) \frac{e^{i\mathbf{q}\cdot\mathbf{r}}}{(2\pi)^{3/2}} g_a \, d^3q$$

$$= \int a(\mathbf{q} - \mathbf{p}) \chi_{a,\mathbf{q}} \, d^3q. \tag{5a}$$

Here $a(\mathbf{q} - \mathbf{p})$ is the amplitude factor with which we weight the various eigenstates $\chi_a$ in $X_{a,\mathbf{p}}$. If we define the wave packet amplitude

$$G(\mathbf{r}) \equiv \int a(\boldsymbol{\rho}) \, e^{i\boldsymbol{\rho}\cdot\mathbf{r}} \, d^3\rho, \tag{6a}$$

we may rewrite Eq. (5) as

$$X_{a,\mathbf{p}} = \frac{e^{i\mathbf{p}\cdot\mathbf{r}}}{(2\pi)^{3/2}} g_a G(\mathbf{r}) = \chi_{a,\mathbf{p}} G(\mathbf{r}). \tag{7}$$

The wave packet $X_{a,\mathbf{p}}$ is normalized to unity. That is,

$$(X_{a,\mathbf{p}}, X_{a,\mathbf{p}}) = 1. \tag{5b}$$

This implies that

$$\int |a(\mathbf{q} - \mathbf{p})|^2 \, d^3q = 1 \tag{5c}$$

and that

$$\int |G(\mathbf{r})|^2 \, d^3r = (2\pi)^3. \tag{6b}$$

We shall be concerned only with wave packets which are *large* on a microscopic scale. That is, we suppose that the wave packets are large compared to the particle wavelength $h/p$,[3] and, of course, large compared to whatever intrinsic "size" the particle may have. Alternatively, we shall interpret this to mean that the spread in $\mathbf{q}$-values in the integrand of Eq. (5) is small compared to the accuracy with which macroscopic measurements of $\mathbf{p}$ are to be made. Therefore, within the wave packet the system described by (7) is in a plane-wave state.

[3] We have agreed to use a system of units in which $\hbar = 1$, except when the value of $\hbar$ is important for our discussion. Because of this, we shall keep $\hbar$ in our equations while discussing the spreading of wave packets.

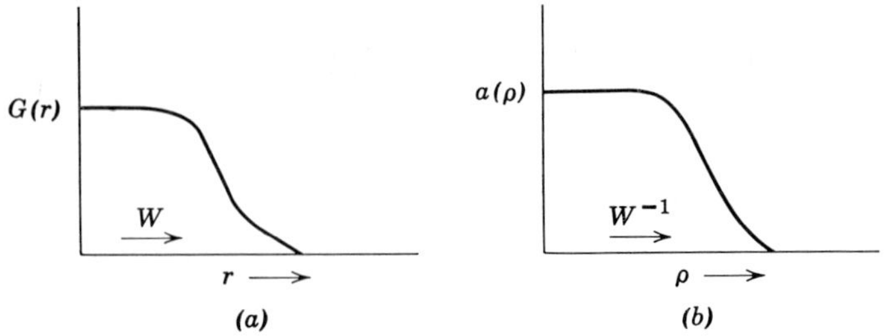


Fig. 3.1 Wave packet amplitudes: (*a*) in coordinate and (*b*) in momentum space.

Let us then suppose that $G(\mathbf{r})$ is characterized by a single parameter $w$, the "width of the packet."[4] By this we mean that $G(\mathbf{r})$ is approximately constant over the interval $0 < r < w$ and falls smoothly to zero for $r$ larger than $w$, although we do not necessarily assume that $G$ represents a spherical packet. To avoid introducing a second parameter, we shall also let the distance over which $G$ falls to zero be of order $w$. This is illustrated in Fig. 3.1*a*. The Fourier amplitude $a(\boldsymbol{\rho})$ is then to be characterized by a width $\hbar/w$ (see Fig. 3.1*b*).

According to Eq. (2.82), the time dependence of the system is described by the wave function

$$\phi_{a,\mathbf{p}}(t) = e^{-(i/\hbar)Kt}X_a = \int d^3q\, a(\mathbf{q}-\mathbf{p})\frac{e^{(i/\hbar)(\mathbf{q}\cdot\mathbf{r}-\epsilon_{a,\mathbf{q}}t)}}{(2\pi\hbar)^{3/2}}\, g_a. \tag{8}$$

Here $\epsilon_{a,\mathbf{q}}$ represents the kinetic energy, plus the internal energy, of the system in state $a$. For a nonrelativistic system of mass $M$

$$\epsilon_{a,q} = \frac{q^2}{2M} + \delta\epsilon_a, \tag{9}$$

where $\delta\epsilon_a$ is its internal energy. For a relativistic system, having rest-mass $M_a$ when in state $a$,

$$\epsilon_{a,q} = c(q^2 + M_a{}^2c^2)^{1/2}. \tag{10}$$

For some applications to solid-state physics or to nuclear physics we may wish to generalize $\epsilon_{a,q}$ to include a portion of the energy of interaction of a particle with the medium through which it is traveling. In this case we often refer to the system as a "quasi-particle," and there is a more general functional relationship than (10) between $\epsilon_{a,q}$ and $\mathbf{q}$.

We anticipate that the wave packet (8) will travel in space and at the same time spread out, in accordance with the indeterminacy principle.

[4] We recognize, of course, that such a wave packet will "contract" under a Lorentz transformation.

Because of the condition that $w \gg \hbar/p$, this spreading will not be of any physical significance for most scattering experiments, as we shall see.

Let us define

$$\boldsymbol{\rho} \equiv \mathbf{q} - \mathbf{p} \tag{11}$$

and expand $\epsilon_{a,\mathbf{q}}$ in a Taylor series in $\boldsymbol{\rho}$.[5]

$$\epsilon_{a,\mathbf{q}} = \epsilon_{a,\mathbf{p}} + \boldsymbol{\rho}\cdot\nabla_p\epsilon_{a,\mathbf{p}} + \tfrac{1}{2}\rho_i\rho_j\frac{\partial^2\epsilon_{a,\mathbf{p}}}{\partial p_i\,\partial p_j} + \cdots. \tag{12}$$

The *group velocity* $\mathbf{v}$ of the packet is defined as

$$\mathbf{v} = \nabla_p\epsilon_{a,\mathbf{p}}. \tag{13}$$

For the case of a particle, with $\epsilon_{a,\mathbf{p}}$ given by one of the expressions (9) or (10), $\mathbf{v}$ is, of course, just the *kinematical* velocity.

We shall now substitute the expansion (12) for $\epsilon_{a,\mathbf{q}}$ in the exponent in Eq. (8). With the assumption that

$$\frac{t}{2\hbar}\rho_i\rho_j\frac{\partial^2\epsilon_{a,\mathbf{p}}}{\partial p_i\,\partial p_j}$$

is small compared to unity, the exponential may be expanded to give

$$\begin{aligned}\phi_{a,\mathbf{p}}(t) &= \frac{e^{(i/\hbar)(\mathbf{p}\cdot\mathbf{r}-\epsilon_{a,\mathbf{p}}t)}}{(2\pi\hbar)^{3/2}}\,g_a\int d^3\rho\,a(\rho)\,e^{(i/\hbar)\boldsymbol{\rho}\cdot(\mathbf{r}-\mathbf{v}t)}\left(1 - \frac{it}{2\hbar}\rho_i\rho_j\frac{\partial^2\epsilon_{a,\mathbf{p}}}{\partial p_i\,\partial p_j} + \cdots\right)\\ &= e^{-i\epsilon_{a,\mathbf{p}}(t/\hbar)}\chi_{a,\mathbf{p}}\left(1 + \frac{it\hbar}{2}\frac{\partial^2\epsilon_{a,\mathbf{p}}}{\partial p_i\,\partial p_j}\nabla_i\nabla_j + \cdots\right)G(\mathbf{r} - \mathbf{v}t).\end{aligned} \tag{14}$$

In a first approximation, for which the parenthetical term may be replaced by unity, we obtain

$$\phi_{a,\mathbf{p}}(t) = e^{-i\epsilon_{a,\mathbf{p}}(t/\hbar)}\chi_{a,\mathbf{p}}G(\mathbf{r} - \mathbf{v}t). \tag{15}$$

The interpretation of Eq. (15) is very simple. In this approximation the wave packet moves with the constant velocity $\mathbf{v}$ and does not spread or change shape. We shall argue later that the approximation (15) can be satisfied with arbitrary accuracy in scattering experiments and in fact holds in practice with negligible error.

### The Spreading of a Wave Packet

To see this, let us consider a simple example for which

$$G(r) = e^{-r^2/2w^2} \tag{16}$$

[5] We are using the summation convention over the repeated vector indices $i$ and $j$ here.

For a nonrelativistic particle of mass $M$

$$\frac{\partial^2 \epsilon_p}{\partial p_i\, \partial p_j} = \delta_{ij}\frac{1}{M},$$

and (14) leads to

$$\phi_{a,\mathbf{p}}(t) = \chi_{a,\mathbf{p}}\, e^{-i\epsilon_{a,\mathbf{p}}(t/\hbar)} G(\mathbf{r} - \mathbf{v}t)\left\{1 + \frac{i\hbar t}{2Mw^2}\left[\frac{1}{w^2}(\mathbf{r} - \mathbf{v}t)^2 - 3\right] + \cdots\right\}. \tag{17}$$

For the case of a relativistic particle traveling parallel to the $z$-axis (or the 3-axis)

$$\frac{\partial^2 \epsilon_p}{\partial p_i\, \partial p_j} = \frac{c^2}{\epsilon_p}\left(\delta_{ij} - \delta_{i3}\,\delta_{j3}\frac{p^2c^2}{\epsilon_p{}^2}\right),$$

and (14) becomes

$$\phi_{a,\mathbf{p}}(t) = \chi_{a,\mathbf{p}} G(\mathbf{x} - \mathbf{v}t)\, e^{-i\epsilon_{a,\mathbf{q}}t/\hbar}$$
$$\times\left\{1 + \frac{i\hbar t}{2}\left(\frac{c^2}{\epsilon_p}\right)\frac{1}{w^2}\left[\frac{x^2 + y^2}{w^2} - 2 + \frac{(Mc^2)^2}{\epsilon_p{}^2}\left(\frac{(z - vt)^2}{w^2} - 1\right)\right] + \cdots\right\} \tag{18}$$

It is characteristic of a relativistic wave packet that spreading occurs more rapidly in directions perpendicular to $\mathbf{v}$.

It is seen from Eq. (17) that the rate of spreading of the packet is determined by the dimensionless parameter

$$\varepsilon \equiv \frac{t}{\hbar}\left(\frac{\hbar}{w}\right)^2 \frac{1}{2M}. \tag{19}$$

If, during the course of the experiment to be performed, the wave packet has traveled a distance $L = vt$, $\varepsilon$ may be reexpressed as

$$\varepsilon = \frac{L}{w}\frac{\hbar}{2wMv}. \tag{20}$$

For the case of a relativistic particle we may either replace $M$ by $\epsilon_p/c^2$ in Eqs. (19) or (20) or replace $Mv$ by the momentum $p$.

For $\varepsilon \ll 1$ the spreading of the wave packet is small and Eq. (15) can be used. When $\varepsilon \gg 1$ the packet will have spread to dimensions large compared to $w$. In the limit as $t \to \infty$, Eq. (8) is easily evaluated by a stationary phase approximation[6,7] to give

$$\phi_{a,\mathbf{p}}(t) = \frac{(d_1 d_2 d_3/\hbar^3)^{-1/2}}{t^{3/2}(2\pi\hbar)^{3/2}} \exp\left[\frac{i}{\hbar}(\mathbf{q}_0 \cdot \mathbf{r} - \epsilon_{a,q_0}t - \frac{\pi}{4}\right] g_a a(\mathbf{q}_0 - \mathbf{p}), \tag{21}$$

[6] See, for example, Ref. [1].

[7] See Eq. (10.289) for a further discussion of wave packet spreading in relativistic theories of particles.

where $d_1$, $d_2$, and $d_3$ are the eigenvalues of the matrix $\partial^2\epsilon_{a,p}/\partial p_i\, \partial p_j$ and $\mathbf{q}_0$ is the value of $\mathbf{q}$ for which the phase

$$P \equiv \hbar^{-1}[-i\hbar \ln a(\mathbf{q} - \mathbf{p}) + \mathbf{q}\cdot\mathbf{r} - \epsilon_{a,q}t]$$

is stationary. Now $\mathbf{q}_0 = m\mathbf{r}/t$ for a nonrelativistic particle and

$$\mathbf{q}_0 = \frac{\varepsilon_{a,\mathbf{q}0}}{c^2}\left(\frac{\mathbf{r}}{t}\right)$$

for a relativistic particle. The argument of $a(\mathbf{q}_0 - \mathbf{p})$ in Eq. (21) is essentially $\mathbf{r}/t - \mathbf{v}$. Thus the packet remains centered at the position of the classical trajectory. The particle density in Eq. (21) decreases as $t^{-3}$ near $\mathbf{r} = \mathbf{v}t$, the center of the packet. We readily verify from Eq. (21) that for conditions applicable to a scattering experiment the spreading rate of the packet never catches up with its kinematical motion. This means that we need not be concerned with the spreading of a packet at very late times.

A few simple examples serve to illustrate typical values of $\mathscr{e}$ and the spreading of the packet:

**Example 1.** A proton, for which $M = 1.6 \times 10^{-24}$ gm, has a velocity $v = 10^9$ cm/sec, or $\epsilon_p = 1$ Mev. The initial wave packet has a width $w = 0.1$ cm. We find that $\mathscr{e} = 1$ when the proton has traveled a distance

$$L = 10^{10}\text{ cm} = 10^5\text{ km},$$

which is comparable to the distance from the earth to the moon! Evidently this packet will not spread back to its point of origin.

**Example 2.** A 1-Mev proton is sent from the sun to the earth. If the initial spread in the wave packet is 7 cm, the packet will roughly have doubled its dimensions at the earth.

**Example 3.** In a laboratory experiment, a beam of 100 Mev electrons travels 10 m from the accelerator to the counters. If the initial beam radius is $10^{-2}$ cm, the value of $\mathscr{e}$ at the counters is about $10^{-6}$.

We have given these examples to emphasize that under the conditions that permit macroscopic observations, the spreading of quantum mechanical wave packets is ordinarily quite negligible. In fact, the experimental techniques for measuring scattering cross sections assume this to be true, since classical mechanics is conventionally used in calculating orbits in accelerators, through focusing magnets, and so on.

We conclude, therefore, that Eq. (15) may be assumed to be correct with *arbitrary* precision in discussing the propagation of wave packets under the macroscopic conditions in which an experiment is done. (Formally, we need only *scale up the dimensions* of $L$ and $w$ to insure ourselves that $\mathscr{e}$ is arbitrarily small.) The final, asymptotic form (21) for the packet will be considered irrelevant for our considerations, since long before appreciable spreading of the packet has occurred the particles being studied will have entered the counter, the shielding walls, or at least have left the laboratory.

**Wave Packets for Two Particles**

A single wave packet such as we have just described is appropriate when the decay of a simple unstable kind of particle is to be studied. A scattering experiment, on the other hand, requires two wave packets; one to describe the *scattered* particle and one to describe the *scattering* particle.

Our analysis of wave packets is easily extended to describe two *noninteracting* wave packets. Setting $\hbar = 1$ in the subsequent equations we introduce

$$\chi_a = \left[\frac{e^{i\mathbf{q}_1\cdot\mathbf{r}_1}}{(2\pi)^{3/2}} g_{a1}\right]\left[\frac{e^{i\mathbf{q}_2\cdot\mathbf{r}_2}}{(2\pi)^{3/2}} g_{a2}\right] \tag{22}$$

as an eigenstate for two noninteracting systems 1 and 2.[8] Equation (22) involves an obvious extension of the earlier notation of this chapter: $\mathbf{q}_1$ and $\mathbf{q}_2$ represent the respective momenta for systems 1 and 2, $g_{a1}$ and $g_{a2}$ the respective internal state functions, and $\mathbf{r}_1$ and $\mathbf{r}_2$ the respective coordinate positions. Now $K$ is taken to be the Hamiltonian for the noninteracting systems, so

$$K\chi_a = E_a\chi_a, \tag{23}$$

where

$$E_a = \epsilon_{a1,\mathbf{q}1} + \epsilon_{a2,\mathbf{q}2}. \tag{24}$$

Wave packets may be constructed for each system with respective amplitudes $a_B(\mathbf{q}_1 - \mathbf{p}_1)$ and $a_T(\mathbf{q}_2 - \mathbf{p}_2)$, so[9]

$$\begin{aligned} G_B(\mathbf{r}_1) &\equiv \int d^3\rho\, a_B(\boldsymbol{\rho})\, e^{i\boldsymbol{\rho}\cdot\mathbf{r}_1}, \\ G_T(\mathbf{r}_2) &\equiv \int d^3\rho\, a_T(\boldsymbol{\rho})\, e^{i\boldsymbol{\rho}\cdot\mathbf{r}_2}. \end{aligned} \tag{25}$$

The "two-particle wave packet" is then

$$\begin{aligned} X_a &= \int d^3q_1\, d^3q_2 a_B(\mathbf{q}_1 - \mathbf{p}_1)a_T(\mathbf{q}_2 - \mathbf{p}_2)\left[\frac{e^{i\mathbf{q}_1\cdot\mathbf{r}_1}}{(2\pi)^{3/2}} g_{a1}\right]\left[\frac{e^{i\mathbf{q}_2\cdot\mathbf{r}_2}}{(2\pi)^{3/2}} g_{a2}\right] \\ &= G_B(\mathbf{r}_1)G_T(\mathbf{r}_2)\chi_{\bar{a}}, \end{aligned} \tag{26a}$$

[8] If systems 1 and 2 are identical particles, or contain identical particles, we must appropriately symmetrize Eq. (22). This is easily done, as will be shown in detail in Chapter 4.

[9] We shall later find it convenient to consider $G_B$ and $G_T$ as describing, respectively, "beam" and "target" wave packets. Each wave packet is now normalized in accordance with Eqs. (5*c*) and (6*b*).

where

$$\chi_{\bar{a}} \equiv \left[\frac{e^{i\mathbf{p}_1\cdot\mathbf{r}_1}}{(2\pi)^{3/2}}\, g_{a\mathbf{p}}\right]\left[\frac{e^{i\mathbf{p}_2\cdot\mathbf{r}_2}}{(2\pi)^{3/2}}\, g_{a2}\right]. \tag{27}$$

Similarly, the behavior of the two systems in time is described by the wave function

$$\phi_a{}^{(t)} = e^{-iKt}X_a = G_B(\mathbf{r}_1 - \mathbf{v}_1 t)G_T(\mathbf{r}_2 - \mathbf{v}_2 t)\chi_{\bar{a}}\, e^{-iE_{\bar{a}}t} \tag{28a}$$

where the group velocities are defined as in Eq. (13),

$$\mathbf{v}_1 = \nabla_{p1}\epsilon_{a1,p1}, \tag{29}$$

$$\mathbf{v}_2 = \nabla_{p2}\epsilon_{a2,p2},$$

and

$$E_{\bar{a}} = \epsilon_{a1,p1} + \epsilon_{a2,p2}.$$

It is evident that under some circumstances we will wish to form packets for variables other than the total momentum of a particle. This can, of course, be included in our description. It is also evidently straightforward to form wave packets, such as that of Eq. (26*a*), for states of any number of particles. For example, if $\chi_a$ is a state of $\mu$ particles, having respective momenta $\mathbf{q}_1, \ldots, \mathbf{q}_\mu$, then

$$X_a \equiv \int d^3q_1 \cdots d^3q_\mu\, a_1(\mathbf{q}_1 - \mathbf{p}_1) \cdots a_\mu(\mathbf{q}_\mu - \mathbf{p}_\mu)\chi_a \tag{26b}$$

is such a wave packet state. The time-dependent wave function $\phi_a(t)$ is in this case again given by the expression

$$\phi_a(t) = e^{-iKt}\, X_a. \tag{28b}$$

## 3.2 INTEGRATION OF THE SCHRÖDINGER EQUATION FOR SCATTERING PROCESSES

### Description of the Interaction

We are now in a position to describe the scattering of two physical systems, designated as 1 and 2, respectively. We shall assume that each of these physical systems is a particle as defined in Section 3.1—but we shall refrain from any implication that they are necessarily "simple" or "elementary." Specifically it will be convenient to refer to particle 1 as the "scattered particle," or "beam particle," and to particle 2 as the "scattering particle," or "target particle."

At an initial time $t_0$ it is supposed that wave packets have been prepared for these two particles, as illustrated in Fig. 3.2. At the time $t_0$ wave packet 1 has its center at a position $\mathbf{R}_1{}^0$ and has a radial extent characterized by the

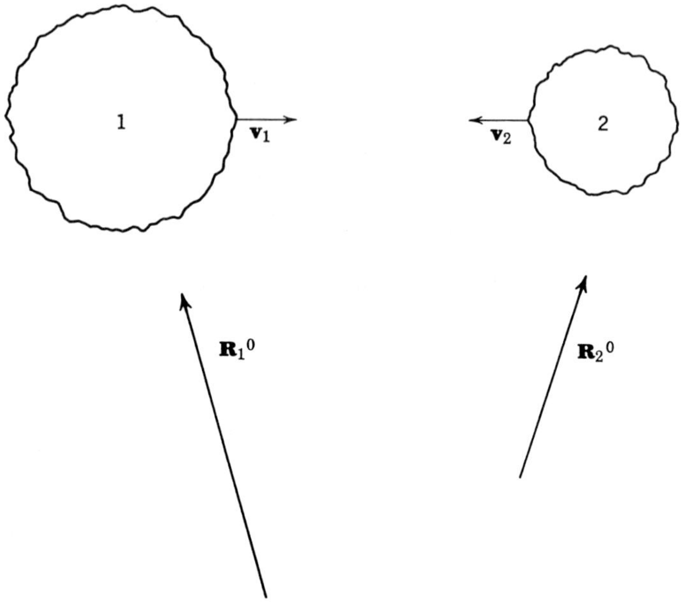


Fig. 3.2 Configuration of the wave packets at initial time $t_0$.

width parameter $w_B$, as described in Section 3.1. Similarly, wave packet 2 has its center at $\mathbf{R}_2{}^0$ and has a radial extent of order $w_T$. The overlap of the two packets at $t = t_0$ is assumed to be arbitrarily small.[10] For some later applications it will be convenient to assume that $w_B \gg w_T$, thus permitting some geometrical simplification without any significant loss of generality.

The respective velocities of the two wave packets are taken as $\mathbf{v}_1$ and $\mathbf{v}_2$, as defined by Eq. (29).[11] We shall suppose that the centers of the two wave packets $\mathbf{R}_1$ and $\mathbf{R}_2$ both coincide with the origin of the coordinate system at the time $t = 0$. Then during the time interval from $t = t_0$ to $t = 0$, we have

$$\mathbf{R}_1 = \mathbf{v}_1 t, \qquad \mathbf{R}_2 = \mathbf{v}_2 t. \tag{30}$$

We shall write $L \equiv |\mathbf{R}_1{}^0 - \mathbf{R}_2{}^0|$, so

$$|t_0| \approx \frac{L}{|\mathbf{v}_1 - \mathbf{v}_2|}\,. \tag{31}$$

[10] In accordance with the discussion in Section 3.1 we shall also assume that the spreading of the two packets is arbitrarily small during the course of the scattering experiment.

[11] In the so-called *laboratory frame of reference* $\mathbf{v}_2 = 0$, so the scattering particle is at rest initially.

Most scattering experiments are performed under conditions such that $L \gg w_1, \; w_2$. At the time $t_c$ defined by

$$|t_c| = \frac{w_1 + w_2}{|\mathbf{v}_1 - \mathbf{v}_2|} \tag{32}$$

the wave packets overlap and the *collision* between the two particles begins to take place. At later times, the wave packets will have moved past each other and the collision will be completed. The interaction between the two particles will give rise to *secondary*, or *scattered waves*, which are detected in the *counting apparatus* (as in Fig. 3.2). The scattered waves describe the particles that have been *scattered* or *created* during the collision. For example, if the incident particles are molecules, an electron from one molecule may be "picked up" by the other molecule, or one of the molecules may be ionized and a free electron ejected. If the initial particles are protons, several pi-mesons may be created during the collision, and so on.

The wave functions $\chi_a$ of Eq. (23) represent a complete set for the two noninteracting particles, and their Hamiltonian is $K$. The wave packet state for the two particles is taken as $X_a$, as given by Eq. (26):

$$X_a = \int d^3q_1 \, d^3q_2 A(\mathbf{q}_1, \mathbf{q}_2)\chi_a, \tag{33}$$

where we use the abbreviation

$$A(\mathbf{q}_1, \mathbf{q}_2) = a_B(\mathbf{q}_1 - \mathbf{p}_1)a_T(\mathbf{q}_2 - \mathbf{p}_2). \tag{34}$$

It imposes no serious restriction on our wave packets to assume that the integrals in Eq. (33) converge uniformly in the following sense: If $\mathcal{O}$ is an operator for which $\mathcal{O}\chi_a$ is defined, then $\mathcal{O}X_a$ is also defined, and furthermore

$$\mathcal{O}X_a = \int d^3q_1 \, d^3q_2 A(\mathbf{q}_1, \mathbf{q}_2)\mathcal{O}\chi_a. \tag{35}$$

The Hamiltonian for the complete interacting system will be written as $H$. $H$ is the operator that generates infinitesimal displacements in time on the system as in Section 1.1. At times long before particles 1 and 2 interact,

$$H = K,$$

the Hamiltonian of Eq. (23). It is convenient to define an *interaction Hamiltonian* $V$ by the equation

$$V \equiv H - K. \tag{36}$$

We shall assume that $V$ has a *range* of interaction, $\mathscr{R}$, such that $V$ leads to a null-state vector when operating on states which correspond to a separation large compared to $\mathscr{R}$ of particles 1 and 2.[12] We may evidently suppose that $\mathscr{R} \ll w_B$ or $w_T$. Then during the interval of time between $t_0$ and $t_c$ in which the wave packets do not overlap we may neglect $V$ and set $H = K$.

In assuming that the momenta $\mathbf{q}_1$ and $\mathbf{q}_2$ of the particles in state $a$ are constants of the motion, we have neglected any interaction of these particles with their environment. Such interactions are of course necessary in preparing the wave packet states prior to the time $t_0$. Ordinarily, however, external forces do not have any significant effect on the collision dynamics. We shall therefore assume that the total momentum $\mathscr{P}$ commutes with the Hamiltonian $H$;[13] that is,

$$[\mathscr{P}, H] = 0.$$

We also assume that the particles emerging from the scattering process travel to the counting apparatus without being influenced by external fields. This is frequently not the case; for example, magnetic fields are often employed in the analysis of an experiment. Such complications may be avoided if we consider these fields as part of the counting apparatus. A more satisfactory description takes the external fields into account in studying the wave packet propagation. We find then that the centers of the packets follow Newton's equations of motion. Such complications are not very relevant for our discussion, however, and will be ignored.

**Time-Dependent Wave Function**

In accordance with our description of the scattering process, the state function $\Psi(t)$ for the complete system must be equal to $\phi_a(t_0) = e^{-iKt_0}X_a$ at $t = t_0$ [see Eqs. (28)]. We may therefore write $\Psi(t)$ as

$$\Psi(t) = e^{-iH(t-t_0)}e^{-iKt_0}X_a, \tag{37}$$

since this is the solution of the Schrödinger equation that reduces to $\phi_a(t_0)$ at $t = t_0$. Now according to Eq. (2.87$a$)

$$e^{iHt_0}e^{-iKt_0} = U(0, t_0), \tag{38}$$

[12] This is not true for gravitional or Coulomb potentials. We do not anticipate applications to scattering by gravitional forces. In practice, Coulomb forces are shielded by the environment at sufficiently large distances, so our assumption of a finite range can be applied to Coulomb scattering.

[13] The reader will easily verify that our discussion requires little modification for the case that $[\mathscr{P}, H] \neq 0$. The effect of the environment on Coulomb scattering is rather trivial (see footnote 12), since this does not violate the condition $[\mathscr{P}, H] \neq 0$.

so Eq. (37) may be rewritten as

$$\Psi(t) = e^{-iHt}\Gamma, \tag{39}$$

where

$$\Gamma \equiv U(0, t_0)X_a = \Psi(0). \tag{40}$$

The possibility that physical systems may run "forward" or "backward" was described in Chapter 2 in connection with a discussion of the time reversal operation. For this reason it is useful to include both the cases that $t_0 < 0$ and $t_0 > 0$ in our analysis. In the former, or "natural," case the scattering develops as described. When $t_0 > 0$, the time ordering of the scattering process is simply reversed. We shall introduce the notation

$$\Gamma = \Gamma^+ \quad \text{for} \quad t_0 < 0$$

and

$$\Gamma = \Gamma^- \quad \text{for} \quad t_0 > 0 \tag{41}$$

when we wish to indicate one of these two specific cases.

Let us first simplify the expression (40). According to the integral equation satisfied by $U(0, t)$, Eq. (2.97),

$$U(0, t_0) = 1 - i\int_{t_0}^{0} dt' U(0, t')V(t') = 1 - i\int_{t_0}^{0} dt'\, e^{iHt'}V\, e^{-iKt'} \tag{42}$$

[since $V(t) = e^{iKt}V\,e^{-iKt}$]. Thus Eq. (40) may be reexpressed as [compare Eq. (23)]

$$\begin{aligned}\Gamma &= X_a - i\int_{t_0}^{0} dt'\, e^{iHt'}V\, e^{-iKt'}X_a \\ &= X_a - i\int d^3q_1\, d^3q_2 A(\mathbf{q}_1, \mathbf{q}_2)\int_{t_0}^{0} dt'\, e^{iHt'}V e^{-iE_at'}\chi_a.\end{aligned} \tag{43}$$

We have here made use of our postulate (35) concerning the wave packets.

According to Eq. (28$a$), the first form for $\Gamma$ here leads to

$$\Gamma = X_a - i\int_{t_0}^{0} dt'\, e^{iHt'}VG_1(\mathbf{r}_1 - \mathbf{v}_1t')G_2(\mathbf{r}_2 - \mathbf{v}_2t')\chi_{\bar{a}}e^{-iE_{\bar{a}}t}. \tag{44}$$

$\Gamma$ is thus independent of $t_0$, since for $t'$ in the interval between $t_0$ and $t_c$

$$VG_1(\mathbf{r}_1 - \mathbf{v}_1t')G_2(\mathbf{r}_2 - \mathbf{v}_2t') = 0. \tag{45}$$

This follows because the wave functions do not overlap and because of our assumption that the range $\mathscr{R}$ of $V$ is small compared to $w_B$ and $w_T$.[14]

[14] Equation (45) is not *rigorously* valid because of the spreading of the wave packets. We saw in Section 3.1, however, that it is possible to choose our wave packets in such a manner that the error in (45) is *arbitrarily* small.

To simplify Eq. (43) let us introduce a complete set of eigenfunctions $\psi_\lambda$ of $H$.

$$H\psi_\lambda = H_\lambda\psi_\lambda, \tag{46}$$

where $H_\lambda$ is the eigenvalue of $H$. We may define

$$\Gamma_\lambda \equiv (\psi_\lambda, \Gamma), \tag{47}$$

$$X_a(\lambda) \equiv (\psi_\lambda, X_a),$$

so

$$\Gamma = \sum_\lambda \psi_\lambda \Gamma_\lambda, \tag{48}$$

$$X_a = \sum_\lambda \psi_\lambda X_a(\lambda).$$

In the $\lambda$-representation, we may write (43) as

$$\Gamma_\lambda - X_a(\lambda) = -i\int d^3q_1\, d^3q_2 A(\mathbf{q}_1, \mathbf{q}_2) \int_{t_0}^{0} dt'\, e^{i(H_\lambda - E_a)t'}(\psi_\lambda, V\chi_a)$$

$$= -i\int d^3q_1\, d^3q_2 A(\mathbf{q}_1, \mathbf{q}_2)\left[\frac{1}{i(H_\lambda - E_a)}(1 - e^{i(H_\lambda - E_a)t_0})\right](\psi_\lambda, V\chi_a). \tag{49}$$

It is evident from Eqs. (39) and (48) that $\Gamma_\lambda$ will appear within an integrand in our final applications. We may always choose the eigenvalue $H_\lambda$ as one of the set of quantum numbers $\lambda$, taking

$$\lambda = (l, H_\lambda), \tag{50}$$

where by $l$ we mean the remaining quantum numbers in the set $\lambda$. Thus we may set

$$\sum_\lambda = \sum_l \int \rho_\lambda\, dH_\lambda, \tag{51}$$

where $\rho_\lambda\, dH_\lambda$ is the number of eigenvalues $H_\lambda$ within the interval $dH_\lambda$. We now assert that within the wave packet integral (49)

$$G^\pm(E_a - H_\lambda) \equiv \frac{1}{E_a - H_\lambda}(1 - e^{i(H_\lambda - E_a)t_0})$$

$$= P\frac{1}{E_a - H_\lambda} \mp i\pi\delta(E_a - H_\lambda). \tag{52}$$

Here $G^+$ refers to the case that $t_0 < 0$ and $G^-$ to the case that $t_0 > 0$, as in Eqs. (41). The symbol $P$ is introduced to indicate that a *principal value* is to be taken when integrating across the singularity of $(E_a - H_\lambda)^{-1}$.

The result (52) is more or less obvious when

$$|H_\lambda - E_{\bar{a}}| \gg \Delta E_a, \tag{53}$$

where $E_a$ is defined by Eq. (29) and $\Delta E_a$ is the "spread" in $E_a$ values over the wave packet. That is,

$$\Delta E_a \approx \frac{\hbar}{w_1} v_1 + \frac{\hbar}{w_2} v_2. \tag{54}$$

Then we may replace $(H_\lambda - E_a)^{-1}$ by $(H_\lambda - E_{\bar{a}})^{-1}$ in Eq. (49). The term involving $e^{i(H_\lambda - E_a)t_0}$ then vanishes, since it is

$$\int \frac{d^3q_1\, d^3q_2}{H_\lambda - E_{\bar{a}}} A(\mathbf{q}_1, \mathbf{q}_2)\, e^{i(H_\lambda - E_a)t_0}(\psi_\lambda, V\chi_a)$$
$$= \frac{e^{i(H_\lambda - E_{\bar{a}})t_0}}{H_\lambda - E_{\bar{a}}} (\psi_\lambda, VG_1(\mathbf{r}_1 - \mathbf{v}_1 t_0)G_2(\mathbf{r}_2 - \mathbf{v}_2 t_0)\chi_{\bar{a}}) = 0,$$

according to Eq. (45).

On integrating $G^\pm$ across the value $H_\lambda = E_a$, we find that the exponential term does give a contribution, however. [We emphasize, again, that the symbolic relation (52) has meaning only when inserted into an integral such as Eq. (49).] To see this, let us write

$$G^\pm = \lim_{\eta \to 0(+)} G^\pm(\eta), \tag{55}$$

where

$$G^\pm(\eta) = \frac{1}{E_a \pm i\eta - H_\lambda} (1 - e^{i(H_\lambda - E_a \mp i\eta)t_0})$$
$$= \frac{1}{E_a \pm i\eta - H_\lambda} - i\int_{t_0}^{\mp\infty} d\tau\, e^{-i(E_a \pm i\eta - H_\lambda)\tau} \tag{56}$$

Now

$$I^\pm(\eta) \equiv \int d^3q_1\, d^3q_2 A(\mathbf{q}_1, \mathbf{q}_2) G^\pm(\eta)(\psi_\lambda, V\chi_a) \tag{57}$$

clearly reduces to $\Gamma_\lambda - X_a(\lambda)$ as $\eta \to 0(+)$, since the expression (49) is not singular in the limit. We observe, that

$$\int d^3q_1\, d^3q_2 A(\mathbf{q}_1, \mathbf{q}_2) \int_{t_0}^{\mp\infty} d\tau\, e^{-i(E_a \pm i\eta - H_\lambda)\tau} (\psi_\lambda, V\chi_a)$$
$$= \int_{t_0}^{\pm\infty} d\tau\, e^{i(H_\lambda \mp i\eta)\tau} (\psi_\lambda, V\phi_a(\tau)) = 0, \tag{58}$$

if we do not go to the limit as $\eta \to 0(+)$ until after integrating over $H_\lambda$. Here $\phi_a(\tau)$ is the wave packet state (28*a*) and according to Eq. (45),

$$V\phi_a(\tau) = 0 \tag{59}$$

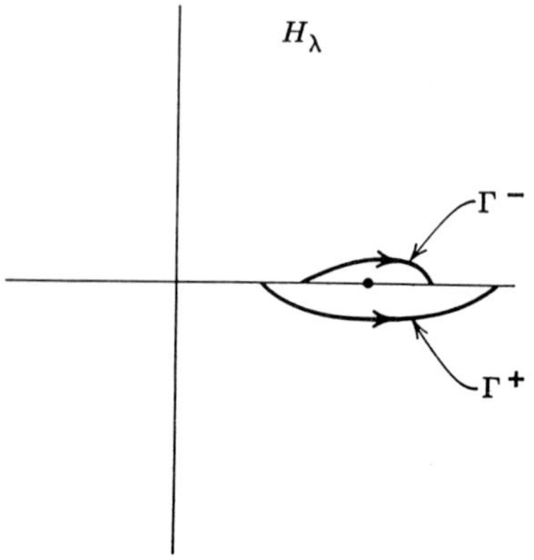


Fig. 3.3 Contours defining integral across the singularity for the expression (63).

during the entire interval of $\tau$-values for the integral (58).[15] Thus

$$I^{\pm}(\eta) = \int d^3q_1\, d^3q_2 A(\mathbf{q}_1, \mathbf{q}_2) \frac{1}{E_a \pm i\eta - H_\lambda} (\psi_\lambda, V\chi_a). \tag{60}$$

We may therefore write

$$\Gamma^{\pm} = \lim_{\eta \to 0(+)} \Gamma^{\pm}(\eta) \tag{61}$$

where

$$\Gamma_\lambda^{\pm}(\eta) = \int d^3q_1\, d^3q_2 A(\mathbf{q}_1, \mathbf{q}_2)\left(\psi_\lambda, \left(1 + \frac{1}{E_a \pm i\eta - H_\lambda} V\right)\chi_a\right). \tag{62}$$

We evidently may not now go to the limit as $\eta \to 0(+)$ until *after* integrating over $H_\lambda$.

The proof of Eq. (52) is completed by the observation that

$$P \frac{1}{E_a - H_\lambda} \mp i\pi\delta(E_a - H_\lambda) = \lim_{\eta \to 0(+)} \frac{1}{E_a \pm i\eta - H_\lambda}, \tag{63}$$

where the limit as $\eta \to 0(+)$ is taken only after integrating over $H_\lambda$. To verify Eq. (63), we consider the integral of a function $F(H_\lambda)$, nonsingular at $H_\lambda = E_a$

$$\lim_{\eta \to 0(+)} \int \frac{F(H_\lambda)\, dH_\lambda}{E_a \pm i\eta - H_\lambda} = -P \int \frac{F(H_\lambda)\, dH_\lambda}{H_\lambda - E_a} \mp \int_C \frac{F(H_\lambda)\, dH_\lambda}{H_\lambda - E_a}.$$

Here $C$ is a semicircular contour of infinitesimal radius, centered at $H_\lambda = E_a$, and above the real axis for $\Gamma^-$ (below it for $\Gamma^+$). This is illustrated in Fig. 3.3.

[15] This is actually a somewhat overly simplified argument. As $t_0 \to \mp\infty$ the wave packet will spread. On following the arguments which led to Eq. (21) we see that the spreading of the packets does not exceed the rate at which the packets recede from each other, so Eq. (59) is valid.

The quantities $\Gamma^{\pm}$ may be reexpressed in a general representation using Eqs. (48):

$$\Gamma^{\pm} = \lim_{\eta \to 0(+)} \Gamma^{\pm}(\eta), \tag{64}$$

where

$$\Gamma^{\pm}(\eta) = \sum_{\lambda} \psi_{\lambda} \Gamma_{\lambda}{}^{\pm}(\eta)$$
$$= \int d^3q_1\, d^3q_2 A(\mathbf{q}_1, \mathbf{q}_2)\left(1 + \frac{1}{E_a \pm i\eta - H} V\right)\chi_a. \tag{65}$$

Here $(E_a \pm i\eta - H)^{-1}$ represents the *Green's function* for the Hamiltonian $H$.

**Time-Independent Wave Functions**

It is convenient to introduce a set of wave functions $\psi_a{}^{\pm}$ defined by

$$\psi_a{}^{\pm} = \lim_{\eta \to 0(+)} \psi_a{}^{\pm}(\eta), \tag{66}$$

where

$$\psi_a{}^{\pm}(\eta) = \left(1 + \frac{1}{E_a \pm i\eta - H} V\right)\chi_a. \tag{67}$$

In terms of these,

$$\Gamma^{\pm} = \int d^3q_1\, d^3q_2 A(\mathbf{q}_1, \mathbf{q}_2)\psi_a{}^{\pm}. \tag{68}$$

We see that $\Gamma^{\pm}$ represents the same wave packet as $X_a$, except that in $\Gamma^{\pm}$ the $\chi_a$ are replaced by the wave functions $\psi_a{}^{\pm}$.

The importance of Eq. (68) will become evident later. We note now, however, that the $\psi_a{}^{\pm}$ are solutions of the Schrödinger equation

$$H\psi_a{}^{\pm} = E_a\psi_a{}^{\pm}. \tag{69}$$

This follows directly from Eqs. (66) and (67):

$$(H - E_a)\psi_a{}^{\pm} = \lim_{\eta \to 0(+)} \left[(H - E_a)\chi_a - (E_a \pm i\eta - H)^{-1}(E_a - H)V\chi_a\right]$$
$$= V\chi_a - V\chi_a = 0,$$

since $H\chi_a = (K + V)\chi_a = (E_a + V)\chi_a$. We shall see in Chapter 5 that the wave functions $\psi_a{}^{+}$ or $\psi_a{}^{-}$ are orthonormal but do not in general form a complete set. Therefore either the $\psi_a{}^{+}$ or the $\psi_a{}^{-}$ may be chosen as part of the complete set $\psi_{\lambda}$ of Eq. (46), and for these $H_{\lambda} = E_a$.

The notation of Eq. (66), by which the limit $\eta \to 0(+)$ is indicated, is a little too cumbersome to be convenient for us. Since we have described the limiting process and have understood that the limit is to be taken after integrating over the singularity of $(E_a \pm i\eta - H)^{-1}$, we can simplify our notation by writing $\psi_a^{\pm}$ for $\psi_a^{\pm}(\eta)$, with the understanding that the limiting procedure is implied. There will be no confusion from this convention; where ambiguities might arise, we shall write out the limit explicitly.

The results obtained here are also applicable to the study of the decay of unstable particles. In this case only a single-particle wave packet, as in Eq. (5), need be prepared at the initial time $t_0$. The interaction $V$ now describes the mechanism responsible for the decay of the particle. We shall discuss the special features of this problem in Chapter 8.

Our results are also directly applicable to a scattering for which any number of particles interact. In the case of $\mu$ particles in the initial state ($\mu > 2$) we would begin with the wave packet (26*b*) at time $t = t_0$. All $\mu$ particles would be supposed to come into simultaneous collision at time $t = 0$.[16] It is evident that the formal consideration leading to the wave function at $t = 0$, $\Psi(0) = \Gamma$ as given by Eq. (68) are unchanged by this generalization.

## 3.3 EVALUATION OF THE SCATTERING CROSS SECTION

In Section 3.2 we considered the preparation of two systems at time $t_0$ in such a manner that they interact at time $t \approx 0$. The wave function $\Psi(t)$, describing the complete system, was determined to be of the form [compare Eqs. (39) and (68)]

$$\begin{aligned}\Psi(t) = e^{-iHt}\Gamma^{\pm} &= \int d^3q_1\, d^3q_2 A(\mathbf{q}_1, \mathbf{q}_2)\, e^{-iHt}\psi_a^{\pm} \\ &= \int d^3q_1\, d^3q_2 A(\mathbf{q}_1, \mathbf{q}_2)\, e^{-iE_a t}\psi_a^{\pm}. \qquad (70)\end{aligned}$$

With Eq. (70) our analysis is half completed. That is, we have described the behavior of the two systems from the time of their preparation to the time of their interaction. We must now describe the final phase of the process, in which the particles emerge from the reaction and travel to the detector. We shall be interested now only in the natural order for a scattering process and so consider that $t_0 = -|t_0|$. This means that we shall use only $\Gamma^{+}$ in Eq. (70).

### The Production of New Particles in a Reaction

In general, the particles reaching the detector may not be the same as those which originally interacted. They may not even represent eigenstates of the Hamiltonian $K$ [compare Eq. (23)] of the original two noninteracting

[16] The case of three particles colliding simultaneously is of interest in molecular reactions. Three-particle collisions are discussed in Appendix B.

particles. This is the case, for example, for such reactions as

$$H + H \rightarrow H^- + p,$$
$$d + d \rightarrow H^3 + p,$$

which are called *pick-up processes*. In the interest of having a uniform terminology, we shall refer to the particles in the initial state as the *incident particles*. The interaction between these particles will be called a *reaction* or *scattering process*. The particles emerging from a reaction will be referred to as *reaction products*. Reactions for which the only reaction products are the incident particles themselves are usually called just "the scattering of two particles."

Let us suppose that the experimental counters used to detect the reaction products are designed to detect those particles and states of those particles which have wave functions $\chi'_b$. The emerging particles are assumed to be sufficiently separated in space that their mutual interactions may be neglected. The Hamiltonian for these noninteracting particles is written as $K'$, so

$$K'\chi'_b = E_b\chi'_b, \tag{72}$$

in analogy to Eq. (23). Since $H$ is the Hamiltonian for the complete system, we may define

$$V' \equiv H - K'. \tag{73}$$

Thus $V'$ is the part of $H$ which describes the interaction between the particles whose eigenstates are the $\chi'_b$.

The states $\chi'_b$, as well as the states $\chi_a$, are members of a complete set of orthonormal functions satisfying *continuum* normalization conditions and the Schrödinger equations (23) and (72).

$$\begin{aligned} (\chi_{a'}, \chi_a) &= \delta_{a',a}, \\ (\chi'_{b'}, \chi'_b) &= \delta_{b',b}. \end{aligned} \tag{74}$$

Thus, for example, we may expand $\Psi(t)$ in either set.

$$\Psi(t) = \sum_a \chi_a(\chi_a, \Psi(t))$$

or

$$\Psi(t) = \sum_b \chi'_b(\chi'_b, \Psi(t)).$$

We have here used the summation symbol $\Sigma_a$ or $\Sigma_b$ as representing a sum over discrete quantum states and an integral over continuous quantum variables.[17]

[17] Strictly speaking, we should use a Stieltjes integral representation, $\int da$, for $\Sigma_a$. The summation symbol is more convenient, however, since we often do not wish to distinguish between *continuum* quantization and quantization in a box.

We emphasize that the representation $\chi'_b$ depends on the conditions set by the experimenter, who measures the cross sections. That is, we choose to use a representation such that $\chi'_b$ is an eigenstate of those particles and the states of the particles that the counting apparatus is supposed to detect. From a purely mathematical point of view, we may use any complete set of wave functions to describe the final state. We should then have to construct the linear combination of states corresponding to the "particles" that the counter can detect. This linear combination would, of course, be just that appropriate to form the $\chi'_b$. Therefore, we might as well use the $\chi'_b$ from the outset. (The case $K' = K$, $V' = V$ is that most often encountered.)

An experimental physicist ordinarily chooses his experimental geometry in such a manner that [compare Eqs. (28)] for the states $\chi'_b$ which are to be observed

$$(\chi'_b, \phi_a) = 0 \tag{75a}$$

to a high degree of accuracy. If he did not, his counters would be *flooded* by the incident beam. The exception to this type of experiment is an "absorption measurement." Then the experimenter first measures the flux of incident particles with the "target particles" removed. Then, with the target in place, he again measures the incident particle flux. The difference in the two counting rates gives the rate at which particles are scattered. In this case

$$\chi'_b = \chi_{a'} \tag{76}$$

where the states $\chi_{a'}$ are those included in the wave packet $X_a$. Equations (75) and (76) may be summarized in one expression.

$$\begin{aligned}(\chi'_b, \phi_a) &= \int d^3q_1\, d^3q_2 A(\mathbf{q}_1, \mathbf{q}_2)\, e^{-iE_at}\delta_{b,a} \\ &= \int d^3q_1\, d^3q_2 A(\mathbf{q}_1, \mathbf{q}_2)\, e^{-iE_bt}\delta_{b,a},\end{aligned} \tag{75b}$$

where $\delta_{b,a}$ vanishes unless $\chi'_b = \chi_{a'}$, in which case it is just the $\delta$-function $\delta_{a',a}$.

### The Scattering Matrix

To evaluate the probability of finding the required *final states* $\chi'_b$ in $\psi_a^+$, we rewrite the defining equations for $\psi_a^+$, Eqs. (66) and (67), in a modified form. To do this we use an operator identity. Let $A$ and $B$ be two operators for which the respective reciprocals $A^{-1}$ and $B^{-1}$ are defined. Then (obviously)

$$\frac{1}{A} - \frac{1}{B} = \frac{1}{A}(B - A)\frac{1}{B} = \frac{1}{B}(B - A)\frac{1}{A}\,. \tag{77}$$

On taking $A = E_a \pm i\eta - H$ and $B = E_a \pm i\eta - K'$, we find

$$\frac{1}{E_a \pm i\eta - H} = \frac{1}{E_a \pm i\eta - K'}\left(1 + V'\,\frac{1}{E_a \pm i\eta - H}\right).$$

Thus we may write Eq. (67) as[18]

$$\psi_a^{\,+} = \chi_a + \frac{1}{E_a + i\eta - K'}\,\mathscr{T}\chi_a, \tag{78}$$

where

$$\mathscr{T} = V + V'\,\frac{1}{E_a + i\eta - H}\,V. \tag{79}$$

The quantity $\mathscr{T}$ is called the *scattering matrix*. We shall give a detailed description of the properties of $\mathscr{T}$ in Chapters 4 and 5. To evaluate the scattering cross section, we need only the expansion (78) and the fact [evident from Eq. (79)] that $\mathscr{T}$ commutes with the total momentum operator $\mathscr{P}$,

$$[\mathscr{P}, \mathscr{T}] = 0. \tag{80}$$

This follows directly from our assumption, made in Section 3.1, that

$$[\mathscr{P}, H] = 0, \qquad [\mathscr{P}, K] = 0, \quad \text{and} \quad [\mathscr{P}, K'] = 0.$$

Equation (78) may be rewritten as

$$\psi_a^{\,+} = \chi_a + \sum_b \frac{\chi'_b}{E_a + i\eta - E_b}\,\langle b|\,\mathscr{T}\,|a\rangle, \tag{81}$$

where

$$\langle b|\,\mathscr{T}\,|a\rangle = \left(\chi'_b, \left(V + V'\,\frac{1}{E_a + i\eta - H}\,V\right)\chi_a\right). \tag{82}$$

Because of momentum conservation, as expressed by Eq. (80), we may reduce this to the form

$$\langle b|\,\mathscr{T}\,|a\rangle = \delta(\mathbf{P}_b - \mathbf{P}_a)T_{ba}, \tag{83}$$

where

$$\begin{aligned} \mathscr{P}\chi_a &= \mathbf{P}_a\chi_a, \\ \mathscr{P}\chi'_b &= \mathbf{P}_b\chi'_b. \end{aligned} \tag{84}$$

Equation (83) provides the definition of the *reduced matrix* $T_{ba}$, which is defined only on the "momentum shell" $\mathbf{P}_b = \mathbf{P}_a$.

[18] We refrain as agreed from writing explicitly $\psi_a^{\pm}(\eta)$ and $\psi_a^{\pm} = \lim_{\eta\to 0(+)} \psi_a^{\pm}(\eta)$. No confusion will result if this limiting procedure is understood as implied. (In this connection see the discussion at the end of Section 3.1.)

**The Transition Probability**

Consider now the unitary operator

$$U'(t, 0) \equiv e^{iK't} e^{-iHt}, \tag{85}$$

which is just $U(t, 0)$ when $K' = K$. In terms of $U'$, we may write the wave function for the system at time $t$ as [compare Eqs. (37) and (39)]

$$\Psi(t) = e^{-iK't}U'(t, 0)U(0, t_0)X_a = e^{-iK't}U'(t, 0)\Gamma^+. \tag{86}$$

We have seen that the scattering interaction occurs during the interval

$$-t_c < t < t_c$$

[compare Eq. (32)]. When $t \gg t_c$, the reaction products separate and cease to interact with each other. We can see the implications of this in the following manner. According to the integral equation for $U(t, 0)$, Eq. (2.98),

$$U'(t, 0) = 1 - i\int_0^t dt'\, e^{iK't'}V'\, e^{-iHt'}. \tag{87}$$

Now $e^{-iHt'}\,\Gamma^+ = \Psi(t')$. For $t' \gg t_c$ the reaction products have separated and therefore[19]

$$V'\Psi(t') = 0, \tag{88}$$

when projected onto the states $\chi'_b$ that are to be observed. It follows, therefore, that $U'(t, 0)\Gamma^+$ is *independent* of $t$ for $t$ large compared to $t_c$. Consequently, we may introduce a *time-independent* operator $S$, defined by

$$SX_a \equiv U'(t, 0)U(0, t_0)X_a \quad \text{for} \quad t \gg t_c. \tag{89}$$

This is the so-called *S-matrix* of Heisenberg. It plays an important role in the theory of collisions among elementary particles.[20]

To obtain a useful expression for the matrix elements of $S$, we write (89) as

$$SX_a = U'(t, 0)\Gamma^+ = e^{iK't}\int d^3q_1\, d^3q_2 A(\mathbf{q}_1, \mathbf{q}_2)\, e^{-iE_at}\psi_a^+, \tag{90}$$

[19] This argument is essentially that which led to Eq. (45), as will be demonstrated in greater detail in Section 3.4.

The case in which very sharp resonances occur does not lead to difficulties, but it warrants a specific description (see Chapter 8).

[20] The $S$-matrix is ordinarily defined for the choice $K' = K$, $V' = V$. Then it is expressed as $S = \lim_{t\to+\infty} U(t, 0)U(0, -t) = \lim_{t\to+\infty} U(t, -t)$, where a suitable limiting process is required (compare Chapter 5).

where use has been made of the expression for $\Gamma$ in terms of the wave packet amplitude $A$, Eq. (68), and also $H\psi_a^+ = E_a\psi_a^+$ together with the definition of $U'(t, 0)$, Eq. (85). Substitution of Eq. (81) for $\psi_a^+$ here gives

$$SX_a = \int d^3q_1\, d^3q_2 A(\mathbf{q}_1, \mathbf{q}_2)\left[e^{i(K'-E_a)t}\chi_a + \sum_b \frac{e^{i(E_b-E_a)t}}{E_a + i\eta - E_b}\chi'_b\langle b|\mathscr{T}|a\rangle\right], \tag{91}$$

where it is understood that $t \gg t_c$.

From Eq. (75*b*), we see that the first term in (91) does not contribute to $(\chi'_b, S\chi_a)$ unless $\chi'_b = \chi_{a'}$, where $\chi_{a'}$ is contained within the packet state $X_a$. In any case, it is evident that we can set $e^{i(K'-E_a)t} = 1$ in evaluating $(\chi'_b, S\chi_a)$ from Eq. (91).

Since Eq. (91) is *independent* of $t$, we may consider $t$ to be very large and set[21]

$$\lim_{t\to+\infty} \frac{e^{i(E_b-E_a)t}}{E_b - (E_a + i\eta)} = 2\pi i\delta(E_b - E_a). \tag{92}$$

Thus

$$(\chi'_b, SX_a) = \int d^3q_1\, d^3q_2\, A(\mathbf{q}_1, \mathbf{q}_2)[\delta_{b,a} - 2\pi i\delta(E_a - E_b)\langle b|\mathscr{T}|a\rangle]. \tag{93}$$

This equation permits us to define the "matrix elements" $(\chi'_b, S\chi_a)$ of $S$. These are

$$(\chi'_b, S\chi_a) \equiv \langle b|\, S\,|a\rangle = \delta_{b,a} - 2\pi i\delta(E_a - E_b)\langle b|\mathscr{T}|a\rangle. \tag{94}$$

It has been emphasized that the representation $\chi'_b$ has been chosen in accordance with the conditions under which the experiment is performed. In the next chapter we shall consider the question of enumerating those classes of final states which correspond to all the observations which may be made on the system having $\phi_a(t_0)$ as the initial state. The various kinds of final states $b$ are called the *channels* from which, or into which, a reaction can proceed. To obtain the cross section for the reaction we are presently considering we need only the submatrix (94).

Our expression for the matrix elements of $S$, Eq. (94), will play an important role in our subsequent consideration. We emphasize that (94) is a matrix element evaluated in a *mixed representation*,[22] unless the $\chi'_b$ happen

[21] To see this, consider

$$I \equiv \lim_{\eta\to 0(+)} \lim_{t\to+\infty} \int_{-a}^{a} \frac{dx\, e^{ixt}}{x - i\eta} f(x),$$

where $f(x)$ is nonsingular near $x = 0$. By the Riemann-Lebesgue theorem (E. C. Titchmarsh [2]), we may replace the contour along the real axis by a small circle about $x = 0$. Then $I = 2\pi i f(0)$.

[22] That is, $\chi'_b$ and $\chi_a$ belong in general to different orthonormal sets of states.

to coincide with the $\chi_a$. The $\delta_{b,a}$ function has been defined in connection with Eq. (75*b*).

To summarize our discussion in slightly different but equivalent terms, let us introduce a projection operator

$$\Lambda_{\bar{b}} = \sum_b \chi'_b(\chi'_b)^\dagger$$

onto those states $\chi'_b$ which we expect to observe. Then

$$\Lambda_{\bar{b}}\Psi(t) = \Lambda_{\bar{b}}\, e^{-iK't}SX_a = \int d^3q_1\, d^3q_2 \sum_b e^{-iE^b t}\chi'_b\langle b|\, S\, |a\rangle A(\mathbf{q}_1, \mathbf{q}_2). \quad (95)$$

Here the sum on $b$ is extended over the states of $\Lambda_{\bar{b}}$.

Now the probability that the system will be found in the state $\chi'_b$ at time $t$ is

$$P_{ba}(t) = |(\phi'_b(t), \Psi(t)|^2, \quad (96)$$

where

$$\phi'_b(t) = e^{-iK't}\chi'_b. \quad (97)$$

When the experimental conditions are such that the states $\chi'_b$ have no overlap with the initial packet at times $t$, Eq. (75), $P_{ba}(t) = 0$ for $t < t_C$. When $t \gg t_C$, $P_{ba}(t)$ approaches a constant value, which we shall call $P_{ba}$. According to Eq. (95), for $t \gg t_C$, we have

$$(\phi'_b(t), \Psi(t)) = (\chi'_b, SX_a). \quad (98)$$

Therefore,

$$P_{ba} = |(\chi'_b, SX_a)|^2. \quad (99)$$

**Lorentz Invariance of the Transition Probability and *S*-Matrix**

For the phenomena we are discussing to be consistent with the special theory of relativity the operator $S$ must be a Lorentz invariant. To see what is implied by this statement, let us consider a wave packet state $X'_b$, formed from the $\chi'_b$, and the transition probability

$$\bar{P}_{ba} = |(X'_b, SX_a)|^2. \quad (100)$$

The physical significance of $\bar{P}_{ba}$ is that it represents the probability that the system makes a transition from $X_a$ to $X'_b$. This probability is a *characteristic of the system* itself, and not of the observer. Thus $\bar{P}_{ba}$ must appear the same to all observers.

Let us suppose that an observer $\mathcal{O}$ has measured the transition probability $\bar{P}_{ba}$. To a second observer $\mathcal{O}'$, in a different Lorentz frame, the states $X_a$ and $X'_b$ will appear as a set of states $X_{t(a)}$ and $X'_{t(b)}$ respectively. These states are obtained from $X_a$ and $X'_b$ by a Lorentz transformation $\mathscr{L}$.

$$X_{t(a)} = \mathscr{L}X_a, \qquad X'_{t(b)} = \mathscr{L}X'_b. \quad (101)$$

Here $\mathscr{L}$ is a unitary operator which effects the Lorentz transformation on the state vectors.[23] We may interpret Eqs. (101) by saying that a state $a$ in the first coordinate system appears as a state $t(a)$ in the second coordinate system, and so on.

The observer $\mathcal{O}'$ will measure the transition probability

$$\bar{P}_{t(b),t(a)} = |(X'_{t(b)}, SX_{t(a)})|^2 = |(X'_{(b)}, \mathscr{L}^{-1}S\mathscr{L}X_a)|^2. \tag{102}$$

Now we have seen that $\bar{P}_{t(b),t(a)} = \bar{P}_{ba}$ is required by the special theory of relativity, so the expressions on the right-hand sides of Eqs. (100) and (102) are equal. Therefore it follows that

$$S = \mathscr{L}^{-1}S\mathscr{L}. \tag{103}$$

Equation (103) is a statement that the operator $S$ is invariant under proper Lorentz transformations.[24]

Let us next obtain the Lorentz transformation for the matrix elements $(\chi'_b, S\chi_a)$ of $S$ between states of definite energy and momentum as given by Eq. (94). Because of the delta-functions for energy and momentum conservation, these matrix elements are singular—in contrast to those of Eqs. (99) or (100). We therefore consider the transition probability, defined in a given Lorentz frame

$$P_{\bar{b}\bar{a}} = \sum_b \sum_a |(\chi'_b, S\chi_a)|^2 = \sum_b \sum_a |\langle b|\, S\, |a\rangle|^2. \tag{104}$$

Here the sums over $b$ and $a$ are carried out over a finite range of states, so chosen that (104) is nonvanishing.[25]

Now an observer in another coordinate system, observing the *same* experiment that led to (104), would measure the transition probability $P_{\overline{t(b)},\overline{t(a)}}$, where $t(b)$ and $t(a)$ are the states resulting when the appropriate Lorentz transformation is made on the states $b$ and $a$ respectively. Since both observers are measuring the same event, we must clearly have

$$P_{\overline{t(b)},\overline{t(a)}} = P_{\bar{b}\bar{a}}.$$

Let us suppose that there are $\nu$ particles in the state $b$. Their respective momenta and energies are taken to be $\mathbf{k}_1, \ldots, \mathbf{k}_\nu$ and $\epsilon_{s1,k1}, \ldots, \epsilon_{s\nu,k\nu}$,

[23] Although the Lorentz group is not unitary, it can be realized by a unitary transformation of the state vectors in an appropriately Lorentz-invariant theory. This follows directly, for example, from a discussion due to Wigner [3, p. 233].

[24] It is to be noted that the argument leading to Eq. (103) need not be restricted to the case that there are only two particles in the state $a$.

[25] The average in Eq. (104) may be considered to result from taking a *statistical ensemble* average over wave packet states.

where $s_1, \ldots, s_\nu$ are sets of quantum numbers describing the internal states of the particles. Then

$$\sum_b = \sum_{s1,\ldots,s\nu} \int d^3k_1 \cdots d^3k_\nu, \tag{105}$$

where the sums and integrals cover an arbitrary range of values (subject to the condition that the right-hand side of (104) be nonvanishing).

We have, until now, supposed that there were just two particles, which are to collide, in the initial states $\chi_a$. It has been mentioned that our formal considerations would be unchanged had we begun with a state of $\mu$ particles ($\mu \geq 2$). If these had the respective momenta $\mathbf{q}_1, \ldots, \mathbf{q}_\mu$, then, as described earlier, we might have formed the wave packet state defined by [compare Eq. (26*a*)],

$$X_a = \int d^3q_1 \cdots d^3q_\mu A(\mathbf{q}_1, \ldots, \mathbf{q}_\mu)\chi_a.$$

The $\mu$ particles can be brought into simultaneous collision at $t = 0$, and our formal considerations leading to Eq. (95) still apply.

For the moment, then, let us suppose there to be $\mu$ particles in the state represented by $\chi_a$. These have the respective momenta and energies

$$\mathbf{q}_1, \ldots, \mathbf{q}_\mu \quad \text{and} \quad \epsilon_{l1,q1}, \ldots, \epsilon_{l\mu,q\mu},$$

where $l_1, \ldots, l_\mu$ are sets of quantum numbers describing the internal states of the $\mu$ particles. Then, in analogy to (105),

$$\sum_a = \sum_{l_1,\ldots,l_\mu} \int d^3q_1 \cdots d^3q_\mu. \tag{106}$$

Let us now consider a second Lorentz frame, moving with a velocity $\mathbf{u}$ with respect to that in which (104) was obtained. The momenta and energies of the particles in states $a$ and $b$ in the new reference system are related to those in the original one by the transformation (1.101). Let these new values of momentum and energy be

$$\mathbf{k}'_1 \cdots \epsilon_{s'\nu,k'\nu}, \qquad \mathbf{q}'_1 \cdots \epsilon_{l'\mu,q'\mu}.$$

Now, for any one particle, we have[26]

$$\begin{aligned} d^3k' &= \gamma\left(1 - \frac{1}{c^2}\mathbf{u}\cdot\mathbf{v}\right) d^3k, \\ \epsilon_{s',k'} &= \gamma\left(1 - \frac{1}{c^2}\mathbf{u}\cdot\mathbf{v}\right)\epsilon_{s,k}, \end{aligned} \tag{107}$$

[26] This is easily seen as follows. Let $\mathbf{u}$ be parallel to the $z$-axis, as in (1.101). Then $dk'_z = \gamma[dk_z - (1/c^2)u\, d\epsilon_{s,k}] = \gamma[dk_z - (1/c^2)u\, d\mathbf{k}\cdot\mathbf{v}]$, $dk'_x = dk_x$, $dk'_y = dk_y$. The Jacobian $J(\partial k'/\partial k) = \gamma[1 - (1/c^2)\mathbf{u}\cdot\mathbf{v}]$.

where $\mathbf{v} = \mathbf{k}/\epsilon_{s,k}c^2$. Thus

$$\frac{d^3k}{\epsilon_{s,k}} \tag{108}$$

is a Lorentz invariant.

Returning to Eq. (104), we may express $P_{\bar{b}\bar{a}}$ as

$$P_{\bar{b}\bar{a}} = \sum_{s1,\ldots,sv} \sum_{l1,\ldots,l\mu} \int \frac{d^3k_1 \cdots d^3k_v}{\epsilon_{s1,k1} \cdots \epsilon_{sv,kv}} \int \frac{d^3q_1 \cdots d^3q_\mu}{\epsilon_{l1,q1} \cdots \epsilon_{l\mu,q\mu}} \times |\sqrt{\epsilon_{s1,k1} \cdots \epsilon_{sv,kv}} \cdot \langle b| S |a\rangle \sqrt{\epsilon_{l1,q1} \cdots \epsilon_{l\mu,q\mu}}|^2. \tag{109}$$

Since $P_{\bar{b}\bar{a}}$ is invariant, and since the sums and integrals are written in invariant form here, it follows that

$$\sqrt{\epsilon_{s1,k1} \cdots \epsilon_{sv,kv}}\, \langle b| S |a\rangle \sqrt{\epsilon_{l1,q1} \cdots \epsilon_{l\mu,q\mu}}$$

is a Lorentz invariant. That is, on evaluating this expression in the two Lorentz frames already introduced, we have

$$\sqrt{\epsilon_{s1,k1} \cdots \epsilon_{sv,kv}}\, \langle b| S |a\rangle \sqrt{\epsilon_{l1,q1} \cdots \epsilon_{l\mu,q\mu}} = \sqrt{\epsilon_{s'1,k'1} \cdots \epsilon_{s'v,k'v}}\, \langle t(b)| S |t(a)\rangle \sqrt{\epsilon_{l'1,q'1} \cdots \epsilon_{l'\mu,q'\mu}}. \tag{110}$$

In general, of course, these two expressions may differ by a phase factor $e^{iw_{ba}}$, which would leave (109) unchanged. In fact, this phase factor can be chosen as independent of the states $a$ and $b$. We could, for example, introduce a new final state

$$C_1\langle b_1| + C_2\langle b_2| \equiv \langle b_{12}|,$$

where $C_1$ and $C_2$ are constants. Application of the Lorentz transformation to

$$\langle b_{12}| S |a\rangle = C_1\langle b_1| S |a\rangle + C_2\langle b_2| S |a\rangle$$

would lead to a dependence of $P_{\bar{b}12,\bar{a}}$ on $e^{iw_{b_1a}}e^{-iw_{b_2a}}$ unless $w_{b1a} = w_{b2a}$. Thus $w_{ba}$ cannot depend on the state $b$. A similar argument shows that $w_{ba}$ cannot depend on the state $a$. It is always possible to choose $e^{iw} = 1$.[27]

From the relation between the matrix elements of $S$ and those of the scattering matrix on the energy momentum shell, Eqs. (83) and (94),

$$\langle b| S |a\rangle = \delta_{b,a} - 2\pi i\delta(E_b - E_a)\delta(\mathbf{P}_b - \mathbf{P}_a)T_{ba}. \tag{111}$$

We readily see that the term $\delta_{b,a}$ is an invariant (*density*), in the sense of Eq. (110), by itself. Thus, we can obtain the transformation of $T_{ba}$ from Eq. (110). The four-dimensional $\delta$-function

$$\delta(E_a - E_b)\delta(\mathbf{P}_b - \mathbf{P}_a)$$

[27] This is shown, for example, by Wigner [3, p. 233].

is a Lorentz invariant.[28] Therefore, $T_{ba}$ transforms just as $\langle b|\, S\, |a\rangle$. More specifically, if we compare $T_{ba}$ in the two Lorentz frames of Eq. (110), we have

$$\sqrt{\epsilon_{s1,k1}\cdots\epsilon_{s\nu,k\nu}}\, T_{ba} \sqrt{\epsilon_{l1,q1}\cdots\epsilon_{l\mu,q\mu}}$$
$$= \sqrt{\epsilon_{s'1,k'1}\cdots\epsilon_{s'\nu,k'\nu}}\, T_{t(b),t(a)} \sqrt{\epsilon_{l'1,q'1}\cdots\epsilon_{l'\mu,q'\mu}}. \quad (112)$$

**Evaluation of the Transition Probability in Terms of the Scattering Matrix**

We are finally ready to evaluate the probability $P_{ba}$ of Eq. (100). Usually the experiment is designed to determine only that the system is within a range of states which is broader than the initial wave packet. That is, the observed probability is

$$\sum_b P_{ba} = \sum_b |(\chi'_b, SX_a)|^2, \quad (113)$$

where the range over which the sum runs is determined experimentally. Now the transition probability $P_{ba}$ has been expressed in terms of the matrix elements of the scattering matrix $\mathscr{T}$ in Eq. (93). In the ordinary experimental situation, where the final states $\chi'_b$ do not overlap the initial wave packet [see Eqs. (75)], we find

$$(\chi'_b, SX_a) = -2\pi i \int d^3q_1\, d^3q_2 a_B(\mathbf{q}_1 - \mathbf{p}_1)$$
$$\times\, a_T(\mathbf{q}_2 - \mathbf{p}_2)\delta(E_a - E_b)\delta(\mathbf{P}_b - \mathbf{P}_a)T_{ba}. \quad (114)$$

Here

$$\mathbf{P}_a = \mathbf{q}_1 + \mathbf{q}_2, \quad (115)$$

and $\mathbf{P}_b$ is the total momentum of all the particles in state $b$. According to Eq. (34), we have set

$$A(\mathbf{q}_1, \mathbf{q}_2) = a_B(\mathbf{q}_1 - \mathbf{p}_1)a_T(\mathbf{q}_2 - \mathbf{p}_2),$$

where $a_B$ and $a_T$ are the beam and target amplitudes, respectively.

It will correspond to the usual experimental circumstance and will considerably simplify our geometry to evaluate Eq. (113) in a coordinate system in which $\mathbf{p}_1$ and $\mathbf{p}_2$ each lies along a given line, say the $z$-axis of a rectangular coordinate system (that is, $\mathbf{p}_1 \times \mathbf{p}_2 = 0$). Since (113) is a Lorentz invariant, the general case may then be obtained from a Lorentz transformation. The two most common coordinates systems used are the *laboratory* coordinate system, in which $\mathbf{p}_2 = 0$, and the *barycentric* coordinate system, in which $\mathbf{p}_1 + \mathbf{p}_2 = 0$. In both cases, we satisfy our restriction that $\mathbf{p}_1 \times \mathbf{p}_2 = 0$.

[28] This can best be seen on writing it in the covariant form $(2\pi)^{-4}\int d^4\lambda\, e^{i(\lambda_\mu p_\mu - p_{0\mu})}$, where $\lambda_\mu$ transforms as a four vector.

The beam and target steady state wave functions are, respectively,

$$X_B = (2\pi)^{-3/2}\, e^{i\mathbf{p}_1\cdot\mathbf{r}_1}\, G_B(\mathbf{r}_1),$$
$$X_T = (2\pi)^{-3/2}\, e^{i\mathbf{p}_2\cdot\mathbf{r}_2} G_T(\mathbf{r}_2), \tag{116}$$

where, as before,

$$G_B(\mathbf{r}) = \int d^3\rho\, a_B(\rho)\, e^{i\boldsymbol{\rho}\cdot\mathbf{r}},$$
$$G_T(\mathbf{r}) = \int d^3\rho\, a_T(\rho)\, e^{i\boldsymbol{\rho}\cdot\mathbf{r}}. \tag{117}$$

It is convenient now (for the first time!) to invoke the assumption made in Section 3.2 that the width $w_T$ of the target wave packet is much less than the width $w_B$ of the beam wave packet. This assumption is in no sense essential for our evaluation; it simplifies the discussion and corresponds to the experimental circumstance that the position of the target particles can ordinarily be measured with greater precision than can that of the beam particles. (It is perhaps unnecessary to call attention to the fact that our analysis is unchanged if we interchange the roles of beam and target particles.) When $w_B \gg w_T$, we may assume that the scattering occurs as long as the target wave packet is inside the beam wave packet.

Since there is one particle in each wave packet, that is, since the wave functions are normalized to unity,

$$(2\pi)^{-3}\int |G_B(\mathbf{r})|^2\, d^3r = 1,$$
$$(2\pi)^{-3}\int |G_T(\mathbf{r})|^2\, d^3r = 1. \tag{118}$$

We have then

$$(2\pi)^{-3}\int |G_T(\mathbf{r})|^2\, d^3r = \int d^3\rho\, a_T(\boldsymbol{\rho}) a_T^*(\boldsymbol{\rho}) = 1. \tag{119}$$

The density of beam particles (per beam particle) is just

$$n_B = (2\pi)^{-3}\, |G_B(\mathbf{r})|^2. \tag{120}$$

We have assumed that both particles are moving parallel to the $z$-axis and have wave packets centered on the coordinate origin at $t = 0$. The flux of beam particles, the number of beam particles per unit area, across the target particle is then

$$F = (2\pi)^{-3}\int_{-\infty}^{+\infty} dz\, |G_B(0, 0, z)|^2, \tag{121}$$

where we have written $G_B(\mathbf{r})$ in terms of the components of $\mathbf{r}$ as $G_B(x, y, z)$.

Now, the transition probability observed, as given by Eq. (113), expressed in terms of the beam and target packets and the reduced matrix elements of $\mathscr{T}$, Eq. (114), becomes

$$\sum_b P_{ba} = (2\pi)^2 \sum_b \int d^3q_1\, d^3q_2\, d^3q'_1\, d^3q'_2$$
$$\times\, a_B(\mathbf{q}_1 - \mathbf{p}_1) a_B^*(\mathbf{q}'_1 - \mathbf{p}_1) a_T(\mathbf{q}_2 - \mathbf{p}_2) a_T^*(\mathbf{q}'_2 - \mathbf{p}_2)$$
$$\times\, \delta(E_b - E_{a'})\delta(E_b - E_a)\delta(\mathbf{P}_b - \mathbf{P}_a)\delta(\mathbf{P}_b - \mathbf{P}_{a'}) T_{ba} T_{ba'}^*, \quad (122)$$

where [see Eqs. (9) and (24)]

$$\begin{aligned} E_a &= \epsilon_{s1,q1} + \epsilon_{s2,q2}, \\ E_{a'} &= \epsilon_{s'1,q'1} + \epsilon_{s'2,q'2}, \\ \mathbf{P}_a &= \mathbf{q}_1 + \mathbf{q}_2, \\ \mathbf{P}_{a'} &= \mathbf{q}'_1 + \mathbf{q}'_2. \end{aligned} \quad (123)$$

We shall now set

$$T_{ba} T_{ba'}^* = |T_{b\bar{a}}|^2, \quad (124)$$

where $\bar{a}$ is the state at the center of the wave packet, customarily so chosen that the error in (124) is negligible. We also set

$$\delta(E_b - E_{a'})\delta(E_b - E_a)$$
$$= \delta(E_b - E_a)\delta(E_{a'} - E_a) \approx \delta(E_b - E_{\bar{a}})\delta(E_{a'} - E_a) \quad (125)$$

and

$$\delta(\mathbf{P}_b - \mathbf{P}_a)\delta(\mathbf{P}_b - \mathbf{P}_{a'})$$
$$= \delta(\mathbf{P}_b - \mathbf{P}_a)\delta(\mathbf{P}_a - \mathbf{P}_{a'}) \approx \delta(\mathbf{P}_b - \mathbf{P}_{\bar{a}})\delta(\mathbf{P}_a - \mathbf{P}_{a'}), \quad (126)$$

where $P_{\bar{a}} = \mathbf{p}_1 + \mathbf{p}_2$. Since the sum over states $b$ in Eq. (122) has been assumed to extend over the wave packet, these approximations are similar to that made in Eq. (124).

Then our transition probability Eq. (122) may be rewritten as

$$\sum_b P_{ba} = \Big[\sum_b (2\pi)^2 \delta(E_b - E_{\bar{a}})\delta(\mathbf{P}_b - \mathbf{P}_{\bar{a}})\, |T_{b\bar{a}}|^2\Big]$$
$$\times \int d^3q_1\, d^3q_2\, d^3q'_1\, d^3q'_2 a_B(\mathbf{q}_1 - \mathbf{p}_1) a_B^*(\mathbf{q}'_1 - \mathbf{p}_1)$$
$$\times\, a_T(\mathbf{q}_2 - \mathbf{p}_2) a_T^*(\mathbf{q}'_2 - \mathbf{p}_2)\delta(E_{a'} - E_a)\delta(\mathbf{P}_a' - \mathbf{P}_a). \quad (127)$$

In the integral here we shall replace $\delta(E_{a'} - E_a)$ by

$$\delta(E_a - E_{a'}) = \frac{1}{2\pi}\int_{-\infty}^{+\infty} d\lambda\, e^{i\lambda(E_{a'} - E_a)}. \quad (128)$$

Now expanding $E_{a'}$ in the neighborhood of $E_a$ according to Eqs. (12), (13), (29), and (123)

$$\begin{aligned} E_{a'} - E_a &\cong (\boldsymbol{\rho}'_1 - \boldsymbol{\rho}_1) \cdot \mathbf{v}_1 + (\boldsymbol{\rho}'_2 - \boldsymbol{\rho}_2) \cdot \mathbf{v}_2 \\ &= (\boldsymbol{\rho}'_1 - \boldsymbol{\rho}_1) \cdot (\mathbf{v}_1 - \mathbf{v}_2) + (\boldsymbol{\rho}'_1 + \boldsymbol{\rho}'_2 - \boldsymbol{\rho}_1 - \boldsymbol{\rho}_2) \cdot \mathbf{v}_2, \end{aligned} \tag{129}$$

where

$$\begin{aligned} \boldsymbol{\rho}_1 &= \mathbf{q}_1 - \mathbf{p}_1, \qquad \boldsymbol{\rho}'_1 = \mathbf{q}'_1 - \mathbf{p}_1, \\ \boldsymbol{\rho}_2 &= \mathbf{q}_2 - \mathbf{p}_2, \qquad \boldsymbol{\rho}'_2 = \mathbf{q}'_2 - \mathbf{p}_2. \end{aligned} \tag{130}$$

Since

$$\delta(\mathbf{P}_{a'} - \mathbf{P}_a) = \delta(\boldsymbol{\rho}'_1 + \boldsymbol{\rho}'_2 - \boldsymbol{\rho}_1 - \boldsymbol{\rho}_2),$$

we may set

$$E_{a'} - E_a = (\boldsymbol{\rho}'_1 - \boldsymbol{\rho}_1) \cdot (\mathbf{v}_1 - \mathbf{v}_2)$$

in the integral in Eq. (127).

The coefficient of the brackets in Eq. (127) is therefore

$$\frac{1}{2\pi} \int_{-\infty}^{+\infty} d\lambda I,$$

where

$$\begin{aligned} I = \int d^3\rho_1\, d^3\rho'_1\, d^3\rho_2 a_B(\boldsymbol{\rho}_1)\, e^{-i\lambda \boldsymbol{\rho}_1 \cdot (\mathbf{v}_1 - \mathbf{v}_2)} \\ \times a_B^*(\boldsymbol{\rho}'_1)\, e^{i\lambda \boldsymbol{\rho}_1' \cdot (\mathbf{v}_1 - \mathbf{v}_2)} a_T(\boldsymbol{\rho}_2) a_T^*(\boldsymbol{\rho}_2 + \boldsymbol{\rho}_1 - \boldsymbol{\rho}'_1). \end{aligned} \tag{131}$$

We have assumed that the width $w_B$ of the beam packet is large compared to $w_T$. This means that we may set

$$a_T^*(\boldsymbol{\rho}_2 + \boldsymbol{\rho}_1 - \boldsymbol{\rho}'_1) \approx a_T^*(\boldsymbol{\rho}_2)$$

in Eq. (131). Now, with the expressions for the wave packet amplitudes in coordinate space, Eq. (117) and the normalization condition, Eq. (119), we find

$$\begin{aligned} I &= |G_B(\lambda(\mathbf{v}_1 - \mathbf{v}_2))|^2 \\ &= |G_B(0, 0, \lambda(v_1 - v_2))|^2. \end{aligned} \tag{132}$$

Finally,

$$\begin{aligned} \frac{1}{2\pi} \int_{-\infty}^{+\infty} d\lambda I &= \frac{1}{2\pi\, |\mathbf{v}_1 - \mathbf{v}_2|} \int dz\, |G_B(0, 0, z)|^2 \\ &= \frac{(2\pi)^2}{|\mathbf{v}_1 - \mathbf{v}_2|} F, \end{aligned} \tag{133}$$

where $F$ is the beam flux (121). Had we not assumed that $w_B \gg w_T$, we would have found (133) to be the average overlap of the two wave packets and would have had essentially the same final result.

Substitution of (133), which is the coefficient of the brackets in Eq. (127), into that equation gives

$$\sum_b P_{ba} = \left[\sum_b \delta(E_b - E_{\bar{a}})\delta(\mathbf{P}_b - \mathbf{P}_{\bar{a}})\,|T_{b\bar{a}}|^2\right]\frac{(2\pi)^4 F}{|\mathbf{v}_1 - \mathbf{v}_2|}. \tag{134}$$

Since the properties of the wave packets appear only in the factor $F$, no confusion will result if we omit the "bar" representing an average for the initial state and simply write $E_{\bar{a}} = E_a$, $\mathbf{p}_{\bar{a}} = \mathbf{p}_a$, and $T_{b\bar{a}} = T_{ba}$.

### The Scattering Cross Section

It is customary to present the experimental measurements of the transition probability (134) as a scattering cross section. This is defined as follows: During an interval of time $\Delta t$ a total number $N_B$ of beam particles are directed against a target containing $N_T$ target particles. The total number of beam particles per unit area striking the target is $F_B$, the beam flux. During the course of the experiment a total number $\Delta N_s$ of scattered particles are detected by the counting apparatus, these being the particles emitted into the range $\Sigma_b$ of states included in the sum (134). The scattering cross section is defined as

$$\Delta\sigma = \frac{\Delta N_s}{N_T F_B}. \tag{135}$$

This expression is related to the transition probability (134), since $\Sigma_b P_{ba}$ is the number of particles scattered into the set of states ($\Sigma_b$) per target particle per beam particle. That is,

$$\sum_b P_{ba} = \frac{\Delta N_s}{N_T N_B}. \tag{136}$$

Thus, since $F_B = N_B F$,

$$\begin{aligned}\Delta\sigma &= \frac{1}{F}\sum_b P_{ba} = \sum_b \delta(E_b - E_a)\delta(\mathbf{P}_b - \mathbf{P}_a)\frac{(2\pi)^4\,|T_{ba}|^2}{|\mathbf{v}_1 - \mathbf{v}_2|} \\ &= \sum_{s'_1,\ldots,s'_v}\int d^3k_1 \cdots d^3k_v \delta(E_b - E_a)\delta(\mathbf{P}_b - \mathbf{P}_a)\frac{(2\pi)^4\,|T_{ba}|^2}{|\mathbf{v}_1 - \mathbf{v}_2|}.\end{aligned} \tag{137}$$

In the last line we have used Eq. (105) to express the sum over final states in specific form. As we have said, the range of variables over which we sum, or integrate, is determined by the resolution of the counting apparatus. In the limit of a very narrow range of values, we write $\Delta\sigma$ as a differential $d\sigma$:

$$\Delta\sigma = d\sigma. \tag{138}$$

When (137) is summed over all the states that can result from the interaction, we write

$$\Delta\sigma = \sigma, \tag{139}$$

which is called the *total* cross section.

For one case of physical importance Eq. (139) has no meaning. It occurs for the *Coulomb scattering* of one charged particle by another. Because of the long range of Coulomb forces, *all* of the beam particles are scattered. That is,

$$\sum_{\text{all final states}} \Delta N_s = N_T N_B,$$

so the right-hand side of Eq. (136) is unity, if we sum over all scattered particles. The arguments leading to Eq. (137) fail to apply because the emerging wave packets *never* get outside the range of their mutual interaction. This difficulty does not really occur in practice, fortunately, since the Coulomb forces are shielded at large distances by matter in the physical environment of the experimental apparatus.[29] We shall see in Chapter 6 that Eq. (137) is valid even for Coulomb scattering if we do not attempt to evaluate the total cross section $\sigma$.

Equation (137) is easily seen to be invariant with respect to Lorentz transformations parallel to the axis along which $\mathbf{v}_1$ and $\mathbf{v}_2$ are directed. We verify that for $\mathbf{v}_1 \times \mathbf{v}_2 = 0$

$$\epsilon_{s1,p1}\epsilon_{s2,p2}\,|\mathbf{v}_1 - \mathbf{v}_2| = \frac{1}{c}\,[(p_{1\lambda}p_{2\lambda})^2 - p_{1\lambda}{}^2 p_{2\rho}{}^2]^{1/2} \tag{140}$$

where $p_{1\lambda}$ and $p_{2\lambda}$ are the respective four-vector representations $(\epsilon_{s1,p1}, \mathbf{p}_1)$ and $(\epsilon_{s2,p2}, \mathbf{p}_2)$. Thus Eq. (137) is equivalent to

$$\Delta\sigma = \sum_{s'_1,\ldots,s'_v} \int \frac{d^3k_1 \cdots d^3k_v}{\epsilon_{s'1,k1} \cdots \epsilon_{s'v,kv}} \delta(E_b - E_a)\delta(\mathbf{P}_b - \mathbf{P}_a)$$
$$\times \frac{(2\pi)^4\,|\sqrt{\epsilon_{s'1,k1} \cdots \epsilon_{s'v,kv}}\, T_{ba} \sqrt{\epsilon_{s1,p1}\epsilon_{s2,p2}}|^2}{(1/c)[(p_{1\lambda}p_{2\lambda})^2 - p_{1\lambda}{}^2 p_{2\rho}{}^2]^{1/2}}. \tag{141}$$

This is *manifestly* an invariant [compare Eqs. (108) and (112)]. Indeed, as Møller [4] has observed, (141) is invariant with respect to a general, proper Lorentz transformation. Thus Eq. (141) provides a definition of the scattering cross section which is invariant with respect to the general

[29] Even in the best vacuum available to us, that of interplanetary space (with a latent density of about 10 free electrons and protons per cm³), a Coulomb field is shielded at distances greater than about 1 m.

(proper) Lorentz transformation. A somewhat more convenient form for Eq. (141) is

$$\Delta\sigma = \sum_{s'_1,\ldots,s'_v} \int d^3k_1 \cdots d^3k_v \delta(E_b - E_a)\delta(\mathbf{P}_b - \mathbf{P}_a) \times \frac{(2\pi)^4 \, |T_{ba}|^2}{[(\mathbf{v}_1 - \mathbf{v}_2)^2 - (1/c^2)(\mathbf{v}_1 \times \mathbf{v}_2)^2]^{1/2}}, \tag{142}$$

which is valid in any Lorentz frame of reference.

**Averages over Spin Orientations**

Ordinarily in scattering experiments the beam and target particles are distributed in a mixture of states before scattering instead of being in specific internal states. The observed scattering cross sections are then appropriately weighted averages of Eqs. (137) or (142). Most frequently, the beam and target particles can be treated as being distributed at random[30] over a set of $N(S_1, S_2)$ degenerate states $S_1$ and $S_2$ since it is relatively easy to separate particles of different energies. The average cross section for such an experiment is then

$$d\bar{\sigma} = \frac{1}{N(S_1, S_2)} \sum_{S_1,S_2} d\sigma^{(S_1,S_2)}. \tag{143a}$$

Here $d\sigma^{(S_1,S_2)}$ is the differential cross section for scattering when particles 1 and 2 are in the specific internal states $S_1$ and $S_2$, respectively.

A second important modification of the expressions for the cross sections results when the internal states of the two incident particles are coupled with definite phase relations. Here the most common example is that for which the spins $S_1$ and $S_2$ of the two particles are coupled to form eigenstates of the total spin $S$. In this case $d\sigma^{(S)}$ is the scattering cross section for an initial state of total spin $S$ of the two particles. The average cross section for a random distribution of such initial spin states is then

$$d\bar{\sigma} = \frac{1}{N(S_1, S_2)} \sum_{S} N(S)\, d\sigma^{(S)}, \tag{143b}$$

where $N(S)$ is the number of states of total spin $S$ and $N(S_1, S_2) = \Sigma_S N(S)$.

We observe that the initial state wave function $X_a$ [Eq. (33)] requires a simple modification when we wish to calculate $d\sigma^{(S)}$. In this case we must replace Eq. (22) by

$$\chi_a = \sum_{a_1,a_2} c_S(a_1, a_2)\left[\frac{e^{i\mathbf{q}_1\cdot\mathbf{r}_1}}{(2\pi)^{3/2}} g_{a1}\right]\left[\frac{e^{i\mathbf{q}_2\cdot\mathbf{r}_2}}{(2\pi)^{3/2}} g_{a2}\right],$$

[30] When the distribution of spin states is not random, the scattering system is said to be *polarized.* Scattering of polarized systems will be discussed in Chapter 7 (see, in particular, Section 7.5).

where $c_S(a_1, a_2)$ is the appropriate coefficient (usually a Wigner coefficient) for constructing the state $S$. The final state wave function $\chi'_b$ may be constructed in a similar manner.

**Examples**

We shall illustrate our results by evaluating the differential cross section for the scattering of two particles, first in the barycentric coordinate system and then in the laboratory system.

In the barycentric system, the initial momenta and internal states of the two particles are written as $(\mathbf{q}_1, s_1)$ and $(\mathbf{q}_2, s_2)$. The $s_1, s_2$ describe spin, or other internal variables. The final momenta and internal states are taken as $(\mathbf{l}_1, s'_1)$ and $(\mathbf{l}_2, s'_2)$, respectively. The corresponding energies are $\epsilon_{s1,q1}$, $\epsilon_{s2,q2}$, $\epsilon_{s'1,l1}$ and $\epsilon_{s'2,l2}$. Now, in the barycentric system $\mathbf{P}_a = \mathbf{q}_1 + \mathbf{q}_2 = 0$, and it is convenient to set $\mathbf{q}_1 \equiv \boldsymbol{\varkappa}$. The initial and final energies are then

$$E_a = \epsilon_{s1,q1} + \epsilon_{s2,q2} = \epsilon_{s1,\kappa} + \epsilon_{s2,\kappa},$$

$$E_b = \epsilon_{s'1,l1} + \epsilon_{s'2,l2}. \tag{144}$$

Since $T_{ba}$ is defined only on the "momentum shell," for which[31]

$$\boldsymbol{\varkappa}_f \equiv \mathbf{l}_1 = -\mathbf{l}_2, \tag{145}$$

it can be chosen to depend on only $\boldsymbol{\varkappa}$, $\boldsymbol{\varkappa}_f$ and the internal variables. Thus

$$T_{ba} = \langle s'_1, s'_2|\, T_B(\boldsymbol{\varkappa}_f, \boldsymbol{\varkappa})\, |s_1, s_2\rangle. \tag{146}$$

The differential cross section, corresponding to Eq. (137) for the case that the internal states are specified, is then

$$d\sigma = \int d^3l_1\, d^3l_2 \delta(E_b - E_a)\delta(\mathbf{l}_1 + \mathbf{l}_2)\frac{(2\pi)^4\,|T_{ba}|^2}{v_{\text{rel},B}}. \tag{147}$$

Here we have written $v_{\text{rel},\,B}$ for $|\mathbf{v}_1 - \mathbf{v}_2|$ in the barycentric system. Integration over $\mathbf{l}_2$ is first performed giving $\mathbf{l}_2 = -\mathbf{l}_1$. Now let us write

$$d^3l_1 = l_1^{\,2}\, dl_1\, d\Omega = \frac{l_1\, d\Omega}{1/\epsilon_{s'1,l1} + 1/\epsilon_{s'2,l1}}\, dE_b, \tag{148}$$

where $d\Omega$ is an element of solid angle into which $\mathbf{l}_1$ is directed. If we hold $d\Omega$ fixed and integrate over $E_b$, Eq. (147) becomes

$$\frac{d\sigma}{d\Omega} = \frac{(2\pi)^4\kappa_f\,|\langle s'_1, s'_2|\, T_B(\boldsymbol{\varkappa}_f, \boldsymbol{\varkappa})\,|s_1, s_2\rangle|^2}{[1/\epsilon_{s'1,\kappa f} + 1/\epsilon_{s'2,\kappa f}]v_{\text{rel},B}} \tag{149}$$

[31] We shall ordinarily use $\boldsymbol{\varkappa}$ and $\boldsymbol{\varkappa}_f$ to designate the initial and final momentum of the beam particle *after* integration is performed over the momentum-conserving $\delta$-function.

where now all variables are defined on the "energy shell," for which

$$\epsilon_{s'1,\kappa f} + \epsilon_{s'2,\kappa f} = \epsilon_{s1,\kappa} + \epsilon_{s2,\kappa}. \tag{150}$$

Let us consider the same cross section in the laboratory frame of reference, for which the target particle is at rest in the initial state. A Lorentz transformation from the barycentric to the laboratory system leads to the transformed set of states

$$\begin{aligned} L\{(\mathbf{q}_1, s_1), (\mathbf{q}_2, s_2)\} &= (\mathbf{p}_1, i_1), (\mathbf{p}_2, i_2); \\ L\{(\mathbf{l}_1, s'_1), (\mathbf{l}_2, s'_2)\} &= (\mathbf{k}_1, i'_1), (\mathbf{k}_2, i'_2). \end{aligned} \tag{151}$$

Here $(\mathbf{p}_1, i_1)$ is the state into which $(\mathbf{q}_1, s_1)$ transforms, and so on. (By $i$ we mean the set of quantum numbers for the transformed internal states.) Now the condition

$$\mathbf{p}_2 = 0, \tag{152}$$

which defines the laboratory coordinate system, specifies the Lorentz transformation $L$. The energies in this coordinate system are $\epsilon_{i1,p1}$, and so on.

The scattering matrix $T_{ba}$ in this coordinate system can be chosen to depend on the momentum variables $\mathbf{k}_1$ and $\mathbf{p}_1$, since on the momentum shell

$$\mathbf{k}_1 + \mathbf{k}_2 = \mathbf{p}_1. \tag{153}$$

Thus according to Eq. (112) we have

$$(T_{ba})_L \equiv \langle i'_1 i'_2 | T_L(\mathbf{k}_1, \mathbf{p}_1) | i_1 i_2 \rangle = \left( \frac{\epsilon_{s'1,\kappa f} \epsilon_{s'2,\kappa f} \epsilon_{s1,\kappa} \epsilon_{s2,\kappa}}{\epsilon_{i'1,k1} \epsilon_{i'2,k2} \epsilon_{i1,p1} \epsilon_{i2,0}} \right)^{1/2} \times \langle s'_1 s'_2 | T_B(\mathbf{k}_f, \boldsymbol{\varkappa}) | s_1 s_2 \rangle \tag{154}$$

for the matrix $T_{ba}$ in the laboratory reference system.

Now Eq. (137) is

$$d\sigma = \int d^3k_1\, d^3k_2\, \delta(E_b - E_a)\, \delta(\mathbf{k}_1 + \mathbf{k}_2 - \mathbf{p}_1) \frac{(2\pi)^4 |(T_{ba})_L|^2}{v_{\text{rel},L}}, \tag{155}$$

where $v_{\text{rel},L}$ is the value of $|\mathbf{v}_1 - \mathbf{v}_2| = v_1$ in the laboratory system. Integrating over $\mathbf{k}_2$ we obtain

$$d\sigma = \int d^3k_1\, \delta(\epsilon_{i'1,k1} + \epsilon_{i'2,|\mathbf{p}1-\mathbf{k}1|} - \epsilon_{i1,p1} - \epsilon_{i2,0}) \frac{(2\pi)^4 |(T_{ba})_L|^2}{v_{\text{rel},L}} \tag{156}$$

or

$$\frac{d\sigma}{d\Omega_L} = \frac{(2\pi)^4 k_1^2 |\langle i'_1, i'_2 | T_L(\mathbf{k}_1, \mathbf{p}_1) | i_1, i_2 \rangle|^2}{\{k_1/\epsilon_{i'1,k1} + [(\mathbf{k}_1 - \mathbf{p}_1) \cdot \hat{\mathbf{k}}_1]/\epsilon_{i'2,|\mathbf{p}1-\mathbf{k}1|}\} v_{\text{rel},L}}. \tag{157}$$

Here $\hat{\mathbf{k}}_1 = \mathbf{k}_1/k_1$ and $d\Omega_L$ is an element of solid angle into which $\mathbf{k}_1$ is directed. Since $d\sigma$ in Eqs. (149) and (157) is the same invariant cross section, these equations provide the appropriate transformation from one coordinate system to the other.

To illustrate the various factors of normalization, let us evaluate a specific cross section in the barycentric coordinate system for the scattering of two simple particles. Let the interaction between the particles be given by the *Yukawa potential*

$$V(r) = \frac{V_0 d}{r} e^{-r/d}, \tag{158}$$

where $V_0$ is the *potential strength* and $d$ is the *range* of the interaction. The scattering matrix is in this case given by Eq. (79) with $V' = V$. For the case that $V_0$ is sufficiently small, we may approximate Eq. (79) as

$$\mathscr{T} \approx V. \tag{159}$$

Then

$$\begin{aligned}\langle \mathbf{l}_1, \mathbf{l}_2 | \mathscr{T} | \mathbf{q}_1, \mathbf{q}_2 \rangle &= (2\pi)^{-6} \int d^3r_1\, d^3r_2\, e^{-i(\mathbf{l}_1\cdot\mathbf{r}_1 + \mathbf{l}_2\cdot\mathbf{r}_2)}\, V(r) e^{i(\mathbf{q}_1\cdot\mathbf{r}_1 + \mathbf{q}_2\cdot\mathbf{r}_2)} \\ &= \delta(\mathbf{l}_1 + \mathbf{l}_2 - \mathbf{q}_1 - \mathbf{q}_2)\bar{V}(|\boldsymbol{\varkappa}_f - \boldsymbol{\varkappa}|),\end{aligned} \tag{160}$$

where

$$\bar{V}(|\boldsymbol{\varkappa}_f - \boldsymbol{\varkappa}|) = \frac{V_0 d}{2\pi^2} \frac{1}{(\boldsymbol{\varkappa}_f - \boldsymbol{\varkappa})^2 + 1/d^2}. \tag{161}$$

Now we see that

$$T_B(\boldsymbol{\varkappa}_f, \boldsymbol{\varkappa}) = \bar{V}(|\boldsymbol{\varkappa}_f - \boldsymbol{\varkappa}|),$$

so Eq. (149) becomes (now $\kappa_f = \kappa$)

$$\begin{aligned}\frac{d\sigma}{d\Omega} &= \frac{4\kappa(V_0 d)^2}{(1/\epsilon_{1,\kappa} + 1/\epsilon_{2,\kappa})v_{\text{rel},B}} \frac{1}{[(\boldsymbol{\varkappa}_f - \boldsymbol{\varkappa})^2 + 1/d^2]^2} \\ &= \frac{4(V_0 d)^2}{(1/\epsilon_{1,\kappa} + 1/\epsilon_{2,\kappa})^2[(\boldsymbol{\varkappa}_f - \boldsymbol{\varkappa})^2 + 1/d^2]^2}.\end{aligned} \tag{162}$$

If we take $d$ as very large, set $V_0\, d = e_1 e_2$, and interpret $e_1$ and $e_2$ as the respective charges of the particles, we obtain the Rutherford cross section for Coulomb scattering,

$$\frac{d\sigma}{d\Omega} = \frac{4\, e_1^{\,2}\, e_2^{\,2}}{[1/\epsilon_{1,\kappa} + 1/\epsilon_{2,\kappa}]^2(\boldsymbol{\varkappa}_f - \boldsymbol{\varkappa})^4}. \tag{163}$$

## 3.4 PHYSICAL INTERPRETATION OF $\Psi(t)$

### The Asymptotic Wave Function in Coordinate Space

Let us first consider a simple two-particle scattering process where one of the particles is much more massive than the other. The heavy particle will be taken as the target and will be assumed to be at rest in the initial state. The wave function for the system as given by Eq. (81) is

$$\psi_a^+ = \chi_a + \sum_b \frac{1}{E_a + i\eta - E_b} \chi_b \langle b | \mathscr{T} | a \rangle. \tag{164}$$

The wave functions $\chi_a$ and $\chi_b$ belong now to the same orthonormal set (since we are treating a pure scattering reaction). We shall take

$$\begin{aligned} \chi_a &= (2\pi)^{-3} e^{i(\mathbf{q}_1 \cdot \mathbf{r}_1 + \mathbf{q}_2 \cdot \mathbf{r}_2)}, \\ \chi_b &= (2\pi)^{-3} e^{i(\mathbf{k}_1 \cdot \mathbf{r}_1 + \mathbf{k}_2 \cdot \mathbf{r}_2)}. \end{aligned} \tag{165}$$

The masses of the two particles are $\mu$ and $M$, where $\mu \ll M$. Then

$$\begin{aligned} E_a &= \epsilon_{q1} + q_2^2/2M \approx \epsilon_{q1} = c(q_1^2 + \mu^2 c^2)^{1/2}, \\ E_b &= \epsilon_{k1} + k_2^2/2M \approx \epsilon_{k1} = c(k_1^2 + \mu^2 c^2)^{1/2}. \end{aligned} \tag{166}$$

The exact time-dependent wave function for the system becomes [see Eqs. (39) and (70)]

$$\Psi(t) = e^{-iHt}\,\Gamma^+ = \int d^3q_1\, d^3q_2 a_B(\mathbf{q}_1 - \mathbf{p}_1) a_T(\mathbf{q}_2)\, e^{-iE_a t} \psi_a^+, \tag{167}$$

since $\mathbf{p}_2 = 0$ for a target particle initially at rest. We shall evaluate this expression explicitly.

We next factor out of the scattering matrix a momentum-conserving $\delta$-function according to Eq. (83),

$$\langle b | \mathscr{T} | a \rangle = \delta(\mathbf{k}_1 + \mathbf{k}_2 - \mathbf{q}_1 - \mathbf{q}_2) T_{ba}. \tag{168}$$

In the present case,

$$\begin{aligned} T_{ba} &\equiv \langle \mathbf{k}_1, \mathbf{k}_2 | T | \mathbf{q}_1, \mathbf{q}_2 \rangle \\ &\approx \langle \mathbf{k}_1, \mathbf{k}_2 | T | \mathbf{q}_1, 0 \rangle, \end{aligned}$$

if $T_{ba}$ varies little over the packet width. [See the discussion in relation to Eq. (124)]. We may eliminate $\mathbf{k}_2$ from this quantity, since $T_{ba}$ is defined only on the momentum shell, $\mathbf{k}_2 = \mathbf{q}_1 - \mathbf{k}_1$. Finally, $T_{ba}$ is invariant with respect to rotations,[32] so it is in general a function of the variables $\hat{\mathbf{k}}_1 \cdot \hat{\mathbf{q}}_1$, $k_1$, and $q_1$ (where $\hat{\mathbf{k}}_1 = \mathbf{k}_1/k_1$, $\hat{\mathbf{q}}_1 = \mathbf{q}_1/q_1$). We write then

$$T_{ba} = T(\hat{\mathbf{k}}_1 \cdot \hat{\mathbf{q}}_1, k_1, q_1). \tag{169}$$

[32] This is a special case of Eq. (112), as the pure rotations are *included* in the Lorentz transformations. The symmetry properties of $T$ will be considered in more detail in Chapters 4 and 5.

The stationary state wave function, Eq. (164), may therefore be re-expressed as

$$\psi_a^+ - \chi_a = (2\pi)^{-3}\int \frac{d^3k_1\, d^3k_2\, e^{i(\mathbf{k}_1\cdot\mathbf{r}_1+\mathbf{k}_2\cdot\mathbf{r}_2)}}{\epsilon_{q1} + i\eta - \epsilon_{k1}}\,\delta(\mathbf{k}_1 + \mathbf{k}_2 - \mathbf{q}_1 - \mathbf{q}_2)T_{ba}$$

$$= (2\pi)^{-3}e^{i(\mathbf{q}_1+\mathbf{q}_2)\cdot\mathbf{r}_2}\int \frac{d^3k_1\, e^{i\mathbf{k}_1\cdot(\mathbf{r}_1-\mathbf{r}_2)}}{\epsilon_{q1} + i\eta - \epsilon_{k1}}\,T(\hat{\mathbf{k}}_1\cdot\hat{\mathbf{q}}_1, k_1, q_1). \qquad (170)$$

Usually this integral may be considered to be finite for all $\mathbf{r}_1$ and $\mathbf{r}_2$. It may, however, be singular at $|\mathbf{r}_1 - \mathbf{r}_2| = 0$.[33] We shall assume that any such singularity is of the form

$$|\mathbf{r}_1 - \mathbf{r}_2|^{-\gamma},$$

where $\gamma < 1$. Then

$$\int^{\infty} \frac{k_1^{\,2}}{\epsilon_{k1}}\, T(\hat{\mathbf{k}}_1\cdot\hat{\mathbf{q}}_1, k_1, q_1)\, dk_1$$

will diverge. We would, indeed, expect (from dimensional arguments) for large $k_1$

$$\frac{k_1^{\,2}}{\epsilon_{k1}}\, T(\hat{\mathbf{k}}_1\cdot\hat{\mathbf{q}}_1, k_1, q_1) \approx k_1^{\gamma-1}. \qquad (171)$$

At present, we are interested in evaluating Eq. (170) only for the case that $|\mathbf{r}_1 - \mathbf{r}_2| \gg \hbar/q_1$. Consider then the integral, as $r \to \infty$,

$$I \equiv \int \frac{d^3k\, e^{i\mathbf{k}\cdot\mathbf{r}}}{\epsilon_q + i\eta - \epsilon_k}\, T(\hat{\mathbf{k}}\cdot\hat{\mathbf{q}}, k), \qquad (172)$$

where we have abbreviated the notation in Eq. (170) and taken $\mathbf{r} \equiv \mathbf{r}_1 - \mathbf{r}_2$. In spherical coordinates, $d^3k = k^2\, dk\, d\phi \sin\theta\, d\theta$, where $\mathbf{k}\cdot\mathbf{r} \equiv kr\cos\theta$. Then,[34] by partial integration,

$$I = \int \frac{k^2\, dk\, d\phi}{\epsilon_q + i\eta - \epsilon_k}\int \frac{d(e^{ikr\cos\theta})}{ikr}\, T(\hat{\mathbf{k}}\cdot\hat{\mathbf{q}}, k)$$

$$= \int \frac{k^2\, dk\, d\phi}{\epsilon_q + i\eta - \epsilon_k}(ikr)^{-1}\left\{[e^{ikr}T(\hat{\mathbf{r}}\cdot\hat{\mathbf{q}}, k) - e^{-ikr}T(-\hat{\mathbf{r}}\cdot\hat{\mathbf{q}}, k)] + \mathcal{O}\left(\frac{T}{kr}\right)\right\}$$

(172)

[33] This happens in the case of the Dirac equation for Coulomb forces, for example. It may very well be that smoothing by radiative corrections, and so on, would lead to nonsingular wave functions for all physical systems.

[34] We use here the notation $\hat{\mathbf{r}}$ for the unit vector parallel to a vector $\mathbf{r}$. This notation will be frequently used henceforth.

for large $r$. We shall drop the terms of order $(kr)^{-2}$. Thus we have

$$I = -\frac{2\pi}{ir}\int_0^\infty \frac{k\,dk}{\epsilon_k - (\epsilon_q + i\eta)}\,[e^{ikr}T(\hat{\mathbf{r}}\cdot\hat{\mathbf{q}}, k) - e^{-ikr}T(-\hat{\mathbf{r}}\cdot\hat{\mathbf{q}}, k)]. \quad (173)$$

Now choose a $\Delta$ sufficiently small that $T(\pm\hat{\mathbf{r}}\cdot\hat{\mathbf{q}}, k) = T(\pm\hat{\mathbf{r}}\cdot\hat{\mathbf{q}}, q)$ to any desired accuracy when $q - \Delta < k < q + \Delta$. Then the integral (173) may be divided into three parts.

$$\int_0^\infty dk \cdots = \int_0^{q-\Delta} dk \cdots + \int_{q-\Delta}^{q+\Delta} dk \cdots + \int_{q+\Delta}^{\infty} dk.$$

In the second integrand we may replace $T(\pm\hat{\mathbf{r}}\cdot\hat{\mathbf{q}}, k)$ by $T(\pm\hat{\mathbf{r}}\cdot\hat{\mathbf{q}}, q)$. The first and third integrals are expected to vanish in the limit that $r \to \infty$. The first integral evidently vanishes, by the Riemann-Lebesgue[35,36] lemma. Now by Eq. (171) we expect the integrand of the third integral to be no larger than $0(k^{\gamma-2})$ for large $k$. Since $\gamma < 1$, the integrand is absolutely convergent for all $r$ and thus (by the Riemann-Lebesgue lemma) vanishes for $r \to \infty$. Finally,

$$I = -\frac{2\pi}{ir}\left[T(\hat{\mathbf{r}}\cdot\hat{\mathbf{q}}, q)\int_{q-\Delta}^{q+\Delta}\frac{dk k\, e^{ikr}}{\epsilon_k - (\epsilon_q + i\eta)} - T(-\hat{\mathbf{r}}\cdot\hat{\mathbf{q}}, q)\int_{q-\Delta}^{q+\Delta}\frac{dk k\, e^{-ikr}}{\epsilon_k - (\epsilon_q + i\eta)}\right].$$

Since $\Delta$ is small, we may set $\epsilon_k - \epsilon_q = (k - q)(\partial\epsilon_q/\partial q) = (k - q)(q/\epsilon_q)$. As $r \to \infty$, we may close the contour of the first integral with a semicircle above the real axis; for the second integral, the semicircle must be below the real axis. The second integral vanishes and the first leads to

$$I = -(2\pi)^2\epsilon_q \frac{e^{iqr}}{r}\, T(\hat{\mathbf{r}}\cdot\hat{\mathbf{q}}, q), \quad (174a)$$

[For the evaluation of $\psi_a^-$ we must replace $+i\eta$ by $-i\eta$ in (172) and the subsequent equation. In this case the first integral in $I$ vanishes and we are led to

$$I^{(-)} = -(2\pi)^2\,\epsilon_q \frac{e^{-iqr}}{r}\, T(-\hat{\mathbf{r}}\cdot\hat{\mathbf{q}}, q). \quad (174b)$$

The factor $e^{-iqr}/r$ is interpreted as an *incoming* scattered wave.]

[35] See, for example, E. C. Titchmarsh [2, Chapter I].
[36] This assumes that $T$ is bounded.

When $|\mathbf{r}_1 - \mathbf{r}_2| \gg \hbar/q_1$, the result for $I$, Eq. (174), may be substituted into Eq. (170) to give

$$\psi_a^+ = \chi_a - (2\pi)^{-3}[(2\pi)^2 \epsilon_{q1} T(\hat{\mathbf{r}} \cdot \hat{\mathbf{q}}_1, q_1, q_1)] \, e^{i(\mathbf{q}_1+\mathbf{q}_2)\cdot \mathbf{r}_2} \frac{e^{iq_1|\mathbf{r}_1-\mathbf{r}_2|}}{|\mathbf{r}_1 - \mathbf{r}_2|}. \quad (175)$$

The scattering matrix $T$ is frequently replaced by the so-called "scattering amplitude" $f(\hat{\mathbf{r}} \cdot \hat{\mathbf{q}})$, defined by

$$f(\hat{\mathbf{r}} \cdot \hat{\mathbf{q}}_1) = -(2\pi)^2 \epsilon_{q1} T(\hat{\mathbf{r}} \cdot \hat{\mathbf{q}}_1, q_1, q_1). \quad (176a)$$

(It is, of course, understood that $f$ is a function of $q_1$, or of the energy of the incident particle.) Then Eq. (175) may also be written as

$$\psi_a^+ = \chi_a + (2\pi)^{-3} \, e^{i(\mathbf{q}_1+\mathbf{q}_2)\cdot \mathbf{r}_2} \frac{e^{iq_1 r}}{r} f(\hat{\mathbf{r}} \cdot \hat{\mathbf{q}}), \quad (177a)$$

valid for $q_1 r \gg 1$.

The corresponding wave function $\psi_a^-$ is given by [compare Eq. (174$a$)]

$$\psi_a^- = \chi_a + (2\pi)^{-3} \, e^{i(\mathbf{q}_1+\mathbf{q}_2)\cdot \mathbf{r}_2} \frac{e^{-iq_1 r}}{r} f^{(-)}(-\hat{\mathbf{r}} \cdot \hat{\mathbf{q}}), \quad (177b)$$

where

$$f^{(-)}(-\hat{\mathbf{r}} \cdot \hat{\mathbf{q}}_1) = -(2\pi)^2 \epsilon_{q1} T^{(-)}(-\hat{\mathbf{r}} \cdot \hat{\mathbf{q}}_1, q_1, q_1). \quad (176b)$$

Here $T^{(-)}$ is the scattering matrix obtained from Eqs. (79) and (83) when $+i\eta$ is replaced by $-i\eta$ in these equations.

The evaluation of the asymptotic form of $\psi_a^\pm$, Eq. (175), or Eq. (177) from Eq. (170) has been given here in some detail. We shall frequently encounter such integrals and shall henceforth not repeat a detailed description of the integration.

The asymptotic form (177) for $\psi_a^+$ may be inserted into Eq. (167) to provide an explicit expression for $\Psi(t)$ valid except in the region $|\mathbf{r}_1 - \mathbf{r}_2| \approx q_1^{-1}$. With the use of the first of Eqs. (166), we obtain[37]

$$\Psi(t) = \phi_a(t) + (2\pi)^{-3} \int d^3q_2 a_T(\mathbf{q}_2) \, e^{i\mathbf{q}_2 \cdot \mathbf{r}_2} \times \int d^3q_1 a_B(\mathbf{q}_1 - \mathbf{p}_1) \, e^{i(q_1 r + \mathbf{q}_1 \cdot \mathbf{r}_2 - \epsilon_{q1} t)} \frac{f(\hat{\mathbf{r}} \cdot \hat{\mathbf{p}}_1)}{r}. \quad (178)$$

The term $\phi_a(t)$ represents the wave packet for the *unscattered* beam particles, and is given by Eq. (28$a$).

[37] We have here replaced $\mathbf{q}_1$ and $q_1$ in $f$ by $\mathbf{p}_1$ and $p_1$, respectively. This is equivalent to the assumption, made previously, that $T_{ba}$ does not vary significantly across the wave packet.

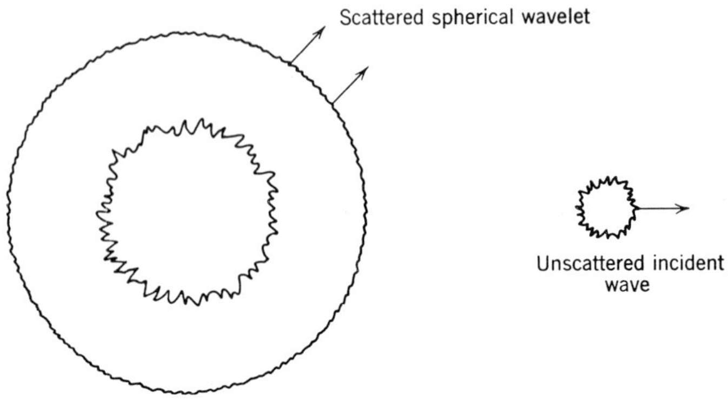


Fig. 3.4 Unscattered and scattered wave packets as described by the wave function (179).

Evaluation of the $\mathbf{q}_2$-integral gives just $G_T(\mathbf{r}_2)$ [see Eq. (117)]. To evaluate the $\mathbf{q}_1$-integral, we set

$$\mathbf{q}_1 = \boldsymbol{\rho}_1 + \mathbf{p}_1,$$

and

$$q_1 r + \mathbf{q}_1 \cdot \mathbf{r}_2 - \epsilon_{q1} t = (p_1 r + \mathbf{p}_1 \cdot \mathbf{r}_2 - \epsilon_{p1} t) + \boldsymbol{\rho}_1 \cdot (\hat{\mathbf{p}}_1 r + \mathbf{r}_2 - \mathbf{v}_1 t),$$

where (as usual) $\hat{\mathbf{p}}_1 = \mathbf{p}_1/p_1$. The $\rho_1$-integral gives just $G_B(\hat{\mathbf{p}}_1 r + \mathbf{r}_2 - \mathbf{v}_1 t)$. Because of the earlier assumption that the width $w_T$ of $G_T$ is much less than that of $G_B$, we may set $\mathbf{r}_2 = 0$ in $G_B$ so that $r \approx r_1$. It is also convenient to define a new momentum variable, $\mathbf{k}_f \equiv p_1 \hat{\mathbf{r}}$. Then

$$\Psi(t) = (2\pi)^{-3} G_T(\mathbf{r}_2) \Big[ e^{(i\mathbf{p}_1 \cdot \mathbf{r}_1 - \epsilon_{p1} t)}\, G_B(\mathbf{r}_1 - \mathbf{v}_1 t) + \frac{f(\hat{\mathbf{k}}_f \cdot \hat{\mathbf{p}}_1)}{r} e^{i[\mathbf{k}_f \cdot (\mathbf{r}_1 - \mathbf{r}_2) + \mathbf{p}_1 \cdot \mathbf{r}_2 - \epsilon_{p1} t]} G_B(0, 0, r_1 - v_1 t) \Big], \quad (179)$$

if $\hat{\mathbf{p}}_1 = \mathbf{v}_1/v_1$ is parallel to the $z$-axis.[38] Note that we have set $r = \hat{\mathbf{r}} \cdot (\mathbf{r}_1 - \mathbf{r}_2)$.

The interpretation of Eq. (179) is quite straightforward. The expression (28$a$) for $\phi_a(t)$ has been used for the case that $\mathbf{v}_2 = 0$. The scattered wave vanishes except for the interval

$$|r_1 - v_1 t| < w_B.$$

Thus the scattered wave is confined to a spherical shell of radius $r = v_1 t$ and thickness $2w_B$. The amplitude of this wave is determined by $f(\hat{\mathbf{k}}_f \cdot \hat{\mathbf{p}}_1) = f(\hat{\mathbf{r}} \cdot \hat{\mathbf{p}}_1)$ at each location $\mathbf{r} = r\hat{\mathbf{r}}$ on the shell. The incident and scattered waves are illustrated in Fig. 3.4.

[38] The notation of Eq. (132) has been used here in writing $G_B$.

The exponential factor

$$e^{i[\mathbf{k}_f \cdot (\mathbf{r}_1 - \mathbf{r}_2) + \mathbf{p}_1 \cdot \mathbf{r}_2]} \tag{180}$$

also has a simple interpretation. The phase of the incident wave $e^{i\mathbf{p}_1 \cdot \mathbf{r}_2}$ at $\mathbf{r}_2$ (the point at which the scattering occurs) is $\mathbf{p}_1 \cdot \mathbf{r}_2$. After scattering, the particle has a new wave vector $\mathbf{k}_f$. When it has traveled from $\mathbf{r}_1$ to $\mathbf{r}_2$, the phase has increased by $\mathbf{k}_f \cdot (\mathbf{r}_1 - \mathbf{r}_2)$. Thus the factor (180) occurs in Eq. (179).

The differential scattering cross section may easily be obtained from Eq. (179). According to Eq. (1.103*b*), the current of scattered particles is obtained from

$$\mathbf{j}(\mathbf{x}) = \operatorname{Re}\left[\left(\Psi, \delta(\mathbf{x} - \mathbf{r}_1)\left(\frac{1}{i}\nabla_{r1}\right)\Psi\right)\epsilon_{p1}^{-1}\right]c^2 \tag{181}$$

Except for a transient effect of order $\hbar/w_B$ near the boundary of the wave packet, we may replace $(1/i)\nabla_{r1}$ by $\mathbf{k}_f$ in this equation. Thus a simple evaluation using Eq. (179) gives

$$\mathbf{j}(\mathbf{x}) = \frac{\mathbf{k}_f c^2}{\epsilon_{p1}} \frac{1}{r^2} (2\pi)^{-6} \int |G_T(\mathbf{r}_2)|^2 \, d^3r_2 \, |G_B(0, 0, x - v_1 t)|^2 \, |f(\hat{\mathbf{k}}_f \cdot \hat{\mathbf{p}}_1)|^2. \tag{182}$$

Now, because of the normalization condition specified by Eqs. (118) and (121),

$$\int \mathbf{j}(\mathbf{x}) \, dt = \hat{\mathbf{r}} F \frac{|f(\hat{\mathbf{k}}_f \cdot \hat{\mathbf{p}}_1)|^2}{r^2}.$$

[Note that $\mathbf{k}_f(\epsilon_{p1}v_1)^{-1}c^2 = \hat{\mathbf{r}}$.] This expression represents the total flux of beam particles per unit area, per incident beam particle, at a point $\mathbf{r}$. If $d\Omega$ is an element of solid angle swept out by $\mathbf{r}$, and if $d\boldsymbol{\Sigma} = \hat{\mathbf{r}} r^2 \, d\Omega$ is the resulting surface element, then

$$\delta N_S = d\boldsymbol{\Sigma} \cdot \int \mathbf{j}(\mathbf{x}) \, dt = d\Omega F \, |f(\hat{\mathbf{k}}_f \cdot \hat{\mathbf{p}}_1)|^2$$

represents the number of scattered particles passing through $\delta\Omega$, per incident beam particle and per target particle.

The differential cross section is now obtained from Eq. (135).[39]

$$\frac{d\sigma}{d\Omega} = \frac{\delta N_S}{d\Omega F} = |f(\hat{\mathbf{k}}_f \cdot \hat{\mathbf{p}}_1)|^2 = (2\pi)^4 \, {\epsilon_{p1}}^2 \, |T(\hat{\mathbf{k}}_f \cdot \hat{\mathbf{p}}_1, p_1, p_1)|^2. \tag{183}$$

Equation (157) reduces to this form for the case that the target particle is very heavy, then

$$[\epsilon_{i'2,|\mathbf{p}_1 - \mathbf{k}_1|}]^{-1} \approx 0 \quad \text{and} \quad v_{\mathrm{rel},L} \approx \frac{k_1}{\epsilon_{i'1k1}} c^2.$$

[39] We have here scaled $N_T$ to unity and $F_B$ to $F$ in Eq. (135).

**The Kinematics of a Reaction**

Let us return to the more general problem for which there are $\nu$ particles emerging from the reaction in the final state $\chi'_b$. The momentum variables for these particles are taken as $\mathbf{k}_1, \ldots, \mathbf{k}_\nu$. In accordance with the notation of Eq. (105),

$$\psi_a^+ - \chi_a = \sum_{s_1,\ldots,s_\nu} \int \frac{d^3k_1 \cdots d^3k_\nu}{E_a + i\eta - E_b} \chi'_b \delta\left(\sum_{j=1}^{\nu} \mathbf{k}_j - \mathbf{q}_1 - \mathbf{q}_2\right) T_{ba}. \quad (184)$$

Here the sum over $s_1, \ldots, s_\nu$ represents a summation over the internal states of the particles in the final states $b$, and $\mathbf{q}_1$, $\mathbf{q}_2$ are the momentum variables of the two particles in the initial state $a$. We then have

$$\chi_a = (2\pi)^{-3} e^{i\mathbf{q}_1\cdot\mathbf{x}_1} e^{i\mathbf{q}_2\cdot\mathbf{x}_2} g_{a1}g_{a2},$$

$$\chi'_b = (2\pi)^{-\frac{3}{2}\nu} \exp\left(i \sum_{j=1}^{\nu} \mathbf{k}_j \cdot \mathbf{r}_j\right)\left(\prod_{\alpha=1}^{\nu} g_{s\alpha}\right). \quad (185)$$

Here, as usual, $g_{a1}, g_{a2}, g_{s1}, \ldots, g_{s\nu}$ are wave functions for the internal states of the particles. Finally,

$$E_a = \epsilon_{1,q1} + \epsilon_{2,q2},$$
$$E_b = \sum_{j=1}^{\nu} \epsilon_{j,kj} \quad (186)$$

in an obvious notation. It is convenient to suppress the internal state index on the energies here. Thus we have written $\epsilon_{j,kj}$ for $\epsilon_{sj,kj}$, and so on.

We shall first find the asymptotic form for $\psi_a^+$ when one of the variables $\mathbf{r}_j$, say $\mathbf{r}_1$, becomes very large. If we integrate Eq. (184) over $\mathbf{k}_2$, for example, the $\delta$-function implies that everywhere in the resulting expression we substitute for $\mathbf{k}_2$

$$\mathbf{k}_2 = \mathbf{q}_1 + \mathbf{q}_2 - \sum_{j(\neq 2)=1}^{\nu} \mathbf{k}_j.$$

Thus

$$\psi_a^+ - \chi_a = \sum_{s_1,\ldots,s_\nu} \prod_{\alpha=1}^{\nu} g_{s\alpha}(2\pi)^{-\frac{3}{2}\nu} \int d^3k_3 \cdots d^3k_\nu \exp\left[i\sum_{j=3}^{\nu} \mathbf{k}_j \cdot (\mathbf{r}_j - \mathbf{r}_2)\right]$$
$$\times\, e^{i(\mathbf{q}_1+\mathbf{q}_2)\cdot\mathbf{r}_2} \int \frac{d^3k_1\, e^{i\mathbf{k}_1\cdot(\mathbf{r}_1-\mathbf{r}_2)}}{E_a + i\eta - E_b} T_{ba}. \quad (187)$$

Consider next the integral

$$I_\nu \equiv \int \frac{d^3k_1\, e^{i\mathbf{k}_1\cdot(\mathbf{r}_1-\mathbf{r}_2)}}{E_a + i\eta - E_b} T_{ba}(\mathbf{k}_1), \quad (188)$$

where we indicate explicitly only the $\mathbf{k}_1$-dependence of $T_{ba}$. This is essentially equivalent to the integral (172), where now

$$E_b = \epsilon_{1,k_1} + \epsilon_{2,|q_1+q_2-\sum_{j(\neq 2)=1}^{\nu}\mathbf{k}_j|} + \epsilon_{3,k_3} + \cdots + \epsilon_{\nu\, k_\nu}. \tag{189}$$

As $|\mathbf{r}_1 - \mathbf{r}_2| \to \infty$, we obtain

$$I_\nu = -\frac{(2\pi)^2 k_1{}^0 T_{ba}(k_1{}^0\hat{\mathbf{r}}_{12})}{[\partial E_b/\partial k_1]_{E_b=E_a}} \frac{e^{ik_1{}^0|\mathbf{r}_1-\mathbf{r}_2|}}{|\mathbf{r}_1 - \mathbf{r}_2|} \tag{190}$$

Here $k_1{}^0$ is the value of $k_1$ "on the energy shell," defined by

$$E_b(k_1{}^0) = E_a. \tag{191}$$

The expression $[\partial E_b/\partial k_1]_{E_b=E_a}$ represents that value of $\partial E_b/\partial k_1$ evaluated at $k_1 = k_1{}^0$. Finally,

$$\hat{\mathbf{r}}_{12} \equiv \frac{\mathbf{r}_1 - \mathbf{r}_2}{|\mathbf{r}_1 - \mathbf{r}_2|}.$$

If we define a *scattering amplitude*

$$f(k_1{}^0\hat{\mathbf{r}}_{12}) \equiv -\frac{(2\pi)^2 k_1{}^0}{[\partial E_b/\partial k_1]_{E_b=E_a}} T_{ba}(k^0\hat{\mathbf{r}}_{12}), \tag{192}$$

Eq. (187) becomes

$$\psi_a{}^+ = \chi_a + \sum_{s_1,\ldots,s_\nu} \prod_{\alpha=1}^{\nu} g_{s\alpha} \int d^3k_3 \cdots d^3k_\nu\, e^{i(\mathbf{q}_1+\mathbf{q}_2)\cdot\mathbf{r}_2}$$
$$\times (2\pi)^{-\frac{3}{2}\nu} \exp\left[i\sum_{j=3}^{\nu}\mathbf{k}_j\cdot(\mathbf{r}_j - \mathbf{r}_2)\right] \frac{e^{ik_1{}^0|\mathbf{r}_1-\mathbf{r}_2|}}{|\mathbf{r}_1 - \mathbf{r}_2|} f. \tag{193}$$

This is valid when $|\mathbf{r}_1 - \mathbf{r}_2| \to \infty$.

We have not specified *which particles* are 1 and 2, so Eq. (193) is a valid asymptotic expression for $\psi_a{}^+$ when the relative separation of any pair of particles is large. In particular, we have not restricted the remaining coordinates in any manner. The asymptotic form (193) is useful for many applications.

Let us finally study the kinematic relations for the $\nu$-particles in the final states $b$. We shall construct a packet of states $\psi_a{}^+$ according to Eqs. (39) and (68). Then the state function is

$$\Psi(t) = \int d^3q_1\, d^3q_2 A(\mathbf{q}_1, \mathbf{q}_2)\, e^{-iE_at}\psi_a{}^+ \equiv \phi_a(t) + \Psi_{sc}(t). \tag{194}$$

Here $\phi_a(t)$ is the wave function for the unscattered beam and target particles, as usual, and $\Psi_{sc}(t)$ represents those particles emitted from the scattering interaction.

We now seek the probability that the $\alpha$th emitted particle ($\alpha = 1, 2, \ldots, \nu$) is to be found near the point $\mathbf{R}_\alpha$ with the probability amplitude $G_\alpha(\mathbf{r}_\alpha - \mathbf{R}_\alpha)$. The $G_\alpha(\alpha = 1, 2, \ldots, \nu)$ will be considered to be wave packets for the particles in the final state $b$. Previous assumptions imply the packets $G_\alpha$ are large compared to the wavelength $\hbar/\boldsymbol{\varkappa}_\alpha$ corresponding to the mean momentum $\boldsymbol{\varkappa}_\alpha$.

The required probability $\bar{P}_{ba}$ is therefore that of finding the system in the final state

$$X'_b = (2\pi)^{-\frac{3}{2}\nu} \prod_{\alpha=1}^{\nu} [G_\alpha(\mathbf{r}_\alpha - \mathbf{R}_\alpha) e^{i\boldsymbol{\varkappa}_\alpha \cdot \mathbf{r}_\alpha} g_{s'\alpha}], \tag{195}$$

where the $g_{s'\alpha}$ are again wave functions for the internal states. The condition

$$(2\pi)^{-3} \int d^3 r_\alpha \, |G_\alpha(\mathbf{r}_\alpha - \mathbf{R}_\alpha)|^2 = 1 \tag{196}$$

ensures the normalization condition

$$(X'_b, X'_b) = 1. \tag{197}$$

Now

$$\bar{P}_{ba} = |(X'_b, \Psi(t)|^2 = |(X'_b, \Psi_{sc}(t)|^2 \equiv |\mathscr{A}|^2 \tag{198}$$

is the probability of finding the system in the state $X'_b$ at time $t$, if it is again assumed that

$$(X'_b, \phi_a(t)) = 0$$

at all times $t$. We shall first evaluate the probability amplitude

$$\mathscr{A} \equiv (X'_b, \Psi_{sc}(t)). \tag{199}$$

An approach which is, of course, equivalent to finding $\mathscr{A}$ is that of projecting $\Psi(t)$ onto the state $X'_b$. This is effected by the projection operator

$$\Lambda_b = |X'_b\rangle \langle X'_b|, \tag{200}$$

so

$$\Psi(t)\,|_b \equiv \Lambda_b \Psi(t). \tag{201}$$

The states $X'_b$ form a complete set if we properly weight the sum over states. The required sum is

$$\sum_{X'_b} = \sum_{s'_1, \ldots, s'_\nu} \int d^3\kappa_1 \cdots d^2\kappa_\nu \int d^3R_1 \cdots d^3R_\nu (2\pi)^{-3\nu}.$$

Thus

$$\sum_{X'_b} \Lambda_b = \int d^3\kappa_1 \cdots d^3\kappa_\nu \int d^3R_1 \cdots d^3R_\nu (2\pi)^{-6\nu}$$

$$\times \left( \sum_{s_1, \ldots, s_\nu} g_{s1} g_{s1}{}^{\dagger} \cdots g_{s\nu} g_{s\nu}{}^{\dagger} \right) \prod_{\alpha=1}^{\nu} e^{i\boldsymbol{\varkappa}_\alpha \cdot (\mathbf{r}_\alpha - \mathbf{r}'_\alpha)} G_\alpha(\mathbf{r}_\alpha - \mathbf{R}_\alpha) G_\alpha^*(\mathbf{r}'_\alpha - \mathbf{R}_\alpha)$$

$$= \prod_{\alpha=1}^{\nu} \left[ \delta(\mathbf{r}_\alpha - \mathbf{r}'_\alpha) \sum_{s_\alpha} g_{s\alpha} g_{s\alpha}{}^{\dagger} \right], \tag{202}$$

which is the unit operator.

Let us write

$$G_\alpha(\mathbf{r}_\alpha - \mathbf{R}_\alpha) = \int d^3l_\alpha a_\alpha(\mathbf{l}_\alpha)\, e^{i(\mathbf{r}_\alpha - \mathbf{R}_\alpha)\cdot \mathbf{l}_\alpha}. \tag{203}$$

If this expansion is introduced into Eq. (195), the state $X'_b$ is expressed as

$$X'_b = (2\pi)^{-\frac{3}{2}\nu} \prod_{\alpha=1}^{\nu} g_{s'\alpha} \int d^3k'_\alpha a_\alpha(\mathbf{k}'_\alpha - \boldsymbol{\varkappa}_\alpha)\, e^{i\mathbf{k}'_\alpha\cdot(\mathbf{r}_\alpha - \mathbf{R}_\alpha)}\, e^{i\boldsymbol{\varkappa}_\alpha\cdot \mathbf{R}_\alpha}. \tag{204}$$

The scattered state function $\Psi_{sc}(t)$ [see Eqs. (184) and (194)] is

$$\Psi_{sc}(t) = (2\pi)^{-\frac{3}{2}\nu} \sum_{s_1,\ldots,s_\nu} \prod_{\alpha=1}^{\nu} g_{s\alpha} \int d^3q_1\, d^3q_2 a_B(\mathbf{q}_1 - \mathbf{p}_1) a_T(\mathbf{q}_2 - \mathbf{p}_2)$$

$$\times \int \frac{d^3k_1 \cdots d^3k_\nu}{E_a + i\eta - E_b}\, e^{-iE_at} \exp\left(i\sum_{j=1}^{\nu} \mathbf{k}_j\cdot\mathbf{r}_j\right)\delta\left(\mathbf{q}_1 + \mathbf{q}_2 - \sum_{j=1}^{\nu}\mathbf{k}_j\right) T_{ba}. \tag{205}$$

Then using (204) the amplitude $\mathscr{A}$ is

$$\mathscr{A} = (X'_b, \Psi_{sc}) = (2\pi)^{-3\nu} \int d^3k'_1 \cdots d^3k'_\nu \int d^3k_1 \cdots d^3k_\nu$$

$$\times \int d^3r_1 \cdots d^3r_\nu \exp\left\{i\sum_{\alpha=1}^{\nu}[(\mathbf{k}_\alpha - \mathbf{k}'_\alpha)\cdot\mathbf{r}_\alpha + i(\mathbf{k}'_\alpha - \boldsymbol{\varkappa}_\alpha)\cdot\mathbf{R}_\alpha]\right\}$$

$$\times a_1^*(\mathbf{k}'_1 - \boldsymbol{\varkappa}_1)\cdots a_\nu^*(\mathbf{k}'_\nu - \boldsymbol{\varkappa}_\nu) \int d^3q_1\, d^3q_2\, e^{-iE_at}$$

$$\times a_B(\mathbf{q}_1 - \mathbf{p}_1) a_T(\mathbf{q}_2 - \mathbf{p}_2)(E_a + i\eta - E_b)^{-1}\,\delta\left(\mathbf{q}_1 + \mathbf{q}_2 - \sum_{j=1}^{\nu}\mathbf{k}_j\right) T_{ba}$$

$$= \int d^3k_1 \cdots d^3k_\nu \int d^3q_1\, d^3q_2 a_1^*(\mathbf{k}_1 - \boldsymbol{\varkappa}_1)\cdots a_\nu^*(\mathbf{k}_\nu - \boldsymbol{\varkappa}_\nu) a_B(\mathbf{q}_1 - \mathbf{p}_1)$$

$$\times a_T(\mathbf{q}_2 - \mathbf{p}_2)\, e^{-iE_at} \exp\left[i\sum_{\alpha=1}^{\nu}(\mathbf{k}_\alpha - \boldsymbol{\varkappa}_\alpha)\cdot\mathbf{R}_\alpha\right] \frac{\delta\left(\mathbf{q}_1 + \mathbf{q}_2 - \sum_{j=1}^{\nu}\mathbf{k}_j\right)}{E_a + i\eta - E_b} T_{ba}. \tag{206}$$

As before, we shall assume that the wave packets are sufficiently narrow (in momentum space) that

$$T_{ba} \approx \langle \mathbf{k}_1 \cdots \mathbf{k}_\nu | T | \mathbf{p}_1, \mathbf{p}_2 \rangle \equiv T_{b\bar{a}}. \tag{207}$$

(The designation of the internal states is suppressed here to simplify the notation.)

To simplify Eq. (206) further we write

$$\frac{1}{E_a + i\eta - E_b} = -i\int_0^\infty d\tau\, e^{i\tau(E_a+i\eta-E_b)},$$

$$\delta\left(\mathbf{q}_1 + \mathbf{q}_2 - \sum_{j=1}^{\nu}\mathbf{k}_j\right) = (2\pi)^{-3}\int d^3\lambda \exp\left[i\boldsymbol{\lambda}\cdot\left(\mathbf{q}_1 + \mathbf{q}_2 - \sum_{j=1}^{\nu}\mathbf{k}_j\right)\right]. \quad (208)$$

These expressions are substituted into Eq. (206) to yield

$$\begin{aligned}\mathscr{A} = T_{b\bar{a}}(-i)\int_0^\infty d\tau(2\pi)^{-3}\int d^3\lambda \int d^3k_1 \cdots d^3k_\nu\, d^3q_1\, d^3q_2 \\ \times e^{iE_a(\tau-t)}\, e^{-i\tau(E_b-i\eta)} \exp\left[i\sum_{\alpha=1}^{\nu}(\mathbf{k}_\alpha - \boldsymbol{\varkappa}_\alpha)\cdot\mathbf{R}_\alpha\right] \\ \times \exp\left[i\boldsymbol{\lambda}\cdot\left(\mathbf{q}_1 + \mathbf{q}_2 - \sum_{j=1}^{\nu}\mathbf{k}_j\right)\right]a_B(\mathbf{q}_1 - \mathbf{p}_1) \\ \times a_T(\mathbf{q}_2 - \mathbf{p}_2)a_1{}^*(\mathbf{k}_1 - \boldsymbol{\varkappa}_1)\cdots a_\nu{}^*(\mathbf{k}_\nu - \boldsymbol{\varkappa}_\nu). \quad (209)\end{aligned}$$

If we introduce $\boldsymbol{\rho}_1 \equiv \mathbf{q}_1 - \mathbf{p}_1$, $\boldsymbol{\rho}_2 = \mathbf{q}_2 - \mathbf{p}_2$, we may expand $E_a$ [compare Eqs. (186) and (29)]

$$\begin{aligned}E_a &= E_{\bar{a}} + \boldsymbol{\rho}_1\cdot\mathbf{v}_1 + \boldsymbol{\rho}_2\cdot\mathbf{v}_2, \\ E_{\bar{a}} &= \epsilon_{1,p1} + \epsilon_{2,p2}.\end{aligned} \quad (210)$$

Integration over $\boldsymbol{\rho}_1$ and $\boldsymbol{\rho}_2$ leads to

$$\begin{aligned}\mathscr{A} = T_{b\bar{a}}(-i)(2\pi)^{-3}\int_0^\infty d\tau\int d^3\lambda\int d^3k_1\cdots d^3k_\nu \\ \times [e^{-iE_{\bar{a}}t}\, e^{i\boldsymbol{\lambda}\cdot(\mathbf{p}_1+\mathbf{p}_2)}\, e^{i\tau(E_{\bar{a}}+i\eta-E_b)}] \\ \times \exp\left[i\sum_{\alpha=1}^{\nu}\mathbf{k}_\alpha\cdot(\mathbf{R}_\alpha - \boldsymbol{\lambda})\right]\exp\left(-i\sum_{\alpha=1}^{\nu}\boldsymbol{\varkappa}_\alpha\cdot\mathbf{R}_\alpha\right) \\ \times a_1{}^*(\mathbf{k}_1 - \boldsymbol{\varkappa}_1)\cdots a_\nu{}^*(\mathbf{k}_\nu - \boldsymbol{\varkappa}_\nu) \\ \times G_B(\boldsymbol{\lambda} - \mathbf{v}_1(t-\tau))G_T(\boldsymbol{\lambda} - \mathbf{v}_2(t-\tau)). \quad (211)\end{aligned}$$

To complete the evaluation, we take [Eq. (186)]

$$\begin{aligned}E_b &= E_{\bar{b}} + \sum_{\alpha=1}^{\nu}\mathbf{v}'_\alpha\cdot\mathbf{l}_\alpha, \\ E_{\bar{b}} &= \sum_\alpha \epsilon_{\alpha,\kappa\alpha}, \\ \mathbf{v}'_\alpha &= \frac{\boldsymbol{\varkappa}_\alpha}{\epsilon_{\alpha,\kappa\alpha}},\end{aligned} \quad (212)$$

where $\mathbf{l}_\alpha \equiv \mathbf{k}_\alpha - \boldsymbol{\varkappa}_\alpha$. With the definitions (203), we obtain

$$\mathscr{A} = T_{b\bar{a}}(-i)(2\pi)^{-3}\int_0^\infty d\tau \int d^3\lambda(e^{-iE_{\bar{a}}t}\, e^{i\boldsymbol{\lambda}\cdot(\mathbf{p}_1+\mathbf{p}_2)})$$
$$\times\, e^{i\tau(E_{\bar{a}}+i\eta-E_b)} \exp\left(-i\boldsymbol{\lambda}\cdot\sum_{\alpha=1}^{\nu}\boldsymbol{\varkappa}_\alpha\right) G_B(\boldsymbol{\lambda} - \mathbf{v}_1(t-\tau))$$
$$\times\, G_T(\boldsymbol{\lambda} - \mathbf{v}_2(t-\tau)) \prod_{\alpha=1}^{\nu} G_\alpha{}^*(\mathbf{R}_\alpha - \boldsymbol{\lambda} - \mathbf{v}'_\alpha\tau). \qquad (213)$$

To interpret Eq. (213) we consider the special case that the target particle is at rest in the initial state, so

$$\mathbf{v}_2 = 0.$$

We also suppose that the $G_\alpha(\alpha = 1, 2, \ldots, \nu)$ are considerably broader than the widths $w_B$ and $w_T$ of $G_B$ and $G_T$. Then Eq. (213) is nonvanishing only for $\boldsymbol{\lambda}$ in the interval

$$|\boldsymbol{\lambda}| \leq w_T,$$

since $G_T(\boldsymbol{\lambda})$ appears in the integrand. Now, since $w_T$ has been assumed to be small compared to $w_B$, we may write

$$G_B(\boldsymbol{\lambda} - \mathbf{v}_1(t-\tau)) \approx G_B(\mathbf{v}_1(\tau - t)). \qquad (214)$$

The $\tau$-variable may be replaced by

$$y \equiv \tau - t,$$

where the integrand in (213) is nonvanishing only for $y$ in the interval

$$|y| \approx \frac{w_B}{v_1}. \qquad (215)$$

Because we have assumed that the $G_\alpha$ have widths large compared to $w_T$ and $w_B$, we may set $\boldsymbol{\lambda} = 0$ and $\tau = t$ in the arguments of these quantities. Then

$$\mathscr{A} = \Bigg\{-i(2\pi)^{-3}\int dy\, d^3\lambda G_T(\boldsymbol{\lambda}) G_B(y\mathbf{v}_1)$$
$$\times \exp\left[i\boldsymbol{\lambda}\cdot\left(\mathbf{p}_1 + \mathbf{p}_2 - \sum_{j=1}^{\nu}\boldsymbol{\varkappa}_j\right)\right] e^{iy(E_{\bar{a}}-E_b)}\Bigg\}$$
$$\times\, T_{b\bar{a}} \prod_{\alpha=1}^{\nu} G_\alpha{}^*(\mathbf{R}_\alpha - \mathbf{v}'_\alpha t)\, e^{-iE_b t} \qquad (216)$$

where we have gone to the limit as $\eta \to 0(+)$. This is possible since the $y$-integral is well defined.

It is evident that $\mathscr{A} = 0$ unless

$$\mathbf{R}_\alpha \approx \mathbf{v}'_\alpha t, \tag{217}$$

which is an obvious kinematical relation. This means that we must place a counter at the point $\mathbf{R}_\alpha = \mathbf{v}'_\alpha t$, if we wish to count particle $\alpha$ with velocity $\mathbf{v}'_\alpha$ at time $t$. A more familiar expression for $\mathscr{A}$ may be given if we introduce "finite $\delta$-functions" $\Delta_E(E_{\bar{a}} - E_{\bar{b}})$ and $\Delta_P\left(\mathbf{p}_1 + \mathbf{p}_2 - \sum_{j=1}^{\nu} \boldsymbol{\varkappa}_j\right)$ according to

$$\bar{G}_B \Delta_E(E_{\bar{a}} - E_{\bar{b}}) \equiv \frac{1}{2\pi} \int dy G_B(\mathbf{v}_1 y)\, e^{iy(E_{\bar{a}} - E_{\bar{b}})}$$

$$\bar{G}_T \Delta_P\left(\mathbf{p}_1 + \mathbf{p}_2 - \sum_{j=1}^{\nu} \boldsymbol{\varkappa}_j\right) \equiv (2\pi)^{-3} \int d^3\lambda\, G_T(\boldsymbol{\lambda}) \exp\left[i\boldsymbol{\lambda} \cdot \left(\mathbf{p}_1 + \mathbf{p}_2 - \sum_{j=1}^{\nu} \boldsymbol{\varkappa}_j\right)\right],$$

where

$$\int dE_{\bar{a}}\, \Delta_E(E_{\bar{a}} - E_{\bar{b}}) = 1, \qquad \int d^3p_1\, \Delta_P(\mathbf{p}_1) = 1. \tag{218}$$

Then

$$\mathscr{A} = [-2\pi i T_{\bar{b}\bar{a}}]\, \Delta_E(E_{\bar{a}} - E_{\bar{b}})\, \Delta_P\left(\mathbf{p}_1 + \mathbf{p}_2 - \sum_{j=1}^{\nu} \boldsymbol{\varkappa}_j\right) \times \bar{G}_T \bar{G}_B \prod_{\alpha=1}^{\nu} G_\alpha^*(\mathbf{R}_\alpha - \mathbf{v}'_\alpha t). \tag{219}$$

This is evidently the wave packet analogue of Eq. (111). There are no singular quantities in Eq. (219), however. It is apparent from this expression that our intuitive expectation for the kinematic behavior of the emitted particles is fulfilled in detail by the wave function $\Psi(t)$. It further shows the universal applicability of the $i\eta$ instruction.

Let us now suppose that experimental counters are arranged to detect each of the particles, the area of each counter being larger than the corresponding wave packet $G_\alpha$. Then, in accordance with Eqs. (198) and (202) the probability of counting all $\nu$-particles is

$$\begin{aligned}\sum_{b} \bar{P}_{ba} &= \sum_{s_1, \ldots, s_\nu} \int d^3\kappa_1 \cdots d^3\kappa_\nu \int d^3R_1 \cdots d^3R_\nu (2\pi)^{-3\nu} \bar{P}_{ba} \\ &= (2\pi)^2 \int d^3\kappa_1 \cdots d^3\kappa_\nu\, |T_{\bar{b}\bar{a}}|^2 \\ &\quad \times \left| \Delta_E(E_{\bar{a}} - E_{\bar{b}})\, \Delta_P\left(\mathbf{p}_1 + \mathbf{p}_2 - \sum_{j=1}^{\nu} \boldsymbol{\varkappa}_\nu\right) \bar{G}_T \bar{G}_B \right|^2,\end{aligned} \tag{220}$$

where the sums and integrals extend over the range of states detected by the counters. If the range of $\boldsymbol{\varkappa}_1, \ldots, \boldsymbol{\varkappa}_\nu$ values to be detected is broad compared to the packet spread, we may write

$$|\Delta_E(E_{\bar{a}} - E_{\bar{b}})\bar{G}_B|^2 = (2\pi)^{-2}\int dy\, dy' G_B(\mathbf{v}_1 y) G_B{}^*(\mathbf{v}_1 y')\, e^{i(y-y')(E_{\bar{a}}-E_{\bar{b}})}$$

$$\approx (2\pi)^{-2}\int dz\, |G_B(0, 0, v_1 z)|^2 \int dx\, e^{ix(E_{\bar{a}}-E_{\bar{b}})} \approx \frac{F}{v_1}(2\pi)^2\, \delta(E_{\bar{a}} - E_{\bar{b}}),$$

valid in the integral of Eq. (220). In addition, in the integrand of Eq. (220) we may set

$$\left|\Delta_P\left(\mathbf{p}_1 + \mathbf{p}_2 - \sum_{j=1}^{\nu} \boldsymbol{\kappa}_j\right)\bar{G}_T\right|^2 \approx \delta\left(\mathbf{p}_1 + \mathbf{p}_2 - \sum_{j=1}^{\nu} \boldsymbol{\kappa}_j\right).$$

In these expressions we have used the normalization condition (118) and (121). Therefore, Eq. (220) is equivalent to

$$\sum_b \bar{P}_{ba} = \sum_{s_1,\ldots,s_\nu} \int d^3\kappa_1 \cdots d^3\kappa_\nu \times \left[\frac{F(2\pi)^4}{v_1}\,\delta(E_{\bar{a}} - E_{\bar{b}})\,\delta\left(\mathbf{p}_1 + \mathbf{p}_2 - \sum_{j=1}^{\nu} \boldsymbol{\kappa}_j\right)|T_{\bar{b}\bar{a}}|^2\right]. \quad (221)$$

This is just Eq. (134) for the case that $\mathbf{v}_2 = 0$.

#### Remarks Concerning the Validity of Our Discussion

We have now completed our general discussion of the kinematical relations concerned with the preparation of the initial state and the observation of final states for a scattering reaction. We have been able to avoid any discussion of the dynamical behavior of the particles during the time that they interact. Instead we have found that a matrix $S$ (or $\mathscr{T}$) which couples initial and final asymptotic states contains a complete specification of possible experimental results.

Interpreted in this manner, our description is expected to have a more general validity than perhaps the Schrödinger equation which has been used to obtain our results. That is, there seems to be little doubt that the Schrödinger equation (or for that matter Newton's equations) provides an adequate description of the propagation of sufficiently large wave packets for noninteracting particles. Therefore, even if an appropriate Schrödinger equation may not really exist for the *evaluation* of $S$, there presumably *does exist* an $S$-matrix in terms of which the cross section for any reaction can be obtained.

The Schrödinger equation has received exhaustive experimental verification only for electromagnetic interactions. Here it has provided both detailed and accurate descriptions for a very wide range of phenomena, including atomic and solid state physics.[40] Applications of the Schrödinger equation to nuclear physics have, of course, been exceedingly useful and have provided considerable insight into the mechanisms of nuclear dynamics. On the other hand, these applications have tended to be either phenomenological or semiquantitative. Applications to elementary particle phenomena (other than electrodynamics) have been crude and never quite convincing—except in the sense we have mentioned, that $S$ is a phenomenological quantity subject to certain symmetry relations.[41]

For these reasons, Heisenberg [5] proposed[42] that $S$ should be introduced as a fundamental quantity in basic theory. In Heisenberg's theory, expressions such as $\Psi(t) \approx e^{-iE_a t}SX_a$, Eqs. (91), (94), and (100) would arise directly from postulates. Unfortunately, no general dynamical principle (other than the Schrödinger equation) has been found for evaluating $S$.[43] Therefore, most efforts to study interactions between elementary systems have been based on quantum field theories, for which the Schrödinger equation provides the dynamical principle.

## REFERENCES AND NOTES

1. C. Chester, B. Friedman, and F. Ursell, *Proc. Cambridge Phil. Soc.* **53,** 599 (1957).
2. E. C. Titchmarsh, *Introduction to the Theory of the Fourier Integral,* Oxford University Press, Oxford, 1937.
3. E. P. Wigner, *Group Theory and its Application to the Quantum Mechanics of Atomic Spectra,* Academic Press, New York, 1959.
4. C. Møller, *Kgl. Danske Videnskab. Selskab., Mat.-fys. Medd.* **23,** 1 (1945).
5. W. Heisenberg, *Z. Physik* **120,** 513, 673 (1943).
6. The importance of using wave packets to develop scattering theory has been emphasized by E. P. Wigner and L. Eisenbud (L. Eisenbud, thesis, Princeton University, 1948) and F. Low (University of Illinois lecture notes, 1953). Subsequent developments using wave packets have been given by K. Watson and K. Brueckner, *Scattering Theory and its Application,* AECU-2765 (1953); H. Sunakawa, *Progr. Theoret. Phys.* (*Kyoto*) **14,** 175 (1955); and T. Sasakawa, *Progr. Theoret. Phys.* (*Kyoto*), Supplement 11, p. 69 (1959).
7. A more complete list of references to the development of scattering theory is given in Chapter 5.

[40] The interaction between relativistic particles may last for a time of the order of $10^{-23}$ sec. The Schrödinger equation was originally proposed to explain atomic reactions, for which the time scales range from $10^{-10}$ to $10^{-15}$ sec.

[41] Successful study of the so-called *weak interactions* does not imply a validity for the Schrödinger description. Since the $S$-matrix, in lowest-order perturbation theory, is $S = 1 - 2\pi i\,\delta(E_b - E_a)V$, where $V$ is the interaction, a correct $V$ can always be (trivially) exhibited if $S$ is known.

[42] See also Møller [4] and Notes [6] and [7].

[43] In recent years dispersion relations have suggested a possible beginning for such a theory. In Chapters 6 and 10 a description of these techniques will be given.

# 4. FURTHER DESCRIPTION OF SCATTERING PROCESSES

In Chapter 3 we discussed the interaction of two (or more) systems that had been prepared initially in isolated wave packet states. The wave function for these states was written as $\phi_a(t_0) = e^{-iKt_0}X_a$ [Eq. (3.33)]. As obtained in Eq. (3.70), the wave function $\Psi(t)$ for the interacting systems has the form

$$\Psi(t) = e^{-iHt}\int d^3q_1\, d^3q_2 A(\mathbf{q}_1, \mathbf{q}_2)\psi_a{}^{\pm}, \tag{1}$$

where the $\psi_a{}^{\pm}$ are wave functions defined by Eqs. (3.66) and (3.67) and satisfying the Schrödinger equation $H\psi_a{}^{\pm} = E_a\psi_a{}^{\pm}$. The choice of $\psi^+$ or $\psi^-$ depends on whether $t_0 = -|t_0|$ or $t_0 = |t_0|$, respectively.

In Section 3.3 we saw that when $t \gg |t_c|$, the collision time, $\Psi(t)$ approaches an asymptotic form when projected onto a particular set of states $\chi'_b$ which are to be observed. If $\Lambda_b$ is the projection operator onto these states, we found that

$$\lim_{t\to\infty} \Lambda_b\Psi(t) = e^{-iK't}\chi'_b S X_a, \tag{2}$$

where $S$ has the matrix elements (3.94). The transition probability and scattering cross sections were expressed in terms of the matrix elements of $S$.

In this chapter we shall extend the description given in Chapter 3. This will include the effect of wave function symmetrization for identical particles. We shall also discuss the implication of several other symmetries of the Schrödinger equation.

### 4.1 DESCRIPTION OF THE CHANNELS OF A REACTION

It was observed in Chapter 3 that the set of final observed states $\chi'_b$ might belong to a different set of complete orthonormal states than the

$\chi_a$. The reason for this is that the "particles" (as defined in Section 3.1) observed in the state $b$ may not be the same as those in the initial state $a$. In this case it was said that the states $a$ and $b$ represent different *channels* for the reaction. As part of the description of a given physical system, it is important to enumerate all possible such channels from which initial scattering states may be formed and into which the system may finally go.

The problem of counting the possible channels for the final asymptotic state is equivalent to enumerating the possible combinations of particles which can be simultaneously observed in that state.[1] To rephrase this, let us suppose two successive scattering experiments are performed, each with initial states prepared in the same manner. If the observed numbers of particles of each type are not the same for the two experiments, we say that the reaction was observed to go into *different channels* in the two experiments. In this connection a question might arise whether two particles are "different," or "different states of the same particle." This requires only a definition of the term "same particle," which in practice is determined by convenience.

**Notation**

Let us suppose that for a given scattering experiment the initial state belongs to a channel $c$. The channel $c$ is specified merely by stating the number of particles of each species which are to interact. A particular channel for the final, asymptotic state will be called $c'$. As we have said, this is specified by enumerating the number of particles of each species which can be simultaneously observed in the final state. For a given physical system, for which the Hamiltonian is $H$, the total number of final channels will be called $C_h$, so $c' = 1, 2, \ldots, C_h$.[2] Evidently, any final state channel may also be used as an initial state channel; we need only reverse the velocity vectors of the receding wave packets. The incident channel $c$ will always be included among the final channels $c'$.[3]

As an example, let us suppose that an experiment involves proton-proton scattering above the energetic threshold for producing a single pi-meson, but below that for producing two pi-mesons. Then the four possible final state channels are

$$(p, p), \qquad (p, p, \pi^0), \qquad (p, n, \pi^+), \qquad (d, \pi^+),$$

[1] A somewhat different definition of channel for a reaction was given by Wigner and Eisenbud, see Ref. [1].

[2] In most applications $C_h$ is finite. When photons are emitted $C_h$ is not defined. This is called the infrared catastrophe, which is, in fact, not a catastrophe, since how to handle the problem is known; see Bloch and Nordsieck [2].

[3] An exception can occur if any of the particles in $c$ is unstable.

the reactions being

$$\begin{aligned} p + p &\rightarrow p + p, \\ p + p &\rightarrow p + p + \pi^0, \\ p + p &\rightarrow p + n + \pi^+, \\ p + p &\rightarrow d + \pi^+. \end{aligned} \tag{3}$$

Any of these final channels might, in principle, be an incident channel, leading to any of the four final channels. For example,

$$\begin{aligned} p + n + \pi^+ &\rightarrow p + p, \\ p + n + \pi^+ &\rightarrow p + p + \pi^0, \\ p + n + \pi^+ &\rightarrow p + n + \pi^+, \\ p + n + \pi^+ &\rightarrow d + \pi^+. \end{aligned} \tag{4}$$

For this system there are seen to be sixteen scattering matrices $(c' |\mathscr{T}| c)$, leading from channel $c$ to channel $c'$.[4]

**Channel Wave Functions and Scattering Operators**

Corresponding to each channel $c$ of a given system the Hamiltonian $H$ may be decomposed into two parts.

$$H = K_c + V_c. \tag{5}$$

Here $V_c$ describes the interaction between the particles of channel $c$ and $K_c$ is the Hamiltonian for these particles when they are *separated* and do *not interact*. In Chapter 3 we wrote $H = K + V$ and $H = K' + V'$, corresponding to two specific choices for $c$ ($c = 1, 2, \ldots, C_h$) in Eq. (5).[5] The wave functions that describe a state $a$ of the system, when it is in channel $c$, are written as $\chi_{c;a}$. These satisfy the equation

$$K_c\chi_{c;a} = E_a\chi_{c;a} \qquad (c = 1, 2, \ldots, C_h). \tag{6a}$$

The $\chi_{c;a}$ do not in general represent a complete set of state vectors for the system. That is, the $\chi_{c;a}$ do not necessarily include all the eigenvectors of $K_c$. These latter functions will be written as $\chi_a{}^c$. They of course *include* the $\chi_{c;a}$, since they satisfy the equation

$$K_c\chi_a{}^c = E_a\chi_a{}^c. \tag{6b}$$

[4] There are of course additional channels, involving $\gamma$-rays, electron-positron pairs, and so on. The matrix elements of $\mathscr{T}$ involving these other channels are much smaller than those described here and may be neglected in a first approximation.

[5] We shall continue to use the notation of Chapter 3 when we do not wish to enumerate channels.

The $\chi_a{}^c$ always form a complete set of state vectors. Examples of the $\chi_a{}^c$, corresponding to two specific choices for $c$, are the $\chi_a$ and $\chi'_b$ of Chapter 3.

The $\chi_{c;a}$ will not form a complete set, for example, when $K_c = K_{c'}$ for two different channels $c$ and $c'$. Then

$$K_c\chi_{c;a} = E_a\chi_{c;a},$$

$$K_c\chi_{c';a'} = E_a\chi_{c';a'},$$

and both channel states must be included in a complete set of states. Again, the $\chi_{c;a}$ will not form a complete set when some of the particles in $c$ are bound states of "elementary" particles.

To illustrate this, we consider a system consisting of a neutron, a proton, and a positron. This system will be found in one of two channels, $c_1 = (d, e^+)$ or $c_2 = (n, p, e^+)$. Let $T_n$, $T_p$, and $T_e$ represent the kinetic energies of the neutron, proton, and positron respectively, and $V_{np}$ and $V_{pe}$ the corresponding interaction energies of the $(n, p)$- and $(p, e^+)$-systems. The Hamiltonians for channels $c_1$ and $c_2$ are then

$$K_{c1} = T_n + T_p + V_{np} + T_e,$$

$$K_{c2} = T_n + T_p + T_e.$$

The eigenfunctions $\chi_{c2;a'}$ of $K_{c2}$ form a complete set. The $\chi_{c1;a}$ evidently do not, since only the states with a bound deuteron are included. The eigenfunctions of $K_{c1}$, including the continuum states of the $(n, p)$-system, do form a complete set, however. (According to our definition of "channel states," those eigenfunctions of $K_c$ for which the neutron and proton are not bound are not channel states, since the interaction $V_{np}$ is included in $K_{c1}$.)

When we wish to imply that any one of the $C_h$ channels is chosen for the initial state of a scattering process, we shall write [see Eqs. (3.26$a$) and (3.70)]

$$\Psi(t) = e^{-iHt}\int d^3q_1 \cdots d^3q_\mu A(\mathbf{q}_1 \cdots \mathbf{q}_\mu)\psi^\pm_{c;a}, \tag{7}$$

where $A(\mathbf{q}_1 \cdots \mathbf{q}_\mu)$ is the incident wave packet amplitude for the $\mu$ particles in channel $c$, and

$$\psi^\pm_{c;a} = \left(1 + \frac{1}{E_a \pm i\eta - H} V_c\right)\chi_{c;a}. \tag{8}$$

We may rewrite this in the form of Eq. (3.81) as

$$\psi^+_{c;a} = \chi_{c;a} + \sum_b \frac{\chi_b{}^{c'}}{E_a + i\eta - E_b}(\chi_b{}^{c'}, \mathscr{T}_{c',c}\chi_{c;a}), \tag{9a}$$

where the $\chi_b^{c'}$ represent the complete set of eigenfunctions of $K_{c'}$ ($c' = 1, 2, \ldots, C_h$), as introduced by Eq. (6*b*). The matrix element of $\mathscr{T}_{c',c}$ is defined by Eq. (3.82). That is,

$$\mathscr{T}_{c',c} = V_c + V_{c'} \frac{1}{E_a + i\eta - H} V_c \tag{10a}$$

defines, for given $c$ and $c'$, the linear operator $\mathscr{T}_{c',c}$.

To obtain the transition probability or scattering cross section, we require the matric elements of $\mathscr{T}_{c',c}$ only near the energy shell, defined by $E_b = E_a$. In this case we may replace the $\chi_b^{c'}$ by the channel states $\chi_{c';b}$ and obtain

$$\begin{aligned}(\chi_{c';b}, \mathscr{T}_{c',c}\chi_{c;a}) &\equiv \langle c'; b|\, \mathscr{T}\, |c; a\rangle \\ &= \left(\chi_{c';b}, \left(V_c + V_{c'} \frac{1}{E_a + i\eta - H} V_c\right)\chi_{c;a}\right).\end{aligned} \tag{11a}$$

In addition to Eqs. (9*a*) and (10*a*), we also have

$$\psi^{-}_{c;a} = \chi_{c;a} + \sum_b \frac{\chi_b^{c'}}{E_a - i\eta - E_b} (\chi_b^{c'}, \mathscr{T}^{(-)}_{c',c}\chi_{c;a}), \tag{9b}$$

where

$$\mathscr{T}^{(-)}_{c',c} = V_c + V_{c'} \frac{1}{E_a - i\eta - H} V_c. \tag{10b}$$

For the channel states $\chi_{c';b}$ we have

$$\begin{aligned}(\chi_{c';b}, \mathscr{T}^{(-)}_{c',c}\chi_{c;a}) &\equiv \langle c'; b|\, \mathscr{T}^{(-)}\, |c; a\rangle \\ &= \left(\chi_{c';b}, \left(V_c + V_{c'} \frac{1}{E_a - i\eta - H} V_c\right)\chi_{c;a}\right).\end{aligned} \tag{11b}$$

On the energy shell defined by $E_b = E_a$, we may evidently write Eqs. (11*a*) and (11*b*) in the form

$$\begin{aligned}\langle c'; b|\, \mathscr{T}\, |c; a\rangle &= (\psi^{-}_{c';b}, V_c\chi_{c;a}), \\ \langle c'; b|\, \mathscr{T}^{(-)}\, |c; a\rangle &= (\psi^{+}_{c';b}, V_c\chi_{c;a}).\end{aligned} \tag{12}$$

Again, because[6]

$$\begin{aligned}(\chi_{c';b}, V_c\chi_{c;a}) &= (\chi_{c';b}, (H - E_a)\chi_{c;a}) \\ &= ((H - E_a)\chi_{c';b}, \chi_{c;a}) \\ &= (V_{c'}\, \chi_{c';b}, \chi_{c;a}) \\ &= (\chi_{c';b}, V_{c'}\chi_{c;a})\end{aligned}$$

[6] This relation is valid, since the $\chi$ describe free particles.

when $E_b = E_a$, we may also write Eqs. (11*a*) and (11*b*) as

$$\begin{aligned} \langle c'; b|\, \mathscr{T}\, |c; a\rangle &= (\chi_{c';b}, V_{c'}\psi^+_{c;a}), \\ \langle c'; b|\, \mathscr{T}^{(-)}\, |c; a\rangle &= (\chi_{c';b}, V_{c'}\psi^-_{c;a}). \end{aligned} \tag{13}$$

The remaining discussion of Chapter 3 may be transcribed to our present notation by merely adding channel labels for the initial and final states. If the eigenvalues of the momentum $\mathscr{P}$ for $\chi_{c;a}$ and $\chi_{c';b}$ are $\mathbf{P}_a$ and $\mathbf{P}_b$ respectively, we may rewrite the matrix elements of the scattering matrix with a momentum-conserving delta-function, Eq. (3.83), as

$$\langle c'; b|\, \mathscr{T}\, |c; a\rangle = \delta(\mathbf{P}_b - \mathbf{P}_a) T_{(c';b),(c;a)}. \tag{14}$$

Equation (3.94) for the *S*-matrix elements is now

$$\langle c'; b|\, S\, |c; a\rangle = \delta_{(c';b),(c;a)} - 2\pi i \delta(E_b - E_a)\langle c'; b|\, \mathscr{T}\, |c; a\rangle. \tag{15}$$

Similarly, the transition probability for the transition $(c; a)$ to $(c'; b)$ is given by Eq. (3.99) as

$$P_{(c';b),(c;a)} = \sum_b |(\chi_{c';b}, SX_{c;a})|^2, \tag{16}$$

where the sum runs over a restricted range of states $b$. The scattering cross section (3.142) may be written with explicit channel labels as

$$\Delta\sigma(c', c) = \sum_{s_1, \ldots, s_\nu} \int d^3k_1 \cdots d^3k_\nu \delta(E_b - E_a)\delta(\mathbf{P}_b - \mathbf{P}_a) \times \frac{(2\pi)^4\, |T_{(c';b),(c;a)}|^2}{[(\mathbf{v}_1 - \mathbf{v}_2)^2 - (1/c^2)(\mathbf{v}_1 \times \mathbf{v}_2)^2]^{1/2}}. \tag{17}$$

It should again be emphasized that we have not generalized the content of the description given in Chapter 3. We have merely added channel labels to equations for which such labels were implicit. These labels are convenient when we wish to indicate an enumeration of several channels. When we wish to consider only one incident and one final channel, we shall continue to use the notation of Chapter 3. When no confusion will result, we shall often simply drop the channel labels.

### Considerations of Orthogonality and Completeness for Channel Wave Functions

We frequently wish to find the transition probability, not into a single channel, but into several channels. This is obtained by summing Eq. (16) over the required channels.

$$P_{(c;a)} = \sum_{c'} P_{(c';b),(c;a)}. \tag{18}$$

In particular, if we sum over all $C_h$ final channels, we obtain the *total* transition probability arising from the initial state $\chi_{c;a}$. In a similar manner we may sum the cross section over channels to obtain

$$\Delta\sigma(c) = \sum_{c'} \Delta\sigma(c', c). \tag{19a}$$

The total cross section, into all channels, is

$$\sigma(c) = \sum_{c'=1}^{C_h} \int d\sigma(c', c). \tag{19b}$$

Equations (18) and (19), although correct, require justification. The reason for this, as mentioned before, is that the wave functions $\chi_{c;a}$ are not necessarily orthogonal for different channels. That is,

$$(\chi_{c';b}, \chi_{c;a}) \neq 0, \tag{20}$$

in general, when $c' \neq c$. This will ordinarily be the case when the Hamiltonians $K_c$ and $K_{c'}$ are different, since then the $\chi_{c;a}$ belong to different sets of complete orthonormal state vectors for different $c$.[7]

To restate this problem, let us introduce the projection operators

$$\Lambda_{c'} = \sum_{b} \chi_{c';b}\chi_{c';b}^{\dagger} \qquad (c' = 1, 2, \ldots, C_h). \tag{21}$$

Here the sum runs over the restricted set of states which are to be observed in an experiment. We may then rewrite Eq. (16) as

$$P_{(c';b),(c;a)} = (\Psi(t), \Lambda_{c'}\Psi(t)), \tag{22a}$$

where $\Psi$ is that state function resulting from the initial state $\phi_{c;a}(t_0)$ and $t \gg |t_c|$, the collision time [Eq. (3.32)]. Then Eq. (18) is

$$P_{(c;a)} = \sum_{c'} (\Psi(t), \Lambda_{c'}, \Psi(t)), \qquad t \gg |t_c| \tag{22b}$$

Because $\Lambda_{c'}\Lambda_{c''} \neq 0$ in general when $c' \neq c''$, it would appear that the sum in Eq. (22*b*) is redundant; that is, it would seem that some final states are being counted more than once. In Appendix C it is shown that this is indeed true, but that

$$\lim_{t\to\infty} (\Psi(t), \Lambda_{c'}\Lambda_{c''}\Psi(t)) = 0 \tag{23a}$$

[7] When the only particles in the $\chi_{c;a}$ are so-called elementary particles, the $\chi_{c;a}$ are all orthogonal. The vector space associated with the eigenvector of each particle represents a subspace (of the space for the entire system) which is orthogonal to the subspace associated with each other particle.

when $c' \neq c''$. Equation (23$a$) implies that the projection operators $\Lambda_{c'}$ and $\Lambda_{c''}$ are orthogonal when operating on the final, asymptotic wave function $\Psi(t)$, as $t \to \infty$. Because of Eq. (23$a$), Eqs. (18), (19), and (23) are valid.

The result expressed by Eq. (23$a$) is of great importance in scattering theory, since it states that we can treat the wave functions for different channels as effectively orthogonal when discussing the final states into which a scattering can go. The physical basis for the validity of Eq. (23$a$) can be easily seen: It was demonstrated in Section 3.4 that each of the $\nu_{c'}$ particles in channel $c'$ emerges from the reaction in a wave packet state, each packet moving with a characteristic velocity $\mathbf{v}_\alpha$ ($\alpha = 1, 2, \ldots, \nu_{c'}$). The nonorthogonality of $\Lambda_{c'}$ and $\Lambda_{c''}$ depends on the extent of the overlap of particles in channel $c''$ with those in channel $c'$. If particle $\beta$ in channel $c''$ has a "size" characterized by $s_\beta$, the fractional overlap with particle $\alpha$ in channel $c'$ is of order

$$\left(\frac{s_\beta}{v_\alpha t}\right)^3.$$

Since $s_\beta \approx 10^{-13}$ cm for nuclear particles ($\approx 10^{-8}$ cm for atoms) and $v_\alpha t \approx 10^2$ cm under most experimental conditions, this overlap is, indeed, negligibly small.

The $S$-matrix and the scattering matrix $\mathscr{T}$ are in general linear operators only in a restricted sense.[8,9] To illustrate this, let us consider an initial scattering wave packet $X_{\bar{a}} = \sum_a A_a \chi_{c;a}$ and the equation

$$SX_{\bar{a}} = \sum_a A_a \left( \sum_{c',b} \chi_{c';b} \langle c'; b|\, S\, |c; a\rangle \right), \tag{24$a$}$$

which would characterize $S$ as a linear operator. Since the $\chi_{c';b}$ are not necessarily orthogonal for different $c'$, it is clear that this equation can be valid only for a restricted class of state vectors $X_{\bar{a}}$. It is also clear from Eq. (23$a$) and Appendix C that this is the class of asymptotically nonoverlapping wave packets. To see this, let us assign to $A_a$ a phase $e^{-iE_a\lambda}$, where $\lambda$ is real. Then Eq. (23$a$) may be restated as

$$\lim_{\lambda \to \pm\infty} (\Lambda_{c'} SX_{\bar{a}}, \Lambda_{c''} SX_{\bar{a}}) = 0. \tag{23$b$}$$

In other words, if we properly choose the phase of the wave packet amplitude, the vectors associated with different channels in Eq. (24$a$) lie in different subspaces of the space of the entire system, and for such

[8] This restriction does not apply to the expressions (10), of course, as long as we hold $c$ and $c'$ fixed.

[9] This point, with a somewhat different emphasis, has been made by Ekstein [3].

packets Eq. (24*a*) is a correct operator equation. Of course, $S$ is a linear operator only in this subspace of nonoverlapping wave packet states. With this understanding and for very narrow wave packets (in momentum space) we may ordinarily just understand that the $A_a$ is implied in Eq. (24*a*) and simply substitute

$$S\chi_{c;a} \leftrightarrow \sum_{c',b} \chi_{c';b}\langle c'; b|\, S\, |c; a\rangle. \tag{24b}$$

These remarks evidently apply also to the submatrix of $\mathscr{T}$ that is defined only on the energy shell.

## 4.2 WAVE FUNCTION SYMMETRIZATION FOR SYSTEMS OF IDENTICAL PARTICLES

The effects of symmetrization for systems containing identical particles were ignored in Chapter 3. We now show how to modify the formulation of the scattering problem to include this possibility. Let us first review, however, the physical restrictions imposed on systems containing several dynamically equivalent particles.

Two particles are said to be *dynamically equivalent* if their physical properties are indistinguishable. In this connection we do not distinguish between two particles which happen only to be in different quantum mechanical states.[10] Thus, two electrons are "dynamically equivalent," even when in different spin states—as are also two $U^{238}$ atoms, even when one may be in an excited state.

### The Permutation Operators

In quantum as well as classical mechanics interchange of two equivalent particles in a physical system leaves its dynamical properties unchanged. For example, let a system consist of $N$ equivalent, or *identical*, particles, each particle being specified by a complete set of variables

$$\xi_i \; (i = 1, 2, \ldots, N).$$

Let the Hamiltonian for this system be $(\xi'_1 \cdots \xi'_N\, |H|\, \xi_1 \cdots \xi_N)$. Then, clearly,

$$(\xi'_1 \cdots \xi'_i \cdots \xi'_j \cdots \xi'_N\, |H|\, \xi_1 \cdots \xi_i \cdots \xi_j \cdots \xi_N)$$
$$= (\xi'_1 \cdots \xi'_j \cdots \xi'_i \cdots \xi'_N\, |H|\, \xi_1 \cdots \xi_j \cdots \xi_i \cdots \xi_N). \tag{25a}$$

[10] In Section 4.1 we noted an ambiguity in the definition of the term "same kind of particle." When the ambiguity is genuine, the definition can be made arbitrarily. In most applications a natural definition is evident. The question arises in a more fundamental form in determining commutation rules for different kinds of fields.

That is, replacing particle $i$ by particle $j$ (and vice versa) in the system does not change its physical properties because of the assumed equivalence of particles $i$ and $j$.

Equation (25$a$) can be expressed more concisely by introducing a permutation operator $Q_{ij}$ which interchanges the variables $\xi_i$ and $\xi_j$.[11] Then Eq. (25$a$) is equivalent to

$$H = Q_{ij} H Q_{ij}^{-1}, \tag{25b}$$

where $Q_{ij}^{-1}$ is the inverse of $Q_{ij}$. Evidently,

$$Q_{ij}^{-1} = Q_{ij} \tag{26a}$$

since two successive interchanges of $\xi_i$ and $\xi_j$ return the variables to their natural order—or are equivalent to the *identity* permutation.

$$Q_{ij}^{\,2} = I. \tag{26b}$$

We may also write Eq. (25$b$) as

$$[Q_{ij}, H] = 0, \tag{27a}$$

so $Q_{ij}$ is a constant of the motion for the system in question.

More generally, there are $N!$ different permutations of the $N$ variable sets $(\xi_1 \cdots \xi_N)$. Let $Q$ be any one of these permutations that replaces $\xi_1$ by $\xi_{Q1}$, $\xi_2$ by $\xi_{Q2}, \ldots, \xi_N$ by $\xi_{QN}$. Then it is evident that[12]

$$[Q, H] = 0, \tag{27b}$$

so $Q$ is a constant of motion for the system.

The result of letting $Q_{ij}$ operate on a wave function $\chi(\xi_1, \xi_2, \ldots, \xi_N)$ is

$$Q_{ij}\chi(\xi_1 \cdots \xi_i \cdots \xi_j \cdots \xi_N) = \chi(\xi_1 \cdots \xi_j \cdots \xi_i \cdots \xi_N). \tag{28a}$$

For the general permutation $Q$, we obtain

$$Q\chi(\xi_1, \xi_2, \ldots, \xi_N) = \chi(\xi_{Q1}, \xi_{Q2}, \ldots, \xi_{QN}). \tag{28b}$$

The permutations $Q$ are linear operators. Let $\chi$ and $\phi$ be two wave functions each depending on the $(\xi_1 \cdots \xi_N)$ and let $a$ and $b$ be two constants. Then

$$Q(a\chi + b\psi) = aQ\chi + bQ\psi.$$

Since the $N!$ permutations $Q$ do not all commute among themselves (except when $N = 2$), they cannot all be diagonalized simultaneously.

[11] We shall refer to $Q_{ij}$ as an "interchange operator" to distinguish it from the general permutations.

[12] Since $Q$ can be expressed as a sequence of interchanges, Eq. (27$b$) follows from Eq. (27$a$).

Thus the eigenfunctions $\psi_\lambda(\xi_1 \cdots \xi_N)$ of $H$ are not in general eigenfunctions of the $Q$'s.

The $N!$ permutations $Q$ of the variable sets $(\xi_1 \cdots \xi_N)$ constitute a group of $N!$ elements, called the symmetry group.[13] To see this, we note: (1) The set of $Q$'s contains the *identity* permutation, which sends $\xi_1 \to \xi_1$, $\xi_2 \to \xi_2, \ldots, \xi_N \to \xi_N$; (2) to each permutation $Q$, which sends $\xi_1 \to \xi_{Q1}, \ldots, \xi_N \to \xi_{QN}$, there exists a unique inverse $Q^{-1}$ which sends $\xi_{Q1} \to \xi_1, \ldots, \xi_{QN} \to \xi_N$; (3) finally, if $P$ and $Q$ are any two specific permutations of the $(\xi_1 \cdots \xi_N)$, then $QP = R$ is another (unique) permutation of the $(\xi_1 \cdots \xi_N)$.[14]

Consider the equation $QP = R$, where $Q$, $P$, and $R$ are permutations. If we hold $P$ fixed and let $Q$ range over the $N!$ elements of the group, $R$ also ranges over the entire group, each element being obtained once and only once. To see this, let $Q_1$ and $Q_2$ be two permutations and consider

$$Q_1P = R_1,$$

$$Q_2P = R_2.$$

Suppose that $R_1 = R_2$ for some $Q_1$ and $Q_2$, but $Q_1 \neq Q_2$. Then $Q_1P = Q_2P$. Operation from the right by $P^{-1}$ would give $Q_1 = Q_2$, which contradicts our hypothesis. Thus $R_1 \neq R_2$, unless $Q_1 = Q_2$. Therefore, as $Q$ runs over the $N!$ group elements, $R\,(= QP)$ runs over $N!$ different group elements and thus spans the group. Similarly we may show that $R' = PQ$ spans the group as $Q$ varies over the $N!$ group elements.

The norm of a wave function $\chi(\xi_1 \cdots \xi_N)$ is unchanged by the permutation $Q$. More generally, if $\phi$ and $\chi$ are two such wave functions for which $(\phi, \chi)$ is defined,

$$(\phi, \chi) = (Q\phi, Q\chi). \tag{29}$$

This may be seen by observing that Eq. (29) is equivalent to[15]

$$\int d\xi_1 \cdots d\xi_N \phi^*(\xi_1 \cdots \xi_N)\chi(\xi_1 \cdots \xi_N)$$
$$= \int d\xi_{Q1} \cdots d\xi_{QN}\phi^*(\xi_{Q1} \cdots {}_{QN})\chi(\xi_{Q1} \cdots \xi_{QN}),$$

where the permutation amounts only to a relabeling of the integration variables.

[13] See, for example, Ref. [4].

[14] That is, the result of applying two permutations in sequence is equivalent to a single permutation. This is called the *group property* of the permutations.

[15] When some of the variables in a variable set $\xi_1$ are discrete, we interpret this as a Stieltjes integral.

The permutations $Q$ represent *unitary* transformations on the

$$\chi(\xi_1 \cdots \xi_N)$$

[Eq. (1.22)]. This follows from Eq. (29). Consider

$$(Q\phi, \chi) = (Q^{-1}Q\phi, Q^{-1}\chi) = (\phi, Q^{-1}\chi), \tag{30}$$

which shows that $Q^\dagger = Q^{-1}$, or that $Q$ is unitary.

Consider now the eigenfunctions $\psi_\lambda(\xi_1 \cdots \xi_N)$ of the Hamiltonian $H$ for a system of $N$ equivalent particles. Then

$$H\psi_\lambda = E_\lambda \psi_\lambda. \tag{31}$$

We may write

$$Q\psi_\lambda = \sum_{\lambda'} (\lambda' |Q| \lambda)\psi_{\lambda'}, \tag{32}$$

where $(\lambda' |Q| \lambda)$ is a matrix element of $Q$ in the $\lambda$-representation. Since $[Q, H] = 0$, it follows from a discussion in Section 1.1 that for all states $\psi_\lambda$ and $\psi_{\lambda'}$, such that $(\lambda' |Q| \lambda) \neq 0$, $E_{\lambda'} = E_\lambda$.

A set of matrices such as $(\lambda' |Q| \lambda)$ is said to provide a *matrix representation* of the group,[16] in that *matrix multiplication* describes the operation of group multiplication. That is, corresponding to $QP = R$, we obtain

$$\sum_{\lambda''} (\lambda' |Q| \lambda'')(\lambda'' |P| \lambda) = (\lambda' |R| \lambda), \tag{33}$$

and so on. Equation (33) follows directly from Eq. (32). Consider first

$$P\psi_\lambda = \sum_{\lambda''} (\lambda'' |P| \lambda)\psi_{\lambda''}.$$

Operation by $Q$ on this equation gives

$$QP\psi_\lambda = R\psi_\lambda = \sum_{\lambda', \lambda''} (\lambda'' |P| \lambda)[(\lambda' |Q| \lambda'')\psi_{\lambda'}],$$

from which Eq. (33) follows. A representation for which $(\lambda' |Q| \lambda)$ is diagonal, for all $Q$, is called a one-dimensional representation.

Our applications will require only the simplest properties of the permutation group. Indeed, according to our present understanding of the particles that occur in nature, all wave functions describing physical systems of identical particles belong to one of two one-dimensional representations of the permutation group.[17] Thus for physical systems of $N$ identical particles the matrices $(\lambda' |Q| \lambda)$ are diagonal; that is,

$$(\lambda' |Q| \lambda) = Q_\lambda \delta_{\lambda'\lambda}. \tag{34}$$

[16] See, for example, Ref. [4].

[17] There are only two one-dimensional representations of the permutation group. See Ref. [4, Chapter 13].

The $Q_\lambda$ may be easily found. Any permutation $Q$ can be expressed (ordinarily in a variety of ways) as a product of simple interchanges of two elements at a time, such as $Q_{ij}$. For a given $Q$, the number of such interchanges is always either *even* or *odd*.[18] Thus each $Q$ can be labeled as *even* or *odd*, depending on whether the number of interchanges in $Q$ is *even* or *odd*.

The simplest representation is obtained by choosing

$$Q_\lambda = 1 \quad \text{for} \quad \text{all } Q. \tag{35a}$$

Wave functions belonging to this representation satisfy

$$\begin{aligned} Q\psi_\lambda(\xi_1 \cdots \xi_N) &= \psi_\lambda(\xi_{Q1} \cdots \xi_{QN}) \\ &= \psi_\lambda(\xi_1 \cdots \xi_N). \end{aligned} \tag{35b}$$

In particular, for a simple interchange $Q_{ij}$, Eq. (35*b*) gives

$$\psi_\lambda(\xi_1 \cdots \xi_i \cdots \xi_j \cdots \xi_N) = \psi_\lambda(\xi_1 \cdots \xi_j \cdots \xi_i \cdots \xi_N). \tag{35c}$$

Wave functions satisfying Eqs. (35) are said to be symmetric, or to satisfy *Bose-Einstein statistics*.

The second representation is obtained by setting

$$Q_\lambda = \epsilon_Q, \tag{36a}$$

where

$$\begin{aligned} \epsilon_Q &= +1 \quad \text{for} \quad \text{an even permutation,} \\ \epsilon_Q &= -1 \quad \text{for} \quad \text{an odd permutation.} \end{aligned} \tag{36b}$$

Thus if $\psi_\lambda$ belongs to this representation,

$$Q\psi_\lambda(\xi_1 \cdots \xi_N) = \psi_\lambda(\xi_{Q1} \cdots \xi_{QN}) = \epsilon_Q \psi_\lambda(\xi_1 \cdots \xi_N). \tag{36c}$$

In particular, for the interchange $Q_{ij}$ Eq. (36*c*) gives

$$\psi_\lambda(\xi_1 \cdots \xi_i \cdots \xi_j \cdots \xi_N) = -\psi_\lambda(\xi_1 \cdots \xi_j \cdots \xi_i \cdots \xi_N). \tag{36d}$$

Such wave functions are called *antisymmetric* and are said to satisfy *Fermi-Dirac statistics*.

It is easily verified that these two representations satisfy the condition (33). This is trivial for the symmetric wave functions (35*b*). It is also evident for the antisymmetric wave functions (36*c*), since $R$ can be expressed as a product of all the interchanges from which $P$ and $Q$ are formed.

We have stated that consistent with our present knowledge all wave functions for systems of identical particles belong to either the symmetric

[18] See, for example, Ref. [4, Chapter 13].

or the antisymmetric representation of the permutation group. Furthermore, systems of identical particles of integral spin are thought to be described by symmetric wave functions (or to obey Bose-Einstein statistics). System of identical particles of half-integral spin are thought to be described by antisymmetric wave functions (or to obey Fermi-Dirac statistics). This is known directly from a variety of experimental phenomena for photons, electrons, protons, and neutrons. For other particles, there are strong theoretical arguments for thinking this to be a general principle.[19]

Our discussion is readily extended to systems containing several different species of particles. Let there be $L$ such different species and let there be $N_i$ particles of species $i$ ($i = 1, 2, \ldots, L$). Permutations that interchange particles of different kinds are not of interest since they do not commute with the Hamiltonian $H$. The $N_i!$ permutations $Q^{(i)}$ on species $i$ do, however, commute with $H$. The wave functions $\psi_\lambda$, representing physical states of the system, belong to one of the representations of the $Q^{(i)}$, for each $i$, as described. If species $i_1$ obey Bose-Einstein statistics,

$$Q^{(i_1)}\psi_\lambda = \psi_\lambda. \tag{37a}$$

If species $i_2$ obeys Fermi-Dirac statistics,

$$Q^{(i_2)}\psi_\lambda = \epsilon_Q \psi_\lambda. \tag{37b}$$

In this discussion we have not found it necessary to distinguish between the so-called "elementary" and "composite" particles. It has therefore not been necessary to define the term "elementary particle," although we have occasionally used it rather loosely. From a formal point of view it is not clear that the term has a meaning other than as given by somewhat arbitrary definitions. We shall adopt a practical definition, saying that a particle is "elementary" if in a particular application we can ignore its possible composite structure as a bound system of other particles.[20] (Thus in most applications involving electronic excitation of atoms we may consider the atomic nuclei to be "elementary.") From a more formal point of view, there is assigned to each elementary particle a vector space which is spanned by the set of state vectors of that particle. This space is a subspace of that for the physical system being considered and is orthogonal to the subspace assigned to each other elementary particle.

According to the terminology we have adopted, we must properly symmetrize a wave function with respect to permutations of identical,

[19] A brief review of these arguments is given in Note [5].

[20] In this connection we recall the definition of particle adopted in Section 3.1.

elementary particles. When composite particles are contained in the system, symmetrization must be effected with respect to the elementary particles from which the composite particles are constructed. For example, in studying the properties of liquid $He^4$ at low temperatures, we often treat the $He^4$ atoms as "elementary" and accordingly symmetrize with respect to the permutation of atoms. On the other hand, for the study of the nuclear scattering of energetic $He^4$ nuclei on $He^4$ nuclei, we must in general symmetrize the wave function with respect to permutations of the constituent neutrons and protons. In Section 4.4 the quantitative criteria will be obtained for determining in which manner symmetrization must be effected.[21]

**Construction of Symmetrized Wave Functions**

It is convenient to have a prescription for constructing symmetrized wave functions from unsymmetrized ones. This is most easily accomplished with the use of projection operators, which project a wave function in the full function space of the variables $(\xi_1 \cdots \xi_N)$ onto the suitably symmetrized subspace. The appropriate operators for a system of $N$ identical particles are

$$\mathscr{S}_A = \frac{1}{N!}\sum_Q \epsilon_Q Q \qquad \text{(antisymmetric, or Fermi-Dirac)}, \qquad (38a)$$

$$\mathscr{S}_S = \frac{1}{N!}\sum_Q Q \qquad \text{(symmetric, or Bose-Einstein)}. \qquad (38b)$$

Here the sum on $Q$ runs over the $N!$ permutations. The function $\epsilon_Q$ is $\pm 1$, depending on whether $Q$ is even or odd, as defined by Eq. (36*b*). It is sometimes convenient to introduce a single symbol to represent either $\mathscr{S}_A$ or $\mathscr{S}_S$. This is

$$\mathscr{S} = \frac{1}{N!}\sum_Q \delta_Q Q \qquad (39)$$

where

$$\delta_Q = 1 \qquad \text{(Bose-Einstein symmetrization)},$$

$$\delta_Q = \epsilon_Q \qquad \text{(Fermi-Dirac symmetrization)}. \qquad (40)$$

Let us first show that $\mathscr{S}$ (that is, $\mathscr{S}_A$ or $\mathscr{S}_S$) is a projection operator [compare Eq. (1.23)]. Letting $P$ and $Q$ be any two permutations on the $(\xi_1 \cdots \xi_N)$, we form

$$\mathscr{S}^2 = \left(\frac{1}{N!}\right)^2 \sum_P \sum_Q \delta_P \delta_Q PQ.$$

[21] This involves the magnitude of certain "exchange" contributions to the scattering matrix.

Now $PQ = R$ is a permutation, and $\delta_P\delta_Q = \delta_R$, since the number of interchanges in $R$ can be chosen as the sum of the respective numbers in $P$ and $Q$. For each $P$, there is a one-to-one correspondence between $Q$ and $R$, so $\Sigma_Q$ may be replaced by $\Sigma_R$ in the expression just given. Thus

$$\mathscr{S}^2 = \frac{1}{(N!)^2}\sum_P\left(\sum_R \delta_R R\right) = \left(\frac{1}{N!}\sum_P\right)\mathscr{S} = \mathscr{S}, \tag{41}$$

as $\Sigma_P = N!$.

The projection operators $\mathscr{S}_A$ and $\mathscr{S}_S$ are Hermitian. To show this, we consider two state vectors $\phi$ and $\chi$ and form the scalar product

$$(\mathscr{S}\phi, \chi) = \frac{1}{N!}\sum_Q \delta_Q(Q\phi, \chi) = \frac{1}{N!}\sum_Q \delta_Q(\phi, Q^{-1}\chi).$$

Since $Q$ and $Q^{-1}$ are simultaneously either even or odd, we have $\delta_Q = \delta_{Q^{-1}}$. Moreover, we may replace $\Sigma_Q$ by $\Sigma_{Q^{-1}}$, as $Q$ and $Q^{-1}$ are uniquely related. Thus the right-hand side of expression (41) is $(\phi, \mathscr{S}\chi)$, and we obtain

$$\mathscr{S} = \mathscr{S}^\dagger. \tag{42}$$

Finally, we must show that $\mathscr{S}_S$ and $\mathscr{S}_A$ are the required projection operators. Consider first

$$\chi_{\mathscr{S}_S} \equiv \mathscr{S}_S\chi(\xi_1 \cdots \xi_N).$$

Now if $P$ is a given permutation,

$$P\chi_{\mathscr{S}_S} = \frac{1}{N!}\sum_Q PQ\chi.$$

On the other hand, $PQ = R$ and $\Sigma_Q$ may be replaced by $\Sigma_R$, so

$$P\chi_{\mathscr{S}_S} = \frac{1}{N!}\sum_R R\chi = \mathscr{S}_S\chi = \chi_{\mathscr{S}_S}. \tag{43}$$

This satisfies the condition (35*b*), as required. Next consider

$$\chi_{\mathscr{S}_A} \equiv \mathscr{S}_A\chi(\xi_1 \cdots \xi_N).$$

Again we operate with $P$ on both sides of this expression to obtain

$$P\chi_{\mathscr{S}_A} = \frac{1}{N!}\sum_Q \epsilon_Q PQ\chi.$$

On setting $PQ = R$, we have $\epsilon_P\epsilon_Q = \epsilon_R$, as was shown in the demonstration of Eq. (41). Since $\epsilon_P^2 = 1$, we may set $\epsilon_Q = \epsilon_P\epsilon_R$ and then obtain

$$P\chi_{\mathscr{S}_A} = \frac{1}{N!}\sum_R \epsilon_P\epsilon_R R\chi = \epsilon_P\mathscr{S}_A\chi = \epsilon_P\chi_{\mathscr{S}_A}. \tag{44}$$

This is the required condition (36*c*).

It follows from Eq. (27*b*) that the projection operator $\mathscr{S}$ commutes with the Hamiltonian $H$ for a system of $N$ identical particles.

$$[\mathscr{S}, H] = 0. \tag{45}$$

Let us illustrate the use of the projection operator $\mathscr{S}$ with a simple example. Consider a wave function $\psi_S$ constructed from a product of $N$ single-particle wave functions $\chi_{S1}(\xi), \ldots, \chi_{SN}(\xi_N)$.

$$\psi_S = \chi_{S1}(\xi_1) \cdots \chi_{SN}(\xi_N). \tag{46}$$

For the moment it will be convenient to suppose that the $\chi_{Si}$ represent a discrete set of orthonormal functions. According to Eqs. (43) and (44), a correctly symmetrized wave function is

$$\psi_{\mathscr{S}(S)} = C\mathscr{S}\psi_S = \frac{C}{N!}\sum_Q \delta_Q \chi_{S1}(\xi_{Q1}) \cdots \chi_{SN}(\xi_{QN}), \tag{47a}$$

in the notation of Eq. (28*b*). Here $C$ is a normalization constant determined from the condition that

$$(\psi_{\mathscr{S}(S)}, \psi_{\mathscr{S}(S)}) = 1. \tag{48}$$

The state (46) is specified by the statement that particle 1 is in state $S_1, \ldots,$ and particle $N$ is in state $S_N$. Because the particles are permuted among the states in Eq. (47), the state described by $\psi_{\mathscr{S}(S)}$ is specified by the statement that there are $n_1$ particles in $S_1$, $n_2$ particles in $S_2$, and so on. It is not meaningful to assign particular particles to given states in $\psi_{\mathscr{S}(S)}$. Alternatively, we may say that the state $S$ of Eq. (46) is $S = (S_1, S_2, \ldots, S_N)$, where the order of the $S_1, S_2, \ldots$ is meaningful. For the state $\mathscr{S}(S) = (S_1, S_2, \ldots, S_N)$, the order in which the $S_1, S_2, \ldots$ are written is irrelevant.

For Fermi-Dirac statistics, for which $\delta_Q = \epsilon_Q$, the wave function $\psi_{\mathscr{S}_A(S)}$ is equivalent to the determinant

$$\psi_{\mathscr{S}_A(S)} = \frac{C}{N!}\begin{vmatrix} \chi_{S1}(\xi_1)\chi_{S2}(\xi_1) \cdots \chi_{SN}(\xi_1) \\ \chi_{S1}(\xi_2)\chi_{S2}(\xi_2) \cdots \chi_{SN}(\xi_2) \\ \chi_{S1}(\xi_N)\chi_{S2}(\xi_N) \cdots \chi_{SN}(\xi_N) \end{vmatrix}. \tag{49}$$

We see that the states $(S_1, S_2, \ldots, S_Q)$ must *all* be different, or $\psi_{\mathscr{S}_A(S)} = 0$. Thus for Fermi-Dirac statistics no two particles "may be in the same state."

To evaluate the constant $C$, we form the scalar product (48) and obtain

$$(\psi_{\mathscr{S}(S)}, \psi_{\mathscr{S}(S)}) = C^2(\mathscr{S}\psi_S, \mathscr{S}\psi_S) = C^2(\psi_S, \mathscr{S}{\psi_S}^2) = C^2(\psi_S, \mathscr{S}\psi_S)$$

$$= \frac{C^2}{N!}\sum_Q \delta_Q(\chi_{S1}(\xi_1) \cdots \chi_{SN}(\xi_N), \chi_{S1}(\xi_{Q1}) \cdots \chi_{SN}(\xi_{QN})). \tag{50}$$

For Fermi-Dirac statistics the $S_1, \ldots, S_N$ are all different, so the scalar products in Eq. (50) vanish for all but the identity permutation $Q = 1$. Thus

$$(\psi_{\mathscr{S}_A(S)}, \psi_{\mathscr{S}_A(S)}) = \frac{C_A^{\,2}}{N!} = 1,$$

or

$$C_A = \sqrt{N!}.$$

Therefore

$$\psi_{\mathscr{S}_A(S)} = \frac{1}{\sqrt{N!}} \sum_Q \epsilon_Q \chi_{S1}(\xi_{Q1}) \cdots \chi_{SN}(\xi_{QN}). \tag{47b}$$

For Bose-Einstein statistics, let us suppose there to be $n_1$ particles in $S_1$, $n_2$ particles in $S_2, \ldots$. Then the permutation, $Q$ for which the terms in Eq. (50) are nonvanishing are those involving the $n_1!$ permutations of particles among the $\chi_{S1}$, the $n_2!$ permutations of particles among the $\chi_{S2}$, and so on. Thus

$$(\psi_{\mathscr{S}_S(S)}, \psi_{\mathscr{S}_S(S)}) = \frac{C_S^{\,2}}{N!} n_1!\, n_2! \cdots = 1,$$

so

$$C_S = \left(\frac{N!}{n_1!\, n_2! \cdots}\right)^{1/2}.$$

Therefore,

$$\psi_{\mathscr{S}_S(S)} = \frac{1}{\sqrt{N!\, n_1!\, n_2! \cdots}} \sum_Q \chi_{S1}(\xi_{Q1}) \cdots \chi_{SN}(\xi_{QN}). \tag{47c}$$

Since $1! = 1$, we need not distinguish between Eqs. (47*b*) and (47*c*) if we write

$$\psi_{\mathscr{S}(S)} = \frac{1}{\sqrt{N!\, n_1!\, n_2! \cdots}} \sum_Q \delta_Q \chi_{S1}(\xi_{Q1}) \cdots \chi_{SN}(\xi\, Q_N). \tag{47d}$$

As an illustration of this equation, we consider two identical particles in states $S_1$ and $S_2$ ($S_1 \neq S_2$). Then $\mathscr{S} = \frac{1}{2}(1 \pm Q_{12})$, and

$$\psi_{\mathscr{S}} = \frac{1}{\sqrt{2}} [\chi_{S1}(\xi_1)\chi_{S2}(\xi_2) \pm \chi_{S1}(\xi_2)\chi_{S2}(\xi_1)], \tag{51}$$

where the plus or minus sign is to be used for Bose-Einstein or Fermi-Dirac statistics, respectively.

Consider now a second state

$$\psi_{\mathscr{S}(S')} = C'\mathscr{S}\chi_{S'1}(\xi_1) \cdots \chi_{S'N}(\xi_N), \tag{52}$$

where $C'$ is the appropriate normalization constant for the state $\mathscr{S}(S')$. Then

$$(\psi_{\mathscr{S}(S')}, \psi_{\mathscr{S}(S)}) = \frac{C'C}{N!} \sum_{Q} \delta_Q(\chi_{S'1}(\xi_1) \cdots \chi_{S'N}(\xi_N), \chi_{S1}(\xi_{Q1}) \cdots \chi_{SN}(\xi_{QN}))$$
$$= \frac{C'C}{N!} \sum_{Q} \delta_Q \delta_{S'Q1,S1} \delta_{S'Q2,S2} \cdots \delta_{S'QN,SN}. \tag{53}$$

This scalar product vanishes unless the numbers of particles (that is, the "occupation numbers") $n_1, n_2, \ldots$ in states $S_1, S_2, \ldots$ are the same in the two states $S$ and $S'$. As described before, we may label the states $\mathscr{S}(S)$ and $\mathscr{S}(S')$ by the occupation number sets $(n_1, n_2, \ldots)$ and $(n'_1, n'_2, \ldots)$, writing

$$\psi_{\mathscr{S}(S)} = |\mathscr{S}(S)\rangle = |n_1, n_2, \ldots\rangle, \tag{54}$$
$$\psi_{\mathscr{S}(S')} = |\mathscr{S}(S')\rangle = |n'_1, n'_2, \ldots\rangle.$$

Equation (53) may then be rewritten as

$$\langle \mathscr{S}(S') \mid \mathscr{S}(S)\rangle = \delta_{n'1,n1}\delta_{n'2,n2} \cdots = I. \tag{55}$$

The sum over states $S$ may be effected by summing over the occupation numbers $n_1, n_2, \ldots$. Alternatively, we may assign an ordering to the single-particle states $S_i$ and write

$$\sum_{\mathscr{S}_A(S)} = \sum_{S_1} \sum_{S_2(>S_1)} \cdots \sum_{S_N(>S_{N-1})}$$
$$\sum_{\mathscr{S}_S(S)} = \sum_{S_1} \sum_{S_2(\geq S_1)} \cdots \sum_{S_N(\geq S_{N-1})} . \tag{56}$$

This is easily verified from Eq. (53) and the requirement that

$$\sum_{\mathscr{S}(S')} (\psi_{\mathscr{S}(S')}, \psi_{\mathscr{S}(S)}) \psi_{\mathscr{S}(S')} = \psi_{\mathscr{S}(S)}. \tag{57}$$

For any function $f(\xi_1 \cdots \xi_N)$ in the symmetrized subspace (that is, $\mathscr{S}f = f$), we verify that

$$If = \int d\xi'_1 \cdots d\xi'_N \left[\frac{1}{N!} \sum_{Q} \delta_Q \delta(\xi'_1 - \xi_{Q1}) \cdots \delta(\xi'_N - \xi_{QN})\right]$$
$$\times f(\xi'_1 \cdots \xi'_N) = f(\xi_1 \cdots \xi_N). \tag{58}$$

The symmetrization of wave functions containing several different kinds of identical particles was described earlier. For species $i$, we have the symmetrization operator [Eq. (39)]

$$\mathscr{S}^{(i)} = \frac{1}{N_i!} \sum_{Q^{(i)}} \delta_{Q^{(i)}} Q^{(i)}, \tag{59}$$

where the $Q^{(i)}$ were described in connection with Eqs. (37). The symmetrization operator for a system having $L$ species of identical particles is

$$\mathscr{S} = \prod_{i=1}^{L} \mathscr{S}^{(i)}, \tag{60a}$$

where the order of the factors is irrelevant. In greater detail, this is

$$\mathscr{S} = \frac{1}{N_1! \cdots N_L!} \sum_{Q^{(1)}} \cdots \sum_{Q^{(L)}} \delta_{Q^{(1)}} \cdots \delta_{Q^{(L)}} Q^{(1)} \cdots Q^{(L)}.$$

This can be conveniently abbreviated by defining a generalized permutation

$$Q \equiv Q^{(1)} \cdots Q^{(L)} \tag{61}$$

and a generalized $\delta_Q$ by

$$\delta_Q \equiv \delta_{Q^{(1)}} \cdots \delta_{Q^{(L)}}.$$

In terms of these, Eq. (60*a*) may be written as

$$\mathscr{S} = \frac{1}{\prod_{i=1}^{L} (N_i!)} \sum_Q \delta_Q Q. \tag{60b}$$

The sum here runs over all $N_1!\, N_2! \cdots N_L!$ permutations of the product $Q = Q^{(1)} \cdots Q^{(L)}$.

The symmetrization operator of Eq. (60) is linear, since the $Q$'s have been seen to be linear operators. It may be applied directly to state functions involving a sum over a number of channel states involving different composite particles but a fixed number of elementary particles. For wave functions involving a sum of terms which describe states with *different* numbers of elementary particles, an extension of Eq. (60) is required, however.

To generalize $\mathscr{S}$ appropriately, let us recall that each elementary particle is described by state vectors that lie in a subspace (of the full system space) which is orthogonal to the subspace of the state vectors of every other elementary particle. Then, for any given channel $c$ of a system, we may define a projection operator $\Lambda^{(\rho)}$ onto the subspace spanned by all the state vectors for the elementary particles in that channel. In particular, in the subspace defined by $\Lambda^{(\rho)}$, let us suppose there to be $N_i^{(\rho)}$ identical particles of species $i$ ($i = 1, 2, \ldots, L_\rho$). The symmetrization operator in this subspace will be written as $\mathscr{S}^{(\rho)}$, where $\mathscr{S}^{(\rho)}$ has the form of Eq. (60*b*). Now let a wave function $\Psi$ describe a given system, and define

$$\begin{aligned} \psi^{(\rho)} &\equiv \Lambda^{(\rho)} \Psi, \\ \Psi &= \sum_\rho C_\rho \psi^{(\rho)}, \end{aligned} \tag{62}$$

where the $C_\rho$ are constants. We may then define the symmetrization operator for the complete system as

$$\mathscr{S} = \sum_\rho \mathscr{S}^{(\rho)}\Lambda^{(\rho)}. \tag{63}$$

For the wave function (62) we obtain

$$\mathscr{S}\Psi = \sum_\rho C_\rho \mathscr{S}^{(\rho)}\psi^{(\rho)}. \tag{64}$$

**Remarks on the "Equivalence" of Particles**

As was mentioned previously, a question often arises as to which particles should be considered "equivalent." One of the best-known examples is that of protons and neutrons. It is customary to consider neutrons and protons as different *isotopic spin* states of a single kind of particle, a nucleon. A wave function containing both neutrons and protons is then *antisymmetrized in all nucleon coordinates.* As we shall observe in Chapter 6, there is good experimental evidence that this is required by the physical properties of the interactions of nucleons. For most problems, it makes no difference, however, whether we antisymmetrize a wave function with respect to neutron and proton coordinates.[22]

We shall illustrate this point by considering a simple example. Let a given physical system contain $N_+$ electrons with spin up and $N_-$ electrons with spin down. We must ask, "Is an electron with spin up distinguishable from an electron with spin down?" Consider the wave function

$$\psi = \psi(\xi_1 \cdots \xi_{N_+}; \xi_{N_++1} \cdots \xi_{N_++N_-}), \tag{65}$$

which is properly symmetrized in the coordinates of the electrons with spin up and in the coordinates of those with spin down, but not symmetrized with respect to exchange of spin up and spin down coordinates. That is,

$$Q_{ij}\psi = -\psi$$

for $i, j = 1, 2, \ldots, N_+$ or for $i, j = N_+ + 1, \ldots, N_+ + N_-$. Because spin up states are orthogonal to spin down states, we have

$$(\psi, Q_{ij}\psi) = 0 \tag{66}$$

[22] There is a rather subtle point involved here, depending on the observability of states with mixed charge. Wick, Wightman, and Wigner [6] have proposed a "superselection rule," according to which states of mixed charge are not measurable. If this is true, antisymmetrization with respect to neutron and proton coordinates is not necessary.

for $i = 1, 2, \ldots, N_+$, $j = N_+ + 1, \ldots, N_+ + N_-$. Now the fully antisymmetrized wave function, obtained by applying $\mathscr{S}_A$ to $\psi$, is

$$\psi\mathscr{S}_A = C\mathscr{S}_A\psi, \tag{67a}$$

where because of Eq. (66)

$$C = \sqrt{(N_+ + N_-)!/N_+!\,N_-!}. \tag{67b}$$

Let us suppose that $\psi$ represents an eigenstate of a Hamiltonian $H$. For this to be possible, $H$ must not contain any "spin-flip" interactions. The results of a physical observation on the system are given as the expectation value of some operator $\mathcal{O}$.

$$\bar{\mathcal{O}}\mathscr{S} = (\psi_{\mathscr{S}}, \mathcal{O}\psi_{\mathscr{S}}) = C^2(\psi, \mathcal{O}\mathscr{S}\psi). \tag{68a}$$

If $\mathcal{O}$ contains no spin-flip terms, *all* permutations $Q$ in $\mathscr{S}$ which interchange coordinates of spin up with coordinates of spin down electrons lead to a vanishing result in Eq. (68*a*) (because of the orthogonality of spin up with spin down states). Only the $(N_+!)$ $(N_-!)$ permutation among the spin up states and the spin down states leads to nonvanishing terms. Thus, an elementary evaluation of (68*a*) gives

$$\bar{\mathcal{O}}_{\mathscr{S}} = (\psi, \mathcal{O}\psi). \tag{68b}$$

In other words, the symmetrization (67) is quite irrelevant for our problem —or electrons with spin up are distinguishable from electrons with spin down. If we generalize the terminology spin up and spin down to any internal state of a particle, Eqs. (68) describe a quite general conclusion.

We now illustrate the symmetrization of neutron and proton states. Let $\chi_{S1,p}(\xi_1)$ represent a proton in state $S_1$ and let $\chi_{S2,n}(\xi_2)$ represent a neutron in state $S_2$. The states $n$ and $p$ may be considered as representing two orthogonal states of a nucleon, so

$$(\chi_{S1,p}(\xi_1), \chi_{S1,n}(\xi_1)) = 0. \tag{69}$$

Let $\tau_{+,1}$ be an operator which acts on the $\xi_1$ to change $\chi_{S1,n}(\xi_1)$ into $\chi_{S1,p}(\xi_1)$. Similarly, $\tau_{-,1}$ is an operator changing $\chi_{S1,p}(\xi_1)$ into $\chi_{S1,n}(\xi_1)$. Most precisely, we define $\tau_{+,1}$ and $\tau_{-,1}$ by

$$\begin{aligned} \tau_{+,1}\chi_{S1,n}(\xi_1) &= \chi_{S1,p}(\xi_1), & \tau_{+,1}\chi_{S1,p}(\xi_1) &= 0, \\ \tau_{-,1}\chi_{S1,n}(\xi_1) &= 0, & \tau_{-,1}\chi_{S1,p}(\xi_1) &= \chi_{S1,n}(\xi_1), \end{aligned} \tag{70}$$

for all $S_1$. Similar operators $\tau_{+,2}$ and $\tau_{-,2}$ operate on the wave function $\chi(\xi_2)$ of the second nucleon. Since $\tau_{\pm,1}$ and $\tau_{\pm,2}$ operate in orthogonal spaces, they commute.

Let two states, describing one neutron and one proton, be represented by

$$\begin{aligned}\psi &= \chi_{S1,p}(\xi_1)\chi_{S2,n}(\xi_2),\\ \psi' &= \chi_{S'1,p}(\xi_1)\chi_{S'2,n}(\xi_2).\end{aligned} \tag{71}$$

The interaction $V$ between the neutron and proton is assumed to have the form

$$V = v + v_x(\tau_{+,1}\tau_{-,2} + \tau_{-,1}\tau_{+,2}), \tag{72}$$

where $v$ and $v_x$ depend in general on the coordinates and spins of the two nucleons.

Because of the orthogonality of neutron and proton states, we find

$$(\psi', V\psi) = (S'_1p, S'_2n\,|v|\,S_1p, S_2n), \tag{73a}$$

where

$$(S'_1p, S'_2n\,|v|\,S_1p, S_2n) = (\chi_{S'1,p}(\xi_1)\chi_{S'2,n}(\xi_2), v\chi_{S1,p}(\xi_1)\chi_{S2,n}(\xi_2)) \tag{73b}$$

Symmetrized wave functions corresponding to Eqs. (71) are $\psi_{\mathscr{S}_A} = \sqrt{2}\mathscr{S}_A\psi$ and $\psi_{\mathscr{S}'_A} = \sqrt{2}\,\mathscr{S}_A\psi'$ [see Eq. (51)]. Now we readily obtain

$$(\psi_{\mathscr{S}'_A}, V\psi_{\mathscr{S}_A}) = (S'_1p, S'_2n\,|v|\,S_1p, S_2n) - (S'_1p, S'_2n\,|v_x|\,S_2p, S_1n), \tag{74}$$

where the matrix element of $v_x$ is defined as in Eq. (73$b$), but with $v$ replaced by $v_x$.

When $v_x = 0$, the expressions (73$a$) and (74) are the same. This represents a special case of the more general conclusion (68$b$). When $v_x \neq 0$, the "flipping" of isotopic spin caused by $\tau_+$ and $\tau_-$ makes Eq. (68$b$) inapplicable. Experimental study of nucleon interactions has shown that "exchange forces" such as that of Eq. (72) do occur in nature. To obtain the result (74) without using antisymmetrized wave functions, we must not use the isotopic spin formalism but must rewrite the potential (72) in terms of space and spin exchange operators.[23]

### 4.3 SYMMETRIZATION OF THE WAVE PACKET STATES FOR A SCATTERING INTERACTION

We now return to a consideration of scattering reactions and to an extension of the discussion of Chapter 3 to include systems containing identical particles. In this section we shall describe the symmetrization of the initial and final asymptotic wave packet states and evaluate the current and particle densities in these states.

#### Time-Dependent Wave Functions

Ignoring, for the moment, wave function symmetrization, we suppose some physical system to have been prepared at time $t = t_0$ in a state

[23] See, for example, Ref. [7, p. 135].

$\phi_a(t_0)$. As was the case in Chapter 3, each particle in this state is localized within a wave packet which does not overlap the wave packet of any other particle. Each composite particle contained in the state $\phi_a(t_0)$ is described by a wave function suitably symmetrized in the coordinates of its identical constituent elementary particles (if any). It would not otherwise be a physical particle. The state $\phi_a(t_0)$ is not, however, considered to be symmetrized with respect to exchange of identical particles in different wave packets. Therefore $\phi_a(t_0)$ is an initial state such as that of Eqs. (3.28).

The (unsymmetrized) time-dependent wave function (3.37) for this system is

$$\Psi(t) = e^{-iH(t-t_0)}\phi_a(t_0), \tag{75}$$

which satisfies the prescribed boundary condition at $t = t_0$,

$$\Psi(t_0) = \phi_a(t_0). \tag{76}$$

For times $t$ prior to the collision time $t_c$, we of course have

$$\Psi(t) = \phi_a(t).$$

To obtain the correct dynamical description of the system, we must project $\Psi(t)$ onto its appropriately symmetrized subspace. The symmetrized incident state, defined for all times $t$, is

$$\phi_{\mathscr{S}(a)}(t) \equiv C\mathscr{S}\phi_a(t), \tag{77}$$

where $\mathscr{S}$ is the symmetrization operator for the complete physical system, as given by Eq. (63). That is, in the notation of Eq. (63),

$$\mathscr{S}\phi_a(t) = \mathscr{S}^{(p)}\phi_a(t),$$

where $\mathscr{S}^{(p)}$ is the symmetrization operator for those particles in the incident channel. The normalization constant $C$ is determined from the condition that

$$(\phi_{\mathscr{S}(a)}(t), \phi_{\mathscr{S}(a)}(t)) = 1 \tag{78}$$

for times $t$ such that the incident wave packets do not overlap.

According to Eq. (3.33) we may write [where $\chi_a$ is given by Eq. (3.23)]

$$\phi_{\mathscr{S}(a)}(t) = \int d^3q_1\, d^3q_2 A(\mathbf{q}_1, \mathbf{q}_2)\, e^{-iE_a t}\chi_{\mathscr{S}(a)}, \tag{79a}$$

$$\chi_{\mathscr{S}(a)} = C\mathscr{S}\chi_a, \tag{79b}$$

and

$$X_{\mathscr{S}(a)} = C\mathscr{S}X_a = \int d^3q_1\, d^3q_2 A(\mathbf{q}_1, \mathbf{q}_2)\chi_{\mathscr{S}(a)}. \tag{79c}$$

More generally, for an initial state containing $\mu$ incident particles, we may symmetrize Eq. (3.26*b*) to obtain

$$\phi_{\mathscr{S}(a)}(t) = \int d^3q_1 \cdots d^3q_\mu A(\mathbf{q}_1 \cdots \mathbf{q}_\mu)\, e^{-iE_at}\chi_{\mathscr{S}(a)}, \tag{80}$$

where now $\chi_a$ is an eigenstate for the $\mu$ noninteracting particles and $\chi_{\mathscr{S}(a)}$ is again given by Eq. (79*b*).

Since the Hamiltonian $H$ commutes with the symmetrization operator $\mathscr{S}$, we may operate on Eq. (75) with $\mathscr{S}$ to obtain

$$\Psi_{\mathscr{S}}(t) \equiv C\mathscr{S}\Psi(t) = e^{-iH(t-t_0)}\phi_{\mathscr{S}(a)}(t_0), \tag{81}$$

for the correctly symmetrized state function $\Psi_{\mathscr{S}}(t)$. From Eq. (81) we have the important conclusion that it is sufficient to symmetrize only the initial state $\phi_a$. In other words, from a symmetrized initial state, the Schrödinger equation will generate only symmetrized wave functions.

### Symmetrization for a System of Elementary Particles

Let us first consider the case of a wave packet state for $N$ identical, elementary particles [see, for example, Eq. (3.7)]:

$$\phi(t) = e^{-iE_at}\prod_{\alpha=1}^{N} [(2\pi)^{-3/2}\, e^{i\mathbf{p}_\alpha \cdot \mathbf{r}_\alpha} g_{s\alpha} G_\alpha(\mathbf{r}_\alpha - \mathbf{v}_\alpha t)], \tag{82}$$

where $g_{s\alpha}$ is a wave function describing the spin, and so on, of particle $\alpha$. The wave packet amplitudes $G_a$ are normalized as usual to

$$(2\pi)^{-3}\int d^3\mathbf{r}\, |G_\alpha(\mathbf{r})|^2 = 1. \tag{83}$$

The symmetrized state $\phi_{\mathscr{S}}$ is, according to Eqs. (77),

$$\phi_{\mathscr{S}}(t) = C\mathscr{S}\phi(t) = \frac{C}{N!}\sum_Q \delta_Q Q\phi(t), \tag{84a}$$

where the sum runs over the $N!$ permutations of the particle coordinates and, according to Eq. (78),

$$C = \sqrt{N!}. \tag{84b}$$

The particle density operator is

$$\rho(\mathbf{x}) = \sum_{\beta=1}^{N} \delta(\mathbf{x} - \mathbf{r}_\beta). \tag{85}$$

For the state (84*a*), at such times that none of the wave packets overlap, we find

$$\bar{\rho}(\mathbf{x}) = (\phi_{\mathscr{S}}, \rho(x)\phi_{\mathscr{S}}) = C^2(\phi, \rho\,\mathscr{S}\phi)$$
$$= \sum_{\alpha=1}^{N} \sum_{Q} \delta_Q(\phi, \delta(\mathbf{x} - \mathbf{r}_\alpha)Q\phi)$$
$$= \sum_{\alpha=1}^{N} (2\pi)^{-3}\, |G_\alpha(\mathbf{x} - \mathbf{v}_\alpha t)|^2. \tag{86}$$

This is just the result which would have been obtained had we *not* symmetrized $\phi(t)$. Thus the particle density is (of course) unchanged by symmetrization.

Similarly, the current operator

$$\mathbf{j}(\mathbf{x}) = \sum_{\beta=1}^{N} \mathbf{v}_\beta \delta(\mathbf{x} - \mathbf{r}_\beta) \tag{87}$$

has an expectation value that is unchanged by symmetrization.

$$\bar{\mathbf{j}}(\mathbf{x}) = (\phi_{\mathscr{S}}, \mathbf{j}(\mathbf{x})\phi_{\mathscr{S}}) = \sum_{\alpha=1}^{N} (2\pi)^{-3}\, \mathbf{v}_\alpha\, |G_\alpha(\mathbf{x} - \mathbf{v}_\alpha t)|^2. \tag{88}$$

We emphasize that Eqs. (86) and (88) are valid only when the wave packets do not overlap. These equations may therefore be used to evaluate the particle, or current, densities for initial or asymptotic final states of a scattering process.

We infer from Eqs. (86) and (88) that we need not reinterpret the evaluation in Chapter 3 of incident and outgoing particle densities for a scattering process involving elementary particles. For example, to evaluate the incident flux of beam particles, we consider some point $\mathbf{x}_0$ in the path of the beam wave packet. If Eq. (88) describes the current in the initial state, only one of the packets will cross $\mathbf{x}_0$. Let this be the packet $\alpha_0$. Then Eq. (88) reduces to

$$\bar{\mathbf{j}}(\mathbf{x}_0) = \mathbf{v}_{\alpha 0}(2\pi)^{-3}\, |G_{\alpha 0}(\mathbf{x}_0 - \mathbf{v}_\alpha t)|^2. \tag{89}$$

If $\mathbf{v}_{\alpha 0}$ is parallel to the $z$-axis, and if $\mathbf{x}_0$ lies on the $z$-axis, the total flux crossing $x_0$ is

$$\int \bar{\mathbf{j}}(x_0)\, dt = (2\pi)^{-3} \int dz\, |G_{\alpha 0}(0, 0, z)|^2 = F. \tag{90}$$

This agrees precisely with the flux obtained in Eq. (3.121).

### Symmetrization for a System of Composite Particles

We may easily extend these conclusions, obtained for elementary particles, to systems of composite particles. Consider next a system of

$N = N_0 + 1$ identical particles, of which $N_0$ are bound to form a composite particle and one is a free particle. Such a state is described by

$$\chi_a(\xi_0, \xi_1, \ldots, \xi_{N0}) = \chi_{s1}(\xi_0)\chi_{s2}(\xi_1 \cdots \xi_{N0}), \tag{91}$$

where $\chi_{s1}$ and $\chi_{s2}$ are considered, for example, to describe beam and target particles, respectively. We suppose that $\chi_{s2}$ is appropriately symmetrized so that $\mathscr{S}\chi_{s2} = \chi_{s2}$, where $\mathscr{S}$ is the operator of Eq. (63).

We shall suppose that an incident wave packet state $\phi_a$ is constructed from the functions (91). Familiar examples are the scattering of an electron by an atom or molecule containing $N_0$ electrons and the scattering of a neutron or proton by a nucleus containing $N_0$ nucleons. The state $\phi_a$ is [compare Eq. (3.28)]

$$\phi_a(t) = e^{-iE_a t}\chi_{\bar{s}1}(\xi_0)\chi_{\bar{s}2}(\xi_1 \cdots \xi_{N0})G_B(\mathbf{r}_1 - \mathbf{v}_1 t)G_T(\mathbf{r}_2 - \mathbf{v}_2 t). \tag{92}$$

It is convenient to choose the "barycentric coordinate" $\bar{\mathbf{r}}_2 = \bar{\mathbf{r}}_2(\xi_1 \cdots \xi_{N0})$ to be symmetric in the variable sets $(\xi_1 \cdots \xi_{N0})$. Thus $Q\bar{\mathbf{r}}_2 = \bar{\mathbf{r}}_2$, where $Q$ is any permutation of the $(\xi_1 \cdots \xi_{N0})$.

The symmetrized state $\phi_{\mathscr{S}(a)}$ may be constructed from $\phi_a$ according to Eqs. (77), (78), and (79). The wave functions $\chi_{\mathscr{S}(a)}$ do not form an orthonormal set in this case, in contrast to the wave functions (47*a*). The reason for this is that

$$Q_{0i}\chi_a = \chi_{s1}(\xi_1)\chi_{s2}(\xi_1 \cdots \xi_{i-1}, \xi_0, \xi_{i+1} \cdots \xi_N) \qquad (i = 1, 2, \ldots, N_0)$$

is not orthogonal to $\chi_a$. Indeed, in the sense of the discussion of Section 4.1, $\chi_a$ and $Q_{0i}\chi_a$ describe different channels for the unsymmetrized wave functions. On the other hand, $Q_{0i}\phi_a(t)$ and $\phi_a(t)$ are orthogonal as long as the wave packets do not overlap. This follows directly from the expression (92) for $\phi_a$.[24]

To evaluate the normalization constant $C$ in Eq. (77), we use Eq. (78) (valid for such times $t$ that the wave packets do not overlap).

$$(\phi_{\mathscr{S}(a)}, \phi_{\mathscr{S}(a)}) = 1 = C^2(\phi_a, \mathscr{S}\phi_a) = \frac{C^2}{N!}\sum_Q \delta_Q(\phi_a, Q\phi_a).$$

The summation here extends over all $N!$ permutations of the $(\xi_0, \xi_1, \ldots, \xi_{N0})$. It is evident that $(\phi_a, Q\phi_a) = 0$, unless $Q$ is one of the $N_0!$ permutations of the $(\xi_1 \cdots \xi_{N0})$, since nonoverlapping of the wave functions

[24] This is similar to the asymptotic orthogonality for different channel states, as was demonstrated in Appendix C.

is assumed for the evaluation of $C$. In this case

$$\delta_Q(\phi_a, Q\phi_a) = (\phi_a, \phi_a) = 1,$$

so

$$\frac{C^2}{N!} N_0! = 1$$

or

$$C = \sqrt{N_0 + 1}. \tag{93}$$

The wave function (79*b*) is therefore

$$\chi_{\mathscr{S}(a)} = \frac{C}{N!} \sum_Q \delta_Q Q \chi_{s1}(\xi_0)\chi_{s2}(\xi_1 \cdots \xi_{N0}). \tag{94}$$

The particle density operator is again given by Eq. (85). A "free-particle" density operator $\rho_f(\mathbf{x})$ is defined by

$$\bar{\rho}_f(\mathbf{x}) = \rho(\mathbf{x})\Lambda_f, \tag{95}$$

where $\rho$ is given by Eq. (85) and $\Lambda_f$ is a projection operator onto the *free*-particle wave packets $G_B$.[25] We then obtain, when the wave packets do not overlap,

$$\bar{\rho}_f(\mathbf{x}) = (\phi_{\mathscr{S}(a)}\rho_f(\mathbf{x})\phi_{\mathscr{S}(a)}) = C^2(\phi_a, \rho_f(\mathbf{x})\mathscr{S}\phi_a)$$
$$= \frac{C^2}{N!} \sum_Q \delta_Q(\phi_a, \rho_f(\mathbf{x})Q\phi_a).$$

Now $\delta_Q(\phi_a, \bar{\rho}_f(\mathbf{x})Q\phi_a)$ vanishes except for the $N_0!$ permutations $Q$ among the $(\xi_1 \cdots \xi_{N0})$. For these permutations the quantity is unity. Thus Eqs. (92) and (93) give

$$\bar{\rho}_f(\mathbf{x}) = (2\pi)^{-3}\,|G_B(\mathbf{x} - \mathbf{v}_1 t)|^2. \tag{96}$$

Similarly, the free-particle current operator is $\mathbf{j}_f(\mathbf{x}) = \mathbf{j}(\mathbf{x})\Lambda_f$, where $\mathrm{j}(\mathbf{x})$ is given by Eq. (87). We obtain, by the arguments leading to (96),

$$\bar{\mathbf{j}}_f(\mathbf{x}) = (\phi_{\mathscr{S}(a)}, \mathbf{j}_f(\mathbf{x})\phi_{\mathscr{S}(a)}) = (2\pi)^{-3}\mathbf{v}_1\,|G_B(\mathbf{x} - \mathbf{v}_1 t)|^2. \tag{97}$$

Again, the wave function symmetrization does not modify the incident current and the incident flux as given by Eq. (3.121).

A useful simplification of Eq. (94) may easily be obtained. The $N!$ permutations $Q$ may be divided into $N$ types. First let $Q_1$ represent any of the $N_0!$ permutations of the $(\xi_1 \cdots \xi_{N0})$. Next let $Q_2(0, j)$ represent

[25] We show in Appendix C that the different channel states can be made orthogonal with wave packets. Thus the projection operator $\Lambda_f$ can be defined as long as the wave packets do not overlap.

the $N_0!$ permutations which result in placing $\xi_j$ into $\chi_{s1}$ and $\xi_0$ into $\chi_{s2}$, for $j = 1, 2, \ldots, N_0$. Now

$$\delta_{Q1} Q_1 \chi_a = \chi_a,$$

$$\delta_{Q2}(0, j) Q_2(0, j) \chi_a = \pm \chi_{s1}(\xi_j) \chi_{s2}(\xi_1 \cdots \xi_{j-1}, \xi_0, \xi_{j+1} \cdots \xi_{N0}),$$

the plus or minus sign referring to Bose-Einstein or Fermi-Dirac statistics respectively. Then using Eq. (93) we reduce Eq. (94) to

$$\chi_{\mathscr{S}(a)} = \frac{1}{\sqrt{N_0 + 1}} \left[ \sum_{j=1}^{N_0} \delta_j Q_{0j} \chi_{s1}(\xi_0) \chi_{s2}(\xi_1 \cdots \xi_{N0}) \right]. \tag{98}$$

Here $Q_{0j}$ interchanges $\xi_0$ and $\xi_j$, and

$$\begin{aligned} &\delta_j = 1 \quad \text{for} \quad \text{Bose-Einstein statistics,} \\ &\left.\begin{aligned} \delta_j &= 1 \quad (j = 0) \\ &= -1 \quad (j = 1, 2, \ldots, N_0) \end{aligned}\right\} \quad \text{for} \quad \text{Fermi-Dirac statistics.} \end{aligned} \tag{99}$$

In addition, of course,

$$\phi_{\mathscr{S}(a)} = \frac{1}{\sqrt{N_0 + 1}} \left[ \sum_{j=1}^{N_0} \delta_j Q_{0j} \phi_a(t) \right]. \tag{100}$$

As a final specific example we consider an initial scattering state involving two composite particles, formed from $N$ identical elementary particles. The first, or beam, particle contains $n$ elementary particles, and the second contains $N - n$ elementary particles. A wave function for such a state is

$$\chi_a = \chi_{s1}(\xi_1 \cdots \xi_n) \chi_{s2}(\xi_{n+1} \cdots \xi_N). \tag{101}$$

We suppose that $\chi_{s1}$ and $\chi_{s2}$ are individually symmetrized, so that $\mathscr{S}\chi_{s1} = \chi_{s1}$ and $\mathscr{S}\chi_{s2} = \chi_{s2}$, where $\mathscr{S}$ is given by Eq. (63). Equation (101) might, for example, represent the initial state for the scattering of two atomic nuclei.

As in Chapter 3, we may write

$$\begin{aligned} \chi_{s1} &= (2\pi)^{-3/2} e^{i\mathbf{q}_1 \cdot \mathbf{r}_1} g_{s1}(\xi_1 \cdots \xi_n), \\ \chi_{s2} &= (2\pi)^{-3/2} e^{i\mathbf{q}_2 \cdot \mathbf{r}_2} g_{s2}(\xi_{n+1} \cdots \xi_N), \end{aligned} \tag{102}$$

whereas

$$\phi_a(t) = e^{-iE_a t} \chi_{\bar{s}1} \chi_{\bar{s}2} G_B(\mathbf{r}_1 - \mathbf{v}_1 t) G_T(\mathbf{r}_2 - \mathbf{v}_2 t).$$

It is again convenient to suppose that $\mathbf{r}_1 = \mathbf{r}_1(\xi_1 \cdots \xi_n)$ and $\mathbf{r}_2 = \mathbf{r}_2(\xi_{n+1} \cdots \xi_N)$ are symmetric functions of the variable sets $(\xi_1 \cdots \xi_n)$ and $(\xi_{n+1} \cdots \xi_N)$. [The advantage of this is that $G_B$ and $G_T$ are unchanged by permutations of the $(\xi_1 \cdots \xi_n)$ or the $(\xi_{n+1} \cdots \xi_N)$ among themselves.]

We now use Eqs. (77), (78), and (79) to construct symmetrized states from that of Eq. (101). When the wave packets do not overlap, we can apply the normalization condition (78) to evaluate $C$. The result is easily seen to be

$$C = \left[\frac{N!}{n!\,(N-n)!}\right]^{1/2}. \tag{103}$$

The symmetrized states $\chi_{\mathscr{S}(a)}$ are given by the equation

$$\chi_{\mathscr{S}(a)} = \frac{C}{N!}\sum_{Q}\delta_Q[Q\chi_{s1}(\xi_1\cdots\xi_n)\chi_{s2}(\xi_{n+1}\cdots\xi_N)]. \tag{104}$$

The density operator for the "beam particles" is (defined for nonoverlapping wave packets)

$$\rho_B(\mathbf{x}) = \left\{\frac{1}{\sqrt{n!\,[(N-n)!]}}\sum_P \delta(\mathbf{x} - P\mathbf{r}_1P^{-1})\right\}\Lambda_B, \tag{105}$$

where the sum runs over all $N!$ permutations $P$ of the $(\xi_1\cdots\xi_N)$ and $P\mathbf{r}_1(\xi_1\cdots\xi_n)P^{-1} = \mathbf{r}_1(\xi_{P1}\cdots\xi_{Pn})$. Finally, $\Lambda_B$ is a projection operator onto the beam particle wave packet states. Again, we find

$$\begin{aligned}\bar{\rho}_B(\mathbf{x}) &= (\phi_{\mathscr{S}(a)}, \rho_B(\mathbf{x})\phi_{\mathscr{S}(a)})\\ &= (2\pi)^{-3}\,|G_B(\mathbf{x}-\mathbf{v}_1t)|^2.\end{aligned} \tag{106}$$

Similarly the current operator for the beam particles is

$$\mathbf{j}_B(\mathbf{x}) = \mathbf{v}_1\rho_B(\mathbf{x}),$$

and

$$\bar{\mathbf{j}}_B(\mathbf{x}) = (2\pi)^{-3}\mathbf{v}_1\,|G_B(\mathbf{x}-\mathbf{v}_1t)|^2. \tag{107}$$

Here again we find that the wave function symmetrization does not modify the particle and current densities, as long as the wave packets do not overlap.

This result is quite general. Consider any incident wave packet state $\phi_{\mathscr{S}(a)}$, representing nonoverlapping wave packets for $\mu$ particles. We may choose one of the wave packets as the "beam particle" packet (defined by the geometry of its path). Current and particle density operators $\mathbf{j}_B(\mathbf{x})$ and $\rho_B(\mathbf{x})$ can be defined with projection operators onto the beam particle wave packet states.[26] Then, for example,

$$\begin{aligned}\bar{\rho}_B(\mathbf{x}) &= (\phi_{\mathscr{S}(a)}, \rho_B(\mathbf{x})\phi_{\mathscr{S}(a)})\\ &= C^2(\phi_a, \rho_B(\mathbf{x})\mathscr{S}\phi_a).\end{aligned} \tag{108}$$

[26] Here again, we recall the orthogonality of different channel states, when constructed as nonoverlapping wave packets.

This should be compared with the normalization condition

$$1 = (\phi_{\mathscr{S}(a)}, \phi_{\mathscr{S}(a)}) = C^2(\phi_a, \mathscr{S}\phi_a). \tag{109}$$

Because of the assumption of nonoverlapping wave packets, only the permutations that do not lead to exchange of elementary particles between different wave packets contribute to either Eq. (108) or Eq. (109). The contribution to $\mathscr{S}$ from these permutations can be designated as $\delta\mathscr{S}$, where $\mathscr{S} - \delta\mathscr{S}$ vanishes in Eqs. (108) and (109). Now $\phi_a$ is an eigenfunction of $\delta\mathscr{S}$; that is, $\delta\mathscr{S}\phi_a = \lambda\phi_a$, where $\lambda$ is a constant. Then Eq. (109) reads

$$1 = C^2(\phi_a, \delta\mathscr{S}\phi_a) = C^2\lambda(\phi_a, \phi_a) = C^2\lambda.$$

Because of this condition, Eq. (108) reduces to

$$\bar{\rho}_B(\mathbf{x}) = C^2(\phi_a, \rho_B(\mathbf{x})\delta\mathscr{S}\phi_a) = C^2\lambda(\phi_a, \rho_B(\mathbf{x})\phi) = (\phi_a, \rho_B(\mathbf{x})\phi_a). \tag{110}$$

A similar argument applies to the current operator. Therefore, wave function symmetrization leaves the evaluation of particle and current densities unchanged for initial, or final, wave packet states. The evaluation of particle fluxes, as made in Chapter 3, is not modified by symmetrization of the wave function.

## 4.4 EVALUATION OF THE SCATTERING CROSS SECTION

Let us now evaluate the transition probability or scattering cross section for a scattering process, as in Section 3.3. The initial state is assumed to be $\phi_{\mathscr{S}(a)}(t_0)$, defined by Eq. (77). This is the symmetrized form of the state $\phi_a$ of Chapter 3. As indicated by Eq. (79), $\phi_{\mathscr{S}(a)}$ is a wave packet state constructed from the functions $\chi_{\mathscr{S}(a)} = C\mathscr{S}\chi_a$, where $\chi_a$ is an eigenfunction of the Hamiltonian $K$.

Let us suppose that the states $\chi_a$ from which $\phi_a$ is formed belong to a channel containing $L$ kinds of elementary particles,[27] there being $N_i$ of type $i$ ($i - 1, 2, \ldots, L$). Then according to Eq. (60$b$)

$$\chi_{\mathscr{S}(a)} = \frac{C}{\prod_{i=1}^{L}(N_i!)} \sum_P \delta_P P\chi_a, \tag{111}$$

where the summation extends over the $\Pi_{i=1}^{L}(N_i!)$ permutations $P$ [see Eq. (61)] among identical particles. The constant $C$ is determined from Eq. (78), as described in Section 4.3. It will prove convenient to adopt the notation

$$P\chi_a \equiv \chi_{P(a)} \equiv |P(a)\rangle. \tag{112}$$

[27] Some of these may be combined to form composite particles, of course.

From the states $\chi'_b$, introduced by Eq. (3.72), we can form symmetrized states

$$\chi'_{\mathscr{S}(b)} \equiv C'\mathscr{S}\chi'_b. \tag{113a}$$

To determine the constant $C'$, we may construct nonoverlapping wave packet states from the $\chi'_{\mathscr{S}(b)}$ and normalize these in accordance with Eq. (78).[28] Let us suppose that those states $\chi'_{\mathscr{S}(b)}$ which are to be observed contain $L'$ kinds of elementary particles,[29] there being $N'_j$ of type $j$ ($j = 1, 2, \ldots, L'$). Then according to Eq. (60$b$)

$$\chi'_{\mathscr{S}(b)} = \frac{C'}{\prod_{j=1}^{L'} (N'_j!)} \sum_Q \delta_Q Q\chi'_b, \tag{113b}$$

where the summation runs over the $\Pi_{j=1}^{L'} (N'_j!)$ permutations $Q$ among identical particles. In analogy to Eq. (112), we introduce the notation

$$Q\chi'_b \equiv \chi'_{Q(b)} \equiv |Q(b')\rangle. \tag{114}$$

**The Scattering of Elementary Particles**

Let us first consider a simple case. We assume that the symmetrization operator $\mathscr{S}$ [Eq. (63)] commutes with both $K$ and $K'$; that is,[30]

$$[\mathscr{S}, K] = [\mathscr{S}, K'] = 0. \tag{115}$$

The conditions (115) will ordinarily be satisfied when there are no composite particles in either the incident or final channel. In this case the incident and emerging particles are elementary and it will usually be possible to choose $K' = K$.

When the conditions (115) are satisfied, the $\chi_{\mathscr{S}(a)}$ are eigenfunctions of $K$ and the $\chi'_{\mathscr{S}(b)}$ are eigenfunctions of $K'$. Thus, the $\chi_{\mathscr{S}(a)}$ and the $\chi'_{\mathscr{S}(b)}$ each represent a complete set of orthonormal wave vectors. In this case we need not modify the development in Chapter 3. It is necessary only to use everywhere the $\chi_{\mathscr{S}(a)}$ in place of the $\chi_a$, and the $\chi'_{\mathscr{S}(b)}$ in place of the $\chi'_b$, in the discussion given in Chapter 3. In this connection we recall that in Section 4.3 it was shown that the asymptotic particle fluxes are unmodified by symmetrization.

[28] For example, we may construct an incident wave packet state from the $\chi'_{\mathscr{S}(b)}$.

[29] Again, some of these may be bound to form composite particles.

[30] The operator $\mathscr{S}$, of course, always commutes with the Hamiltonian $H$ of the complete system.

For example, the wave functions $\psi^{\pm}$ as defined by Eqs. (3.66) and (3.67) are now defined as

$$\psi^{\pm}_{\mathscr{S}(a)}(\eta) = \left(1 + \frac{1}{E_a \pm i\eta - H} V\right)\chi_{\mathscr{S}(a)},$$
$$\psi^{\pm}_{\mathscr{S}(a)} = \lim_{\eta \to 0(+)} \psi^{\pm}_{\mathscr{S}(a)}(\eta). \tag{116}$$

The scattering matrix $\mathscr{T}$ is defined in terms of symmetrized states [compare Eq. (3.82)] as

$$\langle \mathscr{S}(b')| \mathscr{T} |\mathscr{S}(a)\rangle = \left(\chi'_{\mathscr{S}(b)}, \left(V + V' \frac{1}{E_a + i\eta - H} V\right)\chi_{\mathscr{S}(a)}\right), \tag{117}$$

where the operator $\mathscr{T}$ is given by Eq. (3.79). Again, using Eqs. (111) and (113*b*), and the notation of Eqs. (112) and (114), this is

$$\langle \mathscr{S}(b')| \mathscr{T} |\mathscr{S}(a)\rangle = \frac{C'C}{\left[\prod_{i=1}^{L}(N_i!)\right]\left[\prod_{j=1}^{L'}(N'_j!)\right]} \sum_P \delta_P \sum_Q \delta_Q \langle Q(b')| \mathscr{T} |P(a)\rangle, \tag{118a}$$

where

$$\langle Q(b')| \mathscr{T} |P(a)\rangle = \left(\chi'_{Q(b)}, \left(V + V' \frac{1}{E_a + i\eta - H} V\right)\chi_{P(a)}\right). \tag{118b}$$

When the state $\mathscr{S}(a)$ contains $N$ identical particles of only one kind, and the state $\mathscr{S}(b')$ contains $N'$ identical particles of only one kind, Eq. (118*a*) simplifies to

$$\langle \mathscr{S}(b')| \mathscr{T} |\mathscr{S}(a)\rangle = \frac{C'C}{[(N')!](N!)} \sum_P \delta_P \sum_Q \delta_Q \langle Q(b')| \mathscr{T} |P(a)\rangle. \tag{118c}$$

The condition of momentum conservation permits us to simplify Eq. (117), as in Eq. (3.83);

$$\langle \mathscr{S}(b')| \mathscr{T} |\mathscr{S}(a)\rangle = \delta(\mathbf{P}_b - \mathbf{P}_a) T_{\mathscr{S}(b,a)}. \tag{119}$$

Here $\mathbf{P}_b$ and $\mathbf{P}_a$ are the eigenvalues of the momentum operator $\mathscr{P}$ for the states $\mathscr{S}(b')$ and $\mathscr{S}(a)$, respectively, and we have used the condition that $[\mathscr{S}, \mathscr{P}] = 0$.

The $S$-matrix (3.94) is now written as

$$\langle \mathscr{S}(b')| S |\mathscr{S}(a)\rangle = \delta_{ba} - 2\pi i \delta(E_b - E_a) \langle \mathscr{S}(b')| \mathscr{T} |\mathscr{S}(a)\rangle. \tag{120}$$

We use Eq. (3.142) to define the differential scattering cross section

$$d\sigma = \sum_{s_1, \ldots, s_\nu} \int d^3k_1 \cdots d^3k_\nu \delta(E_b - E_a)\delta(\mathbf{P}_b - \mathbf{P}_a) \times \frac{(2\pi)^4 |T_{\mathscr{S}(b,a)}|^2}{[(\mathbf{v}_1 - \mathbf{v}_2)^2 - (1/c^2)(\mathbf{v}_1 \times \mathbf{v}_2)^2]^{1/2}}, \tag{121}$$

where the range of integration covers a small domain which includes the momentum and energy shells.

In evaluating the total cross section into a given channel, we must use the appropriate ordered sum over states, as described by Eqs. (56). That is, we must not sum over redundant states which merely represent interchanges of identical particles.

For example, if the final states in Eq. (121) describe $N'$ particles of a single kind (so $\nu = N$), and if we wish to integrate over all $\mathbf{k}_1, \ldots, \mathbf{k}_\nu$ and sum over all $s_1, \ldots, s_\nu$, we must write

$$\sigma = \frac{1}{N'!} \int d\sigma \tag{122}$$

for the total cross section. Again, if there are $\nu_i$ ($i = 1, 2, \ldots, L'$; $\sum_{\nu_i} = \nu$) identical particles of species $i$ in the final state

$$\sigma = \frac{1}{\prod_{i=1}^{L'} (\nu_i!)} \int d\sigma, \tag{123}$$

where we again sum over all $s_1, \ldots, s_\nu$ and integrate over all $\mathbf{k}_1, \ldots, \mathbf{k}_\nu$.

**The Scattering of Composite Particles**

Let us return to the general case, for which the conditions (115) are not valid. We now symmetrize Eq. (3.70) directly, in accordance with Eq. (81):

$$\Psi_{\mathscr{S}}(t) = C\mathscr{S}\Psi(t) = \int d^3q_1\, d^3q_2 A(\mathbf{q}_1, \mathbf{q}_2) e^{-iE_a t} \Psi^{\pm}_{\mathscr{S}(a)}, \tag{124}$$

where, in place of Eqs. (116), we now define $\psi^{\pm}_{\mathscr{S}(a)}$ by[31]

$$\psi^{\pm}_{\mathscr{S}(a)} = C\mathscr{S}\psi_a^{\pm}. \tag{125}$$

As before, the $\psi_a^{\pm}$ are defined by Eqs. (3.66) and (3.67). The constant $C$ in Eqs. (124) and (125) is determined by the normalization condition (78) on the initial state $\phi_{\mathscr{S}(a)}$. Generalization of Eq. (124) for the initial state (80) is straightforward.

The wave functions $\chi_{P(a)}$ and $\chi'_{Q(b)}$ of Eqs. (112) and (114) now satisfy the equations

$$\begin{aligned} K_P \chi_{P(a)} &= E_a \chi_{P(a)}, \\ K'_Q \chi'_{Q(b)} &= E_b \chi'_{Q(b)}, \end{aligned} \tag{126}$$

where

$$\begin{aligned} K_P &\equiv PKP^{-1}, \\ K'_Q &\equiv QK'Q^{-1}. \end{aligned} \tag{127a}$$

[31] Equations (116) and (125) are equivalent when the conditions (115) are satisfied.

In addition, we may define

$$V_P \equiv H - K_P,\qquad V'_Q = H - K'_Q. \tag{127b}$$

To put Eq. (125) into a more useful form, we write[32] [compare Eq. (3.67)]

$$\begin{aligned}\psi_a^{\pm} &= \left(1 + \frac{1}{E_a \pm i\eta - H} V\right)\chi_a \\ &= \frac{1}{E_a \pm i\eta - H}[(E_a \pm i\eta - H) + V]\chi_a \\ &= \frac{1}{E_a \pm i\eta - H}(E_a - K \pm i\eta)\chi_a \\ &= \frac{\pm i\eta}{E_a \pm i\eta - H}\chi_a,\end{aligned} \tag{128}$$

since $(E_a - K)\chi_a = 0$. Now, $\mathscr{S}$ commutes with $H$, so Eq. (125) is equivalent to

$$\begin{aligned}\psi^{\pm}_{\mathscr{S}(a)} &= \frac{\pm i\eta}{E_a \pm i\eta - H}(C\mathscr{S}\chi_a) = \frac{\pm i\eta}{E_a \pm i\eta - H}\chi_{\mathscr{S}(a)} \\ &= \frac{C}{\prod_{i=1}^{L}(N_i!)}\sum_P \delta_P\left(\frac{\pm i\eta}{E_a \pm i\eta - H}\right)\chi_{P(a)},\end{aligned} \tag{129}$$

because of Eq. (111). On reading Eq. (128) backwards, we obtain

$$\begin{aligned}\frac{\pm i\eta}{E_a \pm i\eta - H}\chi_{P(a)} &= \frac{1}{E_a \pm i\eta - H}(E_a - K_P \pm i\eta)\chi_{P(a)} \\ &= \frac{1}{E_a \pm i\eta - H}[(E_a \pm i\eta - H) + V_P]\chi_{P(a)} \\ &= \left(1 + \frac{1}{E_a \pm i\eta - H}V_P\right)\chi_{P(a)} \\ &= \psi^{\pm}_{P(a)}.\end{aligned} \tag{130}$$

Therefore

$$\psi^{\pm}_{\mathscr{S}(a)} = \frac{C}{\prod_{i=1}^{L}(N_i!)}\sum_P \delta_P \psi^{\pm}_{P(a)}. \tag{131}$$

[32] In accordance with the convention adopted in Chapter 3, we are not using different symbols for $\psi_a^{\pm}$ and $\psi_a^{\pm}(\eta)$.

For a given permutation $Q$ the $\chi'_{Q(b)}$ represent a complete set of state vectors. They may thus be used, instead of the $\chi'_b$, in Eq. (3.81). This permits us to write Eq. (3.81) as

$$\psi^+_{P(a)} = \chi_{P(a)} + \sum_b \frac{\chi'_{Q(b)}}{E_a + i\eta - E_b} \langle Q(b')| \mathcal{T} |P(a)\rangle, \tag{132a}$$

where

$$\langle Q(b')| \mathcal{T} |P(a)\rangle = \left(\chi'_{Q(b)}, \left(V_{P'} + V'_Q \frac{1}{E_a + i\eta - H} V_P\right)\chi_{P(a)}\right). \tag{132b}$$

(For $\psi^-_{P(a)}$, we change the sign of $\eta$ in these equations, as usual.) That is, Eqs. (132) are obtained from Eqs. (3.81) and (3.82) by applying a permutation $P$ to the initial state and a permutation $Q$ to the final state.

If Eq. (132) is inserted into Eq. (131), there results

$$\psi^+_{\mathcal{S}(a)} = \chi_{\mathcal{S}(a)} + \sum_b \frac{\chi'_{Q(b)}}{E_a + i\eta - E_b} \langle Q(b')| \mathcal{T} |\mathcal{S}(a)\rangle \tag{133a}$$

with

$$\langle Q(b')| \mathcal{T} |\mathcal{S}(a)\rangle = \frac{C}{\prod_{i=1}^{L} (N_i!)} \sum_P \delta_P \langle Q(b')| \mathcal{T} |P(a)\rangle, \tag{133b}$$

$\psi^-_{\mathcal{S}(a)}$ being obtained on changing the sign of $\eta$. Finally, Eq. (124) may be expressed as

$$\Psi_{\mathcal{S}}(t) = \int d^3q_1\, d^3q_2 A(\mathbf{q}_1, \mathbf{q}_2) e^{-iE_a t}$$

$$\times \left[\chi_{\mathcal{S}(a)} + \sum_b \frac{\chi'_{Q(b)}}{E_a + i\eta - E_b} \langle Q(b')| \mathcal{T} |\mathcal{S}(a)\rangle\right], \tag{134}$$

for a natural scattering process [that is, when $t_0 = -|t_0|$ in Eq. (76)].

Let it be imagined that we are to observe a limited set of final states

$$\phi'_{\mathcal{S}(b)}(t) = e^{-iE_b t}\chi'_{\mathcal{S}(b)} \tag{135a}$$

at times $t$ much later than the collision time $|t_c|$. The transition probability for observing these states is then

$$P_{\bar{b}a} = \sum_b |(\phi'_{\mathcal{S}(b)}(t), \Psi_{\mathcal{S}}(t))|^2. \tag{136}$$

As has been mentioned in Section 3.4 and Appendix C, it is often better to replace the state (135$a$) by a wave packet

$$\phi'_{\mathcal{S}(\bar{b})}(t) = \sum_b A'_b \phi'_{\mathcal{S}(b)}(t), \tag{135b}$$

having a narrow packet amplitude $A'_b$. The wave functions (135*b*) can be chosen to form an orthonormal set, providing a complete description of asymptotic states.

Now, following the arguments which led to Eq. (3.93), we introduce the expansion (113*b*) to obtain

$$(\phi'_{\mathscr{S}(b)}(t), \Psi_{\mathscr{S}}(t)) = \frac{C'}{\prod_{j=1}^{L'} (N'_j!)} \sum_Q \delta_Q(\chi'_{Q(b)}, e^{iE_bt}\Psi_{\mathscr{S}}(t))$$

$$= \frac{C'}{\prod_{j=1}^{L'} (N'_j!)} \sum_Q \delta_Q \int d^3q_1\, d^3q_2 A(\mathbf{q}_1, \mathbf{q}_2)$$

$$\times \left\{(\chi'_{Q(b)}, \chi_{\mathscr{S}(a)}) + \frac{e^{i(E_b-E_a)t}}{E_a + i\eta - E_b} \langle Q(b)| \mathscr{T} |\mathscr{S}(a)\rangle\right\}. \tag{137}$$

Under the wave packet integral $(\chi'_{Q(b)}, \chi_{\mathscr{S}(b)}) = 0$, unless the $\chi'_b$ belong to the set $\chi_a$—or unless the final channel is the same as the incident channel.[33] Then we may take $\chi'_{\mathscr{S}(b)} = \chi_{\mathscr{S}(a')}$, one of the $\chi_{\mathscr{S}(a)}$, and write $(\chi_{\mathscr{S}(a')}, \chi_{\mathscr{S}(a)}) = \delta_{a',a}$ under the wave packet integral (137). More generally, we may write[34]

$$(\chi'_{\mathscr{S}(b)}, \chi_{\mathscr{S}(a)}) = \delta_{b,a}, \tag{138}$$

as in Eq. (3.75*b*) within the wave packet integral (137).

For the second term in parenthesis, we may use Eq. (3.92). Therefore

$$(\phi'_{\mathscr{S}(b)}, \Psi_{\mathscr{S}}(t)) = \int d^3q_1\, d^3q_2 A(\mathbf{q}_1, \mathbf{q}_2) \langle \mathscr{S}(b')| S |\mathscr{S}(a)\rangle, \tag{139}$$

where

$$\langle \mathscr{S}(b')| S |\mathscr{S}(a)\rangle = \delta_{ba} - 2\pi i \delta(E_b - E_a) \langle \mathscr{S}(b')| \mathscr{T} |\mathscr{S}(a)\rangle \tag{140a}$$

and

$$\langle \mathscr{S}(b')| \mathscr{T} |\mathscr{S}(a)\rangle = \frac{C'}{\prod_{j=1}^{L'} (N'_j!)} \sum_Q \delta_Q \langle Q(b')| \mathscr{T} |\mathscr{S}(a)\rangle$$

$$= \frac{C'C}{\left[\prod_{j=1}^{L'} (N'_j!)\right]\left[\prod_{i=1}^{L} (N_i!)\right]} \sum_P \delta_P \sum_Q \delta_Q \langle Q(b')| \mathscr{T} |P(a)\rangle. \tag{140b}$$

This is a direct generalization of Eq. (118*a*).

[33] The argument here is similar to that of Appendix C.

[34] Although the $\chi_{\mathscr{S}(a)}$ do not form an orthonormal set, they can be treated as such when appearing in nonoverlapping wave packets (see Appendix C).

In a strict sense, Eq. (140) is defined only within a scattering wave packet, or alternatively, when $\mathscr{S}(a)$ and $\mathscr{S}(b')$ are both wave packet states.

To obtain the scattering cross section, we introduce the nonsingular matrix $T_{\mathscr{S}(b,a)}$ as in Eq. (119):

$$\langle \mathscr{S}(b')| \mathscr{T} |\mathscr{S}(a)\rangle = \delta(\mathbf{P}_b - \mathbf{P}_a) T_{\mathscr{S}(b,a)}. \tag{141}$$

Our present Eq. (139) has just the form of Eq. (3.95). The evaluation of $P_{\bar{b}a}$ is therefore identical with that following Eq. (3.113), which we need not repeat. In Eq. (3.134) we must only replace $T_{ba}$ by $T_{\mathscr{S}(b,a)}$. Equation (121) provides us with an expression for the differential scattering cross section, and the remarks associated with Eqs. (122) and (123) are still applicable.

## 4.5 EXAMPLES

In this section we shall discuss several examples to help clarify and illustrate the somewhat formal development of the last section.

### The Case for Which There is Only One Kind of Elementary Particle

Let us first show how to simplify Eq. (140*b*) under conditions that the states $\mathscr{S}(a)$ and $\mathscr{S}(b')$ both contain the same $N$ identical elementary particles and no others. These $N$ elementary particles may be bound in different ways to form composite particles in the incident and final channels; however, we exclude processes leading to the "creation" of new elementary particles. Then Eq. (140*b*) reduces to

$$\langle \mathscr{S}(b')| \mathscr{T} |\mathscr{S}(a)\rangle = \frac{C'C}{(N!)^2} \sum_P \delta_P \sum_Q \delta_Q \langle Q(b')| \mathscr{T} |P(a)\rangle. \tag{142}$$

Here both sums extend over the $N!$ permutations of the coordinates of the $N$ elementary particles. Now, if $R$ is any permutation of the particle coordinates,

$$\langle RQ(b')| \mathscr{T} |RP(a)\rangle = \langle Q(b')| \mathscr{T} |P(a)\rangle, \tag{143}$$

where $|RP(a)\rangle = RP\chi_a$, and so on. That is, we are just relabeling variables of integration in Eq. (143), as was done in Eq. (29). We may, for example, take $R = P^{-1}$, so the matrix element in Eq. (142) is $\langle P^{-1}Q(b')| \mathscr{T} |a\rangle$. Since $\delta_P = \delta_{P^{-1}}$, and thus $\delta_P\delta_Q = \delta_{P^{-1}Q}$, we have

$$\sum_P \delta_P \sum_Q \delta_Q \langle P^{-1}Q(b')| \mathscr{T} |a\rangle = N! \sum_Q \delta_Q \langle Q(b')| \mathscr{T} |a\rangle,$$

where the final sum on $Q$ runs over all $N!$ permutations. This may be substituted into Eq. (142) to give

$$\langle \mathscr{S}(b')| \mathscr{T} |\mathscr{S}(a)\rangle = \frac{C'C}{N!} \sum_Q \delta_Q \langle Q(b')| \mathscr{T} |a\rangle. \tag{144}$$

Equation (144) is ordinarily simpler to use than Eq. (142), since it contains $N!$, rather than $(N!)^2$, terms.

For a more specific example, we consider the simple scattering of two elementary particles. In this case we take

$$\begin{aligned} |a\rangle &= \chi_a = \chi_{s1,\mathbf{q}1}(\xi_1)\chi_{s2,\mathbf{q}2}(\xi_2) = |s_1, \mathbf{q}_1; s_2, \mathbf{q}_2\rangle, \\ |b'\rangle &= \chi_b = \chi_{s'1,\mathbf{k}1}(\xi_1)\chi_{s'2,\mathbf{k}2}(\xi_2) = |s'_1, \mathbf{k}_1; s'_2, \mathbf{k}_2\rangle, \end{aligned} \tag{145}$$

and have $V' = V$, so $\mathscr{T} = V + V(E_a + i\eta - H)^{-1}V$. We may use Eq. (144), with $N = N! = 2$ and $C = C' = \sqrt{2}$. Therefore

$$\begin{aligned} \langle \mathscr{S}(b')| \mathscr{T} |\mathscr{S}(a)\rangle = \langle s'_1, \mathbf{k}_1; s'_2, \mathbf{k}_2| \mathscr{T} |s_1, \mathbf{q}_1; s_2, \mathbf{q}_2\rangle \\ \pm \langle s'_2, \mathbf{k}_2; s'_1, \mathbf{k}_1| \mathscr{T} |s_1, \mathbf{q}_1; s_2, \mathbf{q}_2\rangle, \end{aligned} \tag{146}$$

where the plus or minus sign refers to Bose-Einstein or Fermi-Dirac statistics, respectively. Here

$$\begin{aligned} \langle s'_i, \mathbf{k}_i; s'_j, \mathbf{k}_j| \mathscr{T} |s_1, \mathbf{q}_1; s_2, \mathbf{q}_2\rangle \\ = (\chi_{s'i,\mathbf{k}i}(\xi_1)\chi_{s'j,\mathbf{k}j}(\xi_2), \mathscr{T}(\xi_1, \xi_2)\chi_{s1,\mathbf{q}1}(\xi_1)\chi_{s2,\mathbf{q}2}(\xi_2)) \end{aligned} \tag{147}$$

for $i = 1, j = 2$, or $i = 2, j = 1$. The first term in Eq. (146) is called the "direct" amplitude, while the second is called the "exchange" amplitude. With no wave function symmetrization the exchange term would not appear in Eq. (146).

It is of interest to compare the scattering cross section obtained from Eq. (146) with the corresponding unsymmetrized cross section (3.149). In the barycentric coordinate system [see Eq. (141)] we set $\mathbf{q}_1 = -\mathbf{q}_2 = \boldsymbol{\varkappa}$ and $\mathbf{k}_1 = -\mathbf{k}_2 = \boldsymbol{\varkappa}_f$:

$$T_{\mathscr{S}(b,a)} = \langle s'_1, s'_2| T_B(\boldsymbol{\varkappa}_f, \boldsymbol{\varkappa}) |s_1, s_2\rangle \pm \langle s'_2, s'_1| T_B(-\boldsymbol{\varkappa}_f, \boldsymbol{\varkappa}) |s_1, s_2\rangle, \tag{148}$$

where the quantity $T_B$ was defined in connection with Eq. (3.146). The differential cross section (3.149) is now replaced by

$$\frac{d\sigma}{d\Omega} = \frac{(2\pi)^4\kappa}{(1/\epsilon_{s'1,\kappa} + 1/\epsilon_{s'2,\kappa})v_{\mathrm{rel},B}} \left| \begin{aligned} &\langle s'_1, s'_2| T_B(\boldsymbol{\varkappa}_f, \boldsymbol{\varkappa}) |s_1, s_2\rangle \\ &\qquad \pm \langle s'_2, s'_1| T_B(-\boldsymbol{\varkappa}_f, \boldsymbol{\varkappa}) |s_1, s_2\rangle \end{aligned} \right|^2. \tag{149}$$

For the example of simple potential scattering, as described by Eq. (3.159), we must distinguish several cases. For spin zero particles (satisfying Bose-Einstein statistics, of course) we obtain from Eq. (3.161)

$$T_B(\pm\boldsymbol{\varkappa}_f, \boldsymbol{\varkappa}) = \frac{V_0 d}{2\pi^2} \frac{1}{(\boldsymbol{\varkappa}_f \mp \boldsymbol{\varkappa})^2 + 1/d^2}. \tag{150}$$

We may now set [compare Eq. (3.10)] $\epsilon_{s'1,k} = \epsilon_{s'2,k} = c(\kappa^2 + M^2c^2)^{1/2} \equiv \epsilon_k$. Then Eq. (149) gives

$$\left(\frac{d\sigma}{d\Omega}\right)_{\text{B.E.}} = \frac{\sigma_0}{d^4}\left\{\frac{1}{[(\boldsymbol{\varkappa}_f - \boldsymbol{\varkappa})^2 + 1/d^2]} + \frac{1}{[(\boldsymbol{\varkappa}_f + \boldsymbol{\varkappa})^2 + 1/d^2]}\right\}^2, \quad (151a)$$

where

$$\sigma_0 = (V_0 d)^2 d^4 \epsilon_k^{\ 2}. \quad (151b)$$

For the case of spin one-half particles (Fermi-Dirac statistics), there will be no cross section for spin-flip, since $V(r)$ depends only on the space coordinates of the particles. First, let $s_1 = s_2 = \uparrow$, representing spin up. The only contribution to the cross section comes from states with $s'_1 = s'_2 = \uparrow$, also. Then Eq. (149) gives

$$\frac{d\sigma}{d\Omega}(\uparrow\uparrow) = \frac{\sigma_0}{d^4}\left\{\frac{1}{[(\boldsymbol{\varkappa}_f - \boldsymbol{\varkappa})^2 + 1/d^2]} - \frac{1}{[(\boldsymbol{\varkappa}_f + \boldsymbol{\varkappa})^2 + 1/d^2]}\right\}^2. \quad (152)$$

For the case that $s_1 = \uparrow$, $s_2 = \downarrow$, we sum over the final states $s'_1 = \uparrow$, $s'_2 = \downarrow$, and $s'_1 = \downarrow$, $s'_2 = \uparrow$. The resulting cross section, obtained from Eq. (149), is

$$\frac{d\sigma}{d\Omega}(\uparrow\downarrow) = \frac{\sigma_0}{d^4}\left\{\frac{1}{[(\boldsymbol{\varkappa}_f - \boldsymbol{\varkappa})^2 + 1/d^2]^2} + \frac{1}{[(\boldsymbol{\varkappa}_f + \boldsymbol{\varkappa})^2 + 1/d^2]^2}\right\}. \quad (153)$$

Finally, let $s_1 = \uparrow$, $s_2 = \downarrow$. The cross section corresponding to the scattering for which the emerging particle with momentum $\boldsymbol{\varkappa}_f$ has spin up, while the other has spin down, is

$$\frac{d\sigma}{d\Omega}(\uparrow_{\varkappa_f}) = \frac{\sigma_0}{d^4}\frac{1}{[(\boldsymbol{\varkappa}_f - \boldsymbol{\varkappa})^2 + 1/d^2]^2}. \quad (154)$$

On the other hand, the cross section for which the particle with momentum $\boldsymbol{\varkappa}_f$ has spin down, while the other has spin up, is

$$\frac{d\sigma}{d\Omega}(\downarrow_{\varkappa_f}) = \frac{\sigma_0}{d^4}\frac{1}{[(\boldsymbol{\varkappa}_f + \boldsymbol{\varkappa})^2 + 1/d^2]^2}. \quad (155)$$

Equation (153) is evidently the sum of Eqs. (154) and (155).

The cross sections (151) to (155) should be compared with the original, unsymmetrized cross section (3.162).

The case of Rutherford scattering is of particular importance. This cross section is obtained from those of Eqs. (151) to (155) by setting $d = \infty$ and $(V_0 d)^2 = e_1^{\ 2} e_2^{\ 2}$, as in Eq. (3.163). If we introduce the scattering angle $\theta$, defined by $\cos\theta = (\boldsymbol{\varkappa}_f \cdot \boldsymbol{\varkappa})\boldsymbol{\varkappa}^{-2}$, and

$$\sigma_R = \frac{e_1^{\ 2} e_2^{\ 2}}{(2\kappa^2/\epsilon_\kappa)^2}, \quad (156)$$

Eqs. (151) to (155) reduce to

$$\begin{aligned}
\left(\frac{d\sigma}{d\Omega}\right)_{\text{B.E.}} &= \sigma_R\left(\frac{1}{1-\cos\theta}+\frac{1}{1+\cos\theta}\right)^2,\\
\frac{d\sigma}{d\Omega}(\uparrow\uparrow) &= \sigma_R\left(\frac{1}{1-\cos\theta}-\frac{1}{1+\cos\theta}\right)^2,\\
\frac{d\sigma}{d\Omega}(\uparrow\downarrow) &= \sigma_R\left[\frac{1}{(1-\cos\theta)^2}+\frac{1}{(1+\cos\theta)^2}\right],\\
\frac{d\sigma}{d\Omega}(\uparrow_{\varkappa_f}) &= \sigma_R\left(\frac{1}{1-\cos\theta}\right)^2,\\
\frac{d\sigma}{d\Omega}(\downarrow_{\varkappa_f}) &= \sigma_R\left(\frac{1}{1+\cos\theta}\right)^2.
\end{aligned} \tag{157}$$

The first equation here describes the scattering, for example, of two $He^4$ nuclei at sufficiently low energies that nuclear interactions are not important. The remaining four equations are adequate to describe electron-electron scattering.[35] These cross sections are illustrated in Fig. 4.1.

The total cross section for electron-electron scattering, for a random mixture of initial electron spin, and for any final spins is

$$\begin{aligned}
\frac{d\sigma}{d\Omega}(e, e) &= \frac{1}{2}\left[\frac{d\sigma}{d\Omega}(\uparrow\uparrow)+\frac{d\sigma}{d\Omega}(\uparrow\downarrow)\right]\\
&= \sigma_R\left[\frac{1}{(1-\cos\theta)^2}+\frac{1}{(1+\cos\theta)^2}-\frac{1}{1-\cos^2\theta}\right].
\end{aligned} \tag{158}$$

This equation is also appropriate for describing proton-proton scattering at angles and energies for which nuclear scattering is not important.

### Scattering of an Elementary Particle by a Composite Particle

As a second example, we consider scattering from an incident channel for a system of $N = N_0 + 1$ identical particles. Of these, one is free and the remaining are in a bound state, forming a single composite particle.[36] An unsymmetrized eigenvector $\chi_a$ for this channel is given by Eq. (91). The final channel is chosen to be the same as the initial channel, with (unsymmetrized) eigenvectors

$$\chi_b = \chi_{s'1}(\xi_0)\chi_{s'2}(\xi_1 \cdots \xi_{N0}). \tag{159}$$

(Above the energetic threshold for excitation of the composite particle to

[35] The exact description of nonrelativistic Coulomb scattering will be given in Section 6.4.

[36] As mentioned earlier, two important processes of this type are the scattering of a nucleon by an atomic nucleus and the scattering of an electron by an atom.

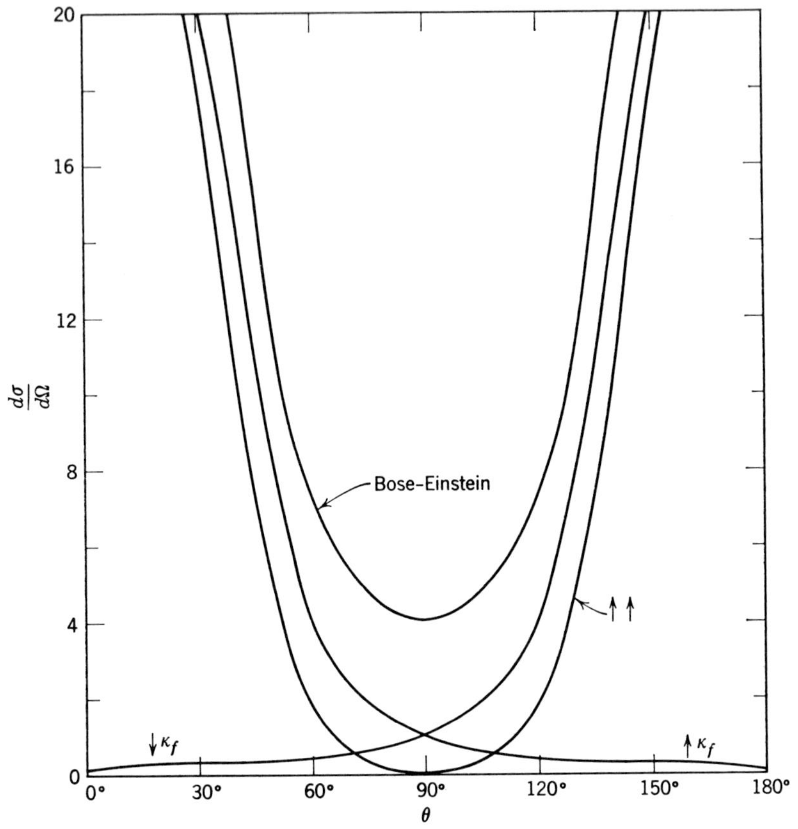


Fig. 4.1 Illustration of the Coulomb scattering cross sections (157), given in units of $\sigma_R$ [Eq. (156)]. The curves are labeled by spin states, as described in the text, and are shown as a function of the scattering angle $\theta$.

the continuum, there are of course other channels into which the reaction may go.) Since the incident and exit channels are the same, $K'_p = K_p$, $V'_p = V_p$ [Eqs. (127)].

We may use Eq. (144) to evaluate the scattering matrix. According to Eq. (93), $C' = C = \sqrt{N_0 + 1}$, so Eq. (144) gives

$$\langle \mathscr{S}(b)| \mathscr{T} |\mathscr{S}(a)\rangle = \frac{1}{N_0!} \sum_Q \delta_Q \langle Q(b)| \mathscr{T} |a\rangle. \tag{160}$$

By precisely the arguments that led from Eq. (94) to Eq. (98), this can be reduced to

$$\langle \mathscr{S}(b)| \mathscr{T} |\mathscr{S}(a)\rangle = \langle b| \mathscr{T} |a\rangle \pm \sum_{j=1}^{N_0} \langle Q_{0j}(b)| \mathscr{T} |a\rangle. \tag{161}$$

Here again the plus sign refers to the case of Bose-Einstein statistics, the minus sign to the Fermi-Dirac statistics, and of course

$$|Q_{0j}(b)\rangle = Q_{0j}\chi_b = \chi_{s'1}(\xi_j)\chi_{s'2}(\xi_1 \cdots \xi_{j-1}, \xi_0, \xi_{j+1} \cdots \xi_{N0}).$$

Now let $Q_{jk}$ ($j, k = 1, 2, \ldots, N_0$) be the operator that interchanges $\xi_i$ and $\xi_j$. According to Eq. (143), then,

$$\langle Q_{0j}(b)| \mathscr{T} |a\rangle = \langle Q_{jk}Q_{0j}(b)| \mathscr{T} |Q_{jk}(a)\rangle = \langle Q_{0k}(b)| \mathscr{T} |a\rangle, \quad (162)$$

since

$$Q_{jk}Q_{0j}\chi_b = \pm Q_{0k}\chi_b,$$
$$Q_{jk}\chi_a = \pm\chi_a.$$

If we call

$$\begin{aligned} \mathscr{T}_d &= \langle b| \mathscr{T} |a\rangle = \delta(\mathbf{P}_b - \mathbf{P}_a)T_d, \\ \mathscr{T}_{\text{ex}} &= \langle Q_{0j}(b)| \mathscr{T} |a\rangle = \delta(\mathbf{P}_b - \mathbf{P}_a)T_{\text{ex}}, \end{aligned} \quad (163)$$

we may rewrite (161) as

$$\langle \mathscr{S}(b)| \mathscr{T} |\mathscr{S}(a)\rangle = \mathscr{T}_d \pm N_0\mathscr{T}_{\text{ex}} = \delta(\mathbf{P}_b - \mathbf{P}_a)(T_d \pm N_0T_{\text{ex}}). \quad (164)$$

The appearance of $T_{\text{ex}}$ here corresponds to the possibility that the incident particle may exchange roles with one of the bound particles. In this case the initially bound particle emerges to be detected as the scattered particle, while the original beam particle is bound in the composite state $\chi_{s'2}$. Because of the identity of the particles, there is no physical means of telling which was the incident particle. The appearance of $N_0$ in Eq. (164) merely reflects the fact that any one of the $N_0$ bound particles can be exchanged for the incident particle.

To evaluate the cross section (121) for this case, let us take the respective momenta for the incident particles to be $\mathbf{q}_1$ and $\mathbf{q}_2$, while those for the final states are $\mathbf{k}_1$ and $\mathbf{k}_2$. We suppose the internal state (other than spin orientation) for particle 2 is unchanged by the scattering. Let the respective angular momenta of the particles be $j_1$ and $j_2$, and let their $Z$-components of angular momentum be $m_1$, $m_2$ and $m'_1$, $m'_2$ for the initial and final states, respectively. Finally, we let $E_a = \epsilon_{q1} + \epsilon_{q2}$ and $E_b = \epsilon_{k1} + \epsilon_{k2}$. Then Eq. (121) may be written as

$$\begin{aligned} d\sigma = \frac{(2\pi)^4}{v_R(2j_1 + 1)(2j_2 + 1)} \sum_{m_1=-j_1}^{j_1} \sum_{m_2=-j_2}^{j_2} \sum_{m'_1=-j_1}^{j_1} \sum_{m'_2=-j_2}^{j_2} \\ \times \int d^3k_1\, d^3k_2\, \delta(\epsilon_{k1} + \epsilon_{k2} - \epsilon_{q1} - \epsilon_{q2}) \\ \times \delta(\mathbf{k}_1 + \mathbf{k}_2 - \mathbf{q}_1 - \mathbf{q}_2)\, |T_d - N_0T_{\text{ex}}|^2. \end{aligned} \quad (165)$$

Here we have taken $\mathbf{v}_1 \times \mathbf{v}_2 = 0$ and $v_R \equiv |\mathbf{v}_1 - \mathbf{v}_2|$. The amplitudes $T_d$ and $T_{\text{ex}}$ are, of course, functions of $m_1$, $m_2$, $m'_1$, $m'_2$, and of *three* of the variables $\mathbf{k}_1$, $\mathbf{k}_2$, $\mathbf{q}_1$, $\mathbf{q}_2$. In Eq. (165) we have averaged over initial and summed over final spin orientations, corresponding to a random distribution of initial spin directions.

**Scattering of an Electron by a Hydrogen Atom**

As an example of the process just discussed, we consider the elastic scattering of an electron by a hydrogen atom; that is,

$$e + \text{H} \rightarrow e + \text{H}. \tag{166}$$

In this case there are two electrons, described by the respective sets of dynamical variables $\xi_1$ and $\xi_2$. The variables $\xi_i$ $(i = 1, 2)$ consist of a space coordinate $\mathbf{x}_i$ and a spin variable $s_i$.

The free electron in the initial state is assumed to have a momentum $\mathbf{q}_1$ and a spin orientation $\nu_1$ $(\nu_1 = \pm\frac{1}{2})$. Its wave function is therefore

$$\chi_{\nu 1,q1}(\xi_1) = (2\pi)^{-3/2}\, e^{i\mathbf{q}_1\cdot\mathbf{x}_1}\, u_{\nu 1}(s_1), \tag{167a}$$

where $u_{\nu 1}$ is the spin eigenvector defined by Eq. (1.69*b*). The second electron is bound in the hydrogen atom. We shall assume that the proton is sufficiently massive compared to the electron that the center-of-mass coordinate of the atom coincides with $\mathbf{r}_2$, the proton coordinate, and that the momentum of the atom is $\mathbf{q}_2$, or that of the proton. It is convenient also to introduce relative coordinates

$$\boldsymbol{\rho}_i = \mathbf{x}_i - \mathbf{r}_2 \qquad (i = 1, 2) \tag{168}$$

for the two electrons. Then the wave function for the hydrogen atom in the initial state is

$$\bar{\chi}_{\nu 2,\mathbf{q}2}(\xi_2) = (2\pi)^{-3/2}\, e^{i\mathbf{q}_2\cdot\mathbf{r}_2}\psi(\rho_2)u_{\nu 2}(s_2). \tag{167b}$$

Here $u_{\nu 2}(s_2)$ is the spin eigenvector for the second electron and

$$\psi(\rho_2) = \frac{1}{\sqrt{\pi a^3}}\, e^{-\rho_2/a} \tag{169}$$

is the ground state hydrogenic wave function, with $a = \hbar^2/me^2$ the Bohr radius. The spin of the proton has been ignored in Eq. (167*b*). This is valid to the extent that its magnetic moment has a negligible effect on the collision dynamics.

The eigenvector for the initial state is then

$$\chi_a = \chi_{\nu 1,\mathbf{q}1}(\xi_1)\bar{\chi}_{\nu 2,\mathbf{q}2}(\xi_2). \tag{167c}$$

A final state eigenvector for elastic scattering is

$$\chi_b = \chi_{v'1,\mathbf{k}1}(\xi_1)\bar{\chi}_{v'2,\mathbf{k}2}(\xi_2), \tag{170a}$$

while that with the two electrons exchanged is

$$\chi_{Q_{12}(b)} = \chi_{v'1,\mathbf{k}1}(\xi_2)\bar{\chi}_{v'2,\mathbf{k}2}(\xi_1). \tag{170b}$$

The kinetic energy of the hydrogen atom will be neglected, consistent with the relatively large mass of the proton. Then the respective energies $E_a$ and $E_b$ for the initial and final states are

$$E_a = \frac{q_1^{\,2}}{2m} + W_0 = E_b = \frac{k_1^{\,2}}{2m} + W_0, \tag{171a}$$

where $m$ is the electron mass and

$$W_0 = -\frac{e^2}{2a} \tag{171b}$$

is the binding energy of the electron in the atom. Again, because of the large mass of the proton, we shall ignore the difference between the barycentric and laboratory coordinate systems. To write down the Hamiltonian, we define

$$K_{ei} = -\frac{1}{2m}\nabla_i^{\,2} \qquad (i = 1, 2) \tag{172}$$

as the kinetic energy operator for electron $i$. The interaction energy of either electron with the proton is (the charge on the electron is written as $-e$)

$$V_i = -\frac{e^2}{\rho_i} \qquad (i = 1, 2) \tag{173}$$

while that of the two electrons is

$$V_{12} = \frac{e^2}{|\boldsymbol{\rho}_1 - \boldsymbol{\rho}_2|}. \tag{174}$$

The Hamiltonian $K$, for which $\chi_a$ and $\chi_b$ are eigenvectors, is

$$K = K_{e1} + K_{e2} + V_2, \tag{175}$$

so $K\chi_a = E_a\chi_a$, and on the energy shell $K\chi_b = E_a\chi_b$. The interaction $V$ is

$$V = H - K = V_1 + V_{12}. \tag{176}$$

Corresponding to Eqs. (127), we have

$$\begin{aligned} K_{Q_{12}} &= K_{e1} + K_{e2} + V_1, \\ V_{Q_{12}} &= V_2 + V_{12}. \end{aligned} \tag{177}$$

To obtain the scattering cross section, we must evaluate [see Eqs. (163) and (132$b$)]

$$\begin{aligned}\mathscr{T}_d &= \left(\chi_b, \left(V + V\frac{1}{E_a + i\eta - H}V\right)\chi_a\right),\\ \mathscr{T}_{\text{ex}} &= \left(\chi_{Q_{12}(b)}, \left(V + V_{Q_{12}}\frac{1}{E_a + i\eta - H}V\right)\chi_a\right).\end{aligned} \tag{178}$$

We shall here discuss the evaluation of these expressions only for "high energies," such that $q_1^2/2m \gg |W_0|$. In this case we may consider the scattering interaction to be "weak" and neglect successive scatterings. Then $V(E_a + i\eta - H)^{-1}V$ may be neglected and

$$\mathscr{T}_d \approx (\chi_b, V\chi_a) = (\chi_b, (V_1 + V_{12})\chi_a), \tag{179}$$

which is valid when $e^2/\hbar v_e \ll 1$, where $v_e = q_1/m$ is the velocity of the incident electron. The validity of Eq. (179), which is called the first Born approximation, will be discussed in detail in Section 6.7. It corresponds to a single impulsive interaction of the incident electron with the atom.

A corresponding approximation for the exchange amplitude $\mathscr{T}_{\text{ex}}$ must be made with some care. This is an example of a so-called rearrangement scattering matrix. Let us first show that $\mathscr{T}_{\text{ex}} = 0$ if the electron-electron interaction $V_{12}$ is neglected. If we set $V_{12} = 0$ in the second of Eqs. (178), there results

$$T_{\text{ex}}^0 = \left(\chi_{Q_{12}(b)}, \left(V_1 + V_2\frac{1}{E_a + i\eta - H_0}V_1\right)\chi_a\right), \tag{180}$$

where

$$H_0 \equiv (K_{e1} + V_1) + (K_{e2} + V_2). \tag{181}$$

Now

$$\begin{aligned}&\left(1 + V_2\frac{1}{E_a + i\eta - H_0}\right)V_1\\ &\qquad = \left[(E_a + i\eta - H_0)\frac{1}{E_a + i\eta - H_0} + V_2\frac{1}{E_a + i\eta - H_0}\right]V_1\\ &\qquad = [E_a + i\eta - (H_0 - V_2)]\frac{1}{E_a + i\eta - H_0}V_1.\end{aligned}$$

On the energy shell, however, $(H_0 - V_2)\chi_{Q_{12}(b)} = E_a\chi_{Q_{12}(b)}$, so Eq. (180) simplifies to

$$\mathscr{T}_{\text{ex}}^0 = \left(\chi_{Q_{12}(b)}, \left(\frac{i\eta}{E_a + i\eta - H_0}\right)V_1\chi_a\right). \tag{182}$$

Since $K_{e1} + V_1$ and $K_{e2} + V_2$ commute, we have

$$\frac{1}{E_a + i\eta - H_0} V_1\chi_a = \frac{1}{E_a + i\eta - H_0} \bar{\chi}_{v2,q2}(\xi_2)V_1\chi_{v1,q1}(\xi_1)$$
$$= \frac{1}{E_a + i\eta - W_0 - (K_{e1} + V_1)} V_1\chi_a. \qquad (183)$$

In addition, when operating to the left in Eq. (182) we have

$$(K_{e1} + V_1)\chi_{Q_{12}(b)} = W_0\chi_{Q_{12}(b)}.$$

Therefore,

$$\mathscr{T}_{\text{ex}}{}^0 = \frac{i\eta}{E_a + i\eta - 2W_0}(\chi_{Q_{12}(b)}, V_1\chi_a). \qquad (184)$$

The quantity $E_a - 2W_0 = q_1{}^2/2m - W_0 = q_1{}^2/2m + |W_0| > 0$, so[37]

$$\mathscr{T}_{\text{ex}}{}^0 = \lim_{\eta\to 0(+)} i\eta\left[\frac{1}{E_a - 2W_0}(\chi_{Q_{12}}, V_1\chi_a)\right] = 0, \qquad (185)$$

$(\chi_{Q_{12}(b)}, V_1\chi_a)$ being independent of $\eta$.

The result expressed by Eq. (185) is of considerable importance for the evaluation of "rearrangement" scattering matrices. Failure to take appropriate account of this when it appears in various guises has led to many inconsistent approximations in scattering theory.

Equation (178) for $\mathscr{T}_{\text{ex}}$ may now be written as

$$\mathscr{T}_{\text{ex}} = \mathscr{T}_{\text{ex}}{}^0 + (\chi_{Q_{12}(b)}, V_{12}\chi_a)$$

plus terms involving multiple scattering interactions. We have just seen that $\mathscr{T}_{\text{ex}}{}^0$ vanishes. The terms involving multiple interactions are of the order of those dropped in Eq. (179), so a consistent approximation for $\mathscr{T}_{\text{ex}}$ is

$$\mathscr{T}_{\text{ex}} \approx (\chi_{Q_{12}(b)}, V_{12}\chi_a). \qquad (186)$$

In Chapter 11 we shall describe more powerful methods for evaluating Eqs. (178).

On inserting the wave functions (167*c*) and (170*a*) into Eq. (179), we obtain

$$\mathscr{T}_d = (u_{v'1}, u_{v1})(u_{v'2}, u_{v2})(2\pi)^{-6}\int d^3x_1\, d^3x_2\, d^3r_2$$
$$\times\; e^{i(\mathbf{q}_1-\mathbf{k}_1)\cdot\mathbf{x}_1}e^{i(\mathbf{q}_2-\mathbf{k}_2)\cdot\mathbf{r}_2}\left(\frac{1}{\pi a^3}e^{-(2/a)\rho_2}\right)\left(-\frac{e^2}{\rho_1} + \frac{e^2}{|\boldsymbol{\rho}_1 - \boldsymbol{\rho}_2|}\right). \qquad (187)$$

[37] Again, let us recall our convention for using a single symbol for $\mathscr{T}_{\text{ex}}{}^0(\eta)$ and $\mathscr{T}_{\text{ex}}{}^0 = \lim_{\eta\to 0(+)} \mathscr{T}_{\text{ex}}{}^0(\eta)$.

An elementary evaluation of this expression leads to

$$\mathscr{T}_d = \delta_{\nu'1,\nu 1}\,\delta_{\nu'2,\nu 2}\,\delta(\mathbf{k}_1 + \mathbf{k}_2 - \mathbf{q}_1 - \mathbf{q}_2) T_d(a; \kappa), \tag{188a}$$

where

$$T_d(a, \kappa) = (2\pi)^{-3} \int d^3\rho_1 \left[ -\frac{e^2}{\rho_1} + e^2 \int \frac{n_e(\rho_2)\, d^3\rho_2}{|\boldsymbol{\rho}_1 - \boldsymbol{\rho}_2|} \right] e^{-i\boldsymbol{\kappa}\cdot\rho_1} \tag{188b}$$

and

$$\boldsymbol{\kappa} \equiv \mathbf{k}_1 - \mathbf{q}_1 \tag{189}$$

is the *momentum transfer* between the electron and the atom. The quantity $n_e$ in Eq. (188*b*) is the electron charge density in the atom, being defined by

$$n_e(\rho_2) = |\psi(\rho_2)|^2.$$

Thus

$$\begin{aligned} V_{(\rho_1)} &\equiv -\frac{e^2}{\rho_1} + e^2 \int \frac{n_e(\rho_2)\, d^3\rho_2}{|\boldsymbol{\rho}_1 - \boldsymbol{\rho}_2|} \\ &= -e^2\left(\frac{1}{\rho_1} + \frac{1}{a}\right) e^{-2(\rho_1/a)}, \end{aligned} \tag{190}$$

which appears in (188*b*), represents the potential associated with the *static* charge density in the atom.[38] The remaining integrations in Eq. (188*b*) are easily performed to give

$$T_d(a; \kappa) = (2\pi)^{-3} \frac{4\pi e^2}{\kappa^2}\left[ -1 + \frac{1}{(1 + a^2\kappa^2/4)^2} \right]. \tag{188c}$$

With the use of the wave function (170*b*) in Eq. (186), we find

$$\mathscr{T}_{ex} = \delta_{\nu'2,\nu 1}\,\delta_{\nu'1,\nu 2}\,\delta(\mathbf{k}_1 + \mathbf{k}_2 - \mathbf{q}_1 - \mathbf{q}_2) T_{ex}(a; \kappa, q_1), \tag{191a}$$

where

$$T_{ex}(a; \kappa, q_1) = (2\pi)^{-3}(\pi a^3)^{-1} \int d^3\rho_1\, d^3\rho_2\, e^{-i\mathbf{k}_1\cdot\rho_2}\, e^{i\mathbf{q}_1\cdot\rho_1} \times e^{-(1/a)(\rho_1+\rho_2)} \frac{e^2}{|\boldsymbol{\rho}_1 - \boldsymbol{\rho}_2|}. \tag{191b}$$

In the high-energy limit that $q_1 \gg 1/a$ [the limit for which Eqs. (179) and (186) are valid] an elementary evaluation gives

$$T_{ex}(a; \kappa, q_1) = (2\pi)^{-3} \frac{4\pi e^2}{{q_1}^2} \frac{1}{(1 + a^2\kappa^2/4)^2}. \tag{191c}$$

[38] As will be seen in Section 11.7, corrections to Eqs. (179) and (186) involve terms which arise because of the *distortion*, or *polarization*, of the atom due to the field of the incident electron.

According to the general Eq. (164), the complete symmetrized amplitude is

$$\langle \mathscr{S}(b)|\, \mathscr{T}\, |\mathscr{S}(a)\rangle = \mathscr{T}_d - \mathscr{T}_{\text{ex}} = \delta(\mathbf{k}_1 + \mathbf{k}_2 - \mathbf{q}_1 - \mathbf{q}_2)T, \quad (192a)$$

where

$$T = \delta_{\nu'1,\nu1}\, \delta_{\nu'2,\nu2} T_d - \delta_{\nu'2,\nu1}\, \delta_{\nu'1,\nu2} T_{\text{ex}}. \quad (192b)$$

To obtain the differential cross sections, we must enumerate the possible combinations of spin orientations.

When the electron spins are parallel, we have $\nu'_1 = \nu'_2 = \nu_1 = \nu_2$ in (192*b*). Then Eqs. (188*c*) and (191*c*) give an amplitude

$$T(\uparrow\uparrow \to \uparrow\uparrow) = (2\pi)^{-3} \frac{4\pi e^2}{\kappa^2}\left[-1 + \frac{1}{(1 + a^2\kappa^2/4)^2}\left(1 - \frac{\kappa^2}{q_1^{\,2}}\right)\right]. \quad (193)$$

When the two electrons have antiparallel spins in the initial state, we must distinguish between two possible final states. First, let the incident and scattered free electron have a fixed spin orientation, so $\nu'_1 = \nu_1$, $\nu'_2 = \nu_2$, but $\nu_1 \neq \nu_2$. Then according to Eq. (192*b*) $T_{\text{ex}}$ gives no contribution and we find an amplitude

$$T(\uparrow\downarrow \to \uparrow\downarrow) = (2\pi)^{-3} \frac{4\pi e^2}{\kappa^2}\left[-1 + \frac{1}{(1 + a^2\kappa^2/4)^2}\right]. \quad (194)$$

Finally, we take $\nu'_2 = \nu_1$, $\nu'_1 = \nu_2$, but $\nu_1 \neq \nu_2$. In this case, the scattered electron has its spin "flipped" by the scattering. The amplitude now has a contribution only from $T_{\text{ex}}$ and is

$$T(\uparrow\downarrow \to \downarrow\uparrow) = -(2\pi)^{-3} \frac{4\pi e^2}{q_1^{\,2}} \frac{1}{(1 + a^2\kappa^2/4)^2}. \quad (195)$$

In giving an interpretation to Eqs. (194) and (195) it is well to recall the discussion associated with Eqs. (68).

Equation (149) provides us with the correct expression for the differential scattering cross section, where our matrices $T$ as given by Eqs. (193), (194), and (195) correspond to the $T_{\mathscr{S}(ba)}$ of Eq. (148). For our present nonrelativistic evaluation, we set $\epsilon_{s'1,k1} = m$ and $\epsilon_{s'2,k2} = \infty$, while $v_{\text{rel},B} = v_e$, the incident electron velocity. Then Eq. (149) simplifies to

$$\frac{d\sigma}{d\Omega} = (2\pi)^4 m^2\, |T|^2. \quad (196)$$

The three cross sections corresponding to Eqs. (193), (194), and (195) are respectively

$$\frac{d\sigma}{d\Omega}(\uparrow\uparrow \to \uparrow\uparrow) = \frac{4e^4m^2}{\kappa^4}\left\{1 - \frac{1}{(1 + a^2\kappa^2/4)^2}\left[1 - \left(\frac{\kappa}{q_1}\right)^2\right]\right\}^2, \quad (197a)$$

$$\frac{d\sigma}{d\Omega}(\uparrow\downarrow \to \uparrow\downarrow) = \frac{4e^4m^2}{\kappa^4}\left[1 - \frac{1}{(1 + a^2\kappa^2/4)^2}\right]^2, \quad (197b)$$

and

$$\frac{d\sigma}{d\Omega}(\uparrow\downarrow \rightarrow \downarrow\uparrow) = \frac{4e^4m^2}{q_1^4} \frac{1}{(1 + a^2\kappa^2/4)^4}. \tag{197c}$$

These are usually considered to be functions of the incident electron energy $q_1^2/2m$ and the scattering angle $\theta$, which is the angle between $\mathbf{k}_1$ and $\mathbf{q}_1$ [so $\kappa^2 = 2q_1^2(1 - \cos\theta)$].

Each of the three differential cross sections (197*a*, *b*, *c*) can be directly measured experimentally. The simplest to measure is, however, that for a random mixture of initial spin states and all final spin states. This is

$$\begin{aligned}\frac{d\bar{\sigma}}{d\Omega} &= \frac{1}{2}\left[\frac{d\sigma}{d\Omega}(\uparrow\uparrow \rightarrow \uparrow\uparrow) + \frac{d\sigma}{d\Omega}(\uparrow\downarrow \rightarrow \uparrow\downarrow) + \frac{d\sigma}{d\Omega}(\uparrow\downarrow \rightarrow \downarrow\uparrow)\right] \\ &= \frac{4e^4m^2}{\kappa^4}\left[1 + \frac{1}{(1 + a^2\kappa^2/4)^4}\left(1 - \frac{\kappa^2}{q_1^2} + \frac{\kappa^4}{q_1^4}\right)\right. \\ &\qquad \left. - \frac{2}{(1 + a^2\kappa^2/4)^2}\left(1 - \frac{1}{2}\frac{\kappa^2}{q_1^2}\right)\right]. \end{aligned} \tag{198}$$

The approximation of high energies by which we were led to the cross sections (197) and (198) is expected to be reasonably accurate for electron energies of about 100 ev or higher. The three cross sections (197) are illustrated in Fig. 4.2 for an electron energy of 110 ev. In comparing these, it is worth noting that (197*b*) arises from $T_d$ only, whereas (197*c*) depends on $T_{ex}$ only.

The total cross section, integrated over all scattering angles, is obtained as usual from the expression

$$\sigma = \int \frac{d\sigma}{d\Omega}\, d\Omega, \tag{199}$$

where $d\sigma/d\Omega$ is any of the differential cross sections of Eqs. (197) or (198). The factor of $(2!)^{-1}$, which appears in Eq. (122), is not required since we are not summing over all final states here. That is, we are including only the ground state of the atom in our sum.

We shall not, at this point, describe the phenomena associated with the process (166) in more detail. This will be done in Chapter 11, where a comprehensive discussion will be given.[39]

[39] For a comparison of the more elementary theories and a discussion of experimental techniques, see Massey and Burhop or Brackmann, Fite, and Neynaber, Refs. [8].

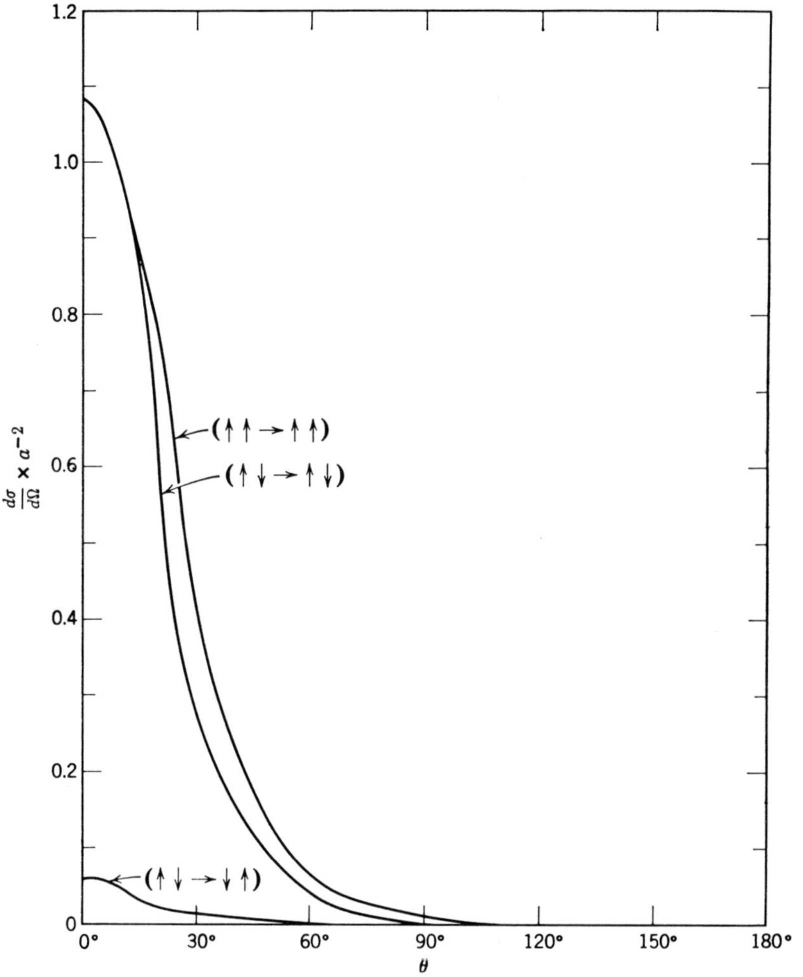


Fig. 4.2 Cross sections for electron-hydrogen scattering, as given by Eqs. (197), in units of the square of the Bohr radius and as a function of the angle $\theta$ through which the electron is deflected.

**Scattering of an Electron by a Helium Atom**

As a slightly more complicated example, we next consider the elastic scattering of electrons by He atoms. This is the process

$$e + \mathrm{He} \rightarrow e + \mathrm{He}. \tag{200}$$

The wave function for the incident electron is again taken as given by

Eq. (167$a$). To avoid an elaborate discussion, we shall use hydrogenic orbital states for the two electrons in the He atom.[40] Then Eq. (167$b$) is replaced by

$$\bar{\chi}_{\mathbf{q}_2}(\xi_2, \xi_3) = (2\pi)^{-3/2} e^{i\mathbf{q}_2 \cdot \mathbf{r}_2} \psi_2(\rho_2)\psi_2(\rho_3)$$

$$\times \frac{1}{\sqrt{2}} [u_{1/2}(s_2)u_{-1/2}(s_3) - u_{1/2}(s_3)u_{-1/2}(s_2)]. \tag{201}$$

Now $\mathbf{r}_2$ is the space coordinate of the He nucleus and $\psi_2(\rho)$ is obtained from Eq. (169) by replacing $a$ by $a/2$ in that equation. The additional variable set $\xi_3$ has been introduced to describe the third electron. (The notation is otherwise that of the previous discussion.) The respective initial and final states are now

$$\chi_a = \chi_{v1,\mathbf{q}1}(\xi_1)\bar{\chi}_{\mathbf{q}2}(\xi_2, \xi_3),$$
$$\chi_b = \chi_{v'1,\mathbf{k}1}(\xi_1)\bar{\chi}_{\mathbf{k}2}(\xi_2, \xi_3), \tag{202}$$

whereas

$$\chi_{Q_{12}(b)} = \chi_{v'1,\mathbf{k}1}(\xi_2)\bar{\chi}_{\mathbf{k}2}(\xi_1, \xi_3). \tag{203}$$

On the following the arguments, which led to the approximate Eqs. (179) and (186), we obtain now

$$\mathscr{T}_d = (\chi_b, (V_1 + V_{12} + V_{13})\chi_a),$$
$$\mathscr{T}_{\text{ex}} = (\chi_{Q_{12}(b)}, V_{12}\chi_a). \tag{204}$$

Here

$$V_1 = -\frac{2e^2}{\rho_1}, \qquad V_{13} = \frac{e^2}{|\boldsymbol{\rho}_1 - \boldsymbol{\rho}_3|},$$

and so on, in the notation of Eqs. (172) and (173).

Evaluation of Eqs. (204), when the incident electron has "high energy" as defined before, leads to

$$\mathscr{T}_d = \delta_{v'1v1}\delta(\mathbf{k}_1 + \mathbf{k}_2 - \mathbf{q}_1 - \mathbf{q}_2)\left[2T_d\left(\frac{a}{2}\,;\kappa\right)\right],$$
$$T_{\text{ex}} = \delta_{v'1v1}\delta(\mathbf{k}_1 + \mathbf{k}_2 - \mathbf{q}_1 - \mathbf{q}_2)\left[\tfrac{1}{2}T_{\text{ex}}\left(\frac{a}{2}\,;\kappa, q_1\right)\right], \tag{205}$$

where $T_d$ and $T_{\text{ex}}$ are defined by Eqs. (188$c$) and (191$c$), respectively.

[40] See, for example, Condon and Shortley [9] for means of obtaining more accurate wave functions.

According to Eq. (164), the symmetrized scattering matrix is

$$\langle \mathscr{S}(b)| \mathscr{T} |\mathscr{S}(a)\rangle = \mathscr{T}_d - 2\mathscr{T}_{\text{ex}}. \tag{206}$$

The cross section is nonvanishing only if the spin of the free electron is unchanged. This is

$$\frac{d\sigma}{d\Omega} = \frac{16e^2m^2}{\kappa^4}\left[1 - \frac{1}{(1 + a^2\kappa^2/16)^2}\left(1 - \frac{1}{4}\frac{\kappa^2}{q_1^{\,2}}\right)\right]^2. \tag{207}$$

**Scattering of Two Composite Particles**

As a final example, let us consider the wave function (101) as describing the initial state for a scattering. The two incident particles are composite, the first containing $n$ and the second $N - n$ bound, identical particles. The final state $\chi_b$ is considered to belong to the same channel as the incident state, so $\chi_b$ also has the form given by Eq. (101). Using Eq. (144),

$$\begin{aligned}\langle \mathscr{S}(b)| \mathscr{T} |\mathscr{S}(a)\rangle &= \frac{C^2}{N!}\sum_Q \delta_Q \langle Q(b)| \mathscr{T} |a\rangle \\ &= \frac{1}{n!\,[(N-n)!]}\sum_Q \delta_Q\langle Q(b)| \mathscr{T} |a\rangle, \end{aligned}\tag{208}$$

where the value (103) has been substituted for $C$. This expression may be simplified in a manner similar to that done for Eq. (164).

First, $n!\,[(N - n)!]$ permutations $Q$ do not exchange elementary particles between the two bound particles. Let us write $\mathscr{T}_d = \delta(\mathbf{P}_b - \mathbf{P}_a)T_d$ for these terms. There are

$$n!\,[(N - n)!]\, n(N - n)$$

permutations for which one elementary particle is exchanged between the bound systems. For these, we set

$$\delta_Q\langle Q(P)| \mathscr{T} |a\rangle = \pm\mathscr{T}_{\text{ex}}(1) = \pm\delta(\mathbf{P}_b - \mathbf{P}_a)T_{\text{ex}}(1), \tag{209}$$

where the plus or minus sign refers to Bose-Einstein or Fermi-Dirac statistics. To continue in this manner, the number of permutations which exchange $l$ elementary particles between the bound systems is

$$[n!][(N - n)!]\frac{n!}{l!\,(n - l)!}\frac{(N - n)!}{l!\,(N - n - l)!}.$$

The corresponding scattering matrices are written as

$$\delta_Q\langle Q(P)| \mathscr{T} |a\rangle = (\pm 1)^l\mathscr{T}_{\text{ex}}(l) = (\pm 1)^l\delta(\mathbf{P}_b - \mathbf{P}_a)T_{\text{ex}}(l). \tag{210}$$

Thus[41] if we choose $n \leq N - n$,

$$\langle \mathscr{S}(b)| \mathscr{T} |\mathscr{S}(a)\rangle = \delta(\mathbf{P}_b - \mathbf{P}_a)$$
$$\times \left[ T_d + \sum_{l=1}^{n} (\pm 1)^l \frac{n!}{l!\,(n-l)!} \frac{(N-n)!}{l!\,(N-n-l)!} T_{\text{ex}}(l) \right]. \quad (211)$$

Under most circumstances it is expected that the $T_{\text{ex}}(l)$ become negligibly small as $l$ becomes large compared to unity. Thus, ordinarily only a few terms need be kept in Eq. (211) even for systems of many bound particles. It may very well happen that $T_{\text{ex}}(l)$ is negligible for all $l = 1, 2, \ldots, n$. In this case the symmetrization of the wave function with respect to exchange of elementary particles between the composite particles may be neglected. When $n = N/2$, the two composite particles are identical. Then $T_{\text{ex}}(N/2)$ corresponds to interchange of the composite particles. When only $T_d$ and $T_{\text{ex}}(N/2)$ need be kept in Eq. (211), we might have ignored the $N$ elementary particles and just have symmetrized the wave function with respect to exchange of the composite particles.

Examples to which Eq. (211) may be applied include the scattering of two atomic nuclei, two atoms, or two molecules.

## 4.6 FURTHER DESCRIPTION AND SIMPLIFICATION OF THE MATRIX ELEMENTS OF $\mathscr{T}$

Having obtained the expressions (132*b*), (140*b*), (141), and (121), we have completed our formal description of the evaluation of the transition probability or scattering cross section. In the next chapter this development will be repeated in a much more condensed form, and a number of useful formal transformations on $\mathscr{T}$ and the $\psi^{\pm}$ will be given. In this section we shall conclude Chapter 4 with a discussion of some of the more simple properties of the scattering matrix.

### Further Discussion of Notation

First it is useful to add channel labels to the initial and final states, in the spirit of Section 4.1. As before, we call the initial channel $c$, so the

[41] As a check on our counting, consider the contour integral that circles the origin once.

$$N! = (N-n)!\,n!\,(2\pi i)^{-1} \oint (1+Z)^N Z^{-N+n-1}\,dZ$$
$$= (N-n)!\,n!\,(2\pi i)^{-1} \oint (1+Z)^n \left(1 + \frac{1}{Z}\right)^{N-n} \frac{dZ}{Z}$$
$$= (n!)^2[(N-n)!]^2 \sum_{l=0}^{n} [l!\,(n-l)!\,l!\,(N-n-l)!]^{-1},$$

on using the binomial theorem.

wave function (111) for the incident state is written as

$$|c;\mathscr{S}(a)\rangle = \chi_{c;\mathscr{S}(a)} = C_c\mathscr{S}\chi_{c;a}. \tag{212}$$

Here $C_c$ is the normalization constant [Eq. (79*b*)] that normalizes Eq. (212) for an asymptotic wave packet state. As $c$ runs over all channels ($c = 1, 2, \ldots, C_h$), and as $a$ runs over all quantum mechanical states corresponding to the channel $c$, we obtain a complete description of the possible initial states which can be prepared for a scattering experiment on the system being considered. When it is used to construct wave packets, a complete orthonormal set of initial states is obtained, as described in Section 4.1 and Appendix C.

As we have seen in Section 4.1, the wave functions (212) also represent the final states into which the system can go. Then we may choose a given channel, say $c'$, and a state $b$ representing a state of the system when in the channel $c'$, to obtain a final state

$$|c';\mathscr{S}(b)\rangle = \chi_{c';\mathscr{S}(b)} = C_{c'}\mathscr{S}\chi_{c';b}. \tag{213}$$

Channel labels may evidently be added (when it is convenient) to the equations of Sections 4.3 and 4.4. For example, Eq. (132*b*) may be written as

$$\langle c';Q(b)|\mathscr{T}|c;P(a)\rangle$$
$$= \left(\chi_{c';Q(b)}, \left(V_{P(c)} + V_{Q(c')}\frac{1}{E_a + i\eta - H}V_{P(c)}\right)\chi_{c;P(a)}\right), \tag{214a}$$

where $V_{P(c)} = PV_cP^{-1}$ and $V_{Q(c')} = QV_{c'}Q^{-1}$. In contrast to the general definition (132*b*), Eq. (214*a*) is defined only for channel states.[42] The corresponding scattering matrix $\mathscr{T}^{(-)}$ is

$$\langle c';Q(b)|\mathscr{T}|c;P(a)\rangle$$
$$= \left(\chi_{c';Q(b)}, \left(V_{P(c)} + V_{Q(c')}\frac{1}{E_a - i\eta - H}V_{P(c)}\right)\chi_{c;P(a)}\right). \tag{215a}$$

The symmetrized scattering matrix of Eq. (140*b*) is now written as

$$\langle c';\mathscr{S}(b)|\mathscr{T}|c;\mathscr{S}(a)\rangle$$
$$= \frac{C_{c'}C_c}{\left[\prod_{j=1}^{L_{c'}}(N_{c',j})!\right]\left[\prod_{i=1}^{L_c}(N_{c,i})!\right]}\sum_Q \delta_Q \sum_P \delta_P\langle c';Q(b)|\mathscr{T}|c;P(a)\rangle, \tag{216}$$

[42] That is, the expression (214*a*) is defined only for initial and final states $\chi_{c;a}$, and so on, but not in general for the states $\chi_a{}^c$, and so on. [See the discussion accompanying Eqs. (6).]

with a corresponding equation for the symmetrized $\mathscr{T}^{(-)}$ in which the expression (215*a*), rather than (214*a*), appears. Similarly, we may write the *S*-matrix [Eq. (140*a*)] as

$$\langle c'; \mathscr{S}(b)| S |c; \mathscr{S}(a)\rangle$$
$$= \delta_{(c';b),c;(a)} - 2\pi i\delta(E_b - E_a)\langle c'; \mathscr{S}(b)| \mathscr{T} |c; \mathscr{S}(a)\rangle. \quad (217)$$

This is evidently nonvanishing only for states $(c'; b)$ and $(c; a)$ on the same energy shell, defined by $E_b = E_a$.

For most applications the scattering matrices $\mathscr{T}$ are needed only on the energy shell, even though our definition (140*b*) is quite general. *On the energy shell* we may use the alternate forms of Eqs. (12) and (13) for the matrix elements of $\mathscr{T}$.

$$\langle c'; Q(b)| \mathscr{T} |c; P(a)\rangle = (\psi^-_{c';Q(b)}, V_{P(c)}\chi_{c;P(a)})$$
$$= (\chi_{c';Q(b)}, V_{Q(c')}\psi^+_{c;P(a)}) \quad (214b)$$

and

$$\langle c; Q(b)| \mathscr{T}^{(-)} |c; P(a)\rangle = (\psi^+_{c';Q(b)}, V_{P(c)}\chi_{c;P(a)})$$
$$= (\chi_{c';Q(b)}, V_{Q(c')}\psi^-_{c;P(a)}), \quad (215b)$$

where the notation of Eq. (130) is used.

Finally, we recall from the discussion in Section 4.1 and will see in Appendix C that $S$ and $\mathscr{T}$ are in general linear operators only in the restricted sense of operating on the asymptotic wave packet states.

### Use of Constants of the Motion to Simplify the Scattering Matrix

The matrices $S$ and $\mathscr{T}$ can in general be greatly simplified by the use of integrals of the motion. Such integrals are most frequently associated with obvious symmetries of the Schrödinger equation. Consider, for example, an operator $\mathscr{D}$ that commutes with the Hamiltonian $H$ and is therefore a constant of motion. It is ordinarily possible and convenient to arrange that $\mathscr{D}$ also commute with $K_c$ $(c = 1, 2, \ldots, C_h)$ as well as with $QK_cQ^{-1}$, where $Q$ is any permutation among the identical elementary particles comprising the system. Then the state $|c; \mathscr{S}(a)\rangle$ may be chosen as an eigenstate, or sum of eigenstates, of $\mathscr{D}$. For example, let us suppose that

$$\mathscr{D}\chi_{c;\mathscr{S}(a)} = D'\chi_{c;\mathscr{S}(a)}, \quad (218)$$

where $D'$ is an eigenvalue of $\mathscr{D}$, for all states $\chi$ in the incident wave packet state $\phi_{\mathscr{S}(a)}(t)$. Then the transition amplitude

$$(\chi_{c';\mathscr{S}(b)}, e^{iE_bt}\Psi(t)) = 0, \quad (219)$$

unless $\chi_{c';\mathscr{S}(b)}$ is also an eigenstate of $\mathscr{D}$ with eigenvalue $D'$.[43] That is,

$$[\mathscr{D}, S] = 0 \tag{220a}$$

$$[\mathscr{D}, \mathscr{T}] = 0. \tag{220b}$$

Therefore, if $D'$ and $D''$ are two eigenvalues of $\mathscr{D}$,

$$\begin{aligned} \langle D''| [\mathscr{D}, \mathscr{T}] |D'\rangle &= (D'' - D')\langle D''| \mathscr{T} |D'\rangle = 0, \\ \langle D''| \mathscr{T} |D'\rangle &= \delta_{D'', D'}\mathscr{T}(D'). \end{aligned} \tag{221}$$

We have already applied this result when $\mathscr{D} = \boldsymbol{\mathscr{P}}$, the operator of total momentum, to obtain

$$\langle c'; \mathscr{S}(b)| \mathscr{T} |c; \mathscr{S}(a)\rangle = \delta(\mathbf{P}_b - \mathbf{P}_a)T_{\mathscr{S}[(c';b)(c;a)]}(\mathbf{P}_a). \tag{222}$$

In Chapter 2 we noted that invariance under rotations implies that the operator $\mathbf{J}$ of total angular momentum will commute with $H$ and $K_c$ (and it also commutes with $\mathscr{S}$). Therefore

$$[\mathbf{J}, S] = 0, \tag{223a}$$

$$[\mathbf{J}, \mathscr{T}] = 0.$$

The angular momentum operator does not in general commute with $\boldsymbol{\mathscr{P}}$, however, so Eq. (222) cannot be further diagonalized in terms of the eigenstates of $J^2$ and $J_Z$ (for example). To see this, we write $\mathbf{J}$ for a system of $\nu$ particles as

$$\mathbf{J} = \sum_{i=1}^{\nu} (\mathbf{r}_i \times \mathbf{p}_i + \mathbf{S}_i), \tag{224}$$

where $\mathbf{r}_i$ and $\mathbf{p}_i$ are the coordinate and momentum operators for the $i$th particle and $\mathbf{S}_i$ is its spin operator.[44] On setting $\boldsymbol{\mathscr{P}} = \Sigma_{i=1}^{\nu} \mathbf{p}_i$, we obtain

$$[\boldsymbol{\mathscr{P}}, \mathbf{J}] = -iI \times \boldsymbol{\mathscr{P}}, \tag{223b}$$

where the $I$ is the unit dyadic.

In the barycentric coordinate system the eigenvalue of $\mathscr{P}$ is zero, and the commutator (223$b$) vanishes. That is, the states of a physical system $|S\rangle$, for which $\mathscr{P}|S\rangle = 0$, define a subspace of the entire space in which $\mathscr{P}$ and $\mathbf{J}$ commute. Within this subspace we may diagonalize $\mathscr{P}$, $J^2$, and $J_Z$ simultaneously. Generality is not lost by restricting the calculation to the barycentric system, however. Equations (3.110) and (3.112) permit us to transform the scattering matrix to a general coordinate system (corresponding to states for which $\mathscr{P}\,|S\rangle \neq 0$), having once evaluated it in the

[43] This follows from the argument given on p. 5, Chapter 1.

[44] When particle $i$ is composite, $\mathbf{S}_i$ is the total angular momentum operator for its internal state.

barycentric system. This is one of several reasons why the theoretical studies of reactions are almost always made in the barycentric coordinate system.

Let us now restrict ourselves to the barycentric system for the initial state, so $\mathbf{P}_a = 0$ in Eq. (222). In this coordinate system the eigenvalues of $J^2$ and $J_z$ will be written as $j(j+1)$ and $m$, respectively [compare Eq. (1.55)]. We may assume that the states $|c; \mathscr{S}(a)\rangle$ are sums of eigenstates of $J^2$ and $J_Z$, so

$$|c; \mathscr{S}(a)\rangle = \sum_{j_a, m_a} B_{j_a}{}^{m_a} \, |c; \mathscr{S}(a, j_a, m_a)\rangle, \tag{225a}$$

where the $B_{j_a}{}^{m_a}$ are numerical coefficients and

$$\begin{aligned} J^2 \, |c; \mathscr{S}(a, j_a, m_a)\rangle &= j_a(j_a+1) |\, c; \mathscr{S}(a, j_a, m_a)\rangle, \\ J_Z \, |c; \mathscr{S}(a, j_a, m_a)\rangle &= m_a \, |c; \mathscr{S}(a, j_a, m_a)\rangle. \end{aligned} \tag{225b}$$

Applying the condition (221) we obtain

$$\langle c'; \mathscr{S}(b, j_b, m_b)| \, \mathscr{T} \, |c; \mathscr{S}(a, j_a, m_a)\rangle$$
$$= \delta(\mathbf{P}_b)\delta_{j_b, j_a}\delta_{m_b, m_a} T_{\mathscr{S}[(c';b), (c;a)]}(0, j_a, m_a). \tag{226a}$$

We can show that $T_{\mathscr{S}}(0, j_a, m_a)$ is independent of $m_a$— an important conclusion in scattering theory. The only nonvanishing matrix elements of $J^{\pm} = J_x \pm iJ_y$ [compare Eq. (1.63)] are of the form $\langle m_a \pm 1|\, J^{\pm} \,|m_a\rangle$. Now because of Eqs. (223*a*) and (226*a*)

$$\langle m_a \pm 1|\, [J^{\pm}, \mathscr{T}] \,|m_a\rangle = \langle m_a \pm 1|\, J^{\pm} \,|m_a\rangle [\mathscr{T}(m_a) - \mathscr{T}(m_a \pm 1)] = 0.$$

[Here $\mathscr{T}(m_a)$ is the submatrix of $\mathscr{T}$ corresponding to states for which $J_Z$ has the eigenvalue $m_a$.] Consequently $\mathscr{T}(m_a) = \mathscr{T}(m_a \pm 1)$,[45] so $\mathscr{T}(m_a)$ is independent of $m_a$. Therefore

$$\langle c'; \mathscr{S}(b, j_b, m_b)| \, \mathscr{T} \, |c; \mathscr{S}(a, j_a, m_a)\rangle$$
$$= \delta(\mathbf{P}_b)\delta_{j_b, j_a}\delta_{m_b, m_a} T_{\mathscr{S}[(c';b), (c;a)]}(0, j_a), \tag{226b}$$

which is valid in the barycentric frame of reference. In applying the result (226*b*) to states like that of Eq. (224), it is important to recall that $\mathscr{T}$ (or $S$) is equivalent to a linear operator when acting on asymptotic scattering states.

As was described in Section 1.2, p. 10, it is thought that the parity operator $\mathscr{P}$ commutes with all Hamiltonians $H$, except those including the so-called weak interactions. In such cases

$$[\mathscr{P}, \mathscr{T}] = 0, \tag{227}$$

[45] When $|m_a \pm 1|$ exceeds $j_a$, the corresponding matrix element of $J^{\pm}$ vanishes, of course.

and the matrix elements vanish between states of different parity. Other approximate symmetry operations lead to "isotopic spin" and "strangeness" conservation. We shall later have occasion to discuss some of these.

**The Barycentric Subspace**

It is convenient to introduce the notion of the barycentric subspace into our considerations. This is the subspace of that space of the entire system in which the matrix $T$ is defined, when the system is in the barycentric coordinate system. That is, in the barycentric subspace those variables describing the translation of the entire system are omitted. In this subspace the $T$-matrix has the same formal definition as does $\mathscr{T}$ in the full space. For example, $T$ is defined by Eq. (214$a$) with the variables of translation being omitted from the $\chi$'s. (This is a somewhat trivial point, but convenient in that we may use the same equations for $\mathscr{T}$ and $T$, once having agreed to this convention.)[46]

**Time Reversal Invariance and the Reciprocity Theorem**

The invariance of physical systems with respect to time reversal was discussed in Chapter 2. This leads to an important symmetry associated with the interchange of initial and final states for a scattering process. The time reversal operator

$$T = WT_0$$

was defined and discussed in connection with Eq. (2.121).

We shall assume that

$$\begin{aligned} THT^{-1} &= H \\ TK_{Q(c)}T^{-1} &= K_{Q(c)} \\ TV_{Q(c)}T^{-1} &= V_{Q(c)} \end{aligned} \tag{228}$$

for all channels $c$ and all permutations $Q$ among identical particles. Then, by Eq. (1.134),

$$T\chi_{c;a} = \mu_{c;-a}\chi_{c;-a}, \tag{229$a$}$$

where $\chi_{c;-a}$ is the time-reversed state obtained from $\chi_{c;a}$. That is, all momenta, angular momenta, and spins are reversed in $\chi_{c;-a}$. In addition to Eq. (229$a$) we have for any permutation $Q$ on the coordinates of identical particles

$$T\chi_{c;Q(a)} = \mu_{c;-a}\chi_{c;Q(-a)}. \tag{229$b$}$$

[46] In Section 7.1 a more complete discussion is given of the use of integrals of the motion to define reduced scattering matrices in a subspace of the original system space.

Consider now the wave function

$$\psi_{c;a}^{\pm}(\eta) = \left(1 + \frac{1}{E_a \pm i\eta - H} V_c\right)\chi_{c;a}. \tag{230}$$

Then

$$T\psi_{c;a}^{\pm}(\eta) = T\left(1 + \frac{1}{E_a \pm i\eta - H} V_c\right)T^{-1}T\chi_{c;a}$$

$$= \mu_{c;-a}\left[\left(1 + \frac{1}{E_a \mp i\eta - H} V_c\right)\chi_{c;-a}\right] = \mu_{c;-a}\psi_{c;-a}^{\mp}(\eta). \tag{231}$$

On going to the limit $\eta \to 0(+)$ we obtain

$$T\psi_{c;a}^{\pm} = \mu_{c;-a}\psi_{c;-a}^{\mp}. \tag{232a}$$

Similarly, for any permutation $Q$ on identical particles [see Eq. (130)]

$$T\psi_{c;Q(a)}^{\pm} = \mu_{c;-a}\psi_{c;Q(-a)}^{\pm}. \tag{232b}$$

Thus for the symmetrized wave functions (131)

$$T\,\psi_{c;\mathscr{S}(a)}^{\pm} = \mu_{c;-a}\,\psi_{c;\mathscr{S}(-a)}^{\pm} \tag{232c}$$

The Eqs. (232) establish an important relationship between the wave functions $\psi^+$ and $\psi^-$.

A symmetry relation arising from time reversal invariance may be derived for $\mathscr{T}$ on the energy shell, defined by $E_b = E_a$. Using Eq. (2.123), we obtain

$$\langle c'; Q(b)|\,\mathscr{T}\,|c; P(a)\rangle$$

$$= \left(x_{c';Q(b)}, \left(V_{P(c)} + V_{Q(c')}\frac{1}{E_a + i\eta - H} V_{P(c)}\right)x_{c;P(a)}\right)$$

$$= \left(T\left(V_{P(c)} + V_{Q(c')}\frac{1}{E_a + i\eta - H} V_{P(c)}\right)\chi_{c\;P(a)}, T\chi_{c';Q(b)}\right)$$

$$= \mu_{c,-a}^{*}\mu_{c',-b}\left(\left(V_{P(c)} + V_{Q(c)}\frac{1}{E_a - i\eta - H} V_{P(c)}\right)\chi_{c;P(-a)}, \chi_{c';Q(-b)}\right)$$

$$= \mu_{c,-a}^{*}\mu_{c',-b}\left(\chi_{c;P(-a)}, \left(V_{P(c)} + V_{P(c)}\frac{1}{E_a + i\eta - H} V_{Q(c')}\right)\chi_{c';Q(-b)}\right). \tag{233}$$

Now according to the discussion preceding Eq. (13)

$$(\chi_{c;P(-a)}, V_{P(c)}\chi_{c';Q(-b)}) = (\chi_{c;P(-a)}, V_{Q(c')}\chi_{c';Q(-b)}),$$

valid *only* on the energy shell. With use of this equality, Eq. (233) reduces to

$$\langle c'; Q(b)|\mathscr{T}\,|c; P(a)\rangle = \mu_{c,-a}^{*}\mu_{c',-b}\langle c; P(-a)|\mathscr{T}\,|c; Q(-b)\rangle. \tag{234}$$

When this is substituted into Eq. (216), there results the general reciprocity relation

$$\langle c'; \mathscr{S}(b)| \mathscr{T} |c; \mathscr{S}(a)\rangle = \mu^{*}_{c,-a}\mu_{c',-b}\langle c; \mathscr{S}(-a)| \mathscr{T} |c'; \mathscr{S}(-b)\rangle. \quad (235)$$

We emphasize again that Eq. (235) is in general correct only on the energy shell. This relation also holds if we replace $\mathscr{T}$ by $\mathscr{T}^{(-)}$ on both sides of Eq. (235).

Equation (235) is, in accordance with present knowledge, considered to be one of the fundamental symmetry relations satisfied by physical systems. It represents the implication for scattering experiments of the property of physical systems to run "backward as well as forward," as was discussed in Section 2.4.

We have purposely used a rather detailed (and therefore sometimes cumbersome) notation in this chapter. This has been done to enhance the clarity of the discussion. In the following chapters it will often be convenient to condense our notation. For example, a single label $\alpha = (c; a)$ may be employed to indicate a state and a channel. The symbol denoting wave function symmetrization will frequently be omitted when it is clear that symmetrized wave functions are being used.

## REFERENCES AND NOTES

1. E. P. Wigner and L. Eisenbud, *Phys. Rev.* **72,** 29 (1947); E. P. Wigner, *Phys. Rev.* **70,** 15 (1946). These authors define channel states in a restricted configuration space in such a manner as to give vanishing overlap integrals for wave functions belonging to different channels. The advantage of having only orthogonal states is obtained at the cost of dealing with rather complex boundary conditions in configuration space. In our treatment the asymptotic wave packet states are orthogonal.
2. F. Bloch and A. Nordsieck, *Phys. Rev.* **52,** 54 (1937).
3. H. Eckstein, *Phys. Rev.* **101,** 880 (1956). See also W. Brenig and R. Haag, *Forts. der Physik* **7,** 183 (1959).
4. E. P. Wigner, *Group Theory*, Academic Press, New York, 1959; F. D. Murghnahan, *The Theory of Group Representations*, Johns Hopkins Press, Baltimore, 1938.
5. The first demonstration of the implied connection between spin and statistics for relativistic field theories is due to Pauli [*Phys. Rev.* **58,** 716 (1940)]. Considerable simplification and clarification of this relationship is achieved by N. Burgoyne [*Nuovo cimento* **8,** 607 (1958)] and by G. Lüders and B. Zumino [*Phys. Rev.* **110,** 1450 (1958)]. These authors have shown that integral spin fields must be quantized according to Bose-Einstein statistics and that half-integral spin fields must be quantized according to Fermi-Dirac statistics if several very general postulates are satisfied by the theory of these fields: (1) There is a unique vacuum state; (2) The theory is invariant for proper Lorentz transformations; (3) the vacuum state is the state of lowest energy; (4) the metric is positive definite; (5) distinct fields either commute or anticommute for spacelike separations. A field theory of spin one-half particles quantized according to modified statistics has recently been exhibited by D. V. Volkov [*J. Exptl. Theoret. Phys. (U.S.S.R.)* **36,** 1560 (1959); *Soviet Phys.—JETP* **9,** 1107 (1959)]. To obtain the modified statistics, Volkov has replaced our

postulate (5) with specified commutation relations for products of three Fermi fields. In this theory there may be zero particles, one particle, or two particles in each state (in contrast to the restriction that there be either zero particles or one particle in each state for Fermi statistics). Dynamical equations may be formulated for these new fields. (Requirements of microcausality would seem to imply that the known hyperons are not of the Volkov type, since they decay by pion (a Boson particle) emission into nucleons, which are known to satisfy Fermi statistics.)

6. G. C. Wick, A. S. Wightman, and E. P. Wigner, *Phys. Rev.* **88,** 101 (1952)
7. J. M. Blatt and V. F. Weisskopf, *Theoretical Nuclear Physics*, John Wiley and Sons, New York, 1952.
8. H. Massey and G. Burhop, *Electronic and Ionic Impact Phenomena*, Oxford University Press, Oxford, 1952, Chapter III. Brackmann, Fite, and Neynaber, *Phys. Rev.* **112,** 1157 (1958).
9. E. Condon and G. Shortley, *Theory of Atomic Spectra*, Cambridge University Press, Cambridge, 1953, Chapter VI.

# 5. FORMAL SCATTERING THEORY

In this chapter we shall develop what has come to be called formal scattering theory. It is to some extent a cavalier (and abbreviated) version of what has been presented already in Chapter 3. There are several reasons for redoing this work. (1) It is possible to handle quite a large class of problems in a uniform manner without repeatedly going through the labor of constructing wave packets; (2) a number of important computation methods may be given for the construction of the scattering matrix; (3) the formalism may be developed without decomposing $H$ into an "asymptotic" part $K$ and a part $V \equiv H - K$. This permits a considerable simplification of much of the notation.

The first systematic discussion of the formal theory was given by Møller [1] in 1945. It was based on the concept of the collision matrix introduced by Wheeler [2] and Heisenberg [3]. Further important developments were made by Lippmann and Schwinger [4] in 1950, who presented powerful variational and algebraic computational methods. Another formulation was given in 1953 by Gell-Mann and Goldberger [5]; the point of view of that formulation will be followed here. A very general treatment, with special attention to field theoretic problems, has been described by Brenig and Haag [6]; several of their results will be incorporated in our treatment. The work of Brenig and Haag has a number of points of similarity with an earlier paper of Ekstein [7] who was the first to show how we may avoid the assumption that the Hamiltonian be split into an unperturbed part and a perturbation.

From a mathematical point of view many of the formal manipulations we shall use have been justified in simple cases (particularly in potential scattering) by Cook [8], Hack [9], and Brenig and Haag [6]. We shall make no pretense at mathematical rigor. Any formal manipulations that are not obviously wrong will be assumed to be correct. Possible ambiguities will be seen below to be easily resolved by recalling the wave packet methods of Chapters 3 and 4.

## 5.1 LIMITING PROCESS AND TRANSITION PROBABILITIES

The physical situation to be described is just the one considered in Chapters 3 and 4. Thus the Hamiltonian $H$ may be split in a variety of ways, giving rise to the notion of channels as described in detail in Chapter 4. Our problem is to calculate the rates of transition between these channel states. We assume that the continuous spectra of $H$ and of the various channel Hamiltonians $K_c$ coincide and that there may be discrete states of $H$ which do not coincide with the discrete states, if any, of the $K_c$. The complete set of eigenstates of $K_c$ is called $\chi_a{}^c$ whereas the channel states are written $\chi_{c;a}$. We do not worry explicitly about symmetrization for identical particles since we have already learned how to deal with this problem in Chapter 4. (Actually at a certain stage in our development symmetrization may easily be incorporated directly into the formalism.)

### Asymptotic Conditions

As in Chapter 3, we imagine that the system is prepared in the remote past in what is essentially an eigenstate of $K_c$, and we follow the subsequent development as it moves into the region of space where the interaction takes place; that is, where $V_c = H - K_c \neq 0$. The difference in the present treatment is that we shall replace the wave packet description by a special limiting process applied to sharp states.

The Schrödinger equation that describes the time development of the system is

$$i\frac{\partial\Psi(t)}{\partial t} = H\Psi(t). \tag{1}$$

The stationary channel eigenstates obtained by setting $V_c = 0$ are

$$\Phi_{c;a}(t) = e^{-iK_ct}\chi_{c;a} = e^{-iE_at}\chi_{c;a}. \tag{2}$$

The latter form is applicable whether or not (because of the identity of particles) we may break $H$ in a unique way into $H = K_c + V_c$. Since we are not going to use wave packets, it is convenient to normalize the $\chi_{c;a}$ to unity in a box of volume $\mathscr{V}$ which will later be allowed to become infinitely large. As we described in Section 4.1, the $\chi_{c;a}$ are *not* in general orthogonal to the basis functions $\chi_{c';a'}$ belonging to other channels. For any eigenstates of $K_c$, say $\chi_{c;a}, \chi_{c;a'}$, we may demand that

$$(\chi_{c;a'}, \chi_{c;a}) = \delta_{a'a}, \sum_{a'}(\chi_{c;a'}, \chi_{c;a}) = 1; \tag{3}$$

but for any finite quantization volume

$$(\chi_{c';a'}, \chi_{c;a}) \neq 0 \tag{4}$$

even if $E_{a'} \neq E_a$. Even if the states were orthogonal for different values of $E$ we could not expect that the value for equal energies is of the appropriate $\delta$-character, so that in the limit $\mathscr{V} \to \infty$, $\Sigma_{a'}(\chi_{c';a'}, \chi_{c;a}) = 1$. These states are to be contrasted with the totality of eigenfunctions of $K_c$, the $\chi_a{}^c$, which form a complete, orthonormal set; the $\chi_{c;a}$ are only a subset of $\chi_a{}^c$.

Now we have seen in Chapter 3 that there exist solutions of the time-dependent Schrödinger equation which reduce in the remote past to a prepared wave packet representing a precollision situation. We shall introduce a limiting process that replaces the wave packet construction and is for most purposes physically equivalent to it. Thus we seek to define a sense in which we may say that

$$\Psi(t) = e^{-iHt}\chi_{c;a} \to \Phi_{c;a}(t) = e^{-iE_a t}\chi_{c;a}$$

in the remote past. An alternative procedure is to imagine an adiabatic switching on an off of interactions at remote time. This is the method used by Lippmann and Schwinger and in many field theoretic papers in the older literature. It seems to be somewhat artificial and unsatisfactory for many applications, however.

We write now $\Psi_{c;a}(t)$ for the state that coincides (in a sense to be defined) in the remote past with $\Phi_{c;a}(t)$. It is most convenient to direct our attention to the state at $t = 0$, called $\psi_{c;a}$, and ask how it evolves from what we shall call the initial configuration $\chi_{c;a}$. Assuming for the moment that we can succeed in defining a limit, we have

$$\begin{aligned} \psi_{c;a} &= e^{iHt}\Psi_{c;a}(t) \underset{t\to-\infty}{\to} e^{iHt}\Phi_{c;a}(t) \\ &= \lim_{t\to-\infty} e^{i(H-E_a)t}\chi_{c;a}. \end{aligned} \tag{5}$$

We must emphasize that this limit will certainly not exist in any simple fashion unless we really construct a precollision packet from the $\chi_{c;a}$. It would be sufficient for our purposes, in fact, to regard the $\Phi_{c;a}(t)$ as just such a packet and write it as $\Sigma_a A_a \chi_{c;a}\, e^{-iE_a t}$. It is customary to omit this precaution, but to do so will cause our limiting process to be somewhat subtle.

If we were to state at a time $\tau$ in the remote past $\Psi_{c;a}(\tau) = \Phi_{c;a}(\tau)$ and thus $\Psi_{c;a}(t) = e^{-iH(t-\tau)}\,\Phi_{c;a}(\tau)$, we would introduce uninteresting transients into the problem; although they are real and physically understandable, we have seen already in Chapter 3 that they play no real role when we start with a packet state at time $\tau$. To eliminate them entirely we imagine feeding in the incident beam over a period of time in the remote past rather than suddenly at a single time $\tau$. The time available is the amount allowed by the precision with which the incident beam is to be specified. In the limit of sharp energy we must take an infinite period. We then define a state $\Psi_{c;a}^{(\eta)}(t)$ by

$$\Psi_{c;a}^{(\eta)}(t) = \eta \int_{-\infty}^{0} d\tau\, e^{\eta\tau}\, e^{-iH(t-\tau)}\Phi_{c;a}(\tau) \tag{6}$$

and ultimately consider the passage to the limit $\eta \to 0(+)$. Clearly this describes smearing of the initial packet over times of $\mathcal{O}(1/\eta)$ before $t = 0$. The particular virtue of this limiting process is that it has no effect on any function $f(\tau)$ which has a proper limit $f(-\infty)$, but it yields zero if $f(\tau)$ oscillates. For example,

$$\lim_{\eta\to 0(+)} \eta \int_{-\infty}^{0} d\tau\, e^{\eta\tau} \sin \alpha\tau = -\lim_{\eta\to 0(+)} \frac{\alpha\eta}{\alpha^2 + \eta^2} \to 0. \tag{7}$$

It is in this way that the transients are eliminated. In the physical problem under consideration there is another limiting process which must be carried out along with $\eta \to 0$. This has to do with the volume of the box inside of which we imagine the system to be confined. Since it makes no sense for the effective duration of time $\eta^{-1}$ of the wave train to exceed the time necessary for the initial systems approaching with velocity $v$ to cross the box (that is, $\mathscr{V}^{1/3}/v$), we must insist that $\mathscr{V}^{1/3}/v \gg 1/\eta$. With this tacit understanding, which will play an important role later, we shall interpret the limit in Eq. (5) to be given by the prescription of Eq. (6), as $\eta \to 0(+)$:

$$\begin{aligned}\psi_{c;a} &= \lim_{\eta\to 0(+)} \Psi_{c;a}^{(\eta)}(0) = \lim_{\eta\to 0(+)} \eta \int_{-\infty}^{0} d\tau\, e^{\eta\tau}\, e^{i(H-E_a)\tau}\chi_{c;a} \\ &= \lim_{\eta\to 0(+)} \frac{i\eta}{E_a - H + i\eta}\chi_{c;a} \equiv \lim_{\eta\to 0(+)} \psi_{c;a}^{(\eta)}.\end{aligned} \tag{8}$$

We shall prove later that $\psi_{c;a}$ is, in fact, an eigenstate of $H$ corresponding to the energy $E_a$. Taking this for granted we may note that in place of Eq. (5) we might have written

$$\begin{aligned}\chi_{c;a} &= e^{iK_ct}\Phi_{c;a}(t) \underset{t\to-\infty}{\longrightarrow} e^{iK_ct}\Psi_{c;a}(t) = e^{iK_ct}\, e^{-iHt}\psi_{c;a} \\ &= \lim_{t\to-\infty} e^{i(K_c-E_a)t}\psi_{c;a}.\end{aligned} \tag{9}$$

The same limiting process as that used in Eq. (5) leads now to

$$\chi_{c;a} = \lim_{\eta\to 0(+)} \frac{i\eta}{E_a - K_c + i\eta}\,\psi_{c;a}. \tag{10}$$

A physical interpretation of the initial configuration may be given in terms of Eq. (9). We start at $t = 0$ with a state $\psi$ and propagate backward in time under the influence of the complete Hamiltonian; this is the factor $e^{-iHt}\psi$. This state is then allowed to propagate forward in time up to $t = 0$, as though there were no interaction, by way of $e^{iKt}$. Thus $\chi$ is the state at $t = 0$ which would have emerged from the remote past if there had been no interaction and if the system had started in the state $\Psi^{\cdot}(t)$.

**The Transition Probability**

To proceed with the computation of the transition rate from a state $\Phi_{c;a}$ to a state $\Phi_{c';a'}$ we ask for the probability at time $t$, $P_{(c;'a)(c;a)}(t)$, of finding the system in the state $\Phi_{c';a'}(t)$ when it evolved in the remote past from $\Phi_{c;a}$. In order to simplify the writing in the many subsequent formal manipulations, let us abbreviate the channel and state labels $(c; a)$ by a single Greek letter $\alpha$; the designation $\alpha'$ will mean in general $(c'; a')$, a different channel and state of that channel. Similarly, we shall as previously write $K$, $K'$ in place of $K_c$, $K_{c'}$ whenever no confusion may arise. Using this notation, we have for the transition probability of interest

$$P^{(\eta)}_{\alpha'\alpha}(t) = |f^{(\eta)}_{\alpha'\alpha}(t)|^2(N^{(\eta)}_{\alpha})^{-1}, \tag{11}$$

where

$$f^{(\eta)}_{\alpha'\alpha}(t) = (\Phi_{\alpha'}(t), \Psi^{\cdot}_{\alpha}(t)), \tag{12}$$

$$N^{(\eta)}_{\alpha} = (\Psi^{\cdot(\eta)}_{\alpha}(t), \Psi^{\cdot(\eta)}_{\alpha}(t)). \tag{13}$$

We do not yet pass to the limit $\eta \to 0(+)$ for a reason that will become apparent in a moment. The fact that $H$ is Hermitian ensures that $N^{(\eta)}_{\alpha}$ is independent of time. It is most convenient to compute the transition rate at $t = 0$. By construction, the "duration" of the interaction is from $-1/\eta$ to $1/\eta$, since the effective length of the wave train is of order $\eta^{-1}$. We shall see later that within this interval the rate $P_{\alpha'\alpha}{}^{(\eta)}$ is essentially independent of time. We note that

$$f^{(\eta)}_{\alpha'\alpha}(t) = (e^{-iE_{\alpha'}t}\chi_{\alpha'}, e^{-iHt}\psi_{\alpha}(\eta)), \tag{14}$$

from which it is obvious that if we pass to the limit $\eta \to 0(+)$ too soon we are in trouble: If, as we claim, $H\psi_{\alpha}(0) = E_{\alpha}\psi_{\alpha}(0)$, the time dependence of $f_{\alpha'\alpha}(t)$ is given by $e^{i(E_{\alpha'}-E_{\alpha})t}$ and $|f^{(0)}_{\alpha'\alpha}|^2 =$ constant; the idea of a transition rate then becomes somewhat absurd. This is the first indication of the subtlety of our limiting process. For the transition rate at $t = 0$ we need only the first derivative of $f^{(\eta)}_{\alpha'\alpha}(t)$.

$$\begin{aligned} \dot{f}^{(\eta)}_{\alpha'\alpha}(0) &= \frac{d}{dt}(e^{i(H-E_{\alpha'})t}\chi_{\alpha'}, \psi_{\alpha}(\eta))\bigg|_{t=0} \\ &= -i((H - E_{\alpha'})\chi_{\alpha'}, \psi_{\alpha}(\eta)) \\ &\equiv -i\mathscr{T}_{\alpha'\alpha}(\eta) \end{aligned} \tag{15}$$

The matrix $\mathscr{T}_{\alpha'\alpha}(\eta)$ is defined[1] by the last two lines of Eq. (15); this form is valid even for identical particles (when the division of $H$ into $K' + V'$ may become meaningless), provided the $\chi_\alpha$ are appropriately symmetrized states. We shall, however, for the present imagine that there is a meaningful separation of $H$ and frequently write

$$\mathscr{T}_{\alpha'\alpha}(\eta) = ((H - E_{\alpha'})\chi_{\alpha'}, \psi_\alpha(\eta)) = (V'\chi_{\alpha'}, \psi_\alpha(\eta))$$
$$= (\chi_{\alpha'}, V'\psi_\alpha(\eta)) \equiv (\chi_{c';a'}, V_{c'}\psi_{c;a}(\eta)), \tag{16}$$

where we momentarily restore the full channel labels to avoid any misunderstanding.

We have for the transition rate at $t = 0$, $\dot{P}^{(\eta)}_{\alpha'\alpha}(0)$,

$$N^{(\eta)}_\alpha \dot{P}^{(\eta)}_{\alpha'\alpha}(0) = [f^{(\eta)}_{\alpha'\alpha}(0)]^* \dot{f}^{(\eta)}_{\alpha'\alpha}(0) + [\dot{f}^{(\eta)}_{\alpha'\alpha}(0)]^* f^{(\eta)}_{\alpha'\alpha}(0); \tag{17}$$

and to complete the evaluation we require $N^{(\eta)}_\alpha$ and $f^{(\eta)}_{\alpha'\alpha}(0)$. We begin by noting various useful forms for $\psi_\alpha(\eta)$ as given in Eq. (8):

$$\psi_\alpha(\eta) \equiv \frac{i\eta}{E_\alpha - H + i\eta}\chi_\alpha = \chi_\alpha + \frac{1}{E_\alpha - H + i\eta}(H - E_\alpha)\chi_\alpha$$
$$= \chi_\alpha + \frac{1}{E_\alpha - H + i\eta}V\chi_\alpha. \tag{18}$$

The operator $(E_\alpha - H + i\eta)^{-1}$ may be rewritten conveniently, as we have done previously, in the form

$$\frac{1}{E_\alpha - H + i\eta} = \frac{1}{E_\alpha - K' + i\eta} + \frac{1}{E_\alpha - K' + i\eta}(H - K')\frac{1}{E_\alpha - H + i\eta} \tag{19}$$

so that from the definition of $\psi_\alpha(\eta)$

$$\psi_\alpha(\eta) = \frac{1}{E_\alpha - K' + i\eta}\chi_\alpha + \frac{1}{E_\alpha - K' + i\eta}(H - K')\psi_\alpha(\eta). \tag{20a}$$

Note that if we had made the division of $H$ with $K$ rather than $K'$ we would have, since $K\chi_\alpha = E_\alpha\chi_\alpha$,

$$\psi_\alpha(\eta) = \chi_\alpha + \frac{1}{E_\alpha - K + i\eta}V\psi_\alpha(\eta) \tag{20b}$$

[1] This defines $\mathscr{T}$ only on the energy shell. For example, compare the expression given here with those in Eqs. (4.13).

where we have written $H - K = V$. It is now a simple matter, using Eq. (20*a*), to evaluate $f^{(\eta)}_{\alpha'\alpha}(0)$:

$$f^{(\eta)}_{\alpha'\alpha}(0) = (\chi_{\alpha'}, \psi_\alpha(\eta)) = \frac{i\eta}{E_\alpha - E_{\alpha'} + i\eta}(\chi_{\alpha'}, \chi_\alpha)$$

$$+ \frac{1}{E_\alpha - E_{\alpha'} + i\eta}(\chi_{\alpha'}, V'\psi_\alpha(\eta))$$

$$= \frac{i\eta}{E_\alpha - E_{\alpha'} + i\eta}(\chi_{\alpha'}, \chi_\alpha) + \frac{1}{E_\alpha - E_{\alpha'} + i\eta}((H - E_{\alpha'})\chi_{\alpha'}, \psi_\alpha(\eta)). \tag{21}$$

Before putting the pieces together and contemplating the passage to the limit $\eta \to 0$, we must discuss the nature of the singularity of $f^{(\eta)}_{\alpha'\alpha}(0)$ near $E_{\alpha'} = E_\alpha$. Because the potential appears in the definition of $\mathcal{T}_{\alpha'\alpha}(\eta)$, there is no reason to expect it to have a singularity at $E_{\alpha'} = E_\alpha$ even when $\eta \to 0$. This may be seen in somewhat greater detail as follows. Using Eq. (18) we may write

$$\mathcal{T}_{\alpha'\alpha}(\eta) = ((H - E_{\alpha'})\chi_{\alpha'}, \psi_\alpha(\eta))$$

$$= ((H - E_{\alpha'})\chi_{\alpha'}, \chi_\alpha) + ((H - E_{\alpha'})\chi_{\alpha'}, \frac{1}{E_\alpha - H + i\eta}(H - E_\alpha)\chi_\alpha). \tag{22}$$

Now we introduce, as in Chapter 3, a complete set of eigenstates of $H$: $H\psi_\lambda = H_\lambda \psi_\lambda$. Introducing these into Eq. (22) we have

$$\mathcal{T}_{\alpha'\alpha}(\eta) = ((H - E_{\alpha'})\chi_{\alpha'}, \chi_\alpha) + \sum_\lambda \frac{((H - E_{\alpha'})\chi_{\alpha'}, \psi_\lambda)(\psi_\lambda, (H - E_\alpha)\chi_\alpha)}{E_\alpha - H_\lambda + i\eta}. \tag{23}$$

As we consider the limit $\eta \to 0$, we must imagine allowing the quantization volume $\mathscr{V}$ to approach infinity; as we do so, the sum over $\lambda$ becomes in part an integration over the continuous spectrum of eigenvalues and the only role played by the $i\eta$ is to give a precise instruction as to how the singularity at $E_\alpha = H_\lambda$ is to be dealt with. Barring a fantastic accident, there is no indication that the point $E_\alpha = E_{\alpha'}$ is at all distinguished, even as $\eta \to 0$. There is the tacit assumption that there are no unexpected $\delta$-functions in the matrix elements, which is certainly true if $(H - E_{\alpha'})\chi_{\alpha'} = V'\chi_{\alpha'}$ with $V'$ a true potential. In certain field theoretic applications the situation is more subtle; we return to this point later.

Finally we must concern ourselves with the simultaneous limit of infinite quantization volume. Because of the normalization of the $\chi$'s there is some definite dependence of the matrix element $\mathcal{T}_{\alpha'\alpha}(\eta)$ on the volume. The precise dependence depends on the number of particles involved in the reaction. For example, if we have two particles in and two particles out, $\chi_\alpha \sim \mathscr{V}^{-1}, \chi_{\alpha'} \sim \mathscr{V}^{-1}$; because of translation invariance one of these factors is canceled when we form $\mathscr{T}_{\alpha'\alpha}(\eta)$, so here $\mathscr{T}_{\alpha'\alpha}(\eta) \sim \mathscr{V}^{-1}$. The complete statement of the regularity hypothesis is then that with the appropriate power $n$

$$\lim_{\substack{\eta\to 0\\ \mathscr{V}\to\infty}} \mathscr{V}^n \mathscr{T}_{\alpha'\alpha}(\eta) = \mathscr{T}_{\alpha'\alpha} \tag{24}$$

is free of singularities as $E_\alpha \to E_{\alpha'}$.

We are now ready to evaluate $N_\alpha^{(\eta)}\dot{P}_{\alpha'\alpha}^{(\eta)}(0)$ from Eq. (17) using Eqs. (15) and (21).

$$\begin{aligned} N_\alpha^{(\eta)}\dot{P}_{\alpha'\alpha}^{(\eta)}(0) &= \frac{-\eta}{E_\alpha - E_{\alpha'} - i\eta}(\chi_{\alpha'}, \chi_\alpha)^* \mathscr{T}_{\alpha'\alpha}(\eta) \\ &\quad - \frac{i\,|\mathscr{T}_{\alpha'\alpha}(\eta)|^2}{E_\alpha - E_{\alpha'} - i\eta} + \text{complex conjugate} \\ &= -\frac{\eta}{(E_\alpha - E_{\alpha'})^2 + \eta^2}[(E_\alpha - E_{\alpha'})(A_{\alpha'\alpha} + A_{\alpha'\alpha}^*) \\ &\quad + i\eta(A_{\alpha'\alpha} - A_{\alpha'\alpha}^*)] \\ &\quad + \frac{2\eta}{(E_\alpha - E_{\alpha'})^2 + \eta^2}|\mathscr{T}_{\alpha'\alpha}(\eta)|^2, \end{aligned} \tag{25}$$

where we have used the abbreviation

$$A_{\alpha'\alpha} = (\chi_{\alpha'}, \chi_\alpha)^* \mathscr{T}_{\alpha'\alpha}(\eta). \tag{26}$$

The complete discussion of Eq. (25) is elementary but somewhat intricate because of the singular nature of the expression. In order to simplify the various limiting processes let us imagine summing over a small band of states $\alpha'$. It is sufficient to regard this as a sum in which the channel label $c'$ in $\alpha' = (c'; a')$ is kept fixed and the sum extends over the $a'$. This amounts almost to a retreat to the wave packet description; however, there is little point in asking for a physically absurd quantity, the transition rate between truly discrete states. As we allow $\eta \to 0$, $\mathscr{V} \to \infty$, the energy levels become continuous, so that the sum over $\alpha'$ involves in part an integration over the continuum $E_{\alpha'}$. We then remark that, if $g(E_{\alpha'})$ is any continuous function of $E_{\alpha'}$,

$$\lim_{\eta\to 0(+)} \int_{E_\alpha - \Delta_1}^{E_\alpha + \Delta_2} dE_{\alpha'} \frac{\eta g(E_{\alpha'})}{(E_{\alpha'} - E_\alpha)^2 + \eta^2} = \pi g(E_\alpha) \tag{27}$$

where $\Delta_1, \Delta_2 > 0$; and if $\Delta_1, \Delta_2$ have opposite signs, we obtain zero.[2] In a formal sense we may write

$$\lim_{\eta\to 0(+)} \frac{\eta}{(E_{\alpha'} - E_\alpha)^2 + \eta^2} = \pi\, \delta(E_{\alpha'} - E_\alpha), \tag{28}$$

which has meaning when the quantity appears under an integral sign. When this result is applied to the sum over $\alpha'$ of $N_\alpha^{(\eta)}\dot{P}_{\alpha'\alpha}^{(\eta)}(0)$, it is clear that the terms involving $A_{\alpha'\alpha}$ make no contribution. Indicating the restricted sum over $\alpha'$ by $(\alpha')$ we have finally

$$\begin{aligned}\sum_{(\alpha')} N_\alpha \dot{P}_{\alpha'\alpha}(0) &= \lim_{\eta\to 0} \sum_{(\alpha')} N_\alpha^{(\eta)} \dot{P}_{\alpha'\alpha}^{(\eta)}(0) \\ &= \lim_{\eta\to 0} \sum_{(\alpha')} 2\pi\, \delta(E_{\alpha'} - E_\alpha)\, |\mathscr{T}_{\alpha'\alpha}(\eta)|^2. \end{aligned} \tag{29}$$

Clearly the transition rate vanishes unless the energy $E_{\alpha'}$ of the set $(\alpha')$ overlaps $E_\alpha$.

To interpret this result and complete the passage to the limit $\eta \to 0$ we must be more explicit about the volume dependence of $\mathscr{T}_{\alpha'\alpha}(\eta)$ and utilize the regularity assumption, Eq. (24). For simplicity, let us imagine that we have two particles in the initial state and $\nu$ particles in the final state. Then $\mathscr{T}_{\alpha'\alpha}(\eta) \sim \mathscr{V}^{-\nu/2}$; superficially we might expect $\mathscr{T}_{\alpha'\alpha}(\eta) \sim \mathscr{V}^{-\nu/2-1}$, but because of momentum conservation (a consequence of translational invariance) one power of $\mathscr{V}$ disappears. In order to convert the sum over $\alpha'$ into an integral over the momenta conjugate to the $\nu$ center-of-mass particle coordinates (that is, for each elementary particle, nucleus, atom, and so on, $\nu$ in all) we introduce the momentum conservation relation and volume dependence of $\mathscr{T}_{\alpha'\alpha}(\eta)$ explicitly according to

$$\mathscr{T}_{\alpha'\alpha}(\eta) = \frac{(2\pi)^{3\nu/2}}{\mathscr{V}^{\nu/2}}\, \delta_{\Sigma \mathbf{k}j,\mathbf{P}\alpha} T_{\alpha'\alpha}(\eta), \tag{30a}$$

where we have put in a factor of $(2\pi)^{3\nu/2}$ in order to make our notation consistent with that of Chapter 3. The momenta of the final particles are called $\mathbf{k}_j$ and the corresponding energies, including internal energies, will be called $\epsilon_j$; $\mathbf{P}_\alpha$ is the total momentum of the state designated $\alpha$. The quantity $T_{\alpha'\alpha}(\eta)$ is seen by Eq. (24) to be regular as $\mathscr{V} \to \infty$, $\eta \to 0$.

[2] This result is almost self-evident. The numbers $\Delta_1, \Delta_2$ may be taken arbitrarily small since it is only where $E_{\alpha'}$ can equal $E_\alpha$ that we get any contribution in the limit as $\eta \to 0$. Thus $g(E_{\alpha'})$ may be evaluated at $E_{\alpha'} = E_\alpha$ and removed from the integral. The substitution $E_{\alpha'} - E_\alpha = \eta x$ transforms the remaining integral into

$$\int_{-\Delta_1/\eta}^{+\Delta_2/\eta} \frac{dx}{x^2 + 1} = \pi \qquad \text{in the limit } \eta \to 0.$$

Using Eq. (30$a$), Eq. (29) becomes

$$\sum_{(\alpha')} N_\alpha \dot{P}_{\alpha'\alpha}(0) = \frac{(2\pi)^4}{\mathscr{V}} \sum_{\alpha'_i} \frac{(2\pi)^{3(\nu-1)}}{\mathscr{V}^{\nu-1}} \sum_{\mathbf{k}_1,\ldots,\mathbf{k}_\nu} \delta(\sum \epsilon_j - E_\alpha)$$

$$\times \delta_{\Sigma \mathbf{k}_j, \mathbf{P}\alpha} |T_{\alpha'\alpha}(\eta)|^2, \tag{31}$$

where the sum over $\alpha'_i$ refers to internal degrees of freedom of the final particles. The sums over the $\nu - 1$ independent momenta are converted to integrations according to the familiar rule

$$\frac{(2\pi)^3}{\mathscr{V}} \sum_{\mathbf{k}_j} \rightarrow \int d^3k_j, \tag{32}$$

and we write $\nu$ such integrals instead of $\nu - 1$ by converting the Kronecker $\delta$ into a Dirac $\delta$-function. Thus we have finally, passing to the limit $\eta \rightarrow 0$,

$$\sum_{(\alpha')} N_\alpha \dot{P}_{\alpha'\alpha}(0) = \frac{(2\pi)^4}{\mathscr{V}} \sum_{\alpha'} \int \prod_{j=1}^{\nu} d^3k_j \, \delta(\sum \mathbf{k}_j - \mathbf{P}_\alpha) \, \delta(\sum \epsilon_j - E_\alpha) |T_{\alpha'\alpha}|^2 \tag{33}$$

The explicit factor of $\mathscr{V}$ will soon disappear.

The only case of physical interest where we might ask for the transition rate between discrete states is where we are concerned with the rate of depletion of the initial state $\alpha$. We will return to this after discussing the evaluation of the normalization factor $N_\alpha^{(\eta)}$ in the limit $\eta \rightarrow 0$.

To compute $N_\alpha^{(\eta)}$, we proceed from the definition of $\psi_\alpha(\eta)$, Eq. (8), and write

$$\begin{aligned}
N_\alpha^{(\eta)} = (\psi_\alpha(\eta), \psi_\alpha(\eta)) &= \left(\frac{i\eta}{E_\alpha - H + i\eta} \chi_\alpha, \frac{i\eta}{E_\alpha - H + i\eta} \chi_\alpha\right) \\
&= (-i\eta)(i\eta)\left(\chi_\alpha, \frac{1}{E_\alpha - H - i\eta} \frac{1}{E_\alpha - H + i\eta} \chi_\alpha\right) \\
&= \frac{-i\eta}{2}\left(\chi_\alpha, \left(\frac{1}{E_\alpha - H - i\eta} - \frac{1}{E_\alpha - H + i\eta}\right)\chi_\alpha\right) \\
&= \tfrac{1}{2}[(\psi_\alpha(\eta), \chi_\alpha) + (\chi_\alpha, \psi_\alpha(\eta))].
\end{aligned} \tag{34}$$

Using Eq. (21) we have immediately

$$\begin{aligned}
N_\alpha^{(\eta)} &= (\chi_\alpha, \chi_\alpha) + \frac{1}{\eta} \operatorname{Im} ((H - E_\alpha)\chi_\alpha, \psi_\alpha(\eta)) \\
&= (\chi_\alpha, \chi_\alpha) + \frac{1}{\eta} \operatorname{Im} \mathscr{T}_{\alpha\alpha}(\eta).
\end{aligned} \tag{35}$$

Now according to the regularity assumption on the $\mathscr{T}_{\alpha\alpha}(\eta)$ embodied in

(24), since $\chi_\alpha$ has been taken to be a two-particle state,

$$\mathscr{T}_{\alpha\alpha}(\eta) = ((H - E_\alpha)\chi_\alpha, \psi_\alpha(\eta)) \sim \mathscr{V}^{-1}.$$

In the double limit $\eta \to 0, \mathscr{V} \to \infty, N_\alpha^{(\eta)} \to N_\alpha = (\chi_\alpha, \chi_\alpha) = 1$. This of course should come as no surprise, since with a Hermitian Hamiltonian we would expect a "packet" growing from the past out of a normalized $\chi_\alpha$ to stay normalized.

The final step in our procedure is to relate the transition rate from state $\alpha$ to a set of states $\alpha'$ to a cross section describing this process. If the relative velocity of the colliding systems is $v$, the incident flux of particles is $v/\mathscr{V}$. We obtain the cross section, $\sigma_{\alpha'\alpha}$, by dividing the transition rate by this quantity.

$$\sigma_{\alpha'\alpha} = \frac{(2\pi)^4}{v} \sum_{\alpha'} \int \prod_{j=1}^{\nu} d^3k_j\, \delta(\sum_j \mathbf{k}_j - \mathbf{P}_\alpha)\, \delta(\sum_j \epsilon_j - E_\alpha)\, |T_{\alpha'\alpha}|^2. \tag{36}$$

We infer from this the differential cross section into the indicated volume in momentum space and into the specified internal states.

$$d\sigma_{\alpha'\alpha} = \frac{(2\pi)^4}{v} \delta(\sum_j \mathbf{k}_j - \mathbf{P}_\alpha)\, \delta(\sum \epsilon_j - E_\alpha)\, |T_{\alpha'\alpha}|^2 \prod_{j=1}^{\nu} d^3k_j \tag{37}$$

This is the same as the cross section obtained from wave packet considerations in Chapter 3.

It should be evident that in arriving at this result we have been able to shorten the labor as compared to that in Chapter 3. As we shall see, many problems can be quickly and concisely treated by these methods. The superficial simplification has been achieved, however, by having on occasion to resort to a certain amount of fast talk. In all cases we can, by using wave packets, make the arguments quite convincing. The most fruitful approach seems to be a combination of these techniques.

We return now to the question of the single state $\alpha$ as a possible final state. From Eq. (25) we have

$$N_\alpha^{(\eta)} \dot{P}_{\alpha\alpha}^{(\eta)}(0) = 2\,\mathrm{Im}\, \mathscr{T}_{\alpha\alpha}(\eta) + \frac{2}{\eta} |\mathscr{T}_{\alpha\alpha}(\eta)|^2; \tag{38}$$

in the double limit $\eta \to 0, \mathscr{V} \to \infty$, the second term is negligibly small compared to the first, since $\mathscr{T}_{\alpha\alpha} \sim \mathscr{V}^{-1}$. We now equate the rate of depletion of the state $\alpha$, $-\dot{P}_{\alpha\alpha}(0)$, to the total transition rate to all other states (which is of course simply a statement of probability conservation).

$$\begin{aligned} -2 \cdot \frac{(2\pi)^3}{\mathscr{V}} \mathrm{Im}\; T_{\alpha\alpha} &= \frac{v}{\mathscr{V}} \sum_{\alpha' \neq \alpha} \sigma_{\alpha'\alpha}, \\ \sum_{\alpha' \neq \alpha} \sigma_{\alpha'\alpha} &= -\frac{2}{v} (2\pi)^3\, \mathrm{Im}\; T_{\alpha\alpha}. \end{aligned} \tag{39}$$

This result, which follows from the fact that the normalization is preserved in time, is called the *optical theorem*. We shall rederive it by formal manipulations shortly.

The one remaining loose end is the justification of our evaluation of the transition rate at time $t = 0$. It is intuitively clear that we should be able to use any time when the systems are interacting, namely the interval $-\eta^{-1} < t < \eta^{-1}$. To show that this is true we compute the second time derivative of $P^{(\eta)}_{\alpha'\alpha}(t)$ at $t = 0$. From the definition, Eq. (11), we find (setting $N_\alpha = 1$)

$$\begin{aligned}\ddot{P}^{(\eta)}_{\alpha'\alpha}(0) = &-((H - E_{\alpha'})^2\chi_{\alpha'}, \psi_\alpha(\eta))(\Psi_\alpha(\eta), \chi_{\alpha'})\\ &+ 2((H - E_{\alpha'})\chi_{\alpha'}, \psi_\alpha(\eta))(\psi_\alpha(\eta), (H - E_{\alpha'})\chi_{\alpha'})\\ &- (\chi_{\alpha'}, \psi_\alpha(\eta))(\psi_\alpha(\eta), (H - E_{\alpha'})^2\chi_{\alpha'}).\end{aligned} \tag{40}$$

This may be rewritten with the aid of the identity

$$\begin{aligned}(\psi_\alpha(\eta), (H - E_{\alpha'})^2\chi_{\alpha'}) &= -i\eta\left(\chi_\alpha, \frac{1}{E_\alpha - H - i\eta}(H - E_{\alpha'})^2\chi_{\alpha'}\right)\\ &= i\eta(\chi_\alpha, (H - E_{\alpha'})\chi_{\alpha'})\\ &\quad - i\eta(E_\alpha - E_{\alpha'} - i\eta)\left(\chi_\alpha, \frac{1}{E_\alpha - H - i\eta}(H - E_{\alpha'})\chi_{\alpha'}\right)\\ &= i\eta(\chi_\alpha, (H - E_{\alpha'})\chi_{\alpha'})\\ &\quad + (E_\alpha - E_{\alpha'} - i\eta)(\psi_\alpha(\eta), (H - E_{\alpha'})\chi_{\alpha'}).\end{aligned} \tag{41}$$

In arriving at Eq. (41) the explicit formula for $\psi_\alpha(\eta)$, Eq. (8), has been used. If we again sum over a band of states $(\alpha')$, we find after uninteresting arithmetic, using Eqs. (21) and (28), the result (to the lowest order in $\eta$)

$$\begin{aligned}\sum_{(\alpha')} \ddot{P}^{(\eta)}_{\alpha'\alpha}(0) = 2\eta \sum_{(\alpha')} \dot{P}_{\alpha'\alpha}(0)&\\ + 2\eta \operatorname{Re} \sum_{(\alpha')} \delta(E_{\alpha'} - E_\alpha)&\mathscr{T}_{\alpha'\alpha}((H - E_{\alpha'})\chi_{\alpha'}, \chi_\alpha)^*.\end{aligned} \tag{42}$$

If we now write

$$\sum_{\alpha'} \dot{P}^{(\eta)}_{\alpha'\alpha}(t) = \sum_{\alpha'} \dot{P}^{(\eta)}_{\alpha'\alpha}(0) + t \sum_{\alpha'} \ddot{P}^{(\eta)}_{\alpha'\alpha}(0) + \cdots,$$

it is clear from Eq. (42) that for any time $t$ such that $\eta t \ll 1$, the transition rate is given accurately by $\dot{P}^{(\eta)}_{\alpha'\alpha}(0)$; in the limit as $\eta \to 0$, this result becomes precise.

**Reactions Involving Several Incident Particles**

Equation (29) permits us to discuss reaction rates involving several incident particles colliding simultaneously. Such processes play an important role in chemical reactions in gases.[3] Let us imagine that in the incident state $\alpha$ there are $\mu$ colliding particles with momenta $\mathbf{p}_1, \ldots, \mathbf{p}_\mu$ and spin (or other internal) variables $s_1, \ldots, s_\mu$. In the final state $\alpha'$ there are $\nu$ particles with momenta $\mathbf{k}_1, \ldots, \mathbf{k}_\nu$ and spin variables $s'_1, \ldots, s'_\nu$.

Equation (30$a$) is now seen to be generalized to the form

$$\mathscr{T}_{\alpha'\alpha}(\eta) = \frac{(2\pi)^{\frac{3}{2}(\nu+\mu)-3}}{\mathscr{V}^{(\nu+\mu)/2} - 1} \delta_{\Sigma \mathbf{k}_j, \mathbf{P}_\alpha} T_{\alpha'\alpha}(\eta), \tag{30b}$$

where $T_{\alpha'\alpha}(\eta)$ is independent of the macroscopic volume $\mathscr{V}$. Now Eq. (29) becomes (we take $N_\alpha = 1$)

$$\sum_{\alpha'} \dot{P}_{\alpha'\alpha}(0) = \lim_{\eta \to 0} 2\pi \sum_{\alpha'} \frac{(2\pi)^{3(\nu+\mu)-6}}{\mathscr{V}^{\nu+\mu-2}} \delta_{\Sigma \mathbf{k}_j, \mathbf{P}_\alpha} \delta\left(\sum \epsilon_j - E_\alpha\right) |T_{\alpha'\alpha}(\eta)|^2.$$

Using Eq. (32) we have

$$\left[\frac{(2\pi)^3}{\mathscr{V}}\right]^\nu \sum_{\alpha'} \to \sum_{s'_1, \ldots, s'_\nu} \int d^3k_1 \cdots d^3k_\nu$$

and

$$\delta_{\Sigma\ j, \mathbf{P}_\alpha} \to \frac{(2\pi)^3}{\mathscr{V}} \delta\left(\sum \mathbf{k}_j - \mathbf{P}_\alpha\right),$$

so

$$\sum_{\alpha'} \dot{P}_{\alpha'\alpha}(0) = \frac{(2\pi)^{3\mu-2}}{\mathscr{V}^{\mu-1}} d^3k_1 \cdots d^3k_\nu \sum_{s'_1, \ldots, s'_\nu}$$
$$\times\ \delta(\mathbf{P}_{\alpha'} - \mathbf{P}_\alpha)\ \delta(E_{\alpha'} - E_\alpha) |T_{\alpha'\alpha}|^2. \tag{43a}$$

This represents the transition rate for one each of the $\mu$ incident particles. If the box of volume $\mathscr{V}$ contains $dN_1, \ldots, dN_\mu$ particles of each species in the momentum ranges $d^3p_1, \ldots, d^3p_\mu$, respectively, the total transition rate per unit volume is

$$dR(\alpha \to \alpha') \equiv \frac{1}{\mathscr{V}} \sum_{\alpha'} \dot{P}_{\alpha'\alpha}(0)\, dN_1 \cdots dN_\mu. \tag{43b}$$

Now

$$\frac{dN_1}{\mathscr{V}} = f_1(\mathbf{p}_1)\, d^3p_1$$

represents the number of particles of species 1 per unit volume in the momentum range $d^3p_1$, and so on. Here $f_1(\mathbf{p}_1)$ (and so on) is the Boltzmann distribution function for particles of species 1. If we average the transition

[3] The additional particles are frequently called catalysts. A wave packet description of three-particle collisions is given in Appendix B.

rate over the $g(\alpha)$ degenerate initial spin states, Eqs. (43*a*) and (43*b*) give[4]

$$dR(\alpha \to \alpha') = [d^3k_1 \cdots d^3k_\nu\, d^3p_1 \cdots d^3p_\mu \times f_1(\mathbf{p}_1) \cdots f_\mu(\mathbf{p}_\mu)]\left[\sum_{s_1,\ldots,s_\mu} \sum_{s'_1,\ldots,s'_\nu} \times \frac{(2\pi)^{3\mu-2}}{g(\alpha)} \delta(\mathbf{P}_{\alpha'} - \mathbf{P}_\alpha)\, \delta(E_{\alpha'} - E_\alpha)\, |T_{\alpha'\alpha}|^2\right]. \tag{43c}$$

Equation (43*c*) gives the mean rate for transitions $\alpha \to \alpha'$, in the indicated momentum ranges, per unit volume. In statistical mechanics it is usually important to compare the rate $dR(\alpha \to \alpha')$ with the inverse transition rate $dR(\alpha' \to \alpha)$, which may be written down immediately from (43*c*) by just interchanging initial and final states. The reciprocity theorem (4.235) provides a fundamental relation between these two rates, since it implies that[5]

$$\sum_{s_1,\ldots,s_\mu} \sum_{s'_1,\ldots,s'_\nu} |T_{\alpha'\alpha}|^2 = \sum_{s_1,\ldots,s_\mu} \sum_{s'_1,\ldots,s'_\nu} |T_{\alpha\alpha'}|^2.$$

This brings to a close the first section. We have not obtained new results, but we have reexpressed much of the content of Chapters 3 and 4 in a new form. For example, the formula for the transition rate written as

$$\mathscr{T}_{\alpha'\alpha} = ((H - E_{\alpha'})\chi_{\alpha'}, \psi_\alpha)$$

rather than $(\chi_{\alpha'}, V'\psi_\alpha)$ is actually correct even if we are dealing with scattering of identical particles, when we cannot write $H = K' + V'$. Then $\chi_{\alpha'}$ is an appropriately symmetrized function made up of all the indistinguishable channel eigenfunctions. In the next section we shall develop the formalism somewhat further and reinterpret some of the formulas in useful ways. The need for splitting the Hamiltonian into perturbed and unperturbed parts will be avoided entirely. We shall also introduce the Heisenberg *S*-matrix from the present point of view and relate it to the more conventional discussions. A number of additional formal developments will be given, such as integral equations for the determination of the scattering matrix, the reaction matrix, and the Heitler integral equation. Some simple examples will be discussed, and finally the modifications of the theory which are necessary in quantum field theory will be briefly treated.

## 5.2 THE S-MATRIX AND FURTHER FORMAL THEORY

We have in the previous section made frequent use of the "packet" functions $\psi_\alpha(\eta)$, which are the state vectors at time zero which develop

[4] Equation (10) of Appendix B represents a special case of Eq. (43*c*).
[5] This relation is valid only if parity is conserved.

from some specified configuration in the remote past. Strictly speaking these initial configurations should be true precollision wave packets like the ones constructed in Chapter 3. When this is so, at least as long as we exclude field theoretic problems involving self-action, the limits as defined actually exist. It is, however, very convenient to use the improper functions defined in terms of very wide (essentially plane-wave) packets for formal manipulations; occasionally we resort to summing over bands of states in order to reduce the singularities encountered and enable the double limit $\eta \to 0$, $\mathscr{V} \to \infty$ to be made more smoothly.

**Precollision Packets**

In the limit $\eta \to 0$, the functions $\psi_\alpha(\eta)$ become eigenfunctions of the Hamiltonian $H$ with eigenvalues $E_\alpha$, as may be easily seen from Eq. (18).

$$\psi_\alpha(\eta) = \chi_\alpha + \frac{1}{E_\alpha - H + i\eta}(H - E_\alpha)\chi_\alpha,$$

$$(E_\alpha - H)\psi_\alpha(\eta) = i\eta(\chi_\alpha - \psi_\alpha(\eta)) \to 0 \quad \text{as} \quad \eta \to 0(+). \tag{44}$$

We shall see later that to insure correct results the actual passage to the limit $\eta \to 0(+)$ should be made only at the end of all calculations. For many purposes, however, we may say that $\psi_\alpha(0+)$ is an eigenfunction of $H$ with a certain boundary condition (that will be made clear in a moment). Let us see how we would be led to the introduction of the $\psi(\eta)$ from the standpoint of wave packets. Suppose at some time, called zero for convenience, we form a precollision packet (more properly, we should refer to a pair of packets) from the channel functions $\chi_\alpha$, $\Sigma A_\alpha \chi_\alpha$; we shall show that such a packet may equally well be represented by $\Sigma A_\alpha \psi_\alpha(\eta)$. This is particularly easy to see when $(H - E_\alpha)\chi_\alpha = V\chi_\alpha$. We have in fact obtained this result already in Chapter 3 where we showed the independence of $\Gamma^+$ on the limit $t_0$. Let us write

$$\begin{aligned}\sum_\alpha A_\alpha \chi_\alpha &= \sum_\alpha A_\alpha \psi_\alpha(\eta) - \sum_\alpha A_\alpha \frac{1}{E_\alpha - H + i\eta} V\chi_\alpha \\ &= \sum_\alpha A_\alpha \psi_\alpha(\eta) + i\int_{-\infty}^{0} dt \sum_\alpha A_\alpha\, e^{\eta t}\, e^{iHt} V\, e^{-iE_\alpha t}\chi_\alpha.\end{aligned} \tag{45}$$

Now since $\Sigma_\alpha A_\alpha \chi_\alpha$ at $t = 0$ represents a pair of precollision packets which presumably move toward collision for positive times, for the negative times encountered in (45) (of the order $-\eta^{-1}$) the packets are receding further from the region where $V \neq 0$. Hence the second term is negligibly small. Actually the restriction that $(H - E_\alpha)\chi_\alpha = V\chi_\alpha$ is not severe. In any problem involving identical particles, the initial packet sum would have to be extended over functions of the appropriate symmetry and for all of these $H - E_\alpha$ would be small in a true precollision situation. Thus we conclude that for a precollision packet we may write either $\Sigma A_\alpha \chi_\alpha$ or $\Sigma A_\alpha \psi_\alpha(\eta)$. Passing now to the limit $\eta \to 0(+)$, we see that our formal limiting process Eq. (6) applied to such a packet leads directly to the old result, Eq. (8). The substance of these remarks is that the term in $\psi_\alpha(\eta)$, which is $(E_\alpha - H + i\eta)^{-1}(H - E_\alpha)\chi_\alpha$, represents only outgoing scattered waves since the $\chi_\alpha$ term represents the entire precollision situation, hence the total incoming flux.

The functions $\psi_\alpha(0+)$ which are eigenfunctions of $H$ are to be obtained from the Schrödinger equation as follows.

$$(H - E_\alpha)[\psi_\alpha(0+) - \chi_\alpha] = -(H - E_\alpha)\chi_\alpha,$$
$$\psi_\alpha(0+) - \chi_\alpha = \lim_{n \to 0(+)} \frac{1}{E_\alpha + i\eta - H}(H - E_\alpha)\chi_\alpha. \tag{46}$$

Thus the $i\eta$ plays the role of giving a precise meaning to the inverse operator. The instruction to use $\eta \to 0(+)$ is the crucial one for our previous considerations because it was just this sign of $\eta$ which led to the integral representation in Eq. (45).

**Ingoing Waves and Postcollision Packets**

We have been talking about states that develop out of the remote past from certain initial configurations. This is the most natural description and the one that corresponds to experience. Actually, the experimental physicist will frequently specify the final configuration of a scattering process. It is of interest therefore to consider states that in the remote future coincide with such prescribed final configurations. By analogy with Eq. (5) we write for the state at $t = 0$, call it $\psi'_{c;a}$,

$$\begin{aligned}\psi'_{c;a} &= e^{iHt}\Psi'_{c;a}(t) \underset{t \to +\infty}{\to} e^{iHt}\Phi_{c;a}(t) \\ &= e^{i(H-E_a)t}\chi_{c;a}.\end{aligned} \tag{47}$$

Here $\Phi_{c;a}(t)$ is a postcollision, that is, final configuration, packet representing separated systems in the indicated channel in the remote future. Again we replace the packet by a sharp energy eigenstate. The limiting process $t \to +\infty$ is defined by an operation similar to that used for $t \to -\infty$ in Eqs. (7) and (8). We write

$$\psi'_{c;a}(\eta) = \eta \int_0^\infty d\tau\, e^{-\eta\tau}\, e^{i(H-E_a)\tau}\chi_{c;a} \tag{48}$$

and contemplate the limit $\eta \to 0(+)$ as before. (Needless to say, this is a substitute limit which could be justified by the use of wave packets.) The earlier remarks about the passage of the quantization volume $\mathscr{V}$ to infinity along with $\eta \to 0$ continue to apply. The alternative possibilities of considering the limits $t \to \pm\infty$ may be summarized in the single formula

$$\psi^{\pm}_{c;a} = \lim_{\eta \to 0(+)} (\mp\eta) \int_0^{\mp\infty} d\tau\, e^{\pm\eta\tau}\, e^{i(H-E_a)\tau} \chi_{c;a}, \tag{49}$$

where we introduce the notation $\psi^{+}_{c;a}$ for the state previously written as $\lim_{\eta \to 0(+)} \psi_{c;a}(\eta)$ and $\psi^{-}_{c;a}$ for $\lim_{\eta \to 0(-)} \psi'_{c;a}(\eta)$. An alternative concise way of writing Eq. (49) is

$$\psi^{\pm}_{c;a} = \lim_{\zeta \to 0(\pm)} (-\zeta) \int_0^{-\zeta/|\zeta|\infty} d\tau\, e^{\zeta\tau}\, e^{i(H-E_a)\tau} \chi_{c;a}. \tag{50}$$

Carrying out the time integration in either Eq. (49) or (50) leads to the representations [see Eq. (8)]

$$\begin{aligned} \psi^{\pm}_{c;a} &= \lim_{\eta \to 0(+)} \frac{\pm i\eta}{E_a - H \pm i\eta} \chi_{c;a} \\ &= \lim_{\eta \to 0(+)} \left[ 1 + \frac{1}{E_a - H \pm i\eta} (H - E_a) \right] \chi_{c;a}. \end{aligned} \tag{51}$$

As might be expected, the ingoing wave states are particularly appropriate for the description of postcollision (that is, after the particles have interacted and are moving apart) packets, and particularly inappropriate for precollision packets. To see this, suppose we had tried to express a precollision packet, Eq. (45), in terms of ingoing wave states [we revert to the simple notation $\alpha$ for $(c; a)$ and again write $(H - E_\alpha)\chi_\alpha = V\chi_\alpha$]:

$$\begin{aligned} \sum_\alpha A_\alpha \chi_\alpha &= \sum_\alpha A_\alpha \psi_\alpha(-\eta) - \sum_\alpha A_\alpha \frac{1}{E_\alpha - H - i\eta} V\chi_\alpha \\ &= \sum_\alpha A_\alpha \psi_\alpha(-\eta) - i \int_0^\infty dt\, e^{-\eta t}\, e^{iHt} V \sum_\alpha A_\alpha \chi_\alpha\, e^{-iE_\alpha t}. \end{aligned} \tag{52}$$

Now, however, we are in the unfortunate situation that we have positive times appearing in the integral in Eq. (51) as contrasted with the negative times of Eq. (45). Thus the packets are moving *into* the region of interaction and not away from it; the second term in Eq. (50) then has no reason to be small. If, however, $\Sigma_\alpha A_\alpha \chi_\alpha$ were a postcollision packet, describing particles receding from the collision the second term in (50) would be negligible. For a postcollision packet, then, we may write

$$\Sigma_\alpha A_\alpha \chi_\alpha = \Sigma_\alpha A_\alpha \psi_\alpha^-.$$

**Properties of the Eigenfunctions $\psi^\pm$**

The functions $\psi_\alpha^\pm$, which are the limits as $\eta \to 0(\pm)$ of the packet functions $\psi_\alpha(\eta)$, are eigenfunctions of the full Hamiltonian operator corresponding to different boundary conditions. The plane-wave states $\chi_\alpha$ from which they originate (in the case of $\psi_\alpha^+$) or to which they converge (in the case of $\psi_\alpha^-$) are not orthogonal in general if the indices $\alpha$ refer, for example, to different channels. Nevertheless the $\psi_\alpha^\pm$ are essentially orthogonal even when they refer to different channels. That they are orthogonal for different values of the energy is a direct consequence of the hermiticity of the Hamiltonian. The fact that at equal energies $\psi_\alpha^+$, $\psi_\beta^+$ for $\alpha \neq \beta$ are effectively orthogonal even for $E_\alpha = E_\beta$ is less obvious. This question has already been discussed in Chapter 4, but we shall consider it again in the present context. For definiteness let us work with the $\psi^+$, the argument for $\psi^-$ is identical. By definition

$$\begin{aligned}(\psi_\beta^+, \psi_\alpha^+) &= \lim_{\substack{\eta \to 0(+) \\ \eta' \to 0(+)}} \left( \frac{i\eta'}{E_\beta - H + i\eta'} \chi_\beta, \frac{i\eta}{E_\alpha - H + i\eta} \chi_\alpha \right) \\ &= \lim\, (-i\eta')(i\eta) \left( \chi_\beta, \frac{1}{E_\beta - H - i\eta'} \frac{1}{E_\alpha - H + i\eta} \chi_\alpha \right) \\ &= \lim \frac{(-i\eta')(i\eta)}{E_\alpha - E_\beta + i(\eta + \eta')} \\ &\quad \times \left( \chi_\beta, \left( \frac{1}{E_\beta - H - i\eta'} - \frac{1}{E_\alpha - H + i\eta} \right) \chi_\alpha \right).\end{aligned} \tag{53}$$

Now it should not matter how we pass to the limits $\eta, \eta' \to 0$, so we do so symmetrically, taking one-half the sum of values obtained by allowing first one and then the other to approach zero. We have then

$$(\psi_\beta^+, \psi_\alpha^+) = \lim_{\eta \to 0(+)} \frac{i\eta}{E_\alpha - E_\beta + i\eta} \frac{1}{2} [(\psi_\beta^+, \chi_\alpha) + (\chi_\beta, \psi_\alpha^+)]. \tag{54}$$

If $E_\beta \neq E_\alpha$, evidently $(\psi_\beta^+, \psi_\alpha^+) = 0$, the factor of $i\eta(E_\alpha - E_\beta + i\eta)^{-1}$ serving as a projection operator for $E_\alpha = E_\beta$. We know from Eq. (21) that *when* $E_\alpha = E_\beta$,

$$(\chi_\beta, \psi_\alpha^+) = \lim_{\eta \to 0(+)} \left[ (\chi_\beta, \chi_\alpha) + \frac{1}{i\eta} ((H - E_\beta)\chi_\beta, \psi_\alpha(\eta)) \right]; \tag{55}$$

because of the double limiting process the second term is small compared to the first. [The argument is just the same as the discussion of volume dependence given in connection with Eq. (35).] Thus we conclude that

$$(\psi_\beta^+, \psi_\alpha^+) = \lim_{\eta\to 0} \frac{i\eta}{E_\alpha - E_\beta + i\eta} (\chi_\beta, \chi_\alpha) \to (\chi_\beta, \chi_\alpha). \tag{56}$$

Unless the configurations $\chi_\beta$, $\chi_\alpha$ correspond to the same channels, the scalar product $(\chi_\beta, \chi_\alpha)$ will have some uncompensated volume factors and the overlap will become negligibly small as the volume approaches infinity. Stated otherwise, consider a sum over a band of states $\beta$ including $E_\beta = E_\alpha$ and pass to the limit of infinite quantization volume. The energy spread may be made arbitrarily small in the limit as $\eta \to 0$ because of the projection operator. The only way such sums can give a finite value is for $(\chi_\beta, \chi_\alpha)$ to be infinite in the interval. This can happen only if the configurations $\chi_\alpha$, $\chi_\beta$ "agree," that is, correspond to the same channels. With this understanding of the symbols, we may write

$$(\psi_\beta^+, \psi_\alpha^+) = \delta_{\beta\alpha}. \tag{57}$$

Another way of arriving at this result is as follows. Construct two precollision packets, one from $\alpha$-components, one from $\beta$-components. We know that for such packets we may write $\Sigma_\alpha A_\alpha \chi_\alpha = \Sigma_\alpha A_\alpha \psi_\alpha^+$ and $\Sigma_\beta A_\beta \chi_\beta = \Sigma_\beta A_\beta \psi_\beta^+$. By definition, precollision packets refer to widely separated configurations of noninteracting systems; if these configurations correspond to different groupings of particles, it is clear that their overlap will in general be negligibly small. Indeed it would require almost unbelievable ingenuity to make it otherwise. We conclude, therefore, for such different configurations $(\Sigma_\beta A_\beta \psi_\beta^+, \Sigma_\alpha A_\alpha \psi_\alpha^+) = 0$. Since the choice of such packets for which the relation is true is almost unlimited to all intents and purposes, we may write $(\psi_\beta^+, \psi_\alpha^+) = \delta_{\beta\alpha}$. Evidently the same considerations apply to the ingoing wave states $\psi_\alpha^-$, so we have finally

$$(\psi_\beta^\pm, \psi_\alpha^\pm) = \delta_{\beta\alpha}. \tag{58}$$

The question naturally arises whether the $\psi_\alpha^\pm$ form a complete set. It is obvious that in general they will not. If there are any states that consist of *all* particles bound, they will surely be orthogonal to the states that arise from (or tend to) initial (final) configurations of the variety we are concerned with in scattering problems. Even aside from these almost trivial situations the fact that the configurations $\chi_{c;a}$, for a given channel $c$ are not complete means that certainly the $\psi_{c;a}^\pm$ are also not complete within the space of scattering states. Another feature of these states is the following:

We may imagine that there is one channel, which we call the all-free channel, in which all the particles are widely separated. The initial and final configurations corresponding to it are simply packets made up from plane wave states, or in the limit, simply a product of plane waves. They form a complete set. Application of our limiting procedure leads to a set of outgoing or ingoing wave states $\psi^{\pm}$ that is trivially *not* complete. Aside from the all-bound configuration, the $\psi^{\pm}$ are clearly orthogonal to eigenstates of the Hamiltonian corresponding to partially bound configurations that have energies below the zero of energy defined by the all-free channel. They are simply eigenstates of $H$ corresponding to different energies. Stated in a different way, no initial configurations corresponding to the all-free channel can describe the partially bound scattering states, at least in a certain energy range. For sufficiently large energies (namely, greater than zero by our definition) the all-free $\psi^{\pm}$ are complete.

It is, of course, a fundamental assumption that the Hamiltonian $H$ has a complete set of eigenstates just as do the various channel Hamiltonians $K_c$. There is every reason to believe that starting from the complete set $\chi_a{}^c$ for a given channel we generate by our limiting process $\psi_a{}^{c\pm}$ that are complete within that portion of the common continuous spectrum of $H$ and $K_c$. For describing physical scattering processes we are interested only in the channel eigenstates $\chi_{c;a}$ corresponding to separated, nonoverlapping channel configurations. Although it is obvious that the corresponding $\psi$ are not complete, at *a given energy* the totality of all channel eigenstates is complete in a well-defined physical sense. In fact the channel configurations are by definition an enumeration at a given energy of all the possible physical situations. We shall therefore assume hereafter that any given physical state can be represented at a given energy by a sum of the form $\Sigma_{c;a} A_{c;a}\psi^{\pm}_{c;a}$, where the sum extends over all channels $c$ and state labels $a$ which are consistent with energy conservation. This basic physical assumption is all that will be required in the way of a completeness relation in our subsequent work.[6]

### The $S$-Matrix in the Heisenberg Representation

In Chapter 2 we discussed the Heisenberg representation, in which the state vectors are independent of time. Since this is an acceptable form of quantum mechanics it must be possible to formulate scattering theory in terms of such states. We ask the following question. A state of the system is characterized by a certain initial configuration; what is the probability amplitude for finding some definite final configuration? In this unusual formulation of the problem there is of course no reference to time. To see

[6] We are restating here, in somewhat different terms, conclusions obtained in Section 4.1 and Appendix C.

how we are led to these considerations we recall the definition of the probability at a time $t$ for finding the system in a state $e^{-iE_\beta t}\chi_\beta$, given that it evolved from the remote past from a precollision packet composed of $\alpha$ [that is, $(c; a)$-] states. This was given by Eqs. (11) and (12).

$$P_{\beta\alpha} = |f_{\beta\alpha}(t)|^2$$

$$f_{\beta\alpha} = (e^{i(H-E_\beta)t}\chi_\beta, \psi_\alpha^+). \tag{59}$$

We have here set $N_\alpha = 1$ and have passed to the limit $\eta \to 0(+)$. Suppose now we consider the limit $t \to +\infty$, that is, we seek the probability of a transition in the remote future after the particles have already separated. Then using, for example, the limiting procedure defined by Eq. (49), we find [compare, for example, Eq. (3.95)]

$$S_{\beta\alpha} \equiv \lim_{t\to\infty} f_{\beta\alpha}(t) = (\psi_\beta^-, \psi_\alpha^+). \tag{60}$$

This same expression of course follows directly from the discussion preceding Eq. (59). We have seen that the state at $t = 0$ which develops from what was a precollision packet $\Sigma_\alpha A_\alpha\chi_\alpha e^{-iE_\alpha t}$ in the remote past takes the form $\Sigma_\alpha A_\alpha\psi_\alpha^+$ at $t = 0$. Alternatively, a postcollision packet $\Sigma_\beta A_\beta\chi_\beta e^{-iE_\beta t}$ evolves in the remote future from a state which at $t = 0$ is $\Sigma_\beta A_\beta\psi_\beta^-$. The probability of finding the state characterized by the final configuration $\chi_\beta$ (allowing the final packet to become sharp) in the state characterized by the initial configuration $\chi_\alpha$ is just $(\psi_\beta^-, \psi_\alpha^+)$. The array $S_{\beta\alpha}$ is called the $S$-matrix, and we shall show that although this may appear to be different from the quantity called by the same name in Chapters 3 and 4, it is really the same set of numbers.

To substantiate this statement we express $\psi_\beta^-$ in terms of $\psi_\beta^+$. [For the rest of this section we shall not write explicitly the limit $\eta \to 0(+)$, in accordance with the convention adopted in Chapter 3. Furthermore the passage of $\mathscr{V}$ to infinity, which must accompany this limit, must be borne in mind.]

$$\psi_\beta^+ = \chi_\beta + \frac{1}{E_\beta - H + i\eta}(H - E_\beta)\chi_\beta,$$

$$\psi_\beta^- = \chi_\beta + \frac{1}{E_\beta - H - i\eta}(H - E_\beta)\chi_\beta, \tag{61}$$

$$\psi_\beta^- = \psi_\beta^+ + \left(\frac{1}{E_\beta - H - i\eta} - \frac{1}{E_\beta - H + i\eta}\right)(H - E_\beta)\chi_\beta.$$

We find then

$$\begin{aligned}
S_{\beta\alpha} &= (\psi_\beta^{-}, \psi_\alpha^{+}) \\
&= (\psi_\beta^{+}, \psi_\alpha^{+}) + \left[(H - E_\beta)\chi_\beta, \left(\frac{1}{E_\beta - H + i\eta} - \frac{1}{E_\beta - H - i\eta}\right)\psi_\alpha^{+}\right] \\
&= (\psi_\beta^{+}, \psi_\alpha^{+}) + \left(\frac{1}{E_\beta - E_\alpha + i\eta} - \frac{1}{E_\beta - E_\alpha - i\eta}\right)((H - E_\beta)\chi_\beta, \psi_\alpha^{+}) \\
&= \lim_{\eta\to 0(+)} \left[\frac{i\eta}{E_\alpha - E_\beta + i\eta}(\chi_\beta, \chi_\alpha)\right. \\
&\qquad \left. - \frac{2\eta i}{(E_\beta - E_\alpha)^2 + \eta^2}((H - E_\beta)\,\chi_\beta, \psi_\alpha^{+})\right]
\end{aligned} \tag{62}$$

In deducing this result we have used Eq. (56) and the fact that $\psi_\alpha^{+}$ is an eigenfunction of $H$. We note that $S_{\beta\alpha}$ is zero unless $E_\beta = E_\alpha$. In the sense of wave packets, or simply summing over a small band of states $\beta$ for which $E_\beta \approx E_\alpha$, we have seen that the first term in Eq. (62) may be replaced by $\delta_{\beta\alpha}$ (that is, all aspects of the initial and final configurations must agree). The $\eta$-dependent factor in the second term is just the one encountered in Eq. (28). We have then, finally,

$$S_{\beta\alpha} = \delta_{\beta\alpha} - 2\pi i\, \delta(E_\beta - E_\alpha)((H - E_\beta)\chi_\beta, \psi_\alpha^{+}). \tag{63}$$

This is precisely Eq. (3.94). We may of course also solve for $\psi_\alpha^{+}$ in terms of $\psi_\alpha^{-}$,

$$\psi_\alpha^{+} = \psi_\alpha^{-} + \left(\frac{1}{E_\alpha - H + i\eta} - \frac{1}{E_\alpha - H - i\eta}\right)(H - E_\alpha)\chi_\alpha, \tag{64}$$

and substitute this into the definition of $S_{\beta\alpha}$. The arithmetic is identical with that of Eq. (62), and we find

$$S_{\beta\alpha} = \delta_{\beta\alpha} - 2\pi i\, \delta(E_\beta - E_\alpha)(\psi_\beta^{-}, (H - E_\alpha)\chi_\alpha). \tag{65}$$

The coefficient of the $\delta$-function appearing in Eq. (63) is just the quantity previously called $\lim_{\eta\to 0(+)}\mathscr{T}_{\beta\alpha}(\eta)$, which we now call $\overline{\mathscr{T}}{}^{(+)}_{\beta\alpha}$; by analogy we call the corresponding coefficient in Eq. (65) $\overline{\mathscr{T}}{}^{(-)}_{\beta\alpha}$:

$$\begin{aligned}
\overline{\mathscr{T}}{}^{(+)}_{\beta\alpha} &= ((H - E_\beta)\chi_\beta, \psi_\alpha^{+}), \\
\overline{\mathscr{T}}{}^{(-)}_{\beta\alpha} &= (\psi_\beta^{-}, (H - E_\alpha)\chi_\alpha).
\end{aligned} \tag{66}$$

For the evaluation of the $S$-matrix these quantities are required only on the energy shell $E_\beta = E_\alpha$; and then (*but only then*) they are equal, by construction. Using the explicit forms for $\psi^{\pm}$ we may define these $\mathscr{T}$-matrix elements quite generally even for unequal energies by[7]

$$\begin{aligned}\overline{\mathscr{T}}^{(+)}_{\beta\alpha} &= ((H - E_\beta)\chi_\beta, \chi_\alpha) + \left((H - E_\beta)\chi_\beta, \frac{1}{E_\alpha - H + i\eta}(H - E_\alpha)\chi_\alpha\right), \\ \overline{\mathscr{T}}^{(-)}_{\beta\alpha} &= (\chi_\beta, (H - E_\alpha)\chi_\alpha) + \left((H - E_\beta)\chi_\beta, \frac{1}{E_\beta - H + i\eta}(H - E_\alpha)\chi_\alpha\right).\end{aligned} \tag{67}$$

We shall return to discuss Eqs. (67) later in this chapter.

In a very cavalier fashion we compute from $S_{\beta\alpha}$ the transition probability per unit time. We have already done this so many times that the illegality of the following operations should not be too unsettling. Assume that $\beta \neq \alpha$, and then

$$\begin{aligned}P_{\beta\alpha} &= |S_{\beta\alpha}|^2 = [2\pi\,\delta(E_\beta - E_\alpha)]^2\,|\overline{\mathscr{T}}^{(\pm)}_{\beta\alpha}|^2 \\ &= [2\pi\,\delta(E_\beta - E_\alpha)]\left(\int_{-\infty}^{\infty} dt\, e^{i(E_\alpha - E_\beta)t}\right)|\overline{\mathscr{T}}^{(\pm)}_{\beta\alpha}|^2 \\ &= \left(\int_{-\infty}^{\infty} dt\right) 2\pi\,\delta(E_\beta - E_\alpha)|\overline{\mathscr{T}}^{(\pm)}_{\beta\alpha}|^2.\end{aligned} \tag{68}$$

The obvious interpretation of the rather large time integral is that it represents the total time of interaction (and that we should make proper use of wave packets, as in Chapter 3). The transition rate is then

$$\dot{P}_{\beta\alpha} = \frac{P_{\beta\alpha}}{\displaystyle\int_{-\infty}^{\infty} dt} = 2\pi\,\delta(E_\beta - E_\alpha)\,|\overline{\mathscr{T}}^{(\pm)}_{\beta\alpha}|^2, \tag{69}$$

which is just Eq. (29) except for the sum over a band of states required to give meaning to the $\delta$-function.

The unitarity of the $S$-matrix in the general case follows in a very simple way from our formalism. An arbitrary scattering state vector may be represented in terms of either precollision or postcollision packets; we may imagine having incident and outgoing packets in all available channels. Thus

$$\Psi = \Psi(t = 0) \equiv \sum_\alpha A_\alpha(-\infty)\psi_\alpha^{+} \equiv \sum_\beta A_\beta(+\infty)\psi_\beta^{-}, \tag{70}$$

where we mean the indices $\alpha$, $\beta$ to refer to both the channel $c$ and the channel eigenvalues $a$ of a state $\chi_{c;a}$. Evidently

$$A_\beta(+\infty) = \sum_\alpha (\psi_\beta^{-}, \psi_\alpha^{+})A_\alpha(-\infty) = \sum_\alpha S_{\beta\alpha}A_\alpha(-\infty), \tag{71}$$

[7] Only on the energy shell do we have $\overline{\mathscr{T}}^{(+)} = \overline{\mathscr{T}}^{(-)} = \mathscr{T}$, where $\mathscr{T}$ is defined by Eq. (3.82). The relations between the various quantities $\overline{\mathscr{T}}^{(+)}$, $\overline{\mathscr{T}}$, and $\overline{\mathscr{T}}^{(-)}$ are described in connection with Eqs. (4.12) and (4.13).

and

$$\sum_{\alpha} A_\alpha^*(-\infty)A_\alpha(-\infty) = \sum_{\beta} A_\beta^*(+\infty)A_\beta(+\infty). \tag{72}$$

These equations are a consequence of the orthonormality of the $\psi^\pm$ in a wave packet sense. If we substitute for $A_\beta(+\infty)$ in Eq. (72) its value as given by Eq. (71) we find

$$\sum_{\beta} S_{\beta\alpha}^* S_{\beta\alpha'} = \delta_{\alpha\alpha'}. \tag{73}$$

Thus the adjoint matrix $S^\dagger$ is the inverse of $S$. Note that if the definition of $S$ is inserted into (73) we have

$$\sum_{\beta} (\psi_\alpha^+, \psi_\beta^-)(\psi_\beta^-, \psi_{\alpha'}^+) = \delta_{\alpha\alpha'}, \tag{74}$$

which implies that although the $\psi_\beta^-$ are not a complete set of states in a true mathematical sense, within the space of physically meaningful asymptotic scattering states (which are restricted by the definition of $S$ to energy shell states) $\Sigma_\beta \psi_\beta^- \psi_\beta^{-\dagger}$ acts like a unit operator.

In closing this section we would like to emphasize that at no point have we needed to split the Hamiltonian into a perturbed and an unperturbed part. The only assumption made isthat the initial and final configurations be well defined in terms of separated noninteracting particle states, and we have made use of the fact that for large (positive or negative) times the state $\Psi(t)$ will correspond to particles outside their mutual ranges of interaction. If we are dealing with identical particles, the appropriate symmetrization of the initial and final configurations must be employed. As an example, we consider rather briefly one discussed in some detail in Chapter 4, namely the elastic scattering of electrons by a hydrogen atom. We refer to Eqs. (4.167*a*) and (4.191*a*) for the notation. The channel states $\chi_\alpha$, $\chi_\beta$ have the form

$$\chi_\alpha = \frac{1}{\sqrt{2}}(\chi_a - \chi_{Q(a)}),$$

$$\chi_\beta = \frac{1}{\sqrt{2}}(\chi_b - \chi_{Q(b)}). \tag{75}$$

(We write simply $Q$ for what was called $Q_{12}$ in Chapter 4.) Calling $E$ the common energy ($E_\alpha = E_\beta = E$), we have

$$(H - E)\chi_{\alpha,\beta} = \frac{1}{\sqrt{2}}(V\chi_{a,b} - V_Q\chi_{Q(a),Q(b)}) \tag{76}$$

and use the second of Eqs. (67) for the $\mathscr{T}$-matrix.

$$\mathscr{T}_{\beta\alpha} = \tfrac{1}{2}[(\chi_b, V\chi_a) - (\chi_b, V_Q\chi_{Q(a)}) - (\chi_{Q(b)}, V\chi_a) + (\chi_{Q(b)}, V_Q\chi_{Q(a)})]$$
$$+ \frac{1}{2}\left[\left((V\chi_b - V_Q\chi_{Q(b)}), \frac{1}{E - H + i\eta}(V\chi_a - V_Q\chi_{Q(a)})\right)\right]. \tag{77}$$

We note that the first and last terms in the first brackets are identical, as are the second and third since the pairs differ only by the labeling of integration variables. In a similar way there is a grouping of the second brackets. The result is

$$\begin{aligned}\mathscr{T}_{\beta\alpha} &= \left(\chi_b, \left(V + V\frac{1}{E - H + i\eta} V\right)\chi_a\right) \\ &\quad - \left(\chi_{Q(b)}, \left(V + V_Q \frac{1}{E - H + i\eta} V\chi_a\right)\right. \\ &= \mathscr{T}_d - \mathscr{T}_{\text{ex}}, \end{aligned} \tag{78}$$

which agrees precisely with Eqs. (4.192). We shall treat a few other examples later in the chapter.

## 5.3 INTEGRAL EQUATIONS FOR THE DETERMINATION OF THE S-MATRIX

Our formal development up to this point has provided us with essentially the kinematical machinery to compute cross sections. The next question is how to go about obtaining either exact values or reliable approximations for the necessary $S$-matrix elements. We have based our formalism largely on the explicit form of the functions $\psi_\alpha^\pm$ which are defined as the limit as $\eta \to 0(\pm)$ of $\psi_\alpha(\eta)$:

$$\psi_\alpha(\eta) = \frac{i\eta}{E_\alpha - H + i\eta}\chi_\alpha = \chi_\alpha + \frac{1}{E_\alpha - H + i\eta}(H - E_\alpha)\chi_\alpha. \tag{79}$$

We shall show that too cavalier a passage to the limit of $\eta \to 0(\pm)$ leads to certain ambiguities.

### The Lippmann-Schwinger Equation

To make the discussion simple, let us imagine that we may write $H = K_c + V_c$ and imagine that $\alpha$ stands for $(c; a)$, $K$ for $K_c$ while $\alpha'$, $K'$ for $(c'; a')$, $K_{c'}$. Then we recall Eq. (20$a$).

$$\psi_\alpha(\eta) = \chi_\alpha + \frac{1}{E_\alpha - K + i\eta} V\psi_\alpha(\eta). \tag{80}$$

We may argue that Eq. (79) is the unique solution of this integral equation for any finite value of $\eta$. The only ambiguity possible is that the homogeneous equation with $\chi_\alpha = 0$ had a solution. The resulting equation may be written as $(E_\alpha - K + i\eta - V)\psi_\alpha(\eta) = 0$, which is just the Schrödinger equation. Now in the asymptotic region where $V \to 0$, we have either exponentially increasing or exponentially decreasing behavior. The increasing solutions are unacceptable physically, and from the decreasing solutions we could in turn prove that the eigenvalues must be real. Thus there is in general no solution, for any finite $\eta$, of the equation

$$(E_\alpha + i\eta - K - V)\psi_\alpha(\eta) = 0.$$

To see the trouble we can encounter,[8] let us recall Eq. (20), which we reproduce here for convenience.

$$\psi_\alpha(\eta) = \frac{i\eta}{E_\alpha - K' + i\eta}\chi_\alpha + \frac{1}{E_\alpha - K' + i\eta}V'\psi_\alpha(\eta). \tag{81}$$

To contemplate the limit $\eta \to 0$, let us introduce the complete set of states belonging to $K'$; call them $\chi'_\gamma$. Then

$$\psi_\alpha(\eta) = \sum_\gamma \frac{i\eta}{E_\alpha - E_\gamma + i\eta}\chi'_\gamma(\chi'_\gamma, \chi_\alpha) + \sum_\gamma \frac{\chi'_\gamma(\chi'_\gamma, V'\psi_\alpha(\eta))}{E_\alpha - E_\gamma + i\eta}. \tag{82}$$

Now as $\eta \to 0$ the first term goes to zero, since $(\chi'_\gamma, \chi_\alpha)$ is finite and only an infinitesimal interval of $E_\gamma \approx E_\alpha$ contributes; in the second term the only role played by the $\eta$ is to give a rule as to how to avoid the singularity at $E_\gamma = E_\alpha$. Thus we would be inclined to write

$$\psi_\alpha^\pm = \left(\lim_{\eta\to 0(\pm)} \frac{1}{E_\alpha - K' + i\eta}\right)V'\psi_\alpha^\pm, \tag{83}$$

where the term in parentheses is to be defined by an expansion like that used in Eq. (82). If we were to write Eq. (80) with $\chi_{\alpha'}$ as the entrance channel, we would naturally write

$$\psi_{\alpha'}(\eta) = \chi_{\alpha'} + \frac{1}{E_{\alpha'} - K' + i\eta}V'\psi_{\alpha'}(\eta); \tag{84}$$

careless passage to the limit $\eta \to 0(\pm)$ leads to

$$\psi_{\alpha'}^\pm = \chi_{\alpha'} + \left(\lim_{\eta\to 0(\pm)} \frac{1}{E_{\alpha'} - K' + i\eta}\right)V'\psi_\alpha^\pm. \tag{85}$$

Evidently to any solution of Eq. (85) we may add a solution of Eq. (83). Thus the solution of Eq. (85) is not unique in contrast to the situation that obtains if $\eta$ is kept finite.

This somewhat unsettling situation might make the reader dubious about some of our earlier manipulations. Such doubts may be dispelled by remarking that we have used the limit $\eta \to 0(\pm)$ only under circumstances where wave packets can be used to verify the results. For example, in computing the overlap of the state (4) with a packet made up of channel states $\chi_{\alpha'}$ (which is essentially what is involved in computing the transition

[8] Attention was called to these difficulties in Refs. [10].

probability $\alpha \to \alpha'$) the first term in Eq. (82) really makes no contribution. Evidently any computation scheme based on Eqs. (83) and (85) is not possible without further information to make the solution unique.[9] The limiting equation, (85) is called the Lippmann-Schwinger equation.[10]

The Lippmann-Schwinger equation, or rather the scalar product of it with some eigenvector, may be derived without making the assumption $H = K_c + V_c$. To see this, let us imagine the limit $\eta \to 0(\pm)$ to be implied in Eq. (79) and take the scalar product with $\psi_\beta^\pm$, using the fact that $\psi_\beta^\pm$ is an eigenfunction of $H$.

$$(\psi_\beta^\pm, \chi_\alpha) = \delta_{\beta\alpha} + \frac{1}{E_\beta - E_\alpha \mp i\eta}(\psi_\beta^\pm, (H - E_\alpha)\chi_\alpha), \tag{86}$$

where we have written

$$(\psi_\beta^\pm, \psi_\alpha^\pm) = \delta_{\beta\alpha}. \tag{87}$$

Actually,

$$(\psi_\beta^{(\pm)}, \psi_\alpha^{(\pm)}) = \lim_{n \to 0(+)} \frac{\pm i\eta}{E_\alpha - E_\beta \pm i\eta}(\chi_\beta, \chi_\alpha), \tag{88}$$

and this is, of course, the source of the difficulty. While it is correct to say that the channel states are orthonormal effectively when their scalar product is taken as weighted with wave packet amplitudes, they are not truly orthonormal in a mathematical sense.

### Nonlinear Equations for the Scattering Matrix

Our difficulty lies in the fact that we appear to be encountering a homogeneous integral equation. There is a way to avoid this, which we now describe. The procedure to be developed is quite general and we discuss it without reference to the examples just considered. We recall Eq. (67) for $\bar{\mathscr{T}}_{\beta\alpha}^{(+)}$:

$$\bar{\mathscr{T}}_{\beta\alpha}^{(+)} = ((H - E_\beta)\chi_\beta, \psi_\alpha^+) = ((H - E_\beta)\chi_\beta, \chi_\alpha) + \left((H - E_\beta)\chi_\beta, \frac{1}{E_\alpha - H + i\eta}(H - E_\alpha)\chi_\alpha\right). \tag{89}$$

To reduce this still further we introduce a complete set of intermediate states which are eigenfunctions of $H$. As we have repeatedly emphasized, the various channel eigenfunctions do not in general form a complete set; rather they are over-complete. The bulk of the spectrum of $H$ may be spanned by the states whose asymptotic configurations are those where all

[9] We emphasize that this lack of uniqueness does not apply to the solutions of Eq. (80). As was shown in the discussion following Eq. (80), $\psi_a^\pm = \lim_{\eta\to 0(\pm)} \psi_a^\pm(\eta)$, where $\psi_a^\pm(\eta)$ is obtained from (80), is unique.

[10] See Refs. [4], where this equation was first obtained.

the particles are free. Aside from true bound states $\psi_d$ corresponding to all particles bound, there is in general a part of the continuous spectrum corresponding to channels in which various numbers of particles are bound together. Let us designate the complete set of continuum states by $\psi_\lambda^{\pm}$, without specifying them further for the moment. We may write, then, choosing $\psi_\lambda^+$, $\psi_d$ for the intermediate states,

$$\mathscr{T}^{(+)}_{\beta\alpha} = ((H - E_\beta)\chi_\beta, \chi_\alpha) + \sum_\lambda \frac{\mathscr{T}^{(+)}_{\beta\lambda}\mathscr{T}^{(+)*}_{\alpha\lambda}}{E_\alpha - E_\lambda + i\eta} + \sum_d \frac{((H - E_\beta)\chi_\beta, \psi_d)(\psi_d, (H - E_\alpha)\chi_\alpha)}{E_\alpha - E_d}. \tag{90}$$

The $i\eta$ is not needed in the last term since the $E_d$ do not coincide with the continuum $E_\alpha$ of interest. In a similar fashion we could consider the corresponding set of equations for $\mathscr{T}^{(-)}_{\beta\alpha}$, also defined by Eq. (89); this time we choose the intermediate states as $\psi_\lambda^-$, $\psi_d$. We then find

$$\mathscr{T}^{(-)}_{\beta\alpha} = (\chi_\beta, (H - E_\alpha)\chi_\alpha) + \sum_\lambda \frac{\mathscr{T}^{(-)}_{\lambda\beta}\,\mathscr{T}^{(-)}_{\lambda\alpha}}{E_\beta - E_\lambda + i\eta} + \sum_d \frac{((H - E_\beta)\chi_\beta, \psi_d)(\psi_d, (H - E_\alpha)\chi_\alpha)}{E_\beta - E_d}. \tag{91}$$

These nonlinear equations for the scattering matrix elements (which are the things of real interest) presumably do not suffer from the lack of uniqueness encountered in the Lippmann-Schwinger equations. They have a very attractive look about them, but they are in fact deceptively complicated. In the first place, in order even to begin we must determine the discrete eigenfunctions (if any) of the entire system. Next, we note that if $\alpha$, $\beta$ correspond to partially bound configurations in the sum over $\lambda$ we need $\mathscr{T}$-matrix elements between the partially bound $\rightarrow$ all-free configurations as well as partially bound $\rightarrow$ partially bound. Similarly, if $\alpha$, $\beta$ both refer to free configurations, the sum over $\lambda$ involves free $\rightarrow$ free as well as free $\rightarrow$ partially bound matrix elements. Finally, if the state described by one index is partially bound and the other is free, the sum involves partially bound $\rightarrow$ free, free $\rightarrow$ free, and partially bound $\rightarrow$ partially bound elements. Using $b$ for partially bound configurations and $f$ for all-free ones we may write the three possible sets of equations symbolically as

$$\begin{aligned} (bb) &= (bf)(fb) + (bb)(bb) + \text{discrete}, \\ (bf) &= (bf)(ff) + (bb)(bf) + \text{discrete}, \\ (ff) &= (ff)(ff) + (fb)(bf) + \text{discrete}. \end{aligned} \tag{92}$$

It is scarcely necessary to remark that this is a rather formidable set of equations in the general situation. One final comment: Even though we ultimately require the $\mathscr{T}$-matrix elements on the energy shell $E_\beta = E_\alpha$ for the evaluation of $S_{\beta\alpha}$ and of the transition probabilities, the integral equations involve these matrix elements off the energy shell. Physically this amounts to the need for knowing the wave function in both the near zone and the wave zone. We note, incidentally, that the Eqs. (90) and (91) provide a very quick derivation of the *optical theorem*, for if we form $\mathscr{T}_{\alpha\alpha} - \mathscr{T}^*_{\alpha\alpha}$, we find

$$2 \operatorname{Im} \mathscr{T}_{\alpha\alpha} = -2\pi \sum_\lambda |\mathscr{T}_{\alpha\lambda}|^2 \, \delta(E_\lambda - E_\alpha), \tag{93}$$

which leads directly to Eq. (39).

In spite of the great complexity of this formalism it has the virtue of complete generality and, as far as we know, has no pathologies. With the appropriate interpretation of the symbols it retains its validity even in quantum field theory. Frequently it may be made the basis of approximation techniques in which we use the operator form of the equations and do not put in a sum over states explicitly. We have already seen in Chapter 4 the power of this approach in our discussion of the exchange scattering of an electron by a hydrogen atom.

Another set of equations is frequently useful, especially in the formulation of variational principles. We assume for the present purpose that we can write the Hamiltonian as $H = K_c + V_c$ for various channels $c$. Indicating only the channel labels we have, as usual,

$$\psi_c^+ = \chi_c + \frac{1}{E_c - H + i\eta} V_c \chi_c. \tag{94}$$

We now use the familiar identity, Eq. (19), to write

$$\psi_c^+ = \chi_c + \frac{1}{E_c - K_{c'} + i\eta}\left(1 + V_{c'} \frac{1}{E_c - H + i\eta}\right) V_c \chi_c \tag{95}$$

and introduce the "operator" [see Eq. (4.10*a*)]

$$\mathscr{T}_{c'c} = V_c + V_{c'} \frac{1}{E_c - H + i\eta} V_c \tag{96}$$

to write $\psi_c^+$ as

$$\psi_c^+ = \chi_c + \frac{1}{E_c - K_{c'} + i\eta} \mathscr{T}_{c'c} \chi_c \,. \tag{97}$$

The quantity $\mathscr{T}_{c'c}$ is to be permitted to act only on states $\chi_c$ which are eigenfunctions of $K_c$ with energy $E_c$, and in practice we are interested in these only as channel states. Similarly, we are interested in final states which are eigenstates of $\chi_{c'}$. Unfortunately, as we shall see in a moment, we cannot confine our attention to channel $\chi_{c'}$ but need the complete set. We may easily derive an integral equation for $\mathscr{T}_{c'c}$ by again using the identity Eq. (19).

$$\mathscr{T}_{c'c} = V_c + V_{c'} \frac{1}{E_c - K_{c'} + i\eta}\left(1 + V_{c'} \frac{1}{E_c - H + i\eta}\right) V_c$$

$$= V_c + V_{c'} \frac{1}{E_c - K_{c'} + i\eta} \mathscr{T}_{c'c}. \tag{98}$$

We observe, once again, that $\mathscr{T}_{c'c}$ is equal to

$$\mathscr{T}^{(-)}_{c'c} = (\Psi_{c'}^{-}, V_c\chi_c) = \left(\chi_{c'}, \left(V_c + V_{c'} \frac{1}{E_{c'} - H + i\eta} V_c\right)\chi_c\right) \tag{99}$$

only if $E_{c'} = E_c$. If we write Eq. (98) in more detail we have

$$(\chi_{c';a'}, \mathscr{T}_{c'c}\chi_{c;a}) = (\chi_{c';a'}, V_c\chi_{c;a})$$
$$+ \sum_{a''} \frac{(\chi_{c';a'}, V_{c'}\chi^{c'}_{a''})(\chi^{c'}_{a''}, \mathscr{T}_{c'c}\chi_{c;a})}{E_{c;a} - E_{c';a''} + i\eta}. \tag{100}$$

We see that we must evidently consider matrix elements of $\mathscr{T}_{c'c}$ not only between channel states $\chi_{c';a'}$ and $\chi_{c;a}$ but also between $\chi^{c'}_{a'}$ (which of course includes the $\chi_{c';a'}$) and $\chi_{c;a}$. As far as we know, this equation has no inherent pathological properties.

## 5.4 SCATTERING FROM TWO POTENTIALS

Our general formulation of scattering theory has provided us with the machinery to discuss the most complicated rearrangement collisions. Needless to say, the evaluation of the general formulas presents severe difficulties in practice, some aspects of which will be explored in later chapters of this book. There is a class of interesting problems which, although included in our development, may be effectively reduced somewhat further by formal manipulations. The situation we have in mind is that of two or more potentials, some of which must for various physical reasons be treated (essentially) exactly, whereas others may be relatively small perturbations. Typical examples are (1) scattering of particles under the combined influence of Coulomb and nuclear forces; (2) bremsstrahlung, the emission of radiation by a particle which is scattered by a strong Coulomb or nuclear force; (3) meson production in a nucleon-nucleon collision; (4) the photoelectric effect, the ejection of a bound electron from an atom or of a bound nucleon from a nucleus by a photon.

The only reason any problem arises is, of course, that physical cross sections are defined in terms of asymptotic configurations in which even the very interactions we want to take into account exactly are no longer effective. Suppose $H = K + U + V$ where $U$ is the potential we wish to take into account exactly. The initial and final configurations are described by eigenfunctions of $K$. We are interested in the scattering matrix elements $\mathscr{T}_{\beta\alpha}$ only on the energy shell $E_\beta = E_\alpha \equiv E$. From the general theory we have

$$\mathscr{T}_{\beta\alpha} = ((H - E)\chi_\beta, \psi_\alpha^+) = (\chi_\beta, (U + V)\psi_\alpha^+), \tag{101}$$

where

$$\begin{aligned}\psi_\alpha^+ &= \chi_\alpha + \frac{1}{E - H + i\eta}(U + V)\chi_\alpha \\ &= \chi_\alpha + \frac{1}{E - K + i\eta}(U + V)\psi_\alpha^+.\end{aligned}$$

Our object is to eliminate the plane-wave state $\chi_\beta$ and the explicit appearance of the strong potential $U$, insofar as it is possible. To do this we introduce the *incoming wave states* $\phi_\beta^-$, which are eigenfunctions (with energy $E$) of the Hamiltonian $K + U$. Thus we write

$$\phi_\beta^- = \chi_\beta + \frac{1}{E - K - i\eta} U\phi_\beta^-. \tag{102}$$

If we substitute the expression (102) for $\chi_\beta$ in Eq. (101) we obtain

$$\begin{aligned}\mathscr{T}_{\beta\alpha} &= (\phi_\beta^-, (U + V)\psi_\alpha^+) - \left(\phi_\beta^-, U\frac{1}{E - K + i\eta}(U + V)\psi_\alpha^+\right) \\ &= (\phi_\beta^-, (U + V)\psi_\alpha^+) - (\phi_\beta^-, U(\psi_\alpha^+ - \chi_\alpha)) \\ &= (\phi_\beta^-, V\psi_\alpha^+) + (\phi_\beta^-, U\chi_\alpha).\end{aligned} \tag{103}$$

Now the second term is just the scattering matrix element that would have been obtained if the potential $V$ had not been present. In several of the examples mentioned it would vanish. In bremsstrahlung $\phi_\beta^-$ contains a photon, $\chi_\alpha$ does not, and $U$ describes either a Coulomb or a nuclear potential; similarly, for a reaction like the meso-disintegration of the deuteron, $\chi_\alpha$ has a meson, whereas $\phi_\beta^-$ does not and $U$ is a two-nucleon potential. Again, for the atomic or nuclear photoeffect the second term vanishes.

The appearance of the ingoing wave state $\phi_\beta^-$ in the first term of equation (103) has been the cause of considerable confusion in the literature. Many papers are simply wrong because they have $\phi_\beta^+$ instead. There is also an extensive literature composed of expositions on the subject of why we should have $\phi_\beta^-$ instead of $\phi_\beta^+$. Actually the problem is completely and correctly discussed in the first edition of the classic book of

Mott and Massey [11]. From any one of several of the standpoints we develop it is inevitable that $\phi_\beta^-$ is the correct quantity. The simplest physical statement is that the $\phi_\beta^-$ are the functions appropriate to the description of postcollision packets, and our wave packet formalism leads directly to this result. Mathematically the result is an elementary consequence of asking the physically relevant question and then simply doing a consistent calculation.

Our derivation would have been even more direct, although essentially equivalent, if we had started with the $\mathscr{T}$-matrix element in the form

$$\mathscr{T}_{\beta\alpha} = (\psi_\beta^-, (H - E)\chi_\alpha) = (\psi_\beta^-, (U + V)\chi_\alpha). \tag{104}$$

We then remark that the state $\psi_\beta^-$, which is defined by

$$\psi_\beta^- = \chi_\beta + \frac{1}{E - H - i\eta}(U + V)\chi_\beta, \tag{105}$$

may be manipulated as follows:

$$\psi_\beta^- = \chi_\beta + \frac{1}{E - K - U - i\eta}U\chi_\beta + \frac{1}{E - H - i\eta}(U + V)\chi_\beta - \frac{1}{E - K - U - i\eta}U\chi_\beta. \tag{106}$$

Here we have merely added and subtracted a term in an obvious way. Recalling the definition of $\phi_\beta^-$, Eq. (102), we have

$$\begin{aligned}\psi_\beta^- &= \phi_\beta^- + \frac{1}{E - H - i\eta}V\chi_\beta + \left(\frac{1}{E - H - i\eta} - \frac{1}{E - K - U - i\eta}\right)U\chi_\beta \\ &= \phi_\beta^- + \frac{1}{E - H - i\eta}V\left(\chi_\beta + \frac{1}{E - K - U - i\eta}U\chi_\beta\right) \qquad (107) \\ &= \phi_\beta^- + \frac{1}{E - H - i\eta}V\phi_\beta^-, \qquad (108)\end{aligned}$$

which is perhaps an obvious result. At any rate, substituting this representation of $\psi_\beta^-$ into Eq. (104) leads to

$$\begin{aligned}\mathscr{T}_{\beta\alpha} &= (\phi_\beta^-, (U + V)\chi_\alpha) + \left(\phi_\beta^-, V\frac{1}{E - H + i\eta}(U + V)\chi_\alpha\right) \\ &= (\phi_\beta^-, (U + V)\chi_\alpha) + (\phi_\beta^-, V(\psi_\alpha^+ - \chi_\alpha)) \qquad (109) \\ &= (\phi_\beta^-, U\chi_\alpha) + (\phi_\beta^-, V\psi_\alpha^+),\end{aligned}$$

which is our previous answer [Eq. (103)]. The state vector $\psi_\alpha^+$ may evidently be expressed as in Eq. (108) by

$$\psi_\alpha^+ = \phi_\alpha^+ + \frac{1}{E_\alpha - H + i\eta} V\phi_\alpha^+, \tag{110}$$

which is the outgoing wave version of Eq. (108). If this is substituted into Eq. (109) we obtain

$$\mathscr{T}_{\beta\alpha} = (\phi_\beta^-, U\chi_\alpha) + \left(\phi_\beta^-, \left(V + V\frac{1}{E_\alpha - H + i\eta} V\right)\phi_\alpha^+\right), \tag{111}$$

in which the $U$-dependence is still partially contained in $H$. To the first order in $V$, however, $U$ appears only in the (presumably) known eigenfunctions $\phi^\pm$.

A third way to do this computation is perhaps worth recording. We use directly the definition of the $S$-matrix element in terms of $\psi^\pm$.

$$\begin{aligned} S_{\beta\alpha} &= (\psi_\beta^-, \psi_\alpha^+) \\ &= \left(\phi_\beta^- + \frac{1}{E_\beta - H - i\eta} V\phi_\beta^-, \phi_\alpha^+ + \frac{1}{E_\alpha - H + i\eta} V\phi_\alpha^+\right). \end{aligned} \tag{112}$$

Strictly speaking we should use different $\eta$ and pass to the limit independently; the result is the same, although more tedious, when obtained carefully. Since we have already done the calculation by two other methods, we shall be a little cavalier here.

$$\begin{aligned} S_{\beta\alpha} = (\phi_\beta^-, \phi_\alpha^+) &+ \left(\phi_\beta^-, \frac{1}{E_\alpha - H + i\eta} V\phi_\alpha^+\right) \\ &+ \left(V\phi_\beta^-, \frac{1}{E_\beta - H + i\eta} \phi_\alpha^+\right) \\ &+ \left(V\phi_\beta^-, \frac{1}{E_\beta - H + i\eta}\frac{1}{E_\alpha - H + i\eta} V\phi_\alpha^+\right). \end{aligned} \tag{113}$$

Now we write

$$\frac{1}{E_\beta - H + i\eta} = \frac{1}{E_\beta - K - U + i\eta} + \frac{1}{E_\beta - H + i\eta} V \frac{1}{E_\beta - K - U + i\eta}, \tag{114}$$

$$\frac{1}{E_\alpha - H + i\eta} = \frac{1}{E_\alpha - K - U + i\eta} + \frac{1}{E_\alpha - K - U + i\eta} V \frac{1}{E_\alpha - H + i\eta}.$$

Substituting these expressions into the second and third terms of Eq. (113) we obtain

$$S_{\beta\alpha} = (\phi_\beta^-, \phi_\alpha^+) + \frac{1}{E_\alpha - E_\beta + i\eta}\left[\left(\phi_\beta^-, \left(1 + V\frac{1}{E_\alpha - H + i\eta}\right)V\phi_\alpha^+\right)\right]$$
$$+ \frac{1}{E_\beta - E_\alpha + i\eta}\left[\left(V\phi_\beta^-, \left(1 + \frac{1}{E_\beta - H + i\eta}V\right)\phi_\alpha^+\right)\right]$$
$$+ \left(V\phi_\beta^-, \frac{1}{E_\beta - H + i\eta}\frac{1}{E_\alpha - H + i\eta}V\phi_\alpha^+\right). \quad (115)$$

Now the last term may be written as

$$\frac{P}{E_\alpha - E_\beta}\left(V\phi_\beta^-, \left(\frac{1}{E_\beta - H + i\eta} - \frac{1}{E_\alpha - H + i\eta}\right)V\phi_\alpha^+\right), \quad (116)$$

where we have arbitrarily chosen to write "principal value" as the instruction for handling the apparent singularity at $E_\alpha = E_\beta$. In fact, there is no real singularity at that point, so any rule is adequate. If now in the second and third terms we write

$$\frac{1}{E_\alpha - E_\beta + i\eta} = \frac{P}{E_\alpha - E_\beta} - i\pi\delta(E_\alpha - E_\beta),$$
$$\frac{1}{E_\beta - E_\alpha + i\eta} = \frac{P}{E_\beta - E_\alpha} - i\pi\delta(E_\alpha - E_\beta), \quad (117)$$

we see that the entire last term in Eq. (115) is canceled and we have

$$S_{\beta\alpha} = (\phi_\beta^-, \phi_\alpha^+) - 2\pi i\delta(E_\alpha - E_\beta)\left(\phi_\beta^-, \left(V + V\frac{1}{E_\alpha - H + i\eta}V\right)\phi_\alpha^+\right). \quad (118)$$

Since $(\phi_\beta^-, \phi_\alpha^+)$ is just the $S$-matrix element that would have obtained if $V = 0$, and since

$$(\phi_\beta^-, \phi_\alpha^+) = \delta_{\beta\alpha} - 2\pi i\delta(E_\alpha - E_\beta)(\chi_\beta, U\phi_\alpha^+)$$
$$= \delta_{\beta\alpha} - 2\pi i\delta(E_\alpha - E_\beta)(\phi_\beta^-, U\chi_\alpha), \quad (119)$$

we see that our earlier result is recovered; that is, the coefficient of $-2\pi i\delta(E_\alpha - E_\beta)$ is just $\mathscr{T}_{\beta\alpha}$.

Actually we have given this argument in perhaps the most difficult possible manner. (The reason for doing so is that the reader might be tempted to try it himself anyway and encounter trouble.) A more elegant method is

$$S_{\beta\alpha} = (\psi_\beta^-, \psi_\alpha^+) = (\phi_\beta^-, \psi_\alpha^+) + \left(\phi_\beta^-, V \frac{1}{E_\beta - H + i\eta} \psi_\alpha^+\right)$$

$$= (\phi_\beta^-, \psi_\alpha^+) + \frac{1}{E_\beta - E_\alpha + i\eta} (\phi_\beta^-, V\psi_\alpha^+), \tag{120}$$

since $\psi_\alpha^+$ is an eigenfunction of $H$ with energy $E_\alpha$. But

$$(\phi_\beta^-, \psi_\alpha^+) = (\phi_\beta^-, \phi_\alpha^+) + \left(\phi_\beta^-, \frac{1}{E_\alpha - H + i\eta} V\phi_\alpha^+\right)$$

$$= (\phi_\beta^-, \phi_\alpha^+) + \left(\phi_\beta^-, \frac{1}{E_\alpha - K - U + i\eta} V\phi_\alpha^+\right)$$

$$+ \left(\phi_\beta^-, \frac{1}{E_\alpha - K - U + i\eta} V \frac{1}{E_\alpha - H + i\eta} V\phi_\alpha^+\right), \tag{121}$$

using the identity, Eq. (114). In the last term we have set $(E_\alpha - H + i\eta)^{-1} V\phi_\alpha^+ = \psi_\alpha^+ - \phi_\alpha^+$, which then cancels the second term and leaves us with

$$(\phi_\beta^-, \psi_\alpha^+) = (\phi_\beta^-, \phi_\alpha^+) + \frac{1}{E_\alpha - E_\beta + i\eta} (\phi_\beta^-, V\psi_\alpha^+). \tag{122}$$

Hence

$$S_{\beta\alpha} = (\phi_\beta^-, \phi_\alpha^+) + \left(\frac{1}{E_\beta - E_\alpha + i\eta} + \frac{1}{E_\alpha - E_\beta + i\eta}\right)(\phi_\beta^-, V\psi_\alpha^+)$$

$$= (\phi_\beta^-, \phi_\alpha^+) - 2\pi i\delta(E_\alpha - E_\beta)(\phi_\beta^-, V\psi_\alpha^+), \tag{123}$$

in the limit as $\eta \to 0$, just as before.

Another example that differs really only in technical detail is afforded by the process $\pi^+ + d \to 2p$. The only slightly different feature here is that the initial configuration, that of a free pi-meson and a deuteron, satisfies a homogeneous integral equation rather than something analogous to the $\phi^\pm$ of the previous example. We imagine that the nucleon-nucleon interaction is called $U$, the pion-nucleon interaction $V$. (The representation of the latter by a "potential" involves some subtleties.) The initial configuration is described by the solution of

$$\phi_\alpha = \frac{1}{E_\alpha - K + i\eta} U\phi_\alpha, \tag{124}$$

where, of course, the meson coordinates are contained only in the kinetic energy operator $K$. The solution to this integral equation consists of just a product of plane waves representing the meson and the deuteron center of mass and the bound-deuteron function.

Now physically we are interested in the $S$-matrix element describing the transition from the initial configuration $\phi_\alpha$ to a final configuration which is a plane-wave state of two outgoing protons. Thus we compute the matrix element $S_{\beta\alpha} = (\psi_\beta^-, \psi_\alpha^+)$ where

$$\begin{aligned}\psi_\alpha^+ &= \phi_\alpha + \frac{1}{E_\alpha - H + i\eta} V\phi_\alpha,\\ \psi_\beta^- &= \chi_\beta + \frac{1}{E_\beta - H - i\eta}(U + V)\chi_\beta,\end{aligned} \tag{125}$$

and $\chi_\beta$ is the two-proton plane-wave state, an eigenfunction of $K$. Then

$$\begin{aligned}S_{\beta\alpha} &= (\chi_\beta, \psi_\alpha^+) + \left(\chi_\beta, (U + V)\frac{1}{E_\beta - H + i\eta}\psi_\alpha^+\right)\\ &= (\chi_\beta, \psi_\alpha^+) + \frac{1}{E_\beta - E_\alpha + i\eta}(\chi_\beta, (U + V)\psi_\alpha^+).\end{aligned} \tag{126}$$

Now

$$\begin{aligned}(\chi_\beta, \psi_\alpha^+) &= (\chi_\beta, \phi_\alpha) + \left(\chi_\beta, \frac{1}{E_\alpha - H + i\eta} V\phi_\alpha\right)\\ &= (\chi_\beta, \phi_\alpha) + \left(\chi_\beta, \frac{1}{E_\alpha - K + i\eta} V\phi_\alpha\right)\\ &\quad + \left(\chi_\beta, \frac{1}{E_\alpha - K + i\eta}(U + V)\frac{1}{E_\alpha - H + i\eta} V\phi_\alpha\right)\\ &= (\chi_\beta, \phi_\alpha) + \left(\chi_\beta, \frac{1}{E_\alpha - K + i\eta} V\phi_\alpha\right)\\ &\quad + \left(\chi_\beta, \frac{1}{E_\alpha - K + i\eta}(U + V)(\psi_\alpha^+ - \phi_\alpha)\right).\end{aligned} \tag{127}$$

The part of the third term involving $\phi_\alpha$ drops out entirely; the $V$ cancels the second term, and the $U$ cancels the first term because of Eq. (124). We are left then with

$$(\chi_\beta, \psi_\alpha^+) = \frac{1}{E_\alpha - E_\beta + i\eta}(\chi_\beta, (U + V)\psi_\alpha^+) \tag{128}$$

and

$$S_{\beta\alpha} = -2\pi i\delta(E_\beta - E_\alpha)(\chi_\beta, (U + V)\psi_\alpha^+). \tag{129}$$

We may again replace the plane wave $\chi_\beta$ in terms of $\phi_\beta^-$, just as we did in the previous example. Calling the common energy of $\chi_\beta$ and $\psi_\alpha^+$ $E$, we have

$$\begin{aligned}(\chi_\beta, (U + V)\psi_\alpha^+) &= (\chi_\beta, U\psi_\alpha^+) + (\phi_\beta^-, V\psi_\alpha^+)\\ &\quad - \left(\chi_\beta, U\frac{1}{E - K - U + i\eta} V\psi_\alpha^+\right),\end{aligned} \tag{130}$$

where we have used $\phi_\beta^-$ in the form

$$\phi_\beta^- = \chi_\beta + \frac{1}{E_\beta - K - U - i\eta} U\chi_\beta. \tag{131}$$

In the last term of Eq. (130) we write $(E - K - U + i\eta)^{-1}V\psi_\alpha^+ = \psi_\alpha^+ - \phi_\alpha$, which follows from Eq. (125) in the standard way.

$$\begin{aligned}\psi_\alpha^+ &= \phi_\alpha + \frac{1}{E_\alpha - K - U + i\eta} V\phi_\alpha + \frac{1}{E_\alpha - K - U + i\eta} V \frac{1}{E_\alpha - H + i\eta} V\phi_\alpha \\ &= \phi_\alpha + \frac{1}{E_\alpha - K - U + i\eta} V\phi_\alpha + \frac{1}{E_\alpha - K - U + i\eta} V(\psi_\alpha^+ - \phi_\alpha) \\ &= \phi_\alpha + \frac{1}{E_\alpha - K - U + i\eta} V\psi_\alpha^+. \end{aligned} \tag{132}$$

Thus we have from Eq. (130)

$$\begin{aligned}(\chi_\beta, (U + V)\psi_\alpha^+) &= (\chi_\beta, U\psi_\alpha^+) + (\phi_\beta^-, V\psi_\alpha^+) - (\chi_\beta, U(\psi_\alpha^+ - \phi_\alpha)) \\ &= (\phi_\beta^-, V\psi_\alpha^+) + (\chi_\beta, U\phi_\alpha). \end{aligned} \tag{133}$$

In our example, since $\phi_\alpha$ contains a meson and $\chi_\beta$ does not, whereas $U$ acts only on nucleons, the second term is zero. The relevant transition probability is given by (where the wave functions are given in the barycentric subspace)

$$\dot{P}_{\beta\alpha} = 2\pi\delta(E_\beta - E_\alpha)\,|(\phi_\beta^-, V\psi_\alpha^+)|^2, \tag{134}$$

as in Eq. (69). Again, *to the first order in* $V$, the nucleon potential $U$ is taken into account *exactly* by replacing $\psi_\alpha^+$ by $\phi_\alpha$.

### 5.5 THE *S*-MATRIX AND THE INTERACTION REPRESENTATION

In order to illustrate most simply the relation between the $S$-matrix in the Heisenberg representation and the $S$-matrix in the interaction representation we consider a situation for which the Hamiltonian may be split in a natural way into precisely two parts $H = K + V$.[11] This covers the example of one-channel scattering processes and, aside from certain technical difficulties to be treated later, many problems in quantum field theory.[12]

The initial and final configurations may be described in terms of the eigenfunctions of $K$, $\chi_\alpha$, which are essentially plane-wave states (or more

[11] We here follow the discussion given in Refs. [5] and [6].

[12] It is no longer very fashionable to use this formalism in field theory but in olden times (circa 1948–1954) it was quite usual.

properly, wave packets composed of them); we assume that these form a complete orthonormal set. We shall also assume that the eigenfunctions $\psi^{\pm}$ of $H$ obtained by our limiting process are complete except for discrete bound states which lie below the continuous spectrum of $H$, which is assumed to coincide with that of $K$. The $S$-matrix elements are written as

$$S_{\alpha'\alpha} = \delta_{\alpha'\alpha} - 2\pi i\delta(E_{\alpha'} - E_{\alpha})\mathscr{T}_{\alpha'\alpha}, \tag{135}$$

where $\mathscr{T}_{\alpha'\alpha}$ is the common value (on the energy shell) of $\bar{\mathscr{T}}^{(\pm)}_{\alpha'\alpha}$ and

$$\bar{\mathscr{T}}^{(+)}_{\alpha'\alpha} = (\chi_{\alpha'}, V\psi_{\alpha}^{+}), \tag{136}$$

$$\bar{\mathscr{T}}^{(-)}_{\alpha'\alpha} = (\psi_{\alpha'}^{-}, V\chi_{\alpha}). \tag{137}$$

Since the $\psi_{\alpha}^{\pm}$ are a complete set that spans the entire continuous spectrum of $H$, the array of numbers $S_{\alpha'\alpha}$ may be regarded as the matrix elements of an operator $S$ which describes the mapping of the basis vectors $\psi^{-}$ onto the corresponding vectors $\psi^{+}$.

$$\psi_{\alpha}^{+} = S\psi_{\alpha}^{-}. \tag{138}$$

We have then

$$\begin{aligned} S_{\alpha'\alpha} &= (\psi_{\alpha'}^{-}, \psi_{\alpha}^{+}) = (\psi_{\alpha'}^{-}, S\psi_{\alpha}^{-}) \\ &= (\psi_{\alpha'}^{+}, S\psi_{\alpha}^{+}). \end{aligned} \tag{139}$$

The unitarity of the operator $S$ has been used to write the last form of this equation. This property must be proved, but it is almost physically obvious. The relation between $\psi_{\alpha}^{\pm}$ and $\chi_{\alpha}$ may be expressed in terms of operators $\Omega^{\pm}$ by

$$\psi_{\alpha}^{\pm} = \Omega^{\pm}\chi_{\alpha}, \tag{140}$$

or

$$\Omega^{\pm} = \sum_{\alpha} |\psi_{\alpha}^{\pm}\rangle\langle\chi_{\alpha}|. \tag{141}$$

An explicit form for these operators may be given in terms of the operator $U(t, t')$ described in Chapter 2, using the method developed earlier in this chapter to take limits of infinite times.

$$\begin{aligned} \Omega^{(\pm)} &= \lim_{t\to\mp\infty} U(0, t) \\ &= \lim_{t\to\mp\infty} e^{iHt}e^{-iKt} \\ &= \lim_{\eta\to 0(+)} \mp\eta\int_0^{\mp\infty} d\tau\, e^{\mp\eta\tau}e^{iH\tau}e^{-iK\tau} \\ &= \lim_{\eta\to 0(+)} \sum_{\alpha} \frac{\pm i\eta}{E_{\alpha} - H \pm i\eta} |\chi_{\alpha}\rangle\langle\chi_{\alpha}|, \end{aligned} \tag{142}$$

where we have used the completeness relation $\Sigma_{\alpha} |\chi_{\alpha}\rangle\langle\chi_{\alpha}| = 1$. Recalling the definition $\psi_\alpha^+ = \lim (i\eta)(E_\alpha - H + i\eta)^{-1}\chi_\alpha$ [Eq. (51)], we see that Eqs. (141) and (142) agree with our previous definition of the $\psi^\pm$. The operators $\Omega^\pm$ are referred to as Møller wave matrices.

$$\langle\chi_{\alpha'}|\, \Omega^\pm \,|\chi_\alpha\rangle = \langle\chi_{\alpha'} \mid \psi_\alpha^\pm\rangle. \tag{143}$$

We note some properties of these operators:

$$\Omega^\pm\dagger\Omega^\pm = \sum_{\alpha\alpha'} |\chi_{\alpha'}\rangle\langle\psi_{\alpha'}^\pm \mid \psi_\alpha^\pm\rangle\langle\chi_\alpha| = \sum_\alpha |\chi_\alpha\rangle\langle\chi_\alpha| = I, \tag{144}$$

where we have used the limiting normalization to unity of the $\psi_\alpha^\pm$ as proved earlier. On the other hand,

$$\Omega^\pm\Omega^\pm\dagger = \sum_{\alpha',\alpha} |\psi_{\alpha'}^\pm\rangle\langle\chi_{\alpha'} \mid \chi_\alpha\rangle\langle\psi_\alpha^\pm|$$

$$= \sum_\alpha |\psi_\alpha^\pm\rangle\langle\psi_\alpha^\pm| = 1 - \Lambda_d \tag{145}$$

where $\Lambda_d$ is the projection operator onto bound states. The $S$-operator may be expressed in terms of the $\Omega$ as follows. From its definition, Eq. (140), we have

$$\Omega^+\chi_\alpha = S\Omega^-\chi_\alpha, \tag{146}$$

or since the $\chi_\alpha$ are complete,

$$\Omega^+ = S\Omega^-, \tag{147}$$

and from Eq. (145) we have

$$\Omega^+\Omega^-\dagger\langle = S(1 - \Lambda_d). \tag{148}$$

Since we are interested in the matrix elements of $S$ only between scattering states, we may drop $\Lambda_d$ and write

$$S = \Omega^+\Omega^-\dagger. \tag{149}$$

Another operator is frequently referred to as the $S$-matrix, or $S$-operator. It is, in fact, the one used earlier in Chapters 3 and 4. This is derived as follows:

$$S_{\alpha'\alpha} = (\psi_{\alpha'}^-, \psi_\alpha^+) = (\Omega^-\chi_{\alpha'}, \Omega^+\chi_\alpha)$$

$$= (\chi_{\alpha'}, \Omega^-\dagger\Omega^+\chi_\alpha). \tag{150}$$

It is natural, therefore, to define an operator which we call $S_I$ (the subscript $I$ refers to the interaction representation for a reason to be revealed shortly),

$$S_I = \Omega^-\dagger\Omega^+. \tag{151}$$

The relation between $S_I$ and the interaction representation is simply derived. We recall from Chapter 2 that the transformation matrix $U(t, t')$ relates the interaction representation states $\Psi_I(t)$ and $\Psi_I(t')$ according to

$$\Psi_I(t) = U(t, t')\Psi_I(t'). \tag{152}$$

Now we remark that $\Psi_s(t)$ is the state in the Schrödinger representation

$$\Psi_I(t) = e^{iKt}\Psi_s(t), \tag{153}$$

so that if $\Psi_s(t)$ approaches $\Phi(t) = e^{-iKt}\chi$, $\Psi_I(t)$ approaches the constant state vector $\chi$. We take the latter to be a pure state, although properly, of course, it is a precollision packet. In the remote future, by the same argument, $\Psi_I(t)$ will approach a constant state vector $\chi'$.

$$\Psi_I(\infty) = \chi' = U(\infty, -\infty)\chi. \tag{154}$$

By definition the $S$-matrix element (between sharp states) $S_{\alpha'\alpha}$ is

$$S_{\alpha',\alpha} = (\chi_{\alpha'}, \chi'_{\alpha}) = (\chi_{\alpha'}, U(\infty, -\infty)\chi_\alpha), \tag{155}$$

and we are led to define an operator $S_I$ acting between plane-wave states (that is, eigenstates of $K$) by

$$S_I = U(\infty, -\infty). \tag{156}$$

We may also arrive at this result by remarking that [see Eq. (2.87)]

$$U(t, t') = e^{iKt}\, e^{-iH(t-t')}\, e^{-iKt'} \tag{157}$$

and

$$\lim_{\substack{t\to+\infty \\ t'\to-\infty}} (\chi_{\alpha'}, U(t, t')\chi_\alpha) = \lim\, (e^{iHt}\, e^{-iKt}\chi_{\alpha'}, e^{iHt'}\, e^{-iKt'}\,\chi_\alpha)$$
$$= (U(0, \infty)\chi_{\alpha'}, U(0, -\infty)\chi_\alpha)$$
$$= (\psi_{\alpha'}^{-}, \psi_\alpha^{+}). \tag{158}$$

From the fact shown in Eq. (2.95), that $U^\dagger(0, \infty) = U(\infty, 0)$, we have immediately the identification $S_I = U(\infty, 0)U(0, -\infty) = U(\infty, -\infty)$. We may also write this as

$$S_I = \Omega^{-\dagger}\Omega^{+}, \tag{159}$$

which is just Eq. (151).

It is clear from the foregoing discussion that the set of numbers $S_{\alpha'\alpha}$ computed either as the matrix elements of $S = \Omega^{+}\Omega^{-\dagger}$ between $\psi_\alpha^{\pm}$ states, or as $S_I = \Omega^{-\dagger}\Omega^{+}$ between $\chi_\alpha$ states, are the same. It is $S_I$ (generalized for many channels) which was used in Chapters 3 and 4. A few further remarks should be made about the relationship between the two operators. We note that

$$\begin{aligned} H\psi_\alpha^{\pm} = H\Omega^{\pm}\chi_\alpha = E_\alpha\psi_\alpha^{\pm} = E_\alpha\Omega^{\pm}\chi_\alpha \\ = \Omega^{\pm}K\chi_\alpha, \end{aligned} \tag{160}$$

so that

$$H\Omega^{\pm} = \Omega^{\pm}K, \qquad \Omega^{\pm\dagger}H = K\Omega^{\pm\dagger}. \tag{161}$$

As a consequence of these equations, we obtain

$$KS_I = K\Omega^{-\dagger}\Omega^{+} = \Omega^{-\dagger}H\Omega^{+}$$
$$= \Omega^{-\dagger}\Omega^{+}K = S_I K \tag{162}$$

and

$$HS = H\Omega^{+}\Omega^{-\dagger} = \Omega^{+}K\Omega^{-\dagger}$$
$$= \Omega^{+}\Omega^{-\dagger}H = SH. \tag{163}$$

Thus $[K, S_I] = 0$ and $[H, S] = 0$.

The proof that $S$ is unitary in the present case is trivial, since the $\psi_\alpha^{\pm}$ are complete except for bound states.

$$S^\dagger S = \sum_{\alpha;\alpha'} |\psi_\alpha^-\rangle\langle\psi_\alpha^+|\psi_{\alpha'}^+\rangle\langle\psi_{\alpha'}^-|$$
$$= \sum_\alpha |\psi_\alpha^+\rangle\langle\psi_\alpha^+| = I - \Lambda_d. \tag{164}$$

Since we are concerned only with the continuous spectrum the $\Lambda_d$ is irrelevant. The proof that $SS^\dagger = I$ is identical. For $S_I$ one point perhaps requires some amplification, and it is connected with the fact that $\Omega^{\pm\dagger}$ acting on a discrete (bound) state gives zero. This is suggested by

$$\Omega^{\pm}\Omega^{\pm\dagger} = 1 - \Lambda_d \tag{165}$$

and also follows from our limiting process on $U(t, t')$.

$$\Omega^{\pm\dagger} = \lim_{t\to\mp\infty} U^\dagger(0, t) = \lim_{t\to\mp\infty} U(t, 0)$$
$$= \lim_{\eta\to 0} \sum_\gamma \frac{\pm i\eta}{E_\gamma - K \pm i\eta} |\psi_\gamma\rangle\langle\psi_\gamma|, \tag{166}$$

where the $|\psi_\gamma\rangle$ represent the complete set, including bound states, of eigenfunctions of $H$. Now allow this operator to act on a particular $|\psi_\gamma\rangle$, a bound state eigenfunction:

$$\Omega^{\pm\dagger}|\psi_d\rangle = \lim_{\eta\to 0} \frac{\pm i\eta}{E_d - K \pm i\eta} |\psi_d\rangle$$
$$= \lim_{\eta\to 0} \sum_\alpha \frac{\pm i\eta}{E_d - E_\alpha \pm i\eta} |\chi_\alpha\rangle\langle\chi_\alpha \mid \psi_d\rangle$$
$$= 0, \tag{167}$$

since $E_d$ lies below the continuous spectrum of eigenvalues of $K$. The demonstration of the unitarity of $S_I$ is now easy.

$$\begin{aligned} S_I{}^\dagger S_I &= \Omega^{+\dagger}\Omega^{-}\Omega^{-\dagger}\Omega^{+} \\ &= \Omega^{+\dagger}(1 - \sum_d |\psi_d\rangle\langle\psi_d|)\Omega^{+} \\ &= \Omega^{+\dagger}\Omega^{+} = I, \end{aligned} \tag{168}$$

since $\Omega^{+\dagger}\,|\psi_d\rangle = 0$. The argument for $S_I S_I{}^\dagger$ is the same. Thus $S_I$ is unitary and defined over the complete set $\chi_\alpha$, whereas $S$ is defined and unitary only in the space spanned by the continuum $\psi_\alpha{}^{\pm}$.

Provided we are willing to introduce a different splitting of $H$ for every channel $c$, we may introduce $\Omega$ operators for every channel, $\Omega_c$, according to the definition

$$\Omega_c{}^{\pm} = \lim_{t\to\mp\infty} e^{iHt}\Lambda_c\, e^{-iK_ct} \tag{169}$$

where $\Lambda_c$ is the channel projection operator. These objects act only on restricted, asymptotic channel spaces, as discussed in Chapter 4. The asymptotic normalization of states discussed there,

$$\lim_{t\to\pm\infty} (\Lambda_{c'}\Psi(t), \Lambda_c\Psi(t)) = 0 \qquad (c' \neq c) \tag{170}$$

implies that since the normalization is preserved in time, if $\Psi(t) \to \Sigma_c\, e^{-iK_ct}X_c(\pm\infty)$ as $t \to \pm\infty$, where $X_c$ is a channel packet state,

$$\lim_{t\to\pm\infty} \sum_c (\Lambda_c\Psi(t), \Lambda_c\Psi(t)) = (\Psi(0), \Psi(0)) = \sum_c (X_c(\pm\infty), X_c(\pm\infty)). \tag{171}$$

Since $X_c(\pm\infty) = \Omega_c{}^{\mp\dagger}\Psi(0)$, we have

$$\sum_c \Omega_c{}^{\pm}\Omega_c{}^{\pm\dagger} = 1. \tag{172}$$

(For simplicity of writing we call the true bound states a "channel," in which $K_c = H$. The fact that such states vanish in the asymptotic region means that there they do not overlap the channel states.) We also have

$$X_c(\pm\infty) = \Omega_c{}^{\mp\dagger}\Psi(0) = \Omega_c{}^{\mp\dagger} \sum_{c'} \Omega_{c'}^{\mp\dagger} X_{c'}(\pm\infty). \tag{173}$$

Now

$$X_{c'}(+\infty) = \Omega_{c'}^{-\dagger}\Psi(0) = \Omega_{c'}^{-\dagger} \sum_c \Omega_c{}^{+} X_c(-\infty), \tag{174}$$

so that the $S$-matrix is defined, acting on packet states, as

$$S_{c'c} = \Omega_{c'}^{-\dagger}\Omega_c{}^{+}. \tag{175}$$

The unitarity of this expression, when operating on the asymptotic channel states, follows from the arguments leading to Eq. (168).

As far as general usefulness is concerned there is not much to choose between $S$ and $S_I$. In the general multichannel case these objects are operators only in a limited sense, as has been discussed in Chapter 4. The $S_I$ is dependent on our willingness to split the Hamiltonian into parts, consequently the $S$-matrix in the Heisenberg representation matrix is (perhaps) more general. It has become very fashionable in recent times to use the Heisenberg representation exclusively, especially in quantum field theory. In spite of this, we are frequently interested in doing actual calculations by what are essentially perturbation methods so that the older interaction representation continues to be important.

## 5.6 THE REACTION MATRIX AND THE HEITLER INTEGRAL EQUATION

In the theory of nuclear reactions as developed primarily by Wigner [12] it is very convenient to introduce an operator called the reaction matrix, which is related to the $\mathcal{T}$-matrix. The particular virtue of this quantity is the fact that on the energy shell it is Hermitian, in contrast to $\mathcal{T}$. We do not propose to consider the general and rather complicated multichannel case treated by Wigner; we rather confine our attention to those problems for which there is only one way of splitting the Hamiltonian. This case is sufficient for many field theoretic applications, where the concept of the reaction matrix was originally introduced by Heitler [13].

There are several ways of obtaining the reaction matrix. One of these is based on the introduction of standing wave eigenstates of the Hamiltonian $H$ which are essentially half the sum of the ingoing and outgoing wave states. Alternatively we may proceed in a purely formal way as follows. From the $\mathcal{T}$-matrix elements

$$\mathcal{T}^{(+)}_{\alpha'\alpha} = \left(\chi_{\alpha'}, \left(V + V\frac{1}{E_\alpha - H + i\eta}V\right)\chi_\alpha\right),$$

$$\bar{\mathcal{T}}^{(-)}_{\alpha'\alpha} = \left(\chi_{\alpha'}, \left(V + V\frac{1}{E_{\alpha'} - H + i\eta}V\right)\chi_\alpha\right), \tag{176}$$

(which are equal to $\mathcal{T}$ on the energy shell) we define an operator $\mathcal{T}(E)$ which is a function of a number $E$:

$$\begin{aligned}\mathcal{T} &= V + V\frac{1}{E - H + i\eta}V \\ &= V + V\frac{1}{E - K + i\eta}\mathcal{T} \\ &= V + \mathcal{T}\frac{1}{E - K + i\eta}V,\end{aligned} \tag{177}$$

where we have used the familiar identity

$$\frac{1}{E - H + i\eta} = \frac{1}{E - K + i\eta} + \frac{1}{E - K + i\eta} V \frac{1}{E - H + i\eta}$$
$$= \frac{1}{E - K + i\eta} + \frac{1}{E - H + i\eta} V \frac{1}{E - K + i\eta}. \tag{178}$$

Next let us introduce $\mathscr{R}(E)$, defined by the equation

$$\mathscr{R}(E) = V + V \frac{P}{E - K} \mathscr{R}(E), \tag{179}$$

where $P/(E - K)$ implies the instruction to take the Cauchy principal value on carrying out an integration over the continuous spectrum of $K$.

$$\frac{P}{E - K} = P \sum_{\lambda} \frac{|\chi_\lambda\rangle\langle\chi_\lambda|}{E - E_\lambda}. \tag{180}$$

Alternatively, the symbol $P$ may be replaced by [this follows from Eqs. (117)]

$$P \frac{1}{E - K} = \frac{1}{2} \lim_{\eta \to 0} \left( \frac{1}{E - K + i\eta} + \frac{1}{E - K - i\eta} \right). \tag{181}$$

If we imagine solving Eq. (179) by iteration, it is clear that the same series is obtained as from the equation

$$\mathscr{R} = V + \mathscr{R} \frac{P}{E - K} V. \tag{182}$$

That this equation is consistent with the form (179) may also be seen by substituting the first expression into Eq. (182).

$$\mathscr{R} = V + V \frac{P}{E - K} V + V \frac{P}{E - K} \mathscr{R} \frac{P}{E - K} V$$
$$= V + V \frac{P}{E - K} \left( V + \mathscr{R} \frac{P}{E - K} V \right), \tag{183}$$

which is just the equation with which we started. Evidently if $E$ is regarded as a number, $\mathscr{R}^\dagger = \mathscr{R}$, so $\mathscr{R}$ is Hermitian. If we identify $E$ with the initial energy in the equation

$$(\chi_{\alpha'}, \mathscr{R}\chi_\alpha) = (\chi_{\alpha'}, V\chi_\alpha) + P \sum_{\lambda} \frac{(\chi_{\alpha'}, V\chi_\lambda)(\chi_\lambda, \mathscr{R}\chi_\alpha)}{E_\alpha - E_\lambda}, \tag{184}$$

$\mathscr{R}_{\alpha'\alpha}(E_\alpha)$ is Hermitian only if $E_\alpha = E_{\alpha'}$.

Now with these definitions we assert that

$$\begin{aligned}\mathcal{T}(E) &= \mathcal{R}(E) - i\pi\mathcal{R}(E)\delta(E-K)\mathcal{T}(E)\\ &= \mathcal{R}(E) - i\pi\mathcal{T}(E)\delta(E-K)\mathcal{R}(E).\end{aligned} \tag{185}$$

To prove these equations we write Eq. (177) for $\mathcal{T}$ as

$$\mathcal{T} = V + V\frac{P}{E-K}\mathcal{T} - i\pi V\delta(E-K)\mathcal{T}, \tag{186}$$

where we use $(E - K + i\eta)^{-1} = P/(E-K) - i\pi\delta(E-K)$. Substitution of the conjectured form (185) for $\mathcal{T}$ in the second term of (186) now gives

$$\begin{aligned}\mathcal{T} &= V + V\frac{P}{E-K}\mathcal{R}[1 - i\pi\delta(E-K)\mathcal{T}] - i\pi V\delta(E-K)\mathcal{T}\\ &= V + (\mathcal{R} - V)[1 - i\pi\delta(E-K)\mathcal{T}] - i\pi V\delta(E-K)\mathcal{T}\\ &= \mathcal{R} - i\pi\mathcal{R}\delta(E-K)\mathcal{T}.\end{aligned} \tag{187}$$

The second line follows from the definition of $\mathcal{R}$. The second equation relating $\mathcal{T}$ and $\mathcal{R}$ may be verified in the same way using the alternative form of the original $\mathcal{T}$-equation. It is customary to write these equations on the energy shell $E_{\alpha'} = E_\alpha = E$ as

$$\begin{aligned}\mathcal{T}_{\alpha'\alpha} &= \mathcal{R}_{\alpha'\alpha} - i\pi\sum_\beta \mathcal{R}_{\alpha'\beta}\delta(E_\beta - E)\mathcal{T}_{\beta\alpha}\\ &= \mathcal{R}_{\alpha'\alpha} - i\pi\sum_\beta \mathcal{T}_{\alpha'\beta}\delta(E_\beta - E)\mathcal{R}_{\beta\alpha}.\end{aligned} \tag{188}$$

In this form they are known as the Heitler integral equations and form the basis of the so-called radiation damping theory (which we do not consider here).

It may appear that very little has been gained by introducing $\mathcal{R}$, since instead of having one equation to solve for $\mathcal{T}$ we must first find $\mathcal{R}$ and then solve yet another to obtain $\mathcal{T}$ from it. The advantage of this separation is that no matter what Hermitian matrix $\mathcal{R}$ is substituted into the Heitler equation we obtain a unitary $S$-matrix. To see this we write out the unitarity relation for the $S$-matrix.

$$\begin{aligned}S_{\alpha'\alpha} &= \delta_{\alpha'\alpha} - 2\pi i\delta(E_{\alpha'} - E_\alpha)\mathcal{T}_{\alpha'\alpha}\\ (S^\dagger S)_{\alpha'\alpha} &= \delta_{\alpha'\alpha} = \delta_{\alpha'\alpha} + 2\pi i\delta(E_{\alpha'} - E_\alpha)\mathcal{T}^\dagger_{\alpha'\alpha}\\ &\quad - 2\pi i\delta(E_{\alpha'} - E_\alpha)\mathcal{T}_{\alpha'\alpha}\\ &\quad - (2\pi i)^2\sum_\beta \delta(E_{\alpha'} - E_\beta)\mathcal{T}^\dagger_{\alpha'\beta}\delta(E_\beta - E_\alpha)\mathcal{T}_{\beta\alpha}.\end{aligned} \tag{189}$$

Factoring out $2\pi i\delta(E_{\alpha'} - E_\alpha)$ and noting that $\delta(E_{\alpha'} - E_\beta)\delta(E_\beta - E_\alpha) = \delta(E_{\alpha'} - E_\alpha)\delta(E_\beta - E_\alpha)$, we have the relation, valid on the energy shell,

$$(\mathscr{T} - \mathscr{T}^\dagger)_{\alpha'\alpha} = -2\pi i \sum_\beta \mathscr{T}^\dagger_{\alpha'\beta}\delta(E_\beta - E)\mathscr{T}_{\beta\alpha}, \tag{190}$$

or in an obvious notation

$$\mathscr{T} - \mathscr{T}^\dagger = -2\pi i \mathscr{T}^\dagger \delta(E - K)\mathscr{T}. \tag{191}$$

[Note, incidentally, that the diagonal matrix element of this equation is equivalent to

$$\operatorname{Im} \mathscr{T}_{\alpha\alpha} = -\pi \sum_\beta |\mathscr{T}_{\beta\alpha}|^2 \,\delta(E_\beta - E_\alpha), \tag{192}$$

which is equivalent to Eq. (39), the optical theorem.] Now form $\mathscr{T} - \mathscr{T}^\dagger$, using the Heitler equations:

$$\begin{aligned} \mathscr{T} - \mathscr{T}^\dagger &= -i\pi \mathscr{R}\delta(E - K)\mathscr{T} - i\pi\mathscr{T}^\dagger\delta(E - K)\mathscr{R} \\ &= -i\pi[\mathscr{T}^\dagger - i\pi\mathscr{T}^\dagger\delta(E - K)\mathscr{R}]\delta(E - K)\mathscr{T} \\ &\quad - i\pi\mathscr{T}^\dagger\delta(E - K)[\mathscr{T} + i\pi\mathscr{R}\delta(E - K)\mathscr{T}] \\ &= -2\pi i\mathscr{T}^\dagger\delta(E - K)\mathscr{T}. \end{aligned} \tag{193}$$

This proves our assertion that the unitarity condition [Eq. (191)] is automatically satisfied for a $\mathscr{T}$ computed from the Heitler equation with any Hermitian $\mathscr{R}$.

## 5.7 APPLICATION TO QUANTUM FIELD THEORY

The problem posed by field theory which distinguishes it from the applications we have considered to this point is that, try as we may, we cannot run away from the interaction. By this we mean the following. A physical electron that has definite properties of mass, spin, and so on as measured in the laboratory is always interacting with the vacuum electromagnetic field. It is customary in the formulation of quantum electrodynamics in the tradition of the canonical formalism of quantum mechanics to speak at least temporarily as though we have a system of electrons, positrons, and photons which may be considered as not interacting. There is a mathematical quantity, call it $m_0$, which is the mass an electron would have if there were no interaction by way of the electromagnetic field. If now the interaction is switched on, the particles interact with the field and thus with each other. As a result of the virtual emission and absorption of photons an electron acquires an electromagnetic mass; and we identify the mass of a physical electron as the sum of $m_0$ and the electromagnetic mass $\delta m$. It so happens that in any known relativistic field theory (to the degree of accuracy that we can carry out computations) $\delta m$

turns out to be infinite. This is a detail for the present purposes, because even if it were finite it would cause a modification of our formulas. There would not be much point in constructing a wave packet state to describe an electron of mass $m_0$ when everyone knows electrons have mass $m = m_0 + \delta m$. Evidently we must generalize our concept of noninteracting states to allow that part of the interaction is always included. It is intuitively quite clear what we would like to do. The asymptotic configuration corresponding to an electron-electron collision must certainly be that of two widely separated packets describing physical particles approaching each other and corresponding to a well-defined current.[13]

A naive application of formal scattering theory, as introduced in this chapter, would lead to at least two severe difficulties. At the most primitive level we might try to describe the initial state in terms of eigenfunctions of the unperturbed Hamiltonian $K$ which contains, say, the free Hamiltonian of an electron-positron field and of the electromagnetic field. This would lead to instant catastrophe because the energy spectra of $K$ and $H$ do not coincide. Evidently this is an unphysical situation and must be corrected at the outset. Formally it is not hard to do [5]. If $\psi_n$ is an eigenstate of $H$ with energy $E_n$, we would like to construct it as a limit of some free eigenfunction $\chi_n$ of $K$, but with the correct energy. If the free configuration corresponding to the desired $\psi_n$ normally satisfies $K\chi_n = \epsilon_n\chi_n$, we write $(K + \Delta)\chi_n = E_n\chi_n$, where $\Delta = \Sigma_n |\chi_n\rangle(E_n - \epsilon_n)\langle\chi_n|$ and then $H = (K + \Delta) + (V - \Delta)$. With $V - \Delta$ in place of $V$ all previous formulas hold except for one feature to be discussed in a moment. It is possible to give a prescription to compute the energy differences $E_n - \epsilon_n$, but it is quite complicated. The second feature that is new has to do with the renormalization of the state vectors caused by the same virtual interactions as those that take place even when we deal with isolated configurations [6, 14]. Even after the energies are appropriately adjusted, as before, the limits $\psi^{\pm} - \pm i\eta(E - H \pm i\eta)^{-1}\chi$ do not lead to normalized functions, that is, $(\psi_\alpha^+, \psi_\alpha^+) \neq 1$ even if $(\chi_\alpha, \chi_\alpha) = 1$. Instead we must write $\sqrt{Z}\psi^{\pm} = \pm i\eta(E - H \pm i\eta)^{-1}\chi$, where we can show that $\sqrt{Z} = (\chi, \psi^{(\pm)})$. Thus $Z$ is the probability of finding $\chi$ in the true normalized eigenstate. This is a circumstance familiar from stationary state perturbation theory. In both cases the interaction lasts over an infinite period and there is a cumulative effect. It is possible to avoid the explicit appearance of the renormalization constant $Z$ by an appropriate definition of the initial configuration [14].

We shall not stop now to explain these remarks in detail. The fact that the formalism developed in this chapter (in particular the discussion in

[13] This is, of course, the point of view taken in Chapter 3. Our general discussion in Chapter 3 is directly applicable to field theories.

which the existence of an unperturbed Hamiltonian was not assumed) may with only trivial modifications be taken over in field theories will be illustrated in Chapter 10. At the same time we shall discuss some of the modern, peculiarly field theoretic techniques, which are somewhat simpler to use in practice.

This chapter marks the end of our largely formal development of collision theory. A number of other topics such as decaying states, resonance scattering, and delay times are rather formal and will be taken up later. It is nevertheless true that all the machinery to treat a general scattering problem has been presented, and the bulk of the remainder of the book will be devoted to applying the results thus far obtained to special situations.

## REFERENCES AND NOTES

1. C. Møller, *Kgl. Danske Videnskab. Selskab, Mat-fys. Medd.* **23,** 1 (1948).
2. J. A. Wheeler, *Phys. Rev.* **52,** 1107 (1937).
3. W. Heisenberg, *Z. Naturforsch.* **1,** 608 (1946).
4. B. Lippmann and J. Schwinger, *Phys. Rev.* **79,** 469 (1950).
5. M. Gell-Mann and M. L. Goldberger, *Phys. Rev.* **91,** 398 (1953).
6. W. Brenig and R. Haag, *Forts. der Physik* **7,** 183 (1959).
7. H. Eckstein, *Phys. Rev.* **101,** 880 (1956).
8. J. Cook, *J. Math. and Phys.* **36,** 82 (1957).
9. M. Hack, *Phys. Rev.* **96,** 196 (1954).
10. L. Foldy and W. Tobocman, *Phys. Rev.* **105,** 1099 (1957); S. Epstein, *Phys. Rev.* **106,** 598 (1957); B. Lippmann, *Phys. Rev.* **102,** 264 (1956).
11. N. F. Mott and H. S. W. Massey, *The Theory of Atomic Collisions*, Clarendon Press, Oxford, 1933, p. 258.
12. E. P. Wigner, *Phys. Rev.* **70,** 15 (1946).
13. W. Heitler, *Proc. Cambridge Phil. Soc.* **37,** 291 (1941). See also M. L. Goldberger, *Phys. Rev.* **84,** 926 (1951).
14. G. C. Wick, *Rev. Modern Phys.* **27,** 339 (1955).

# 6. THE TWO-BODY PROBLEM WITH CENTRAL FORCES

With the conclusion of Chapter 5 we have completed our description of the formal aspects of scattering theory and are now ready to study its applications. In this chapter we shall review the simplest of scattering processes. This is the scattering of two particles interacting by central forces, without the production of new particles or a change of the internal states of the two interacting particles.

In Section 6.1 the transformation of the differential scattering cross section from the laboratory to the barycentric frame of reference will be described in greater detail than was given in Chapter 3. The expressions for the eigenvalues of the reaction matrix in terms of the scattering phase shifts and the solution of the Heitler integral equation are given next. Methods for solution of the nonrelativistic Schrödinger equation are described, and several examples, such as the nuclear scattering of two charged particles, are discussed.

The importance of studying the analytic continuation of scattering amplitudes into the complex energy and momentum planes was emphasized by Heisenberg, Kramers, and Møller [1]. Such studies have enhanced considerably the understanding of elementary particle interactions.[1] In Section 6.5 we shall review the analytic properties of scattering amplitudes for nonrelativistic potential scattering, following the work of Jost, Bargmann, and others [3].

## 6.1 COORDINATE TRANSFORMATIONS FROM THE LABORATORY TO THE BARYCENTRIC FRAME OF REFERENCE

We consider the scattering, in a particular coordinate system $L$, of two particles having masses $M_1$ and $M_2$. Prior to the scattering, the respective momenta and energies of these particles are $\mathbf{p}_1$ and $\mathbf{p}_2$, and $\epsilon_{1,p1} = c(M_1{}^2c^2 + p_1{}^2)^{1/2}$ and $\epsilon_{2,p2} = c(M_2{}^2c^2 + p_2{}^2)^{1/2}$. The total energy of the two

[1] See, for example, M. L. Goldberger or G. F. Chew [2].

particles is written as

$$E_{L,a} = \epsilon_{1,p1} + \epsilon_{2,p2}. \tag{1}$$

In the coordinate system $L$ it is assumed that $\mathbf{p}_1 \times \mathbf{p}_2 = 0$. In the laboratory frame of reference (one of the coordinate systems $L$) particle 2 is at rest, so

$$\mathbf{p}_2 = 0, \qquad \epsilon_{2,p2} = \epsilon_{2,0} = M_2c^2. \tag{2}$$

When the scattering is completed the two particles emerge from the region of interaction with the respective momenta $\mathbf{k}_1$ and $\mathbf{k}_2$ and energies $\epsilon_{1,k1}$ and $\epsilon_{2,k2}$. An element of solid angle into which $\mathbf{k}_1$ is directed is written as $d\Omega_L = d\phi_L \sin\theta_L \, d\theta_L$, where the polar angle $\theta_L$ and the azimuthal angle $\phi_L$ are measured with respect to a polar axis defined by the direction of $\mathbf{p}_1 = \hat{\mathbf{k}}p_1$.[2] The differential cross section $d\sigma/d\Omega_L$ for scattering in the $L$ system is given by Eq. (3.157).

In the barycentric coordinate system, particles 1 and 2 have respective momenta $\boldsymbol{\kappa}$ and $-\boldsymbol{\kappa}$ and energies $\epsilon_{1,\kappa}$ and $\epsilon_{2,\kappa}$ prior to the scattering. The total energy is written as

$$E_a = \epsilon_{1,\kappa} + \epsilon_{2,\kappa}. \tag{3}$$

After the scattering, the two particles will have the respective momenta $\boldsymbol{\kappa}_f$ and $-\boldsymbol{\kappa}_f$. The condition of energy conservation requires that $\kappa_f = \kappa$ in the final asymptotic state. An element of solid angle into which $\boldsymbol{\kappa}_f$ is directed will be expressed as $d\Omega = d\phi \sin\theta \, d\theta$, with the polar axis defined by the direction of $\boldsymbol{\kappa}$. Since $\mathbf{p}_1$ and $\boldsymbol{\kappa}$ are parallel,[3] this axis is the same as that about which $(\theta_L, \phi_L)$ are measured.

We shall assume that the internal states of the two particles are unchanged by the scattering, although their spin orientations may be changed. Thus if $S_1$ and $S_2$ are the respective spins of the particles, we may write $(\nu_1, \nu_2)$ and $(\nu'_1, \nu'_2)$ for the respective initial and final azimuthal components of these spins in the barycentric system. The differential cross section is given by Eq. (3.149). It is

$$\frac{d\sigma}{d\Omega}(\nu'_1, \nu'_2; \nu_1, \nu_2) = \frac{(2\pi)^4}{c^4(1/\epsilon_{1,\kappa} + 1/\epsilon_{2,\kappa})^2} |\langle \nu'_1, \nu'_2| \, T(\boldsymbol{\kappa}_f, \boldsymbol{\kappa}) \, |\nu_1, \nu_2\rangle|^2$$

$$= |f(\hat{\boldsymbol{\kappa}}_f, \hat{\boldsymbol{\kappa}})|^2. \tag{4}$$

[2] For convenience we may consider particle 1 as the beam particle in the initial state and, again, particle 1 as the detected particle in the final state. This notation is particularly useful when particles 1 and 2 are identical.

[3] If $\mathbf{p}_1$ and $\boldsymbol{\kappa}$ should be antiparallel, we may interchange the roles of particles 1 and 2; or we may simply specify that the two polar axes are antiparallel.

The quantity $T(\boldsymbol{\varkappa}_f, \boldsymbol{\varkappa})$ is the scattering matrix (3.146). We have here set

$$v_{\text{rel},B} = \kappa\left(\frac{1}{\epsilon_{1,\kappa}} + \frac{1}{\epsilon_{2,\kappa}}\right)c^2 \tag{5}$$

and in analogy to Eq. (3.176) have introduced the scattering amplitude

$$f(\hat{\boldsymbol{\varkappa}}_f, \hat{\boldsymbol{\varkappa}}) = -\frac{(2\pi)^2}{c^2}\left(\frac{1}{\epsilon_{1,\kappa}} + \frac{1}{\epsilon_{2,\kappa}}\right)^{-1} T(\boldsymbol{\varkappa}_f, \boldsymbol{\varkappa}). \tag{6}$$

For a random distribution of initial spin directions and any final spin orientation, we have the cross section [Eq. (3.143*a*)]

$$\frac{d\bar{\sigma}}{d\Omega} = \frac{1}{(2S_1 + 1)(2S_2 + 1)} \sum_{\nu'_1 \cdot \nu_1} \sum_{\nu'_2 \cdot \nu_2} \frac{d\sigma}{d\Omega}(\nu'_1\nu'_2; \nu_1\nu_2). \tag{7}$$

According to Eq. (3.154) the scattering matrix in the $L$ coordinate system is

$$(T_{ba})_L = \frac{\epsilon_{1,\kappa}\epsilon_{2,\kappa}}{(\epsilon_{1,k1}\epsilon_{2,k2}\epsilon_{1,p1}\epsilon_{2,p2})^{1/2}} \langle \nu'_1, \nu'_2 | \, T(\boldsymbol{\varkappa}_f, \boldsymbol{\varkappa}) \, | \nu_1, \nu_2 \rangle. \tag{8}$$

To obtain the scattering matrix in the laboratory system, we set $\mathbf{p}_2 = 0$ in this equation. The relative velocity of the incident particles in the $L$ system is

$$v_{\text{rel},L} = \left(\frac{p_1}{\epsilon_{1,p1}} - \frac{\hat{\mathbf{p}}_1 \cdot \mathbf{p}_2}{\epsilon_{2,p2}}\right)c^2. \tag{9}$$

In addition, in Eq. (3.157) we may set

$$c^2\left(\frac{k_1}{\epsilon_{1,k1}} - \frac{\hat{k}_1 \cdot \mathbf{k}_2}{\epsilon_{2,k2}}\right) = \hat{\mathbf{k}}_1 \cdot (\mathbf{v}'_1 - \mathbf{v}'_2), \tag{10}$$

where $\mathbf{k}_2 = \mathbf{p}_1 + \mathbf{p}_2 - \mathbf{k}_1$ and

$$\mathbf{v}'_1 = \frac{\mathbf{k}_1}{\epsilon_{1,k1}} c^2, \qquad \mathbf{v}'_2 = \frac{\mathbf{k}_2}{\epsilon_{2,k2}} c^2. \tag{11}$$

With this notation we may combine Eqs. (3.157) and (4) to obtain

$$\frac{d\sigma}{d\Omega_L} = J\left(\frac{\partial\Omega}{\partial\Omega_L}\right)\frac{d\sigma}{d\Omega}, \tag{12a}$$

where

$$J\left(\frac{\partial\Omega}{\partial\Omega_L}\right) = \frac{k_1{}^2c^2}{\epsilon_{1,k1}\epsilon_{2,k2}} \frac{(\epsilon_{1,\kappa} + \epsilon_{2,\kappa})^2}{[\hat{\mathbf{k}}_1 \cdot (\mathbf{v}'_1 - \mathbf{v}'_2)](\mathbf{p}_1\epsilon_{2,p2} - \hat{\mathbf{p}}_1 \cdot \mathbf{p}_2\epsilon_{1,p1})}. \tag{13}$$

On setting $\mathbf{p}_2 = 0$ we specialize to the laboratory coordinate system.

If we sum over final and average over initial spin orientations, as in Eq. (7), we have

$$\frac{d\bar{\sigma}}{d\Omega_L} = J\left(\frac{\partial\Omega}{\partial\Omega_L}\right)\frac{d\bar{\sigma}}{d\Omega}. \tag{12b}$$

The quantity $J$ given by Eq. (13) represents the factor for transforming the differential cross section from the barycentric to the $L$ frame of reference. Scattering cross sections are usually observed in the laboratory frame. The barycentric cross section has the most direct usefulness for theoretical considerations, however, since it does not include the purely kinematical effects associated with a net motion of the entire system.

To complete our discussion, we shall set $\mathbf{p}_2 = 0$ in the preceding equations, specializing to the laboratory coordinate system. Let us suppose that the barycentric system is moving with a uniform velocity $c\beta$ with respect to the laboratory system. Then the Lorentz transformation equations (1.101) provide the relations

$$\begin{aligned}
\epsilon_{1,p1} &= \gamma(c\beta\kappa + \epsilon_{1,\kappa}),\\
M_2c^2 &= \gamma(-c\beta\kappa + \epsilon_{2,\kappa}),\\
p_1 &= \gamma\left(\kappa + \frac{1}{c}\beta\epsilon_{1,\kappa}\right),\\
p_2 &= 0 = \gamma\left(-\kappa + \frac{1}{c}\beta\epsilon_{2,\kappa}\right),
\end{aligned} \tag{14}$$

where $\gamma = (1 - \beta^2)^{-1/2}$. From these equations we may obtain the useful equation

$$E_a^{\,2} = c^2[c^2(M_1^{\,2} - M_2^{\,2}) + 2M_2E_{L,a}] \tag{15}$$

relating the total energies $E_a$ and $E_{L,a} = \gamma E_a$ in the two coordinate systems. The momenta $p_1$ and $\kappa$ are related by the equation

$$p_1 = \kappa\left(\frac{E_a}{M_2c^2}\right), \tag{16}$$

while $\beta$ and $\gamma$ are given by

$$\beta = \frac{c\kappa}{\epsilon_{2,\kappa}}, \qquad \gamma = \frac{\epsilon_{2,\kappa}}{M_2c^2}. \tag{17}$$

Equations (16) and (17) are obtained directly from the general relations (14). In applications it is ordinarily most convenient to evaluate Eq. (15) first, so both $E_{L,a}$ and $E_a$ are known. Next $p_1$ may be obtained from $E_{L,a}$ and then $\kappa$ from Eq. (16).

The transformation for the final states is easily evaluated once $\gamma$, $\beta$, $\kappa$, and $p_1$ are known. From Eqs. (1.101) we find

$$\cos \theta_L = \frac{\gamma}{k_1}\left(\kappa \cos \theta + \frac{\beta}{c} \epsilon_{1,\kappa}\right), \tag{18}$$

relating $\theta$, $\theta_L$, and $k_1$. This expression is supplemented by the equations

$$\begin{aligned} \epsilon_{1,k_1} &= \gamma(c\beta\kappa \cos \theta + \epsilon_{1,\kappa}), \\ \epsilon_{2,k_2} &= \gamma(-c\beta\kappa \cos \theta + \epsilon_{2,\kappa}). \end{aligned} \tag{19}$$

Alternatively, we may write

$$\sin \theta_L k_1 = \sin \theta\kappa, \tag{20}$$

since the transverse components of momentum are unchanged by a Lorentz transformation. The first of Eqs. (19) and Eq. (18) together determine $\theta_L$ and $k_1$ in terms of $\kappa$ and $\cos \theta$. On the other hand, if $\theta_L$ and $k_1$ are given, Eq. (20) may be used to obtain $\theta$ directly, since $\kappa$ is assumed to have been previously determined.

To evaluate $J(\partial\Omega/\partial\Omega_L)$, we may write

$$\frac{1}{c^2}\hat{\mathbf{k}}_1 \cdot (\mathbf{v}'_1 - \mathbf{v}'_2) = k_1\left(\frac{1}{\epsilon_{1,k1}} + \frac{1}{\epsilon_{2,k2}}\right) - \frac{p_1 \cos \theta_L}{\epsilon_{2,k2}}. \tag{21}$$

With this expression we have given a complete set of equations to relate the collision kinematics and differential cross sections in the two coordinate systems.

When $\theta = \theta_L = 0$, the case of *forward scattering*, we have $k_1 = p_1$ and $k_2 = 0$. Then

$$J = \left(\frac{p_1}{\kappa}\right)^2. \tag{22}$$

Inserting this into Eq. (12*a*) and comparing with Eq. (6), we see that the forward scattering amplitude, divided by the momentum, is a Lorentz invariant. That is,

$$\frac{1}{\kappa} f(\hat{\boldsymbol{\kappa}}, \hat{\boldsymbol{\kappa}}) \qquad \text{is a Lorentz invariant.} \tag{23}$$

The discussion given in this section is trivially specialized to the case of nonrelativistic scattering. To illustrate this we consider the scattering of neutrons by protons with a total kinetic energy of 5 Mev in the barycentric system (that is, $E_a - 2M_1c^2 = 5$ Mev, with $M_1 = M_2$). The cross section $d\bar{\sigma}/d\Omega$ is nearly isotropic (independent of $\theta$), as illustrated in Fig. 6.1. The corresponding cross section $d\bar{\sigma}/d\Omega_L$ is also shown in Fig. 6.1. The forward peaking results from the net forward motion of the two particles in the laboratory coordinate system.

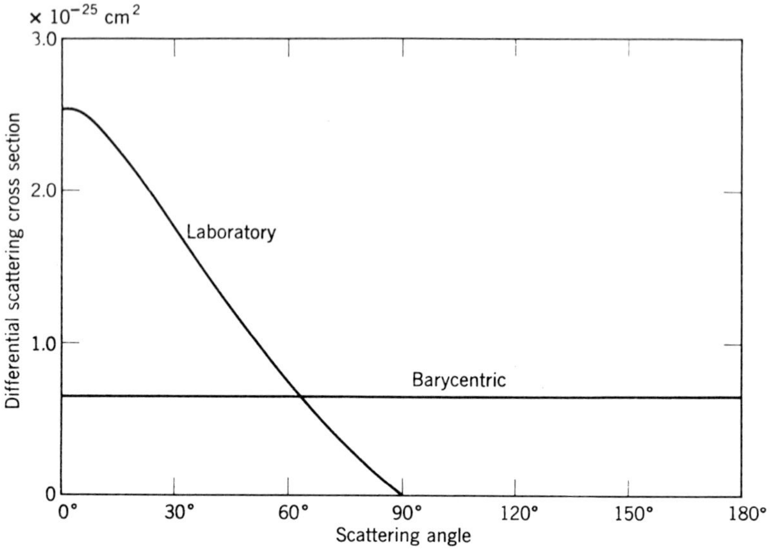


Fig. 6.1 Comparison of differential scattering cross sections for neutron-proton scattering in the laboratory and barycentric systems. The barycentric kinetic energy is 5 MeV. For a given scattering event, the scattering angle in the barycentric system is twice that in the laboratory system.

## 6.2 PARTIAL WAVE ANALYSIS AND THE SCATTERING PHASE SHIFTS

In this section we shall introduce scattering matrices [see Eqs. (4.226)] and wave functions for states of a definite orbital angular momentum $l$. It is often the case that these functions may be evaluated more easily than the corresponding quantities referred to plane-wave states. We shall first show that the Heitler integral equation reduces to a simple algebraic equation in the angular momentum representation. This permits us to express the scattering matrix $T$ in terms of the eigenvalues of the reaction matrix (which are customarily expressed as functions of *scattering phase shifts*). The transformation from the angular momentum to the plane-wave representation enables us to write the complete scattering matrix and the differential scattering cross section in terms of the phase shifts.

### Solution of the Heitler Integral Equation

We have seen that the scattering matrix $\mathscr{T}$ and reaction matrix $\mathscr{R}$ may be written as

$$\begin{aligned} \langle b|\, \mathscr{T}\, |a\rangle &= \delta(\mathbf{P}_b - \mathbf{P}_a)T_{ba}, \\ \langle b|\, \mathscr{R}\, |a\rangle &= \delta(\mathbf{P}_b - \mathbf{P}_a)R_{ba}, \end{aligned} \tag{24}$$

where $\mathbf{P}_a$ and $\mathbf{P}_b$ are the respective total momenta of states $a$ and $b$. According to Eq. (5.188) $\mathscr{T}$ and $\mathscr{R}$ are related by the Heitler integral equation

$$\mathscr{T} = \mathscr{R} - i\pi\mathscr{R}\delta(E_a - K)\mathscr{T}, \tag{25a}$$

in the notation of Section 5.5. For $\mathscr{T}^{(-)}$ there is the corresponding Heitler equation

$$\mathscr{T}^{(-)} = \mathscr{R} + i\pi\mathscr{R}\delta(E_a - K)\mathscr{T}^{(-)}. \tag{25b}$$

We shall henceforth not write separately the equations for $\mathscr{T}^{(-)}$ while solving Eqs. (25), since only the sign of $i$ need be changed to obtain $\mathscr{T}^{(-)}$.

It is convenient to reexpress the Heitler equation (25$a$) in terms of the quantities $T_{ba}$ and $R_{ba}$ on the momentum shell. We first write $E_a = E_a(\mathbf{P}_a)$ and $E_b = E_b(\mathbf{P}_b)$ to indicate that $E_a = E_a(\mathbf{P}_a)$ is the energy associated with the state $a$ having the total momentum $\mathbf{P}_a$, and so on. Then Eqs. (24) may be inserted into Eq. (25$a$) to give

$$\begin{aligned}\delta(\mathbf{P}_b - \mathbf{P}_a)(T_{ba} - R_{ba}) &= -i\pi \sum_{b'}{}' \int d^3P_{b'}\delta(\mathbf{P}_b - \mathbf{P}_{b'})\delta(\mathbf{P}_{b'} - \mathbf{P}_a) \\ &\qquad \times R_{bb'}\delta[E_a(\mathbf{P}_a) - E_{b'}(\mathbf{P}_{b'})]T_{b'a} \\ &= \delta(\mathbf{P}_b - \mathbf{P}_a)\Big\{-i\pi \sum_{b'}{}' R_{bb'}\delta[E_a(\mathbf{P}_a) - E_{b'}(\mathbf{P}_a)]\Big\}T_{b'a}.\end{aligned}$$

[Here $\Sigma_{b'}'$ implies a sum (or integral) over states $b'$, excluding that over $\mathbf{P}_{b'}$.] We thus obtain in place of Eq. (25$a$) the Heitler equation defined on the momentum shell $\mathbf{P}_b = \mathbf{P}_{b'} = \mathbf{P}_a$.

$$T_{ba}' = R_{ba} - i\pi \sum_{b'}{}' R_{bb'}\delta[E_a(\mathbf{P}_a) - E_{b'}(\mathbf{P}_a)]T_{b'a}. \tag{26}$$

The corresponding equation for $T_{ba}^{(-)}$ is obtained by changing the sign of $i$ in the coefficient of the second term in Eq. (26).

At this point we shall specialize our discussion to the scattering in the barycentric system of two particles 1 and 2. As in Section 6.1, we shall write $\boldsymbol{\kappa}$, $-\boldsymbol{\kappa}$ for their respective momenta in state $a$ and $\boldsymbol{\kappa}_f$, $-\boldsymbol{\kappa}_f$ for the corresponding momenta in state $b$. In the barycentric system, since $\mathbf{P}_a = 0$, we may write

$$\begin{aligned}E_a(0) &\equiv \epsilon(\kappa), \\ E_b(0) &\equiv \epsilon(\kappa_f), \\ E_{b'}(0) &\equiv \epsilon(\kappa').\end{aligned} \tag{27}$$

It is convenient, for the present, not to specify the functional forms of $\epsilon(\kappa)$. Thus we include relativistic and nonrelativistic applications, as well as the case of quasi-particles for which $\epsilon$ may be some general function of the momentum $\kappa$.

We shall now write $R_{ba}$ and $T_{ba}$ more specifically as

$$\begin{aligned} R_{ba} &= \langle \boldsymbol{\varkappa}_f, S_f |\, R \,| \boldsymbol{\varkappa}, S\rangle, \\ T_{ba} &= \langle \boldsymbol{\varkappa}_f, S_f |\, T \,| \boldsymbol{\varkappa}, S\rangle. \end{aligned} \tag{28}$$

Here $S$ and $S_f$ label the spin and spin orientation of the states $a$ and $b$ respectively. As we are considering only simple scattering in this chapter, we assume that excitation of internal states may be neglected, or that the two particles are *elementary* in the sense defined in Section 4.2, p. 124.[4] The sum over states in Eq. (26) may be written as

$$\sum_{b'}{}' = \sum_{s'} \int d\Omega_{\kappa'} (\kappa')^2 \, d\kappa',$$

where $d\Omega_{\kappa'}$ is an element of solid angle into which $\boldsymbol{\varkappa}'$ is directed and the sum on $s'$ represents a summation over spin orientations. Then with $\kappa_f = \kappa$

$$\langle \boldsymbol{\varkappa}_f, S_f |\, R\delta[\epsilon(\kappa) - \epsilon(\kappa')]T \,| \boldsymbol{\varkappa}, S\rangle$$
$$= \sum_{S'} \rho_\epsilon \int d\Omega_{\kappa'} \langle \boldsymbol{\varkappa}_f, S_f |\, R \,| \boldsymbol{\varkappa}', S'\rangle \langle \boldsymbol{\varkappa}', S' |\, T \,| \boldsymbol{\varkappa}, S\rangle, \tag{29}$$

where $R$ and $T$ are restricted to the energy shell, so $\kappa_f = \kappa' = \kappa$, and

$$\rho_\epsilon = \frac{\kappa^2}{d\epsilon(\kappa)/d\kappa} \tag{30}$$

is a density of states per unit energy. Equation (29) may be used to put Eq. (26) into the form

$$\langle \boldsymbol{\varkappa}_f, S_f |\, T \,| \boldsymbol{\varkappa}, S\rangle$$
$$= \langle \boldsymbol{\varkappa}_f, S_f |\, R \,| \boldsymbol{\varkappa}, S\rangle - i\pi\rho_\epsilon \sum_{S'} \int d\Omega_{\kappa'} \langle \boldsymbol{\varkappa}_f, S_f |\, R \,| \boldsymbol{\varkappa}', S'\rangle \langle \boldsymbol{\varkappa}', S' |\, T \,| \boldsymbol{\varkappa}, S\rangle \tag{31}$$

where we specify that $\kappa_f = \kappa' = \kappa$.

When the interaction $V$ between the two particles is independent of the spins $S_1$ and $S_2$ of the two particles, $V$ is said to be *spin independent*. In this case $S_1$ and $S_2$ commute with the Hamiltonian $H$. The total angular momentum operator in the barycentric system is

$$\mathbf{J} = \mathbf{S}_1 + \mathbf{S}_2 + \mathbf{L}, \tag{32}$$

where $\mathbf{L}$ is the orbital angular momentum. If the coordinates of the two particles are $\mathbf{r}_1$ and $\mathbf{r}_2$, this is

$$\mathbf{L} = \mathbf{r}_1 \times \boldsymbol{\varkappa} + \mathbf{r}_2 \times (-\boldsymbol{\varkappa}) = \mathbf{r} \times \boldsymbol{\varkappa} = \mathbf{r} \times \left(\frac{1}{i} \nabla_r\right), \tag{33}$$

where $\mathbf{r} = \mathbf{r}_1 - \mathbf{r}_2$ and $\nabla_r$ is the gradient with respect to $\mathbf{r}$.

[4] For example, if either particle were a molecule or atomic nucleus, we assume that excitation to internal states other than the initial state may be neglected.

Since $[\mathbf{J}, H] = 0$ and since we have assumed that $[\mathbf{S}_1, H] = [\mathbf{S}_2, H] = 0$, it follows that

$$[\mathbf{L}, H] = 0. \tag{34}$$

Therefore $\mathbf{S}_1$, $\mathbf{S}_2$, and $\mathbf{L}$ all commute with $T$ [and with $R$, as is easily seen from Eq. (5.179)], as was shown in Section 4.6, p. 167. This has several important consequences. Because $\mathbf{L}$ generates the infinitesimal rotations in coordinate (or in momentum) space (compare Section 2.1) it follows that $T$ and $R$ will be functions of only scalar products of $\boldsymbol{\varkappa}_f$ and $\boldsymbol{\varkappa}$, or of $\hat{\boldsymbol{\varkappa}}_f \cdot \hat{\boldsymbol{\varkappa}}$, $\kappa_f$, and $\kappa$. Moreover, as was shown in Section 4.6, $T$ and $R$ will be diagonal in and independent of $\nu_1$ and $\nu_2$, the eigenvalues of the azimuthal components of $S_1$ and $S_2$. We may therefore write the expressions in Eq. (28) as

$$\begin{aligned}\langle \boldsymbol{\varkappa}_f, \nu'_1, \nu'_2 | \, T \, | \boldsymbol{\varkappa}, \nu_1, \nu_2 \rangle &= \delta_{\nu'1,\nu1}\delta_{\nu'2,\nu2} T(\hat{\boldsymbol{\varkappa}}_f \cdot \hat{\boldsymbol{\varkappa}}, \kappa_f, \kappa),\\ \langle \boldsymbol{\varkappa}_f, \nu'_1, \nu'_2 | \, R \, | \boldsymbol{\varkappa}, \nu_1, \nu_2 \rangle &= \delta_{\nu'1,\nu1}\delta_{\nu'2,\nu2} R(\hat{\boldsymbol{\varkappa}}_f \cdot \hat{\boldsymbol{\varkappa}}, \kappa_f, \kappa),\end{aligned} \tag{35a}$$

where the coefficients on the right-hand side are functions of $\hat{\boldsymbol{\varkappa}}_f \cdot \hat{\boldsymbol{\varkappa}}$, $\kappa_f$, and $\kappa$ only, as described before.

The total spin of the two particles is $\mathbf{S} = \mathbf{S}_1 + \mathbf{S}_2$. Their spin states may, rather than as in Eqs. (35*a*), be labeled by the eigenvalues of $\mathbf{S}^2$ and $\mathbf{S} \cdot \hat{\mathbf{k}}$, where $\hat{\mathbf{k}}$ defines a convenient axis. These will be written as $S(S+1)$ and $\nu$ respectively. If the interaction $V$ depends on $S$, but not on the orientation of $S$, it is said to represent a *central force*. That is, for a central force $V$ has the general form

$$\langle S' | \, V \, | S \rangle = \delta_{S',S} V^{(S)}, \tag{36}$$

where $V^{(S)}$ is the interaction in a state of total spin $S$. The condition of *spin independence*, which led to Eqs. (35*a*), implies that $V^{(S)}$ is independent of $S$. A spin-independent force evidently represents a special case of central forces.

For a central force we have $[\mathbf{S}, H] = [\mathbf{L}, H] = 0$, although not in general the more restrictive conditions $[\mathbf{S}_1, H] = [\mathbf{S}_2, H] = 0$. Since Eq. (34) is valid, we see that for central forces $T$ and $R$ are again functions of $\hat{\boldsymbol{\varkappa}}_f \cdot \hat{\boldsymbol{\varkappa}}$, $\kappa_f$, and $\kappa$. The discussion in Section 4.6, p. 168 now permits us to conclude that $T$ and $R$ are diagonal in $S$ and $\nu$ and are independent of $\nu$. In this case we may therefore write the expressions (28) in the form

$$\begin{aligned}\langle \boldsymbol{\varkappa}_f, S', \nu' | \, T \, | \boldsymbol{\varkappa}, S, \nu \rangle &= \delta_{S',S}\delta_{\nu',\nu} T(\hat{\boldsymbol{\varkappa}}_f \cdot \hat{\boldsymbol{\varkappa}}, \kappa_f, \kappa, S),\\ \langle \boldsymbol{\varkappa}_f, S', \nu' | \, R \, | \boldsymbol{\varkappa}, S, \nu \rangle &= \delta_{S',S}\delta_{\nu',\nu} R(\hat{\boldsymbol{\varkappa}}_f \cdot \hat{\boldsymbol{\varkappa}}, \kappa_f, \kappa, S),\end{aligned} \tag{35b}$$

recalling the arguments leading to Eq. (4.226*a*).

For the case of spin-independent forces the Hamiltonian $H$ is diagonal in $\nu_1$ and $\nu_2$, and we may evidently write the Schrödinger wave function $\Psi(t)$ for the system as (we shall later take account of wave function symmetrization for particles 1 and 2 that are identical)

$$\Psi(t) = \Psi_r(t) u_{\nu 1}(s_1) u_{\nu 2}(s_2). \tag{37a}$$

Here $u_{\nu 1}(s_1)$ and $u_{\nu 2}(s_2)$ are the respective spin eigenvectors for the two particles and $\Psi_r(t)$ depends on only their space coordinates. For central forces the Hamiltonian is diagonal in $S$ and $\nu$, so we may write

$$\Psi(t) = \Psi_r(t) u_{S,\nu}(s_1, s_2), \tag{37b}$$

where $\Psi_r(t)$ again depends on the space coordinates and

$$u_{S,\nu}(s_1, s_2) = \sum_{\nu_1} (S_1, S_2; \nu_1, \nu_2 \mid S, \nu) u_{\nu 1}(s_1) u_{\nu 2}(s_2). \tag{38}$$

Here $(S_1, S_2; \nu_1, \nu_2 \mid S, \nu)$ is the appropriate vector addition coefficient.[5]

It is perhaps useful to recall once again the special sense in which we use the term *elementary particle*. Because we have assumed that $V$ has no matrix elements which lead to a transition between internal states of the particles,[6] we describe them as "elementary." This means that we need not distinguish between internal wave functions $g_i(\xi)$ and the spin functions $u_{S,\nu}$.

The wave functions $\Psi_r(t)$ may be written as

$$\Psi_r(t) = \sum_a A_a \psi_a^{\pm}, \tag{39}$$

where $A_a$ is a wave packet amplitude and the $\psi_a^{\pm}$ are eigenfunctions of $H$, as defined by Eq. (3.67) or (5.51). The initial state wave functions in the barycentric system are

$$\chi_\varkappa(\mathbf{r}) = (2\pi)^{-3/2} e^{i\varkappa \cdot \mathbf{r}}. \tag{40}$$

These form a complete set on the barycentric momentum shell in the sense that [see Eqs. (24)]

$$\mathscr{T}\chi_\varkappa = \int d^3\kappa' \chi_{\varkappa'} T(\hat{\varkappa}' \cdot \hat{\varkappa}; \kappa', \kappa), \tag{41}$$

applicable for the case of Eq. (35*a*) or of Eq. (35*b*). [To avoid having to write duplicate equations for the $(\nu_1, \nu_2)$ and the $(S, \nu)$ representations, we shall now omit the spin label $S$, it being understood that a specific value of

[5] See Section 1.2.

[6] We are not here concerned with the validity of this as an approximation. See, for example, Chapters 4 and 11 for a discussion of some special cases.

$S$ is implied if $R$ and $T$ depend on $S$.] Then according to Eq. (3.78) or Eq. (5.51)

$$\psi_a^+ \equiv \psi_{\boldsymbol{\varkappa}}^+(\mathbf{r}) = \left[1 + \frac{1}{\epsilon(\kappa) + i\eta - K}\mathscr{T}\right]\chi_{\boldsymbol{\varkappa}}(\mathbf{r})$$

$$= \chi_{\boldsymbol{\varkappa}}(\mathbf{r}) + \int \frac{d^3\kappa'\chi_{\boldsymbol{\varkappa}'}(\mathbf{r})}{\epsilon(\kappa) + i\eta - \epsilon(\kappa')} T(\hat{\boldsymbol{\varkappa}}' \cdot \hat{\boldsymbol{\varkappa}}, \kappa', \kappa). \tag{42}$$

Substitution of either of the Eqs. (35*a*) or (35*b*) into the Heitler equation (31) gives us

$$T(\hat{\boldsymbol{\varkappa}}_f \cdot \hat{\boldsymbol{\varkappa}}, \kappa, \kappa)$$
$$= R(\hat{\boldsymbol{\varkappa}}_f \cdot \hat{\boldsymbol{\varkappa}}, \kappa, \kappa) - i\pi\rho_\epsilon \int d\Omega_{\boldsymbol{\varkappa}'} R(\hat{\boldsymbol{\varkappa}}_f \cdot \hat{\boldsymbol{\varkappa}}', \kappa, \kappa) T(\hat{\boldsymbol{\varkappa}}' \cdot \hat{\boldsymbol{\varkappa}}, \kappa, \kappa), \tag{43}$$

since $\kappa_f = \kappa' = \kappa$. To solve this equation we first expand $R$ and $T$ in a series of Legendre polynomials[7] $P_l(\cos\theta)$, where $\cos\theta \equiv \hat{\boldsymbol{\varkappa}}_f \cdot \hat{\boldsymbol{\varkappa}}$.

$$R(\hat{\boldsymbol{\varkappa}}_f \cdot \hat{\boldsymbol{\varkappa}}, \kappa_f, \kappa) = \sum_{l=0}^{\infty} \frac{2l+1}{4\pi} P_l(\cos\theta) R_l(\kappa_f; \kappa)$$
$$= \sum_{l=0}^{\infty} \sum_{m=-l}^{+l} Y_l^m(\hat{\boldsymbol{\varkappa}}_f) Y_l^{m*}(\hat{\boldsymbol{\varkappa}}) R_l(\kappa_f; \kappa), \tag{44a}$$

and

$$T(\hat{\boldsymbol{\varkappa}}_f \cdot \hat{\boldsymbol{\varkappa}}, \kappa_f, \kappa) = \sum_{l=0}^{\infty} \frac{2l+1}{4\pi} P_l(\cos\theta) T_l(\kappa_f; \kappa)$$
$$= \sum_{l=0}^{\infty} \sum_{m=-l}^{+l} Y_l^m(\hat{\boldsymbol{\varkappa}}_f) Y_l^{m*}(\hat{\boldsymbol{\varkappa}}) T_l(\kappa_f; \kappa). \tag{44b}$$

The coefficients $R_l$ and $T_l$ are defined by Eqs. (44*a*) and (44*b*). In the alternate forms of these equations we have used the addition theorem [Eq. (1.78)] to express the Legendre polynomials in terms of the spherical harmonics $Y_l^m$, for which we have used the notation of Eq. (1.86).

On the energy shell, since $\kappa_f = \kappa$, we may write

$$R_l(\kappa; \kappa) \equiv R_l(\kappa), \tag{45}$$
$$T_l(\kappa; \kappa) = T_l(\kappa).$$

The function $R_l(\kappa)$ is *real*. To see this, we recall from Section 5.6 that $\mathscr{R}$ is Hermitian. Therefore on the momentum shell $R$ is Hermitian. This means that

$$R(\hat{\boldsymbol{\varkappa}}_f \cdot \hat{\boldsymbol{\varkappa}}, \kappa_f, \kappa) = R^*(\hat{\boldsymbol{\varkappa}}_f \cdot \hat{\boldsymbol{\varkappa}}, \kappa, \kappa_f),$$

[7] See Section 1.3.

so $R(\hat{\boldsymbol{\varkappa}}_f \cdot \hat{\boldsymbol{\varkappa}}, \kappa, \kappa)$ is real. Since the Legendre polynomials are real functions, we conclude from the first form of Eq. (44*a*) that[8]

$$R_l(\kappa) = R_l^*(\kappa). \tag{46}$$

The spherical harmonic expansions (44) permit us to reduce the Heitler equation (43) to a set of algebraic equations. Using the orthogonality relation (1.77) for the spherical harmonics and the expansions (44), we easily show that on the energy shell

$$\int d\Omega_{\varkappa'} R(\hat{\boldsymbol{\varkappa}}_f \cdot \hat{\boldsymbol{\varkappa}}', \kappa, \kappa) T(\hat{\boldsymbol{\varkappa}}' \cdot \hat{\boldsymbol{\varkappa}}, \kappa, \kappa) = \sum_{l,m} Y_l^m(\hat{\boldsymbol{\varkappa}}_f) Y_l^{m*}(\hat{\boldsymbol{\varkappa}}) R_l(\kappa) T_l(\kappa). \tag{47}$$

Therefore Eq. (43) reduces to

$$\sum_{l,m} Y_l^m(\hat{\boldsymbol{\varkappa}}_f) Y_l^{m*}(\hat{\boldsymbol{\varkappa}})[T_l(\kappa) - R_l(\kappa) + i\pi\rho_\epsilon R_l(\kappa) T_l(\kappa)] = 0. \tag{48}$$

Since Eq. (48) must hold as an identity in $\hat{\boldsymbol{\varkappa}}_f$ and $\hat{\boldsymbol{\varkappa}}$, it follows that[9]

$$T_l(\kappa) = R_l(\kappa) - i\pi\rho_\epsilon R_l(\kappa) T_l(\kappa) \tag{49a}$$

for $l = 0, 1, 2, \ldots$. This may be solved for $T_l$, giving

$$T_l(\kappa) = \frac{R_l(\kappa)}{1 + i\pi\rho_\epsilon R_l(\kappa)}. \tag{49b}$$

This expression is conveniently simplified by writing

$$R_l(\kappa) = -\frac{1}{\pi\rho_\epsilon} \tan \delta_l(\kappa) = -\left(\frac{\pi\kappa^2}{d\epsilon/d\kappa}\right)^{-1} \tan \delta_l(\kappa) \tag{50}$$

on substituting the value (30) for $\rho_\epsilon$. Since $R_l$ is a real number, we may evidently always represent it in the form (50) with a real $\delta_l(\kappa)$. The quantity $\delta_l(\kappa)$ is the *scattering phase shift* for the $l$th partial wave in $\psi^+$. It is defined only mod $\pi$ by Eq. (50). A convention to determine $\delta_l$ uniquely will be adopted in Sections 6.3 and 6.5. Insertion of Eq. (50) into (49*b*) leads to

$$T_l(\kappa) = -\left(\frac{\pi\kappa^2}{d\epsilon/d\kappa}\right)^{-1} e^{i\delta_l} \sin \delta_l. \tag{51a}$$

The scattering matrix $T^{(-)}$ may be expanded in spherical harmonics, as in Eq. (44*b*). We have seen that the Heitler equation for $T^{(-)}$ differs from that for $T$ only in the sign of $i$. Therefore the coefficient $T_l^{(-)}(\kappa)$ corresponding to $T_l(\kappa)$ is given by the expression

$$T_l^{(-)}(\kappa) = T_0 T_l(\kappa) T_0^{-1} = -\left(\frac{\pi\kappa^2}{d\epsilon/d\kappa}\right)^{-1} e^{-i\delta_l} \sin \delta_l. \tag{51b}$$

[8] A more general reality condition is easily derived. See Note [4].

[9] To show this, we may multiply Eq. (48) by $Y_l^{\mu*}(\hat{\boldsymbol{\varkappa}}_f)$, for example, and integrate over the angles of $\hat{\boldsymbol{\varkappa}}_f$.

[Here $T_0$ is the operator of complex conjugation, as introduced in Eq. (2.121).]

It is useful to make the spherical harmonic expansion (44) also for the $S$-matrix [compare Eqs. (3.94) or (5.63)]. We first write

$$S_{ba} = \delta(\mathbf{P}_b - 0)\langle \boldsymbol{\kappa}_f | \, S \, | \boldsymbol{\kappa} \rangle \delta_{S',S} \delta_{\nu'\nu} \tag{52}$$

in analogy to Eqs. (35*a*) or (35*b*). The coefficient $\langle \boldsymbol{\kappa}_f | \, S \, | \boldsymbol{\kappa} \rangle$ is defined here only on the momentum shell, $\mathbf{P}_b = \mathbf{P}_a = 0$, and may depend on the total spin $S$. Then

$$\langle \boldsymbol{\kappa}_f | \, S \, | \boldsymbol{\kappa} \rangle = \delta(\boldsymbol{\kappa}_f - \boldsymbol{\kappa}) - 2\pi i \delta[\epsilon(\kappa) - \epsilon(\kappa_f)] T(\hat{\boldsymbol{\kappa}}_f \cdot \hat{\boldsymbol{\kappa}}, \kappa, \kappa). \tag{53}$$

Now [see Eq. (30)][10]

$$\delta(\boldsymbol{\kappa}_f - \boldsymbol{\kappa}) = \sum_{l,m} Y_l^m(\hat{\boldsymbol{\kappa}}_f) Y_l^{m*}(\hat{\boldsymbol{\kappa}}) \rho_\epsilon^{-1} \delta[\epsilon(\kappa) - \epsilon(\kappa_f)]. \tag{54}$$

If we insert this expression and the expansion (44*b*) into Eq. (53), we find

$$\langle \boldsymbol{\kappa}_f | \, S \, | \boldsymbol{\kappa} \rangle = \sum_{l,m} Y_l^m(\hat{\boldsymbol{\kappa}}_f) Y_l^{m*}(\hat{\boldsymbol{\kappa}}) \rho_\epsilon^{-1} \delta[\epsilon(\kappa) - \epsilon(\kappa_f)] \times [1 - 2\pi i \rho_\epsilon T_l(\kappa)]. \tag{55}$$

Therefore $\langle \boldsymbol{\kappa}_f | \, S \, | \boldsymbol{\kappa} \rangle$ has the partial wave expansion

$$\langle \boldsymbol{\kappa}_f | \, S \, | \boldsymbol{\kappa} \rangle = \sum_{l,m} Y_l^m(\hat{\boldsymbol{\kappa}}_f) Y_l^{m*}(\hat{\boldsymbol{\kappa}}) \rho_\epsilon^{-1} \delta[\epsilon(\kappa) - \epsilon(\kappa_f)] S_l(\kappa), \tag{56}$$

where

$$S_l(\kappa) = 1 - 2\pi i \rho_\epsilon T_l(\kappa) = 1 + 2i \, e^{i\delta_l} \sin \delta_l = e^{2i\delta_l(\kappa)}. \tag{57}$$

The quantity $S_l(\kappa)$ represents an eigenvalue of the $S$-matrix.

### Partial Wave Expansion of the Wave Function

Let us return to the wave function (42). We may expand Eq. (40) [according to Eq. (1.85*a*)] as

$$\chi_{\boldsymbol{\kappa}}(\mathbf{r}) = (2\pi)^{-3/2} e^{i\boldsymbol{\kappa}\cdot\mathbf{r}} = (2\pi)^{-3/2} \sum_{l=0}^{\infty} (2l+1) i^l P_l(\cos\theta) j_l(\kappa r)$$

$$= \sum_{l=0}^{\infty} \sum_{m=-l}^{+l} Y_l^{m*}(\hat{\boldsymbol{\kappa}}) Y_l^m(\hat{\mathbf{r}}) \chi_{\kappa,l}(r), \tag{58a}$$

[10] This is easily verified on introducing the expansion

$$f(\boldsymbol{\kappa}) = \sum_{l,m} Y_l^m(\hat{\boldsymbol{\kappa}}) f_{l,m}(\kappa)$$

for some function $f(\boldsymbol{\kappa})$ into the integrand of the expression

$$\int \delta(\boldsymbol{\kappa}_f - \boldsymbol{\kappa}) f(\boldsymbol{\kappa}) \, d^3\kappa = f(\boldsymbol{\kappa}_f).$$

where

$$\chi_{\kappa,l}(r) \equiv i^l\sqrt{2/\pi}\, j_l(\kappa r) \tag{58b}$$

and $\cos\theta = \hat{\boldsymbol{\varkappa}} \cdot \hat{\mathbf{r}}$. The addition theorem (1.78) has been used again here. Similarly we may expand $\psi_{\boldsymbol{\varkappa}}^{+}(\mathbf{r})$ in the form

$$\psi_{\boldsymbol{\varkappa}}^{+}(\mathbf{r}) = \sum_{l=0}^{\infty} \sum_{m=-l}^{+l} Y_l^{m*}(\hat{\boldsymbol{\varkappa}}) Y_l^{m}(\hat{\mathbf{r}}) \psi_{\kappa,l}^{+}(r). \tag{59}$$

To evaluate $\psi_{\kappa,l}^{+}(r)$, we introduce the expansions (44*b*) and (58) into the second term on the right side of Eq. (42) to obtain

$$\int \frac{d^3\kappa' \chi_{\boldsymbol{\varkappa}'}}{\epsilon(\kappa) + i\eta - \epsilon(\kappa')}\, T(\hat{\boldsymbol{\varkappa}}' \cdot \hat{\boldsymbol{\varkappa}}, \kappa', \kappa) = \sum_{l,m} Y_l^{m*}(\hat{\boldsymbol{\varkappa}}) Y_l^{m}(\hat{r}) i^l \sqrt{2/\pi} \int \frac{(\kappa')^2\, d\kappa' j_l(\kappa' r) T_l(\kappa'; \kappa)}{\epsilon(\kappa) + i\eta - \epsilon(\kappa')}. \tag{60}$$

Now we may insert Eqs. (58), (59), and (60) into Eq. (42) to obtain an explicit expression for $\psi_{\kappa,l}^{+}(r)$.

$$\psi_{\kappa,l}^{+}(r) = i^l \sqrt{2/\pi} \left[ j_l(\kappa r) + \int \frac{(\kappa')^2\, d\kappa' j_l(\kappa' r) T_l(\kappa'; \kappa)}{\epsilon(\kappa) + i\eta - \epsilon(\kappa')} \right]. \tag{61}$$

It follows directly from the definition of $\mathscr{T}$ [see Eqs. (4.13)] that when $\kappa_f = \kappa$

$$T(\hat{\boldsymbol{\varkappa}}_f \cdot \hat{\boldsymbol{\varkappa}}, \kappa, \kappa) = (\chi_{\boldsymbol{\varkappa}_f}, V\psi_{\kappa}^{+}). \tag{62a}$$

If we insert the expansions (44*b*), (58*b*), and (59) into Eq. (62*b*), we find the useful relation

$$T_l(\kappa) = (\chi_{\kappa,l}, V\psi_{\kappa,l}^{+}). \tag{62b}$$

This expression permits us to calculate the $T_l(\kappa)$, and thus the scattering cross section, when we have evaluated $\psi_{\kappa,l}^{+}$ only at small distances (within the "range" of $V$).

It follows from Eq. (5.55) that the normalization of $\psi_{\boldsymbol{\varkappa}}^{+}$ is

$$(\psi_{\boldsymbol{\varkappa}'}^{+}, \psi_{\boldsymbol{\varkappa}}^{+}) = (\chi_{\boldsymbol{\varkappa}'}, \chi_{\boldsymbol{\varkappa}}) = \delta(\boldsymbol{\varkappa}' - \boldsymbol{\varkappa}).$$

From this and the expansion (59) we easily obtain the important normalization condition

$$\int [\psi_{\kappa',l}^{+}(r)]^* [\psi_{\kappa,l}^{+}(r)] r^2\, dr = \frac{1}{\kappa^2}\, \delta(\kappa' - \kappa). \tag{63a}$$

When there are no bound states of the two particles, for a given value of $l$, the $\psi_{\kappa,l}^{+}$ represent a complete set of states (as described in Section 5.2, p. 191). In this case

$$\int [\psi_{\kappa,l}^{+}(r')]^* \psi_{\kappa,l}^{+}(r) \kappa^2\, d\kappa = \frac{1}{r^2}\, \delta(r' - r). \tag{63b}$$

The asymptotic form of $\psi^+_{\kappa,l}(r)$, as $r \to \infty$, is important for our later applications. Since

$$\lim_{r\to\infty} j_l(\kappa r) = \frac{\sin(\kappa r - \pi l/2)}{\kappa r}, \tag{64}$$

we may write

$$\lim_{r\to\infty} \int \frac{(\kappa')^2\, d\kappa'}{\epsilon(\kappa) + i\eta - \epsilon(\kappa')} j_l(\kappa' r) T_l(\kappa'; \kappa)$$

$$= -\int \frac{(\kappa')^2}{d\epsilon(\kappa')/d\kappa'} \frac{d\epsilon(\kappa') T_l(\kappa'; \kappa)}{\epsilon(\kappa') - [\epsilon(\kappa) + i\eta]} \frac{e^{i(\kappa' r - \pi l/2)} - e^{-i(\kappa' r - \pi l/2)}}{2i\kappa' r}$$

$$= -\frac{\pi\kappa^2}{d\epsilon/d\kappa} \frac{e^{i\kappa r}}{\kappa r} e^{-i(\pi l/2)} T_l(\kappa). \tag{65}$$

The evaluation of Eq. (65) follows that of Eq. (3.174), which was described in detail in Section 3.4. The result (65) and the asymptotic form (64) may be inserted into Eq. (61) to give

$$\begin{aligned} \lim_{r\to\infty} \psi^+_{\kappa,l}(r) &= \sqrt{2/\pi}(2i\kappa r)^{-1}[-(-1)^l e^{-i\kappa r} + S_l(\kappa)\, e^{i\kappa r}] \\ &= i^l \sqrt{2/\pi}\, e^{i\delta_l} \frac{\sin(\kappa r - \pi l/2 + \delta_l)}{\kappa r}. \end{aligned} \tag{66a}$$

In obtaining this result, we have used the relation between $S_l$ and the phase shift $\delta_l$, Eqs. (51$a$) and (57), in Eq. (65).

It follows from Eq. (4.232$a$) that the wave function $\psi_{\boldsymbol{\kappa}}^-$ can be obtained by application of the time reversal operator $T_0$ to $\psi_{\boldsymbol{\kappa}}^+$. Thus

$$\psi_{\boldsymbol{\kappa}}^- = T_0 \psi_{-\boldsymbol{\kappa}}^+ = \sum_{l=0}^{\infty} \sum_{m=-l}^{+l} Y_l^{m*}(-\hat{\boldsymbol{\kappa}}) Y_l^m(\hat{\mathbf{r}}) \psi^-_{\kappa,l}(r), \tag{67}$$

where we have introduced an expansion analogous to that of Eq. (59) for $\psi_\kappa^+$. Now since $Y_l^m(-\hat{\boldsymbol{\kappa}}) = (-1)^l Y_l^m(\hat{\boldsymbol{\kappa}})$ and $Y_l^{m*} = (-1)^m Y_l^{-m}$, we may change $m$ to $-m$ in the sum over $m$ to obtain

$$\begin{aligned} \psi^-_{\kappa,l}(r) &= (-1)^l T_0 \psi^+_{\kappa,l} = (-1)^l [\psi^+_{\kappa,l}(r)]^* \\ \lim_{r\to\infty} &= (i)^l \sqrt{2/\pi}\, e^{-i\delta_l} \frac{\sin(\kappa r - \pi l/2 + \delta_l)}{\kappa r}. \end{aligned} \tag{66b}$$

The asymptotic form here differs only by a phase factor from that for $\psi^+_{\kappa,l}$; in fact $\psi^{(+)}_{\kappa,l} = S_l \psi^{(-)}_{\kappa,l}$, a quite general result [see, for example, Eq. (5.138)]. It may also be anticipated from the fact that the solution of the Schrödinger equation must be unique if we specify as definite eigenstate—in this case defined by ($\kappa$, $l$, $m$). The result (66$b$) might, of course, have been obtained directly from the definition of the $\psi^-$ functions and Eq. (51$b$).

The series (59) may be summed in a convenient form when $r \to \infty$. To do this, we first rewrite Eq. (66$a$) as

$$\lim_{r\to\infty} \psi^+_{\kappa,l}(r) = \sqrt{2/\pi}\left[ i^l j_l(\kappa r) + \frac{S_l(\kappa) - 1}{2i\kappa r}\, e^{i\kappa r}\right]. \tag{68}$$

When this is substituted into Eq. (59) we may use Eq. (58) to obtain

$$\lim_{r\to\infty} \psi_{\boldsymbol{\kappa}}^+(\mathbf{r}) = (2\pi)^{-3/2}\left[ e^{i\boldsymbol{\kappa}\cdot\mathbf{r}} + \frac{e^{i\kappa r}}{r} f(\hat{\mathbf{r}}\cdot\hat{\boldsymbol{\kappa}})\right], \tag{69}$$

where

$$f(\hat{\mathbf{r}}\cdot\hat{\boldsymbol{\kappa}}) = \sum_{l=0}^{\infty} (2l+1)[S_l(\kappa) - 1](2i\kappa)^{-1} P_l(\hat{\mathbf{r}}\cdot\hat{\boldsymbol{\kappa}})$$

$$= \sum_{l=0}^{\infty} \frac{2l+1}{\kappa}\, e^{i\delta_l} \sin\delta_l P_l(\hat{\mathbf{r}}\cdot\hat{\boldsymbol{\kappa}}). \tag{70a}$$

On comparison with Eqs. (44$b$) and (51$a$) it is seen that this may also be written as

$$f(\hat{\mathbf{r}}\cdot\hat{\boldsymbol{\kappa}}) = -(2\pi)^2\left(\frac{\kappa}{d\epsilon/d\kappa}\right) T(\hat{\mathbf{r}}\cdot\hat{\boldsymbol{\kappa}}, \kappa, \kappa). \tag{70b}$$

For spin-dependent central forces it will often be desirable to add a spin label $S$ in Eqs. (70$a$) and (70$b$). In this case, we write, using the notation of Eqs. (35$b$),

$$f^{(S)}(\hat{\mathbf{r}}\cdot\hat{\boldsymbol{\kappa}}) = -(2\pi)^2\left(\frac{\kappa}{d\epsilon/d\kappa}\right) T(\hat{\mathbf{r}}\cdot\hat{\boldsymbol{\kappa}}, \kappa, \kappa, S). \tag{70c}$$

The phase shift $\delta_l$ is then written as $\delta_l^{(S)}$ in Eq. (70$a$).

On comparing this with our previous notation, we note that here the unit vector $\hat{\mathbf{r}}$, rather than $\hat{\boldsymbol{\kappa}}_f$, appears in the scattering matrix and in $f$. There is no discrepancy, however, since $\hat{\mathbf{r}} = \hat{\boldsymbol{\kappa}}_f$ for a final asymptotic

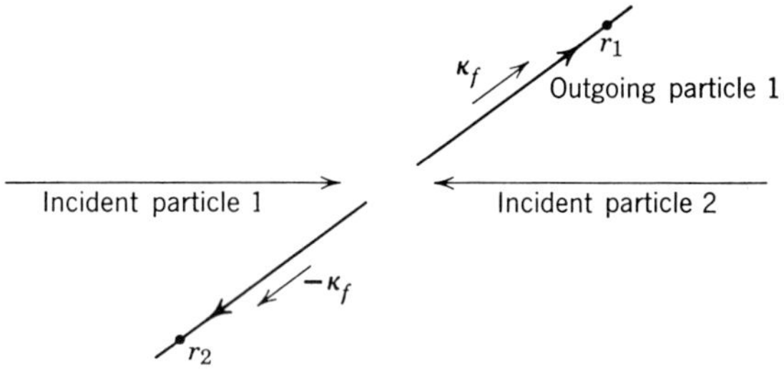


Fig. 6.2 Illustration of the vectors $\hat{\mathbf{r}}$ and $\boldsymbol{\kappa}_f$ for the scattering of two particles.

scattering state. That is, the relative momentum vector $\hat{\boldsymbol{\kappa}}_f$ is directed along the vector $\mathbf{r}$ joining the two particles,[11] as is illustrated in Fig. 6.2.

**The Scattering Cross Section and the Scattering Amplitude**

The quantity $f(\hat{\mathbf{r}} \cdot \hat{\boldsymbol{\kappa}})$ is the scattering amplitude. It has been encountered previously.[12] For the case of scattering by two physical particles we have [see Eq. (3)]

$$\epsilon(\kappa) = \epsilon_{1,\kappa} + \epsilon_{2,\kappa},$$

and Eq. (70$b$) agrees precisely with Eq. (6). The differential scattering cross section is given by Eq. (4).

Equation (4) may be easily generalized to the case that $\epsilon$ is a general function of $\kappa$, as for quasi-particles. According to Eq. (3.147) we have[13]

$$d\sigma = \frac{(2\pi)^4}{v_{\text{rel},B}} \int d^3P_b \, d^3\kappa_f \, \delta(\mathbf{P}_b) \, \delta[\epsilon(\kappa) - \epsilon(\kappa_f)] \, |T|^2$$

$$= (2\pi)^4 \left( \frac{\kappa^2}{d\epsilon/d\kappa} \frac{1}{v_{\text{rel},B}} |T|^2 \right) d\Omega. \tag{71}$$

Now according to Eq. (3.13)

$$v_{\text{rel},B} = \frac{d\epsilon}{d\kappa}, \tag{72}$$

so Eq. (71) simplifies to

$$\frac{d\sigma}{d\Omega} = |f(\hat{\mathbf{r}} \cdot \hat{\boldsymbol{\kappa}})|^2, \tag{73$a$}$$

using the definition (70$b$) for $f(\hat{\mathbf{r}} \cdot \hat{\boldsymbol{\kappa}})$. This agrees with the second form of Eq. (4) since $\hat{\boldsymbol{\kappa}}_f = \hat{\mathbf{r}}$.

For spin-independent forces, the initial and final spin orientations are the same according to Eq. (35$a$), so $\nu'_1 = \nu_1$ and $\nu'_2 = \nu_2$ is implied in Eq. (73$a$). In this case the cross section $d\bar{\sigma}/d\Omega$, obtained by averaging over initial and summing over final spin orientations, is equal to $d\sigma/d\Omega$. That is, Eq. (7) becomes

$$\frac{d\bar{\sigma}}{d\Omega} = \frac{d\sigma}{d\Omega} \tag{74$a$}$$

for spin-independent forces.

For spin-dependent central forces, on the other hand, Eq. (73$a$) must be labeled by the total spin $S$ of the two particles [see Eq. (35$b$)]. Then for a

[11] This is clear from the wave packet description of scattering given in Chapter 3.

[12] Equation (69) may be compared with Eq. (3.177). The additional factor of $(2\pi)^{-3/2} e^{i\mathbf{P}_a \cdot \mathbf{r}_2}$ in Chapter 3 describes the motion of the center-of-mass.

[13] The Jacobian of the transformation to variables $\mathbf{P}_b$ and $\boldsymbol{\kappa}_f$ is unity.

given state of spin $S$ we have [Eq. (70$c$)]

$$\frac{d\sigma^{(S)}}{d\Omega} = |f^{(S)}(\hat{\mathbf{r}} \cdot \hat{\boldsymbol{\varkappa}})|^2. \qquad (73b)$$

For a random distribution of initial spin orientation the state $S$ occurs with a probability

$$\frac{2S+1}{(2S_1+1)(2S_2+1)}.$$

The differential cross section for any final spin orientation and a random distribution of initial spin orientations is therefore

$$\frac{d\bar{\sigma}}{d\Omega} = \frac{1}{(2S_1+1)(2S_2+1)} \sum_{S=|S_1-S_2|}^{|S_1+S_2|} (2S+1) \frac{d\sigma^{(S)}}{d\Omega}. \qquad (74b)$$

This reduces to Eq. (73$b$) when $f^{(S)}(\hat{\mathbf{r}} \cdot \hat{\boldsymbol{\varkappa}})$ is independent of $S$.

The total cross section $\bar{\sigma}$ is obtained by integrating the differential cross section over all scattering angles. For the case of spin-independent forces, we obtain from Eqs. (70$a$) and (73$a$) and the orthogonality relations (1.77)

$$\bar{\sigma} = \sigma = \int \frac{d\bar{\sigma}}{d\Omega}\, d\Omega = \frac{4\pi}{\kappa^2} \sum_{l=0}^{\infty} (2l+1) \sin^2 \delta_l. \qquad (74c)$$

On comparing Eqs. (70$a$) and (74$c$), we obtain a special case of the important "optical theorem" in scattering theory.

$$\bar{\sigma} = \frac{4\pi}{\kappa} \operatorname{Im}\,[f(1)]. \qquad (74d)$$

For scattering when the forces are spin dependent we find from Eq. (74$b$)

$$\bar{\sigma} = \frac{4\pi}{(2S_1+1)(2S_2+1)\kappa^2} \sum_{S=|S_1-S_2|}^{|S_1+S_2|} (2S+1) \sum_{l=0}^{\infty} (2l+1) \sin^2 \delta_l^{(S)}. \qquad (74e)$$

The maximum scattering possible from a given state $l$ (and $S$) is seen to occur when $\sin^2 \delta_l = 1$.

It is instructive to rederive Eq. (73$a$) directly from Eq. (69). The incident beam current associated with [see Eq. (40)] $\chi_\kappa(\mathbf{r})$ is

$$\mathbf{j}_{\text{in}} = (2\pi)^{-3}\, v_{\text{rel},B} \hat{\boldsymbol{\varkappa}}, \qquad (75a)$$

where $v_{\text{rel},B}$ has been given in Eq. (72). The asymptotic flux of scattered particles is, as obtained from the second term in Eq. (69),

$$\mathbf{j}_{\text{out}} = \frac{\hat{\mathbf{r}}}{r^2}\, v_{\text{rel},B} (2\pi)^{-3}\, |f(\hat{\mathbf{r}} \cdot \hat{\boldsymbol{\varkappa}})|^2. \qquad (75b)$$

The differential cross section is then [see the arguments leading to Eq. (3.183)]

$$\frac{d\sigma}{d\Omega} = r^2 \frac{j_{\text{out}}}{j_{\text{in}}} = |f(\hat{\mathbf{r}} \cdot \hat{\boldsymbol{\varkappa}})|^2,$$

which is just Eq. (73*a*).

**The Scattering of Two Identical Particles**

We have yet to correct our discussion for wave function symmetrization when the two particles are identical. According to Eqs. (37*a*) and (42) we may write

$$\psi^{\pm}_{\varkappa,\nu 1,\nu 2} = \psi_{\varkappa}^{\pm}(\mathbf{r}) u_{\nu 1}(s_1) u_{\nu 2}(s_2) \tag{76a}$$

for the continuum eigenfunctions of $H$ when the force is spin independent. More generally, for central forces [see Eqs. (37*b*) and (38)] we have

$$\psi^{\pm}_{\varkappa,S,\nu} = \psi^{\pm}_{\varkappa,S}(r) u_{S,\nu}(s_1, s_2). \tag{76b}$$

The symmetrization operator $\mathscr{S}$ [Eq. (4.39)] is now

$$\mathscr{S} = \tfrac{1}{2}(1 + \delta_{Q12} Q_{12}), \tag{77}$$

where $Q_{12}$ interchanges the coordinates of particles 1 and 2. For example, $Q_{12}\mathbf{r}Q_{12}^{-1} = -\mathbf{r}$. Therefore the symmetrized eigenfunctions (76*a*) and (76*b*) are, according to Eq. (4.131),

$$\begin{aligned}\psi^{\pm}_{\mathscr{S}(\varkappa,\nu 1,\nu 2)} &= \sqrt{2}\,\mathscr{S}\psi^{\pm}_{\varkappa,\nu 1,\nu 2} \\ &= \frac{1}{\sqrt{2}}[\psi_{\kappa}^{\pm}(\mathbf{r}) u_{\nu 1}(s_1) u_{\nu 2}(s_2) + \delta_{Q12}\psi_{\varkappa}^{\pm}(-\mathbf{r}) u_{\nu 1}(s_2) u_{\nu 2}(s_1)]\end{aligned} \tag{78a}$$

and

$$\psi^{\pm}_{\mathscr{S}(\kappa,S,\nu)} = \frac{1}{\sqrt{2}}[\psi^{\pm}_{\varkappa,S}(\mathbf{r}) u_{S,\nu}(s_1, s_2) + \delta_{Q12}\psi^{\pm}_{\varkappa,S}(-\mathbf{r}) u_{S,\nu}(s_1, s_2)]. \tag{78b}$$

The asymptotic form of $\psi_{\kappa}^{+}(-\mathbf{r})$ [or of $\psi^{+}_{\varkappa,S}(-\mathbf{r})$] is obtained from Eq. (69) as

$$\lim_{r\to\infty} \psi_{\varkappa}^{+}(-\mathbf{r}) = \chi_{\varkappa}(-\mathbf{r}) + (2\pi)^{-3/2} \frac{e^{i\kappa r}}{r} f(-\hat{\mathbf{r}} \cdot \hat{\boldsymbol{\varkappa}}) \tag{79}$$

where the label $S$ is implied for spin-dependent forces.

As a first example, let us consider the scattering of two spinless Bose-Einstein particles. In this case $\delta_{Q12} = 1$ and $u_{\nu 1}, u_{\nu 2}$ do not appear in Eq. (78*a*). Thus

$$\lim_{r\to\infty} \psi^{+}_{\mathscr{S}(\varkappa)} = \frac{1}{\sqrt{2}}\left\{\chi_{\varkappa}(\mathbf{r}) + \chi_{\varkappa}(-\mathbf{r}) + (2\pi)^{-3/2} \frac{e^{\,i\kappa r}}{r} [f(\hat{\mathbf{r}} \cdot \hat{\boldsymbol{\varkappa}}) + f(-\hat{\mathbf{r}} \cdot \hat{\boldsymbol{\varkappa}})]\right\}. \tag{80}$$

To evaluate the differential cross section, we recall from Section 4.3 that the incident current is unaffected by symmetrization. The outgoing current is given by Eq. (75*b*), with $f(\hat{\mathbf{r}} \cdot \hat{\boldsymbol{\varkappa}})$ replaced by

$$f_{\mathscr{S}S}(\hat{\mathbf{r}} \cdot \hat{\boldsymbol{\varkappa}}) \equiv f(\hat{\mathbf{r}} \cdot \hat{\boldsymbol{\varkappa}}) + f(-\hat{\mathbf{r}} \cdot \hat{\boldsymbol{\varkappa}}). \tag{81a}$$

To understand this, we observe that $\mathbf{j}_{\text{out}}$ first gets a factor of $\frac{1}{2} = (1/\sqrt{2})^2$ because of the $1/\sqrt{2}$ in Eq. (80). A compensating factor of 2 appears because $\mathbf{j}_{\text{out}} = \mathbf{j}_{\text{out}}(1) + \mathbf{j}_{\text{out}}(2)$, the sum of the identical currents associated with particles 1 and 2. Therefore the differential cross section is

$$\frac{d\sigma}{d\Omega} = |f_{\mathscr{S}S}|^2 = |f(\hat{\mathbf{r}} \cdot \hat{\boldsymbol{\varkappa}}) + f(-\hat{\mathbf{r}} \cdot \hat{\boldsymbol{\varkappa}})|^2. \tag{82a}$$

The total cross section, as obtained from Eq. (4.122), is

$$\sigma = \frac{1}{2} \int \frac{d\sigma}{d\Omega}\, d\Omega. \tag{82b}$$

Using Eq. (70*a*) we see that

$$f_{\mathscr{S}S}(\hat{\mathbf{r}} \cdot \hat{\boldsymbol{\varkappa}}) = 2 \sum_{l \text{ even}} \frac{2l+1}{\kappa}\, e^{i\delta_l} \sin \delta_l P_l(\hat{\mathbf{r}} \cdot \hat{\boldsymbol{\varkappa}}) \tag{81b}$$

where the sum runs over all *even* values of $l$.[14]

Equations (80), (81), and (82) are applicable, for example, for the scattering of two He atoms or of two pi-mesons having the same charge.

For two Fermi-Dirac particles in the *same* spin state, Eq. (78*a*) gives

$$\lim_{r\to\infty} \psi^{+}_{\mathscr{S}A(\varkappa,\nu1,\nu2)} = \frac{1}{\sqrt{2}} \left\{ \chi_{\varkappa}(\mathbf{r}) - \chi_{\varkappa}(-\mathbf{r}) + \frac{e^{i\kappa r}}{r} \left[ f(\hat{\mathbf{r}} \cdot \hat{\boldsymbol{\varkappa}}) - f(-\hat{\mathbf{r}} \cdot \hat{\boldsymbol{\varkappa}}) \right] \right\} u_{\nu 1}(s_1) u_{\nu 1}(s_2). \tag{83}$$

As was the case before, the study in Section 4.3 implies that $\mathbf{j}_{\text{in}}$ is unaffected by antisymmetrization of the wave function (83). Evaluation of the outgoing current is similar to that for Eq. (80), except that we are now led to

$$f_{\mathscr{S}A}(\hat{\mathbf{r}} \cdot \hat{\boldsymbol{\varkappa}}) \equiv f(\hat{\mathbf{r}} \cdot \hat{\boldsymbol{\varkappa}}) - f(-\hat{\mathbf{r}} \cdot \hat{\boldsymbol{\varkappa}}) \tag{84a}$$

and

$$\frac{d\sigma}{d\Omega} = |f_{\mathscr{S}A}|^2 = |f(\hat{\mathbf{r}} \cdot \hat{\boldsymbol{\varkappa}}) - f(-\hat{\mathbf{r}} \cdot \hat{\boldsymbol{\varkappa}})|^2, \tag{85a}$$

with

$$\sigma = \frac{1}{2} \int \frac{d\sigma}{d\Omega}\, d\Omega. \tag{85b}$$

[14] If the argument leading to Eqs. (81) and (82) appears a little too brief, the reader may reassure himself by constructing wave packets. We shall verify these equations in what follows by use of the general theory of Section 4.4.

According to Eq. (70*a*) we have

$$f_{\mathscr{S}A}(\hat{\mathbf{r}} \cdot \hat{\boldsymbol{\varkappa}}) = 2 \sum_{l \text{ odd}} \frac{2l+1}{\kappa} e^{i\delta_l} \sin \delta_l P_l(\hat{\mathbf{r}} \cdot \hat{\boldsymbol{\varkappa}}), \tag{84b}$$

where the sum now runs over all *odd* values of $l$.

To describe generally the scattering of identical particles, let us first consider the case of spin-independent forces and write, according to Eq. (4.146),

$$T_{\mathscr{S}} = \langle \boldsymbol{\varkappa}_f, \nu'_1, \nu'_2 | \, T \, | \boldsymbol{\varkappa}, \nu_1, \nu_2 \rangle + \delta_{Q12} \langle Q_{12}(\boldsymbol{\varkappa}_f, \nu'_1, \nu'_2) | \, T \, | \boldsymbol{\varkappa}, \nu_1, \nu_2 \rangle, \tag{86a}$$

in the notation of Eq. (35*a*). Now since

$$Q_{12}[\chi_{\varkappa f}(\mathbf{r}) u_{\nu'1}(s_1) u_{\nu'2}(s_2)] = \chi_{-\varkappa f}(\mathbf{r}) u_{\nu'1}(s_2) u_{\nu'2}(s_1)$$

we have [on removing the factor $\delta(\mathbf{P}_b)$ describing momentum conservation in Eq. (4.118*b*)]

$$\langle Q_{12}(\boldsymbol{\varkappa}_f, \nu'_1, \nu'_2) | \, T \, | \boldsymbol{\varkappa}, \nu_1, \nu_2 \rangle$$

$$= \left( \chi_{-\varkappa f}(\mathbf{r}) u_{\nu'1}(s_2) u_{\nu'2}(s_1), \left[ V + V \frac{1}{\epsilon(\kappa) + i\eta - H} V \right] \chi_{\varkappa}(\mathbf{r}) u_{\nu 1}(s_1) u_{\nu 2}(s_2) \right)$$

$$= \delta_{\nu'2,\nu 1} \delta_{\nu'1,\nu 2} T(-\hat{\boldsymbol{\varkappa}}_f \cdot \hat{\boldsymbol{\varkappa}}, \kappa_f, \kappa). \tag{87}$$

Therefore using Eqs. (35*a*) and (86*a*) we find that

$$T_{\mathscr{S}} = T(\hat{\boldsymbol{\varkappa}}_f \cdot \hat{\boldsymbol{\varkappa}}, \kappa_f, \kappa) \, \delta_{\nu'1,\nu 1} \, \delta_{\nu'2,\nu 2} + \delta_{Q12} T(-\hat{\boldsymbol{\varkappa}}_f \cdot \hat{\boldsymbol{\varkappa}}, \kappa_f, \kappa) \, \delta_{\nu'2,\nu 1} \, \delta_{\nu'1,\nu 2}. \tag{88a}$$

For the more general case of central forces we may again use Eq. (4.146) for the $T$-matrix of Eq. (35*b*). Thus

$$T_{\mathscr{S}} = \langle \boldsymbol{\varkappa}_f, S', \nu' | \, T \, | \boldsymbol{\varkappa}, S, \nu \rangle + \delta_{Q12} \langle Q_{12}(\boldsymbol{\varkappa}_f, S', \nu') | \, T \, | \boldsymbol{\varkappa}, S, \nu \rangle. \tag{86b}$$

Because of the symmetry relation of Eq. (1.60*a*) for the vector addition coefficients, we find from Eq. (38) that

$$Q_{12} u_{S,\nu}(s_1, s_2) = (-1)^{2S_1 - S} u_{S,\nu}(s_1, s_2), \tag{89}$$

since for identical particles $S_1 = S_2$. Therefore

$$\langle Q_{12}(\boldsymbol{\varkappa}_f, S', \nu') | \, T \, | \boldsymbol{\varkappa}, S, \nu \rangle = (-1)^{2S_1 - S} \langle -\boldsymbol{\varkappa}_f, S', \nu' | \, T \, | \boldsymbol{\varkappa}, S, \nu \rangle,$$

and Eq. (86*b*) simplifies to

$$T_{\mathscr{S}} = [T(\hat{\boldsymbol{\varkappa}}_f \cdot \hat{\boldsymbol{\varkappa}}, \kappa_f, \kappa, S) + (-1)^{2S_1 - S} \delta_{Q12} T(-\hat{\boldsymbol{\varkappa}}_f \cdot \hat{\boldsymbol{\varkappa}}, \kappa_f, \kappa, S)] \, \delta_{S' S} \delta_{\nu',\nu}. \tag{88b}$$

The general expression (4.121) for the differential cross section is, for the present example,

$$d\sigma = \frac{(2\pi)^4}{v_{\mathrm{rel},B}} \int d^3P_b\, d^3\kappa_f\, \delta(\mathbf{P}_b)\, \delta[\epsilon(\kappa) - \epsilon(\kappa_f)]\, |T_{\mathscr{S}}|^2 \equiv d\Omega\, |f_{\mathscr{S}}|^2, \quad (90)$$

where

$$f_{\mathscr{S}} = -(2\pi)^2 \frac{\kappa}{d\epsilon/d\kappa} T_{\mathscr{S}}. \quad (91)$$

The sequence of arguments here is essentially that of Eqs. (71), (72), and (73). The scattering matrix $T$ in these equations is restricted to the energy shell, so $\kappa_f = \kappa$.

For spin-independent forces, we may write Eq. (91) as [see Eq. (88*a*)]

$$\langle \nu'_1, \nu'_2\, | f_{\mathscr{S}} | \nu_1, \nu_2 \rangle = f(\hat{\boldsymbol{\kappa}}_f \cdot \hat{\boldsymbol{\kappa}})\, \delta_{\nu'1,\nu1}\, \delta_{\nu'2,\nu2} \pm f(-\hat{\boldsymbol{\kappa}}_f \cdot \hat{\boldsymbol{\kappa}})\, \delta_{\nu'2,\nu1}\, \delta_{\nu'1,\nu2}, \quad (92a)$$

where $f(\hat{\boldsymbol{\kappa}}_f \cdot \hat{\boldsymbol{\kappa}})$ is defined by Eqs. (70). For spin-dependent forces, we have instead [from Eq. (88*b*)]

$$\langle \mathrm{S}', v' | f_{\mathscr{S}} | \mathrm{S}, v \rangle = [f^{(S)}(\hat{\boldsymbol{\kappa}}_f \cdot \hat{\boldsymbol{\kappa}}) \pm (-1)^{S2_1 - S} f^{(S)}(-\hat{\boldsymbol{\kappa}}_f \cdot \hat{\boldsymbol{\kappa}})]\, \delta_{S',S}\, \delta_{v',v} \quad (92b)$$

where $f^{(S)}(\hat{\boldsymbol{\kappa}}_f \cdot \hat{\boldsymbol{\kappa}})$ is again defined by Eqs. (70). In Eqs. (92*a*) and (92*b*) the plus or minus sign pertains, as usual, to Bose-Einstein or Fermi-Dirac statistics.

For spin-independent forces, we may use Eqs. (88*a*), (90), and (92*a*) to write

$$\begin{aligned} \frac{d\sigma}{d\Omega}(\nu'_1, \nu'_2; \nu_1, \nu_2) &= |f_{\mathscr{S}}|^2 = |f(\hat{\boldsymbol{\kappa}}_f \cdot \hat{\boldsymbol{\kappa}})|^2\, \delta_{\nu'1,\nu1}\, \delta_{\nu'2,\nu2} \\ &\quad + |f(-\hat{\boldsymbol{\kappa}}_f \cdot \hat{\boldsymbol{\kappa}})|^2\, \delta_{\nu'2,\nu1}\, \delta_{\nu'1,\nu2} \\ &\quad \pm 2\, \mathrm{Re}\, [f^*(\hat{\boldsymbol{\kappa}}_f \cdot \hat{\boldsymbol{\kappa}}) f(-\hat{\boldsymbol{\kappa}}_f \cdot \hat{\boldsymbol{\kappa}})]\, \delta_{\nu'1,\nu1}\, \delta_{\nu'2,\nu2}\, \delta_{\nu'1,\nu2}\, \delta_{\nu'2,\nu1}, \quad (93a) \end{aligned}$$

where the plus or minus sign refers to the case of Bose-Einstein or Fermi-Dirac statistics, respectively.

The barycentric cross section (92*a*) describes the scattering of particle 1 through an angle $\theta$, where $\cos\theta = \hat{\boldsymbol{\kappa}}_f \cdot \hat{\boldsymbol{\kappa}}$ [$= \hat{\mathbf{r}} \cdot \hat{\boldsymbol{\kappa}}$ in the notation of Eq. (73)]. The initial spin orientations $(\nu_1, \nu_2)$ becomes $(\nu'_1, \nu'_2)$ in the final state. The contrast with Eq. (73*a*) which applies for the case of distinguishable particles and for which $\nu_1 = \nu'_1$, $\nu_2 = \nu'_2$, should be noted. The difference arises not from a dynamical spin interaction but from our inability to determine whether particle 1 in the final state (that is, our detected particle) is the same as particle 1 in the initial state (the beam particle). If we set $\nu_1 = \nu_2$ in Eq. (92*a*), we immediately recover the appropriate Eq. (82*a*) or (85*a*).

For a random distribution of initial spin orientations, we obtain, instead of Eq. (74$a$),

$$\frac{d\bar{\sigma}}{d\Omega} = \frac{1}{(2S_1 + 1)^2} \sum_{\nu_1,\nu_2,\nu'_1,\nu'_2} \frac{d\sigma}{d\Omega}(\nu'_1, \nu'_2; \nu_1, \nu_2)$$

$$= |f(\hat{\varkappa}_f \cdot \hat{\varkappa})|^2 + |f(-\hat{\varkappa}_f \cdot \hat{\varkappa})|^2$$

$$\pm \frac{2}{2S_1 + 1} \operatorname{Re}\,[f^*(\hat{\varkappa}_f \cdot \hat{\varkappa}) f(-\hat{\varkappa}_f \cdot \hat{\varkappa})], \tag{94a}$$

where, we recall, $S_1$ is the spin of either particle. Again, the plus or minus sign refers to the case of Bose-Einstein or Fermi-Dirac particles.

Equations (92$a$) and (94$a$) apply to the scattering of any two identical particles (for example, two atoms, nuclei, or electrons) provided that we may adequately describe their interaction by a spin-independent potential $V$ which does not lead to excitation of internal states of the particles or to the production of new particles.[15]

More generally, for central forces, we must describe the differential cross section (90) separately for the various eigenstates $S$ of the total spin. In this case we must use the expression (88$b$) in Eqs. (90), (91), and (92$b$). Then the differential cross section for a total spin $S$ is

$$\frac{d\sigma^{(S)}}{d\Omega} = |f_{\mathscr{S}}|^2 = |f^{(S)}(\hat{\varkappa}_f \cdot \hat{\varkappa}) \pm (-1)^{2S_1 - S} f^{(S)}(-\hat{\varkappa}_f \cdot \hat{\varkappa})|^2, \tag{92c}$$

where the scattering amplitude $f^{(S)}$ is given by Eq. (70$c$) and the plus or minus sign refers to Bose-Einstein or Fermi-Dirac statistics, respectively.

Equation (92$b$) gives the appropriate modification of Eq. (73$b$) when the two particles are identical. Here Eq. (74$b$) becomes

$$\frac{d\bar{\sigma}}{d\Omega} = \frac{1}{(2S_1 + 1)^2} \sum_{S=0}^{2S_1} (2S + 1) \frac{d\sigma^{(S)}}{d\Omega} \tag{94b}$$

for a random distribution of initial spin orientations.[16]

## 6.3 NONRELATIVISTIC POTENTIAL SCATTERING

### The Schrödinger Equation

Scattering by a central potential is most frequently encountered for non-relativistic energies of the two interacting particles. In this case the kinetic energy operator for the two particles is

$$K = K_1 + K_2, \tag{95a}$$

[15] It is easily verified that Eqs. (92$a$) and (94$a$) are consistent with the examples discussed in Section 4.5.

[16] When $d\sigma^{(S)}/d\Omega$ is independent of $S$, it is easy to show that Eq. (94$b$) reduces to Eq. (94$a$).

where

$$K_1 = -\frac{1}{2M_1}\nabla_1^{\,2},$$

$$K_2 = -\frac{1}{2M_2}\nabla_2^{\,2}, \tag{95b}$$

and the respective masses are $M_1$ and $M_2$, as before. Hence $\nabla_1$ and $\nabla_2$ are the gradient operators acting on the respective particle coordinates $\mathbf{r}_1$ and $\mathbf{r}_2$. In addition to the relative coordinate $\mathbf{r} = \mathbf{r}_1 - \mathbf{r}_2$ used in Section 6.2, we may now profitably introduce the center-of-mass coordinate

$$\mathbf{R} = \frac{M_1\mathbf{r}_1 + M_2\mathbf{r}_2}{M_1 + M_2}. \tag{96}$$

In terms of the gradients $\nabla_r$ and $\nabla_R$ acting on $\mathbf{r}$ and $\mathbf{R}$ respectively, the kinetic energy operator is

$$K = -\frac{1}{2M_r}\nabla_r^{\,2} - \frac{1}{2(M_1 + M_2)}\nabla_R^{\,2}, \tag{97}$$

where

$$M_r = \frac{M_1M_2}{M_1 + M_2} \tag{98}$$

is the reduced mass.

For spin-independent forces the interaction operator will be assumed to be a function of $r$ only. That is,

$$V = V(r). \tag{99a}$$

For central forces we have more generally

$$V = \sum_S \Lambda^{(S)}V^{(S)}(r), \tag{99b}$$

in the notation of Eq. (36). Here $\Lambda^{(S)}$ is a projection operator onto states of total spin $S$ of the two particles and $V^{(S)}(r)$ depends on only $S$ and $r$. We may assume without loss of generality that $V(r) \to 0$ as $r \to \infty$.[17]

The Schrödinger equation for the continuum wave functions $\psi_a^{\,\pm}$ is therefore

$$(K + V)\psi_a^{\,\pm} = E_a\psi_a^{\,\pm}, \tag{100}$$

using our conventional notation. For spin-independent forces, we may evidently take [see Eq. (76*a*)]

$$\psi_a^{\,\pm} = \frac{e^{i\mathbf{P}_a\cdot\mathbf{R}}}{(2\pi)^{3/2}}\,\psi^{\pm}_{\boldsymbol{\varkappa},\nu 1,\nu 2} = \frac{e^{i\mathbf{P}_a\cdot\mathbf{R}}}{(2\pi)^{3/2}}\,\psi_{\boldsymbol{\varkappa}}^{\,\pm}(\mathbf{r})u_{\nu 1}(s_1)u_{\nu 2}(s_2), \tag{101a}$$

[17] We shall not concern ourselves here with the problem of determining under what conditions a *local* potential, such as that of Eq. (99*a*), is adequate to describe the interaction between two particles.

whereas for general central potentials we may write [see Eqs. (38) and (76*b*)]

$$\psi_a^{\pm} = \frac{e^{i\mathbf{P}_a\cdot\mathbf{R}}}{(2\pi)^{3/2}}\,\psi^{\pm}_{\varkappa,S,\nu_2} = \frac{e^{i\mathbf{P}_a\cdot\mathbf{R}}}{(2\pi)^{3/2}}\quad \psi^{\pm}_{\varkappa\, S}\,(\mathbf{r})u_{S,\nu}(s_1,s_2). \qquad (101b)$$

If we set

$$E_a = \epsilon(\kappa) + \frac{\mathbf{P}_a^{\,2}}{2(M_1 + M_2)},$$
$$\epsilon(\kappa) = \frac{\kappa^2}{2M_r}, \qquad (102)$$

and substitute Eqs. (101*a*) and (101*b*) into Eq. (100), we obtain, respectively,

$$\left[-\frac{1}{2M_r}\nabla_r^{\,2} + V(r)\right]\psi_\varkappa^{\pm} = \epsilon(\kappa)\psi_\varkappa^{\pm}, \qquad (103a)$$

and

$$\left[-\frac{1}{2M_r}\nabla_r^{\,2} + V^{(S)}(r)\right]\psi^{\pm}_{\varkappa,S} = \epsilon(\kappa)\psi^{\pm}_{\varkappa,S}. \qquad (103b)$$

Since these equations are independent of the total momentum $\mathbf{P}_a$, and therefore valid when $\mathbf{P}_a = 0$, we see that the $\psi_\kappa^{\pm}$, or $\psi^{\pm}_{\varkappa,S}$, are just the barycentric wave functions discussed in detail in Section 6.2.

It is convenient to drop the label $S$ again in Eq. (103*b*), with the understanding that it is implied when the potential is spin dependent. It is also convenient to introduce the "reduced potential"

$$v(r) = 2M_rV(r). \qquad (104)$$

Then both Eqs. (103*a*) and (103*b*) may be written as

$$[\nabla_r^{\,2} + \kappa^2 - v(r)]\psi_\varkappa^{\pm} = 0. \qquad (105)$$

If we introduce spherical coordinates $(r, \theta, \phi)$, Eq. (105) becomes

$$\left[\frac{1}{r^2}\frac{\partial}{\partial r}\left(r^2\frac{\partial}{\partial r}\right) - \frac{L^2}{r^2} + \kappa^2 - v(r)\right]\psi_\kappa^{\pm} = 0, \qquad (106)$$

where $\mathbf{L}$ is the orbital angular momentum operator (33) and

$$-L^2 = \frac{1}{\sin\theta}\frac{\partial}{\partial\theta}\left(\sin\theta\frac{\partial}{\partial\theta}\right) + \frac{1}{\sin^2\theta}\frac{\partial^2}{\partial\phi^2}. \qquad (107)$$

The spherical harmonic expansions (59) and (67) into partial waves for $\psi_\kappa^{\pm}$ may be substituted into Eq. (106) to give

$$[\nabla_r^{\,2} + \kappa^2 - v(r)]\psi_\varkappa^{\pm} = \sum_{l=0}^{\infty}\sum_{m=-l}^{+l} Y_l^{m^*}(\hat{\boldsymbol{\varkappa}})Y_l^m(\hat{\mathbf{r}})$$
$$\times\left[\frac{1}{r^2}\frac{\partial}{\partial r}\left(r^2\frac{\partial}{\partial r}\right) - \frac{l(l+1)}{r^2} + \kappa^2 - v(r)\right]\psi^{\pm}_{\kappa,l}(r) = 0.$$

Since this is an identity in $\hat{\mathbf{r}}$, we obtain

$$\left[\frac{1}{r^2}\frac{\partial}{\partial r}\left(r^2\frac{\partial}{\partial r}\right) - \frac{l(l+1)}{r^2} + \kappa^2 - v(r)\right]\psi^{\pm}_{\kappa,l}(r) = 0 \tag{108}$$

for each $l$. This equation can be further simplified if we set

$$\psi^{+}_{\kappa,l} = i^l\, e^{i\delta_l(\kappa)}\,\frac{w_l(\kappa; r)}{\kappa r}, \tag{109a}$$

where the asymptotic form (66$a$) has suggested the numerical factor. Substitution of Eq. (109$a$) into Eq. (108) leads to the differential equation for $w_l(\kappa; r)$,

$$w''_l - \left[\frac{l(l+1)}{r^2} + v(r) - \kappa^2\right]w_l = 0. \tag{110}$$

We have here written $w' \equiv dw/dr$, and so on. There is clearly no loss of generality in assuming that $w_l$ is real, since the real and imaginary parts of a complex $w_l$ would separately satisfy Eq. (110). The choice of coefficient in Eq. (109$a$) ensures us that $w_l$ is indeed real, if $\psi^{+}_{\kappa,l}$ is to have the asymptotic form (66$a$). Since $w_l$ is real, we obtain from Eq. (66$b$)

$$\psi^{-}_{\kappa,l}(r) = i^l\, e^{-i\delta_l}\,\frac{w_l(\kappa; r)}{\kappa r}. \tag{109b}$$

There is therefore only a single radial wave function $w_l(\kappa; r)$ to be found for each $l$.

We have already observed that $v(r)$ must vanish for large $r$. If

$$v(r) > 0 \quad \text{for} \quad \text{all } r,$$

the potential is said to be repulsive.[18] On the other hand, if

$$v(r) < 0 \quad \text{for} \quad \text{all } r,$$

the potential is said to be attractive. Let us consider a domain $0 < r < d$ such that

$$v(r) > 0, \qquad r \text{ in domain},$$

or

$$v(r) < 0, \qquad r \text{ in domain}.$$

In the first case $v$ is "repulsive at the origin"; in the second $v$ is "attractive at the origin." When $v(r) < 0$ for any interval $r_1 < r < r_2$, $v$ is said to be

[18] Conventional terminology in quantum mechanics is somewhat unfortunate here. In classical mechanics we would say that the force ($= -\nabla V$) is repulsive. Since "forces" are rarely used directly in quantum mechanics, it has become customary to refer to potentials as "repulsive" or "attractive."

"attractive" in that interval. If $v$ is attractive in any interval, there may be a certain number of solutions of Eq. (110) corresponding to bound states of the two particles. These may be designated as $w_{n,l}(r)$, where $n$ ($n = 1, 2, \ldots, n_b$) labels the bound states corresponding to an orbital angular momentum $l$. For each $n$ there is a value of $\kappa^2$ in Eq. (110) such that

$$\kappa^2 \equiv -\kappa_n^{\ 2} = -2M_r\epsilon_n, \tag{111a}$$

where $\epsilon_n(\epsilon_n > 0)$ is the binding energy. As $r$ becomes large, we verify from Eq. (110) that

$$w_{n,l} \approx \beta_n\, e^{-|\kappa_n|r}, \tag{111b}$$

where $\beta_n$ is a constant.

Physical considerations imply that $v(r)$ will have a singularity, if anywhere, only at $r = 0$.[19] When $v(r)$ is repulsive at the origin no general restriction need be placed on this singularity at $r = 0$. Indeed, an essential singularity is often assumed for atomic and nuclear scattering. The nature of the singularity is limited, however, if $v(r)$ is attractive at the origin. In this case let

$$v(r) \approx r^q b_0, \tag{112}$$

where $b_0 < 0$, in a small neighborhood of $r = 0$. If $q < -2$, there is no lower bound on the energy of the system. Since this is unphysical we must exclude the case $q < -2$. If $q = -2$, there is a restricted range of values for $b_0$ for which Eq. (110) has physically acceptable solutions.[20] Since this case does not seem to have applications, we shall assume that $q > -2$.

**Boundary Conditions**

To solve Eq. (110) for the wave functions $w_l$ it is necessary to specify the appropriate boundary conditions on $w_l$. The boundary condition at $r = \infty$ is obtained from Eq. (66$a$). This is

$$\lim_{r\to\infty} w_l(\kappa; r) = \sqrt{2/\pi}\,\sin\left(\kappa r - \frac{\pi l}{2} + \delta_l\right). \tag{113}$$

This condition restricts the form of $v(r)$ for large $r$, as will be seen later. The boundary condition at $r = 0$ is determined from the general requirement that the possible physical states are described by a complete orthonormal set of wave functions.[21] According to Eqs. (63$a$) and (109$a$), the $w_l$ must satisfy the orthogonality condition

$$\int_0^\infty w_l(\kappa_1; r) w_l(\kappa_2; r)\, dr = \delta(\kappa_1 - \kappa_2) \tag{114}$$

[19] Any assumed form for $V(r)$ near $r = 0$ represents an idealization of the actual interaction between two particles, of course.

[20] These occur for $b_0 > -\frac{1}{4}$. See Mott and Massey [5, Section 3.3].

[21] This statement seems to be due to von Neumann; see K. M. Case [6].

for two solutions $w_l(\kappa_1; r)$ and $w_l(\kappa_2; r)$ of Eq. (110). When there are bound states, the completeness relation (63*b*) becomes

$$\int_0^\infty w_l(\kappa; r')w_l(\kappa; r)\,d\kappa + \sum_{n=1}^{n_b} w_{n,l}(r')w_{n,l}(r) = \delta(r' - r). \tag{115}$$

To discuss the condition (114), we consider the integral [and use Eq. (110)]

$$\begin{aligned} I_X(\kappa_1, \kappa_2) - I_0(\kappa_1, \kappa_2) &\equiv (\kappa_1^2 - \kappa_2^2)\int_0^X w_l(\kappa_1; r)w_l(\kappa_2; r)\,dr \\ &= \int_0^X \left\{\left[\kappa_1^2 - v - \frac{l(l+1)}{r^2}\right]w_l(\kappa_1; r)w_l(\kappa_2; r)\right. \\ &\qquad \left. - \left[\kappa_2^2 - v - \frac{l(l+1)}{r^2}\right]w_l(\kappa_2; r)w_l(\kappa_1; r)\right\} dr \\ &= \int_0^X [w_l(\kappa_1; r)w''_l(\kappa_2; r) - w''_l(\kappa_1; r)w_l(\kappa_2; r)]\,dr \\ &= \int_0^X dr\,\frac{d}{dr}[w_l(\kappa_1; r)w'_l(\kappa_2; r) - w'_l(\kappa_1; r)w_l(\kappa_2; r)] \\ &= [w_l(\kappa_1; r)w'_l(\kappa_2; r) - w'_l(\kappa_1; r)w_l(\kappa_2; r)]\Big|_0^X. \end{aligned} \tag{116}$$

[If one of the $w$ describes a bound state, this will vanish at the upper limit when $X \to \infty$, because of Eq. (111*b*).]

From Eq. (116) we obtain

$$\int_0^\infty w_l(\kappa_1; r)w_l(\kappa_2; r)\,dr = \lim_{X\to\infty} \frac{1}{\kappa_1^2 - \kappa_2^2}[I_X(\kappa_1, \kappa_2) - I_0(\kappa_1, \kappa_2)]. \tag{117}$$

Now $I_X(\kappa_1, \kappa_2)$ oscillates as $\sin[(\kappa_1 \pm \kappa_2)X]$ and will effectively vanish under an integral as $X \to \infty$, except near $\kappa_1 = \kappa_2$. For large $X$ we may use the asymptotic form (113) to evaluate $I_X(\kappa_1, \kappa_2)$. When $\kappa_1 \approx \kappa_2$ we obtain

$$\begin{aligned} I_X(\kappa_1, \kappa_2) &\equiv [w_l(\kappa_1; X)w'_l(\kappa_2; X) - w'_l(\kappa_1; X)w_l(\kappa_2; X)] \\ &\approx \frac{1}{\pi}\frac{\sin[(\kappa_1 - \kappa_2)X]}{\kappa_1 - \kappa_2}[\kappa_1^2 - \kappa_2^2] \end{aligned}$$

Therefore[22]

$$\lim_{X\to\infty} \frac{I_X(\kappa_1, \kappa_2)}{\kappa_1^2 - \kappa_2^2} = \delta(\kappa_1 - \kappa_2). \tag{118}$$

[22] We recall the representation

$$\delta(y) = \lim_{X\to\infty} \frac{1}{\pi}\frac{\sin Xy}{y}$$

for the delta-function.

This permits us to write Eq. (117) as

$$\int_0^\infty w_l(\kappa_1; r) w_l(\kappa_2; r)\, dr = \delta(\kappa_1 - \kappa_2) - \frac{1}{\kappa_1^{\,2} - \kappa_2^{\,2}} I_0(\kappa_1, \kappa_2). \quad (119)$$

On comparing this with Eq. (114), we see that the orthogonality condition requires that

$$I_0(\kappa_1, \kappa_2) = 0,$$

or that

$$w_l(\kappa_1; 0) w'_l(\kappa_2; 0) - w'_l(\kappa_1; 0) w_l(\kappa_2; 0) = 0. \quad (120)$$

Equation (120) imposes the boundary condition on the functions $w_l$ at $r = 0$.

Before discussing further the asymptotic boundary condition (113), let us review the solutions of Eq. (110) for $v(r) = 0$. In this case we have the equation

$$y''_l + \left[\kappa^2 - \frac{l(l+1)}{r^2}\right] y_l = 0. \quad (121)$$

This equation is satisfied by

$$y_l^{\pm} = \sqrt{\kappa r}\, J_{\pm(l+1/2)}(\kappa r), \quad (122)$$

where the $J_{\pm(l+1/2)}(x)$ are Bessel functions of order $\pm(l + \frac{1}{2})$.[23] Several auxiliary functions constructed in terms of these quantities are useful in scattering theory. First, if $J_\nu(x)$ is the Bessel function of order $\nu$, we may define the Neumann function[24]

$$N_\nu(x) = \frac{J_\nu(x) \cos \nu\pi - J_{-\nu}(x)}{\sin \nu\pi}. \quad (123)$$

In addition, two complex quantities, the Hankel functions, are defined by

$$\begin{aligned} H_\nu^{(1)}(x) &= J_\nu(x) + i N_\nu(x), \\ H_\nu^{(2)}(x) &= J_\nu(x) - i N_\nu(x). \end{aligned} \quad (124)$$

The wave functions $\psi_{\kappa,l}$ are proportional to linear combinations of $(1/\sqrt{r}) J_{\pm(l+1/2)}$ in regions of $r$ for which $v(r)$ is negligibly small. It is therefore convenient to replace the Bessel functions by the so-called spherical Bessel functions, defined as

$$\begin{aligned} j_l(x) &= \sqrt{\pi/2x}\, J_{(l+1/2)}(x), \\ n_l(x) &= \sqrt{\pi/2x}\, N_{(l+1/2)}(x), \\ h_l^{\pm}(x) &= j_l(x) \pm i n_l(x). \end{aligned} \quad (125)$$

[23] See, for example, Whittaker and Watson [7, Chapter 17], or Morse and Feshbach [8, p. 619].

[24] This quantity is denoted by $Y_\nu(x)$ in Ref. [7].

(The $h_l^{\pm}$ are, of course, proportional to the Hankel functions.) The functions $j_l(x)$ have already been used in Chapter 1 and in Section 6.2.

For small $x$ we have the approximate forms[25] [valid when $x \ll 2\sqrt{l+\frac{3}{2}}$; see Eq. (1.74)]

$$
\begin{aligned}
i_l(x) &\approx \frac{x^l}{(2l+1)!!} = \frac{\sqrt{\pi}}{\Gamma(l+\frac{3}{2})}\frac{x^l}{2^{l+1}}, \\
n_l(x) &\approx (-1)^{l+1}\frac{2^l}{x^{l+1}}\frac{\Gamma(l+\frac{1}{2})}{\sqrt{\pi}}.
\end{aligned} \tag{126}
$$

On the other hand, as $x \to \infty$ we have [to $\mathcal{O}((l/x)^2)$]

$$
\begin{aligned}
j_l(x) &= \frac{\sin(x - \pi l/2)}{x}, \\
n_l(x) &= -\frac{\cos(x - \pi l/2)}{x}, \\
h_l^{\pm}(x) &= \mp i\frac{e^{\pm i(x-\pi l/2)}}{x}.
\end{aligned} \tag{127}
$$

We note that $\psi_{\kappa,l}^{\pm} = j_l(\kappa r)$ times a constant when $\delta_l = 0$ [Eqs. (66*a*, *b*)].

As we have remarked before, it is expected that $w_l/\kappa r$ will asymptotically approach a linear combination of the functions (125) as $r \to \infty$. To investigate the condition under which this will occur, we write

$$
w_l = y_l(r)F_l(r), \tag{128}
$$

where either

$$
y_l(r) = rh_l^{+}(\kappa r),
$$

or

$$
y_l(r) = rh_l^{-}(\kappa r).
$$

On substituting the expression (128) into Eq. (110), and using Eq. (121), we obtain

$$
\frac{F''_l}{F_l} + \frac{2y'_l}{y_l}\frac{F'_l}{F_l} = v(r). \tag{129}
$$

For large $r$, we may use the third of Eqs. (127) to write

$$
\frac{2y'_l}{y_l} \approx \pm 2\kappa i.
$$

We may also neglect the $F''_l$ term in Eq. (129), since $F''_l = \mathcal{O}(F'_l/r)$. Thus

$$
\frac{F'_l}{F_l} = \pm\frac{v(r)}{2\kappa i},
$$

[25] See, for example, Refs. [7] or [8]. $\Gamma(x)$ is the gamma-function; Whittaker and Watson [7, Chapter 12].

or

$$\ln F_l = \pm \frac{1}{2\kappa i} \int^r v(r')\, dr', \tag{130}$$

valid in the limit of large $r$. Now if

$$\lim_{r\to\infty} |v(r)| < \frac{M}{r^{1+\epsilon}} \tag{131}$$

for some $M$ and $\epsilon$ greater than zero, it follows from Eq. (130) that

$$\lim_{r\to\infty} F_l(r) = \text{constant}.$$

Therefore when the condition (131) is satisfied $w_l$ approaches a linear combination of the $y_l$ for large $r$, and

$$\psi_{\kappa,l}^{+} = \alpha_+ h_l^+ + \alpha_- h_l^-,$$

where $\alpha_+$ and $\alpha_-$ are constants. Reference to Eq. (66*a*) shows that

$$\begin{aligned}\alpha_+ &= \frac{i^l}{2}\sqrt{2/\pi}\, S_l(\kappa),\\ \alpha_- &= \frac{i^l}{2}\sqrt{2/\pi}.\end{aligned} \tag{132}$$

The asymptotic form for $\psi_{\kappa,l}^{+}$ when $r$ is sufficiently large that $v(r)$ is negligible, is then

$$\psi_{\kappa,l}^{+} = \frac{i^l}{\sqrt{2\pi}} [h_l^-(\kappa r) + S_l(\kappa) h_l^+(\kappa r)]. \tag{133a}$$

Alternatively, we may use Eqs. (57) and (125) to reexpress this as

$$\psi_{\kappa,l}^{+} = \frac{i^l}{\sqrt{2\pi}} [1 + S_l(\kappa)][j_l(\kappa r) - \tan \delta_l n_l(\kappa r)]. \tag{133b}$$

Equations (133*a*) and (133*b*) of course approach the general asymptotic form (66*a*) as $r \to \infty$. When $v(r)$ falls off faster than $r^{-2}$ when $r$ is large, Eqs. (133*a*) and (133*b*) will be valid for a greater range of values of $r$ than Eq. (66*a*).

Let us now return to the boundary condition at $r = 0$, as given by Eq. (120). First let it be supposed that in a neighborhood of $r = 0$ we may write

$$v(r) = r^q(b_0 + b_1 r + \cdots), \tag{134}$$

where $q$ is an integer and $q \geq -1$ if $b_0 < 0$. Then $w_l$ may be expanded in the form

$$w_l = r^s(a_0 + a_1 r + \cdots). \tag{135}$$

Substitution of Eqs. (134) and (135) into Eq. (110) determines the coefficients $a_0, a_1, \ldots$. The indicial equation[26] gives $s = l + 1$ or $s = -l$. The choice $s = l + 1$ corresponds to the *regular*, and physically acceptable, solutions. The alternate class of solutions are the *irregular* solutions. These fail to satisfy the boundary condition (120) and are therefore to be discarded as unacceptable.

An application of Eq. (120) to the study of Eqs. (110) and (134) for $q < -2$ and $b_0 < 0$ has been made by Case [6].

The discussion of Eq. (110) when the exponent $q$ in Eq. (134) is not an integer is more difficult.[27] For the special case that $q$ is rational, however, a simple transformation of the independent variable reduces the equation to a form that can be solved by conventional methods. Let $q \equiv -(m/n)$, where $m$ and $n$ are integers, and suppose that $n > 0$, $m < 2n$. Then we may define a new variable $y$ as $r = y^n$ and reduce Eq. (110) to the form

$$\frac{d^2w_l}{dy^2} - \frac{n-1}{y}\frac{dw_l}{dy} - n^2y^{2n-2-m}(b_0 + b_1y^n + \cdots)w_l + n^2\left[\kappa^2y^{2n-2} - \frac{l(l+1)}{y^2}\right]w_l = 0. \quad (136)$$

## 6.4 EXAMPLES

In general we must resort to numerical integration of Eq. (110). This is quite straightforward, involving no difficulties either in principle or in practice. On the other hand, it is evidently advantageous to pursue analytical methods for the discussion of this equation when possible.

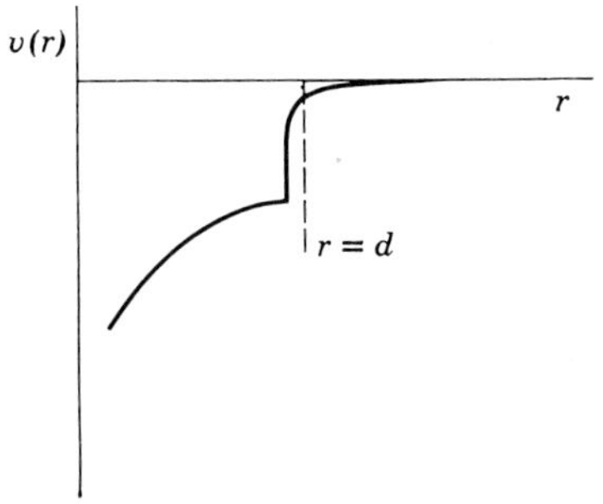


Fig. 6.3 Illustration of a potential having a sharp boundary for which the jump conditions (138) may be used.

Analytical approaches fall into several categories. First, we may discuss general properties of the solutions of Eq. (110). This will be done in Section 6.5. Second, we may develop approximation methods for the integration of this equation. Such methods, some of which will be reviewed in Section 6.7, have been of great practical usefulness. Finally, for a very restricted class of potentials, Eq. (110) may be integrated in terms of known functions. Some solutions of this type will be given here.

[26] See, for example, Whittaker and Watson [7, Chapter 10, and especially Section 10.32].

[27] See, for example, E. L. Ince [9].

**Integration across a Discontinuity in the Potential**

Let us first, however, consider the integration of Eq. (110) across a discontinuity of the potential $V(r)$. We do not anticipate that such a discontinuity can occur (we now exclude the point $r = 0$) for physical potentials. On the other hand, it is often convenient to suppose that $v(r)$ varies between two values over a very small interval of distance. For example, let us suppose that $v(r)$ varies rapidly, but remains finite, over the interval $d - \epsilon < r < d + \epsilon$. For $r$ outside this interval $v(r)$ is assumed to vary slowly. We shall set $v(d - \epsilon) = v_1$ and $v(d + \epsilon) = v_2$ and suppose that $\epsilon$ is *small* in a sense to be defined later. Such a potential is illustrated in Fig. 6.3.

Integration of Eq. (110) across this interval yields

$$\int_{d-\epsilon}^{d+\epsilon} \left\{ w''_l - \left[\frac{l(l+1)}{r^2} + v(r) - \kappa^2\right] w_l \right\} dr = w'_l \Big|_{d-\epsilon}^{d+\epsilon} - \mathcal{O}(\epsilon) = 0,$$

where

$$\mathcal{O}(\epsilon) \equiv \int_{d-\epsilon}^{d+\epsilon} \left[\frac{l(l+1)}{r^2} + v(r) - \kappa^2\right] w_l \, dr. \tag{137a}$$

Now if $\epsilon$ is sufficiently small that

$$|w'_l(d)| \gg \mathcal{O}(\epsilon), \tag{137b}$$

we obtain

$$w'_l(d + \epsilon) = w'_l(d - \epsilon). \tag{138a}$$

This is generally expressed as a "jump condition" by going formally to the limit $\epsilon = 0$.

$$\lim_{\epsilon \to 0} [w'_l(d + \epsilon) - w'_l(d - \epsilon)] = 0. \tag{138b}$$

In other words, $w'_l(r)$ is continuous near $r = d$. This, of course, implies that $w_l(r)$ is also continuous, since

$$\int_{d-\epsilon}^{d+\epsilon} w'_l(r)\, dr = w_l(d + \epsilon) - w_l(d - \epsilon) = \mathcal{O}\, |\epsilon w'_l(d)| \underset{\substack{\lim \\ \epsilon \to 0}}{=} 0.$$

Thus the second "jump condition" is

$$\lim_{\epsilon \to 0} [w_l(d + \epsilon) - w_l(d - \epsilon)] = 0. \tag{138c}$$

Because Eq. (110) is of the second order, the two boundary conditions (138*b*) and (138*c*) permit unique integration across the discontinuity at $r = d$.

Since $r^{-1}$ is continuous at $r = d$, we may rewrite Eqs. (138) in terms of the wave functions $\psi^{\pm}_{\kappa,l}$ of Eqs. (109):

$$\lim_{\epsilon\to 0} \psi^{\pm}_{\kappa,l}(r)\Big|_{r=d-\epsilon}^{r=d+\epsilon} = \lim_{\epsilon\to 0} \frac{d}{dr}\psi^{\pm}_{\kappa,l}(r)\Big|_{r=d-\epsilon}^{r=d+\epsilon} = 0. \tag{139}$$

**The Square-Well Potential**

As an application of the connection formulas just obtained we consider the square-well potential. This is defined by the equations

$$\begin{aligned} v(r) &= v_0, \text{ a constant,} \quad \text{for} \quad r < d; \\ v(r) &= 0 \quad \text{for} \quad r > d. \end{aligned} \tag{140}$$

Let us first suppose that $v_0 < \kappa^2$ and define

$$\kappa_0^{\,2} = \kappa^2 - v_0 > 0. \tag{141}$$

Then in the interval $0 > r > d$, Eq. (110) becomes

$$w''_l - \left[\frac{l(l+1)}{r^2} - \kappa_0^{\,2}\right] w_l = 0. \tag{142}$$

This has the form of Eq. (121), so

$$\frac{w_l(r)}{(\kappa_0 r)} = Cj_l(\kappa_0 r), \qquad 0 < r < d, \tag{143}$$

where $C$ is a constant. We have kept only the *regular* solution here, as described in Section 6.3.

For $r > d$ we see from Eq. (133*b*) that

$$\frac{w_l}{(\kappa r)} = \sqrt{2/\pi}\cos\delta_l[j_l(\kappa r) - \tan\delta_l n_l(\kappa r)]. \tag{144}$$

The quantities $\delta_l$ and $C$ may be determined from the jump conditions (139). We find

$$\begin{aligned} Cj_l(\kappa_0 d) &= \sqrt{2/\pi}\cos\delta_l[j_l(\kappa d) - \tan\delta_l n_l(\kappa d)], \\ \kappa_0 Cj'_l(\kappa_0 d) &= \kappa\sqrt{2/\pi}\cos\delta_l[j'_l(\kappa d) - \tan\delta_l n'_l(\kappa d)], \end{aligned} \tag{145}$$

where $j'_l(x) = (d/dx)j_l(x)$, and so on. Eliminating $C$ from these equations and solving for $\tan\delta_l$ gives

$$\tan\delta_l = \frac{j'_l(\kappa d) - Dj_l(\kappa d)}{n_l'(\kappa d) - Dn_l(\kappa d)}, \tag{146a}$$

where

$$D = \frac{\kappa_0}{\kappa}\frac{j'_l(\kappa_0 d)}{j_l(\kappa_0 d)}. \tag{146b}$$

The scattering amplitude and cross section may then be obtained from Eqs. (70*a*) and (73*a*) if the two interacting particles are not identical. When the two particles are identical, we must use Eqs. (70*a*), (70*c*), and (92*b*).

At sufficiently low energies, we need consider scattering only in the $l = 0$ or $S$-state.[28] On setting

$$j_0(x) = \frac{\sin x}{x},$$

$$n_0(x) = -\frac{\cos x}{x},$$

we obtain from Eqs. (145)

$$\kappa \cot(\kappa d + \delta_0) = \kappa_0 \cot \kappa_0 d. \tag{147}$$

At such energies that $\kappa^2 \ll |v_0|$ and $\kappa d \ll |\delta_0|$, we find

$$\kappa \cot \delta_0 \approx \frac{1}{a_0}, \tag{148a}$$

where

$$\frac{1}{a_0} \equiv \sqrt{-v_0} \cot \sqrt{-v_0}\, d^2. \tag{148b}$$

The scattering amplitude $f(\hat{\boldsymbol{\kappa}}_f \cdot \hat{\boldsymbol{\kappa}})$ is now [Eq. (70*a*)]

$$f(\hat{\boldsymbol{\kappa}}_f \cdot \hat{\boldsymbol{\kappa}}) = \frac{1}{\kappa} e^{i\delta_0} \sin \delta_0. \tag{149}$$

For particles that are not identical, the barycentric differential cross section is

$$\frac{d\sigma}{d\Omega} = \frac{1}{\kappa^2} \sin^2 \delta_0, \tag{150a}$$

while the total cross section (in any coordinate system) is

$$\sigma = \frac{4\pi}{\kappa^2} \sin^2 \delta_0. \tag{150b}$$

Now according to Eq. (148*a*)

$$\sin^2 \delta_0 = \frac{(\kappa a_0)^2}{1 + (\kappa a_0)^2},$$

so

$$\sigma = \frac{4\pi a_0^{\,2}}{1 + (\kappa a_0)^2}. \tag{150c}$$

For identical particles of spin $S_1$ and spin-independent forces, Eq. (94*a*) gives now

$$\frac{d\bar{\sigma}}{d\Omega} = 2\left(1 \pm \frac{1}{2S_1 + 1}\right)\frac{1}{\kappa^2} \sin^2 \delta_0,$$

[28] This will be shown in detail on p. 259.

and[29]

$$\bar{\sigma} = \left(1 \pm \frac{1}{2S_1 + 1}\right) \frac{4\pi a_0^2}{1 + (\kappa a_0)^2}. \tag{150d}$$

Equations (146) are valid for repulsive potentials as long as the condition (141) is satisfied. When $v_0 > \kappa^2$, we may define

$$\kappa_1^2 = v_0 - \kappa^2 > 0. \tag{151}$$

For $S$-wave scattering, and in the interval $0 < r < d$, Eq. (110) is now

$$w''_0 - \kappa_1^2 w_0 = 0,$$

so

$$w_0 = C(e^{\kappa_1 r} - e^{-\kappa_1 r}), \tag{152}$$

where $C$ is a constant. For $r > d$ we have [see Eq. (113)]

$$w_0 = \sqrt{2/\pi}\,\sin(\kappa r + \delta_0).$$

The connection formulas (138*b*) and (138*c*) therefore lead to the equation

$$\kappa \cot(\kappa d + \delta_0) = \kappa_1 \coth \kappa_1 d. \tag{153}$$

In the limit when $v_0 \to \infty$, the case of hard-sphere scattering, we obtain

$$\delta_0 = -\kappa d. \tag{154}$$

For energies sufficiently low that $\kappa d \ll 1$, the scattering cross section is

$$\sigma = 4\pi d^2 \tag{155a}$$

if the particles are not identical. For spinless Bose-Einstein particles (He atoms, for example), we obtain from Eq. (94*a*)

$$\bar{\sigma} = 8\pi d^2. \tag{155b}$$

On the other hand, for Fermi-Dirac particles having spin one-half, Eq. (94*a*) gives[30]

$$\bar{\sigma} = 2\pi d^2. \tag{155c}$$

**The Exponential Well**

Let us consider only $S$-wave scattering for an attractive potential of the form

$$v(r) = -v_0\, e^{-r/d}. \tag{156}$$

[29] Note that Eq. (4.122) implies that we must multiply the above expression by one-half when evaluating the total cross section.

[30] On integrating Eq. (94*a*) over solid angles to obtain the total cross section, we must again divide by $2! = 2$ to take account of the interchange of identical particles [see Eq. (4.122)].

If we make a change of independent variable from $r$ to

$$y = \alpha\, e^{-r/2d}, \tag{157a}$$

where

$$\alpha = 2dv_0^{1/2}, \tag{157b}$$

Eq. (110) simplifies to (for $l = 0$)

$$\frac{d^2w_0}{dy^2} + \frac{1}{y}\frac{dw_0}{dy} + \left(1 - \frac{\nu^2}{y^2}\right)w_0 = 0, \tag{158}$$

where

$$\nu = i\,2\kappa d. \tag{159}$$

The general solution of Eq. (158) is

$$w_0 = C_1J_\nu(y) + C_2J_{-\nu}(y), \tag{160a}$$

where $C_1$ and $C_2$ are constants and $J_\nu(y)$ is a Bessel function. At the point $r = 0$ we have $y = \alpha$. Therefore the regular solution of Eq. (158) is

$$w_0(\kappa; r) = C[J_{-\nu}(\alpha)J_\nu(y) - J_\nu(\alpha)J_{-\nu}(y)], \tag{160b}$$

which evidently vanishes at $r = 0$.

As $r \to \infty$, $y \to 0$, and

$$J_\nu(y) \approx \frac{y^\nu}{2^\nu\Gamma(\nu + 1)} = \left(\frac{\alpha}{2}\right)^\nu \frac{1}{\Gamma(\nu + 1)}\, e^{-i\kappa r}.$$

Therefore

$$\lim_{r\to\infty} w_0(\kappa; r) = -C\left(\frac{\alpha}{2}\right)^\nu \frac{J_{-\nu}}{\Gamma(\nu + 1)}(\alpha)\left[e^{-i\kappa r} - \left(\frac{\alpha}{2}\right)^{-2\nu}\frac{J_\nu(\alpha)\Gamma(\nu + 1)}{J_{-\nu}(\alpha)\Gamma(-\nu + 1)}\, e^{i\kappa r}\right]. \tag{161}$$

On comparing this with the standard asymptotic form (113) for $w$, we see that

$$C = -\sqrt{2/\pi}\left[\left(\frac{\alpha}{2}\right)^\nu J_{-\nu}(\alpha)\right]^{-1} \frac{e^{-i\delta_0}}{2i}\,\Gamma(\nu + 1), \tag{162a}$$

and

$$S_0(\kappa) = e^{2i\delta_0(\kappa)} = \left(\frac{\alpha}{2}\right)^{-2\nu} \frac{J_\nu(\alpha)\Gamma(\nu + 1)}{J_{-\nu}(\alpha)\Gamma(-\nu + 1)}. \tag{162b}$$

This provides an equation for the $S$-wave phase shift as a function of the energy of the colliding particles.[31]

[31] Equation (162$b$) was obtained by S. T. Ma [10].

There are other potentials for which the wave equation (110) may be integrated for $S$-waves. A review of this subject has been given by Bargmann [11].

**Phase Shifts for Large Angular Momentum**

We next consider a potential which has a "range" $d$ in the sense that $v(r)$ is negligibly small when $r > d$. In classical mechanics we would expect little scattering of the particles when their relative angular momentum $l \gg \kappa d$. The reason for this is that the classical impact parameter is $l/\kappa$, which is larger than the range of the force when $l \gg \kappa d$. A similar conclusion may be obtained from quantum mechanical considerations.

Integration of Eq. (110) in the interval $0 < r < d$ will give a wave function $w_l(\kappa; r)$. For $r > d$, we may use Eq. (133*b*) to write

$$\frac{w_l(\kappa; r)}{\kappa r} = \sqrt{2/\pi} \cos \delta_l [j_l(\kappa r) - \tan \delta_l n_l(\kappa r)]. \tag{163}$$

If we define, using the $w_l$ obtained for the interval $0 < r < d$,

$$\kappa D \equiv \frac{d}{dr}\left(\ln \frac{w_l(\kappa; r)}{\kappa r}\right)_{r=d}, \tag{164}$$

the connection formulas (139) give

$$\tan \delta_l = \frac{j'_l(\kappa d) - D j_l(\kappa d)}{n'_l(\kappa d) - D n_l(\kappa d)}. \tag{165}$$

We may evidently set

$$D \equiv \frac{j'_l(\kappa d)}{j_l(\kappa d)} p, \tag{166}$$

where $p$ is a parameter depending on the strength of the potential. Clearly, if $v(r)$ were negligibly small, we would have $p = 1$.

When $l$ is large and nearly equal to $\kappa d$, $n_l$ and $j_l$ are approximately equal in magnitude. For $l \gg \kappa d$, however,[32]

$$\begin{aligned} j_l(\kappa d) &\approx C_l \left(\frac{\kappa d}{l + \frac{1}{2}}\right)^{l+\frac{1}{2}}, \\ n_l(\kappa d) &\approx -2C_l \left(\frac{\kappa d}{l + \frac{1}{2}}\right)^{-(l+\frac{1}{2})}, \end{aligned} \tag{167}$$

where $C_l$ is independent of $\kappa d$.

If the expressions (166) and (167) are inserted into Eq. (165), we obtain

$$\tan \delta_l \approx \left(\frac{\kappa d}{l + \frac{1}{2}}\right)^{2l+1} \frac{1 - p}{1 + p}. \tag{168}$$

When $\kappa d \ll l$, the phase shift $\delta_l$ will be negligibly small, and thus will not

[32] Morse and Feshbach [8, p. 631, Vol. I].

contribute to the scattering amplitude unless $(1 - p)/(1 + p)$ happens to be very large. This can happen only for those values of $\kappa$ for which $p \approx -1$.

Potentials for which this can occur must be attractive and *very strong*. In this case $\delta_l$ ordinarily passes through $\pi/2$ at a given energy, giving rise to a so-called resonance in the scattering cross section.[33] It should be emphasized that quite strong interactions are needed to produce a resonance when $\kappa d \ll l$, however.

With the possible exception just mentioned, we may conclude from Eq. (168) that the scattering phase shift is negligibly small when $l \gg \kappa d$. This has the important practical consequence that such $l$-values need not be included in the sum (70*a*) for the scattering amplitude $f(\hat{\boldsymbol{\kappa}}_f \cdot \hat{\boldsymbol{\kappa}})$. To illustrate this, we consider again the repulsive square-well for which $v_0 = \infty$. When $\kappa d \ll l$, we easily obtain [using the expansions (126)]

$$\tan \delta_l = (-1)^{l+1}\left(\frac{\kappa d}{2}\right)^{2l+1} \frac{\pi}{(l + \frac{1}{2})[\Gamma(l + \frac{1}{2})]^2}. \tag{169}$$

With the use of Stirling's approximation for the gamma-function, we can put this into a form similar to that of Eq. (168).

**Wave Functions for a Coulomb Field**

The Coulomb potential for two particles having the respective electric charges $e_1$ and $e_2$ is $V(r) = e_1e_2/r$. The quantity $v(r)$ given by Eq. (104) is then

$$v_c(r) = \frac{2M_r e_1 e_2}{r}. \tag{170a}$$

This potential does not satisfy the condition (131). Therefore, the Coulomb wave functions will not have the asymptotic form (113). Our formulation of the scattering problem is consequently not applicable for the interaction (170*a*).

Fortunately this difficulty does not occur in practice. The Coulomb potential describing the interaction of two charged particles is always "shielded" at large distances by the physical environment within which a scattering experiment is conducted.[34] We may therefore replace $v_c(r)$ by a "screened potential"

$$v_c(r)g_c(r), \tag{170b}$$

where $g_c(r)$ corrects Eq. (170*a*) for the shielding of the Coulomb interaction. We shall here assume that

$$\begin{aligned} g_c(r) &= 1 \quad \text{for} \quad r < \mathscr{R}, \\ &= 0 \quad \text{for} \quad r \gg \mathscr{R}, \end{aligned} \tag{170c}$$

[33] Resonance scattering will be discussed in Section 8.6.
[34] See, for example, footnote 29, Chapter 3.

where $\mathscr{R}$ represents the screening radius, or distance at which the two charges are effectively shielded from each other. We shall also assume here that $\mathscr{R}$ is so large that its precise value is irrelevant for the applications to be considered.

For some applications, such as to the scattering of low-energy electrons by atoms or molecules, the function $g_c(r)$ must be determined with care. On the other hand, the simple expression (170$c$) is often adequate. This is the case for most experiments in nuclear and elementary particle physics, for which the Coulomb interaction is incidental. In such situations the relevant collision distances are of the order of $10^{-13}$ cm, while $\mathscr{R} \approx 10^{-8}$ cm, since screening is then usually due to orbital electrons. As we shall see, the actual form of $g_c(r)$ and the actual value of $\mathscr{R}$ are ordinarily irrelevant for these applications.

Let us first, however, discuss the solutions to Eq. (110) for the potential $v_c(r)$, as given by Eq. (170$a$). We must then solve the differential equation

$$w''_l + \left[\kappa^2 - v_c(r) - \frac{l(l+1)}{r^2}\right] w_l = 0. \tag{171}$$

If we define[35]

$$z = 2i\kappa r, \quad \mu = l + \tfrac{1}{2}, \quad \xi = \frac{M_r e_1 e_2}{\kappa}, \tag{172}$$

Eq. (171) becomes

$$\frac{d^2 w_l}{dz^2} + \left(-\frac{1}{4} + \frac{i\xi}{z} + \frac{\frac{1}{4} - \mu^2}{z^2}\right) w_l = 0. \tag{173}$$

This equation is satisfied by either of the two confluent hypergeometric functions[36] $W_{k,\mu}(z)$ or $W_{-k,\mu}(-z)$, where

$$k \equiv i\xi. \tag{174}$$

The general solution of Eq. (173) is therefore

$$w_l = C_+ W_{k,\mu}(z) + C_- W_{-k,\mu}(-z), \tag{175}$$

where $C_+$ and $C_-$ are constants. The functions $W_{k,\mu}(z)$ are singular at

[35] The quantity $\xi$ is dimensionless. In centimeter-gram-second units it has the form $\xi = (M_r/\kappa)(e_1 e_2/\hbar)$.

[36] See, for example, Whittaker and Watson [7, Chapter 16].

$z = 0$. Equation (173) has the regular solution[37,38]

$$M_{k,\mu}(z) = \frac{\Gamma(2\mu + 1)}{\Gamma(\frac{1}{2} + \mu - k)} e^{-k\pi i} W_{-k,\mu}(-z)$$
$$+ \frac{\Gamma(2\mu + 1)}{\Gamma(\frac{1}{2} + \mu + k)} e^{i\pi(\frac{1}{2}+\mu-k)} W_{k,\mu}(z). \qquad (176a)$$

Near $z = 0$ this function has the expansion

$$M_{k,\mu}(z) = z^{\frac{1}{2}+\mu} e^{-z/2}\left[1 + \frac{\frac{1}{2} + \mu - k}{1!\,(2\mu + 1)} z + \cdots\right]. \qquad (176b)$$

The asymptotic forms of the $W$, when $z$ is large, are

$$W_{k,\mu}(z) \approx e^{-z/2} z^k = e^{-i\kappa r} e^{i\xi \ln (2\kappa r)} e^{-\xi(\pi/2)},$$
$$W_{-k,\mu}(-z) \simeq e^{z/2}(-z)^{-k} = e^{i\kappa r} e^{-i\xi \ln (2\kappa r)} e^{-\xi(\pi/2)}. \qquad (177)$$

Following Yost, Wheeler, and Breit [12] we define

$$Y_l \equiv i\, e^{-i\sigma_l} e^{i(\pi/2)l} e^{\xi(\pi/2)} W_{k,\mu}(z),$$
$$Y_l^* \equiv (-i)\, e^{i\sigma_l} e^{-i(\pi/2)l} e^{\xi(\pi/2)} W_{-k,\mu}(-z), \qquad (178)$$

where

$$e^{2i\sigma_l} = \frac{\Gamma(l + 1 + i\xi)}{\Gamma(l + 1 - i\xi)}. \qquad (179)$$

It follows from Eqs. (177) and (178) that the asymptotic expressions for $Y_l$ and $Y_l^*$ are, when $\kappa r \to \infty$,

$$Y_l \approx i\, e^{-i[\kappa r - \xi \ln (2\kappa r) - \pi l/2 + \sigma_l]},$$
$$Y_l^* \approx (-i)\, e^{+i[\kappa r - \xi \ln (2\kappa r) - \pi l/2 + \sigma_l]}. \qquad (180)$$

In terms of the regular solution of Eq. (173), we may define the normalized solution

$$F_l = \frac{1}{2}\sqrt{2/\pi}\, \frac{[\Gamma(\frac{1}{2} + \mu - k)\Gamma(\frac{1}{2} + \mu + k)]^{\frac{1}{2}}}{\Gamma(2\mu + 1)} e^{-i(\pi/2)(\frac{1}{2}+\mu-k)} M_{k,\mu}(z)$$
$$= \frac{1}{2}\sqrt{2/\pi}(Y_l + Y_l^*), \qquad (181)$$

the second form following directly from Eqs. (176*a*) and (178). A second,

[37] The function $M_{k,\mu}$ is also defined by Whittaker and Watson [7, Chapter 16].

[38] We are following here the discussion of Yost, Wheeler, and Breit [12]. Note that the sign of $k$ in the exponentials in Eq. (176*a*), as given by these authors, differs from that given in Whittaker and Watson [7].

irregular solution of Eq. (173) was chosen by Yost, Wheeler, and Breit to be[39]

$$G_l = \sqrt{2/\pi}\left[\frac{1}{2i}(Y_l^* - Y_l)\right]. \tag{182}$$

The asymptotic forms of $F_l$ and $G_l$ are then [compare Eqs. (180)]

$$\begin{aligned}\lim_{\kappa r\to\infty} F_l &= \sqrt{2/\pi}\,\sin\left[\kappa r - \frac{\pi l}{2} - \xi\ln(2\kappa r) + \sigma_l\right],\\ \lim_{\kappa r\to\infty} G_l &= -\sqrt{2/\pi}\,\cos\left[\kappa r - \frac{\pi l}{2} - \xi\ln(2\kappa r) + \sigma_l\right].\end{aligned} \tag{183}$$

Except for the term $\xi \ln(2\kappa r)$, we see that $F_l$ agrees with the asymptotic form (113) for $w_l$, as $\kappa r \to \infty$. The quantity $\sigma_l$ is identified as the Coulomb scattering phase shift. The function $G_l$ is chosen to have a form analogous to that of $n_l(r)$ [Eq. (127)] when $\kappa r$ is large.

Near $\kappa r = 0$, $F_l$ and $G_l$ have the expansions given by Yost, Wheeler, and Breit;

$$\begin{aligned}F_l &= \sqrt{2/\pi}C_l\,e^{-i\kappa r}(\kappa r)^{l+1}\left[1 + \frac{l+1-i\xi}{1!\,(2l+2)}2i\kappa r + \cdots\right],\\ G_l &= -\,\mathrm{Re}\left\{\sqrt{2/\pi}\left[D_l\,e^{-i\kappa r}(\kappa r)^{-l}\left(1 + \frac{l+i\xi}{2l}\frac{2i\kappa r}{1!} + \cdots\right)\right.\right.\\ &\qquad\left.\left. + \sqrt{\pi/2}\,pD_l(\ln(2\kappa r) + \text{constant})\frac{F_l}{C_l} + \mathcal{O}((\kappa r)^{l+1})\right]\right\}.\end{aligned} \tag{184}$$

Here [13]

$$\begin{aligned}C_l^2 &= \left[\frac{2^l}{\Gamma(2l+2)}\right]^2[\Gamma(l+1-i\xi)\Gamma(l+1+i\xi)]\,e^{\pi\xi}\\ &= \left[\frac{2^l}{(2l+1)!}\right]^2(l^2+\xi^2)[(l-1)^2+\xi^2]\cdots(1+\xi^2)\frac{2\pi\xi}{e^{2\pi\xi}-1},\\ D_l &= [(2l+1)C_l]^{-1},\\ p &= \frac{2l+1}{\pi}(e^{2\pi\xi}-1)C_l^2.\end{aligned} \tag{185}$$

For the frequently encountered case that $l = 0$, we write in more detail

$$C_0^2 = \frac{2\pi\xi}{e^{2\pi\xi}-1} \tag{186a}$$

[39] These authors defined $F_l = \frac{1}{2}(Y_l + Y_l^*)$, $G_l = (1/2i)(Y_l - Y_l^*)$. Our definitions are chosen to be consistent with the asymptotic normalization (113).

and

$$F_0 = (\sqrt{2/\pi}C_0)\kappa r(1 + \xi\kappa r - \tfrac{1}{6}\kappa^2r^2 + \tfrac{1}{3}\xi^2\kappa^2r^2 + \cdots),$$

$$G_0 = -\left(\sqrt{2/\pi}\,\frac{1}{C_0}\right)\Big\{1 + 2\kappa r\xi[\ln(2\kappa r) + 2\gamma - 1 + h(\xi) + \ln\xi] - \frac{\kappa^2r^2}{2} + 3\xi^2\kappa^2r^2 + \cdots\Big\}. \quad (186b)$$

Here[40]

$$h(\xi) + \ln\xi = \mathrm{Re}\left[\frac{\Gamma'(-i\xi)}{\Gamma(-i\xi)}\right] = \xi^2\sum_{j=1}^{\infty}\frac{1}{j(j^2+\xi^2)} - \gamma, \quad (186c)$$

and $\gamma = 0.5772$ is Euler's constant.

**Scattering by a Coulomb plus Short-Range Force**

Coulomb scattering is most frequently encountered as being incidental (and ordinarily not welcome) to experiments studying other interactions between charged particles. Classic examples include the scattering of protons by protons and by atomic nuclei. Here we are usually interested in the nuclear interaction but must untangle the effects of the long-range Coulomb field. We shall now show how to do this, following the work of Breit and his collaborators [12, 15]. There is no difficulty of principle involved, but numerical applications may be tedious.

Let us write $v_n(r)$ for the potential associated with the nuclear interaction. We shall suppose that $v_n(r)$ has a range $d$ decreasing rapidly to zero[41] when $r > d$. Then Eq. (110) becomes

$$w''_l - \left[v_c(r)g_c(r) + v_n(r) + \frac{l(l+1)}{r^2} - \kappa^2\right]w_l = 0. \quad (187)$$

We have included here the cutoff $g_c(r)$, as defined by Eq. (170$c$).

Three intervals for the range of $r$ are to be distinguished in Eq. (187). For $r < d$, the interaction by both nuclear and Coulomb forces is of importance (although often it is permissible to neglect $v_c$ in this interval, because of the relatively greater strength of the nuclear forces). Now $\mathscr{R}$ has been assumed to be very large, so in the interval $d \ll r < \mathscr{R}$ the nuclear

[40] This notation is used by Jackson and Blatt [14].

[41] For nuclear forces we may assume that $v_n(r)$ decreases exponentially to zero. For our general theoretical considerations, we need only assume that $v_n = \mathcal{O}(1/r^{1+\epsilon})$, where $\epsilon > 0$.

force $v_n$ may be neglected. In this interval $w_l$ will be a linear combination of $F_l$ and $G_l$, as defined by Eqs. (181) and (182), having the form

$$w_l = B_1(F_l - \tan \nu_l G_l), \tag{188}$$

where $B_1$ and $\nu_l$ are constants. According to Eqs. (183), this will have the asymptotic form for large $\kappa r$;

$$w_l \approx B_2 \sin (\phi_l + \nu_l), \tag{189a}$$

where

$$\phi_l \equiv \kappa r - \frac{\pi l}{2} - \xi \ln (2\kappa r) + \sigma_l \tag{189b}$$

and $B_2$ ($B_2 = \sqrt{2/\pi}$) is a second constant. The quantity $\nu_l$ is called the nuclear scattering phase shift. It is, of course, determined by integration of Eq. (187) from the origin out to $r \gg d$.

In the third interval for $r$, for which $r \gg \mathcal{R}$, there is no interaction and $w_l$ has the standard asymptotic form (113).

$$w_l = \sqrt{2/\pi} \sin \left(\kappa r - \frac{\pi l}{2} + \delta_l\right). \tag{190}$$

We must now evaluate the phase shift $\delta_l$. It will prove convenient to write $\delta_l$ in the form

$$\delta_l = \sigma_l + \nu_l + \lambda_l, \tag{191}$$

where $\lambda_l$ is to be determined. From the discussion on pp. 258–259, we can conclude that[42]

$$\nu_l \approx 0 \quad \text{for} \quad l \gg \kappa d \tag{192a}$$

and

$$\delta_l \approx 0 \quad \text{for} \quad l \gg \kappa \mathcal{R}, \tag{192b}$$

so

$$\lambda_l \approx -\sigma_l \quad \text{for} \quad l \gg \kappa \mathcal{R}. \tag{192c}$$

To determine $\lambda_l$ when $l \ll \kappa \mathcal{R}$, we consider two solutions $h_{c,l}^{\pm}$ of the equation

$$h_{c,l}^{\pm\,''} - \left[v_c + v_n + \frac{l(l+1)}{r^2} - \kappa^2\right] h_{c,l}^{\pm} = 0, \tag{193}$$

which have the asymptotic forms [see Eqs. (189)]

$$\lim_{r \to \infty} h_{c,l}^{\pm} = e^{\pm i(\phi_l + \nu_l)}. \tag{194}$$

[42] The condition (192$a$) may be shown using the arguments leading to Eq. (168), but replacing $j_l$, $n_l$ by the Coulomb functions $F_l$, $G_l$. For most applications, the Coulomb functions are well approximated by $j_l$ and $n_l$ near $r = d$, however.

Then a pair of functions $w_l^{\pm}$ that are solutions of Eq. (187) and are equal to $h_{c,l}^{\pm}$ when $r < \mathscr{R}$ may be defined. Following the discussion associated with Eqs. (128) and (129), we set

$$w_l^{\pm} \equiv h_{c,l}^{\pm} F^{\pm} \tag{195}$$

and substitute this into Eq. (187) to obtain an equation for $F^{\pm}$. Equation (130) is now replaced by

$$\ln F^{\pm} = \pm \frac{1}{2i\kappa} \int_{\mathscr{R}}^{r} v_c(r')[g_c(r') - 1]\, dr', \tag{196a}$$

since $F^{\pm} = 1$ for $r < \mathscr{R}$. When $r \gg \mathscr{R}$, this may be integrated to give

$$\ln F^{\pm} = \text{constant} \pm i\xi \ln (2\kappa r). \tag{196b}$$

Since $\ln (2\kappa r)$ varies slowly with $r$, and since $\ln F^{\pm} = 0$ for $r = \mathscr{R}$, the constant may be evaluated to give

$$\ln F^{\pm} = \mp i\xi[\ln (2\kappa\mathscr{R}) - \ln (2\kappa r)].$$

Therefore

$$\lim_{r\to\infty} w_l^{\pm} = e^{\pm i[\kappa r - \pi l/2 + \sigma_l + \nu_l - \xi \ln (2\kappa\mathscr{R})]}. \tag{197}$$

From these we can construct the asymptotic function (190) and obtain $\delta_l$. We see then that [Eq. (191)]

$$\lambda_l = -\xi \ln (2\kappa\mathscr{R}) \equiv \lambda, \tag{198}$$

when $l \ll \kappa\mathscr{R}$.

The scattering amplitude (70*a*) is

$$f(\hat{\boldsymbol{\kappa}}_f \cdot \hat{\boldsymbol{\kappa}}) = \sum_{l=0}^{\infty} \frac{2l+1}{2i\kappa} (e^{2i\delta_l} - 1) P_l (\cos \theta), \tag{199}$$

where $\cos \theta = \hat{\boldsymbol{\kappa}}_f \cdot \hat{\boldsymbol{\kappa}}$. To simplify this, we use Eq. (191) to write

$$e^{2i\delta_l} - 1 = [(e^{2i\lambda} - 1) + e^{2i\sigma_l}(e^{2i\lambda_l} - e^{2i\lambda})] + e^{2i\lambda}(e^{2i\sigma_l} - 1) + e^{2i\sigma_l} e^{2i\lambda_l}(e^{2i\nu_l} - 1). \tag{200}$$

Now because of Eqs. (192*a*) and (198) we may set

$$e^{2i\lambda_l}(e^{2i\nu_l} - 1) = e^{2i\lambda}(e^{2i\nu_l} - 1). \tag{201}$$

Therefore Eq. (199) may be reexpressed as

$$f(\hat{\boldsymbol{\kappa}}_f \cdot \hat{\boldsymbol{\kappa}}) = f_{\text{sp}}(\hat{\boldsymbol{\kappa}}_f \cdot \hat{\boldsymbol{\kappa}}) + e^{2i\lambda}[f_{\text{Coul}}(\hat{\boldsymbol{\kappa}}_f \cdot \hat{\boldsymbol{\kappa}}) + f_{\text{nuc}}(\hat{\boldsymbol{\kappa}}_f \cdot \hat{\boldsymbol{\kappa}})], \tag{202a}$$

where

$$f_{\text{Coul}} = \sum_{l=0}^{\infty} \frac{2l+1}{2i\kappa} (e^{2i\sigma_l} - 1) P_l(\cos \theta), \tag{202b}$$

$$f_{\text{nuc}} = \sum_{l=0}^{\infty} \frac{2l+1}{2i\kappa} e^{2i\sigma_l}(e^{2i\nu_l} - 1) P_l(\cos \theta), \tag{202c}$$

and

$$f_{\rm sp} = \sum_{l=0}^{\infty} \frac{2l+1}{2i\kappa}[(e^{2i\lambda} - 1) + e^{2i\sigma_l}(e^{2i\lambda_l} - e^{2i\lambda})]P_l(\cos\theta). \quad (202d)$$

The quantity $f_{\rm Coul}$ is the usual Coulomb scattering amplitude[43]

$$f_{\rm Coul}(\cos\theta) = -\frac{e_1e_2M_r}{2\kappa^2}\left(\sin\frac{\theta}{2}\right)^{-2} e^{i\{2\sigma_0 - 2\xi\ln[\sin(\theta/2)]\}}. \quad (203)$$

The term $f_{\rm nuc}$ in Eq. (202*a*) is the nuclear scattering amplitude. It describes the scattering by the potential $v_n(r)$ (as modified by the Coulomb force).

For most scattering processes involving nuclear and elementary particle interactions the term $f_{\rm sp}$ may be neglected in Eq. (202*a*). To see this, we approximate $f_{\rm sp}$ as

$$f_{\rm sp} \approx f_< + f_>, \quad (204a)$$

where

$$f_< = \sum_{l=0}^{L} \frac{2l+1}{2i\kappa}(e^{2i\lambda} - 1)P_l(\cos\theta),$$
$$f_> = -\sum_{l=L}^{\infty} \frac{2l+1}{2i\kappa} e^{2i\lambda}(e^{2i\sigma_l} - 1)P_l(\cos\theta), \quad (204b)$$

and $L \approx \kappa\mathscr{R}$. To obtain these equations, we have set $\lambda_l = \lambda$ for $l < L$ and $\lambda_l = -\sigma_l$ for $l > L$ in Eq. (202*d*). The effect of $f_>$ is to provide a cutoff at $l > L$ in Eq. (202*b*), as is required by the condition (192*b*). If evaluated, this cutoff would remove the singularity at $\theta = 0$ in $f_{\rm Coul}$, when Eqs. (202*b*) and (204*a*) are substituted into Eq. (202*a*). The quantity $f_<$ describes diffraction at the boundary of the sphere at which $v_c$ is cutoff.

To evaluate $f_<$, we note that[44]

$$\sum_{l=0}^{L}(2l+1)P_l(\cos\theta) = \frac{L+1}{1-\cos\theta}[P_L(\cos\theta) - P_{L+1}(\cos\theta)].$$

Now for $L \gg 1$ and $L\theta \approx \mathscr{O}(1)$

$$P_L(\cos\theta) \approx J_0(L\theta), \quad (205)$$

so

$$P_L(\cos\theta) - P_{L+1}(\cos\theta) \approx -\frac{dJ_0(L\theta)}{dL} = \theta J_1(L\theta).$$

[43] Mott and Massey [5, Chapter 3]. We are using the notation of Jackson and Blatt [14].

[44] Magnus and Oberhettinger [16, Sections 2 and 5]. This identity is easily verified by multiplying both sides by $1 - \cos\theta$ and using the recursion relation for Legendre polynomials $(l+1)P_{l+1}(x) = (2l+1)xP_l(x) - lP_{l-1}(x)$.

Substitution into the first of Eqs. (204*b*) then gives, since $L \approx \kappa\mathscr{R}$,

$$f_<(\cos\theta) \approx \frac{e^{2i\lambda}-1}{2i\kappa}\,\frac{2\kappa\mathscr{R}}{\theta}\,J_1(\kappa\mathscr{R}\theta), \tag{206}$$

which is valid for $\theta = \mathscr{O}((\kappa\mathscr{R})^{-1})$.

To simplify $f_>$ we use Stirling's approximation[45] for the gamma-function to write Eq. (179) in the form

$$e^{2i\sigma_l} \approx (l+1)^{2i\xi}. \tag{207}$$

Then with the approximation (205) the second of Eqs. (204*b*) becomes

$$f_> \approx -\frac{e^{2i\lambda}}{i\kappa}\int_L^\infty dl\; l[(l+1)^{2i\xi}-1]J_0(l\theta). \tag{208}$$

We see from Eqs. (206) and (208) that $f_{sp}$ $(\cos\theta)$ varies as

$$f_{sp} \approx \sin(\kappa\mathscr{R}\theta + \text{constant}), \tag{209}$$

when $\theta$ is larger than $(\kappa\mathscr{R})^{-1}$. For nearly all nuclear scattering experiments $\kappa\mathscr{R}\theta \gg 1$: We have seen that for most scattering experiments $\mathscr{R} \approx 10^{-8}$ cm, or larger; for nucleons or mesons of energies greater than a few Mev, $\kappa > 10^{13}$ cm$^{-1}$. Therefore we ordinarily anticipate that $\kappa\mathscr{R} > 10^5$ for nuclear or elementary particle interactions.

To see the significance of this, we return to the wave packet description of scattering given in Chapter 3. To obtain the transition probability, or scattering cross section, in Eqs. (3.124) or (3.207) we assumed that the scattering matrix (and thus the scattering amplitude) varied by only a small amount over the spread of the wave packet. This will not in general be true for the amplitude $f_{sp}$ when $\kappa\mathscr{R}\theta \gg 1$. Unless the wave packet spread in momentum space is very small, the expression (209) will oscillate and contribute little to Eqs. (3.122) or (3.206). For example, with a Gaussian wave packet of width $\Delta\kappa$, we find

$$\bar{f}_{sp} \approx e^{-[\theta(\Delta\kappa\mathscr{R})]^2} \tag{210}$$

for the mean value of $f_{sp}$, averaged over the wave packet.

We conclude from this that the contribution of $f_{sp}(\cos\theta)$ to Eq. (202*a*) is effectively negligible if the wave packet has a width $\Delta(\kappa\theta)$, corresponding to the range of $\kappa\theta$ values, such that $\Delta(\kappa\theta) \gg 1/\mathscr{R}$. This condition is usually well satisfied for the energy and angular resolution encountered in nuclear scattering experiments. In such cases it is customary simply to ignore the term $f_{sp}$ in Eq. (202*a*). We must recognize, however, that the effects of $f_{sp}$ can be observed in an experiment for which the energy and

[45] Whittaker and Watson [7, Chapter 12].

angle collimation are very good for both the incident beam and for the counting apparatus.

In accordance with the discussion just given, we shall assume that $f_{sp}$ may be neglected in Eq. (202$a$) and write

$$f(\hat{\boldsymbol{\kappa}}_f \cdot \hat{\boldsymbol{\kappa}}) = f_{\text{Coul}}(\hat{\boldsymbol{\kappa}}_f \cdot \hat{\boldsymbol{\kappa}}) + f_{\text{nuc}}(\hat{\boldsymbol{\kappa}}_f \cdot \hat{\boldsymbol{\kappa}}), \tag{211}$$

dropping the unobservable phase factor $e^{2i\lambda}$. Integration of Eq. (187) to obtain the nuclear phase shift $\nu_l$ and thus the scattering amplitude (211) is straightforward. We may set $g_c(r) = 1$ and integrate the differential equation only from the point $r = 0$ out to some point $r_e$ sufficiently greater than $d$ that $v_n$ is negligibly small. The $w_l$ so obtained may then be matched to the function (188), using the matching conditions (138), to evaluate $\nu_l$. The differential scattering cross section is then

$$\frac{d\sigma}{d\Omega} = |f_{\text{Coul}}(\cos\theta) + f_{\text{nuc}}(\cos\theta)|^2, \tag{212}$$

when $\theta \gg (\kappa\mathscr{R})^{-1}$. Equation (212) is applicable when the two interacting particles are not identical. For identical particles we must use Eqs. (94$a$) or (94$b$).

The application of this analysis to the scattering of protons by protons has been of great importance for the understanding of nuclear forces.[46] Since the proton has a spin $S_1 = \frac{1}{2}$, the total spin for two protons may be $S = 0$ or $S = 1$. It is known that the force between nucleons is not central; however, at low energies (say $\epsilon < 25$ Mev) we may adequately describe the scattering by spin-dependent central forces.[47] Thus a spin index $S$ must be attached to the scattering amplitude $f_{\text{nuc}}$ and to the phase shift $\nu_l$, as in Eq. (70$c$).

For the case that $S = 0$, or for singlet scattering, we have according to Eq. (92$b$)

$$f_{\mathscr{S}}^{(0)}(\hat{\boldsymbol{\kappa}}_f \cdot \hat{\boldsymbol{\kappa}}) = f^{(0)}(\hat{\boldsymbol{\kappa}}_f \cdot \hat{\boldsymbol{\kappa}}) + f^{(0)}(-\hat{\boldsymbol{\kappa}}_f \cdot \hat{\boldsymbol{\kappa}}). \tag{213}$$

For triplet scattering, with $S = 1$, we have

$$f_{\mathscr{S}}^{(1)}(\hat{\boldsymbol{\kappa}}_f \cdot \hat{\boldsymbol{\kappa}}) = f^{(1)}(\hat{\boldsymbol{\kappa}}_f \cdot \hat{\boldsymbol{\kappa}}) - f^{(1)}(-\hat{\boldsymbol{\kappa}}_f \cdot \hat{\boldsymbol{\kappa}}), \tag{214}$$

where $f^{(S)}$ represents the amplitude (211) with an appropriate spin label. Then, according to Eq. (202$c$), the symmetrized nuclear scattering amplitudes are

$$\begin{aligned} f_{\text{nuc}}^{(0)}(\hat{\kappa}_f \cdot \hat{\kappa}) &= 2 \sum_{l\,\text{even}=0}^{\infty} \frac{2l+1}{2i\kappa} e^{2i\sigma_l}(e^{2i\nu_l^{(0)}} - 1)P_l(\cos\theta), \\ f_{\text{nuc}}^{(1)}(\hat{\kappa}_f \cdot \hat{\kappa}) &= 2 \sum_{l\,\text{odd}=0}^{\infty} \frac{2l+1}{2i\kappa} e^{2i\sigma_l}(e^{2i\nu_l^{(1)}} - 1)P_l(\cos\theta). \end{aligned} \tag{215}$$

[46] See Refs. [12], [14], and [15] for example.

[47] See, for example, Ref. [14].

Here only even values of $l$ are kept in the first summation, and only odd values are kept in the second equation. This is, of course, a direct consequence of Eqs. (213) and (214) and the fact that $P_l(-\cos\theta) = (-1)^l P_l(\cos\theta)$. This result may also be seen directly from Eqs. (78*b*) and (89).

The differential scattering cross sections (in the barycentric coordinate system) may now be obtained from Eqs. (92*b*), (211), (213), (214), and (215). These are

$$\begin{aligned}\frac{d\sigma^{(0)}}{d\Omega} &= |f_{\mathscr{S}}^{(0)}|^2 \\ &= |f_{\text{Coul}}(\hat{\boldsymbol{\kappa}}_f\cdot\hat{\boldsymbol{\kappa}}) + f_{\text{Coul}}(-\hat{\boldsymbol{\kappa}}_f\cdot\hat{\boldsymbol{\kappa}}) + f_{\mathscr{S}\,\text{nuc}}^{(0)}(\hat{\boldsymbol{\kappa}}_f\cdot\hat{\boldsymbol{\kappa}})|^2 \\ \frac{d\sigma^{(1)}}{d\Omega} &= |f_{\mathscr{S}}^{(1)}|^2 \\ &= |f_{\text{Coul}}(\hat{\boldsymbol{\kappa}}_f\cdot\hat{\boldsymbol{\kappa}}) - f_{\text{Coul}}(-\hat{\boldsymbol{\kappa}}_f\cdot\hat{\boldsymbol{\kappa}}) + f_{\mathscr{S}\,\text{nuc}}^{(1)}(\hat{\boldsymbol{\kappa}}_f\cdot\hat{\boldsymbol{\kappa}})|^2.\end{aligned} \tag{216}$$

For convenient reference we give

$$f_{\text{Coul}}(-\hat{\boldsymbol{\kappa}}_f\cdot\hat{\boldsymbol{\kappa}}) = -\frac{e_1 e_2 M_r}{2\kappa^2}\left(\cos\frac{\theta}{2}\right)^{-2} e^{2i\{\sigma_0 - \xi\ln[\cos(\theta/2)]\}},$$

which can be immediately obtained from Eq. (203).

Finally, for a random distribution of initial spins we may now write Eq. (94*b*) as

$$\frac{d\sigma}{d\Omega} = \frac{1}{4}\left(\frac{d\sigma^{(0)}}{d\Omega} + 3\,\frac{d\sigma^{(1)}}{d\Omega}\right). \tag{217}$$

The transformation to the laboratory frame of reference, which must be done when comparing these cross sections with experimental measurements, has been described in Section 6.1.

## 6.5 ANALYTIC PROPERTIES OF PARTIAL WAVE AMPLITUDES

As remarked previously, the problem of determining the phase shifts $\delta_l$, for particles interacting by way of a nonrelativistic central potential, is essentially trivial. For example, numerical integration of Eq. (110) requires only a few hours with a desk computer and only a few minutes with a computing machine. It is, however, of considerable theoretical interest to see what general statements may be made for (almost) arbitrary potentials. Typical questions are: Given the phase shift $\delta_l(\kappa)$ for any angular momentum and for all moments $\kappa$, is the potential $V(r)$ uniquely defined? Is there any relation between the phase shift at zero energy and at infinite energy? What is the relation between the scattering amplitude and bound states?

A powerful method for answering these questions depends on the study of the analytic properties of the scattering amplitude as a function of complex energies or wave numbers. These same techniques also serve as guides to similar problems which arise in quantum field theory. Historically, the possibility of relating bound states to analytic properties of scattering amplitudes was first emphasized by Heisenberg and developed further by Møller and Kramers [1]. The first careful study was carried out by Jost [3], who made a number of very important contributions to the problem. Further work by Bargmann [3], Kohn [17, 18], Newton [19], and Gel'fand and Levitan [20] has clarified most of the remaining questions. Recent work in dispersion theory has also provided deep insight into the analytic structure of the scattering amplitude.[48]

A detailed, rigorous description of this subject will not be attempted here. Instead, we shall discuss a few topics carefully and present the rest with what we hope are convincing arguments. Many of the results emerge quite simply from the dispersion theoretic standpoint and will be rederived in Chapter 10.

### The Jost Function

The bulk of our treatment will be confined to $S$-wave scattering. Higher angular momenta are more complicated in an uninteresting way, simply because of the singularity at $r = 0$ of the "centrifugal term" $l(l + 1)/r^2$ in Eq. (110). We shall therefore merely quote many of the results which pertain to the case that the angular momentum $l \neq 0$.

For $S$-wave scattering, Eq. (110) is

$$\frac{d^2 w_0}{dr^2} + [\kappa^2 - v(r)]w_0 = 0. \tag{218}$$

We shall assume in this section that the reduced potential $v(r)$ satisfies the conditions

$$\begin{aligned} \int_0^\infty dr r\, |v(r)| &< \infty, \\ \int_0^\infty dr r^2\, |v(r)| &< \infty. \end{aligned} \tag{219}$$

These inequalities imply that $v(r)$ is less singular than $r^{-2}$ at the origin and that it falls off faster than $1/r^3$ at infinity. It will be convenient to introduce a new function

$$\phi(\kappa, r) = \alpha^{-1} w_0(\kappa, r), \tag{220}$$

[48] See, for example, Refs. [2] and [21].

where $\alpha$ is a constant. This function, which is also evidently a solution of Eq. (218), is defined by the boundary conditions

$$\begin{aligned} \phi(\kappa, 0) &= 0, \\ \phi'(\kappa, 0) &= 1. \end{aligned} \tag{221}$$

These boundary conditions imply that not only is $\phi(\kappa, r)$ a real function for real $\kappa$ (as was shown to be the case for $w_0$), but that it is an even function of $\kappa$. The boundary conditions (221) also imply that $\phi(\kappa, r)$ is an entire function of $\kappa$; that is, $\phi$ is analytic everywhere in the open complex $\kappa$-plane. This follows from a theorem of Poincaré [22], which states that a solution of Eq. (218), subject to boundary conditions independent of $\kappa$, is an entire function of $\kappa$ for any value of $r$.

Another useful solution of Eq. (218) is the Jost function [3] $\mathbf{f}(\kappa, r)$, defined by the boundary condition

$$\lim_{r\to\infty} e^{i\kappa r}\mathbf{f}(\kappa, r) = 1. \tag{222}$$

Since the differential equation (218) is real for real $\kappa$, we see from the boundary condition (222) that

$$\mathbf{f}^*(-\kappa, r) = \mathbf{f}(\kappa, r) \tag{223a}$$

For complex values of $\kappa$, we noted that $\phi(\kappa, r)$ is an entire function of $\kappa$. The boundary condition (222) does not define $\mathbf{f}(\kappa, r)$ in the upper half-$\kappa$-plane. We are able to conclude, however, that $\mathbf{f}(\kappa, r)$ is analytic at least in the lower half-plane. In a region of analyticity connected to the real axis, we see from Eq. (222) that

$$\mathbf{f}^*(-\kappa^*, r) = \mathbf{f}(\kappa, r), \tag{223b}$$

a generalization of Eq. (223*a*).

The proof of the analyticity of $\mathbf{f}(\kappa, r)$ may be demonstrated by a method given by Bargmann [3]. We see from Eq. (218) that the function $g(\kappa, r) \equiv e^{i\kappa r}\mathbf{f}(\kappa, r)$ satisfies the differential equation

$$g'' - 2i\kappa g' = v(r)g, \tag{224}$$

with the boundary condition $g(\kappa, r) \to 1$ as $r \to \infty$. This differential equation is conveniently replaced by the integral equation

$$g(\kappa, r) = 1 + \int_r^\infty dr' G_\kappa(r' - r)v(r')g(\kappa, r'), \tag{225a}$$

where

$$G_\kappa(x) = \frac{1 - e^{-2i\kappa x}}{2i\kappa} = \int_0^x dy\, e^{-2i\kappa y}. \tag{225b}$$

Equation (225*a*) may be solved in the form

$$g(\kappa, r) = \sum_{n=0}^{\infty} g_n(\kappa, r),$$

$$g_0(\kappa, r) = 1, \tag{226}$$

$$g_{n+1}(\kappa, r) = \int_r^{\infty} dr' G_\kappa(r' - r) v(r') g_n(\kappa, r').$$

We can prove that the series converges when Im $\kappa < 0$.[49] Then

$$|G_\kappa(r' - r)| \le (r' - r) \le r',$$

so by induction [23]

$$|g_n(\kappa, r)| \le \frac{\left(\int_r^{\infty} dr' r' \, |v(r')|\right)^n}{n!}$$

$$\le \frac{\left(\int_0^{\infty} dr' r' \, |v(r')|\right)^n}{n!} \tag{227}$$

The series (226) therefore converges as a consequence of our first inequality (219). If we differentiate the series (226) with respect to $\kappa$, we see by the same argument that the sum exists because of the second inequality (219). Therefore, $\mathbf{f}(\kappa, r)$ exists and is analytic everywhere in the lower half-$\kappa$-plane.

It is possible to extend the domain of analyticity of **f** into the upper half-plane under some circumstances. If the integral

$$\int_0^{\infty} dr \, e^{\mu r} \, |v(r)| < \infty, \tag{228}$$

for $\mu$ real and greater than zero, then $f(\kappa, r)$ is analytic for Im $\kappa < \mu/2$. The previous argument is modified only in that $|G_\kappa(x)| \le x \, e^{\mu x}$.

The functions $\mathbf{f}(\kappa, r)$ and $\mathbf{f}(-\kappa, r)$ are evidently independent solutions of Eq. (218). We may therefore construct $\phi(\kappa, r)$ from them. Since $\phi(\kappa, r)$ is real, even in $\kappa$, and subject to the conditions (221), we have

$$\phi(\kappa, r) = \frac{1}{2i\kappa} [\mathbf{f}(\kappa)\mathbf{f}(-\kappa, r) - \mathbf{f}(-\kappa)\mathbf{f}(\kappa, r)], \tag{229}$$

where

$$\mathbf{f}(\pm\kappa) \equiv \mathbf{f}(\pm\kappa, 0). \tag{230}$$

To check the numerical factor in Eq. (229) we evaluate $\phi'(\kappa, 0)$. This is

[49] By Im $(\cdots)$ we mean the imaginary part of $(\cdots)$, as usual.

easily done, since the Wronskian of two independent solutions $\mathbf{f}(\pm\kappa, r)$ is independent of $r$.

$$W[\mathbf{f}(\kappa, r), \mathbf{f}(-\kappa, r)] \equiv \mathbf{f}(\kappa, r)\mathbf{f}'(-\kappa, r) - \mathbf{f}'(\kappa, r)\mathbf{f}(-\kappa, r)$$
$$= 2i\kappa, \tag{231}$$

which follows from the asymptotic forms $\mathbf{f}(\pm\kappa, r) = e^{\mp i\kappa r}$. It is precisely the Wronskian at $r = 0$ that we encounter in evaluating $\phi'(\kappa, 0)$. For future reference, we note that the equation

$$\mathbf{f}(\kappa) = W[\mathbf{f}(\kappa, r), \phi(\kappa, r)] \tag{232}$$

follows directly from Eq. (229).

To relate $\mathbf{f}(\kappa)$ to the $S$-matrix, we need the asymptotic form of $\phi$ for large $r$.

$$\phi \to \frac{1}{2i\kappa}[\mathbf{f}(\kappa)\, e^{i\kappa r} - \mathbf{f}(-\kappa)\, e^{-i\kappa r}]$$
$$= \frac{\mathbf{f}(-\kappa)}{2i\kappa}\left[-e^{-i\kappa r} + \frac{\mathbf{f}(\kappa)}{\mathbf{f}(-\kappa)}\, e^{i\kappa r}\right]. \tag{233}$$

Comparison with Eq. (66$a$) leads to the identification

$$S_0(\kappa) = \frac{\mathbf{f}(\kappa)}{\mathbf{f}(-\kappa)},$$
$$S_0(\kappa) = \frac{1}{S_0(-\kappa)}. \tag{234}$$

Then according to Eq. (57)

$$\mathbf{f}(\kappa) = |\mathbf{f}(\kappa)|e^{i\delta_0(\kappa)}, \tag{235}$$

where $\delta_0$ is the $S$-wave scattering phase shift. Evidently

$$\delta_0(-\kappa) = -\delta_0(\kappa). \tag{236}$$

The relation of $\phi$ to the wave function $w_0$ may be easily obtained using Eqs. (233) and (235). The constant $\alpha$ in Eq. (220) is seen to be

$$\alpha = \left[\sqrt{\pi/2}\,\frac{|\mathbf{f}(-\kappa)|}{\kappa}\right]^{-1}. \tag{237}$$

Equation (237) permits us to derive a result which has been useful in describing reactions of elementary particles. The ratio

$$\mathscr{F}' \equiv \left.\frac{w_0(\kappa, r)}{\sqrt{\pi/2}\,\sin\kappa r}\right|_{r=0} \tag{238$a$}$$

may be evaluated using the boundary conditions (221). Evidently,

$$\mathscr{F}' = \alpha\sqrt{\pi/2}\,\frac{\phi'(\kappa, 0)}{\kappa} = \frac{1}{|\mathbf{f}(\kappa)|}. \tag{238b}$$

In terms of the "physical wave function" $\psi^{+}_{\kappa,0}$ [Eq. (109*a*)] we then have

$$\mathscr{F} \equiv \left.\frac{\psi^{+}_{\kappa,0}}{\sqrt{2/\pi}\,(\sin\kappa r/\kappa r)}\right|_{r=0} = \frac{e^{i\delta_0(\kappa)}}{|\mathbf{f}(\kappa)|}. \tag{238c}$$

The quantity $|\mathscr{F}|^2$ is the so-called *enhancement factor*.[50] It is the ratio of the probability of finding the two particles at $r = 0$, when their interaction is $V(r)$, to that for no interaction between the particles.

We have seen that $\mathbf{f}(\kappa)$ is analytic in the lower half-plane. It is, furthermore, analytic for $\operatorname{Im}\kappa < \mu_0/2$, where $\mu_0$ is the largest value of $\mu$ for which (228) is satisfied. Actually, it is possible to say something about the properties of $\mathbf{f}(\kappa)$ for $\operatorname{Im}\kappa > \mu_0/2$. To do this, we write

$$\mathbf{f}(\kappa, r) = e^{-i\kappa r} + \mathbf{f}^{(1)}(\kappa, r) \tag{239}$$

and note that because of Eqs. (225)

$$\begin{aligned}\mathbf{f}(\kappa, r) &= e^{-i\kappa r} + \int_r^\infty dr'\,\frac{\sin[\kappa(r'-r)]}{\kappa}\,v(r')\mathbf{f}(\kappa, r').\\ \mathbf{f}^{(1)}(\kappa, r) &= \mathbf{f}_B^{(1)}(\kappa, r) + \int_r^\infty dr'\,\frac{\sin[\kappa(r'-r)]}{\kappa}\,v(r')\mathbf{f}^{(1)}(\kappa, r'),\end{aligned} \tag{240}$$

where

$$\mathbf{f}_B^{(1)}(\kappa, r) = \int_r^\infty dr'\,\frac{\sin[\kappa(r'-r)]}{\kappa}\,v(r')\,e^{-i\kappa r'}. \tag{241}$$

The explicit evaluation of this integral will have an analytic continuation into some region of the upper half-plane with $\operatorname{Im}\kappa > \mu_0/2$. If, for example, $|\mathbf{f}_B^{(1)}| < M\,e^{\nu r}$, where $M > 0$ and $\nu > 0$, by the method used in connection with Eq. (225), we can show that $\mathbf{f}^{(1)}(\kappa, r)$ is analytic for $\operatorname{Im}\kappa < \mu_0 - \nu$ (previously $\nu$ was $\operatorname{Im}\kappa$, which led to the condition $\operatorname{Im}\kappa < \mu_0/2$). If $\nu$ is $\operatorname{Im}\kappa - \mu_0$, as it indeed is for $\operatorname{Im}\kappa < \mu_0/2$, we can analytically continue up to $\operatorname{Im}\kappa < \mu_0$. Repeating this procedure, introducing $\mathbf{f}_B^{(2)}$, $\mathbf{f}_B^{(3)}$, and so on, this process can then be continued further into the positive imaginary half-plane, except for specific, isolated singularities.[51] When the condition (228) is satisfied for any value of $\mu$, so $v(r)$ decreases faster than an exponential, $\mathbf{f}(\kappa)$ is evidently analytic in the entire complex plane, except at $\kappa = \infty$.

[50] This was introduced by Fermi and by Watson et al. in describing the production and absorption of pi-mesons. See Refs. [24].

[51] This argument is given by Newton [19].

**Bound States**

The zeros of the function $\mathbf{f}(\kappa)$ in the lower half-plane correspond to bound states. This may be seen by remarking that if $\mathbf{f}(\kappa) = 0$ for $\kappa = \bar{\kappa}$, then by Eq. (232) the Wronskian is zero and $\mathbf{f}(\bar{\kappa}, r)$ is thus a multiple of $\phi(\bar{\kappa}, r)$.

$$\mathbf{f}(\bar{\kappa}, r) = \gamma_n \phi(\bar{\kappa}, r), \tag{242}$$

where $\gamma_n$ is a constant and $\bar{\kappa} = -i\kappa_n$ $(n = 1, 2, \ldots, n_b)$ in the notation of Eq. (111*a*). Now, since the left side of this equation vanishes exponentially for $r \to \infty$, and by definition $\phi(\kappa, r)$ vanishes at $r = 0$, we see that $\phi(\bar{\kappa}, r)$ is a quadratically integrable eigenfunction of the Schrödinger equation with $\bar{\kappa}^2$ a discrete eigenvalue. We proved in Section 1.1 that $\bar{\kappa}^2$ is real. It is instructive to show this directly, however. By virtue of the relation (223*b*), if $\mathbf{f}(\bar{\kappa}) = 0$, it follows that $\mathbf{f}(-\bar{\kappa}^*) = 0$. Therefore $\phi(-\bar{\kappa}^*, r) = \phi(\bar{\kappa}^*, r)$ (since $\phi$ is even in $\bar{\kappa}$) is an integrable eigenfunction of Eq. (218). Now a straightforward calculation shows that

$$\begin{aligned} \frac{d}{dr} W[\phi(\bar{\kappa}, r), \phi(\bar{\kappa}^*, r)] &= 2i \operatorname{Im} \bar{\kappa}^2 \, |\phi(\bar{\kappa}, r)|^2 \\ &= 0, \end{aligned} \tag{243}$$

since the Wronskian is a constant. Therefore $\operatorname{Im} \bar{\kappa}^2 = 0$, and $\kappa_n$ $(\kappa_n = i\bar{\kappa})$ is real and positive.

When the point $\kappa = i\kappa_n$ lies within the region of analyticity of $\mathbf{f}(\kappa)$ connected with the real axis, the constant $\gamma_n$ may be evaluated. From Eqs. (242) and (229) we obtain

$$\gamma_n = -2 \frac{\kappa_n}{\mathbf{f}(i\kappa_n)}, \tag{244}$$

and thus

$$\lim_{r \to \infty} \phi(-i\kappa_n, r) = \frac{1}{\gamma_n} e^{-\kappa_n r} \tag{245}$$

It is important to note that $\mathbf{f}(\kappa)$ cannot vanish for real $\varkappa \neq 0$. If it did, so would $\mathbf{f}(-\kappa)$, which because of Eq. (229) would imply that $\phi(\kappa, r) \equiv 0$. This is impossible, since it violates the boundary conditions (221). The question arises whether $\kappa = 0$ corresponds to a bound state. If $\mathbf{f}(\kappa) = 0$ for $\kappa = 0$, we conclude from Eq. (242) that

$$\mathbf{f}(0, r) = \gamma_0 \phi(0, r), \tag{246}$$

where $\gamma_0$ is an appropriate constant. But since $\mathbf{f}(0, r) \to 1$ as $r \to \infty$, we see that $\phi(0, r)$ is not quadratically integrable. (This is the first place that there arises a difference between $l = 0$ and $l \geq 1$; in the latter case $\kappa = 0$ can correspond to a discrete eigenvalue.[52])

It is not hard to show that the zeros of $\mathbf{f}$ at the points $\kappa = \bar{\kappa} = -i\kappa_n$, $\kappa_n > 0$, are simple. The method used is essentially an adaptation of one used by Kramers [25] to normalize wave functions. On differentiating Eq. (232) with respect to $\kappa^2$, and then setting $\kappa = \bar{\kappa}$, we find

$$\begin{aligned}\frac{d\mathbf{f}}{d\kappa^2} \equiv \dot{\mathbf{f}}(\bar{\kappa}) &= W[\dot{\mathbf{f}}(\bar{\kappa}, r), \phi(\bar{\kappa}, r)] + W[\mathbf{f}(\bar{\kappa}, r), \dot{\phi}(\bar{\kappa}, r)] \\ &= \gamma_n^{-1} W[\dot{\mathbf{f}}(\bar{\kappa}, r), \mathbf{f}(\bar{\kappa}, r) \ + \gamma_n W[\phi(\bar{\kappa}, r), \dot{\phi}(\bar{\kappa}, r)]. \end{aligned} \tag{247}$$

Since $\mathbf{f}$ and $\phi$ both satisfy Eq. (218), we deduce the Wronskian conditions[53]

$$\begin{aligned} W[\phi(\kappa, r), \phi(\bar{\kappa}, r)] &= (\kappa^2 - \bar{\kappa}^2) \int_0^r dr' \, \phi(\kappa, r')\phi(\bar{\kappa}, r'), \\ W[\mathbf{f}(\kappa, r), \mathbf{f}(\bar{\kappa}, r)] &= -(\kappa^2 - \bar{\kappa}^2) \int_r^\infty dr' \, \mathbf{f}(\kappa, r')\mathbf{f}(\bar{\kappa}, r'). \end{aligned} \tag{248}$$

Differentiating these expressions with respect to $\kappa^2$ and setting $\kappa^2 = \bar{\kappa}^2$ then yields the expressions needed for evaluating Eq. (247). We find

$$\begin{aligned} \dot{\mathbf{f}}(\bar{\kappa}) &= -\left[\gamma_n \int_0^r dr' \, \phi^2(\bar{\kappa}, r') + \gamma_n^{-1} \int_r^\infty dr' \, \mathbf{f}^2(\bar{\kappa}, r)\right] \\ &= -\gamma_n \int_0^\infty dr \phi^2(\bar{\kappa}, r). \end{aligned}$$

Since $\gamma_n$ is not zero and since $\phi(\bar{\kappa}, r)$ is real, we conclude that

$$\frac{\partial \mathbf{f}(\bar{\kappa})}{\partial \bar{\kappa}} = -2\bar{\kappa}\gamma_n \int_0^\infty dr \phi^2(\bar{\kappa}, r) \tag{249}$$

is different from zero. Thus the zeros of $\mathbf{f}(\kappa)$, Im $\kappa < 0$, are simple.

The bound-state wave functions $w_{n,0}(r)$ [see Eq. (111*b*)] may now be normalized, since for these

$$\int_0^\infty w_{n,0}^2(r) \, dr = 1. \tag{250}$$

In the notation of Eq. (220), we have

$$w_{n,0}(r) = \alpha_n \phi(\bar{\kappa}, r), \tag{251}$$

[52] See Ref. [19, Eq. (4.18′)].

[53] These fcllow, for example, from Eq. (116).

where $\bar{\kappa} = -i\kappa_n$. Because of Eq. (249),

$$1 = \int_0^\infty w_{n,0}^2(r)\,dr = \alpha_n^{\ 2} \int_0^\infty dr \phi^2(\bar{\kappa}, r)$$

$$= \alpha_n^{\ 2}\left[-\frac{d\mathbf{f}(\bar{\kappa})}{2\bar{\kappa}\,d\bar{\kappa}}\right]\frac{1}{\gamma_n},$$

so

$$\alpha_n^{\ 2} = \frac{2\kappa_n\gamma_n}{d\mathbf{f}(-i\kappa_n)/d\kappa_n} = -\frac{4\kappa_n^{\ 2}}{\mathbf{f}(i\kappa_n)[d\mathbf{f}(-i\kappa_n)/d\kappa_n]}. \tag{252}$$

It should be emphasized again that this result is valid only when $\kappa = i\kappa_n$ lies within the domain of analyticity of $\mathbf{f}(\kappa)$ which is connected to the real axis. We observe that in this case $\mathbf{f}(i\kappa_n)$ cannot vanish, for this would imply the vanishing of $\phi(-i\kappa_n, r)$.[54]

A sometimes more convenient expression for $w_{n,0}(r)$ is

$$w_{n,0}(r) = \beta_n \mathbf{f}(-i\kappa_n, r), \tag{253}$$

so[55]

$$\lim_{r\to\infty} w_{n,0}(r) = \beta_n\, e^{-\kappa_n r}. \tag{254}$$

Because of Eqs. (251) and (242), we have

$$\beta_n^{\ 2} = \frac{\alpha_n^{\ 2}}{\gamma_n^{\ 2}} = -i\frac{\mathbf{f}(-\bar{\kappa})}{d\mathbf{f}(\bar{\kappa})/d\bar{\kappa}} = \frac{\mathbf{f}(i\kappa_n)}{d\mathbf{f}(-i\kappa_n)/d\kappa_n}. \tag{255a}$$

Since $S_0(\kappa) = \mathbf{f}(\kappa)/\mathbf{f}(-\kappa)$, this may be rewritten in the form

$$\beta_n^{\ 2} = \frac{1}{2\pi}\oint d\kappa S_0(\kappa). \tag{255b}$$

Here the contour is a small circle, in the positive sense, about the point $\kappa = i\kappa_n$. This equation relates the bound-state wave function normalization directly to the scattering phase shift. It has been very useful in discussing phenomena associated with the production and absorption of low-energy mesons.[56]

When $\mathbf{f}(0) = 0$, Eq. (249) requires special consideration. Since $\phi(\bar{\kappa}, r) \to (1/\gamma_0)\, e^{-i\bar{\kappa}r}$ [see Eq. (246)], when $r \to \infty$,

$$\int_0^\infty dr \phi^2(\bar{\kappa}, r) = \frac{1}{2i\bar{\kappa}\gamma_0^{\ 2}} + \text{finite term},$$

[54] Unless $\mathbf{f}(-i\kappa_n, r)$ is singular. See Ref. [19].
[55] This is just Eq. (111*b*).
[56] See, for example, Refs. [24].

as $\bar{\kappa} \to 0$. Thus Eq. (249) becomes

$$\left.\frac{d\mathbf{f}(\bar{\kappa})}{d\bar{\kappa}}\right|_{\bar{\kappa}\to 0} = \frac{i}{\gamma_0}, \tag{256}$$

so $\mathbf{f}(\kappa)$ has a simple zero at $\kappa = 0$ when $f(0) = 0$.

**Higher Angular Momenta**

The case that $l \geq 1$ has been discussed by Newton [19]. We shall quote, without derivation, some of his results. For any $l$, solutions $\phi_l(\kappa, r)$ of Eq. (110) subject to the boundary conditions (221) are entire functions of $\kappa$.[57] Additional solutions $\mathbf{f}_l(\pm\kappa, r)$ may be defined by the boundary condition

$$\lim_{r\to\infty} e^{i\kappa r}\mathbf{f}_l(\kappa, r) = i^l. \tag{257}$$

The amplitudes $\mathbf{f}_l(\kappa)$ are then defined as[58]

$$\mathbf{f}_l(\pm\kappa) = \lim_{r\to 0} \frac{(\pm\kappa r)^l}{(2l-1)!!}\mathbf{f}_l(\pm\kappa, r). \tag{258}$$

The relation between $\phi_l$ and the $\mathbf{f}_l$ is seen to be[59]

$$\phi_l(\kappa, r) = \frac{i}{2\kappa^{l+1}}[\mathbf{f}_l(-\kappa)\mathbf{f}_l(\kappa, r) - (-1)^l\mathbf{f}_l(\kappa)\mathbf{f}_l(-\kappa, r)]. \tag{259}$$

Comparison of the asymptotic form of Eq. (259) with Eq. (66*a*) shows that the $S$-matrix elements are

$$S_l(\kappa) = \frac{\mathbf{f}_l(\kappa)}{\mathbf{f}_l(-\kappa)}. \tag{260}$$

When the conditions (219) are satisfied by $v(r)$, $\mathbf{f}_l(\kappa, r)$ is analytic in the lower half-$\kappa$-plane. If the condition (228) is satisfied, the domain of analyticity may be extended to Im $\kappa < \mu/2$. Most of the remaining discussion of the case $l = 0$ is unchanged except for occasional numerical factors. An exception occurs if $\mathbf{f}_l(\kappa) = 0$ for $\kappa = 0$. Then $\mathbf{f}_l(\kappa) = \mathcal{O}(\kappa^2)$ near $\kappa = 0$ for $l \geq 1$.

An important result due to Bargmann [26] states that the number of bound states $n_{b,l}$ of angular momentum $l$ is subject to

$$(2l+1)n_{b,l} < \int_0^\infty dr\, r\,|v(r)|, \tag{261}$$

if the conditions (219) are satisfied.

[57] This follows from the theorem of Poincaré [22].

[58] The double factoral notation is defined in connection with Eq. (1.74).

[59] The function (259) satisfies the boundary condition

$$\lim_{r\to 0} (2l+1)!!\, r^{-(l+1)}\phi_l(\kappa, r) = 1.$$

**Analytic Properties of the $S$-Matrix**

We turn now to a discussion of the analytic properties of $S_l(\kappa)$ which can be deduced from those of $\mathbf{f}_l(\kappa)$. Because $S_l(\kappa) = \mathbf{f}_l(\kappa)/\mathbf{f}_l(-\kappa)$, the analyticity of $S_l$ is rather more complicated. On the real axis we can conclude that $S_l(\kappa) \to 1$ as $\kappa \to \pm\infty$, but leaving the real axis is an ugly business. Either $\mathbf{f}_l(\kappa)$ or $\mathbf{f}_l(-\kappa)$ may have singularities in general. We might expect that at $\kappa = i\kappa_n$, where $-\kappa_n^2$ is an eigenvalue, $S_l(\kappa)$ will have a pole; this need not be the case because $\mathbf{f}_l(i\kappa_n)$ might vanish. If, however, Eq. (228) is valid, $S_l(\kappa)$ is analytic in strip $0 < |\text{Im}\ \kappa| < \mu/2$, except for poles at $\kappa = i\kappa_n$. [In this case $\mathbf{f}_l(i\kappa_n)$ does not vanish, because of Eqs. (242) and (244).]

In the early days of $S$-matrix theory it was noted that $S_0(\kappa) = 0$ at the bound states $\kappa = -i\kappa_n$. Evidently, we could also have $S_0 = 0$ if for some value of $\kappa$, $\mathbf{f}(-\kappa) = \infty$ while $\mathbf{f}(\kappa)$ were finite. This was indeed shown to be the case for the exponential potential $v(r) = -v_0 e^{-r/d}$ [see Eq. (156)].[60]

These zeros of $S_0$, called *redundant zeros*, have been seen to disappear when the potential is cut off at any arbitrarily large distance. It may at first seem strange that a cutoff at a distance, say of 1 km, would play any role in elementary particle interactions.[61]

It is perhaps worth going into this point in a little more detail here. Imagine cutting off the exponential tail of a potential at some large $r = R$. If we write $\delta v(r)$ for the part of $v(r)$ that was cut off, then according to Eq. (240) the effect of adding back $\delta v$ is to modify $\mathbf{f}(\kappa, r)$ by $\delta\mathbf{f}(\kappa, r)$, where

$$\delta\mathbf{f}(\kappa, r) = \frac{1}{\kappa}\int_r^\infty dr' \sin [\kappa(r' - r)]\ \delta v(r')\mathbf{f}(\kappa, r').$$

Here $\mathbf{f}(\kappa, r)$ is the appropriate solution to Eq. (240) for the cutoff potential and $\delta v$ has been treated as a small perturbation. Since $\delta v = 0$ for $r < R$, we have

$$\delta\mathbf{f}(\kappa) = \frac{1}{\kappa}\int_R^\infty dr' \sin (\kappa r')\, \delta v(r')\mathbf{f}(\kappa, r')$$

$$\approx \frac{1}{2i\kappa}\int_R^\infty dr(1 - e^{-2i\kappa r})\ \delta v(r), \tag{262}$$

where we have set $\mathbf{f}(\kappa, r)$ equal to its asymptotic value.

Clearly, for any real $\kappa$, taking $R$ sufficiently large will produce an arbitrarily small effect on $\mathbf{f}$; the same conclusion holds for Im $\kappa < 0$. In the

[60] See Eq. (162*b*). These zeros occur for $\nu = 1, 2, \ldots,$ for which $\Gamma(-\nu + 1) = \infty$. The existence of redundant zeros was noted by S. T. Ma [10].

[61] We of course encountered an example in discussing the Coulomb field in Section 6.4.

upper half-plane, however, we have difficulties. If we take $\delta v = -v_0\, e^{-\mu r}$, we find for real $\kappa$

$$\delta\mathbf{f}(\kappa) = -\frac{v_0}{2i\kappa}\left(\frac{e^{-\mu R}}{\mu} - \frac{e^{-(\mu+2i\kappa)R}}{\mu + 2i\kappa}\right). \tag{263}$$

If this is continued analytically into the upper half-plane we see that $\delta\mathbf{f}(\kappa)$ changes enormously when $\text{Im}\,\kappa > \mu/2 - 1/R$. In addition, $\delta\mathbf{f}(\kappa) = \infty$ at $\kappa = i(\mu/2)$, which is, in fact, the first redundant zero of Ma[62,63] for $\mathbf{f}(-\kappa)$.

The residues of the poles of $S_0(\kappa)$ that lie within the strip of analyticity have been evaluated in Eq. (255*b*) in terms of the normalization of the bound-state wave functions. A similar result may be obtained for higher angular momenta.

**An Example**

To illustrate our discussion, we study the example of an attractive square-well potential discussed in Section 6.4, p. 254 and defined by Eq. (140), with $v_0 = -|v_0|$. According to the boundary conditions (222), the solution to Eq. (218) is

$$\mathbf{f}(\kappa, r) = e^{-i\kappa r} \quad \text{for} \quad r > d. \tag{264}$$

When $r < d$, we have

$$\mathbf{f}(\kappa, r) = C_1\, e^{i\kappa_0 r} + C_2\, e^{-i\kappa_0 r}, \tag{265}$$

where $C_1$ and $C_2$ are constants and $\kappa_0^{\,2} = \kappa^2 + |v_0|$, as defined by Eq. (141). The "jump" conditions (138) determine the constants $C_1$ and $C_2$. We find on evaluating them that

$$\mathbf{f}(\kappa) = C_1 + C_2 = e^{-i\kappa d}\left(\cos \kappa_0 d + \frac{i\kappa}{\kappa_0}\sin \kappa_0 d\right). \tag{266}$$

This is an entire function of $\kappa$, as expected from our general considerations. The $S$-matrix is

$$S_0(\kappa) = e^{-2i\kappa d}\,\frac{\cos \kappa_0 d + (i\kappa/k_0)\sin \kappa_0 d}{\cos \kappa_0 d - i(\kappa/\kappa_0)\sin \kappa_0 d}, \tag{267}$$

which is also analytic, except for isolated poles, for all finite $\kappa$, but has an essential singularity at $|\kappa| = \infty$.

For $\kappa \approx 0$, if we set $\kappa_0 d$ equal to an odd multiple of $\pi/2$, we find that

$$\mathbf{f}(\kappa) \approx \pm i\,\frac{\kappa}{\kappa_0}, \tag{268}$$

consistent with Eq. (256). The zeros of $\mathbf{f}(\kappa)$ are found by setting $\kappa = -i\kappa_n$ and solving the equation

$$\tan \kappa_0 d = -\frac{\kappa_0}{\kappa_n}, \tag{269}$$

[62] Reference [10]. This is verified from Eq. (162*b*).

[63] The discussion leading to Eq. (263) was suggested by G. C. Wick (unpublished).

where $\kappa_0 = (|v_0| - \kappa_n^2)^{1/2}$. We may easily see that Eq. (269) is equivalent to the bound-state eigenvalue problem for the potential (140).[64]

**Integral Relations and Dispersion Equations**

The analyticity of the $\mathbf{f}_l(\kappa)$ in the lower half-$\kappa$-plane permits us to derive some useful integral relations. We have remarked that $\mathbf{f}_l(\kappa) - 1$ approaches zero as $\kappa \to \infty$ anywhere in the lower half-plane, including the real axis. By Cauchy's theorem, then, for any value of $\kappa$ in the lower half-plane,

$$\mathbf{f}_l(\kappa) - 1 = \frac{1}{2\pi i}\int_C d\kappa' \frac{\mathbf{f}_l(\kappa') - 1}{\kappa' - \kappa}. \tag{270}$$

Here $C$ is the closed contour consisting of a path on the real axis from $+\kappa_c$ to $-\kappa_c$ and a semicircle of radius $\kappa_c$ in the lower half-plane, with $\kappa_c$ sufficiently large that the point $\kappa$ lies within the contour. In the limit as $\kappa_c \to \infty$, the contribution from the semicircle vanishes and we have

$$\mathbf{f}_l(\kappa) = 1 - \frac{1}{2\pi i}\int_{-\infty}^{+\infty} d\kappa' \frac{\mathbf{f}_l(\kappa') - 1}{\kappa' - \kappa}. \tag{271}$$

Allowing $\kappa$ to approach the real axis (from below, of course), we find from (271)

$$\mathbf{f}_l(\kappa) = 1 - \frac{P}{i\pi}\int_{-\infty}^{+\infty} d\kappa' \frac{\mathbf{f}_l(\kappa') - 1}{\kappa' - \kappa}, \tag{272}$$

where the symbol $P$ implies (as usual) that a principal value integral be taken. The real part of Eq. (272) is

$$\mathrm{Re}\,\mathbf{f}_l(\kappa) = 1 - \frac{P}{\pi}\int_{-\infty}^{\infty} d\kappa' \frac{\mathrm{Im}\,\mathbf{f}_l(\kappa')}{\kappa' - \kappa}, \tag{273a}$$

or

$$\mathbf{f}_l(\kappa) = 1 - \frac{1}{\pi}\int_{-\infty}^{\infty} d\kappa' \frac{\mathrm{Im}\,\mathbf{f}_l(\kappa')}{\kappa' - \kappa + i\eta}, \tag{273b}$$

where the limit $\eta \to 0(+)$ is understood. Equation (273*b*) follows from Eq. (273*a*) since

$$\lim_{\eta \to 0(+)} \frac{1}{x + i\eta} = P\left(\frac{1}{x}\right) - i\pi\delta(x).$$

Equation (273*a*) is called a *dispersion relation* for the function $\mathbf{f}_l(\kappa)$. It may be regarded as an integral equation for $\mathbf{f}_l(\kappa)$ in a certain sense. Since $\mathbf{f}_l(\kappa) = |\mathbf{f}_l(\kappa)|\, e^{i\delta_l(\kappa)}$ and $\mathrm{Im}\,\mathbf{f}_l(\kappa) = |\mathbf{f}_l(\kappa)| \sin\delta_l(\kappa) = \mathbf{f}_l(\kappa) \sin\delta_l(\kappa)\, e^{-i\delta_l(\kappa)}$,

[64] It is instructive to construct the function $\phi(\kappa, r)$ explicitly according to Eq. (229) and to check such relations as (244), (252), and (255*b*) for this example.

we may write Eq. (273*b*) as

$$\mathbf{f}_l(\kappa) = 1 - \frac{1}{\pi}\int_{-\infty}^{\infty} d\kappa' \, \frac{\mathbf{f}_l(\kappa')\, e^{-i\delta_l(\kappa')} \sin \delta_l(\kappa')}{\kappa' - \kappa + i\eta}. \tag{274}$$

We ask, given $\delta_l(\kappa)$, is it possible to solve this equation for $\mathbf{f}_l(\kappa)$? The answer is, in general, no.[65] We must give also the location of the zeros of $\mathbf{f}_l(\kappa)$ in the lower half-plane, corresponding to the bound-state energies. In Chapter 10 we shall return to such equations and a discussion of their solutions. We note that Eq. (274) simply states the fact that $\mathbf{f}_l(\kappa)$ may be extended to a function analytic in the lower half-$\kappa$-plane, given its value on the real axis.

We know that $\mathbf{f}_l(\kappa)$ has zeros at $\kappa = -i\kappa_n$, $\kappa_n > 0$, corresponding to bound states. Therefore the function $\bar{\mathbf{f}}_l(\kappa)$ defined by

$$\bar{\mathbf{f}}_l(\kappa) \equiv \left(\prod_{n=1}^{n_{b,l}} \frac{\kappa - i\kappa_n}{\kappa + i\kappa_n}\right) \mathbf{f}_l(\kappa) \tag{275}$$

is analytic in the lower half-plane, has no zeros there, and approaches unity at infinity. Following the arguments that led to Eq. (272), we deduce

$$\ln \bar{\mathbf{f}}_l(\kappa) = -\frac{P}{\pi i}\int_{-\infty}^{\infty} d\kappa' \, \frac{\ln \bar{\mathbf{f}}_l(\kappa')}{\kappa' - \kappa}.$$

Taking the real part of this and noting that $|\bar{\mathbf{f}}_l| = |\mathbf{f}_l|$, we have

$$\ln |\mathbf{f}_l| = -\frac{P}{\pi}\int_{-\infty}^{\infty} d\kappa' \, \frac{\operatorname{Im} \ln \bar{\mathbf{f}}_l(\kappa')}{\kappa' - \kappa}. \tag{276}$$

Now,

$$\operatorname{Im} \ln \bar{\mathbf{f}}_l = \delta_l(\kappa) - i \sum_n [\ln (\kappa - i\kappa_n) - \ln (\kappa + i\kappa_n)].$$

This may be inserted into Eq. (276) and the integral over the sum evaluated by contour integration. Writing $\epsilon_n = -\kappa_n^2/2M_r$ and $\epsilon = \kappa^2/2M_r$, we have

$$\ln |\mathbf{f}_l(\kappa)| = -\frac{P}{\pi}\int_{-\infty}^{\infty} d\kappa' \, \frac{\delta_l(\kappa')}{\kappa' - \kappa} + \prod_{n=1}^{n_{b,l}} \ln \frac{\epsilon - \epsilon_n}{\epsilon}, \tag{277}$$

where to give this equation meaning we adopt the convention that

$$\lim_{\kappa \to \pm\infty} \delta_l(\kappa) = 0. \tag{278}$$

This may evidently be done since $\lim_{\kappa \to \pm\infty} \mathbf{f}_l(\kappa) = 1$.

[65] Bargmann [3] has shown that there exists whole classes of potentials which all lead to the same phase shift but differ in bound states.

Since $\ln |\mathbf{f}_l| = \ln \mathbf{f}_l - i\delta_l$, we may write Eq. (277) as

$$\ln \mathbf{f}_l = -\frac{1}{\pi}\int_{-\infty}^{\infty} d\kappa' \frac{\delta_l(\kappa')}{\kappa' - \kappa + i\eta} + \sum_n \ln \frac{\epsilon - \epsilon_n}{\epsilon},$$

or[66]

$$\mathbf{f}_l(\kappa) = \prod_{n=1}^{n_{b,l}} \left(1 - \frac{\epsilon_n}{\epsilon}\right) \exp\left[-\frac{1}{\pi}\int_{-\infty}^{\infty} d\kappa' \frac{\delta_l(\kappa')}{\kappa' - \kappa + i\eta}\right]. \tag{279}$$

This is, in fact, the solution to the integral equation (274).

It is often convenient to discuss the function $\mathbf{f}_l(-\kappa)$ in the energy plane. By the relation $\epsilon = \kappa^2/2M_r$, we see that the upper half-$\kappa$-plane is mapped into the entire $\epsilon$-plane while the lower half-$\kappa$-plane is mapped into the second sheet of the Riemann surface of the function $\epsilon^{1/2}$, conveniently cut along the positive real axis. We refer to the first sheet as the *physical sheet* (Im $\kappa > 0$). (In this connection we recall that the wave functions $\psi^{\pm}$ and the scattering matrix $\mathscr{T}$ have discontinuities on the real energy axis. The choice $\psi^{+}$ corresponds to $\epsilon + i\eta = \kappa^2/2M_r$; thus the term *physical sheet*.)

Our representation for $\mathbf{f}_l(\kappa)$ yields for $\mathbf{f}_l(-\kappa)$

$$\mathbf{f}_l(-\kappa) = 1 + \frac{1}{\pi}\int_{-\infty}^{\infty} d\kappa' \frac{(\kappa' + \kappa)\,\mathrm{Im}\,\mathbf{f}_l(-\kappa')}{(\kappa')^2 - (\kappa + i\eta)^2}. \tag{280}$$

Since $\mathbf{f}_l(-\kappa) = \mathbf{f}_l^*(\kappa)$, Im $\mathbf{f}_l(\kappa)$ is an odd function of $\kappa$[67] and we may write[68] in terms of the variable $\epsilon$,

$$D_l(\epsilon) \equiv \mathbf{f}_l(-\kappa) = 1 + \frac{1}{\pi}\int_0^{\infty} d\epsilon' \frac{\mathrm{Im}\, D_l(\epsilon')}{\epsilon' - \epsilon - i\eta}. \tag{281}$$

$D_l(\epsilon)$ may be extended by this representation to a function analytic in the entire complex energy plane, cut from zero to infinity along the positive real axis. The physical value of $D_l$ is obtained by approaching the real axis from above. By construction, the phase of $D_l(\epsilon)$ along the real axis is $-\delta_l(\epsilon)$.[69] Evidently,

$$\lim_{\eta \to 0(+)} \frac{D_l(\epsilon + i\eta)}{D_l(\epsilon - i\eta)} = e^{-2i\delta_l(\epsilon)} = S_l^{-1}(\epsilon), \tag{282}$$

[66] This expression was obtained by Jost and Kohn [27] and by Newton [19].

[67] As is the phase shift $\delta_l(\kappa)$.

[68] The quantity $D(\epsilon)$ represents the Fredholm determinant which arises in solving the integral equation (225). See Ref. [28].

[69] To avoid introducing a new symbol, we have adopted a rather loose notation here, writing $\delta_l(\kappa) = \delta_l(\epsilon)$.

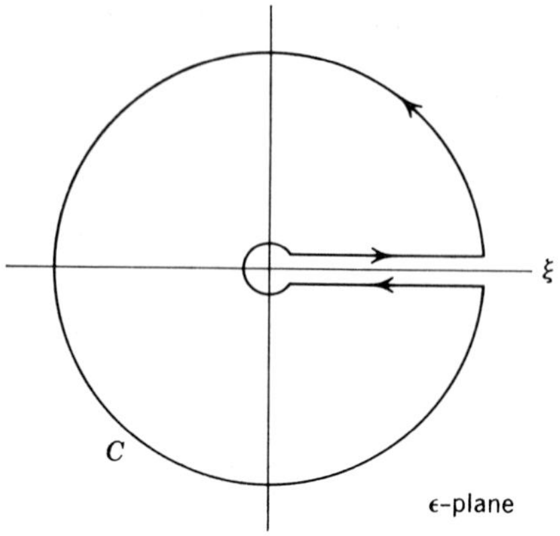


Fig. 6.4 Contour of integration for Eq. (284).

where $S_l(\epsilon) \equiv S_l(\kappa)$, and $D_l{}^*(Z) = D_l(Z^*)$. From Eq. (279) we obtain

$$D_l(\epsilon) = \prod_{n=1}^{n_{b,l}} \left(1 - \frac{\epsilon_n}{\epsilon}\right) e^{-\Delta(\epsilon)},$$
$$\Delta(\epsilon) = \frac{1}{\pi}\int_0^\infty d\epsilon' \frac{\delta_l(\epsilon')}{\epsilon' - \epsilon - i\eta}. \tag{283}$$

**The Levinson Theorem**

The Levinson theorem [29] relates the difference between the phase shift at zero and infinite energy to the number of bound states. Consider the quantity $n$ defined by

$$n = \frac{1}{2\pi i}\int_C dZ \frac{d}{dZ} \ln D_l(Z), \tag{284}$$

where the contour consists of a path leading from $+\xi$ below the real axis to the origin, which is avoided by a circle of radius $\epsilon$, then above the real axis to $+\xi$ and closed by a circle of radius $\xi$. It is supposed that all zeros of $D_l(\epsilon) = f_l(-\kappa)$ are enclosed within the circle of radius $\xi$. This contour is illustrated in Fig. 6.4.

Since $D_l(\epsilon)$ is analytic in the region enclosed by the contour $C$ in Eq. (284), the only contribution comes from the simple, discrete zeros at the points $\epsilon_n = -\kappa_n{}^2/2M_r$. This may, indeed, be seen directly from the expression (283) for $D_l$, since $\Delta(\epsilon)$ is analytic in the region enclosed by the contour. Thus Eq. (284) gives us $n = n_{b,l}$, the number of bound states. The case that $\mathbf{f}_l(\kappa) = 0$ for $\kappa = 0$ requires separate attention. It was observed that then $\mathbf{f}_0 \approx \kappa$ and $\mathbf{f}_l \approx \kappa^2$ for $l \geq 1$. Writing $D_l(Z) = Z^\beta$, where $\beta = \frac{1}{2}$ for $l = 0$ and $\beta = 1$ for $l \geq 1$, we obtain a contribution $-\beta$ to (284) from the integral around the small circle.

Letting the radius of the large circle approach infinity in (284), we then obtain

$$(n_{b,l} + \beta) = \frac{1}{2\pi i}\left[\int_{+\infty}^{0} d\epsilon \frac{d}{d\epsilon} \ln D_l(\epsilon - i\eta) + \int_0^{\infty} d\epsilon \frac{d}{d\epsilon} \ln D_l(\epsilon + i\eta)\right]$$

$$= \frac{1}{2\pi i}\int_0^{\infty} d\epsilon \frac{d}{d\epsilon} \ln \frac{D_l(\epsilon + i\eta)}{D_l(\epsilon - i\eta)}$$

$$= \frac{1}{2\pi i}\int_0^{\infty} d\epsilon \frac{b}{d\epsilon} \ln e^{-2i\delta_l(\epsilon)}$$

$$= \frac{1}{\pi}[\delta_l(0) - \delta_l(\infty)]. \tag{285}$$

Thus since we have agreed to the convention that $\delta_l(\infty) = 0$,

$$\begin{aligned} \delta_l(0) &= (n_{b,l} + \tfrac{1}{2})\pi \quad \text{for} \quad l = 0, \quad D_0(0) = 0, \\ &= n_{b,l}\pi \quad \text{for} \quad \text{all other cases.} \end{aligned} \tag{286}$$

[It is unnecessary to write $n_{b,l} + 1$ in case $D_l(0) = 0$ for $l \geq 1$, since such a zero describes a true bound state.] Equation (286) is the statement of the Levinson theorem.

### Behavior of $S_l(\kappa)$ at Low Energies

Because of Eq. (286) we have

$$\begin{aligned} S_l(0) = e^{2i\delta_l(0)} &= -1 \quad \text{for} \quad l = 0, \quad f_0(0) = 0, \\ &= 1 \qquad \text{otherwise.} \end{aligned} \tag{287}$$

Therefore the scattering matrix [see Eq. (57)]

$$T_l(\kappa) = -\frac{S_l(\kappa) - 1}{2\pi i\kappa M_r} \tag{288}$$

is finite at $\kappa = 0$ except for the case that $l = 0$, $\mathbf{f}_0(0) = 0$, for which it is infinite. This singular case is called a *zero-energy resonance*.

To obtain the $\kappa$-dependence of $T_l(\kappa)$ near $\kappa = 0$, we write [see Eq. (44*b*)]

$$T(\hat{\boldsymbol{\kappa}}_f \cdot \hat{\boldsymbol{\kappa}}, \kappa_f, \kappa) = (\chi_{\kappa_f}, T\chi_{\kappa}), \tag{289a}$$

where

$$T = V + V\frac{1}{\epsilon(\kappa) - H + i\eta}V. \tag{289b}$$

For $\epsilon(\kappa) \to 0$, we define

$$T_0 = \lim_{\epsilon(\kappa)\to 0} T = \langle \mathbf{r}' | T_0 | \mathbf{r} \rangle. \tag{290}$$

If we substitute the expansion (58) for $\chi_{\varkappa}$ into (289$a$), there results in the neighborhood of $\boldsymbol{\varkappa} = \boldsymbol{\varkappa}_f = 0$

$$T(\hat{\boldsymbol{\varkappa}}_f \cdot \hat{\boldsymbol{\varkappa}}, \kappa_f, \kappa) = \frac{2}{\pi} \sum_{l=0}^{\infty} \sum_{m=-l}^{l} Y_l^m(\hat{\boldsymbol{\varkappa}}_f) Y_l^{m*}(\hat{\boldsymbol{\varkappa}}) (j_l(\kappa_f r'), T_0^l j_l(\kappa r)), \tag{291}$$

where

$$\langle r' \,|T_0^l|\, r\rangle = \int d\Omega_{\hat{\mathbf{r}}'} \int d\Omega_{\hat{\mathbf{r}}} Y_l^{m*}(\hat{\mathbf{r}}')\langle \mathbf{r}' \,|T_0|\, \mathbf{r}\rangle Y_l^m(\hat{\mathbf{r}}). \tag{292}$$

Comparison of Eqs. (44$b$) and (291) leads to

$$T_l(\kappa) = \frac{2}{\pi} \int_0^{\infty} (r')^2\, dr' \int_0^{\infty} r^2\, dr j_l(\kappa r') j_l(\kappa r) \langle r' \,|T_0^{(l)}|\, r\rangle. \tag{293}$$

Now for sufficiently small $\kappa$,

$$j_l(\kappa r) \approx (\kappa r)^l$$

according to Eq. (126), so

$$T_l(\kappa) = \text{constant} \times \kappa^{2l} \tag{294}$$

near $\kappa = 0$, if the integral

$$\int_0^{\infty} (r')^{l+2}\, dr' \int_0^{\infty} r^{l+2} \langle r' \,|T_0^{(l)}|\, r\rangle\, dr$$

converges. This will be the case for all $l$ if the condition (228) is satisfied for any finite positive value of $\mu$.[70]

Therefore we see from Eq. (288) that

$$S_l(\kappa) - 1 = \text{constant} \times \kappa^{2l+1}, \tag{295}$$

except for the singular case that $f_0(0) = 0$ and $l = 0$.

### 6.6 EFFECTIVE-RANGE THEORY

In the early days of nuclear physics many attempts were made to fit experimental data on low-energy scattering phase shifts for the nucleon-nucleon system by solving the Schrödinger equation with various potential shapes. It was found that there was a peculiar insensitivity to the precise potential shape and that the data could be fit by almost any shape. In particular it was found by Breit and his collaborators [15] in their analysis of the $S$-wave phase shift in proton-proton scattering that a particular function of the phase shift was almost energy independent. This fact was somehow overlooked for many years and was rediscovered much later by

[70] More precisely, D. S. Carter (quoted by Newton [19]) has shown that Eq. (294) is valid if $\int_0^{\infty} dr r^{2l+2} \,|V(r)| < \infty$.

Landau and Smorodinsky and by Schwinger [30, 31]. The essential result was (for neutron-proton scattering, where the Coulomb effects do not complicate the issue) that the function $\kappa \cot \delta$ (where $\kappa$ is the center-of-mass wave number and $\delta$ the $S$-wave phase shift) is to an excellent approximation a linear function of $\kappa^2$.

$$\kappa \cot \delta = \frac{1}{a} + \frac{1}{2} r_e \kappa^2. \tag{296}$$

Any potential, therefore, that gave the same values of the parameters $a$ and $r_e$ would fit the data. Since all the old trial potentials had adjustable range and depth, all could be made to work. The parameter $a$ is called the scattering length, whereas $r_e$ is called the effective range, and the approximation to $\kappa \cot \delta$ given in Eq. (296) is referred to as the effective-range approximation. The name scattering length comes from the fact that the asymptotic form of the two-particle scattering wave function at very low energies takes the form $1 + a/r$, where $a$ is the constant value of the scattering amplitude $f(\cos \theta)$ at zero kinetic energy. (Scattering in all angular momentum states, except $l = 0$, vanishes at zero energy.) The effective range will be seen to measure essentially the range of the interaction, the distance beyond which the potential is negligible. Actually, $r_e$ is a rather more complicated quantity; for example, if the potential changes sign, $r_e$ can even be negative, although in most cases it has just the simple interpretation first given.

The first systematic investigation of effective-range theory, as it has come to be called, was carried out by Schwinger, using the variational method described in Section 6.7. A much simpler derivation was given a few years later by Chew and Goldberger; their method was independently proposed by Bethe and by Barker and Peierls [32–34]. An exhaustive analysis of the experimental data was made, using this theory, by Blatt and Jackson [14, 35]. We shall not enter into the latter aspect of the program but content ourselves with the development of the formalism. It is perhaps worth mentioning that the existence of an expansion of $\kappa \cot \delta$ in powers of $\kappa^2$ is reasonable on the basis of investigations of the analytic properties of scattering amplitudes which were studied in Section 6.5.

**Convergence of the Low-Energy Expansion**

To see how this comes about, let us express the scattering amplitude $T_l(\kappa)$ in terms of $\mathbf{f}_l(\kappa)$, using Eqs. (51$a$) and (260).

$$T_l(\kappa) = -\frac{e^{i\delta_l} \sin \delta_l}{\pi \kappa M_r} = -\frac{\mathbf{f}_l(\kappa) - \mathbf{f}_l(-\kappa)}{2\pi i \kappa M_r \mathbf{f}_l(-\kappa)}. \tag{297}$$

In the general case we cannot extend this function off the real axis, since we know nothing about the analyticity of $\mathbf{f}_l(\kappa)$ in the upper half-$\kappa$-plane. This precludes any meaningful discussion of an expansion of $T_l(\kappa)$, or a function thereof, in powers of $\kappa$. We therefore restrict ourselves to the case of a potential satisfying the condition (228) and thus decreasing at least as fast as $e^{-\mu r}$ for large $r$. As we saw in the preceding section, this implies that $\mathbf{f}_l(\kappa)$ will be analytic for $\text{Im }\kappa < \mu/2$.

The energy plane will be convenient for our discussion. We write, then,

$$-\pi M_r T_l(\kappa) = \frac{N_l(\epsilon)}{D_l(\epsilon)}, \tag{298}$$

where $D_l(\epsilon)$ was defined in Eq. (281) and

$$N_l(\epsilon) = \frac{\mathbf{f}_l(\kappa) - \mathbf{f}_l(-\kappa)}{2i\kappa}. \tag{299}$$

Now when $\kappa \to \kappa e^{i\pi}$, we see that $N_l(\epsilon)$ is unchanged. Since $\mathbf{f}(\kappa)$ is analytic for $\text{Im }\kappa < \mu/2$, it therefore follows that $N_l(\epsilon)$ is analytic in the $\epsilon$-plane within the circle defined by $|\epsilon| < (\mu/2)^2/2M_r$. For real $\epsilon$, we obtain

$$-\frac{1}{\pi M_r T_l(\kappa)} = \frac{D_l(\epsilon)}{N_l(\epsilon)} = \kappa \cot \delta_l(\kappa) - i\kappa \tag{300}$$

and observe that $N_l(\epsilon)$ is real. Thus it follows from Eq. (281) that

$$\kappa N_l(\epsilon) \cot \delta_l = \text{Re } D_l(\epsilon) = 1 + \frac{P}{\pi}\int_0^\infty d\epsilon' \frac{\text{Im } D_l(\epsilon')}{\epsilon' - \epsilon}, \tag{301}$$

for real $\epsilon$. In addition, in the integrand here we have

$$\text{Im } D_l(\epsilon) = N_l(\epsilon) \text{ Im } \frac{1}{-\pi M_r T_l} = -\kappa N_l(\epsilon), \tag{302}$$

so

$$\begin{aligned}\kappa N_l(\epsilon) \cot \delta_l &= 1 - \frac{P}{\pi}\int_0^\infty d\epsilon \frac{\kappa' N_l(\epsilon')}{\epsilon' - \epsilon} \\ &= 1 - \frac{1}{\pi}\int_0^\infty d\epsilon' \frac{\kappa' N_l(\epsilon')}{\epsilon' - \epsilon - i\eta} + i\kappa N_l(\epsilon).\end{aligned} \tag{303}$$

The purpose of the second writing is to see explicitly what is the domain of analyticity of the left-hand side. The integral is analytic everywhere except along the real axis. Now, as $\epsilon + i\eta \to \epsilon\, e^{2\pi i} + i\eta$, the integral term increases by $i[2\kappa N_l(\epsilon)]$. At the same time, $\kappa N_l(\epsilon) \to -\kappa N_l(\epsilon)$, so the right-hand side of (303) is unchanged. Therefore $\kappa N_l \cot \delta_l$ is analytic within the domain $|\epsilon| < (\mu/2)^2/2M_r$ and may be expanded in a power series in $\epsilon$, convergent within that domain:

$$\kappa N_l \cot \delta_l = a_0 + a_1\epsilon + a_2\epsilon^2 + \cdots, \tag{304a}$$

where the $a$'s are numerical coefficients. For example,

$$a_0 = 1 - \frac{1}{\pi}\int_0^\infty d\epsilon' \frac{\kappa'}{\epsilon'} N_l(\epsilon'). \tag{304b}$$

Within the same domain we may expand $N_l$ in a series in $\epsilon$. According to Eqs. (294) and (298), this has the form

$$N_l(\epsilon) = \kappa^{2l}(b_0 + b_1\epsilon + b_2\epsilon^2 + \cdots), \tag{305}$$

except for the exceptional case of a zero-energy resonance [for which $D_l(0) = 0$]. On substituting Eq. (305) into (304$a$) and reexpanding, we obtain

$$\kappa^{2l+1} \cot \delta_l = \alpha_0 + \alpha_1\epsilon + \alpha_2\epsilon^2 + \cdots, \tag{306}$$

where $\alpha_0, \alpha_1, \ldots$ are constants and the expansion converges for $|\epsilon| < (\mu/2)^2(1/2M_r)$.

Equation (306) has been of great use in the analysis of low-energy nuclear and elementary particle interactions. The special significance of the function $\kappa^{2l+1} \cot \delta_l$ is seen to depend on its being analytic near $\epsilon = 0$. From simple dimensional considerations we would be inclined to conclude that the coefficients $\alpha_i$ are approximately related by

$$\frac{\alpha_i}{\alpha_{i+1}} = \mathcal{O}\left(\frac{(\mu/2)^2}{2M_r}\right),$$

for $i \geq 1$. (The coefficient $\alpha_0$ depends sensitively on details of the interaction.)

### Evaluation of the Effective Range

The series (306) may be studied empirically, using observed scattering cross sections to deduce the coefficients $\alpha_i$;[71] and it may, of course, be evaluated directly from the solutions of Eq. (110) when we have chosen a potential $v(r)$. To illustrate the evaluation we shall discuss in some detail the case for $l = 0$, which is the one most frequently encountered.[72] Let us first define a new function

$$\bar{w}(r) = \sqrt{\pi/2}[\sin \delta_0(\kappa)]^{-1} w_0(\kappa, r), \tag{307a}$$

which [see Eq. (113)] has the asymptotic form

$$\bar{w} \to \frac{\sin(\kappa r + \delta_0)}{\sin \delta_0} \tag{307b}$$

[71] Careful studies of neutron-proton and proton-proton scattering have been made in terms of the expansion (306). See, for example, Refs. [14] and [35].

[72] We shall here follow the discussion of Chew and Goldberger [32].

as $r \to \infty$. Therefore

$$\left[\frac{d^2}{dr^2} - v(r)\right]\bar{w} = -\kappa^2\bar{w}. \tag{308}$$

Consider a second solution $\bar{w}_1$ of Eq. (308), corresponding to a momentum $\kappa_1$.

$$\bar{w}_1(r) = \sqrt{\pi/2}[\sin \delta_0(\kappa_1)]^{-1}w_0(\kappa_1, r).$$

Then

$$\frac{d}{dr} W_r[\bar{w}_1, \bar{w}] = (\kappa_1{}^2 - \kappa^2)\bar{w}_1\bar{w}, \tag{309}$$

where the Wronskian $W_r$ is $\bar{w}_1\bar{w}' - \bar{w}'_1\bar{w}$, evaluated at $r$. Since $\bar{w}$ is the regular solution of Eq. (308), $\bar{w}$ and $\bar{w}_1$ vanish at $r = 0$, and we may write

$$W_r[\bar{w}_1, \bar{w}] = (\kappa_1{}^2 - \kappa^2)\int_0^r dr'\bar{w}_1(r')\bar{w}(r'). \tag{310}$$

Now let us introduce solutions $\bar{u}$ and $\bar{u}_1$ of the equation

$$\frac{d^2}{dr^2}\bar{u} = -\kappa^2\bar{u}, \tag{311}$$

such that $\bar{w} \to \bar{u}$ and $\bar{w}_1 \to \bar{u}_1$ as $r \to \infty$. That is,

$$\bar{u}(r) = \frac{\sin[\kappa r + \delta_0(\kappa)]}{\sin \delta_0(\kappa)}, \tag{312a}$$

with a similar expression for $\bar{u}_1$ in terms of $\kappa_1$. Evidently,

$$\begin{aligned} \bar{u}(0) &= 1, \\ \bar{u}'(0) &= \kappa \cot \delta_0(\kappa). \end{aligned} \tag{312b}$$

On following the procedure which led to Eq. (310), we find

$$W_r[\bar{u}_1, \bar{u}] - W_0[\bar{u}_1, \bar{u}] = (\kappa_1{}^2 - \kappa^2)\int^r dr'\bar{u}_1(r')\bar{u}(r'). \tag{313}$$

If we subtract Eq. (313) from Eq. (310) and set $r = \infty$, there results

$$W_0[\bar{u}_1, \bar{u}] = (\kappa_1{}^2 - \kappa^2)\int_0^\infty dr[\bar{w}_1(r)\bar{w}(r) - \bar{u}_1(r)\bar{u}(r)], \tag{314}$$

since $W_r[\bar{w}_1, \bar{w}] - W_r[\bar{u}_1, \bar{u}] = 0$ as $r \to \infty$. The Wronskian $W_0[\bar{u}_1, \bar{u}]$ may be evaluated using Eqs. (312*b*), and Eq. (314) reduces to

$$\kappa \cot \delta_0(\kappa) - \kappa_1 \cot \delta_0(\kappa_1) = (\kappa^2 - \kappa_1{}^2)\int_0^\infty dr[\bar{u}_1(r)\bar{u}(r) - \bar{w}_1(r)\bar{w}(r)]. \tag{315}$$

From this equation we obtain two rigorous and important formulas. Differentiate with respect to $\kappa^2$ and then set $\kappa_1 = \kappa$.

$$\frac{d}{d\kappa^2}\,[\kappa \cot \delta_0(\kappa)] = \int_0^\infty dr[\bar{u}^2(r) - \bar{w}^2(r)]. \tag{316}$$

Next set $\kappa_1 = 0$ and define

$$\lim_{\kappa_1 \to 0} \kappa_1 \cot \delta_0(\kappa_1) \equiv \frac{1}{a}\,. \tag{317}$$

From Eq. (315) we then find

$$\kappa \cot \delta_0(\kappa) = \frac{1}{a} + \kappa^2 \int_0^\infty dr[\bar{u}_0(r)\bar{u}(r) - \bar{w}_0(r)\bar{w}(r)], \tag{318}$$

where $\bar{u}_0$ is the value of $\bar{u}_1$ and $\bar{w}_0$ is the value of $\bar{w}_1$ for $\kappa_1 = 0$.

The quantity $a$ is called the *scattering length*. It is just the value of the scattering amplitude $f(\hat{\boldsymbol{\kappa}}_f \cdot \hat{\boldsymbol{\kappa}})$ at zero energy. To see this we must evaluate Eq. (70$a$) at $\kappa = 0$. Because of the condition (295), $f = (\sin \delta_0)/\kappa = a$ at $\kappa = 0$, and the differential cross section (73$a$) is

$$\frac{d\sigma}{d\Omega} = a^2. \tag{319}$$

It follows directly from the Levinson theorem that $a$ is positive if the system has an even number of bound states for $l = 0$ (including the case of no bound states) and that $a$ is negative for an odd number of bound states. That is, in the notation of Eq. (286),

$$a = (-1)^{n_{b,0}}|a|, \tag{320}$$

unless there is a zero energy resonance, in which case $a = \infty$.

In retrospect we see why the choice (307$a$) for the wave function normalization was made. With Eq. (306) we obtained the result that $\kappa \cot \delta_0$ may be expanded in a series in $\epsilon$ (that is, in $\kappa^2$). With the function (307$a$) we obtain an expression for $\kappa \cot \delta_0$ directly, without the necessity of carrying along a numerical factor in our equations. The relation of $\bar{w}$ to the function (220) may, for example, be obtained directly using Eq. (237) and (307$a$). The asymptotic form of $\bar{w}_0$ is seen to be

$$\lim_{r \to \infty} \bar{w}_0 = 1 + \frac{r}{a}\,. \tag{321}$$

If we carry out the differentiation indicated in Eq. (316), we obtain

$$\frac{d\delta_0}{d\kappa} = \frac{\sin 2\delta_0}{2\kappa} + 2\sin^2 \delta_0 \int_0^\infty dr[\bar{w}^2 - \bar{u}^2]. \tag{322a}$$

Let us suppose that the potential $v(r)$ is negligibly small outside the range $r = d$. Then $\bar{w} \approx \bar{u}$ for $r > d$, and we may set the upper limit of the integral in Eq. (322*a*) equal to $d$. The integral over $\bar{u}^2$ may be evaluated using Eq. (312*a*), and we are led to

$$\begin{aligned} \frac{d\delta_0}{d\kappa} &= -d + \frac{\sin [2(\kappa d + \delta_0)]}{2\kappa} + 2 \sin^2 \delta_0 \int_0^d dr \bar{w}^2(r) \\ &\geq -d + \frac{\sin [2(\kappa d + \delta_0)]}{2\kappa} \geq -\left(d + \frac{1}{2\kappa}\right), \end{aligned} \tag{322b}$$

a relation first given by Wigner [36].[73] Wigner's derivation is much more general, since it assumes only that the interaction vanishes outside $r = d$ and does not even require the existence of a potential in the sense in which we use it.

If we define[74]

$$r_e \equiv 2 \int_0^\infty dr[\bar{u}_0^2(r) - \bar{w}_0^2(r)], \tag{323}$$

we see from Eq. (318) that

$$\kappa \cot \delta_0(\kappa) = \frac{1}{a} + \frac{1}{2} r_e \kappa^2 + \mathcal{O}(\kappa^4). \tag{324}$$

The quantity $r_e$ is called the *effective range*. It may be obtained, along with the scattering length $a$, by integrating the zero-energy Schrödinger equation, subject to the boundary condition (321).

Once the zero-energy solution $\bar{w}_0$ is known, the computation of the higher-order terms in the expansion of $\kappa \cot \delta_0$ may be done systematically. Consistent power series expansions for both $\bar{w}$ and $\bar{u}$ are required. This may be done for $\bar{u}$ explicitly in terms of the assumed expansion of $\kappa \cot \delta_0$. Let us assume, then, that we have a series of the form[75]

$$\begin{aligned} \bar{u} &= \sum_{n=0}^{\infty} (\kappa^2)^n \bar{u}_n, \\ \bar{w} &= \sum_{n=0}^{\infty} (\kappa^2)^n \bar{w}_n. \end{aligned} \tag{325}$$

[73] We shall return in Chapter 8 to this result and to a description of its physical significance.

[74] It follows from Eq. (306) that this integral exists.

[75] The existence of these expansions may be established from that of $\phi(\kappa, r)$ [Eq. (220)], which is an entire function of $\kappa$. The expansions (325) are not expected to converge outside the domain of analyticity of the Jost function.

If these expressions are substituted in Eq. (318), there results

$$\kappa \cot \delta_0(\kappa) = \frac{1}{a} + \sum_{n=0}^{\infty} (\kappa^2)^{n+1} X_{0,n}, \tag{326}$$

where we define the integrals

$$X_{l,m} \equiv \int_0^\infty dr(\bar{u}_l \bar{u}_m - \bar{w}_l \bar{w}_m) = X_{m,l}, \tag{327}$$

for $l, m = 0, 1, 2, \ldots$. Evidently $2X_{0,0} = r_e$, as defined by Eq. (323).

The computation of Eq. (326) may be simplified with the use of a set of identities which may be obtained from Eq. (316). If we insert the expansion (326) into the left side of (316) and the expansions (325) into the right side of (316) and equate corresponding coefficients of $\kappa^2$, we find

$$(n + 1)X_{0,n} = \sum_{l=0}^{n} X_{l,n-l}. \tag{328}$$

Because of the symmetry of $X_{l,m} = X_{m,l}$, this sum may be simplified. For $n$ odd, we obtain

$$X_{0,2m+1} = \frac{1}{m} \sum_{l=1}^{m} X_{l,2m+1-l}. \tag{329a}$$

For $n$ even, we find

$$X_{0,2m} = \frac{1}{2m - 1}\left(X_{m,m} + 2\sum_{l=1}^{m-1} X_{l,2m-l}\right). \tag{329b}$$

Therefore, if we know $\bar{u}_m$ and $\bar{w}_m$ for $m < m_0$, we may obtain $X_{0,m_0+1}$ if $m_0 \geq 1$.

If there happens to be a bound state[76] of the system with its binding energy $\epsilon_b = -|\epsilon_b|$, within the domain of convergence of the expansion (306) we may relate the scattering parameters to the binding energy. The Schrödinger equation is then

$$\left[\frac{d^2}{dr^2} - v(r)\right]\bar{w}_b = \alpha^2 \bar{w}_b, \tag{330}$$

where $\alpha^2 = 2M_r|\epsilon_b|$. This is formally the same as Eq. (308) with $\kappa^2 = -\alpha^2$. The corresponding solution of Eq. (311) will be called $\bar{u}_b$. Now, the wave functions $\bar{w}_b$ and $\bar{u}_b$ may be obtained by analytic continuation of $\bar{w}$ and $\bar{u}$ into the positive imaginary $\kappa$-plane to $\kappa = i\alpha$ (we have assumed that the point $\epsilon = -\alpha^2/2M_r$ lies within the domain of analyticity, or that $\alpha < \mu/2$).

[76] An important example is that of the triplet $S$-state of the neutron-proton system, for which the deutron represents the bound state.

To do this we rewrite Eq. (312$a$) as

$$\bar{u} = \frac{e^{i\kappa r} - e^{-i\kappa r}(1/S_0)}{1 - (1/S_0)}, \tag{331}$$

where $S_0(\kappa) = e^{2i\delta_0}$ is the $S$-matrix element. Now, in the last section we saw that the point $\kappa = i\alpha$ corresponds to a pole of the $S$-matrix, so we set $\kappa = i\alpha$ in Eq. (331) to obtain

$$\bar{u}_b = e^{-\alpha r}. \tag{332a}$$

Then since $\bar{w} = \bar{u}$, as $r \to \infty$, we see that

$$\lim_{r\to\infty} \bar{w}_b = e^{-\alpha r}. \tag{332b}$$

If we now set $\kappa_1 = 0$ and $\kappa = i\alpha$ in Eq. (314), there results [see Eq. (321)]

$$W_0\left(1 + \frac{r}{a}, e^{-\alpha r}\right) = \alpha^2 \int_0^\infty dr[\bar{w}_0(r)\bar{w}_b(r) - \bar{u}_0(r)\, e^{-\alpha r}].$$

Evaluation of the Wronskian then gives directly

$$\frac{1}{a} = -\alpha - \alpha^2 \int_0^\infty dr[\bar{w}_0(r)\bar{w}_b(r) - \bar{u}_0(r)\, e^{-\alpha r}]. \tag{333}$$

Substitution of the expansions (325) for $\bar{w}_b$ and $e^{-\alpha r}$ into this expression and insertion of the result into Eq. (326) gives us the relation

$$\kappa \cot \delta_0(\kappa) = -\alpha + \sum_{n=0}^{\infty} [(\kappa^2)^{n+1} - (-\alpha^2)^{n+1}]X_{0,n}. \tag{334}$$

It is instructive to obtain the expansion (334) using the general theory of Section 6.5. There we showed that the bound states correspond to poles of the $S$-matrix in the positive imaginary $\kappa$-plane (giving the physical sheet in the $\epsilon$-plane). Now

$$S_0(\kappa) = \frac{\kappa \cot \delta_0 + i\kappa}{\kappa \cot \delta_0 - i\kappa},$$

so the positions of the bound states are determined by setting $\kappa \cot \delta_0 = i\kappa$ and solving this equation for $\kappa$. If we insert the expansion (326) into this equation and set $\kappa = i\alpha$, we obtain

$$\frac{1}{a} = -\alpha - \sum_{n=0}^{\infty} (-\alpha^2)^{n+1} X_{0,n}. \tag{335}$$

This expression for $a^{-1}$, when inserted into Eq. (326), leads immediately to Eq. (334).

**Low-Energy Scattering in the Presence of a Coulomb Potential**

The corresponding treatment of the low-energy scattering for charged particles is slightly more involved because of the long-range Coulomb potential. We shall illustrate this with the description of $S$-wave proton-proton scattering, following the analysis of Chew and Goldberger [32].[77] The wave equation (187) is now

$$\left[\frac{d^2}{dr^2} - \frac{\beta}{r} - v_n(r)\right]\bar{w}(r) = -\kappa^2\bar{w}(r), \tag{336}$$

where

$$\beta = \frac{Me^2}{\hbar^2}, \tag{337a}$$

and $M$ and $e$ are the mass and charge, respectively, of a proton. We note that $\beta^{-1}$ is the proton "Bohr radius" and that $\beta$ is related to the parameter $\xi$ of Eq. (172) by

$$\beta = 2\kappa\xi. \tag{337b}$$

In accordance with the discussion of Section 6.4, p. 259, we have dropped the Coulomb cutoff factor $g_c(r)$, since we need integrate Eq. (336) only out beyond the range $d$ of the nuclear force to find the nuclear phase shift $\nu_0$ [using the expression (188)].

The normalization of the wave function $\bar{w}$ is defined by the equation

$$\bar{w}(r) = \sqrt{\pi/2}\,\frac{C_0}{\sin \nu_0}\, w_0(\kappa, r), \tag{338a}$$

where $w_0$ is the normalized wave function described by Eqs. (189) and (190). The constant $C_0$ is defined by Eq. (186$a$). Using the asymptotic forms (188), (189), and (190) for $w_0$, we see that

$$\bar{w} \to \bar{u}(r) \equiv \sqrt{\pi/2}\left[C_0^{\,2}\kappa \cot \nu_0\left(\frac{F_0}{C_0\kappa}\right) - C_0G_0\right], \tag{338b}$$

for $r \gg d$, the range of the nuclear potential $v_n(r)$. Here $F_0$ and $G_0$ are the Coulomb functions as defined by Eqs. (186$b$). The function $\bar{u}(r)$ is that solution of Eq. (336), with $v_n$ equal to zero, which agrees asymptotically with $\bar{w}$.

On introducing a second set of solutions $\bar{w}_1$ and $\bar{u}_1$, corresponding to a momentum $\kappa_1$, we may again derive the Wronskian relation (314). On setting $\kappa_1 = 0$, we obtain from Eq. (338$b$)

$$\bar{u}_1 \to \bar{u}_0 = \bar{K}(r) + \frac{\bar{I}(r)}{a}, \tag{339a}$$

[77] See also Refs. [14] and [15].

where

$$\bar{I}(r) = \beta^{-1}(\beta r)^{1/2} I_1[2(\beta r)^{1/2}],$$
$$\bar{K}(r) = 2(\beta r)^{1/2} K_1[2(\beta r)^{1/2}]. \qquad (339b)$$

Here $a$ is a constant and $I_1$ and $K_1$ are modified Bessel functions of the second kind of pure imaginary argument.[78] Near $r = 0$, $\bar{I}$ and $\bar{K}$ have the expansions

$$\bar{I}(r) = r + \beta \frac{r^2}{2} + \cdots,$$
$$\bar{K}(r) = 1 + \beta r[\ln(\beta r) + 2\gamma - 1] + \cdots. \qquad (339c)$$

Equation (339*a*) may be verified by inserting the expansions (186*b*) into (338*b*), going to the limit $\kappa = 0$, and comparing with the expansions (339*c*). The constant $a$ is then seen to be given by

$$\frac{1}{a} = \lim_{\kappa \to 0} [C_0{}^2 \kappa \cot \nu_0 + \beta h(\xi)], \qquad (340)$$

where $h(\xi)$ is defined by Eq. (186*c*).

The Wronskian $W_0[\bar{u}_0, \bar{u}]$ may be evaluated from Eqs. (338*b*) and (339*a*). This gives

$$W_0[\bar{u}_0, \bar{u}] = C_0{}^2 \kappa \cot \nu_0(\kappa) - \frac{1}{a} + \beta h(\xi). \qquad (341)$$

Thus in place of Eq. (318) we have

$$C_0{}^2 \kappa \cot \nu_0 + \beta h(\xi) = \frac{1}{a} + \kappa^2 \int_0^\infty dr[\bar{u}_0(r)\bar{u}(r) - \bar{w}_0(r)\bar{w}(r)]. \qquad (342)$$

In the lowest approximation, we obtain the relation analogous to Eq. (324).

$$C_0{}^2 \kappa \cot \nu_0 + \beta h(\xi) \approx \frac{1}{a} + \frac{1}{2} r_e \kappa^2, \qquad (343)$$

where

$$r_e = 2\int_0^\infty dr[\bar{u}_0{}^2(r) - \bar{w}_0{}^2(r)]. \qquad (344)$$

This is a remarkable result in the sense that $C_0{}^2$ and $h$ are quite complicated functions of the energy. The left-hand side of Eq. (343) is, in fact, precisely the expression that Breit and his collaborators first observed to be energy independent for a zero-range interaction. The presence of the Coulomb potential precludes an appeal to the Jost function to state precisely the radius of convergence of the series, but in all probability it is

[78] More specifically, $I_\nu(Z) = e^{-i\pi\nu/2} J_\nu(Z\, e^{\pi i/2})$, for $-\pi < \arg Z < \pi/2$, and $K_\nu(Z) = (\pi i/2)\, e^{i\pi\nu/2} H_\nu^{(1)}(iZ)$, where $J_\nu$ and $H_\nu^{(1)}$ were defined in Section 6.3. See G. N. Watson [7].

the same for the non-Coulomb case. That is, if the nuclear potential has an exponential range $d = \mu^{-1}$, the critical value is expected to be given by $\kappa^2 = (\mu/2)^2$. In any case, the expression (343) has been of great practical value in studying proton-proton scattering.

Let us next consider the possibility of treating the Coulomb field as a perturbation. This may seem silly because in the asymptotic region the Coulomb potential is of course the dominant interaction. Nevertheless, it is possible to arrange the calculation in such a way that only the Coulomb potential within the range of the nuclear potential need be considered. For $r < d$, the Coulomb potential is about two orders of magnitude weaker than the nuclear potential.

We rewrite Eq. (336) as

$$\left[\frac{d^2}{dr^2} + \kappa^2 - v_n(r)\right]\bar{w} = \frac{\beta}{r}\,\bar{w}, \tag{345}$$

to indicate that the $\beta/r$ term is to be considered as a perturbation. Now, if we set $\beta = 0$, Eq. (345) has a solution which we call $\bar{w}^{(0)}$. The normalization of $\bar{w}^{(0)}$ is so chosen that

$$\lim_{r\to\infty} \bar{w}^{(0)} = \bar{u}^{(0)} \equiv \kappa \cot \nu_0^{(0)} \frac{\sin \kappa r}{\kappa} + \cos \kappa r, \tag{346}$$

which has the form of Eq. (307*b*). The quantity $\nu_0^{(0)}(\kappa)$ is, of course, just the nuclear phase shift, obtained by dropping the Coulomb interaction.

The same procedure which led to Eq. (314) now yields

$$W_{r_0}[\bar{u}^{(0)}, \bar{u}] = -\beta \int_{r_0}^{\infty} \frac{dr}{r}\,[\bar{u}^{(0)}(r)\bar{u}(r) - \bar{w}^{(0)}(r)\bar{w}(r)], \tag{347}$$

where the singularity of the perturbation at the origin requires a finite lower limit $r_0$, which may ultimately be allowed to go to zero. Evaluation of the Wronskian for small $r_0$ leads to

$$\begin{aligned} C_0{}^2\kappa \cot \nu_0 + \beta h = \kappa \cot \nu_0^{(0)} &- \beta \int_{r_0}^{\infty} \frac{dr}{r}\,(\bar{u}^{(0)}\bar{u} - \bar{w}^{(0)}\bar{w}) \\ &- \beta[\ln \beta r_0 + 2\gamma]. \end{aligned} \tag{348}$$

The limit as $r_0 \to 0$ may now be taken, since the logarithmic singularity of the integral is canceled by that of the term $-\beta \ln \beta r_0$.

We repeat that $\cot \nu_0^{(0)}$ is obtained from the nuclear potential *only* and is consequently given by the expansion (326). The explicit modification of this expansion by the Coulomb interaction is given by the last two terms on the right-hand side of Eq. (348), since $\bar{w}$ and $\bar{u}$ may be expanded in powers of $\beta$.

As an application of our discussion, we consider the scattering of low-energy neutrons by protons. When only $S$-wave scattering need be considered (that is, $l = 0$),[79] the scattering amplitude is, according to Eq. (70$a$),

$$f(\hat{\boldsymbol{\varkappa}}_f \cdot \hat{\boldsymbol{\varkappa}}) = \frac{1}{\kappa} e^{i\delta_0} \sin \delta_0. \tag{349}$$

Since this is independent of the scattering angle, the total cross section is

$$\sigma = 4\pi \, |f|^2 = \frac{4\pi}{\kappa^2} \sin^2 \delta_0 = \frac{4\pi}{\kappa^2 \cot^2 \delta_0 + \kappa^2} \, . \tag{350a}$$

If we consider $\kappa^2$ to be sufficiently small, we may drop terms of order $\kappa^4$ in Eq. (324) and substitute the resulting expression for $\cot \delta_0$ into Eq. (350$a$) to obtain

$$\sigma = \frac{4\pi a^2}{(1 + r_e a\kappa^2/2)^2 + (a\kappa)^2} \, . \tag{350b}$$

Because of the spin dependence of nuclear forces, we must distinguish between singlet ($S = 0$) and triplet ($S = 1$) scattering. For the former we write $a^{(0)}$, $r_e^{(0)}$, and $\sigma^{(0)}$; for the latter we write $a^{(1)}$, $r_e^{(1)}$, and $\sigma^{(1)}$, in the notation of Section 6.2, p. 238. Now the triplet $S$-states include the bound (deuteron) state. Therefore, we may use Eq. (335) to write

$$\frac{1}{a^{(1)}} \cong -\alpha + \tfrac{1}{2} r_e^{(1)} \alpha^2, \tag{351}$$

where $\alpha = [(M/\hbar^2)\epsilon_D]^{1/2}$ and $\epsilon_D = 2.23$ Mev is the binding energy of the deuteron. For scattering with a random mixture of initial spins, we have

$$\bar{\sigma} = \tfrac{1}{4}\sigma^{(0)} + \tfrac{3}{4}\sigma^{(1)},$$

in the notation of Eq. (94$b$).

Numerical values for the scattering lengths and effective ranges have been deduced[80] from experimental studies of the scattering. Approximate values are

$$\begin{aligned} a^{(0)} &= 2.4 \times 10^{-12} \text{ cm}, \\ a^{(1)} &= -5.4 \times 10^{-13} \text{ cm}, \\ r_e^{(0)} &\approx 2.5 \times 10^{-13} \text{ cm}, \\ r_e^{(1)} &= 1.7 \times 10^{-13} \text{ cm}. \end{aligned} \tag{352}$$

[79] This is a good approximation when the barycentric kinetic energy is less than 5 Mev.

[80] Determination of the low-energy scattering parameters is reviewed, for example, by Blatt and Weisskopf [37]. See also Refs. [14] and [35].

[The negative value of $a^{(1)}$ is expected from the Levinson theorem, as is clear from Eqs. (320) and (351).]

The term of order $\kappa^4$ in the expansion (326) has been studied by Blatt and Jackson [35]. They write

$$\kappa \cot \delta_0 = \frac{1}{a} + \frac{1}{2} r_e \kappa^2 - P r_e^{\,3} \kappa^4 + \mathcal{O}(\kappa^6). \tag{353}$$

The dimensionless parameter $P$ was found to vary between $-0.05$ and 0.15. Now our basic expansion (306) is expected to converge only for $\epsilon < 5$ Mev. At the limit of convergence of the expansion, the $\kappa^4$ term appears therefore to be surprisingly small ($\approx 0.1$ the preceding term).

For proton-proton scattering, only the singlet state occurs for $l = 0$, because of the Pauli principle. The symmetrized scattering amplitude, as given by Eq. (215), is thus

$$f^{(0)}_{\mathscr{S}\,\mathrm{nuc}} = \frac{2}{\kappa} e^{i(2\sigma_0 + \nu_0^{(0)})} \sin \nu_0^{(0)}, \tag{354}$$

where $\nu_0^{(0)}$ is the nuclear scattering phase shift for the singlet $S$-state. This may be evaluated from Eq. (343) in terms of the proton-proton "scattering length" $a(pp)$ and effective range $r_e(pp)$. The differential scattering cross section may then be obtained from Eqs. (216) and (217). (The triplet amplitude $f^{(1)}_{\mathscr{S}\,\mathrm{nuc}}$ may be neglected at low energies, say less than 5 Mev.)

According to the postulate of *charge independence*, the nuclear interaction between any two nucleons is the same for a given state of spin and angular momentum, independently of whether the particles are neutrons or protons. Therefore, the proton-proton "phase shift" $\nu_0^{(0)}$ in Eq. (348) should be identical with the neutron-proton singlet $S$-state phase shift $\delta_0^{(0)}$. Since the last two terms on the right in Eq. (348) are rather small, we expect the conditions $r_e(pp) \approx r_e^{(0)}$ and $a(pp) \approx a^{(0)}$ to be valid. Furthermore, the two correction terms in Eq. (348) may be evaluated to give a quantitative relation between $r_e(pp)$ and $r_e^{(0)}$ and between $a(pp)$ and $a^{(0)}$. This has permitted a direct test of the charge independence hypothesis.[81,82]

### 6.7 APPROXIMATION METHODS

For the study of most scattering processes of physical interest approximate treatments are necessary. As we have indicated, this is not the case when solving Eq. (110), which can always be integrated by numerical methods. On the other hand, approximate solutions may often be obtained

[81] See, for example, Refs. [14] and [37].

[82] Our expressions for the scattering cross sections are unmodified if we employ the isotopic spin formalism and antisymmetrize the neutron-proton wave functions (see Chapter 7).

much more easily than exact solutions and may be as accurate as the physical situation warrants. Furthermore, the various approximate treatments of Eq. (110) serve as a convenient introduction to similar methods for handling much more difficult problems.

In this section we shall discuss three classes of approximations used in scattering theory. The most often encountered of these is the Born expansion. The study of this expansion provides a well-defined mathematical problem, which has been investigated in some detail. A second class of approximation methods is that making use of variational principles. They are often very effective (especially for more complex problems than those now being considered), but they require care in their application, since the error incurred is often difficult to estimate. Finally, we shall describe the WKB and eikonal approximations, which have been useful in nuclear and atomic physics.

It may be well to emphasize again that each of these methods may be applied (under suitable conditions) to much more complex problems than that of simple potential scattering.

**Use of Green's Functions**

The Green's function (or resolvent) for the Hamiltonian operator is important for both theoretical studies and the actual solution of the Schrödinger equation. Although we have frequently used Green's functions in previous chapters to define scattering operators and wave functions, it is appropriate to give a somewhat more detailed description in this section.

Continuing to use our accepted notation, we consider two interacting elementary particles and let $H = K + V$ be the Hamiltonian describing their dynamics. Then, for complex $E$, we have the two familiar Green's functions

$$\mathscr{G}_0(E) = \frac{1}{E - K}, \qquad \mathscr{G}(E) = \frac{1}{E - H}. \tag{355a}$$

With the use of the algebraic relation (3.77), we may relate these by the integral equation

$$\frac{1}{E-H} = \frac{1}{E-K} + \frac{1}{E-K} V \frac{1}{E-H},$$

or

$$\mathscr{G}(E) = \mathscr{G}_0(E) + \mathscr{G}_0(E) V \mathscr{G}(E). \tag{356}$$

Now, if $E$ is considered to approach the real axis, so $E \rightarrow E \pm i\eta$, where $E$

is now real and $\eta$ is an infinitesimal positive number, the expressions (355*a*) become

$$\mathscr{G}_0^{\pm}(E) = \frac{1}{E \pm i\eta - K},$$
$$\mathscr{G}^{\pm}(E) = \frac{1}{E \pm i\eta - H}. \tag{355b}$$

(The notation $\pm$ is understood to imply that $E$ is real.) These are, of course, just the operators which have been used in Chapters 3 and 5 to define the wave functions $\psi^{\pm}$.

From Eq. (3.63) we have

$$\frac{1}{E \pm i\eta - K} = P\left(\frac{1}{E - K}\right) \mp i\pi\delta(E - K),$$

where $P[1/(E - K)]$ means that a principal value is to be taken when evaluating an integral involving $1/(E - K)$. We may therefore define a third Green's function (again for real $E$) as

$$\mathscr{G}_0^{0}(E) = P\left(\frac{1}{E - K}\right) = \frac{1}{2}[\mathscr{G}_0^{+}(E) + \mathscr{G}_0^{-}(E)]. \tag{355c}$$

It will be recalled that this operator appeared in the integral equation defining the reaction matrix.

It follows from the definitions (355*b*) and (355*c*) that

$$[\mathscr{G}_0^{+}(E)]^{\dagger} = \mathscr{G}_0^{-}(E),$$
$$[\mathscr{G}^{+}(E)]^{\dagger} = \mathscr{G}^{-}(E), \tag{357}$$
$$[\mathscr{G}_0^{0}(E)]^{\dagger} = \mathscr{G}_0^{0}(E).$$

Moreover, if $H$ and $K$ are invariant under the operation of time reversal, it is easily seen that

$$\mathscr{G}_0^{-}(E) = T\mathscr{G}_0^{+}(E)T^{-1},$$
$$\mathscr{G}^{-}(E) = T\mathscr{G}^{+}(E)T^{-1}, \tag{358}$$

where $T$ is the time reversal operator [see Eq. (2.122)].

For systems not interacting with external fields, the operator of total momentum $\mathscr{P}$ commutes with both $H$ and $K$, and therefore commutes with the corresponding Green's functions. This permits us to define sub-matrices $G$ on the momentum shell. Let $a \equiv (\mathbf{P}_a, i_a)$ and $b \equiv (\mathbf{P}_b, i_b)$ label two eigenstates of $K$, or of $H$. Here $\mathbf{P}_a$ and $\mathbf{P}_b$ are the respective eigenvalues of $\mathscr{P}$ and $i_a$, $i_b$ label other eigenvariables. Then we may evidently write

$$\langle \mathbf{P}_b, i_b|\, \mathscr{G}_0(E)\, |\mathbf{P}_a, i_a\rangle = \delta(\mathbf{P}_b - \mathbf{P}_a)\langle i_b|\, G_0(E)\, |i_a\rangle,$$
$$\langle \mathbf{P}_b, i_b|\, \mathscr{G}(E)\, |\mathbf{P}_a, i_a\rangle = \delta(\mathbf{P}_b - \mathbf{P}_a)\langle i_b|\, G(E)\, |i_a\rangle. \tag{359}$$

By letting $E$ approach the real axis, we define Green's functions $G^{\pm}$, $G^0$, $G_0^{\pm}$, and $G_0^0$ on the momentum shell. These are frequently more useful than the full Green's functions.

In the remainder of this section we shall consider the $G$'s as operators only in the *barycentric subspace* (as defined in Section 4.6). In this subspace, corresponding to $\mathbf{P}_a = 0$, $K$ has the eigenvalues $\epsilon(\kappa)$ [see Eq. (27)] and the eigenfunctions $\chi_{\varkappa}(\mathbf{r})$ [see Eq. (40)]:[83]

$$K\chi_{\varkappa} = \epsilon(\kappa)\chi_{\varkappa}.$$

Then, if $\epsilon$ is a complex variable, we have

$$\begin{aligned} G_0(\epsilon) &= \frac{1}{\epsilon - K}, \\ G(\epsilon) &= \frac{1}{\epsilon - H}, \end{aligned} \tag{360a}$$

defined as operators in the barycentric subspace. For example,

$$\begin{aligned} \langle \boldsymbol{\varkappa}_2 | \, G_0(\epsilon) \, | \boldsymbol{\varkappa}_1 \rangle &= (\chi_{\varkappa_2}, G_0\chi_{\varkappa_1}) = \delta(\boldsymbol{\varkappa}_2 - \boldsymbol{\varkappa}_1)\frac{1}{\epsilon - \epsilon(\kappa_1)}, \\ \langle \boldsymbol{\varkappa}_2 | \, G(\epsilon) \, | \boldsymbol{\varkappa}_1 \rangle &= \left(\chi_{\varkappa_2}, \frac{1}{\epsilon - H}\chi_{\varkappa_1}\right). \end{aligned} \tag{360b}$$

Several useful relations may be expressed in terms of these operators. It follows from Eqs. (360*a*) that, in analogy to Eq. (356), for real $\epsilon$

$$G^{\pm}(\epsilon) = G_0^{\pm}(\epsilon) + G_0^{\pm}(\epsilon)VG^{\pm}(\epsilon). \tag{361}$$

Again we evidently have

$$\mathscr{T} = V + V\mathscr{G}^+V = \delta(\mathbf{P}_b - \mathbf{P}_a)T = \delta(\mathbf{P}_b - \mathbf{P}_a)(V + VG^+V),$$

so the scattering operator $T$ is given by either of the equations

$$T = V + VG^+V = V + VG_0^+T. \tag{362a}$$

The reaction matrix $R$ [see Eq. (5.179)] may also be written in the barycentric subspace as

$$R = V + VG_0^0R. \tag{362b}$$

The first of Eqs. (362*a*) takes rather a simple form in a coordinate representation, in which $V(r)$ has been assumed (in this chapter) to be diagonal.

$$\langle \mathbf{r} | \, T \, | \mathbf{r}' \rangle = V(r)\delta(\mathbf{r} - \mathbf{r}') + V(r)\langle \mathbf{r} | \, G^+ \, | \mathbf{r}' \rangle V(r'). \tag{363}$$

[83] Since the particle spin, if any, has been assumed to commute with $H$ and $K$, we may simply omit spin labels and spin eigenfunctions without loss of generality.

We shall restrict ourselves to the scattering of nonrelativistic particles in the remainder of this section. Then [see Eq. (98)]

$$\epsilon = \frac{1}{2M_r}\,\kappa^2,$$

$$K\,|\boldsymbol{\varkappa}_1\rangle = \frac{1}{2M_r}\,\kappa_1^{\,2}\,|\boldsymbol{\varkappa}_1\rangle,$$

and

$$\langle\boldsymbol{\varkappa}_2|\,G_0^{\,\pm}\,|\boldsymbol{\varkappa}_1\rangle = 2M_r\delta(\boldsymbol{\varkappa}_2 - \boldsymbol{\varkappa}_1)\,\frac{1}{\kappa^2 \pm i\eta - \kappa_1^{\,2}}\,. \tag{364a}$$

Transformed to coordinate space, this becomes

$$\langle\mathbf{r}|\,G_0^{\pm}\,|\mathbf{r}'\rangle = (2\pi)^{-3}\int d^3\kappa_1\,d^3\kappa_2\,e^{i(\boldsymbol{\varkappa}_2\cdot\mathbf{r} - \boldsymbol{\varkappa}_1\cdot\mathbf{r}')}\langle\boldsymbol{\varkappa}_2|\,G_0^{\pm}\,|\boldsymbol{\varkappa}_1\rangle$$

$$= -\,\frac{2M_r}{4\pi}\,\frac{e^{\pm i\kappa|\mathbf{r}-\mathbf{r}'|}}{|\mathbf{r}-\mathbf{r}'|}\,. \tag{364b}$$

A Green's function may also be constructed for the radial wave equation (110), which we write as

$$\mathscr{D}_l(r)w_l \equiv \left[\frac{d^2}{dr^2} - \frac{l(l+1)}{r^2} + \kappa^2\right]w_l = vw_l, \tag{365}$$

where $v$ is the reduced potential defined by Eq. (104). The Green's function $G_{0l}$ is defined as

$$G_{0l} = \mathscr{D}_l^{-1}, \tag{366a}$$

or by the equation

$$\mathscr{D}_l(r)\langle r|\,G_{0l}\,|r'\rangle = \delta(r - r'). \tag{366b}$$

This quantity may be used to construct a regular solution, call it $w_l^{(0)}(\kappa; r)$, of Eq. (365). We first recall the function $y_l(\kappa r)$ defined by Eq. (122),

$$y_l(\kappa r) = \sqrt{2/\pi}(\kappa r)j_l(\kappa r), \tag{367a}$$

and satisfying the homogeneous equation $\mathscr{D}_l y_l = 0$. Then using Eq. (366$b$) we verify that

$$w_l^{(0)}(\kappa; r) = y_l(\kappa r) + \int_0^\infty \langle r|\,G_{0l}\,|r'\rangle v(r')w_l^{(0)}(\kappa; r')\,dr' \tag{368}$$

is a solution of Eq. (365). Since $w_l^{(0)}$ is specified to be the regular solution of Eq. (365), it must vanish at $r = 0$. This, in turn, implies the boundary condition

$$\langle 0|\,G_{0l}\,|r'\rangle = 0. \tag{369a}$$

In addition to $y_l$ it is convenient to introduce the function

$$z_l(r) = \sqrt{2/\pi}(\kappa r)n_l(\kappa r), \tag{367b}$$

which is an irregular solution of the equation $\mathscr{D}_l z_l = 0$.

Now, since $w_l$ and $w_l^{(0)}$ are both regular solutions of the second-order differential equation (110), they can differ only by a constant factor. For reasons that will become clear, it is convenient to choose this as

$$w_l^{(0)}(\kappa; r) = \frac{w_l(\kappa; r)}{\cos \delta_l}, \tag{370}$$

where $\delta_l$ is the scattering phase shift, as usual. It follows then from Eq. (133*b*) that

$$\lim_{r\to\infty} w_l^{(0)} = y_l - \tan \delta_l z_l. \tag{371}$$

According to Eqs. (50) and (51*a*), the reaction matrix $R_l(\kappa)$ is, for non-relativistic scattering,

$$R_l(\kappa) = -\frac{1}{\pi\kappa M_r} \tan \delta_l = \frac{e^{-i\delta_l}}{\cos \delta_l} T_l(\kappa). \tag{372a}$$

On using Eq. (62*b*) for $T_l$, Eq. (109*a*) for $\psi^+_{\kappa,l}$, and Eq. (370), we may rewrite (372*a*) as

$$R_l(\kappa) = \frac{1}{\kappa^2}\int_0^\infty y_l(\kappa r)V(r)w_l^{(0)}(\kappa; r)\, dr. \tag{372b}$$

An alternate form of this is

$$\kappa \tan \delta_l = -\frac{\pi}{2}\int_0^\infty y_l(\kappa r)v(r)w_l^{(0)}(\kappa; r)\, dr. \tag{372c}$$

Let us now construct the Green's function $\langle r|\, G_{0l}\,|r'\rangle$.[84] First, since $\langle r|\, G_{0l}\,|r'\rangle$ satisfies the equation $\mathscr{D}_l(r)\langle r|\, G_{0l}\,|r'\rangle = 0$, except at $r = r'$, it must be a linear combination of $y_l(\kappa r)$ and $z_l(\kappa r)$. Because of the condition (369), it must be proportional to $y_l(\kappa r)$ only near $r = 0$. On the other hand, it follows from Eqs. (368) and (371) that, for sufficiently large $r$, $\langle r|\, G_{0l}\,|r'\rangle$ must be proportional to $z_l(\kappa r)$ only. Finally, $\langle r|\, G_{0l}\,|r'\rangle$ can fail to satisfy the equation $\mathscr{D}_l G_{0l} = 0$ only at $r = r'$. Therefore the most general form $G_{0l}$ can have is

$$\begin{aligned}\langle r|\, G_{0l}\,|r'\rangle &= y_l(\kappa r)F_>(r') \quad \text{for} \quad r < r' \\ &= z_l(\kappa r)F_<(r') \quad \text{for} \quad r > r',\end{aligned} \tag{373a}$$

where $F_>$ and $F_<$ are two functions of $r'$.

[84] For a general discussion pertaining to the construction of Green's functions, see Courant and Hilbert [30, p. 315], or Morse and Feshbach [8, p. 807, Vol. I].

We next establish that $G_{0l}$ is symmetric:

$$\langle r|\, G_{0l}\,|r'\rangle = \langle r'|\, G_{0l}\,|r\rangle. \tag{369b}$$

To see this we consider the equation

$$\int_0^\infty [\langle r|\, G_{0l}\,|r''\rangle \mathscr{D}_l(r)\langle r|\, G_{0l}\,|r'\rangle - \langle r|\, G_{0l}\,|r'\rangle \mathscr{D}_l(r)\langle r|\, G_{0l}\,|r''\rangle]\, dr$$
$$= \langle r'|\, G_{0l}\,|r''\rangle - \langle r''|\, G_{0l}\,|r'\rangle,$$

which follows from Eq. (366*b*). Now the left-hand side of this equation is

$$\int_0^\infty dr\, \frac{d}{dr}\left(\langle r|\, G_{0l}\,|r''\rangle \frac{d}{dr}\langle r|\, G_{0l}\,|r'\rangle - \langle r|\, G_{0l}\,|r'\rangle \frac{d}{dr}\langle r|\, G_{0l}\,|r''\rangle\right)$$
$$= \left[\langle r|\, G_{0l}\,|r''\rangle \frac{d}{dr}\langle r|\, G_{0l}\,|r'\rangle - \langle r|\, G_{0l}\,|r'\rangle \frac{d}{dr}\langle r|\, G_{0l}\,|r''\rangle\right]_{r=0}^{r=\infty} = 0,$$

which follows directly from Eq. (373*a*). We thus establish the validity of Eq. (369*b*).

It follows from Eq. (369*b*) that

$$\begin{aligned} F_>(r') &= bz_l(\kappa r'), \\ F_<(r') &= by_l(\kappa r'), \end{aligned} \tag{373b}$$

where $b$ is a constant. This constant may be evaluated from Eq. (366*b*). If we integrate this equation over the *very* small interval $r' - \delta < r < r' + \delta$, there results

$$\int_{r'-\delta}^{r'+\delta}\left[\frac{d^2}{dr^2} - \frac{l(l+1)}{r^2} + \kappa^2\right]\langle r|\, G_{0l}\,|r'\rangle\, dr = \int_{r'-\delta}^{r'+\delta}\delta(r - r')\, dr = 1$$
$$= \frac{d}{dr}\langle r|\, G_{0l}\,|r'\rangle\Bigg|_{r=r'-\delta}^{r=r'+\delta} = bW[y_l(\kappa r'), z_l(\kappa r')], \tag{374}$$

where $W[y_l, z_l]$ is the Wronskian of the two solutions $y_l$ and $z_l$. Now the Wronskian is a constant, independent of $r'$. From the asymptotic values of $y_l$ and $z_l$, as $r' \to \infty$, we obtain $W = 2\kappa/\pi$. Therefore we obtain from Eq. (374) $b = \pi/2\kappa$. This permits us to write Eq. (373*a*) in explicit form as

$$\begin{aligned} \langle r|\, G_{0l}\,|r'\rangle &= \frac{\pi}{2\kappa} y_l(\kappa r) z_l(\kappa r') \quad \text{for} \quad r < r', \\ &= \frac{\pi}{2\kappa} z_l(\kappa r) y_l(\kappa r') \quad \text{for} \quad r > r'. \end{aligned} \tag{375}$$

The Green's function (375), when substituted into Eq. (368), leads to the solution $w_l^{(0)}$ which is normalized according to Eq. (370). This normalization is useful when we are using Eq. (372*c*) to evaluate $\tan \delta_l$. For most formal considerations, the $w_l$ are more useful because of the normalization (114).

**The Born Approximation**

The most frequently encountered approximation for solving the Schrödinger equation is that of Born [39]. Here a series expansion is developed in "powers" of the interaction potential. Such expansions may be given in several forms, one of the most simple being obtained on iterating the integral equation (362*a*). For $T(\hat{\boldsymbol{\varkappa}}_f \cdot \hat{\kappa}, \kappa_f, \kappa)$ [see Eq. (35*a*) or (35*b*)] we find

$$T = V + VG_0^+V + VG_0^+VG_0^+V + \cdots = T^{(1)} + T^{(2)} + \cdots, \tag{376a}$$

where

$$\begin{aligned} T^{(1)} &= (\chi_{\varkappa_f}, V\chi_{\varkappa}), \\ T^{(2)} &= \left(\chi_{\varkappa_f}, V\frac{1}{\epsilon + i\eta - K} V\chi_{\varkappa}\right), \end{aligned} \tag{376b}$$

and so on. Here $\epsilon = \kappa^2/2M_r$ and $M_r$ is the reduced mass, as before. A corresponding expansion for the scattering amplitude $f(\hat{\boldsymbol{\varkappa}}_f \cdot \hat{\boldsymbol{\varkappa}})$ is then obtained on substituting the series (376) into [see Eqs. (70)] the equation

$$f(\hat{\boldsymbol{\varkappa}}_f \cdot \hat{\boldsymbol{\varkappa}}) = -(2\pi)^2 M_r T(\hat{\boldsymbol{\varkappa}}_f \cdot \hat{\boldsymbol{\varkappa}}, \kappa, \kappa). \tag{377}$$

When only the first term $T^{(1)}$ is kept in (376*a*), we obtain the "first Born approximation"—often simply called the Born approximation.

To obtain an estimate for the validity of the first Born approximation at *high energies*, we make a somewhat crude approximation to $T^{(2)}$, setting

$$T^{(2)} \approx \frac{1}{\epsilon}(\chi_{\varkappa_f}, V^2\chi_{\varkappa}). \tag{378}$$

Except for very small scattering angles, this expression should be of the correct order of magnitude if $(\chi_{\varkappa_f}, V^2\chi_{\kappa})$ is finite. Then the ratio

$$\left|\frac{T^{(2)}}{T^{(1)}}\right| \approx \frac{1}{\epsilon}\left|\frac{(\chi_{\varkappa_f}, V^2\chi_{\varkappa})}{(\chi_{\varkappa_f}, V\chi_{\varkappa})}\right| \tag{379}$$

provides a rough estimate for the error in the first Born approximation. In particular, we might anticipate that at sufficiently high energies this ratio will become small, implying that the first Born approximation becomes adequate.

To illustrate this, let us consider as an example the Yukawa potential

$$V = \frac{V_0}{\mu}\frac{1}{r}e^{-\mu r}, \tag{380}$$

for which [on using Eq. (378)]

$$T^{(1)} = \frac{V_0}{\mu}\frac{1}{2\pi^2}\frac{1}{(\Delta\kappa)^2 + \mu^2},$$
$$T^{(2)} \approx \frac{1}{\epsilon}\left(\frac{V_0}{\mu}\right)^2 \frac{1}{2\pi^2}\left\{\frac{1}{\Delta\kappa}\sin^{-1}\frac{\Delta\kappa}{[\Delta\kappa^2 + (2\mu)^2]^{1/2}}\right\}, \tag{381a}$$

where $\Delta\kappa \equiv |\boldsymbol{\kappa}_f - \boldsymbol{\kappa}|$. When $\Delta\kappa \gg \mu$, we find that

$$\left|\frac{T^{(2)}}{T^{(1)}}\right| \approx \frac{V_0}{\epsilon}\frac{\Delta\kappa}{\mu} \approx \frac{M_r V_0}{\mu\kappa}. \tag{381b}$$

For the case of nucleon-nucleon scattering, $V_0 \approx 100$ Mev, $2M_r \approx 1000$ Mev, and $\mu \approx 140$ Mev. The expression (381$b$) is not small unless $\epsilon$ is several hundreds of Mev; for such energies a nonrelativistic interaction potential is probably quite inadequate to describe the scattering. The Born approximation is not therefore expected to be useful in studying nucleon-nucleon scattering.[85]

An investigation of the convergence of the series (376) was made by Zemach and Klein [40]. They studied the convergence of the series expansion of Eq. (361) for the Green's function:

$$G^+ = G_0^+ + G_0^+VG_0^+ + G_0^+VG_0^+VG_0^+ + \cdots. \tag{382}$$

On substituting this expansion into Eq. (363), we obtain the series (376$a$)

A set of quantities $\langle\mathbf{r}|\, G_n\, |\mathbf{r}'\rangle$ ($n = 0, 1, 2, \ldots$) may be defined by the equations

$$G_{n+1}^+ = G_0^+VG_n^+, \tag{383a}$$

or, equivalently,

$$G_{n+m+1}^+ = G_n^+VG_m^+ \qquad (n, m = 0, 1, \ldots). \tag{383b}$$

In a coordinate representation this has the form

$$\langle\mathbf{r}|\, G_{n+m+1}^+\, |\mathbf{r}'\rangle = \int \langle\mathbf{r}|\, G_n^+\, |\mathbf{x}\rangle V(x) \langle\mathbf{x}|\, G_m^+\, |\mathbf{r}'\rangle\, d^3x. \tag{383c}$$

Following Zemach and Klein, we next define

$$\langle\mathbf{r}|\, g_n\, |\mathbf{r}'\rangle = \frac{\langle\mathbf{r}|\, G_n^+\, |\mathbf{r}'\rangle}{\langle\mathbf{r}|\, G_0^+\, |\mathbf{r}'\rangle},$$
$$g_0 = 1, \tag{384}$$

and

$$D_{r,r'}(\mathbf{x}) = -\frac{1}{4\pi}\frac{|\mathbf{r} - \mathbf{r}'|}{|\mathbf{r} - \mathbf{x}|\,|\mathbf{r}' - \mathbf{x}|}\, e^{i\kappa[|\mathbf{r}-\mathbf{x}|+|\mathbf{r}'-\mathbf{x}|-|\mathbf{r}'-\mathbf{r}|]}. \tag{385}$$

[85] This conclusion was reached by Jost and Pais [28].

Then using the expression (364$b$) for $G_0^+$ and Eq. (383$c$) we obtain [recalling that $v(x) = 2M_rV(x)$]

$$\langle \mathbf{r}| g_{n+m+1} |\mathbf{r}'\rangle = \int \langle \mathbf{r}| g_n |\mathbf{x}\rangle D_{r,r'}(\mathbf{x}) v(x) \langle \mathbf{x}| g_m |\mathbf{r}'\rangle \, d^3x. \tag{386}$$

Consider now the convergence of the series

$$g = \sum_{n=0}^{\infty} g_n, \tag{387}$$

which is just the algebraic ratio of $G^+$ to $G_0^+$:

$$\langle \mathbf{r}| g |\mathbf{r}'\rangle = \frac{\langle \mathbf{r}| G^+ |\mathbf{r}'\rangle}{\langle \mathbf{r}| G_0^+ |\mathbf{r}'\rangle}.$$

Now let

$$B \equiv \max_{(r)} \left( \frac{1}{2\pi} \int \frac{|v(x)|}{|\mathbf{r} - \mathbf{x}|} \, d^3x \right), \tag{388$a$}$$

maximized by varying $r$, and suppose $B < \infty$. In addition, let

$$Q_n \equiv \max_{(r,r')} (|\langle \mathbf{r}| g_n |\mathbf{r}'\rangle|), \tag{388$b$}$$

defined by varying $\mathbf{r}$ and $\mathbf{r}'$ to obtain the maximum value of $|\langle \mathbf{r}| g_n |\mathbf{r}'\rangle|$. Finally we use the triangle inequality,

$$|\mathbf{r} - \mathbf{r}'| < |\mathbf{r} - \mathbf{x}| + |\mathbf{r}' - \mathbf{x}|,$$

to observe that [see Eq. (385)]

$$|D_{\mathbf{r},\mathbf{r}'}(\mathbf{x})| \le \frac{1}{4\pi} \left( \frac{1}{|\mathbf{r} - \mathbf{x}|} + \frac{1}{|\mathbf{r}' - \mathbf{x}|} \right). \tag{389}$$

The Eqs. (388$a$) and (388$b$) and the inequality (389) permit us to deduce from Eq. (386) the inequality

$$Q_{n+m+1} < Q_n Q_m B. \tag{390}$$

A fundamental step in the study of Zemach and Klein is the argument that

$$\lim_{\kappa \to \infty} Q_1 \to 0$$

for a wide class of potentials.[86] This conclusion is obtained from Eq. (386) for the case that $n = m = 0$.

$$Q_1 = \max_{(\mathbf{r},\mathbf{r}')} \left| \int D_{\mathbf{r},\mathbf{r}'}(\mathbf{x}) v(x) \, d^3x \right|.$$

[86] See the Appendix to Ref. [40].

Because of the exponential in the expression (385), the condition (391) is seen to depend on a three-dimensional analogue of the Reimann-Lebesgue theorem in Fourier analysis.[87] For the validity of (391), we require that $v(x)$ be less singular than $x^{-2}$ near $x = 0$ and decrease to zero faster than $x^{-3}$ as $x \to \infty$.[88]

If we set $m = 1$ in the inequality (390), we obtain

$$Q_{n+2} < Q_n Q_1 B.$$

From this it is seen that

$$Q_{2n} < (Q_1 B)^n, \tag{392}$$
$$Q_{2n+1} < (Q_1 B)^n Q_1.$$

The series (387) therefore converges if the series

$$|g| \leq \sum_{n=0}^{\infty} (Q_1 B)^n (1 + Q_1) = 1 + Q_1 + Q_1 B + \cdots \tag{393}$$

converges. This will be the case if

$$Q_1 B < 1. \tag{394a}$$

Convergence of the expansion (387) implies that $G^+$ is defined by substitution of the series into

$$\langle \mathbf{r} | \, G^+ \, | \mathbf{r}' \rangle = \langle \mathbf{r} | \, G_0{}^+ \, | \mathbf{r}' \rangle \langle \mathbf{r} | \, g \, | \mathbf{r}' \rangle.$$

Finally, Eq. (363) permits us to obtain the Born expansion for the scattering matrix $T$ in terms of that for $G^+$. The condition (394*a*) is therefore sufficient to imply the convergence of the Born series for the scattering amplitude.

Because of Eq. (391), we see that (394*a*) implies the convergence of the Born series for sufficiently high energies.[89] This is an important conclusion. [It represents a precise statement of the related conjecture made following Eq. (379).]

A condition for the convergence of the Born series at all energies may also be obtained from our considerations. On setting $n = m = 0$ in Eq. (390), we see that

$$Q_1 \leq B.$$

Therefore, if

$$B < 1, \tag{394b}$$

the condition (394*a*) is satisfied and the Born series converges for all energies.

[87] See Ref. [53, Chapter III].

[88] More precise conditions are given in Ref. [40].

[89] This conclusion is not necessarily valid for relativistic scattering.

Let us first apply these results to the case of scattering by the Yukawa potential (380). We find that

$$B = \frac{2M_r\,|V_0|}{\mu^2}, \tag{395a}$$

and that[90]

$$Q_1 \approx \frac{M_r\,|V_0|}{\mu\kappa}, \tag{395b}$$

which has the form of the expression (381*b*). At high energies $Q_1 \ll B$, so the condition (394*a*) is more stringent than was the estimate made in connection with Eq. (381*b*). This difference is somewhat clarified if we refer to Eq. (393). The term $Q_1$ in Eq. (393), corresponding to the expression (381*b*), is smaller than the following term when $B > 1$. Equation (394*a*) represents a sufficient, rather than a necessary, condition for convergence of the Born series, however, so it is possible that there may be weaker convergence conditions.

As a second illustration of the conditions (394), let us consider the elastic scattering of electrons by hydrogen atoms, using the potential (4.190). In this case we find that

$$B = \frac{me^2}{\pi}\int \frac{1}{x}\left(\frac{1}{x} + \frac{1}{a}\right)e^{-2x/a}\,d^3x = 3 > 1. \tag{396a}$$

The condition (394*b*) is not satisfied, so the Born series is not expected to converge at all energies. At high energies we obtain the estimate,

$$Q_1 \approx \frac{1}{\kappa a}, \tag{396b}$$

so the condition (394*a*) implies that the (approximate) inequality

$$\epsilon(\kappa) = \frac{\kappa^2}{2m} > 10\,\frac{e^2}{a} \tag{397}$$

must be satisfied if the Born series is to converge. In other words, the Born approximation is not expected to be useful for electron energies below several hundred electron volts.

The Born series (376*a*) provides an expansion for the scattering amplitude $f(\hat{\boldsymbol{\varkappa}}_f \cdot \hat{\boldsymbol{\varkappa}})$. A corresponding Born expansion for the partial wave amplitudes may be obtained just by iterating the integral equation (368),

$$w_l^{(0)}(\kappa; r) = (1 + G_{0l}v + G_{0l}vG_{0l}v + \cdots)y_l. \tag{398a}$$

[90] See the Appendix to Ref. [40] for an evaluation of the integrals.

This may be substituted into Eq. (372*b*) to give an expansion for the reaction matrix $R_l$.

$$2M_r\kappa^2 R_l = \int y_l(\kappa r)[v(r)\delta(r - r') + \langle r|\, vG_{0l}v\, |r'\rangle + \cdots]y_l(\kappa r')\, dr\, dr'. \quad (398b)$$

The convergence of this series has been discussed in some detail by Kohn [17].[91] The differential equation (365) is rewritten by Kohn in the form

$$\left[\frac{d^2}{dr^2} - \frac{l(l+1)}{r^2} - \lambda v(r) + \kappa^2\right] w_l^{(0)} = 0, \quad (399)$$

where $\lambda$ is a dimensionless parameter. The choice $\lambda = 1$ corresponds to the scattering problem of physical interest. On using Eq. (399) to evaluate $R_l$, it is clear that the series (398*b*) is modified only in that each $v$ is multiplied by a factor of $\lambda$, so this equation becomes a power series in $\lambda$. This series will in general have a finite radius of convergence, which, let us say, occurs at $\lambda = |\lambda_r|$. If $|\lambda_r| > 1$, the series (398*b*) is convergent for the actual scattering potential, corresponding to $\lambda = 1$.

To find $\lambda_r$, a pair of functions $\mathbf{f}_l(\lambda, \pm\kappa, r)$ are introduced, which are solutions of Eq. (399) and which satisfy the boundary conditions

$$\lim_{r\to\infty} e^{i\kappa r}\, \mathbf{f}_l(\lambda, \kappa, r) = i^l. \quad (400)$$

For $\lambda = 1$, this is, of course, just the function defined by Eq. (257). In analogy to Eq. (258), we may now define

$$\mathbf{f}_l(\lambda, \kappa) = \lim_{r\to 0} \frac{(\kappa r)^l}{(2l-1)!!}\, \mathbf{f}_l(\lambda, \kappa, r). \quad (401)$$

According to the theorem of Poincaré quoted previously [22], $\mathbf{f}_l(\lambda, \pm\kappa, r)$ and $\mathbf{f}_l(\lambda, \pm\kappa)$ are entire functions of $\lambda$.

We may use Eq. (260) to write

$$\tan \delta_l = -i\, \frac{\mathbf{f}_l(\lambda, \kappa) - \mathbf{f}_l(\lambda, -\kappa)}{\mathbf{f}_l(\lambda, \kappa) + \mathbf{f}_l(\lambda, -\kappa)}, \quad (402)$$

where $\delta_l$ is now the phase shift for scattering by the potential $\lambda v$. Since

$$R_l = -\frac{1}{\pi M_r \kappa} \tan \delta_l,$$

the convergence of the series (398*b*) (with the factors of $\lambda$ inserted) is restricted to the domain $|\lambda| < |\lambda_r|$, where $\lambda_r$ is that singularity of (402)

[91] See also Jost and Pais [28].

nearest $\lambda = 0$. Because the $\mathbf{f}_l(\lambda, \pm\kappa)$ are entire functions of $\lambda$, the quantity $\lambda_r$ is in general that root of the equation

$$\mathbf{f}_l(\lambda, \kappa) + \mathbf{f}_l(\lambda, -\kappa) = 0 \tag{403}$$

which is nearest the origin at $\lambda = 0$. As has been observed before, if $|\lambda_r| > 1$ the series (398*b*) converges for the "physical potential."

Kohn's investigation of Eq. (403) is a little too lengthy to reproduce here, so we shall merely quote some of his conclusions. For the case of zero-energy scattering, that is, for $\kappa = 0$, the Born series converges if

$$\int_0^\infty r\,|v(r)|\,dr < 2l + 1. \tag{404}$$

This result is not surprising in view of Bargmann's condition (261) that there be bound states. That is,

$$\int_0^\infty r\,|v(r)|\,dr = 2l + 1$$

is a necessary condition that there be a bound state at zero energy.

For scattering at high energies, we distinguish two cases: If the potential satisfies the condition that

$$\int_0^1 |v(r)|\,dr < \infty,$$

the Born series (398*b*) will converge if

$$\left|\int_0^\infty v(r)\,dr\right| < \pi\kappa. \tag{405}$$

If the potential has an $r^{-1}$ singularity at $r = 0$, we write it in the form

$$v(r) = \frac{2M_r V_0}{\mu r} N(r),$$

where $N(r)$ is a dimensionless function of $r$ satisfying the equation

$$N(0) = 1.$$

A sufficient condition for convergence of the Born series (398*b*) is that

$$\left(\frac{2}{\pi}\ln\frac{\kappa}{\mu}\right)\frac{M_r\,|V_0|}{\mu\kappa} < 1. \tag{406}$$

Except for the factor $(2/\pi)\ln(\kappa/\mu)$, this is equivalent to the condition that the expressions (381*b*) and (395*b*) be less than unity.

The terms beyond the lowest order in the Born series are ordinarily cumbersome to evaluate, so the usefulness of the method is generally restricted to conditions under which the first term provides an adequate approximation for the scattering amplitude or phase shifts. For most applications, the validity of the first Born approximation may be easily estimated from one or more of the conditions given in this section. It is perhaps worth mentioning that the approximation is somewhat capricious; it often works much better than would be expected. It is also true that many loose statements as to when we would imagine it to be valid are wrong. For example, if the phase shifts are small, it is frequently said to be good. This is manifestly false in the case of a very strong repulsive short-range potential, where for $S$-states $\delta_0 = -\kappa a$, with an arbitrarily small range. The Born approximation leads to nonsense in this case.

**Variational Principles**

Variational principles are of great use in obtaining approximate numerical solutions to many kinds of differential and integral equations. In this section we shall describe two such methods for solving Eq. (110).

The variational principles to be described are applicable to more general problems than that of simple potential scattering. They have been widely used in field theory and, in particular, in studying scattering by composite systems. The rather elementary application to Eq. (110), or to Eqs. (103), provides a convenient introduction to these techniques and helps to illustrate possible pitfalls in their use. It is often difficult to estimate the error associated with the use of a variational solution, since we may be restricted to a limited class of trial functions. Indeed, an ill-considered application of a variational principle may lead to absurd, or at least very inaccurate, results.

We first consider the method of Kohn [41] and Hulthén [42] as applied to Eq. (110). Let $w(\kappa; r)$ be a function of $r$, depending parametrically on the momentum $\kappa$, and consider the integral

$$I(w) \equiv \int_0^\infty w\left[\frac{d^2}{dr^2} + \kappa^2 - v - \frac{l(l+1)}{r^2}\right] w\,dr. \tag{407}$$

Now if we set $w = w_l^{(0)}$ [see Eq. (370) or (371)], we have

$$I(w_l^{(0)}) \equiv I_0 = 0, \tag{408}$$

since $w_l^{(0)}$ is a solution of (110).

Let us next suppose that we have chosen some "trial function" $w_t$, which we believe to be a reasonable approximation to $w_l^{(0)}$, and set $w = w_t$ in Eq. (407). The new quantity $I(w_t)$ will not in general vanish.

It is convenient from the outset to restrict the class of allowed trial functions $w_t$ to satisfy the same boundary conditions as do the $w_l^{(0)}$. These are [see Eqs. (370) and (371)]

$$\begin{aligned} w_l^{(0)} &= 0 \quad \text{for} \quad r = 0, \\ w_t &= 0 \quad \text{for} \quad r = 0, \end{aligned} \tag{409a}$$

and

$$\lim_{r\to\infty} w_l^{(0)} = \sqrt{2/\pi}\left[\sin\left(\kappa r - \frac{\pi l}{2}\right) + \tan\delta_l \cos\left(\kappa r - \frac{\pi l}{2}\right)\right],$$

$$\lim_{r\to\infty} w_t = \sqrt{2/\pi}\left[\sin\left(\kappa r - \frac{\pi l}{2}\right) + \tan\delta_{l,t} \cos\left(\kappa r - \frac{\pi l}{2}\right)\right]. \tag{409b}$$

Here $\delta_{l,t}$ is the "trial value" of the phase shift $\delta_l$. The "variation" in $w$ is then defined as

$$\delta w = w_t - w_l^{(0)}. \tag{410}$$

It follows from Eqs. (409*a*) and (409*b*) that

$$\delta w = 0 \quad \text{for} \quad r = 0 \tag{409c}$$

and that

$$\lim_{r\to\infty} \delta w = \sqrt{2/\pi}\, \delta(\tan\delta_l) \cos\left(\kappa r - \frac{\pi l}{2}\right), \tag{409d}$$

where

$$\delta(\tan\delta_l) = \tan\delta_{l,t} - \tan\delta_l. \tag{411}$$

We now simplify the difference

$$\delta I \equiv I(w_t) - I(w_l^{(0)}), \tag{412}$$

retaining only first-order terms in $\delta w$. Since $w_l^{(0)}$ satisfies Eq. (110), we find immediately that

$$\delta I = \int_0^\infty w_l^{(0)} \left[\frac{d^2}{dr^2} + \kappa^2 - v(r) - \frac{l(l+1)}{r^2}\right] \delta w\, dr. \tag{413}$$

To simplify this further, we set

$$w_l^{(0)} \frac{d^2}{dr^2} \delta w \equiv w_l^{(0)}\, \delta w'' = \delta w\, w_l^{(0)''} + \frac{d}{dr}(w_l^{(0)}\, \delta w' - \delta w\, w_l^{(0)'})$$

in Eq. (413). This gives us

$$\begin{aligned} \delta I &= \int_0^\infty \delta w \left\{ w_l^{(0)''} + \left[\kappa^2 - v - \frac{l(l+1)}{r^2}\right] w_l^{(0)} \right\} dr \\ &\quad + (w_l^{(0)}\, \delta w' - \delta w\, w_l^{(0)'})_{r=0}^{r=\infty} \\ &= (w_l^{(0)}\, \delta w' - \delta w\, w_l^{(0)'})_{r=0}^{r=\infty} = -\kappa \frac{2}{\pi} \delta(\tan\delta_l). \end{aligned} \tag{414}$$

The last step follows directly on using the limiting expressions (409*a*, *b*, *c*, and *d*).

Equation (414) may then be written as

$$\delta\left(I + \frac{2}{\pi}\kappa \tan \delta_l\right) = 0, \tag{415}$$

valid to the first order in $\delta w$. That is, the quantity

$$\Gamma \equiv I + \frac{2}{\pi}\kappa \tan \delta_l \tag{416}$$

is stationary with respect to small variations of $w_t$ about $w_l^{(0)}$, the exact wave function, if we restrict the variations to satisfy the boundary conditions (409). For example, let us suppose that $\Gamma$ has been evaluated for a trial wave function $w_t$, giving

$$\Gamma_t \equiv I(w_t) + \frac{2}{\pi}\kappa \tan \delta_{l,t}. \tag{417}$$

Then the exact value of $\Gamma$ is equal to $\Gamma_t$ to within the first order in $\delta w$; that is,

$$\Gamma = \Gamma_t + \mathcal{O}(\delta w^2).$$

Now because of Eq. (408) the exact $\Gamma$ is just $\Gamma = (2/\pi)\kappa \tan \delta_l$. Thus, to the first order, we have

$$\tan \delta_l = \tan \delta_{l,t} + \frac{\pi}{2\kappa} I(w_t). \tag{418}$$

Equation (418) then provides the variational estimate for $\tan \delta_l$. Having obtained by some other method an approximate wave function and phase shift, this gives us an improved phase shift if terms of $\mathcal{O}(\delta w^2)$ may be neglected.

The variational principle (415) permits us to improve the trial wave functions $w_t$. For example, let $w_t$ depend parametrically on $\nu$ parameters $\lambda_1, \lambda_2, \ldots, \lambda_\nu$ and consider

$$\lambda_\delta \equiv \tan \delta_{l,t} \tag{419}$$

to be an additional parameter. The quantity $\Gamma_t(\lambda_1, \lambda_2, \ldots, \lambda_\delta)$ then also depends on the $\nu + 1$ parameters. If we vary the $\lambda$, we have, to the first order, according to Eq. (415),

$$\delta\Gamma_t = \sum_{i=1}^{\nu} \frac{\partial \Gamma_t}{\partial \lambda_i}\delta\lambda_i + \frac{\partial \Gamma_t}{\partial \lambda_\delta}\delta\lambda_\delta = 0,$$

or

$$\frac{\partial I(w_t)}{\partial \lambda_i} = 0, \qquad i = 1, 2, \ldots, \nu,$$

$$\frac{\partial I(w_t)}{\partial \lambda_\delta} + \frac{2}{\pi}\kappa = 0. \tag{420}$$

This set of $\nu + 1$ algebraic equations determines the parameters $\lambda_1, \ldots, \lambda_\nu$ and $\lambda_\delta = \tan \delta_{l,t}$. Equation (418) may then be used to give the final value for $\tan \delta_l$. The use of Eqs. (418) and (420) to find $\tan \delta$ represents the formulation due to Kohn [41] of the variational principle.

Hulthén [42] has used the variational principle (415) in a somewhat modified form. He determines the parameters $\lambda_1, \ldots, \lambda_\nu$ and $\lambda_\delta$ from the equations

$$\begin{aligned} I_t(\lambda_1, \ldots, \lambda_\nu, \lambda_\delta) &= 0, \\ \frac{\partial I(w_t)}{\partial \lambda_i} &= 0, \qquad i = 1, 2, \ldots, \nu. \end{aligned} \tag{421}$$

The first of these equations is quadratic in $\lambda_\delta$, so an extraneous value for $\tan \delta_{l,t}$ is obtained in Hulthén's method. It is also somewhat more complicated to use in practice than is Kohn's; consequently the conditions (420) have had more extensive application than have (421).

We shall illustrate the methods just described with two examples. First we consider $S$-wave scattering by the square-well potential (140) for zero energy (that is, $\kappa = 0$).[92] The potential is assumed to have the form

$$\begin{aligned} v(r) &= \frac{-4.01}{d^2} \quad \text{for} \quad r < d, \\ &= 0 \qquad\quad \text{for} \quad r > d, \end{aligned} \tag{422a}$$

and

$$d = 2.8 \times 10^{-13} \text{ cm}. \tag{422b}$$

The exact solution $w_0^{(0)}(0; r)$ satisfies the equation

$$\frac{d^2 w_0^{(0)}}{dr^2} + v(r) w_0^{(0)} = 0.$$

The exact phase shift is given by Eqs. (148*a*, *b*). For the parameters of Eqs. (422*a*, *b*) we obtain

$$\lim_{\kappa \to 0} \frac{1}{\kappa} \tan \delta_0 = -2.08d. \tag{423}$$

[92] This example is taken from Ref. [41].

To apply the Kohn variational principle, a simple choice for $w_t$ is

$$\begin{aligned} \frac{1}{\kappa} w_t &= \sum_{j=1}^{\nu+1} \lambda_j r^j \quad \text{for} \quad r < d, \\ &= Cr + D \quad \text{for} \quad r > d. \end{aligned} \tag{424}$$

This choice assures us that the trial wave function has the correct boundary conditions (409*a*, *b*) in the limit of small $\kappa$, if we set $C = \sqrt{2/\pi}$ and $D = \sqrt{2/\pi}(1/\kappa) \tan \delta_{l,t}$.

**Table 6.1**

| Trial Function | $\lim_{\kappa \to 0} \frac{1}{\kappa d} \tan \delta_l$ |
|---|---|
| Exact | $-2.08$ |
| $\lambda_1 r$ | $+1.34$ |
| $\lambda_1 r + \lambda_2 r^2$ | $-2.13$ |
| $\lambda_1 r + \lambda_2 r^2 + \lambda_3 r^3$ | $-2.08$ |

Values are indicated for the scattering phase shift using the Kohn variational method and the indicated trial functions for the square-well potential (422*a*, *b*). Values are taken from Ref. [41].

An additional parameter is used to equate the two forms for $w_t$ at $r = d$. The values obtained from Eqs. (418) and (420) for

$$\lim_{\kappa \to 0} \frac{1}{\kappa} \tan \delta_l$$

for several choices for $\nu$ are given in Table 6.1. It is seen that a very accurate value for the phase shift is obtained for quite simple trial functions.

As a second example, we illustrate the use of the variational principle for *S*-state electron-hydrogen atom scattering, using the potential given in Eq. (4.190):[93]

$$v(r) = -2me^2\left(\frac{1}{r} + \frac{1}{a}\right)e^{-2r/a}, \tag{425}$$

where $m$ is the electron mass and $a$ the Bohr radius. A convenient trial wave function, satisfying the boundary conditions (409), is

$$w_t(\kappa; r_\alpha) = \sqrt{2/\pi}[\sin \kappa r + (\tan \delta_{l,t} + \lambda_1 e^{-r/a})(1 - e^{-r/a}) \cos \kappa r]. \tag{426}$$

For this choice of trial wave function the Kohn and Hulthén formulations

[93] This example is taken from calculations of Massey and Moiseiwitsch [43].

give essentially the same results. These are illustrated in Table 6.2 for several values of $\kappa a$, where they are compared with the results of "exact" (numerical) integration of Eq. (110). The excellent agreement is reassuring—but it is undoubtedly dependent on the evidently fortunate choice (426) for $w_t$.

**Table 6.2**

| $\kappa a$ | $\delta_0$ Numerical Integration | $\delta_0$ Variational Evaluation |
|---|---|---|
| 0.1 | 0.730 | 0.721 |
| 0.2 | 0.973 | 0.972 |
| 0.5 | 1.045 | 1.044 |
| 1.0 | 0.906 | 0.904 |

The scattering phase shift for $S$-wave scattering of an electron by an hydrogen atom, using the potential (425), is illustrated for several electron energies. Comparison is made of the results of the Kohn-Hulthén variational method with those of numerical integration. Taken from Ref. [43].

The Kohn variational principle may also be formulated for the scattering amplitude $f(\hat{\boldsymbol{\varkappa}}_f \cdot \hat{\boldsymbol{\varkappa}})$. In this case we consider either of the Eqs. (103) in the form

$$[\nabla^2 + \kappa^2 - v(r)]\psi_{\varkappa}^{+}(\mathbf{r}) = 0. \tag{427}$$

The wave function $\psi_{\varkappa}^{+}(r)$ has the asymptotic form (69):

$$\lim_{r\to\infty} \psi_{\varkappa}^{+}(\mathbf{r}_\alpha) = (2\pi)^{-3/2}\left[e^{i\varkappa\cdot\mathbf{r}} + \frac{e^{i\kappa r}}{r} f(\hat{\boldsymbol{\varkappa}}\cdot\hat{\mathbf{r}})\right]. \tag{428}$$

We now consider the integral, analogous to that of Eq. (407),

$$I(\psi_{\varkappa_f}^{+}, \psi_{\varkappa}^{+}) \equiv \int \psi_{\varkappa_f}^{+}(\mathbf{r})[\nabla^2 + \kappa^2 - v(r)]\psi_{\varkappa}^{+}(\mathbf{r})\, d^3r, \tag{429}$$

where $\kappa_f = \kappa$. On introducing the variations $\delta\psi_{\varkappa}^{+}$ and $\delta\psi_{\varkappa_f}^{+}$, where

$$\begin{aligned} \lim_{r\to\infty} \delta\psi_{\varkappa}^{+} &= (2\pi)^{-3/2} \frac{e^{i\kappa r}}{r}\, \delta[f(\hat{\boldsymbol{\varkappa}}\cdot\hat{\mathbf{r}})], \\ \lim_{r\to\infty} \delta\psi_{\varkappa_f}^{+} &= (2\pi)^{-3/2} \frac{e^{i\kappa r}}{r}\, \delta[f(\hat{\boldsymbol{\varkappa}}_f\cdot\hat{\mathbf{r}})], \end{aligned} \tag{430}$$

we obtain [by the arguments leading to Eq. (414)]

$$\delta I = \lim_{r\to\infty} \int r^2\left(\psi_{\varkappa_f} + \frac{\partial \delta\psi_{\varkappa}^{+}}{\partial r} - \delta\psi_{\varkappa}\frac{\partial \psi_{\varkappa_f}{}^{+}}{\partial r}\right) d\Omega$$
$$= \lim_{r\to\infty} (2\pi)^{-3} \int \delta[f(\hat{\boldsymbol{\varkappa}}\cdot\hat{\mathbf{r}})] r^2\left[e^{i\boldsymbol{\varkappa}_f\cdot\mathbf{r}}\frac{\partial}{\partial r}\left(\frac{e^{i\kappa r}}{r}\right) - \frac{e^{i\kappa r}}{r}\frac{\partial}{\partial r}(e^{i\boldsymbol{\varkappa}_f\cdot\mathbf{r}})\right] d\Omega. \tag{431}$$

Here the integral extends over all directions of the vector **r**. In further simplifying Eq. (431), we note, for example, that

$$\int e^{i\boldsymbol{\varkappa}_f\cdot\mathbf{r}}\delta[f(\hat{\boldsymbol{\varkappa}}\cdot\hat{\mathbf{r}})]\, d\Omega = \frac{2\pi}{i\kappa r}\{e^{i\kappa r}\delta[f(\hat{\boldsymbol{\varkappa}}_f\cdot\hat{\boldsymbol{\varkappa}})] - e^{-i\kappa r}\delta[f(-\hat{\boldsymbol{\varkappa}}_f\cdot\hat{\boldsymbol{\varkappa}})]\} + \mathcal{O}\left(\frac{1}{r^2}\right). \tag{432}$$

After a little algebra and the use of Eq. (432), we find that Eq. (431) reduces to

$$\delta I = -\frac{1}{2\pi^2}\delta[f(-\hat{\boldsymbol{\varkappa}}_f\cdot\hat{\boldsymbol{\varkappa}})]. \tag{433}$$

In other words, the quantity

$$\Gamma \equiv I(\psi_{\varkappa_f}{}^{+}, \psi_{\varkappa}{}^{+}) + \frac{1}{2\pi^2}f(-\hat{\boldsymbol{\varkappa}}_f\cdot\hat{\boldsymbol{\varkappa}}) \tag{434a}$$

is stationary, to the first order, in variations of $\psi_{\varkappa}{}^{+}$ and $\psi_{\varkappa_f}{}^{+}$ about their exact values; that is,

$$\delta\Gamma = 0. \tag{435a}$$

Let us suppose that trial functions $\psi_{\varkappa,t}^{+}$ and $\psi_{\varkappa_f,t}^{+}$, having trial scattering amplitudes $f_t(\hat{\boldsymbol{\varkappa}}\cdot\hat{\mathbf{r}}), f_t(\hat{\boldsymbol{\varkappa}}_f\cdot\hat{\mathbf{r}})$, are used to evaluate $I$. Since $I = 0$ for the exact wave functions, we obtain from Eqs. (434*a*) and (435*a*) the result that

$$f(\hat{\boldsymbol{\varkappa}}_f\cdot\hat{\boldsymbol{\varkappa}}) = f_t(\hat{\boldsymbol{\varkappa}}_f\cdot\hat{\boldsymbol{\varkappa}}) + 2\pi^2 I(\psi_{-\varkappa_f,t}^{+}, \psi_{\varkappa,t}^{+}). \tag{434b}$$

This provides an improved value for $f(\hat{\boldsymbol{\varkappa}}_f\cdot\hat{\boldsymbol{\varkappa}})$ if terms of $\mathcal{O}(\delta\psi^2)$ may be neglected. Equation (434*b*) is, of course, a direct analogue of Eq. (418).

To obtain a trial wave function, we may let $\psi_{\varkappa,t}^{+}$ have some assumed functional dependence on parameters $\lambda_1, \ldots, \lambda_\nu$. From Eq. (435*a*), we then obtain the first-order equation

$$\frac{\partial\Gamma}{\partial\lambda_i} = 0 \qquad i = 1, 2, \ldots, \nu, \tag{435b}$$

to determine the $\nu$ parameters $\lambda$. We here suppose the trial amplitudes $f_t(\hat{\boldsymbol{\varkappa}}_f \cdot \hat{\mathbf{r}})$ to depend also in some specified manner on the $\lambda_1, \ldots, \lambda_\nu$.

As a trivial illustration of Eqs. (434) and (435), let us choose

$$\psi^+_{\boldsymbol{\varkappa},t} = (2\pi)^{-3/2} e^{i\boldsymbol{\varkappa}\cdot\mathbf{r}},$$

$$\psi^+_{\boldsymbol{\varkappa}_f,t} = (2\pi)^{-3/2} e^{i\boldsymbol{\varkappa}_f\cdot\mathbf{r}},$$

so

$$f_t(\hat{\boldsymbol{\varkappa}} \cdot \hat{\mathbf{r}}) = 0.$$

In this case

$$I = -(2\pi)^{-3} \int e^{-i(\boldsymbol{\varkappa}_f - \boldsymbol{\varkappa})\cdot\mathbf{r}}\, v(r)\, d^3r,$$

and [since $v(r) = 2M_r V(r)$] Eq. (434*b*) leads to

$$f(\hat{\boldsymbol{\varkappa}}_f \cdot \hat{\boldsymbol{\varkappa}}) = -\frac{M_r}{2\pi} \int V(r)\, e^{-i(\boldsymbol{\varkappa}_f - \hat{\boldsymbol{\varkappa}})\cdot\mathbf{r}}\, d^3r. \tag{436}$$

This is just the first Born approximation. [Had we set $w_t = y_l$ and $\delta_{l,t} = 0$ in Eq. (418), we would then have obtained just the first Born approximation for $\tan \delta_l$.]

Schwinger[94] has given another variational principle which has been widely used in scattering theory. We shall first discuss this as applied to the reaction matrix, satisfying the integral equation (362*b*):

$$R = V + VG_0{}^0R.$$

A related wave function $\psi_a$ may be defined by the integral equation

$$\psi_a = \chi_a + G_0{}^0 V \psi_a, \tag{437}$$

where $\chi_a$ is, as usual, an "unperturbed" wave function satisfying the equation $K\chi_a = \epsilon_a \chi_a$.

Now a matrix element of $R$ may evidently be written as

$$R_{ba} = (\chi_b, R\chi_a) = (\chi_b, V\psi_a) = (\psi_b, V\chi_a), \tag{438}$$

where $\psi_b$ also satisfies Eq. (437), but with $\chi_a$ replaced by $\chi_b$. We may again express $R_{ba}$ in the form

$$R_{ba} = (\psi_b, (V - VG_0{}^0V)\psi_a). \tag{439}$$

This follows directly from Eq. (437), since

$$(\psi_b, (V - VG_0{}^0V)\psi_a) = (\psi_b, V(1 - G_0{}^0V)\psi_a) = (\psi_b, V(\psi_a - \psi_a + \chi_a)).$$

[94] J. Schwinger, unpublished lectures. See also Ref. [35].

Equations (438) and (439) permit us to set

$$R_{ba} = \frac{(\psi_b, V\chi_a)(\chi_b, V\psi_a)}{(\psi_b, (V - VG_0{}^0V)\psi_a)}. \tag{440}$$

The significance of this expression lies in the fact that it is stationary with respect to small variations of either $\psi_a$ or $\psi_b$ about its correct form. To see this, let us first consider the variation $\psi_a \to \psi_a + \delta\psi_a$. The corresponding variation in (440) is, to the first order in $\delta\psi_a$,

$$\begin{aligned} \delta P_{ba|\delta\psi_a} &= \frac{(\psi_b, V\chi_a)(\chi_b, V\,\delta\psi_a)}{(\psi_b, (V - VG_0{}^0V)\psi_a)} \\ &\quad - \frac{(\psi_b, V\chi_a)(\chi_b, V\psi_a)}{(\psi_b, (V - VG_0{}^0V)\psi_a)^2}(\psi_b, (V - VG_0{}^0V)\,\delta\psi_a) \\ &= (\chi_b, V\,\delta\psi_a) - (\psi_b, (V - VG_0{}^0V)\,\delta\psi_a) = 0, \end{aligned} \tag{441a}$$

since

$$(\psi_b, (V - VG_0{}^0V)\,\delta\psi_a) = ((1 - G_0{}^0V)\psi_b, V\,\delta\psi_a) = (\chi_b, V\,\delta\psi_a).$$

In a similar manner, we may show that if $\psi_b$ is varied about its exact value,

$$\delta R_{ba|\delta\psi_b} = 0, \tag{441b}$$

to the first order in $\delta\psi_b$. (Note that our stationary expression for $R_{ba}$ is independent of the normalization of $\psi_a$ and $\psi_b$.)

This result may be generalized without difficulty. Consider the $T$-matrix, as defined by Eq. (362*a*), and the wave functions

$$\psi_a{}^{\pm} = \chi_a + G_0{}^{\pm}V\psi_a{}^{\pm}. \tag{442}$$

Now

$$T_{ba} = (\chi_b, T\chi_a) = (\chi_b, V\psi_a{}^+) = (\psi_b{}^-, V\chi_a) \tag{443a}$$

$$= \frac{(\psi_b{}^-, V\chi_a)(\chi_b, V\psi_a{}^+)}{(\psi_b{}^-, (V - VG_0{}^+V)\psi_a{}^+)}. \tag{443b}$$

The arguments leading to Eqs. (441*a*) and (441*b*) are directly applicable to the expression (443*b*). If we perform the variation $\psi_a{}^+ \to \psi_a{}^+ + \delta\psi_a{}^+$ or $\psi_b{}^- \to \psi_b{}^- + \delta\psi_b{}^-$, respectively, we therefore obtain

$$\begin{aligned} \delta T_{ba|\delta\psi_a{}^+} &= 0, \\ \delta T_{ba|\delta\psi_b{}^-} &= 0, \end{aligned} \tag{444}$$

to the first order in the variations.

It should be noted that the derivation of the variational principles (441) and (444) is quite general. This derivation depended only on the assumption that the Hamiltonian $H$ can be decomposed into $K + V$ and that $T$, or $R$, satisfies integral equations of the form (362). Our results are then directly applicable to the scattering of composite particles or to field theory.[95]

As a first illustration, let us replace $\psi_a^+$ and $\psi_b^-$ by the trial functions $\chi_a$ and $\chi_b$, respectively, in the expression (443$b$) for $T_{ba}$. If we expand this in powers of $V$, there results

$$T_{ba} = (\chi_b, V\chi_a) + (\chi_b, VG_0^+V\chi_a) + \cdots. \tag{445}$$

This agrees with the Born series (376$a$) through the second order [but disagrees in the third order]. The corresponding approximation using the Kohn-Hulthén method gave just the first Born approximation, we recall.

In applications of Eq. (443$b$) we would ordinarily not expand the denominator as was done in Eq. (445). We may illustrate this with the case of simple potential scattering with $V = V(r)$ and the choice of trial functions

$$\psi_a^+ = (2\pi)^{-3/2}\, e^{i\boldsymbol{\kappa}\cdot\mathbf{r}},$$
$$\psi_b^- = (2\pi)^{-3/2}\, e^{i\boldsymbol{\kappa}_f\cdot\mathbf{r}}.$$

Then Eq. (443$b$) gives [in the notation of Eq. (44$b$)]

$$T(\hat{\boldsymbol{\kappa}}_f \cdot \hat{\boldsymbol{\kappa}}, \kappa, \kappa) = (2\pi)^{-3}\left[\int e^{-i(\boldsymbol{\kappa}_f-\boldsymbol{\kappa})\cdot\mathbf{r}}\, V(r)\, d^3r\right]^2 \left[\int e^{-i(\boldsymbol{\kappa}_f-\boldsymbol{\kappa})\cdot\mathbf{r}}\, V(r)\, d^3r \right.$$
$$\left. + \int d^3r\, d^3r'\, e^{-i\boldsymbol{\kappa}_f\cdot\mathbf{r}'}\, V(r')\left(\frac{2M_r\, e^{i\kappa|\mathbf{r}-\mathbf{r}'|}}{4\pi\,|\mathbf{r}-\mathbf{r}'|}\right) V(r)\, e^{i\boldsymbol{\kappa}\cdot\mathbf{r}}\right]^{-1}, \tag{446}$$

where Eq. (364$b$) has been used for $G_0^+$.

An evaluation of this expression for nucleon-nucleon scattering by a Yukawa potential has been given by Gerjuoy and Saxon [44] for a potential of realistic strength and for an energy of about 10 Mev. They compared the variational results and the results of the first Born approximation with the exact scattering amplitudes. This comparison showed the first Born approximation to be actually less inaccurate than the variational approximation! Such examples illustrate the difficulties that may arise in using variational principles with poor trial wave functions.

It was indicated in Section 6.7, pp. 306–307, that nuclear potentials are much too strong for plane-wave approximations to be considered as reliable. Applications of the approximation (446) to weaker interactions—as, for example, that of Eq. (4.190)—may provide sufficiently accurate scattering amplitudes.

Equation (440) provides also a variational principle for the partial wave

[95] A slightly modified form of the variational principle is required for rearrangement collisions, however.

scattering amplitudes. We need only identify $R_l$ with $-(2/\pi)\kappa \tan \delta_l$, as in Eq. (372c), $V$ with the reduced potentially $v$, and $G_0{}^0$ with $G_{0l}$ [see Eq. (366a)]. Then, on setting $\psi_a = \psi_b = w_l^{(0)}$, we obtain

$$-\frac{2}{\pi}\kappa \tan \delta_l = \frac{(y_l, vw_l^{(0)})(w_l^{(0)}, vy_l)}{(w_l^{(0)}, (v - vG_{0l}v)w_l^{(0)})} . \tag{447}$$

This is stationary with respect to small variations of $w_l^{(0)}$ about its correct value.

It is perhaps worth writing Eq. (447) out in detail, using the expressions (375) for $G_{0l}$. Let us first define

$$B_{l1} \equiv \int_0^\infty y_l(\kappa r)v(r)w_l^{(0)}(\kappa; r)\, dr,$$

$$B_{l2} = \frac{\pi}{2\kappa}\int_0^\infty dr' \left\{\int_0^{r'} dr[w_l^{(0)}(\kappa; r)v(r)y_l(\kappa r)][z_l(\kappa r')v(r')w_l^{(0)}(\kappa; r')] + \int_{r'}^\infty dr[w_l^{(0)}(\kappa; r)v(r)z_l(\kappa r)][y_l(\kappa r')v(r')w_l^{(0)}(\kappa; r')]\right\}. \tag{448}$$

Then Eq. (447) is just

$$-\frac{2}{\pi}\kappa \tan \delta_l = \frac{B_{l1}}{1 - B_{l2}/B_{l1}} . \tag{449}$$

As an application of this expression let us try the approximation

$$w_l^{(0)}(\kappa; r) = y_l(\kappa r), \tag{450}$$

and consider $S$-wave scattering at zero energy by the square-well potential (140). This example has already been discussed by the use of the Kohn method for the choice (422a) of the parameters for the potential. As we saw previously, the choice (450) for the trial function in the Kohn variational principle leads precisely to the first Born approximation:

$$\frac{2}{\pi}\kappa \tan \delta_l = -B_{l1},$$

in the notation of Eq. (449).

To evaluate the integrals in Eqs. (448) for this example, we set

$$y_0 = \sqrt{2/\pi}\kappa r,$$
$$z_0 = -\sqrt{2/\pi},$$

and use Eq. (140) for $v(r)$, as specified. The integration is elementary, and we are led to (valid as $\kappa \to 0$)

$$\frac{1}{\kappa d}\tan \delta_0 = -\frac{1}{3}\frac{v_0 d^2}{1 + \frac{2}{5}(v_0 d^2)} . \tag{451}$$

As mentioned before, it is obvious that this reduces to the first Born approximation when the term $\frac{2}{5}(v_0 d^2)$ may be neglected in the denominator. It is instructive to compare this conclusion with that obtained from Eq. (404), the condition for convergence of the Born series. For the potential (140), Eq. (404) reads

$$|v_0| \frac{d^2}{2} < 1, \tag{452}$$

which is sufficient to ensure that Eq. (451) is nonsingular.

We have seen that a bound state of zero energy would correspond to a potential of such a strength that $\tan \delta_0 = \infty$ for $\kappa = 0$. If we take Eq. (451) literally, this occurs for

$$v_0 d^2 = -2.5. \tag{453a}$$

On the other hand, the exact condition that there be a bound state of zero energy is that $1/a_0 = 0$ in Eq. (148*b*). This corresponds to

$$v_0 d^2 = -\left(\frac{\pi}{2}\right)^2 = -2.46, \tag{453b}$$

which agrees remarkably well with (453*a*).

The potential (422*a*) corresponds to $v_0 d^2 = -4.01$. It is evidently somewhat dangerous to continue to use (451) beyond its singularity. If we do evaluate (451) for the potential (422*a*), we find, however,

$$\frac{1}{\kappa d} \tan \delta_0 = -2.23. \tag{454}$$

This agrees with the correct value (423) to within about 10 percent.

### The WKB Approximation

One of the more useful of the approximate methods for solving the Schrödinger equation is the semiclassical, or so-called WKB, method.[96] This approximation is applicable when the scattering potential changes slowly over a distance comparable to the wavelength $1/\kappa$. Such conditions are frequently encountered in discussing the high-energy elastic scattering of particles by atoms or by atomic nuclei.

Let us return to Eq. (110) and describe the WKB method for determining the radial wave function $w_l(\kappa; r)$. It is convenient first to make a change of variable, replacing the coordinate $r$ by $x$, defined implicitly by[97]

$$r = \kappa^{-1} e^x. \tag{455}$$

As $r$ varies over the range $0 \leq r < \infty$, $x$ varies over the range $-\infty < x < \infty$.

[96] See Refs. [45].

[97] See R. E. Langer [46].

On making the transformation (455) and on setting

$$w_l(\kappa; r) \equiv e^{x/2}u(x), \tag{456a}$$

the wave equation (110) is transformed into

$$\frac{d^2u}{dx^2} + Q^2(x)u = 0, \tag{456b}$$

where

$$Q^2(x) \equiv e^{2x}\left(1 - \frac{v}{\kappa^2}\right) - (l + \tfrac{1}{2})^2. \tag{456c}$$

It will be convenient to suppose that $v$ falls to zero faster than $r^{-1}$ as $r \to \infty$ and that it is less singular than $r^{-2}$ at $r = 0$.[98]

We see that for $x$ sufficiently large and positive, $Q^2 > 0$; on the other hand, when $x$ is sufficiently large in magnitude and negative, $Q^2 < 0$. Let us then suppose that $Q^2(x) = 0$ at the point $x = x_0$. For convenience, we shall suppose that $Q^2$ vanishes for no other value of $x$.[99] Therefore,

$$\begin{aligned} Q^2(x) &> 0 \quad \text{for} \quad x > x_0, \\ Q^2(x) &< 0 \quad \text{for} \quad x < x_0. \end{aligned} \tag{457}$$

An approximate solution of Eq. (456*b*) may now be introduced in the form

$$u(x) = \frac{1}{\sqrt{q(x)}} \exp\left[\pm i \int_{x_0}^{x} q(s)\, ds\right], \tag{458}$$

where $q(x)$ is to be determined. If this expression is substituted into Eq. (456*b*), there results the equation

$$q^2(x) + \epsilon(x) = Q^2(x), \tag{459a}$$

where

$$\epsilon(x) = \frac{1}{2}\frac{q''}{q} - \frac{3}{4}\frac{(q')^2}{q^2}, \tag{459b}$$

(with $q' \equiv dq/dx$). In the *WKB approximation* we neglect the term $\epsilon$ in Eq. (459*a*) to obtain

$$q^2(x) = Q^2(x). \tag{460}$$

When this is substituted into Eq. (458), two independent approximate solutions of Eq. (456*b*) result.

[98] The WKB method is directly applicable to Coulomb scattering if the phase shift is evaluated as in Section 6.4, p. 264.

[99] The WKB method is applicable when there are several roots of the equation $Q^2(x) = 0$. In this case we must repeat the analysis given for obtaining the "connection formula" at each of the zeros of $Q^2$.

The condition for these to be valid solutions is evidently expressed by the inequality

$$|\epsilon(x)| \ll |Q^2(x)|, \tag{461}$$

where $\epsilon(x)$ is evaluated using Eq. (460). Roughly speaking, the inequality (461) requires that the distance $d$ over which the potential changes significantly must be large compared with the wavelength $1/\kappa$.

In the WKB approximation, and in the domain $x > x_0$, the general solution of Eq. (456$b$) is of the form

$$u = \frac{1}{\sqrt{Q}}\left\{C_+ \exp\left[+i\int_{x_0}^{x} Q(s)\,ds\right] + C_- \exp\left[-i\int_{x_0}^{x} Q(s)\,ds\right]\right\}, \tag{462a}$$

where $C_+$ and $C_-$ are constants and $Q > 0$. In the domain $x < x_0$, the corresponding general solution is evidently of the form

$$u = \frac{1}{\sqrt{|Q|}}\left\{D_+ \exp\left[-\int_{x_0}^{x} |Q(s)|\,ds\right] + D_- \exp\left[+\int_{x_0}^{x} |Q(s)|\,ds\right]\right\}, \tag{462b}$$

where $D_+$ and $D_-$ are two more constants. The boundary condition that $|w_l| \leq \mathcal{O}(r)$ near $r = 0$ implies that $u(x) = 0$ at $x = -\infty$. We must therefore set $D_+ = 0$ in Eq. (462$b$), giving

$$u(x) = \frac{1}{\sqrt{|Q(x)|}}\, D_- \exp\left[-\int_{x}^{x_0} |Q(s)|\,ds\right] \tag{463}$$

in the domain $x < x_0$.

The point $x = x_0$ is not a singular point of the differential equation (456$b$). There is consequently no difficulty in principle of integrating Eq. (456$b$) past the point $x = x_0$ to determine the coefficients $C_+$ and $C_-$ [Eq. (462$a$)] in terms of $D_-$. We see from Eq. (461), however, that the WKB approximation fails at $x = x_0$, the "turning point."[100] Thus we cannot simply match Eqs. (462$a$) and (463) at $x = x_0$, since neither expression is valid at this point. It is necessary, then, to return to the integration of the original equation (456$b$) in the neighborhood of $x = x_0$.

Let us introduce the new variable

$$y \equiv x - x_0 \tag{464}$$

and assume that near $x = x_0$, $Q^2$ has the form

$$Q^2(x) = \alpha y, \tag{465}$$

[100] In classical mechanics a particle cannot penetrate past the turning point—thus its name.

where $\alpha$ is a constant.[101,102] Then in the neighborhood of $x = x_0$, Eq. (456*b*) assumes the form

$$\frac{d^2u}{dy^2} + \alpha y u = 0. \tag{466}$$

As we shall see, this equation may be integrated explicitly. The solution of Eq. (466) may then be matched to Eqs. (462*a*) and (463), permitting us to determine the constants $C_+$ and $C_-$. In order that we can do this, however, the "linear turning point" approximation (465) must remain valid for sufficiently large values of $|y|$ that the condition (461) is satisfied. Using (465), we find that (461) is equivalent to

$$|y|^3 \gg \frac{1}{\alpha}. \tag{467}$$

If there exist positive and negative values of $y$ for which (465) is satisfied and for which the Eq. (465) is still valid, our treatment of the connection formula, relating $C_+$ and $C_-$ to $D_-$, will be valid.

The solution of Eq. (466) may be expressed in terms of the Airy integral[103]

$$u = \int_C e^{\phi(t)}\, dt, \tag{468a}$$

where

$$\phi(t) = yt + \frac{1}{3\alpha} t^3, \tag{468b}$$

and the contour $C$ is chosen so that $\phi = -\infty$ at the endpoints of the integration. [That the expression (468) then satisfies Eq. (466) is easily verified by substitution.] Since $\phi$ is regular in the open $t$-plane, this means that the endpoints of $C$ must lie at $|t| = \infty$ in regions for which $\mathrm{Re}(t^3) < 0$. If we write $t = |t|\, e^{i\omega}$, $\omega$ real, this implies that the contour must come from, or go to, infinity in one of the regions

$$\begin{aligned} \frac{\pi}{6} &< \omega < \frac{\pi}{2}, \\ \frac{5\pi}{6} &< \omega < \frac{7\pi}{6}, \\ \frac{3\pi}{2} &< \omega < \frac{11\pi}{6}. \end{aligned} \tag{469}$$

These allowed regions are indicated as unshaded areas in Fig. 6.5. There

[101] We neglect the possibility that for isolated values of $\kappa$ and some potentials $\alpha$ might vanish.

[102] For other discussions of the WKB connection formula, see, for example, Langer [46] and Kemble [47].

[103] See, for example, G. N. Watson [7, p. 188].

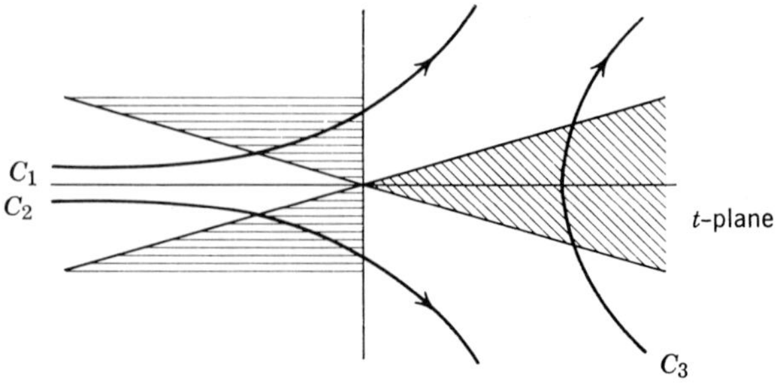


Fig. 6.5 Possible contours of integration for the integral (468*a*).

are three contours, then, joining these three regions and labeled as $C_1$, $C_2$, and $C_3$ in Fig. 6.5. Corresponding to each of these contours there is a function $u$ satisfying Eq. (466):

$$u_i = \int_{C_i} e^{\phi(t)}\, dt \qquad (i = 1, 2, 3). \tag{470}$$

Since $y = 0$ is not a singular point of Eq. (466), each of these functions is defined for positive and negative values of $y$. Only two of these functions are independent, however, since

$$u_1 = u_2 + u_3. \tag{471}$$

That is, $C_2 + C_3 - C_1$ is equivalent to a closed contour about the origin.

Let us first consider the case that $y > 0$. The integrand in Eq. (468*a*) has two saddle points at (here $\phi' \equiv d\phi/dt$)

$$\phi'(t^{\pm}) = 0,$$

or

$$t^{\pm} = \pm i\sqrt{\alpha y}. \tag{472}$$

Near these points, $\phi(t)$ has the form

$$\phi(t) = \phi(t^{\pm}) + \tfrac{1}{2}(t - t^{\pm})^2\phi''(t^{\pm}) + \mathcal{O}\left[\frac{1}{3!}(t - t^{\pm})^3\phi'''(t^{\pm})\right]. \tag{473}$$

When the condition (467) is satisfied, we may drop the last contribution here and evaluate (470) by a saddle point approximation. That is, we set

$$t - t^{\pm} \equiv e^{\pm i\pi/4}\tau,$$

and obtain for $u$, whose contour is made to pass through the saddle point at $t^+$,

$$\begin{aligned} u_1 &\approx e^{i\pi/4}\, e^{\phi(t^+)} \int_{-\infty}^{\infty} e^{-(y/\alpha)^{1/2}\tau^2}\, d\tau \\ &= \left[\frac{\pi}{(y/\alpha)^{1/2}}\right]^{1/2} e^{i\pi/4}\, e^{\phi(t^+)}, \end{aligned} \tag{474a}$$

where $\phi(t^+) = \frac{2}{3}i\sqrt{\alpha}y^{3/2}$. Similarly, the contour of $u_2$ is deformed to pass through $t^-$, showing that

$$u_2 \approx \left[\frac{\pi}{(y/a)^{1/2}}\right]^{1/2} e^{-i\pi/4} e^{\phi(t^-)}, \tag{474b}$$

where $\phi(t^-) = -\frac{2}{3}i\sqrt{\alpha}\, y^{3/2}$.

Now according to Eq. (465)

$$\phi(t^\pm) = \pm i \int_{x_0}^{x} Q(s)\, ds. \tag{475}$$

Therefore we see that near $x = x_0$ Eq. (462$a$) becomes equal to the function

$$u = \frac{1}{\sqrt{\pi\alpha}}(C_+ e^{-i\pi/4}u_1 + C_- e^{i\pi/4}u_2). \tag{476}$$

When $y < 0$, the saddle points in Eq. (468$a$) occur at

$$t = t^\pm = \pm\sqrt{(-y)\alpha}. \tag{477}$$

Now, near $x = x_0$

$$\phi(t^+) = -\tfrac{2}{3}\alpha^{1/2}(-y)^{3/2} = -\int_{x}^{x_0} |Q(s)|\, ds. \tag{478}$$

The solution that decays exponentially for $y < 0$ is seen to be $u_3$. On setting $t - t^+ = i\tau$, this becomes

$$\begin{aligned} u_3 &\approx i\, e^{\phi(t^+)} \int_{-\infty}^{\infty} e^{-(-y/\alpha)^{1/2}\tau^2}\, d\tau \\ &= i\, e^{\phi(t^+)} \left[\frac{\pi}{(-y/a)^{1/2}}\right]^{1/2} \\ &= i\sqrt{\pi\alpha}\, \frac{1}{\sqrt{|Q|}} \exp\left[-\int_{x}^{x_0} |Q(s)|\, ds\right]. \end{aligned} \tag{479}$$

Therefore, as $x$ approaches $x_0$, Eq. (463) reads

$$u = \frac{1}{i\sqrt{\pi\alpha}} D_- u_3 = \frac{D_-}{i\sqrt{\pi\alpha}}(u_1 - u_2), \tag{480}$$

the last step following from Eq. (471). This equation is valid on both sides of the turning point, since (as we have seen) the differential Eq. (466) is not singular at $y = 0$. Comparison of Eqs. (480) and (476) then leads immediately to the connection formula

$$\begin{aligned} C_+ &= e^{i\pi/4} \frac{D_-}{i}, \\ C_- &= -e^{-i\pi/4} \frac{D_-}{i}. \end{aligned} \tag{481}$$

We have now joined Eqs. (462$a$) and (463) across the region $x \approx x_0$. The wave function (462$a$) is therefore for $x > x_0$

$$u = \frac{2D_-}{\sqrt{(Qx)}} \sin\left[\frac{\pi}{4} + \int_{x_0}^{x} Q(s)\,ds\right]. \tag{482}$$

To find the scattering phase shift, we transform back to the variable $r$ and take the limit $r \to \infty$. Using Eq. (455), we set

$$Q^2(x) = r^2 p^2(r), \tag{483a}$$

where

$$p^2(r) = \kappa^2 - v(r) - \frac{(l+\frac{1}{2})^2}{r^2}. \tag{483b}$$

Then with Eq. (456$a$) we find that for $r > r_0 \equiv (1/\kappa)\,e^{x_0}$

$$w_l(\kappa; r) = \sqrt{\kappa r}\, u = 2\sqrt{\kappa/p}\, D_- \sin\left[\frac{\pi}{4} + \int_{r_0}^{r} p(r')\,dr'\right]. \tag{484}$$

When $r \to \infty$, $p \to \kappa$, and Eq. (484) becomes

$$\lim_{r\to\infty} w_l(\kappa; r) = 2D_- \sin\left\{\frac{\pi}{4} + \int_{r_0}^{\infty} [p(r') - \kappa]\,dr' + \kappa(r - r_0)\right\}. \tag{485}$$

On comparing this with the standard asymptotic form (113) for $w_l$, we see that

$$D_- = \tfrac{1}{2}\sqrt{2/\pi}, \tag{486a}$$

and

$$\delta_l = \left(l + \frac{1}{2}\right)\frac{\pi}{2} - \kappa r_0 + \int_{r_0}^{\infty} [p(r') - \kappa]\,dr'. \tag{486b}$$

This is the scattering phase shift $\delta_l$, given in the WKB approximation. It is easily verified that if we set $v(r) = 0$ in Eq. (486$b$), the phase shift $\delta_l$ vanishes.

Let us review once again the approximations leading to Eq. (486$b$). First, we require the validity of the inequality (461) except near the turning point at $x = x_0$. We require next that the expression (465) be an acceptable approximation near $x = x_0$ over a range overlapping that for which the condition (461) is valid. If these conditions are met [implying the inequality (467)], the conditions for validity of the saddle point evaluation of the integrals (470) are automatically satisfied.

### The Eikonal Approximation

Of greater antiquity, but closely related to the WKB approximation, is the eikonal approximation. It is by means of the eikonal method that geometrical optics is obtained as a limiting approximation to physical optics.[104] We shall describe a somewhat simplified form of the eikonal

[104] See, for example, Goldstein [48].

which has been quite useful in discussing the scattering of very fast particles by smooth potentials.[105] The presentation we shall give follows an application made to nuclear scattering.[106]

The Green's function (361) satisfies the differential equation

$$[\nabla_r^2 + \kappa^2 - v(r)]\langle \mathbf{r}|\, G^+ \,|\mathbf{r}'\rangle = 2M_r\, \delta(\mathbf{r} - \mathbf{r}'). \tag{487}$$

To obtain an approximate solution to this equation, we shall assume that

$$|v(r)| \ll \kappa^2 \quad \text{for} \quad \text{all } r, \tag{488a}$$

and that

$$|\nabla v(r)| \ll \kappa\, |v(r)| \quad \text{for} \quad \text{all } r. \tag{488b}$$

Guided by the form of Eq. (364*b*), we look for a solution to (487) of the form

$$\langle \mathbf{r}|\, G^+ \,|\mathbf{r}'\rangle = -\frac{2M_r}{4\pi}\frac{e^{i\phi}}{|\mathbf{r}' - \mathbf{r}|}, \tag{489a}$$

where

$$\phi = \int_0^{|\mathbf{r}-\mathbf{r}'|} p(\mathbf{x})\, ds. \tag{489b}$$

Here $p(\mathbf{x})$ is a function yet to be determined and $\mathbf{x}$ is a point lying on the straight line joining $\mathbf{r}$ to $\mathbf{r}'$. Finally, $ds$ is an element of path lying along this line. These coordinates are illustrated in Fig. 6.6.

Now, on inserting the expressions (489*a*) and (489*b*) into Eq. (487), we find that

$$[\nabla_r^2 + \kappa^2 - v(r)]\langle \mathbf{r}|\, G^+ \,|\mathbf{r}'\rangle$$
$$= \{-p^2 + \kappa^2 - v(r) + i[\hat{\mathbf{s}} \cdot \nabla p(\mathbf{r})]\}\langle \mathbf{r}|\, G^+ \,|\mathbf{r}'\rangle, \tag{490}$$

where $\hat{\mathbf{s}}$ is a unit vector parallel to $\mathbf{r} - \mathbf{r}'$. If we invoke the conditions (488) to drop the term $\hat{\mathbf{s}} \cdot \nabla p$ here, we see that Eq. (487) is satisfied, for $|\mathbf{r} - \mathbf{r}'| \neq 0$, with the choice

$$p(\mathbf{r}) = [\kappa^2 - v(r)]^{1/2}. \tag{491}$$

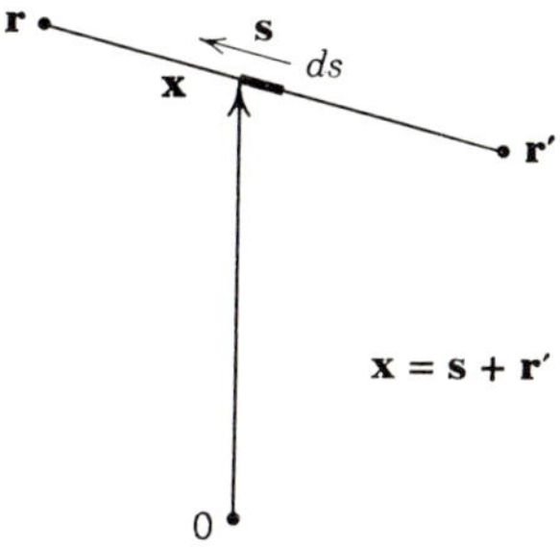

Fig. 6.6 Illustration of the coordinates used in Eq. (489*b*).

When $\mathbf{r} \approx \mathbf{r}'$, the phase $\phi$ is irrelevant and the factor $(2M_r/4\pi)(|\mathbf{r} - \mathbf{r}'|)^{-1}$ is appropriate to give the $\delta$-function on the right side of Eq. (487). Equations (489) and (491) therefore provide an explicit expression for the Green's function $G^+$ when the term $\hat{\mathbf{s}} \cdot \nabla p$ may be neglected in Eq. (490).

[105] The importance of this method was emphasized by the work of Fernbach, Serber, and Taylor [49].

[106] K. M. Watson [50]. For other developments see Saxon and Schiff [51].

The Green's function $G^-$ may be obtained from $G^+$ with the use of the operator of complex conjugation $T_0$, as in Eqs. (358):

$$\langle \mathbf{r}|\, G^- \,|\mathbf{r}'\rangle = T_0 \langle \mathbf{r}|\, G^+ \,|\mathbf{r}'\rangle T_0^{-1} = -\frac{2M_r}{4\pi}\frac{e^{-i\phi}}{|\mathbf{r}-\mathbf{r}'|}. \tag{492}$$

The relatively simple expressions (489) and (492) for the Green's functions for interacting particles permit us to evaluate the wave functions $\psi^\pm$ and the scattering matrix $T$ without the tedious use of partial waves. Indeed, in situations for which the conditions (488) are satisfied, a very large number of partial waves would ordinarily be required.

A strict eikonal approximation would involve defining the path of integration in Eq. (489*b*) along a curved trajectory—the path of classical motion of the particles. This more accurate evaluation would permit us to relax the condition (488*a*). The approximation used in Eq. (489*b*) is valid when the classical trajectory is curved only slightly by the potential. This will tend to be the case when Eq. (488*a*) is satisfied.

The evaluation given for $G^\pm$ is valid even for *non-Hermitian* potentials such as those encountered in the optical model of nuclear scattering. For such potentials we have

$$v(r) = v_R(r) - iv_I(r), \tag{493a}$$

where $v_R$ and $v_I$ are real, and

$$v_I(r) \geq 0 \quad \text{for} \quad \text{all } r. \tag{493b}$$

The wave describing the propagation of a particle through such a potential is exponentially attenuated.[107]

The argument leading to Eqs. (489) and (491) is unchanged for the potential (493), so these equations are still valid.[108] Equation (492) is modified in an obvious manner, however. Now we have

$$\langle \mathbf{r}|\, G^- \,|\mathbf{r}'\rangle = T_0 \langle \mathbf{r}|\, G^+ \,|\mathbf{r}'\rangle T_0^{-1} = -\frac{2M_r}{4\pi}\frac{e^{-i\phi^*}}{|\mathbf{r}-\mathbf{r}'|}, \tag{494a}$$

where

$$\phi^* = \int_0^{|\mathbf{r}-\mathbf{r}'|} p^*(\mathbf{x})\, ds. \tag{494b}$$

The scattering wave function $\psi_\varkappa^+(\mathbf{r})$ may easily be evaluated in the eikonal approximation. We have

$$\psi_\varkappa^+(\mathbf{r}) = \chi_\varkappa(\mathbf{r}) + \int d^3r' \langle \mathbf{r}|\, G^+ \,|\mathbf{r}'\rangle V(r')\chi_\varkappa(\mathbf{r}') \equiv \chi_\varkappa(\mathbf{r}) + \psi_{\text{sc}}(\mathbf{r}), \tag{495}$$

[107] A detailed discussion of the physical basis for this kind of potential will be given in Section 11.4.

[108] Because of the condition (488*a*), no problem of turning points arises here.

where, as usual,

$$\chi_{\varkappa}(\mathbf{r}) = (2\pi)^{-3/2}\, e^{i\varkappa\cdot\mathbf{r}}.$$

Equation (489) may now be inserted into Eq. (495), giving

$$\psi_{\text{sc}}(\mathbf{r}) = -\frac{1}{4\pi}\int\frac{d^3r'}{|\mathbf{r}-\mathbf{r}'|}\exp\left[i\int_0^{|\mathbf{r}-\mathbf{r}'|}p(\mathbf{x})\,ds\right]v(r')\frac{e^{i\varkappa\cdot\mathbf{r}'}}{(2\pi)^{3/2}}. \tag{496a}$$

With the substitutions $\mathbf{y} = \mathbf{r}' - \mathbf{r}$ and $\cos\theta \equiv \hat{\mathbf{y}}\cdot\hat{\varkappa}$, this becomes

$$(2\pi)^{3/2}\psi_{\text{sc}}(\mathbf{r}) = -\frac{1}{4\pi}\int\frac{d^3y}{y}\,e^{i\phi}v(r')\,e^{i\kappa y\cos\theta}\,e^{i\varkappa\cdot\mathbf{r}}. \tag{496b}$$

Figure 6.7 illustrates the relevant vectors in this integrand.

By partial integration with respect to $\theta$, Eq. (496*b*) is transformed into

$$(2\pi)^{3/2}\psi_{\text{sc}}(\mathbf{r}) = -\frac{1}{2i\kappa}\int y\,dy\,e^{i\varkappa\cdot\mathbf{r}}\left\{\left[v(r')\,e^{i\phi}\frac{e^{i\kappa y}}{y}\right]_{\hat{\mathbf{y}}\|\hat{\varkappa}} - \left[v(r')\,e^{i\phi}\frac{e^{-i\kappa y}}{y}\right]_{\hat{\mathbf{y}}\|(-\hat{\varkappa})} + \mathcal{O}\left(\frac{v\,e^{i\phi}}{\kappa y^2}\right)\right\}. \tag{496c}$$

The first term in the integrand oscillates rapidly when the integral over $y$ is performed, leading to a small term which we shall neglect. The last term of $\mathcal{O}(v\,e^{i\phi}/\kappa y^2)$ will also be neglected, since in our semiclassical approximation we consider $\kappa y$ to be large. Then, because we must consider $\phi - \kappa y$ to be small, the principal contribution to Eq. (496*c*) must come from the second term. Thus

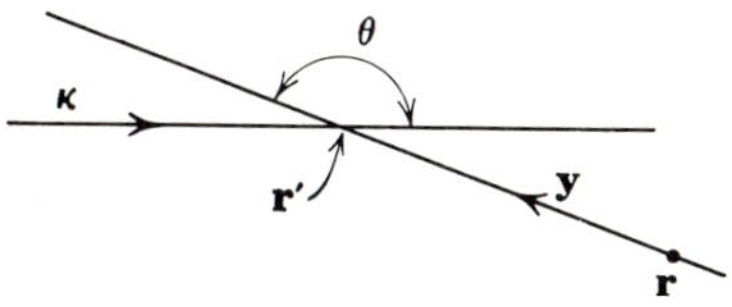

Fig. 6.7 Illustration of the vectors appearing in Eq. (496*b*).

$$\psi_{\text{sc}}(\mathbf{r}) \approx \left[\frac{e^{i\varkappa\cdot\mathbf{r}}}{(2\pi)^{3/2}}\right]\frac{1}{2i\kappa}\int dy\,v(r')\,e^{i(\phi-\kappa y)}. \tag{496d}$$

Here, because of the restrictions (488),

$$\phi - \kappa y \approx \int_{\mathbf{r}'}^{\mathbf{r}}\left[-\frac{1}{2\kappa}v(x)\right]ds \equiv \Delta\phi(\mathbf{r},\mathbf{r}'), \tag{497}$$

if we write $p(x) \approx \kappa - (1/2\kappa)v(x)$. The integral is taken [as specified in connection with Eq. (489*b*)] along the straight line joining $\mathbf{r}'$ to $\mathbf{r}$. Because of the condition in Eq. (496*c*) that $y$ be parallel to $-\hat{\varkappa}$, the vector $\mathbf{r} - \mathbf{r}'$ is here parallel to $\hat{\varkappa}$.

The $y$-integral in Eq. (496*d*) is easily evaluated, since

$$\frac{v(x)}{2\kappa}dy = d(\Delta\phi).$$

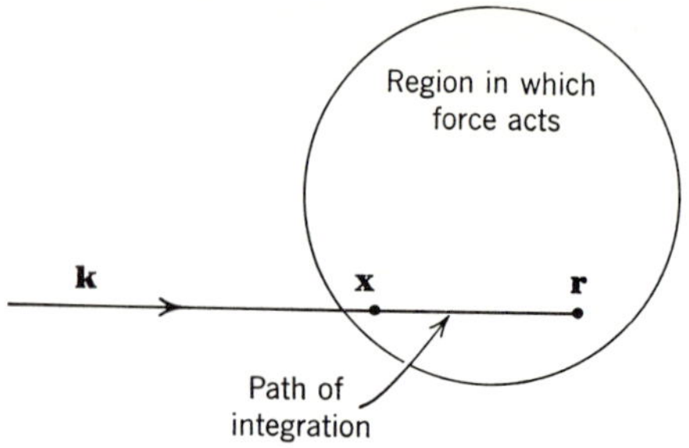


Fig. 6.8 Path of integration for the integral in Eq. (498).

We find then

$$\psi_{sc} = \frac{e^{i\boldsymbol{\varkappa}\cdot\mathbf{r}}}{(2\pi)^{3/2}}\left[-\frac{1}{i}\int_{\Delta\phi=0}^{\Delta\phi(\mathbf{r})} d(\Delta\phi)\, e^{i\Delta\phi}\right] = (e^{i\Delta\phi(\mathbf{r})} - 1)\frac{e^{i\boldsymbol{\varkappa}\cdot\mathbf{r}}}{(2\pi)^{3/2}}, \quad (496e)$$

where

$$\Delta\phi(\mathbf{r}) = \int_{-\infty}^{\mathbf{r}} [p(\mathbf{x}) - \kappa]\, ds. \quad (498)$$

The path of integration in Eq. (498) is taken along points $\mathbf{x}$ on the *straight line* parallel to $\boldsymbol{\varkappa}$ and passing through the point $\mathbf{r}$. This is illustrated in Fig. 6.8.

On inserting the expression (496*e*) into Eq. (495), we obtain

$$\psi_{\varkappa}^{+}(\mathbf{r}) = (2\pi)^{-3/2}\, e^{i\boldsymbol{\varkappa}\cdot\mathbf{r}}\, e^{i\Delta\phi(\mathbf{r})}. \quad (499a)$$

The canceling of the incident wave $\chi_{\varkappa}(\mathbf{r})$ here is an example of the extinction theorem in optics. That is, the phase of the incoming wave is modified by a refractive index in Eq. (499*a*). We can see this perhaps more clearly if we rewrite (499*a*) as

$$\psi_{\varkappa}^{+}(\mathbf{r}) = (2\pi)^{-3/2} \exp\left[i\int^{\mathbf{r}} p(\mathbf{x})\, ds\right]. \quad (499b)$$

Here the path of integration is that used in Eq. (498), but with a finite lower limit. [Thus Eq. (499*b*) differs by a constant irrelevant phase factor from Eq. (499*a*).] By following once again the steps leading to Eq. (499*a*), we find that

$$\psi_{\varkappa}^{-}(\mathbf{r}) = \chi_{\varkappa}(\mathbf{r}) + \int d^3r' \langle\mathbf{r}|\, G^{-}\, |\mathbf{r}'\rangle\, V(r')\psi_{\varkappa}(\mathbf{r}') = (2\pi)^{-3/2}\, e^{i\boldsymbol{\varkappa}\cdot\mathbf{r}}\, e^{-i\Delta\phi^{-}(\mathbf{r})}, \quad (500)$$

where

$$\Delta\phi^{-}(\mathbf{r}) = \int_{\mathbf{r}}^{\infty} [p^*(\mathbf{x}) - \kappa]\, ds \quad (501)$$

[see Eq. (494*b*)]. Here the path of integration is taken over those points **x** on the straight line running from **r** to $+\infty$ and parallel to $\boldsymbol{\varkappa}$, as in Fig. 6.5.

To find the scattering cross section we cannot use Eq. (499*a*) directly, because the simple eikonal approximation used does not permit the particle to be deflected. Instead, we use the expression

$$T(\hat{\boldsymbol{\varkappa}}_f \cdot \hat{\boldsymbol{\varkappa}}, \kappa, \kappa) = (\chi_{\boldsymbol{\varkappa}_f}, V\psi_{\boldsymbol{\varkappa}}^{+}) = (2\pi)^{-3}\int d^3r\, e^{-i(\boldsymbol{\varkappa}_f-\boldsymbol{\varkappa})\cdot\mathbf{r}}V(r)\, e^{i\Delta\phi(\mathbf{r})}. \tag{502}$$

The corresponding expression for the scattering amplitude $f(\hat{\boldsymbol{\varkappa}}_f \cdot \hat{\boldsymbol{\varkappa}})$ is given as [see Eq. (70*b*)]

$$f(\hat{\boldsymbol{\varkappa}}_f \cdot \hat{\boldsymbol{\varkappa}}) = -\frac{1}{4\pi}\int d^3r\, e^{-i(\boldsymbol{\varkappa}_f-\boldsymbol{\varkappa})\cdot\mathbf{r}}v(r)\, e^{i\Delta\phi(\mathbf{r})}. \tag{503}$$

The approximation involved in deriving Eqs. (502) and (503) may be understood as follows. First, the semiclassical approximation is equivalent to the assumption that secondary wavelets, arising at any point **x** due to scattering by $V(x)$, average to zero except along the orbit of the classical trajectory. Second, the classical trajectory is replaced by a straight line. The phase of the wave arriving at **r** is then $\boldsymbol{\varkappa}\cdot\mathbf{r} + \Delta\phi(\mathbf{r})$. The actual phase should be evaluated along the curved trajectory, however. For slightly curved trajectories, the error in the phase is of second order in the reciprocal radius of curvature.

If it is desired to estimate the error in Eqs. (502) and (503), we can recalculate the phase in Eq. (499) for a curved trajectory. We can also estimate the error in Eq. (491) by considering $\hat{\mathbf{s}}\cdot\nabla p$ to be a small perturbation in Eq. (490).

The total cross section for scattering may be obtained from the optical theorem. [See Eq. (74*d*).] In Chapter 11 it is shown that this theorem is valid even for potentials like that given by Eqs. (493). That is,

$$\sigma_{\text{tot}} = \frac{4\pi}{\kappa}\,\text{Im}\, f(\hat{\boldsymbol{\varkappa}}\cdot\hat{\boldsymbol{\varkappa}}) = -\frac{1}{\kappa}\,\text{Im}\left[\int d^3r v(r)\, e^{i\Delta\phi(\mathbf{r})}\right]. \tag{504}$$

The expressions (502), (503), and (504) are often relatively easy to evaluate and have had extensive use in describing the elastic scattering of high-energy particles by complex systems [such as atomic nuclei].

If we make explicit use of our approximation (488*a*), we may reduce Eq. (503) somewhat further. We write it as

$$f(\hat{\boldsymbol{\varkappa}}_f \cdot \hat{\boldsymbol{\varkappa}}) = -\frac{1}{4\pi}\int d^3r\, e^{-i(\boldsymbol{\varkappa}_f-\boldsymbol{\varkappa})\cdot\mathbf{r}}v(r) \times \exp\left\{-\frac{i}{2\kappa}\int_{-\infty}^{\mathbf{r}\cdot\hat{\boldsymbol{\varkappa}}} dz'v[(\rho^2+z'^2)^{1/2}]\right\}, \tag{505}$$

where we have written $r^2 = \rho^2 + z^2$, in a cylindrical coordinate system, and have written $p - \kappa \approx -v(r)/2\kappa$. As we have commented, one way to improve the approximation is to evaluate the integral in $\Delta\phi$ along the actual curved classical trajectory; we make a partial concession to this method by integrating from $-\infty$ to $\mathbf{r}$ along a line parallel to $\boldsymbol{\kappa} + \boldsymbol{\kappa}_f$. Choosing our axes for the integration so that $z$ lies along $\boldsymbol{\kappa} + \boldsymbol{\kappa}_f$ and letting $\boldsymbol{\rho}$ be a vector perpendicular to this, we have

$$\begin{aligned} f(\hat{\boldsymbol{\kappa}}_f \cdot \hat{\boldsymbol{\kappa}}) &= -\frac{1}{4\pi}\int d^2\rho \int_{-\infty}^{\infty} dz\, e^{-i(\boldsymbol{\kappa}_f - \boldsymbol{\kappa})\cdot \boldsymbol{\rho}} v(\rho, z) \exp\left[-\frac{i}{2\kappa}\int_{-\infty}^{z} dz' v(\rho, z')\right] \\ &= -\frac{i\kappa}{2\pi}\int d^2\rho\, e^{-i(\boldsymbol{\kappa}_f - \boldsymbol{\kappa})\cdot \boldsymbol{\rho}}\left\{\exp\left[-\frac{i}{2\kappa}\int_{-\infty}^{\infty} dz v(\rho, z)\right] - 1\right\} \\ &= -i\kappa\int_0^{\infty} \rho\, d\rho J_0\left(2\kappa\rho \sin\frac{\theta}{2}\right)\left\{\exp\left[-\frac{i}{2\kappa}\int_{-\infty}^{\infty} dz v(\rho, z)\right] - 1\right\}, \end{aligned} \tag{506}$$

where we have written

$$|\boldsymbol{\kappa}_f - \boldsymbol{\kappa}| = 2\kappa \sin\frac{\theta}{2}, \tag{507}$$

$$J_0(x) = \frac{1}{2\pi}\int_0^{2\pi} d\phi\, e^{ix\cos\phi}$$

[$J_0(x)$ is the zero-order Bessel function].

This result is closely related to the familiar expression for the scattering amplitude summed over phase shifts in the case that the number of angular momenta $l$ is so large that the sum over $l$ may be replaced by an integral. We recall the approximate formula [see Eq. (205)]

$$P_l(\cos\theta) \cong J_0[(l + \tfrac{1}{2})\theta]. \tag{508}$$

Then

$$\begin{aligned} f(\hat{\boldsymbol{\kappa}}_f \cdot \hat{\boldsymbol{\kappa}}) &= \sum_l (2l + 1)\left[\frac{e^{2i\delta_l} - 1}{2i\kappa}\right] P_l(cos\,\theta) \\ &\approx -\frac{i}{k}\sum_l (l + \tfrac{1}{2}) J_0[(l + \tfrac{1}{2})\theta](e^{2i\delta_l} - 1). \end{aligned} \tag{509}$$

Introducing the "impact parameter" $\rho$ by

$$l + \tfrac{1}{2} = \kappa\rho, \tag{510}$$

and replacing the sum over $l$ by an integration over $\rho$, we have

$$f(\hat{\boldsymbol{\kappa}}_f \cdot \hat{\boldsymbol{\kappa}}) = -i\kappa\int_0^{\infty} \rho\, d\rho J_0(\kappa\rho\theta)(e^{2i\delta(\rho,\kappa)} - 1), \tag{511}$$

where

$$\delta(\rho, \kappa) \equiv \delta_l(\kappa) \qquad (l \cong \kappa\rho). \tag{512}$$

To the extent that our approximate formula for $P_l(\cos\theta)$ is valid, namely, small angles, we may replace $\kappa\rho\theta$ by $2\kappa\rho\sin(\theta/2)$, in which case we recover Eq. (506), with $2\delta(\rho, \kappa)$ given by the explicit formula

$$2\delta(\rho, \kappa) = -\frac{1}{2\kappa}\int_{-\infty}^{\infty} dz v(\rho, z). \tag{513}$$

[It is perhaps worth mentioning that the Bessel function approximation for $P_l(\cos\theta)$ is very accurate; even for $l = 0$ it is good to less than 10 percent for $\theta = \pi/3$.]

We may explicitly verify the optical theorem, Eq. (504), in the limit of high energies using Eq. (506), which we write in a self-explanatory notation as

$$f(\hat{\boldsymbol{\kappa}}_f \cdot \hat{\boldsymbol{\kappa}}) = -i\kappa\int_0^{\infty} \rho\, d\rho J_0\left(2\kappa\rho\sin\frac{\theta}{2}\right)(e^{i\chi(\rho,\kappa)} - 1). \tag{514}$$

We may imagine that $v$ is complex and hence so is $\chi$. The total scattering cross section is obtained in the usual way.

$$\begin{aligned}\sigma_s &= \int d\Omega_f\, |f(\hat{\boldsymbol{\kappa}}_f \cdot \hat{\boldsymbol{\kappa}})|^2 = \kappa^2\int_0^{\infty}\rho\, d\rho\int_0^{\infty}\rho'\, d\rho'(e^{-i\chi^*(\rho,\kappa)} - 1)\\ &\times (e^{i\chi(\rho',\kappa)} - 1)\int d\Omega J_0\left(2\kappa\rho\sin\frac{\theta}{2}\right)J_0\left(2\kappa\rho'\sin\frac{\theta}{2}\right)\\ &\cong 2\pi\int_0^{\infty}\rho\, d\rho\int_0^{\infty}\rho'\, d\rho'(e^{-i\chi^*(\rho,\kappa)} - 1)(e^{i\chi(\rho',\kappa)} - 1)\\ &\times\int_0^{2\kappa} x\, dx J_0(x\rho)J_0(x\rho'). \end{aligned}\tag{515}$$

To the extent that the upper limit of the $x$-integral may be taken to be infinite, since

$$\int_0^{\infty} x\, dx J_0(x\rho)J_0(x\rho') = \frac{1}{\rho}\,\delta(\rho - \rho'), \tag{516}$$

we find

$$\sigma_s = 4\pi\int_0^{\infty}\rho\, d\rho(1 - e^{-\chi_i}\cos\chi_r) - 2\pi\int_0^{\infty}\rho\, d\rho(1 - e^{-2\chi_i}), \tag{517}$$

where we have written $\chi = \chi_r + i\chi_i$, with $\chi_r$ and $\chi_i$ both real. Now,

$$\frac{4\pi}{\kappa}\operatorname{Im} f(\hat{\boldsymbol{\kappa}} \cdot \hat{\boldsymbol{\kappa}}) = 4\pi\int_0^{\infty}\rho\, d\rho(1 - e^{-\chi_i}\cos\chi_r). \tag{518}$$

If we tentatively identify this with the total cross section $\sigma_t$ [Eq. (504)], we may write

$$\sigma_t \equiv \frac{4\pi}{\kappa}\operatorname{Im} f(\hat{\boldsymbol{\kappa}} \cdot \hat{\boldsymbol{\kappa}}) = \sigma_s + 2\pi\int_0^{\infty}\rho\, d\rho(1 - e^{-2\chi_i}). \tag{519}$$

The second term has an obvious physical significance. It is simply the total absorption cross section $\sigma_{ab}$, since the intensity of a wave of unit amplitude passing at impact parameter $\rho$ is, after penetrating the potential, $e^{-2\chi_i}$; $1 - e^{-2\chi_i}$ represents the absorption by the annular area $2\pi\rho\, d\rho$. Evidently if $\chi_i = 0$, our representation automatically satisfies the optical theorem (a consequence of the unitarity of the $S$-matrix) at high energies. The error is of $\mathcal{O}(1/\kappa d)$, where $d$ is the range of the potential. (See the example at the end of this section.)

We shall return in Chapter 11 to a more extensive discussion of applications of the approximations just described. Now we consider two simple examples. First we note that from Eq. (506), in the high-energy limit where the exponent is very small, we have

$$\begin{aligned} f(\hat{\boldsymbol{\kappa}}_f \cdot \hat{\boldsymbol{\kappa}}) &\approx -\frac{1}{4\pi}\int d^2\rho \int_{-\infty}^{\infty} dz\, e^{-i(\boldsymbol{\kappa}_f - \boldsymbol{\kappa})\cdot \boldsymbol{\rho}}\, v(\rho, z) \\ &= -\frac{1}{4\pi}\int d^3 r\, e^{-i(\boldsymbol{\kappa}_f - \boldsymbol{\kappa})\cdot \mathbf{r}}\, v(r). \end{aligned} \tag{520}$$

This is just the first Born approximation, which we know to be exact at high energies.

As a second example, consider the scattering by a complex square-well potential,

$$\begin{aligned} v(r) &= v_0 \quad \text{for} \quad r < d, \\ &= 0 \quad \text{for} \quad r > d, \end{aligned} \tag{521}$$

with $v_0$ a complex constant.

$$v_0 = v_r - iv_i \qquad (v_i > 0). \tag{522}$$

Now we have

$$\begin{aligned} \exp\left[-\frac{i}{2\kappa}\int_{-\infty}^{\infty} dz v(r)\right] &= e^{-(iv_0/\kappa)(d^2-\rho^2)^{1/2}} \qquad (0 < \rho < d), \\ &= 1 \qquad (\rho > d), \end{aligned} \tag{523}$$

and note that

$$\left|\exp\left[-\frac{i}{2\kappa}\int_{-\infty}^{\infty} dz v(r)\right]\right|^2 = e^{-(v_i/\kappa)2(d^2-\rho^2)^{1/2}} \qquad (0 < \rho < d). \tag{524}$$

Since $2(d^2 - \rho^2)^{1/2}$ is the distance traveled through the potential at impact parameter $\rho$, we interpret $\kappa/v_i$ as a mean free path through the potential. The scattering amplitude becomes

$$f(\cos\theta) = i\kappa \int_0^d \rho\, d\rho J_0\left(2\kappa\rho \sin\frac{\theta}{2}\right)(1 - e^{-i(v_0/\kappa)(d^2-\kappa^2)^{1/2}}). \tag{525}$$

In the limit of very small mean free path, $v_i d/\kappa \gg 1$, we neglect the second term and obtain

$$f(\cos\theta) = i\kappa \int_0^d \rho\, d\rho J_0\left(2\kappa\rho \sin\frac{\theta}{2}\right)$$
$$= id\,\frac{J_1[2\kappa d \sin(\theta/2)]}{2\sin(\theta/2)}. \tag{526}$$

The differential cross section for scattering by the potential is therefore (in this small mean free path limit)

$$\frac{d\sigma}{d\Omega} = |f(\cos\theta)|^2 = (\pi d^2)\frac{1}{\pi}\left[\frac{J_1[2\kappa d\sin(\theta/2)]}{2\sin(\theta/2)}\right]^2, \tag{527}$$

which is familiar in optics as characterizing the diffraction scattering from a black sphere. The total cross section is obtained by integrating

$$\sigma = 2\pi d^2 \int_0^\pi \sin\theta\, d\theta\, \frac{J_1^2[2\kappa d\sin(\theta/2)]}{4\sin^2(\theta/2)}$$
$$= 2\pi d^2 \int_0^{2\kappa d} dx\, \frac{J_1^2(x)}{x}$$
$$= \pi d^2[1 - J_0^2(2\kappa d) - J_1^2(2\kappa d)]$$
$$\approx \pi d^2\left(1 - \frac{1}{\pi\kappa d}\right) \qquad (\kappa d \gg 1). \tag{528}$$

Finally, we evaluate the absorption cross section $\sigma_{ab}$. We have from the second term in Eq. (519)

$$\sigma_{ab} = 2\pi \int_0^d \rho\, d\rho\,(1 - e^{-(2v_i/\kappa)(d^2-\rho^2)^{1/2}})$$
$$= \pi d^2\left\{1 - \frac{\lambda^2}{2d^2}\left[1 - \left(1 + \frac{2d}{\lambda}\right)e^{-2d/\lambda}\right]\right\}, \tag{529}$$

where we have introduced $\kappa/v_i = \lambda$, the mean free path as described earlier. In the limit $d/\lambda \gg 1$, we find

$$\sigma_{ab} \approx \pi d^2. \tag{530}$$

Thus the total cross section $\sigma_t = \sigma_s + \sigma_{ab}$ in this limit is $2\pi d^2$. The physical basis for the diffraction scattering $\pi d^2$ by a black sphere (that is, $d/\lambda \gg 1$) will be described in more detail in Chapter 11.[109]

## 6.8 THE POTENTIAL SCATTERING OF RELATIVISTIC PARTICLES

In some situations of physical interest it is possible to describe the interaction of two relativistic particles by a simple, local potential $V(r)$. One of

[109] An elementary discussion is given by Francis and Watson [51].

the most common examples is that of the elastic scattering of nucleons or mesons by atomic nuclei. As we shall see, for many applications of this kind we can apply the considerations of this chapter with only trivial modifications.

Let us consider such a scattering in the barycentric coordinate system.[110] The beam particle is assigned a rest-mass $M_1$ and is, in general, relativistic. It will also be supposed to have zero spin.[111] The rest-mass $M_2$ of the target particle is, however, considered to be so large that the motion of this particle is nonrelativistic. In the barycentric subspace, the respective momentum operators for the beam and target particles are taken to be $\mathbf{p} = -i\nabla$ and $-\mathbf{p} = i\nabla$. The beam particle is supposed to have a momentum $\boldsymbol{\varkappa}$ before the scattering has occurred.

Let us now return to Eq. (1.102*a*), describing nonrelativistic potential scattering. We must add a kinetic energy term $p^2/2M_2$ for the target particle to the Hamiltonian. Then for a stationary state wave function $\psi_{\varkappa}^{+}(r)$, Eq. (1.102*a*) becomes

$$\left[\omega + \frac{p^2}{2M_2} + V(r)\right]\psi_{\varkappa}^{+}(r) = \epsilon\psi_{\varkappa}^{+}(r), \tag{531}$$

where $\epsilon = \epsilon(\kappa)$ is the energy. Here $\omega = c(M_1{}^2c^2 + p^2)^{1/2}$, as in Chapter 1. Our assumptions imply that $M_2$ is sufficiently large that $\langle p^2/2M_2\rangle \ll \langle\omega\rangle$.

On following the arguments leading to the second of Eqs. (1.102*b*), we obtain

$$(c^2p^2 + M_1{}^2c^4 + [\omega, V])\psi_{\varkappa}^{+} = \left(\epsilon - V - \frac{p^2}{2M_2}\right)^2\psi_{\varkappa}^{+}. \tag{532}$$

Now, as remarked in Chapter 1, it is only when $|V| \ll M_1c^2$ and $|\nabla V|/|V| \ll \kappa$ that a one-particle relativistic theory is expected to be sensible. Assuming that these conditions are satisfied, we may simplify (532) to

$$[\nabla^2 + \kappa^2 - v(r)]\psi_{\varkappa}^{+}(\mathbf{r}) = 0. \tag{533}$$

Here we have introduced

$$\begin{aligned} v(r) &\equiv \frac{2\epsilon}{c^2}\left(1 + \frac{\epsilon}{M_2c^2}\right)^{-1} V(r), \\ \kappa^2 &\equiv \left(\frac{\epsilon^2}{c^2} - M_1{}^2c^2\right)\left(1 + \frac{\epsilon}{M_2c^2}\right)^{-1}. \end{aligned} \tag{534}$$

Equation (533) is identical in structure with Eq. (105). This means that our partial wave expansion leading to Eq. (110) may be used without

[110] We have shown in Section 6.1 how to transform the differential cross section to other coordinate frames.

[111] The corresponding problem for spin one-half particles is treated in Chapter 9.

modification. Furthermore, the methods for solving Eqs. (105) and (110) and the properties of the solutions as described in this chapter are applicable to Eq. (533).

Indeed, the only modifications required when we apply the conclusions of previous sections to relativistic particle scattering [when Eq. (533) is applicable] are rather trivial and obvious. For example, the conclusion that the Born expansion converges at sufficiently high energies, as obtained in Section 6.7, p. 309, is no longer applicable, since $v(r)$ itself now increases with increasing energy. At the same time, however, the general conditions for convergence are not changed.

For evaluation of the scattering cross section, we must return to the general expressions of Section 6.2. For example, the relation

$$\frac{d\epsilon}{d\kappa} = \frac{\kappa}{\epsilon} \frac{(1 + \epsilon/M_2c^2)^2}{1 + (\epsilon^2 + M_1^2c^4)/2\epsilon M_2c^2} \tag{535}$$

must be used in Eqs. (70*b*) and (70*c*) to evaluate the scattering amplitude. The differential scattering cross sections may then be obtained using Eqs. (73*a*) and (73*b*).

## REFERENCES AND NOTES

1. W. Heisenberg, *Z. Naturforsch.* **1,** 608 (1946); H. A. Kramers, *Hand- und Jahrbuch der Chemischer Physik* **1,** 312 (1938); C. Møller, *Kgl. Danske Videnskab. Selskab, Mat-fys. Medd.* **23,** 1 (1945).
2. This subject is reviewed by M. L. Goldberger, *Proc. Ann. Intern. Conf. High-Energy Phys., Geneva,* Interscience Publishers, New York, 1959, p. 208; G. F. Chew, *Ann. Rev. Nuclear Sci.* **9,** 29 (1959).
3. R. Jost, *Helv. Phys. Acta* **20,** 356 (1947); V. Bargmann, *Rev. Modern Phys.* **21,** 488 (1949).
4. When the Hamiltonian is invariant under the time reversal operation, so is $\mathscr{R}$. To see this, we recall the integral equation defining $\mathscr{R}$ [Eq. (5.179)],

$$\mathscr{R} = V + V\frac{P}{E_a - K}\mathscr{R},$$

and suppose that $TVT^{-1} = V$, $TKT^{-1} = K$, where $T$ is the time reversal operator. We readily verify from the given equation for $\mathscr{R}$ that $T\mathscr{R}T^{-1} = \mathscr{R}$. Then

$$\langle b|\, \mathscr{R}\, |a\rangle = (\chi_b, \mathscr{R}\chi_a) = (\chi_b, T^{-1}\mathscr{R}T\chi_a) = (T\chi_b, \mathscr{R}T\chi_a)^*.$$

Now we have seen that $T\chi_a = \mu_{-a}\chi_{-a}$, and so on, where $\mu_{-a}$ is a phase factor. Thus

$$\langle b|\, \mathscr{R}\, |a\rangle = \mu_{-b}\mu_{-a}^*(\chi_{-a}, \mathscr{R}\chi_{-b}) = \mu_{-b}\mu_{-a}^*\langle -a|\, \mathscr{R}\, |-b\rangle = \langle a|\, \mathscr{R}\, |b\rangle^*.$$

[This last relation follows since $\mathscr{R}$ is Hermitian.] When $\mathscr{R}$ does not involve spins, but commutes with the parity operator, $\langle -a|\, \mathscr{R}\, |-b\rangle = \langle a|\, \mathscr{R}\, |b\rangle$. In this case $\mu_{-a}$ and $\mu_{-b}$ may be chosen to be unity; then it follows that $\langle a|\, \mathscr{R}\, |b\rangle$ is real. This result also applies when spin interactions occur, if the appropriate states are introduced.

5. N. F. Mott and H. M. Massey, *The Theory of Atomic Collisions*, Clarendon Press, Oxford, 1933.
6. K. M. Case, *Phys. Rev.* **80,** 797 (1950).
7. E. T. Whittaker and G. N. Watson, *Modern Analysis*, Cambridge University Press, Cambridge, 1943; G. N. Watson, *Treatise on the Theory of Bessel Functions*, Cambridge University Press, Cambridge, 1944.
8. P. M. Morse and H. Feshbach, *Methods of Theoretical Physics*, McGraw-Hill Book Co., New York, 1953.
9. E. L. Ince, *Ordinary Differential Equations*, Dover Publications, New York, 1949.
10. S. T. Ma, *Phys. Rev.* **71,** 195 (1947).
11. V. Bargmann, *Rev. Modern Phys.* **21,** 488 (1949).
12. R. Yost, J. Wheeler, and G. Breit, *Phys. Rev.* **49,** 174 (1936).
13. The expansion for $F_l$ follows immediately from Eqs. (176*b*) and (181), using the relations $\Gamma(1 + z) = z\Gamma(z)$ and $\Gamma(z)\Gamma(-z) = -\pi(z \sin \pi z)^{-1}$. See Ref. [7, Chapter 12]. On setting $z = i\xi$, we have
$$\Gamma(i\xi)\Gamma(-i\xi) = \frac{2\pi}{\xi} e^{-\pi\xi}(e^{2\pi\xi} - 1)^{-1}.$$
For the evaluation of $G_l$, see Ref. [12].
14. J. D. Jackson and J. M. Blatt, *Rev. Modern Phys.* **22,** 77 (1950).
15. G. Breit, E. Condon, and R. Present, *Phys. Rev.* **50,** 825 (1936); G. Breit, B. Thaxton, and L. Eisenbud, *Phys. Rev.* **55,** 1018 (1939).
16. W. Magnus and F. Oberhettinger, *Special Functions of Mathematical Physics*, Chelsea Publishing Co., New York, 1949.
17. W. Kohn, *Rev. Modern Phys.* **26,** 292 (1954).
18. R. Jost and W. Kohn, *Danske Videnskab. Selskab, Mat-fys. Medd.* **27,** 9 (1953).
19. R. G. Newton, *J. Math. Phys.* **1,** 319 (1960).
20. I. M. Gel'fand and B. M. Levitan, *Doklady Akad. Nauk S.S.S.R.* **77,** 557 (1951), and *Izvest. Akad. Nauk S.S.S.R.* **15,** 309 (1951).
21. M. L. Goldberger and R. Blankenbecler, *Phys. Rev.* **126,** 766 (1962).
22. H. Poincaré, *Acta Math.* **4,** 215 (1884).
23. The condition (227) is evidently satisfied by $g_0(\kappa, r)$. Let us assume that it is true for $g_n$ and show that it is valid for $g_{n+1}$. Define $j(r) = \int_r^\infty dr'r' \, |v(r')|$, so by hypothesis
$$|g_n(\kappa, r)| \leq \frac{j^n(r)}{n!}.$$
Then from the recurrence relation (226) we have
$$|g_{n+1}| \leq \int_r^\infty dr'r' \, |v(r')| \frac{j^n(r')}{n!} = \int_0^{j(r)} dj \frac{j^n}{n!} = \frac{j^{n+1}(r)}{(n + 1)!}.$$
We thus see that the inequality (227) is valid for $|g_{n+1}|$.
24. E. Fermi, unpublished; K. M. Watson, *Phys. Rev.* **88,** 1163 (1952) and **95,** 228 (1954); M. Gell-Mann and K. Watson, *Ann. Rev. Nuclear Sci.* **4,** 219 (1954); K. Watson, J. Keck, A. Tollestrup, and R. Walker, *Phys. Rev.* **101,** 1159 (1956).
25. H. A. Kramers, *Quantum Mechanics*, North-Holland Publishing Co., Amsterdam, 1957, p. 306.
26. V. Bargmann, *Proc. Natl. Acad. Sci. U.S.* **38,** 961 (1952).
27. R. Jost and W. Kohn, *Phys. Rev.* **87,** 977 (1952).

28. R. Jost and A. Pais, *Phys. Rev.* **82,** 840 (1951).
29. N. Levinson, *Danske Videnskab. Selskab, Mat.-fys. Medd.* **25,** No. 9 (1949).
30. L. Landau and J. Smorodinsky, *J. Phys.* (*U.S.S.R.*) **8,** 154 (1944); J. Smorodinsky, *J. Phys.* (*U.S.S.R.*) **8,** 219 (1944) and **11,** 195 (1947).
31. J. Schwinger, hectographed notes on nuclear physics, Harvard University, 1947.
32. G. F. Chew and M. L. Goldberger, *Phys. Rev.* **75,** 1637 (1949).
33. H. A. Bethe, *Phys. Rev.* **76,** 38 (1949).
34. F. C. Barker and R. E. Peierls, *Phys. Rev.* **75,** 3122 (1949).
35. J. M. Blatt and J. D. Jackson, *Phys. Rev.* **76,** 18 (1949).
36. E. P. Wigner, *Phys. Rev.* **98,** 145 (1955).
37. J. M. Blatt and V. F. Weisskopf, *Theoretical Nuclear Physics*, John Wiley and Sons, New York, 1952.
38. R. Courant and D. Hilbert, *Methods of Mathematical Physics*, Interscience Publishers, New York, 1957.
39. M. Born, *Z. Physik.* **37,** 863 (1926); **38,** 803 (1926).
40. A. C. Zemach and A. Klein, *Nuovo cimento* **10,** 1078 (1958).
41. W. Kohn, *Phys. Rev.* **74,** 1763 (1948).
42. L. Hulthén, *Kgl. Fys. Sällskapets Lund. Förhandl.* **14,** 1 (1944).
43. H. S. W. Massey and B. L. Moiseiwitsch, *Proc. Roy. Soc.* (*London*) **205A,** 483 (1950).
44. E. Gerjuoy and D. S. Saxon, *Phys. Rev.* **94,** 478 (1954).
45. H. Jeffreys, *Proc. Lond. Math. Soc.* **23,** 428 (1924); G. Wentzel, *Z. Physik.* **38,** 518 (1926); H. A. Kramers, *Z. Physik.* **39,** 828 (1926); M. Brillouin, *J. de physique* **3,** 65 (1922).
46. R. E. Langer, *Phys. Rev.* **51,** 669 (1937).
47. E. C. Kemble, *Principles of Quantum Mechanics*, McGraw-Hill Book Co., New York, 1937.
48. H. Goldstein, *Classical Mechanics*, Addison-Wesley Publishing Co., New York, 1953.
49. S. Fernbach, R. Serber, and T. Taylor, *Phys. Rev.* **75,** 1352 (1949).
50. K. M. Watson, *Phys. Rev.* **89,** 575 (1953).
51. L. I. Schiff, *Phys. Rev.* **103,** 443 (1956) and **104,** 1481 (1956); D. S. Saxon and L. I. Schiff, *Nuovo cimento* **6,** 614 (1957).
52. N. Francis and K. Watson, *Am. J. Phys.* **21,** 659 (1953).
53. E. C. Titchmarsh, *Introduction to the Theory of the Fourier Integral*, Oxford University Press, Oxford, 1937.

# 7. SCATTERING BY NONCENTRAL FORCES

In Chapter 6 we have given a detailed discussion of the scattering of two particles by central forces. In this case there is no coupling between the spin and orbital angular momentum, so the spin appears just parametrically in the scattering matrix, which commutes with the orbital angular momentum. In the present chapter we shall extend our discussion to the case that two particles scatter by way of noncentral interactions. Such interactions couple the spin orientation to the orbital angular momentum.

Much of our discussion in Chapter 6 requires little modification here. For example, our analysis leading to Eq. (6-31), relating the reaction and scattering matrices, is quite general. Equation (6-34) will not generally be correct, however, for noncentral forces. This means that the scattering matrix $T$ is not a function only of the scalar product $\hat{\boldsymbol{\varkappa}}_f \cdot \hat{\boldsymbol{\varkappa}}$, but may depend on both $\hat{\boldsymbol{\varkappa}}_f$ and $\hat{\boldsymbol{\varkappa}}$. This makes it necessary to generalize the eigenstate decomposition of Eqs. (6-44) or (6-56). Much of the resulting additional complexity is, however, of a rather trivial and purely kinematical nature. Indeed, our formal considerations for noncentral forces are not significantly more complicated than those of Chapter 6, although the specific calculations tend to be more tedious.

The special methods for solving the Schrödinger equation which were discussed in Chapter 6 may in most cases be easily adapted to the study of noncentral interactions. Furthermore, the description given in Section 6.5 of the analytic properties of scattering amplitudes is not significantly changed [1].

In Section 7.1 we give a detailed kinematical description of the scattering of two particles. This will be illustrated in Section 7.2 with applications to the scattering of spin zero by spin one-half particles and in Section 7.4 to nucleon-nucleon scattering. In Section 7.3 a discussion will be given of reaction cross sections for the special case that there are just two particles in the initial state and two in the final state. This will be illustrated in Chapter 9 with a description of the photoproduction of pi-mesons.

Finally, in Section 7.5 we shall describe the use of the density matrix for studying scattering by polarized particles.

## 7.1 KINEMATICAL DESCRIPTION OF SCATTERING BY NONCENTRAL FORCES

In this section we describe the scattering of two particles by noncentral forces, restricting ourselves to simple scattering for which the same particles that interacted in the initial state emerge in the final state. Our discussion is also restricted to the barycentric coordinate system, the particles having respective momenta $\boldsymbol{\kappa}$, $-\boldsymbol{\kappa}$ in the initial state and $\boldsymbol{\kappa}_f$, $-\boldsymbol{\kappa}_f$ in the final state, as was assumed in Chapter 6. We also continue to suppose that the two particles have spins $S_1$ and $S_2$.

### Review of Central Force Scattering

We begin our discussion with a brief review of the treatment of central force scattering, as given in Section 6.2. In the barycentric subspace the $S$-matrix has the form [see Eqs. (6-53) and (6-55)]

$$\langle \boldsymbol{\kappa}_f | S | \boldsymbol{\kappa} \rangle = \frac{1}{\rho_\epsilon} \delta[\epsilon(\kappa_f) - \epsilon(\kappa)] \langle \hat{\boldsymbol{\kappa}}_f | S(\kappa) | \hat{\boldsymbol{\kappa}} \rangle. \tag{1a}$$

Here $\epsilon(\kappa)$ is the energy of the two particles when their respective momenta are $\boldsymbol{\kappa}$ and $-\boldsymbol{\kappa}$, and

$$\rho_\epsilon = \frac{\kappa^2}{d\epsilon/d\kappa}$$

is the density of states per unit energy [as defined by Eq. (6-30)]. The quantity $\langle \hat{\boldsymbol{\kappa}}_f | S(\kappa) | \hat{\boldsymbol{\kappa}} \rangle$ represents a matrix element of $S$ on the energy and momentum shells.

The normalization in Eq. ($1a$) is chosen to preserve the unitarity of the matrix $\langle \hat{\boldsymbol{\kappa}}_f | S(\kappa) | \hat{\boldsymbol{\kappa}} \rangle$ in the subspace of unit vectors $\hat{\boldsymbol{\kappa}}$. To see this, we note first the general expression of unitarity:

$$\begin{aligned} \langle \boldsymbol{\kappa}' | S^\dagger S | \boldsymbol{\kappa} \rangle &= \int \langle \boldsymbol{\kappa}' | S^\dagger | \boldsymbol{\kappa}_f \rangle \, d^3\boldsymbol{\kappa}_f \langle \boldsymbol{\kappa}_f | S | \boldsymbol{\kappa} \rangle \\ &= \delta(\boldsymbol{\kappa}' - \boldsymbol{\kappa}). \end{aligned} \tag{2}$$

Now

$$d^3\kappa_f = \rho_{\epsilon_f} \, d\epsilon_f \, d\Omega_{\hat{\kappa}_f},$$

so the use of ($1a$) in Eq. (2) leads to

$$\begin{aligned} \langle \boldsymbol{\kappa}' | S^\dagger S | \boldsymbol{\kappa} \rangle &= \frac{1}{\rho_\epsilon} \delta[\epsilon(\kappa') - \epsilon(\kappa)] \\ &\quad \times \int \langle \hat{\boldsymbol{\kappa}}' | S^\dagger(\kappa) | \hat{\boldsymbol{\kappa}}_f \rangle \, d\Omega_{\hat{\kappa}_f} \langle \hat{\boldsymbol{\kappa}}_f | S(\kappa) | \hat{\boldsymbol{\kappa}} \rangle \\ &= \delta(\boldsymbol{\kappa}' - \boldsymbol{\kappa}). \end{aligned} \tag{3}$$

By Eq. (6-54), however,

$$\delta(\boldsymbol{\varkappa}' - \boldsymbol{\varkappa}) = \frac{1}{\rho_\epsilon}\,\delta[\epsilon(\kappa') - \epsilon(\kappa)]\delta_{\hat{\boldsymbol{\varkappa}}',\hat{\boldsymbol{\varkappa}}}, \tag{4a}$$

where[1]

$$\delta_{\hat{\boldsymbol{\varkappa}}',\hat{\boldsymbol{\varkappa}}} \equiv \sum_{l=0}^{\infty}\sum_{m=-l}^{+l} Y_l^m(\hat{\boldsymbol{\varkappa}}')Y_l^{m*}(\hat{\boldsymbol{\varkappa}}). \tag{4b}$$

On inserting the expression (4*a*) into Eq. (3), we obtain

$$\begin{aligned}\langle\hat{\boldsymbol{\varkappa}}'|\, S^\dagger S\,|\hat{\boldsymbol{\varkappa}}\rangle &= \int \langle\hat{\boldsymbol{\varkappa}}'|\, S^\dagger(\kappa)\,|\hat{\boldsymbol{\varkappa}}_f\rangle\, d\Omega_{\hat{\boldsymbol{\varkappa}}_f}\langle\hat{\boldsymbol{\varkappa}}_f|\, S(\kappa)\,|\hat{\boldsymbol{\varkappa}}\rangle \\ &= \delta_{\hat{\boldsymbol{\varkappa}}',\hat{\boldsymbol{\varkappa}}}.\end{aligned} \tag{5}$$

This equation represents an expression of the unitarity of $\langle\hat{\boldsymbol{\varkappa}}_f|\, S(\kappa)\,|\hat{\boldsymbol{\varkappa}}\rangle$ in the subspace of the unit vectors $\hat{\boldsymbol{\varkappa}}$.

It is sometimes useful to replace the states $|\boldsymbol{\varkappa}\rangle$ by wave functions

$$|\epsilon, \hat{\boldsymbol{\varkappa}}\rangle \equiv |\boldsymbol{\varkappa}\rangle\, \rho_\epsilon^{1/2} \tag{6}$$

normalized per unit energy interval. In the representation (6) Eq. (1*a*) takes the simpler form

$$\langle\hat{\boldsymbol{\varkappa}}_f, \epsilon_f|\, S\,|\hat{\boldsymbol{\varkappa}}, \epsilon\rangle = \delta(\epsilon_f - \epsilon)\,\langle\hat{\boldsymbol{\varkappa}}_f|\, S(\kappa)\,|\hat{\boldsymbol{\varkappa}}\rangle. \tag{1b}$$

The partial wave decomposition of $\langle\hat{\boldsymbol{\varkappa}}_f|\, S(\kappa)\,|\hat{\boldsymbol{\varkappa}}\rangle$ [see Eq. (6-55)] is

$$\langle\hat{\boldsymbol{\varkappa}}_f|\, S(\kappa)\,|\hat{\boldsymbol{\varkappa}}\rangle = \sum_{l'm'lm}\langle\hat{\boldsymbol{\varkappa}}_f \mid l'm'\rangle\langle l'm'|\, S\,|lm\rangle\langle lm \mid \hat{\boldsymbol{\varkappa}}\rangle, \tag{7a}$$

where, of course,

$$\langle\hat{\boldsymbol{\varkappa}} \mid lm\rangle \equiv Y_l^m(\hat{\boldsymbol{\varkappa}}) \tag{8}$$

is the transformation function from the $(l, m)$ to the $\hat{\boldsymbol{\varkappa}}$ representation.[2] Now, because of Eq. (4-226*b*), $\langle l'm'|\, S\,|lm\rangle$ has the general form

$$\langle l'm'|\, S\,|lm\rangle = S_l(\kappa)\delta_{l',l}\delta_{m',m}. \tag{9}$$

Since the transformation (7) is unitary, the $S_l(\kappa)$ form a diagonal unitary matrix in the $(l, m)$ representation. They may therefore be written as $S_l = e^{2i\delta_l}$, where $\delta_l$ is real, as in Eq. (6-57).

### Description of Scattering by Noncentral Forces

We now repeat what is essentially the argument just given, but for scattering with noncentral interactions. We emphasize again that our

[1] Here $\int\delta_{\hat{\boldsymbol{\varkappa}}',\hat{\boldsymbol{\varkappa}}} f(\hat{\boldsymbol{\varkappa}})\, d\Omega_{\hat{\boldsymbol{\varkappa}}} = f(\hat{\boldsymbol{\varkappa}}')$, where $f(\hat{\boldsymbol{\varkappa}})$ is some function of the unit vector $\hat{\boldsymbol{\varkappa}}$.

[2] See the footnote to Eq. (6-54).

description is restricted to the barycentric coordinate system.[3] Two particles having respective momenta $\boldsymbol{\kappa}$, $-\boldsymbol{\kappa}$ and spins $S_1$, $S_2$ scatter. They emerge into the final state with respective momenta $\boldsymbol{\kappa}_f$, $-\boldsymbol{\kappa}_f$. The spins $S_1$ and $S_2$ are unchanged, although the spin orientations may be different. Retaining the notation of Eqs. (6-27), the total energies of the particles in the initial and final states are written as $\epsilon(\kappa)$ and $\epsilon(\kappa_f)$, respectively.

It is convenient to add the spins of the two particles, as in Eq. (6-38), to give a total spin $S$, with component $\nu$ along the axis of quantization.[4] The initial state prior to scattering is then [see Eq. (6-40)]

$$|\boldsymbol{\kappa}, S, \nu\rangle = \chi_{\boldsymbol{\kappa},S,\nu} = (2\pi)^{-3/2} e^{i\boldsymbol{\kappa}\cdot\mathbf{r}} u_{S,\nu}(s_1, s_2), \tag{10a}$$

where $u_{S,\nu}$ is the spin eigenfunction (6-38).[5] A final state, after the scattering has occurred, will then have the form

$$|\boldsymbol{\kappa}_f, S', \nu'\rangle = \chi_{\boldsymbol{\kappa}_f,S',\nu'}(\mathbf{r}) = (2\pi)^{-3/2} e^{i\boldsymbol{\kappa}_f\cdot\mathbf{r}} u_{S',\nu'}(s_1, s_2). \tag{10b}$$

Here $(S', \nu')$ represents the total spin and spin orientation for the two particles in the final state.

Following Eq. (6-58), we may expand $|\boldsymbol{\kappa}, S, \nu\rangle$ into partial waves:

$$\chi_{\boldsymbol{\kappa},S,\nu}(\mathbf{r}) = \sqrt{2/\pi} \sum_{l,m} i^l Y_l^{m*}(\hat{\boldsymbol{\kappa}}) Y_l^m(\hat{\mathbf{r}}) j_l(\kappa r) u_{S,\nu}. \tag{11a}$$

This may in turn be resolved into eigenstates of the total angular momentum operator $\mathbf{J}$, using a representation $(J, M)$ for the magnitude and orientation of $\mathbf{J}$ along the axis of quantization. Let us define

$$\langle \hat{\mathbf{r}}, l'S'\nu' |\, \mathscr{Y}^J \,| \hat{\boldsymbol{\kappa}}, lS\nu\rangle \equiv \sum_{m'mM} Y_{l'}^{m'}(\hat{\mathbf{r}}) Y_l^{m*}(\hat{\boldsymbol{\kappa}}) \langle l', S'; m', \nu' \mid J, M\rangle \times \langle l, S; m, \nu \mid J, M\rangle, \tag{12a}$$

so that

$$\sum_{\nu'} \langle \hat{\mathbf{r}}, l'S'\nu' |\, \mathscr{Y}^J \,| \hat{\boldsymbol{\kappa}}, lS\nu\rangle u_{S',\nu'} \tag{12b}$$

is an eigenstate of $\mathbf{J}^2$, corresponding to the eigenvalue $J(J+1)$ in the $(l'S'J)$ representation. The expression (12$a$) permits us to write Eq. (11$a$) as

$$\chi_{\boldsymbol{\kappa},S,\nu}(\mathbf{r}) = \sqrt{2/\pi} \sum_{l,J} i^l \langle \hat{\mathbf{r}}, lS\nu |\, \mathscr{Y}^J \,| \hat{\boldsymbol{\kappa}}, lS\nu\rangle u_{S,\nu} j_l(\kappa r). \tag{11b}$$

[3] As we saw in Chapter 3, this involves no loss of generality since a Lorentz transformation can always be made to describe the scattering in a general coordinate system.

[4] It is often (but not always) convenient to choose the axis of quantization parallel to $\boldsymbol{\kappa}$.

[5] If we wish to consider the two scattering particles to be composite systems (as will be done in Section 7.3), we may replace the $u_{S,\nu}$ by internal state wave functions $g_{i,S,\nu}(\xi)$.

Now, the $S$-matrix which describes the scattering in the barycentric coordinate system has the general form

$$\langle b| S |a\rangle = \delta(\mathbf{P}_b - 0)\langle \boldsymbol{\kappa}_f, S', \nu'| S |\boldsymbol{\kappa}, S, \nu\rangle, \tag{13}$$

since the total initial momentum $\mathbf{P}_a$ vanishes. Here the quantity $\langle \boldsymbol{\kappa}_f, S', \nu'| S |\boldsymbol{\kappa}, S, \nu\rangle$ represents an element of the $S$-matrix in the barycentric subspace. Now, just as was done in connection with Eq. (1$a$), we can define a submatrix $\langle \hat{\boldsymbol{\kappa}}_f, S', \nu'| S(\kappa) |\hat{\boldsymbol{\kappa}}, S, \nu\rangle$ on the energy shell:

$$\langle \boldsymbol{\kappa}_f, S', \nu'| S |\boldsymbol{\kappa}, S, \nu\rangle = \frac{1}{\rho_\epsilon}\,\delta[\epsilon(\kappa_f) - \epsilon(\kappa)]\langle \hat{\boldsymbol{\kappa}}_f, S', \nu'| S(\kappa) |\hat{\boldsymbol{\kappa}}, S, \nu\rangle. \tag{14a}$$

Again, the normalization here is chosen to give the unitarity relation

$$\sum_{S',\nu'} \int \langle \hat{\boldsymbol{\kappa}}'', S'', \nu''| S^\dagger(\kappa) |\hat{\boldsymbol{\kappa}}_f, S', \nu'\rangle \, d\Omega_{\hat{\kappa}_f} \langle \hat{\boldsymbol{\kappa}}_f, S', \nu'| S(\kappa) |\hat{\boldsymbol{\kappa}}, S, \nu\rangle$$
$$= \delta_{\hat{\kappa}'',\hat{\kappa}}\delta_{S'',S}\delta_{\nu'',\nu}. \tag{15}$$

[The proof of Eq. (15) follows precisely that of Eq. (5).]

The density factor $\rho_\epsilon$ may be eliminated from (14$a$) if, as in Eq. (6), we define states

$$|\epsilon, \boldsymbol{\kappa}, S, \nu\rangle \equiv |\boldsymbol{\kappa}, S, \nu\rangle \rho_\epsilon^{\frac{1}{2}}.$$

In this representation

$$\langle \epsilon_f, \hat{\boldsymbol{\kappa}}_f, S', \nu'| S |\epsilon, \hat{\boldsymbol{\kappa}}, S, \nu\rangle = \delta(\epsilon_f - \epsilon)\langle \hat{\boldsymbol{\kappa}}_f, S', \nu'| S(\kappa) |\hat{\boldsymbol{\kappa}}, S, \nu\rangle. \tag{14b}$$

The $T$-matrix in the barycentric subspace may be written as

$$\langle \boldsymbol{\kappa}_f, S', \nu'| T |\boldsymbol{\kappa}, S, \nu\rangle.$$

On the energy shell this is expressed as

$$\langle \hat{\boldsymbol{\kappa}}_f, S', \nu'| T(\kappa) |\hat{\boldsymbol{\kappa}}, S, \nu\rangle = \langle \boldsymbol{\kappa}_f, S', \nu'| T |\boldsymbol{\kappa}, S, \nu\rangle|_{\kappa_f=\kappa}. \tag{16}$$

Then, using Eq. (3-94),

$$\langle b| S |a\rangle = \delta(\mathbf{P}_b - 0)\{\delta_{S',S}\delta_{\nu',\nu}\delta(\boldsymbol{\kappa}_f - \boldsymbol{\kappa}) - 2\pi i\delta[\epsilon(\kappa_f) - \epsilon(\kappa)]$$
$$\times \langle \hat{\boldsymbol{\kappa}}_f, S', \nu'| T(\kappa) |\hat{\boldsymbol{\kappa}}, S, \nu\rangle\}. \tag{17}$$

On comparing this with Eq. (14$a$), using the representation (4$a$) for $\delta(\boldsymbol{\kappa}_f - \boldsymbol{\kappa})$, we obtain

$$\langle \hat{\boldsymbol{\kappa}}_f, S', \nu'| S(\kappa) |\hat{\boldsymbol{\kappa}}, S, \nu\rangle$$
$$= \delta_{S',S}\delta_{\nu',\nu}\delta_{\hat{\kappa},\hat{\kappa}_f} - 2\pi i\rho_\epsilon\langle \hat{\boldsymbol{\kappa}}_f, S', \nu'| T(\kappa) |\hat{\boldsymbol{\kappa}}, S, \nu\rangle. \tag{18}$$

The differential cross section (3-147) may now be written as (in the barycentric coordinate system)

$$d\sigma(\hat{\boldsymbol{\kappa}}_f, S', \nu'; \hat{\boldsymbol{\kappa}}, S, \nu) = \frac{(2\pi)^4}{v_{\text{rel},B}} \int d^3\kappa_f \delta[\epsilon(\kappa_f) - \epsilon(\kappa)] |\langle \hat{\boldsymbol{\kappa}}_f, S', \nu'| T |\hat{\boldsymbol{\kappa}}, S, \nu\rangle|^2, \tag{19}$$

or

$$\frac{d\sigma}{d\Omega} = \frac{(2\pi)^4}{v_{\text{rel},B}} \rho_\epsilon |\langle \hat{\boldsymbol{\kappa}}_f, S', \nu'| T(\kappa) |\hat{\boldsymbol{\kappa}}, S, \nu\rangle|^2. \tag{20a}$$

The relative velocity is

$$v_{\text{rel},B} = \frac{d\epsilon(\kappa)}{d\kappa} = \frac{\kappa^2}{\rho_\epsilon}, \tag{21}$$

so

$$\frac{d\sigma}{d\Omega} = \frac{(2\pi)^4}{\kappa^2} \rho_\epsilon^2 |\langle \hat{\boldsymbol{\kappa}}_f, S', \nu'| T(\kappa) |\hat{\boldsymbol{\kappa}}, S, \nu\rangle|^2. \tag{20b}$$

Again, using Eq. (18) we have

$$\frac{d\sigma}{d\Omega} = \frac{(2\pi)^2}{\kappa^2} |\langle \hat{\boldsymbol{\kappa}}_f, S', \nu'| S(\kappa) |\hat{\boldsymbol{\kappa}}, S, \nu\rangle|^2, \tag{20c}$$

valid when $\delta_{S',S}\delta_{\nu',\nu}\delta_{\hat{\boldsymbol{\kappa}}_f,\hat{\boldsymbol{\kappa}}} = 0$. We shall describe later the modification of Eqs. (20) when the two particles are identical.

More frequently than not the incident beam and target contain a random mixture of spin orientations. For such experiments, we must sum Eqs. (20) over all $S$ and $\nu$ and divide by $(2S_1 + 1)(2S_2 + 1)$, the total number of spin orientations. If all possible states $(S', \nu')$ are counted by the experimentor, we must also sum Eqs. (20) over $(S', \nu')$. Thus the average cross section, for an *unpolarized* beam and target and for *any* final spins, is [see Eq. (3-143*b*)]

$$\frac{d\bar{\sigma}}{d\Omega} = \frac{1}{(2S_1 + 1)(2S_2 + 1)} \sum_{S',\nu',S,\nu} \frac{d\sigma}{d\Omega} (\hat{\boldsymbol{\kappa}}_f, S', \nu'; \hat{\boldsymbol{\kappa}}, S, \nu). \tag{22}$$

We shall return in Section 7.5 to a discussion of Eqs. (20) for the case that the spin orientations are observed (and not random).

Let us now turn to the problem of resolving the matrix

$$\langle \boldsymbol{\kappa}_f, S', \nu'| S |\boldsymbol{\kappa}, S, \nu\rangle$$

into its submatrices for states of given total angular momentum $J$, as was done in Eqs. (7) and (9). To do this we first resolve the plane-wave states into partial waves with the transformation matrix [see Eq. (8)].

$$\langle \hat{\boldsymbol{\kappa}}_f, S' \nu' \mid l'm'S'\nu'\rangle = Y_{l'}^{m'}(\hat{\boldsymbol{\kappa}}_f). \tag{23}$$

Next, the state $(l'm'S'\nu')$ is resolved into states of total angular momentum $(J', M')$, using the Wigner coefficient $\langle l', S'; m', \nu' \mid J', M' \rangle$ [see Eq. (1-59*a*)]. [This has already been done, of course, in Eq. (11*b*).] For the $S$-matrix we then have the generalization of Eq. (7):

$$\langle \hat{\boldsymbol{\kappa}}_f, S', \nu' | \, S(\kappa) \, | \hat{\boldsymbol{\kappa}}, S, \nu \rangle = \sum_{\substack{l'm'J'M' \\ lmJM}} \langle \hat{\boldsymbol{\kappa}}_f, S'\nu' \mid l'm'S'\nu' \rangle$$
$$\times \langle l', S'; m', \nu' \mid J', M' \rangle \langle l'S'J'M' | \, S(\kappa) \, | lSJM \rangle$$
$$\times \langle l, S; m, \nu \mid J, M \rangle \langle lmS\nu \mid \hat{\boldsymbol{\kappa}}, S\nu \rangle. \tag{24}$$

Now according to Eq. (4-226*b*)

$$\langle l'S'J'M' | \, S(\kappa) \, | lSJM \rangle = \delta_{J',J} \delta_{M',M} S^J_{l'S',lS}(\kappa). \tag{25}$$

Here $S^J_{l'S',lS}$ is the $S$-matrix in the $(lSJM)$ representation. It is a unitary matrix, since the transformation (24) is unitary.[6]

Equations (23) and (25) permit us to write Eq. (24) in the form

$$\langle \hat{\boldsymbol{\kappa}}_f, S', \nu' | \, S(\kappa) \, | \hat{\boldsymbol{\kappa}}, S, \nu \rangle = \sum_{l'm'lmJM} Y_{l'}^{m'}(\hat{\boldsymbol{\kappa}}_f)$$
$$\times \langle l'S'; m', \nu' \mid J, M \rangle S^J_{l'S',lS}(\kappa)$$
$$\times \langle l, S; m, \nu \mid J, M \rangle Y_l^{m*}(\hat{\boldsymbol{\kappa}})$$
$$= \sum_{l'lJ} \langle \hat{\boldsymbol{\kappa}}_f, l'S'\nu' | \, \mathscr{Y}^J \, | \hat{\boldsymbol{\kappa}}, lS\nu \rangle S^J_{l'S',lS}(\kappa), \tag{26}$$

if the expression (12*a*) is used. This is an important equation; it permits us to calculate the differential scattering cross section in terms of the set of complex numbers $S^J_{l'S',lS}$, which are functions of the momentum $\kappa$ (or energy $\epsilon$).

The scattering matrix may also be decomposed into its matrix elements for states of given total angular momentum. On transforming to the $(lSJM)$ representation we have, in analogy to Eq. (26),

$$\langle \hat{\boldsymbol{\kappa}}', S', \nu' | \, T \, | \boldsymbol{\kappa}, S, \nu \rangle = \sum_{l'lJ} \langle \hat{\boldsymbol{\kappa}}', l'S'\nu' | \, \mathscr{Y}^J \, | \hat{\boldsymbol{\kappa}}, lS\nu \rangle T^J_{l'S',lS}(\kappa', \kappa). \tag{27}$$

Here the $T^J_{l'S',lS}$ represent $T$ in the $(\kappa lSJ)$ representation. On the energy shell, with $\kappa' = \kappa$, we have

$$T^J_{l'S',lS}(\kappa) \equiv T^J_{l'S',lS}(\kappa, \kappa). \tag{28}$$

If we substitute Eqs. (26) and (27) into Eq. (18) and use the expansion (4*b*), we obtain the simple relation

$$S^J_{l'S',lS}(\kappa) = \delta_{l',l} \delta_{S',S} - 2\pi i \rho_\epsilon T^J_{l'S',lS}(\kappa). \tag{29}$$

[6] This can be explicitly verified by substitution of the expansion (24) into Eq. (15), to obtain $\sum_{l''S''} [S^J_{l''S'',lS}]^* \; [S^J_{l''S'',l'S'}] = \delta_{l'l} \delta_{S',S}$.

We have obtained Eq. (26) using only the symmetry of invariance with respect to rotations. Ordinarily, there will be other symmetries to simplify the $S$-matrix further. For example, if parity is conserved the orbital angular momenta $l$ and $l'$ must be both even or both odd. That is, the scattering will not lead from a state of even (odd) to a state of odd (even) orbital angular momentum. Again, when the two particles are identical, Eq. (26) must be symmetrized as prescribed by Eq. (4-217). This will in general mean that certain states are missing from the sums in Eq. (26).

An important simplification of Eq. (26) results from the condition of invariance under time reversal. To study this, let us suppose the phase of the spin function to be so chosen that [see Eq. (2-144)]

$$Tu_{S,\nu} = (-1)^{S+\nu} u_{S,-\nu}, \tag{30}$$

where $T$ is the time reversal operator. Then according to Eqs. (4-232)

$$T\psi^{\pm}_{\boldsymbol{\varkappa},S,\nu} = (-1)^{S+\nu} \psi^{\mp}_{-\boldsymbol{\varkappa},S,-\nu}, \tag{31}$$

where $\psi^{\pm}_{\boldsymbol{\varkappa},S,\nu}$ are the stationary-state wave functions for the scattering process being considered (and $T$ is again the time reversal operator). With the representation (5-60) for the $S$-matrix, we may use Eq. (31) to write

$$\begin{aligned}\langle \boldsymbol{\varkappa}_f, S', \nu'|\, S\, |\boldsymbol{\varkappa}, S, \nu\rangle &= (\psi^{-}_{\boldsymbol{\varkappa}_f,S',\nu'}, \psi^{+}_{\boldsymbol{\varkappa},S,\nu}) \\ &= (T\psi^{-}_{\boldsymbol{\varkappa}_f,S',\nu'}, T\psi^{+}_{\boldsymbol{\varkappa},S,\nu})^* \\ &= (-1)^{S'+\nu'-S-\nu}(\psi^{-}_{-\boldsymbol{\varkappa},S,-\nu}, \psi^{+}_{-\boldsymbol{\varkappa}',S',-\nu'}) \\ &= (-1)^{S'+\nu'-S-\nu}\langle -\boldsymbol{\varkappa}, S, -\nu|\, S\, |-\boldsymbol{\varkappa}_f, S', -\nu'\rangle. \end{aligned} \tag{32}$$

[This equation is exactly equivalent, of course, to Eq. (4-235).] On comparing this with Eq. (14*a*), we conclude that

$$\langle \hat{\boldsymbol{\varkappa}}_f, S', \nu'|\, S(\kappa)\, |\hat{\boldsymbol{\varkappa}}, S, \nu\rangle = (-1)^{S'+\nu'-S-\nu}\langle -\hat{\boldsymbol{\varkappa}}, S, -\nu|\, S(\kappa)\, |-\hat{\boldsymbol{\varkappa}}_f, S', -\nu'\rangle. \tag{33}$$

Finally, if we substitute the appropriate expansion (26) into both sides of this equation, we obtain the important relation[7]

$$S^J_{l'S',lS} = S^J_{lS,l'S'}. \tag{34}$$

That is, the matrix $S^J_{l'S',lS}$ is symmetric.

Now, the symmetric, unitary matrix $S^J_{l'S',lS}$ may be diagonalized by a *real*, orthogonal matrix $U$.[8] That is,

$$S^J_{l'S',lS} = \sum_\alpha U^J_{l'S';\alpha}\, e^{2i\delta^J_\alpha}\, U^J_{\alpha;lS}. \tag{35}$$

[7] In obtaining Eq. (34) we use the relation $Y_l^m(\hat{\boldsymbol{\varkappa}}) = (-1)^{l+m} Y_l^{m*}(-\hat{\boldsymbol{\varkappa}})$ and Eqs. (1-61) for the Wigner coefficients.

[8] See, for example, Wigner [2, p. 29].

Here the quantities $e^{2i\delta_\alpha^J}$ are the diagonal elements of the $S$-matrix. The unitarity of $S$ implies that the $\delta_\alpha{}^J$, called the *eigenphase shifts*, are *real.*

The expression (35) permits a simple counting of the number of parameters required to characterize the submatrix $S^J_{l'S',lS}$. Let us suppose that there are $n$ combinations of $(l, S)$ which contribute to $S^J$, so this is an $n \times n$ matrix. Since it is complex, $2n^2$ real parameters would in general be required to define it. However, a real orthogonal matrix (which describes a rotation) is characterized by $(n/2)(n-1)$ real numbers. To these we must add the $n$ eigenphases $\delta_\alpha{}^J$ to obtain the result that $S^J_{l'S',lS}$ depends on $(n/2)(n+1)$, real parameters.

Nucleon-nucleon scattering provides an important illustration of this. For scattering in the triplet state (that is, $S = 1$), we may have (among other possibilities) $l', l = J+1$ and $l', l = J-1$. Then $U$ and $S^J_{l'S',lS}$ are $2 \times 2$ matrices. The general real orthogonal $2 \times 2$ matrix is of the form

$$U_{l'1;\alpha} = \begin{pmatrix} \cos\epsilon & \sin\epsilon \\ -\sin\epsilon & \cos\epsilon \end{pmatrix}, \tag{36a}$$

where $\epsilon$ is some real parameter. For this case Eq. (35) becomes

$$S = \begin{pmatrix} \cos^2\epsilon\, e^{2i\delta_1} + \sin^2\epsilon\, e^{2i\delta_2} & \frac{1}{2}\sin 2\epsilon(e^{2i\delta_1} - e^{2i\delta_2} \\ \frac{1}{2}\sin 2\epsilon(e^{2i\delta_1} - e^{2i\delta_2}) & \sin^2\epsilon\, e^{2i\delta_1} + \cos^2\epsilon\, e^{2i\delta_2} \end{pmatrix}. \tag{36b}$$

For convenience we have here omitted the symbol $J$. The quantity $\epsilon$ is called the "mixing parameter".[9]

**The Wave Functions**

The wave functions $\psi^+_{\varkappa,S,\nu}$ in the barycentric subspace may also be resolved into substates of given total angular momentum $J$, just as was done in Eq. (11*b*) for the incident wave. According to Eq. (3-78) we have

$$\psi^+_{\varkappa,S,\nu}(\mathbf{r}) = \chi_{\varkappa,S,\nu} + \sum_{S',\nu'} \int \frac{d^3\kappa'\chi_{\varkappa',S',\nu'}}{\epsilon(\kappa) + i\eta - \epsilon(\kappa')} \langle \varkappa', S', \nu' |\, T \,| \varkappa, S, \nu\rangle. \tag{37}$$

On inserting the expansions (11*a*) and (27) into this equation we obtain

$$\begin{aligned} \psi^+_{\varkappa,S,\nu} = \chi_{\varkappa,S,\nu} + &\sum_{l'm'S'\nu'lmJM} \sqrt{2/\pi}\, i^{l'} \\ &\times Y_{l'}^{m'}(\hat{\mathbf{r}}) \langle l', S'; m', \nu' |\, J, M\rangle \langle l, S; m, \nu \,|\, J, M\rangle \\ &\times Y_l^{m*}(\hat{\varkappa}) u_{S',\nu'} \int \frac{(\kappa')^2\, d\kappa' j_l(\kappa' r)}{\epsilon(\kappa) + i\eta - \epsilon(\kappa')} T^J_{l'S',lS}(\kappa', \kappa). \end{aligned} \tag{38}$$

[9] When $\epsilon = 0$, the scattering reduces to that from central forces, as discussed in Chapter 6.

Finally, use of Eqs. (11*b*) and (12*a*) permits us to write this as

$$\psi^{+}_{\boldsymbol{\kappa},S,\nu}(\mathbf{r}) = \sum_{l'S'lJ\nu'} \langle \hat{\mathbf{r}}, l'S'\nu' |\, \mathscr{Y}^{J}\, | \hat{\boldsymbol{\kappa}}, lS\nu\rangle u_{S',\nu'} \psi^{+}_{l'S';\kappa JlS}(r), \tag{39a}$$

where

$$\psi^{+}_{l'S';\kappa JlS}(r) = i^{l'} \sqrt{2/\pi}$$
$$\times \left\{\delta_{l',l}\delta_{S',S}\, j_l(\kappa r) + \int \frac{(\kappa')^2\, d\kappa'\, j_l(\kappa' r)}{\epsilon(\kappa) + i\eta - \epsilon(\kappa')}\, T^{J}_{l'S',lS}(\kappa', \kappa)\right\}. \tag{39b}$$

The wave functions $\psi^{-}$ may of course be obtained from applying the time reversal operation to $\psi^{+}$. For example,

$$\psi^{-}_{l'S';\kappa JlS}(r) = (-1)^{l}[\psi^{+}_{l'S';\kappa JlS}(r)]^{*}. \tag{39c}$$

The close similarity of the expressions (39) to those of Eqs. (6-59) should be noted. The radial wave function (39*b*) is a direct generalization of the wave function $\psi^{+}_{\kappa,l}(r)$ to which we devoted so much attention in Chapter 6.

The normalization conditions of Chapter 6 [see Eq. (6-63*a*)] are now replaced by

$$(\psi^{+}_{\bar{\boldsymbol{\kappa}},\bar{S}\bar{\nu}}, \psi^{+}_{\boldsymbol{\kappa},S,\nu}) = \delta(\bar{\boldsymbol{\kappa}} - \boldsymbol{\kappa})\delta_{\bar{S},S}\delta_{\bar{\nu},\nu} \tag{40a}$$

and

$$\sum_{l'S'} \int [\psi^{+}_{l'S';\bar{\kappa}J\bar{l}\bar{S}}]^{*} \psi^{+}_{l'S';\kappa JlS}\, r^2\, dr = \frac{1}{\kappa^2}\, \delta(\bar{\kappa} - \kappa)\delta_{\bar{l},l}\delta_{\bar{S},S}. \tag{40b}$$

Equation (40*b*) follows on inserting the expansion (39*a*) into (40*a*).

The radial wave functions assume a simple form in the limit that $r$ becomes large. Following the argument which led to Eq. (6-65), we find that

$$\lim_{r\to\infty} \psi^{+}_{l'S';\kappa JlS}(r) = \sqrt{2/\pi}\, i^{l'} \left\{ j_l(\kappa r)\delta_{l',l}\delta_{S',S} \right.$$
$$\left. - \frac{e^{i\kappa r}}{\kappa r}\, [\pi \rho_{\epsilon}(-i)^{l'} T^{J}_{l'S',lS}(\kappa)] \right\}. \tag{41a}$$

With the use of Eqs. (6-64) and (29) this simplifies to the form

$$\lim_{r\to\infty} \psi^{+}_{l'S';\kappa JlS}(r) = (i\sqrt{2\pi}\kappa r)^{-1}$$
$$\times \{-(-1)^{l}\delta_{l',l}\delta_{S',S}\, e^{-i\kappa r} + S^{J}_{l'S',lS}\, e^{i\kappa r}\}, \tag{41b}$$

which bears a close resemblance to the first of Eqs. (6-66*a*).

Direct evaluation of Eq. (37) as $r \to \infty$ leads, on the other hand, to [see, for example, Eq. (3-174*a*)]

$$\lim_{r\to\infty} \psi^{+}_{\boldsymbol{\kappa},S,\nu}(\mathbf{r}) = \chi_{\boldsymbol{\kappa},S,\nu} + (2\pi)^{-3/2}\, \frac{e^{i\kappa r}}{r} \sum_{S',\nu'} \langle S', \nu' |\, \mathbf{f}(\hat{\mathbf{r}}, \hat{\boldsymbol{\kappa}})\, | S, \nu\rangle u_{S',\nu'}, \tag{42}$$

where we have introduced the generalized scattering amplitude

$$\langle S', \nu'|\, \mathbf{f}(\hat{\mathbf{r}}, \hat{\boldsymbol{\varkappa}})\, |S, \nu\rangle \equiv -\frac{(2\pi)^2 \rho_\epsilon}{\kappa} \langle \hat{\mathbf{r}}, S', \nu'|\, T(\kappa)\, |\hat{\boldsymbol{\varkappa}}, S, \nu\rangle. \tag{43}$$

[Again, we note the similarity of Eqs. (42) and (43) to Eqs. (6-69) and (6-70*b*).]

The differential scattering cross section may be obtained directly from Eq. (42). It is

$$\frac{d\sigma}{d\Omega}(\hat{\boldsymbol{\varkappa}}_f, S', \nu'; \hat{\boldsymbol{\varkappa}}, S, \nu) = |\langle S', \nu'|\, \mathbf{f}(\hat{\boldsymbol{\varkappa}}_f, \hat{\boldsymbol{\varkappa}})\, |S, \nu\rangle^2. \tag{44}$$

With the definition (43) this is seen to be identical with Eq. (20*b*).

For the case of nonrelativistic potential scattering the wave functions $\psi^+_{\varkappa,S,\nu}$ satisfy a Schrödinger equation analogous to that given by Eq. (6-105). The interaction potential $V$ is now in general a function of the coordinate $\mathbf{r}$, the angular momentum $\mathbf{L}$, and the spin operators $\mathbf{S}_1$ and $\mathbf{S}_2$. Invariance under rotations requires of course that $[\mathbf{J}, V] = 0$, where $\mathbf{J} = \mathbf{S}_1 + \mathbf{S}_2 + \mathbf{L}$ is the total angular momentum operator.

Then, in the notation of Section 6.3 [see Eq. (6-105)]

$$(\nabla_r^2 + \kappa^2 - v)\psi^+_{\varkappa,S,\nu} = 0, \tag{45}$$

where $v = 2M_r V$, and $\nabla_r$ is the gradient operator acting on $\mathbf{r}$.

To obtain an equation for the radial wave function (39*b*) we first define the matrix

$$v^J_{l'',S'';l'S'}(r) = i^{l'-l''}\frac{2S+1}{2J+1}\sum_{\nu'',\nu'}(u_{S'',\nu''}, \int d\Omega_{\hat{\varkappa}}\, d\Omega_{\hat{\mathbf{r}}}$$
$$\times \langle \hat{\mathbf{r}}, l''S''\nu''|\, \mathscr{Y}^J\, |\hat{\boldsymbol{\varkappa}}, lS\nu\rangle^*\, v(\hat{\mathbf{r}}, l'S'\nu'|\, \mathscr{Y}^J\, |\hat{\boldsymbol{\varkappa}}, lS\nu\rangle u_{S',\nu'}). \tag{46a}$$

If $v$ is invariant under time reversal, this expression is symmetric in the $(lS)$ indices; that is,

$$v^J_{l'',S''\ l'S'} = v^J_{l',S'\ l''S''}. \tag{46b}$$

The proof of Eq. (46*b*) is analogous to that which led to Eq. (34). Finally, since $v$ is Hermitian, it follows from Eq. (46*b*) that $v^J_{l'',S'';l',S'}$ is real:

$$v^J_{l'',S'';l',S'}(r) = [v^J_{l'',S;''l',S'}(r)]^*. \tag{46c}$$

On simplifying the expression

$$\sum_{\nu'}(u_{S',\nu'}, \int d\Omega_{\hat{\varkappa}}\, d\Omega_{\hat{\mathbf{r}}}\langle \hat{\mathbf{r}}, l'S'\nu'|\, \mathscr{Y}^J\, |\hat{\boldsymbol{\varkappa}}, lS\nu\rangle^*(\nabla_r^2 + \kappa^2 - v)\psi^+_{\varkappa,S,\nu}) = 0,$$

using the expansion (39*a*) and the definition (46), we obtain the radial wave equations [compare Eq. (6-108)]

$$\left[\frac{1}{r^2}\frac{\partial}{\partial r}\left(r^2\frac{\partial}{\partial r}\right) - \frac{l'(l'+1)}{r^2} + \kappa^2\right]\psi^+_{l'S';\kappa JlS}(r)$$
$$= \sum_{l'',S''} i^{l'-l''} v^J_{l',S';l'',S''}(r)\psi^+_{l''S'';\kappa JlS}(r). \quad (47a)$$

In contrast to the situation encountered in Chapter 6, we must now solve a set of coupled differential equations to obtain the wave functions and the scattering amplitudes.[10] These equations are to be solved subject to the asymptotic boundary conditions implied by Eq. (41*a*) and to regularity conditions at $r = 0$.

These latter conditions may easily be obtained on following the argument which led to Eq. (6-120): Let us for the moment suppress the indices $(JlS)$ on $\psi^+_{l'S';\kappa JlS}$, writing

$$\psi^+_{l'S'\kappa;JlS}(r) \equiv i^{l'}\frac{w_{l'S'}(\kappa; r)}{\kappa r}. \quad (48)$$

Then Eq. (47*a*) may be rewritten as [compare Eq. (6-110)]

$$w''_{l'S'} - \left[\frac{l'(l'+1)}{r^2} - \kappa^2\right]w_{l'S'} - \sum_{l''S''} v^J_{l'S';l''S''} w_{l''S''} = 0, \quad (47b)$$

where $w' \equiv dw/dr$, and so on. This may be put into a more concise matrix form if we consider $w$ to be a column matrix. Then (47*b*) is equivalent to

$$w'' - \left(\frac{L^2}{r^2} + v - \kappa^2\right)w = 0. \quad (47c)$$

Here $L^2$ is the diagonal matrix with elements $l'(l'+1)$ and $v$ represents the interaction appearing in Eq. (47*b*). Equation (47*c*) is a real differential equation in the sense that it is satisfied by $w^*$, the complex conjugate of $w$:

$$w^{*''} - \left(\frac{L^2}{r^2} + v - \kappa^2\right)w^* = 0 \quad (47d)$$

Let us now consider two wave functions $w(\kappa_1; r)$ and $w(\kappa_2; r)$ both of which are defined for the same values of $(JlS)$ [see Eq. (48)]. The orthogonality relation (40*b*) may be expressed in terms of these as

$$\int_0^\infty \tilde{w}^*(\kappa_2; r)w(\kappa_1; r)\,dr \equiv \sum_{l'S'}\int_0^\infty w^*_{l'S'}(\kappa_2; r)w_{l'S'}(\kappa_1; r)\,dr$$
$$= \delta(\kappa_2 - \kappa_1). \quad (49)$$

[10] For central forces, we recall that $v^J_{l'S',lS} = \delta_{l',l}\delta_{S',S}v^{(S)}(r)$. Because of the boundary condition on the incident wave implied by Eq. (41*a*), we see that Eq. (47) becomes equivalent to (6-108).

Equation (6-116) may be expressed in terms of the column matrices $w$. Taking account of the fact that the $w$ are now complex, and using Eqs. (47$c$) and (47$d$), we obtain

$$
\begin{aligned}
I_x(\kappa_1, \kappa_2) - I_0(\kappa_1, \kappa_2) &\equiv (\kappa_1^2 - \kappa_2^2) \int_0^x \tilde{w}^*(\kappa_2; r) w(\kappa_1; r)\, dr \\
&= \int_0^x \left[ \tilde{w}^*(\kappa_2; r)\left(\kappa_1^2 - v - \frac{L^2}{r^2}\right) w(\kappa_1; r) \right. \\
&\qquad \left. - \tilde{w}(\kappa_1; r)\left(\kappa_2^2 - v - \frac{L^2}{r^2}\right) w^*(\kappa_2; r) \right] dr \\
&= \int_0^x [\tilde{w}(\kappa_1; r) w^{*\prime\prime}(\kappa_2; r) - \tilde{w}^*(\kappa_2; r) w''(\kappa_1; r)]\, dr \\
&= \int_0^x dr \frac{d}{dr} [\tilde{w}(\kappa_1; r) w^{*\prime}(\kappa_2; r) - \tilde{w}^*(\kappa_2; r) w'(\kappa_1; r)] \\
&= [\tilde{w}(\kappa_1; r) w^{*\prime}(\kappa_2; r) - \tilde{w}^*(\kappa_2; r) w'(\kappa_1; r)]\Big|_0^x.
\end{aligned}
$$

On using the asymptotic form (41$b$) and the unitarity of the $S$-matrix, we find that

$$
\lim_{x \to \infty} I_x(\kappa_1, \kappa_2) = (\kappa_1^2 - \kappa_2^2)\delta(\kappa_1 - \kappa_2).
$$

Then, because of the relation (49), we obtain

$$
\begin{aligned}
I_0(\kappa_1, \kappa_2) &\equiv [\tilde{w}(\kappa_1; 0) w^{*\prime}(\kappa_2; 0) - \tilde{w}^*(\kappa_2; 0) w'(\kappa_1; 0)] \\
&= \sum_{l'S'} [w_{l'S';JlS}(\kappa_1; 0) w^{*\prime}_{l'S';JlS}(\kappa_2; 0) - w^{*}_{l'S';JlS}(\kappa_2; 0) \\
&\times w_{l'S';JlS}(\kappa_1; 0)] = 0
\end{aligned}
\tag{50}
$$

[where we have now reinserted the ($JlS$) label in Eq. (48)]. Equation (50) [a generalization of (6-120)] provides the boundary condition at $r = 0$ which is required for integrating the differential equations (47). The remaining boundary conditions are determined by the asymptotic expression (41$b$).

It is clear from the form of (41$b$) that the $w_{l'S'}(\kappa; r)$ are complex functions. It follows from Eqs. (47$c$) and (47$d$), however, that these functions may be expressed in terms of real functions satisfying Eq. (47$c$). These real functions correspond to linear combinations of the $w_{l'S'}$ for different initial values of ($l$, $S$)—which appear in our equations only because of the boundary conditions (41$b$).[11] [When Eqs. (47$c$) are integrated numerically, the real functions are usually first obtained, since

[11] These real functions are exhibited in Section 7.4, p. 387.

complex equations are somewhat cumbersome.] The physical interpretation of the $\psi^{+}_{l'S';\kappa JlS}$ is straightforward. We see from Eq. (41*b*) that this describes the scattering from an initial state ($\kappa JlS$) to a final state ($\kappa Jl'S'$). Once these wave functions have been evaluated, the $S$-matrix elements may be obtained on comparison with Eq. (41*b*).

Alternatively, we may use Eqs. (4-13) or (5-66) to write

$$\langle \hat{\boldsymbol{\varkappa}}_f, S', \nu' |\, T(\kappa)\, | \hat{\boldsymbol{\varkappa}}, S, \nu \rangle = (\chi_{\varkappa_f, S', \nu'}, V \psi^{+}_{\varkappa, S\nu}), \tag{51a}$$

where $\kappa_f = \kappa$. If we insert the expansions (11*b*), (27) and (39*a*) into Eq. (51), we obtain

$$T^{J}_{l'S', lS}(\kappa) = \frac{1}{2M_r} \sum_{l'', S''} \int_0^\infty \chi^{*}_{\kappa, l'}(r) v^{J}_{l'S'; l''S''}(r) i^{l'-l''} \psi^{+}_{l''S''; \kappa JlS}(r) r^2\, dr, \tag{51b}$$

where [see Eq. (58*b*)]

$$\chi_{\kappa, l} \equiv \sqrt{2/\pi}\, i^l j_l(\kappa r).$$

Most of the discussion given in Section 6.5 of the analytic properties of scattering amplitudes may be directly generalized to scattering by noncentral forces.[12]

### The Case of Identical Particles

When the two particles are identical we must take proper account of wave function symmetrization, as described in Chapter 4. In fact, we can now use the corresponding discussion of Chapter 6 with only minor changes.

Since we assume that the particles are identical, they both have the same spin $S_1$. The symmetrization operator

$$\mathscr{S} = \tfrac{1}{2}[1 + \delta_{Q12} Q_{12}]$$

is just that of Eq. (6-77). The symmetrized scattering wave function is then

$$\psi^{+}_{\mathscr{S}(\varkappa, S, \nu)} = \sqrt{2}\, \mathscr{S} \psi^{+}_{\varkappa, S, \nu}. \tag{52a}$$

Using Eq. (6-89), we find from Eq. (42) that

$$\lim_{r \to \infty} \psi^{+}_{\mathscr{S}(\varkappa, S, \nu)} = \frac{1}{\sqrt{2}} \Big[ \chi_{\varkappa, S, \nu} + \delta_{Q12}(-1)^{2S_1 - S} \chi_{-\varkappa, S, \nu} + (2\pi)^{-3/2} \sum_{S', \nu'} \frac{e^{i\kappa r}}{r} \langle S', \nu' |\, \mathbf{f}_{\mathscr{S}}(\hat{\mathbf{r}}, \hat{\boldsymbol{\varkappa}})\, | S, \nu \rangle u_{S', \nu'} \Big], \tag{52b}$$

where

$$\langle S', \nu' |\, \mathbf{f}_{\mathscr{S}}(\hat{\mathbf{r}}, \hat{\boldsymbol{\varkappa}})\, | S, \nu \rangle \equiv \langle S', \nu' |\, \mathbf{f}(\hat{\mathbf{r}}, \hat{\boldsymbol{\varkappa}})\, | S, \nu \rangle + \delta_{Q12}(-1)^{2S_1 - S'} \langle S', \nu' |\, \mathbf{f}(-\hat{\mathbf{r}}, \hat{\boldsymbol{\varkappa}})\, | S, \nu \rangle. \tag{52c}$$

[12] See, for example, R. G. Newton [1].

From Eq. (52*b*) we see that the differential cross section is

$$\frac{d\sigma}{d\Omega}(\hat{\boldsymbol{\varkappa}}_f, S', \nu'; \hat{\boldsymbol{\varkappa}}, S, \nu) = |\langle S', \nu'|\, \mathbf{f}_{\mathscr{S}}(\hat{\boldsymbol{\varkappa}}_f, \hat{\boldsymbol{\varkappa}})\, |S, \nu\rangle|^2. \tag{53a}$$

Alternatively, according to Eq. (6-86*b*), the symmetrized *T*-matrix is

$$\begin{aligned}(\hat{\boldsymbol{\varkappa}}_f, S', \nu'|\, T_{\mathscr{S}}(\kappa)\, |\hat{\boldsymbol{\varkappa}}, S, \nu\rangle &\equiv \langle \hat{\boldsymbol{\varkappa}}_f, S', \nu'|\, T(\kappa)\, |\hat{\boldsymbol{\varkappa}}, S, \nu\rangle \\ &+ \delta_{Q12}(-1)^{2S_1-S'}\langle -\hat{\boldsymbol{\varkappa}}_f, S', \nu'|\, T(\kappa)\, |\hat{\boldsymbol{\varkappa}}, S, \nu\rangle. \end{aligned} \tag{54}$$

The differential cross section (4-121), specialized to our problem, is

$$\frac{d\sigma}{d\Omega}(\hat{\boldsymbol{\varkappa}}_f, S', \nu'; \hat{\boldsymbol{\varkappa}}, S, \nu) = \frac{(2\pi)^4}{\kappa^2}\, \rho_\epsilon^{\,2} |\langle \hat{\boldsymbol{\varkappa}}_f, S', \nu'|\, T_{\mathscr{S}}(\kappa)\, |\hat{\boldsymbol{\varkappa}}, S, \nu\rangle|^2. \tag{53b}$$

This is, of course, identical with (53*a*).

## 7.2 APPLICATION TO THE SCATTERING OF SPIN ZERO BY SPIN ONE-HALF PARTICLES

It is perhaps well to interrupt our general discussion at this point with a specific application of the results obtained in Section 7.1. We shall describe the scattering of spin zero by spin one-half particles, which provides the simplest example of scattering by noncentral forces. Applications include the scattering of nucleons by spin zero nuclei and pi-meson scattering by nucleons.[13]

We continue to describe the scattering in the barycentric coordinate system. Then in a state in which the total angular momentum is $J$ the orbital angular momentum may be $l = J + \frac{1}{2}$ or $l = J - \frac{1}{2}$, since the spin in Eqs. (10) and (24) is $S = S' = \frac{1}{2}$.

To within a common phase factor, the respective parities of the states $|lSJM\rangle$ and $|l'SJM\rangle$ in Eq. (24) are $(-1)^l$ and $(-1)^{l'}$. We shall assume for the present that parity is conserved by the interactions that lead to the scattering. Then [see the discussion following Eq. (4-221)]

$$(-1)^l = (-1)^{l'}.$$

Because $l, l'$ must be $J \pm \frac{1}{2}$, we see that $l = l'$. This conclusion permits us to write [see Eq. (25)]

$$S^J_{l'\frac{1}{2}, l\frac{1}{2}} \equiv \delta_{l'l} S_l^{\,J}, \tag{55a}$$

and Eq. (35) reduces trivially to

$$S_l^{\,J} = e^{2i\delta_l^J}. \tag{55b}$$

Here $\delta_l^{\,J}$ is the scattering phase shift in the state $(l, J)$.

[13] In this section we shall neglect complications associated with such charge exchange reactions as $\pi^- + p \rightarrow \pi^0 + n$. (See Chapter 9 for a more detailed description of pi-meson nucleon scattering.)

The fact that for each $l$ there are only two $J$ values, that is, $J = l \pm \frac{1}{2}$, suggests a more convenient notation for the quantities appearing in Eqs. (55). We write for

$$\begin{aligned} J &= l + \tfrac{1}{2}, \qquad S_l^{l+1/2} \equiv S_{l+} = e^{2i\delta_{l+}}, \\ J &= l - \tfrac{1}{2}, \qquad S_l^{l-1/2} \equiv S_{l-} = e^{2i\delta_{l-}}. \end{aligned} \tag{55c}$$

We may now specialize Eqs. (26) and (27) to the form

$$\langle \hat{\boldsymbol{\varkappa}}_f, \tfrac{1}{2}, \nu' | \, S(\kappa) \, | \hat{\boldsymbol{\varkappa}}, \tfrac{1}{2}, \nu \rangle = \sum_{lm'mJM} Y_l^{m'}(\hat{\boldsymbol{\varkappa}}_f) Y_l^{m*}(\hat{\boldsymbol{\varkappa}}) \times \langle l, \tfrac{1}{2}; m', \nu' \,|\, J, M \rangle \langle l, \tfrac{1}{2}; m, \nu \,|\, J, M \rangle S_l^J, \tag{56a}$$

and

$$\langle \hat{\boldsymbol{\varkappa}}_f, \tfrac{1}{2}, \nu' | \, T(\kappa) \, | \hat{\boldsymbol{\varkappa}}, \tfrac{1}{2}, \nu \rangle = \sum_{lm'mJM} Y_l^{m'}(\hat{\boldsymbol{\varkappa}}_f) Y_l^{m*}(\hat{\boldsymbol{\varkappa}}) \times \langle l, \tfrac{1}{2}; m', \nu' \,|\, J, M \rangle \langle l, \tfrac{1}{2}; m, \nu \,|\, J, M \rangle T_l^J. \tag{56b}$$

Here [see Eq. (29)]

$$T_l^J = \frac{1}{2\pi i \rho_\epsilon} (1 - S_l^J). \tag{57a}$$

In the notation of Eq. (55) this may be written as

$$T_{l\pm} = \frac{1}{2\pi i \rho_\epsilon} (1 - e^{2i\delta_{l\pm}}) = - \frac{1}{\pi \rho_\epsilon} e^{i\delta_{l\pm}} \sin \delta_{l\pm}, \tag{57b}$$

which should be compared with Eq. (6-51$a$).

The expressions (56$a$) and (56$b$) are easily evaluated using the expressions given in Appendix A for the Wigner coefficients.[14] The appropriate cross section may then be obtained from the expressions (20) or (22).

### The Schrödinger Equation for Nonrelativistic Scattering

In this section we specialize the results just obtained for the scattering of a spin zero by a spin one-half particle to the case that both particles have nonrelativistic energies and that they interact through a known spin orbit potential. We are therefore discussing a special case of Eqs. (45) and (47) and shall continue to use the representation and coordinates of Eqs. (6-95), (6-96), and (6-97).

The reduced potential $v = 2M_r V$ in Eq. (45) is now taken to have the familiar form[15]

$$v \equiv v_c(r) + v_s(r) \, \boldsymbol{\sigma} \cdot \mathbf{L}. \tag{58}$$

Here $v_c(r)$ and $v_s(r)$ are, respectively, the "central potential" and the "spin orbit potential," $\boldsymbol{\sigma}$ is the Pauli spin operator and $\mathbf{L}$ is the operator

[14] We shall defer an explicit evaluation of these expressions to Section 7.2, p. 363.
[15] See Ref. [3, p. 276], for example.

of orbital angular momentum for the two particles in the barycentric subspace. We shall assume that $v_c$ and $v_s$ in Eq. (58) are functions only of $r$, the separation of the two particles.

Potentials of the form (58) were first used to describe the interaction of an electron's spin with its orbital angular momentum in an atom. In nuclear physics such interactions are used to describe the elastic scattering of nucleons by atomic nuclei.

The wave functions $\psi^{\pm}_{\varkappa,S,\nu}$ [see Eq. (37)] may now be written in the form $\psi^{\pm}_{\varkappa,1/2,\nu}$; they satisfy the equation

$$(\nabla_r^{\,2} + \kappa^2 - v)\psi^{\pm}_{\varkappa,1/2,\nu} = 0, \tag{59}$$

a special case of Eq. (45). Rather than to use the general equation (47), we shall simplify this directly. Using Eq. (6-106), we may write (59) as

$$\left[\frac{1}{r^2}\frac{\partial}{\partial r}\left(r^2\frac{\partial}{\partial r}\right) - \frac{L^2}{r^2} + \kappa^2 - v_c(r) - v_s(r)\boldsymbol{\sigma}\cdot\mathbf{L}\right]\psi^{\pm}_{\varkappa,1/2,\nu} = 0. \tag{60}$$

The expansion (39*a*) may be inserted here for $\psi^{\pm}$. We now have only terms corresponding to $l' = l$ and $S' = S = \frac{1}{2}$. We also note [see the discussion accompanying the expression (12*b*)] that terms corresponding to a given value of $l$ and $J$ in the expansion (39*a*) are eigenvectors of $\boldsymbol{\sigma}\cdot\mathbf{L}$. In particular,[16]

$$\boldsymbol{\sigma}\cdot\mathbf{L}\,|l\tfrac{1}{2}JM\rangle = l\,|l\tfrac{1}{2}JM\rangle \quad \text{for} \quad J = l + \tfrac{1}{2}, \tag{61a}$$

$$\boldsymbol{\sigma}\cdot\mathbf{L}\,|l\tfrac{1}{2}JM\rangle = -(l+1)\,|l\tfrac{1}{2}JM\rangle \quad \text{for} \quad J = l - \tfrac{1}{2}, \tag{61b}$$

where $|l\frac{1}{2}JM\rangle$ is an eigenstate of total angular momentum $(J, M)$ constructed from $Y_l^{\,m}(\hat{\mathbf{r}})$ and $u_{1/2,\nu}$.

For the present application we may write the radial wave functions (39*b*) in the form

$$\begin{aligned} \psi^{\pm}_{l'1/2;\kappa J l 1/2} &\equiv \delta_{l',l}\psi^{\pm}_{\kappa J l} \\ &\equiv \delta_{l',l}\psi^{\pm}_{\kappa l+} \quad \text{for} \quad J = l + \tfrac{1}{2} \\ &\equiv \delta_{l',l}\psi^{\pm}_{\kappa l-}, \quad \text{for} \quad J = l - \tfrac{1}{2}. \end{aligned} \tag{62}$$

Equation (39*a*) is then

$$\psi^{\pm}_{\varkappa,1/2,\nu} = \sum_{l,J}\left(\sum_{\nu'}\langle\hat{\mathbf{r}},\, l\tfrac{1}{2}\nu'|\,\mathscr{Y}^J\,|\hat{\boldsymbol{\varkappa}},\, l\tfrac{1}{2}\nu\rangle u_{1/2,\nu'}\right)\psi^{\pm}_{\kappa J l}(r). \tag{63}$$

We have remarked that the coefficient of $\psi^{\pm}_{\varkappa J l}$ here is an eigenfunction of $\mathbf{J}^2$, corresponding to the eigenvalue $J(J+1)$. Therefore, substitution

[16] Equations (61) are easily obtained with the use of the operator identity $\boldsymbol{\sigma}\cdot\mathbf{L} = \mathbf{J}^2 - \frac{1}{4}\boldsymbol{\sigma}^2 - \mathbf{L}^2$, where $\mathbf{J} = \frac{1}{2}\boldsymbol{\sigma} + \mathbf{L}$.

of Eq. (63) into Eq. (60), with use of Eqs. (61), leads to the set of radial wave equations [compare Eq. (6-108)]

$$\left[\frac{1}{r^2}\frac{\partial}{\partial r}\left(r^2\frac{\partial}{\partial r}\right) - \frac{l(l+1)}{r^2} + \kappa^2 - v_{l+}(r)\right]\psi^{\pm}_{\kappa l+} = 0, \quad \text{for} \quad J = l + \tfrac{1}{2} \tag{64a}$$

and

$$\left[\frac{1}{r^2}\frac{\partial}{\partial r}\left(r^2\frac{\partial}{\partial r}\right) - \frac{l(l+1)}{r^2} + \kappa^2 - v_{l-}(r)\right]\psi^{\pm}_{\kappa l-} = 0, \quad \text{for} \quad J = l - \tfrac{1}{2}. \tag{64b}$$

Here

$$\begin{aligned} v_{l+}(r) &\equiv v_c(r) + l v_s(r), \\ v_{l-}(r) &\equiv v_c(r) - (l+1) v_s(r). \end{aligned} \tag{65}$$

Finally, we may introduce a pair of radial wave functions $w_{l\pm}(\kappa; r)$, analogous to that of Eq. (6-109*a*).

$$\begin{aligned} \psi^{+}_{\kappa l\pm}(r) &= i^l\, e^{i\delta_{l\pm}(\kappa)}\, \frac{w_{l\pm}(\kappa; r)}{\kappa r}, \\ \psi^{-}_{\kappa l\pm}(r) &= i^l\, e^{-i\delta_{l\pm}(\kappa)}\, \frac{w_{l\pm}(\kappa; r)}{\kappa r}, \end{aligned} \tag{66}$$

where the phase shifts $\delta_{l\pm}$ were introduced in Eqs. (55*c*). Substitution of these expressions into Eqs. (64) leads to the set of radial equations

$$w''_{l\pm} - \left[\frac{l(l+1)}{r^2} + v_{l\pm}(r) - \kappa^2\right] w_{l\pm} = 0. \tag{67}$$

Equations (67) are formally equivalent to Eq. (6-110). Therefore, our extensive discussion of Eq. (6-110) given in Chapter 6 is directly applicable to the present case. This includes the study of the analytic properties of the wave functions and amplitudes, as well as the special methods described for solving Eq. (6-110).

### Explicit Evaluation of the Scattering Amplitudes

As we have remarked, the evaluation of the amplitudes (56) is quite straightforward. In practice this is often most easily accomplished using projection operators, as we shall now demonstrate. We first note that Eq. (56*b*) contains implicitly the projection operator (see Section 1.1, p. 6)

$$\Lambda_J \equiv \sum_M |l\tfrac{1}{2}JM\rangle\langle l\tfrac{1}{2}JM|, \tag{68}$$

which when acting on the product state $|l\tfrac{1}{2}m\nu\rangle = Y_l^{\,m} u_{\frac{1}{2},\nu}$ yields an eigenstate of total angular momentum ($J$, $M = m + \nu$). Equations (61) permit us to write the operators $\Lambda_J$ in the form

$$\Lambda_{l+} \equiv \Lambda_{J=l+\frac{1}{2}} = \frac{l + 1 + \boldsymbol{\sigma} \cdot \mathbf{L}}{2l + 1}$$
$$\Lambda_{l-} \equiv \Lambda_{J=l-\frac{1}{2}} = \frac{l - \boldsymbol{\sigma} \cdot \mathbf{L}}{2l + 1}. \tag{69}$$

We may now use Eq. (68) to write (56*b*) in the form

$$\begin{aligned}\langle \hat{\boldsymbol{\varkappa}}_f, \tfrac{1}{2}, \nu' | T | \hat{\boldsymbol{\varkappa}}, \tfrac{1}{2}, \nu \rangle &= \sum_{lm'm} Y_l^{m'}(\hat{\boldsymbol{\varkappa}}_f)(\langle l\tfrac{1}{2}m'\nu' | \Lambda_{l+} | l\tfrac{1}{2}m\nu \rangle T_{l+} \\ &\quad + \langle l\tfrac{1}{2}m'\nu' | \Lambda_{l-} | l\tfrac{1}{2}m\nu \rangle T_{l-}) Y_l^{m*}(\hat{\boldsymbol{\varkappa}}) \\ &= \sum_{l'lm'm} Y_{l'}^{m'}(\hat{\boldsymbol{\varkappa}}_f)(\langle l'\tfrac{1}{2}m'\nu' | \Lambda_{l+} | l\tfrac{1}{2}m\nu \rangle T_{l+} \\ &\quad + \langle l'\tfrac{1}{2}m'\nu' | \Lambda_{l-} | l\tfrac{1}{2}m\nu \rangle T_{l-}) Y_l^{m*}(\hat{\boldsymbol{\varkappa}}).\end{aligned} \tag{70}$$

The second form given here is seen to be equivalent to the first form, since it is evident from Eqs. (69) that the matrix elements of $\Lambda_{l\pm}$ vanish unless $l' = l$.

To simplify Eq. (70), we note, for example, that

$$\langle l'\tfrac{1}{2}m'\nu' | \Lambda_{l+} | l\tfrac{1}{2}m\nu \rangle = \left(u_{\frac{1}{2},\nu'}, \int d\Omega_{\hat{\mathbf{q}}} Y_{l'}^{m'*}(\hat{\mathbf{q}}) \left( \frac{l + 1 + \frac{1}{i}\, \boldsymbol{\sigma} \cdot \mathbf{q} \times \nabla_{\mathbf{q}}}{2l + 1} \right) \times Y_l^{m}(\hat{\mathbf{q}}) u_{\frac{1}{2},\nu} \right). \tag{71}$$

We have here written $\mathbf{L} = -i\mathbf{q} \times \nabla_{\mathbf{q}}$, valid in a momentum representation.[17] We now multiply this expression by $Y_l^{m*}(\hat{\boldsymbol{\varkappa}})$ and sum over $m$, as in Eq. (70). There results a factor [see Eq. (1-78)]

$$\sum_{m=-l}^{+l} Y_l^{m}(\hat{\mathbf{q}}) Y_l^{m*}(\hat{\boldsymbol{\varkappa}}) = \frac{2l + 1}{4\pi} P_l(\hat{\mathbf{q}} \cdot \hat{\boldsymbol{\varkappa}}) \tag{72}$$

in the integrand. Next, we multiply the expression (71) by $Y_{l'}^{m'}(\hat{\boldsymbol{\varkappa}}_f)$ and sum over $l'$ and $m'$. Using Eq. (4*b*), we obtain a factor $\delta_{\hat{\boldsymbol{\varkappa}}_f, \hat{\mathbf{q}}}$ in the integrand. If we repeat this procedure for the $\Lambda_{l-}$ term and integrate over the solid angles $d\Omega_{\hat{\mathbf{q}}}$, we finally obtain

$$\begin{aligned}4\pi \langle \hat{\boldsymbol{\varkappa}}_f, \tfrac{1}{2}, \nu' | T | \hat{\boldsymbol{\varkappa}}, \tfrac{1}{2}, \nu \rangle &= \Bigg( u_{\frac{1}{2},\nu'}, \sum_l \Bigg[ T_{l+}\left( l + 1 + \frac{1}{i}\, \boldsymbol{\sigma} \cdot \boldsymbol{\varkappa}_f \times \nabla_{\boldsymbol{\varkappa}_f} \right) \\ &\quad + T_{l-}\left( l - \frac{1}{i}\, \boldsymbol{\sigma} \cdot \boldsymbol{\varkappa}_f \times \nabla_{\boldsymbol{\varkappa}_f} \right) \Bigg] P_l(\hat{\boldsymbol{\varkappa}}_f \cdot \hat{\boldsymbol{\varkappa}}) u_{\frac{1}{2},\nu} \Bigg).\end{aligned} \tag{73}$$

[17] That is, $\mathbf{L} = \mathbf{r} \times \mathbf{q}$, where $\mathbf{r} = i\nabla_{\mathbf{q}}$.

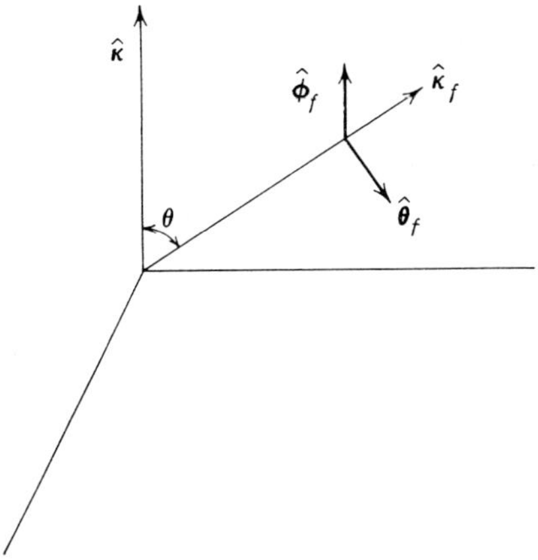


Fig. 7.1 Illustration of the unit vectors in Eq. (74).

This may be further simplified if we carry out the indicated differentiation. To do this we introduce a spherical coordinate system with coordinates $(\kappa_f, \theta, \phi)$, where $\cos\theta \equiv \hat{\boldsymbol{\varkappa}}_f \cdot \hat{\boldsymbol{\varkappa}}$ and $\phi$ is measured by the orientation of the plane $\hat{\boldsymbol{\varkappa}} \times \hat{\boldsymbol{\varkappa}}_f$. Unit vectors $\hat{\boldsymbol{\theta}}_f$ and $\hat{\boldsymbol{\phi}}_f$ are defined by the equations $\sin\theta\, \hat{\boldsymbol{\phi}}_f = \boldsymbol{\varkappa} \times \boldsymbol{\varkappa}_f$ and $\hat{\boldsymbol{\theta}}_f = \hat{\boldsymbol{\phi}}_f \times \hat{\boldsymbol{\varkappa}}_f$. These are illustrated in Fig. 7.1. Then

$$\nabla_{\varkappa_f} = \hat{\boldsymbol{\varkappa}}_f \frac{\partial}{\partial \kappa_f} + \hat{\boldsymbol{\theta}}_f \frac{1}{\kappa_f}\frac{\partial}{\partial\theta} + \hat{\boldsymbol{\phi}}_f \frac{1}{\kappa_f \sin\theta}\frac{\partial}{\partial\phi} \tag{74}$$

and

$$\nabla_{\varkappa_f} P_l(\hat{\boldsymbol{\varkappa}}_f \cdot \hat{\boldsymbol{\varkappa}}) = \frac{\hat{\boldsymbol{\theta}}_f}{\kappa_f}\frac{\partial P_l(\cos\theta)}{\partial\theta} = -\frac{\sin\theta}{\kappa_f}\hat{\boldsymbol{\theta}}_f P'_l(\cos\theta),$$

where $P'_l(z) \equiv \dfrac{dP_l(z)}{dz}$. Therefore

$$\boldsymbol{\sigma} \cdot \boldsymbol{\varkappa}_f \times \nabla_{\varkappa_f} P_l(\hat{\boldsymbol{\varkappa}}_f \cdot \hat{\boldsymbol{\varkappa}}) = -\boldsymbol{\sigma} \cdot (\hat{\boldsymbol{\varkappa}} \times \hat{\boldsymbol{\varkappa}}_f) P'_l(\cos\theta). \tag{75}$$

We obtain then, for Eq. (73),

$$4\pi \langle \hat{\boldsymbol{\varkappa}}_f, \tfrac{1}{2}, \nu' |\, T \,| \hat{\boldsymbol{\varkappa}}, \tfrac{1}{2}, \nu \rangle = \Big( u_{\frac{1}{2},\nu'}, \sum_{l=0}^{\infty} \{[(l+1)T_{l+} + lT_{l-}]P_l(\cos\theta)$$
$$+ (T_{l+} - T_{l-}) i\boldsymbol{\sigma} \cdot \hat{\boldsymbol{\varkappa}} \times \hat{\boldsymbol{\varkappa}}_f P'_l(\cos\theta)\} u_{\frac{1}{2},\nu} \Big). \tag{76}$$

The scattering amplitude (43) is often more convenient to use than is $T$. This is

$$\langle \nu' |\, \mathbf{f}(\hat{\boldsymbol{\varkappa}}_f, \hat{\boldsymbol{\varkappa}}) \,| \nu \rangle = -\frac{(2\pi)^2}{\kappa} \rho_\epsilon \langle \hat{\boldsymbol{\varkappa}}_f, \tfrac{1}{2}, \nu' |\, T \,| \hat{\boldsymbol{\varkappa}}, \tfrac{1}{2}, \nu \rangle$$
$$= (u_{\frac{1}{2},\nu'}, (f + i\boldsymbol{\sigma} \cdot \hat{\boldsymbol{\varkappa}} \times \hat{\boldsymbol{\varkappa}}_f g) u_{\frac{1}{2},\nu}), \tag{77}$$

where we have omitted the spin index $S$ and written

$$f(\cos\theta) = -\frac{\pi\rho_\epsilon}{\kappa}\sum_{l=0}^{\infty}[(l+1)T_{l+} + lT_{l-}]P_l(\cos\theta),$$
$$g(\cos\theta) = -\frac{\pi\rho_\epsilon}{\kappa}\sum_{l=1}^{\infty}(T_{l+} - T_{l-})P'_l(\cos\theta). \tag{78}$$

The amplitude $\mathbf{f}$ is a $2 \times 2$ matrix in spin space. We may therefore write it simply as

$$\mathbf{f} = f + i\boldsymbol{\sigma}\cdot\hat{\boldsymbol{\kappa}}\times\hat{\boldsymbol{\kappa}}_f g, \tag{79}$$

where it is understood that the first term is multiplied by the unit $2 \times 2$ matrix.

The differential cross section (22) for an unpolarized beam (or target) is easily evaluated from Eq. (77). We find that

$$\frac{d\bar{\sigma}}{d\Omega} = \tfrac{1}{2}\sum_{\nu,\nu'}|\langle\nu'|\,\mathbf{f}\,|\nu\rangle|^2$$
$$= \tfrac{1}{2}\,\mathrm{Tr}\,[(f^* - i\boldsymbol{\sigma}\cdot\hat{\boldsymbol{\kappa}}\times\hat{\boldsymbol{\kappa}}_f g^*)(f + i\boldsymbol{\sigma}\cdot\hat{\boldsymbol{\kappa}}\times\hat{\boldsymbol{\kappa}}_f g)]$$
$$= |f(\cos\theta)|^2 + \sin^2\theta\,|g(\cos\theta)|^2. \tag{80}$$

The "spin flip" term $g$ does not therefore contribute to the scattering for $\theta \cong 0$ (unless $g$ is singular at $\theta = 0$).[18]

When the scattering depends only on $l$, and not on $J$, so $\delta_{l+} = \delta_{l-}$ and $T_{l+} = T_{l-}$, we verify [using Eq. (57*b*)] that the scattering amplitude $\mathbf{f}$ as given by Eq. (79) reduces to that of Eq. (6-70*a*), valid for central forces. This is easily seen for the special case of potential scattering as determined from Eqs. (67). If we set the spin orbit potential $v_s = 0$, then $v_{l+} = v_{l-} = v_c$, which in turn implies that $w_{l+} = w_{l-}$.

It is frequently the case that the scattering is negligible except for $S$-waves and $P$-waves ($l = 0$ and 1). In this case we obtain from Eqs. (57*b*) and (78) the explicit expressions

$$f(\cos\theta) = \frac{1}{\kappa}\,[e^{i\delta_{0+}}\sin\delta_{0+} + (2e^{i\delta_{1+}}\sin\delta_{1+} + e^{i\delta_{1-}}\sin\delta_1)\cos\theta],$$
$$g(\cos\theta) = \frac{1}{\kappa}\,(e^{i\delta_{1+}}\sin\delta_{1+} - e^{i\delta_{1-}}\sin\delta_{1-}). \tag{81}$$

The expressions (81) have, for example, been found adequate to give a good description of pion-nucleon scattering for energies of less than 200 Mev.

[18] We saw in Section 6.4 that only a limited number of partial waves need be kept in sums such as (78) for forces of finite range. For such forces $g$ is not singular at $\theta = 0$.

**Comments Concerning the Physical Interpretation of the Scattering**

Let us return to Eq. (77). The vector $\hat{\boldsymbol{\kappa}} \times \hat{\boldsymbol{\kappa}}_f$ determines the *plane of the scattering*. This is illustrated in Fig. 7.2*a* for the case that scattering occurs to the left, so the plane of the scattering is directed *out* of the paper. Let us first suppose that the axis of quantization ($z$-axis) is directed out of the paper. Then Eq. (77) simplifies to the form

$$\begin{aligned}\langle\nu'|\,\mathbf{f}\,|\nu\rangle &= (u_{\frac{1}{2},\nu'}, (f + i\sin\theta g\sigma_z)u_{\frac{1}{2},\nu}) \\ &= \delta_{\nu',\nu}\,[f + i(-1)^{\frac{1}{2}-\nu}\sin\theta g].\end{aligned} \tag{82}$$

Here $\sin\theta > 0$ if the scattering is to the left, as in Fig. 7.2*a*, and $\sin\theta < 0$ if the scattering is to the right, as in Fig. 7.2*b*.[19]

The amplitude (82) describes the scattering of particles polarized parallel to the $z$-axis for $\nu = \frac{1}{2}$, antiparallel for $\nu = -\frac{1}{2}$. For such orientations there is no spin flip. The differential scattering cross section [see Eq. (44)] is

$$\begin{aligned}\frac{d\sigma}{d\Omega} &= |\langle\nu|\,\mathbf{f}\,|\nu\rangle|^2 \\ &= |f|^2 + \sin^2\theta\,|g|^2 + i(-1)^{\frac{1}{2}-\nu}\sin\theta(f^*g - fg^*)\end{aligned} \tag{83}$$

Unless $f^*g - fg^* = 0$, we see that the scattering shows a left-right asymmetry. This kind of asymmetry is characteristic of scattering with *polarized* particles and has been of great importance in experimental studies of scattering for noncentral interactions.

We may illustrate the dynamical origin of this asymmetry by considering a semiclassical model. Let us suppose that the spin one-half particle is scattered by a very massive spin zero particle which is at rest and that the interaction between the particles varies so slowly that a classical orbit may

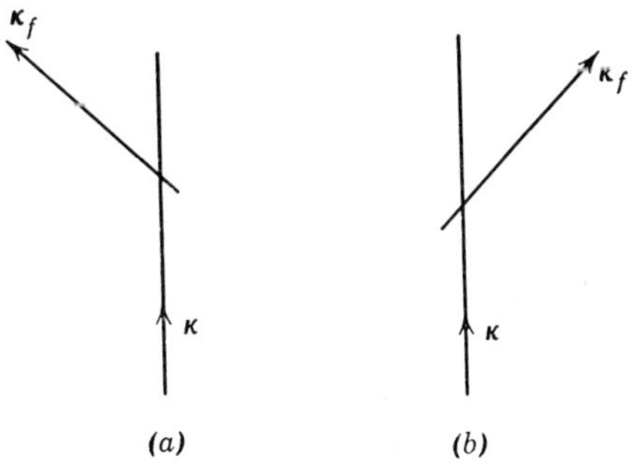


Fig. 7.2 (*a*) Scattering to the left. (*b*) Scattering to the right.

[19] We suppose the $z$-axis to be again directed out of the paper in Fig. 7.2*b*.

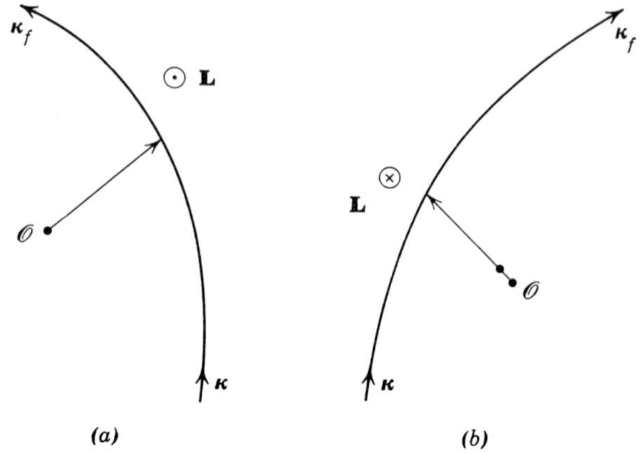


Fig. 7.3 Illustration of scattering by an attractive force. (*a*) Orbital angular momentum directed out of the page. (*b*) Orbital angular momentum directed into the page.

be defined. This is illustrated in Figs. 7.3, where the heavy particle is at point $\mathcal{O}$. The scattering is described by Eq. (59).

We shall suppose that $v_c$ and $v_s$ [Eq. (58)] are both negative, but that $|v_c|$ is sufficiently greater than $|v_s|$ that the total potential is attractive along the trajectory of the spin one-half particle. It will be assumed that the spin one-half particle has its spin directed out of the paper. Because the potential has been chosen to be attractive, the lighter particle will be deflected toward the point $\mathcal{O}$. Thus, if it passes to the right of $\mathcal{O}$, it will be deflected to the left; if it passes to the left of $\mathcal{O}$, however, it will be deflected to the right. This is illustrated in Figs. 7.3*a* and 7.3*b*, respectively. Now, for the scattering illustrated in Fig. 7.3*a* the orbital angular momentum **L** is directed out of the paper, so $\boldsymbol{\sigma}$ and **L** are "parallel." On the other hand, for the scattering illustrated in Fig. 7.3*b* the spin and orbital angular momentum are "antiparallel."

Referring to Eq. (58) [or to Eqs. (65)] we see that in the first case the interaction is *stronger* than it is in the second case. This means that for random-impact parameters there will be a net tendency for a beam of spin one-half particles to be scattered to the left. On the other hand, had we scattered particles with spin oriented into the paper, the effect just described would have been reversed and there would have been a net tendency to scatter to the right. For an experiment conducted with equal numbers of particles having spins up and down in the incident beam the asymmetry is just canceled. This may be seen from Eq. (80) or from Eq. (83).

Now, let us suppose that we do the foregoing experiment with an *unpolarized beam* (that is, one containing equal numbers of particles with spin out and spin in), but that we detect only particles scattered to the left. If we measure the spin orientations of the detected particles, we will find that more have spin out than have spin into the paper (in the geometry of Fig. 7.3). This provides a method for obtaining polarized beams of particles.[20]

Let us next consider another kind of experiment. We shall suppose that the axis of quantization ($z$-axis) is parallel to $\hat{\boldsymbol{\varkappa}}$ and that the $y$-axis is directed out of the paper (see Fig. 7.2). Now Eq. (79) reads

$$\mathbf{f} = f + i \sin \theta g \sigma_y. \tag{84a}$$

This may be written in the form of a $2 \times 2$ matrix as

$$\mathbf{f} = \begin{pmatrix} f & \sin \theta g \\ -\sin \theta g & f \end{pmatrix}. \tag{84b}$$

The four amplitudes (77) are therefore

$$\begin{aligned} \langle \tfrac{1}{2} | \, \mathbf{f} \, | \tfrac{1}{2} \rangle &= \langle -\tfrac{1}{2} | \, \mathbf{f} \, | -\tfrac{1}{2} \rangle = f, \\ \langle \tfrac{1}{2} | \, \mathbf{f} \, | -\tfrac{1}{2} \rangle &= -\langle -\tfrac{1}{2} | \, \mathbf{f} \, | \tfrac{1}{2} \rangle = \sin \theta g. \end{aligned} \tag{85}$$

In this case the scattering cross section shows no left-right asymmetry. The cross section for scattering without spin flip is

$$\frac{d\sigma}{d\Omega}(\uparrow \to \uparrow) = |f(\cos \theta)|^2. \tag{86a}$$

That for spin flip is

$$\frac{d\sigma}{d\Omega}(\uparrow \to \downarrow) = \frac{d\sigma}{d\Omega}(\downarrow \to \uparrow) = \sin^2 \theta \, |g(\cos \theta)|^2. \tag{86b}$$

In Section 7.5 we shall give a much more general description of the scattering of polarized particles.

### Some Formal Considerations

The $2 \times 2$ matrix (79) may be written in the alternative form

$$\mathbf{f} = f_1 + \boldsymbol{\sigma} \cdot \hat{\boldsymbol{\varkappa}}_f \boldsymbol{\sigma} \cdot \hat{\boldsymbol{\varkappa}} f_2. \tag{87}$$

Since $\boldsymbol{\sigma} \cdot \hat{\boldsymbol{\varkappa}}_f \boldsymbol{\sigma} \cdot \hat{\boldsymbol{\varkappa}} = \hat{\boldsymbol{\varkappa}}_f \cdot \hat{\boldsymbol{\varkappa}} - i\boldsymbol{\sigma} \cdot \hat{\boldsymbol{\varkappa}} \times \hat{\boldsymbol{\varkappa}}_f$, we see that

$$f_1 = f + \cos \theta g, \qquad f_2 = -g \tag{88}$$

[20] We shall give a more detailed discussion of scattering with polarized beams in Section 7.5.

in the notation of Eq. (78). Using the recurrence relations for Legendre polynomials,

$$\begin{aligned} (l+1)P_l(\cos\theta) + \cos\theta P'_l(\cos\theta) &= P'_{l+1}(\cos\theta) \\ lP_l(\cos\theta) - \cos\theta P'_l(\cos\theta) &= -P'_{l-1}(\cos\theta), \end{aligned} \tag{89}$$

we may write

$$\begin{aligned} f_1(\cos\theta) &= \frac{-\pi\rho_\epsilon}{\kappa}\left[\sum_{l=0}^{\infty} T_{l+}P'_{l+1}(\cos\theta) - \sum_{l=2}^{\infty} T_{l-}P'_{l-1}(\cos\theta)\right] \\ f_2(\cos\theta) &= \frac{-\pi\rho_\epsilon}{\kappa}\left[\sum_{l=1}^{\infty}(T_{l-} - T_{l+})P'_l(\cos\theta)\right]. \end{aligned} \tag{90}$$

Referring again to Eq. (77), we may ask what is the most general form for the scattering matrix. The most general operator in the two-component spin space is a linear combination of 1, the unit $2 \times 2$ matrix, and the three Pauli spin matrices.[21] We may therefore conclude quite generally that the scattering matrix on the energy shell must have the form

$$\langle \hat{\boldsymbol{\kappa}}_f, \tfrac{1}{2}, \nu' |\, T \,| \hat{\boldsymbol{\kappa}}, \tfrac{1}{2}, \nu \rangle = (u_{\frac{1}{2},\nu'}, (A + i\boldsymbol{\sigma}\cdot\mathbf{B})u_{\frac{1}{2},\nu}). \tag{91}$$

Here, since $T$ is a rotational invariant, $A = A(\hat{\boldsymbol{\kappa}}_f \cdot \hat{\boldsymbol{\kappa}})$ can be a function of $\hat{\boldsymbol{\kappa}}_f \cdot \hat{\boldsymbol{\kappa}}$ (and of course the energy), while $\mathbf{B}$ will be of the form

$$\mathbf{B} = b_1\hat{\boldsymbol{\kappa}} + b_2\hat{\boldsymbol{\kappa}}_f + b_3\hat{\boldsymbol{\kappa}} \times \hat{\boldsymbol{\kappa}}_f, \tag{92}$$

where $b_1$, $b_2$, and $b_3$ are functions of $\hat{\boldsymbol{\kappa}}_f \cdot \hat{\boldsymbol{\kappa}}$ (and the total energy).[22]

Equations (91) and (92) represent the most general form for $T$ consistent with rotational invariance. The unitarity of the $S$-matrix places further restrictions on Eq. (91). This condition follows directly from the optical theorem [Eq. (5-39)], or on inserting the expression (18) into Eq. (15). The unitarity condition is then expressed as an integral equation which must be satisfied by $T$. [Our Eq. (77) of course satisfies this condition automatically, since the $S$-matrix (54*b*) is manifestly unitary.]

If the interaction is invariant under the time reversal operation, an additional constraint is imposed on Eq. (91). From Eqs. (18) and (33) we see that

$$\langle \hat{\boldsymbol{\kappa}}_f, \tfrac{1}{2}, \nu' |\, T \,| \hat{\boldsymbol{\kappa}}, \tfrac{1}{2}, \nu \rangle = (-1)^{\frac{1}{2}+\nu-\frac{1}{2}-\nu'} \langle -\hat{\boldsymbol{\kappa}}, \tfrac{1}{2}, -\nu |\, T \,| -\hat{\boldsymbol{\kappa}}_f, \tfrac{1}{2}, -\nu' \rangle. \tag{93}$$

To see the implications of this, let us write

$$A + i\boldsymbol{\sigma}\cdot\mathbf{B} \equiv T(\hat{\boldsymbol{\kappa}}_f, \hat{\boldsymbol{\kappa}}, \boldsymbol{\sigma}) \tag{94}$$

[21] That is, an arbitrary $2 \times 2$ matrix may be written as $a \cdot 1 + \mathbf{a}\cdot\boldsymbol{\sigma}$, where $a$, $a_x$, $a_y$, and $a_z$ are four complex constants.

[22] This is fairly obvious. A proof is given in Note 4.

in Eq. (91). Because of Eq. (30), which now reads [see Eq. (1-130)]

$$i\sigma_2 u^*_{1/2,\nu} = (-1)^{1/2+\nu} u_{1/2,-\nu}, \tag{95}$$

we may rewrite the right-hand side of Eq. (93) as

$$\begin{aligned}(i\sigma_2 u^*_{1/2,\nu}, T(-\hat{\boldsymbol{\varkappa}}, -\hat{\boldsymbol{\varkappa}}_f, \boldsymbol{\sigma}) i\sigma_2 u^*_{1/2,\nu'}) &= (T^*(-\hat{\boldsymbol{\varkappa}}, -\hat{\boldsymbol{\varkappa}}_f, \boldsymbol{\sigma})\sigma_2 u_{1/2,\nu'}, \sigma_2 u_{1/2,\nu}) \\ &= (u_{1/2,\nu'}, \sigma_2 T(-\hat{\boldsymbol{\varkappa}}, -\hat{\boldsymbol{\varkappa}}_f, \tilde{\boldsymbol{\sigma}})\sigma_2 u_{1/2,\nu}) \\ &= (u_{1/2,\nu'}, T(-\hat{\boldsymbol{\varkappa}}, -\hat{\boldsymbol{\varkappa}}_f, -\boldsymbol{\sigma}) u_{1/2,\nu}),\end{aligned} \tag{96}$$

since $\sigma_2 \tilde{\boldsymbol{\sigma}} \sigma_2 = -\boldsymbol{\sigma}$. We obtain then from Eqs. (93) and (96) the general implication of time reversal invariance

$$T(\hat{\boldsymbol{\varkappa}}_f, \hat{\boldsymbol{\varkappa}}, \boldsymbol{\sigma}) = T(-\hat{\boldsymbol{\varkappa}}, -\hat{\boldsymbol{\varkappa}}_f, -\boldsymbol{\sigma}). \tag{97}$$

Now, on comparing Eqs. (94) and (97) we find that

$$A(\hat{\boldsymbol{\varkappa}}_f \cdot \hat{\boldsymbol{\varkappa}}) + i\boldsymbol{\sigma} \cdot \mathbf{B}(\hat{\boldsymbol{\varkappa}}_f, \hat{\boldsymbol{\varkappa}}) = A(\hat{\boldsymbol{\varkappa}}_f \cdot \hat{\boldsymbol{\varkappa}}) - i\boldsymbol{\sigma} \cdot \mathbf{B}(-\hat{\boldsymbol{\varkappa}}, -\hat{\boldsymbol{\varkappa}}_f). \tag{98}$$

Time reversal invariance therefore imposes no restriction on the form of $A$.[23] On the other hand, inserting the expression (92) into Eq. (98) leads us to the conclusion that $b_1 = b_2$, while $b_3$ is not subject to restrictions.[23] We thus have

$$\begin{aligned}\langle \hat{\boldsymbol{\varkappa}}_f, \tfrac{1}{2}, \nu' | T | \hat{\boldsymbol{\varkappa}}, \tfrac{1}{2}, \nu \rangle = (u_{1/2,\nu'}, [A &+ ib_1 \boldsymbol{\sigma} \cdot (\hat{\boldsymbol{\varkappa}} + \hat{\boldsymbol{\varkappa}}_f) \\ &+ ib_3 \boldsymbol{\sigma} \cdot \hat{\boldsymbol{\varkappa}} \times \hat{\boldsymbol{\varkappa}}_f] u_{1/2,\nu}),\end{aligned} \tag{99}$$

as the most general expression consistent with invariance under rotations and time reversal.

Let us next suppose that parity is conserved, so $\mathscr{P} T \mathscr{P}^{-1} = T$, where $\mathscr{P}$ is the parity operator and $T$ the scattering matrix. We verify from Eq. (10*a*) that

$$\mathscr{P} \chi_{\varkappa,1/2,\nu} = \chi_{-\varkappa,1/2,\nu}, \tag{100}$$

to within a phase factor which is independent of $\boldsymbol{\varkappa}$ and $\nu$.[24] Therefore,

$$\begin{aligned}\langle \hat{\boldsymbol{\varkappa}}_f, \tfrac{1}{2}, \nu' | T | \hat{\boldsymbol{\varkappa}}, \tfrac{1}{2}, \nu \rangle &= (\chi_{\varkappa_f,1/2,\nu'}, T\chi_{\varkappa,1/2,\nu}) \\ &= (\chi_{\varkappa_f,1/2,\nu'}, \mathscr{P}^{-1} T \mathscr{P} \chi_{\varkappa,1/2,\nu}) \\ &= \langle -\hat{\boldsymbol{\varkappa}}_f, \tfrac{1}{2}, \nu' | T | -\hat{\boldsymbol{\varkappa}}, \tfrac{1}{2}, \nu \rangle \\ &= (u_{1/2,\nu'}, T(-\hat{\boldsymbol{\varkappa}}_f, -\hat{\boldsymbol{\varkappa}}, \boldsymbol{\sigma}) u_{1/2,\nu}),\end{aligned} \tag{101}$$

in the notation of Eqs. (91) and (94). Reference to Eq. (99) shows that now $b_1 = 0$. Therefore Eq. (77) represents the most general expression for the scattering matrix which is consistent with time reversal and rotational invariance and with parity conservation.

[23] This would not have been the case had we not first imposed the condition of invariance under rotations.

[24] See the discussion in Section 1.2.

**The Case That Parity Is Not Conserved**

We conclude our discussion of the scattering of a spinless particle by a spin one-half particle by describing the scattering for the case that parity is not conserved. Our results are also applicable to a reaction in which one of the two particles changes its intrinsic parity during the scattering process. Some useful operator techniques will be illustrated in the course of our discussion.

When parity is not conserved, we have the condition

$$(-1)^l = -(-1)^{l'}$$

on the nonvanishing matrix elements of $T^J_{l'\frac{1}{2},l\frac{1}{2}}$. In this case we need consider only the amplitudes $T^{l+\frac{1}{2}}_{(l+1)\frac{1}{2},l\frac{1}{2}}$ and $T^{l-\frac{1}{2}}_{(l-1)\frac{1}{2},l\frac{1}{2}}$. Now Eq. (27) takes the form

$$\begin{aligned}\langle \hat{\boldsymbol{\varkappa}}_f, \tfrac{1}{2}, \nu' |\, T\, | \hat{\boldsymbol{\varkappa}}, \tfrac{1}{2}, \nu\rangle = \sum_{lm'mM} \Big[ & Y^{m'}_{l+1}(\hat{\boldsymbol{\varkappa}}_f) \\ & \times \langle l+1, \tfrac{1}{2}; m', \nu' \,|\, l+\tfrac{1}{2}, M\rangle T^{l+\frac{1}{2}}_{(l+1)\frac{1}{2},l\frac{1}{2}} \\ & \times \langle l, \tfrac{1}{2}; m, \nu \,|\, l+\tfrac{1}{2}, M\rangle Y_l^{m*}(\hat{\boldsymbol{\varkappa}}) \\ & + Y^{m'}_{l-1}(\hat{\boldsymbol{\varkappa}}_f)\langle l-1, \tfrac{1}{2}; m', \nu' \,|\, l-\tfrac{1}{2}, M\rangle T^{l-\frac{1}{2}}_{(l-1)\frac{1}{2},l\frac{1}{2}} \\ & \times \langle l, \tfrac{1}{2}; m, \nu \,|\, l-\tfrac{1}{2}, M\rangle Y_l^{m*}(\hat{\boldsymbol{\varkappa}}) \Big]. \end{aligned} \tag{102}$$

To simplify this we must modify our use of projection operators as made in Section 7.2, p. 362. To do this, we first consider the eigenfunction of total angular momentum

$$\begin{aligned}\mathscr{Y}^M_{Jl\frac{1}{2}} &\equiv \langle \hat{\mathbf{q}}, s \,|\, l, \tfrac{1}{2}, J, M\rangle \\ &= \sum_{m,\nu} \langle l, \tfrac{1}{2}; m, \nu \,|\, J, M\rangle Y_l^m(\hat{\mathbf{q}}) u_{\frac{1}{2},\nu}(s),\end{aligned} \tag{103}$$

where $s$ represents a spin variable. Making use of Eq. (30), we see that

$$T\mathscr{Y}^M_{Jl\frac{1}{2}} = (-1)^{\frac{1}{2}+M}\mathscr{Y}^{-M}_{Jl\frac{1}{2}}, \tag{104}$$

where $T$ is the time reversal operator.

Now,

$$\boldsymbol{\sigma} \cdot \hat{\mathbf{q}}\, \mathscr{Y}^M_{J(J\mp\frac{1}{2})\frac{1}{2}} = \alpha_\pm \mathscr{Y}^M_{J(J\pm\frac{1}{2})\frac{1}{2}}, \tag{105}$$

where the $\alpha_\pm$ are two constants. To show this, we first note that $\boldsymbol{\sigma} \cdot \hat{\mathbf{q}}$ is a rotational invariant and consequently commutes with the total angular momentum operator $\mathbf{J} = \frac{1}{2}\boldsymbol{\sigma} + \mathbf{L}$. Therefore, multiplication by $\boldsymbol{\sigma} \cdot \hat{\mathbf{q}}$ *does not* change the total angular momentum. Since $\boldsymbol{\sigma} \cdot \hat{\mathbf{q}}$ changes sign as $\hat{\mathbf{q}} \to -\hat{\mathbf{q}}$, however, multiplication of $\mathscr{Y}$ by $\boldsymbol{\sigma} \cdot \hat{\mathbf{q}}$ must change the orbital angular momentum by one unit. Equation (105) follows directly from these two observations.

Since $(\boldsymbol{\sigma}\cdot\hat{\mathbf{q}})^2 = 1$, and since the $\mathscr{Y}$'s are normalized to unity, we have

$$(\sigma\cdot q\mathscr{Y},\, \sigma\cdot q\mathscr{Y}) = (\mathscr{Y},\mathscr{Y}) = 1 = |\alpha_{\pm}|^2.$$

Thus the $\alpha_{\pm}$ are of modulus unity. On applying $\boldsymbol{\sigma}\cdot\hat{\mathbf{q}}$ a second time to Eq. (105), we obtain $\alpha_{+}\alpha_{-} = 1$. Finally, we may apply the time reversal operator to both sides of Eq. (105) and use Eq. (104) to show that $\alpha_{\pm} = (\alpha_{\pm})^*$, since $\boldsymbol{\sigma}$ changes sign under time reversal. We conclude then that either $\alpha_{+} = \alpha_{-} = 1$, or $\alpha_{+} = \alpha_{-} = -1$. We shall take $\alpha_{+} = \alpha_{-} = 1$, the sign ambiguity being irrelevant for our considerations.

In anticipation that we shall use the $(\hat{\mathbf{q}}, s)$ representation, we may therefore write

$$\sum_M |l \pm 1, \tfrac{1}{2}, l \pm \tfrac{1}{2}, M\rangle\langle l, \tfrac{1}{2}, l \pm \tfrac{1}{2}, M|$$
$$= \boldsymbol{\sigma}\cdot\hat{\mathbf{q}} \sum_M |l, \tfrac{1}{2}, l \pm \tfrac{1}{2}, M\rangle\langle l, \tfrac{1}{2}, l \pm \tfrac{1}{2}, M| = \boldsymbol{\sigma}\cdot\hat{\mathbf{q}}\,\Lambda_{l\pm}, \quad (106)$$

where the $\Lambda_{l\pm}$ are the projection operators introduced in Eqs. (69).

The argument now goes through just as it did for Eq. (70). For example,

$$\sum_{m'm} Y_{l+1}^{m'}(\hat{\boldsymbol{\varkappa}}_f)\langle l+1, \tfrac{1}{2}; m', \nu'|\ \boldsymbol{\sigma}\cdot\hat{\mathbf{q}}\,\Lambda_{l+}\,|l, \tfrac{1}{2}; m, \nu\rangle Y_l^{m*}(\hat{\boldsymbol{\varkappa}})$$
$$= \left(u_{\frac{1}{2},\nu'}, \sum_{m'} Y_{l+1}^{m'}(\hat{\boldsymbol{\varkappa}}_f) \int d\Omega_{\hat{\mathbf{q}}}\, Y_{l+1}^{m'*}(\hat{\mathbf{q}})\boldsymbol{\sigma}\cdot\hat{\mathbf{q}}\,\Lambda_{l+}(\hat{\mathbf{q}})\,\frac{2l+1}{4\pi}\,P_l(\hat{\mathbf{q}}\cdot\hat{\boldsymbol{\varkappa}})u_{\frac{1}{2},\nu}\right)$$
$$= \left(u_{\frac{1}{2},\nu'}, \sum_{l',m'} Y_{l'}^{m'}(\hat{\boldsymbol{\varkappa}}_f) \int d\Omega_{\hat{\mathbf{q}}}\, Y_{l'}^{m'*}(\hat{\mathbf{q}})\boldsymbol{\sigma}\cdot\hat{\mathbf{q}}\,\Lambda_{l+}(\hat{\mathbf{q}})\,\frac{2l+1}{4\pi}\,P_l(\hat{\mathbf{q}}\cdot\hat{\boldsymbol{\varkappa}})u_{\frac{1}{2},\nu}\right)$$
$$= (u_{\frac{1}{2},\nu'}, \boldsymbol{\sigma}\cdot\hat{\boldsymbol{\varkappa}}_f\,\Lambda_{l+}(\hat{\boldsymbol{\varkappa}}_f)P_l(\hat{\boldsymbol{\varkappa}}_f\cdot\hat{\boldsymbol{\varkappa}})u_{\frac{1}{2},\nu})\,\frac{2l+1}{4\pi}, \quad (107)$$

where [see Eq. (71)]

$$\Lambda_{l+}(\hat{\mathbf{q}}) = \frac{l+1+(1/i)\boldsymbol{\sigma}\cdot\mathbf{q}\times\nabla_{\mathbf{q}}}{2l+1}. \quad (108)$$

Just as was the case in Section 7.2, p. 362, we are able to sum over $l'$ here since only the term with $l' = l+1$ is nonvanishing.

On completing the evaluation of Eq. (102), we obtain (here $\cos\theta \equiv \hat{\boldsymbol{\varkappa}}_f\cdot\hat{\boldsymbol{\varkappa}}$)

$$4\pi\langle\hat{\boldsymbol{\varkappa}}_f, \tfrac{1}{2}, \nu'|\,T\,|\hat{\boldsymbol{\varkappa}}, \tfrac{1}{2}, \nu\rangle = (u_{\frac{1}{2},\nu'}, \sum_l \{T^{l+\frac{1}{2}}_{(l+1)\frac{1}{2},l\frac{1}{2}}\boldsymbol{\sigma}\cdot\hat{\boldsymbol{\varkappa}}_f$$
$$\times [(l+1)P_l(\cos\theta) + i\boldsymbol{\sigma}\cdot\hat{\boldsymbol{\varkappa}}\times\hat{\boldsymbol{\varkappa}}_f P'_l(\cos\theta)]$$
$$+ T^{l-\frac{1}{2}}_{(l-1)\frac{1}{2},l\frac{1}{2}}\boldsymbol{\sigma}\cdot\hat{\boldsymbol{\varkappa}}_f[lP_l(\cos\theta)$$
$$- i\boldsymbol{\sigma}\cdot\hat{\boldsymbol{\varkappa}}\times\hat{\boldsymbol{\varkappa}}_f P'_l(\cos\theta)]\}u_{\frac{1}{2},\nu}). \quad (109)$$

If we simplify the spin products and use Eqs. (89), this becomes

$$4\pi, \langle \hat{\boldsymbol{\varkappa}}_f, \tfrac{1}{2}, \nu' |\, T\, | \hat{\boldsymbol{\varkappa}}, \tfrac{1}{2}, \nu \rangle = (u_{\frac{1}{2},\nu'}, (a\boldsymbol{\sigma}\cdot\hat{\boldsymbol{\varkappa}}_f + b\boldsymbol{\sigma}\cdot\hat{\boldsymbol{\varkappa}})\, u_{\frac{1}{2},\nu}), \tag{110}$$

where

$$\begin{aligned} a &= \sum_l [T^{l+\frac{1}{2}}_{(l+1)\frac{1}{2},l\frac{1}{2}}\, P'_{l+1}(\cos\theta) - T^{l-\frac{1}{2}}_{(l-1)\frac{1}{2},l\frac{1}{2}}\, P'_{l-1}(\cos\theta)], \\ b &= \sum_l [-T^{l+\frac{1}{2}}_{(l+1)\frac{1}{2},l\frac{1}{2}}\, P'_l(\cos\theta) + T^{l-\frac{1}{2}}_{(l-1)\frac{1}{2},l\frac{1}{2}}\, P'_l(\cos\theta)]. \end{aligned} \tag{111}$$

We have seen that time reversal invariance implies that

$$T^{l+\frac{1}{2}}_{(l+1)\frac{1}{2},l\frac{1}{2}} = T^{l+\frac{1}{2}}_{l\frac{1}{2},(l+1)\frac{1}{2}}$$

and

$$T^{l-\frac{1}{2}}_{(l-1)\frac{1}{2},l\frac{1}{2}} = T^{l-\frac{1}{2}}_{l\frac{1}{2},(l-1)\frac{1}{2}}.$$

When these conditions are satisfied, $a = b$, consistent with Eq. (99).

When the scattering interaction contains a parity-conserving part and a parity-violating part, the scattering matrix will be the sum of the expressions (77) and (110).

### 7.3 SCATTERING REACTIONS

In this section we consider scattering reactions, restricting ourselves, however, to the case that there are two particles in the initial and two in the final state. Using the notation of Section 4.1, we designate the incident (or entrance) channel by $c$ and the exit channel by $c'$. In channel $c$ the two particles have respective spins $S_1$ and $S_2$; in channel $c'$ the particles have spins $S'_1$ and $S'_2$.

As before, we describe the scattering in the barycentric coordinate system. We shall continue to write $\boldsymbol{\varkappa}$ and $-\boldsymbol{\varkappa}$ for the momenta of the two particles in the initial state, $\boldsymbol{\varkappa}_f$ and $-\boldsymbol{\varkappa}_f$ for the momenta of the two particles in the final state. The respective total energies are then expressed as

$$\begin{aligned} \epsilon_c(\kappa) &= \text{energy in initial state}, \\ \epsilon_{c'}(\kappa_f) &= \text{energy in final state}. \end{aligned} \tag{112}$$

The expression (11$a$) for a wave function in the initial channel is now (in the barycentric subspace)

$$|c; \boldsymbol{\varkappa}, S, \nu\rangle = \chi_{c;\boldsymbol{\varkappa},S,\nu} = (2\pi)^{-\frac{3}{2}}\, e^{i\boldsymbol{\varkappa}\cdot\mathbf{r}_c} g_{c;S,\nu}(\xi), \tag{113}$$

with corresponding wave functions $|c'; \boldsymbol{\varkappa}_f, S, \nu\rangle$ for the final channel. Here $\mathbf{r}_c$, $\mathbf{r}_{c'}$, and so on, represent relative coordinates for the particles in the respective channels $c$ and $c'$, and so on. To indicate that the particles are in general *composite*, we have written $g_{c;S,\nu}(\xi)$ [and $g_{c';S,\nu}(\xi)$, and so on] to designate the internal-state wave functions for the two particles, where the spins $S_1$ and $S_2$ (or $S'_1$, $S'_2$) have been combined to give a resultant spin $S$ with orientation $\nu$ (or $S'$, $\nu'$).[25]

We shall illustrate our present discussion in Chapter 9 with an application to the photoproduction of pi-mesons, the reaction being

$$\gamma + n \rightarrow \pi + n,$$

where $n$ represents a nucleon.

**Eigenstate Analysis of the Scattering Matrix**

We shall now transform the $S$-matrix to a representation in which it is diagonal, as was done in Section 7.1. Equation (13) is generalized for two-particle reactions to read

$$\langle c'; b|\, S\, |c; a\rangle = \delta(\mathbf{P}_b - 0)\langle c'; \boldsymbol{\varkappa}_f, S', \nu'|\, S\, |c; \boldsymbol{\varkappa}, S, \nu\rangle. \tag{114}$$

The coefficient of the $\delta$-function on the right-hand side of this equation is unitary in the barycentric subspace.

Instead of Eq. (14$a$), we now write

$$\langle c'; \boldsymbol{\varkappa}_f, S', \nu'|\, S\, |c; \boldsymbol{\varkappa}, S, \nu\rangle = \frac{1}{\sqrt{\rho_{\epsilon_{c'}}\rho_{\epsilon_c}}}\,\delta[\epsilon_{c'}(\kappa_f) - \epsilon_c(\kappa)] \times \langle c'; \hat{\boldsymbol{\varkappa}}_f, S', \nu'|\, S(\kappa)\, |c; \hat{\boldsymbol{\varkappa}}, S, \nu\rangle \tag{115$a$}$$

which defines $\langle c'; \hat{\boldsymbol{\varkappa}}_f, S', \nu'|\, S(\kappa)\, |c; \hat{\boldsymbol{\varkappa}}, S, \nu\rangle$ on the energy and momentum shells. Here we have generalized Eq. (6-30) to define

$$\rho_{\epsilon_c}(\kappa) = \frac{\kappa^2}{d\epsilon_c/d\kappa} \tag{116}$$

for all channels $c$.

The expression of unitarity for the $S$-matrix is

$$\begin{aligned}\delta_{c',c}\delta(\boldsymbol{\varkappa}' - \boldsymbol{\varkappa})\delta_{S',S}\delta_{\nu',\nu} &= \sum_{c'',S'',\nu''}\int \langle c' : \boldsymbol{\varkappa}', S', \nu'|\, S^\dagger\, |c''; \boldsymbol{\varkappa}'', S'', \nu''\rangle\, d^3\boldsymbol{\varkappa}'' \\ &\quad \times \langle c''; \boldsymbol{\varkappa}'', S'', \nu''|\, S\, |c; \boldsymbol{\varkappa}, S, \nu\rangle \\ &= \frac{1}{\sqrt{\rho_{\epsilon_{c'}}\rho_{\epsilon_c}}}\,\delta[\epsilon_{c'}(\kappa') - \epsilon_c(\kappa)] \\ &\quad \times \sum_{c'',S'',\nu''}\int \langle c'; \hat{\boldsymbol{\varkappa}}', S', \nu'|\, S^\dagger(\kappa)\, |c''; \hat{\boldsymbol{\varkappa}}'', S'', \nu''\rangle \\ &\quad \times d\Omega_{\hat{\varkappa}''}\langle c''; \hat{\boldsymbol{\varkappa}}'', S'', \nu''|\, S(\kappa)\, |c; \hat{\boldsymbol{\varkappa}}, S, \nu\rangle.\end{aligned} \tag{117}$$

[25] We recall the convention adopted in Chapter 6 that we write $u_{S,\nu}$ for the internal-state wave function when we prefer to treat the particles as "elementary." We emphasize again that this distinction is considered as one of convenience, rather than as being fundamental.

With use of the relation [see Eq. (4*a*)]

$$\delta_{c',c}\delta(\boldsymbol{\varkappa}'-\boldsymbol{\varkappa})=\frac{1}{\sqrt{\rho_{\epsilon_{c'}}\rho_{\epsilon_c}}}\,\delta[\epsilon_{c'}(\kappa')-\epsilon_c(\kappa)]\delta_{\hat{\boldsymbol{\varkappa}}',\hat{\boldsymbol{\varkappa}}}\delta_{c',c}, \tag{118}$$

we may conclude from Eq. (117) that

$$\sum_{c'',S'',\nu''}\int\langle c';\hat{\boldsymbol{\varkappa}}',S',\nu'|\,S^\dagger(\kappa)\,|c'';\hat{\boldsymbol{\varkappa}}'',S'',\nu''\rangle \times d\Omega_{\hat{\boldsymbol{\varkappa}}''}\langle c'';\hat{\boldsymbol{\varkappa}}'',S'',\nu''|\,S(\kappa)\,|c;\hat{\boldsymbol{\varkappa}},S,\nu\rangle=\delta_{\hat{\boldsymbol{\varkappa}}',\hat{\boldsymbol{\varkappa}}}\delta_{c',c}\delta_{S',S}\delta_{\nu',\nu}. \tag{119}$$

This equation represents an explicit statement of the unitarity of the $S$-matrix on the energy and momentum shells. It is a direct generalization of Eq. (15) to multichannel scattering.

If in analogy to Eq. (14*b*) we normalize the wave functions per unit energy interval, Eq. (115*a*) is replaced by

$$\langle c';\epsilon_{c'},\hat{\boldsymbol{\varkappa}}_f,S',\nu'|\,S\,|c;\epsilon_c,\hat{\boldsymbol{\varkappa}},S,\nu\rangle =\delta(\epsilon_{c'}-\epsilon_c)\langle c';\hat{\boldsymbol{\varkappa}}_f,S',\nu'|\,S\,|c;\hat{\boldsymbol{\varkappa}},S,\nu\rangle. \tag{115b}$$

The $S$-matrix may be related to the $T$-matrix in the barycentric subspace by the equation [see Eq. (4-15)]

$$\langle c';b|\,S\,|c;a\rangle=\delta(\mathbf{P}_b-0)\{\delta_{c',c}\delta_{S',S}\delta_{\nu',\nu}\delta(\boldsymbol{\varkappa}_f-\boldsymbol{\varkappa}) -2\pi i\delta[\epsilon_{c'}(\kappa_f)-\epsilon_c(\kappa)]\langle c';\hat{\boldsymbol{\varkappa}}_f,S',\nu'|\,T(\kappa)\,|c;\hat{\boldsymbol{\varkappa}},S,\nu\rangle\}, \tag{120}$$

where $\langle c';\hat{\boldsymbol{\varkappa}}_f,S',\nu'|\,T(\kappa)\,|c;\hat{\boldsymbol{\varkappa}},S,\nu\rangle$ is the $T$-matrix on the energy and momentum shells in the barycentric coordinate system. The relation (118) and Eqs. (114) and (115*a*) permit us to rewrite Eq. (120) as

$$\langle c';\hat{\boldsymbol{\varkappa}}_f,S',\nu'|\,S(\kappa)\,|c;\hat{\boldsymbol{\varkappa}},S,\nu\rangle=\delta_{c',c}\delta_{S',S}\delta_{\nu',\nu}\delta_{\hat{\boldsymbol{\varkappa}}_f,\hat{\boldsymbol{\varkappa}}} -2\pi i\sqrt{\rho_{\epsilon_{c'}}\rho_{\epsilon_c}}\langle c';\hat{\boldsymbol{\varkappa}}_f,S',\nu'|\,T(\kappa)\,|c;\hat{\boldsymbol{\varkappa}},S,\nu\rangle. \tag{121}$$

The scattering cross section is [see Eq. (19)]

$$d\sigma(c';\boldsymbol{\varkappa}_f,S',\nu';c;\boldsymbol{\varkappa},S,\nu)=\frac{(2\pi)^4}{v_{\text{rel},B}}\int d^3\kappa_f\delta[\epsilon_{c'}(\kappa_f)-\epsilon_c(\kappa)] \times|\langle c';\hat{\boldsymbol{\varkappa}}_f,S',\nu'|\,T(\kappa)\,|c;\hat{\boldsymbol{\varkappa}},S,\nu\rangle|^2. \tag{122}$$

Now, the incident relative velocity of the colliding particles is

$$v_{\text{rel},B}=\frac{d\epsilon_c}{d\kappa}=\frac{\kappa^2}{\rho_{\epsilon_c}},$$

so

$$\begin{aligned}\frac{d\sigma}{d\Omega}&=\frac{(2\pi)^4}{\kappa^2}(\rho_{\epsilon_{c'}}\rho_{\epsilon_c})\,|\langle c';\hat{\boldsymbol{\varkappa}}_f,S',\nu'|\,T(\kappa)\,|c;\hat{\boldsymbol{\varkappa}},S,\nu\rangle|^2\\&=\frac{(2\pi)^2}{\kappa^2}\,|\langle c';\hat{\boldsymbol{\varkappa}}_f,S',\nu'|\,S(\kappa)\,|c;\hat{\boldsymbol{\varkappa}},S,\nu\rangle|^2.\end{aligned} \tag{123}$$

[The last form is valid, according to Eq. (121), when $\delta_{c'c}\delta_{S',S}\delta_{\nu',\nu}\delta_{\hat{\boldsymbol{\varkappa}}_f,\hat{\boldsymbol{\varkappa}}} = 0$.] The differential scattering cross section for an unpolarized beam target and any final spin orientation and spin is given by Eq. (22).

Equation (24) is now generalized to the form [see also Eq. (25)]

$$\begin{aligned}\langle c'; \hat{\boldsymbol{\varkappa}}_f, S', \nu' |\, S(\kappa)\, | c; \hat{\boldsymbol{\varkappa}}, S, \nu\rangle &= \sum_{l'm'lmJM} Y_{l'}^{m'}(\hat{\boldsymbol{\varkappa}}_f)\langle l', S'; m', \nu' \mid J, M\rangle \\ &\qquad \times S^J_{c'l'S',clS}\langle l, S; m, \nu \mid J, M\rangle Y_l^{m*}(\hat{\boldsymbol{\varkappa}}) \\ &= \sum_{l'lJ} \langle \hat{\boldsymbol{\varkappa}}_f, l'S'\nu' |\, \mathscr{Y}^J \,| \hat{\boldsymbol{\varkappa}}, lS\nu\rangle S^J_{c'l'S',clS}, \end{aligned} \tag{124}$$

where the notation of Eq. (12*a*) has been used. The quantities $S^J_{c'l'S',clS}$ represent the matrix elements of the $S$-matrix in the $(clSJM)$ representation. Since the transformation (124) is unitary we have

$$\sum_{c''l''S''} (S^J_{c''l''S'',c'l'S'})^*(S^J_{c''l''S'',clS}) = \delta_{c',c}\delta_{l',l}\delta_{S',S}. \tag{125}$$

To obtain the consequence of time reversal invariance we generalize Eq. (30) to read

$$Tg_{c;S,\nu} = (-1)^{S+\nu} g_{c;S,-\nu} \tag{126}$$

for all channels $c$. On applying the time reversal operator $T$ to the steady-state scattering wave functions, we have then instead of Eq. (31)

$$T\psi^{\pm}_{c;\hat{\boldsymbol{\varkappa}},S,\nu} = (-1)^{S+\nu}\psi^{\pm}_{c;-\hat{\boldsymbol{\varkappa}},S,-\nu} \tag{127}$$

for each incident channel $c$. Now, since by Eq. (5-60)

$$\langle c'; \boldsymbol{\varkappa}_f, S', \nu' |\, S\, | c; \boldsymbol{\varkappa}, S, \nu\rangle = (\psi^-_{c';\hat{\boldsymbol{\varkappa}}_f,S',\nu'}, \psi^+_{c;\hat{\boldsymbol{\varkappa}}_f,S,\nu}), \tag{128}$$

we obtain, by an argument identical with that leading to Eq. (34),

$$S^J_{c'l'S',clS} = S^J_{clS,c'l'S'}. \tag{129}$$

The symmetric, unitary matrix $S^J_{c'l'S',clS}$ may be diagonalized by a real, orthogonal matrix $U$. Thus, Eq. (35) can be generalized to the form

$$S^J_{c'l'S',clS} = \sum_\alpha U^J_{c'l'S';\alpha}\, e^{2i\delta_\alpha{}^J} U^J_{\alpha;clS}, \tag{130}$$

where the $\delta_\alpha{}^J$ are again called *eigenphase* shifts.

With Eq. (130) we have accomplished our objective of diagonalizing the $S$-matrix. This is an important relation used in specific evaluations of the $S$-matrix. As described in connection with Eq. (35), it also permits a simple counting of the number of parameters required to describe a given reaction.[26]

The scattering matrix $T$ may also be easily transformed to the $(clSJ)$

[26] An alternative formulation of this problem is described in Appendix E.

representation. On following the arguments leading to Eq. (27), we obtain now

$$\langle c'; \hat{\boldsymbol{\kappa}}_f, S', \nu' |\, T(\kappa)\,| c; \hat{\boldsymbol{\kappa}}, S, \nu\rangle = \sum_{l'lJ} (\hat{\boldsymbol{\kappa}}_f, l'S'\nu' |\, \mathcal{Y}^J\, | \hat{\boldsymbol{\kappa}}, lS\nu\rangle T^J_{c'l'S',clS}(\kappa). \quad (131)$$

Here the $T^J_{c'l'S',clS}(\kappa)$ are the matrix elements of $T$ on the energy shell and in the $(clSJ)$ representation. [Off the energy shell these are written as $T^J_{c'l'S',clS}(\kappa', \kappa)$, as was done in Eq. (27).]

The expansions (124) and (131) may be inserted into Eq. (121) to give the important relation

$$S^J_{c'l'S',clS} = \delta_{c',c}\delta_{l',l}\delta_{S',S} - 2\pi i\sqrt{\rho_{\epsilon_{c'}}\rho_{\epsilon_c}}\, T^J_{c'l'S',clS}(\kappa). \quad (132)$$

The expansions (124) and (131) are generally valid, even for interacting charged particles (with the understanding that the Coulomb force is shielded at large distances). Since the partial wave expansion for Coulomb scattering converges slowly, it is ordinarily convenient to sum this contribution in closed form. To do this we proceed as in Section 6.4, p. 265, writing

$$S^J_{c'l'S',clS} = \delta_{c',c}\delta_{l',l}\delta_{S',S}\, e^{2i\sigma_{l,c}} + (S^J_{c'l'S',clS} - \delta_{c',c}\delta_{l',l}\delta_{S',S}\, e^{2i\sigma_{l,c}}).$$

Here $\sigma_{l,c}$ is the phase shift in channel $c$ for Coulomb scattering, defined by Eq. (6-179). When this expression is inserted into Eq. (124) the first term may be summed to give the Coulomb scattering amplitude. The second term describes the effects of other interactions. This can be put into a simpler form by defining a reduced $S$-matrix $s^J_{c'l'S,clS}$.

$$S^J_{c'l'S',clS} \equiv e^{i\sigma_{l',c'}} s^J_{c'l'S',clS}\, e^{i\sigma_{l,c}}. \quad (133a)$$

We then have

$$S^J_{c'l'S',clS} = \delta_{c',c}\delta_{l',l}\delta_{S',S}\, e^{2i\sigma_{l,c}} + e^{i\sigma_{l',c'}}(s^J_{c'l'S',clS} - \delta_{c',c}\delta_{l',l}\delta_{S',S})e^{i\sigma_{l,c}}. \quad (133b)$$

When the interaction arises from Coulomb plus short-range forces, of range $d$, this form is convenient since then

$$s^J_{c'l'S',clS} \cong \delta_{c',c}\delta_{l',l}\delta_{S',S}$$

for $l' \gg \kappa_f d$ or $l \gg \kappa d$ (see the discussion on p. 258).

### The Wave Function

The wave function describing scattering from the initial state $|c; \boldsymbol{\kappa}, S, \nu\rangle$ is defined by Eq. (4-8) as

$$\psi^+_{c;\hat{\boldsymbol{\kappa}},S,\nu} = \left[1 + \frac{1}{\epsilon_c(\kappa) + i\eta - H}\, V_c\right]\chi_{c;\hat{\boldsymbol{\kappa}},S,\nu}, \quad (134a)$$

where $H$ is the Hamiltonian for the system and $V_c$ is the interaction in channel $c$. This wave function satisfies the Schrödinger equation

$$[H - \epsilon_c(\kappa)]\psi^+_{c;\hat{\boldsymbol{\kappa}},S,\nu} = 0. \tag{134b}$$

The multichannel wave functions $\psi^\pm$ are usually quite complicated. In the asymptotic region corresponding to a large separation of the particles they are relatively simple in structure, however. To discuss the asymptotic form of these wave functions it is convenient to adopt a more symmetric notation for the momenta $\boldsymbol{\kappa}$, writing $\boldsymbol{\kappa}_c$, $\boldsymbol{\kappa}_{c'}$ for the relative momenta in channels $c$, $c'$, respectively.

Then as the relative coordinate $r_{c'}$ (in channel $c'$) becomes very large, we have

$$\begin{aligned}\lim_{r_{c'}\to\infty} \psi^+_{c;\boldsymbol{\kappa}_c,S,\nu} &= \chi_{c;\boldsymbol{\kappa}_c,S,\nu} + \lim_{r_{c'}\to\infty} \sum_{S',\nu'} \int \frac{d^3\kappa' \chi_{c';\boldsymbol{\kappa}',S',\nu'}}{\epsilon_c(\kappa_c) + i\eta - \epsilon_{c'}(\kappa')} \\ &\quad \times \langle c'; \boldsymbol{\kappa}', S', \nu' |\, T \,| c; \boldsymbol{\kappa}, S, \nu\rangle \\ &= \chi_{c;\boldsymbol{\kappa}_c,S,\nu} - (2\pi)^{-3/2} \frac{(2\pi)^2}{\kappa_{c'}} \rho_{\epsilon_{c'}} \frac{e^{i\kappa_{c'}r_{c'}}}{r_{c'}} \\ &\quad \times \sum_{S',\nu'} \langle c'; \hat{\mathbf{r}}_{c'}, S', \nu' |\, T(\kappa) \,| c; \hat{\boldsymbol{\kappa}}, S, \nu\rangle g_{c';S',\nu'}. \end{aligned} \tag{135}$$

In obtaining this result we have introduced the expression (113) for the wave function $\chi_{c';\boldsymbol{\kappa}',S',\nu'}$ and evaluated the integral in the manner described in Section 3.4. The momentum $\kappa_{c'}$ in channel $c'$ is here defined on the energy shell as

$$\epsilon_{c'}(\kappa_{c'}) = \epsilon_c(\kappa_c). \tag{136}$$

The expansions (11$b$) and (131) permit us to write Eq. (135) as

$$\lim_{r_{c'}\to\infty} \psi^+_{c;\boldsymbol{\kappa}_c,S,\nu} = \sum_{l'lS'J\nu'} \langle \hat{\mathbf{r}}_{c'}, l', S', \nu' |\, \mathscr{Y}^J \,| \hat{\boldsymbol{\kappa}}_c, l, S, \nu\rangle \times g_{c';S',\nu'} \psi^+_{c'l'S';\kappa JlS}, \tag{137}$$

where

$$\psi^+_{c'l'S';\kappa JlS} = -\frac{i}{\sqrt{2\pi\rho_{\epsilon_c}}} [-(-1)^{l'} I_{c'} \delta_{l',l} \delta_{S',S} \delta_{c',c} + S^J_{c'l'S',clS} O_{c'}]. \tag{138}$$

To obtain this expression, we have used Eq. (132) and have introduced the useful set of wave functions

$$\begin{aligned} I_c &\equiv \frac{e^{-i\kappa_c r_c}}{r_c\sqrt{d\epsilon_c/d\kappa_c}}. \\ O_c &\equiv \frac{e^{i\kappa_c r_c}}{r_c\sqrt{d\epsilon_c/d\kappa_c}}. \end{aligned} \tag{139}$$

It is clear from our discussion of wave packets in Section 3.1 that $I_c$ represents a unit flux per steradian of *incoming* particles in channel $c$, while $O_c$ represents a unit flux per steradian of *outgoing* particles in channel $c$. More precisely, if $\mathbf{j}(\mathbf{x})$ is the particle current operator, we have

$$
\begin{aligned}
(I_c, \mathbf{j}(\mathbf{x}) I_c) &= -\frac{\hat{\mathbf{x}}}{x^2}, \\
(O_c, \mathbf{j}(\mathbf{x}) O_c) &= \frac{\hat{\mathbf{x}}}{x^2}.
\end{aligned}
\tag{140}
$$

**The Total Cross Section**

The total cross section $\bar{\sigma}$, averaged over spin orientations and integrated over scattering angles, is relatively simple. According to Eq. (22),

$$
\begin{aligned}
\bar{\sigma}(c', c) &= \frac{1}{(2S_1 + 1)(2S_2 + 1)} \sum_{\nu', \nu} \int d\Omega_{\hat{\boldsymbol{\kappa}}_f} \frac{d\sigma}{d\Omega}(c'; \hat{\boldsymbol{\kappa}}_f, S', \nu'; c; \hat{\boldsymbol{\kappa}}, S, \nu) \\
&= \frac{1}{4\pi(2S_1 + 1)(2S_2 + 1)} \sum_{\nu', \nu} \int d\Omega_{\hat{\boldsymbol{\kappa}}_f}\, d\Omega_{\hat{\boldsymbol{\kappa}}} \frac{d\sigma}{d\Omega}(c'; \hat{\boldsymbol{\kappa}}_f, S', \nu'; c; \hat{\boldsymbol{\kappa}}, S, \nu).
\end{aligned}
\tag{141}
$$

The second form here follows directly from the first, since $\bar{\sigma}(c', c)$ is independent of the direction of $\hat{\boldsymbol{\kappa}}$.

If we substitute the first form of Eq. (123) into this, use the expansion (131), and perform the angular integrations, there results

$$
\begin{aligned}
\bar{\sigma}(c', c) = {} & \frac{(2\pi)^4 \rho_{\epsilon_{c'}} \rho_{\epsilon_c}}{4\pi(2S_1 + 1)(2S_2 + 1)\kappa_c^{\,2}} \sum_{\nu', \nu} \sum_{ll'mm'MJ\bar{J}} \\
& \times \langle l', S'; m', \nu' \mid J, M \rangle \langle l', S'; m', \nu' \mid \bar{J}, M \rangle \\
& \times \langle l, S; m, \nu \mid J, M \rangle \langle l, S; m, \nu \mid \bar{J}, M \rangle T^{J}_{c'l'S', clS} \left(T^{\bar{J}}_{c'l'S', clS}\right)^{*}.
\end{aligned}
\tag{142}
$$

With the use of Eqs. (1-60*b*) and (1-60*d*), this becomes

$$
\bar{\sigma}(c', c) = \sum_J \frac{(2J + 1)\pi}{(2S_1 + 1)(2S_2 + 1)\kappa_c^{\,2}} \sum_{l, l'} |S^J_{c'l'S', clS} - \delta_{c', c}\delta_{l', l}\delta_{S', S}|^2. \tag{143}
$$

If often happens that in studying a given reaction we find a predominant contribution to the cross section from only a single state $J$ and given $l$, $l'$. When this occurs, we may replace Eq. (143) by

$$
\bar{\sigma}(c', c) \cong \frac{(2J + 1)\pi}{(2S_1 + 1)(2S_2 + 1)\kappa_c^{\,2}} |S^J_{c'l'S', clS} - \delta_{c', c}\delta_{l', l}\delta_{S', S}|^2. \tag{144}
$$

**Reactions Involving Identical Particles**

When the two particles in the initial or final channel are identical, we must symmetrize the scattering amplitudes as described in Chapter 4. Since the exit channel may not be the same as the entrance channel, we may not use the simple expression (4-144), but must use the general Eqs. (4-216) or (4-217).

If the two particles in the incident channel $c$ are identical, we write $Q_{12}$

for the operator which interchanges their coordinates. Then [see Eq. (6-89)]

$$|c; Q_{12}(\boldsymbol{\varkappa}, S, \nu)\rangle = (-1)^{2S_1 - S}\,|c; -\boldsymbol{\varkappa}, S, \nu\rangle, \tag{145a}$$

where $S_1$ is the spin of either particle. If the two particles in the exit channel $c'$ are identical, we write $Q'_{12}$ for the operator which interchanges their coordinates. Then

$$|c'; Q'_{12}(\boldsymbol{\varkappa}_f, S', \nu')\rangle = (-1)^{2S'_1 - S'}\,|c'; -\boldsymbol{\varkappa}_f, S, \nu\rangle, \tag{145b}$$

where $S'_1$ is now the spin of either particle in channel $c'$.

We must distinguish three cases:

*Case* (*a*). The two particles in channel $c$ are identical. Those in channel $c'$ are not.

*Case* (*b*). The two particles in channel $c'$ are identical. Those in channel $c$ are not.

*Case* (*c*). The two particles in channel $c$ are identical, and the two in channel $c'$ are identical.

If we remove the $\delta$-function describing momentum conservation from Eq. (4-216), we obtain directly the symmetrized scattering matrix $T_{\mathscr{S}}$ on the momentum shell. It may be recalled that

$$\begin{aligned} \delta_{Q12} &= +1 \quad \text{if the particles in channel } c \text{ satisfy Bose-Einstein statistics,} \\ &= -1 \quad \text{if the particles in channel } c \text{ satisfy Fermi-Dirac statistics.} \end{aligned}$$

The quantity $\delta_{Q'12}$ is defined in a similar manner for the particles in channel $c'$. Then according to Eq. (4-216) we have

$$\begin{aligned}
&\langle c'; \hat{\boldsymbol{\varkappa}}_f, S', \nu'|\, T_{\mathscr{S}}\,|c; \hat{\boldsymbol{\varkappa}}, S, \nu\rangle \\
&\quad = \frac{1}{\sqrt{2}}\,[\langle c'; \hat{\boldsymbol{\varkappa}}_f, S', \nu'|\, T\,|c; \hat{\boldsymbol{\varkappa}}, S, \nu\rangle \\
&\qquad + \delta_{Q12}(-1)^{2S_1 - S}\langle c'; \hat{\boldsymbol{\varkappa}}_f, S', \nu'|\, T\,|c; -\hat{\boldsymbol{\varkappa}}, S, \nu\rangle] \\
&\qquad\qquad (\text{case } a) \\
&\quad = \frac{1}{\sqrt{2}}\,[\langle c'; \hat{\boldsymbol{\varkappa}}_f, S', \nu'|\, T\,|c; \hat{\boldsymbol{\varkappa}}, S, \nu\rangle \\
&\qquad + \delta_{Q'12}(-1)^{2S'_1 - S'}\langle c'; -\hat{\boldsymbol{\varkappa}}_f, S', \nu'|\, T\,|c; \hat{\boldsymbol{\varkappa}}, S, \nu\rangle] \\
&\qquad\qquad (\text{case } b) \\
&\quad = \tfrac{1}{2}[\langle c'; \hat{\boldsymbol{\varkappa}}_f, S', \nu'|\, T\,|c; \hat{\boldsymbol{\varkappa}}, S, \nu\rangle \\
&\qquad + \delta_{Q'12}(-1)^{2S'_1 - S'}\langle c'; -\hat{\boldsymbol{\varkappa}}_f, S', \nu'|\, T\,|c; \hat{\boldsymbol{\varkappa}}, S, \nu\rangle \\
&\qquad + \delta_{Q12}(-1)^{2S_1 - S}\langle c'; \hat{\boldsymbol{\varkappa}}_f, S', \nu'|\, T\,|c; -\hat{\boldsymbol{\varkappa}}, S, \nu\rangle \\
&\qquad + \delta_{Q'12}\delta_{Q12}(-1)^{2S_1 + 2S'_1 - S - S'}\langle c'; -\hat{\boldsymbol{\varkappa}}_f, S', \nu'|\, T\,|c; -\hat{\boldsymbol{\varkappa}}, S, \nu\rangle]. \qquad (146) \\
&\qquad\qquad (\text{case } c)
\end{aligned}$$

For any one of these three cases the differential scattering cross section is, of course,

$$\frac{d\sigma}{d\Omega} = \frac{(2\pi)^4}{\kappa^2}(\rho_{\epsilon_{c'}}\,\rho_{\epsilon_c})\,|\langle c';\hat{\boldsymbol{\varkappa}}_f, S', \nu'|\,T_{\mathscr{S}}\,|c;\hat{\boldsymbol{\varkappa}}, S, \nu\rangle|^2. \tag{147}$$

**Energy Dependence of the Scattering Matrix at Low Energy**

When the forces responsible for the scattering have a finite range $d$, such that they are very small for particles separated by a distance greater than $d$, the scattering matrix has several simple limiting properties at "low energies." To discuss these we first consider the expressions [see Eqs. (4-12) and (4-13)]

$$\langle c';\hat{\boldsymbol{\varkappa}}_f, S', \nu'|\,T(\kappa)\,|c;\hat{\boldsymbol{\varkappa}}, S, \nu) = (\chi_{c';\varkappa_f,S',\nu'}, V_{c'}\psi^+_{c;\varkappa,S,\nu}) \tag{148a}$$

$$= (\psi^-_{c';\varkappa_f,S',\nu'}, V_c\chi_{c;\varkappa,S,\nu}). \tag{148b}$$

Here $V_c$ and $V_{c'}$ are the interactions in channels $c$ and $c'$, respectively, and [see Eqs. (113), (11$b$), and (12$a$)]

$$\chi_{c'\varkappa_f,S',\nu'} = \sum_{l',J}\langle\hat{\mathbf{r}}_{c'}, l'S'\nu'|\,\mathscr{Y}^J\,|\hat{\boldsymbol{\varkappa}}_f, l'S'\nu'\rangle\chi_{\kappa_f,l'}(r_{c'})g_{c';S',\nu'}, \tag{149}$$

where $\chi_{\varkappa,l'}$ is defined by Eq. (6-58$b$). We require, of course, that $\epsilon_{c'}(\kappa_f) = \epsilon_c(\kappa)$ in Eqs. (148$a$) and (148$b$).

If we "solve" Eq. (131) for $T^J_{c'l'S',clS}$, we may write Eq. (148$a$) as

$$\begin{aligned} T^J_{c'l'S',clS} = \frac{1}{2J+1}\sum_{mm'm''\nu\nu'MJ'}&\int d\Omega_{\hat{\varkappa}}\,Y_l^m(\hat{\boldsymbol{\varkappa}}) \\ &\times (Y_{l'}^{m''}(\hat{\mathbf{r}}_{c'})\chi_{\kappa_f,l'}(r_{c'})g_{c';S',\nu'}, V_{c'}\psi^+_{c;\varkappa,S,\nu}) \\ &\times \langle l', S'; m', \nu' \mid J, M\rangle\langle l', S'; m', \nu' \mid J', M\rangle \\ &\times \langle l', S'; m'', \nu' \mid J', M\rangle\langle l, S; m, \nu \mid J, M\rangle. \end{aligned} \tag{150}$$

A similar expression may be obtained using Eq. (148$b$).

Let us now suppose that the energy of the scattering system is such that

$$\kappa_f d \ll 2\sqrt{l' + \tfrac{3}{2}}. \tag{151}$$

Then, according to Eq. (6-126),

$$\chi_{\kappa f,l'} \approx (\kappa_f r_{c'})^{l'} \times \text{constant}.$$

If, for the moment, we ignore the relation between $\kappa_f$ and $\kappa$ implied by energy conservation, we see that the right-hand side of Eq. (150) depends on $\kappa_f$ only through a factor of $\kappa_f^{\,l'}$. Therefore, when the condition (151) is valid, we may set

$$T^J_{c'l'S',clS} \cong (\kappa_f d)^{l'} \times C_f, \tag{152}$$

where $C_f$ is independent of $\kappa_f$ (but depends on $\kappa$).

The expression (152) is very useful when $\kappa_f$ is sufficiently small that the equation $\epsilon_{c'}(\kappa_f) = \epsilon_c(\kappa)$ can be approximated as $\epsilon_{c'}(0) = \epsilon_c(\kappa)$ with a real, positive value of $\kappa$.[27] Then $C_f$ can be treated as a true constant in Eq. (152), and the energy dependence of $T^J_{c'l'S',clS}$ is known explicitly. If, for example, only a single value of $l'$ is of significance in the scattering cross section, we see from Eqs. (152), (144), and (132) that

$$\bar{\sigma}(c';c) \approx \rho_{\epsilon_{c'}}\kappa_f^{2l'} \times \text{constant}. \tag{153}$$

This is an important relation, since it provides the energy dependence of a reaction cross section at its energetic threshold.[28]

When the energy is such that

$$\kappa d \ll 2\sqrt{l + \tfrac{3}{2}}, \tag{154}$$

we may use the expression (148*b*) to conclude that

$$T^J_{c'l'S',clS} \cong (\kappa d)^l \times C, \tag{155}$$

where $C$ is independent of $\kappa$ (but depends upon $\kappa_f$). Then if $\kappa$ is sufficiently small that the condition for energy conservation may be approximated as $\epsilon_{c'}(\kappa_f) = \epsilon_c(0)$, with a real, positive $\kappa_f$,[29] we may treat $C$ as a genuine constant in Eq. (155). In this case also the explicit energy dependence of $T^J_{c'l'S',clS}$ is determined.

When only a single $l$-value contributes to the scattering cross section, we may use Eqs. (155) and (144) to write

$$\bar{\sigma}(c';c) \cong \rho_{\epsilon_c}\kappa^{2l-2} \times \text{constant}. \tag{156}$$

Finally, if both conditions (151) and (154) are simultaneously satisfied, we see that

$$T^J_{c'l'S',clS} \cong (\kappa_f d)^{l'}(\kappa d)^l \times \text{constant}. \tag{157}$$

Again, if only a single value of $l$ and a single value of $l'$ give a significant contribution to the cross section, we may use Eqs. (157) and (144) to write

$$\bar{\sigma}(c',c) \approx \rho_{\epsilon_c}\rho_{\epsilon_{c'}}\kappa_f^{2l'}\kappa^{2l-2} \times \text{constant}. \tag{158}$$

In Section 6.4, p. 258, we showed that for scattering by central forces we can ordinarily neglect scattering in states of orbital angular $l$ such that $l \gg \kappa d$. This conclusion can easily be generalized to our present discussion of reactions. Reference to Eq. (6-167) shows that for $l \gg \kappa d$,

$$T^J_{c'l'S',clS} \cong \left(\frac{\kappa d}{l + \frac{1}{2}}\right)^{l+\frac{1}{2}} \times \text{function independent of } \kappa. \tag{159}$$

[27] This, of course, requires that the reaction be endothermic.

[28] A general discussion of the threshold energy dependence of reaction cross sections was given by Wigner [5].

[29] In this case the reaction is exothermic.

When $\kappa d \ll l$, this expression becomes quite small (compared to its value for $\kappa d \approx l$). This strongly suggests that we can ordinarily neglect scattering in $l$ states such that

$$l \gg \kappa d. \tag{160a}$$

Similarly, when $l' \gg \kappa_f d$, we see that

$$T^J_{c'l'S',clS} \approx \left(\frac{\kappa_f d}{l' + \frac{1}{2}}\right)^{l'+\frac{1}{2}}.$$

From this, we conclude that scattering may be neglected in states $l'$ for which

$$l' \gg \kappa_f d. \tag{160b}$$

The conditions (160*a*) and 160*b*) permit us to estimate the $l$ and $l'$ values at which the series (124) and (131) may be terminated. Experience has usually shown these conditions to be very reliable—and certainly very useful. It should be emphasized, however, that these criteria cannot be applied to scattering by long-range forces, such as Coulomb interactions.[30]

The expressions (152), (155), and (157) have been of considerable importance in studying reactions at low energy, or near energetic thresholds, in nuclear and elementary particle physics. Perhaps the best-known example is that of the "$(1/v)$-law" for exothermic capture cross sections. In this case we assume that $\kappa d$ is sufficiently small that only the state $l = 0$ need be considered [Eq. (160*a*)]. Then taking

$$\epsilon_c(\kappa) = \frac{\kappa^2}{2M_c},$$

where $M_c$ is the reduced mass of the colliding particles in channel $c$, we obtain

$$\rho_{\epsilon_c} = M_c\kappa.$$

Equation (156) then reduces to

$$\bar{\sigma}(c', c) \simeq \frac{1}{\kappa} \times \text{constant}. \tag{161}$$

For an endothermic reaction very near threshold $(\kappa_f d) \ll 1$, and we need consider only states for which $l' = 0$. Taking $\rho_{\epsilon_{c'}} \approx \kappa_f$, we find from Eq. (153) that

$$\bar{\sigma}(c', c) \approx \kappa_f \times \text{constant}. \tag{162}$$

The energy dependence implied by Eqs. (152), (155), and (157) has been studied extensively in low-energy pi-meson interactions. Here much of the observed phenomena appears to follow from simple and general arguments like those given in this section. For example, it has been found

[30] L. Fonda and R. G. Newton have discussed the threshold behavior of charged-particle reactions [6].

that the pion-nucleon scattering cross section, for pion energies below 100 Mev, is adequately described by Eq. (158). Here we take $c' = c$, $\kappa_f = \kappa$, $l' = l = 1$, and

$$\epsilon_c = \frac{\kappa^2}{2M} + c(\mu^2c^2 + \kappa^2)^{1/2}, \tag{163}$$

where $M$ is the nucleon rest-mass and $\mu$ the pion rest-mass. Then Eq. (158) leads to the expression

$$\bar{\sigma} = \kappa^4 \Big/ \left[\frac{1}{M} + \frac{c}{(\mu^2c^2 + \kappa^2)^{1/2}}\right]^2 \times \text{constant}. \tag{164}$$

This expression is compared in Fig. 7.4 with the experimental cross section for scattering $\pi^+$-mesons by protons.

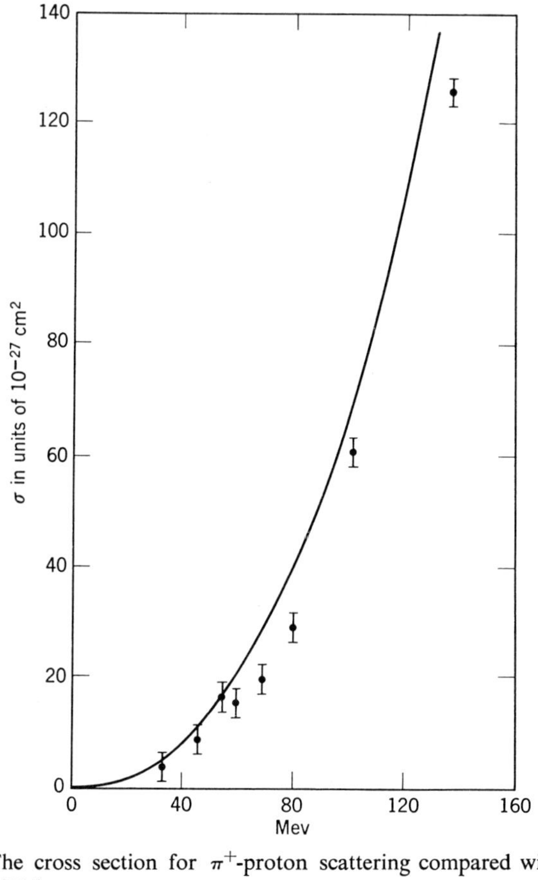


Fig. 7.4 The cross section for $\pi^+$-proton scattering compared with the predicted cross section (164).

## 7.4 THE SCATTERING OF TWO SPIN ONE-HALF PARTICLES

The most frequently encountered example of the scattering of two spin one-half particles is that of nucleon-nucleon scattering, to which our discussion in this section is primarily devoted. Here experimental study has clearly shown that the interaction is noncentral; thus the general techniques developed in Section 7.1 are required to describe the scattering of one nucleon by another.[31]

The coupling of the two spins leads in this case to one singlet state ($S = 0, \nu = 0$) and three triplet states ($S = 1, \nu = 1, 0, -1$). For the singlet state and for a given value $J$ of the total angular momentum, there is of course only a single state $l = J$ of orbital angular momentum. For the triplet state, on the other hand, we may have $l = J, J \pm 1$. The state with $l = J$ has a parity (defined with respect to the choice of intrinsic parities for the two particles) $\mathscr{P} = (-1)^J$. The parity of the two states with $l = J \pm 1$ is then $\mathscr{P} = -(-1)^J$. These eigenstates may be expressed as usual in the form

$$|lSJM\rangle \equiv \mathscr{Y}^M_{JlS} = \sum_{m,\nu} \langle l,S;\, m,\nu \mid J, M\rangle\, |lSm\nu\rangle, \tag{165}$$

where $|lSm\nu\rangle$ describes a state of uncoupled orbital and spin angular momentum.

If parity is conserved, as we shall assume to be the case in this section, the $S$-matrix $S^J_{l'S',lS}$ will vanish for the transition $l = J,\ l' = J \pm 1$ or $l = J \pm 1,\ l' = J$.

For the scattering of two identical particles there is an additional selection rule forbidding singlet-triplet transitions. That is, $S^J_{l'S',lS} = 0$ unless $S' = S$. To see this we note that $\mathscr{P}Q_{12}$, where $\mathscr{P}$ is the operator for coordinate reflection and $Q_{12}$ is the operator which interchanges space and spin coordinates for the two particles, is equivalent to an operator which interchanges just the spin coordinates of the two particles. Therefore [see Eq. (6-89)]

$$\mathscr{P}Q_{12}\, |lSJM\rangle = (-1)^{1-S}\, |lSJM\rangle. \tag{166}$$

Now by hypothesis $[\mathscr{P}Q_{12}, T] = 0$, where $T$ is the scattering matrix, which with (166) implies that $S$ is a constant of the motion.

For the scattering of two neutrons or two protons transitions between singlet and triplet states do not, therefore, occur.[32] Experience has shown that to a very good approximation singlet-triplet spin transitions do not

[31] At sufficiently low energies the noncentral interactions can be represented by equivalent central interactions, however. See, for example, the discussion of Christian and Hart [7].

[32] We repeat that the very small effects associated with the nonconservation of parity are being neglected in this section.

occur for neutron-proton scattering. This follows from the principle of "charge symmetry," which states that nuclear interactions commute with the operator $Q_{np}$ which interchanges neutron and proton space and spin coordinates.[33] Then $\mathscr{P}Q_{np}$ is a constant of the motion and has the eigenvalue $(-1)^{1-S}$ when operating on the state $|lSJM\rangle$ [see Eq. (166)]. The argument that $S' = S$ then follows that given previously for the case of identical particles.

In the remainder of this section we shall assume that $S$, as well as $\mathscr{P}$, is a constant of the motion. Our discussion is then not applicable, for example, to the scattering of hyperons by nucleons. The extension to the general case is, however, straightforward with the use of the more general equations obtained in Section 7.1.

### The Schrödinger Equation for Tensor Forces

The most general interaction potential for two spin one-half particles that depends on only the relative coordinate $\mathbf{r}$ and spins $\boldsymbol{\sigma}_1$ and $\boldsymbol{\sigma}_2$ of the particles and which conserves parity and angular momentum is of the form

$$V = V_{ct}(r)\Lambda^{(1)} + V_{cs}(r)\Lambda^{(0)} + V_t(r)S_{12}. \tag{167}$$

Here $V_{ct}$, $V_{cs}$, and $V_t$ are functions of $r$ and

$$\begin{aligned} \Lambda^{(1)} &\equiv \tfrac{1}{4}(3 + \boldsymbol{\sigma}_1 \cdot \boldsymbol{\sigma}_2) \\ \Lambda^{(0)} &\equiv \tfrac{1}{4}(1 - \boldsymbol{\sigma}_1 \cdot \boldsymbol{\sigma}_2) \end{aligned} \tag{168}$$

$$S_{12} \equiv 3\boldsymbol{\sigma}_1 \cdot \hat{\mathbf{r}}\boldsymbol{\sigma}_2 \cdot \hat{\mathbf{r}} - \boldsymbol{\sigma}_1 \cdot \boldsymbol{\sigma}_2. \tag{169}$$

The quantities $\Lambda^{(1)}$ and $\Lambda^{(0)}$ are, respectively, projection operators on the triplet and singlet spin states. The operator $S_{12}$ describes a noncentral or "tensor interaction," since it couples states of different orbital angular momentum. It is easy to verify that

$$\begin{aligned} \Lambda^{(1)}u_{0,0} &= S_{12}u_{0,0} = \Lambda^{(0)}u_{1,\nu} = 0, \\ \Lambda^{(1)}u_{1,\nu} &= u_{1,\nu} \\ \Lambda^{(0)}u_{0,0} &= u_{0,0}, \end{aligned}$$

where $u_{s,\nu}$ are the spin functions for the two particles.

The interaction (167) is said to arise from "tensor forces." It has been

[33] This is a special case of the more restrictive principle of "charge independence" (see Section 9.2). Since electromagnetic interactions violate both these symmetry principles, they can be correct only to within terms of the relative order of the fine structure constant.

studied extensively in connection with nucleon-nucleon scattering.[34] The scattering due to the interaction $V$ is described by the Schrödinger equation (45), where now

$$v \equiv 2M_r V \equiv v_{ct}(r)\Lambda^{(1)} + v_{cs}(r)\Lambda^{(0)} + v_t(r)S_{12}. \tag{171}$$

The wave function $\psi^+_{\boldsymbol{\varkappa},S,\nu}(\mathbf{r})$ may be decomposed into partial waves using Eq. (39$a$). For the singlet states ($S = S' = 0$) we have from (39$a$)

$$\psi^+_{\boldsymbol{\varkappa},0,0}(r) = \sum_J \sum_{m=-J}^{J} Y_J{}^m(\hat{\mathbf{r}})Y_J{}^{m*}(\hat{\boldsymbol{\varkappa}})u_{0,0}\psi^+_{J0;\kappa JJ0}(r). \tag{172}$$

We see immediately, therefore, that scattering in the singlet spin state is entirely equivalent to scattering by central forces, so the techniques developed in Chapter 6 are applicable. Indeed, we easily verify from Eq. (46$a$) that

$$V^J_{l'',0;l',0} = 0 \quad \text{unless} \quad l'' = l' = J, \tag{173$a$}$$

and

$$v^J_{J,0,J,0} = v_{cs}(r). \tag{173$b$}$$

Equations (173) imply that Eq. (47$a$) is equivalent to Eq. (6-108).

For scattering in triplet states we have [see Eq. (39$a$)]

$$\psi^+_{\boldsymbol{\varkappa},1,\nu} = \sum_{l'lJ\nu'} \langle \hat{\mathbf{r}}, l'1\nu'|\ \mathscr{Y}^J\ |\hat{\boldsymbol{\varkappa}}, l1\nu\rangle u_{1,\nu'}\psi^+_{l'1;\kappa Jl1}(r). \tag{174}$$

The matrix elements (46$a$) of $v$ are seen to reduce to the form

$$\begin{aligned} v^J_{l'',1;l',1} &= \delta_{l'',l'}v_{ct}(r) + \frac{3}{2J+1}\, i^{l'-l''}v_t(r)\sum_M (\mathscr{Y}^M_{Jl''1}, S_{12}\mathscr{Y}^M_{Jl'1}) \\ &\qquad\qquad \times \sum_m (\langle l, 1; m, \nu \mid J, M\rangle)^2 \qquad (175) \\ &= \delta_{l'',l'}v_{ct}(r) + i^{l'-l''}v_t(r)(\mathscr{Y}^M_{Jl''1}, S_{12}\mathscr{Y}^M_{Jl'1}), \end{aligned}$$

where the $\mathscr{Y}^M_{Jls}$ are defined by Eq. (165). The straightforward, but somewhat tedious, evaluation of the matrix elements of $S_{12}$ is described in Appendix F. The only nonvanishing matrix elements of $v$ for triplet

[34] See, for example, Refs. [7–10]. It is of course necessary to include the Coulomb interaction when studying the scattering of two protons.

To describe the observed nucleon-nucleon scattering at a few hundred Mev energy, an additional term of the form [see Eq. (58)] $V_s(r)(\boldsymbol{\sigma}_1 + \boldsymbol{\sigma}_2)\cdot\mathbf{L}$ seems to be required (Refs. [11]). Such an interaction is easily included, as described in Section 7.2, in the radial Schrödinger equations.

states are obtained directly from Eq. (175) and from Eq. (16) of Appendix F. These are[35]

$$\begin{aligned}
v^J_{J-1,1;J-1,1} &= v_{ct} - 2\frac{J-1}{2J+1}v_t \\
v^J_{J+1,1;J-1,1} &= -6\frac{[J(J+1)]^{1/2}}{2J+1}v_t \\
v^J_{J-1,1;J+1,1} &= -6\frac{[J(J+1)]^{1/2}}{2J+1}v_t \\
v^J_{J+1,1;J+1,1} &= v_{ct} - 2\frac{J+2}{2J+1}v_t \\
v^J_{J,1;J,1} &= v_{ct} + 2v_t.
\end{aligned} \tag{176}$$

These expressions permit us to write out the explicit radial equations (47*a*) or (47*b*) for the wave functions. It is clear from Eqs. (176) that scattering in the state with $l = J$ is equivalent to scattering by central forces, since this state is not coupled to any other states.

For further discussion of these equations it is useful to transform the $w_{l'S';JlS}(\kappa; r)$ of Eq. (48) (the label $JlS$ is now reinserted) into a more convenient form. The asymptotic expression for $w_{l'S';JlS}$ is, according to Eq. (41*b*),

$$\begin{aligned}
\lim_{r\to\infty} w_{l'S';JlS}(\kappa; r) &= \frac{i^{-l'}}{i\sqrt{2\pi}}\left[-(-1)^{l'}\delta_{l',l}\,\delta_{S',S}\,e^{-i\kappa r} + S^J_{l'S',lS}\,e^{i\kappa r}\right] \\
&= \sqrt{2/\pi}\sum_\alpha e^{i\delta_\alpha^J}\,U^J_{l'S':\alpha}U^J_{\alpha;lS}\sin\left(\kappa r - \frac{\pi l'}{2} + \delta_\alpha{}^J\right),
\end{aligned} \tag{177}$$

where the $U^J_{lS;\alpha}$ are the real, orthogonal transformation matrices of Eq. (35) and the $\delta_\alpha{}^J$ are the eigenphase shifts. Let us now define the new wave functions,

$$\bar{w}_{l'S'J\alpha}(\kappa; r) \equiv e^{-i\delta_\alpha^J}\sum_{l,S} w_{l'S',JlS}(\kappa; r)U^J_{lS;\alpha}, \tag{178}$$

which because of Eq. (177) have the asymptotic form

$$\lim_{r\to\infty} \bar{w}_{l'S';J\alpha}(\kappa; r) = \sqrt{2/\pi}\,U^J_{l'S';\alpha}\sin\left(\kappa r - \frac{\pi l'}{2} + \delta_\alpha{}^J\right) \tag{179}$$

Since this expression is real and the differential equations (47*b*) are also real, we see that the $\bar{w}$ are real functions; that is,

$$\bar{w}_{l'S';J\alpha} = \bar{w}^*_{l'S';J\alpha}. \tag{180}$$

[35] The sign convention here is determined by our phase convention for time reversal and our insistence on conditions (46*b*) and (46*c*).

These functions are direct generalizations of the $w_l$ of Chapter 6 [compare Eq. (6-113) with (179)].

Equations (178) and (179) are quite general. For the case of scattering of two spin one-half particles, they are trivial for $S = S' = 0$ and for $l' = l = J$, $S = 1$. For singlet scattering we use the label $\alpha = 0$ and have

$$U^J_{l'S';0} = \delta_{l',J}\delta_{S'0}.$$

Then Eq. (179) gives

$$\lim_{r\to\infty} \bar{w}_{J0;J0} = \sqrt{2/\pi}\,\sin\left(\kappa r - \frac{\pi J}{2} + \delta_0{}^J\right), \tag{181a}$$

where $\delta_0{}^J$ is the phase shift for singlet scattering. For triplet scattering with $l' = l = J$, we use the label $\alpha = 3$, write $\delta_3{}^J$ for the scattering phase shift, and obtain

$$U^J_{l'S';3} = \delta_{l',J}\delta_{S',1}$$

and

$$\lim_{r\to\infty} \bar{w}_{J1;J3} = \sqrt{2/\pi}\,\sin\left(\kappa r - \frac{\pi J}{2} + \delta_3{}^J\right). \tag{181b}$$

For scattering in the states $S = S' = 1$, $l' = J \pm 1$, the $2 \times 2$ $S$-matrix is expressed by Eqs. (36*a*, *b*) in terms of two eigenphase shifts $\delta_\alpha{}^J(\alpha = 1, 2)$ and a mixing parameter $\epsilon_J$. We shall choose the label $\alpha$ in such a manner that as the mixing parameter $\epsilon_J \to 0$,[36] the states labeled as $\alpha = 1$ and $\alpha = 2$ describe scattering in only the $l' = J - 1$ and the $l' = J + 1$ state, respectively.[37] Then, using Eq. (36*a*) for the matrix elements of $U$, we find from Eq. (179) that for $\alpha = 1$

$$\begin{aligned}\lim_{r\to\infty} \bar{w}_{J-1,1;J1} &= \sqrt{2/\pi}\,\cos\epsilon_J\,\sin\left[\kappa r - \frac{\pi(J-1)}{2} + \delta_1{}^J\right]\\ \lim_{r\to\infty} \bar{w}_{J+1,1;J1} &= \sqrt{2/\pi}\,\sin\epsilon_J\,\sin\left[\kappa r - \frac{\pi(J+1)}{2} + \delta_1{}^J\right],\end{aligned} \tag{181c}$$

and for $\alpha = 2$

$$\begin{aligned}\lim_{r\to\infty} \bar{w}_{J-1,1;J2} &= -\sqrt{2/\pi}\,\sin\epsilon_J\,\sin\left[\kappa r - \frac{\pi(J-1)}{2} + \delta_2{}^J\right]\\ \lim_{r\to\infty} \bar{w}_{J+1,1;J2} &= \sqrt{2/\pi}\,\cos\epsilon_J\,\sin\left[\kappa r - \frac{\pi(J+1)}{2} + \delta_2{}^J\right].\end{aligned} \tag{181d}$$

The radial differential equations satisfied by the $\bar{w}_{l'S',J\alpha}$ are obtained from Eqs. (47*b*), (173), and (176). For $\alpha = 0$,

$$\left[\frac{d^2}{dr^2} + \kappa^2 - \frac{J(J+1)}{r^2}\right]\bar{w}_{J0;J0} = v_{cs}(r)\bar{w}_{J0;J0}. \tag{182a}$$

[36] This has a precise meaning, since $\epsilon_J$ vanishes as $\kappa \to 0$. It is for this reason that tensor forces may be represented by equivalent central forces at sufficiently low energies.

[37] This corresponds to the choice made by Blatt and Biedenharn [12].

For $\alpha = 3$,

$$\left[\frac{d^2}{dr^2} + \kappa^2 - \frac{J(J+1)}{r^2}\right]\bar{w}_{J1;\,J3} = [v_{ct}(r) + 2v_t(r)]\bar{w}_{J1;\,J3}. \quad (182b)$$

Finally, for $\alpha = 1$ or 2,

$$\left[\frac{d^2}{dr^2} + \kappa^2 - \frac{J(J-1)}{r^2}\right]\bar{w}_{J-1,1;\,J\alpha}$$
$$= v^J_{J-1,1;\,J-1,1}\bar{w}_{J-1,1;\,J\alpha} + v^J_{J-1,1;J+1,1}\bar{w}_{J+1,1;\,J\alpha},$$
$$\left[\frac{d^2}{dr^2} + \kappa^2 - \frac{(J+1)(J+2)}{r^2}\right]\bar{w}_{J+1,1;\,J\alpha} \quad (182c)$$
$$= v^J_{J+1,1;\,J-1,1}\bar{w}_{J-1,1;\,J\alpha} + v^J_{J+1,1;\,J+1,1}\bar{w}_{J+1,1;\,J\alpha}.$$

We emphasize once again, as is clear from Eqs. (182), that the equations for the states $S = 0$ and $S = 1$, $l' = l = J$ are uncoupled. To obtain the scattering in the states $S = S' = 1$, $l' = J \pm 1$, we must solve the two coupled differential equations (182*c*), subject to the boundary conditions (50) and (181*c*, *d*).

The wave functions satisfying Eqs. (182) may be expressed as solutions to a set of integral equations, similar to (6-368). To obtain these we introduce the wave functions [analogous to those of Eq. (6-370)]

$$\bar{w}^{(0)}_{l'S';\,JlS}(\kappa;\,r) \equiv \sum_\alpha \frac{U^J_{\alpha;\,lS}}{\cos\delta_\alpha{}^J}\,\bar{w}_{l'S';\,J\alpha}(\kappa;\,r), \quad (183)$$

which satisfy Eq. (47*b*):

$$\left[\frac{d^2}{dr^2} + \kappa^2 - \frac{l'(l'+1)}{r^2}\right]\bar{w}^{(0)}_{l'S';\,JlS} = \sum_{l''S''} v^J_{l',S';\,l''S''}\bar{w}^{(0)}_{l''S'';\,JlS}. \quad (184)$$

To obtain the integral equations satisfied by the functions (183), we note from Eq. (29) that

$$\sum_{l'S'lS} U^J_{\alpha;\,l'S'}T^J_{l'S',lS}U^J_{lS;\,\beta} \equiv \delta_{\alpha\beta}T_\alpha{}^J = \delta_{\alpha\beta}\left[-\frac{1}{2\pi i M_r\kappa}(e^{2i\delta_\alpha^J} - 1)\right] \quad (185)$$

and from Eq. (51*b*) that

$$T^J_{l'S',lS} = \frac{1}{2M_r\kappa^2}\sum_{l''S''}\int_0^\infty y_{l'}(\kappa r)v^J_{l',S';\,l'',S''}\bar{w}_{l''S'';\,JlS}(\kappa;\,r)\,dr, \quad (186)$$

where $y_l(\kappa r)$ is defined by Eq. (6-367*a*). Using Eqs. (183), (185), and (186) we may show that

$$R^J_{l'S',lS} \equiv \frac{1}{2M_r\kappa^2}\sum_{l''S''}\int_0^\infty y_{l'}v^J_{l',S';\,l'',S''}\bar{w}^{(0)}_{l''S'';\,JlS}\,dr$$
$$= -\frac{1}{\pi M_r\kappa}\sum_\alpha U^J_{l'S';\,\alpha}\tan\delta_\alpha{}^J U^J_{\alpha;\,lS}, \quad (187)$$

which should be compared with Eqs. (6-372). It is evident that $R^J_{l'S',lS} = R^J_{lS,l'S'}$ and that $R^J_{l'S',lS} = R^{J*}_{l'S',lS}$.

We may now easily verify that the wave functions (183), satisfying Eqs. (184), are the solutions to the integral equations

$$\bar{w}^{(0)}_{l'S';\,JlS}(\kappa; r) = \delta_{l',l}\,y_l(\kappa r) + \sum_{l''S''} \int_0^\infty \langle r|\, G_{0l'}\, |r'\rangle v^J_{l'S';l'',S''}(r')\bar{w}^{(0)}_{l''S'';\,JlS}(\kappa; r')\, dr'. \quad (188)$$

Here the $\langle r|\, G_{0l}\, |r'\rangle$ are the radial Green's functions defined by Eq. (6-375). The functions defined by (188) obviously represent a regular solution to the coupled Eqs. (184). That they agree with the definition (183) follows from a comparison of the asymptotic forms, as $r \to \infty$.

The expression (187) may evidently be written in the form

$$R^J_{l'S',lS} = \sum_\alpha U^J_{l'S';\,\alpha} R_\alpha{}^J U^J_{\alpha;\,lS} \quad (189a)$$

with [see Eq. (6-372*a*)]

$$R_\alpha{}^J = -\frac{1}{\pi M_r \kappa} \tan \delta_\alpha{}^J. \quad (189b)$$

These are evidently the eigenvalues of the reaction matrix [Eq. (5-182)]. This is related to the $T$-matrix by Heitler's integral equation (5-188). We have

$$T_\alpha{}^J = R_\alpha{}^J - i\pi M_r \kappa R_\alpha{}^J T_\alpha{}^J, \quad (190a)$$

or

$$T^J_{l'S',lS} = R^J_{l'S',lS} - i\pi M_r \kappa \sum_{l''S''} R^J_{l'S',l''S''} T^J_{l''S'',lS}. \quad (190b)$$

Variational principles such as those described in Section 6.7, p. 313, are easily obtained for the expressions (187), using Eqs. (188). Such an expression is, for example, obtained directly from Eq. (6-440) for the function $2M_r\kappa^2 R^J_{l'S',lS}$.

**Formal Considerations**

The scattering matrix $\langle \hat{\boldsymbol{\kappa}}_f, S', \nu'|\, T(\kappa)\, |\hat{\boldsymbol{\kappa}}, S, \nu\rangle$, considered as a $4 \times 4$ spin matrix, may evidently be written in terms of the two Pauli spin matrices $\boldsymbol{\sigma}_1$ and $\boldsymbol{\sigma}_2$ for the interacting particles. In analogy to Eqs. (91) and (94), we write

$$\langle \hat{\boldsymbol{\kappa}}_f, S', \nu'|\, T(\kappa)\, |\hat{\boldsymbol{\kappa}}, S, \nu\rangle = (u_{S',\nu'}, T(\hat{\boldsymbol{\kappa}}_f, \hat{\boldsymbol{\kappa}}, \boldsymbol{\sigma}_1, \boldsymbol{\sigma}_2) u_{S,\nu}), \quad (191)$$

where $T(\hat{\boldsymbol{\kappa}}_f, \hat{\boldsymbol{\kappa}}, \boldsymbol{\sigma}_1, \boldsymbol{\sigma}_2)$ is some function of $\hat{\boldsymbol{\kappa}}_f$, $\hat{\boldsymbol{\kappa}}$, $\boldsymbol{\sigma}_1$, and $\boldsymbol{\sigma}_2$ (and of course of the energy of the particles).

The functional form of $T(\hat{\boldsymbol{\kappa}}_f, \hat{\boldsymbol{\kappa}}, \boldsymbol{\sigma}_1, \boldsymbol{\sigma}_2)$ is ordinarily limited by symmetry restrictions. First, invariance with respect to rotations implies that $T$ is a scalar function of $\hat{\boldsymbol{\kappa}}_f$, $\hat{\boldsymbol{\kappa}}$, $\boldsymbol{\sigma}_1$ and $\boldsymbol{\sigma}_2$.[38] The assumption that the interaction is invariant with respect to space inversion implies that [see Eq. (101)]

$$T(-\hat{\boldsymbol{\kappa}}_f, -\hat{\boldsymbol{\kappa}}, \boldsymbol{\sigma}_1, \boldsymbol{\sigma}_2) = T(\hat{\boldsymbol{\kappa}}_f, \hat{\boldsymbol{\kappa}}, \boldsymbol{\sigma}_1, \boldsymbol{\sigma}_2). \tag{192}$$

Invariance with respect to time reversal leads, as in Eq. (97), to the condition that

$$T(\hat{\boldsymbol{\kappa}}_f, \hat{\boldsymbol{\kappa}}, \boldsymbol{\sigma}_1, \boldsymbol{\sigma}_2) = T(-\hat{\boldsymbol{\kappa}}, -\hat{\boldsymbol{\kappa}}_f, -\boldsymbol{\sigma}_1, -\boldsymbol{\sigma}_2). \tag{193}$$

These relations permit us to give $T$ a more specific form. First, we may write quite generally

$$T = A_2 + \mathbf{B}_2 \cdot \boldsymbol{\sigma}_1, \tag{194}$$

where $A_2$ and $B_2$ may be linear functions of $\boldsymbol{\sigma}_2$. They may also depend on the three mutually orthogonal unit vectors

$$\hat{\mathbf{l}} \equiv \frac{\hat{\boldsymbol{\kappa}} + \hat{\boldsymbol{\kappa}}_f}{2\cos(\theta/2)}, \qquad \hat{\mathbf{m}} \equiv \frac{\hat{\boldsymbol{\kappa}} - \hat{\boldsymbol{\kappa}}_f}{2\sin(\theta/2)}, \qquad \hat{\mathbf{n}} \equiv \frac{\hat{\boldsymbol{\kappa}} \times \hat{\boldsymbol{\kappa}}_f}{\sin\theta}, \tag{195}$$

where $\theta$ is the angle between $\hat{\boldsymbol{\kappa}}$ and $\hat{\boldsymbol{\kappa}}_f$. Because $T$ must be invariant under rotations, we may write $\mathbf{B}_2$ in the general form

$$\mathbf{B}_2 = b\hat{\mathbf{l}} + c\hat{\mathbf{m}} + d\hat{\mathbf{n}} \tag{196}$$

The requirement (192) of invariance under spatial reflections implies that

$$\begin{aligned} &A_2 \text{ is a scalar,} \\ &b, c \text{ are pseudoscalars,} \\ &d \text{ is a scalar.} \end{aligned} \tag{197}$$

The most general form for $A_2$ is therefore

$$A_2 = \alpha + a_3 \boldsymbol{\sigma}_2 \cdot \hat{\mathbf{n}}, \tag{198}$$

where $\alpha$ and $a_3$ are functions only of $\kappa$ and $\hat{\boldsymbol{\kappa}} \cdot \hat{\boldsymbol{\kappa}}_f$. The expression (198) is seen to be consistent with Eq. (193).

The most general expressions for $b$, $c$, and $d$, taking account of the conditions (193) and (197), are

$$\begin{aligned} b &= \epsilon \boldsymbol{\sigma}_2 \cdot \hat{\mathbf{l}} \\ c &= \delta \boldsymbol{\sigma}_2 \cdot \hat{\mathbf{m}} \\ d &= d_3 + \beta \boldsymbol{\sigma}_2 \cdot \hat{\mathbf{n}}, \end{aligned} \tag{199}$$

where $\epsilon$, $\delta$, $d_3$, and $\beta$ may depend on $\kappa$ and $\hat{\boldsymbol{\kappa}} \cdot \hat{\boldsymbol{\kappa}}_f$ only.

Combining Eqs. (194) and (199), we obtain

$$\begin{aligned} T(\hat{\boldsymbol{\kappa}}_f, \hat{\boldsymbol{\kappa}}, \boldsymbol{\sigma}_1, \boldsymbol{\sigma}_2) = \alpha &+ \beta \boldsymbol{\sigma}_1 \cdot \hat{\mathbf{n}} \boldsymbol{\sigma}_2 \cdot \hat{\mathbf{n}} + i\gamma(\boldsymbol{\sigma}_1 + \boldsymbol{\sigma}_2) \cdot \hat{\mathbf{n}} \\ &+ i\gamma'(\boldsymbol{\sigma}_1 - \boldsymbol{\sigma}_2) \cdot \hat{\mathbf{n}} + \delta \boldsymbol{\sigma}_1 \cdot \hat{\mathbf{m}} \boldsymbol{\sigma}_2 \cdot \hat{\mathbf{m}} + \epsilon \boldsymbol{\sigma}_1 \cdot \hat{\mathbf{l}} \boldsymbol{\sigma}_2 \cdot \hat{\mathbf{l}}, \end{aligned} \tag{200}$$

[38] The proof follows that given in Note 4.

where

$$d_3 + a_3 \equiv 2i\gamma,$$
$$d_3 - a_3 \equiv 2i\gamma'. \tag{201}$$

Equation (200) expresses $T$ in the most general form consistent with parity and time reversal invariance. Additional symmetry conditions may of course further restrict the form of $T$. For example, if the two interacting particles are identical, we have [see Eq. (166)]

$$T(\hat{\boldsymbol{\varkappa}}_f, \hat{\boldsymbol{\varkappa}}, \boldsymbol{\sigma}_1, \boldsymbol{\sigma}_2) = (\mathscr{P}Q_{12})T(\mathscr{P}Q_{12})^{-1} = T(\hat{\boldsymbol{\varkappa}}_f, \hat{\boldsymbol{\varkappa}}, \boldsymbol{\sigma}_2, \boldsymbol{\sigma}_1). \tag{202}$$

This condition implies that $\gamma' = 0$, so Eq. (200) becomes

$$T(\boldsymbol{\varkappa}_f, \boldsymbol{\varkappa}, \boldsymbol{\sigma}_1, \boldsymbol{\sigma}_2) = \alpha + \beta\boldsymbol{\sigma}_1 \cdot \hat{\mathbf{n}}\boldsymbol{\sigma}_2 \cdot \hat{\mathbf{n}} + i\gamma(\boldsymbol{\sigma}_1 + \boldsymbol{\sigma}_2) \cdot \hat{\mathbf{n}} + \delta\boldsymbol{\sigma}_1 \cdot \hat{\mathbf{m}}\boldsymbol{\sigma}_2 \cdot \hat{\mathbf{m}} + \epsilon\boldsymbol{\sigma}_1 \cdot \hat{\mathbf{l}}\boldsymbol{\sigma}_2 \cdot \hat{\mathbf{l}} \tag{203}$$

Since this expression is symmetric in $\boldsymbol{\sigma}_1$ and $\boldsymbol{\sigma}_2$, it has no nonvanishing matrix elements connecting singlet and triplet states. This conclusion is of course just that obtained in the discussion following Eq. (166). For nucleon-nucleon scattering the assumption of charge symmetry also implies that $\gamma' = 0$. To show this, we need only replace $Q_{12}$ in Eq. (202) by $Q_{np}$, as was argued on p. 385.

We see that $T$ depends on at most five different functions $\alpha$, $\beta$, $\gamma$, $\delta$, $\epsilon$ when $\gamma' = 0$. This may also be seen as follows. For the singlet state, to obtain the total angular momentum $J$ a single orbital momentum $l = J$ is required, and thus there is a single amplitude. For the triplet states with $l = J$ again a single amplitude is needed. For the triplet states with $l = J \pm 1$ there are four amplitudes. Two of these are equal, however, because of the condition (34). We obtain then just five amplitudes. (If singlet-triplet transitions may occur, there is one more amplitude associated with the $S$-matrix elements $S^J_{J0;J1} = S^J_{J1;J0}$.)[39]

It is worth noting that the $T$-matrix may be easily transformed to a representation in which the spins of the two particles are uncoupled. Letting $u_{\frac{1}{2},\nu 1}(s_1)$ and $u_{\frac{1}{2},\nu 2}(s_2)$ be the respective spin wave functions for particles 1 and 2, we obtain, using Eq. (191),

$$\begin{aligned}\langle\hat{\boldsymbol{\varkappa}}_f, \tfrac{1}{2}\nu'_1, \tfrac{1}{2}\nu'_2|\, T(\kappa)\,|\hat{\boldsymbol{\varkappa}}, \tfrac{1}{2}\nu_1, \tfrac{1}{2}\nu_2\rangle &= \sum_{S',S}\langle\tfrac{1}{2}, \tfrac{1}{2}; \nu'_1, \nu'_2 \mid S', \nu'\rangle\langle\hat{\boldsymbol{\varkappa}}_f, S', \nu'|\, T(\kappa)\,|\hat{\boldsymbol{\varkappa}}, S, \nu\rangle\langle\tfrac{1}{2}, \tfrac{1}{2}; \nu_1, \nu_2 \mid S, \nu\rangle \\ &= (u_{\frac{1}{2},\nu'1}u_{\frac{1}{2},\nu'2}, T(\hat{\boldsymbol{\varkappa}}_f, \hat{\boldsymbol{\varkappa}}, \boldsymbol{\sigma}_1, \boldsymbol{\sigma}_2)u_{\frac{1}{2},\nu 1}u_{\frac{1}{2},\nu 2}).\end{aligned} \tag{204}$$

[39] As another example, consider nucleon-deuteron scattering. Now $S_1 = \frac{1}{2}$, $S_2 = 1$, and there are doublet and quartet spin states. For the doublet states, with a given $J$, we have $l = J \pm \frac{1}{2}$ and hence two amplitudes. For the quartet states the pair $l = J + \frac{3}{2}$ and $l = J - \frac{1}{2}$ require three amplitudes, as does the pair $l = J + \frac{1}{2}$ and $l = J - \frac{3}{2}$. Finally, there are two $(J \pm \frac{1}{2})_{\text{doubl}} \leftrightarrow (J \pm \frac{1}{2})_{\text{quart}}$ and two $(J \pm \frac{1}{2})_{\text{doubl}} \leftrightarrow (J \mp \frac{3}{2})_{\text{quart}}$ amplitudes. Therefore twelve amplitudes are required.

The representation (200) [or (203)] is seen to be particularly convenient for this transformation.

### Representation of the Scattering Matrix in Terms of the Eigenphase Shifts

The general expressions (26), (27), and (35) permit the scattering matrix to be expressed explicitly in terms of the eigenphase shifts $\delta_\alpha^J$ and the mixing parameters $\epsilon_J$. This is a straightforward, although somewhat tedious, calculation. It is also of interest to express the parametrized form (203) for $T$ in terms of the phase shifts and mixing parameters.

Let us now return to the general expressions (26) and (27) and choose a special rectangular coordinate system with the $z$-axis (polar axis) parallel to the vector $\hat{\boldsymbol{\kappa}}$. Then, taking $\cos\theta \equiv \hat{\boldsymbol{\kappa}}_f \cdot \hat{\boldsymbol{\kappa}}$ and $\phi$ to be the azimuthal angle of the vector $\hat{\boldsymbol{\kappa}}_f$, the scattering matrix can be written in the form

$$\langle \hat{\boldsymbol{\kappa}}_f, 1, \nu'|\, T(\kappa)\, |\hat{\boldsymbol{\kappa}}, 1, \nu\rangle = T_{\nu',\nu}(\cos\theta)e^{i(\nu-\nu')\phi}, \tag{205}$$

$$\langle \hat{\boldsymbol{\kappa}}_f, 0, 0|\, T(\kappa)\, |\hat{\boldsymbol{\kappa}}, 0, 0\rangle = T_{s,s}(\cos\theta). \tag{206}$$

The quantities $T_{\nu',\nu}$ and $T_{s,s}$ are functions of only $\cos\theta$ and $\kappa$.

It is also evident from Eq. (27) that[40]

$$\begin{aligned} T_{0,-1} &= -T_{0,1}, \qquad & T_{-1,1} &= T_{1,-1}, \\ T_{-1,0} &= -T_{1,0}, \qquad & T_{1,-1} &= T_{1,1}. \end{aligned} \tag{207}$$

The condition of time reversal invariance implies an additional relation among the $T_{\nu',\nu}$. As we shall show later, this is

$$T_{1,1} - T_{1,-1} - T_{0,0} = \sqrt{2}\cot\theta\,(T_{1,0} + T_{0,1}). \tag{208}$$

The five relations (207) and (208) among the ten $T_{\nu',\nu}$ and $T_{s,s}$ leave five independent amplitudes, as was shown to be required in the last section.[41]

We next seek the relations which express the coefficients $\alpha$, $\beta$, $\gamma$, $\delta$, and $\epsilon$ of Eq. (203) in terms of the $T_{\nu',\nu}$. This may be done using the equations

$$\begin{aligned} \alpha &= \tfrac{1}{4}\,\mathrm{Tr}\, T, \\ \beta &= \tfrac{1}{4}\,\mathrm{Tr}\,(T\boldsymbol{\sigma}_1 \cdot \hat{\mathbf{n}}\boldsymbol{\sigma}_2 \cdot \hat{\mathbf{n}}), \\ \gamma &= \frac{1}{8i}\,\mathrm{Tr}\,[T(\boldsymbol{\sigma}_1 + \boldsymbol{\sigma}_2)\cdot \hat{\mathbf{n}}], \\ \delta &= \tfrac{1}{4}\,\mathrm{Tr}\,(T\boldsymbol{\sigma}_1 \cdot \hat{\mathbf{m}}\boldsymbol{\sigma}_2 \cdot \hat{\mathbf{m}}), \\ \epsilon &= \tfrac{1}{4}\,\mathrm{Tr}\,(T\boldsymbol{\sigma}_1 \cdot \hat{\mathbf{l}}\boldsymbol{\sigma}_2 \cdot \hat{\mathbf{l}}). \end{aligned} \tag{209}$$

Here the trace of the $4 \times 4$ spin matrices is implied.

[40] This is demonstrated directly in Eq. (212).

[41] We are following here the treatment of Ref. [13].

To evaluate these expressions we require the matrix elements of $\boldsymbol{\sigma}_1 \cdot \hat{\mathbf{n}}\boldsymbol{\sigma}_2 \cdot \hat{\mathbf{n}}$, and so on, in the $(S, \nu)$ representation. To find these we first need the rectangular components of $\hat{\mathbf{n}}$, $\hat{\mathbf{l}}$, and $\hat{\mathbf{m}}$, which are

$$\begin{aligned}\hat{\mathbf{n}} &= -\sin\phi, \quad \cos\phi, \quad 0,\\ \hat{\mathbf{l}} &= \sin\frac{\theta}{2}\cos\phi, \quad \sin\frac{\theta}{2}\sin\phi, \quad \cos\frac{\theta}{2},\\ \hat{\mathbf{m}} &= -\cos\frac{\theta}{2}\cos\phi, \quad -\cos\frac{\theta}{2}\sin\phi, \quad \sin\frac{\theta}{2}.\end{aligned} \tag{210}$$

Now, a direct calculation gives the results shown in Tables 7.1–7.4, which together constitute Eqs. (211).

These matrices, along with Eqs. (203) and (209), lead to the expression

$$\begin{aligned}\alpha &= \tfrac{1}{4}(2T_{1,1} + T_{0,0} + T_{s,s}),\\ \beta &= \tfrac{1}{4}(-2T_{1,-1} + T_{0,0} - T_{s,s}),\\ \gamma &= \frac{\sqrt{2}}{4}(T_{1,0} - T_{0,1}),\\ \delta &= \tfrac{1}{4}[(T_{1,1} + T_{1,-1} - T_{s,s}) - \sec\theta\,(T_{1,1} - T_{1,-1} - T_{0,0})]\\ \epsilon &= \tfrac{1}{4}[(T_{1,1} + T_{1,-1} - T_{s,s}) + \sec\theta\,(T_{1,1} - T_{1,-1} - T_{0,0})].\end{aligned} \tag{211}$$

On the other hand, we may compare Eqs. (203), (205), and (206), using the matrices of Tables 7.1 to 7.4, to obtain

$$\begin{aligned}T_{s,s} &= \alpha - \beta - \delta - \epsilon,\\ T_{1,1} &= T_{-1,-1} = \alpha + \delta\sin^2\frac{\theta}{2} + \epsilon\cos^2\frac{\theta}{2},\\ T_{1,0} &= -T_{-1,0} = \sqrt{2}\,\gamma - \frac{1}{\sqrt{2}}\delta\sin\theta + \frac{1}{\sqrt{2}}\epsilon\sin\theta,\\ T_{1,-1} &= T_{-1,1} = -\beta + \delta\cos^2\frac{\theta}{2} + \epsilon\sin^2\frac{\theta}{2},\\ T_{0,1} &= -T_{0,-1} = -\sqrt{2}\,\gamma - \frac{\delta}{\sqrt{2}}\sin\theta + \frac{\epsilon}{\sqrt{2}}\sin\theta,\\ T_{0,0} &= \alpha + \beta + \delta\cos\theta - \epsilon\cos\theta.\end{aligned} \tag{212}$$

Equation (208) follows directly from these expressions. We may also derive Eq. (208) from the fact that Eq. (203) contains no term of the form $\boldsymbol{\sigma}_1 \cdot \hat{\mathbf{m}}\boldsymbol{\sigma}_2 \cdot \hat{\mathbf{l}} + \boldsymbol{\sigma}_1 \cdot \hat{\mathbf{l}}\boldsymbol{\sigma}_2 \cdot \hat{\mathbf{m}}$, a consequence of time reversal invariance.

**Table 7.1**

$\boldsymbol{\sigma}_1 \cdot \hat{\mathbf{n}} \boldsymbol{\sigma}_2 \cdot \hat{\mathbf{n}} =$

| $\nu' \backslash \nu$ | $S = 1$ | | | $S = 0$ |
|---|---|---|---|---|
| | 1 | 0 | $-1$ | 0 |
| 1 | 0 | 0 | $-e^{-2i\phi}$ | 0 |
| 0 | 0 | 1 | 0 | 0 |
| $-1$ | $-e^{2i\phi}$ | 0 | 0 | 0 |
| 0 | 0 | 0 | 0 | $-1$ |

**Table 7.2**

$\dfrac{1}{i\sqrt{2}}(\boldsymbol{\sigma}_1 + \boldsymbol{\sigma}_2) \cdot \hat{\mathbf{n}} =$

| $\nu' \backslash \nu$ | $S = 1$ | | | $S = 0$ |
|---|---|---|---|---|
| | 1 | 0 | $-1$ | 0 |
| 1 | 0 | $-e^{-i\phi}$ | 0 | 0 |
| 0 | $e^{i\phi}$ | 0 | $-e^{-i\phi}$ | 0 |
| $-1$ | 0 | $e^{i\phi}$ | 0 | 0 |
| 0 | 0 | 0 | 0 | 0 |

**Table 7.3**

$\boldsymbol{\sigma}_1 \cdot \hat{\mathbf{m}} \boldsymbol{\sigma}_2 \cdot \hat{\mathbf{m}} =$

| $\nu' \backslash \nu$ | $S = 1$ | | | $S = 0$ |
|---|---|---|---|---|
| | 1 | 0 | $-1$ | 0 |
| 1 | $\sin^2 \dfrac{\theta}{2}$ | $-\dfrac{1}{\sqrt{2}} \sin\theta\, e^{-i\phi}$ | $\cos^2 \dfrac{\theta}{2}\, e^{-2i\phi}$ | 0 |
| 0 | $-\dfrac{1}{\sqrt{2}} \sin\theta\, e^{i\phi}$ | $\cos\theta$ | $\dfrac{1}{\sqrt{2}} \sin\theta\, e^{-i\phi}$ | 0 |
| $-1$ | $\cos^2 \dfrac{\theta}{2}\, e^{2i\phi}$ | $\dfrac{1}{\sqrt{2}} \sin\theta\, e^{i\phi}$ | $\sin^2 \dfrac{\theta}{2}$ | 0 |
| 0 | 0 | 0 | 0 | $-1$ |

**Table 7.4**

$\boldsymbol{\sigma}_1 \cdot \hat{\mathbf{l}} \boldsymbol{\sigma}_2 \cdot \hat{\mathbf{l}} =$

| $\nu' \backslash \nu$ | $S = 1$ | | | $S = 0$ |
|---|---|---|---|---|
| | 1 | 0 | $-1$ | 0 |
| 1 | $\cos^2 \dfrac{\theta}{2}$ | $\dfrac{1}{\sqrt{2}} \sin\theta\, e^{-i\phi}$ | $\sin^2 \dfrac{\theta}{2}\, e^{-2i\phi}$ | 0 |
| 0 | $\dfrac{1}{\sqrt{2}} \sin\theta\, e^{i\phi}$ | $-\cos\theta$ | $-\dfrac{1}{\sqrt{2}} \sin\theta\, e^{-i\phi}$ | 0 |
| $-1$ | $\sin^2 \dfrac{\theta}{2}\, e^{2i\phi}$ | $-\dfrac{1}{\sqrt{2}} \sin\theta\, e^{i\phi}$ | $\cos^2 \dfrac{\theta}{2}$ | 0 |
| 0 | 0 | 0 | 0 | $-1$ |

This implies that

$$\mathrm{Tr}\,[T(\boldsymbol{\sigma}_1 \cdot \hat{\mathbf{m}}\boldsymbol{\sigma}_2 \cdot \hat{\mathbf{l}} + \boldsymbol{\sigma}_1 \cdot \hat{\mathbf{l}}\boldsymbol{\sigma}_2 \cdot \hat{\mathbf{m}})] = 0, \tag{213}$$

from which (208) follows.

Equations (211) and (212) depend upon only the scattering angle $\theta$ and not on the azimuthal angle $\phi$. They are therefore independent of the orientation of the coordinate system and generally correct, even though Eq. (205) was specialized to the coordinate system with $\hat{\boldsymbol{\varkappa}}$ parallel to the $z$-axis.

A scattering amplitude analogous to that of Eq. (77) is frequently used. This is defined as

$$\mathbf{f}(\hat{\boldsymbol{\varkappa}}_f, \hat{\boldsymbol{\varkappa}}, \boldsymbol{\sigma}_1, \boldsymbol{\sigma}_2) \equiv -\frac{(2\pi)^2}{\kappa}\rho_\epsilon T(\hat{\boldsymbol{\varkappa}}_f, \hat{\boldsymbol{\varkappa}}, \boldsymbol{\sigma}_1, \boldsymbol{\sigma}_2). \tag{214}$$

It is convenient to project this into its singlet and triplet subspaces, using the singlet and triplet projection operators $\Lambda^{(0)}$ and $\Lambda^{(1)}$ defined by Eqs. (168). In terms of these we may write

$$\begin{aligned}\mathbf{f}(\hat{\boldsymbol{\varkappa}}_f, \hat{\boldsymbol{\varkappa}}, \boldsymbol{\sigma}_1, \boldsymbol{\sigma}_2) &= A\Lambda^{(0)} + Ci(\boldsymbol{\sigma}_1 + \boldsymbol{\sigma}_2) \cdot \hat{\mathbf{n}} \\ &\quad + (B\boldsymbol{\sigma}_1 \cdot \hat{\mathbf{n}}\boldsymbol{\sigma}_2 \cdot \hat{\mathbf{n}} + D\boldsymbol{\sigma}_1 \cdot \hat{\mathbf{m}}\boldsymbol{\sigma}_2 \cdot \hat{\mathbf{m}} + E\boldsymbol{\sigma}_1 \cdot \hat{\mathbf{l}}\boldsymbol{\sigma}_2 \cdot \hat{\mathbf{l}})\Lambda^{(1)}, \end{aligned} \tag{215}$$

where $A$, $B$, $C$, $D$, and $E$ are functions of $\kappa$ and $\hat{\boldsymbol{\varkappa}}_f \cdot \hat{\boldsymbol{\varkappa}} = \cos\theta$. The relations between the coefficients in Eqs. (203) and (215) are easily found, using Eq. (214)[42]

$$\begin{aligned}A &= -\frac{(2\pi)^2}{\kappa}\rho_\epsilon(\alpha - \beta - \delta - \epsilon),\\ B &= -\frac{(2\pi)^2}{\kappa}\rho_\epsilon(\alpha + \beta),\\ C &= -\frac{(2\pi)^2}{\kappa}\rho_\epsilon\gamma,\\ D &= -\frac{(2\pi)^2}{\kappa}\rho_\epsilon(\alpha + \delta),\\ E &= -\frac{(2\pi)^2}{\kappa}\rho_\epsilon(\alpha + \epsilon).\end{aligned} \tag{216}$$

The coefficients $A, B, \ldots, E$ or $\alpha, \beta, \ldots, \epsilon$ may be expressed in terms of the eigenphase shifts and mixing parameters, using Eqs. (36), (211), and (216). Equations (177) and (178) define these quantities in terms of the asymptotic wave functions, while Eqs. (36) relate them to the $S$-matrix

[42] To obtain the expressions (216), we set $T = T(\Lambda^{(0)} + \Lambda^{(1)})$ and observe that $\boldsymbol{\sigma}_1 \cdot \hat{\mathbf{n}}\boldsymbol{\sigma}_2 \cdot \hat{\mathbf{n}}\Lambda^{(0)} = -\Lambda^{(0)}$, and so on, and that

$$\Lambda^{(1)} = \boldsymbol{\sigma}_1 \cdot \boldsymbol{\sigma}_2\Lambda^{(1)} = (\boldsymbol{\sigma}_1 \cdot \hat{\mathbf{n}}\boldsymbol{\sigma}_2 \cdot \hat{\mathbf{n}} + \boldsymbol{\sigma}_1 \cdot \hat{\mathbf{l}}\boldsymbol{\sigma}_2 \cdot \hat{\mathbf{l}} + \boldsymbol{\sigma}_1 \cdot \hat{\mathbf{m}}\boldsymbol{\sigma}_2 \cdot \hat{\mathbf{m}})\Lambda^{(1)}.$$

elements. Using the notation of Eqs. (181), we first define (for $S = 1$)

$$a_J^J \equiv \sin \delta_3{}^J e^{i\delta_3{}^J} \qquad (\text{for } l' = J),$$

$$a_{J-1}^J \equiv \alpha^J \cos^2 \epsilon_J + \beta^J \sin^2 \epsilon_J - \left(\frac{J}{J+1}\right)^{1/2} \frac{1}{2}(\alpha^J - \beta^J) \sin^2 \epsilon_J,$$

$$b_{J-1}^J \equiv \alpha^J \cos^2 \epsilon_J + \beta^J \sin^2 \epsilon_J + \left(\frac{J+1}{J}\right)^{1/2} \frac{1}{2}(\alpha^J - \beta^J) \sin^2 \epsilon_J \qquad (\text{for } l' = J - 1), \qquad (217a)$$

$$a_{J+1}^J \equiv \alpha^J \sin^2 \epsilon_J + \beta^J \cos^2 \epsilon_J - \left(\frac{J+1}{J}\right)^{1/2} \frac{1}{2}(\alpha^J - \beta^J) \sin^2 \epsilon_J,$$

$$b_{J+1}^J \equiv \alpha^J \sin^2 \epsilon_J + \beta^J \cos^2 \epsilon_J + \left(\frac{J}{J+1}\right)^{1/2} \frac{1}{2}(\alpha^J - \beta^J) \sin^2 \epsilon_J \qquad (\text{for } l' = J + 1).$$

Here

$$\alpha^J \equiv \sin \delta_1{}^J e^{i\delta_1{}^J}$$
$$\beta^J \equiv \sin \delta_2{}^J e^{i\delta_2{}^J},$$

and when $J = 0$, $\epsilon_0 = 0$, $\delta_1{}^0 = 0$, $\delta_2{}^0 = \delta(^3P_0)$, the phase shift for $S = 1$, $l' = 1$. For the singlet states we introduce the amplitudes

$$C_J \equiv \sin \delta_0{}^J e^{i\delta_0{}^J}. \qquad (217b)$$

After a somewhat lengthy calculation, there results[43]

$$\frac{(2\pi)^2}{\kappa} \rho_\epsilon T_{1,1} = -\frac{1}{2\kappa} \sum_{L=0}^{\infty} [(L + 2(a_L^{L+1} + (2L + 1)a_L{}^L + (L - 1)a_L^{L-1}]P_L,$$

$$\frac{(2\pi)^2}{\kappa} \rho_\epsilon T_{0,0} = -\frac{1}{\kappa} \sum_{L=0}^{\infty} [(L + 1)b_L^{L+1} + Lb_L^{L-1}]P_L,$$

$$\frac{(2\pi)^2}{\kappa} \rho_\epsilon T_{1,-1} = -\frac{1}{2\kappa} \sum_{L=2}^{\infty} \left[\frac{1}{L+1} a_L^{L+1} - \frac{2L+1}{L(L+1)} a_L^L + \frac{1}{L} a_L^{L-1}\right] P_L{}^2,$$

$$\frac{(2\pi)^2}{\kappa} \rho_\epsilon T_{1,0} = -\frac{\sqrt{2}}{2\kappa} \sum_{L=1}^{\infty} (b_L^{L+1} - b_L^{L-1})P_L{}^1,$$

$$\frac{(2\pi)^2}{\kappa} \rho_\epsilon T_{0,1} = -\frac{\sqrt{2}}{2\kappa} \sum_{L=1}^{\infty} \times \left[-\frac{L+2}{L+1} a_L^{L+1} + \frac{2L+1}{L(L+1)} a_L^L + \frac{L-1}{L} a_L^{L-1}\right] P_L{}^1,$$

$$\frac{(2\pi)^2}{\kappa} \rho_\epsilon T_{s,s} = -\frac{1}{\kappa} \sum_{L=0}^{\infty} (2L + 1)C_L P_L.$$

[43] The evaluation of Eqs. (218) is outlined in Appendix F.
These expressions have been given in Refs. [13] and [14] (where modifications for Coulomb scattering in the case of the proton-proton interaction have been included).
Here $P_l \equiv P_l(\cos\theta)$ and $P_l{}^m \equiv \sin^m \theta\, [d^m/d(\cos\theta)^m]P_l(\cos\theta)$.

For the scattering of two neutrons or two protons the scattering amplitude must be appropriately symmetrized because of the Pauli principle. The required modification in Eq. (215) may be made easily. Using Eqs. (214) and (52), we find the symmetrized scattering amplitude

$$\langle S, \nu'|\, \mathbf{f}_{\mathscr{S}}(\hat{\boldsymbol{\varkappa}}_f, \hat{\boldsymbol{\varkappa}}, \boldsymbol{\sigma}_1, \boldsymbol{\sigma}_2)\,|S, \nu\rangle = \langle S, \nu'|\, \mathbf{f}(\hat{\boldsymbol{\varkappa}}_f, \hat{\boldsymbol{\varkappa}}, \boldsymbol{\sigma}_1, \boldsymbol{\sigma}_2)\,|S, \nu\rangle + (-1)^S \langle S, \nu'|\, \mathbf{f}(-\hat{\boldsymbol{\varkappa}}_f, \hat{\boldsymbol{\varkappa}}, \boldsymbol{\sigma}_1, \boldsymbol{\sigma}_2)\,|S, \nu\rangle. \quad (219)$$

Now the transformation $\boldsymbol{\varkappa}_f \to -\boldsymbol{\varkappa}_f$ implies that $\theta \to \pi - \theta$, $\hat{\mathbf{n}} \to -\hat{\mathbf{n}}$, $\hat{\mathbf{l}} \to \hat{\mathbf{m}}$, $\hat{\mathbf{m}} \to \hat{\mathbf{l}}$. Therefore,

$$\begin{aligned}\mathbf{f}_{\mathscr{S}}(\hat{\boldsymbol{\varkappa}}_f, \hat{\boldsymbol{\varkappa}}, \boldsymbol{\sigma}_1, \boldsymbol{\sigma}_2) = {} & [A(\theta) + A(\pi - \theta)]\Lambda^{(0)} + [C(\theta) - C(\pi - \theta)] \\ & \times i(\boldsymbol{\sigma}_1 + \boldsymbol{\sigma}_2)\cdot\hat{\mathbf{n}} + \{[B(\theta) - B(\pi - \theta)]\boldsymbol{\sigma}_1\cdot\hat{\mathbf{n}}\boldsymbol{\sigma}_2\cdot\hat{\mathbf{n}} \\ & + [D(\theta) - E(\pi - \theta)]\boldsymbol{\sigma}_1\cdot\hat{\mathbf{m}}\boldsymbol{\sigma}_2\cdot\hat{\mathbf{m}} \\ & + [E(\theta) - D(\pi - \theta)]\boldsymbol{\sigma}_1\cdot\hat{\mathbf{l}}\boldsymbol{\sigma}_2\cdot\hat{\mathbf{l}}\}\Lambda^{(1)}. \end{aligned} \quad (220)$$

### The Scattering Cross Section

The differential cross section for nonidentical particles may be obtained from Eqs. (215) and (44), while for identical particles it is given by Eqs. (220) and (53*a*). For any spins and unpolarized initial states the differential cross section is

$$\frac{d\bar{\sigma}}{d\Omega} = \frac{1}{4}\,\mathrm{Tr}(\mathbf{f}^\dagger\,\mathbf{f}) = |A|^2 + 2\,|C|^2 + |B|^2 + |D|^2 + |E|^2, \quad (221a)$$

if the two particles are not equivalent. For identical particles, we find

$$\frac{d\bar{\sigma}}{d\Omega} = \frac{1}{4}\,\mathrm{Tr}(\mathbf{f}_{\mathscr{S}}{}^\dagger\,\mathbf{f}_{\mathscr{S}}) = |A(\theta) + A(\pi - \theta)|^2 + 2\,|C(\theta) + C(\pi - \theta)|^2 + |B(\theta) - B(\pi - \theta)|^2 + |D(\theta) - E(\pi - \theta)|^2 + |E(\theta) - D(\pi - \theta)|^2. \quad (221b)$$

(In these expressions $\mathbf{f}^\dagger$ represents the adjoint of the $4 \times 4$ matrix $\mathbf{f}$.)

## 7.5 THE DENSITY MATRIX AND THE SCATTERING OF POLARIZED PARTICLES

We have seen that the scattering amplitude **f** involves two independent functions of energy [Eq. (79)] and scattering angle for spin zero–spin one-half scattering and five (or six) such functions [Eq. (215)] for spin one-half–spin one-half scattering. Clearly a simple scattering experiment to measure $d\bar{\sigma}/d\Omega$ as a function of energy and scattering angle cannot determine more than one relation between these functions. It is for this reason that experiments are performed with polarized beams, or targets, since such experiments give further relations among the parameters of the scattering amplitude. In principle, these parameters may be completely determined—in practice, however, this usually involves quite elaborate experiments.

In this section we shall describe special techniques for the description of scattering with beams and targets which may be polarized. It will be assumed that the interactions causing the scattering are invariant with respect to coordinate reflection and time reversal. Extension to processes for which these symmetry relations may not hold is straightforward, but it requires that several theorems developed in this section be modified.

**Illustration for the Scattering of Spin Zero and Spin One-half Particles**

In order to explain these phenomena in the simplest way, we begin with an extension of the description given in Section 7.2 of the scattering of spin one-half by spin zero particles (examples include pion-nucleon and nucleon-nucleus scattering). We saw from Eq. (77) that the scattering amplitude describing scattering from a spin state $u_\nu$ to $u_{\nu'}$ is given by[44]

$$\langle \nu' | \mathbf{f} | \nu \rangle = (u_{\nu'}, (f + i\boldsymbol{\sigma} \cdot \hat{\mathbf{n}} \sin \theta g) u_\nu), \tag{222}$$

where $\hat{\mathbf{n}} \sin \theta \equiv \hat{\boldsymbol{\varkappa}} \times \hat{\boldsymbol{\varkappa}}_f$, as before.

For a given initial spin orientation $\nu$, the amplitude of the state produced by the scattering is

$$A^\nu \equiv \mathbf{f} u_\nu = (f + i\boldsymbol{\sigma} \cdot \hat{\mathbf{n}} \sin \theta g) u_\nu. \tag{223}$$

Since $\mathbf{f}$ is a $2 \times 2$ matrix, $A^\nu$ is a two component spinor, as is $u_\nu$. The differential cross section is obtained by forming the absolute square of $A^\nu$:

$$\begin{aligned} \frac{d\sigma_\nu}{d\Omega} &= \sum_{\nu'} |\langle \nu' | \mathbf{f} | \nu \rangle|^2 = (u_\nu, \mathbf{f}^\dagger \mathbf{f} u_\nu) \\ &= (u_\nu, [|\mathbf{f}|^2 + |g|^2 \sin^2 \theta + 2\boldsymbol{\sigma} \cdot \hat{\mathbf{n}} \sin \theta \operatorname{Im} (\mathbf{f} g^*)] u_\nu). \end{aligned} \tag{224}$$

On averaging over the initial spin orientations $\nu$, we obtain the cross section (80):

$$\frac{d\bar{\sigma}}{d\Omega} = |\mathbf{f}|^2 + |g|^2 \sin^2 \theta. \tag{225}$$

It is evident that the differential cross section depends on the spin orientation of the initial beam provided $g^* f$ is not real (which would be an accident). Clearly if the spin is quantized along $\hat{\boldsymbol{\varkappa}}$, $d\sigma_{\hat{\varkappa}}/d\Omega = d\bar{\sigma}/d\Omega$, whereas if it is quantized parallel or antiparallel to $\hat{\mathbf{n}}$,

$$\frac{d\sigma_{\pm\hat{\mathbf{n}}}}{d\Omega} = |f \pm ig \sin \theta|^2. \tag{226}$$

In general, for a beam polarized in some direction $\mathbf{P}_i$, say, we have $(\chi, \boldsymbol{\sigma}\chi) = \mathbf{P}_i$ for some spinor $\chi$, provided $\chi$ is normalized. This is easily

[44] For simplicity of notation we now write the spin one-half wave functions, which were expressed as $u_{1/2,\nu}$ in Section 7.2, as $u_\nu$.

seen by noting that if $\boldsymbol{\sigma} \cdot \mathbf{P}_i\chi = \chi$ (which is what we mean when we say a particle has a spin in a certain direction),

$$(\chi, \boldsymbol{\sigma}\chi) = \tfrac{1}{2}(\chi, \boldsymbol{\sigma}\boldsymbol{\sigma} \cdot \mathbf{P}_i\chi) + \tfrac{1}{2}(\boldsymbol{\sigma} \cdot \mathbf{P}_i\chi, \boldsymbol{\sigma}\chi),$$

which (using $\boldsymbol{\sigma}\boldsymbol{\sigma} \cdot \mathbf{P}_i = \mathbf{P}_i - i\boldsymbol{\sigma} \times \mathbf{P}_i$, $\boldsymbol{\sigma} \cdot \mathbf{P}_i\boldsymbol{\sigma} = \mathbf{P}_i + i\boldsymbol{\sigma} \times \mathbf{P}_i$) leads immediately to the stated result. Thus we have

$$\frac{d\sigma(\mathbf{P}_i)}{d\Omega} = |f|^2 + |g|^2 \sin^2\theta + 2\mathbf{P}_i \cdot \hat{\mathbf{n}} \sin\theta \operatorname{Im}(g^*f), \tag{227}$$

showing that the differential cross section for an initially polarized beam has in general a dependence on the cosine of the angle between the normal to the scattering plane $\hat{\mathbf{n}}$ and the direction of polarization $\mathbf{P}_i$ (note that $\mathbf{P}_i^2 = 1$).

Since the scattering cross section depends upon the orientation of the initial spin vector with respect to the scattering plane, we may expect that after the scattering the beam will have a definite degree of polarization. It is natural to define this quantity as the net fraction of the scattered beam whose spin is oriented in a given direction. It is in fact convenient to define a vector $\mathbf{P}_\nu$ which expresses this by

$$\mathbf{P}_\nu = \frac{(A^\nu, \boldsymbol{\sigma}A^\nu)}{(A^\nu, A^\nu)} = \frac{(u_\nu, \mathbf{f}^\dagger\boldsymbol{\sigma}\mathbf{f}u_\nu)}{(u_\nu, \mathbf{f}^\dagger\mathbf{f}u_\nu)}. \tag{228}$$

As described before, we have

$$\mathbf{P}_i = (u_\nu, \boldsymbol{\sigma}u_\nu), \tag{229}$$

which is the polarization of the initial beam. Substituting Eq. (223) for $\mathbf{f}u_\nu$ in Eq. (228), we find

$$\begin{aligned}\mathbf{P}_\nu = \{&\hat{\mathbf{n}}[2\operatorname{Im}(g^*f)\sin\theta + 2|g|^2\sin^2\theta\,\hat{\mathbf{n}} \cdot \mathbf{P}_i] \\ &+ (|f|^2 - |g|^2\sin^2\theta)\mathbf{P}_i + 2\operatorname{Re}(g^*f)\sin\theta\,\mathbf{P}_i \times \hat{\mathbf{n}}\} \\ &\times [|f|^2 + |g|^2\sin^2\theta + 2\hat{\mathbf{n}} \cdot \mathbf{P}_i \sin\theta \operatorname{Im}(g^*f)]^{-1}.\end{aligned} \tag{230}$$

The general structure of $\mathbf{P}_\nu$ is what would be expected, since there are only the axial vectors $\hat{\mathbf{n}}$, $\mathbf{P}_i$, and $\mathbf{P}_i \times \hat{\mathbf{n}}$ upon which the axial vector $\mathbf{P}_\nu$ can possibly depend.

In an unpolarized incident beam, one-half of the spin one-half particles have $\nu = +\frac{1}{2}$, and one-half have $\nu = -\frac{1}{2}$. The average polarization $\mathbf{P}$ following the scattering of an unpolarized beam is then given by

$$\frac{d\bar{\sigma}}{d\Omega}\mathbf{P} = \frac{1}{2}\sum_\nu \frac{d\sigma_\nu}{d\Omega}\mathbf{P}_\nu,$$

or

$$\mathbf{P} = \hat{\mathbf{n}}\,\frac{2\operatorname{Im}(g^*f)\sin\theta}{|f|^2 + |g|^2\sin^2\theta}, \tag{231}$$

which follows directly from Eq. (230).

As an illustration of Eq. (231) let us consider scattering for $\theta = 90°$ and under such conditions that only interaction in the states $l = 0, J = \frac{1}{2}$ and $l = 1, J = \frac{3}{2}$ need be considered. Then from Eqs. (81) and (231) we obtain

$$\mathbf{P} = \hat{\mathbf{n}} \frac{2 \sin \delta_{0+} \sin \delta_{1+} \sin (\delta_{0+} - \delta_{1+})}{\sin^2 \delta_{0+} + \sin^2 \delta_{1+}}. \tag{232}$$

When the phase shifts are small so is the polarization $\mathbf{P}$. This conclusion represents a special case of a general result to be obtained later.

Suppose that we now imagine letting an unpolarized beam undergo two successive scatterings and that we ask for the differential cross section for this event. We may obtain the answer directly from our previous calculations by substituting into Eq. (224) for $u_\nu$ the wave function $A^\nu = \mathbf{f}u_\nu$ and normalizing the result by dividing by $(A^\nu, A^\nu)$. The first scattering goes through an angle $\theta_1$, and the normal to the scattering plane is $\hat{\mathbf{n}}_1$; the second scattering is through an angle $\theta_2$ and the normal to its plane is $\hat{\mathbf{n}}_2$. Then we obtain simply

$$\frac{d\bar{\sigma}_2}{d\Omega_2} = [|f(\theta_2)|^2 + |g(\theta_2)|^2 \sin^2 \theta_2][1 + P_2(\theta_2)P_1(\theta_1)\hat{\mathbf{n}}_2 \cdot \hat{\mathbf{n}}_1], \tag{233}$$

where $\mathbf{P}_1$, $\mathbf{P}_2$ are the polarizations which would be obtained from the scattering of an unpolarized beam by scattering through angles $\theta_1$ and $\theta_2$. Evidently we have an explicit dependence on the azimuthal angle between the two scattering planes. If the two scatterings take place in the same plane, then $\hat{\mathbf{n}}_1 \cdot \hat{\mathbf{n}}_2 = \pm 1$.

To see the geometrical significance of this, we let $\hat{\boldsymbol{\varkappa}}$, $\hat{\boldsymbol{\varkappa}}_1$, and $\hat{\boldsymbol{\varkappa}}_2$ be the respective unit vectors parallel to the relative momentum vectors prior to the first scattering, following the first scattering, and following the second scattering. Now imagine an observer standing parallel to the unit vector $\hat{\mathbf{n}}_1 = (\hat{\boldsymbol{\varkappa}} \times \hat{\boldsymbol{\varkappa}}_1)/\sin \theta_1$. If the second scattering is to his left, we have $\hat{\mathbf{n}}_1 \cdot \hat{\mathbf{n}}_2 = 1$; if to his right, we have $\hat{\mathbf{n}}_1 \cdot \hat{\mathbf{n}}_2 = -1$. (This is illustrated in Fig. 7.5.) The corresponding differential cross sections may be written as $d\sigma_2/d\Omega_2$ (left) and $d\sigma_2/d\Omega_2$ (right). In terms of these it is customary to define an asymmetry parameter $e$ by

$$\begin{aligned} e &= \left[\frac{d\sigma_2}{d\Omega_2}(\text{left}) - \frac{d\sigma_2}{d\Omega_2}(\text{right})\right] \bigg/ \left[\frac{d\sigma_2}{d\Omega_2}(\text{left}) + \frac{d\sigma_2}{d\Omega_2}(\text{right})\right] \\ &= P_2(\theta_2)P_1(\theta_1). \end{aligned} \tag{234}$$

The measurement of $d\sigma_2/d\Omega_2$ (left) and $d\sigma_2/d\Omega_2$ (right) therefore gives the product $P_1(\theta_1)P_2(\theta_2)$. To determine, say $P_1(\theta_1)$, it is necessary to arrange that the two scatterings be equivalent, so that $P_2(\theta_2) = P_1(\theta_1)$. Then the

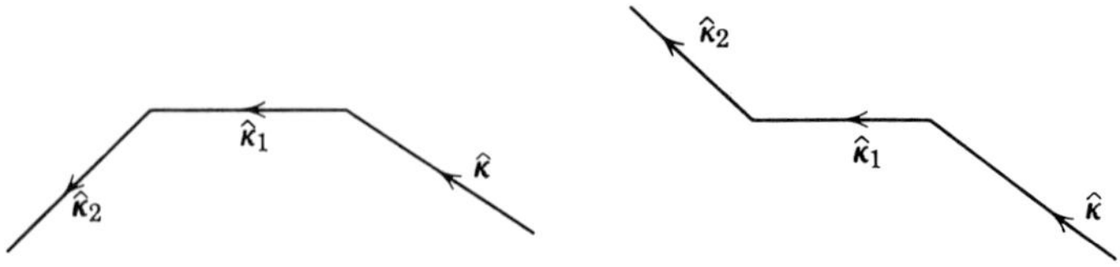


Fig. 7.5 The two coplanar scattering experiments from which the asymmetry parameter [Eq. (234)] is deduced.

experimental measurement of $\epsilon$ leads directly to $P_1(\theta_1)$:

$$P_1(\theta_1) = \sqrt{\epsilon}. \tag{235}$$

Of course, if $P_2(\theta_2)$ is already known, then $P_1(\theta_1)$ may be obtained directly from $\epsilon$.

**Use of the Density Matrix for Describing Partially Polarized Beams**

For beams which are neither completely polarized or unpolarized the description we have used is somewhat cumbersome. For describing partially polarized beams and/or targets the density matrix of statistical mechanics is very useful.[45]

It is evident that a statistical description is appropriate for scattering experiments. The beam and target do not in fact consist of a single particle each. Instead, the target and beam contain a very large number of particles and the experimenter measures the *average* scattering of a large number of pairs of beam and target particles. This suggests that we imagine the target and beam to constitute a *statistical ensemble*[46] of systems, each consisting of a pair of interacting particles. It is then various averages over the ensemble which correspond to experimentally measured physical quantities.

To illustrate this in a general way, let us suppose the ensemble to consist of $N$ systems, each representing a pair of interacting particles. Let $\psi(\alpha)$ be the wave function for system $\alpha$ ($\alpha = 1, 2, \ldots, N$) in the ensemble. Then the mean value of a physical quantity represented by an operator $G$ is

$$\langle G \rangle = \frac{1}{N} \sum_{\alpha=1}^{N} (\psi(\alpha), G\psi(\alpha)). \tag{236}$$

If the $\psi(\alpha)$ are expressed in terms of an orthonormal set $|\lambda\rangle$ of basis functions, we have

$$\psi(\alpha) = \sum_{\lambda} a_\lambda(\alpha) \, |\lambda\rangle, \tag{237}$$

[45] See, for example, Refs. [15] and [16].

[46] The notion of an ensemble of systems and its statistical properties are admirably described by Tolman [17, Chapter 9].

where the $a_\lambda(\alpha)$ are appropriate expansion amplitudes—satisfying the normalization condition $\sum_\lambda |a_\lambda(\alpha)|^2 = 1$. The expansion (237) may be inserted into Eq. (236) to give

$$\langle G \rangle = \frac{1}{N} \sum_{\alpha=1}^{N} \sum_{\lambda',\lambda} a_{\lambda'}(\alpha)^* \langle \lambda' | \, G \, | \lambda \rangle a_\lambda(\alpha).$$

This suggests that we define the Hermitian matrix, called the *density matrix*,

$$\langle \lambda | \, \rho \, | \lambda' \rangle \equiv \sum_{\alpha=1}^{N} \frac{a_\lambda(\alpha) a_{\lambda'}(\alpha)^*}{N}, \tag{238}$$

in terms of which

$$\begin{aligned} \langle G \rangle &= \sum_{\lambda,\lambda'} \langle \lambda | \, \rho \, | \lambda' \rangle \langle \lambda' | \, G \, | \lambda \rangle \\ &= \operatorname{Tr} (\rho G). \end{aligned} \tag{239}$$

It is evident that

$$\operatorname{Tr} \rho = 1. \tag{240}$$

The density matrix $\rho$ is very useful, since with it we may calculate the mean value over the ensemble of any physical quantity, using Eq. (239). If the members of the ensemble are all in a single, pure state $\psi$, then

$$\langle \lambda | \, \rho \, | \lambda' \rangle = a_\lambda a_{\lambda'}^*.$$

In this case $\rho$ is a projection operator, since

$$\langle \lambda | \, \rho^2 \, | \lambda' \rangle = \langle \lambda | \, \rho \, | \lambda' \rangle.$$

In general, however, we may show that

$$\operatorname{Tr} \rho^2 \leq \operatorname{Tr} \rho = 1.$$

The diagonal matrix elements of $\rho$,

$$\langle \lambda | \, \rho \, | \lambda \rangle = \frac{1}{N} \sum_{\alpha=1}^{N} |a_\lambda(\alpha)|^2 = P(\lambda),$$

give the probability of finding a member of the ensemble in the state $\lambda$. Since $\rho$ is Hermitian, it may be diagonalized by a unitary transformation. If we write $|\gamma\rangle$ for the states in the representation in which $\rho$ is diagonal, we evidently have

$$\langle \gamma | \, \rho \, | \gamma' \rangle = \delta_{\gamma,\gamma'} P(\gamma), \tag{241}$$

where $P(\gamma)$ is the fraction of the members of the ensemble in the state $\gamma$.

To express $\rho$ as an operator, we transform Eq. (238) to a general representation:

$$\rho = \sum_{\lambda,\lambda'} |\lambda\rangle \langle \lambda | \, \rho \, | \lambda' \rangle \langle \lambda' |. \tag{242a}$$

In the representation $|\gamma\rangle$ of Eq. (241), this becomes

$$\rho = \sum_\gamma |\gamma\rangle P(\gamma) \langle \gamma |. \tag{242b}$$

This may also be written directly as an ensemble average. It follows from Eq. (238) that $P(\gamma) = N(\gamma)/N$, where $N(\gamma)$ is the number of systems in the ensemble in the state $\gamma$. Therefore, Eq. (242*b*) is equivalent to

$$\rho = \frac{1}{N} \sum_{\alpha=1}^{N} |\gamma(\alpha)\rangle \langle\gamma(\alpha)|, \tag{242c}$$

where $\gamma(\alpha)$ describes the state of system $\alpha$.

To apply these considerations to a scattering experiment, let us return to the interaction described in Section 7.1. Two particles having the respective spins and spin orientations $(S_1, \nu_1)$ and $(S_2, \nu_2)$ scatter in the barycentric reference system. The basis spin functions [see Eq. (6-38)] are then

$$u_{S1,\nu1}(s_1)u_{S2,\nu2}(s_2) \equiv |S_1S_2\nu_1\nu_2\rangle. \tag{243}$$

The final spin orientation following the scattering of the two particles is assumed to have been observed to be $(\nu'_1, \nu'_2)$.

In an experimental study of this interaction, many such scatterings are observed. We have agreed to represent the beam and target particles by an ensemble, so it is possible to define a density matrix for the ensemble spins. This will have the form $\langle\nu'_1, \nu'_2|\, \rho \,|\nu_1, \nu_2\rangle$. In the $(S, \nu)$ representation [see Eqs. (10)] it is transformed into $\langle S', \nu'|\, \rho \,|S, \nu\rangle$. As we have seen, this density matrix can be used to evaluate the average value over the ensemble of any function of the spins $\mathbf{S}_1$ and $\mathbf{S}_2$.

The density matrix for the incident, or initial, state in a scattering experiment will be called $\rho_i$. We emphasize that this describes only the mean distribution of spin orientations (or *polarization*) for the beam and target and not the spin orientation of individual particles. (For the sake of brevity, we shall refer to the "polarization of the beam or target" when we mean the "average polarization over the ensemble of beam and target systems.") The appropriate density matrix $\rho_i$ for a given experiment of course depends on the manner in which the beam and target are prepared. One of the most frequently encountered cases is that of an *unpolarized* beam and target. In this case we have

$$\langle\nu'_1, \nu'_2|\, \rho_i \,|\nu_1, \nu_2\rangle = \frac{1}{(2S_1+1)(2S_2+1)}\, \delta_{\nu'1,\nu1}\delta_{\nu'2,\nu2}, \tag{244a}$$

or in a general representation, we may use Eqs. (244) to write

$$\begin{aligned}\rho_i &= \frac{1}{(2S_1+1)(2S_2+1)} \sum_{\nu1,\nu2} |S_1S_2\nu_1\nu_2\rangle \langle S_1S_2\nu_1\nu_2| \\ &= \frac{1}{(2S_1+1)(2S_2+1)}\, 1,\end{aligned} \tag{244b}$$

where 1 is the unit matrix. It is evident from Eq. (241) that Eqs. (244)

describe an ensemble with no preferred spin orientations. Since the unit matrix remains the unit matrix in any representation, this conclusion is independent of representation.[47] For example, we may rewrite (244$a$) as

$$\langle S', \nu'| \rho_i |S, \nu\rangle = \frac{1}{(2S_1 + 1)(2S_2 + 1)} \delta_{S',S}\delta_{\nu',\nu}. \tag{244c}$$

Let us now express the scattering cross section in terms of $\rho_i$. The scattering amplitude (43) may be considered to be a $(2S_1 + 1)(2S_2 + 1) \times (2S_1 + 1)(2S_2 + 1)$ matrix in spin space. In a general representation we may write it as a function of the spin operators $\mathbf{S}_1$ and $\mathbf{S}_2$:

$$\mathbf{f}(\hat{\boldsymbol{\varkappa}}_f, \hat{\boldsymbol{\varkappa}}, \mathbf{S}_1, \mathbf{S}_2). \tag{245}$$

[The expressions (79) and (214) are special cases of (245).] Then [see Eq. (43)]

$$\langle S', \nu'| \mathbf{f} |S, \nu\rangle = (u_{S',\nu'}, \mathbf{f}(\hat{\boldsymbol{\varkappa}}_f, \hat{\boldsymbol{\varkappa}}, \mathbf{S}_1, \mathbf{S}_2)u_{S,\nu}),$$

or

$$\langle \nu'_1, \nu'_2| \mathbf{f} |\nu_1, \nu_2\rangle = (u_{S1,\nu'1}u_{S2,\nu'2}, \mathbf{f}(\hat{\boldsymbol{\varkappa}}_f, \hat{\boldsymbol{\varkappa}}, \mathbf{S}_1, \mathbf{S}_2)u_{S1,\nu1}u_{S2,\nu2}). \tag{246}$$

It is desirable at this point to agree on several notational conventions for the $(2S_1 + 1)(2S_2 + 1) \times (2S_1 + 1)(2S_2 + 1)$ matrix $\mathbf{f}$. First, the complex conjugate of $\mathbf{f}$ will be written as

$$[\mathbf{f}(\hat{\boldsymbol{\varkappa}}_f, \hat{\boldsymbol{\varkappa}}, \mathbf{S}_1, \mathbf{S}_2)]^* \equiv \mathbf{f}^*(\hat{\boldsymbol{\varkappa}}_f, \hat{\boldsymbol{\varkappa}}, \mathbf{S}_1^*, \mathbf{S}_2^*). \tag{247}$$

This then indicates the complex conjugate function of the complex conjugate operators $\mathbf{S}_1^* = \tilde{\mathbf{S}}_1^{\dagger}$, and so on.[48] We shall write

$$\widetilde{\mathbf{f}(\hat{\boldsymbol{\varkappa}}_f, \hat{\boldsymbol{\varkappa}}, \mathbf{S}_1, \mathbf{S}_2)}$$

for the transpose of the $(2S_1 + 1)(2S_2 + 1) \times (2S_1 + 1)(2S_2 + 1)$ matrix $\mathbf{f}$. Finally, the adjoint of $\mathbf{f}$ is

$$\mathbf{f}^{\dagger}(\hat{\boldsymbol{\varkappa}}_f, \hat{\boldsymbol{\varkappa}}, \mathbf{S}_1, \mathbf{S}_2) = [\widetilde{\mathbf{f}(\hat{\boldsymbol{\varkappa}}_f, \hat{\boldsymbol{\varkappa}}, \mathbf{S}_1, \mathbf{S}_2)}]^*. \tag{248}$$

The differential cross section for a final spin orientation $(\nu'_1, \nu'_2)$ and an unpolarized incident beam is obtained from Eq. (44):

$$\begin{aligned} \frac{d\sigma}{d\Omega}(\nu'_1, \nu'_2) &= \frac{1}{(2S_1 + 1)(2S_2 + 1)} \sum_{\nu_1,\nu_2} |\langle \nu'_1, \nu'_2| \mathbf{f} |\nu_1, \nu_2\rangle|^2 \\ &= \langle \nu'_1, \nu'_2| \mathbf{f}\rho_i\mathbf{f}^{\dagger} |\nu'_1, \nu'_2\rangle, \end{aligned} \tag{249}$$

[47] Equations (244) properly define the term "unpolarized beam and target." It is thus an experimental question as to whether the beam and target in a given experiment are unpolarized. The definition (244) is evidently equivalent to that which we have been using for unpolarized systems.

[48] To restate this, let us replace $\mathbf{S}_1$ and $\mathbf{S}_2$ by two *real* vectors $\mathbf{A}$ and $\mathbf{B}$. Then $[\mathbf{f}(\hat{\boldsymbol{\varkappa}}_f, \hat{\boldsymbol{\varkappa}}, \mathbf{A}, \mathbf{B})]^* = \mathbf{f}^*(\hat{\boldsymbol{\varkappa}}_f, \hat{\boldsymbol{\varkappa}}, \mathbf{A}, \mathbf{B})$ which defines $f^*(\hat{\boldsymbol{\varkappa}}_f, \hat{\boldsymbol{\varkappa}}, \mathbf{A}, \mathbf{B})$ as a function of $\hat{\boldsymbol{\varkappa}}_f$, $\hat{\boldsymbol{\varkappa}}$, $\mathbf{A}$, and $\mathbf{B}$. If we now set $\mathbf{A} = \mathbf{S}_1^*$, $\mathbf{B} = \mathbf{S}_2^*$ in this function, we obtain the right-hand side of Eq. (247).

where $\rho_i$ is given by Eq. (244). The cross section for any final spin is then given in the concise form

$$\frac{d\bar{\sigma}}{d\Omega} = \sum_{\nu'_1, \nu'_2} \frac{d\sigma}{d\Omega}(\nu'_1, \nu'_2) = \mathrm{Tr}\,(\mathbf{f}\rho_i\mathbf{f}^\dagger).$$

The trace here is evaluated in the $(2S_1 + 1)(2S_2 + 1)$-dimensional spin space.

These expressions are actually valid for arbitrary polarization of the beam and target. To see this let us express the matrix elements of $\mathbf{f}$ in the representation $\gamma$ which diagonalizes the initial density matrix $\rho_i$. The cross section for any pair $\alpha$ of particles in the ensemble is then

$$\frac{d\sigma(\alpha)}{d\Omega} = \sum_{\gamma'} |\langle\gamma'|\,\mathbf{f}\,|\gamma(\alpha)\rangle|^2,$$

where $\gamma(\alpha)$ is the initial spin state of the particles. The average cross section for the ensemble is therefore

$$\begin{aligned}\frac{d\bar{\sigma}}{d\Omega} &= \frac{1}{N}\sum_{\alpha=1}^{N} \frac{d\sigma(\alpha)}{d\Omega}\\ &= \frac{1}{N}\sum_{\alpha}\sum_{\gamma'} \langle\gamma'|\,\mathbf{f}\,|\gamma(\alpha)\rangle\langle\gamma(\alpha)|\,\mathbf{f}^\dagger\,|\gamma'\rangle \qquad (250)\\ &= \mathrm{Tr}\,(\mathbf{f}\rho_i\mathbf{f}^\dagger),\end{aligned}$$

on using Eqs. (238) and (242*c*). This is a very important expression, for it permits us to calculate the average cross section for a beam and target of any polarization.

Before continuing with the general discussion, let us return to the case of spin zero–spin one-half scattering to illustrate our remarks. The scattering amplitude is $\mathbf{f} = f + i\boldsymbol{\sigma}\cdot\hat{\mathbf{n}}\sin\theta g$. To determine $\rho_i$, we use the fact that the unit $2 \times 2$ matrix 1 and the three Pauli spin matrices $\sigma_x$, $\sigma_y$, and $\sigma_z$ form a complete, orthogonal basis set[49] of $2 \times 2$ matrices. We may therefore write $\rho_i$ in the general form

$$\rho_i = a_0 \cdot 1 + \mathbf{a}\cdot\boldsymbol{\sigma}, \qquad (251)$$

where $a_0$ and $a_x$, $a_y$, and $a_z$ are four parameters. Since

$$\mathrm{Tr}\,\rho_i = 1 = 2a_0,$$

we obtain $a_0 = \frac{1}{2}$. The mean polarization of the incident beam can [as follows from the general Eq. (239)] be assumed to be

$$\mathbf{P}_i = \mathrm{Tr}\,(\rho_i\boldsymbol{\sigma}). \qquad (252)$$

Substitution of the expression (251) into this equation gives

$$\mathbf{P}_i = 2\mathbf{a}.$$

[49] Recall that an arbitrary $2 \times 2$ matrix may be written as $a \cdot 1 + \mathbf{a}\cdot\boldsymbol{\sigma}$, where $a$, $a_x$, $a_y$, and $a_z$ are four complex constants.

Summarizing these results, we may rewrite (251) as

$$\rho_i = \tfrac{1}{2}(1 + \mathbf{P}_i \cdot \boldsymbol{\sigma}). \tag{253}$$

We may now substitute Eq. (253) into Eq. (250) to obtain

$$\frac{d\bar{\sigma}}{d\Omega} = |f|^2 + |g|^2 \sin^2\theta + 2\mathbf{P}_i \cdot \hat{\mathbf{n}} \sin\theta \, \mathrm{Im}\,(g^*f). \tag{254}$$

This has the form of Eq. (227), but is more general, since now $P_i$ may have any value in the range $0 \leq P_i \leq 1$.

Let us next construct the density matrix $\rho_f$ for the spins of an ensemble of two particle systems following scattering from some initial state. Let us choose spin functions $u_\gamma$ such that the density matrix $\rho_i$ is diagonal in the $\gamma$-representation. Then the asymptotic wave function (42) may be transformed into

$$\psi^{+}_{\varkappa,S_1,S_2,\gamma} = (2\pi)^{-3/2}\left(e^{i\varkappa\cdot\mathbf{r}} + \frac{e^{i\kappa r}}{r}\mathbf{f}\right)u_\gamma. \tag{255}$$

This suggests defining an amplitude $A^\gamma$, analogous to that of Eq. (223), as

$$A^\gamma \equiv \mathbf{f}u_\gamma. \tag{256}$$

Then $A^\gamma$ is the (unnormalized) spin wave function of the scattered wave when the two particles had the "spin orientation" $\gamma$ in the initial state.

We now suppose that $N$ scatterings have been observed, all into the same increment of solid angle. If we let $\gamma(\alpha)$ ($\alpha = 1, 2, \ldots, N$) be the initial spin state for the pair $\alpha$ of scattering particles, the density function $\rho_f$ is seen by Eqs. (242) to be given by the equation

$$\begin{aligned} A\rho_f &= \frac{1}{N}\sum_\alpha A^{\gamma(\alpha)}A^{\gamma(\alpha)\dagger} \\ &= \mathbf{f}\left(\frac{1}{N}\sum_\alpha u_{\gamma(\alpha)}u^\dagger_{\gamma(\alpha)}\right)\mathbf{f}^\dagger \\ &= \mathbf{f}\rho_i\mathbf{f}^\dagger. \end{aligned} \tag{257}$$

Since the density matrix is so normalized that $\mathrm{Tr}\,\rho_f = 1$, the quantity $A$ is just

$$A = \mathrm{Tr}\,(\mathbf{f}\rho_i\mathbf{f}^\dagger) = \frac{d\bar{\sigma}}{d\Omega}, \tag{258}$$

according to Eq. (250). To find the average polarization in the ensemble following the scattering, we may multiply $\rho_f$ by products of the components of $\mathbf{S}_1$ and $\mathbf{S}_2$ and form the trace. For the special case of spin zero–spin one-half scattering, we may easily show that

$$\mathbf{P} = \mathrm{Tr}\,(\rho_f\boldsymbol{\sigma}) \tag{259}$$

leads to precisely Eq. (231).

Let us next consider a sequence of two scatterings for the beam particles, which are assumed to have a spin $S_1$. The target particles for the first and second scatterings have, respectively, spins $S_2$ and $S_3$. The relative momenta prior to the first scattering, following the first scattering, and following the second scattering, will be taken as $\boldsymbol{\varkappa}$, $\boldsymbol{\varkappa}_1$, and $\boldsymbol{\varkappa}_2$. Prior to the first scattering the beam and both targets will be considered to be unpolarized. Then, letting $\mathbf{f}_1$ and $\mathbf{f}_2$ be the respective amplitudes for the two scatterings, we see from Eq. (256) that the spin amplitude following the first scattering is

$$A_1^{\nu_1\nu_2\nu_3} = f_1(\hat{\boldsymbol{\varkappa}}_1, \hat{\boldsymbol{\varkappa}}, S_1, S_2)u_{S1,\nu1}u_{S2,\nu2}u_{S3,\nu3}. \tag{260}$$

To describe the second scattering, we may use Eq. (255), replacing $\mathbf{f}$ by $\mathbf{f}_2$ and the wave functions $u_{S1;\nu1}\ u_{S2,\nu2}$ by $A_1^{\nu_1\nu_2\nu_3}$. The differential cross section following the second scattering is then[50]

$$\begin{aligned}\frac{d\bar{\sigma}_2}{d\Omega_2} &= \frac{1}{(2S_1+1)(2S_2+1)(2S_3+1)} \sum_{\nu_1,\nu_2,\nu_3} \frac{(A^{\nu_1\nu_2\nu_3}, \mathbf{f}_2{}^{\dagger}\mathbf{f}_2 A^{\nu_1\nu_2\nu_3})}{A_1} \\ &= \operatorname{Tr}(\mathbf{f}_2\rho_1\mathbf{f}_2{}^{\dagger}) = \operatorname{Tr}\frac{\mathbf{f}_2\mathbf{f}_1\rho_i\mathbf{f}_1{}^{\dagger}\mathbf{f}_2{}^{\dagger}}{A_1}\end{aligned} \tag{261}$$

Here

$$A_1\rho_1 \equiv f_1\rho_i f_1{}^{\dagger}, \qquad \operatorname{Tr}\rho_1 = 1 \tag{262}$$

and

$$\rho_i = \frac{1}{(2S_1+1)(2S_2+1)(2S_3+1)}\,1, \tag{263}$$

where 1 is the unit matrix acting on the three sets of spin functions. We may easily show that Eq. (261) agrees with Eq. (233) for spin one-half–spin zero scattering. [It is also easy to show that the last two forms of Eq. (261) are valid for arbitrary polarization of the initial beam and target.]

The cross section and density matrix following $n$ scatterings of a beam of particles having spin $S_1$ may now be obtained by induction. The respective target spins are assumed to be $S_2, S_3, \ldots, S_{n+1}$. The scattering amplitude for the $j$th scattering ($j = 1, 2, \ldots, n$) is written as $\mathbf{f}_j$. The initial density matrix is taken to be

$$\rho_i = \frac{1}{(2S_1+1)(2S_2+1)\cdots(2S_{n+1}+1)}\,1, \tag{264}$$

where 1 is the unit matrix in the space of the $n+1$ spins. Equation (264) describes an ensemble for which the beam and targets are unpolarized in the initial state. By reasoning analogous to that which led to Eq. (262),

[50] For particles with relativistic energies, a Lorentz transformation of the spins to the new barycentric coordinate system must be made unless we define all scattering amplitudes in the laboratory frame of reference.

we see that the density matrix $\rho_j$, following the $j$th scattering, is defined by the set of equations

$$A_j\rho_j = \mathbf{f}_j\rho_{j-1}\mathbf{f}_j^{\dagger}, \qquad (265)$$

$$\mathrm{Tr}\ \rho_j = 1,$$

using Eq. (262) as a starting condition.

The differential cross section for the $n$th scattering is

$$\frac{d\bar{\sigma}_n}{d\Omega_n} = \mathrm{Tr}\ (\mathbf{f}_n\rho_{n-1}\mathbf{f}_n^{\dagger})$$

$$= \frac{\mathrm{Tr}\ (\mathbf{f}_n\mathbf{f}_{n-1}\cdots\mathbf{f}_1\rho_1\mathbf{f}_1^{\dagger}\cdots\mathbf{f}_n^{\dagger})}{A_1A_2\cdots A_{n-1}}. \qquad (266)$$

The polarization of the beam particles after the $n$th scattering is given by

$$\langle\mathbf{S}_1\rangle = \mathrm{Tr}\ (\rho_n\mathbf{S}_1), \qquad (267)$$

and so on.

**Special Representations for the Density Matrix and Scattering Cross Sections**

It is often convenient to expand the density matrix $\rho$ for a pair of interacting particles in terms of a complete orthogonal basis set of Hermitian matrices. This can be done in a manner analogous to that by which we obtained Eq. (253).

Let us first consider a general $(2S + 1) \times (2S + 1)$ matrix $M$ in the space of a spin operator $\mathbf{S}$. In this space we may introduce a complete orthogonal basis of $(2S + 1) \times (2S + 1)$ Hermitian matrices $\omega^{\lambda}(S)$ $[\lambda = 1, 2, \ldots, (2S + 1)^2]$. These are considered to be normalized in such a way that for any pair $\omega^{\lambda}(S)$, $\omega^{\mu}(S)$,

$$\mathrm{Tr}\ [\omega^{\lambda}(S)\omega^{\mu}(S)] = (2S + 1)\delta_{\lambda,\mu}. \qquad (268)$$

It is also convenient to define these so that one of them, say $\omega^1(S)$, is just the unit matrix.

$$\omega^1(S) = 1. \qquad (269)$$

The general matrix $M$ may then be expanded in the form

$$M = \sum_{\lambda=1}^{(2S+1)^2} b_{\lambda}\omega^{\lambda}(S), \qquad (270)$$

where the $b_{\lambda}$ are numerical coefficients. It follows from Eqs. (268) and (270) that

$$b_{\lambda} = \frac{1}{2S + 1}\mathrm{Tr}\ [M\omega^{\lambda}(S)]. \qquad (271)$$

Such a set of basis matrices can be constructed in the $(2S_1 + 1) \times (2S_2 + 1) \cdots (2S_n + 1)$ space of $n$ spins, $\mathbf{S}_1, \mathbf{S}_2, \ldots, \mathbf{S}_n$. These are

$$\Omega^\lambda = \omega^{\lambda_1}(S_1) \times \omega^{\lambda_2}(S_2) \times \cdots \times \omega^{\lambda_n}(S_n), \tag{272a}$$

where the direct (or Kronecker) product is implied.[51] Here the index $\lambda = (\lambda_1, \lambda_2, \ldots, \lambda_n)$ may be chosen in such a way as to run over the values $\lambda = 1, 2, \ldots, (2S_1 + 1)^2(2S_2 + 1)^2, \ldots, (2S_n + 1)^2$. We now choose, as in Eq. (269),

$$\Omega^1 = 1, \tag{273}$$

where 1 is the unit $(2S_1 + 1)(2S_2 + 1) \cdots (2S_n + 1) \times (2S_1 + 1) \cdots (2S_n + 1)$ matrix.

For a system of two particles with spins $S_1$ and $S_2$, we have

$$\Omega^\lambda = \omega^{\lambda_1}(S_1) \times \omega^{\lambda_2}(S_2). \tag{272b}$$

The density matrix $\rho$ for this system is therefore of the form

$$\rho = \sum_{\lambda=1}^{(2S_1+1)^2(2S_2+1)^2} a_\lambda \Omega^\lambda, \tag{274}$$

where

$$a_1 = \frac{1}{(2S_1 + 1)(2S_2 + 1)}, \tag{275a}$$

and for any $\lambda$,

$$\begin{aligned} a_\lambda &= \frac{1}{(2S_1 + 1)(2S_2 + 1)} \operatorname{Tr}(\rho\Omega^\lambda) \\ &\equiv \frac{1}{(2S_1 + 1)(2S_2 + 1)} \langle \Omega^\lambda \rangle. \end{aligned} \tag{275b}$$

For the initial ensemble of beam and target particles prior to scattering, the density matrix may then be written as

$$\rho_i = \frac{1}{(2S_1 + 1)(2S_2 + 1)} \left( \sum_{\lambda=1}^{(2S_1+1)^2(2S_2+1)^2} \langle \Omega^\lambda \rangle_i \Omega^\lambda \right). \tag{276}$$

Here $\langle \Omega^\lambda \rangle_i$ is the average over the ensemble of expectation values of $\Omega^\lambda$. The complete set of quantities $\langle \Omega^\lambda \rangle_i$ are seen to be required if the initial system is to be specified.

An unpolarized beam and target correspond to [see Eq. (273)]

$$\begin{aligned} \langle \Omega^1 \rangle_i &= 1 \\ \langle \Omega^\lambda \rangle_i &= 0, \qquad i \neq 1. \end{aligned} \tag{277}$$

[51] See, for example, Murnaghan [18, Chapter 3].

In this case we see that Eq. (276) reduces to Eq. (244*b*). For a polarized beam, but unpolarized target, of respective spins $S_1$ and $S_2$, we see that

$$\langle\Omega^{\lambda}\rangle_i = \langle\omega^{\lambda_1}(S_1) \times \omega^{\lambda_2}(S_2)\rangle_i = 0$$

unless $\lambda_2 = 1$. Then Eq. (276) simplifies to

$$\rho_i = \frac{1}{(2S_1 + 1)(2S_2 + 1)} \left\{ \sum_{\lambda_1=1}^{(2S_1+1)^2} \langle\omega^{\lambda_1}(S_1)\rangle_i \omega^{\lambda_1}(S_1) \right\}, \qquad (278)$$

where it is implied that the unit $(2S_2 + 1) \times (2S_2 + 1)$ matrix appears as a factor.

The matrices $\omega^{\lambda}(S)$ may be constructed from products of the components of the spin operator **S**. For spin zero there is only the single matrix $\omega^1(0) = 1$. In the case of spin one-half we have seen that the four matrices $\omega^{\lambda}(\frac{1}{2})$ are 1, $\sigma_x$, $\sigma_y$, and $\sigma_z$. For a particle with spin 1 there are $3^2 = 9$ matrices $\omega^{\lambda}(1)$. If we write $S_{\alpha}$ ($\alpha = 1, 2, 3$) for the three components of **S**, these are

$$\omega^1(1) = 1,$$

$$\omega^2(1) = \frac{1}{\sqrt{2}} S_1,$$

$$\omega^3(1) = \frac{1}{\sqrt{2}} S_2,$$

$$\omega^4(1) = \frac{1}{\sqrt{2}} S_3,$$

$$\omega^5(1) = \frac{1}{\sqrt{2}} (S_1S_2 + S_2S_1),$$

$$\omega^6(1) = \frac{1}{\sqrt{2}} (S_2S_3 + S_3S_2),$$

$$\omega^7(1) = \frac{1}{\sqrt{2}} (S_3S_1 + S_1S_3),$$

$$\omega^8(1) = \frac{1}{2}\left(S_1^2 - S_2^2 - \sqrt{3}S_3^2 + \frac{2}{\sqrt{3}}\right),$$

$$\omega^9(1) = \frac{1}{2}\left(S_2^2 - S_1^2 - \sqrt{3}S_3^2 + \frac{2}{\sqrt{3}}\right). \qquad (279)$$

That these satisfy the relations (268) may be verified using Eqs. (14) in Appendix F.

Using Eq. (276), we may write the differential cross section (250) in the form[52]

$$\frac{d\bar{\sigma}}{d\Omega} = \frac{1}{(2S_1+1)(2S_2+1)} \sum_{\lambda=1}^{(2S_1+1)^2(2S_2+1)^2} \langle\Omega^\lambda\rangle_i \operatorname{Tr}(\mathbf{f}\Omega^\lambda\mathbf{f}^\dagger). \qquad (280)$$

The polarization following the scattering is given by the set of $(2S_1 + 1)^2 \times (2S_2 + 1)^2$ quantities

$$\begin{aligned}\langle\Omega^\mu\rangle_f &= \operatorname{Tr}(\rho_f\Omega^\mu)\\ &= \frac{\operatorname{Tr}(\mathbf{f}\rho_i\mathbf{f}^\dagger\Omega^\mu)}{\operatorname{Tr}(\mathbf{f}\rho_i\mathbf{f}^\dagger)}\\ &= \frac{\sum_\lambda \langle\Omega^\lambda\rangle_i \operatorname{Tr}(\mathbf{f}\Omega^\lambda\mathbf{f}^\dagger\Omega^\mu)}{\sum_\lambda \langle\Omega^\lambda\rangle_i \operatorname{Tr}(\mathbf{f}\Omega^\lambda\mathbf{f}^\dagger)}.\end{aligned} \qquad (281)$$

The expressions (280) and (281), along with the recurrence relations (262), (264), and (265), permit us to describe a sequence of any number of scatterings of beam (and of recoil target) particles.

As a final illustration of our general results, let us describe the scattering of polarized spin one-half particles by an unpolarized target of any spin, say $S_2$. Then Eq. (278) gives

$$\rho_i = \frac{1}{2(2S_2+1)}(1 + \mathbf{P}_i \cdot \boldsymbol{\sigma}), \qquad (282)$$

where $\mathbf{P}_i = \langle\boldsymbol{\sigma}\rangle_i$ is the polarization of the incident beam. Equation (282) is the direct product of the expression (253) with the unit $(2S_2 + 1) \times (2S_2 + 1)$ matrix, divided by the normalizing factor $2S_2 + 1$. The scattering cross section (250) is now

$$\begin{aligned}\left.\frac{d\bar{\sigma}}{d\Omega}\right)_{P_i} &= \operatorname{Tr}(\mathbf{f}\rho_i\mathbf{f}^\dagger)\\ &= \frac{1}{2(2S_2+1)}[\operatorname{Tr}(\mathbf{f}^\dagger\mathbf{f}) + \operatorname{Tr}(\mathbf{f}\boldsymbol{\sigma}\mathbf{f}^\dagger)\cdot\mathbf{P}_i].\end{aligned} \qquad (283)$$

Evidently the first term,

$$\left.\frac{d\bar{\sigma}}{d\Omega}\right)_u = \frac{1}{2(2S_2+1)}\operatorname{Tr}(\mathbf{f}^\dagger\mathbf{f}) \qquad (284)$$

is the cross section for an unpolarized target. The polarization of the spin one-half particle following the scattering is

$$\begin{aligned}\mathbf{P}(P_i) &= \operatorname{Tr}(\rho_f\boldsymbol{\sigma})\\ &= \frac{[\operatorname{Tr}(\mathbf{f}^\dagger\boldsymbol{\sigma}\mathbf{f}) + \operatorname{Tr}(\boldsymbol{\sigma}\mathbf{f}\boldsymbol{\sigma}\mathbf{f}^\dagger)\cdot\mathbf{P}_i]}{[\operatorname{Tr}(\mathbf{f}^\dagger\mathbf{f}) + \operatorname{Tr}(\mathbf{f}\boldsymbol{\sigma}\mathbf{f}^\dagger)\cdot\mathbf{P}_i]}.\end{aligned} \qquad (285)$$

[52] On p. 419 we shall show how to simplify Eq. (280).

For an initially unpolarized beam, on the other hand, the polarization would be

$$P_u = \frac{\mathrm{Tr}\,(\mathbf{f}^\dagger \boldsymbol{\sigma} \mathbf{f})}{\mathrm{Tr}\,(\mathbf{f}^\dagger \mathbf{f})}. \tag{286}$$

Now, in the next section we shall prove that

$$\mathrm{Tr}\,(\mathbf{f}^\dagger \boldsymbol{\sigma} \mathbf{f}) = \mathrm{Tr}\,(\mathbf{f} \boldsymbol{\sigma} \mathbf{f}^\dagger), \tag{287}$$

which follows from time reversal invariance. Using the identity (287), we may write Eq. (283) as

$$\left.\frac{d\bar{\sigma}}{d\Omega}\right)_{P_i} = \left.\frac{d\bar{\sigma}}{d\Omega}\right)_u (1 + \mathbf{P}_u \cdot \mathbf{P}_i), \tag{288}$$

and (285) as

$$\mathbf{P}(P_i) = \frac{\mathbf{P}_u + \mathscr{P} \cdot \mathbf{P}_i}{1 + \mathbf{P}_u \cdot \mathbf{P}_i} \tag{289}$$

where

$$\mathscr{P} \equiv \frac{\mathrm{Tr}\,(\boldsymbol{\sigma} \mathbf{f} \boldsymbol{\sigma} \mathbf{f}^\dagger)}{\mathrm{Tr}\,(\mathbf{f}\mathbf{f}^\dagger)}. \tag{290}$$

Equations (288) and (289) are seen to be generalizations of Eqs. (233) and (230), respectively.

**Two General Theorems**

Let us first prove a generalized form of Eq. (287). To do this we require an explicit statement of the time reversal invariance of $\mathbf{f}(\hat{\boldsymbol{\varkappa}}_f, \hat{\boldsymbol{\varkappa}}, \mathbf{S}_1, \mathbf{S}_2)$, the matrix amplitude for the scattering of two particles of respective spins $S_1$ and $S_2$. Transcribing Eq. (32) to our present notation, we have [see Eq. (43)]

$$\begin{aligned}\langle S', \nu' |\, \mathbf{f}(\hat{\boldsymbol{\varkappa}}_f, \hat{\boldsymbol{\varkappa}})\, | S, \nu\rangle &\equiv (u_{S',\nu'}, \mathbf{f}(\hat{\boldsymbol{\varkappa}}_f, \hat{\boldsymbol{\varkappa}}, \mathbf{S}_1, \mathbf{S}_2) u_{S,\nu}) \\ &= (-1)^{S'+\nu'-S-\nu} \langle S, -\nu |\, \mathbf{f}(-\hat{\boldsymbol{\varkappa}}, -\hat{\boldsymbol{\varkappa}}_f)\, | S', -\nu'\rangle.\end{aligned} \tag{291}$$

Here we have used the phase convention of Eq. (30). Therefore,

$$W u^*_{S,\nu} = (-1)^{S+\nu} u_{S,-\nu}, \tag{292}$$

and so on, where the time reversal operator is $T = WT_0$ [see Eq. (2-121)]. Similarly, from Eq. (2-129), we see that

$$W S_\alpha^* W^{-1} = -S_\alpha, \qquad \alpha = 1, 2. \tag{293}$$

Now using Eqs. (292) and (293) and the notation of Eqs. (247) and (248), we may write Eq. (291) as

$$
\begin{aligned}
(-1)^{S'+\nu'-S-\nu}&(u_{S,-\nu}, \mathbf{f}(-\hat{\boldsymbol{\varkappa}}, -\hat{\boldsymbol{\varkappa}}_f, \mathbf{S}_1, \mathbf{S}_2)u_{S',-\nu'}) \\
&= (Wu^*_{S,\nu}, \mathbf{f}(-\hat{\boldsymbol{\varkappa}}, -\hat{\boldsymbol{\varkappa}}_f, \mathbf{S}_1, \mathbf{S}_2)Wu^*_{S',\nu'}) \\
&= (u^*_{S,\nu}, \mathbf{f}(-\hat{\boldsymbol{\varkappa}}, -\hat{\boldsymbol{\varkappa}}_f, W^{-1}\mathbf{S}_1W, W^{-1}\mathbf{S}_2W)u^*_{S',\nu'}) \\
&= (u^*_{S,\nu}, f(-\hat{\boldsymbol{\varkappa}}, -\hat{\boldsymbol{\varkappa}}_f, -\mathbf{S}_1{}^*, -\mathbf{S}_2{}^*)u^*_{S',\nu'}) \\
&= (\overgroup{u_{S',\nu'}, \mathbf{f}(-\hat{\boldsymbol{\varkappa}}, -\hat{\boldsymbol{\varkappa}}_f, -\mathbf{S}_1{}^*, -\mathbf{S}_2{}^*})u_{S,\nu}) \\
&= (u_{S',\nu'}, \mathbf{f}(\hat{\boldsymbol{\varkappa}}_f, \hat{\boldsymbol{\varkappa}}, \mathbf{S}_1, \mathbf{S}_2)u_{S,\nu}).
\end{aligned}
$$

Then we have the general condition of time reversal invariance.

$$
\mathbf{f}(\hat{\boldsymbol{\varkappa}}_f, \hat{\boldsymbol{\varkappa}}, \mathbf{S}_1, \mathbf{S}_2) = \mathbf{f}\overgroup{(-\hat{\boldsymbol{\varkappa}}, -\hat{\boldsymbol{\varkappa}}_f, -\mathbf{S}_1{}^*, -\mathbf{S}_2{}^*}). \tag{294}
$$

In the special case that $\mathbf{f}$ is a linear function of $\mathbf{S}_1$ and of $\mathbf{S}_2$, this simplifies to

$$
\mathbf{f}(\hat{\boldsymbol{\varkappa}}_f, \hat{\boldsymbol{\varkappa}}, \mathbf{S}_1, \mathbf{S}_2) = \mathbf{f}(-\hat{\boldsymbol{\varkappa}}, -\hat{\boldsymbol{\varkappa}}_f, -\mathbf{S}_1, -\mathbf{S}_2),
$$

of which Eqs. (97) and (193) are special cases.

We may now show the validity of the following equation.

$$
\begin{aligned}
\mathrm{Tr}\,&[\mathbf{f}(\hat{\boldsymbol{\varkappa}}_f, \hat{\boldsymbol{\varkappa}}, \mathbf{S}_1, \mathbf{S}_2)\mathbf{f}^\dagger(\hat{\boldsymbol{\varkappa}}_f, \hat{\boldsymbol{\varkappa}}, \mathbf{S}_1, \mathbf{S}_2)\mathbf{S}_\alpha] \\
&= \mathrm{Tr}\,[\mathbf{f}^\dagger(\hat{\boldsymbol{\varkappa}}_f, \hat{\boldsymbol{\varkappa}}, \mathbf{S}_1, \mathbf{S}_2)\mathbf{f}(\hat{\boldsymbol{\varkappa}}_f, \hat{\boldsymbol{\varkappa}}, \mathbf{S}_1, \mathbf{S}_2)\mathbf{S}_\alpha],
\end{aligned} \tag{295}
$$

where $\alpha = 1$ or 2. Equation (287) corresponds to the special case that $\mathbf{S}_1 = \boldsymbol{\sigma}$, $\alpha = 1$.

Consider the quantity (which is of course an axial vector)

$$
\hat{\boldsymbol{\varkappa}} \times \hat{\boldsymbol{\varkappa}}_f F(\hat{\boldsymbol{\varkappa}}_f \cdot \hat{\boldsymbol{\varkappa}}) \equiv \mathrm{Tr}\,[\mathbf{f}(\hat{\boldsymbol{\varkappa}}_f, \hat{\boldsymbol{\varkappa}}, \mathbf{S}_1, \mathbf{S}_2)\mathbf{f}^\dagger(\hat{\boldsymbol{\varkappa}}_f, \hat{\boldsymbol{\varkappa}}, \mathbf{S}_1, \mathbf{S}_2)\mathbf{S}_\alpha]. \tag{296}
$$

It is obvious that $F$ is real, or that

$$
F(\hat{\boldsymbol{\varkappa}}_f \cdot \hat{\boldsymbol{\varkappa}}) = F(\hat{\boldsymbol{\varkappa}}_f \cdot \hat{\boldsymbol{\varkappa}})^*. \tag{297a}
$$

Also, $F$ is invariant under either of the transformations

$$
\begin{aligned}
\hat{\boldsymbol{\varkappa}} &\to \hat{\boldsymbol{\varkappa}}_f, \qquad & \hat{\boldsymbol{\varkappa}}_f &\to \hat{\boldsymbol{\varkappa}}, \\
\hat{\boldsymbol{\varkappa}} &\to -\hat{\boldsymbol{\varkappa}}, \qquad & \hat{\boldsymbol{\varkappa}}_f &\to -\hat{\boldsymbol{\varkappa}}_f,
\end{aligned} \tag{297b}
$$

which follows trivially from the fact that $F$ can depend on only $\hat{\boldsymbol{\varkappa}}_f \cdot \hat{\boldsymbol{\varkappa}}$ (and the energy of the particles, which is irrelevant for the present argument). We may now use the relations (293) and (294) and the symmetry conditions

(297) to write

$$\begin{aligned}
\hat{\boldsymbol{\kappa}} \times \hat{\boldsymbol{\kappa}}_f F(\hat{\boldsymbol{\kappa}}_f \cdot \hat{\boldsymbol{\kappa}}) \\
&= -\mathrm{Tr}\,[\mathbf{f}(\hat{\boldsymbol{\kappa}}_f, \hat{\boldsymbol{\kappa}}, \mathbf{S}_1, \mathbf{S}_2)\mathbf{f}^\dagger(\hat{\boldsymbol{\kappa}}_f, \hat{\boldsymbol{\kappa}}, \mathbf{S}_1, \mathbf{S}_2) W \mathbf{S}_\alpha{}^* W^{-1}] \\
&= -\mathrm{Tr}\,[\mathbf{f}(\hat{\boldsymbol{\kappa}}_f, \hat{\boldsymbol{\kappa}}, -\mathbf{S}_1{}^*, -\mathbf{S}_2{}^*)\mathbf{f}^\dagger(\hat{\boldsymbol{\kappa}}_f, \hat{\boldsymbol{\kappa}}, -\mathbf{S}_1{}^*, -\mathbf{S}_2{}^*)\mathbf{S}_\alpha{}^*] \\
&= \mathrm{Tr}\,[\mathbf{f}(-\hat{\boldsymbol{\kappa}}, -\hat{\boldsymbol{\kappa}}_f, -\mathbf{S}_1{}^*, -\mathbf{S}_2{}^*)\mathbf{f}^\dagger(-\hat{\boldsymbol{\kappa}}, -\hat{\boldsymbol{\kappa}}_f, -\mathbf{S}_1{}^*, -\mathbf{S}_2{}^*)\mathbf{S}_\alpha{}^*] \\
&= \mathrm{Tr}\,\widetilde{\mathbf{f}(\hat{\boldsymbol{\kappa}}_f, \hat{\boldsymbol{\kappa}}, \mathbf{S}_1, \mathbf{S}_2})[\mathbf{f}(\hat{\boldsymbol{\kappa}}_f, \hat{\boldsymbol{\kappa}}, \mathbf{S}_1, \mathbf{S}_2)]^* \mathbf{S}_\alpha{}^*\} \\
&= \mathrm{Tr}\,[\mathbf{f}^\dagger(\hat{\boldsymbol{\kappa}}_f, \hat{\boldsymbol{\kappa}}, \mathbf{S}_1, \mathbf{S}_2)\mathbf{f}(\hat{\boldsymbol{\kappa}}_f, \hat{\boldsymbol{\kappa}}, \mathbf{S}_1, \mathbf{S}_2)\mathbf{S}_\alpha]. \qquad (298)
\end{aligned}$$

This is precisely the relation (295) which we set out to prove.

The equality (295) was adequate for the derivation of Eq. (288). To make a similar simplification of Eq. (280), a generalized form of Eq. (295) is required. In the next section we shall prove a more general form of (295), applicable to reaction cross sections.

We may use Eq. (295) to prove another very useful result. That is, we can show that in the first Born approximation the polarization *vanishes* following scattering with an unpolarized beam and target. We recall that in the first Born approximation the scattering matrix $T$ is equal to the interaction potential $V$. Since $V$ is Hermitian,[53] so are $T$ and the scattering amplitude **f** [see Eq. (43)]. That is,

$$\langle S', \nu'|\, \mathbf{f}(\hat{\boldsymbol{\kappa}}_f, \hat{\boldsymbol{\kappa}})\, |S, \nu\rangle = [\langle S, \nu|\, \mathbf{f}(\hat{\boldsymbol{\kappa}}, \hat{\boldsymbol{\kappa}}_f)\, |S', \nu'\rangle]^*. \qquad (299)$$

We may rewrite this in the form

$$\begin{aligned}
(u_{S',\nu'}, \mathbf{f}(\hat{\boldsymbol{\kappa}}_f, \hat{\boldsymbol{\kappa}}, \mathbf{S}_1, \mathbf{S}_2)u_{S,\nu}) &= (u_{S,\nu}, \mathbf{f}(\hat{\boldsymbol{\kappa}}, \hat{\boldsymbol{\kappa}}_f, \mathbf{S}_1, \mathbf{S}_2)u_{S',\nu'})^* \\
&= (u_{S',\nu'}, \mathbf{f}^\dagger(\hat{\boldsymbol{\kappa}}, \hat{\boldsymbol{\kappa}}_f, \mathbf{S}_1, \mathbf{S}_2)u_{S,\nu}).
\end{aligned}$$

Therefore,

$$\mathbf{f}^\dagger(\hat{\boldsymbol{\kappa}}_f, \hat{\boldsymbol{\kappa}}, \mathbf{S}_1, \mathbf{S}_2) = \mathbf{f}(\hat{\boldsymbol{\kappa}}, \hat{\boldsymbol{\kappa}}_f, \mathbf{S}_1, \mathbf{S}_2). \qquad (300)$$

This expression is, we emphasize, valid only in the first Born approximation [whereas Eq. (295) is generally valid if the system is invariant under time reversal].

Now, the polarization of particle $\alpha$ ($\alpha = 1, 2$) following scattering by an unpolarized beam is

$$P_\alpha = \frac{\mathrm{Tr}\,(\mathbf{f}\mathbf{f}^\dagger \mathbf{S}_\alpha)}{\mathrm{Tr}\,(\mathbf{f}\mathbf{f}^\dagger)}. \qquad (301)$$

We must consider then the expression

$$\begin{aligned}
\mathrm{Tr}\,[\mathbf{f}(\hat{\boldsymbol{\kappa}}_f, \hat{\boldsymbol{\kappa}}, \mathbf{S}_1, \mathbf{S}_2)\mathbf{f}^\dagger(\hat{\boldsymbol{\kappa}}_f, \hat{\boldsymbol{\kappa}}, \mathbf{S}_1, \mathbf{S}_2)\mathbf{S}_\alpha] \\
&= \mathrm{Tr}\,[\mathbf{f}^\dagger(\hat{\boldsymbol{\kappa}}, \hat{\boldsymbol{\kappa}}_f, \mathbf{S}_1, \mathbf{S}_2)\mathbf{f}(\hat{\boldsymbol{\kappa}}, \hat{\boldsymbol{\kappa}}_f, \mathbf{S}_1, \mathbf{S}_2)\mathbf{S}_\alpha] \\
&= -\mathrm{Tr}\,[\mathbf{f}^\dagger(\hat{\boldsymbol{\kappa}}_f, \hat{\boldsymbol{\kappa}}, \mathbf{S}_1, \mathbf{S}_2)\mathbf{f}(\hat{\boldsymbol{\kappa}}_f, \hat{\boldsymbol{\kappa}}, \mathbf{S}_1, \mathbf{S}_2)\mathbf{S}_\alpha]. \qquad (302)
\end{aligned}$$

[53] Our theorem does not apply to scattering with a non-Hermitian potential.

The first equality follows from Eq. (300) while the second follows from the fact that the right-hand side of Eq. (296) changes sign when $\hat{\boldsymbol{\varkappa}}_f$ and $\hat{\boldsymbol{\varkappa}}$ are interchanged. Comparing Eqs. (295) and (302) we see that the expression (301) vanishes. This completes our demonstration that particles cannot be polarized by scattering when the first Born approximation is valid.

**Polarization Resulting from Reactions**

The description of polarization given in the last section requires little modification for application to reaction cross sections, as described in Section 7.3. Let us consider two particles of respective spins $S_1$ and $S_2$ and relative momentum $\boldsymbol{\varkappa}_c$ in the entrance channel $c$. In the exit channel $c'$ there will be two particles with spins $S'_1$ and $S'_2$ and having a relative momentum $\boldsymbol{\varkappa}_{c'}$. A scattering experiment may be considered as determining the average properties of an ensemble of such interacting systems.

We may define a scattering amplitude for the reaction as

$$\langle S', \nu'| \mathbf{f}(\hat{\boldsymbol{\varkappa}}_{c'}, c'; \hat{\boldsymbol{\varkappa}}_c, c) |S, \nu\rangle$$
$$= -(2\pi)^2 \sqrt{\frac{\rho_{\epsilon_c'}\rho_{\epsilon_c}}{\kappa_{c'}\kappa_c}} \langle c'; \hat{\boldsymbol{\varkappa}}_{c'}, S', \nu'| T |c; \hat{\boldsymbol{\varkappa}}_c, S, \nu\rangle. \quad (303)$$

Unless the spins $S'_1$ and $S'_2$ in channel $c'$ are the same as $S_1$ and $S_2$ in channel $c$, the matrix $\mathbf{f}$ will not ordinarily be a square matrix. In general, it will be a rectangular $(2S'_1 + 1)(2S'_2 + 1) \times (2S_1 + 1)(2S_2 + 1)$ matrix. This causes no real complication, however. For example, we see from Eq. (123) that the differential cross section for an unpolarized beam and target in the initial state will be

$$\frac{d\bar{\sigma}(c', c)}{d\Omega} = \frac{\kappa_{c'}}{\kappa_c(2S_1 + 1)(2S_2 + 1)} \operatorname{Tr}(\mathbf{f}^\dagger\mathbf{f}). \quad (304)$$

For an initial-state ensemble with some degree of polarization we introduce the square $(2S_1 + 1)(2S_2 + 1) \times (2S_1 + 1)(2S_2 + 1)$ matrix $\rho_c$, as given by Eq. (276). The density matrix $\rho_{c'}$ in the final channel $c'$ will be a $(2S'_1 + 1)(2S'_2 + 1) \times (2S'_1 + 1)(2S'_2 + 1)$ matrix, defined as in Eq. (257):

$$A\rho_{c'} = \mathbf{f}\rho_c\mathbf{f}^\dagger. \quad (305)$$

Since $\operatorname{Tr} \rho_{c'} = 1$, we see that

$$\frac{\kappa_{c'}}{\kappa_c} A = \frac{\kappa_{c'}}{\kappa_c} \operatorname{Tr}(\mathbf{f}\rho_c\mathbf{f}^\dagger)$$
$$= \left.\frac{d\bar{\sigma}(c', c)}{d\Omega}\right|_{P}, \quad (306)$$

which is the average differential cross section for scattering of the particles in the initial ensemble.

Complete, orthogonal sets of basis matrices, $\Omega_{c'}^{\mu}$ and $\Omega_c^{\lambda}$, may be defined as in Eq. (272*b*) for channels $c'$ and $c$, respectively. Then the density matrix in the channel $c$ has the expansion (276):

$$\rho_c = \frac{1}{(2S_1 + 1)(2S_2 + 1)} \left( \sum_{\lambda=1}^{(2S_1+1)^2(2S_1+1)^2} \langle \Omega_c^{\lambda} \rangle_i \Omega_c^{\lambda} \right), \tag{307}$$

where $\langle \Omega_c^{\lambda} \rangle_i$ are the averages of the $\Omega_c^{\lambda}$ over the initial ensemble. The polarization in the final state is specified by the $(2S'_1 + 1)^2(2S'_2 + 1)^2$ expectation values

$$\langle \Omega_{c'}^{\mu} \rangle = \mathrm{Tr}\,(\rho_{c'} \Omega_{c'}^{\mu}). \tag{308}$$

Equation (307) permits us to rewrite Eq. (306) as

$$\frac{d\bar{\sigma}(c', c)}{d\Omega} = \frac{\kappa_{c'}}{\kappa_c} \frac{1}{(2S_1 + 1)(2S_2 + 1)} \sum_{\lambda} \langle \Omega_c^{\lambda} \rangle_i \,\mathrm{Tr}\,(\mathbf{f}^{\dagger}\mathbf{f}\Omega_c^{\lambda}). \tag{309}$$

To simplify this, we must first generalize Eq. (295).

For this purpose, we shall introduce (for the moment) the abbreviations

$$\begin{aligned} \mathbf{f} &\equiv \mathbf{f}(\hat{\boldsymbol{\varkappa}}_{c'}, c'; \hat{\boldsymbol{\varkappa}}_c, c) \\ \mathbf{f}(r) &\equiv \mathbf{f}(-\hat{\boldsymbol{\varkappa}}_c, c; -\hat{\boldsymbol{\varkappa}}_{c'}, c'). \end{aligned} \tag{310}$$

Here $\mathbf{f}$ is a $(2S'_1 + 1)(2S'_2 + 1) \times (2S_1 + 1)(2S_2 + 1)$ matrix, while $\mathbf{f}(r)$ is a $(2S_1 + 1)(2S_2 + 1) \times (2S'_1 + 1)(2S'_2 + 1)$ matrix. The expression $\mathbf{f}^*$ will be written for the matrix whose elements are the complex conjugates of those of $\mathbf{f}$, $\tilde{\mathbf{f}}$ for the transpose of $\mathbf{f}$, and $\mathbf{f}^{\dagger}$ for the matrix $\tilde{\mathbf{f}}^*$.

If the particles in channels $c$ or $c'$ are composite, proper account of internal coordinates must have been taken in evaluating $\mathbf{f}$. We can thus introduce just simple spin eigenfunctions $u_{S,\nu}$ and $u_{S',\nu'}$ in channels $c$ and $c'$, respectively, to evaluate (303). These are assumed to satisfy the time reversal relations (30):

$$\begin{aligned} W_c u_{S,\nu}^* &= (-1)^{S+\nu} u_{S,-\nu}, \\ W_{c'} u_{S',\nu'}^* &= (-1)^{S'+\nu'} u_{S',-\nu'}, \end{aligned} \tag{311}$$

where $W_c$ and $W_{c'}$ are the appropriate time reversal matrices in channels $c$ and $c'$. Operating on Eqs. (311) once again with the time reversal operator, we obtain

$$\begin{aligned} W_c^* &= (-1)^{2S} W_c^{-1} \\ W_{c'}^* &= (-1)^{2S'} W_{c'}^{-1}. \end{aligned} \tag{312}$$

Now the assumption of time reversal invariance gives us the general relation

$$\begin{aligned}(u_{S',\nu'}, \mathbf{f}u_{S,\nu}) &= (-1)^{S'+\nu'-S-\nu}(u_{S,-\nu}, \mathbf{f}(r)u_{S',-\nu'}) \\ &= (u^*_{S,\nu}, W_c^{-1}\mathbf{f}(r)W_{c'}u^*_{S',\nu'}) \\ &= (u_{S',\nu'}, W^{*-1}_{c'}\tilde{\mathbf{f}}(r)W_c^* u_{S,\nu}).\end{aligned}$$

Therefore,

$$W_{c'}^{-1}\,\mathbf{f}^\dagger(r)W_c = \mathbf{f}^*. \tag{313$a$}$$

This may also be written as [with the use of Eqs. (311)]

$$W_{c'}^{-1}\mathbf{f}W_c = (-1)^{2(S'-S)}\tilde{\mathbf{f}}(r) \tag{313$b$}$$

or

$$W_c^{-1}\mathbf{f}^\dagger W_{c'} = (-1)^{2(S'-S)}\mathbf{f}^*(r). \tag{313$c$}$$

The basis matrices $\Omega_c^\lambda$ may be constructed as tensors formed from products of the components of $\mathbf{S}_1$ and $\mathbf{S}_2$. We thus set $\Omega_c^\lambda = \Omega_c^\lambda(\mathbf{S}_1, \mathbf{S}_2)$. Let us now assume that the $\Omega_c^\lambda$ have been so defined that

$$\Omega_c^\lambda(-\mathbf{S}_1, -\mathbf{S}_2) = e_\lambda \Omega_c^\lambda(\mathbf{S}_1, \mathbf{S}_2), \tag{314}$$

where $e_\lambda = \pm 1$. Then, since $W_c\mathbf{S}_1^* W_c^{-1} = -\mathbf{S}_1$ and $W_c\mathbf{S}_2^* W_c^{-1} = -\mathbf{S}_2$ [see Eq. (293)], we have

$$W_c\Omega_c^{\lambda^*} W_c^{-1} = e_\lambda \Omega_c^\lambda. \tag{315}$$

Suppose that we have done the inverse of the scattering experiment described by Eq. (303). That is, the two particles in channel $c'$ collide with initial relative momentum $-\boldsymbol{\varkappa}_{c'}$. The reaction then leads to a final relative momentum $-\boldsymbol{\varkappa}_c$ in channel $c$. If the spins are unpolarized in the channel $c'$, the final state of polarization will be specified by the averages [compare Eq. (308)]

$$\langle \Omega_c^\lambda \rangle_f \equiv \frac{\mathrm{Tr}\,[\mathbf{f}(r)\mathbf{f}^\dagger(r)\Omega_c^\lambda]}{\mathrm{Tr}\,[\mathbf{f}(r)\mathbf{f}^\dagger(r)]}. \tag{316}$$

The expression can be related to the traces appearing in Eq. (309). We have

$$\begin{aligned}\mathrm{Tr}\,(\mathbf{f}^\dagger\mathbf{f}\Omega_c^\lambda) &= e_\lambda\,\mathrm{Tr}\,(\mathbf{f}^\dagger\mathbf{f}W_c\Omega_c^{\lambda^*}W_c^{-1}) \\ &= e_\lambda\,\mathrm{Tr}\,(W_c^{-1}\mathbf{f}^\dagger W_{c'}W_{c'}^{-1}\mathbf{f}W_c\Omega_c^{\lambda^*}) \\ &= e_\lambda\,\mathrm{Tr}\,[\mathbf{f}^*(r)\tilde{\mathbf{f}}(r)\Omega_c^{\lambda^*}] \\ &= e_\lambda\,\mathrm{Tr}\,[\mathbf{f}(r)\mathbf{f}^\dagger(r)\Omega_c^\lambda].\end{aligned} \tag{317}$$

Here we have used Eqs. (313$b$) and (313$c$) and the fact that the traces in Eqs. (317) are real. Equation (317) is seen to be a generalized form of (295).

A special case of (317) corresponds to $\Omega_c^1 = 1$, the unit operator. Then we have

$$\mathrm{Tr}\,(\mathbf{f}^\dagger\mathbf{f}) = \mathrm{Tr}\,[\mathbf{f}^\dagger(r)\mathbf{f}(r)]. \tag{318}$$

Equations (316), (317), and (318) permit us to write (309) as

$$\left.\frac{d\bar{\sigma}(c', c)}{d\Omega}\right|_P = \left.\frac{d\bar{\sigma}(c', c)}{d\Omega}\right|_u \left\{\sum_\lambda e_\lambda \langle \Omega_c{}^\lambda \rangle_i \langle \Omega_c{}^\lambda \rangle_f \right\}. \tag{319}$$

Here $[d\bar{\sigma}(c', c)/d\Omega]_u$ is the differential cross section for an unpolarized initial state.

Since $\langle \Omega_c{}^\lambda \rangle_f$ is unchanged if $\boldsymbol{\varkappa}_c \to -\boldsymbol{\varkappa}_c$, $\boldsymbol{\varkappa}_{c'} \to -\boldsymbol{\varkappa}_{c'}$, we may also use the reaction $(\boldsymbol{\varkappa}_c, c) \to (\boldsymbol{\varkappa}_{c'}, c')$ to determine $\langle \Omega_c{}^\lambda \rangle_f$. Equation (288) is therefore seen to be a special case of Eq. (319).

There are several useful general rules relating to the polarization of a beam which can result from the scattering of an unpolarized beam on an unpolarized target. These apply to either reaction or scattering cross sections. We state them without proof.[54]

(*a*) If there are only *S*-waves in the initial and in the final channels, there will be no polarization.

(*b*) If a reaction goes through a single intermediate state which has a total angular momentum $J = \frac{1}{2}$ and a single parity (or $J = 0$ and mixed parity), there will be no polarization.

(*c*) For the channel spin $S' = 0$ in the final channel, there will be no polarization.

(*d*) Polarization can result only from interference of different spins or partial waves.

(*e*) Polarization will not occur unless there are noncentral interactions.

(*f*) The polarization vector **P** is always in one of the directions $\pm(\hat{\boldsymbol{\varkappa}}_c \times \hat{\boldsymbol{\varkappa}}_{c'})$.

(*g*) If $l_{\max}$ is the greatest value of the orbital angular momentum which need be considered, then

$$\frac{d\bar{\sigma}}{d\Omega} P = \sum_{n=0}^{2l_{\max}-1} d_n \cos^n \theta \sin \theta, \tag{320a}$$

where the $d_n$ are constants, $\theta$ is the barycentric scattering angle, and $P$ is the polarization. Equation (320*a*) should be supplemented by a theorem of Yang [20], which states that the scattering cross section has the general form

$$\frac{d\bar{\sigma}}{d\Omega} = \sum_{n=0}^{2l_{\max}} c_n \cos^n \theta, \tag{320b}$$

where the $c_n$ are another set of constants.

[54] These are proved in Refs. [19].

## 7.6 THE SOLUTION OF THE DIRAC EQUATION IN A CENTRAL FIELD

The problem of separating the Dirac equation into radial and angular parts is quite straightforward. For the sake of completeness we shall describe the procedure in a more or less conventional fashion.

Our Hamiltonian is (see Section 1.4)[55]

$$H = \boldsymbol{\alpha} \cdot \mathbf{p} + \beta M + V(r)$$

where $\mathbf{p}$ is the differential operator $-i\nabla$. It is an elementary exercise to verify that the total angular momentum operator $\mathbf{J} = \mathbf{L} + \boldsymbol{\sigma}/2$ commutes with the Hamiltonian, since $\boldsymbol{\sigma}$ commutes with $\beta$ and [see the discussion following Eq. (1-105*b*) for a definition of $\rho_1$]

$$\begin{aligned} [\boldsymbol{\alpha} \cdot \mathbf{p}, \mathbf{L}] &= i\mathbf{p} \times \boldsymbol{\alpha} \\ [\boldsymbol{\alpha} \cdot \mathbf{p}, \boldsymbol{\sigma}] = \rho_1[\boldsymbol{\sigma} \cdot \mathbf{p}, \boldsymbol{\sigma}] &= 2i\rho_1\boldsymbol{\sigma} \times \mathbf{p} = 2i\boldsymbol{\alpha} \times \mathbf{p}. \end{aligned} \tag{321}$$

Thus since $J^2$ also commutes with $H$ as well as with $J_z$, we may seek eigenfunctions of $H$ which are simultaneously eigenfunctions of $J^2$ and $J_z$. The latter are, of course, functions of angle and spin variables alone. We have already effectively found these functions in the course of our general discussion of the scattering of spin one-half particles. They are proportional to the projection operators $\Lambda_{l+}$ and $\Lambda_{l-}$ [see Eq. (69)] acting on the product $Y_l^m(\hat{\mathbf{r}})u_\nu(s)$, where $u_\nu(s)$ is a spin eigenfunction. The method for carrying out the calculation in detail has been described before quite explicitly.

$$\chi_{jm} = \sqrt{(l + m + \tfrac{1}{2})/(2l + 1)}\,Y_l^{m-\frac{1}{2}}(\hat{\mathbf{r}})u_{\frac{1}{2}}(s) + \sqrt{(l - m + \tfrac{1}{2})/(2l + 1)}\,Y_l^{m+\frac{1}{2}}(\hat{\mathbf{r}})u_{-\frac{1}{2}}(s)$$

for $j = l + \frac{1}{2}$, and

$$\chi_{jm} = \sqrt{(l - m + \tfrac{1}{2})/(2l + 1)}\,Y_l^{m-\frac{1}{2}}(\hat{\mathbf{r}})u_{\frac{1}{2}}(s) - \sqrt{(l + m + \tfrac{1}{2})/(2l + 1)}\,Y_l^{m+\frac{1}{2}}(\hat{\mathbf{r}})u_{-\frac{1}{2}}(s) \tag{322}$$

for $j = l - \frac{1}{2}$. It is convenient to regard these $\chi_{jm}$'s as two component spinors, and we write $\chi_{jm}^{(+)}$ for $j = l + \frac{1}{2}$, $\chi_{jm}^{(-)}$ for $j = l - \frac{1}{2}$.

$$\begin{aligned} \chi_{jm}^{(+)} &= \begin{pmatrix} \sqrt{(j + m)/2j} & Y_{j-\frac{1}{2}}^{m-\frac{1}{2}} \\ \sqrt{(j - m)/2j} & Y_{j-\frac{1}{2}}^{m+\frac{1}{2}} \end{pmatrix}, \\ \chi_{jm}^{(-)} &= \begin{pmatrix} \sqrt{(j - m + 1)/2(j + 1)} & Y_{j+\frac{1}{2}}^{m-\frac{1}{2}} \\ -\sqrt{(j + m + 1)/2(j + 1)} & Y_{j+\frac{1}{2}}^{m+\frac{1}{2}} \end{pmatrix}. \end{aligned} \tag{323}$$

These two component spinors have opposite parity, for if $\mathbf{r} \to -\mathbf{r}$,

[55] For convenience of notation, we shall choose our units in this section in such a manner that $\hbar = c = 1$.

$Y_l^m(\hat{\mathbf{r}}) = (-1)^l Y_l^m(-\hat{\mathbf{r}})$ and the $l$ values of the spinors $\chi^{(+)}$, $\chi^{(-)}$ differ by one unit.

It is useful to classify the stationary states according to rigorously conserved quantities. One such is the total angular momentum **J**. If there is no degeneracy we may also use the parity operator, which has the form $\beta \times$ (space exchange) and which has eigenvalues $\pm 1$. A possible choice of eigenfunction might be a radial function times a four-component spinor, the upper two components of which are $\chi^{(\pm)}$ and the lower ones zero. Substitution into the Dirac equation shows instantly that such an attempt will not work. The precise way in which it fails is illuminating and shows what modification must be made if we are, in fact, to succeed. We note that

$$\boldsymbol{\alpha} \cdot \mathbf{p} = \frac{\boldsymbol{\alpha} \cdot \mathbf{r}\boldsymbol{\alpha} \cdot \mathbf{r}\boldsymbol{\alpha} \cdot \mathbf{p}}{r^2} = \frac{\boldsymbol{\alpha} \cdot \mathbf{r}}{r^2}(\mathbf{r} \cdot \mathbf{p} + i\boldsymbol{\sigma} \cdot \mathbf{r} \times \mathbf{p})$$

$$= \frac{\boldsymbol{\alpha} \cdot \mathbf{r}}{r^2}\left(-ir\frac{\partial}{\partial r} + i\boldsymbol{\sigma} \cdot \mathbf{L}\right). \tag{324}$$

It is clear now that there can be no solution of the conjectured variety; starting out with upper components leads inevitably to a function with lower components and vice versa. Since $\boldsymbol{\sigma} \cdot \mathbf{L}$ operating on one of our $\chi$'s is just a number times the same $\chi$ $[\boldsymbol{\sigma} \cdot \mathbf{L}\chi^{(+)} = (j - \frac{1}{2})\chi^{(+)}, \boldsymbol{\sigma} \cdot \mathbf{L}\chi^{(-)} = -(j + \frac{3}{2})\chi^{(-)}]$,

$$HR(r)\begin{pmatrix} \chi_{jm}^{(\pm)} \\ 0 \end{pmatrix} \simeq R_1(r)\begin{pmatrix} \chi_{jm}^{(\pm)} \\ 0 \end{pmatrix} + R_2(r)\boldsymbol{\alpha} \cdot \hat{\mathbf{r}}\begin{pmatrix} \chi_{jm}^{(\pm)} \\ 0 \end{pmatrix},$$

where $R_1(r)$, $R_2(r)$ are radial functions. This suggests that our general eigenfunctions have the form

$$\psi_{jm}^{(\pm)}(\mathbf{r}) = F(r)\begin{pmatrix} \chi_{jm}^{(\pm)} \\ 0 \end{pmatrix} - i\boldsymbol{\alpha} \cdot \hat{\mathbf{r}}G(r)\begin{pmatrix} \chi_{jm}^{(\pm)} \\ 0 \end{pmatrix}. \tag{325}$$

Needless to say, our choice of coefficient in the second term is purely for later convenience. Note that under the parity operation

$$\psi_{jm}^{(\pm)}(\mathbf{r}) \to \beta\psi_{jm}^{(\pm)}(-\mathbf{r}) = (-1)^{j\mp 1/2}\psi_{jm}^{(\pm)}(\mathbf{r}), \tag{326}$$

so these are indeed states of definite parity. If we substitute the general form (325) into the Dirac equation and identify the coefficients of the terms which do not involve $\boldsymbol{\alpha} \cdot \hat{\mathbf{r}}$, we find immediately two scalar equations for $F$ and $G$. For $\psi_{jm}^{(+)}$ we have

$$\frac{dG}{dr} + \left(j + \frac{3}{2}\right)\frac{G}{r} = -(E - M - V)F, \qquad (j = l + \tfrac{1}{2})$$

$$\frac{dF}{dr} - \left(j - \frac{1}{2}\right)\frac{F}{r} = (E + M - V)G; \tag{327a}$$

for $\psi_{jm}^{(-)}$ we find

$$\frac{dG}{dr} - \left(j - \frac{1}{2}\right)\frac{G}{r} = -(E - M - V)F,$$
$$\frac{dF}{dr} + \left(j + \frac{3}{2}\right)\frac{F}{r} = (E + M - V)G, \qquad (j = l - \tfrac{1}{2}) \tag{327b}$$

where $E$ is the energy of the particle.

These two sets of equations may be written as a single pair in terms of a new quantum number $k = -(j + \frac{1}{2})$ referring to $\psi_{jm}^{(+)}$ and $k = j + \frac{1}{2}$ for $\psi_{jm}^{(-)}$.

$$\frac{dF}{dr} + (1 + k)\frac{F}{r} = (E + M - V)G,$$
$$\frac{dG}{dr} + (1 - k)\frac{G}{r} = -(E - M - V)F. \tag{328}$$

We shall not discuss the solution of these equations for any case except that of $V = 0$, the free-particle problem. If we introduce $f = rF$ and $g = rG$, we find in general a second-order equation for $f$ which is

$$\frac{d^2f}{dr^2} + \frac{dV/dr}{E - V + M}\frac{df}{dr} + \left[(E - V)^2 - M^2 - \frac{k(k+1)}{r^2} + \frac{k}{r}\frac{dV/dr}{E - V + M}\right]f = 0,$$

and which reduces to (here $\kappa^2 = E^2 - M^2$)

$$\frac{d^2f}{dr^2} + \left[\kappa^2 - \frac{k(k + 1)}{r^2}\right]f = 0 \tag{329}$$

when $V = 0$. The regular solution of this equation is

$$f = rj_l(\kappa r),$$

where $j_l$ is the spherical Bessel function [see Eq. (6-125)]. The small component $g$ may be obtained from the first-order differential equation for $f$, which is

$$\left(\frac{d}{dr} + \frac{k}{r}\right)f = (E + M)g.$$

We find

$$g = \frac{k}{|k|}\frac{\kappa r}{E + M}j_l(\kappa r). \tag{330}$$

The extension of these methods to the solution of the Dirac equation for noncentral interactions is straightforward.

## REFERENCES AND NOTES


1. R. G. Newton, *J. Math. Phys.* **1,** 319 (1960).
2. E. P. Wigner, *Group Theory*, Academic Press, New York, 1959.
3. L. I. Schiff, *Quantum Mechanics*, McGraw-Hill Book Co., New York, 1949.
4. We wish to show that the expression $T(\hat{\boldsymbol{\kappa}}_f, \hat{\boldsymbol{\kappa}}, \boldsymbol{\sigma})$, Eq. (94), is invariant under rotations of the vectors $\hat{\boldsymbol{\kappa}}_f$, $\hat{\boldsymbol{\kappa}}$ and $\boldsymbol{\sigma}$. The rotation operator (2–42) may be written in the form $\mathscr{R} = \mathscr{R}_c\mathscr{R}_s$, where $\mathscr{R}_c$ acts on the space coordinates and $\mathscr{R}_s$ on spin coordinates. Let a transformation send $(\hat{\boldsymbol{\kappa}}, \nu)$ into $(\bar{\hat{\boldsymbol{\kappa}}}, \bar{\nu})$ and $(\hat{\boldsymbol{\kappa}}_f, \nu')$ into $(\bar{\hat{\boldsymbol{\kappa}}}_f, \bar{\nu}')$. Then invariance under coordinate rotations implies that (with an appropriate choice of phases)

$$\langle \bar{\hat{\boldsymbol{\kappa}}}_f, \tfrac{1}{2}, \bar{\nu}' |\, T\, | \bar{\hat{\boldsymbol{\kappa}}}, \tfrac{1}{2}, \bar{\nu} \rangle = \langle \hat{\boldsymbol{\kappa}}_f, \tfrac{1}{2}, \nu' |\, T\, | \hat{\boldsymbol{\kappa}}, \tfrac{1}{2}, \nu \rangle.$$

Now $u_{1/2,\bar{\nu}} = \mathscr{R}_s u_{1/2,\nu}$, and so on, so in the notation of Eq. (94),

$$\begin{aligned}(u_{1/2,\bar{\nu}'}, T(\bar{\hat{\boldsymbol{\kappa}}}_f, \bar{\hat{\boldsymbol{\kappa}}}, \boldsymbol{\sigma}) u_{1/2,\bar{\nu}}) &= (u_{1/2,\nu'}, \mathscr{R}_s^{-1} T(\bar{\hat{\boldsymbol{\kappa}}}_f, \bar{\hat{\boldsymbol{\kappa}}}, \boldsymbol{\sigma}) \mathscr{R}_s u_{1/2,\nu}) \\ &= (u_{1/2,\nu'}, T(\hat{\boldsymbol{\kappa}}_f, \hat{\boldsymbol{\kappa}}, \boldsymbol{\sigma}) u_{1/2,\nu}).\end{aligned}$$

Therefore

$$\begin{aligned}\mathscr{R}_s^{-1} T(\bar{\hat{\boldsymbol{\kappa}}}_f, \bar{\hat{\boldsymbol{\kappa}}}, \boldsymbol{\sigma}) \mathscr{R}_s &\equiv T(\bar{\hat{\boldsymbol{\kappa}}}_f, \bar{\hat{\boldsymbol{\kappa}}}, \bar{\boldsymbol{\sigma}}) \\ &= T(\hat{\boldsymbol{\kappa}}_f, \hat{\boldsymbol{\kappa}}, \boldsymbol{\sigma}),\end{aligned}$$

which establishes the desired result.
5. E. P. Wigner, *Phys. Rev.* **73,** 1002 (1948).
6. L. Fonda and R. G. Newton, *Ann. Phys.* **7,** 133 (1959).
7. R. S. Christian and E. W. Hart, *Phys. Rev.* **77,** 441 (1950).
8. W. Rarita and J. Schwinger, *Phys. Rev.* **59,** 436 (1941).
9. R. S. Christian and H. P. Noyes, *Phys. Rev.* **79,** 85 (1950).
10. F. Rohrlich and J. Eisenstein, *Phys. Rev.* **75,** 705 (1949).
11. J. L. Gammel and R. M. Thaler, *Phys. Rev.* **107,** 291 (1957); P. S. Signell and R. E. Marshak, *Phys. Rev.* **106,** 832 (1957).
12. J. M. Blatt and L. C. Biedenharn, *Phys. Rev.* **86,** 399 (1952).
13. M. L. Goldberger, Y. Nambu, and R. Oehme, *Ann. Phys.* **2,** 226 (1957).
14. S. C. Wright, *Phys. Rev.* **99,** 996 (1955).
15. L. Wolfenstein and J. Ashkin, *Phys. Rev.* **85,** 947 (1952).
16. L. Wolfenstein, *Ann. Rev. Nuclear Sci.* **6,** 43 (1956).
17. R. C. Tolman, *The Principles of Statistical Mechanics*, Oxford University Press, London, 1938.
18. F. D. Murnaghan, *The Theory of Group Representations*, The Johns Hopkins University Press, Baltimore, 1938.
19. A. Simon and T. Welton, *Phys. Rev.* **90,** 1036 (1953); A. Simon, *Phys. Rev.* **92,** 1050 (1953). [See also the correction of several errors in these papers: *Phys. Rev.* **93,** 1435 (1954).]
20. C. N. Yang, *Phys. Rev.* **74,** 764 (1948).

# 8. THE LIFETIME AND DECAY OF VIRTUAL STATES

When we described in Chapter 3 the sense in which we use the term "particle," we included as a supplementary concept that of "unstable particle." It was observed that an unstable particle does not represent a true stationary state of the system described and that the term "particle" may therefore be applied to such systems as a matter of convenience. For example, one does not usually hesitate to call a free neutron, which has a mean life of about 17 min, a "particle." On the other hand, the composite system that occurs for "resonant" pion-nucleon scattering has a life expectancy of about $10^{-23}$ secs. Using the term "particle" for this system does not obviously add greatly to one's understanding of pion-nucleon scattering.

In this chapter we shall discuss in some detail, and by several methods, the property of states which are not stationary—formally, states which are not eigenstates of the operator of time translation [see Eq. (1.20*a*)]. We shall see that the "lifetime" of such states is physically well defined; it may be observed experimentally or calculated from a knowledge of the dynamics of the system being considered.

In addition to the study of the decay of "unstable particles," the discussion of unstable states has many applications. This includes such familiar examples as the decay by radiative transitions of excited atomic and nuclear states. It includes also the phenomenon of "resonances" in scattering and reaction cross sections, which will be seen to arise as a consequence of a long-lived virtual state of the system.

Preliminary to the principal discussion of this chapter, we shall give a brief review of stationary state perturbation theory. As we shall see, this bears a close formal relationship to both scattering theory, as described in earlier chapters, and the theory of decaying states, to be described in this chapter. Study of the analytic structure of the scattering matrix will be seen to provide a very useful basis for discussing resonance and decay phenomena.

## 8.1 STATIONARY STATE PERTURBATION THEORY

Many aspects of the problem posed by stationary state perturbation theory are similar to those of scattering theory. We consider a system described by the Hamiltonian $H$ and imagine this to be decomposed into an "unperturbed" part $K$ and a "perturbation" $V$. It is supposed that the eigenstates $|a\rangle$ of $K$ are explicitly known and that they form a complete set of states for the system. Thus

$$K\,|a\rangle = \epsilon_a\,|a\rangle, \tag{1}$$

where $\epsilon_a$ is the eigenvalue of $K$ for the state $a$.

Although our formal considerations will not be so limited, it is convenient to require that the system be confined to a box of finite volume $\mathscr{V}$. This implies that the eigenfunctions $|a\rangle$ and eigenvalues $\epsilon_a$ are discrete.

The eigenfunctions and eigenvalues of $H$ will be written as $|\lambda\rangle$ and $E_\lambda$, respectively, so

$$H\,|\lambda\rangle = E_\lambda\,|\lambda\rangle. \tag{2}$$

In the finite volume $\mathscr{V}$ these states and eigenvalues will also be discrete. The central problem of perturbation theory is to evaluate the $|\lambda\rangle$ and $E_\lambda$ directly in terms of the states $|a\rangle$ and energies $\epsilon_a$ of the "unperturbed system."

It will be convenient to suppose that the states $|a\rangle$ and $|\lambda\rangle$ are defined in the barycentric subspace of the system (see Section 4.6), this means that $\delta$-functions specifying momentum conservation and center-of-mass coordinates need not be carried through our analysis. Then the operator for the Green's function for the Hamiltonian $H$ is (in the notation of Section 6.7, p. 302)

$$G(E) = (E - H)^{-1}. \tag{3a}$$

In the representation $|a\rangle$ this is

$$\langle a'|\,G(E)\,|a\rangle = \sum_\lambda \frac{\langle a' \mid \lambda\rangle\langle\lambda \mid a\rangle}{E - E_\lambda}. \tag{3b}$$

The task of finding the eigenvalues $E_\lambda$ of Eq. (2) is seen to be equivalent to that of finding the poles at $E = E_\lambda$ of the matrix elements of $G(E)$. Alternatively, we may attempt to solve the algebraic equation

$$[\langle a'|\,G(E)\,|a\rangle]^{-1} = 0 \tag{4a}$$

for the eigenvalues $E = E_\lambda$. The diagonal matrix elements of $G$ are particularly convenient for this purpose. We therefore define

$$G_a(E) \equiv \langle a|\,G(E)\,|a\rangle \tag{5}$$

and have, as a special case of Eq. (4*a*),[1]

$$[G_a(E)]^{-1} = 0. \tag{4b}$$

To evaluate $G_a(E)$ it is convenient to begin with the matrix equation

$$(E - K)G = 1 + VG \tag{6}$$

(here 1 is the unit matrix) obtained from (3*a*). This may be put into a more convenient form by introducing a new operator $F$, defined by the equation

$$\langle a'|\, G(E)\, |a\rangle \equiv \langle a'|\, F\, |a\rangle G_a, \tag{7a}$$

where $G_a$ is defined by Eq. (5). We may also write Eq. (7*a*) as

$$G = F\mathcal{G}, \tag{7b}$$

where $\mathcal{G}$ is the diagonal matrix

$$\langle a'|\, \mathcal{G}\, |a\rangle = \delta_{a'a} G_a(E). \tag{8}$$

It follows from Eqs. (5) and (7) that the diagonal matrix elements of $F$ are unity, that is,

$$\langle a|\, F\, |a\rangle = 1. \tag{9}$$

Substitution of Eq. (7*b*) into Eq. (6) now gives

$$(E - K)F\mathcal{G} = 1 + VF\mathcal{G}. \tag{10}$$

The diagonal matrix element of this equation for a given state $a$ is

$$(E - \epsilon_a)G_a = 1 + R_a G_a, \tag{11}$$

where Eq. (9) has been used and

$$R_a(E) \equiv \langle a|\, VF\, |a\rangle \tag{12a}$$

is the diagonal matrix element of the operator

$$R = VF. \tag{12b}$$

Equation (11) is just a simple algebraic equation (not a matrix equation) and may be directly solved for $G_a$.

$$G_a = \frac{1}{E - \epsilon_a - R_a(E)}\,. \tag{13}$$

[1] We are here following a formulation of stationary state perturbation theory given by Riesenfeld and Watson [1]. Related operator techniques were previously used in this context by Brueckner and his collaborators [2], by Eden and Francis [3], and by Van Hove [4].

Then, according to Eq. (4*b*), the eigenvalues $E_\lambda$ of $H$ are solutions of the algebraic equation

$$E_\lambda = \epsilon_a + R_a(E_\lambda). \tag{14}$$

Equation (14) is one of the fundamental equations of perturbation theory since it may be explicitly solved for the eigenvalue $E_\lambda$. One may, however, very well ask what role the unperturbed state $a$ plays here. Formally, it is irrelevant since Eq. (14) is satisfied by many eigenvalues $E_\lambda$ for each state $a$.[2] In practice the state $a$ is important in solving Eq. (14) since this equation is expected to be much simpler to handle if $|a\rangle$ is a reasonably good approximation to the state $|\lambda\rangle$. Indeed, Eq. (14) is ordinarily evaluated by a sequence of approximations, of which the first is $E_\lambda = \epsilon_a$.[3] Because of this, $R \equiv VF$ is called the "level shift" operator. That is, its diagonal matrix elements represent the shift between the "unperturbed" energy $\epsilon_a$ and the "perturbed" energy $E$.

It should be noted that when the energy $E_\lambda$ is degenerate, Eq. (14) does not by itself specify each of the states $|\lambda\rangle$ having the eigenvalue $E_\lambda$ since it provides only the numerical value of $E_\lambda$. In many applications the degenerate states may be resolved by symmetry operations.[4] If this cannot be done, the degenerate states $|\lambda\rangle$ may be obtained from the equation for $F$ [see Eqs. (19)].

To obtain the level shift $R_a$, we must first evaluate $F$ from Eq. (10). This equation may evidently be rewritten in the form

$$(E - K - \mathcal{O})F\mathcal{g} = 1 + (V - \mathcal{O})F\mathcal{g}, \tag{15}$$

where $\mathcal{O}$ is (essentially) an arbitrary matrix. For our purposes, it will be both sufficient, and convenient, to assume that $\mathcal{O}$ is diagonal in the $|a\rangle$ representation; so

$$\langle a'|\, \mathcal{O}\, |a\rangle = \delta_{a'a}\mathcal{O}_a. \tag{16}$$

Equation (15) may next be written as

$$F = \frac{1}{E - K - \mathcal{O}}\frac{1}{\mathcal{g}} + \frac{1}{E - K - \mathcal{O}}(V - \mathcal{O})F, \tag{17a}$$

an explicit equation for $F$. For the vector $F\,|a\rangle$ this takes the form

$$F\,|a\rangle = \frac{1}{E - K - \mathcal{O}_a}(E - K - R_a)\,|a\rangle + \frac{1}{E - K - \mathcal{O}}(V - \mathcal{O})F\,|a\rangle. \tag{17b}$$

[2] It might appear from Eqs. (3*b*) and (4*a*) that Eq. (14) is satisfied by all $E_\lambda$ for any $a$. This is not generally true, however, since the states $|a\rangle$ and $|\lambda\rangle$ may have common constants of the motion (such as angular momentum).

[3] This will be illustrated with some examples in our subsequent discussion.

[4] In principle, one can always remove the degeneracy by means of a suitable perturbation. This is usually unnecessarily cumbersome, however.

An obvious simplification of this equation results if we choose

$$\mathscr{O}_a = R_a(E) \tag{18}$$

for the single matrix element $\mathscr{O}_a$ of $\mathscr{O}$. We shall accept Eq. (18) as a restriction on the form of $\mathscr{O}$, but we shall not now impose other conditions [in addition to Eq. (16)] on the remaining matrix elements of $\mathscr{O}$. Now Eq. (17*b*) becomes

$$F\,|a\rangle = |a\rangle + \frac{1}{E - K - \mathscr{O}}(V - \mathscr{O})F\,|a\rangle, \tag{19a}$$

from which we may evaluate $F\,|a\rangle$ and, in turn, $R_a = \langle a|\,VF\,|a\rangle$. [It should be recalled that $F\,|a\rangle$ does not depend on the choice made for $\mathscr{O}$, except for the restrictions imposed by Eqs. (16) and (18).]

In matrix form, Eq. (19*a*) is the equation

$$\langle a'|F|a\rangle = \delta_{a'a} + \sum_{a''}\frac{1}{E - \epsilon_a - \mathscr{O}_{a'}}[\langle a'|\,V|a''\rangle - \delta_{a'a''}\mathscr{O}_{a'}]\langle a''|\,F\,|a\rangle. \tag{19b}$$

A third form may be given. It depends on the observation that

$$\langle a|\,(V - \mathscr{O})F\,|a\rangle = 0,$$

which follows from Eqs. (19), (12*a*), and (18). This permits us to write Eq. (19*a*) as

$$F\,|a\rangle = |a\rangle + \frac{1}{E - K - \mathscr{O}}(1 - \Lambda_a)(V - \mathscr{O})F\,|a\rangle, \tag{19c}$$

where

$$\Lambda_a = |a\rangle\,\langle a| \tag{20}$$

is the projection operator onto the state $|a\rangle$.

Equations (19) are similar in structure to the equations encountered in previous chapters for the scattering wave functions $\psi_a^{\pm}$. Indeed, if we let $E = E_a \pm i\eta$ and set $\mathscr{O}_{a'} = 0$ for all $a' \neq a$, the equation (19*c*) for $F\,|a\rangle$ is identical with that for $\psi_a^{\pm}$ when the limit $\mathscr{V} \to \infty$ is taken before $\eta \to 0(+)$. The matrix $R$ is then just the scattering matrix $T$. On the other hand, if we let $\eta \to 0(+)$ and then let $\mathscr{V} \to \infty$, the matrix $R$ is the reaction matrix [see Eq. (5.179)]. In this case, we can easily show that for a finite volume $\mathscr{V}$, and for $E = E_\lambda$, the vector $F\,|a\rangle$ is equal to the eigenfunction $|\lambda\rangle$ of $H$ (to within a normalizing constant). That is,

$$F(E_\lambda)\,|a\rangle = C\,|\lambda\rangle, \tag{21}$$

where $C$ is a constant. To see this, let us multiply both sides of Eq. (19*a*) from the left by $(E - K - \mathscr{O})$ to obtain

$$(E - K - \mathscr{O})F\,|a\rangle = (E - \epsilon_a - \mathscr{O}_a)\,|a\rangle + (V - \mathscr{O})F\,|a\rangle.$$

Now, since $\mathcal{O}_a(E_\lambda) = E_\lambda - \epsilon_a$, the first term on the right side of this equation vanishes and we are left with

$$[E_\lambda - K - \mathcal{O} - (V - \mathcal{O})]F\,|a\rangle = 0,$$

which is equivalent to Eq. (2).

The matrix $\mathcal{O}$ is normally chosen in such a manner as to simplify the evaluation of Eqs. (19). We shall illustrate this with two examples. First, let

$$\mathcal{O} = R_a(E)\Lambda_a, \tag{22}$$

where $\Lambda_a$ is the projection operator (20) onto the state $|a\rangle$. This permits us to write Eq. (19*c*) as

$$F\,|a\rangle = |a\rangle + \frac{1}{E - K}(1 - \Lambda_a)VF\,|a\rangle. \tag{23}$$

The level shift operator $R$ [see Eqs. (12)] is now given by the equation

$$R = V + V\frac{1 - \Lambda_a}{E - K}VF = V + V\frac{1 - \Lambda_a}{E - K}R, \tag{24}$$

where it is understood that $R$ is to operate on the state $|a\rangle$.

Equation (24) provides an integral equation for $R$. If we iterate this equation by successive substitution into the right-hand side, an expansion is obtained for $R$ in powers of $V$. This is

$$R = V + V\frac{1 - \Lambda_a}{E - K}V + V\frac{1 - \Lambda_a}{E - K}V\frac{1 - \Lambda_a}{E - K}V + \cdots. \tag{25}$$

According to Eq. (14), we may obtain $E_\lambda$ from the expansion

$$E_\lambda = \epsilon_a + \langle a|\,V|a\rangle + \sum_{a'(\neq a)} \langle a|\,V|a'\rangle \frac{1}{E_\lambda - \epsilon_{a'}} \langle a'|\,V|a\rangle + \cdots. \tag{26}$$

This is the Wigner-Brillouin [5] form of the perturbation expansion for $E_\lambda$. When the series converges, it provides a correspondence between the levels $E_\lambda$ of the perturbed system and the unperturbed energies $\epsilon_a$. The sense of this correspondence may be understood as follows: If we reduce the strength of the interaction $V$ to zero, $E_\lambda \to \epsilon_a$.[5]

As a second illustration of Eqs. (19), let us choose

$$\mathcal{O} = R_a(E)\Lambda_a + (E - \epsilon_a)(1 - \Lambda_a). \tag{27}$$

Then, on writing

$$\delta E = E - \epsilon_a,$$

[5] It is understood that extraneous solutions not satisfying this condition are to be discarded.

the equation (19$c$) for $F$ becomes

$$F\,|a\rangle = |a\rangle + \frac{1}{\epsilon_a - K}(1 - \Lambda_a)(V - \delta E)F\,|a\rangle, \tag{28$a$}$$

and Eq. (14) becomes

$$\delta E = R_a(\delta E + \epsilon_a). \tag{28$b$}$$

A systematic expansion of Eqs. (28) in powers of $V$ gives the Rayleigh-Schrödinger perturbation series for the eigenenergies $E_\lambda$.

As a final example of our equations for $R$, let us consider two particles confined in a "box" of such properties that their wave function vanishes whenever the separation of the particles is greater than $R_0$, the "radius of the box." We suppose that the particles have nonrelativistic energies and that they interact by a central potential. In Section 6.2 we wrote the reaction matrix for two such particles as [see Eq. (6.50)]

$$R_l(\kappa) = (\chi_{\kappa,l}, R\chi_{\kappa,l}) = -\frac{1}{\pi M_r\kappa}\tan\delta_l, \tag{29$a$}$$

when the relative orbital angular momentum is $l$ and the volume of the box is infinite. For a large, but finite, spherical box of volume $\mathscr{V} = (4\pi/3)R_0^3$, the normalization of Eq. (29) must be modified. Instead of choosing

$$\chi_{\kappa,l} = i^l\sqrt{2/\pi}\,j_l(\kappa r),$$

as in Eq. (6.58$b$), we now take

$$\chi_{\kappa,l} = \sqrt{2/R_0}\,\kappa i^l j_l(\kappa r), \tag{30}$$

as in Eq. (1.84). This change in normalization modifies the right-hand side of Eq. (29$a$) by a factor $\pi\kappa^2/R_0$, so

$$R_l(\kappa) = -\frac{\kappa}{M_r R_0}\tan\delta_l(\kappa), \tag{29$b$}$$

valid for particles in a spherical box of radius $R_0$.

The complete reaction matrix is obtained from this using the transformation (1.85):

$$R(\hat{\boldsymbol{\kappa}}_f\cdot\hat{\boldsymbol{\kappa}}, \kappa, \kappa) = \sum_{l,m}\langle\boldsymbol{\kappa}_f\,|\,\kappa, l, m\rangle R_l(\kappa)\langle\kappa, l, m\,|\,\boldsymbol{\kappa}\rangle. \tag{30$a$}$$

We verify that this differs from the expression (6.44$a$) by a factor $(2\pi)^3/\mathscr{V}$. Comparison with Eqs. (1.87) and (1.88) shows that this is indeed just the appropriate factor to convert from "continuum normalization" to "box normalization."

We recall from Chapter 5 that Eqs. (29$b$) and (30) are in a strict sense valid only in the limit as $\mathscr{V} \to \infty$. For a finite volume the levels are discrete. In this case Eqs. (29) are not valid. The correct $R_l$ may be evaluated in a straightforward manner using Eqs. (24) and (14). If we set

$$E = \epsilon(\kappa) + \delta E,$$

where $\epsilon(\kappa)$ corresponds to the perturbed energy $\epsilon_a$, then Eq. (14) implies that

$$\delta E = R_l(\kappa, E) \equiv (\chi_{\kappa,l}, R(\epsilon)\chi_{\kappa,l}). \tag{31}$$

This equation may be solved for $\delta E$ to give[6]

$$\delta E = R_l(\kappa, \delta E) = -\frac{\kappa}{M_r R_0}\,\delta_l(\kappa). \tag{32}$$

This differs from the form (29*b*) in that tan $\delta_l(\kappa)$ is here replaced by just the phase shift $\delta_l(\kappa)$.

The physical basis for the difference between Eqs. (29*b*) and (32) may be easily seen. If we were to construct scattering wave packets in a sufficiently large finite box, we would observe scattering as described in Chapters 3 and 5.[7] If the experimenter persisted, however, in continuing to observe the system even after the wave packets struck the walls and were reflected back many times, he would require the discrete states of this section to describe the system.

## 8.2 THE DECAY OF A PREPARED STATE

In this section we shall develop the general theory describing the decay of an unstable physical system which has been prepared at some given time, say $t = 0$. Examples of such "decay" processes include the radiative transitions to lower states of excited atoms, molecules, and atomic nuclei. Most of the so-called elementary particles are not stable, but they decay spontaneously into other particles.

In a strict sense, the process of forming an unstable system and its subsequent decay constitutes a scattering experiment. We are therefore not required to give a special formulation of the description of decaying

[6] The evaluation of Eq. (32) is straightforward, but a little too long to give here. It is done in detail in Ref. [1] and, in a different form in Refs. [6] and [7]. An elementary derivation of Eq. (32) is based on the observation that the radial wave function must vanish at the walls of the box; that is, sin $[\kappa R_0 + \delta_l(\kappa) - l\pi/2] = 0$. For the unperturbed system, we have instead sin $(\kappa_0 R_0 - l\pi/2) = 0$. Equating the arguments of the two sine functions gives $(\kappa - \kappa_0)R_0 + \delta_l(\kappa) = 0$, so

$$\delta E \simeq \frac{\kappa}{M_r}\,\delta\kappa = \frac{\kappa}{M_r}(\kappa - \kappa_0) = -\frac{\kappa}{M_r R_0}\,\delta_l(\kappa).$$

[7] It will be recalled that the description of scattering given in Chapter 3 did not require an infinite volume, but only one so large that the wave packets did not come near the walls during the course of the experiment.

states. On the other hand, if such an unstable system has a sufficiently long lifetime that experimental observations may be made on it before its decay, the mode of formation is irrelevant and the decay is conveniently studied separately.[8] The actual preparation of such states is studied in Section 8.4.

**Mathematical Description of the Decay**

As described in Section 3.1, we suppose the unstable system to be represented by an approximate Hamiltonian $K$. Certain eigenstates $\chi_a$, satisfying the equation

$$K\chi_a = E_a\chi_a, \tag{33}$$

are considered to provide an acceptable description of the "states" of the unstable system. These states $\chi_a$ may be written in the form

$$\chi_a = (2\pi)^{-3/2} e^{i\mathbf{p}_a\cdot\mathbf{r}} g_a(\xi), \tag{34}$$

where $\mathbf{P}_a$ is the total momentum and $\mathbf{r}$ is the corresponding canonical "barycentric" coordinate of the system. The wave function $g_a(\xi)$ describes the internal state $a$ of the system. For our discussion in this section (and in most of this chapter),[9] we shall assume that the states $a$ are discrete, so

$$(g_a, g_a) = 1.$$

The exact Hamiltonian for the system will be written as $H$. The difference

$$V \equiv H - K \tag{35}$$

will be called the "interaction" responsible for the decay.

We emphasize that the approximate description (33) is dictated by the conditions under which experimental observations are to be made on the system. To make this more precise, let us suppose that at the time $t = 0$ the system has been prepared in the wave packet state

$$X_a = \sum_{\mathbf{P}_a} A(\mathbf{P}_a)\chi_a, \tag{36a}$$

where $A(\mathbf{P}_a)$ is the wave packet amplitude for a momentum $\mathbf{P}_a$, and the $X_a$ are normalized to unity:

$$(X_a, X_a) = \sum_{\mathbf{P}_a} |A(\mathbf{P}_a)|^2 = 1. \tag{36b}$$

As in Chapter 3, we shall suppose that the wave packet contains a very small range of values of $\mathbf{P}_a$ and that these are centered at a mean momentum

[8] This reflects the fact that the Schrödinger equation is of first order.

[9] In Section 8.5 we shall apply our present considerations to a scattering experiment taking $g_a$ to be an incident plane-wave state.

$\mathbf{P}_{\bar{a}}$. (Our treatment is easily generalized to include a sum over internal states $a$ in the packet.)

Our central problem is now the study of the dynamical behavior of the system, as determined by the exact Hamiltonian $H$. The time-dependent wave function $\Psi(t)$ has the explicit form

$$\Psi(t) = e^{-iHt}X_a, \tag{37}$$

which satisfies the specified boundary condition

$$\Psi(0) = X_a \tag{38}$$

at $t = 0$.

It will prove convenient to take a Laplace transform of Eq. (37). We then obtain

$$L(\Psi) \equiv \int_0^\infty e^{-pt}\Psi(t)\, dt = \frac{1}{p + iH} X_a. \tag{39}$$

The inverse transform then gives us $\Psi(t)$ again.

$$\Psi(t) = \frac{1}{2\pi i}\int_{\mathscr{C}_1} dp\, e^{pt}\frac{1}{p + iH} X_a, \tag{40$a$}$$

where $\mathscr{C}_1$ is the usual Laplace transform contour extending from $c - i\infty$ to $c + i\infty$ with $c$ a positive constant sufficiently large that no singularities of $(p + iH)^{-1}$ lie to the right of $\mathscr{C}_1$. This is illustrated in Fig. 8.1*a*. Let us next make a change in the variable of integration in Eq. (40*a*), setting $p = -iE$. Then the contour $\mathscr{C}_1$ is rotated through 90° in the positive

Fig. 8.1 Contours of integration for Eqs. (40*a*) and (40*b*) are illustrated in (*a*) and (*b*).

sense as illustrated in Fig. 8.1$b$, and Eq. (40$a$) is modified to read

$$\Psi(t) = \frac{1}{2\pi i}\int_{\mathscr{C}_2} \frac{dE\, e^{-iEt}}{E - H} X_a. \tag{40b}$$

The contour $\mathscr{C}_2$ runs from $ic + \infty$ to $ic - \infty$.

We recognize $(E - H)^{-1}$ as being the Green's function

$$\mathscr{G}(E) = \frac{1}{E - H}, \tag{41}$$

so we may write Eq. (40$b$) as

$$\Psi(t) = \frac{1}{2\pi i}\int_{\mathscr{C}_2} dE\, e^{-iEt}\mathscr{G}(E)X_a. \tag{40c}$$

This may be further simplified if we note that the singularities of $\mathscr{G}(E)$ lie on the real $E$-axis[10] since the eigenvalues of $H$ are real. This permits us to lower the contour $\mathscr{C}_2$ to within an infinitesimal distance $\eta$ from the real axis. Then, in the notation of Eq. (6.355$b$), we set

$$\mathscr{G}^+(E) = \lim_{\eta \to 0(+)} \mathscr{G}(E + i\eta),$$

with $E$ now real, and rewrite Eq. (40$c$) as

$$\Psi(t) = \frac{1}{2\pi i}\int_{-\infty}^{\infty} dE\, e^{-iEt}\mathscr{G}^+(E)X_a. \tag{40d}$$

The wave function $\Psi(t)$ contains a complete description of the system for $t > 0$. Since it has been assumed that $H \neq K$, it is not expected that the system will remain in the initial state $g_a$. Indeed, measurements made at times $t > 0$ will in general indicate a finite probability that the system is in other channels than the initial one. The wave functions for one of these decay channels may, for example, be

$$\chi'_b = (2\pi)^{-3/2} e^{i\mathbf{p}_b\cdot\mathbf{r}'} g'_b(\xi). \tag{42}$$

Here $g'_b(\xi)$ is the wave function describing the internal state of the system in the new channel and $\mathbf{P}_b$ is the total momentum. To illustrate this, we may imagine that the initial state $g_a$ describes an atom $A^*$ in an excited state. The new channel will then ordinarily describe the atom in a lower state and a photon emitted by the atom. When the atom is left in its lowest state, this transition would normally be written as

$$A^* \to A + \gamma.$$

Let us first ask for the probability of finding the system in a restricted range of states $g'_b$ but having any final momentum $\mathbf{P}_b$. The number of

[10] A proof of this statement will be given later.

states to be observed is $\Sigma_b$, which we shall consider to be an infinitesimal quantity. The required probability is then

$$dP_b(t) = \sum_{\mathbf{P}b} \sum_{b} | \mathscr{A}_b |^2, \tag{43}$$

where $\mathscr{A}_b$ is the probability amplitude

$$\mathscr{A}_b = (\chi'_b, \Psi(t)) = \sum_{\mathbf{P}a} A(\mathbf{P}_a) \frac{1}{2\pi i} \int_{\mathscr{C}2} dE\, e^{-iEt} (\chi'_b, \mathscr{G}(E)\chi_a). \tag{44a}$$

To find the probability that the system remain in the initial state $g_a$, we write

$$\chi_{a'} = (2\pi)^{-3/2} e^{i\mathbf{P}'_a \cdot \mathbf{r}} g_a(\xi) \tag{45}$$

and introduce the amplitude

$$\mathscr{A}_a = (\chi_{a'}, \Psi(t)) = \sum_{\mathbf{P}a} A(\mathbf{P}_a) \frac{1}{2\pi i} \int_{\mathscr{C}2} dE\, e^{-iEt} (\chi_{a'}, \mathscr{G}(E)\chi_a). \tag{46a}$$

The probability that the system is in the state $g_a$ at time $t$ is then

$$P_a(t) = \sum_{\mathbf{P}'a} |\mathscr{A}_a|^2. \tag{47}$$

Here we sum over all final total momenta $\mathbf{P}'_a$.

Equations (44*a*) and (46*a*) may be simplified by removing the kinematical factor associated with conservation of total momentum. As described in connection with Eq. (6.359), we set

$$\begin{aligned} (\chi_{a'}, \mathscr{G}(E)\chi_a) &= \delta(\mathbf{P}'_a - \mathbf{P}_a)(g_a, G(E)g_a) \\ (\chi'_b, \mathscr{G}(E)\chi_a) &= \delta(\mathbf{P}_b - \mathbf{P}_a)(g'_b, G(E)g_a). \end{aligned} \tag{48}$$

Here, as before,

$$G(E) = \frac{1}{E - H}, \tag{49}$$

but this now operates only on the internal state vectors of the system. The diagonal matrix element

$$G_a(E) = (g_a, G(E)g_a) \tag{50}$$

was encountered in Eq. (5) of Section 8.1, where it was seen to play an essential role in stationary state perturbation theory. Its properties will also be of great importance here in studying decay processes.

To continue, we may use the first of Eqs. (48) and Eq. (50) to write Eq. (46*a*) as

$$\mathscr{A}_a = \sum_{\mathbf{P}a} A(\mathbf{P}_a)\delta(\mathbf{P}'_a - \mathbf{P}_a)\mathscr{I}_a, \tag{46b}$$

where

$$\mathscr{I}_a = \frac{1}{2\pi i}\int_{\mathscr{C}2} dE\, e^{-iEt} G_a(E). \tag{51}$$

Similarly, Eq. (44*a*) becomes

$$\mathscr{A}_b = \sum_{\mathbf{P}a} A(\mathbf{P}_a)\delta(\mathbf{P}_b - \mathbf{P}_a)\mathscr{I}_b, \tag{44b}$$

where

$$\mathscr{I}_b \equiv \frac{1}{2\pi i}\int_{\mathscr{C}2} dE\, e^{-iEt}(g'_b, G(E)g_a). \tag{52}$$

To simplify Eq. (47), we insert into it the expression (46*b*). We obtain

$$P_a(t) = \sum_{\mathbf{P}'a} |A(\mathbf{P}'_a)|^2\, |\mathscr{I}_a|^2 = |\mathscr{I}_a|^2. \tag{53}$$

Here we have first used the condition that the wave packet is very narrow to remove $|\mathscr{I}_a|^2$ from the sum over $\mathbf{P}'_a$ and then used the normalization condition (36*b*). In the final form of Eq. (53), $\mathscr{I}_a$ is evaluated at the mean momentum $\mathbf{P}_{\bar{a}}$ and all kinematical factors describing center-of-mass motion have been removed.

In a similar manner, we may simplify Eq. (43), giving the probability of finding the system in the range $\Sigma_b$ of states $g'_b$. Using Eq. (44*b*) and the fact that the wave packet is very narrow, we easily obtain

$$dP_b(t) = \sum_{\mathbf{P}b}\sum_{b}\delta(\mathbf{P}_b - \mathbf{P}_a)\, |\mathscr{I}_b|^2 = \sum_{b} |\mathscr{I}_b|^2. \tag{54a}$$

In most applications the states $g'_b$ have a continuous spectrum (in contrast to the $g_a$ which are usually discrete). This means that we should replace the sum over $b$ by an integral, using

$$\sum_b = d\rho'(\epsilon_b)\, d\epsilon_b, \tag{55}$$

where $\epsilon_b$ is the energy of the state $g'_b$ and $d\rho'$ is the number of states per unit energy. For example, if the system were to decay into two non-relativistic particles with relative momentum $\boldsymbol{\varkappa}_f$,

$$\sum_b = d^3\boldsymbol{\varkappa}_f = M_r\kappa_f\, d\Omega_f\, d\epsilon_f, \tag{56}$$

where $M_r$ is the reduced mass of the two particles and $d\Omega_f$ is an element of solid angle into which $\boldsymbol{\varkappa}_f$ is directed. When Eq. (55) is applicable, we have

$$dP_b(t) = d\rho'(\epsilon_b)\, d\epsilon_b\, |\mathscr{I}_b|^2. \tag{54b}$$

It will often be true that the system can decay into any one of several different channels. For such systems it may be useful to add channel labels to our expressions. Then, in the notation of Chapter 4, the probability of finding the system in the state $g_{c;b}$, corresponding to channel $c$, at time $t$ is

$$dP_{c;b}(t) = d\rho_c(\epsilon_b)\, d\epsilon_b\, |\mathscr{I}_{c;b}|^2. \tag{54c}$$

Here $d\rho_c$ is the density of states in channel $c$, and $\mathscr{I}_{c;b}$ is just the quantity defined by Eq. (52), but with channel labels added.[11]

It involves no loss of generality to now specialize our discussion to the barycentric coordinate system so that the mean momentum $\mathbf{P}_{\bar{a}}$ in the packet (36*a*) vanishes. Then the Green's function (49) is an operator in the "barycentric subspace," as defined in Section 4.7. In this subspace, we have

$$Kg_a = \epsilon_a g_a, \tag{57}$$

where $\epsilon_a$ is the energy of the state $a$.

We shall also limit our discussion now to such processes that the states $g'_b$ in the decay channel are eigenstates of $K$;[12] thus

$$Kg'_b = \epsilon_b g'_b. \tag{58}$$

This permits us to conclude that

$$(g'_b, g_a) = 0. \tag{59}$$

Consistent with our use of $\epsilon$ to represent a barycentric energy, we shall replace $E$ by $\epsilon$ as a variable of integration in Eqs. (51) and (52).

The Green's function $G(\epsilon)$ may be put into the form of Eqs. (7). That is, we write

$$G(\epsilon) = F\mathscr{g}, \tag{60}$$

where $\mathscr{g}$ is diagonal and

$$\langle a|\, \mathscr{g}\, |a\rangle = (g_a, Gg_a) = G_a(\epsilon).$$

Similarly, $F$ satisfies Eq. (23).

$$Fg_a = g_a + \frac{1}{\epsilon - K}(1 - \Lambda_a)VFg_a, \tag{61}$$

where $\Lambda_a = g_a g_a^{\dagger}$. The operator

$$R = VF, \tag{62a}$$

[11] Since we are discussing only a single channel for the initial state $g_a$, it seems unnecessary to add a channel label to $g_a$.

[12] The general case may be treated in a straightforward manner, following the lines of the analysis given here. In Section 8.6 we shall return to the general case, giving a different description of decay processes.

defined by Eq. (12*b*), has diagonal matrix elements [see Eq. (12*a*)]

$$R_a(\epsilon) = (g_a, VFg_a). \tag{62b}$$

This permits us to adapt Eq. (13) to Eq. (50), which is equivalent to

$$G_a(\epsilon) = \frac{1}{\epsilon - \epsilon_a - R_a(\epsilon)}. \tag{63}$$

We may also use Eqs. (60) and (61) to write

$$G_{ba}(\epsilon) = (g'_b, G(\epsilon)g_a) = \frac{1}{\epsilon - \epsilon_b} R_{ba}(\epsilon) \frac{1}{\epsilon - \epsilon_a - R_a(\epsilon)}, \tag{64a}$$

where

$$R_{ba}(\epsilon) = (g'_b, VFg_a). \tag{62c}$$

If several decay channels are possible, we may add channel labels to Eq. (64) by inserting

$$R_{(c;b)a} = (g_{c;b}, VFg_a) \tag{62d}$$

for $R_{ba}$ and replacing $G_{ba}$ by $G_{(c;b)a}$, with,

$$G_{(c;b)a} = \frac{1}{\epsilon - \epsilon_b} R_{(c;b)a}(\epsilon) \frac{1}{\epsilon - \epsilon_a - R_a(\epsilon)}. \tag{64b}$$

**Some Formal Properties of *G* and *R***

When the eigenvalues of $H$ are discrete, as was the case in Section 8.1, the singularities of $G_a$ are poles on the real $\epsilon$-axis at the position of these eigenvalues. Here Eq. (51) may be evaluated in terms of the residues at these poles. Indeed, we see from Eq. (3*b*) that

$$\mathscr{I}_a = \sum_\lambda e^{-iE_\lambda t} |\langle a \mid \lambda \rangle|^2. \tag{65}$$

The system therefore distributes itself among eigenstates $|\lambda\rangle$ of $H$, weighted by the overlap $|\langle a \mid \lambda \rangle|^2$ between the initial state $g_a \equiv |a\rangle$ and the eigenstates $|\lambda\rangle$. Since this is a discrete sum of simple oscillating exponentials, a true transition out of the state $a$ does not occur in the sense that $\mathscr{I}_a \to 0$ as $t \to \infty$. On the other hand, when the energies $\epsilon$ are closely spaced, so that a large number of terms is important in Eq. (65), the behavior of the system may closely approximate a transition (as described later) for a long period of time.[13]

When $H$ has a continuous spectrum of eigenvalues, the situation is quite

[13] The situation here is similar to that encountered in connection with Eqs. (29*b*) and (32). In a finite enclosure the system will keep "rebounding from the walls" and there is always a finite chance that it will again be found in the initial state.

different. Here let us suppose that the eigenvalues $E$ of $H$ (in the barycentric system) are continuous over the range

$$\epsilon_M < E_\lambda < \infty \tag{66}$$

and that there are no discrete states corresponding to $E_\lambda < \epsilon_M$.[14] We shall also assume that $\epsilon_a$, the energy of the initial state $a$, is greater than $\epsilon_M$. When these conditions are satisfied, it is easy to show that $G_a(\epsilon)$ is analytic everywhere in the complex $\epsilon$-plane, except along a cut extending from $\epsilon_M$ to plus infinity on the real axis.

To show this, rewrite Eq. (3*b*) for a continuous spectrum and for the diagonal matrix element $G_a$ as

$$G_a(\epsilon) = \int \rho_H(\lambda)\, d\lambda\, \frac{|\langle a \mid \lambda\rangle|^2}{\epsilon - E_\lambda}, \tag{67}$$

where $\rho_H(\lambda)$ is the density of states $\lambda$, defined so that

$$\int |\lambda\rangle \rho_H(\lambda)\, d\lambda \langle\lambda| = I,$$

the identity operator. We now choose any point $\epsilon$ not on the cut. There clearly exists a positive, nonvanishing number $\Delta$ such that

$$|(\epsilon - E_\lambda)^{-1}| \geq \Delta^{-1} > 0$$

in the integrand of Eq. (67). Therefore, by the comparison theorem,[15]

$$|G_a(\epsilon)| \leq \int \frac{\rho_H(\lambda)\, d\lambda}{\Delta} |\langle a \mid \lambda\rangle|^2 = \frac{1}{\Delta}\langle a \mid a\rangle = \frac{1}{\Delta}. \tag{68}$$

Similarly,

$$\frac{dG_a}{d\epsilon} = -\int \rho_H(\lambda)\, d\lambda\, \frac{|\langle a \mid \lambda\rangle|^2}{(\epsilon - E_\lambda)^2}$$

and thus

$$\left|\frac{dG_a}{d\epsilon}\right| \leq \frac{1}{\Delta^2}. \tag{69}$$

It follows from Eqs. (68) and (69) that $G_a$ and its first derivative exist everywhere except along the cut and that $G_a(\epsilon)$ is therefore analytic in $\epsilon$ everywhere except on the cut extending over the interval $\epsilon_M < \epsilon < \infty$.[16]

[14] This corresponds to the situation of greatest physical interest. There is no difficulty in principle involved in our analysis if $H$ has discrete eigenvalues below $\epsilon_M$. The contribution from such states is given by Eq. (65), where now the sum would extend just over the discrete levels.

[15] See, for example, Whittaker and Watson [8], Chapters 2 to 4.

[16] Because the states $g_a = |a\rangle$ and $|\lambda\rangle$ may have common constants of motion, the cut may actually begin at some energy $\epsilon'_M > \epsilon_M$. This would correspond to the vanishing of $|\langle a \mid \lambda\rangle|^2$ over the range $\epsilon_M < E_\lambda < \epsilon'_M$. Here we may simply replace $\epsilon_M$ by $\epsilon'_M$ in our subsequent discussion.

We may show by similar arguments that $G_{ba}(\epsilon)$ [see Eq. (64$a$)] is also analytic, except on the same cut, subject to some reasonable restrictions on the system. We suppose first that the integral (here $|b\rangle \equiv g'_b$)

$$I\,dE_\lambda \equiv \int_D \langle b \mid \lambda\rangle \rho_H(\lambda)\, d\lambda \langle \lambda \mid a\rangle$$

extends over the domain of states for which the eigenvalues of $H$ lie between $E_\lambda$ and $E_\lambda + dE_\lambda$. Then let us assume that

$$\int_{\epsilon_M}^{U} I\,dE_\lambda$$

remains bounded as $U \to \infty$, and that

$$\int_{\epsilon_M}^{\infty} I\,dE_\lambda = \langle b \mid a\rangle$$

is well defined [we have, in fact, assumed that this vanishes, see Eq. (59)]. Then

$$G_{ba}(\epsilon) = \int \rho_H(\lambda)\, d\lambda \frac{\langle b \mid \lambda\rangle\langle \lambda \mid a\rangle}{\epsilon - E_\lambda} = \int_{\epsilon_M}^{\infty} \frac{I\,dE_\lambda}{\epsilon - E_\lambda} \tag{70}$$

exists for any $\epsilon$ not on the cut extending from $\epsilon_M$ to $\infty$. This follows from Chartier's test for convergence. [8, page 72]. In a similar manner we may show that $dG_{ba}/d\epsilon$ exists for $\epsilon$ not on the cut. We may therefore conclude that $G_{ba}$ is analytic except on the cut.

These conclusions concerning the domain of analyticity of $G_a(\epsilon)$ and $G_{ba}(\epsilon)$ are evidently important for the evaluation of Eqs. (51) and (52).

We shall find it useful to study also the analytic properties of $R_a(\epsilon)$ [Eq. (62$b$)] for the case just discussed that $H$ has the eigenvalue spectrum (66). For this purpose it is convenient to begin with the formal solution

$$R = V + V(1 - \Lambda_a)\frac{1}{\epsilon - H_a}(1 - \Lambda_a)V \tag{71}$$

of Eq. (24). Here $H_a$ is the Hermitian operator

$$H_a = K + (1 - \Lambda_a)V(1 - \Lambda_a). \tag{72}$$

To verify Eq. (71), we substitute the identity

$$\frac{1}{\epsilon - H_a} = \frac{1}{\epsilon - K} + \frac{1}{\epsilon - K}(1 - \Lambda_a)V(1 - \Lambda_a)\frac{1}{\epsilon - H_a}$$

into Eq. (71). This gives

$$R = V + V(1 - \Lambda_a)\frac{1}{\epsilon - K}\left[V + V(1 - \Lambda_a)\frac{1}{\epsilon - H_a}(1 - \Lambda_a)V\right]$$

$$= V + V(1 - \Lambda_a)\frac{1}{\epsilon - K}R.$$

Therefore the expression (71) satisfies the integral equation (24) for $R$ and is consequently the solution of that equation.

A simple property of $R(\epsilon)$, which follows directly from Eq. (71), is

$$R^{\dagger}(\epsilon) = R(\epsilon^*). \tag{73}$$

We shall now show that, subject to certain general restrictions on the system, the quantity $R_a(\epsilon)$ is analytic in the entire complex $\epsilon$-plane, except on the cut defined by $\epsilon_M < \epsilon < \infty$. The first condition which we impose on the system is that

$$\mathscr{M} \equiv \langle a| V(1 - \Lambda_a)V |a\rangle < \infty. \tag{74}$$

We next assume that $H_a$ has a complete set of eigenfunctions $|\gamma\rangle$ with corresponding eigenvalues $\epsilon_\gamma$.

$$H_a |\gamma\rangle = \epsilon_\gamma |\gamma\rangle. \tag{75}$$

The eigenvalue spectrum of $H_a$ will in general be continuous over some interval

$$\epsilon_L \leq \epsilon_\gamma < \infty. \tag{76a}$$

We shall see that $\epsilon = \epsilon_L$ corresponds to a branch point of the function $R_a(\epsilon)$. It follows from Eq. (63) that this must coincide with the branch point at $\epsilon = \epsilon_M$ of $G_a(\epsilon)$. Therefore

$$\epsilon_L = \epsilon_M. \tag{77}$$

We can also see that $H_a$ has no discrete eigenstates corresponding to $\epsilon < \epsilon_L = \epsilon_M$. It follows from Eq. (71) that $R_a(\epsilon)$ would have a pole at each such discrete eigenenergy. Such a pole would correspond to a zero of $G_a(\epsilon)$. The expression (67) for $G_a$ is seen to be negative definite for $\epsilon < \epsilon_M$, however, and so cannot vanish. We conclude, then, that the eigenvalue spectrum of $H_a$ is continuous and extends over the interval

$$\epsilon_M \leq \epsilon_\gamma < \infty. \tag{76b}$$

It coincides therefore with the spectrum (66) of the eigenvalues of $H$.

To continue, we introduce the density $\rho_{H_a}(\gamma)$ of states $\gamma$, defined so that

$$\int |\gamma\rangle \rho_{H_a}(\gamma)\, d\gamma \langle\gamma| = \mathrm{I},$$

the unit operator. Then, according to Eq. (71),

$$R_a(\epsilon) = \langle a| V |a\rangle + \int \rho_{H_a}(\gamma)\, d\gamma \, \frac{\langle a| V(1 - \Lambda_a) |\gamma\rangle\langle\gamma| (1 - \Lambda_a)V |a\rangle}{\epsilon - \epsilon_\gamma}. \tag{78}$$

We may now follow the argument leading to Eq. (68). Let us choose an $\epsilon$ not on the cut $\epsilon_M < \epsilon < \infty$ and a real number $\Delta$ such that

$$|(\epsilon - \epsilon_\gamma)^{-1}| \geq \Delta^{-1} > 0. \tag{79}$$

Then [see Eq. (74)]

$$|R_a(\epsilon) - \langle a| V |a\rangle| < \frac{1}{\Delta} \mathscr{M}. \tag{80}$$

Similarly, we may show that

$$\left| \frac{dR_a(\epsilon)}{d\epsilon} \right| < \frac{1}{\Delta^2} \mathscr{M} \tag{81}$$

when $\epsilon$ is not on the cut. Now, if $R_a$ is to be defined under any circumstances $\langle a| V |a\rangle$ must be finite. It follows then from Eqs. (80) and (81) that $R_a(\epsilon)$ is analytic in the complex $\epsilon$-plane, except on the cut defined by $\epsilon_M < \epsilon < \infty$.

As $\epsilon$ approaches the real axis from above (below), we obtain the functions

$$R_a^{\pm}(\epsilon) = \lim_{\eta \to 0(+)} R_a(\epsilon \pm i\eta), \tag{82}$$

where $\epsilon$ is now real.[17,18] Now, it follows by familiar arguments from Eqs. (71) and (72) that (for real $\epsilon$)

$$\begin{aligned} &R(\epsilon \pm i\eta) - R^\dagger(\epsilon \pm i\eta) \\ &= V(1 - \Lambda_a)\left(\frac{1}{\epsilon \pm i\eta - H_a} - \frac{1}{\epsilon \mp i\eta - H_a}\right)(1 - \Lambda_a)V \\ &= (\mp 2i\eta)R^\dagger \frac{1 - \Lambda_a}{(\epsilon \pm i\eta - K)(\epsilon \mp i\eta - K)} R. \end{aligned} \tag{83}$$

When $\eta$ is very small (and positive), we recall Eq. (5.28) to set

$$\frac{2i\eta}{(\epsilon - K)^2 + \eta^2} = 2\pi i \delta(\epsilon - K).$$

This permits us to write Eq. (83) in the form (valid for very small $\eta$)

$$R(\epsilon \pm i\eta) - R^\dagger(\epsilon \pm i\eta) = \mp 2\pi i R^\dagger(\epsilon \pm i\eta)\, \delta(\epsilon - K)(1 - \Lambda_a) R(\epsilon \pm i\eta). \tag{84}$$

On taking the diagonal matrix element corresponding to the state $a$, we

[17] It is clear from Eq. (78) that the limit (82) will ordinarily exist if the condition (74) is satisfied.

[18] The notation $R_a^{\pm}(\epsilon)$ will be understood to imply that $\epsilon$ is real.

obtain from Eqs. (82) and (84)

$$\operatorname{Im}\,[R_a^{\pm}(\epsilon)] = \lim_{\eta\to 0(+)} \operatorname{Im}\,[R_a(\epsilon \pm i\eta)] \equiv \mp I(\epsilon), \tag{85a}$$

$$\begin{aligned} I(\epsilon) &= \lim_{\eta\to 0(+)} \pi\langle a \mid R^{\dagger}(\epsilon \pm i\eta)\delta(\epsilon - K)(1-\Lambda_a)R(\epsilon \pm i\eta)\,|a\rangle \\ &= \pi\langle a|\, R^{+\dagger}(\epsilon)\delta(\epsilon - K)(1-\Lambda_a)R^{+}(\epsilon)\,|a\rangle. \end{aligned} \tag{85b}$$

That $I(\epsilon)$ is the same for either choice ($\pm i\eta$) follows from Eq. (73). We emphasize that $\epsilon$ is real in Eqs. (85).

When there is only one decay channel, we may use the notation of Eq. (55) to rewrite (85*b*) as

$$\begin{aligned} I(\epsilon) &= \pi \int d\rho'(\epsilon_b)\, d\epsilon_b \delta(\epsilon - \epsilon_b)\,|R_{ba}{}^{+}(\epsilon)|^2 \\ &= \pi \int d\rho'(\epsilon)\,|R_{ba}{}^{+}(\epsilon)|^2. \end{aligned} \tag{86a}$$

When there are several decay channels, we must write instead

$$I(\epsilon) = \pi \sum_c \int d\rho_c(\epsilon)\,|R^{+}_{(c;b)a}(\epsilon)|^2, \tag{86b}$$

where the notation of Eqs. (54*c*) and (62*d*) has been used.

Since $R_a(\epsilon)$ is analytic except for a cut extending from $\epsilon_M$ to infinity on the real axis, a dispersion relation[19] may be derived for it. To obtain this, let us first define

$$\tilde{R}_a(\epsilon) \equiv R_a(\epsilon) - \langle a|\, V\,|a\rangle \tag{87a}$$

and suppose that[20]

$$\lim_{|\epsilon|\to\infty} \tilde{R}_a(\epsilon) = \mathcal{O}(\epsilon^{-q}), \qquad q > 0. \tag{87b}$$

Then, if $\operatorname{Im}(\epsilon) > 0$,

$$\tilde{R}_a(\epsilon) = \frac{1}{2\pi i}\int_{\mathscr{C}} d\epsilon' \,\frac{\tilde{R}_a(\epsilon')}{\epsilon' - \epsilon}, \tag{88}$$

where the contour $\mathscr{C}$ runs from $-\infty$ to $+\infty$ above the real axis, but below the point $\epsilon$. This is illustrated in Fig. 8.2. Equation (88) defines $\tilde{R}_a(\epsilon)$ for $\operatorname{Im}(\epsilon) < 0$ by analytic continuation around the branch cut.

If we replace $\epsilon$ by $\epsilon + i\eta$ in Eq. (88), where $\epsilon$ is now real and $\eta$ is small and positive, and move the contour $\mathscr{C}$ down to the real axis, we obtain

$$\begin{aligned} \tilde{R}_a{}^{+}(\epsilon) &= \frac{1}{2\pi i}\int_{-\infty}^{\infty} d\epsilon' \,\frac{\tilde{R}_a{}^{+}(\epsilon')}{\epsilon' - \epsilon - i\eta} \\ &= \frac{P}{\pi i}\int_{-\infty}^{\infty} d\epsilon' \,\frac{\tilde{R}_a{}^{+}(\epsilon')}{\epsilon' - \epsilon}. \end{aligned} \tag{89}$$

[19] Equivalent relations were obtained in Section 6.5. See also Chapter 10 for a more comprehensive discussion of methods by which dispersion relations may be derived.

[20] If the condition (87*b*) were not satisfied, further "subtractions" such as that of Eq. (87*a*) would be necessary.

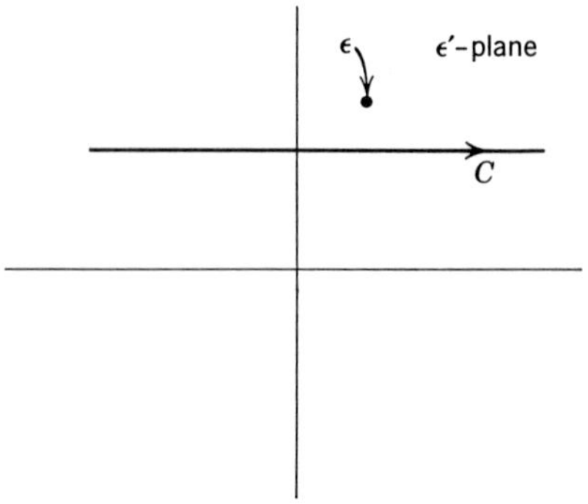


Fig. 8.2 Contour of integration for the integral in Eq. (88).

Here $\tilde{R}_a^+ = R_a^+ - \langle a|\,V\,|a\rangle$ and the symbol $P$ means that a principal value integral is to be taken. The argument by which the second form of Eq. (89) is obtained is just that by which we were led from Eq. (6.271) to Eq. (6.272).

Let us write

$$R_a^+(\epsilon) = D(\epsilon) - iI(\epsilon), \tag{90}$$

where $D(\epsilon)$ is the real part of $R_a^+(\epsilon)$. Then the real part of Eq. (89) reads [see Eqs. (85)]

$$D(\epsilon) = \langle a|\,V\,|a\rangle - \frac{P}{\pi}\int_{-\infty}^{\infty}\frac{I(\epsilon')\,d\epsilon'}{\epsilon' - \epsilon}. \tag{91}$$

This may evidently be reexpressed as

$$R_a^+(\epsilon) = \langle a|\,V\,|a\rangle - \frac{1}{\pi}\int_{-\infty}^{\infty}\frac{I(\epsilon')}{\epsilon' - \epsilon - i\eta}\,d\epsilon', \tag{92a}$$

where $\eta$ is (as usual) very small and positive.

This equation may be further simplified if we note from Eq. (78) that $R_a(\epsilon)$ is real for $\epsilon < \epsilon_M$ and on the real axis. Therefore

$$I(\epsilon) = 0 \quad \text{for} \quad \epsilon < \epsilon_M. \tag{93}$$

This permits us to rewrite Eq. (92*a*) as

$$R_a^+(\epsilon) = \langle a|\,V\,|a\rangle - \frac{1}{\pi}\int_{\epsilon_M}^{\infty}\frac{I(\epsilon')\,d\epsilon'}{\epsilon' - \epsilon - i\eta}. \tag{92b}$$

Equation (92*b*) permits us to define $R_a(\epsilon)$ in the entire $\epsilon$-plane, except on the cut. We need only consider $\epsilon + i\eta$ to be a complex variable and continue the function (92*b*) analytically throughout the complex plane. Thus the function $R_a(\epsilon)$ for complex $\epsilon$ is defined in terms of $I(\epsilon)$ over the real interval $\epsilon_M < \epsilon < \infty$. This is a useful conclusion, since $I(\epsilon)$ is ordinarily easier to evaluate directly than is $R_a(\epsilon)$.

If we analytically continue Eq. (90) to points $\epsilon - i\eta$ just below the real axis, we obtain

$$R_a^-(\epsilon) = \langle a| V |a\rangle - \frac{1}{\pi}\int_{\epsilon_M}^{\infty} \frac{I(\epsilon')\, d\epsilon'}{\epsilon' - \epsilon + i\eta}. \tag{94}$$

It is evident from this equation that $\mathrm{Re}[R_a^-(\epsilon)] = \mathrm{Re}[R_a^+(\epsilon)]$, so we have in general

$$R_a^{\pm}(\epsilon) = D(\epsilon) \mp iI(\epsilon). \tag{95}$$

According to Eqs. (91) and (93),

$$D(\epsilon) = \langle a| V |a\rangle - \frac{P}{\pi}\int_{\epsilon_M}^{\infty} \frac{I(\epsilon')\, d\epsilon'}{\epsilon' - \epsilon}. \tag{96}$$

Before continuing with our description of decay processes, it is perhaps worth observing that the operator $R$ used here satisfies the Schwinger variational principle (6.441). This follows from the fact that the integral equation (24) is formally equivalent to that for the reaction matrix.

**Evaluation of the Decay Probability**

To obtain the probability $[1 - P_a(t)]$ that the state $a$ has decayed, we must evaluate the integral (51).

$$\mathscr{I}_a = \frac{1}{2\pi i}\int_{\mathscr{C}2} d\epsilon \frac{e^{-i\epsilon t}}{\epsilon - \epsilon_a - R_a(\epsilon)}. \tag{97a}$$

If we lower the contour to the real axis, this becomes

$$\mathscr{I}_a = \frac{1}{2\pi i}\int_{+\infty}^{-\infty} d\epsilon \frac{e^{-i\epsilon t}}{\epsilon - \epsilon_a - D(\epsilon) + iI(\epsilon)}. \tag{97b}$$

Since $I(\epsilon) > 0$ for $\epsilon > \epsilon_M$, the integrand here has no poles on the real axis[21] (or anywhere else in the complex plane).

The significance of this can be best understood in terms of our description of the eigenvalue problem given in Section 8.1. When there are discrete eigenvalues of $H$, the operator $R_a$ represents the "level shift," that is, the difference between the actual eigenvalues of $H$ and the unperturbed eigenvalue $\epsilon_a$. For a true decay process, the initial state $g_a$ is not simply related to eigenstates of $H$ in the sense of perturbation theory, but it decays *irreversibly* into one or more channels. This is to be contrasted with the situation described by Eq. (65) for which the eigenstates of $H$ were discrete.

To anticipate our more detailed results, let us imagine that $|D(\epsilon)|$ and $I(\epsilon)$ are numerically much smaller than $\epsilon_a$ in Eq. (97$b$). Then we expect that

$$|\mathscr{I}_a| \approx e^{-I(\epsilon_a)t}. \tag{98a}$$

[21] We are still considering the case that $H$ has the eigenvalue spectrum (66).

This would imply that the lifetime of the state $a$ is roughly

$$\Delta t \approx \frac{\hbar}{2I(\epsilon_a)} . \tag{98b}$$

Now, as discussed in the introduction to this chapter, unless $\Delta t \gg \hbar/\epsilon_a$ the state $g_a$ will be difficult to observe and will not ordinarily provide a useful representation for the system being studied. We shall consequently find it most useful to restrict our evaluation of Eq. (97*a*) to situations for which

$$\frac{I(\epsilon_a)}{|\epsilon_a|} \ll 1 . \tag{99}$$

The condition (99) is ordinarily interpreted as implying that both $|D(\epsilon)|$ and $I(\epsilon)$ are very small quantities, as compared to $|\epsilon_a|$, in the integrand (97*b*).

To give a more quantitative discussion, we first recall that Eq. (92*b*) defines $R_a(\epsilon)$ by analytic continuation in the complex $\epsilon$-plane for[22]

$$0 < \text{Arg}\,[\epsilon - \epsilon_M] < 2\pi, \tag{100a}$$

as

$$R_a(\epsilon) = \langle a|\, V\,|a\rangle - \frac{1}{\pi}\int_{\epsilon_M}^{\infty} \frac{I(\epsilon')\,d\epsilon'}{\epsilon' - \epsilon} . \tag{100b}$$

We may restate this by saying that Eqs. (100) define $R_a(\epsilon) = R_a{}^{\text{I}}(\epsilon)$ on the *first Riemann sheet*. The function $R_a(\epsilon)$ may be continued analytically onto the second Riemann sheet [8, Chapter 5], where we write $R_a(\epsilon) = R_a{}^{\text{II}}(\epsilon)$. The continuation onto the second sheet is effected be defining, for real $\epsilon$,

$$R_a{}^{\text{II}}(\epsilon - i\eta) = R_a{}^{\text{I}}(\epsilon + i\eta) = D(\epsilon) - iI(\epsilon), \tag{101}$$

where $\eta$ is as usual an infinitesimal positive number. The function $R_a{}^{\text{II}}(\epsilon)$ is defined for $0 > \text{Arg}\,[\epsilon - \epsilon_M] > -2\pi$.

We have seen that $[\epsilon - \epsilon_a - R_a(\epsilon)]^{-1}$ is analytic on the first sheet. This will not in general be true on the second sheet, however. Indeed, when the condition (99) is satisfied, we expect $G_a(\epsilon)$ to have a pole at the point

$$\epsilon = \epsilon_0 \cong \epsilon_a + R_a(\epsilon_a) = \epsilon_a + D(\epsilon_a) - iI(\epsilon_a) \tag{102a}$$

on the second sheet. We introduce a special symbol for $I(\epsilon_a)$, writing [see Eq. (86*b*)]

$$\frac{\Gamma}{2} \equiv I(\epsilon_a) = \pi \sum_c \int d\rho_c(\epsilon_a)\,|R^+_{(c;b)a}(\epsilon_a)|^2. \tag{103}$$

[22] Here $\text{Arg}\,[Z] \equiv \text{Im}[\ln Z]$.

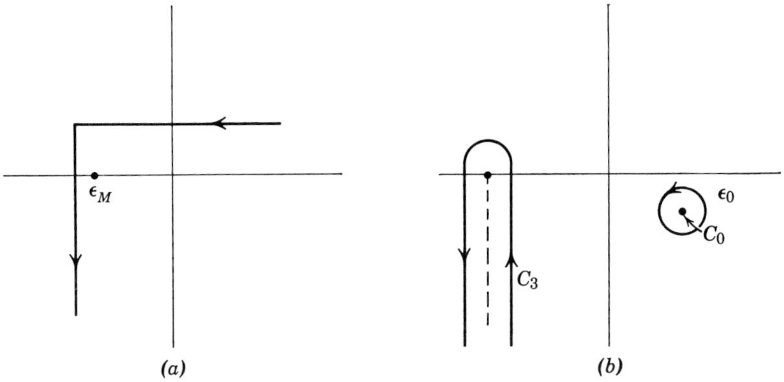


Fig. 8.3 Modification of the contour of Fig. 8.1*b* to give the integral (104).

Then, we may take

$$\epsilon_0 \cong \epsilon_a + D(\epsilon_a) - i\frac{\Gamma}{2}. \tag{102b}$$

The existence of a pole at $\epsilon_0$ suggests that the integral (97*a*) may be most easily evaluated by deforming the contour $\mathscr{C}_2$ onto the second sheet. To do this, we first deform the contour to the left of $\epsilon_M$, rotating it by 90 degrees so that it extends from $\epsilon = \epsilon_M$ to $\epsilon = \epsilon_M - i\infty$ as illustrated in Fig. 8.3*a*. We next deform the contour to the right of $\epsilon_M$ into the negative imaginary plane on the second sheet. If the pole at $\epsilon = \epsilon_0$ were the only singularity of $G_a(\epsilon)$, we would obtain the contours shown in Fig. 8.3*b*. The new contour $\mathscr{C}_3$ would run from $\epsilon_M - i\infty$ to $\epsilon_M$ on the second sheet and return to $\epsilon_M - i\infty$ on the first sheet. In addition, a small circular contour $\mathscr{C}_0$ encloses the pole at $\epsilon_0$.

In general, there may be other singularities of $G_a(\epsilon)$ in the second sheet within the domain $0 > \mathrm{Arg}\,|\epsilon - \epsilon_M| > -\pi/2$, and suitable contours must enclose these singularities. When the condition (99) is met well enough, we can expect that these singularities will occur at points $\epsilon$ for which $|\mathrm{Im}\,(\epsilon)| > |\mathrm{Im}\,(\epsilon_0)|$. Thus their contribution (which will be of $\mathscr{O}[e^{-|\mathrm{Im}(\epsilon)|t}]$) will decrease with time more rapidly than does that from $\epsilon_0$. In this case a suitable approximation for $\mathscr{I}_a$ is

$$\mathscr{I}_a = \frac{1}{2\pi i}\left(\int_{\mathscr{C}_0} + \int_{\mathscr{C}_3}\right)\frac{d\epsilon\, e^{-i\epsilon t}}{\epsilon - \epsilon_a - R_a(\epsilon)}. \tag{104}$$

Since the contours lie in the negative imaginary half-plane, we see that

$$\lim_{t\to\infty} |\mathscr{I}_a| = 0. \tag{105}$$

That is, the probability of finding the system in the state $a$ at time $t$ vanishes as $t \to \infty$.

On evaluating the residue at the pole within $\mathscr{C}_0$, we obtain from Eq. (104)

$$\begin{aligned} \mathscr{I}_a \cong e^{-i\epsilon_0 t} &+ \frac{1}{2\pi i}\int_{\epsilon_M - i\infty}^{\epsilon_M} \frac{e^{-i\epsilon t}\,d\epsilon}{\epsilon - \epsilon_a - R_a^{\text{II}}(\epsilon)} \\ &+ \frac{1}{2\pi i}\int_{\epsilon_M}^{\epsilon_M - i\infty} \frac{e^{-i\epsilon t}\,d\epsilon}{\epsilon - \epsilon_a - R_a^{\text{I}}(\epsilon)}, \end{aligned} \tag{106}$$

where we have used the condition (99) to replace $(1 - dR_a/d\epsilon)_{\epsilon_a}$ by unity. This may be further simplified if we define

$$\begin{aligned} U^{\text{I}}(Z) &\equiv \epsilon - \epsilon_a - R_a^{\text{I}}(\epsilon) \\ U^{\text{II}}(Z) &\equiv \epsilon - \epsilon_a - R_a^{\text{II}}(\epsilon), \end{aligned} \tag{107}$$

where $Z = \epsilon - \epsilon_M$. Then on setting $Z = ye^{\frac{3}{2}\pi i}$ on the first sheet and $Z = ye^{-\pi/2i}$ on the second sheet, we may write $\mathscr{I}_a$ as

$$\mathscr{I}_a = e^{-i\epsilon_0 t} + \frac{e^{-i\epsilon_0 t}}{2\pi}\int_0^\infty dy\, e^{-yt}\left[\frac{1}{U^{\text{II}}(-iy)} - \frac{1}{U^{\text{I}}(-iy)}\right]. \tag{108}$$

Ordinarily, when the condition (99) is satisfied, the first term $e^{-i\epsilon_0 t}$ will give a very good approximation to $\mathscr{I}_a$ until $t$ becomes quite large. That is, from dimensional arguments we expect the integral in Eq. (108) to be of order $\Gamma/\epsilon_a$, or less. For large values of $t$, however, the integral usually gives the dominant contribution to $\mathscr{I}_a$. At such times $\mathscr{I}_a$ is already so small that the decay is for practical purposes completed. Therefore it is usually justified to simply drop the second term in (108) and write

$$\mathscr{I}_a \cong e^{-i\epsilon_0 t} \cong e^{-i[\epsilon_a + D(\epsilon_a)]t} e^{-(\Gamma/2)t} \tag{109}$$

[when the inequality (99) is valid].

It is instructive to evaluate the second term in Eq. (108) with a specific model. Let us suppose that the system has only one decay channel and that the condition (99) is satisfied. We may also, without loss of generality, choose the energy scale in such a manner that $\epsilon_M = 0$. Then as $t$ becomes very large (more precisely, for $\epsilon_a t \gg 1$), we see that the important contributions to the integral in Eq. (108) come from $y \approx 0$, or from $\epsilon \cong 0$. This suggests that we shall need $I(\epsilon)$ only for $\epsilon \approx 0$.

We shall therefore set

$$|R_{ba}^{+}(\epsilon)|^2 = \text{constant}$$

in Eq. (86*a*). This corresponds, for example, to the decay into two particles emitted into an $S$-state. We shall also take

$$\rho'(\epsilon) = \int d\rho'(\epsilon) = \epsilon^{\frac{1}{2}} \times \text{constant},$$

corresponding to a state $g'_b$ describing decay into two nonrelativistic particles. Combining these expressions, we obtain from Eq. (86*a*)

$$I(\epsilon) \approx \alpha\sqrt{\epsilon}, \tag{110}$$

where $\alpha$ is a constant. The expression (110) is, of course, valid only in the neighborhood of $\epsilon = 0$.

We shall now use the Eqs. (110) and (100*b*) to evaluate $R_a(\epsilon)$. Because (110) is valid only for $\epsilon \cong 0$, we set $\epsilon = 0$ in Eq. (100*b*) and subtract the resulting expression from Eq. (100*b*) to obtain

$$\begin{aligned} R_a(\epsilon) - R_a(0) &= -\frac{1}{\pi}\int_0^\infty d\epsilon'\, I(\epsilon')\left(\frac{1}{\epsilon' - \epsilon} - \frac{1}{\epsilon'}\right) \\ &= -\frac{\epsilon}{\pi}\int_0^\infty \frac{I(\epsilon')\, d\epsilon'}{\epsilon'(\epsilon' - \epsilon)} \\ &\cong -\frac{\alpha}{\pi}\epsilon\int_0^\infty \frac{d\epsilon'}{\sqrt{\epsilon'}(\epsilon' - \epsilon)} \\ &= \alpha\sqrt{-\epsilon}. \end{aligned} \tag{111}$$

Unless $I(\epsilon)$ deviates significantly from the approximate form (110), this should be a good approximation for $R_a$. Because $I(0) = 0$, the quantity $R_a(0)$ is real.

If we set $\epsilon = ye^{i\theta}$, we have on the first sheet

$$R_a^{\mathrm{I}}(\epsilon) - R_a^{\mathrm{I}}(0) = \alpha\sqrt{y}\, e^{(i/2)(\theta-\pi)}, \qquad 0 < \theta < 2\pi. \tag{112a}$$

On the second sheet we find that

$$R_a^{\mathrm{II}}(\epsilon) - R_a^{\mathrm{II}}(0) = \alpha\sqrt{y}\, e^{(i/2)(\theta-\pi)}, \qquad 0 > \theta > -2\pi. \tag{112b}$$

Therefore [see Eqs. (107)]

$$\begin{aligned} U^{\mathrm{I}}(-iy) &= +\alpha\sqrt{y}\, e^{i(\pi/4)} - iy - \epsilon_a - R_a(0), \\ U^{\mathrm{II}}(-iy) &= -\alpha\sqrt{y}\, e^{i(\pi/4)} - iy - \epsilon_a - R_a(0). \end{aligned} \tag{113}$$

Then, because of the condition (99), we have when $t$ is large,

$$\begin{aligned} \int_0^\infty dy\, e^{-yt}\left[\frac{1}{U^{\mathrm{II}}(-iy)} - \frac{1}{U^{\mathrm{I}}(-iy)}\right] &\approx \frac{2\alpha\, e^{i(\pi/4)}}{\epsilon_a^{\,2}}\int_0^\infty y^{1/2}\, e^{-yt}\, dy \\ &= +\frac{\alpha\sqrt{\pi}\, e^{i(\pi/4)}}{t^{3/2}\epsilon_a^{\,2}}. \end{aligned} \tag{114}$$

Comparison of Eqs. (103) and (110) suggests that it will be convenient to write

$$\alpha \equiv \frac{\Gamma}{2\sqrt{\epsilon_c}}, \tag{115}$$

where $\epsilon_c$ is a parameter having the dimensions of energy, and is presumably of the order of $\epsilon_a$. Then insertion of the expression (114) into Eq. (108) gives (on inserting factors of $\hbar$)

$$\mathscr{I}_a \simeq e^{-i[\epsilon_a+D(\epsilon_a)]t/\hbar}\, e^{-\Gamma t/2\hbar} + \frac{\Gamma}{4\sqrt{\pi}}\, e^{i(\pi/4)}\, \frac{\hbar^{3/2}}{t^{3/2}(\epsilon_a{}^2\sqrt{\epsilon_c})}\,. \tag{116}$$

This expression is valid when $\epsilon_a t \gg \hbar$.

As we have remarked, when the condition (99) is satisfied, the second term here is rarely large enough to be of physical importance. We may therefore ordinarily neglect it to obtain [see Eq. (53)]

$$P_a(t) = e^{-\Gamma/t\hbar} \tag{117}$$

This represents the probability at time $t$ of finding the system in the state $g_a$.[23]

The quantity $D(\epsilon_a)$ in Eq. (116) represents a "level shift" for the energy $\epsilon_a$, playing a role somewhat like that of the level shift operator in Section 8.1. Its appearance here does not, of course, violate energy conservation since the energy of the initial state $\Psi(0) = X_a$ is not precisely defined. Straightforward perturbation calculations of $D(\epsilon_a)$ in field theories (other than "cutoff" theories) ordinarily lead to an infinite value for $D(\epsilon_a)$, which includes a "self-energy." The reason for this is that the quantity $\mathscr{M}$ defined by Eq. (74) is infinite in such theories. The quantity $I(\epsilon)$, on the other hand, is finite in its lowest nonvanishing order, so when the condition (99) is satisfied one can ordinarily evaluate the decay rate $\Gamma/\hbar$ and Eq. (117) without complication.

In the renormalizable theory the infinite terms in $R_a(\epsilon)$ may be removed as mass and coupling constant renormalizations, as has been shown by Low [12]. In using dispersion integrals such as that of Eq. (100*b*), this is accomplished by making "subtractions" such as was done in Eq. (111).

Let us next study the decay into the state $g'_b$, using Eq. (54*b*) and assuming the validity of (99). To do this, we must first evaluate the integral (52). Using the expression (64*a*), we have

$$\mathscr{I}_b = \frac{1}{2\pi i}\int_{\mathscr{C}2} d\epsilon\, e^{-i\epsilon t}\frac{R_{ba}(\epsilon)}{(\epsilon-\epsilon_b)[\epsilon-\epsilon_a-R_a(\epsilon)]}\,. \tag{118}$$

This integral is evaluated most easily by deforming the contour onto the second sheet as was done in Eq. (104). In addition to the pole at $\epsilon = \epsilon_0$, the integrand in Eq. (118) has a pole on the real axis at $\epsilon = \epsilon_b$. For large $t$

[23] Equation (117) was first obtained by Weisskopf and Wigner [9]. For further discussion see Refs. [11] and [12], in particular the paper by Zumino.

this gives the dominant contribution to Eq. (118). If we neglect the contribution from the contour $\mathscr{C}_3$ [of $\mathscr{O}(R/\epsilon_a)$] then

$$\mathscr{I}_b \approx e^{-i\epsilon_b t} \frac{R^+{}_{ba}(\epsilon_b)}{\epsilon_b - \epsilon_a - R_a{}^+(\epsilon_b)} - \frac{e^{-i[\epsilon_a + D(\epsilon_a)]t}\, e^{-(\Gamma/2)t}\, R_{ba}{}^+(\epsilon_a)}{[\epsilon_b - \epsilon_a - R_a{}^+(\epsilon_a)]}. \tag{119a}$$

Now this expression is small, except when $\epsilon_b \cong \epsilon_a$, so we shall set

$$\begin{aligned} R^+{}_{ba}(\epsilon_b) &= R^+{}_{ba}(\epsilon_a), \\ R_a{}^+(\epsilon_b) &= R_a{}^+(\epsilon_a). \end{aligned} \tag{120}$$

Thus

$$\mathscr{I}_b \cong \frac{R^+{}_{ba}(\epsilon_b)\, e^{-i\epsilon_b t}}{[\epsilon_b - \epsilon_a - R_a{}^+(\epsilon_a)]} \left\{1 - \exp\left[i\left(\epsilon_b - \epsilon_a - D(\epsilon_a) + i\frac{\Gamma}{2}\right)t\right]\right\}. \tag{119b}$$

When $t \gg 1$, we may neglect the exponential factor in brackets here to obtain from Eq. (54*b*)

$$d^2P_b = \lim_{t\to\infty} dP_b(t) = \frac{d\rho'(\epsilon_b)\, |R^+{}_{ba}(\epsilon_b)|^2\, d\epsilon_b}{[\epsilon_b - \epsilon_a - D(\epsilon_a)]^2 + (\Gamma^2/4)}. \tag{120a}$$

If there are several decay channels, we may use the notation of Eq. (54*c*) to write this as

$$d^2P_{c;\,b} = \frac{d\rho_c(\epsilon_b)\, |R_{(c;b)a}{}^+(\epsilon_b)|^2\, d\epsilon_b}{[\epsilon_b - \epsilon_a - D(\epsilon_a)]^2 + (\Gamma^2/4)}. \tag{120b}$$

If we desire only the decay probability into an energy range $d\epsilon_b$, we may integrate Eqs. (120) over the remaining coordinates. According to Eqs. (86),

$$\int d\rho'(\epsilon_b)\, |R^+{}_{ba}(\epsilon_b)|^2 = \frac{1}{\pi}\, I(\epsilon_a) = \frac{1}{2\pi}\, \Gamma, \tag{121a}$$

or

$$\int d\rho_c(\epsilon_b)\, |R_{(c;b)a^+}(\epsilon_b)|^2 \equiv \frac{1}{2\pi}\, \Gamma_c, \tag{121b}$$

where

$$\sum_c \Gamma_c = \Gamma \tag{121c}$$

is the total "width." These expressions permit us to write Eqs. (120) in integrated form as

$$dP_b = \frac{\Gamma}{2\pi} \frac{d\epsilon_b}{(\epsilon_b - \epsilon_a - D)^2 + (\Gamma^2/4)}, \tag{122a}$$

or

$$dP_{c;b} = \frac{\Gamma_c}{2\pi} \frac{d\epsilon_b}{(\epsilon_b - \epsilon_a - D)^2 + (\Gamma^2/4)}. \tag{122b}$$

Integration of these expressions over the energy finally gives

$$P_b = \frac{\Gamma}{2\pi} \int \frac{d\epsilon_b}{(\epsilon_b - \epsilon_a - D)^2 + (\Gamma^2/4)} = 1, \tag{123a}$$

or

$$P_{c;b} = \frac{\Gamma_c}{\Gamma} \tag{123b}$$

with

$$\sum_c P_{c;b} = 1. \tag{123c}$$

Equations (123) merely confirm our earlier observation that the system will have undergone complete decay from the state $a$ after a long period of time.[24]

We may easily evaluate the probability (54$b$) that the system is in the state $g'_b$ at a finite time $t$. Substitution of Eq. (119$b$) into (54$b$) gives, with the approximation (120),

$$d^2P_b(t) = d\rho'(\epsilon_b)\, d\epsilon_b\, |\mathscr{I}_b|^2 = \frac{d\rho'(\epsilon_b)\, d\epsilon_b\, |R^+{}_{ba}(\epsilon_b)|^2}{(\epsilon_b + \epsilon_a - D)^2 + (\Gamma^2/4)}$$
$$\times \{1 - 2e^{-(\Gamma/2)t} \cos [(\epsilon_b - \epsilon_a - D)t] + e^{-\Gamma t}\}. \tag{124}$$

We shall express this in several forms. First, let us integrate over the energy $\epsilon_b$, using the assumed smallness of $\Gamma$ to set [see Eqs. (120)]

$$d\rho'(\epsilon_b)\, |R_{ba}{}^+(\epsilon_b)|^2 = d\rho'(\epsilon_a)\, |R_{ba}{}^+(\epsilon_a)|^2.$$

Now, since

$$\int \frac{d\epsilon_b}{(\epsilon_b - \epsilon_a - D)^2 + (\Gamma^2/4)} = \frac{2\pi}{\Gamma}$$

and

$$\int \frac{d\epsilon_b \cos [(\epsilon_b - \epsilon_a - D)t]}{(\epsilon_b - \epsilon_a - D)^2 + (\Gamma^2/4)} = \frac{2\pi}{\Gamma} e^{-(\Gamma/2)t},$$

we obtain

$$dP_b(t) = \frac{2\pi}{\Gamma} d\rho'(\epsilon_a)\, |R^+{}_{ba}(\epsilon_a)|^2\, [1 - e^{-\Gamma t}]. \tag{125a}$$

An important quantity obtained from this is the *initial decay rate*. This is

$$d\dot{P}_b \equiv \frac{d}{dt} |(dP_b)|_{t=0} = 2\pi\, d\rho'(\epsilon_a)\, |R_{ba}{}^+(\epsilon_a)^2|. \tag{126a}$$

[24] Equations (123) are generally valid. They follow directly from our demonstration that $P_a(t) \to 0$ as $t \to \infty$ and the normalization of $\Psi(t)$.

If channel labels are added to these equations, they have the respective forms

$$dP_{c;b}(t) = \frac{2\pi}{\Gamma}\, d\rho_c(\epsilon_a)\, |R^+_{(c;b)a}(\epsilon_a)|^2\, [1 - e^{-\Gamma t}] \tag{125b}$$

and

$$d\dot{P}_{c;b} = 2\pi\, d\rho_c(\epsilon_a)\, |R^+_{(c;b)a}(\epsilon_a)|^2. \tag{126b}$$

Again, if we integrate Eqs. (125) and (126) over the remaining variables, they become

$$P_b(t) = [1 - e^{-\Gamma(t/\hbar)}], \tag{127a}$$

$$P_{c;b}(t) = \frac{\Gamma_c}{\Gamma}\, [1 - e^{-\Gamma(t/\hbar)}], \tag{127b}$$

and

$$\dot{P}_b = \frac{\Gamma}{\hbar}, \tag{128a}$$

$$\dot{P}_{c;b} = \frac{\Gamma_c}{\hbar}, \tag{128b}$$

where

$$\hbar \sum_c \dot{P}_{c;b} = \sum \Gamma_c = \Gamma. \tag{128c}$$

The decay rate at $t = 0$ is usually used in the following physical context. Let us suppose that a large number of identical unstable systems are being studied. A series of observations are made to determine the number in state $a$, say $N_a(t_i)$, at a sequence of times $t_i$ ($i = 1, 2, \ldots, n$). Since the Schrödinger equation is of first order, we may conclude that the decay probability per unit time following each observation is just $\Gamma/\hbar$. Then, if $\Gamma(t_{i+1} - t_i) \ll 1$, the change $\delta N_a$ during the interval $(t_{i+1} - t_i)$ is just

$$\delta N_a = -\Gamma N_a(t_i)(t_{i+1} - t_i).$$

On taking the time intervals to be arbitrarily small, we may write this as the differential equation

$$\frac{dN_a(t)}{dt} = -\Gamma N_a(t). \tag{129}$$

In a similar manner, we may let

$$dN_b(t) = N_b(t)\, d\rho'(\epsilon_b)\, d\epsilon_b$$

be the number of systems in the range $d\rho'(\epsilon_b)\, d\epsilon_b$ of states $g'_b$. Then, in analogy to Eq. (129) we find, for example,

$$\frac{dN_b(t)}{dt} = \frac{|R^+{}_{ba}(\epsilon_b)|^2}{(\epsilon_b - \epsilon_a - D)^2 + (\Gamma^2/4)}\, N_a(t). \tag{130}$$

**Example of an Unstable Decay Product**

The discussion just given is easily extended to include situations for which the states represented by the $g'_b$ also contain unstable particles which, in turn, decay. As an illustration of this, we observe that a $\pi^+$-meson decays into a $\mu^+$-meson and a neutrino:

$$\pi^+ \to \mu^+ + \nu. \tag{131a}$$

The decay rate (for example, see [13]) for this process is $\Gamma/\hbar \approx 0.4 \times 10^8\ \text{sec}^{-1}$. The $\mu^+$-meson then decays into a positron and two neutrinos with a rate $\Gamma/\hbar \approx 0.4 \times 10^6\ \text{sec}^{-1}$.

$$\mu^+ \to e^+ + \nu + \bar{\nu}. \tag{131b}$$

We shall write the Hamiltonian for such a two-step decay process as

$$H = K + V + V'. \tag{132}$$

Here it is supposed that $V$ is responsible for the decay $g_a \to g'_b$. The interaction $V'$ is considered to vanish when acting on the state $g_a$[25] (that is, $V'g_a = 0$), but it is responsible for the decay of the state $g'_b$ to a new channel, say with wave functions $g''_d$.

To describe the two-step decay, we first simply replace $K$ by $K + V'$ in the formal equations of Section 8.2, p. 438. Then, Eq. (64*a*) becomes

$$G_{ba}(\epsilon) = \left(g'_b, \frac{1}{\epsilon - K - V'}\, g'_b\right) R_{ba} \frac{1}{\epsilon - \epsilon_a - R_a(\epsilon)}. \tag{133}$$

Now, $(g'_b, (\epsilon - K - V')^{-1} g'_b)$ is formally of the same structure as was $G_a$ in the previous discussion, so we may write[26]

$$\left(g'_b, \frac{1}{\epsilon - K - V'}\, g'_b\right) = \frac{1}{\epsilon - \epsilon_b - R'_b(\epsilon)}, \tag{134}$$

where $R'_b(\epsilon)$ is defined in the same manner as $R_a(\epsilon)$, except with $V$ replaced by $V'$ and the state $g_a$ replaced by $g'_b$.

Using the expressions (133) and (134), we see that Eq. (118) takes the form

$$\mathscr{I}_b = \frac{1}{2\pi i} \int_{\mathscr{C}_2} d\epsilon\, \frac{e^{-i\epsilon t} R_{ba}(\epsilon)}{[\epsilon - \epsilon_b - R'_b(\epsilon)][\epsilon - \epsilon_a - R_a(\epsilon)]}. \tag{135}$$

[25] The argument is easily extended to more general cases.

[26] Strictly speaking, Eq. (134) will contain a momentum conserving $\delta$-function as did Eqs. (48). Removal of these is straightforward by the method used in Eq. (53). For simplicity, we shall not do this in detail now.

Without repeating the details, we follow once again the steps which led to Eq. (119*b*). Let

$$R'_b(\epsilon_b) \approx R'_b(\epsilon_a) \equiv D' - iI' \equiv D' - i\frac{\Gamma'}{2}. \tag{136}$$

Then

$$\mathscr{I}_b \approx \frac{R_{ba}}{\left(\epsilon_b - \epsilon_a + D' - D + i\dfrac{\Gamma - \Gamma'}{2}\right)} \times \left\{1 - \exp i\left[\epsilon_b - \epsilon_a + D' - D + \frac{i}{2}(\Gamma - \Gamma')\right]t\right\}. \tag{137}$$

Furthermore, also as in Eq. (124)

$$\begin{aligned} P_b(t) &= \int \frac{d\rho'(\epsilon_b)\, d\epsilon_b\, |R_{ba}|^2\, e^{-\Gamma' t}}{(\epsilon_b - \epsilon_a + D' - D)^2 + (1/4)(\Gamma - \Gamma')^2} \\ &\quad \times \{1 - 2e^{-(\Gamma-\Gamma')t/2} \cos[(\epsilon_b + D' - \epsilon_a - D)t] + e^{-(\Gamma-\Gamma')t}\} \\ &= \frac{1}{\Gamma - \Gamma'}(e^{-\Gamma' t} - e^{-\Gamma t})2\pi \int d\rho'\, |R_{ba}|^2 \\ &= \frac{\Gamma}{\Gamma - \Gamma'}(e^{-\Gamma' t} - e^{-\Gamma t}). \end{aligned} \tag{138}$$

Equation (138) is just that expected from classical considerations. By simple arguments from probability theory we have

$$\frac{dP_b}{dt} = \Gamma P_a - \Gamma' P_b, \tag{139}$$

where [by Eq. (117)]

$$P_a = e^{-\Gamma t}.$$

Integration of the differential equation (139) leads directly to Eq. (138). The state $b$ is "fed" by the decay from state $a$, but it itself decays into the state $d$. Thus $P_b(t)$ also vanishes as $t \to \infty$.

**Decay of Excitations in a Liquid or a Gas**

The methods we have described for studying perturbation theory and the decay of states may be applied to a variety of problems. These include, for example, the study of the partition function in statistical mechanics[27] which may be written as

$$Z = \sum_a \frac{1}{2\pi i} \int dE e^{-\beta E} \langle a|\, G(E)\, |a\rangle, \tag{140}$$

where $G(E)$ is the Green's function for the Hamiltonian of a macroscopic

[27] See, for example, references [14], [15], and [16].

system. Related to this are a class of methods developed for treating the perturbation theory of many particle systems, following the work of Brueckner [17].[28] More closely related to our description of decay processes is the work of Van Hove on irreversible processes in the statistical theory of nonequilibrium processes (Ref. [4]).

We shall illustrate these methods with a rather simple application of Eq. (40*c*) to the description of the decay of an initially prepared state of a macroscopic system. The system is supposed to contain a very large number $N$ of atoms or molecules and will be imagined to correspond to either a liquid or a gas. If such a system is disturbed from a state of thermal equilibrium, the disturbance will propagate from its point of origin as sound waves (or in some other form). Because of the effects of viscosity (or other dissipative mechanisms), these waves will eventually decay, their energy being converted to thermal energy. This decay represents, of course, the irreversible trend toward thermal equilibrium of any macroscopic physical system (a trend which is, in fact, a characteristic of the physicist observing the system rather than the system itself, which we expect to satisfy the condition of dynamic reversibility as discussed in Chapter 2).[29]

To describe the transient response of the system, we shall use an approximate method involving the introduction of so-called quasi-particles.[30] These represent *collective excitations* of the system and have certain particle-like properties. They are not true eigenstates, however, and so decay after a sufficient period of time, reflecting an ergodic mixing into the true (but exceedingly complicated) eigenstates.

Let the $N$-particle system be confined to a box of volume $\mathscr{V}$ and let $H$ be the Hamiltonian describing its dynamical properties. Let us suppose that $\psi_0$ represents a true eigenstate of energy $E_0$ of the system, so

$$H\psi_0 = E_0\psi_0. \tag{141}$$

It is also convenient to suppose that the system is at rest in the state $\psi_0$, so that

$$D^{-1}(\boldsymbol{\lambda})\psi_0 = \psi_0, \tag{142}$$

[28] These depend on "cluster" expansions similar to those of classical statistical mechanics and are derived by techniques developed in scattering theory. Several derivations which are essentially equivalent, but differing considerably in detail, are available (see Refs. [18]).

[29] "It is conceivable that quantum dynamics was created by God, but thermodynamics was certainly created by Man."

[30] See, for example, Landau and Lifschitz [19]. Quasi-particle excitations have been used for example by Bogoliubov [20] in describing $He^4$ at low temperatures and by Sawada [21] in describing the electron gas. Our discussion of the dynamic response of the system follows Glassgold, Heckrotte, and Watson [22].

where $D(\boldsymbol{\lambda})$ is the displacement operator defined by Eq. (1.40). Later in our description, we shall suppose that a statistical average is to be made over states $\psi_0$.

We next suppose that an approximate Hamiltonian $K$ has been found to describe the system and that the eigenstates $\phi_{\mathbf{q}}$ of $K$ are known.

$$K\phi_{\mathbf{q}} = (E_0 + \epsilon_{\mathbf{q}})\phi_{\mathbf{q}}. \tag{143a}$$

The states $\phi_{\mathbf{q}}$ are considered to be eigenstates of the momentum operator for the system, corresponding to the eigenmomentum $\mathbf{q}$. Thus

$$D^{-1}(\boldsymbol{\lambda})\phi_{\mathbf{q}} = e^{i\cdot\lambda\mathbf{q}}\phi_{\mathbf{q}}. \tag{144a}$$

The states $\phi_{\mathbf{q}}$ are said to represent quasi-particle excitations of the system. If multiple excitations occur, these are assumed to have eigenstates $\phi_{\mathbf{q}_1,\ldots,\mathbf{q}_\nu}$ satisfying the equations

$$K\phi_{\mathbf{q}_1,\ldots,\mathbf{q}_\nu} = (E_0 + \epsilon_{q_1} + \cdots + \epsilon_{q_\nu})\phi_{\mathbf{q}_1,\ldots,\mathbf{q}_\nu} \tag{143b}$$

and

$$D^{-1}(\boldsymbol{\lambda})\phi_{\mathbf{q}_1,\ldots,\mathbf{q}_\nu} = \exp\left(i\boldsymbol{\lambda}\cdot\sum_{j=1}^{\nu}\mathbf{q}_j\right)\phi_{\mathbf{q}_1,\ldots,\mathbf{q}_\nu}. \tag{144b}$$

It is because Eqs. (143) and (144) associate many "particle-like" qualities with these states that the term quasi-particle is applied to them. We shall not here discuss the detailed construction of the operator $K$ for specific systems, however,[31] since we require only the property that they are associated with an energy $\epsilon_q$ and a momentum $\mathbf{q}$. In this context, we recall that in Chapter 3 it was observed that the scattering theory developed there can be applied to the scattering of quasi-particles. The interaction responsible for the scattering is then

$$V \equiv H - K. \tag{145}$$

In order to speak of eigenstates of a definite momentum in a box, we must, of course, assume that the box is very large (on a microscopic scale) and that the eigenstates are obtained in the limit of large wave packets, which are, however, much smaller than the dimensions of the box. This is, of course, equivalent to the problem of discussing scattering theory in a box as was done in Chapter 5.

At the time $t = 0$ let it be supposed that the system has been slightly disturbed from the state $\psi_0$, its wave function being

$$\Psi(0) = \psi_0 + \sum_{\mathbf{q}} A_{\mathbf{q}}\phi_{\mathbf{q}}. \tag{146}$$

Here $A_q$ represents a wave packet amplitude for quasi-particle excitation,

[31] This is done, for example, in Refs. [20], [21], and [22].

and the statement that the system is "slightly disturbed" is interpreted to mean that

$$\sum_q |A_{\mathbf{q}}|^2 \ll 1. \tag{147}$$

Now, according to Eq. (40$c$), the wave function at time $t$ is

$$\Psi(t) = e^{-iE_0t}\psi_0 + \sum_{\mathbf{q}} A_{\mathbf{q}}\left(\frac{1}{2\pi i}\int_{\mathscr{C}_2} e^{-i\epsilon t}\frac{1}{\epsilon - K - V}\phi_{\mathbf{q}}\, d\epsilon\right), \tag{148}$$

where $V$ is defined by Eq. (145) and represents the "error" incurred in the use of $K$ rather than $H$ as the Hamiltonian. For a sufficiently large box we may assume that $H = K + V$ commutes with $D^{-1}(\boldsymbol{\lambda})$, so

$$\frac{1}{\epsilon - K - V}\phi_{\mathbf{q}} = \phi_{\mathbf{q}}\left[\frac{1}{\epsilon - E_0 - \epsilon_{\mathbf{q}} - R_{\mathbf{q}}(\epsilon)}\right] \tag{149}$$

plus terms involving multiple quasi-particle excitations. Here $R_{\mathbf{q}}(\epsilon)$ is, of course, just the diagonal matrix element with respect to the $\phi_{\mathbf{q}}$ of the operator (62$a$). Now, if $K$ is indeed a suitable approximation to $H$, we may consider $R$ to be small and neglect the multiple quasi-particle excitations to write Eq. (148) as

$$\begin{aligned} \Psi(t) &= e^{-iE_0t}\left(\psi_0 + \sum_{\mathbf{q}} A_{\mathbf{q}}\phi_{\mathbf{q}}\frac{1}{2\pi i}\int_{\mathscr{C}_2}\frac{e^{-i\epsilon t}\, d\epsilon}{\epsilon - \epsilon_q - R_q(\epsilon)}\right) \\ &= e^{-iE_0t}\left(\psi_0 + \sum_{\mathbf{q}} e^{-i[\epsilon_q + R_q(\epsilon_q)]t}A_{\mathbf{q}}\phi_{\mathbf{q}}\right). \end{aligned} \tag{150}$$

Here we have followed the arguments leading to Eq. (109), omitting all but the contributions from the pole at $\epsilon \approx \epsilon_q + R_q(\epsilon_q)$. This requires the assumption equivalent to (99) that $\dfrac{|R|}{\epsilon_q} \ll 1$—an assumption that is often difficult to justify in practice because of the great complexity of many-particle systems.

We now use the condition that the volume $V$ is arbitrarily large[32] to write [see Eq. (102)]

$$R_a(\epsilon_q) = D_q - \frac{i}{2}\Gamma_q, \tag{151}$$

where $\Gamma_q > 0$. The decay rate of the quasi-particle states is seen to be $\Gamma_q/\hbar$. The energy of these states goes into more complex excitations which are generally called "thermal energy."

To describe the physical significance of the state $\Psi(t)$, let us evaluate the

[32] Equation (151) is valid in a large finite box when $\Delta Et \ll 1$, where $\Delta E$ is the level spacing. This is, once again, the problem of "scattering" in a finite box.

mean value of the actual particle density at a point $R$ in the box. The particle density operator is

$$n(\mathbf{R}) = \sum_{i=1}^{N} \delta(\mathbf{R} - \mathbf{r}_i), \tag{152a}$$

where $\mathbf{r}_i$ is the coordinate of the $i$th particle (not quasi-particle) in the system. Fourier expansion of the $\delta$-functions in Eq. (152$a$) permits us to write $n(\mathbf{R})$ as

$$n(R) = \sum_{\mathbf{k}} \frac{e^{i\mathbf{k}\cdot\mathbf{R}}}{\mathscr{V}} \sum_{i=1}^{N} e^{-i\mathbf{k}\cdot\mathbf{r}_i}. \tag{152b}$$

In the steady state $\psi_0$ we suppose the density to have the constant value

$$n_0 = (\psi_0, n(\mathbf{R})\psi_0) = \frac{N}{\mathscr{V}} \tag{153}$$

(although strictly speaking we should take an ensemble average of this quantity). Then for the perturbed state (150), we have to first order in the excitations

$$\begin{aligned} \bar{n}(\mathbf{R}, t) &= (\Psi(t), n(\mathbf{R})\Psi(t)) \\ &\approx n_0 + \frac{2\,\mathrm{Re}}{\mathscr{V}}\left[\int \sum_{\mathbf{q}} e^{-i(\epsilon_q + R_q)t} A_{\mathbf{q}} \sum_{\mathbf{k}} e^{i\mathbf{k}\cdot\mathbf{R}}\left(\psi_0, \sum_{i=1}^{N} e^{-i\mathbf{k}\cdot\mathbf{r}_i}\phi_{\mathbf{q}}\right)\right]. \end{aligned} \tag{154}$$

Now, because of Eqs. (142) and (144$a$),

$$\left(\psi_0, \sum_{i=1}^{N} e^{-i\mathbf{k}\cdot\mathbf{r}_i}\phi_{\mathbf{q}}\right) = \delta(\mathbf{k} - \mathbf{q})C_{\mathbf{q}}, \tag{155}$$

where $C_q$ is some complex number depending on $\mathbf{q}$. It follows then that

$$\bar{n}(\mathbf{R}, t) = n_0 + \frac{2\,\mathrm{Re}}{\mathscr{V}}\left[\sum_{\mathbf{q}} C_{\mathbf{q}} A_{\mathbf{q}} e^{i(\mathbf{q}\cdot\mathbf{R} - \epsilon_q t - R_q t)}\right]. \tag{156}$$

In practice, one would ordinarily take an ensemble average of this equation. This would replace $C_{\mathbf{q}}$ by its ensemble average $\langle C_{\mathbf{q}}\rangle$, but it would not otherwise modify the equation.

Equation (156) describes the propagation of waves in the system. These waves decay in a time depending on $\Gamma_q$; the decay is usually described in transport theory as an "irreversible process." When $R_q$ is small, we may conveniently neglect it in Eq. (156) for times of the order of $\hbar/\Gamma_q$. In one of the cases of greatest physical interest, the excitation energy $\epsilon_q$ varies linearly with $q$.

$$\epsilon_q = vq, \tag{157}$$

where $v$ is a constant having the dimensions of a velocity. For times such that $R_q$ may be neglected, we may then write Eq. (156) as

$$\delta\bar{n}(\mathbf{R}, t) \equiv \bar{n}(\mathbf{R}, t) - n_0$$
$$= \frac{2\,\mathrm{Re}}{\mathscr{V}}\left[\sum_{\mathbf{q}} C_{\mathbf{q}} A_{\mathbf{q}} e^{i(\mathbf{q}\cdot\mathbf{R}-qvt)}\right]. \tag{158}$$

This satisfies the conventional wave equation

$$\left(\frac{1}{v^2}\frac{\partial^2}{\partial t^2} - \nabla^2\right)\delta\bar{n} = 0, \tag{159}$$

which describes the propagation of ordinary sound waves with a sound speed $v$.

Other examples of quasi-particle excitations include phonons and rotons in $He^4$, plasma waves, and the excitations of the theory of superconductivity. For example, in a plasma the energy of the excitation is described by the equation

$$\epsilon_q^{\,2} = \alpha + \beta q^2, \tag{160}$$

where $\alpha$ and $\beta$ are constants.

## 8.3 EXAMPLES OF THE DECAY OF A PREPARED STATE

In this section we shall consider in some detail two examples of decay processes. The first is the familiar example of the radiative transition from an excited state to a lower state in an atom or molecule. As a second example, we shall describe the decay of a pi-meson into a mu-meson and a neutrino. The relation of the decay of a prepared state to the scattering reaction by which the unstable state is prepared will be studied in Section 8.4.

### Radiative Transitions between Atomic States

As an application of the general theory developed in Section 8.2, we discuss the decay by radiative transitions of the excited states of an atom or molecule. In the notation of Eqs. (33) and (34), we now set

$$K = \frac{P_a^{\,2}}{2M} + h + K_r. \tag{161}$$

Here $M$ is the mass of the atom,[33] assumed to be very large, and $\mathbf{P}_a$ is its momentum in the initial state $\chi_a$. The operator $h$ is the Hamiltonian for the internal electronic states of the atom (for a molecule these include

[33] For brevity we shall from now on refer to the particle as an atom. Our general considerations apply to either atoms or molecules, however.

rotational, vibrational, and electronic states). Thus

$$hg_a = W_a g_a, \tag{162}$$

where $g_a(\xi)$ is the internal state wave function and $W_a$ is the energy of the atom in the state $a$. Finally, $K_r$ is the energy of the radiation field. It may be expressed as[34]

$$K_r = \sum_{\mathbf{k},\hat{e}} (kc) n_{\mathbf{k},\hat{e}} \tag{163}$$

where $n_{\mathbf{k},\hat{e}}$ is the number of photons in a momentum state $\mathbf{k}$ and having polarization parallel to $\hat{e}$ (and, of course, $\mathbf{k} \cdot \hat{e} = 0$ since the field is transverse).

We shall suppose that the possible states of the system are quantized in a large box of volume $\mathscr{V}$. This simplifies somewhat the evaluation of the photon flux and density. It is also consistent with our considerations of decay in Section 8.2 if (and only if) we pass to the limit $\mathscr{V} = \infty$ before evaluating sums over virtual states. The initial state $\chi_a$ is then [see Eq. (34)] taken as

$$\chi_a = \frac{1}{\sqrt{\mathscr{V}}} e^{i\mathbf{P}_a \cdot \mathbf{r}} g_a(\xi)\omega_0. \tag{164}$$

Here $\omega_0$ describes a state of the radiation field containing various number of photons. This is an eigenstate of $K_r$,

$$K_r \omega_0 = \epsilon_r \omega_0, \tag{165a}$$

where $\epsilon_r$ is the energy in the radiation field.

The initial wave packet [see Eqs. (36*a*,*b*)] describing an atom in the internal state $a$ and localized within some region of $\mathscr{V}$ is

$$X_a = \sum_{\mathbf{P}_a} A(\mathbf{P}_a)\chi_a. \tag{166}$$

Let us now suppose that the atom may radiate to a lower atomic state $g_b(\xi)$, having energy $W_b$, accompanied by the emission of a photon of momentum $\mathbf{k}$ and polarization $\hat{e}$. Here the final state $\chi'_b$ [Eq. (42)] is

$$\chi'_b = \frac{1}{\sqrt{\mathscr{V}}} e^{i\mathbf{P}_b \cdot \mathbf{r}} g_b(\xi)\omega(\mathbf{k}, \hat{e}), \tag{167}$$

where $\omega(\mathbf{k}, \hat{e})$ describes a state of the radiation field which corresponds to adding a single photon of momentum $\mathbf{k}$ and polarization $\hat{e}$ to the state described by $\omega_0$. Thus

$$K_r \omega(\mathbf{k}, \hat{e}) = (kc + \epsilon_r)\omega(\mathbf{k}, \hat{e}). \tag{165b}$$

The states $\omega(\mathbf{k}, \hat{e})$ and $\omega_0$ are, of course, orthogonal.[35]

[34] See, for example, Heitler [10].

[35] See the discussion on page 124 in Chapter 4.

The transition from state $a$ to state $b$ of an atom will involve a change of electronic states in that atom. For a molecule the transition may involve an electronic transition (characteristically in the optical or ultraviolet spectrum), a vibrational transition (characteristically in the infrared spectrum), or a rotational transition (which is likely to be in the radio frequency spectrum). The energies $W_a$ and $W_b$ are usually degenerate, corresponding to several states of the particle. We shall let $m_a$ and $m_b$, respectively, label these degenerate states, there being $\mu(a)$ and $\mu(b)$ of them. Ordinarily, $m_a$ and $m_b$ label the orientation of the spins $S_a$ and $S_b$ of the states $a$ and $b$, and then

$$\begin{aligned} \mu(a) &= 2S_a + 1, \\ \mu(b) &= 2S_b + 1. \end{aligned} \tag{168}$$

The notion of the "interaction" $V$ responsible for the radioactive transition was introduced by Eq. (35). For this example this interaction is taken as

$$V_r = -\sum_{j=1}^{Z} \frac{e}{mc} \mathbf{p}_j \cdot \mathbf{A}(\mathbf{r}_j), \tag{169}$$

where $e$ and $m$ are the charge and mass, respectively, of an electron. The summation in Eq. (169) extends over the coordinates $\mathbf{r}_j$ and momenta $\mathbf{p}_j$ of the $Z$ electrons in the atom. Finally, the quantity $\mathbf{A}(\mathbf{x})$ is the vector potential of the radiation field evaluated at a point $\mathbf{x}$.

To describe radiative transitions in a molecule, we replace $e$ by $e_j$ and $m$ by $m_j$ in Eq. (169) and extend the sum over all electrons and ions.

The vector potential $\mathbf{A}(\mathbf{x})$ may be expanded in a series of plane waves. This expansion is[36,37]

$$\mathbf{A}(\mathbf{x}) = \sum_{\mathbf{k},\hat{e}} \left(\frac{2\pi}{\mathscr{V}} \frac{\hbar^2 c}{k}\right)^{1/2} \hat{e}[a_{\mathbf{k},\hat{e}} e^{(i\mathbf{k}\cdot\mathbf{x})/\hbar} + a^{\dagger}_{\mathbf{k},\hat{e}} e^{-i\mathbf{k}\cdot\mathbf{x}/\hbar}]. \tag{170}$$

The sum on $\mathbf{k}$ here extends over all plane wave momentum states $\mathbf{k}$ in the box $\mathscr{V}$. The sum on $\hat{e}$ extends over the two allowed directions of polarization of a photon having a momentum $\mathbf{k}$. The quantities $a_{\mathbf{k},\hat{e}}$ and $a^{\dagger}_{\mathbf{k},\hat{e}}$ are the respective operators for absorption and creation of photons in the state $(\mathbf{k}, \hat{e})$. Their only nonvanishing matrix elements may be written as

$$\begin{aligned} \langle n_{\mathbf{k},\hat{e}} - 1 |\, a_{\mathbf{k},\hat{e}} \,| n_{\mathbf{k},\hat{e}} \rangle &= \sqrt{n_{\mathbf{k},\hat{e}}}, \\ \langle n_{\mathbf{k},\hat{e}} + 1 |\, a^{\dagger}_{\mathbf{k},\hat{e}} \,| n_{\mathbf{k},\hat{e}} \rangle &= \sqrt{n_{\mathbf{k},\hat{e}} + 1}. \end{aligned} \tag{171}$$

[36] See Heitler [10].

[37] Because of the existence of a standardized notation in connection with radiative transitions, we shall attempt to follow this and keep $\hbar$ and $h = 2\pi\hbar$ in our equations here.

The normalization of the expression (170) is easily determined from an evaluation of the photon flux in a given state of the radiation field.[38]

Now, the electromagnetic interaction of charged particles with the transverse field is "weak," in the sense that virtual transitions involving the emission and reabsorption of photons contribute very little to scattering cross sections and decay rates. This means that we may evaluate the operator $R_{ba}(\epsilon)$ in lowest order perturbation theory. (We shall return later to an estimate for the validity of this approximation.) Then according to Eqs. (61) and (62),

$$\begin{aligned} R_{ba}(\epsilon) &= (\omega(\mathbf{k}, \hat{e})g_b, V_r \omega_0 g_a) \\ &= \langle \mathbf{k}, \hat{e}, b | \, V_r \, | a \rangle. \end{aligned} \tag{172}$$

The differential decay rate for the transition from state $a$ to state $b$ is obtained from Eq. (126$a$) as

$$d\dot{P}_{ba} = \frac{2\pi}{\hbar} d\rho' \sum_{m_a, m_b} \frac{1}{\mu(a)} \sum_{\hat{e}} |\langle \mathbf{k}, \hat{e}, b | \, V_r \, | a \rangle|^2, \tag{173}$$

where

$$kc \equiv h\nu_{ab} \equiv W_a - W_b \tag{174}$$

is the energy of the emitted photon and

$$d\rho' = \frac{k^2}{h^3 c} d\Omega_{\hat{k}} \tag{175}$$

is the number of photon states of a given polarization in $\mathscr{V}$, per unit energy, directed into the solid angle $d\Omega_{\hat{k}}$. In Eq. (173) we have summed over all final states of energy $W_b$ and averaged over the initial states of energy $W_a$. We have also summed over the two states of polarization for each $\mathbf{k}$.

If we integrate Eq. (173) over the angles of the emitted photon, we obtain

[38] We may see this as follows. Consider the Poynting flux $\mathbf{S} = (c/4\pi)\mathbf{E} \times \mathbf{B}$ at the point $\mathbf{x} = 0$. Now, for $n_{\mathbf{k},\hat{e}}$ photons in a single state $(\mathbf{k}, \hat{e})$, we have

$$\begin{aligned} \mathbf{B} &= \nabla \times \mathbf{A} \\ &= i\left(\frac{2\pi c}{\mathscr{V}k}\right)^{1/2} \mathbf{k} \times \hat{e}(a_{\mathbf{k},\hat{e}} - a^\dagger_{\mathbf{k},\hat{e}}) \end{aligned}$$

and

$$\mathbf{E} = -\frac{1}{c}\dot{\mathbf{A}} = \frac{ik}{\hbar}\mathbf{A} = i\left(\frac{2\pi c}{\mathscr{V}k}\right)^{1/2} k\hat{e}(a_{\mathbf{k},\hat{e}} - a^\dagger_{\mathbf{k},\hat{e}}).$$

On dropping terms which would vanish on averaging over the angles of $\mathbf{k}$ (that is, vacuum fluctuations) we obtain for the expectation value of $\mathbf{S}$: $\langle \mathbf{S} \rangle = c(c\mathbf{k})n_{\mathbf{k},\hat{e}}/\mathscr{V}$.

the total transition rate [see Eqs. (103) and (128*a*)]

$$\dot{P}_{ba} = \frac{\Gamma_{ba}}{\hbar} = \frac{2\pi}{\hbar} \sum_{m_a, m_b} \frac{1}{\mu(a)} \int d\rho' \sum_{\hat{e}} |\langle \mathbf{k}, \hat{e}, b | V_r | a \rangle|^2. \tag{176}$$

Using Eqs. (169), (170), and (171), we find that

$$\langle \mathbf{k}, \hat{e}, b | V_r | a \rangle = -e\left(\frac{2\pi\hbar^2 c}{\mathscr{V} k}\right)^{1/2} \sqrt{n_{\mathbf{k},\hat{e}} + 1}\left(g_b, \sum_{j=1}^{Z} \frac{\hat{e} \cdot \mathbf{p}_j}{mc} e^{i\mathbf{k}\cdot\mathbf{r}_j} g_a\right). \tag{177}$$

Here $n_{\mathbf{k},\hat{e}}$ represents the number of photons already in the state $(\mathbf{k}, \hat{e})$ before the transition has occurred. We shall assume that an *electric dipole transition* is allowed for the decay from state $a$ to state $b$ and set

$$e^{i\mathbf{k}\cdot\mathbf{r}_j} = 1$$

in Eq. (177). This is an excellent approximation for atomic transitions since $k\langle r_j \rangle \approx 10^{-3}$.

The expression (177) may be inserted into Eq. (173) to give

$$d\dot{P}_{ba} = \frac{e^2 k}{2\pi\hbar^2} d\Omega_{\hat{k}} \sum_{m_a\, m_b, \hat{e}} \frac{1}{\mu(a)} |\langle b| \frac{\mathscr{P} \cdot \hat{e}}{mc} |a\rangle|^2 (1 + n_{\mathbf{k},\hat{e}}), \tag{178}$$

where

$$\mathscr{P} = \sum_{j=1}^{Z} \mathbf{p}_j \tag{179}$$

is the total momentum operator for the electrons.

Because of the sum and average over spin substates of the atom in Eq. (178), the only possible dependence on $\mathbf{k}$ and $\hat{e}$ comes from $n_{\mathbf{k},\hat{e}}$. We may thus introduce the "oscillator strength," defined as

$$f_{ba} \equiv \frac{mc^2}{h\nu_{ba}} \int \frac{d\Omega_{\hat{k}}}{4\pi} \sum_{\hat{e}} \sum_{m_a, m_b} \frac{1}{\mu(a)} |\langle b| \frac{\mathscr{P} \cdot \hat{e}}{mc} |a\rangle|^2, \tag{180}$$

where $h\nu_{ba} = W_b - W_a$, and write Eq. (178) as

$$d\dot{P}_{ba} = \frac{2\pi e^2}{mc^3} \nu_{ab}{}^2 \, d\Omega_{\hat{k}} \, |f_{ba}| \left[\frac{1}{2} \sum_{\hat{e}} (1 + n_{\mathbf{k},\hat{e}})\right]. \tag{181}$$

When there are no photons initially in the state $(\mathbf{k}, \hat{e})$, we may integrate Eq. (181) to obtain the total spontaneous transition rate

$$\dot{P}_{ba} = \frac{\Gamma_{ba}}{\hbar} = \frac{8\pi^2 e^2}{mc^3} \nu_{ab}{}^2 \, |f_{ba}|. \tag{182}$$

Before continuing we shall simplify the expression (180) for the oscillator strength. Let

$$\mathbf{d} \equiv \sum_{j=1}^{Z} e\mathbf{r}_j, \tag{183}$$

be the dipole moment operator for the atom. Now,

$$(W_b - W_a)\langle b|\, \mathbf{d} \cdot \hat{e}\, |a\rangle = \langle b|\, [h, \mathbf{d} \cdot \hat{e}]\, |a\rangle = -\frac{ie}{m}\hbar\langle b|\, \mathscr{P} \cdot \hat{e}\, |a\rangle. \quad (184)$$

[For evaluating the commutator, we need keep only the kinetic energy $\left(-\frac{\hbar^2}{2m}\sum_{j=1}^{Z} \nabla_j^2\right)$ in $h$.] Thus

$$\begin{aligned} f_{ba} &= \frac{4\pi^2 \nu_{ba} m}{he^2} \int \frac{d\Omega_{\hat{k}}}{4\pi} \sum_{m_a, m_b} \frac{1}{\mu(a)} \sum_{\hat{e}} |\langle b|\, \mathbf{d} \cdot \hat{e}\, |a\rangle|^2 \\ &= \frac{4\pi^2 \nu_{ba} m}{he^2} \int \frac{d\Omega_{\hat{k}}}{4\pi} \sum_{m_a, m_b} \frac{1}{\mu(a)} \left( |\langle b|\, \mathbf{d}\, |a\rangle|^2 - \frac{1}{k^2} |\langle b|\, \mathbf{d} \cdot \mathbf{k}\, |a\rangle|^2 \right) \\ &= \frac{8\pi^2 \nu_{ba} m}{3he^2} \sum_{m_a, m_b} \frac{1}{\mu(a)} |\langle b|\, \mathbf{d}\, |a\rangle|^2. \end{aligned} \quad (185)$$

We note the Thomas-Kuhn-Reiche sum rule

$$\sum_b f_{ba} = Z,$$

which is easily derived using the expressions (180), (184), and (185).

If we wish the rate at which photons are emitted into a frequency range between $\nu$ and $\nu + d\nu$, we must multiply Eq. (181) by the probability $b_{ba}(\nu)\, d\nu$ for emission into this frequency range [see Eq. (122*a*)].[39] This is

$$b_{ba}(\nu)\, d\nu = \frac{\Gamma}{2\pi} \frac{h\, d\nu}{(W_b - W_a + h\nu)^2 + \dfrac{\Gamma^2}{4}}, \quad (186a)$$

$$\int b_{ba}(\nu)\, d\nu = 1, \quad (186b)$$

where [by Eq. (121*c*)]

$$\Gamma = \sum_b \Gamma_{ba}, \quad (187)$$

the sum extending over all states $b$ into which the transition may go (that is, over states for which $W_b < W_a$). The level shift $D$ has been omitted in Eq. (186) since (as we shall see) it is very small. The transition rate into a solid angle $d\Omega_{\hat{k}}$ and into the frequency interval $d\nu$ is then

$$d^2\dot{P}_{ba} = \frac{2\pi e^2 \nu^2}{mc^3} |f_{ba}|\, b_{ba}(\nu) \left[\frac{1}{2} \sum_{\hat{e}} (1 + n_{\mathbf{k},\hat{e}})\right] d\Omega_{\hat{k}}\, d\nu. \quad (188a)$$

[39] In practice, the "natural line width" (186) may be modified by Stark or Doppler broadening (see Ref. [23]).

When $n_{\mathbf{k},\hat{e}} = n_{\mathbf{k}}$ is independent of $\hat{e}$, this reduces to

$$d^2\dot{P}_{ba} = \frac{2\pi e^2}{mc^3}\nu^2 |f_{ba}|\, b_{ba}(\nu)(1 + n_{\mathbf{k}})\, d\Omega_{\hat{\mathbf{k}}}\, d\nu$$

$$\equiv w_{ba}(\nu)\, d\Omega_{\hat{\mathbf{k}}}\, d\nu. \tag{188b}$$

The quantity $w_{ba}(\nu)$ represents the transition rate per atom per unit solid angle and per unit frequency range. Thus, if there are $n(a)$ atoms per unit volume in the state $a$, the change in time $dt$ in $n(a)$ due to transitions to the state $b$ and the emission of photons into $d\Omega_{\hat{\mathbf{k}}}\, d\nu$ is [see Eq. (129)]

$$dn(a)|_{\mathrm{rad}} = -w_{ba}(\nu)\, d\nu\, d\Omega_{\hat{\mathbf{k}}} n(a)\, dt. \tag{189}$$

In addition to the contribution (181) to $dn(a)$ due to radiation by atoms in the state $a$, there will be a change due to absorption of radiation of frequency $\nu$ coming from $d\Omega_{\hat{\mathbf{k}}}$ by atoms in the state $b$. This change is written as

$$dn(a)|_{ab} = w_{ab}^{(\mathrm{ab})}(\nu) n(b)\, d\nu\, d\Omega_{\hat{\mathbf{k}}}\, dt, \tag{190}$$

where $n(b)$ is the number of atoms per unit volume in state $b$ and $w_{ab}^{(\mathrm{ab})}$ is an absorption coefficient per atom.

We might calculate $w_{ab}^{(\mathrm{ab})}$ directly. It is simpler, however, to evaluate it using the statistical principle of "detailed balance," which states that when the atoms and radiation field are in thermal equilibrium

$$dn(a)|_{ab} + dn(a)|_{\mathrm{rad}} = 0. \tag{191}$$

Now in thermal equilibrium at temperature $\beta^{-1}$ we have[40]

$$\begin{aligned} n(a) &= n^0\mu(a)e^{-\beta W_a}, \\ n(b) &= n^0\mu(b)e^{-\beta W_b}, \\ n_{\mathbf{k}} &= \frac{1}{e^{\beta\hbar\nu} - 1}. \end{aligned} \tag{192}$$

Substitution of Eqs. (189) and (190) into Eq. (191) gives the relation

$$w_{ab}^{(\mathrm{ab})}\, d\nu\, d\Omega_{\hat{\mathbf{k}}} n(b) = w_{ba}\, d\nu\, d\Omega_{\hat{\mathbf{k}}} n(a),$$

which [with Eqs. (188$b$) and (192)] yields

$$w_{ab}^{(\mathrm{ab})}(\nu) = \frac{2\pi e^2}{mc^3}\nu^2\left[\frac{\mu(a)}{\mu(b)}|f_{ba}|\right] b_{ba}(\nu) n_{\mathbf{k}}. \tag{192a}$$

From Eq. (185) we see that

$$f_{ab} = \frac{\mu(a)}{\mu(b)}|f_{ba}|, \tag{193}$$

so

$$w_{ab}^{(\mathrm{ab})}(\nu) = \frac{2\pi e^2}{mc^3}\nu^2 f_{ab} b_{ba}(\nu) n_{\mathbf{k}}. \tag{192b}$$

[40] See, for example, Ref. [19].

This is a purely dynamical relation, independent of the distribution of particles, so it is generally valid, whether or not the atoms and radiation are in thermal equilibrium.

The expression for the absorption rate may be put into the convenient form of an absorption cross section. To do this, we note from Eq. (175) that the flux of photons per unit frequency interval per steradian is

$$\frac{I(\nu, \hat{\mathbf{k}})}{h\nu} = \frac{2\nu^2}{c^2} n_{\mathbf{k}}.$$

The absorption cross section is the transition rate per atom per unit flux of photons. This is then

$$\sigma_{ab}^{(\mathrm{ab})}(\nu) = \frac{w_{ab}^{(\mathrm{ab})}}{[I(\nu, \hat{\mathbf{k}})/h\nu]} = \frac{\pi e^2}{mc} f_{ab} b_{ba}(\nu). \tag{194a}$$

This expression may be put into a simpler form if we use Eqs. (182), (186), and (193).

$$\sigma_{ab}^{(\mathrm{ab})}(\nu) = \pi \left(\frac{\hbar}{k}\right)^2 \frac{\mu(a)}{2\mu(b)} \frac{\Gamma_{ba}\Gamma}{[h\nu - (W_a - W_b)]^2 + \Gamma^2/4}, \tag{194b}$$

where $k = h\nu_{ab}/c$ as before. When the expressions (168) for $\mu(a)$ and $\mu(b)$ may be used, this takes the more conventional form

$$\sigma_{ab}^{(\mathrm{ab})}(\nu) = \frac{(2S_a + 1)}{2(2S_b + 1)} \pi \left(\frac{\hbar}{k}\right)^2 \frac{\Gamma_{ba}\Gamma}{[h\nu - (W_a - W_b)]^2 + \Gamma^2/4}. \tag{195}$$

The factor $2S_a + 1$ is the statistical weight of the state $a$, whereas $[2(2S_b + 1)]$ is the statistical weight of the initial state—the factor of 2 arises from the two polarization states of the photon. The cross section (195) represents the total cross section for absorption of a photon by an atom in the state $b$, leading to an excited atom in the state $a$. The subsequent mode of reradiation is not specified in this equation, however.

Equations (194) and (195) provide our first illustration of a *resonance* cross section. In the next section we shall rederive this result directly from scattering theory. Finally, in Section 8.5 we shall discuss such "resonances" in scattering from a more fundamental point of view.

In writing the general equation (122*a*) in the form (186) for $b_{ba}(\nu)$, we neglected the energy shift $D(\epsilon)$. Let us now estimate $D$ using Eq. (96). We first require the function $I(\epsilon)$. Now, in the event that only one decay state $b$ need be considered

$$I(\epsilon_a) = \frac{\Gamma_{ba}}{2}.$$

We see from Eq. (172) that $R_{ba}^{+}(\epsilon) = \langle \mathbf{k}, \hat{e}, b | V_r | a \rangle$ is, to the order kept, independent of $\epsilon$. Now, $d\rho' \simeq \epsilon^2$ [see Eq. (175)] and therefore $I(\epsilon) \simeq \epsilon$, taking account of the factor of $(kc)^{-1/2} = \epsilon^{-1/2}$ in the vector potential $A$. This permits us to write

$$I(\epsilon) = \frac{\nu}{\nu_{ab}} \frac{\Gamma_{ba}}{2}, \tag{196}$$

where $\epsilon = h\nu$.

If this expression is inserted into Eq. (96), there results

$$\begin{aligned} D(h\nu) &= -\frac{P}{\pi} \int_0^\infty \frac{I(h\nu')}{\nu' - \nu - \nu_{ab}}\, d\nu' \\ &= -\frac{\Gamma_{ba}}{2\pi\nu_{ab}} \int_0^\infty \frac{\nu'\, d\nu'}{\nu' - \nu - \nu_{ab}}. \end{aligned} \tag{197a}$$

The linear divergence of this integral for large $\nu'$ provides a familiar example of "self-energy" divergences in field theories. [Had we used relativistic wave functions for the electrons and evaluated $D$ keeping all decay transitions, we would have found only a logarithmic divergence in Eq. (197*a*).] The origin of such divergences is well understood, in our case it results from the failure of $V_r$, as defined by Eq. (169), to satisfy the condition (74).

It is known that for the "renormalizable" field theories such divergences may be removed.[41] This may be done in electrodynamics by modifying $V_r$ in that an electron self-mass $\delta mc^2$ is explicitly subtracted from $V_r$ and a corresponding "renormalization" of the charge $e$ is made. In this case, we would modify Eq. (197*a*) to read

$$D(h\nu) = -\delta mc^2 - \frac{\Gamma_{ba}}{2\pi\nu_{ab}} P \int_0^\infty \frac{\nu'\, d\nu'}{\nu' - \nu - \nu_{ab}}, \tag{197b}$$

and choose $\delta mc^2$ so as to make $D(0)$ vanish.[42] Alternatively, we may formally subtract $D(0)$ from both sides of Eq. (197*a*), obtaining

$$D(h\nu) - D(0) = -\frac{\Gamma_{ba}\nu}{2\pi\nu_{ab}} P \int_0^\infty \frac{\nu'\, d\nu'}{(\nu' - \nu_{ab})(\nu' - \nu - \nu_{ab})}. \tag{197c}$$

The subtraction constant $D(0)$ would be treated as a mass renormalization in a thorough discussion of this problem.

The integral (197*c*) is still divergent, although it would be finite had we treated the electrons relativistically. We can obtain a correct order-of-magnitude estimate, however, if we cut off the integral at an upper limit $\nu' = mc^2/h$.[43] This then gives

$$D(h\nu) - D(0) = \mathcal{O}\left(\frac{\Gamma_{ba}}{2\pi}\right). \tag{198}$$

[41] See, for example, Jauch and Rohrlich [24].

[42] This has been done explicitly by Low [12].

[43] See Ref. [12].

Now, this is seen to have the value

$$\frac{\Gamma_{ba}}{2\pi W_a} \cong 7\left(\frac{e^2}{\hbar c}\right)^3 \cong 10^{-6}. \tag{199}$$

Although our estimate is very crude, it suffices to show that the real level shift $D$ gives an exceedingly small contribution to the line-shape function $b_{ba}(\nu)$. This means that $D$ may ordinarily be neglected in studies of the line shape.

**Pi-Meson Decay**

As a second example of our description of the decay of a prepared state, we shall give a phenomenological discussion of pi-meson decay. As we have mentioned earlier, a charged pi-meson decays into a mu-meson and a neutrino with a rate

$$\frac{\Gamma}{\hbar} = 0.4 \times 10^8 \text{ sec}^{-1}. \tag{200}$$

If we accept the principle of "lepton conservation," by which (e, $\mu$, $\nu$) are leptons and ($e^+$, $\mu^+$, $\bar{\nu}$) are antileptons,[44] the reactions for $\pi - \mu$ decay are

$$\pi^+ \rightarrow \mu^+ + \nu \tag{201a}$$

$$\pi^- \rightarrow \mu^- + \bar{\nu}. \tag{201b}$$

We shall specifically discuss the second reaction, that is, the decay of a $\pi^-$-meson into a negative mu-meson and an antineutrino.[45]

To describe the decay (201*b*), we require the Dirac wave functions for the $\mu^-$ and $\bar{\nu}$ particles. It will be recalled that the Dirac spinor $u_\lambda(p)$ defined by Eq. (2.63) describes a particle of a four-momentum[46] ($\mathbf{p}$, $\epsilon_p$) for $\lambda = 1, 2$ and describes a particle of negative energy in a state ($-\mathbf{p}$, $-\epsilon_p$) for $\lambda = -1, -2$. Associated with the $u_\lambda(p)$ are the wave functions $v_\lambda(p)$, where by Eq. (2.67)

$$v_{-\lambda}(p) = C\tilde{\bar{u}}_\lambda(p). \tag{202a}$$

Here $C$ is the matrix for charge conjugation. Equation (202*a*) may be inverted to read

$$u_{-\lambda}(p) = C\tilde{\bar{v}}_\lambda(p), \tag{202b}$$

which shows the physical equivalence of the states described by $u_\lambda$ and by $v_\lambda$.

[44] For a review of the "weak interactions" see, for example, S. B. Treiman [25].
[45] The two reactions (201*a*) and (201*b*) are related by charge conjugation.
[46] In this section we shall find it convenient to use a system of units in which $\hbar = c = 1$.

The wave function $\psi$ for a free Dirac particle is

$$\psi(\mathbf{x}) = u_\lambda(p)\frac{e^{i\dot{\mathbf{p}}\cdot\mathbf{x}}}{(2\pi)^{3/2}}, \tag{203a}$$

where we have introduced the notation

$$\dot{\mathbf{p}} \equiv \mathbf{p}\frac{\lambda}{|\lambda|}. \tag{204}$$

The charge conjugate wave function corresponding to (203$a$) is

$$\psi^C(\mathbf{x}) = C\tilde{\bar{\psi}} = v_{-\lambda}(p)\frac{e^{-i\dot{\mathbf{p}}\cdot\mathbf{x}}}{(2\pi)^{3/2}}. \tag{203b}$$

The Dirac field operator $\Psi$ is then[47]

$$\Psi = \sum_{\mathbf{p},\lambda} a_{\dot{\mathbf{p}},\lambda} u_\lambda(p)\frac{e^{i\dot{\mathbf{p}}\cdot\mathbf{x}}}{(2\pi)^{3/2}}. \tag{205a}$$

Here $a_{\mathbf{p},\lambda}$ is the operator which annihilates the Dirac particle in a state of momentum $\mathbf{p}$, having spin and energy labeled by $\lambda$. The corresponding operator which creates a particle in this state is $a^\dagger_{\mathbf{p},\lambda}$. We may write the charge conjugate field operator corresponding to $\Psi$ as

$$\begin{aligned}\Psi^C = C\tilde{\bar{\Psi}} &= \sum_{\mathbf{p},\lambda} a^\dagger_{\dot{\mathbf{p}},\lambda} v_{-\lambda}(p)\frac{e^{-i\dot{\mathbf{p}}\cdot\mathbf{x}}}{(2\pi)^{3/2}} \\ &= \sum_{\mathbf{p},\lambda} a^\dagger_{-\dot{\mathbf{p}},-\lambda} v_\lambda(p)\frac{e^{i\dot{\mathbf{p}}\cdot\mathbf{x}}}{(2\pi)^{3/2}}.\end{aligned} \tag{205b}$$

We recall that in the charge conjugate states the particles and antiparticles just change roles. This permits us to define

$$\begin{aligned} b_{\mathbf{p},\lambda} &\equiv a^\dagger_{-\mathbf{p},-\lambda}, \\ b^\dagger_{\mathbf{p},\lambda} &\equiv a_{-\mathbf{p},-\lambda} \end{aligned} \tag{206}$$

where $b_{\mathbf{p},\lambda}$ *annihilates* an antiparticle of momentum $\mathbf{p}$ in the spinor state $\lambda$ and $b^\dagger_{\mathbf{p},\lambda}$ creates such a particle. (See Eqs. (2.73) for a specific definition of these states.)

Let us consider an operator of the form (here $\theta$ is some $4 \times 4$ Dirac matrix)

$$\mathcal{M} = (\bar{\Psi}\theta\Psi) = (2\pi)^{-3}\sum_{\mathbf{p}',\lambda'}\sum_{\mathbf{p},\lambda} a^\dagger_{\dot{\mathbf{p}}',\lambda'} a_{\dot{\mathbf{p}},\lambda}[\bar{u}_{\lambda'}(p')\theta u_\lambda(p)], \tag{207}$$

where the coordinate $\mathbf{x}$ has been set equal to zero. In field theories involving the interactions of Fermi particles, products of operators such as $\mathcal{M}$ with other similar operators are normally encountered. In our example,

[47] See, for example, Refs. [10] and [24].

we shall suppose that one of the operators $\Psi$ represents a neutrino field, whereas the other represents a mu-meson field.

To obtain the matrix element of $\mathscr{M}$ corresponding to the creation of a particle and an antiparticle, we set $a_{\mathbf{p},\lambda} = b^{\dagger}_{-\mathbf{p},-\lambda}$ in (207). The required matrix element for creating a particle in the state $\mathbf{p}'$, $\lambda'$ and an antiparticle in the state $\mathbf{p}$, $\lambda$ is then

$$\langle \mathbf{p}', \lambda', \overline{\mathbf{p}, \lambda} | \mathscr{M} | 0 \rangle = (2\pi)^{-3} [\bar{u}_{\lambda'}(p') \theta u_{-\lambda}(p)]. \tag{208}$$

For the specific case of the reaction (201*b*), this becomes

$$\langle \mu^-, \bar{\nu} | \mathscr{M} | 0 \rangle = (2\pi)^{-3} [\bar{u}_{\lambda',\mu}(p') \theta u_{-\lambda,\nu}(p)], \tag{209}$$

where $u_{\lambda',\mu}$ and $u_{\lambda,\nu}$ are the respective Dirac spinors for a mu-meson and a neutrino.

Considerations of Lorentz invariance severely limit the choice of matrices $\theta$ in the expression (207). The possible transformation properties for such expressions have been enumerated in Eq. (1.114). Because of the failure of weak interactions to conserve parity, however, the matrix $\theta$ may consist of an arbitrarily linear combination of an operator and its *dual*. Lee and Yang have proposed the "two-component theory" of the neutrino as a means of removing this arbitrariness.[48] To make our discussion definite, we shall accept the two-component neutrino theory, which is compatible with present experimental knowledge.

Let us now turn to a description of, say, the process (201*b*). We shall make a simple phenomenological choice for the interaction $V$ of Eq. (35) responsible for the $\pi^-$-decay. This will be written as

$$V = (2\pi)^{3/2} G_\pi \Phi(0)(\bar{\Psi}_\mu a^+ \Psi_\nu) + (2\pi)^{3/2} G_\pi \Phi^\dagger(0)(\bar{\Psi}_\nu a^- \Psi_\mu). \tag{210}$$

Here $a^{\pm}$ are the $4 \times 4$ Dirac matrices

$$a^{\pm} \equiv \tfrac{1}{2}(1 \pm \gamma_5). \tag{211}$$

The appearance of this operator is a result of our choice of the two-component neutrino theory. It assures us that a neutrino produced can have its spin orientation only antiparallel to its momentum vector, whereas an antineutrino can have its spin oriented only in the direction of its momentum vector. (This will be demonstrated later.) The quantities $\Psi_\nu$ and $\Psi_\mu$ are the Dirac field operators [see Eqs. (205)] for the neutrino and mu-meson, respectively. The field operator responsible for annihilation of the pi-meson has been written as $\Phi(0)$ in Eq. (210). Its matrix element corresponding to the annihilation of a pi-meson at rest is[49]

$$\langle 0 | \Phi(0) | \pi \rangle = [2(2\pi)^3 m_\pi]^{-1/2}, \tag{212}$$

[48] For an excellent review of this subject, see Ref. [13].

[49] See, for example, Bethe and de Hoffman [20] for an evaluation of matrix elements of meson field operators.

where $m_\pi$ is the rest mass of the meson. Finally, $G_\pi$ is a coupling constant describing the effective strength of the decay interaction.

We shall accept the experimental fact that the decay interaction is weak and write $R = V$, as was done in Eq. (172). Then, according to Eq. (103) [we insert a $\delta$-function specifying momentum conservation, as in Eq. (54*a*)], the decay rate for a pi-meson at rest is

$$\Gamma = 2\pi \sum_{\text{spins}} \int d^3p\, d^3q\, \delta(\epsilon + q - m_\pi)\, \delta(\mathbf{p} + \mathbf{q}) \times |\langle \mu^-, \bar{\nu}|\, V\, |\pi\rangle|^2, \quad (213)$$

where we have written $\mathbf{p}$ for the momentum of the mu-meson, $\epsilon \equiv (m_\mu^{\ 2} + p^2)^{1/2}$ for its energy and $\mathbf{q}$ for the momentum of the antineutrino. According to Eqs. (209), (210), and (212),

$$\langle \mu^-, \pi|\, V\, |\pi\rangle = \frac{G_\pi}{\sqrt{2m_\pi}} (2\pi)^{-3} [\bar{u}_{\lambda',\mu}(p) a^+ u_{-\lambda,\nu}(q)]. \quad (214)$$

A summation over the spin orientations of the mu-meson and the neutrino is implied in Eq. (213).

To simplify Eq. (214), we define

$$\begin{aligned} \gamma &\equiv \sum_{\text{spins}} |\langle \mu^-, \bar{\nu}|\, V\, |\pi\rangle|^2 \\ &= \frac{G_\pi^{\ 2}}{2m_\pi} (2\pi)^{-6} \sum_{\text{spins}} [u^\dagger_{-\lambda,\nu}(q) a^+ \gamma_4 u_{\lambda',\mu}(p)] \\ &\quad \times u^+_{\lambda',\mu}(p) \gamma_4 a^+ u_{-\lambda,\nu}(q). \end{aligned} \quad (215)$$

This is most easily evaluated with the help of the Casimir projection operators

$$\begin{aligned} \Lambda_\mu^{\ +}(\mathbf{p}) &\equiv i\beta \frac{\gamma_\alpha Q_\alpha^{\ +} - im_\mu}{2\epsilon}, \\ \Lambda_\nu^{\ -}(-\mathbf{q}) &\equiv -i\beta \frac{\gamma_\sigma Q_\sigma^{\ -}}{2|q|}, \end{aligned} \quad (216a)$$

where $m_\mu$ is the rest mass of the mu-meson,

$$\begin{aligned} Q_\alpha^{\ +} &= (\mathbf{p}, -i\epsilon), \\ Q_\sigma^{\ -} &= (-\mathbf{q}, iq) = (\mathbf{p}, ip), \end{aligned} \quad (216b)$$

and $\epsilon = (p^2 + m_\mu^{\ 2})^{1/2}$. That is, $\Lambda_\mu^{\ +}(\mathbf{p})$ is the projection operator onto positive energy states of momentum $\mathbf{p}$ of the mu-meson, and $\Lambda_\nu^{\ -}(-\mathbf{q})$ is the projection onto negative energy neutrino states of momentum $(-\mathbf{q})$.[50]

[50] This result, which will be shown in detail in Section 9.1, is easily verified.

With the aid of Eqs. (216), we may write Eq. (215) in the form

$$\gamma = (2\pi)^{-6}\frac{G_\pi^{\ 2}}{2m_\pi}\,\mathrm{Tr}\,\{a^+\gamma_4\Lambda_\mu^{\ +}(\mathbf{p})\gamma_4 a^+\Lambda_\nu^{\ -}(-\mathbf{q})\}$$

$$= (2\pi)^{-6}\frac{G_\pi^{\ 2}}{2m_\pi}\left(\frac{p}{\epsilon}+1\right). \tag{217}$$

Here the symbol Tr $\{\cdots\}$ implies that the trace of the Dirac matrix $\{\cdots\}$ is to be evaluated. The result expressed by Eq. (217) may finally be inserted into Eq. (213) to give

$$\Gamma = 2\pi\int d^3p\, d^3q\, \delta(\epsilon + q - m_\pi)\, \delta(\mathbf{p}+\mathbf{q})\gamma$$

$$= (2\pi)^{-3}\frac{G_\pi^{\ 2}}{4\pi}\frac{p^2}{m_\pi}, \tag{218}$$

with

$$(p^2 + m_\mu^{\ 2})^{1/2} + p = m_\pi. \tag{219}$$

Comparison of Eq. (218) with the experimental decay rate (200) permits evaluation of the effective coupling strength $G_\pi$.

It is instructive to evaluate the average spin orientation of the mu-meson. This is given by

$$\Gamma\langle\boldsymbol{\sigma}\cdot\hat{\mathbf{p}}\rangle = 2\pi\int d^3p\, d^3q\, \delta(\epsilon + q - m_\pi)\, \delta(\mathbf{p}+\mathbf{q})\langle\pi|\, V\boldsymbol{\sigma}\cdot\hat{\mathbf{p}}V\,|\pi\rangle$$

$$= 2\pi\int d^3p\, d^3q\, \delta(\epsilon + q - m_\pi)\, \delta(\mathbf{p}+\mathbf{q})$$

$$\times\left[(2\pi)^{-6}\frac{G_\pi^{\ 2}}{2m_\pi}\right]\mathrm{Tr}\,\{a^+\gamma_4\Lambda_\mu^{\ +}(\mathbf{p})\boldsymbol{\sigma}\cdot\hat{\mathbf{p}}\gamma_4 a^+\Lambda_\nu^{\ -}(-\mathbf{q})\}$$

$$= \Gamma. \tag{220}$$

Therefore

$$\langle\boldsymbol{\sigma}\cdot\hat{\mathbf{p}}\rangle = 1, \tag{221}$$

or the spin of the mu-meson is parallel to its momentum. This was to have been expected from our description of the two-component neutrino.

Were we to follow the discussion which led to Eq. (194), we would obtain an expression for the scattering of neutrinos by mu-mesons.

$$\mu^- + \bar{\nu} \rightarrow \mu^- + \bar{\nu}, \tag{222}$$

valid very near the energy $\epsilon + p = m_\pi$. Considering the neutrino as the "atom," we see that $\mu(a) = \mu(b) = 1$. This follows since the neutrino has only one spin orientation in the two-component theory. The factor of $\frac{1}{2}$ arising from the two polarization direction of the photon remains now, however, since the mu-meson has two spin orientations. Therefore we see that

$$\sigma(\mu^- + \bar{\nu} \rightarrow \mu^- + \bar{\nu}) = \frac{\pi}{2p^2} \frac{\Gamma^2}{(\epsilon + p - m_\pi)^2 + \Gamma^2/4}. \tag{223}$$

Equation (223) is, of course, valid only for $\epsilon + p \cong m_\pi$.

In Section 8.4 we shall show how to obtain such a resonance cross section by direct calculation, whereas in Section 8.6 a much more general discussion will be given.

The resonance described by Eq. (223) is *exceedingly narrow* since by Eq. (200)

$$\Gamma \cong 10^{-6} \text{ ev}. \tag{224}$$

Thus even if mu-mesons and neutrinos were available in macroscopic quantities, it would be difficult to observe this resonance.

## 8.4 PREPARATION OF AN UNSTABLE STATE BY SCATTERING INTERACTIONS

In the previous two sections we have described the decay of a system prepared in an unstable state, but we have not discussed the scattering mechanism by which this state is prepared. It is to this latter problem that we now turn, illustrating it with two examples. First, the cross section for resonant scattering of light by an atom will be obtained by direct calculation, the expression (195) being obtained again. Next, we shall discuss the collisional excitation of an atom by electron impact, noting specifically the subsequent deexcitation by a radiative transition.

These two examples are given to illustrate general techniques for the study of reactions involving long-lived virtual states.

### Resonant Scattering of Light by an Atom

Let us now consider the scattering as illustrated in Fig. 8.4 of a photon having momentum **k** and polarization $\hat{e}$ by an atom in a discrete state $b$

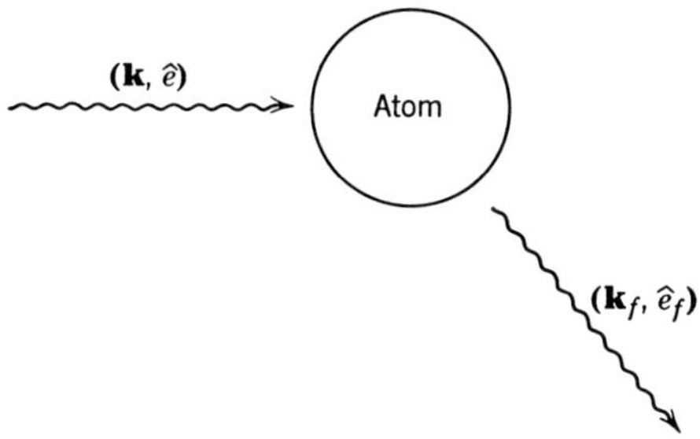


Fig. 8.4 Illustration of the scattering of light by an atom.

(we shall assume this to be its ground state). The photon will be scattered into a final state $(\mathbf{k}_f, \hat{e}_f)$, the atom being left in a state $b'$. When the energies $W_b$ and $W_{b'}$ of the atomic states $b$ and $b'$ are equal, that is, when

$$W_b = W_{b'}, \tag{225a}$$

the scattering is elastic and

$$k_f = k. \tag{225b}$$

Even when the condition (225*a*) is satisfied, however, the spin orientations[51] $m_b$ and $m_{b'}$ of the states $b$ and $b'$ may differ.

The Hamiltonian for the system being considered is just that of Section 8.3, p. 460.

$$H = K + V_r, \tag{226}$$

where $K$ and $V_r$ are defined by Eqs. (161) and (169). We agreed in Section 8.3, p. 460 to consider the atom as being very heavy, so the laboratory and barycentric frames of reference are equivalent. Then, in the barycentric subspace, the scattering matrix $T$ is [see Eq. (3.79)]

$$T = V_r + V_r \frac{1}{E + i\eta - H} V_r, \tag{227a}$$

where

$$E = kc + W_b. \tag{228}$$

It is seen from Eqs. (169) and (170) that the interaction $V_r$ has no non-vanishing matrix elements connecting the states $(\mathbf{k}, \hat{e})$ and $(\mathbf{k}_f, \hat{e}_f)$; this permits us to rewrite Eq. (227*a*) as

$$T = V_r G(E + i\eta) V_r \tag{227b}$$

where, as before,

$$G(E + i\eta) \equiv \frac{1}{E + i\eta - H}.$$

An exact evaluation of the expression (227*b*) is not feasible. Fortunately, the weakness of the coupling of the electrons to the radiation field may once again be invoked [see Eq. (199)] to provide a simple and quite accurate approximation for $T$. To be specific, we may assume that one of the factors of $V_r$ in (227*b*) emits a photon into the state $(\mathbf{k}_f, \hat{e}_f)$, whereas the other absorbs the incident photon from the state $(\mathbf{k}, \hat{e})$. Other matrix elements of these factors correspond to emission and absorption of "virtual quanta" and contribute terms of the relative order of $10^{-3}$.

[51] We shall continue to follow the notation of Section 8.3, p. 462, except for returning to our convention by which we set $\hbar = 1$.

We may therefore replace the exact expression (227*b*) by the approximate expression

$$\langle \mathbf{k}_f, \hat{e}_f, b'|\, T\,|\mathbf{k}, \hat{e}, b\rangle = \sum_{a',a} \langle \mathbf{k}_f, \hat{e}_f, b'|\, V_r\,|a'\rangle\langle a'|\, G\,|a\rangle$$
$$\times \langle a|\, V_r\,|\mathbf{k}, \hat{e}, b\rangle + \sum_{a',a} \langle \mathbf{k}_f, \hat{e}_f, b'|\, V_r\,|\mathbf{k}, \hat{e}, \mathbf{k}_f, \hat{e}_f, a'\rangle$$
$$\times \langle \mathbf{k}, \hat{e}, \mathbf{k}_f, \hat{e}_f, a'|\, G\,|\mathbf{k}, \hat{e}, \mathbf{k}_f, \hat{e}_f, a\rangle\langle \mathbf{k}, \hat{e}, \mathbf{k}_f, \hat{e}_f, a|\, V_r\,|\mathbf{k}, \hat{e}, b\rangle. \quad (229)$$

Here $a$ and $a'$ represent excited states of the atom. The scattering cross section is obtained by using the expression (229) in the general equation (3.142). Since the relative velocity of the photon and atom is now just $c$, the speed of light, we have

$$\sigma = \frac{(2\pi)^4}{c}\frac{1}{2\mu(b)}\sum_{m_b,m_{b'}}\sum_{\hat{e}_f,\hat{e}}\int d^3k_f\, d^3P_f$$
$$\times\, \delta(\mathbf{P}_f + \mathbf{k}_f - \mathbf{k})\, \delta(k_f c - kc)\, |\langle \mathbf{k}_f, \hat{e}_f, b'|\, T\,|\mathbf{k}, \hat{e}, b\rangle|^2. \quad (230)$$

Here $\mathbf{P}_f$ is the final recoil momentum of the atom (which was at rest initially) and the notation of Section 8.3, p. 463 has been used for the substates $m_b$ and $m_{b'}$ describing spin orientation. The factor $[2\mu(b)]^{-1}$ arises in averaging over spin and polarization directions in the initial state.

If we neglect $V_r$ in the Green's functions $G$ in Eq. (229), these are diagonal in the atomic states $a$ and have the values

$$\langle a|\, G\,|a\rangle = \frac{1}{kc + W_b - W_a},$$
$$\langle \mathbf{k}, \hat{e}, \mathbf{k}_f, \hat{e}_f, a|\, G\,|\mathbf{k}, \hat{e}, \mathbf{k}_f, \hat{e}_f, a\rangle = \frac{1}{W_b - W_a - k_f c}, \quad (231)$$

where $W_a$ is the energy of the state $a$. Now, since $W_a > W_b$, the second expression here is never singular and may be used as it is in Eq. (229). The first expression (231) is, however, singular when $kc = W_a - W_b$ for any discrete atomic state $a$. This, of course, corresponds just to the energy of a photon which would be emitted in the radiative transition from state $a$ to state $b$.

When $kc$ does not lie near the energy $W_a - W_b$, corresponding to some discrete state $a$, we may use the expressions (231) in Eq. (229). (As we have already mentioned, the error resulting from this approximation is only of the order of a tenth of a percent.) The evaluation is relatively straightforward and one obtains a cross section of the order of the Rayleigh cross section for elastic scattering[52] of long wavelength radiation.

$$\sigma_{\text{Ray}} = 8\pi\left(\frac{k}{\hbar}\right)^4 \alpha_p^{\ 2}, \tag{232}$$

where $\alpha_p$ is the atomic polarizability.[53]

When the energy of the incident photon is such that for some discrete state $a$

$$|kc + W_b - W_a| \cong \Gamma, \tag{233}$$

[see Eq. (187)] we must use the correct expression (63) for

$$\langle a|\, G\, |a\rangle = G_a = \frac{1}{kc + i\eta + W_b - W_a - R_a^{\ +}(E)}\,. \tag{234}$$

In this case $G_a$ is very large and we may write[54] Eq. (229) as

$$\langle \mathbf{k}_f, \hat{e}_f, b'|\, T\, |\mathbf{k}, \hat{e}, b\rangle$$
$$\cong \sum_{m_a} \langle \mathbf{k}_f, \hat{e}_f, b'|\, V_r\, |a\rangle \frac{1}{E + i\eta - W_a - R_a^{\ +}(E)} \langle a|\, V_r\, |\mathbf{k}, \hat{e}, b\rangle. \tag{235}$$

Here the sum extends only over the spin orientations of the virtual states $a$. Roughly speaking, we are justified in keeping only states of energy $W_a$ in (235) when this amplitude is large compared to that for the Rayleigh scattering.

Now,

$$R_a^{\ +}(E) \approx R_a^{\ +}(W_a) = D(E) - i\frac{\Gamma}{2} \cong -i\frac{\Gamma}{2}, \tag{236}$$

where $\Gamma$ is defined by Eq. (187) and the real level shift may be neglected for the reasons discussed in Section 8.3, p. 469. This permits us to write the cross section (230) as

$$\sigma_{\text{res}} = \frac{(2\pi)^4}{2c^2\mu(b)} \sum_{m_b, m_{b'}} \sum_{\hat{e}_f, \hat{e}} k_f^{\ 2} \int d\Omega_{\hat{\mathbf{k}}_f}$$
$$\times \left|\sum_{m_a} \langle \mathbf{k}_f, \hat{e}_f, b'|\, V_r\, |a\rangle\langle a|\, V_r\, |\mathbf{k}, \hat{e}, b\rangle\right|^2$$
$$\times \frac{1}{[kc - (W_a - W_b)]^2 + \Gamma^2/4}\,. \tag{237}$$

[52] See, for example, Heitler [10].

[53] The polarizability $\alpha_p$ has the dimensions of a volume and tends to be of the order of $a_0^{\ 3}$, where $a_0$ is the Bohr radius.

[54] We here follow a treatment of this problem given by Brueckner and Watson [27].

Using arguments of rotational invariance, we may remove the sum over $m_a$ from the squared quantity. That is,

$$\sum_{m_b, m_b'} \sum_{\hat{e}_f, \hat{e}} \int d\Omega_{\hat{\mathbf{k}}_f} \left| \sum_{m_a} \langle \mathbf{k}_f, \hat{e}_f, b' | V_r | a \rangle \langle a | V_r | \mathbf{k}, \hat{e}, b \rangle \right|^2$$
$$= \sum_{m_a} \sum_{m_b, m_b'} \sum_{\hat{e}_f, \hat{e}} \int d\Omega_{\hat{\mathbf{k}}_f} |\langle \mathbf{k}_f, \hat{e}, b' | V_r | a \rangle|^2 \, |\langle \mathbf{k}, \hat{e}, b | V_r | a \rangle|^2. \quad (238)$$

Finally, we may use the expression (176) to set

$$\sum_{\hat{e}_f} \sum_{m_b'} 2\pi \frac{k_f^2}{c} \int d\Omega_{\hat{\mathbf{k}}_f} |\langle \mathbf{k}_f, \hat{e}_f, b' | V_r | a \rangle|^2 = \Gamma_{b'a} \quad (239a)$$

and

$$\sum_{\hat{e}} \sum_{m_a, m_b} |\langle \mathbf{k}, \hat{e}, b | V_r | a \rangle|^2 = \frac{\mu(a)\Gamma_{ba}}{(k^2/c)(2\pi)(4\pi)}. \quad (239b)$$

To see this, we note that the left side of Eq. (239*b*) is independent of $\hat{\mathbf{k}}$. The factor $(4\pi)^{-1}$ on the right then arises to compensate for the "missing" integral over the directions of $\mathbf{k}$. On combining Eqs. (237), (238), and (239), we are led to

$$\sigma_{\text{res}} = \frac{\pi}{k^2} \frac{\mu(a)}{2\mu(b)} \frac{\Gamma_{b'a}\Gamma_{ba}}{[kc - (W_a - W_b)]^2 + \Gamma^2/4}. \quad (240a)$$

If $\mu(a) = (2S_a + 1)$ and $\mu(b) = (2S_b + 1)$, we obtain [see Eq. (195)]

$$\sigma_{\text{res}} = \frac{\pi}{k^2} \frac{(2S_a + 1)}{2(2S_b + 1)} \frac{\Gamma_{ba}\Gamma_{b'a}}{[kc - (W_a - W_b)]^2 + \Gamma^2/4}. \quad (240b)$$

Finally, if we sum over all final states $b'$ to which the atom can radiate, the cross section is

$$\sigma_{\text{res}} = \frac{\pi}{k^2} \frac{(2S_a + 1)}{2(2S_b + 1)} \frac{\Gamma_{ba}\Gamma}{[kc - (W_a - W_b)]^2 + \Gamma^2/4}. \quad (240c)$$

This agrees precisely with Eq. (195).

Equation (240*b*) is a specific example of a "one level" resonance scattering cross section. Such resonances are often encountered in atomic, nuclear, and elementary particle physics. Our derivation, although for a simple example, is of considerable generality and may be applied to many such phenomena. Generally speaking, a resonance will occur in scattering when the energy of the initial state lies close to the energy of a long-lived virtual state of the composite system.[55] To see how this comes about, let us review the assumptions made in our derivation of Eq. (240*b*).

[55] A formal proof of this statement will be given in Section 8.6.

First, the assumption of a "long life" for the virtual state $a$ (which follows from the observation that $\Gamma \ll W_a$) permitted us to write Eq. (235) when the condition (233) is satisfied. This essential step and the general arguments leading to Eqs. (239) led immediately to the cross sections (240).

When the atom is left in an excited state $b'$ after emitting the photon of momentum $\mathbf{k}_f$, it will, of course, radiate again, the process ending only when the atom has reached its lowest state. A detailed description of the subsequent radiative transitions may be made easily following the method to be given in the next section [see also Section 8.2].

**Excitation of an Atom by Collision**

Let us now consider the preparation of an excited atomic (or molecular) state by impact with, say, an electron. As before, the atom will be considered to be very heavy and at rest before being struck. The electron will be assumed to have an initial momentum $\boldsymbol{\varkappa}$[56] and the atom to be initially in its ground state $b$. The electron is scattered into a momentum state $\boldsymbol{\varkappa}_f$, leaving the atom in an excited state $a$. This subsequently makes a radiative transition to a state $b'$, emitting a photon of momentum $\mathbf{k}$ and polarization $\hat{e}$. Such a process is illustrated in Fig. 8.5.

The Hamiltonian for the system is

$$H = K + V_r + V, \tag{241}$$

where $V_r$ is defined by Eq. (169), $V$ represents the interaction of the electron with the atom, and

$$K = h + K_r + T_e. \tag{242}$$

The notation here is that of Eq. (161), except that we have now neglected

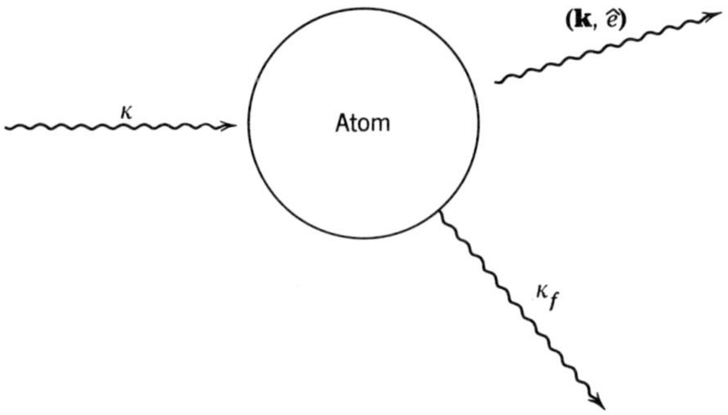


Fig. 8.5 Collisional excitation of an atom by an electron with the subsequent emission of a photon.

[56] For brevity of notation, we shall not write explicitly the label for the electron spin orientation.

the kinetic energy of the atom and have added the kinetic energy operator $T_e$ of the incident electron.

We shall in due course take proper account of wave function symmetrization for the electrons as described in Chapter 4. As implied before, the atomic wave functions are assumed to be correctly symmetrized from the outset.

In the barycentric subspace we have the scattering matrix

$$T = (V_r + V) + (V_r + V)\frac{1}{E + i\eta - H}(V_r + V), \tag{243}$$

where

$$E = \epsilon(\kappa) + W_b, \tag{244a}$$

and

$$\epsilon(\kappa) = \frac{\kappa^2}{2m} \tag{244b}$$

is the kinetic energy of the incident electron.

We recall once again that $V_r$ represents a weak interaction and might very well be neglected in Eq. (243) except for one reason: if we wait a long enough time, the atom will always radiate back to its ground state. To evaluate $T$ we need only find that particular occurrence of $V_r$ in (243) which effects this transition. Our physical intuition tells us that the scattering will be complete before the radiation is emitted, so the "last" interaction (on the left) must be $V_r$. That is, we may replace Eq. (243) by

$$T = V_r \frac{1}{E + i\eta - H} V, \tag{245}$$

the error resulting being that characteristic of radiative corrections ($\approx 0.1\%$). As we shall see in retrospect, only the term (245) among those in (243) has a nearly singular Green's function to compensate for the smallness of $V_r$. This is essentially equivalent to the condition by which we were led to Eq. (235).

To express this in more physical terms, the electron is expected to scatter and be a great distance from the atom long before the atom radiates.

Now, with the algebraic relation (3.77), we can write

$$\frac{1}{E + i\eta - H} = \frac{1}{E + i\eta - K - V_r}\left(1 + V\frac{1}{E + i\eta - H}\right). \tag{246}$$

Thus Eq. (245) is equivalent to

$$T = V_r \frac{1}{E + i\eta - K - V_r}\left(V + V\frac{1}{E + i\eta - H}V\right). \tag{247}$$

The weak interaction $V_r$ can contribute only small radiative corrections (of the relative order $10^{-3}$) to the electron-atom collision, so we may set

$$V + V\frac{1}{E + i\eta - H}V \cong V + V\frac{1}{E + i\eta - K - V}V \equiv T_{\text{sc}}. \quad (248)$$

We recognize $T_{\text{sc}}$ as being just the scattering matrix for electron-atom scattering, ignoring the coupling to the radiation field. Therefore

$$T = V_r\frac{1}{E + i\eta - K - V_r}T_{\text{sc}}. \quad (249)$$

As we have already supposed, let us imagine that the atom is excited to a state $a$ by the collision. For this process, we have the scattering matrix

$$\langle \boldsymbol{\varkappa}_f, a|\, T_{\text{sc}}\, |\boldsymbol{\varkappa}, b\rangle.$$

If we were to neglect the radiative interaction $V_r$ in the Green's function in (249), this would be

$$\frac{1}{E + i\eta - K} = [\epsilon(\kappa) + W_b + i\eta - \epsilon(\kappa_f) - W_a]^{-1}. \quad (250)$$

When we set $\eta = 0$, this is singular on the energy shell. This singularity is removed, however, if we make the substitution

$$\frac{1}{E + i\eta - K - V_r} \to G_a(E) = [\epsilon(\kappa) + i\eta + W_b - \epsilon(\kappa_f) - W_a - R_a^{+}(E)]^{-1}$$

$$\cong \left[\epsilon(\kappa) + W_b - \epsilon(\kappa_f) - W_a + i\frac{\Gamma}{2}\right]^{-1}. \quad (251)$$

The width $\Gamma$ here is recognized as being precisely the quantity defined by Eq. (187).

Using (251) in Eq. (249), we see that

$$T = \sum_{m_a} \frac{\langle \mathbf{k}, \hat{e}, b'|\, V_r\, |a\rangle\langle \boldsymbol{\varkappa}_f, a|\, T_{\text{sc}}\, |\boldsymbol{\varkappa}, b\rangle}{\epsilon(\kappa) + W_b - \epsilon(\kappa_f) - W_a + i(\Gamma/2)}, \quad (252)$$

where the sum over spin orientations of the atom has been explained in connection with Eq. (235). This expression must now be appropriately symmetrized since the incident electron is indistinguishable from that in the atom. As described in connection with Eq. (4.118), the symmetrized $T$-matrix is just

$$T_{\mathscr{S}} = \sum_{m_a} \frac{\langle \mathbf{k}, \hat{e}, b'|\, V_r\, |a\rangle\langle \mathscr{S}(\boldsymbol{\varkappa}_f, a)|\, T_{\text{sc}}\, |\mathscr{S}(\boldsymbol{\varkappa}, b)\rangle}{\epsilon(\kappa) + W_b - \epsilon(\kappa_f) - W_a + i(\Gamma/2)}. \quad (253)$$

Here

$$\langle \mathscr{S}(\boldsymbol{\varkappa}_f, a)|\, T_{\text{sc}}\, |\mathscr{S}(\boldsymbol{\varkappa}, b)\rangle$$

is the symmetrized scattering matrix for electron-atom scattering, written in the notation of Eq. (4.118). (The matrix elements of $V_r$ are, of course, unaffected by the symmetrization since we have assumed that the atomic states were symmetrized from the outset.)

It is appropriate at this point to review our derivation of Eq. (253). In obtaining the expression (245) from (243), we anticipated the singular denominator (250), which makes (245) the dominant term in the scattering matrix. Even the nonsingular Green's function (251) is very large since we have seen that $\dfrac{\Gamma}{|W_b|} \approx 10^{-6}$. Had we been tempted to keep matrix elements of $(\epsilon + i\eta - K - V_r)^{-1}$ in Eq. (249) which connect different atomic states, we would again have found nonsingular denominators and thus only radiative corrections (of the order of 0.1%). We conclude therefore that Eq. (253) becomes exact in the limit that $\Gamma$ is very small, or that the virtual state $a$ has a very long lifetime.

If $\mathbf{P}_f$ is the recoil momentum of the atom, the scattering cross section is (here $v = \kappa/m$ is the velocity of the incident electron and the remaining notation is that of the preceding section)

$$\begin{aligned} d\sigma &= \frac{(2\pi)^4}{v\mu(b)} \sum_{\hat{e}} \sum_{m_b, m_{b'}} \int d^3k\, d^3\kappa_f\, d^3P_f \\ &\quad \times \delta(\mathbf{P}_f + \mathbf{k} + \boldsymbol{\varkappa}_f - \boldsymbol{\varkappa})\, \delta[\epsilon(\kappa_f) + kc - \epsilon(\kappa) + W_{b'} - W_b]\, |T_{\mathscr{S}}|^2 \\ &= \frac{(2\pi)^4}{v} \frac{\kappa_f m}{\mu(b)} d\Omega_{\hat{\varkappa}_f} k^2\, dk \int d\Omega_{\hat{\varkappa}} \sum_{\hat{e}} \sum_{m_b, m_{b'}, m_a} \\ &\quad \times \frac{|\langle \mathbf{k}, \hat{e}, b' |\, V_r\, | a \rangle|^2}{[kc - (W_a - W_{b'})]^2 + (\Gamma^2/4)} |\langle \mathscr{S}(\boldsymbol{\varkappa}_f, a)|\, T_{\mathrm{SC}}\, |\mathscr{S}(\boldsymbol{\varkappa}, b)\rangle|^2, \end{aligned} \tag{254}$$

with $\kappa_f$ now defined by the condition of energy conservation (that is, the vanishing of the $\delta$-function on energy above). Now,

$$d\sigma_{\mathrm{SC}} = \frac{(2\pi)^4}{v} \sum_{m_a, m_b} \frac{1}{\mu(b)} \kappa_f m\, d\Omega_{\hat{\varkappa}_f} |\langle \mathscr{S}(\boldsymbol{\varkappa}_f, a)|\, T_{\mathrm{SC}}\, |\mathscr{S}(\boldsymbol{\varkappa}, b)\rangle|^2 \tag{255}$$

is just the scattering cross section (ignoring radiation) for exciting the atomic state $a$. Similarly, by Eq. (176),

$$k^2 \int d\Omega_{\hat{\mathbf{k}}} \sum_{\hat{e}} \sum_{m_{b'}} |\langle \mathbf{k}, \hat{e}, b' |\, V_r\, | a \rangle|^2 = \frac{\Gamma_{b'a}}{2\pi} c. \tag{256}$$

Therefore Eq. (254) simplifies to

$$d^2\sigma(b') = \left[ \frac{(\Gamma_{b'a}/2\pi) d(kc)}{[kc - (W_a - W_{b'})]^2 + \Gamma^2/4} \right] d\sigma_{\mathrm{SC}} = \frac{\Gamma_{b'a}}{\Gamma} [b_{b'a}(\nu)\, d\nu]\, d\sigma_{\mathrm{SC}}. \tag{257}$$

Here $b_{b'a}(\nu)$ is the natural line width, as defined by Eqs. (186).

Since

$$\int b_{b'a}(\nu)\, d\nu = 1,$$

the cross section (257) integrated over all frequencies is

$$d\sigma(b') = \frac{\Gamma_{b'a}}{\Gamma}\, d\sigma_{\text{sc}}. \tag{258}$$

If this is summed over all states $b'$ to which the atom can radiate, Eq. (258) gives

$$d\sigma = \sum_{b'} d\sigma(b') = d\sigma_{\text{sc}}, \tag{259}$$

using Eq. (187).

Equation (259) has a clear physical significance. The cross section for excitation followed by radiative decay to any state is the same as the excitation cross section, since the atom will *always* eventually radiate back to a lower state.

It is instructive to give a symbolic derivation of the result expressed by Eq. (259). The total cross section $\sigma$ is just

$$\sigma = \frac{(2\pi)^4}{v} \langle \boldsymbol{\varkappa}, b|\, T^\dagger\, \delta(E - K) T\, |\boldsymbol{\varkappa}, b\rangle, \tag{260}$$

where the operators are defined in the barycentric subspace. Now, by Eq. (252),

$$T = V_r \frac{1}{E - K + i(\Gamma/2)}\, T_{\text{sc}},$$

so

$$\sigma = \frac{(2\pi)^4}{v} \langle \boldsymbol{\varkappa}, b|\, T_{\text{sc}}^\dagger \frac{1}{E - K - i(\Gamma/2)}$$
$$\times\, V_r\, \delta(E - K) V_r \frac{1}{E - K + i(\Gamma/2)}\, T_{\text{sc}}\, |\boldsymbol{\varkappa}, b\rangle. \tag{261}$$

In the spirit of our preceding discussion, we shall suppose that excitation only to the atomic state $a$ is important (although this may have degeneracy with respect to spin orientation). Then we may factor out of Eq. (261) the expression

$$\langle a|\, V_r\, \delta(E - K) V_r\, |a\rangle = \frac{\Gamma}{2\pi}, \tag{262}$$

which is just Eqs. (176) and (187) in an abbreviated notation. This permits us to write Eq. (261) as

$$\sigma = \frac{(2\pi)^4}{v} \langle \boldsymbol{\varkappa}, b|\, T^{\dagger}_{\mathrm{sc}} \frac{\Gamma/2\pi}{(E - K)^2 + \Gamma^2/4} T_{\mathrm{sc}} \,|\boldsymbol{\varkappa}, b\rangle$$

$$= \frac{(2\pi)^4}{v} \langle \boldsymbol{\varkappa}, b|\, T^{\dagger}_{\mathrm{sc}}\, \delta(E - K) T_{\mathrm{sc}} \,|\boldsymbol{\varkappa}, b\rangle = \sigma_{\mathrm{sc}}. \tag{263}$$

We have here used Eq. (5.28) to set

$$\frac{\Gamma}{2\pi} \frac{1}{(E - K)^2 + \Gamma^2/4} = \delta(E - K),$$

since $\Gamma$ is considered to be very small compared to $E$.

Equation (263) implies, as did Eq. (259), that the total scattering cross section is unaffected by the delayed decay of states excited by the scattering.

**An Alternate Discussion of Radiation Damping**

The term "radiation damping" is often applied to the decay of an initial state. That is, as the system radiates outgoing waves into its decay channels, the original state becomes depleted—and this, of course, leads to a self-damping of the radiative process. We have seen that it is the imaginary part of the operator $R_a$ which determines the damping, which occurs in a manner similar to that of the decay of electrical current in a tuned circuit having a small amount of resistance.

It is intuitively helpful to think of the operator $(E + i\eta - K - V_r)^{-1}$ in Eqs. (227*b*) and (249) as describing the propagation of a wave which is decaying exponentially (in space or time, as the case may be). This is to be contrasted with the propagator

$$e^{-1} \equiv (E + i\eta - K)^{-1} \tag{264}$$

which describes a freely propagating wave. A useful relation between these operators is established by the algebraic identity

$$\frac{1}{E + i\eta - K - V_r} = \frac{1}{e - V_r}$$

$$= \frac{1}{e - V_r(1/e)V_r} + \frac{1}{e} V_r \frac{1}{e - V_r(1/e)V_r}. \tag{265}$$

This expression may be easily derived by expanding both sides of the equation in powers of $V_r$ or, more elegantly, by showing that both sides satisfy the same integral equations.[57] Now, by Eq. (3.63),

$$V_r \frac{1}{e} V_r = V_r \frac{P}{E - K} V_r - i\pi V_r \, \delta(E - K) V_r. \tag{266}$$

The first term here corresponds to the level shift $D(\epsilon)$, whereas the second corresponds to $-i(\Gamma/2)$ as seen from Eq. (262). We thus obtain immediately the approximate relation

$$\frac{1}{E + i\eta - K - V_r} \cong \frac{1}{E - K + i(\Gamma/2)}, \tag{267}$$

if the real level shift $D$ and the second term in Eq. (265) are neglected.

As was observed earlier, the occurrence of $i(\Gamma/2)$ here implies the damping of a propagating wave. This phenomenon is encountered in many contexts. Perhaps the most common is that of the multiple scattering of particles by an extended, absorbing medium. For example, this might represent the scattering of neutrons in a gas or the scattering of pi-mesons by atomic nuclei. In describing such phenomena, we encounter the operator $R_a$ in the guise of a pseudo-potential, frequently called the "optical model potential." We shall return to this in Chapter 11 where scattering by composite systems will be discussed.

## 8.5 GENERAL THEORY OF THE LIFETIME OF SCATTERING STATES

It has been observed by Wigner [29] that the lifetime of the composite system in a scattering event may be given a precise and general definition. This concept has been generalized by Smith [30], who has introduced a "lifetime matrix." In this section we shall first discuss the lifetime by means of the Schrödinger equation. This will permit us to demonstrate the relation between the Wigner lifetime and the decay period of a prepared state as discussed in Section 8.2. We shall next obtain the lifetime of a scattering state directly from the asymptotic wave functions, following Wigner's original treatment. This description has the advantage of defining the lifetime independently of the assumption that a Schrödinger equation describes the dynamics of the collision. Obtained in this manner, we shall see that the notion of a lifetime provides a kind of dynamical description directly in terms of the $S$-matrix.

### Definition of the Lifetime for a Scattering Interaction

Let us first construct an initial wave packet state for two particles which are to collide as was done in Chapter 3. The scattering will be supposed to

[57] Equation (265) provides an example of a number of such relations derived in references (28).

occur in the barycentric coordinate system and the variables describing the motion of the center-of-mass will be removed. That is, we shall describe the interaction in the barycentric subspace as introduced in Chapter 4.

We shall now use the energy representation introduced in Eq. (7.6). The eigenfunctions for the noninteracting particles will be written as $\chi_{s,\epsilon}$; these will be assumed to satisfy the Schrödinger equation

$$K\chi_{s,\epsilon} = \epsilon\chi_{s,\epsilon}. \tag{268}$$

That is, the energy $\epsilon$ is chosen as one of the quantum variables labeling the state, the remaining state labels being written as $s$. We shall suppose that the operators whose eigevalues are the quantities $s$ commute with the complete Hamiltonian $H = K + V$. Then the scattering eigenfunctions may be written as

$$H\psi_{s,\epsilon}^{+} = \epsilon\psi_{s,\epsilon}^{+}. \tag{269}$$

The initial state wave packet is written as

$$\phi_s(t) = \sum_{\epsilon} A_\epsilon \chi_{s,\epsilon} e^{-i\epsilon t}, \tag{270a}$$

where $A_\epsilon$ is a suitable wave packet amplitude. The wave function $\varphi_s$ is normalized to unity, so

$$(\varphi_s, \varphi_s) = 1 \tag{270b}$$

and

$$\sum_{\epsilon} |A_\epsilon|^2 = 1. \tag{270c}$$

The time-dependent wave function for the system $\Psi_s(t)$ is chosen to coincide with $\varphi_s(t)$ at the time $t = -T$ at which the experiment is begun. According to Eq. (3.68), this is

$$\Psi_s(t) = \sum_{\epsilon} A_\epsilon \psi_{s,\epsilon} e^{-i\epsilon t} \tag{271a}$$

and is normalized to

$$(\Psi_s(t), \Psi_s(t)) = 1. \tag{271b}$$

Finally, we shall make the special assumption that the representation $s$ has been (and can be) so chosen that in it the $S$-matrix is diagonal. Then clearly our experiment is one in which the incident wave is a spherical packet collapsing on the point at which the interaction is to occur. The scattered wave will be an outgoing spherical packet expanding away from the local region of interaction. This is illustrated in Fig. 8.6.

Consider now a large volume $S_R$ in the barycentric configuration space. This volume is characterized by a radius parameter $\mathscr{R}$ and is sufficiently large that the particles do not interact when lying outside $S_R$. We also suppose that $T$ is so large that the wave packets lie outside $S_R$ with

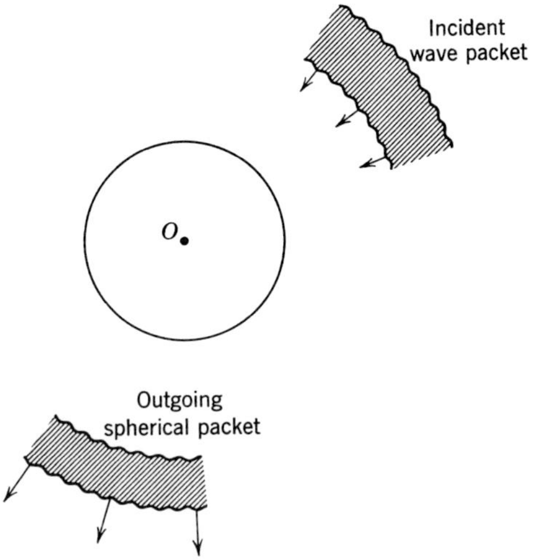


Fig. 8.6 Scattering of a spherical wave packet.

arbitrary precision when $t < -T$ or $t > T$. Then the lifetime $Q$ of the scattering state is defined by the equation

$$Q = \lim_{\mathscr{R}\to\infty} \lim_{T\to\infty} \int_{-T}^{T} dt \int_{S_R} d\tau[\Psi_s^*(t)\Psi_s(t) - \varphi_s^*(t)\varphi_s(t)]. \qquad (272)$$

Here $d\tau$ is a volume element in the configuration space and the integral is confined to the domain $S_R$. The limit $\mathscr{R} \to \infty$ is understood to imply that in the limit the integral extends over the entire configuration space.

Equation (272) has a simple physical interpretation. The integral

$$I \equiv \int_{S_R} d\tau \Psi_s^*(t)\Psi_s(t)$$

is unity when the wave packets are entirely contained within $S_R$ and it vanishes when the wave packets lie outside $S_R$. Therefore

$$\int_{-\infty}^{+\infty} I\, dt$$

represents the total time during which the particles lie within $S_R$. Now, in Eq. (272) we consider the difference in the times spent in $S_R$ by the interacting and the noninteracting particles. As long as $S_R$ is large compared to the domain in which the particles interact, this difference should be independent of the actual size of $S_R$. The limit $\mathscr{R} \to \infty$ is seen then to be irrelevant, but mathematically convenient.[58]

[58] Had we not chosen the $s$-representation in such a manner that the $S$-matrix is diagonal, the limit would not in general exist.

We see that $Q$ represents the additional time, over and above the "free flight time," that the particles spend in the range of their interaction. If $Q > 0$, the particles tend to "stick together," whereas if $Q < 0$, the interaction tends to force them apart.

Let us now integrate Eq. (272) over time. Using the wave packet expansions (270$a$) and (271$a$), we have

$$\begin{aligned} Q &= \lim_{\mathscr{R}\to\infty} 2\pi \left\{ \sum_{\epsilon',\epsilon} A_{\epsilon'}{}^* A_\epsilon \int_{S_R} d\tau \right. \\ &\qquad \left. \times \left[\psi^{+*}_{s,\epsilon'}\, \delta(\epsilon' - \epsilon)\psi^+_{s,\epsilon} - \chi^*_{s,\epsilon'}\, \delta(\epsilon' - \epsilon)\chi_{s,\epsilon}\right] \right\} \\ &= \lim_{\mathscr{R}\to\infty} 2\pi \left\{ \sum_{\epsilon} |A_\epsilon|^2 \int_{S_R} d\tau \right. \\ &\qquad \left. \times \sum_{\epsilon'} \left[\psi^{+*}_{s,\epsilon'}\, \delta(\epsilon - H)\psi^+_{s,\epsilon'} - \chi_{s,\epsilon'}\, \delta(\epsilon - K)\chi_{s,\epsilon'}\right] \right\}. \end{aligned} \tag{273}$$

According to Eq. (3.63), we have the relations

$$\begin{aligned} \delta(\epsilon - K) &= \frac{i}{2\pi}\,[G_0{}^+(\epsilon) - G_0{}^-(\epsilon)], \\ \delta(\epsilon - H) &= \frac{i}{2\pi}\,[G^+(\epsilon) - G^-(\epsilon)], \end{aligned} \tag{274}$$

where, as previously, we have set

$$\begin{aligned} G_0{}^\pm(\epsilon) &\equiv \frac{1}{\epsilon \pm i\eta - K}, \\ G^\pm(\epsilon) &\equiv \frac{1}{\epsilon \pm i\eta - H}. \end{aligned} \tag{275}$$

If the relations (274) are introduced into Eq. (273) and the limit is taken, there results

$$\begin{aligned} Q &= i \sum_\epsilon |A_\epsilon|^2 \sum_{\epsilon'} \langle s, \epsilon'|\, [G^+(\epsilon) - G^-(\epsilon) - G_0{}^+(\epsilon) + G_0{}^-(\epsilon)]\, |s, \epsilon'\rangle \\ &= i \sum_\epsilon |A_\epsilon|^2\, \mathrm{Tr}\{G^+(\epsilon) - G^-(\epsilon) - G_0{}^+(\epsilon) + G_0{}^-(\epsilon)\}. \end{aligned} \tag{276}$$

By the symbol $\mathrm{Tr}\{\cdots\}$ here we mean only a partial trace since the variables $s$ are held constant and the sum runs over only the energy $\epsilon'$.

Evaluation of the trace in Eq. (276) evidently requires some care since the individual terms diverge. This is a purely mathematical problem independent of the context in which we are using it and is most easily accomplished by first quantizing the system in a box and then passing to the limit of an infinite volume. Let us then write $E_\lambda$ for the discrete eigenvalues of $H$ (we suppress the index $s$ since this is held constant in our problem). Then, for example, a typical term in Eq. (276) is

$$\begin{aligned}\operatorname{Tr} G^+(\epsilon) &= \sum_\lambda \frac{1}{\epsilon + i\eta - E_\lambda} \\ &= \frac{d}{d\epsilon} \ln\left[\prod_\lambda (\epsilon + i\eta - E_\lambda)\right] \\ &= \frac{d}{d\epsilon} \ln \det\,[G^+(\epsilon)]^{-1} = -\frac{d}{d\epsilon} \ln \det G^+(\epsilon).\end{aligned} \tag{277}$$

Here det $G^+$ means the determinant of the matrix $G^+(\epsilon)$. It is evident, of course, that for the convergence of these expressions the four traces must be combined as implied in Eq. (276).

If we combine the four traces, each having the form implied by Eq. (277), we may simplify Eq. (276) to the form

$$Q = i \sum_\epsilon |A_\epsilon|^2 \frac{d}{d\epsilon} \left\{\ln \det \left[\frac{G^-(\epsilon)G_0^+(\epsilon)}{G^+(\epsilon)G_0^-(\epsilon)}\right]\right\}. \tag{278}$$

The determinant appearing here has been evaluated by DeWitt [31] whose analysis we shall follow. Goldberger [32] has shown that the expression on the right of Eq. (278) appears in the evaluation of the second virial coefficient in statistical mechanics. This is very plausible since the partition function is essentially an integral over allowed regions of phase space, weighted by the time spent in each region (that is, by the lifetime $Q$ in the first approximation beyond that of an ideal gas).

We may use Eqs. (275) to write

$$\begin{aligned} G_0^+[G^+]^{-1} &= 1 - G_0^+V, \\ G_0^-[G^-]^{-1} &= 1 - G_0^-V, \end{aligned} \tag{279}$$

where $V \equiv H - K$, as usual. Then

$$\begin{aligned} D_G &\equiv \det\left[\frac{G_0^-(G^-)^{-1}}{G_0^+(G^+)^{-1}}\right] \\ &= \det\left[(1 - G_0^-V)\frac{1}{1 - G_0^+V}\right]. \end{aligned} \tag{280}$$

Now,

$$\frac{1}{1 - G_0^+V} = 1 + G_0^+V\frac{1}{1 - G_0^+V} = 1 + G_0^+T, \tag{281}$$

where

$$T(\epsilon) = V\frac{1}{1 - G_0^+(\epsilon)V}$$

is the scattering matrix. This is easily verified since the integral equation (5.177) for the $T$-matrix is obtained if we insert here the expression (281) for $(1 - G_0{}^+V)^{-1}$.

Equation (281) permits us to write (280) as

$$D_G = \det\left[1 + (G_0{}^+ - G_0{}^-)V\frac{1}{1 - G_0{}^+V}\right]$$
$$= \det\left[1 - 2\pi i\,\delta(\epsilon - K)T(\epsilon)\right], \tag{282a}$$

where the first of Eqs. (274) has been used. Matrix labels inserted in Eq. (282$a$) make this appear in a more specific form as

$$D_G = \det\left[\delta_{\epsilon'',\epsilon'} - 2\pi i\,\delta_{\epsilon'',\epsilon}T_{\epsilon'',\epsilon'}(\epsilon)\right]. \tag{282b}$$

Now, the matrix from which the determinant is formed vanishes except along the principal diagonal and along the row $\epsilon'' = \epsilon$. The value of the determinant is therefore

$$D_G = 1 - 2\pi i T_{\epsilon\epsilon}(\epsilon) = S(\epsilon), \tag{283}$$

where $S(\epsilon)$ is the eigenvalue of the $S$-matrix for the state $(s, \epsilon)$.[59,60]

The value (283) for the expression (280) permits us to write Eq. (278) as

$$Q = -\,i\sum_\epsilon |A_\epsilon|^2\frac{d}{d\epsilon}\ln S(\epsilon). \tag{284}$$

For a wave packet of sufficiently small energy spread we may remove $(d/d\epsilon)\ln S(\epsilon)$ from the summand and use the condition (270$c$) to write

$$Q(\epsilon) = -\,i\frac{d}{d\epsilon}\ln S(\epsilon). \tag{285a}$$

This is the "lifetime" of a narrow wave packet of mean energy $\epsilon$ (it being sufficiently narrow that its time of passage across the region of interaction is much larger than $Q$).

Equation (285$a$) may be written in several alternate forms. First, we take $S = e^{2i\delta(\epsilon)}$, where $\delta(\epsilon)$ is a phase shift. Then

$$Q = 2\frac{d\,\delta(\epsilon)}{d\epsilon}, \tag{285b}$$

[59] We repeat again that since the labels $s$ have been held constant, $s$ appears only parametrically in our equations.

[60] Compare Eq. (283) with Eqs. (5.63) and (5.66), or Eq. (6.57) for the $S$-matrix elements. The density function $\rho_\epsilon$ in Eq. (6.57) is now unity since our states are labeled with $\epsilon$ as a quantum number, as described in connection with Eq. (7.115$b$).

which is Wigner's original result [29]. Again, we may rewrite Eq. (285*a*) as

$$Q = -\, iS^{-1}\frac{d}{d\epsilon}S. \tag{285c}$$

This takes a more appealing form if we set

$$-\, i\frac{d}{d\epsilon} \equiv \mathcal{t} \tag{286}$$

a kind of "time operator." Then

$$Q = S^{-1}\mathcal{t}S, \tag{285d}$$

valid in any representation for which the transformation matrix commutes with $\mathcal{t}$.

Our derivation of Eqs. (285) is not completely general. We have assumed that the eigenfunctions $\chi_{s,\epsilon}$, which define a representation in which the $S$-matrix is diagonal, are eigenfunctions of an Hamiltonian $K$ [Eq. (268)]. We therefore have not included cases which require a different Hamiltonian $K$ for each channel. We have also not included many particle channels for which the labels $s$ are not discrete. A more general derivation of $Q$ will be given on p. 493, however.

As a simple application of Eq. (285*b*), let us rederive Wigner's bound (6.322*b*) on the derivative of the phase shift for scattering by a force which vanishes outside a range $d$. If the particles did not interact, the total time spent within the distance $d$ of each other would be $[2d + 1/\kappa]1/v$, where $1/\kappa$ is their wavelength and $v$ is their relative velocity. When they interact, the total time spent within the range $d$ is

$$\Delta t = Q + \frac{1}{v}\left(2d + \frac{1}{\kappa}\right) \geq 0,$$

or

$$\frac{d\,\delta(\epsilon)}{d\epsilon} \geq -\frac{1}{v}\left(d + \frac{1}{2\kappa}\right), \tag{287}$$

which is equivalent to Eq. (6.322*b*).

### Relation to the Decay of a Prepared State

Let us now establish the relation between the decay of a prepared state, as discussed in Section 8.2, and the time delay $Q$ for a scattering reaction. To do this, we return to the scattering described by the wave function (271*a*). It is convenient to suppose that at the time $t = 0$ the systems which are to scatter are both within the large volume $S_R$, but that the scattering has not yet occurred.

First, we construct the wave packet

$$X_{S_R} = \sum_{\epsilon} B_\epsilon \chi_{s,\epsilon} \tag{288}$$

from the eigenvectors (268). Here the amplitude $B_\epsilon$ is so chosen that

$$\begin{aligned} \Lambda_{S_R} \equiv X_{S_R} X_{S_R}^{\dagger} &= 1 \qquad \text{for systems in } S_R, \\ &= 0 \qquad \text{for systems not in } S_R. \end{aligned} \tag{289}$$

Then [see Eq. (271*a*)]

$$\mathscr{A}_{S_R} = (X_{S_R}, \Psi_s(t)) \tag{290}$$

is the amplitude for the probability

$$P_{S_R}(t) = |\mathscr{A}_{S_R}|^2 \tag{291}$$

of finding the system in $S_R$. This, however, is just the probability (47) that the initial condition that the system is in $S_R$ has not "decayed." The time spent in $S_R$ is then

$$t_{S_R} = \int_0^\infty P_{S_R}(t)\, dt. \tag{292}$$

If the colliding systems did not interact, the time spent in $S_R$ would be the "free flight time"

$$t_{S_R}{}^0 = \int_0^\infty P_{S_R}{}^0(t)\, dt, \tag{293}$$

where [see Eq. (270*a*)]

$$P_{S_R}{}^0 = |(X_{S_R}, \varphi_{S_R}(t))|^2. \tag{294}$$

The difference between $t_{S_R}$ and $t_{S_R}{}^0$ is just the time delay $Q$.

$$\begin{aligned} Q &= \int_0^\infty dt\, [P_{S_R}(t) - P_{S_R}{}^0(t)] \\ &= \int_0^\infty dt \int_{S_R} d\tau\, (\Psi_s{}^*\Psi_s - \varphi_s{}^*\varphi_s), \end{aligned} \tag{295}$$

where the last step follows from Eq. (289). Since $\Psi_s(t) = \varphi_s(t)$ until the collision begins, the integral over time in Eq. (295) is independent of its lower limit, which may be replaced by $(-\infty)$. We see therefore that Eq. (295) is precisely equivalent to Eq. (272), which defines Wigner's time delay.

With this observation, we have established a quantitative relation between the decay rate of a prepared unstable state and the lifetime of a virtual state formed by colliding systems.

### A More General Derivation of the Lifetime of an Interacting System

Let us consider the most general conditions under which the scattering wave function $\psi_{s,\epsilon}^+$ will have the asymptotic form

$$\psi_{s,\epsilon}^+ = [\mathscr{I}_{s,\epsilon} - S_s(\epsilon)\mathscr{O}_{s,\epsilon}] \tag{296}$$

when the interacting systems are separated at great distances. Here $\mathscr{I}_{s,\epsilon}$ represents an incoming wave and $\mathscr{O}_{s,\epsilon}$ an outgoing wave [see Eq. (7.139)].[61] As before, we assume that the $S$-matrix is diagonal in the $(s, \epsilon)$ representation. We do not necessarily suppose that $\psi^{+}_{s,\epsilon}$ satisfies a Schrödinger equation, however.

We next construct a wave packet of the states (296) to describe a scattering experiment. The wave packet amplitude $A(\epsilon - \bar{\epsilon})$ will be considered to be centered about a mean energy $\bar{\epsilon}$ and to extend over only a narrow range of energy. Then the time-dependent wave function is

$$\Psi_s(t) = \sum_{\epsilon} A(\epsilon - \bar{\epsilon})\psi^{+}_{s,\epsilon} e^{-i\epsilon t}. \tag{297}$$

The asymptotic, outgoing wave part of this is obtained from Eq. (296) as

$$\Psi_{\text{out}}(t) = -\sum_{\epsilon} A(\epsilon - \bar{\epsilon})S(\epsilon)\mathscr{O}_{s,\epsilon} e^{-i\epsilon t}. \tag{298}$$

(Since the state label $s$ is held constant, we shall again omit this when writing $Q$ and $S$.)

We have supposed the wave packet to be very narrow. This permits us to write

$$\begin{aligned} S(\epsilon) = e^{\ln S(\epsilon)} &= \exp\left[\ln S(\bar{\epsilon}) + \Delta\epsilon \frac{d}{d\bar{\epsilon}} \ln S(\bar{\epsilon})\right] \\ &= S(\bar{\epsilon}) \exp\left[\Delta\epsilon \frac{d}{d\bar{\epsilon}} \ln S(\bar{\epsilon})\right], \end{aligned} \tag{299}$$

where

$$\Delta\epsilon \equiv \epsilon - \bar{\epsilon}$$

in Eq. (298). We thus have

$$\Psi_{\text{out}}(t) = -S(\bar{\epsilon})e^{-i\bar{\epsilon}t} \sum_{\Delta\epsilon} A(\Delta\epsilon)\mathscr{O}_{s,\bar{\epsilon}+\Delta\epsilon} \exp\left\{-i\left[t + i\frac{d}{d\bar{\epsilon}} \ln S(\bar{\epsilon})\right]\right\}\Delta\epsilon. \tag{300}$$

It is seen then that the interaction between the particles (recalling that $S = 1$ for noninteracting systems) leads to a time delay

$$Q = -i\frac{d}{d\bar{\epsilon}} \ln S(\bar{\epsilon}) \tag{301a}$$

in the appearance of the asymptotic wave packet. This general conclusion[62] agrees with Eq. (285$a$), which was obtained under more restrictive assumptions.

[61] We assume here that the wave functions describing the internal states of the system are contained within $\mathscr{I}_{s,\epsilon}$ and $\mathscr{O}_{s,\epsilon}$.

[62] The derivation just given follows the spirit of Wigner's original evaluation of $Q$ (see Refs. [29]).

As has been shown by Froissart, Goldberger, and Watson [33], it is also possible to define the time delay for scattering of a plane wave. For a wave packet amplitude $A(\boldsymbol{\varkappa} - \mathbf{p})$ the wave function at points $\mathbf{x}$ far from the scatterer is

$$\Psi(t) = \int d^3\kappa A(\boldsymbol{\varkappa} - \mathbf{p})\left[e^{i\boldsymbol{\varkappa}\cdot\mathbf{x}} + \frac{e^{i\kappa x}}{x} f(\kappa, \hat{\mathbf{x}}\cdot\hat{\boldsymbol{\varkappa}})\right]e^{-i\epsilon_\kappa t}$$

where $\epsilon_\kappa$ is the energy of the scattered particle when its momentum is $\boldsymbol{\varkappa}$. When the amplitude $A(\boldsymbol{\varkappa} - \mathbf{p})$ is sharply peaked near $\boldsymbol{\varkappa} = \mathbf{p}$, the scattering amplitude $f$ may be expanded as in Eq. (299). In this case the time delay is found to be

$$Q = \frac{\partial}{\partial \epsilon_p} \operatorname{Arg} f(p, \hat{\mathbf{x}}\cdot\hat{\mathbf{p}}). \tag{301b}$$

The expression (301$b$) may evidently be generalized to apply in the case of multichannel scattering.

**Dynamical Implications of the Time Delay**

If, independently of the Schrödinger equation, a set of "rules" were available for evaluating the $S$-matrix for physical systems (and if this were extended to give bound state energies),[63] one would be able to calculate all properties of all scattering processes. By themselves such a set of "rules" would be inadequate, however, since they would not describe the evolution in time of physical processes.

The Schrödinger equation (1.20$a$) provides just such a dynamical description of physical systems.[64] This equation prescribes the change in the state vector of a system for arbitrarily short intervals of time. A question naturally arises, however, as to whether physical experience requires a knowledge of the variation of a system over infinitesimal time intervals. Indeed, it may very well be asked to what extent Wigner's time delay $Q$, which gives the time lapse between asymptotic states, provides an adequate dynamical description of physical systems, assuming that the $S$-matrix may be obtained by some means.

Such questions cannot be answered at present. We shall show, however, that Wigner's time delay $Q$ permits us to obtain a kind of "coarse-grained" Schrödinger equation expressed entirely in terms of the $S$-matrix. We shall also see that this equation is adequate for obtaining classical dynamics.

Let us begin by considering a particle which undergoes a sequence of scattering interactions, as illustrated in Fig. 8.7. These scatterings are

[63] In this connection, we recall the problems raised in Chapter 6, p. 279.

[64] In this section we shall apply the term "Schrödinger equation" only to the time-dependent Eq. (1.20).

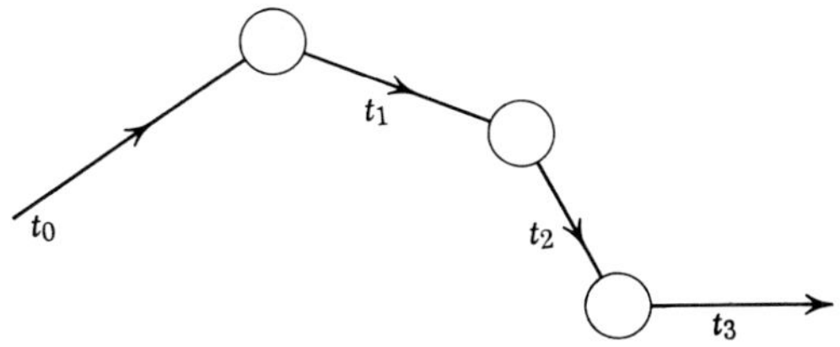


Fig. 8.7 Illustration of a sequence of scatterings.

supposed to be sufficiently well separated in space and time that each interaction is described by an $S$-matrix. A sequence of times $t_0, t_1, t_2 \cdots$ is chosen such that $t_0$ precedes the first scattering, which is completed by $t_1$; $t_1$ precedes the second scattering, which is completed by $t_2$; etc. Following our conventional notation, we write the Hamiltonian for the particle as $H = K + V$, where $K$ represents its kinetic energy and $V$ its interaction energy.

Now at the time $t_{n-1}$ we suppose the wave function of the particle to be $\Psi_{n-1}$. Then at the time $t_n$ its wave function will be [see Eq. (1.87$a$)]

$$\Psi_n = e^{-iK(t_n - t_{n-1})} U(t_n, t_{n-1})\Psi_{n-1}. \tag{302}$$

This provides a recurrence relation for the wave function of the system at the various times $t_1, t_2, \ldots$ in terms of $\Psi_0$, the initial wave function.

With the assumption that the interval

$$\delta t_n \equiv t_n - t_{n-1} \tag{303}$$

is large enough that energy of the particle is sharply defined at the beginning and end of the interval, we may set

$$U(t_n, t_{n-1}) = S_n, \tag{304}$$

the $S$-matrix for the $n$th interval. This permits us to rewrite Eq. (302) as

$$\Psi_n = s_n \Psi_{n-1}, \tag{305}$$

where

$$\begin{aligned} s_n &\equiv e^{-iK(t_n - t_{n-1})} S_n \\ &= e^{-iE(t_n - t_{n-1})} S_n(E), \end{aligned} \tag{306}$$

where $E$ is the energy of the particle.

Now, according to Eq. (301), the time interval $\delta t_n$ is

$$\begin{aligned} \delta t_n &= \frac{d_n}{v} + Q_n \\ &= \frac{d_n}{v} - i\frac{d}{dE}\ln S_n, \end{aligned} \tag{307$a$}$$

where $d_n$ is the distance traveled by the particle in the $n$th interval and $v$ is is asymptotic velocity. Since $dE = v\,dp$, where $p$ is the particle momentum, we may write Eq. (307$a$) as

$$\delta t_n = \frac{d_n}{v}\left(1 - i\,\frac{\hbar}{d_n}\,\frac{d \ln S_n}{dp}\right). \tag{307b}$$

This result may be generalized. Let us imagine the regions of interaction to overlap slightly so that the particle is always moving in a potential, but that this varies slowly over a wavelength (at least in the overlap region). Then the regions of interaction, labeled by $n$, define a coarse-grained decomposition of coordinate space and $S_n$ may be replaced by $S(\mathbf{r}, \mathbf{p})$, where $\mathbf{r}$ is the coarse-grained coordinate.[65] We may then write Eq. (307$b$) as

$$\delta t_n = (\delta t_n)_{\text{free}}(1 - i\hbar\nabla_{\mathbf{r}} \cdot \nabla_{\mathbf{p}} \ln S). \tag{307c}$$

Here $(\delta t_n)_{\text{free}}$ is the "free flight time" across the $n$th region and $\nabla_{\mathbf{r}}$ and $\nabla_{\mathbf{p}}$ are the respective gradient operators in coordinate and momentum space.

Returning to Eq. (305), we introduce

$$\delta\Psi_n \equiv \Psi_n - \Psi_{n-1} = (s_n - 1)\Psi_{n-1}, \tag{308}$$

and have

$$i\,\frac{\delta\Psi_n}{\delta t_n} = \frac{i(s_n - 1)}{\delta t_n}\Psi_{n-1}. \tag{309}$$

This is a kind of "coarse-grained" analog of the Schrödinger equation. It permits us to determine the wave function of a system as a function of time, except during short intervals of time when scattering is actually occurring (within small regions of space). The significant feature of this equation is that it involves only the $S$-matrix.

Equation (309) may formally be replaced by a differential equation. Recalling that $S_n$ is just a constant in each interval of time, we write

$$\frac{s_n - 1}{\delta t_n} = \left[\frac{1}{\delta t_n}(e^{-iE\delta t_n} - 1) + \frac{1}{\delta t_n}(1 - S_n^{-1})\right] S_n. \tag{310}$$

This expression permits us to write (309) as[66] (on setting $E = K$)

$$\hbar i\,\frac{\partial\Psi}{\partial t} = \left(K + i\hbar\,\frac{\ln S_n}{\delta t_n}\right)\Psi, \tag{311}$$

which has the appearance of a Schrödinger equation with a "potential"

$$[i\hbar\,(\ln S_n/\delta t_n)].$$

[65] For sufficiently large regions this does not, of course, violate the principle of complementarity.

[66] Integration of Eq. (311) over a single-time interval leads us back to Eq. (309).

Equation (311) contains a much more limited description of the system than does the Schrödinger equation. It will, however, agree with the Schrödinger equation for observations made on the system in macroscopic time intervals. Expressed otherwise, Eq. (311) is equivalent to the Schrödinger equation in "asymptotic regions" where interactions vary slowly over a wavelength $\hbar/p$. It is important that the time interval $\delta t_n$ can itself be expressed in terms of the $S$-matrix.

To illustrate these remarks, let us apply Eq. (311) to the scattering of a particle by a potential which varies so slowly that the eikonal description is valid. The volume in which the particle interacts is divided into coarse-grained cellular regions. In the $n$th such region the mean energy of the particle is written as $E = (1/2M)p^2$. The variation in potential energy about its mean value is then taken to be $V$. Because of the assumption that $V$ varies slowly over a wavelength, the cellular structure may be so chosen that

$$|V| \ll E$$

in the $n$th cell.

Now, according to Eqs. (6.499),

$$S_n = \exp\left\{i\int_{x_{n-1}}^{x_n} [\kappa(x') - p]\, dx'\right\}, \tag{312}$$

where $x_n - x_{n-1} = d_n$ is the path length in the $n$th cell and

$$\kappa^2 = p^2 - 2MV = 2M(E - V). \tag{313}$$

According to Eqs. (307*a*) and (312), we may choose the cell sizes so that

$$\begin{aligned}\delta t_n &= \frac{d_n}{v} + d_n \frac{d}{dE}(\kappa - p)\\ &\cong \frac{M}{\kappa}\, d_n.\end{aligned} \tag{314}$$

Similarly, we may arrange that

$$\ln S_n \cong \frac{id_n}{2\kappa}(-2MV) = \frac{-iM}{\kappa}\, d_n V. \tag{315}$$

Substitution into Eq. (311) gives the equation

$$i\frac{\partial \Psi}{\partial t} = (K + V)\Psi, \tag{316}$$

which is just the Schrödinger equation. Therefore Eq. (311) and the Schrödinger equation will give the same result for scattering in the semi-classical limit.

We see then that Eq. (311) contains a description of the semiclassical limit of quantum dynamics. It is thus adequate to describe gross macroscopic phenomena, the passage of wave packets between asymptotic limits, and Newtonian mechanics. For example, it will give a correct description of the scattering of elementary particles; it will not, however, permit us to study rapid fluctuations of the system while these particles are interacting. Whether Eq. (311), or a similar equation, may represent a limit on the observability of small time intervals is not known.[67]

## 8.6 THE GENERAL THEORY OF RESONANT SCATTERING

The scattering cross section (240) provides a specific example of the Breit-Wigner [34] one-level resonance formula. As we have noted, such resonances are frequently encountered in experimental studies of scattering. These have been of special importance in nuclear physics for which the theory has been developed in elaborate detail particularly by Wigner and his collaborators [35, 36].[68] Interactions among "elementary" particles have also shown resonances.

Our derivation of Eqs. (240) is of limited generality, since it was assumed that a weak interaction $V_r$ was responsible for the formation and decay of the long-lived virtual state. In this section we shall describe much more generally the physical basis for resonant scattering in terms of the lifetime $Q$ of virtual states.

A necessary condition that the one-level resonance formula provide an adequate description of the scattering cross section is that a single, long-lived virtual state contribute predominantly to the interaction. This was manifestly the case in our derivation of Eqs. (240). We shall now see that it is in general necessary (and, subject to weak restrictions, sufficient) for there to be "resonant scattering."

In anticipation of our requirements, let us put the $S$-matrix into a useful form. We shall restrict our considerations to the case for which there are just two particles in the initial state and two particles in the final state.[69] We may also, without loss of generality, discuss only a single state of angular momentum, parity,[70] etc. This permits us to label $S$ by the channel $c$ and energy $\epsilon$ only. Then, in the barycentric subspace, the $S$-matrix has the usual form [see Eq. (7.115$b$)]

$$\begin{aligned}\langle c'; \epsilon'| S |c; \epsilon\rangle &= \delta_{(c';\epsilon'),(c;\epsilon)} - 2\pi i\delta(\epsilon' - \epsilon)T_{c'c}(\epsilon)\\ &= \delta_{\epsilon',\epsilon}[\delta_{c',c} - 2\pi i T_{c'c}(\epsilon)] \equiv \delta_{\epsilon',\epsilon}S_{c'c}(\epsilon),\end{aligned} \tag{317}$$

[67] Some further discussion of this point is given in Refs. [33].

[68] Extensive applications are made in Refs. [37] and [38].

[69] When there are more than two particles in either the initial or final state, it is sometimes possible to treat formally several particles as a single "particle" in order to use two-particle scattering theory.

[70] This is possible only if parity is conserved, of course.

where $T_{c'c}$ is the element of the scattering matrix connecting channels $c$ and $c'$.

Since $S$ is unitary, it may be diagonalized. Let it then be diagonal in a representation for which the states are labeled as $(s, \epsilon)$. That is,

$$\begin{aligned}\langle s', \epsilon'|\, S\, |s, \epsilon\rangle &= \delta_{\epsilon',\epsilon}\delta_{s',s}[1 - 2\pi i T_s(\epsilon)] \\ &\equiv \delta_{\epsilon',\epsilon}\delta_{s',s}S_s(\epsilon).\end{aligned} \tag{318}$$

If we write $U_{cs}$ [and $U_{sc}{}^{-1} = U_{cs}{}^*$] for the unitary matrix which transforms from the $s$ to the $c$ representation, then

$$S_{c'c}(\epsilon) = \sum_s U_{c's}S_s(\epsilon)U^{-1}{}_{sc}, \tag{319a}$$

or

$$\delta_{s's}S_s(\epsilon) = \sum_{c,c'} U^{-1}{}_{s'c'}S_{c'c}(\epsilon)U_{cs}. \tag{319b}$$

Finally, if the scattering in the state under consideration is sufficiently larger than that in the other states, the cross section for scattering from channel $c$ to channel $c'$ is [see Eq. (7.144)]

$$\sigma_{c'c} = \frac{\pi}{\kappa_c{}^2}\frac{2J+1}{(2S_1+1)(2S_2+1)}|S_{c'c}(\epsilon) - \delta_{c'c}|^2. \tag{320}$$

Here $\kappa_c$ is the momentum of either of the two particles in the incident channel $c$, whereas $S_1$ and $S_2$ are their respective spins.

**The Consequences of a Long-Lived Virtual State**

Let us now ask what will be the consequence if the lifetime $Q$ in a particular eigenstate $(s, \epsilon, j, \ldots)$ has a maximum value at a given energy $\epsilon = \epsilon_0$, but that strong scattering does not occur except near $\epsilon = \epsilon_0$ in this state. We shall assume that $Q(\epsilon)$ is large and positive near $\epsilon = \epsilon_0$, but decreases to a value of the order of the "free flight time" when $\epsilon$ is not close to $\epsilon_0$. A large value of $Q$ for $\epsilon \approx \epsilon_0$ means, of course, that the "particles stick together" for an extended time. Such a phenomena is closely analogous to the slow decay of electrical currents in a resonant circuit when the frequency is near the resonance frequency.

The assertion that $Q(\epsilon)$ has a maximum near $\epsilon = \epsilon_0$ suggests that $Q^{-1}$ be expanded in a Taylor series of the form

$$\frac{1}{Q} = \alpha + \beta(\epsilon - \epsilon_0)^2 + \gamma(\epsilon - \epsilon_0)^3 + \cdots, \tag{321}$$

where $\alpha$, $\beta$, $\gamma$ are constants and $\alpha, \beta > 0$.[71] Normally, we might expect

[71] It is clear from our physical requirement that $Q$ have a sharp maximum near $\epsilon = \epsilon_0$ that a Taylor expansion of $Q$, instead of $Q^{-1}$, will not be very useful because of the implied narrow interval of convergence of the series. Because of the condition (323$a$) the first zero of $Q^{-1}$ (which limits the radius of convergence of the series for $Q$) occurs in the complex $\epsilon$-plane very near the energy $\epsilon = \epsilon_0$.

that

$$\beta = \mathcal{O}\left(\frac{\alpha}{\epsilon_0{}^2}\right),$$
$$\gamma = \mathcal{O}\left(\frac{\beta}{\epsilon_0}\right), \ldots. \tag{322}$$

Our assumption of a long life-time near the energy $\epsilon_0$ will be interpreted more precisely to mean that

$$\alpha \ll \epsilon_0{}^2\beta, \tag{323a}$$

but that

$$\gamma = \mathcal{O}\left(\frac{\beta}{\epsilon_0}\right). \tag{323b}$$

The conditions (323) imply that

$$\frac{1}{Q} \approx \alpha + \beta(\epsilon - \epsilon_0)^2 \tag{324}$$

for a range of energies such that $|\epsilon - \epsilon_0| \ll \mathcal{O}(\epsilon_0)$.[72]

Let us return to Eq. (321) and set $Z \equiv \epsilon - \epsilon_0$. We then write

$$F(Z) \equiv \gamma Z^3 + \cdots, \tag{325}$$

so [by Eq. (301)]

$$\frac{1}{Q} = i\,\frac{dZ}{d\ln S} = \alpha + \beta Z^2 + F(Z). \tag{326}$$

(For convenience, we shall drop the state label $s$ on the $S$-matrix.) The term $F(Z)$ will be considered as small over the range of $Z$ used. This suggests that we integrate Eq. (326) in the form

$$\begin{aligned}\ln S &= i\int \frac{dZ}{\alpha + \beta Z^2 + F(Z)}\\ &= i\int \frac{dZ}{\alpha + \beta Z^2} - i\int \frac{F(Z)\,dZ}{(\alpha + \beta Z^2)[\alpha + \beta Z^2 + F(Z)]}\\ &= \frac{i}{\sqrt{\alpha\beta}}\tan^{-1}\left(\frac{\sqrt{\alpha/\beta}}{\epsilon_0 - \epsilon}\right) + 2i\nu(\epsilon).\end{aligned} \tag{327}$$

Here

$$2\nu(\epsilon) = -\int^{\epsilon-\epsilon_0} \frac{F(Z)\,dZ}{(\alpha + \beta Z^2)[\alpha + \beta Z^2 + F(Z)]}, \tag{328}$$

the lower limit on the integral being chosen in a manner consistent with Eq. (327).

[72] Our argument is unchanged if some other energy than $\epsilon_0$ is required to characterize the size of the higher order terms in the series (321).

It is convenient to replace $\alpha$ and $\beta$ by two new parameters $r$ and $\Gamma$, defined by the equations

$$r \equiv \tfrac{1}{2}(\alpha\beta)^{-1/2}$$
$$\Gamma \equiv 2\sqrt{\alpha/\beta}. \tag{329}$$

Then Eq. (327) takes the form

$$S = \exp\left[2ri\tan^{-1}\left(\frac{\Gamma/2}{\epsilon_0 - \epsilon}\right) + 2i\nu(\epsilon)\right]. \tag{330}$$

The condition (323*a*) implies that $\Gamma/\epsilon_0 \ll 1$, so the term involving $\tan^{-1}[\Gamma/2/(\epsilon_0 - \epsilon)]$ in Eq. (330) will be negligible (unless $r$ is very large) except for $\epsilon \cong \epsilon_0$. We have specified that the scattering in the state being considered is small except (possibly) near the energy $\epsilon_0$. This implies that $\nu(\epsilon)$ will not give a very important contribution to $S$ when $\epsilon \cong \epsilon_0$. To estimate the contribution from $\nu(\epsilon)$, we take $r = 1$, so

$$\alpha = \frac{\Gamma}{4}, \qquad \beta = \frac{1}{\Gamma}.$$

Then for $\epsilon - \epsilon_0 = \Gamma/2$, the series (321) is

$$\frac{1}{Q} = \frac{\Gamma}{4} + \frac{\Gamma}{4} + \frac{\Gamma}{4}\left(\frac{\Gamma}{2\epsilon_0}\right) + \cdots, \tag{331}$$

using Eq. (323*b*). The condition (323*a*) implies that the higher terms in the series (331) are negligible.

We must now ask what may be said concerning the value of the parameter $r$. For this it is important to know the analytic properties of $S(\epsilon)$ in the complex $\epsilon$-plane. The scattering matrix $T(\epsilon)$ is defined and analytic when $\epsilon$ has a sufficiently small positive imaginary part, since $T$ was obtained as a function of $\epsilon + i\eta$ in the limit $\eta \to 0(+)$. From this we see that $S(\epsilon)$ is analytic in a domain above and bounded by the real axis. If $\Gamma$ is sufficiently small, the point

$$\epsilon = \epsilon_0 + i\frac{\Gamma}{2} \tag{332}$$

will lie in this domain of analyticity. In order to understand the implications of this, let us write

$$\tan^{-1}\left(\frac{\Gamma/2}{\epsilon_0 - \epsilon}\right) = \frac{i}{2}\left[\ln\left(\epsilon - \epsilon_0 + i\frac{\Gamma}{2}\right) - \ln\left(\epsilon - \epsilon_0 - i\frac{\Gamma}{2}\right)\right]. \tag{333}$$

Now, restricting $\epsilon$ to the domain of analyticity of $S(\epsilon)$, let the point $\epsilon$ move up from the real axis, describe a circle about the point $\epsilon_0 + i(\Gamma/2)$, and then return to its initial point on the real axis (as illustrated in Fig. 8.8).

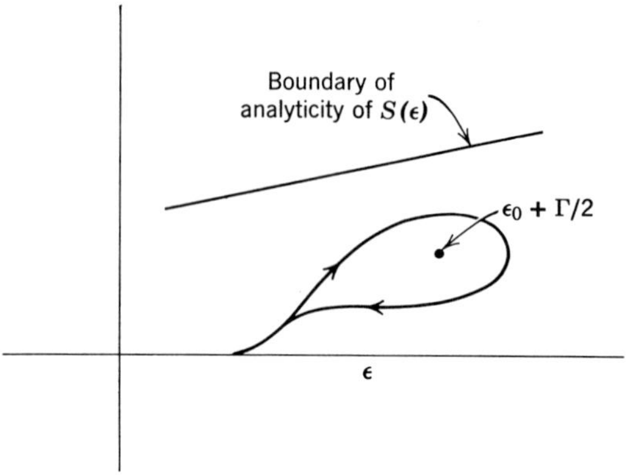


Fig. 8.8 A closed contour about the point $\epsilon_0 + i(\Gamma/2)$ in the complex $\epsilon$-plane.

The expression (333) does not return to its initial value, however, but acquires an additive term.

$$\tan^{-1}\left(\frac{\Gamma/2}{\epsilon_0 - \epsilon}\right) \to \tan^{-1}\left(\frac{\Gamma/2}{\epsilon_0 - \epsilon}\right) + \pi. \tag{334}$$

Then moving $\epsilon$ around the closed contour, as in Fig. 8.8, transforms $S$ according to [see Eq. (330)][73]

$$S \to Se^{2\pi ir}. \tag{335}$$

This violates the condition that $S(\epsilon)$ be analytic in the domain in which the contour lies, unless $r$ is an integer. We therefore must have

$$r = 0, 1, 2, \ldots, \tag{336}$$

which is the principal result of this section.

The case $r = 0$ is trivial and uninteresting. Of principal physical interest is the case that

$$r = 1. \tag{337}$$

As we shall see, this leads to the Breit-Wigner one-level resonance cross section. To our knowledge, scattering corresponding to $r \geq 2$ has not been observed, although this does not seem to be excluded by general considerations.

Our derivation of the condition (336) on $r$ depended on the assumption that the point $\epsilon_0 + i(\Gamma/2)$ lies in a region of analyticity for $S(\epsilon)$ connected to the real axis. In Section 8.6, p. 505, we shall return to a quantitative study of the validity of this assumption. For the moment, we shall assume it to be correct and investigate the consequences of Eq. (336).

[73] We have seen from Eq. (331) that $\nu(\epsilon)$ may be neglected when $\Gamma$ is sufficiently small.

Let us first suppose that $r = 1$. Then from Eq. (330) we obtain, writing $\tan\delta = (\Gamma/2)/(\epsilon_0 - \epsilon)$,

$$\begin{aligned} S(\epsilon) &\equiv e^{2i\delta(\epsilon)}\, e^{2i\nu(\epsilon)} \\ &= \frac{1 + i\tan\delta}{1 - i\tan\delta}\, e^{2i\nu} \\ &= 1 + \frac{i\Gamma}{(\epsilon_0 - \epsilon) - i(\Gamma/2)} + \frac{\epsilon_0 - \epsilon + i(\Gamma/2)}{\epsilon_0 - \epsilon - i(\Gamma/2)}(e^{2i\nu} - 1). \end{aligned} \tag{338}$$

The second term here represents "resonance scattering," whereas the third term is called "potential scattering" in the literature. By hypothesis, it gives only a small contribution to $S$ for $\epsilon \cong \epsilon_0$.

To discuss the scattering cross section, it is convenient to reinsert the state label $s$ on $S(\epsilon)$, writing $S_s(\epsilon)$ as in Eq. (318). Let the state for which $S$ has the form (338) be $s_0$. For simplicity, let us neglect the scattering in other states and the potential scattering in the state $s_0$. Then, we may take

$$S_s = 1 \qquad \text{for } s \neq s_0$$

$$S_{s_0} = 1 + \frac{i\Gamma}{\epsilon_0 - \epsilon - i(\Gamma/2)}\,. \tag{339}$$

This permits us to write $S_{c'c}$ in the channel representation as [see Eq. (319*a*)]

$$S_{c'c} = \delta_{c',c} + i\,\frac{U_{c's_0}\overset{*}{U}_{cs_0}\Gamma}{\epsilon_0 - \epsilon - i(\Gamma/2)}\,. \tag{340}$$

The scattering cross section (320) is then

$$\sigma_{c'c} = \frac{\pi}{\kappa_c^{\,2}}\,\frac{2J + 1}{(2S_1 + 1)(2S_2 + 1)}\,\frac{\Gamma_{c'}\Gamma_c}{(\epsilon_0 - \epsilon)^2 + \Gamma^2/4}\,. \tag{341}$$

Here we have defined

$$\begin{aligned} \Gamma_c &\equiv |U_{cs_0}|^2\,\Gamma, \\ \Gamma_{c'} &\equiv |U_{c's_0}|^2\,\Gamma. \end{aligned} \tag{342}$$

It follows from Eqs. (342) and the unitarity of $U_{cs}$ that

$$\sum_c \Gamma_c = \Gamma. \tag{343}$$

Equation (341) has the precise form of Eq. (240*b*), which was obtained for a specific model. The expression (341) was obtained under very general conditions, however, from the assumption that the composite state of the system has a long lifetime $Q$ at the energy $\epsilon_0$. This makes it clear why the model leading to Eq. (240*b*) also gives a resonant scattering cross section;

this model was based on the assumption that a virtual state, which decays slowly, was formed.

We verify also that Eq. (187) is a special case of Eq. (343).

The physical basis for our derivation of Eq. (341) is very satisfying. The notion that resonance occurs when a slowly decaying state is excited is familiar from classical mechanics and from electrical circuit theory.

When the potential scattering term in Eq. (338) cannot be neglected, Eq. (341) is modified in a straightforward manner. Also, when scattering in states other than $s_0$ must be included, it is again straightforward (but often tedious) to evaluate $\sigma_{c'c}$ from Eqs. (319*a*) and (320). If the differential scattering cross section is required, we may evaluate it from the eigenvalues of the $S$-matrix in the manner described in Chapter 7.

**Scattering for $r = 2$**

When the potential scattering can be neglected, so $v \simeq 0$, Eq. (330) takes the form

$$S = \left[\frac{\epsilon_0 - \epsilon + i(\Gamma/2)}{\epsilon_0 - \epsilon - i(\Gamma/2)}\right]^r, \tag{344}$$

where $r = 0, 1, 2, \ldots$. When $r = 2$, we have

$$S - 1 = \frac{2i(\epsilon - \epsilon_0)\Gamma}{[(\epsilon - \epsilon_0)^2 - \Gamma^2/4] - i(\epsilon - \epsilon_0)\Gamma}. \tag{345}$$

In the simple case of a single channel, Eq. (320) then leads to the cross section

$$\sigma = \frac{4\pi}{\kappa^2}\frac{2J + 1}{(2S_1 + 1)(2S_2 + 1)}\frac{\Gamma^2(\epsilon - \epsilon_0)^2}{[(\epsilon - \epsilon_0)^2 - \Gamma^2/4]^2 + \Gamma^2(\epsilon - \epsilon_0)^2}. \tag{346}$$

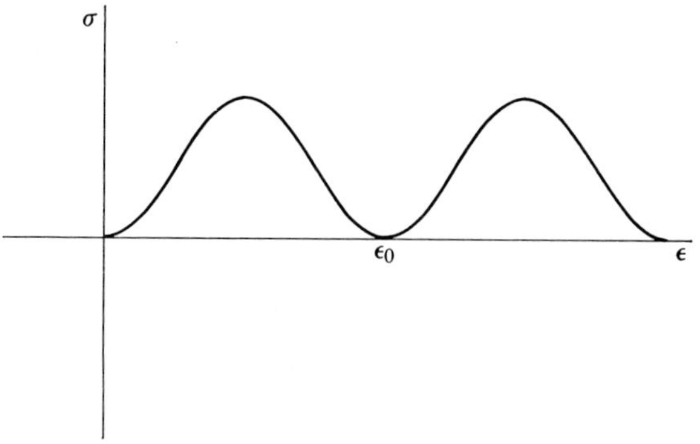


Fig. 8.9 Illustration of the energy dependence of the scattering cross section (346).

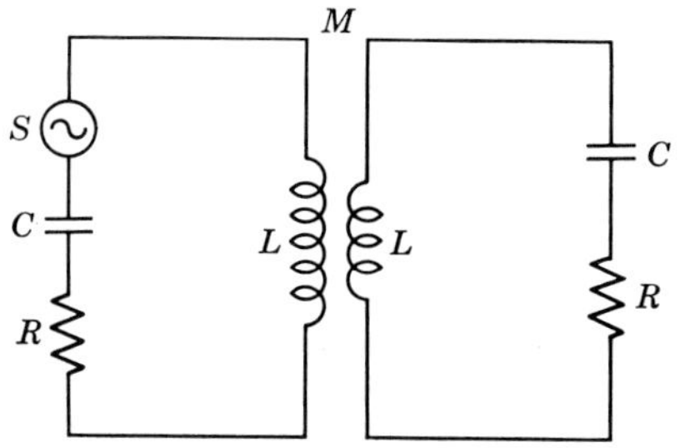


Fig. 8.10 Electric circuit illustrating Eq. (347).

This describes a "double-humped" resonance and is illustrated in Fig. 8.9. It vanishes, curiously enough, for $\epsilon = \epsilon_0$.[74]

Equation (345) bears some resemblance to the flow of current in coupled, resonant electrical circuits. Such a circuit is illustrated in Fig. 8.10. Simple series circuits each containing an inductance $L$, resistance $R$, and capacitance $C$ are coupled with a mutual inductance $M$. On one of the circuits there is an alternating voltage source $S$ of frequency $\omega$. The current $I$ flowing in this circuit is proportional to

$$I \cong \frac{\epsilon_0 - \epsilon + i\gamma}{(\epsilon_0 - \epsilon + i\gamma)^2 - \epsilon M^2}, \tag{347}$$

where

$$\epsilon_0 \equiv \frac{1}{LC},$$
$$\gamma \equiv \omega \frac{R}{L}, \tag{348}$$
$$\epsilon \equiv \omega^2.$$

**Analytic Properties of the *S*-Matrix**

We have yet to discuss the conditions under which $S(\epsilon)$ is analytic in a strip above the real axis containing the point $\epsilon = \epsilon_0 + i(\Gamma/2)$. This was, we recall, an essential assumption in the derivation of the cross section for resonant scattering.

The analytic properties of the $S$-matrix were described in Section 6.5 for the scattering of two simple particles by a potential. We must now generalize this to multichannel scattering of (possibly) composite particles. According to Eq. (317) or Eq. (7.115*b*)

$$S_{c'c}(\epsilon) = \delta_{c'c} - 2\pi i T_{c'c}(\epsilon). \tag{349}$$

[74] Scattering in other states precludes the actual vanishing of the total cross section at $\epsilon = \epsilon_0$.

It therefore suffices to discuss the analytic properties of $T_{c'c}(\epsilon)$.[75] This may be written in the form

$$T_{c'c} = \left(\chi_{c';\epsilon}, \left(V_c + V_{c'} \frac{1}{\epsilon + i\eta - H} V_c\right)\chi_{c;\epsilon}\right). \tag{350}$$

Here $H$ is the Hamiltonian of the system and $V_c$, $V_{c'}$ are the appropriate channel interactions in the notation of Eq. (4.11$a$). If the scattering involves identical particles, the symmetrized scattering matrix is a sum of such expressions as (350). The $\chi_{c;\epsilon}$ are the channel wave functions for the noninteracting particles.

As in Section 8.2, we suppose that wave functions $\psi_\lambda$ represent a complete set of eigenfunctions of $H$, corresponding to eigenvalues $E_\lambda$. Then Eq. (350) may be written as

$$T_{c'c} = X_1 + X_2, \tag{351$a$}$$

where

$$X_1 \equiv (\chi_{c';\epsilon}, V_c\chi_{c;\epsilon}) \tag{351$b$}$$

and

$$\begin{aligned} X_2 &\equiv \left(\chi_{c';\epsilon}, V_{c'} \frac{1}{\epsilon + i\eta - H} V_c\chi_{c;\epsilon}\right) \\ &= \sum_\lambda \frac{(\chi_{c';\epsilon}, V_{c'}\psi_\lambda)(\psi_\lambda V_c\chi_{c;\epsilon})}{\epsilon + i\eta - E_\lambda} \end{aligned} \tag{351$c$}$$

Now, following the argument given in connection with Eq. (70), we define

$$I' dE_\lambda \equiv \int \rho(\lambda)\, d\lambda (\chi_{c';\epsilon}, V_{c'}\psi_\lambda)(\psi_\lambda, V_c\chi_{c;\epsilon}), \tag{352}$$

where $\rho(\lambda)$ is the appropriate density function for the states $\lambda$ so that

$$\int I' dE_\lambda = (\chi_{c';\epsilon}, V_{c'}V_c\chi_{c;\epsilon}) \tag{353}$$

Here the integral extends over the complete eigenvalue spectrum of $H$ and is interpreted as a Stielje integral if there are discrete states.

Let us now suppose that Im $(\epsilon) \geq 0$. Then by Chartier's test for convergence, [see the discussion in connection with Eq. (70)] if the integral (353) is finite, the integral

$$X_2 = \int \frac{I' dE_\lambda}{\epsilon + i\eta - E_\lambda} \tag{354}$$

exists. To complete our demonstration that $T_{c'c}$ is defined, we must show that the expressions (351$b$) and (353) exist.

[75] We see from Eq. (340) that the transformation matrix $U_{cs}$ may be taken to be independent of the energy in the neighborhood of the resonance energy.

We shall show that this is the case in a strip above the real axis in the $\epsilon$-plane if we assume that the interactions $V_c$ decrease at least as fast an an exponential when the particles are separated to large distances in any channel. Let us consider $\mathbf{r}$ to be a coordinate describing the separation of the particles (when they are at large distances from each other) in whatever channel is involved. Then, not explicitly writing other variables than $\mathbf{r}$, we assume that

$$\begin{aligned} \lim_{r\to\infty} \langle \mathbf{r}'| V_c |\mathbf{r}\rangle &\leq \mathcal{O}(e^{-(\mu/2)r}), \\ \lim_{r'\to\infty} \langle \mathbf{r}'| V_c |\mathbf{r}\rangle &\leq \mathcal{O}(e^{-(\mu/2)r'}), \\ \lim_{r,r'\to\infty} \langle \mathbf{r}'| V_c |\mathbf{r}\rangle &\leq \mathcal{O}(e^{-(\mu/2)r'} e^{-(\mu/2)r}), \end{aligned} \tag{355}$$

for any pair of channels. Here $d = \mu^{-1}$ represents the range of the interaction. The conditions (355) are thought to be satisfied by all interactions except those associated with electromagnetic and gravitational forces.

Now, the channel wave functions $\chi_{c;\epsilon}$ have the asymptotic forms for large separation of the particles,

$$\chi_{c;\epsilon} \cong \mathcal{O}(e^{\pm i\kappa_c r}) \times (\text{a function of internal coordinates}). \tag{356}$$

Here the $(\pm)$ signs in the exponential refer to outgoing, or incoming waves, respectively—both of which are contained in $\chi_{s,\epsilon}$. We have written $\kappa_c$ for the relative momentum of the particles in channel $c$. In the complex $\epsilon$-plane when $\text{Im}(\epsilon) > 0$, we have

$$\kappa_c = k_c + il_c, \tag{357}$$

where $k_c$ and $l_c$ are real and $k_c, l_c > 0$.

Then, for $\text{Im}(\epsilon) > 0$,

$$X_1 = \mathcal{O}\left[\int_0^\infty dr'\, dr\, e^{i\kappa_{c'}^{*} r'} \langle r'| V_c |r\rangle e^{-i\kappa_c r}\right]. \tag{358}$$

This integral converges,[76] then, if for all channels $c$,

$$l_c < \frac{\mu}{2}. \tag{359}$$

In a similar manner we see that the expression (353) exists if the conditions (359) are satisfied and if the product $V_{c'}V_c$ is not infinite because of singularities at small separations of the particles.

The existence of $dX_1/d\epsilon$ and $dX_2/d\epsilon$ is established by similar arguments. We conclude therefore that $T_{c'c}(\epsilon)$ is analytic for complex $\epsilon$ such that the

[76] It is assumed that the integral is defined for small $r$.

condition (359) is met, if: (1) The interactions $V_c$ decrease at least exponentially for large separations of the particles; (2) the interactions are not so singular at small distances that the product $V_{c'}V_c$ fails to be defined.

The condition (359) is evidently a generalization of the corresponding result, obtained for simple potential scattering, found in Section 6.5, p.272.

To interpret the physical significance of the condition (359), we observe that

$$\text{Im}\,(\epsilon) \approx \frac{k_c}{M_c}\, l_c \approx v_c l_c. \tag{360}$$

Here $M_c$ is the reduced mass in channel $c$, and $v_c$ is the corresponding relative velocity of the particles. The derivation of Eq. (336) required that

$$\frac{\Gamma}{2} < \text{Im}\,(\epsilon) = v_c l_c < v_c \frac{\mu}{2} \approx \frac{v_c}{2d}, \tag{361}$$

where $d = \mu^{-1}$ is the range of the forces between the particles. Now, near resonance, $\Gamma/2 \approx 1/Q$, so Eq. (360) implies that for all $c$

$$Q > \frac{2d}{v_c}. \tag{362}$$

The quantity $2d/v_c$ is just the free flight time across the range of interaction. Therefore $S(\epsilon)$ is analytic at $\epsilon = \epsilon_0 + i(\Gamma/2)$, and our discussion of resonance is valid, if the lifetime $Q$ is *greater* than the free flight time (in any channel) across the range of the forces.

## REFERENCES


1. W. B. Riesenfeld and K. M. Watson, *Phys. Rev.* **104,** 492 (1956).
2. Brueckner, Levinson, and Mahmond, *Phys. Rev.* **95,** 217 (1954); K. A. Brueckner, *Phys. Rev.* **96,** 508 (1954); **97,** 1353 (1955).
3. R. J. Eden and N. C. Francis, *Phys. Rev.* **97,** 1366 (1955).
4. L. Van Hove, *Physica* **21,** 901 (1955); **22,** 343 (1956).
5. E. P. Wigner, *Math u. Naturw. Ang. Ungar. Akad. Wiss.* **53,** 475 (1935); L. Brillouin, *J. Phys. Radium* **4,** 1 (1933).
6. N. Fukuda and R. Newton, *Phys. Rev.* **103,** 1558 (1956).
7. B. S. DeWitt, *Phys. Rev.* **103,** 1565 (1956).
8. E. T. Whittaker and G. N. Watson, *Modern Analysis,* (The MacMillan Co., New York, 1943).
9. V. F. Weisskopf and E. P. Wigner, *Zeit. f. Phys.* **63,** 54 (1930); **65,** 18 (1930).
10. W. Heitler, "Quantum Theory of Radiation" (Oxford, Clarendon Press, Third Edition, 1954).
11. B. Zumino, "On the Formal Theory of Collision and Reaction Processes" (New York University Research Report No. CX-23); E. J. Hellund, *Phys. Rev.* **89,** 919 (1953); G. Höhler, *Zeit. F. Phys.* **152,** 546 (1958); B. S. DeWitt, Univeristy of California Radiation Laboratory Report 2884 (1954).
12. F. E. Low, *Phys. Rev.* **88,** 53 (1952).

13. M. Gell-Mann and A. H. Rosenfeld, *Annual Review of Nuclear Science* (Annual Reviews, Inc., Stanford, California, 1957).
14. K. M. Watson, *Phys. Rev.* **104,** 492 (1956).
15. M. L. Goldberger, *Phys. of Fluids* **2,** 252 (1959).
16. T. Matsubara, *Prog. Theoret. Phys.* (*Kyoto*) **14,** 351 (1955); C. Bloch and C. de Dominicis, *Nuclear Phys.* **7,** 459 (1958); D. J. Thouless, *Phys. Rev.* **107,** 1162 (1957); Glassgold, Heckrotte, and Watson, *Phys. Rev.* **115,** 1374 (1959).
17. K. A. Brueckner, *Phys. Rev.* **100,** 36 (1955).
18. W. B. Riesenfeld and K. M. Watson, *Phys. Rev.* **104,** 492 (1956); J. Goldstone, *Proc. Roy. Soc.* (*London*) **A239,** 267 (1957); N. Hugenholtz, Physica **23,** 481 (1957); C. Bloch, *Nuclear Phys.* **7,** 451 (1958).
19. L. Landau and E. Lifschitz *Statistical Physics* (Addison-Wesley Publishing Co., Inc. Reading, Mass., 1958).
20. N. N. Bogolyubov, *J. Phys.* (*U.S.S.R.*) **11,** 23 (1947).
21. K. Sawada, *Phys. Rev.* **106,** 372 (1957).
22. Glassgold, Heckrotte, and Watson, *Annals of Phys.* **6,** 1 (1959).
23. H. Margenau and M. Lewis, *Rev. Mod. Phys.* **31,** 594 (1959).
24. J. M. Jauch and F. Rohrlich, *Theory of Photons and Electrons* (Addison-Wesley Publishing Co., Inc. Reading, Mass., 1955).
25. S. B. Trieman, *Dispersion Relations and Elementary Particles* (Hermann, Paris (1960)).
26. H. Bethe and F. deHoffman, *Mesons and Fields* (Row, Peterson and Co., Evanston, Ill., 1955).
27. K. A. Brueckner and K. M. Watson (Los Alamos Lectures, 1952).
28. K. M. Watson, *Phys. Rev.* **89,** 575 (1953); K. A. Brueckner and K. M. Watson, *Phys. Rev.* **90,** 699 (1953).
29. E. P. Wigner, *Phys. Rev.* **98,** 145 (1955). See also L. Eisenbud, Thesis (1948), Princeton University.
30. F. T. Smith, *Phys. Rev.* **118,** 349 (1960).
31. B. S. DeWitt, *Phys. Rev.* **105,** 1565 (1956).
32. M. L. Goldberger, *Phys. Fluids* **2,** 252 (1959).
33. M. L. Goldberger and K. M. Watson, *Phys. Rev.* **127,** 2284 (1962). M. Froissart, M. L. Goldberger, and K. M. Watson, *Phys. Rev.* **131,** 2820 (1963).
34. G. Breit and E. P. Wigner, *Phys. Rev.* **49,** 519, 642 (1936).
35. E. P. Wigner, *Phys. Rev.* **70,** 15, 606 (1946); E. P. Wigner and L. Eisenbud, *Phys. Rev.* **72,** 29 (1947).
36. T. Teichmann, *Phys. Rev.* **77,** 506 (1950).
37. J. M. Blatt and V. F. Weisskopf, *Theoretical Nuclear Physics* (John Wiley and Sons, New York, 1952).
38. R. G. Sachs, Nuclear Physics (Addison-Wesley Publishing Co., Inc., Reading, Mass., 1953).

# 9. MISCELLANEOUS APPLICATIONS

In this chapter a number of rather disconnected topics are discussed by way of application of the general techniques developed in earlier chapters.

Despite the fact that most of our formal development has been applicable to the scattering of relativistic particles, much of it has had a nonrelativistic appearance. To dispel any misconceptions on this point, we treat here the relativistic theory of the scattering of spin one-half particles by a force center using previously developed techniques. As an illustration of the general formalism developed in Chapter 7, we study in considerable detail the problem of photoproduction of mesons from nucleons. We take up briefly the theory of final state interactions. The object here is to treat the strong interaction between particles created in some sort of production process. Finally, we illustrate that the formalism of Chapter 5 can cope with simple field theoretical problems such as those of static meson theory.

## 9.1 RELATIVISTIC DESCRIPTION OF THE SCATTERING OF SPIN ONE-HALF PARTICLES BY A FORCE CENTER

The scattering of high energy electrons from nuclei is a process which is both historically and practically of very great importance. Much of the formal description of the scattering of spin one-half particles by a force center can be taken over directly from the theory developed in Chapter 7. This may at first sight seem surprising since we never specifically mentioned the Dirac equation there. We must dispel any apprehensions that the earlier description in terms of two-component (Pauli) wave functions rather than four component (Dirac) wave functions involved making any kind of nonrelativistic approximation. Stated briefly, it does not! The point is simply that the kinematical description of the scattering process is essentially a geometrical problem. We are concerned with only asymptotic free-particle states in which the small components of the Dirac wave function are expressible in terms of the large components, the ones which are nonzero in the rest frame of the particle. In the particle's rest frame, a two-component description is rigorously correct; the use of the usual four-component functions is just an easy way to avoid thinking about Lorentz transformations. The formalism developed in Appendix E to Chapter 7 requires no explicit use of a wave equation for a particle and provides a completely covariant description of the scattering process. The Dirac theory is so familiar that in practice it is frequently worthwhile to use it, however.

The general theory of scattering and reaction processes developed in Chapters 3 and 5 has a noncovariant look about it. Nevertheless, as we have repeatedly emphasized, there is no implication that the formalism is a nonrelativistic one, and we shall show in this case that it is directly applicable to the problem at hand.

We consider the elastic scattering of spin one-half particles by an electrostatic scalar spherically symmetric potential $V(r)$. The Dirac Hamiltonian $H$ is

$$H = \boldsymbol{\alpha} \cdot \mathbf{p} + \beta m + V(r). \tag{1}$$

(The notation here is as in Section 1.4 and we have set $\hbar = c = 1$.) The assumption that $V$ be an electrostatic scalar will not play much of a role in our subsequent discussion; we could equally well imagine $V$ to be multiplied by $\beta$, where then we would say that the interaction was a *world scalar*, or by $\rho_2$, where it would be a *pseudoscalar*.

The scattering eigenstates of the Hamiltonian evolve from wave packets built up by superposing positive energy plane-wave solutions of the Dirac equation with $V = 0$. These plane wave solutions were discussed in detail in Chapters 1 and 2. The same "$E + i\eta$" prescription for constructing outgoing wave states used throughout the book continues to apply, of course. In the limit of a sharp momentum state we are led to the following formal solution to the equation $H\psi = \epsilon\psi$, with the outgoing wave boundary condition

$$\psi_{\boldsymbol{\varkappa},\lambda}^{+}(\mathbf{r}) = \chi_{\boldsymbol{\varkappa},\lambda}(\mathbf{r}) + \frac{1}{\epsilon + i\eta - H} V\chi_{\boldsymbol{\varkappa},\lambda}(\mathbf{r}), \tag{2}$$

where

$$\chi_{\boldsymbol{\varkappa},\lambda}(\mathbf{r}) = \frac{e^{i\boldsymbol{\varkappa}\cdot\mathbf{r}}}{(2\pi)^{3/2}} u^{\lambda}(\boldsymbol{\varkappa}), \tag{3}$$

and $u^{\lambda}(\boldsymbol{\varkappa})$ is a four-component spinor,[1] a solution of

$$(\epsilon_{\kappa} - \boldsymbol{\alpha} \cdot \boldsymbol{\varkappa} - \beta m)u^{\lambda}(\boldsymbol{\varkappa}) = 0 \tag{4}$$

with $\epsilon_{\varkappa} = +(m^2 + \kappa^2)^{1/2}$. The superscript $\lambda$ on $u^{\lambda}(\boldsymbol{\varkappa})$ takes on the values

[1] We note that the function $u^{\lambda}$ introduced here differs in normalization from that of Eq. (2.64).

1, 2 and labels the spin state of the incident electron; the latter specification is made more precise in a moment. It is convenient to normalize the spinors according to

$$u^{\lambda'\dagger}(\boldsymbol{\varkappa})u^{\lambda}(\boldsymbol{\varkappa}) = \sum_{\alpha=1}^{4} u_{\alpha}^{\lambda'}{}^{*}u_{\alpha}{}^{\lambda} = \delta_{\lambda'\lambda}. \tag{5}$$

Quite explicitly we have

$$u^{\lambda}(\boldsymbol{\varkappa}) = \frac{\epsilon_{\kappa} + (\boldsymbol{\alpha}\cdot\boldsymbol{\varkappa} + \beta m)\beta}{[2\epsilon_{\kappa}(\epsilon_{\kappa} + m)]^{1/2}}\begin{pmatrix}\chi^{\lambda}\\ 0\end{pmatrix}, \tag{6}$$

where $\chi^{\lambda}$ is a two-component Pauli spinor and 0 means the two-component spinor (0, 0). The normalization of the $\chi^{\lambda}$'s is taken to be $\chi^{\lambda'\dagger}\chi^{\lambda} = \delta_{\lambda'\lambda}$. The $\chi^{\lambda}$ might perhaps be eigenfunctions of the Pauli matrix $\sigma_z$ or eigenfunctions of $\boldsymbol{\sigma}\cdot\hat{\mathbf{u}}$, where $\hat{u}$ is a unit vector in a direction characterized by polar angles $\theta_0$, $\phi_0$.

$$\chi^{\lambda} = \frac{1 + \boldsymbol{\sigma}\cdot\hat{\boldsymbol{u}}\sigma_z}{\sqrt{2}(1 + \cos\theta_0)^{1/2}}\, w^{\lambda} \tag{7}$$

with $w^1 = (1, 0)$ and $w^2 = (0, 1)$. The matrix appearing in Eq. (6) is just that which describes a Lorentz transformation to a system which is moving along the direction $-\hat{\boldsymbol{\varkappa}}$ with velocity $\kappa/\epsilon_{\kappa}$, thus imparting to a particle at rest momentum $\boldsymbol{\varkappa}$.

From our general theory, the $T$-matrix element describing the elastic scattering of a particle of momentum $\boldsymbol{\varkappa}$ in spin state $u^{\lambda}(\boldsymbol{\varkappa})$ into a state with momentum $\boldsymbol{\varkappa}_f$ and spin state $u^{\lambda'}(\boldsymbol{\varkappa}_f)$ is given by

$$\begin{aligned} T^{\lambda'\lambda}_{\boldsymbol{\varkappa}_f,\boldsymbol{\varkappa}} &= (\chi_{\boldsymbol{\varkappa}_f,\lambda'}, V\psi^{+}_{\boldsymbol{\varkappa},\lambda}) \\ &= \int d^3r\, \frac{e^{-i\boldsymbol{\varkappa}_f\cdot\mathbf{r}}}{(2\pi)^{3/2}}\, u^{\lambda'*}(\boldsymbol{\varkappa}_f)V\psi^{+}_{\boldsymbol{\varkappa},\lambda}(\mathbf{r}). \end{aligned} \tag{8}$$

The current density of the incident wave is $(2\pi)^{-3}\kappa/\epsilon_{\kappa}$, so from the general equation (5.37) for the differential cross section

$$[(2\pi)^{-3}\kappa/\epsilon_{\kappa}]\, d\sigma^{\lambda'\lambda} = 2\pi\, |T^{\lambda'\lambda}_{\boldsymbol{\varkappa}_f,\boldsymbol{\varkappa}}|^2\, \delta[(\epsilon_{\kappa_f} - \epsilon_{\kappa})\, d^3\kappa_f, \tag{9a}$$

we obtain

$$\frac{d\sigma^{\lambda'\lambda}}{d\Omega} = (2\pi)^4\epsilon^2 \left| \int d^3r\, \frac{e^{-i\boldsymbol{\varkappa}_f\cdot\mathbf{r}}}{(2\pi)^{3/2}}\, u^{\lambda'*}(\boldsymbol{\varkappa}_f)V\psi^{+}_{\boldsymbol{\varkappa},\lambda}\right|^2; \tag{9b}$$

we have written $\epsilon_{\kappa} = \epsilon_{\kappa_f} = \epsilon$ and $d\Omega$ for the element of solid angle about $\boldsymbol{\varkappa}_f$.

The total differential cross section for scattering into any final spin state is obtained by summing over $\lambda' = 1, 2$. This operation may be carried out quite explicitly by using the expression (6) for the spinor $u^{\lambda'}(\boldsymbol{\varkappa}_f)$. Consider the matrix in Dirac spinor indices defined by

$$\begin{aligned}\Lambda_{\mu\mu'}(\boldsymbol{\varkappa}) &\equiv \sum_{\lambda=1,2} u_\mu{}^\lambda(\boldsymbol{\varkappa})u_{\mu'}^{\lambda*}(\boldsymbol{\varkappa}) \\ &= \sum_{\lambda=1,2}\left\{\frac{\epsilon + H_0(\boldsymbol{\varkappa})\beta}{[2\epsilon(\epsilon+m)]^{1/2}}\begin{pmatrix}\chi^\lambda\\0\end{pmatrix}\right\}_\mu \\ &\quad\times\left\{\frac{\epsilon + H_0(\boldsymbol{\varkappa})\beta}{[2\epsilon(\epsilon+m)]^{1/2}}\begin{pmatrix}\chi^\lambda\\0\end{pmatrix}_{\mu'}\right\}^\dagger,\end{aligned} \tag{10}$$

where $H_0(\boldsymbol{\varkappa}) = \boldsymbol{\alpha}\cdot\boldsymbol{\varkappa} + \beta m$. Now the complete set of basic spinors has, in addition to $(\chi^1, 0)$ and $(\chi^2, 0)$, also $(0, \chi^1)$ and $(0, \chi^2)$ corresponding to the negative energy solutions discussed in Chapter 2 and labeled by $\lambda = -1$, $-2$. In order to utilize the completeness relation for the basic spinors we must be able to sum over the full set of four. This may be easily arranged by the insertion into Eq. (10) of a projection operator which eliminates the unwanted contributions. We merely insert in $\{\ \}_\mu$ in (10) a factor $(1+\beta)/2$ in front of $(\chi_\lambda, 0)$ which gives unity when acting on $(\chi^1, 0)$ and $(\chi^2, 0)$, but zero when applied to $(0, \chi^1)$ or $(0, \chi^2)$. Adding such a factor enables us to use the relation

$$\sum_{\lambda=1,2}\left\{\begin{pmatrix}\chi^\lambda\\0\end{pmatrix}_\nu\begin{pmatrix}\chi^\lambda\\0\end{pmatrix}_{\nu'}^\dagger + \begin{pmatrix}0\\\chi^\lambda\end{pmatrix}_\nu\begin{pmatrix}0\\\chi^\lambda\end{pmatrix}_{\nu'}^\dagger\right\} = \delta_{\nu\nu'}, \tag{11}$$

and then to obtain for $\Lambda_{\mu\mu'}(\boldsymbol{\varkappa})$

$$\Lambda_{\mu\mu'}(\boldsymbol{\varkappa}) = \frac{1}{2\epsilon(\epsilon+m)}\left\{[\epsilon + H_0(\boldsymbol{\varkappa})\beta]\left(\frac{1+\beta}{2}\right)[\epsilon + \beta H_0(\boldsymbol{\varkappa})]\right\}_{\mu\mu'}; \tag{12}$$

using the fact that $\beta(1+\beta)\beta = 1+\beta$ we may write (as a Dirac matrix relation)

$$\Lambda(\boldsymbol{\varkappa}) = \frac{1}{2\epsilon(\epsilon+m)}[\epsilon + H_0(\boldsymbol{\varkappa})]\frac{1+\beta}{2}[\epsilon + H_0(\boldsymbol{\varkappa})]. \tag{13}$$

Finally, we note that $H_0\beta + \beta H_0 = 2m$ and $H_0\beta H_0 = -\epsilon^2\beta + 2mH_0$ so that

$$\Lambda(\boldsymbol{\varkappa}) = \frac{\epsilon + H_0(\boldsymbol{\varkappa})}{2\epsilon}. \tag{14}$$

The educated reader will at this point remark that we have taken a singularly obtuse way of obtaining $\Lambda(\boldsymbol{\varkappa})$, which is called a *Casimir projection operator*.[2] Obviously, $\Lambda(\boldsymbol{\varkappa})$ gives unity when applied to a positive

[2] The form given here for these projection operators is easily seen to be equivalent to that used in Eqs. (8.216).

energy spinor belonging to momentum $\boldsymbol{\varkappa}$ and zero when applied to a negative energy one. We could then have written immediately

$$\Lambda_{\mu\mu'}(\boldsymbol{\varkappa}) = \sum_{\substack{\lambda=1,2\\-1,-2}} \left[\frac{\epsilon(\boldsymbol{\varkappa}) + H_0(\boldsymbol{\varkappa})}{2\epsilon} u^\lambda(\boldsymbol{\varkappa})\right]_\mu u_{\mu'}^{\lambda *}(\boldsymbol{\varkappa}), \tag{15}$$

which leads directly to Eq. (14) when we use the ordinary completeness relation

$$\sum_{\substack{\lambda=1,2\\-1,-2}} u_\nu{}^\lambda(\boldsymbol{\varkappa}) u_{\nu'}^{\lambda *}(\boldsymbol{\varkappa}) = \delta_{\nu\nu'}. \tag{16}$$

The reason for our circuitous route is that it is easily extended to the case where, in addition to the restriction of the sum over $\lambda$ to positive values in the definition of $\Lambda$, we might want the further condition that the spin in the rest-frame point in a given direction, $\hat{\mathbf{u}}$. This corresponds to the $\chi^\lambda$ being given by Eq. (7). We would be led then to consider the rather awkward object defined by

$$\Lambda_{\mu\mu'}(\boldsymbol{\varkappa}, \hat{\mathbf{u}}) = u_\mu{}^\lambda(\boldsymbol{\varkappa}) u_{\mu'}^{\lambda *}(\boldsymbol{\varkappa}) \tag{17}$$

with *no* sum over $\lambda$ but only the restriction that it refer to a positive energy state. The sum over $\lambda$ may be reinstated by introducing the operator $(1 + \boldsymbol{\sigma} \cdot \hat{\mathbf{u}})/2$, where $\boldsymbol{\sigma}$ is the $4 \times 4$ matrix

$$\begin{pmatrix} \boldsymbol{\sigma} & 0 \\ 0 & \boldsymbol{\sigma} \end{pmatrix}, \tag{18}$$

as follows:

$$\Delta_{\mu\mu'}(\boldsymbol{\varkappa}, \hat{\mathbf{u}}) = \sum_{\lambda=1,2} \left\{\frac{\epsilon + H_0(\boldsymbol{\varkappa})\beta}{[2\epsilon(\epsilon + m)]^{1/2}} \frac{1 + \boldsymbol{\sigma} \cdot \hat{\mathbf{u}}}{2} \begin{pmatrix} \chi^\lambda \\ 0 \end{pmatrix}\right\}_\mu \left\{\frac{\epsilon + H_0(\boldsymbol{\varkappa})\beta}{[2\epsilon(\epsilon + m)]^{1/2}} \begin{pmatrix} \chi^\lambda \\ 0 \end{pmatrix}\right\}_{\mu'}^\dagger. \tag{19}$$

Next, we insert a factor of $(1 + \beta)/2$ as before and use the completeness relation (11) for the basic spinors, Eq. (11). Recalling that $\beta$ and $\sigma$ commute, we find

$$\Lambda(\boldsymbol{\varkappa}, \hat{\mathbf{u}}) = \frac{2\epsilon}{\epsilon + m} \Lambda(\boldsymbol{\varkappa}) \frac{1 + \boldsymbol{\sigma} \cdot \hat{\mathbf{u}}}{2} \frac{1 + \beta}{2} \Lambda(\boldsymbol{\varkappa}). \tag{20}$$

Finally, we remark that if we were concerned with a projection operator corresponding to a statistical mixture of states with a polarization $\mathbf{P}_0$, we would have

$$\Lambda(\boldsymbol{\varkappa}, \mathbf{P}_0) = \frac{2\epsilon}{\epsilon + m} \Lambda(\boldsymbol{\varkappa}) \frac{1 + \boldsymbol{\sigma} \cdot \mathbf{P}_0}{2} \frac{1 + \beta}{2} \Lambda(\boldsymbol{\varkappa}). \tag{21}$$

It is clear that $\Lambda(\boldsymbol{\varkappa}, \hat{\mathbf{u}})$ may be interpreted as the density matrix for a spin one-half particle with momentum $\boldsymbol{\varkappa}$ and which in its rest-frame has its spin along $\hat{\mathbf{u}}$. Similarly, $\Lambda(\boldsymbol{\varkappa}, \mathbf{P}_0)$ is the density matrix for a particle which in its rest-frame is characterized by the density matrix $(1 + \boldsymbol{\sigma} \cdot \mathbf{P}_0)/2$ appropriate for a particle with degree of polarization $\mathbf{P}_0$.

We return now to Eq. (9*b*) for the differential cross section and carry out the sum over $\lambda'$ using Eq. (14); we find

$$d\sigma^\lambda = (2\pi)^4\epsilon^2 \int d^3r \int d^3r' \, \psi^{+*}_{\boldsymbol{\varkappa},\lambda}(\mathbf{r}')V(r')\frac{e^{i\boldsymbol{\varkappa}_f\cdot\mathbf{r}'}}{(2\pi)^{3/2}} \times \frac{\epsilon + H_0(\boldsymbol{\varkappa}_f)}{2\epsilon}\frac{e^{-i\boldsymbol{\varkappa}_f\cdot\mathbf{r}}}{(2\pi)^{3/2}} V(r)\psi^+_{\kappa,\lambda}(\mathbf{r}). \quad (22a)$$

Since

$$\frac{\epsilon + H_0(\boldsymbol{\varkappa}_f)}{2\epsilon} = \left[\frac{\epsilon + H_0(\boldsymbol{\varkappa}_f)}{2\epsilon}\right]^2, \quad (23)$$

we see that $d\sigma^\lambda$ may be written in terms of a scattering amplitude $f^\lambda$ (which is itself a Dirac spinor):

$$d\sigma^\lambda = f^{\lambda\dagger}f^\lambda = \sum_{\mu=1}^{4} f_\mu^{\lambda*}f_\mu^{\lambda}. \quad (22b)$$

Here

$$f^\lambda = \frac{\epsilon + H_0(\boldsymbol{\varkappa}_f)}{2\epsilon}\left[-(2\pi)^2\epsilon \int d^3r \frac{e^{-i\boldsymbol{\varkappa}_f\cdot\mathbf{r}}}{(2\pi)^{3/2}} V\psi^+_{\boldsymbol{\varkappa},\lambda}(\mathbf{r})\right].^{3} \quad (24)$$

The quantity in the brackets is precisely what is met in nonrelativistic theory, except then $\epsilon = m$. As a matter of fact, it is just the quantity introduced in Eq. (7.77) since $\epsilon = \kappa/(d\epsilon/d\kappa) = \rho_\varepsilon/\kappa$). [See also Eq. (6.70*b*).]

It is important to note that the spinor $f^\lambda$, which is, of course, the amplitude of the scattered wave in coordinate space, is a solution of the free-particle Dirac equation $[\epsilon - H_0(\boldsymbol{\varkappa}_f)]f^\lambda = 0$, which should come as no surprise. The small components of $f^\lambda$ are in a definite and known ratio to the large ones. This then shows that a knowledge of the asymptotic form of the large components of the solution of the Dirac equation in the presence of $V$ is sufficient to determine the scattering cross section. It is essentially for this reason that our earlier description in terms of two-component quantities is complete and involves no nonrelativistic approximation.

To pursue this matter a little further we remark that we may define a $T$-matrix in Dirac spinor space which relates the amplitudes of the initial and final spinors. Thus from Eq. (8) we write

$$T^{\lambda'\lambda}_{\boldsymbol{\varkappa}_f,\boldsymbol{\varkappa}} = u^{\lambda'\dagger}(\boldsymbol{\varkappa}_f)T(\hat{\boldsymbol{\varkappa}}_f \cdot \hat{\boldsymbol{\varkappa}}, \kappa_f, \kappa)u^\lambda(\kappa) \quad (25)$$

[3] We have nowhere made use of the fact that $V$ is an electrostatic scalar, and since we have kept the order of factors straight, Eq. (24) is valid for any form of interaction.

in much the same manner as we introduced a $T$-matrix acting on plane wave states in Chapter 6. Let us see how $T$ can be related to an amplitude $\bar{\mathbf{f}}$ which involves only the large components and hence may be regarded as a $2 \times 2$ matrix. This is trivial since we need use only our explicit formulas for the $u$'s. We restrict our attention to elastic scattering from one positive energy state to another, with $\kappa_f = \kappa$; then

$$\begin{aligned} T^{\lambda'\lambda}_{\boldsymbol{\varkappa}_f,\boldsymbol{\varkappa}} &= \frac{2\epsilon}{\epsilon + m}\begin{pmatrix}\chi^{\lambda'}\\ 0\end{pmatrix}^{\dagger} \frac{\epsilon + H_0(\boldsymbol{\varkappa}_f)}{2\epsilon}\, T(\hat{\boldsymbol{\varkappa}}_f\cdot\hat{\boldsymbol{\varkappa}}, \kappa, \kappa)\, \frac{\epsilon + H_0(\boldsymbol{\varkappa})}{2\epsilon}\begin{pmatrix}\chi^{\lambda}\\ 0\end{pmatrix} \\ &= \frac{2\epsilon}{\epsilon + m}\begin{pmatrix}\chi^{\lambda'}\\ 0\end{pmatrix}^{\dagger}\left(\frac{1+\beta}{2}\right)\Lambda(\boldsymbol{\varkappa}_f)T(\hat{\boldsymbol{\varkappa}}_f\cdot\hat{\boldsymbol{\varkappa}}, \kappa, \kappa)\Lambda(\boldsymbol{\varkappa})\left(\frac{1+\beta}{2}\right)\begin{pmatrix}\chi^{\lambda}\\ 0\end{pmatrix}, \end{aligned} \tag{26}$$

where the factors of $(1 + \beta)/2$ introduced in the second line are unity when acting on our basic spinors. The desired $2 \times 2$ quantity $\bar{\mathbf{f}}$ which is to be sandwiched between $\chi^{\lambda'*}$ and $\chi^{\lambda}$ to yield the $T$-matrix is clearly

$$\bar{\mathbf{f}} = \frac{2\epsilon}{\epsilon + m}\left(\frac{1+\beta}{2}\right)\Lambda(\boldsymbol{\varkappa}_f)T(\hat{\boldsymbol{\varkappa}}_f\cdot\hat{\boldsymbol{\varkappa}}, \kappa, \kappa)\Lambda(\boldsymbol{\varkappa})\frac{1+\beta}{2}. \tag{27}$$

Now consider the spinor $g^{\lambda}(\boldsymbol{\varkappa}_f)$ which results from the action of $T(\hat{\boldsymbol{\varkappa}}_f\cdot\hat{\boldsymbol{\varkappa}}, \kappa, \kappa)$ on $u^{\lambda}(\boldsymbol{\varkappa})$. Evidently $g^{\lambda}(\boldsymbol{\varkappa}_f)$ is, to within a factor, the same as the $f^{\lambda}$ introduced in Eq. (24). As such, it also is a solution of the equation $[\epsilon - H_0(\boldsymbol{\varkappa}_f)]g^{\lambda} = 0$. Hence we may write

$$g^{\lambda}(\boldsymbol{\varkappa}_f) = T(\hat{\boldsymbol{\varkappa}}_f\cdot\hat{\boldsymbol{\varkappa}}, \kappa, \kappa)u^{\lambda}(\boldsymbol{\varkappa}) = \Lambda(\boldsymbol{\varkappa}_f)T(\hat{\boldsymbol{\varkappa}}_f\cdot\hat{\boldsymbol{\varkappa}}, \kappa, \kappa)\Lambda(\boldsymbol{\varkappa})u^{\lambda}(\boldsymbol{\varkappa}), \tag{28}$$

since $\Lambda(\boldsymbol{\varkappa})u^{\lambda}(\boldsymbol{\varkappa}) = u^{\lambda}(\boldsymbol{\varkappa})$. Dropping the arguments of $T$, we insert into Eq. (28) the expression (13) for $\Lambda(\boldsymbol{\varkappa})$ in the form

$$\Lambda(\boldsymbol{\varkappa}) = \frac{2\epsilon}{\epsilon + m}\Lambda(\boldsymbol{\varkappa})\frac{1+\beta}{2}\cdot\frac{1+\beta}{2}\Lambda(\boldsymbol{\varkappa}), \tag{29}$$

[since $(1 + \beta)/2^2 = (1 + \beta)/2$] and obtain

$$g^{\lambda}(\boldsymbol{\varkappa}_f) = \frac{2\epsilon}{\epsilon + m}\Lambda(\boldsymbol{\varkappa}_f)T\Lambda(\boldsymbol{\varkappa})\frac{1+\beta}{2}\cdot\frac{1+\beta}{2}\Lambda(\boldsymbol{\varkappa})u^{\lambda}(\boldsymbol{\varkappa}). \tag{30}$$

Finally multiplying by $(1 + \beta)/2$ and using the definition of $\bar{\mathbf{f}}$ together with the fact that $\Lambda(\boldsymbol{\varkappa})u^{\lambda}(\boldsymbol{\varkappa}) = u^{\lambda}(\boldsymbol{\varkappa})$, we find

$$\frac{1+\beta}{2}g^{\lambda}(\boldsymbol{\varkappa}_f) = \bar{\mathbf{f}}\left(\frac{1+\beta}{2}\right)u^{\lambda}(\boldsymbol{\varkappa}). \tag{31}$$

Thus we see that $\bar{\mathbf{f}}$ transforms the large components of the initial spinor into the large components of the final spinor.

If we were interested in the density matrix $\rho^{\lambda}_{\mu\mu'}$ associated with the scattered beam originating from some specified initial density matrix, we would obtain

$$\begin{aligned}\rho^{\lambda}_{\mu\mu'} &= \overline{g_{\mu}{}^{\lambda} g^{\lambda *}_{\mu'}} \\ &= \overline{[\Lambda(\boldsymbol{\varkappa}_f) T u^{\lambda}(\boldsymbol{\varkappa})]_{\mu} [u^{\lambda *}(\boldsymbol{\varkappa}) T^{\dagger} \Lambda(\boldsymbol{\varkappa}_f)]_{\mu'}} \\ &= [\Lambda(\boldsymbol{\varkappa}_f) T\, \Lambda(\boldsymbol{\varkappa}, \mathbf{P}_0) T^{\dagger}\, \Lambda(\boldsymbol{\varkappa}_f)]_{\mu\mu'},\end{aligned} \tag{32}$$

where $\Lambda(\boldsymbol{\varkappa}, \mathbf{P}_0)$ is defined by Eq. (21). Alternatively, we could work with the $2 \times 2$ quantity obtained from $\bar{\mathbf{f}}$ just as we did in Chapter 7. Again, in spite of its nonrelativistic appearance, the formalism developed there is not so restricted.

The general structure of the matrix $T(\hat{\boldsymbol{\varkappa}}_f \cdot \hat{\boldsymbol{\varkappa}}, \kappa, \kappa)$ is easy to exhibit. It must behave like a true scalar under both rotations and reflections (if the interaction is either an electrostatic or world scalar). Since it is to be "sandwiched" between free-particle spinors for the initial and final states, any terms like $\boldsymbol{\alpha} \cdot \boldsymbol{\varkappa}$ or $\boldsymbol{\alpha} \cdot \boldsymbol{\varkappa}_f$, which we might be tempted to write in the general structure, may be eliminated. No pure pseudoscalar like $\beta \rho_2$ can appear, since one cannot make a pseudoscalar from the two vectors $\boldsymbol{\varkappa}$, $\boldsymbol{\varkappa}_f$. An apparently legitimate term of the form $\beta \boldsymbol{\sigma} \cdot \boldsymbol{\varkappa} \times \boldsymbol{\varkappa}_f$ can be eliminated because $\boldsymbol{\sigma} \cdot (\boldsymbol{\varkappa} \times \boldsymbol{\varkappa}_f) = -i(\boldsymbol{\alpha} \cdot \boldsymbol{\varkappa}, \boldsymbol{\alpha} \cdot \boldsymbol{\varkappa}_f)/2$ and the Dirac equation can again be used to dispose of the $\boldsymbol{\alpha}$'s. It is clear then that the most general form is

$$T = A + B\beta, \tag{33}$$

where $A$, $B$ are scalar functions of $\boldsymbol{\varkappa}_f \cdot \boldsymbol{\varkappa}$ and $\kappa$. If this is substituted into Eq. (27), we find that $\bar{\mathbf{f}}$ has the familiar form

$$\bar{\mathbf{f}} = g_1 + i\boldsymbol{\sigma} \cdot \boldsymbol{\varkappa} \times \boldsymbol{\varkappa}_f g_2, \tag{34}$$

where

$$\begin{aligned}g_1 &= \frac{\epsilon + m + (\epsilon - m)\hat{\boldsymbol{\varkappa}}_f \cdot \hat{\boldsymbol{\varkappa}}}{2\epsilon}(A + B), \\ g_2 &= -\frac{\epsilon - m}{2\epsilon}(A - B).\end{aligned} \tag{35}$$

We shall not discuss the problem of actually integrating the Dirac equation for special cases. This subject is exhaustively treated in a book by Rose devoted entirely to relativistic electron theory.[4] As a very simple illustration of the techniques just developed, we consider the Born approximation for scattering in a screened Coulomb potential.

$$V(r) = \frac{e_1 e_2}{r} e^{-\mu r}, \tag{36}$$

[4] Reference [1]. Further discussions of formal operations with spinors are given by Rose.

where $e_1$, $e_2$ are the charges of projectile and target. The Born approximation consists of replacing $\psi^{(+)}_{\varkappa,\lambda}$ by $\chi_{\varkappa,\lambda}$ in the scattering amplitude, Eq. (24). Thus

$$f^\lambda \simeq \frac{\epsilon + H_0(\varkappa_f)}{2m}\left(-\frac{2m}{4\pi}\int d^3r e^{-i\varkappa_f\cdot\mathbf{r}}\,\frac{e_1e_2}{r}\,e^{-\mu r}e^{i\varkappa\cdot\mathbf{r}}\right)u^\lambda(\varkappa). \tag{37}$$

The bracketed quantity is just the nonrelativistic amplitude. (Note that the $B$ term in Eq. (33) is missing in the present approximation.) We find

$$u^{\lambda'\dagger}(\varkappa_f)f^\lambda = -2\epsilon\frac{e_1e_2}{\mu^2 + (\varkappa - \varkappa_f)^2}\,u^{\lambda'\dagger}(\varkappa_f)u^\lambda(\varkappa), \tag{38}$$

so the $A$ part of the scattering amplitude matrix (the coefficient of $u^{\lambda'\dagger}u^\lambda$ in (38), which differs from the $T$-matrix only by a factor), is

$$A = -2\epsilon\frac{e_1e_2}{\mu^2 + (\varkappa - \varkappa_f)^2}. \tag{39}$$

The counterparts of the $g_1$, $g_2$ (call them $f$, $g$) in Eq. (35) for the scattering amplitude are real, so there are no interesting polarization phenomena. We content ourselves with giving just the differential cross section averaged over initial spin states (see Eq. 7.80).

$$\begin{aligned}\frac{d\bar{\sigma}}{d\Omega} &= f^2 + g^2\sin^2\theta \\ &= \left(\frac{e_1e_2}{2\kappa v}\right)^2\frac{1 - v^2\sin^2\theta/2}{\sin^4\theta/2}.\end{aligned} \tag{40}$$

Here $v$ is the velocity $\kappa/\epsilon$ and we have set $\mu = 0$. In the limit $v^2 \ll 1$, $\kappa \approx mv$ and we recover the nonrelativistic result.

## 9.2 THE SCATTERING AND PHOTOPRODUCTION OF PI-MESONS FROM NUCLEONS

As an illustration of the general methods of Section 7.3 we consider the physically interesting problem of the scattering of pi-mesons (called pions hereafter) by nucleons and the production of pions by photons incident on nucleons. What we shall be concerned with is the diagonalization of the $S$-matrix for these processes in terms of eigenstates labeled with appropriate quantum numbers. We restrict our attention to energies so low that production of another pion is forbidden. Furthermore, we exploit the smallness of the coupling constant between the charged particles and the electromagnetic field to restrict attention to states where no more than a single photon appears. From a practical standpoint this is an entirely harmless approximation. Ignoring for a moment different charge states (as well as all weak interactions which couple pions and nucleons to electrons, mesons, and neutrinos), we evidently have to do with a two-channel process: $\gamma$-ray plus nucleon and pion plus nucleon. Schematically, the $S$-matrix has the form

$$S = \begin{pmatrix} S_{\pi\pi} & S_{\pi\gamma} \\ S_{\gamma\pi} & S_{\gamma\gamma} \end{pmatrix}. \tag{41}$$

Here $S_{\pi\pi}$ describes pion-nucleon scattering, $S_{\pi\gamma}$ is the matrix element for $\gamma$ plus nucleon $\rightarrow$ pion plus nucleon, $S_{\gamma\pi}$ describes the inverse reaction of pion absorption with $\gamma$-ray emission, and finally $S_{\gamma\gamma}$ is the matrix element for Compton scattering from protons.

Our assumption about the weakness of the electromagnetic interaction allows us to simplify the $S$-matrix somewhat. Since $S_{\gamma\gamma}$ differs from unity by terms of the order of the square of the electric charge $e$, (since a photon has to be absorbed and emitted), we set $S_{\gamma\gamma}$ equal to the unit matrix in the $\gamma$-ray nucleon subspace. The matrix elements $S_{\pi\gamma}$ and $S_{\gamma\pi}$ are proportional to the first power of $e$ and are small compared to $S_{\pi\pi}$. Furthermore, we shall assume in treating $S_{\pi\pi}$ that we may ignore the radiative corrections ($\sim e^2$) to pion scattering.

Unfortunately, the problem of real physical interest is considerably more complicated. Pions are known to occur in three charge states ($\pi^+$, $\pi^-$, $\pi^\circ$) and nucleons in two charge states (neutrons and protons); consequently, there are in all six pion-nucleon channels. There are four $\gamma$-ray channels corresponding to the two possible circular polarizations of photons and the two nucleon charge states. Thus $S$ is a $10 \times 10$ matrix in "channel space."

We shall work entirely in the barycentric subspace as was done in Chapter 7. Then in addition to pion and nucleon charge labels, $S$ must also have matrix labels giving the relative momentum of the pion and nucleon or $\gamma$-ray and nucleon, nucleon spin, and $\gamma$-ray polarization.

The submatrix $S_{\pi\pi}$ is treated first. We assume that electric charge, parity, and angular momentum are conserved and that all interactions are invariant under time reversal. Furthermore, we shall also accept the principle of charge independence (to be explained later). Our procedure is to follow very closely the results of Chapter 7 and not to use many of the special tricks which can always be devised to solve any given problem. A much more compact treatment, using such techniques, is given in Chapter 10, Section 10.6, p. 652.

**Pion-Nucleon Scattering**

The $6 \times 6$ submatrix $S_{\pi\pi}$ may be diagonalized using the general expressions (7.124) and (7.130). To effect the final diagonalization, (7.130), explicitly we shall assume the principle of charge independence to which we have alluded previously and which is now described.

We introduce three pion "charge-state" wave functions $\pi^\alpha(\alpha = \pm 1, 0)$, defined so that

$\pi^{+1}$ corresponds to a pion of positive charge,
$\pi^0$ corresponds to a pion of neutral charge,
$\pi^{-1}$ corresponds to a pion of negative charge.

The two nucleon charge-state wave functions may be written as $n^\lambda$ ($\lambda = \pm\frac{1}{2}$), where

$n^{1/2}$ corresponds to a proton,
$n^{-1/2}$ corresponds to a neutron.

From a purely mathematical point of view the $\pi^\alpha$ may be considered as transforming as first-order spherical harmonics under the three-dimensional rotations in "charge space."[5] The $n^\lambda$ transform as spinors. It is sometimes convenient to use in place of the $\pi^\alpha$ quantities related to the $\pi^\alpha$ in the same way the components of the vector $\mathbf{r}$ are related to the first-order spherical harmonics $Y_1{}^m$. At the (small) risk of confusion we refer to these as $\pi_\alpha$, where the subscript $\alpha$ takes on the values 1, 2, 3. The precise relation will be taken to be

$$\pi^+ = -\frac{\pi_1 + i\pi_2}{(2)^{1/2}},$$

$$\pi^0 = \pi_3,$$

$$\pi^- = \frac{\pi_1 - i\pi_2}{(2)^{1/2}} \quad . \tag{42}$$

Pursuing the obvious analogy with angular momenta, we define a set of operators $\tau_\alpha$ ($\alpha = 1, 2, 3$) which describe the behavior of the spinors $n^\lambda$ under rotations and which are numerically indistinguishable from the Pauli spin matrices

$$\tau_1 = \begin{pmatrix} 0 & 1 \\ 1 & 0 \end{pmatrix} \qquad \tau_2 = \begin{pmatrix} 0 & -i \\ i & 0 \end{pmatrix} \qquad \tau_3 = \begin{pmatrix} 1 & 0 \\ 0 & -1 \end{pmatrix}. \tag{43}$$

The corresponding set which generates rotations among the $\pi_\alpha$ may be taken to be the spin one matrices (2.37)

$$(T_j)_{\alpha\beta} = i\epsilon_{\alpha j\beta}, \qquad j = 1, 2, 3. \tag{44}$$

[5] This was first observed by Kemmer [2]. That charge independence leads to restrictions on pion-nucleon scattering was pointed out by Heitler (Ref. [3]). Charge independence as a general symmetry principle (formulated independently of field theory) for elementary particle reactions was suggested by Brueckner and Watson [4].

We shall not have much cause to use these operators in this chapter, but we will return to them in Section 10.6. We may, of course, choose a representation in which $T_3$ is diagonal, and in fact, it is just the combination of states given in (42) which does this. Then $T_3\pi^+ = \pi^+$, $T_3\pi^0 = 0$, $T_3\pi^- = -\pi^-$. The operator $\tau_3$ is already diagonal and the nucleon charge operator may evidently be taken to be $Q = (1 + \tau_3)/2$, so $Qn^{1/2} = n^{1/2}$, $Qn^{-1/2} = 0$. For the pions, $Q = T_3$, so the total charge operator is $Q = \tau_3/2 + T_3 + \frac{1}{2}$.

We may form a total isotopic spin operator $\mathbf{I} = \boldsymbol{\tau}/2 + \mathbf{T}$ and construct eigenstates of $\mathbf{I}$, $\mathbf{I}_z$ with eigenvalues $I^2 = i(i + 1)$, $I_z = \mu$ in the familiar fashion using Wigner coefficients. According to Eq. (1.59*a*), the eigenstates corresponding to the eigenvalues $i$, $\mu$ are of the form

$$|i, \mu\rangle = \sum_{\lambda,\alpha} \langle 1, \tfrac{1}{2}; \alpha, \lambda \mid i, \mu\rangle \, |\pi^\alpha n^\lambda\rangle. \tag{45}$$

The possible values of $i$ and $\mu$ are $i = \frac{3}{2}$, $\mu = \frac{3}{2}, \frac{1}{2}, -\frac{1}{2}, -\frac{3}{2}$ or $i = \frac{1}{2}$, $\mu = \frac{1}{2}, -\frac{1}{2}$.

The principle of charge independence is the statement that nuclear interactions are invariant under rotations in isotopic space. Since these rotations are formally similar to rotations in coordinate space, the diagonalization of the scattering matrix for isotropic states is formally equivalent to that for angular momentum. In particular, the $S$-matrix (or the scattering matrix $T$) is diagonal in $i$ and independent of the charge quantum number $\mu$. Charge independence is clearly not exact. It is violated by the electromagnetic interactions and the weak interactions. Nevertheless, it has been of great use in the analysis of the strong nuclear interactions.

To describe pion-nucleon scattering, we may use Eq. (7.124) as it stands. The initial and final relative momenta of pion and nucleon are written as $\boldsymbol{\varkappa}$ and $\boldsymbol{\varkappa}_f$, respectively. (The detailed kinematics are taken up in Section 10.6.) The pion has spin zero, so $s = s' = \frac{1}{2}$ is just the nucleon spin in (7.124). The channel labels are written as

$$\begin{aligned} c &= \pi(\alpha\lambda), \\ c' &= \pi(\alpha'\lambda') \end{aligned} \tag{46}$$

to designate a pion-nucleon state with charges $\alpha(\alpha')$ and $\lambda(\lambda')$ for the pion and nucleon, respectively. In order to save writing, with the understanding that we are concerned with the submatrix $S_{\pi\pi}$, we write the channel labels simply as $(\alpha\lambda)$, $(\alpha'\lambda')$. For a given value of $J$, we may have the orbital angular momentum $l = J \pm \frac{1}{2}$; since parity is conserved, $l' = l$. The $S$-matrix elements in (7.124) take the form

$$S^J_{c'l's',cls} \equiv S^{Jl}_{\alpha'\lambda',\alpha\lambda}\delta_{l',l}\delta_{\alpha'+\lambda',\alpha+\lambda}. \tag{47}$$

The condition $\alpha' + \lambda' = \alpha + \lambda$ is just a statement of charge conservation.

The differential scattering cross section may be written down immediately using Eqs. (7.78) and (7.80). For the scattering corresponding to $(\alpha\lambda) \to (\alpha'\lambda')$, where we always understand the restriction $\alpha + \lambda = \alpha' + \lambda'$, we have

$$\frac{d\bar{\sigma}}{d\Omega}(\alpha\lambda \to \alpha'\lambda') = |f|^2 + \sin^2\theta\,|g|^2. \tag{48}$$

Here $f$ and $g$ are given by Eq. (7.78) provided we generalize the expressions (7.576) for $T_{l\pm}$ to take into account the charge labels. We make the substitution[6] [see Eq. (7.132)]

$$T_{l\pm} \to \frac{1}{2\pi i\rho_\epsilon}(\delta_{\alpha',\alpha}\delta_{\lambda',\lambda} - S^{(l\pm\frac{1}{2})l}_{\alpha'\lambda',\alpha\lambda}) \tag{49}$$

in Eq. (7.78).

To illustrate our results we use the experimental fact that only states with $l = 0, 1$ seem to be required for a good description of pion-nucleon scattering in the (kinetic) energy region 0 to 200 Mev. Keeping only $l = 0, 1$, we find

$$\begin{aligned} f &= \frac{1}{2i\kappa}\{[S^{\frac{1}{2}\,0}_{\alpha'\lambda',\alpha\lambda} - \delta_{\alpha'\alpha}\delta_{\lambda'\lambda}] + [2(S^{\frac{3}{2}\,1}_{\alpha'\lambda',\alpha\lambda} - \delta_{\alpha',\alpha}\delta_{\lambda',\lambda}) \\ &\quad + S^{\frac{1}{2}\,1}_{\alpha'\lambda',\alpha\lambda} - \delta_{\alpha',\alpha}\delta_{\lambda',\lambda}]\cos\theta\}, \\ g &= \frac{1}{2i\kappa}\{S^{\frac{3}{2}\,1}_{\alpha'\lambda',\alpha\lambda} - S^{\frac{1}{2}\,1}_{\alpha'\lambda',\alpha\lambda}\}. \end{aligned} \tag{50}$$

We note that these expressions are obtained from Eq. (7.81) with the substitution

$$e^{i\delta}\sin\delta \to \frac{1}{2i}(S_{\alpha'\lambda'\cdot\alpha\lambda} - \delta_{\alpha',\alpha}\delta_{\lambda',\lambda}). \tag{51}$$

We have done nothing more than to transcribe Eq. (7.124) into a form appropriate to pion-nucleon scattering. To complete the diagonalization of $S$ we turn to Eq. (7.130) and use the principle of charge independence. Since the matrix elements $S^J_{c'l's',cls}$ are already diagonal in all except the charge labels, the general matrix $U^J_{c'l's';a}$ will depend only on these. Indeed, by definition, it can be only the Wigner coefficient giving the transformation from the $\alpha\lambda$ to the $i\mu$ representation

$$U^J_{c'ls;\alpha} = \langle 1, \tfrac{1}{2}; \alpha, \lambda \mid i, \mu\rangle. \tag{52}$$

[6] We are here neglecting the neutron-proton mass difference as well as the $\pi^\pm - \pi^0$, mass difference; these are (presumably) electromagnetic in origin and, of course, necessarily violate the charge independence hypothesis. Thus in (7.132) we have $\rho_{\epsilon_c'} = \rho_{\epsilon_c}$ for the six pion-nucleon channels.

Therefore Eq. (7.130) takes the form

$$S^{Jl}_{\alpha'\lambda',\alpha\lambda} = \sum_i \langle 1, \tfrac{1}{2}; \alpha', \lambda' \mid i, \alpha + \lambda\rangle e^{2i\delta(Jli)} \langle i, \lambda + \alpha \mid 1, \tfrac{1}{2}; \alpha, \lambda\rangle. \qquad (53)$$

The $\delta(Jli)$ are the eigenphase shifts for pion-nucleon scattering. Since most analyses of experimental data have included only $s$- and $p$-wave scattering, a special notation has come into general use. This is

$$\begin{aligned}
s\text{-wave: } & J = \tfrac{1}{2};\ l = 0;\ i = \tfrac{1}{2}, \tfrac{3}{2}; \\
& \delta(\tfrac{1}{2}, 0, \tfrac{1}{2}) \equiv \delta_1, \\
& \delta(\tfrac{1}{2}, 0, \tfrac{3}{2}) \equiv \delta_3. \\
p\text{-wave: } & J = \tfrac{1}{2}, \tfrac{3}{2};\ l = 1;\ i = \tfrac{1}{2}, \tfrac{3}{2}; \\
& \delta(\tfrac{1}{2}, 1, \tfrac{1}{2}) \equiv \delta_{11}, \\
& \delta(\tfrac{1}{2}, 1, \tfrac{3}{2}) \equiv \delta_{31}, \\
& \delta(\tfrac{3}{2}, 1, \tfrac{1}{2}) \equiv \delta_{13}, \\
& \delta(\tfrac{3}{2}, 1, \tfrac{3}{2}) \equiv \delta_{33}.
\end{aligned} \qquad (54)$$

It has been found experimentally that there is a resonance in the $J = \frac{3}{2}$, $i = \frac{3}{2}$ state at a barycentric kinetic energy $\epsilon_0 \simeq 160$ Mev.[7] This suggests an approximation in which we neglect scattering in all but this state. From Eqs. (48) and (50), we obtain

$$\bar{\sigma}(\alpha\lambda \to \alpha'\lambda') = \int d\Omega \frac{d\sigma}{d\Omega}(\alpha\lambda - \alpha'\lambda') = \frac{2\pi}{\kappa^2} |S^{3/2 1}_{\alpha'\lambda',\alpha\lambda} - \delta_{\alpha'\alpha}\delta_{\lambda'\lambda}|^2, \qquad (55)$$

and using the explicit values for the Wigner coefficients in Eq. (53), we find

$$\begin{aligned}
\pi^+ p \to \pi^+ p: &\quad S^{3/2 1}_{1 1/2, 1 1/2} = e^{2i\delta_{33}}, \\
\pi^- p \to \pi^- p: &\quad S^{3/2 1}_{-1 1/2, -1 1/2} = \tfrac{1}{3}(e^{2i\delta_{33}} - 1) + 1, \\
\pi^- p \to \pi^0 n: &\quad S^{3/2 1}_{0 -1/2, -1 1/2} = \frac{\sqrt{2}}{3}(e^{2i\delta_{33}} - 1).
\end{aligned} \qquad (56)$$

Therefore

$$\begin{aligned}
\bar{\sigma}(\pi^+ p \to \pi^+ p) &= \frac{8\pi}{2} \sin^2 \delta_{33}, \\
\bar{\sigma}(\pi^- p \to \pi^- p) &= \tfrac{1}{9}\,\bar{\sigma}(\pi^+ p - \pi^+ p), \\
\bar{\sigma}(\pi^- p \to \pi^0 n) &= \tfrac{2}{9}\,\bar{\sigma}(\pi^+ p - \pi^+ p).
\end{aligned} \qquad (57)$$

Actually, the preceding ratios 9:1:2 obtain under a less restrictive assumption, namely, that the scattering takes place only in the state of isotopic

[7] See, for example, Ref. [5].

spin for which $i = \frac{3}{2}$. This may be seen by carrying out the diagonalization of the $S$-matrix in isotopic space before even discussing the angular dependence of the matrix elements. In this way we find, in an abbreviated notation,

$$\begin{aligned} \langle \pi^+ p | \, T \, | \pi^+ p \rangle &= T_{3/2}, \\ \langle \pi^- p | \, T \, | \pi^- p \rangle &= \frac{T_{3/2} + 2T_{1/2}}{3}, \\ \langle \pi^0 n | \, T \, | \pi^- p \rangle &= \frac{\sqrt{2}\,(T_{3/2} - T_{1/2})}{3}, \end{aligned} \tag{58}$$

where $T_{3/2}$, $T_{1/2}$ are the scattering matrices for the pure isotopic spin states $i = \frac{3}{2}$, $i = \frac{1}{2}$. Since the differential cross sections are proportional to the absolute squares of these $T$-matrix elements, we see that

$$\begin{aligned} &\frac{d\sigma}{d\Omega}(\pi^+ + p \rightarrow \pi^+ + p): \\ &\frac{d\sigma}{d\Omega}(\pi^- + p \rightarrow \pi^- + p): \\ &\frac{d\sigma}{d\Omega}(\pi^- + p \rightarrow \pi^0 + n) = 9:1:2 \end{aligned} \tag{59}$$

provided $T_{1/2} = 0$. Note also the general relation

$$\frac{d\sigma}{d\Omega}(\pi^- + p - \pi^- + p) + \frac{d\sigma}{d\Omega}(\pi^- + p - \pi^0 + p) = \frac{1}{3}\frac{d\sigma_{3/2}}{d\Omega} + \frac{2}{3}\frac{d\sigma_{1/2}}{d\Omega}, \tag{60}$$

where $(d\sigma_{3/2}/d\Omega)$, $(d\sigma_{1/2}/d\Omega)$ are the differential cross sections for the indicated pure isotopic spin states. Of course, $d\sigma_{3/2}/d\Omega$ is just the cross section for $(\pi^+ p)$ scattering. Hence the isotopic spin one-half cross section may be found from experimental measurements of $(\pi^+ p)$ scattering and the total differential $\pi^- p$ cross section.

We have commented on the fact that the scattering in the $J = \frac{3}{2}$, $i = \frac{3}{2}$ state corresponds to a resonance. We may therefore use Eq. (8.341) to describe this. Since the $\pi^+ p$ state is pure $i = \frac{3}{2}$, we take $\Gamma_{c'} = \Gamma_c = \Gamma$ in (8.341) and obtain

$$\bar{\sigma}(\pi^+ + p \rightarrow \pi^+ + p) = \frac{2\pi}{\kappa^2} \frac{\Gamma^2}{(\epsilon - \epsilon_0)^2 + \Gamma^2/4}. \tag{61}$$

A comparison of this expression with experiment is shown in Fig. 9.1[8].

[8] This is taken from Fig. 5 of Ref. [6].

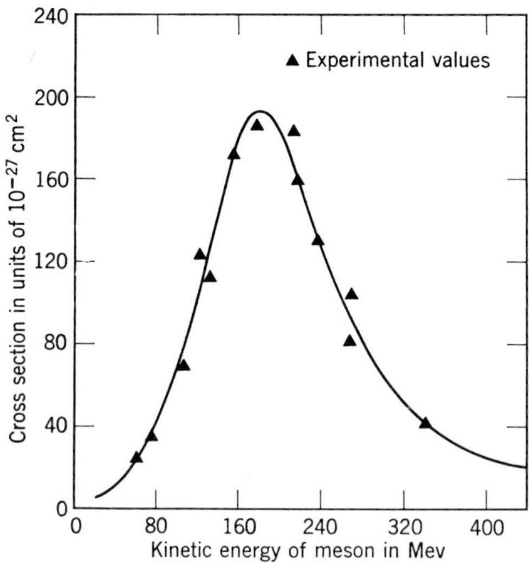


Fig. 9.1. The cross section of Eq. (61) compared with experimental $\pi^+$-proton cross sections.

**The Eigenstates for Photons**

Before discussing the $S$-matrix for photomeson production from nucleons, we must describe how to decompose the wave function for a photon into states of given angular momentum. This problem may be handled in an elegant manner using the techniques of Jacob and Wick, described in Appendix E. We choose, however, to discuss it in a more pedestrian fashion, and then later give a very simple projection operator method for describing the photoproduction process.

Since photons have spin one, we may introduce a spin function $\mathscr{Y}^\rho(\rho = 1, 0, -1)$. If we call $\mathbf{S}_\gamma$ the photon spin operator and quantize along the $z$-direction, we have[9]

$$\mathbf{S}_\gamma^2 \mathscr{Y}^\rho = 2\mathscr{Y}^\rho,$$
$$S_{\gamma z} \mathscr{Y}^\rho = \rho \mathscr{Y}^\rho. \tag{62}$$

Corresponding to an orbital angular momentum $l$ with $z$-projection $l_z = m$, we may construct product states $|l, 1;\ m\rho\rangle$ and eigenstates of total angular momentum $j$ by writing

$$\mathscr{Y}_{jl}{}^\mu = \sum_\rho |l, 1; \mu - \rho, \rho\rangle\langle l, 1; \mu - \rho, \rho \mid j, \mu\rangle, \tag{63}$$

where $j$ takes on the values $j = l, l \pm 1$.

[9] Let the nervous reader be assured that we will ultimately recognize what he already knows, that photons have only two polarizations, not three.

Because of the Lorentz condition, which implies that the radiation field is transverse, these states $\mathscr{Y}_{jl}{}^{\mu}$ are not, in general, physically allowed photon states. The physical states $\Phi$ must satisfy the condition

$$\nabla \cdot \mathbf{A}(\mathbf{r})\Phi = 0 \tag{64}$$

where $\mathbf{A}(\mathbf{r})$ is the vector potential field operator. This has the form [See Eq. (8.170)][10,11]

$$\mathbf{A}(\mathbf{r}) = \sum_{\hat{e}} \int \frac{d^3k}{[(2\pi)^2 k]^{1/2}}\, \hat{e}[a_{\mathbf{k},\hat{e}} e^{i\mathbf{k}\cdot\mathbf{r}} + a^{\dagger}_{\mathbf{k},\hat{e}} e^{-i\mathbf{k}\cdot\mathbf{r}}]. \tag{65}$$

As in Chapter 8, $\hat{e}$ represents the photon polarization vector (which depends on $\mathbf{k}$ even though we suppress this label). The annihilation ($a_{\mathbf{k},\hat{e}}$) and creation ($a^{\dagger}_{\mathbf{k},\hat{e}}$) operators now satisfy the commutation rules

$$\begin{aligned} [a_{\mathbf{k},\hat{e}}, a_{\mathbf{k}',\hat{e}'}] &= [a^{\dagger}_{\mathbf{k},\hat{e}}, a^{\dagger}_{\mathbf{k}',\hat{e}'}] = 0, \\ [a_{\mathbf{k},\hat{e}}, a^{\dagger}_{\mathbf{k}',\hat{e}'}] &= \delta(\mathbf{k} - \mathbf{k}')\delta_{\hat{e},\hat{e}'}. \end{aligned} \tag{66}$$

The polarization vectors $\hat{e}$ are restricted to be orthogonal to $\mathbf{k}$. We may extend the sum to any three orthogonal unit vectors as long as we confine our attention to states satisfying $\nabla \cdot \mathbf{A}\Phi = 0$. That is, as long as the physical states have no "longitudinal" photons, the extra dynamical degree of freedom has no significance.

With this proviso, we will henceforth include the longitudinal polarization in our sums over $\hat{e}$. It is convenient to define three new field operators $A^{\rho}(\mathbf{r})[\rho = 1, 0, -1]$ by

$$\begin{aligned} A^{\pm}(\mathbf{r}) &= \mp \frac{1}{\sqrt{2}}(A_x \pm iA_y), \\ A^{0}(\mathbf{r}) &= A_z, \end{aligned} \tag{67}$$

expressed in terms of rectangular components of $A$. Similarly, a gradient operator $\nabla^{\rho}$ is defined by

$$\begin{aligned} \nabla^{\pm} &= \mp \frac{1}{\sqrt{2}}\left(\frac{\partial}{\partial x} \pm i\frac{\partial}{\partial y}\right), \\ \nabla^{0} &= \frac{\partial}{\partial z}. \end{aligned} \tag{68}$$

The Lorentz condition on physical states becomes

$$\sum_{\rho} (\nabla^{\rho})^{*} A^{\rho}(\mathbf{r})\Phi = 0. \tag{69}$$

[10] We continue to use units in which $\hbar = c = 1$.

[11] In obtaining this form from Eq. (8.170), we have gone to continuum normalization with the substitutions $\sum_{\mathbf{k}} \to \frac{\mathscr{V}}{(2\pi)^3}\int d^3k$ and $a_{\mathbf{k},\hat{e}} \to \sqrt{(2\pi)^3/\mathscr{V}}\, a_{\mathbf{k},\hat{e}}$.

Using these new definitions, we may write

$$A^\rho(\mathbf{r}) = \int \frac{d^3k}{2\pi k^{1/2}} [(-1)^\rho a_{\mathbf{k},-\rho} e^{i\mathbf{k}\cdot\mathbf{r}} + a^\dagger_{\mathbf{k},\rho} e^{-i\mathbf{k}\cdot\mathbf{r}}], \tag{70}$$

where

$$\begin{aligned} a^\dagger_{\mathbf{k},\rho} &= \sum_{\hat{e}} a^\dagger_{\mathbf{k},\hat{e}} e^\rho(\hat{e}), \\ a_{\mathbf{k},\rho} &= [a^\dagger_{\mathbf{k},\rho}]^\dagger, \\ e^\pm(\hat{e}) &= \mp \frac{1}{\sqrt{2}} (\hat{e}_x \pm i\hat{e}_y), \\ e^0(\hat{e}) &= \hat{e}_z. \end{aligned} \tag{71}$$

Since our sum over $\hat{e}$ extends over three orthogonal unit vectors for every $\mathbf{k}$, we obtain

$$[a_{\mathbf{k},\rho}, a^\dagger_{\mathbf{k}',\rho'}] = \delta(\mathbf{k} - \mathbf{k}')\delta_{\rho,\rho'} \tag{72}$$

and all other commutators vanish.

To discuss the scattering matrix, we require only the asymptotic photon states at large distances from regions of interaction. We consider a plane photon wave corresponding to a photon with propagation vector $\mathbf{k}$ and "polarization" $\rho'$, say. Such a state is obtained by operating on the vacuum state $\Phi_0$ with $a^\dagger_{\mathbf{k},\rho'}$.

$$\Phi(\mathbf{k}, \rho') = a^\dagger_{\mathbf{k},\rho'}\Phi_0. \tag{73}$$

We imagine $\mathbf{k}$ to be along the $z$-direction, the axis of quantization. Then the "wave function" for such a state is defined as

$$(\Phi_0, A^\rho(\mathbf{r})\Phi(\mathbf{k}, \rho')) = \frac{(-1)^{-\rho}}{2\pi k^{1/2}} e^{i\mathbf{k}\cdot\mathbf{r}} \delta_{-\rho,\rho'}. \tag{74}$$

The Lorentz condition becomes

$$(\Phi_0, \sum_\rho (\nabla^\rho)^* A^\rho(\mathbf{r})\Phi(\mathbf{k}, \rho')) = \frac{(-1)^{-\rho'}}{2\pi k^{1/2}} (\nabla^{-\rho'})^* e^{i\mathbf{k}\cdot\mathbf{r}}. \tag{75}$$

Since $\mathbf{k}$ is parallel to $z$, this vanishes for $\rho' = \pm 1$, but not for $\rho' = 0$. The states $\Phi(k, \rho)$ for which $\rho = \pm 1$ are therefore physically admissible, but that corresponding to $\rho = 0$ is not. The allowed states are conventionally referred to as left-circularly polarized, $\rho = +1$, and right-circularly polarized, $\rho = -1$.

We turn now to the rather involved problem of finding eigenstates of the total angular momentum $j$ for photons at large distances from any region of interaction. The general form of such states will be

$$\Phi_j{}^\mu = \sum_l \alpha_l \mathscr{Y}_{jl}{}^\mu, \tag{76}$$

where $l$ runs over $-1, j$, and $j + 1$; the $\alpha_l$ are constants to be determined. We normalize the $\Phi^{\mu}$ to unity.

$$(\Phi_j^{\mu}, \Phi_j^{\mu}) = \sum_l |\alpha_l|^2 = 1. \tag{77}$$

It is useful to classify the states $\Phi_j^{\mu}$ according to their parity. Since **A** may be taken as a polar vector the one-photon spin states may be taken to have negative parity (measured with respect to the vacuum). Thus we take

$$\mathscr{P}\mathscr{Y}^{\rho} = -\mathscr{Y}^{\rho}, \tag{78}$$

where $\mathscr{P}$ is the parity operator. The states $\Phi_j^{\mu}$ are therefore of two classes: those with parity $-(-1)^j$ are uniquely defined and written as

$$\Phi_j^{\mu}(M) = \mathscr{Y}_{jj}^{\mu}; \tag{79}$$

these are called magnetic multipoles of order $j$. The angular distribution (and parity) characterized by $l = 1$, $j = 1$ is called a magnetic dipole radiation pattern, that for which $l = 2, j = 2$ is called magnetic quadrupole and so on.

The remaining allowed states have parity $(-1)^j$ and the general form

$$\Phi_j^{\mu}(E) = \alpha_+\mathscr{Y}^{\mu}_{j,j+1} + \alpha_-\mathscr{Y}^{\mu}_{j,j-1}, \tag{80}$$

where $\alpha_+$ and $\alpha_-$ are two constants satisfying

$$|\alpha_+|^2 + |\alpha_-|^2 = 1. \tag{81}$$

The allowed states must satisfy the Lorentz condition

$$(\Phi_0, (\nabla^{\rho})^* A^{\rho}(\mathbf{r})\Phi_j^{\mu}) = 0, \tag{82}$$

which together with the normalization condition on the $\alpha$'s can be used to find $\alpha_+$, $\alpha_-$.

It is convenient to express the operator **A** in terms of spherical waves using Eq. (1.85*a*):

$$e^{i\mathbf{k}\cdot\mathbf{r}} = \sum_{l=0} \sum_{m=-l}^{l} 4\pi i^l j_l(kr) Y_l^m(\hat{\mathbf{r}}) Y_l^{m*}(\hat{\mathbf{k}}). \tag{83}$$

Then

$$A^{\rho}(\hat{\mathbf{r}}) = 2\int k^{3/2}\, dk \sum_{l,m} i^l j_l(kr)[(-1)^{\rho} a_{k,l,m,-\rho} Y_l^m(\hat{\mathbf{r}}) + (-1)^l a^{\dagger}_{k,l,m,\rho} Y_l^{m*}(\hat{\mathbf{r}})], \tag{84}$$

where we have defined

$$a^{\dagger}_{\mathbf{k},l,m,\rho} \equiv \int d\Omega_{\mathbf{k}} a^{\dagger}_{\mathbf{k},\rho} Y_l^m(\hat{\mathbf{k}}), \tag{85}$$

$$a_{k,l,m,\rho} \equiv [a^{\dagger}_{k,l,m,\rho}]^{\dagger}.$$

These operators correspond to creation and destruction operators for photons in states $|l1;\ m\rho\rangle$. The commutation relations they satisfy are [See Eq. (6.54)]

$$[a_{k,lm,\rho}, a^\dagger_{k',l',m',\rho'}] = \frac{1}{k^2}\,\delta(k-k')\delta_{l,l'}\delta_{m,m'}\delta_{\rho,\rho'}. \tag{86}$$

Using this new expansion for $A^\rho(\mathbf{r})$, the Lorentz condition becomes

$$\int k^{3/2}\, dk \sum_{l,m,\rho} (-1)^\rho i^l[(\nabla^{-\rho})^* j_l(kr) Y_l^{\,m}(\hat{\mathbf{r}})](\Phi_0, a_{k,l,m,\rho}\Phi_j^{\,\mu}) = 0. \tag{87}$$

We need concern ourselves only with asymptotic states, so we use

$$\lim_{r\to\infty} (-1)^\rho i^l[(\nabla^{-\rho})^* j_l(kr) Y_l^{\,m}(\hat{\mathbf{r}})] = F(r) Y_1^{\,\rho}(\hat{\mathbf{r}}) Y_l^{\,m}(\hat{\mathbf{r}}). \tag{88}$$

Here $F(r)$ is a function of $r$ independent of $l$. With this result, the Lorentz condition reduces to the requirement

$$\sum_{l,m,\rho} Y_1^{\,\rho}(\hat{\mathbf{r}}) Y_l^{\,m}(\hat{\mathbf{r}})(\Phi_0, a_{k,l,m,\rho}\Phi_j^{\,\mu}) = 0. \tag{89}$$

Now from Eq. (1.79) we have

$$Y_1^{\,\rho}(\hat{\mathbf{r}}) Y_l^{M-\rho}(\hat{\mathbf{r}}) = \sum_{L=|l-1|}^{L=l+1} \left[\frac{3(2l+1)}{4\pi(2L+1)}\right]^{1/2} \langle L, M \mid l, 1; M-\rho, \rho\rangle$$
$$\times\ \langle l, 1; 0, 0 \mid L, 0\rangle Y_L^{\,M}(\hat{\mathbf{r}}), \tag{90}$$

and this is to be substituted into Eq. (89). The coefficient of each $Y_L^{\,M}(\hat{\mathbf{r}})$ must vanish identically since (89) must hold for all $\mathbf{r}$. We find the set of conditions

$$\sum_l (2l+1)^{1/2} \langle l, 1; 0, 0 \mid L, 0\rangle(\Phi_0, a^L_{k,l,M}\Phi_j^{\,\mu}) = 0, \tag{91}$$

where

$$a^L_{k,l,M} = \sum_\rho \langle L, M \mid l, 1; M-\rho, \rho\rangle a_{k,l,m,\rho}. \tag{92}$$

This is evidently the destruction operator for a photon in the state $\mathscr{Y}^M_{L,l}$.

This is our final statement of the Lorentz condition. The magnetic multipole states $\Phi_j^{\,\mu}(M)$ satisfy it automatically since

$$\langle j, 1;\ 0, 0 \mid j, 0\rangle = 0. \tag{93}$$

This should come as no surprise, since there was only a single state of the required parity available.

The requirement on the electric multipole state $\Phi_j^{\,\mu}(E)$ becomes

$$(2j+3)^{1/2}\langle j+1, 1;\ 0, 0 \mid j, 0\rangle\alpha_+$$
$$+\ (2j-1)^{1/2}\langle j-1, 1;\ 0, 0 \mid j, 0\rangle\alpha_- = 0. \tag{94}$$

Using the explicit formulas for the Wigner coefficients and the normalization condition on the $\alpha$'s, we find

$$\Phi_j^{\mu}(E) = \sqrt{j/(2j+1)}\mathscr{Y}^{\mu}_{j,j+1} + \sqrt{(j+1)/(2j+1)}\mathscr{Y}^{\mu}_{j,j-1}. \tag{95}$$

The expressions for $\Phi_j^{\mu}(M)$ and $\Phi_j^{\mu}(E)$ represent the physically admissible photon states. It is worth noting that there is no state corresponding to $j = 0$, which is not unexpected in view of the vector character of the electromagnetic field.

There is a simple operatorial representation of these multipole wave functions which is quite useful; for the magnetic multipoles,

$$\langle j\mu \mid \mathbf{k}, \hat{e}\rangle_{\text{mag}} = \frac{\hat{e}\cdot\mathbf{L}_k}{\sqrt{j(j+1)}}\, Y_j^{\mu}(\hat{\mathbf{k}})^* \tag{96}$$

where $\mathbf{L}_k = -i\mathbf{k}\times\nabla_{\mathbf{k}}$. Note that under a reflection in which $\mathbf{k}\to-\mathbf{k}$, $\hat{e}\to-\hat{e}$, $\Phi_j^{\mu}\to(-1)^{j+1}\Phi_j^{\mu}$ as it should. The factor $[j(j+1)]^{-1/2}$ insures the normalization to unity, as required. For the electric multipoles we have

$$\langle j\mu \mid \mathbf{k}, \hat{e}\rangle_{\text{elec}} = -i\,\frac{\hat{e}\cdot(\mathbf{k}\times\mathbf{L}_k)}{\sqrt{j(j+1)}}\, Y_j^{\mu *}(\hat{\mathbf{k}}). \tag{97}$$

The factor $(-i)$ is purely conventional. Note that the parity here is $(-1)^j$, which is correct.

### Photomeson Production

We turn now to the discussion of the submatrix $S_{\pi\gamma}$ which describes the photoproduction of pi-mesons from nucleons. The initial channel is designated as $c = (\gamma, \lambda)$, implying that it contains a photon and a nucleon in charge state $\lambda$. The incident plane wave state is written as

$$|(\nu, \lambda);\ \hat{\mathbf{k}}, \rho, \tfrac{1}{2}, \nu\rangle \tag{98}$$

where $\rho(=\pm1)$ describes the photon polarization and $\nu(=\pm\frac{1}{2})$ is the spin orientation of the nucleon, both spins being quantized along the direction $k$ of the incident photon in the barycentric system. The final state is written as

$$|(\pi\alpha'\lambda');\ \hat{\boldsymbol{\varkappa}}_f, \tfrac{1}{2}, \nu'\rangle \tag{99}$$

where $\alpha'$ is the pion charge state, $\lambda'$ that of the nucleon, $\hat{\boldsymbol{\varkappa}}_f$ the direction of the relative momentum $\boldsymbol{\varkappa}_f$ of pion and nucleon, and $\nu'$ is the nucleon spin projection.

In order to express the relevant $S$-matrix elements, $\langle(\pi\alpha'\lambda');\ \hat{\boldsymbol{\varkappa}}_f, \frac{1}{2}, \nu'|\, S\, |(\gamma, \lambda);\ \hat{\mathbf{k}}, \rho, \frac{1}{2}, \nu\rangle$ in terms of the multipole amplitudes discussed on p. 528, we proceed just as in connection with Eq. (7.124). Our procedure must be modified slightly to allow for the two kinds of photon states, that is, electric and magnetic multipoles. The latter are designated by $|j\phi\mu\rangle$, corresponding to total angular momentum $j$, $z$-projection $\mu$, and multipole-type $\phi(\phi = E$ for electric, $\phi = M$ for magnetic). The total angular momentum and $z$-projection are called $J$, $M$. Note also that conservation of charge requires that $\lambda = \alpha' + \lambda'$.

We have

$$\langle(\pi a'\lambda'); \hat{\boldsymbol{\varkappa}}_f, \tfrac{1}{2}, \nu'|\, S\, |(\gamma, \lambda); \hat{\mathbf{k}}, \rho, \tfrac{1}{2}, \nu\rangle = \sum_{\substack{l'm'jJM\\ \mu\phi}} \times Y_{l'}^{m'}(\hat{\boldsymbol{\varkappa}}_f)\langle 1', \tfrac{1}{2}; m', \nu' \mid J, M\rangle S^J_{(\pi\alpha'\lambda')l';(\gamma\lambda)J\phi}\langle J, M \mid j, \tfrac{1}{2}; \mu, \nu\rangle \times \langle j\phi\mu \mid \hat{\mathbf{k}}\rho\rangle. \tag{100}$$

In this expression

$$\langle j\phi\mu \mid \hat{\mathbf{k}}, \rho\rangle = (\Phi_j^{\mu}(\phi), \mathscr{Y}^{\rho}), \tag{101}$$

where for $\phi = M$ we use Eq. (79), whereas for $\phi = E$ we use Eq. (95). The photon spin function was defined in Eq. (62). Using these equations and the definition of the $\mathscr{Y}_{jl}^{\mu}$, Eq. (63), we obtain

$$\begin{aligned}(\Phi_j^{\mu}(M), \mathscr{Y}^{\rho}) &= \langle j, 1; \mu - \rho, \rho \mid j, \mu\rangle Y_j^{\mu-\rho}(\hat{\mathbf{k}})^*,\\ (\Phi_j^{\mu}(E), \mathscr{Y}^{\rho}) &= \sqrt{j/(2j+1)}\langle j+1, 1; \mu - \rho, \rho \mid j, \mu\rangle Y_{j+1}^{\mu-\rho^*}(\hat{\mathbf{k}})\\ &\quad + \sqrt{(j+1)/(2j+1)}\langle j-1, 1; \mu - \rho, \rho \mid j, \mu\rangle Y_{j-1}^{\mu-\rho^*}(\hat{\mathbf{k}}).\end{aligned} \tag{102}$$

These formulas may be simplified by choosing the axis of quantization along $\mathbf{k}$, and then $Y_j^{\mu-\rho}(\hat{\mathbf{k}}) = \delta_{\mu,\rho}[(2j+1)/4\pi]^{1/2}$. Then

$$\begin{aligned}(\Phi_j^{\mu}(M), \mathscr{Y}^{\rho}) &= \delta_{\mu,\rho}\langle j, 1; 0, \rho \mid j, \rho\rangle \left(\frac{2j+1}{4\pi}\right)^{1/2},\\ (\Phi_j^{\mu}(E), \mathscr{Y}^{\rho}) &= \delta_{\mu,\rho}\left\{\left[\frac{j(2j+3)}{4\pi(2j+1)}\right]^{1/2}\langle j+1, 1; 0, \rho \mid j, \rho\rangle\right.\\ &\quad \left. + \left[\frac{(j+1)(2j-1)}{4\pi(2j+1)}\right]^{1/2}\langle j-1, 1; 0, \rho \mid j, \rho\rangle\right\}.\end{aligned} \tag{103}$$

For energies below about 400 Mev, we may neglect pion states other than those corresponding to $l' = 0$ and $l' = 1$. Then only electric dipole ($E_d$), electric quadrupole ($E_q$), and magnetic dipole ($M_d$) radiations contribute to photoproduction. The relevant transitions and an abbreviated notation for the $S$-matrix elements are summarized in Table 9.1. (Recall that by our conventions, the intrinsic parity of the pion is odd.) We have suppressed all indices and written, for example,

$$S^J_{(\pi\alpha'\lambda')l',(\gamma\lambda)j\phi} = S(M_d^{3/2}), \tag{104}$$

**Table 9.1 Electromagnetic Multipole Transitions for Photomeson Production**

| $\gamma$-Ray State | Total Angular Momentum | | Pion-Nucleon State |
|---|---|---|---|
| $E_d$ ($\mathscr{P}$ odd) | $J = \frac{1}{2}$ | $\xrightarrow[S(E_d{}^{1/2})]{}$ | $S_{1/2}$ |
| | $J = \frac{3}{2}$ | $\longrightarrow$ | $D_{3/2}$ |
| $M_d$ ($\mathscr{P}$ even) | $J = \frac{1}{2}$ | $\xrightarrow[S(M_d{}^{1/2})]{}$ | $P_{1/2}$ |
| | $J = \frac{3}{2}$ | $\xrightarrow[S(M_d{}^{3/2})]{}$ | $P_{3/2}$ |
| $E_q$ ($\mathscr{P}$ even) | $J = \frac{3}{2}$ | $\xrightarrow[S(E_q{}^{3/2})]{}$ | $P_{3/2}$ |
| | $J = \frac{5}{2}$ | $\longrightarrow$ | $F_{5/2}$ |

for $J = \frac{3}{2}, j = 1, \phi = M, l' = 1$; the appropriate charge indices must be supplied.

The differential cross section may be obtained by substituting the expression (100) into Eq. (7.123). Carrying out an average over the two values of $\rho(\pm 1)$ and $\nu(\pm\frac{1}{2})$ and a sum over $\nu'(\pm\frac{1}{2})$, we obtain for the differential cross section for an unpolarized $\gamma$-ray on an unpolarized nucleon (after much algebra)

$$\frac{d\bar{\sigma}_{\pi\gamma}}{d\Omega}(\lambda \to \alpha'\lambda') = A + B\cos\theta + C\cos^2\theta \tag{105}$$

In this equation $\cos\theta = \hat{k}\cdot\kappa_f$ and $A$, $B$, $C$ are given by

$$\begin{aligned} A &= |X|^2 + |Y|^2 + |Z|^2, \\ B &= -2\,\mathrm{Re}\, Z^*K, \\ C &= |K|^2 - |X|^2 - |Y|^2, \end{aligned} \tag{106}$$

with

$$\begin{aligned} X &= \frac{1}{4\sqrt{2}\,k}[3S(M_d^{3/2}) + \sqrt{3}\,S(E_q^{3/2})], \\ Y &= \frac{1}{4\sqrt{2}\,k}[S(M_d^{3/2}) - \sqrt{3}\,S(E_q^{3/2}) + 2S(M_d^{1/2})], \\ K &= \frac{1}{2\sqrt{2}\,k}[S(M_d^{3/2}) - \sqrt{3}\,S(E_q^{3/2}) - S(M_d^{1/2})], \\ Z &= \frac{1}{2\sqrt{2}\,k}[S(E_d^{1/2})]. \end{aligned} \tag{107}$$

The appropriate charge indices must be supplied to the abbreviated $S$-matrix elements in terms of which the cross section has been expressed.

The principle of charge independence is not valid in the presence of electromagnetic interactions, so that isotopic spin $i$ is not conserved in the photopion process. Nevertheless, a certain simplification is achieved when we consider $S$-matrix elements between eigenstates of the isotopic spin. Our initial state is already an eigenstate corresponding to $i = \frac{1}{2}$ with $z$-projection $\lambda$, $\lambda = +\frac{1}{2}$ for proton, $\lambda = -\frac{1}{2}$ for neutron. The final state composed of a pion and nucleon may be expressed in terms of eigenstates $i' = \frac{3}{2}, \frac{1}{2}$, with $z$-projection $\alpha' + \lambda' = \lambda$, and the transformation matrix to such states was given earlier Eq. (52). We have then

$$S^J_{(\pi\alpha'\lambda')l',(\gamma\lambda)j\phi} \equiv \sum_{i'} \langle 1, \tfrac{1}{2}; \alpha', \lambda' | i', \lambda\rangle S^J_{(\pi i')l',(\gamma\lambda)j\phi}, \tag{108}$$

where the $S$-matrix element on the right describes the transition to an eigenstate $i'$ of the isotopic spin.

A further simplification of $S$-matrix elements can be made if we use the condition of time reversal invariance. In discussing the $S$-matrix for radiative absorption of pions, the inverse process, call it $S_{\gamma\pi}$, we would have encountered $S$-matrix elements $S^J_{(\gamma\lambda)j\phi,(\pi i')l'}$. The condition expressed by Eq. (7.129) for time reversal invariance leads in this case to

$$S^J_{(\gamma\lambda)j\phi,(\pi i')l'} = S^J_{(\pi i')l',(\gamma\lambda)j\phi}. \tag{109}$$

The fact that $S$ is unitary, that is, $S^\dagger S = 1$, leads in this case of a reaction process to

$$S^\dagger{}_{\pi\pi}S_{\pi\gamma} + S^\dagger{}_{\pi\gamma}S_{\gamma\gamma} = 0, \tag{110}$$

or

$$S^*{}_{\pi\pi}S_{\pi\gamma} + S^*{}_{\gamma\pi}S_{\gamma\gamma} = 0, \tag{111}$$

where we have assumed that no other than the indicated intermediate states can be present because the energy is too low for multiparticle production. Now we have agreed to set $S_{\gamma\gamma} = 1$, so that using the time reversal invariance condition, Eq. (109), we obtain

$$S^*{}_{\gamma\pi} = S^*{}_{\pi\gamma} = -S^*{}_{\pi\pi}S_{\pi\gamma}. \tag{112}$$

In the isotopic spin representation [see Eq. (53)],

$$S_{\pi\pi} = e^{2i\delta(Jl'i')}, \tag{113}$$

so that

$$S^{J*}_{(\pi i')l',(\gamma\lambda)j\phi} = -e^{-2i\delta(Jl'i')}S^J_{(\pi i')l',(\gamma\lambda)j\phi}. \tag{114}$$

This statement may be reexpressed in the form

$$S^J_{(\pi i')l',(\gamma\lambda)j\phi} = ie^{i\delta(Jl'i')}s^J_{(\pi i')l',(\gamma\lambda)j\phi}, \tag{115}$$

where the matrix elements of $s$ are purely *real* quantities.

Equation (115) is an important result and has been very useful in understanding the experimental studies of photopion production. It states that the phases of the $S$-matrix elements for photopion production are given by the eigenphases for pion-nucleon scattering.[12]

There is an additional simplification possible if we make use of what is known as the principle of "minimal electromagnetic interaction."[13] This means that the only way the electromagnetic interaction is introduced by the replacement of the differential operator $\partial/\partial x_\mu$ in the Lagrangian for the system by $\partial/\partial x_\mu - iQA_\mu$, where $A_\mu$ is the vector potential operator and $Q$ is a linear operator in the particle isotopic spin space. For example, for the nucleon field, $Q = e(1 + \tau_3)/2$; for the pion field, $Q = eT_3$, where $T_3$ is the diagonal spin one matrix used earlier in a representation using charged and neutral pions. The general consequence of the minimal interaction hypothesis is that the current density operator $j_\mu$ is composed of a scalar and third component of a vector in isotopic spin space.

The $T$-matrix element for photoproduction may be shown by the methods of Chapter 10 to be given quite generally by the following:

$$\langle \pi N_{\text{out}} \mid \gamma, N_{\text{in}} \rangle = \langle \pi N_{\text{out}} | \, j_\mu \, | N \rangle e_\mu, \tag{116}$$

where we have factored out momentum conserving $\delta$-functions; $e_\mu$ is the polarization vector of the photon, $|N\rangle$ is a one nucleon state, $|\pi N_{\text{out}}\rangle$ is an interacting pion-nucleon state with the ingoing wave boundary condition (that is, the $-i\eta$ prescription of Chapter 5). Since we agree to treat the electromagnetic field only to lowest order, having exhibited the $j_\mu$ which is proportional to $e$, the final $\pi$-$N$ state may be described accurately in terms of isotopic spin eigenstates.

The one nucleon state is a pure $i = \frac{1}{2}$ state, whereas the pion-nucleon system may be in a linear combination of $i' = \frac{1}{2}$, $i' = \frac{3}{2}$ states. If we write

$$j_\mu = j_\mu^{\,s} + j_\mu^{\,v}, \tag{117}$$

where $j_\mu^{\,s}$ is a scalar operator in isotopic spin space and $j_\mu^{\,v}$ is a vector (actually the third component of a vector operator), we note that

$$\begin{aligned} \langle \tfrac{3}{2}, \lambda | \, j_\mu \, | \tfrac{1}{2}, \lambda \rangle &= \langle \tfrac{3}{2}, \lambda | \, j_\mu^{\,v} \, | \tfrac{1}{2}, \lambda \rangle \\ \langle \tfrac{1}{2}, \lambda | \, j_\mu \, | \tfrac{1}{2}, \lambda \rangle &= \langle \tfrac{1}{2}, \lambda | \, j_\mu^{\,s} \, | \tfrac{1}{2}, \lambda \rangle + \langle \tfrac{1}{2}, \lambda | \, j_\mu^{\,v} \, | \tfrac{1}{2}, \lambda \rangle, \end{aligned} \tag{118}$$

since only the isotopic vector current can cause the transition from $i = \frac{1}{2}$ to $i' = \frac{3}{2}$. Furthermore, it is easy to show, using the commutation relation between the vector operator $j_\mu^{\,v}$ and the raising and lowering operators in isotopic spin space $I_\pm$, that $\langle \frac{3}{2}, \lambda | \, j_\mu^{\,v} \, | \frac{1}{2}, \lambda \rangle$ is *independent* of $\lambda$. The $\lambda$

[12] A detailed treatment of the subject has been given in Refs. [6] and [7].
[13] See Ref. [4].

dependence of the $\frac{1}{2} \rightarrow \frac{1}{2}$ transitions may be explicitly exhibited, but we shall not stop to do so [4]. We remark only that since different combinations of $j_\mu^s$ and $j_\mu^v$ emerge, $\langle \frac{1}{2}, \lambda | j_\mu | \frac{1}{2}, \lambda \rangle$ is, in general, different corresponding to $\lambda = \pm\frac{1}{2}$.

We may now make the following notational changes using these facts:

$$S^J_{(\pi i')l',(\gamma\lambda)j\phi} = \begin{cases} s(E_3) & \text{electric dipole} \rightarrow i' = \frac{3}{2}, j = \frac{1}{2} \\ s(E_1\lambda) & \text{electric dipole} \rightarrow i' = \frac{1}{2}, j = \frac{1}{2} \\ s(E_{33}) & \text{electric quadrupole} \rightarrow i' = \frac{3}{2}, j = \frac{3}{2} \\ s(E_{13}\lambda) & \text{electric quadrupole} \rightarrow i' = \frac{1}{2}, j = \frac{3}{2} \\ s(M_{33}) & \text{magnetic dipole} \rightarrow i' = \frac{3}{2}, j = \frac{3}{2} \\ s(M_{31}) & \text{magnetic dipole} \rightarrow i' = \frac{3}{2}, j = \frac{1}{2} \\ s(M_{13}\lambda) & \text{magnetic dipole} \rightarrow i' = \frac{1}{2}, j = \frac{3}{2} \\ s(M_{11}\lambda) & \text{magnetic dipole} \rightarrow i' = \frac{1}{2}, j = \frac{1}{2}. \end{cases} \tag{119}$$

Finally, using the vector addition coefficients and the notation of Table 1, we have

For $\gamma + p \rightarrow \pi^+ + n(\lambda = +\frac{1}{2})$

$$S(E_d\,\tfrac{1}{2}) = \frac{i}{\sqrt{3}}\{e^{i\delta_3}s(E_3) + \sqrt{2}\,e^{i\delta_1}s(E_1\,\tfrac{1}{2})\}$$

$$S(M_d\,\tfrac{3}{2}) = \frac{i}{3}\{e^{i\delta_{33}}s(M_{33}) + \sqrt{2}\,e^{i\delta_{13}}s(M_{13}\,\tfrac{1}{2})\}$$

$$S(M_d\,\tfrac{1}{2}) = \frac{i}{\sqrt{3}}\{e^{i\delta_{31}}s(M_{31}) + \sqrt{2}\,e^{i\delta_{11}}s(M_{11}\,\tfrac{1}{2})\}$$

$$S(E_q\,\tfrac{3}{2}) = \frac{i}{\sqrt{3}}\{e^{i\delta_{33}}s(E_{33}) + \sqrt{2}\,e^{i\delta_{13}}s(E_{13}\,\tfrac{1}{2})\} \tag{120a}$$

For $\gamma + p \rightarrow \pi^0 + n(\lambda = +\frac{1}{2})$

$$S(E_d\,\tfrac{1}{2}) = \frac{i}{\sqrt{3}}\{\sqrt{2}\,e^{i\delta_3}s(E_3) - e^{i\delta_1}s(E_1\,\tfrac{1}{2})\}$$

$$S(M_d\,\tfrac{3}{2}) = \frac{i}{\sqrt{3}}\{\sqrt{2}\,e^{i\delta_{33}}s(M_{33}) - e^{i\delta_{13}}s(M_{13}\,\tfrac{1}{2})\}$$

$$S(M_d\,\tfrac{1}{2}) = \frac{i}{\sqrt{3}}\{\sqrt{2}\,e^{i\delta_{31}}s(M_{31}) - e^{i\delta_{11}}s(M_{11}\,\tfrac{1}{2})\}$$

$$S(E_q\,\tfrac{3}{2}) = \frac{i}{\sqrt{3}}\{\sqrt{2}\,e^{i\delta_{33}}s(M_{33}) - e^{i\delta_{13}}s(E_{13}\,\tfrac{1}{2})\}. \tag{120b}$$

The corresponding expressions for $\gamma + n \rightarrow \pi^- + p$ are obtained from those in Eq. (120*a*) by replacing $\lambda = \frac{1}{2}$ by $\lambda = -\frac{1}{2}$. Similarly, the *S*-matrix elements for $\gamma + n \to \pi^0 + n$ are obtained from Eq. (120*b*) by replacing $\lambda = +\frac{1}{2}$ by $\lambda = -\frac{1}{2}$. There are then twelve real coefficients *s* to determine the cross sections for the four physical processes in the present approximation.

We have up to this point treated the photoproduction process in an entirely straightforward way, using the general formalism we have developed. It is possible to carry out the calculation in a very general and more compact way using projection operator techniques, analogous to those used in Section 7.2. First we express the isotopic spin structure of the *T*-matrix element as a matrix element in the isotopic spin space of the nucleon. For this purpose it is convenient to use isotopic spin notation for the pion field also, namely, the $\pi_\alpha$ designation used at the beginning of this chapter. In the process $\gamma$ + nucleon $\to$ nucleon + $\pi_\alpha$, it is clear that as a matrix in nucleon isotopic spin space we must have a $\tau$-operator with the label $\alpha$, but this may, in addition, also involve $\tau_3$ as well, since the current operator contains a part which is the three-component of an isotopic vector. Thus we may write (suppressing all other labels) for the *T*-matrix element corresponding to the production of $\pi_\alpha$,

$$T_\alpha = T^{(+)}\mathscr{I}_\alpha^{(+)} + T^{(-)}\mathscr{I}_\alpha^{(-)} + T^{(0)}\mathscr{I}_\alpha^{(0)}, \tag{121}$$

where

$$\begin{aligned} \mathscr{I}_\alpha^{(+)} &= \tfrac{1}{2}(\tau_\alpha\tau_3 + \tau_3\tau_\alpha) = \delta_{\alpha 3}, \\ \mathscr{I}_\alpha^{(-)} &= \tfrac{1}{2}(\tau_\alpha\tau_3 - \tau_3\tau_\alpha) = \tfrac{1}{2}[\tau_\alpha, \tau_3], \\ \mathscr{I}_\alpha^{(0)} &= \tau_\alpha. \end{aligned} \tag{122}$$

The amplitudes for producing pions of a given charge may be taken to be

$$\begin{aligned} T_{\pi^+} &= \frac{T_1 - iT_2}{\sqrt{2}} = \sqrt{2}\,(T^{(-)} + T^{(0)})\frac{\tau_1 - i\tau_2}{2}, \\ T_{\pi^-} &= \frac{T_1 + iT_2}{\sqrt{2}} = \sqrt{2}\,(-T^{(-)} + T^{(0)})\frac{\tau_1 + i\tau_2}{2}, \\ T_{\pi^0} &= T_3 = T^{(+)} + \tau_3 T^{(0)}. \end{aligned} \tag{123}$$

We have dropped the minus sign in the definition of the $\pi^+$ wave function Eq. (42) (which is only a convention anyway) so that $\pi^+ = (\pi^-)^*$; the algebraic signs then follow if we remember that the pion occurs in the final state; hence it is the complex conjugate of the wave functions in Eq. (42) that give the relevant combinations. Clearly, to produce a positive meson, we must start with proton, and the combination $(\tau_1 - i\tau_2)/2$ on a proton state $\to$ a neutron state. Similarly, to make a $\pi^-$, we must start with a neutron and $(\tau_1 + i\tau_2)/2$ converts a neutron into a proton. A $\pi^0$, of course, may be produced by either a neutron or a proton.

Having disposed of the isotopic spin dependence, we ask now for the dependence of the matrix element on nucleon spin. Since we have to do with the matrix element between two one nucleon states, we know that the general form of any of the $T^{(+,-,0)}$ must be

$$T = A + \mathbf{B} \cdot \boldsymbol{\sigma}. \tag{124}$$

Since we are dealing with the absorption of a single photon, we know that $A$, $B$ must involve the photon polarization vector $\hat{e}$ linearly. Finally, the gauge invariance condition demands that if we replace $\hat{e}$ by $\mathbf{k}$, $T = 0$. We may span the three-dimensional space conveniently with the vectors $\mathbf{k}$, $\boldsymbol{\varkappa}_\perp$, and $\mathbf{k} \times \boldsymbol{\varkappa}$, where $\boldsymbol{\varkappa}$ is the barycentric pion momentum and $\boldsymbol{\varkappa}_\perp = \boldsymbol{\varkappa} - \boldsymbol{\varkappa} \cdot \mathbf{k} \times \mathbf{k}/k^2$. Since we are producing a pseudoscalar pion, $A$ must be a pseudoscalar and $B$ a pseudovector. Clearly, $A$ must be proportional to $\hat{e} \cdot \mathbf{k} \times \boldsymbol{\varkappa}$. If we write

$$\mathbf{B} = b_1\mathbf{k} + b_2\boldsymbol{\varkappa}_\perp + b_3\mathbf{k} \times \boldsymbol{\varkappa}, \tag{125}$$

we see that $b_1$, $b_2$ are scalars, whereas $b_3$ is a pseudoscalar. Then of the two scalars linearly proportional to $\hat{e}$, namely $\hat{e} \cdot \mathbf{k}$, $\hat{e} \cdot \boldsymbol{\varkappa}_\perp$, we must reject the $\hat{e} \cdot \mathbf{k}$ since it does not vanish if $\hat{e} \to \mathbf{k}$. Thus $b_1$ and $b_2$ must be proportional to $\hat{e} \cdot \boldsymbol{\varkappa}_\perp$. Finally, $b_3$ can involve only $\hat{e} \cdot \mathbf{k} \times \boldsymbol{\varkappa}$. The general form then becomes

$$T = A\hat{e} \cdot (\mathbf{k} \times \boldsymbol{\varkappa}) + b_1\hat{e} \cdot \boldsymbol{\varkappa}_\perp \boldsymbol{\sigma} \cdot \mathbf{k} + b_2\hat{e} \cdot \boldsymbol{\varkappa}_\perp \sigma \cdot \boldsymbol{\varkappa}_\perp + b_3\hat{e} \cdot (\mathbf{k} \times \boldsymbol{\varkappa})\boldsymbol{\sigma} \cdot \mathbf{k} \times \boldsymbol{\varkappa}. \tag{126}$$

There are therefore for each of $T^{(+,-,0)}$ four scalar functions of energy and angle, hence twelve functions to describe the four physical processes $\gamma_+ : p \to \pi^+ + n, n \to \pi^- + p, p \to \pi^0 + p, n \to \pi^0 + n$. If we *now* choose a special gauge, conventionally taken as $\hat{e} \cdot \mathbf{k} = 0$, we may, using elementary vector identities and properties of the $\boldsymbol{\sigma}$-matrices, write $T$ in the somewhat more convenient form

$$T = i\boldsymbol{\sigma} \cdot \hat{e}\mathscr{F}_1 + \frac{\boldsymbol{\sigma} \cdot \boldsymbol{\varkappa}\boldsymbol{\sigma} \cdot (\mathbf{k} \times \hat{e})}{\kappa k}\mathscr{F}_2 + \frac{i\boldsymbol{\sigma} \cdot \mathbf{k}\boldsymbol{\varkappa} \cdot \hat{e}}{\kappa k}\mathscr{F}_3 + \frac{i\boldsymbol{\sigma} \cdot \boldsymbol{\varkappa}\boldsymbol{\varkappa} \cdot \hat{e}}{\kappa^2}\mathscr{F}_4. \tag{127}$$

Naturally, having chosen a special gauge, we can no longer expect this to vanish when $\hat{e}$ is replaced by $\mathbf{k}$.

It is, of course, possible to deduce the expressions for the $\mathscr{F}_i$ in terms of the multipole amplitudes by comparing matrix elements of the preceding form taken between specific spin states with our general expression, Eq. (100). Our purpose, however, was to avoid Wigner coefficients and so on and to arrive directly at the desired relations. We write as in Section 7.3, suppressing isotopic labels,

$$\langle \hat{\boldsymbol{\varkappa}}, \nu' | T | \hat{\mathbf{k}}, \hat{e}, \nu \rangle = \sum_{\substack{j,\mu,\phi \\ l',m'}} \langle \hat{\boldsymbol{\varkappa}} | l'm' \rangle \langle l', \tfrac{1}{2}; m', \nu' | JM \rangle T^{J}_{l',j\phi}$$
$$\times \langle J, M | j, \tfrac{1}{2}; \mu\nu \rangle \langle j, \mu; \phi | \hat{\mathbf{k}}, \hat{e} \rangle. \quad (128)$$

The general structure of the nonzero terms is shown in Table 9.2. The amplitudes indicated in Table 9.2 are proportional to the $T^{J}_{l',j\phi}$ and will

**Table 9.2 Allowed Transitions in Photomeson Production**

| Photon Angular Momentum $j$ | Total Angular $J$ | Parity | Meson Angular Momentum $l'$ | Amplitude |
|---|---|---|---|---|
| $j$ | $j + \frac{1}{2}$ | $(-1)^j$ | $j + 1$ | $E^{-}_{j+1}$ |
| $j$ | $j - \frac{1}{2}$ | $(-1)^j$ | $j - 1$ | $E^{+}_{j-1}$ |
| $j$ | $j + \frac{1}{2}$ | $-(-1)^j$ | $j$ | $M_j^{+}$ |
| $j$ | $j - \frac{1}{2}$ | $-(-1)^j$ | $j$ | $M_j^{-}$ |

be defined precisely in a moment. The superscript indicates whether the total angular momentum is obtained by adding (+) or subtracting (−) one-half from the meson orbital angular momentum.

For the magnetic multipoles, where $l' = j$ we encounter just the projection operator met in (7.68):

$$\sum_M |j, \tfrac{1}{2}, j \pm \tfrac{1}{2}, M\rangle\langle j, \tfrac{1}{2}, j \pm \tfrac{1}{2}, M| = \Lambda_{j\pm}. \quad (129)$$

With the electric multipoles, we require

$$\sum_M |j \pm 1, \tfrac{1}{2}, j \pm \tfrac{1}{2}, M\rangle\langle j, \tfrac{1}{2}, j \pm \tfrac{1}{2}, M| = \boldsymbol{\sigma} \cdot \hat{\boldsymbol{\varkappa}} \Lambda_{j\pm}, \quad (130)$$

a relation established in Chapter 7 (7.106). We recall the following equations for $\langle j\mu\phi | \hat{\mathbf{k}}, \hat{e}\rangle$ [Eqs. (96), (97)](arbitrarily multiplied by $+i$):

$$\langle j, \mu; M | \hat{\mathbf{k}}, \hat{e} \rangle = \frac{i\hat{e} \cdot \mathbf{L}_{\mathbf{k}}}{[j(j+1)]^{1/2}} Y_j^{\mu *}(\hat{\mathbf{k}}),$$
$$\langle j, \mu; E | \hat{\mathbf{k}}, \hat{e} \rangle = \frac{\hat{e} \cdot \mathbf{k} \times \mathbf{L}_{\mathbf{k}}}{[j(j+1)]^{1/2}} Y_j^{\mu *}(\hat{\mathbf{k}}), \quad (131)$$

where $\mathbf{L}_{\mathbf{k}} = -i\mathbf{k} \times \nabla_{\mathbf{k}}$. By following the procedure that led to Eqs. (7.70) and (7.107), we obtain for the desired $T$-matrix element, regarding it as a matrix in the nucleon spin space,

$$4\pi T = \sum_{j=1}^{\infty} \{M_j^{+}(j + 1 + \boldsymbol{\sigma} \cdot \mathbf{L}_{\varkappa}) i\hat{e} \cdot \mathbf{L}_{\mathbf{k}} + M_j^{-}(j - \boldsymbol{\sigma} \cdot \mathbf{L}_{\varkappa}) i\hat{e} \cdot \mathbf{L}_{\mathbf{k}}$$
$$+ E^{-}_{j+1} \boldsymbol{\sigma} \cdot \hat{\boldsymbol{\varkappa}}(j + 1 + \boldsymbol{\sigma} \cdot \mathbf{L}_{\varkappa}) \hat{e} \cdot \mathbf{k} \times \mathbf{L}_{\mathbf{k}}$$
$$+ E^{+}_{j-1} \boldsymbol{\sigma} \cdot \hat{\boldsymbol{\varkappa}}(j - \boldsymbol{\sigma} \cdot \mathbf{L}_{\varkappa}) \hat{e} \cdot \mathbf{k} \times \mathbf{L}_{\mathbf{k}}\} P_j(\hat{\varkappa} \cdot \hat{\mathbf{k}}), \quad (132)$$

where

$$M_j^+ = \frac{T^{j+\frac{1}{2}}_{j,jM}}{[j(j+1)]^{\frac{1}{2}}}, \qquad M_j^- = \frac{T^{j-\frac{1}{2}}_{j,jM}}{[j(j+1)]^{\frac{1}{2}}},$$

$$E^-_{j+1} = \frac{T^{j+\frac{1}{2}}_{j+1,jE}}{[j(j+1)]^{\frac{1}{2}}}, \qquad E^+_{j-1} = \frac{T^{j-\frac{1}{2}}_{j-1,jM}}{[j(j+1)]^{\frac{1}{2}}}. \tag{133}$$

We have not yet made a commitment to a special gauge; the matrix element vanishes if $\hat{e}$ is replaced by $\mathbf{k}$. If we make use of our previous choice $\hat{e} \cdot \mathbf{k} = 0$ and the following identities, we obtain an answer in our standard form, Eq. (127).

$$\begin{aligned}
\hat{e} \cdot \hat{\mathbf{k}} \times \hat{\boldsymbol{\varkappa}} &= -\boldsymbol{\sigma} \cdot \hat{\boldsymbol{\varkappa}}\boldsymbol{\sigma} \cdot (\hat{\mathbf{k}} \times \hat{e}) - i\boldsymbol{\sigma} \cdot \hat{e}\hat{\boldsymbol{\varkappa}} \cdot \hat{\mathbf{k}} + i\boldsymbol{\sigma} \cdot \hat{\mathbf{k}}\hat{\boldsymbol{\varkappa}} \cdot \hat{e} \\
&= \boldsymbol{\sigma} \cdot \hat{e}\boldsymbol{\sigma} \cdot (\hat{\mathbf{k}} \times \hat{\boldsymbol{\varkappa}}) - i\boldsymbol{\sigma} \cdot \hat{\mathbf{k}}\hat{e} \cdot \hat{\boldsymbol{\varkappa}}, \\
(1 - x^2)P''_j(x) &= 2xP'_j(x) - j(j+1)P_j(x) \\
xP'_j(x) &= P'_{j+1}(x) - (j+1)P_j(x), \\
xP'_j(x) &= P'_{j-1}(x) + jP_j(x), \\
i\hat{e} \cdot \mathbf{L}_{\mathbf{k}}P_j(\hat{\boldsymbol{\varkappa}} \cdot \hat{\mathbf{k}}) &= \hat{e} \cdot \hat{\mathbf{k}} \times \hat{\boldsymbol{\varkappa}}P'_j(\hat{\boldsymbol{\varkappa}} \cdot \hat{\mathbf{k}}), \\
\boldsymbol{\sigma} \cdot \mathbf{L}_{\varkappa} i\hat{e} \cdot \mathbf{L}_{\mathbf{k}}P_j(\hat{\boldsymbol{\varkappa}} \cdot \hat{\mathbf{k}}) &= -i\boldsymbol{\sigma} \cdot \hat{\boldsymbol{\varkappa}} \times (\hat{e} \times \hat{\mathbf{k}})P'_j(\hat{\boldsymbol{\varkappa}} \cdot \hat{\mathbf{k}}) \qquad (134) \\
&\quad + i\hat{e} \cdot \hat{\mathbf{k}} \times \hat{\boldsymbol{\varkappa}}\boldsymbol{\sigma} \cdot \hat{k} \times \hat{\boldsymbol{\varkappa}}P''_j(\hat{\boldsymbol{\varkappa}} \cdot \hat{\mathbf{k}}), \\
\hat{e} \cdot \hat{\mathbf{k}} \times \mathbf{L}_{\mathbf{k}}P_j(\hat{\boldsymbol{\varkappa}} \cdot \hat{\mathbf{k}}) &= i\hat{e} \cdot \hat{\boldsymbol{\varkappa}}P'_j(\hat{\boldsymbol{\varkappa}} \cdot \hat{\mathbf{k}}), \\
\boldsymbol{\sigma} \cdot \mathbf{L}_{\varkappa}\hat{e} \cdot \hat{\mathbf{k}} \times \mathbf{L}_{\mathbf{k}}P_j(\hat{\boldsymbol{\varkappa}} \cdot \hat{\mathbf{k}}) &= \boldsymbol{\sigma} \cdot \hat{\boldsymbol{\varkappa}} \times \hat{e}P'_j(\hat{\boldsymbol{\varkappa}} \cdot \hat{\mathbf{k}}) + \boldsymbol{\sigma} \cdot \hat{\boldsymbol{\varkappa}} \times \hat{\mathbf{k}}\hat{e} \cdot \boldsymbol{\varkappa}P''_j(\hat{\boldsymbol{\varkappa}} \cdot \hat{\mathbf{k}}).
\end{aligned}$$

We write the final answer in terms of a sum over the final meson angular momentum which we now call $l$ (rather than $l'$ as before).

$$\begin{aligned}
4\pi\mathscr{F}_1 &= -\sum_{l=0} \{[lM_l^+ - E_l^-]P'_{l+1}(x) + [(l+1)M_l^- + E_l^-]P'_{l-1}(x), \\
4\pi\mathscr{F}_2 &= -\sum_{l=1} [(l+1)M_l^+ + lM_l^-]P'_l(x), \\
4\pi\mathscr{F}_3 &= \sum_{l=1} \{[M_l^+ + E_l^+]P''_{l+1}(x) - [M_l^- + E_l^-]P''_{l-1}(x)\}, \qquad (135) \\
4\pi\mathscr{F}_4 &= \sum_{l=2} \{-M_l^+ - E_l^+ + M_l^- + E_l^-\}P''_l(x),
\end{aligned}$$

where $x = \hat{\boldsymbol{\varkappa}} \cdot \hat{\mathbf{k}}$ is the cosine of the angle between the photon and meson momenta. These formulas may be inverted to yield the $M_l$, $E_l$ in terms of integrals over $\mathscr{F}$'s and Legendre polynomials.

With conventions on the pion wave functions which we have adopted in this section the relation between eigenstates of isotopic spin designated by

$|i, \lambda\rangle$ and the various combinations $(\pi^0 n)$, $(\pi - p)(\lambda = -\frac{1}{2})$ and $(\pi^0 p)$ $(\pi^+ n)(\lambda = +\frac{1}{2})$ becomes

$$(\pi^0 n) = \frac{\sqrt{2}\,|\frac{3}{2}, -\frac{1}{2}\rangle + |\frac{1}{2}, -\frac{1}{2}\rangle}{\sqrt{3}},$$

$$(\pi^- p) = \frac{|\frac{3}{2}, -\frac{1}{2}\rangle - \sqrt{2}\,|\frac{1}{2}, -\frac{1}{2}\rangle}{\sqrt{3}},$$

$$(\pi^0 p) = \frac{\sqrt{2}\,|\frac{3}{2}, \frac{1}{2}\rangle + |\frac{1}{2}, \frac{1}{2}\rangle}{\sqrt{3}},$$

$$(\pi^+ n) = \frac{-|\frac{3}{2}, \frac{1}{2}\rangle + \sqrt{2}\,|\frac{1}{2}, \frac{1}{2}\rangle}{\sqrt{3}}. \tag{136}$$

The $T$-matrix elements describing the transitions from $i = \frac{1}{2}$, $\lambda = \pm\frac{1}{2}$ to $i' = \frac{3}{2}, \frac{1}{2}$, $\lambda' = \pm\frac{1}{2}$ are related to $T^{+,-,0}$ according to

$$\begin{aligned}
\langle\tfrac{3}{2}, \tfrac{1}{2}|\, T\, |\tfrac{1}{2}, \tfrac{1}{2}\rangle &= \langle\tfrac{3}{2}, -\tfrac{1}{2}|\, T\, |\tfrac{1}{2}, -\tfrac{1}{2}\rangle = \sqrt{\tfrac{2}{3}}(T^+ - T^-) \\
\langle\tfrac{1}{2}, \tfrac{1}{2}|\, T\, |\tfrac{1}{2}, \tfrac{1}{2}\rangle &= \frac{1}{\sqrt{3}}(T^+ + 2T^- + 3T^0) \\
\langle\tfrac{1}{2}, -\tfrac{1}{2}|\, T\, |\tfrac{1}{2}, -\tfrac{1}{2}\rangle &= \frac{1}{\sqrt{3}}(T^+ + 2T^- - 3T^0)
\end{aligned} \tag{137}$$

We have, of course, rediscovered the fact that the matrix elements to isotopic spin $\frac{3}{2}$ states are independent of the initial charge states.

## 9.3 THE THEORY OF FINAL STATE INTERACTIONS

In Chapter 5 we studied a class of problems loosely referred to as "two potential problems." The general characterization was that one part of the interaction was in some sense "responsible" for the transition of interest, whereas the other described a type of internal reaction among some of the constituents which was so strong that we wished to treat it exactly. The situation of greatest interest is in connection with particles produced in elementary particle reactions, although there are many much more ancient examples. In the photoelectric effect in atoms, the Coulomb attraction between the electron and the residual ion drastically affects the energy spectrum of the photoelectrons; the same sort of Coulomb effect similarly modifies the electron spectra in beta-decay in heavy elements. In particle physics we encounter processes such as photomeson production, where the interaction between the produced pion and nucleon is very important; the capture of a negative pion by a deuteron leading to two neutrons and a photon—the force between the two neutrons has a profound effect on the photon spectrum; similarly, the nucleon-nucleon interaction has a very strong influence on the spectrum of mesons produced in nucleon-nucleon collisions.

It is evident that if a really clean separation of the primary interaction (for example, the production process) and what we shall term final state interactions can be made, much useful information about the nature of this interaction can be deduced from the analysis of such experiments. This information is especially welcome if we wish to study the forces between those unstable elementary particles on which direct experiments are impossible. For example, to learn about the $\pi$-$\pi$ interaction we would consider the reaction $\pi + N \rightarrow \pi + \pi + N$. This reaction serves to illustrate the idea and also some of the difficulties. Clearly, we expect the $\pi$-$\pi$ interaction to be the most important when the pion momenta are such that the $\pi$-$N$ forces are the least effective. An adequate theory of final state effects in the presence of three or more strongly interacting particles does not, to our knowledge, exist, although there are some plausible conjectures to which we shall refer. We shall therefore confine our quantitative discussion to the case where the strong interaction between only two particles needs be considered. For example, in the reaction leading to two nucleons and a meson, the high-energy end of the meson spectrum will be most influenced by the nucleon-nucleon interaction, since the nuclear forces are most effective at low energies. Barring unusual circumstances, we expect that final state modifications will be most profound when the relative energies of particles are low; although we shall be able to give a quite adequate treatment for any energies whenever the "primary interaction volume," call it $a^3$, is such that $a \ll$ "range" of the force of the interacting particles in the final state. Finally, we shall see that the effects are much more pronounced when we have to do with attractive rather than repulsive interactions. It is intuitively clear that an attractive force will tend to enhance the *magnitude* of the cross section relative to that for no interaction, whereas a repulsive one will tend to depress it. It does not seem quite so obvious that the spectral modifications would be so much less effective.

It is possible and quite fashionable nowadays to discuss the theory of final state interactions with the language of dispersion theory. We shall in fact make some reference to this approach and anticipate some of the things treated in Chapter 10. The complete dispersion theoretic formulation of the class of problems of interest to us is so involved that we shall largely use the more familiar Hamiltonian methods of Chapters 3 and 5.

We develop the general formalism along exactly the lines given in Section 5.4. There are two interactions, $U$ and $V$; we regard $V$ as the primary one, such that if it were zero, the process in question would not occur. The remaining part of the potential $U$ is the one responsible for the final state interactions of interest. For definiteness it is useful to keep in mind an example such as $p + p \to \pi^+ + n + p$, production of a positive meson in a proton-proton collision. It is the forces between nucleons that we are interested in taking into account. We recall the expression for the relevant $T$-matrix element, $\mathscr{T}_{\beta\alpha}$ [see Eq. (5.109)].

$$\mathscr{T}_{\beta\alpha} = (\phi_\beta^-, V\psi_\alpha^+) + (\phi_\beta^-, U\chi_\alpha), \tag{138}$$

where $\chi_\alpha$ characterizes the initial state of energy $E_\alpha = E$ and satisfies the equation

$$K\chi_\alpha = E\chi_\alpha, \tag{139}$$

and $\psi_\alpha^+$ is the complete wave function given by

$$\psi_\alpha^+ = \left[1 + \frac{1}{E + i\eta - K - V - U}(V + U)\right]\chi_\alpha. \tag{140}$$

The final state is described by $\chi_\beta$, which has been eliminated in favor of $\phi_\beta^-$, where

$$\phi_\beta^- = \left(1 + \frac{1}{E - i\eta - K - U}U\right)\chi_\beta \tag{141}$$

($E_\beta = E$ for the physical process). In the example mentioned, $U$ contains the interactions between all the final particles and may be regarded as also acting between the initial particles. The remaining part $V$ creates the final meson contained in $\phi_\beta^-$. It is consistent with this description to drop the second term in $\mathscr{T}_{\beta\alpha}$, since we assume $U$ cannot create real mesons; $\phi_\beta^-$ contains a meson, whereas $\chi_\alpha$ has none.

Now the bulk of the final state interaction of interest is contained in $\phi_\beta^-$ in the surviving term in $\mathscr{T}_{\beta\alpha}$.

$$\mathscr{T}_{\beta\alpha} = (\phi_\beta^-, V\psi_\alpha^+), \tag{142}$$

although $\psi_\alpha^+$ still "knows" about it. The part of $U$ that affects the initial particles can be easily taken into account fairly accurately by first writing for $\psi_\alpha^+$

$$\psi_\alpha^+ = \left(1 + \frac{1}{E + i\eta - K - U - V}V\right)\phi_\alpha^+, \tag{143}$$

where

$$\phi_\alpha^+ = \left(1 + \frac{1}{E + i\eta - K - U}U\right)\chi_\alpha, \tag{144}$$

and then approximating $\psi_\alpha^+$ in $\mathscr{T}_{\beta\alpha}$ by $\phi_\alpha^+$. This will certainly be a good approximation if $V$ is in some sense weak. Unless it is possible to think of the "last" interaction with $V$ as being somehow distinguished, there is little hope of making any simple discussion.[14]

The situation of greatest physical importance is one where we have three or more particles in the final state. In order to orient ourselves, however, imagine that we are dealing with a two-particle reaction such as the photoelectric effect in a hydrogen atom near threshold. Rather than treat this specific problem we consider first a model calculation which contains all the elements of interest without such incidental complications as a Coulomb field. Suppose the basic interaction can be treated in lowest order so that the $\mathscr{T}$-matrix element has the form

$$\mathscr{T} = \int d^3r \phi_{\boldsymbol{\varkappa}}^{-}(\mathbf{r})^* V(\mathbf{r}, \mathbf{k}) \phi_0(\mathbf{r}), \tag{145}$$

where we assume that $\phi_{\boldsymbol{\varkappa}}^{-}(\mathbf{r})$ belongs to the continuous spectrum and describes a scattering state characterized by momentum $\boldsymbol{\varkappa}$. The basic interaction $V(\mathbf{r}, \mathbf{k})$ depends parametrically on the incident particle momentum $\mathbf{k}$ which we regard as a parameter, and $\phi_0(\mathbf{r})$ is some sort of bound state wave function that has a relatively small extension. Of course, $|\mathbf{k}|$ and $|\boldsymbol{\varkappa}|$ are connected by energy conservation, but we ignore this for the present. Suppose further that we are sufficiently near threshold that only $S$-waves need be considered in $\phi_{\boldsymbol{\varkappa}}^{-}(\mathbf{r})$, so

$$\phi_{\boldsymbol{\varkappa}}^{-}(\mathbf{r}) = e^{-i\delta(\kappa)} \frac{w_0(\kappa, r)}{4\pi\kappa r} \tag{146}$$

[See Eq. (6.109$b$)]. Here $w_0(\kappa, r)$ is the regular solution of

$$\left[\frac{d^2}{dr^2} + \kappa^2 - 2mU(r)\right] w_0(\kappa, r) = 0, \tag{147}$$

which is asymptotic to [see Eq. (6.113)].

$$w_0(\kappa, r) \to \sqrt{2/\pi} \sin(\kappa r + \delta). \tag{148}$$

Thus the relevant $S$-wave part of $\mathscr{T}$, call it $\mathscr{T}_s$, becomes

$$4\pi \mathscr{T}_s = \int d^3r \phi_{\kappa,0}^{-*}(r) V(\mathbf{r}, \mathbf{k}) \phi_0(\mathbf{r}). \tag{149}$$

Now we make use of the fact that the interaction takes place over a very small region of radius $a$, say. This could be because the interaction emphasizes small $r$ (such as would be the case if we were creating a particle which would require a high momentum transfer), or because $\phi_0$ is effectively nonzero over a small region. As long as the wave length $1/\kappa$ is large compared to $a$, the precise $r$ dependence of $\phi_{\boldsymbol{\varkappa},0}^{-}(r)$ should be unimportant.

[14] This point is discussed in more detail in Ref. [8].

In Chapter 6, in connection with the analytic properties of partial wave amplitudes, we deduced the result [see Eq. (6.238$c$)]

$$\left.\frac{\phi^{+}_{\kappa,0}(r)}{\sqrt{2/\pi}\,\dfrac{\sin \kappa r}{\kappa r}}\right|_{r=0} = \frac{1}{\mathbf{f}(-\kappa)} \equiv \frac{1}{D(E)}, \tag{150}$$

where $E = \kappa^2/2m$ and $\mathbf{f}(-\kappa) = D(E)$ [Eq. (6.281)]. The function $D(E)$ is the Jost function which for positive $E$ has just the phase $-\delta_0(E)$, where $\delta_0(E)$ is the $S$-wave phase shift, and which has zeros at the negative values of $E$ corresponding to bound states. We shall therefore, in the spirit of the preceding remarks, replace $\phi_{\varkappa,0}(r)$ in $\mathscr{T}_s$ by

$$\phi^{-}_{\kappa,0}(r) = [\phi^{+}_{\kappa,0}(r)]^* \approx \frac{1}{[D(E)]^*}\sqrt{2/\pi}\,\frac{\sin \kappa r}{\kappa r} \tag{151}$$

and obtain for $\mathscr{T}_s$ the result

$$\mathscr{T}_s \approx \frac{1}{D(E)}\,\mathscr{T}_s(\text{Born}), \tag{152}$$

where

$$4\pi\mathscr{T}_s(\text{Born}) = \int d^3r \sqrt{2/\pi}\,\frac{\sin \kappa r}{\kappa r}\;V(\mathbf{r}, \mathbf{k})\phi_0(\mathbf{r}). \tag{153}$$

Clearly $\mathscr{T}_s$ (Born) is just the result that would have been obtained if we had neglected the final state interaction effects entirely. Note also that $\mathscr{T}_s \sim e^{+i\delta_0(E)}$ since $D(E) = |D(E)|\, e^{-i\delta_0(E)}$, just the result we found in connection with photo-pion production [see Eq. (115)].

To the extent that the relevant $r$ values are so small that the factor sin $\kappa r/\kappa r$ may be replaced by unity, the only $\kappa$-dependence of the matrix is that given by the factor $1/D(E)$. Of course, the initial momentum $\mathbf{k}$ and $\boldsymbol{\varkappa}$ are not independent, but in a typical instance of an endothermic reaction, $|\mathbf{k}|$ may be set at the threshold value, in which case the principal energy variation of the cross section will be given by phase space factors times $|1/D(E)|^2$. Thus, with $S$-waves,

$$\sigma \sim \frac{\kappa}{|D(E)|^2}, \tag{154}$$

where, of course, $E = \kappa^2/2m = E_0$, the initial energy and the variation of $\mathscr{T}_s$ (Born) with $E_0$ has been neglected.

This simple modification of a Born approximation calculation by the insertion of what is just the value of the wave function at the origin is an ancient procedure which is periodically rediscovered.

In the first "modern" paper on final state effects by Watson [8], a somewhat different formula was obtained for $\mathscr{T}_s$, namely

$$\mathscr{T}_s^{\,w} \approx e^{i\delta}\frac{\sin\delta}{\kappa} \times \text{constant}. \tag{155}$$

In many practical situations, as we shall see in a moment, the numerical agreement between $\mathscr{T}_s^{\,w}$ and $\mathscr{T}_s$ as given here is quite close. It seems clear that the present derivation is more satisfactory for two reasons: (1) the result we have obtained evidently becomes exact in the limit of a point interaction, and (2) for very large $E$ it has the property that $\mathscr{T}_s \to \mathscr{T}_s$ (Born), since $D(E) \to 1$ for high energies [see Eq. (6.281)]. This is, of course, to be expected, since final state effects should disappear for large $E$.

It is possible to make a somewhat more ambitious calculation which could be of use in some simple problems. We recall that the wave function $\phi_{\kappa,0}^{-}(r)$ may be related to the function extensively studied in Section 6.5, called there $\phi(\kappa, r)$, by

$$r\phi_{\kappa,0}^{-}(r)^* = \sqrt{2/\pi}\,\frac{\phi(\kappa, r)}{\mathbf{f}(-\kappa)} = \sqrt{2/\pi}\,\frac{\phi(\kappa, r)}{D(E)}. \tag{156}$$

[See Eqs. (6.220, 221) for the definitions.] The important thing for this discussion is the fact that $\phi(\kappa, r)$ is an entire function of $\kappa^2$. Our purpose is to study the analytic properties of $\mathscr{T}_s$ as a function of $\kappa$. We see that $\mathscr{T}_s$ can be written as

$$\mathscr{T}_s = \frac{\mathscr{F}}{D(E)}, \tag{157}$$

where

$$4\pi\mathscr{F}(\kappa^2) = \sqrt{2/\pi}\int \frac{d^3r}{r}\,\phi(\kappa, r)V\phi_0(r), \tag{158}$$

and all the singularities associated with the final state interaction are exhibited in the factor $1/D(E)$. $\mathscr{F}(\kappa^2)$, of course, may be expected to have singularities associated with the structure of $V$ and $\phi_0$, but these have nothing to do with the final state effects of interest.

The actual domain of analyticity of $\mathscr{F}$ cannot be given precisely unless $V$ and $\phi_0$ are given. To see what they might be in a typical case, consider something like photodisintegration of a deuteron. Then $\phi_0$ is the ground-state wave function of the deuteron and $V \sim \sin kr/kr$. Clearly, the non-analyticity of $\mathscr{F}$ shows up as a failure of the integral to converge for large $r$, so we may use the asymptotic form of $\phi(\kappa, r)$ and of $\phi_0$ to study this:

$$\mathscr{F} \sim \int r^2\,dr\,\frac{\sin\kappa r}{\kappa r}\,\frac{\sin kr}{kr}\,\frac{e^{-\alpha r}}{r}, \tag{159}$$

where $\alpha$ is related to the deuteron binding energy. It is easily seen that $\mathscr{F}$ has cuts running from $\kappa = i\alpha \pm k$ to $\kappa = i\infty \pm k$. The true singularities of $\mathscr{F}$ have this same general character. The shorter the range of the final state interaction, the farther away from the physical region ($\kappa$ real, $>0$) will the remaining singularities be. It is reasonable to make the approximation of neglecting all but these Born effects which can be easily taken into account.

It is always possible to choose phases in such a way that $\mathscr{F}(\kappa^2)$ is real. We note then that since $D(E) = f(-\kappa) = |f(-\kappa)|\, e^{-i\delta(\kappa)}$, along the real $\kappa$-axis,

$$\operatorname{Im} \mathscr{T}_s = \frac{\mathscr{F}(\kappa^2)}{|f(-\kappa)|} \sin\delta = e^{-i\delta} \sin\delta \mathscr{T}_s. \tag{160}$$

Then with the Born approximation to $\mathscr{T}_s$ given by $\mathscr{F}_B(\kappa^2)$, where

$$4\pi \mathscr{F}_B(\kappa^2) = \sqrt{2/\pi} \int d^3r \, \frac{\sin \kappa r}{\kappa r} \, V \phi_0, \tag{161}$$

we may say (regarding now the energy $E = \kappa^2/2m$ as the variable) $\mathscr{T}_s - \mathscr{F}_B$ is an analytic function in the entire $E$-plane except for (1) a cut along the positive real axis with discontinuity given by Eq. (160), (2) poles arising from the zeros of $D(E)$ along the negative real axis (the bound states of the final state system), and (3) the cuts arising from the remainder of $\phi(\kappa, r)$, that is, from $\phi(\kappa, r) - \sqrt{2/\pi} \times \sin \kappa r/\kappa r$. We propose to neglect the latter. We make the additional assumption that $\mathscr{T}_s(E) \to \mathscr{F}_B(E)$ for large $E$. It follows then that

$$\mathscr{T}_s(E) - \mathscr{F}_B(E) = \sum_i \frac{\Gamma_i}{E - E_i} + \int_0^\infty \frac{dE'}{\pi} \frac{\operatorname{Im}[\mathscr{T}_s(E') - \mathscr{F}_B(E')]}{E' - E - i\eta}, \tag{162}$$

where

$$\Gamma_i = \frac{\mathscr{F}(E_i)}{D'(E_i)}, \tag{163}$$

and the $E_i$ are the bound-state energies ($E_i < 0$) for which $D(E_i) = 0$.

We shall return to the evaluation of the $\Gamma_i$ in a moment. Using the fact that $\mathscr{F}_B(E)$ is real and Eq. (160) for $\operatorname{Im} \mathscr{T}_s(E)$, we have the following integral equation or $\mathscr{T}_s(E)$:

$$\mathscr{T}_s(E) = \mathscr{F}_B(E) + \sum_i \frac{\Gamma_i}{E - E_i} + \int_0^\infty \frac{dE'}{\pi} \frac{e^{-i\delta(E')} \sin \delta(E') \mathscr{T}_s(E')}{E' - E - i\eta}. \tag{164}$$

The solution of this type of equation is discussed in detail in Appendix G. We find

$$\mathscr{T}_s(E) = \mathscr{F}_B(E) + \sum_i \frac{\Gamma_i e^{v(E) - v(E_i)}}{E - E_i} + e^{v(E)} \int_0^\infty \frac{dE'}{\pi} \frac{\mathscr{F}_B(E') e^{-p(E')} \sin \delta(E')}{E' - E - i\eta}, \tag{165}$$

where

$$v(E) = \int_0^\infty \frac{dE'}{\pi} \frac{\delta(E')}{E' - E - i\eta}$$
$$= P\int_0^\infty \frac{dE'}{\pi} \frac{\delta(E')}{E' - E} + i\delta(E) \tag{166}$$
$$\equiv p(E) + i\delta(E),$$

and $P$ means the integral is to be taken in the sense of principal values. If we break the integral over $\mathscr{F}_B$ into its real and imaginary parts, we may also write

$$\mathscr{T}_s(E) = \mathscr{F}_B(E)e^{i\delta(E)} \cos \delta(E) + \sum_i \frac{\Gamma_i e^{v(E)-v(E_i)}}{E - E_i}$$
$$+ e^{v(E)} P\int_0^\infty \frac{dE'}{\pi} \frac{\mathscr{F}_B(E')e^{-p(E')}\sin \delta(E')}{E' - E}. \tag{167}$$

The residues $\Gamma_i$ may under some circumstances be expressed directly in terms of the bound-state wave functions and the Jost function $f(\kappa)$. Provided the bound-state energies lie inside the domain of analyticity of $f(\kappa)$ connected to the real $\kappa$-axis, we may use Eq. (6.251) which relates $\phi(\kappa, r)$ with $\kappa = -i\kappa_i$, $E_i = -\kappa_i^2/2m$ to the normalized bound state wave functions.

$$r\phi_i(r) = \alpha_i \phi(-i\kappa_i, r), \tag{168}$$

and [see Eq. (6.252)]

$$\alpha_i^2 = -\frac{4\kappa_i^2}{\mathbf{f}(i\kappa_i)\dfrac{d\mathbf{f}(-i\kappa_i)}{d\kappa_i}}. \tag{169}$$

The residues $\Gamma_i$ may now be easily calculated.

In order to get some insight into the qualitative nature of the final state interaction effects, let us return to the simplest approximation which corresponds to neglecting the energy dependence of $\mathscr{F}_B(E)$, which is correct for a point interaction or for sufficiently small $\kappa$. Here

$$\mathscr{T}_s(E) = \frac{\mathscr{F}_B(0)}{D(E)}, \tag{170}$$

where

$$4\pi\mathscr{F}_B(0) = \sqrt{2/\pi}\int d^3r V(\mathbf{r}, \mathbf{k})\phi_0(\mathbf{r}). \tag{171}$$

We recall that $D(E)$ may be represented by [see Eq. (6.279)]

$$D(E) = \prod_i \left(1 - \frac{E_i}{E}\right) \exp\left[-\frac{1}{\pi}\int_0^\infty dE' \frac{\delta(E')}{E' - E - i\eta}\right]. \tag{172}$$

In practice, of course, we will not usually know $\delta(E)$ for large energies, but provided we are interested in small $E$, the high-energy behavior of $\delta(E)$ should not be too important. We may, of course, avoid the explicit calculation of the preceding integral if we can find an expression for the $S$-matrix in the Jost form

$$S(E) = \frac{\mathbf{f}(\kappa)}{\mathbf{f}(-\kappa)} \tag{173}$$

since $D(E) = \mathbf{f}(-\kappa)$. The approximation corresponding to a scattering length interaction is frequently made. That is, we set $\kappa \cot \delta = 1/a$, in which case

$$S(E) = \frac{\cot \delta + i}{\cot \delta - i} = \frac{1 + i\kappa a}{1 - i\kappa a}, \tag{174}$$

from which we read $\mathbf{f}(-\kappa)$:

$$\mathbf{f}(-\kappa) = 1 - i\kappa a \tag{175}$$

and then from Eq. (154)

$$\sigma \sim \frac{\kappa}{1 + \kappa^2 a^2}. \tag{176}$$

This shows the characteristic peaking at small $\kappa^2$. Actually, the scattering length approximation violates the conditions under which the Jost function theory applies, since $\mathbf{f}(-\kappa)$ does not approach unity for large $\kappa$. Stated otherwise, the phase shift approaches $\pi/2$ rather than zero. It is worth noting that the Watson approximation (155) in this case leads to the same numerical answer since $e^{i\delta} \sin \delta/\kappa = (1 - i\kappa a)^{-1}a$.

A somewhat more legitimate $S$-matrix is provided by the effective range approximation: $\kappa \cot \delta = 1/a + \frac{1}{2}r_e\kappa^2$ [see Eq. (6.296)]. We can actually carry out the required integration over the phase shift and compute $D(E)$ directly or indirectly as follows. We write the $S$-matrix as

$$S = \frac{1/a + \frac{1}{2}r_e\kappa^2 + i\kappa}{1/a + \frac{1}{2}r_e\kappa^2 - i\kappa} = \frac{(\kappa + i\alpha)(\kappa - i\beta)}{(\kappa - i\alpha)(\kappa + i\beta)}, \tag{177}$$

where the parameters $\alpha$, $\beta$ are given by

$$\begin{aligned} \tfrac{1}{2}r_e(\alpha - \beta) &= 1 \\ \tfrac{1}{2}r_e\alpha\beta &= \frac{1}{a}. \end{aligned} \tag{178}$$

Suppose we are interested in the case where we have a bound state near zero energy. Then $a < 0$ and $r_e/|a| \ll 1$. We choose $\alpha > 0$, $\beta < 0$ and write

$$\delta(\kappa) = \tan^{-1}\left[\frac{\kappa}{1/a + (r_e/2)\kappa^2}\right]$$

$$= \frac{i}{2}\ln\left[\frac{1/a + (r_e/2)\kappa^2 - i\kappa}{1/a + (r_e/2)\kappa^2 + i\kappa}\right] = \frac{i}{2}\ln\left[\frac{(\kappa - i\alpha)(\kappa + i\beta)}{(\kappa + i\alpha)(\kappa - i\beta)}\right].$$

The representation (6.279) may now be used to obtain

$$\mathbf{f}(\kappa) = \frac{\kappa - i\beta}{\kappa - i\alpha}, \qquad \mathbf{f}(-\kappa) = \frac{\kappa + i\beta}{\kappa + i\alpha}. \tag{179}$$

Since $\alpha > 0$, we see that $\mathbf{f}(\kappa)$ is analytic in the lower half $\kappa$-plane, as required by the general considerations of Section 6.5, p. 271.

According to Eq. (170), the enhancement factor is now

$$\frac{1}{D(E)} = \frac{1}{\mathbf{f}(-\kappa)} = \frac{(\kappa^2 + \alpha^2)r_e/2}{1/a + (r_e/2)\kappa^2 - i\kappa}. \tag{180}$$

This is to be compared with the Watson factor

$$\frac{e^{i\delta}\sin\delta}{\kappa} = \frac{S - 1}{2i\kappa} = \frac{\alpha - \beta}{1/a + (r_e/2)\kappa^2 - i\kappa}. \tag{181}$$

For $\kappa < \alpha$ the two expressions will give essentially the same energy dependence. The factor (180) contains the correct normalization, however, and in the limit $\kappa \to \infty$ approaches unity (as it should).

To study the relative sensitivity of the final state effect to the sign of the interaction, let us compute $D(E)$ by perturbation theory. We recall that according to Eq. (6.281)

$$D(E) = 1 + \frac{1}{\pi}\int_0^\infty dE' \frac{\operatorname{Im} D(E')}{E' - E - i\eta}. \tag{182}$$

From Eq. (6.240), to first order in $V$,

$$\mathbf{f}(-\kappa) = D(E) = 1 + \int_0^\infty dr \frac{\sin \kappa r}{\kappa} \cdot 2mV(r) \cdot e^{i\kappa r}, \tag{183}$$

which leads to

$$\operatorname{Im} D(E) = 2m \int_0^\infty dr \frac{\sin^2 \kappa r}{\kappa} V(r). \tag{184}$$

Now if $V(r)$ is positive (repulsive potential), $\operatorname{Im} D(E) > 0$ and furthermore, $\operatorname{Im} D(E) \sim \sqrt{E}$ for small $E$. The effect of these remarks is that for small $E$, one may expect $\operatorname{Re} D(E)$ to be positive and $D(E)$ itself can never get very small; there is also then little tendency for $D(E)$ to vary much with $E$. For the example $V(r) = V_0 e^{-\mu r/\mu r}$, we find

$$\mathbf{f}(-\kappa) = 1 + \frac{mV_0}{\kappa\mu}\tan^{-1}\frac{2\mu}{\kappa} + \frac{imV_0}{2\kappa\mu}\ln\left(1 + \frac{4\kappa^2}{\mu^2}\right). \tag{185}$$

The cross section enhancement,

$$\frac{1}{|D(E)|^2}$$

is illustrated as a function of pion momentum in Figs. 9.2 for a final state consisting of two nucleons and one pion. The potential $V(r)$ is assumed to be a square-well of radius $10^{-13}$ cm, the Jost function then being obtained from Eq. (6.266). The case of an attractive potential is illustrated in Fig. 9.2*a*, whereas that of a repulsive interaction is shown in Fig. 9.2*b*.

We must now consider what happens when we have more than two particles in the final state. Here the theory is on rather less firm ground. To the extent that we may assume that among the final particles only interactions between a particular pair is of importance, we may more or less take over the treatment we have just given. In the general formula for $\mathscr{T}_{\beta\alpha}$ we imagine the final state wave function may be approximated by

$$\phi_\beta^- \approx \phi_{\kappa,0}^-(r)h_{\beta'}^-(\xi), \tag{186}$$

where we imagine that the label $\beta'$ does not depend on the relative momentum $\kappa$ of the chosen pair and the coordinates $\xi$ do not contain the relative coordinate $r$. We again restrict attention to final $S$-states for the interacting pair. If we make the approximation of replacing $\phi_{\kappa,0}^-(r)$ by $\phi_{\kappa,0}^-(0)$ a factor $1/D(E)$ may again be extracted. The $\kappa$-dependence of the cross section will be proportional to $|D(E)|^{-2}$ together with a conventional phase space factors. Thus

$$d\sigma \sim \frac{1}{v_r}\int \prod_i d^3k_i\, \delta\left(\sum_i E(k_i) - E_\alpha\right)\delta(\textstyle\sum \mathbf{k}_i)\,|D(E)|^{-2}\, G(\mathbf{k}_i), \tag{187}$$

where $v_r$ is the relative velocity in the center-of-mass system, and $G(\mathbf{k}_i)$ stands for the dependence of the matrix element on all the momenta other than the chosen relative momentum $\kappa$, where $E = \kappa^2/2m_r$ and $m_r$ is the reduced mass of the interacting pair. If we imagine that $\boldsymbol{\kappa} = \frac{1}{2}(\mathbf{k}_1 - \mathbf{k}_2)$, $\mathbf{p} = \mathbf{k}_1 + \mathbf{k}_2$, there will appear a factor of $d^3\kappa$ in the preceding integral. For small $\kappa$ the energy conservation condition will not restrict the values of $\kappa$ and we have

$$d\sigma \sim \frac{\kappa^2\, d\kappa}{|D(E)|^2}. \tag{188}$$

We can, of course, do the phase space exactly if we choose. If there were

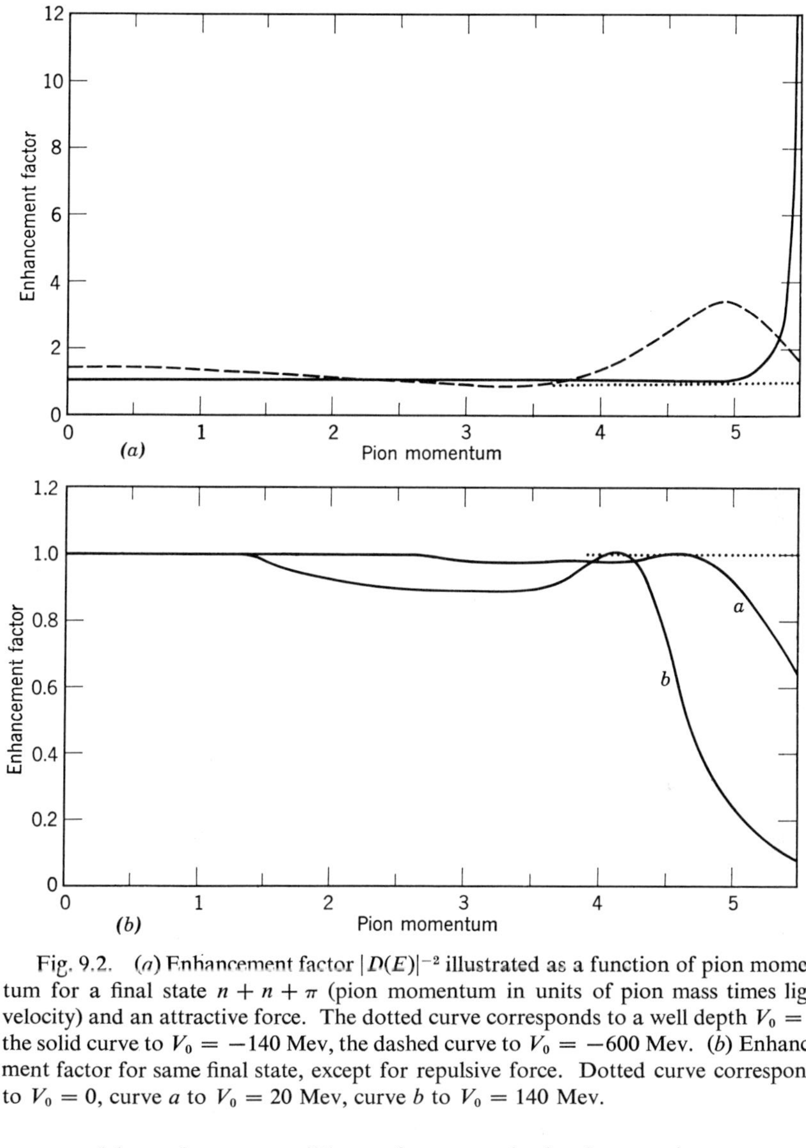


Fig. 9.2. (*a*) Enhancement factor $|D(E)|^{-2}$ illustrated as a function of pion momentum for a final state $n + n + \pi$ (pion momentum in units of pion mass times light velocity) and an attractive force. The dotted curve corresponds to a well depth $V_0 = 0$, the solid curve to $V_0 = -140$ Mev, the dashed curve to $V_0 = -600$ Mev. (*b*) Enhancement factor for same final state, except for repulsive force. Dotted curve corresponds to $V_0 = 0$, curve *a* to $V_0 = 20$ Mev, curve *b* to $V_0 = 140$ Mev.

an actual bound state possible as, for example, in the reactions $p + p \to n + p + \pi^+$ or $p + p \to \pi^+ + d$, where $d$ is a deuteron, we may handle the latter situation by replacing $\phi^-_{\kappa,0}(0)$ by $\phi_i(0) = \alpha_i$ according to Eq. (168). The computation of the cross sections for the two cases involves then merely evaluation of the relevant phase space factors.

This formulation of final state interaction theory is very close to what emerges in a dispersion theoretic treatment of elementary particle reactions.[15] The situation is somewhat more complicated and we cannot discuss the matter in detail. A very simple case is treated in Chapter 10. One important fact that emerges is that there is no need to limit the discussion to nonrelativistic kinematics for the interacting particles. If the four-momenta of the particles are called $k_1$, $k_2$, we define the square of the total energy $s = -(k_1 + k_2)^2$ and in terms of this variable we find an enhancement factor (assuming no bound states)

$$\frac{1}{D(s)} = \exp\left[\int_{(m_1+m_2)^2}^{\infty} \frac{ds'}{\pi} \frac{\delta(s')}{s' - s - i\eta}\right] \tag{189}$$

assuming $\delta \to 0$ for large $s$. If this is not the case, we write

$$\frac{D(s_0)}{D(s)} = \exp\left[(s - s_0)\int_{(m_1+m_2)^2}^{\infty} \frac{ds'}{\pi} \frac{\delta(s')}{(s' - s - i\eta)(s' - s_0 - i\eta)}\right], \tag{190}$$

where $(s_0)^{1/2}$ is some conveniently chosen energy.

In a very schematic way we would like to outline a semi-quantitative treatment of three particle final state effects.[16] In terms of a wave function description we imagine using the following approximation for the final state:

$$\phi_\beta^- = (\tilde{\phi}_{12} + \phi'_{12})(\tilde{\phi}_{23} + \phi'_{23})(\tilde{\phi}_{31} + \phi'_{31}) \tag{191}$$

where, for example, $\tilde{\phi}_{12}$ means take the Born approximation wave function $\Phi_{12}{}^{B}$ and subtract off that angular momentum state whose strong interaction is to be taken into account. Then we add the interacting state $\phi'_{12}$. We write this out as

$$\begin{aligned} \tilde{\phi}_{12} &= \Phi_{12}{}^{B} - \phi_{12}{}^{B}, \\ \phi'_{12} &= \frac{1}{D_{12}}\phi_{12}{}^{B}, \end{aligned} \tag{192}$$

where the superscript B means Born approximation, $D_{12}{}^{-1}$ is the enhancement factor for particles 1 and 2. Now define

$$p_{12} = \frac{\phi_{12}{}^{B}}{\Phi_{12}{}^{B}} \tag{193}$$

as a kind of probability of finding the angular momentum state of interest. The matrix element would then have the rough form

$$\mathcal{T} \sim \left(1 - \bar{p}_{12} + \frac{\bar{p}_{12}}{D_{12}}\right)\left(1 - \bar{p}_{23} + \frac{\bar{p}_{23}}{D_{23}}\right)\left(1 - \bar{p}_{31} + \frac{\bar{p}_{31}}{D_{31}}\right) \times (\text{Born}) \tag{194}$$

[15] See Refs. [9], for example.

[16] This procedure was suggested by Dr. Richard Blankenbecler, unpublished.

where the $\bar{p}$'s are defined as a kind of mean of the $p$'s. If we define $s_{ij} = -(k_i + k_j)^2$, where the $k_i$ are the four-momenta of the produced particles, the $D_{ij}$ may be taken as

$$D_{ij}(s_{ij}) = \exp\left[-\int_{(m_i+m_j)^2}^{\infty} \frac{ds'}{\pi} \frac{\delta_{ij}(s_{ij})}{s' - s_{ij} - i\eta}\right] \tag{195}$$

If it is kinematically possible for two pairs of the final particles to resonate simultaneously (roughly speaking this means Re $D_{ij} = 0$), there should be a sort of superenhancement, and if all three pairs can resonate, almost an explosion. It is perhaps worth remarking that if one of the final particles may be regarded as infinitely massive and only interactions between it and the other two taken into account, the form (194) becomes exact.

## 9.4 APPLICATION OF FORMAL SCATTERING THEORY TO FIELD THEORY

In Chapter 5 we remarked that the formalism developed there could be applied in quantum field theory. We shall indicate only very briefly how this can be done. The importance of so doing is largely historical since there are superior methods available, some of which are described in Chapter 10. Detailed expositions are to be found in a review article by Wick [10] and in the book of Schweber [11].

We consider the so-called static meson theory in which a meson field interacts with an infinitely heavy nucleon which may, however, have spin and charge degrees of freedom. The Hamiltonian of the system may be taken as

$$H = \sum_{\mathbf{k}} \alpha_k{}^{\dagger} \alpha_{\mathbf{k}} \omega_k + \sum_{\mathbf{k}} (a_{\mathbf{k}} V_{\mathbf{k}} + a_{\mathbf{k}}{}^{\dagger} V_{\mathbf{k}}{}^{\dagger}), \tag{196}$$

where $\omega_{\mathbf{k}} = (\mu^2 + \mathbf{k}^2)$ with $\mu$ the meson mass, and the operators $a_{\mathbf{k}}$ are essentially the same as those introduced in connection with the quantization of the electromagnetic field. The nucleon has been placed at the origin of the coordinate system and the $V_{\mathbf{k}}$ may contain operators which act on nucleon charge and spin variables. The subscript $\mathbf{k}$ may be taken to include a meson charge label as well as the momentum $k$. The $a_k$ satisfy the commutation relation

$$[a_{\mathbf{k}}, a_{\mathbf{k}'}^{\dagger}] = \delta_{\mathbf{k}\mathbf{k}'} \tag{197}$$

and all others are zero. For convenience we use quantization in a box of unit volume, so that the momenta are discrete. We assume the existence of a complete set of states describing various numbers of mesons interacting with our nucleon.

We confine our attention to the problem of meson-nucleon scattering.

The only problem we need to discuss is that of characterizing the "plane-wave" states $\chi$ of Chapter 5. We are evidently interested in studying the scattering of mesons by physical nucleons which have an energy $E_0$, where $E_0$ is the lowest eigenvalue of $H$, corresponding to having no mesons in the field. We would like to claim that the initial state is characterized by an incoming packet which is built up from the states

$$\chi_{\mathbf{k}} = a_{\mathbf{k}}^{\dagger}\Psi_0, \tag{198}$$

where

$$H\Psi_0 = E_0\Psi_0. \tag{199}$$

What one must show is that the state

$$X(t) = \sum_{\mathbf{k}} C_{\mathbf{k}} e^{-i(E_0+\omega_k)t} a_{\mathbf{k}}^{\dagger}\Psi_0 \tag{200}$$

satisfies the Schrödinger equation

$$i\frac{\partial X}{\partial t}(t) = HX(t) \tag{201}$$

provided that the meson packet (the $C_{\mathbf{k}}$) is chosen so that it does not overlap the nucleon for $t$ in the remote past. What this means is that the packet

$$g(\mathbf{r}, t) = \sum_{\mathbf{k}} C_{\mathbf{k}}\, e^{i(\mathbf{k}\cdot\mathbf{r}-\omega_k t)} \tag{202}$$

has the property $g(0, t) = 0$ for $t \ll 0$. Using the elementary relation

$$[H, a_{\mathbf{k}}^{\dagger}]\Phi_0 = (E_0 + \omega_k)a_{\mathbf{k}}^{\dagger}\Psi_0 + V_{\mathbf{k}}\Psi_0, \tag{203}$$

we find

$$\begin{aligned} i\frac{\partial X}{\partial t} - HX &= \sum_{\mathbf{k}} C_{\mathbf{k}} V_{\mathbf{k}} e^{-i(E_0+\omega_k)t}\Psi_0 \\ &\approx V_{\bar{\mathbf{k}}} e^{-iE_0 t} \sum C_{\mathbf{k}} e^{-i\omega_k t}\Psi_0 \\ &= 0, \end{aligned} \tag{204}$$

where we have taken out the operator $V_{\mathbf{k}}$ at the mean momentum of the packet $\bar{\mathbf{k}}$. The last line then follows from $g(0, t) = 0$. This argument is exactly the one given by Wick.

We now recall our general formula for the $\mathscr{T}$-matrix element given in Chapter 5, Eq. (67).

$$\begin{aligned} \mathscr{T}_{\beta\alpha}^{-} &= (\psi_\beta^{-}, (H - E_\alpha)\chi_\alpha) \\ &= (\chi_\beta, (H - E_\alpha)\chi_\alpha) + \left((H - E_\beta)\chi_\beta, \frac{1}{E_\beta - H + i\eta}(H - E_\alpha)\chi_\alpha\right). \end{aligned} \tag{205}$$

We write

$$\chi_\beta \equiv a_{\mathbf{k}}^\dagger \Psi'_0, \qquad \chi_\alpha \equiv a_{\mathbf{k}}^\dagger \Psi_0, \tag{206}$$

where $\Psi'_0$ is written for the final state to indicate that the final nucleon might have a different charge or spin from the initial one. Now $E_\alpha = E_0 + \omega_k$, $E_\beta = E_0 + \omega_{k'}$, and

$$(H - E_0 - \omega_k) a_{\mathbf{k}}^\dagger \Psi_0 = V_{\mathbf{k}} \Psi_0, \tag{207}$$

so we find

$$\mathscr{T}^-_{\mathbf{k}'\mathbf{k}} = (\Psi'_0, a_{\mathbf{k}'} V_{\mathbf{k}} \Psi_0) + \left(\Psi'_0, V_{\mathbf{k}'}^\dagger \frac{1}{E_0 + \omega_{k'} - H + i\eta} V_{\mathbf{k}} \Psi_0\right). \tag{208}$$

To get rid of the operator $a_{k'}$, we remark that it commutes with $V_{\mathbf{k}}$ (which depends only on nucleon coordinates) and

$$\begin{aligned} H a_{\mathbf{k}'} \Psi_0 &= a_{\mathbf{k}'} H \Psi_0 + [H, a_{\mathbf{k}'}] \Psi_0 \\ &= E_0 a_{\mathbf{k}'} \Psi_0 + (-\omega_{k'} a_{\mathbf{k}'} - V_{\mathbf{k}'}^\dagger) \Psi_0. \end{aligned} \tag{209}$$

Thus

$$(\omega_{k'} - E_0 + H) a_{\mathbf{k}'} \Psi_0 = -V_{\mathbf{k}'}^\dagger \Psi_0. \tag{210}$$

Since the eigenvalues of $H - E_0$ are all greater than or equal to zero, we may solve for $a_{\mathbf{k}'} \Psi_0$.

$$a_{\mathbf{k}'} \Psi_0 = -\left(\frac{1}{\omega_{k'} - E_0 + H}\right) V_{\mathbf{k}'}^\dagger \Psi_0. \tag{211}$$

Substituting this into $\mathscr{T}^-_{\mathbf{k}'\mathbf{k}}$ we find

$$\begin{aligned} \mathscr{T}^-_{\mathbf{k}'\mathbf{k}} = &\left(\Psi'_0, V_{\mathbf{k}} \frac{1}{E_0 - \omega_{k'} - H} V_{\mathbf{k}'}^\dagger \Psi_0\right) \\ &+ \left(\Psi'_0, V_{\mathbf{k}'}^\dagger \frac{1}{E_0 + \omega_{k'} - H + i\eta} V_{\mathbf{k}} \Psi_0\right). \end{aligned} \tag{212}$$

It is possible to regard Eq. (212) as a kind of integral equation by noting that for any state $\beta$,

$$\mathscr{T}_{\beta\mathbf{k}}^- = (\psi_\beta^-, V_{\mathbf{k}} \Psi_0), \mathscr{T}_{\beta\mathbf{k}}^{-'} = (\psi_\beta^-, V_{\mathbf{k}} \Psi'_0). \tag{213}$$

If we now introduce a complete set of "$(-)$" states into Eq. (212) we find (assuming either $V_{\mathbf{k}} = V_{\mathbf{k}}^\dagger$ or $V_{\mathbf{k}} = -V_{\mathbf{k}}^\dagger$)

$$\mathscr{T}^-_{\mathbf{k}'\mathbf{k}} = -\sum_\gamma \frac{\mathscr{T}^{-'*}_{\gamma\mathbf{k}} \mathscr{T}^-_{\gamma\mathbf{k}'}}{\omega_\gamma + \omega_{\mathbf{k}'}} + \frac{\mathscr{T}^{-'*}_{\gamma\mathbf{k}'} \mathscr{T}^-_{\gamma\mathbf{k}}}{\omega_\gamma - \omega_{\mathbf{k}'} - i\eta} \tag{214}$$

If we restrict the sum over $\gamma$ to states with zero and one meson, then this becomes a true nonlinear integral equation for $\mathscr{T}^-$ with the zero meson term as an inhomogeneity. We shall not pursue this question any further (see Refs. 10, 11).

As one trivial application of our general result, Eq. (212), consider neutral scalar mesons interacting with nucleons. In this case the $V_{\mathbf{k}}$'s are just numbers—there is no spin or charge dependence. We see immediately that $\mathscr{T}^{-}_{\mathbf{k}'\mathbf{k}} \equiv 0$, a well-known result.

## REFERENCES

1. M. E. Rose, *Relativistic Electron Theory* (John Wiley and Sons, New York, 1961).
2. N. Kemmer, *Proc. Cambridge Phil. Soc.* **34,** 354 (1938).
3. W. Heitler, *Proc. Roy. Irish Acad.* **51,** 33 (1946).
4. K. A. Brueckner and K. M. Watson, *Phys. Rev.* **83,** 1 (1951) and K. M. Watson, *Phys. Rev.* **85,** 852 (1952).
5. H. Anderson, E. Fermi, R. Martin and D. Nagle, *Phys. Rev.* **91,** 155 (1953).
6. M. Gell-Mann and K. M. Watson, *Ann. Rev. of Nuclear Sci.* **4,** 219 (1954).
7. K. Watson, J. Keck, A. Tollestrup, and R. Walker, *Phys. Rev.* **101,** 1159 (1956).
8. K. M. Watson, *Phys. Rev.* **88,** 1163 (1952).
9. W. R. Frazer and J. R. Fulco, *Phys. Rev. Letters* **2,** 365 (1959).
10. G. C. Wick, *Rev. Mod. Phys.* **27,** 339 (1955).
11. S. S. Schweber, *An Introduction to Relativistic Quantum Field Theory* (Row, Peterson, Evanston, Illinois 1961).

# 10. INTRODUCTION TO DISPERSION THEORY[1]

In our treatment of scattering theory we have repeatedly placed considerable emphasis on the unitarity of the $S$-matrix and on relativistic invariance. There is another general concept which has not yet been exploited: the principle of microscopic causality. It is impossible to do full justice to this important concept without some excursion into quantum field theory. This in turn would require almost a book in itself. We shall therefore have to make some sort of compromise and try to explain the important consequences of the causality condition within the framework of a formalism in which it plays little, if any, explicit role, that is, nonrelativistic scattering theory. As we shall see, this is not as idiotic as it sounds. In order to give some feeling for the field theoretic applications, however, we shall discuss a little of the general theory at the end of this chapter.

## 10.1 HISTORICAL BACKGROUND

It was Kronig [2], in 1946, who first drew attention to the possibility that the requirement that signals not propagate at supervedic speeds might impose limitations on the $S$-matrix. It is this prohibition which is generally referred to as the *principle of microcausality*. In practice the technical use of the statement is rather more poorly founded in the theory of measurement than one would like it to be. Nonetheless there is a formalism which appears to be related to microcausality and which has very far reaching consequences.

At the time of Kronig's remark (which incidentally consisted of little more than the first sentence of the preceding paragraph) no one had a very clear idea of how to implement it. Actually, Kronig himself and Kramers [3] some time ago had done work which had considerable bearing on the subject, the importance of which was not recognized for twenty

[1] The contents of this chapter are based largely on lectures of M. L. Goldberger given at Princeton and Les Houches [1].

years. The subject of their investigation was the propagation of electromagnetic radiation in optical media. They showed that the requirement that light shall not propagate with the speeds exceeding $c$ leads to a relation between the real and imaginary parts of the index of refraction of the medium—a relation long known from the study of special models in classical electrodynamics. If we write the index of refraction for light of frequency $\omega$ as $n(\omega) = n_r(\omega) + in_i(\omega)$, the imaginary part is related to the absorption coefficient $\alpha(\omega)$ by $n_i(\omega) = c\alpha(\omega)/2\omega$ (in a distance $x$ the *intensity* decreases by $e^{-\alpha x}$). The result of classical electrodynamics is[2]

$$n_r(\omega) = 1 + \frac{c}{\pi} P \int_0^\infty d\omega' \frac{\alpha(\omega')}{\omega'^2 - \omega^2}, \tag{1}$$

where the $P$ implies that the principal value of the integral is to be taken at the singularity $\omega' = \omega$. Following Kramers, we define the index of refraction for negative $\omega$ by the relation $n(\omega) = n^*(-\omega)$; this means that $n_r(\omega)$ is an even function of $\omega$, $n_i(\omega)$ is odd, and so $\alpha(\omega)$ is even. We may then rewrite Eq. (1) in the self-evident forms

$$\begin{aligned} n_r(\omega) - 1 &= \frac{c}{\pi} P \int_{-\infty}^\infty d\omega' \frac{\alpha(\omega')/2\omega'}{\omega' - \omega}, \\ \operatorname{Re}[n(\omega) - 1] &= \frac{P}{\pi} \int_{-\infty}^\infty d\omega' \frac{\operatorname{Im}[n(\omega') - 1]}{\omega' - \omega}. \end{aligned} \tag{2}$$

The second of Eqs. (2) may be recognized as the real part of the equation

$$\begin{aligned} n(\omega) - 1 &= \frac{P}{i\pi} \int_{-\infty}^\infty d\omega' \frac{n(\omega') - 1}{\omega' - \omega}, \\ &= \frac{1}{2\pi i} \lim_{\eta \to 0(+)} \int_{-\infty}^\infty d\omega' \frac{n(\omega') - 1}{\omega' - \omega - i\eta}. \end{aligned} \tag{3}$$

In writing the second form of Eq. (3) we have used the familiar relation $(x - i\eta)^{-1} = P(1/x) + i\pi\,\delta(x)$ in the limit $\eta \to 0(+)$.[3] It should also be noted that we are tacitly assuming that $n(\omega) - 1$ approaches zero at least faster than $1/\ln \omega$ for large $\omega$, so that the integral will exist.

Next we observe that the representation (3) enables us to define $n(\omega)$ as a function of a complex variable $z$ which is analytic in the entire upper half complex $z$-plane.

$$n(z) - 1 = \frac{1}{2\pi i} \int_{-\infty}^\infty d\omega' \frac{n(\omega') - 1}{\omega' - z}; \tag{4}$$

[2] See Ref. [4].

[3] Frequently we shall not write explicitly the limit $\eta \to 0$, as agreed in Section 3.2.

the physical $n$ is obtained as the boundary value $z \to \omega + i\eta$, $\eta \to 0(+)$. Conversely, if we simply assume that $n$ may be defined as a function of complex frequency, call it $z$, which is analytic in the entire upper half $z$-plane, and which approaches unity at infinity, we find that Eq. (4) is a direct consequence of Cauchy's theorem. The steps we have outlined are Kramer's argument; he went on to show that if we consider the propagation of light described by an index of refraction with the preceding characteristics, no signals could be transmitted with speeds greater than $c$.

The equations relating the real and imaginary parts of $n(\omega)$, Eqs. (2) and (3), are generally referred to as *dispersion relations*.[4] It is very useful to relate the macroscopic concept of index of refraction to the scattering properties of individual atoms. This is done by means of the Lorentz relation

$$n(\omega) = 1 + \frac{2\pi c^2}{\omega^2} Nf(\omega), \tag{5}$$

where $N$ is the number of scattering centers per cubic centimeter in the medium and $f(\omega)$ is the coherent forward scattering amplitude for light of frequency $\omega$. This relation is easily derived by considering the transmission of a wave through a thin slab of randomly distributed scatterers.[5] We may verify that it yields our well-known relation between total cross section and the forward scattering amplitude: The intensity of a wave in passing through a medium with refractive index $n(\omega)$ is reduced in a distance $x$ by a factor $|\exp\{i(\omega/c)n(\omega)x\}|^2 = \exp\{-(2\omega/c)n_i x\}$; by definition of the total cross section, this same reduction of intensity is given by $e^{\{-N\sigma x\}}$. Using Eq. (5) for $n(\omega)$, we find

$$\sigma(\omega) = \frac{4\pi \operatorname{Im} f(\omega)}{\omega/c}, \tag{6}$$

which is of the form of Eq. (6.74*d*). If we now combine our dispersion relation in Eq. (2) with Eq. (5), we obtain

$$\begin{aligned} \operatorname{Re} f(\omega) &= \frac{\omega^2}{\pi} P \int_{-\infty}^{\infty} d\omega' \frac{\operatorname{Im} f(\omega')}{\omega'^2(\omega' - \omega)} \\ &= \frac{2\omega^2}{\pi} P \int_0^{\infty} d\omega' \frac{\operatorname{Im} f(\omega')}{\omega'(\omega'^2 - \omega^2)}. \end{aligned} \tag{7}$$

The second form of Eq. (7) follows from the fact that $\operatorname{Im} f(\omega)$ is an odd function of $\omega$.

[4] We have already encountered examples of dispersion relations in Sections 6.5, p. 281 and 8.2, p. 443.

[5] A derivation of Eq. (5) will be given in Section 11.4. See also Ref. [5], for example.

The result expressed by Eq. (7) is not correct in all cases because we have tacitly assumed that $f(\omega) = 0$ for $\omega = 0$; otherwise $n(\omega) - 1$ could have a pole of some order at $\omega = 0$. In case $f(0)$ is finite we must modify Eq. (7) to read

$$\begin{aligned} \operatorname{Re} f(\omega) - \operatorname{Re} f(0) &= \frac{2\omega^2}{\pi} P \int_0^\infty d\omega' \frac{\operatorname{Im} f(\omega')}{\omega'(\omega'^2 - \omega^2)} \\ &= \frac{\omega^2}{2\pi^2 c} P \int_0^\infty d\omega' \frac{\sigma(\omega')}{\omega'^2 - \omega^2}, \end{aligned} \tag{8}$$

where we have used Eq. (6). This is generally referred to as the Kramers-Kronig dispersion relation. It is useful to note that Eq. (8) would also result from the formula

$$\operatorname{Re} f(\omega) = \frac{1}{2\pi^2 c} P \int_0^\infty d\omega' \frac{\omega'^2 \sigma(\omega')}{\omega'^2 - \omega^2}, \tag{9}$$

under the assumption that the integral converges, which is a stronger requirement than that implied by Eq. (8). Quite generally we refer to the procedure of passing from Eq. (9) to Eq. (8) as making a *subtraction*. Obviously, Eq. (9) is a stronger statement than Eq. (8) since it enables us to *compute* $\operatorname{Re} f(0)$ according to

$$\operatorname{Re} f(0) = \frac{1}{2\pi c} \int_0^\infty d\omega \sigma(\omega). \tag{10}$$

Unfortunately, this relation often fails and we must resort to the weaker Kramers-Kronig formula. A representative example is that of the scattering of light by a free electron; the zero frequency amplitude is given rigorously by the Thompson amplitude,[6] $f(0) = -e^2/mc^2$, whereas the right-hand side of Eq. (10) is manifestly positive.

There are at least two reasons to be unhappy about the derivation of Eq. (8) if we are interested in elementary scattering processes. One is that we have used classical electrodynamic considerations (this is not so serious with light since the quantum theory really does have a classical limit) and the other is that the connection with the scattering is made by way of Eq. (5), the circumstances for the validity of which are quite restricted.

What we would like is a purely quantum mechanical, straightforward approach. Unfortunately, this is not so easy. The only really general principle which is involved in the preceding discussion is that signals should not propagate with speeds exceeding that of light. We shall refer to this as a causality condition. Such a condition certainly has no place

[6] This can be proved quite rigorously (see M. Gell-Mann and M. L. Goldberger [6]), but it is really obvious. On what else could the amplitude depend at zero frequency beside the total charge and mass, and then how could it disagree with the classical limit?

in nonrelativistic quantum mechanics, but we shall find nevertheless that relations such as that of Kramers and Kronig indeed hold in the non-relativistic domain. In 1951, at the suggestion of Wigner, Schutzer and Tiomno [7] studied the problem of the scattering of a nonrelativistic particle by a force center which had a definite range. Their causality condition was the statement that the amplitude of the outgoing (that is, scattered) wave must be zero for all times before the ingoing (that is, incident) wave hits the scattering center, which is assumed to have a finite radius of interaction. This approach was criticized by van Kampen [8] who pointed out that one cannot construct wave packets which are zero prior to a particular time using positive energy states only. He proposed substituting a weaker condition namely, that if an incident wave packet is normalized at $t = -\infty$ to represent a single particle, the total probability of finding the particle at any later time outside the scattering region cannot exceed unity. This requirement seems to be intimately connected with the completeness relation. Within the framework of classical electrodynamics van Kampen [9] showed that the Schutzer-Tiomno causality condition is valid and with it he derived dispersion relations of the Kramers-Kronig variety for $S$-matrix elements corresponding to the scattering of waves of specified multipolarity. These relations involve explicitly the size of the scattering region.

A very general derivation of dispersion relations for the scattering of particles by finite range scatterers was given by Wigner [10]. It is very interesting in that no mention is made of causality (that is, the time sequence of events). The completeness relation, unitarity of the $S$-matrix, and the existence of a local current density form the basis of the proof which is conceptually simple but which involves a rather difficult and little known mathematical theorem. Wigner obtains dispersion relations for partial wave amplitudes where again the size of the scatterer appears explicitly.

There were a number of applications of the Kramers-Kronig relation to problems in quantum electrodynamics such as the scattering of light by a Coulomb field and the scattering of light by light [11]. The possibility of such calculations stems from the observation that, for example, the *amplitude* for light scattering from a Coulomb field is proportional to the square of the electric charge $e^2$, and hence that contribution to the cross section is proportional to $e^4$; on the other hand, there is an $e^2$ contribution to the total cross section corresponding to pair creation in the Coulomb field. Thus, given the total cross section for pair creation (to order $e^2$), one can compute by the dispersion relation the forward light scattering amplitude to order $e^2$.

The first attempt at a complete formulation of the causality condition and a derivation of the Kramers-Kronig relation within the framework of quantum field theory was made by Gell-Mann, Goldberger, and Thirring in 1954 [6]. The statement of causality proposed is as follows: The measurement of two observable quantities should not interfere if the points of measurement have a spacelike separation. Unfortunately, this entirely reasonable sounding requirement must be formulated mathematically in a way which is very often quite remote from any discussion of measurement. In terms of Heisenberg field operators (see Section 10.3) describing particles (photons, mesons, electrons, etc.) we demand that the commutator (or anticommutator for particles obeying Fermi statistics) of two such operators taken at spacelike points vanish. For the special case of electrodynamics, where we are concerned with a description of light scattering, our causality condition (which we shall frequently refer to as the *commutator condition*, or as *local commutativity*) has a direct relation with measurement, since the field strengths are measurable quantities. For massive bosons and even more so for massive fermions (by massive we do not mean big, but simply that the rest-mass is finite!) the fields customarily used have no direct physical meaning. Thus, although the local commutativity assumption plays a fundamental role in all derivations of dispersion relations, it has no simple connection with microscopic causality (to say nothing of macroscopic causality in the sense of measurement).

It turns out,[7] as we shall see, that the derivation of the dispersion relation for forward light scattering is exceedingly simple and of very great generality. If it were to be found to disagree with experience, the shock to physics would be far more profound than that suffered when nonconservation of parity was discovered. The situation is essentially the same for the corresponding relations describing the scattering of mesons by nucleons. The deduction of the latter relation is a little more involved, but it is quite rigorous insofar as any derivations in quantum field theory can be regarded as convincing at present.

Since 1954, dispersion theory has become a very highly developed art. The Kramers-Kronig relations have been generalized to describe nonforward scattering. More precisely, if the scattering process is characterized by the center-of-mass energy and momentum transfer rather than, as is more familiar, by energy and scattering angle, we find a dispersion relation for fixed momentum transfer which, of course, reduces to the usual Kramers-Kronig form as the momentum transfer goes to zero. (We shall see later why it is advantageous to hold the momentum transfer, rather than the scattering angle fixed.) The derivation of these generalized relations is exceedingly involved and in many important processes has not yet been given, except in perturbation theory. Since it is impossible for

[7] See Ref. [12].

us to go into this case in detail, we shall ultimately content ourselves with quoting the results and some of the applications.

The second and even more fundamental advance has been the conjecture by Mandelstam [13] of a two-dimensional or double dispersion representation in which not only the energy but also the momentum transfer dependence of the scattering amplitude is displayed explicitly. Accepting this representation we have the beginnings of truly dynamical theory, which in a sense replaces the conventional formalism of quantum field theory. The derivation of the double dispersion representation from field theory presents problems which have not yet been entirely overcome even within the framework of perturbation theory. Most people believe that the representation is nevertheless true.

In order to explain these developments we shall confine our attention largely to nonrelativistic potential scattering where things may be proved with considerable rigor and without the use of involved mathematics. We shall develop the theory in such a way that the transition to the corresponding field theoretic problem may be made rather easily. It is also our feeling that establishing the formalism in a familiar setting will help to understand the rather considerable body of literature on the subject.

## 10.2 FORMAL CONSIDERATIONS

Before discussing the nonrelativistic theory in detail, it is useful to review some formal manipulations with Fourier integrals which serve to illustrate in further detail the relation between dispersion relations and causality.

Suppose we are given a function of frequency $\omega, f(\omega)$, and the information that it is the Fourier transform of a function of time $t$, $g(t)$, which is zero for negative values of $t$. Thus

$$f(\omega) = \int_{-\infty}^{\infty} dt\, g(t)e^{i\omega t} = \int_{0}^{\infty} dt\, g(t)e^{i\omega t}. \tag{11}$$

It is clear that if the integral converges for real $\omega$, it certainly will do so for $\omega \to \omega + i\gamma$, $\gamma > 0$. This representation may evidently be used to continue $f(\omega)$ into the upper half complex $\omega$-plane, as we shall see shortly. It is convenient to extend $g(t)$ to negative values of $t$, where it is arbitrary and write the obvious statement

$$g(t) = \theta(t)g(t), \tag{12}$$

where $\theta(t)$ is the unit step function defined by

$$\begin{aligned} \theta(t) &= 1, && t > 0, \\ &= \tfrac{1}{2} && t = 0, \\ &= 0 && t < 0. \end{aligned} \tag{13}$$

This may be written in terms of the sign function $t/|t| = \epsilon(t)$ as

$$\theta(t) = \frac{1 + \epsilon(t)}{2}. \tag{14}$$

Provided that $g(t)$ has no $\delta$-function singularity at $t = 0$, corresponding to a constant in $f(\omega)$, there is no difficulty in this procedure. The point is that $\epsilon(t)\,\delta(t)$ is not well defined. We shall see in a moment how this contingency may be handled. It is convenient to imagine that $g(t)$ is extended to negative $t$ as an odd function of $t$.

We write now, using Eq. (12), $f(\omega) = d(\omega) + ia(\omega)$, where

$$\begin{aligned} d(\omega) &\equiv \frac{1}{2}\int_{-\infty}^{\infty} dt\, \epsilon(t)g(t)e^{i\omega t}, \\ ia(\omega) &\equiv \frac{1}{2}\int_{-\infty}^{\infty} dt\, g(t)e^{i\omega t}. \end{aligned} \tag{15}$$

With our assumption that $g(t) = -g(-t)$ and if $g$ is real, we note that $d(\omega)$ and $a(\omega)$ are real, although this plays no role in what we are about to do. We shall refer to $d(\omega)$ as the dispersive part of $f(\omega)$ and $a(\omega)$ as the absorptive part of $f(\omega)$. Next, we note that

$$\epsilon(t)e^{i\omega t} = \frac{P}{i\pi}\int_{-\infty}^{\infty} d\omega' \frac{e^{i\omega' t}}{\omega' - \omega}, \tag{16}$$

where $P$ implies, as before, that we take the Cauchy principal value at $\omega' = \omega$. This formula is trivially verified by contour integration. Substituting Eq. (16) into the first of Eqs. (15), we find

$$\begin{aligned} d(\omega) &= \frac{P}{2\pi i}\int_{-\infty}^{\infty} dt\, g(t)\int_{-\infty}^{\infty} d\omega' \frac{e^{i\omega' t}}{\omega' - \omega} \\ &= \frac{P}{\pi}\int_{-\infty}^{\infty} d\omega' \frac{1}{\omega' - \omega}\frac{1}{2i}\int_{-\infty}^{\infty} dt\, g(t)e^{i\omega' t}. \end{aligned}$$

or

$$d(\omega) = \frac{P}{\pi}\int_{-\infty}^{\infty} d\omega' \frac{a(\omega')}{\omega' - \omega}, \tag{17}$$

provided the interchange of order of the integration as made can be justified. (This is not a very involved point and depends only on whether $f(\omega)$ approaches zero sufficiently rapidly for large $\omega$. For example, if $f(\omega)$ is quadratically integrable, there is no difficulty. This is an unrealistically strong assumption for problems in physics and such a restriction will be removed later.) We may "invert" Eq. (17) and express $a(\omega)$ as an integral over $d(\omega)$ as follows: We write

$$1 = \epsilon(t) \cdot \epsilon(t) = \epsilon(t) \frac{P}{i\pi} \int_{-\infty}^{\infty} d\omega' \frac{e^{i(\omega'-\omega)t}}{\omega' - \omega}$$

and substitute into the second of Eqs. (15)

$$\begin{aligned} ia(\omega) &= \frac{1}{2} \int_{-\infty}^{\infty} dt\, g(t)\epsilon(t) \frac{P}{i\pi} \int_{-\infty}^{\infty} d\omega' \frac{e^{i\omega' t}}{\omega' - \omega} \\ &= \frac{P}{\pi i} \int_{-\infty}^{\infty} d\omega' \frac{1}{\omega' - \omega} \frac{1}{2} \int_{-\infty}^{\infty} dt\, g(t) e^{i\omega' t} \epsilon(t) \\ &= \frac{P}{i\pi} \int_{-\infty}^{\infty} d\omega' \frac{d(\omega')}{\omega' - \omega}, \end{aligned}$$

or

$$a(\omega) = -\frac{P}{\pi} \int_{-\infty}^{\infty} d\omega' \frac{d(\omega')}{\omega' - \omega}. \tag{18}$$

Then, finally, Eq. (17) may be written as

$$f(\omega) = d(\omega) + ia(\omega) = \lim_{\eta \to 0(+)} \frac{1}{\pi} \int_{-\infty}^{\infty} d\omega' \frac{a(\omega')}{\omega' - \omega - i\eta} \tag{19}$$

[as agreed above, we shall usually omit the instruction lim $\eta \to 0(+)$].

The pair of relations, Eqs. (17) and (18), are known as Hilbert transforms. Our rather cavalier manipulations may be made entirely rigorous on the basis of the so-called Titchmarsh theorem which we now quote:[8] If $f(\omega)$ is the Fourier transform of a function which vanishes for negative values of $t$ [as in Eq. (11)] and

$$\int_{-\infty}^{\infty} d\omega\, |f(\omega)|^2 < \infty,$$

then $f(\omega)$ is the boundary value of a function $f(\omega + i\gamma)$, $\gamma \to 0$ [the boundary is $\gamma \to 0(+)$], which is analytic in the upper half plane. Furthermore,

$$\int_{-\infty}^{\infty} d\omega\, |f(\omega + i\gamma)|^2 < \infty \qquad (\text{all } \gamma > 0).$$

These are necessary and sufficient conditions for the Hilbert transform relation

$$f(\omega) = \frac{P}{i\pi} \int_{-\infty}^{\infty} d\omega' \frac{f(\omega')}{\omega' - \omega}, \tag{20}$$

or

$$f(\omega) = \frac{1}{2\pi i} \int_{-\infty}^{\infty} d\omega' \frac{f(\omega')}{\omega' - \omega - i\eta}. \tag{21}$$

[8] See Chapter 5 of Ref. [2] in Chapter 3.

The real and imaginary parts of Eq. (20) are just our relations (17) and (18) between $d$ and $a$

$$\operatorname{Re} f(\omega) = \frac{P}{\pi} \int_{-\infty}^{\infty} d\omega' \frac{\operatorname{Im} f(\omega')}{\omega' - \omega}, \qquad \operatorname{Im} f(\omega) = -\frac{P}{\pi} \int_{-\infty}^{\infty} d\omega' \frac{\operatorname{Re} f(\omega')}{\omega' - \omega}. \tag{22}$$

It is very easy to free ourselves of the restriction to quadratically integrable functions $f(\omega)$. Suppose we know that $f(\omega)$ approaches a constant $f(\infty)$ as $\omega \to \infty$ (in any direction in the upper half $\omega$-plane). Then if we integrate $[f(\omega') - f(\infty)]/(\omega' - \omega - i\eta)$ over a contour consisting of the real axis and an infinite semicircle in the upper half $\omega'$-plane, we find

$$f(\omega + i\eta) - f(\infty) = \frac{1}{2\pi i} \int_{-\infty}^{\infty} d\omega' \frac{f(\omega') - f(\infty)}{\omega' - \omega - i\eta}, \tag{23}$$

and as $\eta \to 0(+)$

$$f(\omega) - f(\infty) = \frac{P}{i\pi} \int_{-\infty}^{\infty} d\omega' \frac{f(\omega') - f(\infty)}{\omega' - \omega}. \tag{24}$$

We may in practice not know $f(\infty)$, but instead know $f(\omega)$ at some finite frequency $\omega_0$. If we express the latter by Eq. (24) and subtract it from Eq. (24), we have

$$f(\omega) - f(\omega_0) = \frac{P}{\pi i} (\omega - \omega_0) \int_{-\infty}^{\infty} d\omega' \frac{f(\omega')}{(\omega' - \omega)(\omega' - \omega_0)} - \frac{P}{\pi i} (\omega - \omega_0) f(\infty) \int_{-\infty}^{\infty} d\omega' \frac{1}{\omega' - \omega} \frac{1}{\omega' - \omega_0}. \tag{25}$$

Now

$$P \int_{-\infty}^{\infty} d\omega' \frac{1}{\omega' - \omega} \frac{1}{\omega' - \omega_0} = \pi^2 \delta(\omega - \omega_0), \tag{26}$$

as can be verified by using

$$P \frac{1}{x} = \frac{i}{2} \int_{-\infty}^{\infty} d\lambda\, e^{i\lambda x} \epsilon(\lambda),$$

so that the second term on the right-hand side of Eq. (25) may be omitted. We have then what we call a *subtracted dispersion relation*,

$$f(\omega) - f(\omega_0) = \frac{P}{\pi i} (\omega - \omega_0) \int_{-\infty}^{\infty} d\omega' \frac{f(\omega')}{(\omega' - \omega)(\omega' - \omega_0)}, \tag{27}$$

or

$$\operatorname{Re} f(\omega) - \operatorname{Re} f(\omega_0) = \frac{P}{\pi} (\omega - \omega_0) \int_{-\infty}^{\infty} d\omega' \frac{\operatorname{Im} f(\omega')}{(\omega' - \omega)(\omega' - \omega_0)}. \tag{28}$$

If we interpret the $f(\omega)$ in this equation as the forward light scattering amplitude, take $\operatorname{Im} f(\omega) = \omega/4\pi c\sigma(\omega)$ (with $\sigma(\omega)$ the total cross section) to be an odd function of $\omega$, and set $\omega_0 = 0$, we recover the Kramers-Kronig relation

$$\operatorname{Re} f(\omega) - \operatorname{Re} f(0) = \frac{\omega^2}{2\pi^2 c} P \int_0^\infty d\omega' \frac{\sigma(\omega')}{\omega'^2 - \omega^2}. \tag{29}$$

In passing, we notice that Eq. (21) may be obtained from Eq. (12) by taking the Fourier transform of Eq. (12) and using the representation

$$\theta(t) = \frac{1}{2\pi i} \int_{-\infty}^{\infty} d\omega' \frac{e^{i\omega' t}}{\omega' - i\eta}, \tag{30}$$

we obtain

$$\begin{aligned} f(\omega) &= \int_{-\infty}^{\infty} dt\, g(t) e^{i\omega t} = \int_{-\infty}^{\infty} dt\, g(t) e^{i\omega t} \frac{1}{2\pi i} \int_{-\infty}^{\infty} d\omega' \frac{e^{i\omega' t}}{\omega' - i\eta} \\ &= \frac{1}{2\pi i} \int_{-\infty}^{\infty} d\omega' \frac{f(\omega' + \omega)}{\omega' - i\eta} = \frac{1}{2\pi i} \int_{-\infty}^{\infty} d\omega' \frac{f(\omega')}{\omega' - \omega - i\eta}. \end{aligned} \tag{31}$$

Interchanges of the order of integration must, of course, be scrutinized; again quadratic integrability of $f(\omega)$ is quite adequate.

We turn now to the nonrelativistic theory.

## 10.3 NONRELATIVISTIC THEORY

The absence of a limiting velocity in nonrelativistic theory might lead one to suspect that there would be nothing like a dispersion relation for the scattering amplitude. Nevertheless, as we shall see, this is not the case. Historically, the rigorous derivation of such relations followed their development in the relativistic theory, although the nonrelativistic result was obtained in perturbation theory as early as 1953 by Gell-Mann and Goldberger (unpublished). It was obtained first in full generality using the Fredholm theory of integral equations in 1957 by N. Khuri [14]. We shall use a simpler procedure here after first reexpressing some of the results of Chapters 3 and 5 in a rather highbrow manner, which is closely related to what is now considered good form in quantum field theory.

We shall treat the scattering of spinless particles of mass $m$ by a time-independent local potential $V(r)$. The particles will be taken to be bosons, although this plays no role here. Since the potential is independent of time, there is no particle creation. According to Eqs. (5.60), (5.63), and (5.65), the $S$-matrix element describing the scattering by a potential $V(r)$ of a particle of momentum $\mathbf{k}$ into momentum $\mathbf{k}'$ is given by

$$\begin{aligned} S_{\mathbf{k}'\mathbf{k}} = (\psi_{\mathbf{k}'}^-, \psi_{\mathbf{k}}^+) &= \delta_{\mathbf{k}'\mathbf{k}} - 2\pi i\, \delta(E_{k'} - E_k)(\psi_{\mathbf{k}'}^-, V\chi_{\mathbf{k}}) \\ &= \delta_{\mathbf{k}'\mathbf{k}} - 2\pi i\, \delta(E_{k'} - E_k)(\chi_{\mathbf{k}'}, V\psi_{\mathbf{k}}^+), \end{aligned} \tag{32}$$

where we are using normalization in a box of unit volume. Here [see Eq. (6.364*b*)]

$$E_k = \frac{k^2}{2m},$$
$$\chi_{\mathbf{k}} = e^{i\mathbf{k}\cdot\mathbf{r}}, \tag{33}$$
$$\psi_{\mathbf{k}}^{\pm} = e^{i\mathbf{k}\cdot\mathbf{r}} - 2m\int d^3r' \frac{e^{\pm ik|\mathbf{r}-\mathbf{r}'|}}{4\pi|\mathbf{r}-\mathbf{r}'|} V(\mathbf{r}')\psi_{\mathbf{k}}^{\pm}(\mathbf{r}').$$

**Review of Second Quantization**

We could proceed directly with Eq. (32), but instead shall reformulate the problem using the formalism of second quantization in order to ease the way into the corresponding field theoretic case.[9] To do this, we first recall some of the pertinent formulas: One introduces a wave field $\boldsymbol{\psi}(\mathbf{r})$, which we write in boldface form in order to distinguish it from the Schrödinger wave function. In terms of this wave field (which is an *operator*) the Hamiltonian for a system containing an arbitrary number of particles is written as

$$H = \int d^3r \boldsymbol{\psi}^\dagger(\mathbf{r})\left[-\frac{1}{2m}\nabla^2 + V(\mathbf{r})\right]\boldsymbol{\psi}(\mathbf{r}), \tag{34}$$

where $\boldsymbol{\psi}(\mathbf{r})$ and its adjoint $\boldsymbol{\psi}^\dagger(\mathbf{r})$ satisfy the commutation relation

$$[\boldsymbol{\psi}(\mathbf{r}), \boldsymbol{\psi}^\dagger(\mathbf{r}')] = \delta(\mathbf{r}-\mathbf{r}') \tag{35}$$

and all other combinations of operators commute. If we had assumed that our particles were fermions, we would have written

$$\{\boldsymbol{\psi}(\mathbf{r}), \boldsymbol{\psi}^\dagger(\mathbf{r}')\} = \delta(\mathbf{r}-\mathbf{r}'), \tag{36}$$

where $\{\cdot\cdot\cdot\}$ represents an *anti*commutator [see Eq. (1.25*b*)].

Consider now a set of basis vectors which are eigenvectors of the coordinate operator corresponding to having particles at points $\mathbf{r}_1$, $\mathbf{r}_2, \ldots \mathbf{r}_n$. Acting on such basis vectors, we have[10]

$$\begin{aligned}\boldsymbol{\psi}^\dagger(\mathbf{r})\,|\mathbf{r}_1, \mathbf{r}_2, \ldots, \mathbf{r}_n\rangle &= (n+1)^{1/2}\,|\mathbf{r}_1, \mathbf{r}_2, \ldots, \mathbf{r}_n\mathbf{r}\rangle,\\ \boldsymbol{\psi}(\mathbf{r})\,|\mathbf{r}_1, \mathbf{r}_2, \ldots, \mathbf{r}_n\rangle &= \frac{1}{\sqrt{n}}\sum_{s=1}^{n}\delta(\mathbf{r}-\mathbf{r}_s)\,|\mathbf{r}_1\cdots\mathbf{r}_{s-1}\mathbf{r}_{s+1}\cdots\mathbf{r}_n\rangle.\end{aligned} \tag{37}$$

[9] Second quantization is described in many texts. An especially concise and neat treatment is given by Becker and Leibfried [15].

[10] For Fermi-Dirac statistics, Eqs. (37) are replaced by the equations

$$\boldsymbol{\psi}^\dagger(\mathbf{r})\,|\mathbf{r}_1, \mathbf{r}_2, \ldots, \mathbf{r}_n\rangle = (n+1)^{1/2}\,|\mathbf{r}_1, \mathbf{r}_2, \ldots, \mathbf{r}_n\mathbf{r}\rangle,$$

$$\boldsymbol{\psi}(\mathbf{r})\,|\mathbf{r}_1, \mathbf{r}_2, \ldots, \mathbf{r}_n\rangle = \frac{1}{\sqrt{n}}\sum_{s=1}^{n}(-1)^{s-1}\delta(\mathbf{r}-\mathbf{r}_s)\,|\mathbf{r}_1\cdots\mathbf{r}_{s-1}\mathbf{r}_{s+1}\cdots\mathbf{r}_n\rangle.$$

The commutation relation Eq. (35) is readily seen to be compatible with these formulas. Note also that

$$|\mathbf{r}_1\rangle = \psi^\dagger(\mathbf{r}_1)\,|0\rangle,$$
$$|\mathbf{r}_1\mathbf{r}_2\rangle = \frac{1}{\sqrt{2!}}\psi^\dagger(\mathbf{r}_1)\psi^\dagger(\mathbf{r}_2)\,|0\rangle, \text{ and so on,} \tag{38}$$

where $|0\rangle$ represents the state containing no particles or the vacuum state; since the $\psi^\dagger$'s all commute, the basis functions are evidently symmetric functions of their arguments. The orthonormality of the basis vectors is easily established from Eqs. (38). For example,

$$\begin{aligned}\langle \mathbf{r}_1 \mid \mathbf{r}'_1\rangle &= \langle 0|\,\Psi(\mathbf{r}_1)\psi^\dagger(\mathbf{r}'_1)\,|0\rangle \\ &= \delta(\mathbf{r}_1 - \mathbf{r}'_1) + \langle 0|\,\psi^\dagger(\mathbf{r}'_1)\psi(\mathbf{r}_1)\,|0\rangle,\end{aligned} \tag{39}$$

where we have used Eq. (35) and the fact that $\langle 0 \mid 0\rangle = 1$. Now from the second of Eqs. (37) we see that we must have $\psi(\mathbf{r})\,|0\rangle = 0$, so from Eq. (39) we conclude that

$$\langle \mathbf{r}_1 \mid \mathbf{r}'_1\rangle = \delta(\mathbf{r}_1 - \mathbf{r}'_1). \tag{40}$$

Similarly, we find

$$\langle \mathbf{r}_1\mathbf{r}_2 \mid \mathbf{r}'_1\mathbf{r}'_2\rangle = \frac{1}{2!}\,[\delta(\mathbf{r}_1 - \mathbf{r}'_1)\,\delta(\mathbf{r}_2 - \mathbf{r}'_2) + \delta(\mathbf{r}_1 - \mathbf{r}'_2)\,\delta(\mathbf{r}_2 - \mathbf{r}'_1)] \tag{41}$$

and in general

$$\langle \mathbf{r}_1, \mathbf{r}_2, \ldots, \mathbf{r}_n \mid \mathbf{r}'_1, \mathbf{r}'_2, \ldots, \mathbf{r}'_{n'}\rangle = \frac{\delta_{n,n'}}{n!}\sum_{\alpha,\beta,\ldots,\nu=1}^{n} \delta(\mathbf{r}_1 - \mathbf{r}'_\alpha)\,\delta(\mathbf{r}_2 - \mathbf{r}'_\beta)\cdots\delta(\mathbf{r}_n - \mathbf{r}'_\nu). \tag{42}$$

A general state vector may be written as

$$|\Psi\rangle = \phi_0\,|0\rangle + \sum_{n=1}^{\infty}\int d^3r_1\,d^3r_2\cdots d^3r_n\phi_n(\mathbf{r}_1, \mathbf{r}_2, \ldots, \mathbf{r}_n)\,|\mathbf{r}_1, \mathbf{r}_2, \ldots, \mathbf{r}_n\rangle, \tag{43}$$

where the $\phi_i$, the "wave functions," are symmetric functions of their arguments. The normalization of the state vector is given by

$$\langle\Psi \mid \Psi\rangle = (\phi_0, \phi_0) + \sum_{n=1}^{\infty}\int d^3r_1\cdots d^3r_n\,|\phi_n(\mathbf{r}_1, \ldots, \mathbf{r}_n)|^2. \tag{44}$$

Time-dependent (Heisenberg representation) operators are formed according to the usual rules

$$\psi(\mathbf{r}, t) = e^{iHt}\psi(\mathbf{r})e^{-iHt} \tag{45}$$

and the equation of motion for $\psi(\mathbf{r}, t)$ is found to be

$$i\dot{\psi}(\mathbf{r}, t) = [\psi(\mathbf{r}, t), H] = \left[-\frac{1}{2m}\nabla^2 + V(\mathbf{r})\right]\psi(\mathbf{r}, t), \tag{46}$$

as follows immediately from the Hamiltonian (34) and the commutation relation (35). To the casual reader this might appear to be just the Schrödinger equation for a particle moving in the potential $V(\mathbf{r})$. Whereas this is, of course, formally true in this very simple situation of particles interacting with $V(\mathbf{r})$ but not with each other, the meaning of Eq. (46) is quite different. It is the analog of the equations of motion for the coordinate operator $\mathbf{r}_{\text{op}}(t)$ and $\mathbf{p}_{\text{op}}(t)$ in ordinary particle quantum mechanics

$$\begin{aligned} \dot{\mathbf{r}}_{\text{op}}(t) &= \frac{\mathbf{p}_{\text{op}}(t)}{m}, \\ \dot{\mathbf{p}}_{\text{op}}(t) &= -\nabla V[\mathbf{r}_{\text{op}}(t)], \end{aligned} \tag{47}$$

which are to be solved subject to the boundary conditions $[x(0), p_x(0)] = i$, and so on.

In our particular case it is possible in principle to solve Eq. (46) for $\psi$ in a very simple fashion. If we can find a complete orthonormal set of eigenstates for the Schrödinger equation, call them $\psi_\alpha(\mathbf{r})\, e^{-iE_\alpha t}$, then we may write

$$\boldsymbol{\psi}(\mathbf{r}, t) = \sum_\alpha a_\alpha \psi_\alpha(\mathbf{r}) e^{-iE_\alpha t}, \tag{48}$$

where the $a_\alpha$ are, of course, *operators.* In order to satisfy the basic commutation relation (35) [which, of course, holds for our time-dependent operators $\boldsymbol{\psi}(\mathbf{r}, t)$ and $\boldsymbol{\psi}^\dagger(\mathbf{r}, t)$, provided they are taken at the same time], we must demand that

$$[a_\alpha, a_\beta^\dagger] = \delta_{\alpha\beta}, \; [a_\alpha, \alpha_\beta] = 0, \qquad [a_\alpha^\dagger, a_\beta^\dagger] = 0; \tag{49}$$

here $a_\beta^\dagger$, the adjoint of $a_\beta$, is the quantity that appears in the expansion $\boldsymbol{\psi}^\dagger(\mathbf{r}, t)$:

$$\boldsymbol{\psi}^\dagger(\mathbf{r}, t) = \sum_\alpha a_\alpha^\dagger \psi_\alpha^*(\mathbf{r}) e^{iE_\alpha t}. \tag{50}$$

To see the significance of the operators $a_\alpha$, let us solve Eq. (50) for $a_\alpha^\dagger$. In doing this, it is convenient to set $t = 0$. We have then

$$a_\alpha^\dagger = \int d^3r \boldsymbol{\psi}^\dagger(\mathbf{r}) \psi_\alpha(\mathbf{r}). \tag{51}$$

Now imagine allowing $a_\alpha^\dagger$ to operate on the no-particle state $|0\rangle$

$$\begin{aligned} a_\alpha^\dagger\, |0\rangle \equiv |\alpha\rangle &= \int d^3r \boldsymbol{\psi}^\dagger(\mathbf{r})\, |0\rangle \cdot \psi_\alpha(\mathbf{r}) \\ &= \int d^3r\, |\mathbf{r}\rangle \psi_\alpha(\mathbf{r}). \end{aligned} \tag{52}$$

The state $|\alpha\rangle$ evidently corresponds to one where we have a single particle described by the wave function $\psi_\alpha(\mathbf{r})$. Since this state has been created by the action of $a_\alpha^\dagger$ on the no-particle, or vacuum state, it is called a creation operator. We verify immediately that

$$\begin{aligned}
a_\beta\,|\alpha\rangle &= \int d^3r'\,\psi_\beta^*(\mathbf{r}')\psi(\mathbf{r}')\,|\alpha\rangle \\
&= \int d^3r' \int d^3r\,\psi_\beta^*(\mathbf{r}')\psi_\alpha(\mathbf{r})\boldsymbol{\psi}(\mathbf{r}')\,|\mathbf{r}\rangle \\
&= \int d^3r' \int d^3r\,\psi_\beta^*(\mathbf{r}')\psi_\alpha(\mathbf{r})\,\delta(\mathbf{r}'-\mathbf{r})\,|0\rangle \\
&= \delta_{\beta\alpha}\,|0\rangle,
\end{aligned} \tag{53}$$

so that $a_\beta$ either gives zero when acting on $|\alpha\rangle$ or it annihilates the particle that was present to yield the vacuum state. Hence the $a_\beta$ are called annihilation operators.

These annihilation and creation operators are, of course, quite familiar in ordinary quantum mechanics. They are usually introduced in connection with the study of the harmonic oscillator. All their properties may however, be deduced from those of the operators $\boldsymbol{\psi}$ and $\boldsymbol{\psi}^\dagger$, which have been previously given. For example, consider $a_\beta^\dagger\,|\alpha\rangle$, where $\beta \neq \alpha$:

$$\begin{aligned}
a_\beta^\dagger\,|\alpha\rangle &= \int d^3r' \int d^3r\,\boldsymbol{\psi}^\dagger(\mathbf{r}')\,|\mathbf{r}\rangle\,\psi_\beta(\mathbf{r}')\psi_\alpha(\mathbf{r}) \\
&= \sqrt{2}\int d^3r' \int d^3r\,|\mathbf{r}\mathbf{r}'\rangle\psi_\beta(\mathbf{r}')\psi_\alpha(\mathbf{r}) \\
&= \int d^3r' \int d^3r\,|\mathbf{r}\mathbf{r}'\rangle\,\frac{1}{\sqrt{2}}\,\{\psi_\beta(\mathbf{r}')\psi_\alpha(\mathbf{r}) + \psi_\alpha(\mathbf{r}')\psi_\beta(\mathbf{r})\},
\end{aligned} \tag{54}$$

where the third line has been obtained from the second by using the symmetry of $|\mathbf{r}\mathbf{r}'\rangle$. Now the coefficient of $|\mathbf{r}\mathbf{r}'\rangle$ is just the normalized, appropriately symmetric wave function for a system of two identical noninteracting particles, one in state $\alpha$ and one in state $\beta$. If the states $\alpha$ and $\beta$ were the same, we see that a factor of $\sqrt{2}$ would appear explicitly, again familiar from the properties of particles obeying Bose statistics.

In general, we classify the states of the system in terms of the number of particles in particular single particle eigenstates. Thus the state with $n_\alpha$ particles in $\psi_\alpha$, $n_\beta$ in $\psi_\beta$, and so on is given by

$$|n_\alpha n_\beta \cdots n_\nu\rangle = \frac{1}{\sqrt{n_\alpha!\,n_\beta!\cdots n_\nu!}}\,(a_\alpha^\dagger)^{n_\alpha}(a_\beta^\dagger)^{n_\beta}\cdots(a_\nu^\dagger)^{n_\nu}\,|0\rangle, \tag{55}$$

where all numerical factors are properly taken into account.

We have chosen to satisfy the basic commutation relation (35) in terms of the eigenfunctions of the single particle Hamiltonian. The outstanding virtue of this choice is that the time-dependent operators are trivially constructed from the values at $t = 0$. This follows from the fact that the Hamiltonian $H$ becomes

$$H = \sum_\alpha a_\alpha^\dagger a_\alpha E_\alpha \tag{56}$$

and

$$\begin{aligned} a_\alpha(t) &= e^{iHt} a_\alpha e^{-iHt} = a_\alpha e^{-iE_\alpha t}, \\ a_\alpha^\dagger(t) &= e^{iHt} a_\alpha^\dagger e^{-iHt} = a_\alpha^\dagger e^{iE_\alpha t}. \end{aligned} \tag{57}$$

In spite of the fact that the Hamiltonian has been diagonalized by our choice of functions $\psi_\alpha$, scattering by the potential does take place. The point is that we must construct out of the $\psi_\alpha$'s wave packets appropriate to physical boundary conditions, as discussed in Chapter 5.

**Scattering Theory**

It is frequently convenient to satisfy Eq. (35) at $t = 0$ in terms of some other convenient complete set of functions, say the eigenfunctions $\chi_\alpha$ of $-\nabla^2/2m$. (We use the same labeling as before, since we shall be putting the $\psi_\alpha$ into one-to-one correspondence with the $\chi_\alpha$, aside from possible bound states which must be handled separately.) The formulation then goes through as shown, with the exception that the time dependence of the corresponding $a_\alpha$ operators is no longer so simple. We consider, in addition to the field operators $\psi(\mathbf{r}, t)$, two other ones called "in" and "out" operators, $\psi_i(\mathbf{r}, t), \psi_o(\mathbf{r}, t)$, which satisfy the "field equation"

$$i\dot{\psi}_{i,o}(\mathbf{r}, t) = -\frac{1}{2m} \nabla^2 \psi_{i,o}(\mathbf{r}, t). \tag{58}$$

These are required to satisfy the commutation relation (35)

$$[\psi_{i,o}(\mathbf{r}, t), \psi_{i,o}^\dagger(\mathbf{r}', t)] = \delta(\mathbf{r} - \mathbf{r}'), \tag{59}$$

and the property shared with $\psi(\mathbf{r}, t)$ (in our simple case)

$$\psi_i(\mathbf{r}, t)\,|0\rangle = \psi_o(\mathbf{r}, t)\,|0\rangle \equiv 0. \tag{60}$$

Since the operators $\psi_i$ and $\psi_o$ satisfy the same commutation relations and field equations, they must be related by a unitary transformation. The operator $S$ which effects this will be found to be the $S$-operator of collision theory.

Now consider the operators $\psi^\alpha(t)$ defined by

$$\psi^\alpha(t) = \int d^3r \chi_\alpha^*(\mathbf{r}, t)\psi(\mathbf{r}, t). \tag{61}$$

These are nothing more than the operators $a_\alpha(t)$, defined with respect to the complete set of functions $\chi_\alpha(\mathbf{r}, t)$, which are normalized solutions of

$$i\dot{\chi}_\alpha(\mathbf{r}, t) = -\frac{1}{2m}\nabla^2\chi_\alpha(\mathbf{r}, t). \tag{62}$$

We *must* imagine that the $\chi_\alpha$ are wave packet functions, although in most applications we shall be able ultimately to allow the $\chi_\alpha$ to approach plane waves of the form[11]

$$\chi_\alpha \to e^{i(\mathbf{k}\cdot\mathbf{r} - tk^2/2m)} \tag{63}$$

In precisely the same way we define operators $\psi_i^{\alpha}, \psi_o^{\alpha}$ according to

$$\psi_{i,o}^{\alpha}(t) = \int d^3r\chi_\alpha^*(\mathbf{r}, t)\psi_{i,o}(\mathbf{r}, t), \tag{64}$$

which are easily seen to be independent of time. That is,

$$i\dot{\psi}_{i,o}^{\alpha} = \int d^3r\left\{\chi_\alpha^*(\mathbf{r}, t)\left(-\frac{1}{2m}\nabla^2\right)\psi_{i,o}(\mathbf{r}, t) - \left[-\frac{1}{2m}\nabla^2\chi_\alpha^*(\mathbf{r}, t)\right]\psi_{i,o}(\mathbf{r}, t)\right\}$$

$$= -\frac{1}{2m}\int d^3r\nabla\cdot\{\chi_\alpha^*(\mathbf{r}, t)\nabla\psi_{i,o}(\mathbf{r}, t) - [\nabla\chi_\alpha^*(\mathbf{r}, t)]\psi_{i,o}(\mathbf{r}, t)\} = 0, \tag{65}$$

since this may be converted into a surface integral which vanishes by virtue of the fact that $\chi_\alpha(\mathbf{r}, t)$ is a localized packet. By comparison with our earlier discussion, we see that operators $\psi_{i,o}^{\alpha}$ annihilate states with one particle in the state $\chi_\alpha$. It can be shown under quite mild restrictions on the potential $V(r)$ that, for any states $|\Phi_1\rangle$ and $|\Phi_2\rangle$, the operators $\psi_i$ and $\psi_o$ may be so chosen that[12]

$$\begin{aligned}\lim_{t\to-\infty}\langle\Phi_1|\,\psi^\alpha(t)\,|\Phi_2\rangle &= \langle\Phi_1|\,\psi_i^{\alpha}\,|\Phi_2\rangle,\\ \lim_{t\to+\infty}\langle\Phi_1|\,\psi^\alpha(t)\,|\Phi_2\rangle &= \langle\Phi_1|\,\psi_o^{\alpha}\,|\Phi_2\rangle.\end{aligned} \tag{66}$$

We also note the relation

$$[\psi_{i,o}^{\alpha}, \psi_{i,o}^{\beta\dagger}] = \delta_{\alpha\beta}, \tag{67}$$

which follows from Eq. (60) and the definition (64).

In our very simple problem in which particles do not interact with each other but only with $V(r)$, it is not very hard to show that the limits (66), in fact, exist. It is sufficient to consider the matrix elements of the operators between one particle and no particle states, since these are the only relevant ones for us. We construct a one-particle "in" state according to the

[11] It will be recalled that we agreed to take the quantization volume as unity.

[12] The operators $\psi_i$ and $\psi_o$ are given explicit definitions in Eq. (75).

definition

$$|\Phi_{\text{in}}{}^{\alpha}\rangle = \psi_i^{\alpha\dagger}\,|0\rangle. \tag{68}$$

We note then

$$\langle 0|\,\psi_i{}^{\beta}\,|\Phi_{\text{in}}{}^{\alpha}\rangle = \langle 0|\,\psi_i{}^{\beta}\psi_i{}^{\alpha\dagger}\,|0\rangle = \delta_{\beta\alpha}. \tag{69}$$

Now consider

$$\langle 0|\,\psi^{\alpha}(t)\,|\Phi_{\text{in}}{}^{\alpha}\rangle = \int d^3r\chi_{\alpha}{}^*(\mathbf{r}, t)\,\langle 0|\,\psi(\mathbf{r}, t)\,|\Phi_{\text{in}}{}^{\alpha}\rangle. \tag{70}$$

The "wave function" $\varphi_{\alpha}(\mathbf{r}, t) \equiv \langle 0|\,\psi(\mathbf{r}, t)\,|\Phi_{\text{in}}{}^{\alpha}\rangle$ clearly satisfies the one-particle Schrödinger equation with interaction. We know from Chapter 5 that there exist solutions which coincide in the remote past with an incoming packet solution of the free Schrödinger equation. In other words, we have

$$\phi_{\alpha}(\mathbf{r}, t) \underset{t\to-\infty}{\longrightarrow} \chi_{\alpha}(\mathbf{r}, t), \tag{71}$$

which is somewhat more than is really required to establish Eq. (66). Furthermore, we know that there are also solutions which coincide in the remote future with a packet moving out from the scattering center.[13]

It is convenient to introduce two complete sets of states by allowing the in and out operators to act on the vacuum state:

$$\begin{aligned} |\Phi_{\text{in}}^{\alpha_1\alpha_2\cdots\alpha_\nu}\rangle &= \frac{1}{\sqrt{n_{\alpha_1}!\,n_{\alpha_2}!\cdots n_{\alpha_\nu}!}}(\psi_i^{\alpha_1\dagger})^{n_{\alpha_1}}\cdots(\psi_i^{\alpha_\nu\dagger})^{n_{\alpha_\nu}}\,|0\rangle,\\ |\Phi_{\text{out}}^{\alpha_1\alpha_2\cdots\alpha_\nu}\rangle &= \frac{1}{\sqrt{n_{\alpha_1}!\,n_{\alpha_2}!\cdots n_{\alpha_\nu}!}}(\psi_o^{\alpha_1\dagger})^{n_{\alpha_1}}\cdots(\psi_o^{\alpha_\nu\dagger})^{n_{\alpha_\nu}}\,|0\rangle. \end{aligned} \tag{72}$$

These "in" and "out" states are not to be interpreted as plane-wave states. For example, $|\Phi_{\text{in}}^{\alpha}\rangle$ is that eigenstate of the full Hamiltonian which in the remote past was described by an incoming packet $\chi_{\alpha}$. The relation (66), which is called the asymptotic condition, insures that $\psi_i{}^{\alpha}$ has information about the interaction $V$. In a similar fashion $|\Phi_{\text{out}}^{\alpha}\rangle$ is a state which in the remote future has an outgoing packet $\chi_{\alpha}$. Consequently, $\langle\Phi_{\text{out}}^{\alpha'}\mid\Phi_{\text{in}}^{\alpha}\rangle$ is a nonvanishing quantity and is in fact just an $S$-matrix element, as we shall shortly verify.

Some of these statements may appear paradoxical: How can the states created by the action of the "in" and "out" operators be eigenstates of the full Hamiltonian with interaction? Does this imply that in some sense the unperturbed Hamiltonian (with $V = 0$) expressed in terms of $\psi_i$ or $\psi_o$ is equal to the full Hamiltonian expressed in terms of $\psi$? In order to clarify these questions we construct in a reasonably precise (but admittedly far from rigorous) fashion the "in" and "out" operators quite explicitly.[14]

[13] A proof of Eqs. (66) is obtained with Eq. (93), given later.

[14] We are indebted to Professor A. S. Wightman for this treatment. It has been carried out by him in a mathematically satisfactory fashion which is a little too involved to be given in detail here.

To do this, we introduce an operator $\boldsymbol{\phi}(\mathbf{r}, t; t')$, which coincides with the Heisenberg operator $\boldsymbol{\psi}(\mathbf{r}, t')$ for $t = t'$

$$\begin{aligned}\boldsymbol{\phi}(\mathbf{r}, t; t') &= e^{-i(t-t')(-\nabla^2/2m)}\boldsymbol{\psi}(\mathbf{r}, t') \\ &= \int d^3r' e^{-i(t-t')(-\nabla^2/2m)}\,\delta(\mathbf{r}-\mathbf{r}')\boldsymbol{\psi}(\mathbf{r}', t') \\ &= \int d^3r' G_0(\mathbf{r}-\mathbf{r}'; t-t')\boldsymbol{\psi}(\mathbf{r}', t'),\end{aligned} \tag{73}$$

where

$$\begin{aligned}G_0(\mathbf{r}-\mathbf{r}'; t-t') &= e^{-i(t-t')(-\nabla^2/2m)}\,\delta(\mathbf{r}-\mathbf{r}') \\ &= \frac{e^{im(\mathbf{r}-\mathbf{r}')^2/2(t-t')}}{[2\pi(t-t')/mi]^{3/2}}\cdot\end{aligned} \tag{74}$$

By construction we see that $G_0(\mathbf{r}-\mathbf{r}'; 0) = \delta(\mathbf{r}-\mathbf{r}')$. We now define $\boldsymbol{\psi}_i(\mathbf{r}, t)$ and $\boldsymbol{\psi}_o(\mathbf{r}, t)$ by the requirement that

$$\begin{aligned}\boldsymbol{\psi}_i(\mathbf{r}, t) &= \lim_{t'\to-\infty} \boldsymbol{\phi}(\mathbf{r}, t; t'), \\ \boldsymbol{\psi}_o(r, t) &= \lim_{t'\to+\infty} \boldsymbol{\phi}(\mathbf{r}, t; t').\end{aligned} \tag{75}$$

$\boldsymbol{\psi}_i(\mathbf{r}, t)$ is the field operator which would exist at $\mathbf{r}, t$ if we start in the remote past with some configuration and follow its development in time with the free Hamiltonian; similarly, $\boldsymbol{\psi}_o(\mathbf{r}, t)$ is that configuration at $\mathbf{r}, t$ which coincides in the remote future with some specified one. The parallel between this discussion and that of Chapter 5 should be apparent.

Next we consider the role of the full Hamiltonian, including interaction, as the generator of time translations. We have already used this property in connection with the construction of $\boldsymbol{\psi}(\mathbf{r}, t)$ from $\boldsymbol{\psi}(\mathbf{r}, 0)$. It may appear somewhat less than obvious that

$$e^{iH\tau}\boldsymbol{\psi}_i^\dagger(\mathbf{r}, t)e^{-iH\tau} = \boldsymbol{\psi}_i^\dagger(\mathbf{r}, t+\tau), \tag{76}$$

with similar equations for $\boldsymbol{\psi}_i(\mathbf{r}, t)$, $\boldsymbol{\psi}_o(\mathbf{r}, t)$ and $\boldsymbol{\psi}_o^\dagger(\mathbf{r}, t)$. To prove this we note

$$\begin{aligned}e^{iH\tau}\boldsymbol{\phi}^\dagger(\mathbf{r}, t; t')e^{-iH\tau} &= \int d^3r' G_0(\mathbf{r}-\mathbf{r}'; t-t')e^{iH\tau}\boldsymbol{\psi}^\dagger(\mathbf{r}, t')e^{-iH\tau} \\ &= \int d^3r' G_0(\mathbf{r}-\mathbf{r}'; t-t')\boldsymbol{\psi}^\dagger(\mathbf{r}, t'+\tau) \\ &= \int d^3r' G_0(\mathbf{r}-\mathbf{r}'; t+\tau-t'-\tau)\boldsymbol{\psi}^\dagger(\mathbf{r}, t'+\tau) \\ &= \boldsymbol{\phi}^\dagger(\mathbf{r}, t+\tau; t'+\tau).\end{aligned} \tag{77}$$

In the limit as $t' \to -\infty$, for any finite $\tau$, we have

$$e^{iH\tau}\boldsymbol{\psi}_i^\dagger(\mathbf{r}, t)e^{-iH\tau} = \boldsymbol{\psi}_i^\dagger(\mathbf{r}, t+\tau). \tag{78}$$

Suppose now we allow this operator, integrated over one of our packet functions $\chi_\alpha(\mathbf{r}, t)$, to act on the vacuum state:[15]

$$e^{iH\tau}\int d^3r\psi_i^\dagger(\mathbf{r}, t)\,|0\rangle\,\chi_\alpha(\mathbf{r}, t) = \int d^3r\psi_i^\dagger(\mathbf{r}, t+\tau)\,|0\rangle\,\chi_\alpha(\mathbf{r}, t). \tag{79}$$

To the extent that we may assign a definite energy to the packet state, we may write

$$\chi_\alpha(\mathbf{r}, t) = \chi_\alpha(\mathbf{r}, t+\tau)e^{iE_\alpha\tau} \tag{80}$$

and obtain

$$e^{iH\tau}\,|\Phi_{\text{in}}{}^\alpha\rangle = e^{iE_\alpha\tau}\,|\Phi_{\text{in}}{}^\alpha\rangle, \tag{81}$$

which shows that the "in" states are eigenstates of the full Hamiltonian operator.

It is easily shown that $H_0(\psi_i) = H(\psi)$, where

$$\begin{aligned} H_0(\psi_i) &= \int d^3r\psi_i^\dagger(\mathbf{r}, t)\left(-\frac{\nabla^2}{2m}\right)\psi_i(\mathbf{r}, t), \\ H(\psi) &= \int d^3r\psi^\dagger(\mathbf{r}, t)\left[-\frac{\nabla^2}{2m} + V(r)\right]\psi(\mathbf{r}, t). \end{aligned} \tag{82}$$

We first note that by direct use of the commutation relations for $\psi_i(\mathbf{r}, t)$ and $\psi_i^\dagger(\mathbf{r}, t)$ and Eq. (58),

$$e^{iH_0(\psi_i)\tau}\psi_i(\mathbf{r}, t)e^{-iH_0(\psi_i)\tau} = \psi_i(\mathbf{r}, t+\tau). \tag{83}$$

But we also have shown in Eq. (78) that this same time translation is effected by $H(\psi)$. We may write then

$$e^{-iH(\psi)\tau}e^{iH_0(\psi_i)\tau}\psi_i(\mathbf{r}, t)e^{-iH_0(\psi_i)\tau}e^{iH(\psi)\tau} = \psi_i(\mathbf{r}, t), \tag{84}$$

or

$$[e^{-iH(\psi)\tau}e^{iH_0(\psi_i)\tau}, \psi_i(\mathbf{r}, t)] = 0. \tag{85}$$

Precisely the same argument may be given for $\psi_i^\dagger(\mathbf{r}, t)$, and since all operators in our theory may be made up of these, we conclude that $e^{-iH(\psi)\tau} \times e^{iH_0(\psi_i)\tau}$ is (to within a phase factor) a multiple of the unit operator:

$$e^{-iH(\psi)\tau}e^{iH_0(\psi_i)\tau} = e^{i\phi(\tau)}1, \tag{86}$$

with $\phi$ real. If we take the vacuum expectation value of this equation, we conclude that $\phi = 0$. Thus we conclude

$$H_0(\psi_i) = H(\psi), \tag{87}$$

which was to be shown.

[15] It follows from Eq. (34) that the vacuum state is an eigenstate of $H$, corresponding to the eigenvalue zero.

Now let us consider the problem of formally solving the operator Schrödinger equation for $\psi(\mathbf{r}, t)$. Equation (73) permits us to write (here it is assumed that $t_2 \geq t \geq t_1$)

$$\begin{aligned}\int_{t_1}^{t_2} dt' \frac{\partial \phi(\mathbf{r}, t; t')}{\partial t'} &= \phi(\mathbf{r}, t; t_2) - \phi(\mathbf{r}, t; t_1)\\ &= i \int_{t_1}^{t_2} dt' \int d^3r' \left\{ \left[ -i \frac{\partial}{\partial t'} G_0(\mathbf{r} - \mathbf{r}'; t - t') \right] \psi(\mathbf{r}', t') \right.\\ &\qquad \left. - iG_0(\mathbf{r} - \mathbf{r}'; t - t') \frac{\partial \psi}{\partial t'}(\mathbf{r}', t') \right\}\\ &= -i \int_{t_1}^{t_2} dt' \int d^3r' G_0(\mathbf{r} - \mathbf{r}'; t - t') \left[ i \frac{\partial}{\partial t'} - \left( -\frac{\nabla'^2}{2m} \right) \right]\\ &\qquad \times \psi(\mathbf{r}', t')\\ &= -i \int_{t_1}^{t_2} dt' \int d^3r' G_0(\mathbf{r} - \mathbf{r}'; t - t') V(r') \psi(\mathbf{r}', t'), \end{aligned} \tag{88}$$

where we have used the equation satisfied by $G_0$, namely,

$$\left[ -i \frac{\partial}{\partial t'} - \left( \frac{\nabla'^2}{2m} \right) \right] G_0(\mathbf{r} - \mathbf{r}'; t - t') = 0, \tag{89}$$

and integrated spatially by parts twice. The validity of the latter operation is essentially insured by the oscillatory character of $G_0$.

If in Eq. (88) we take $t_2 = t$ and $t_1 = -\infty$, we have

$$\psi(\mathbf{r}, t) = \psi_i(\mathbf{r}, t) - i \int_{-\infty}^{t} dt' \int d^3r' G_0(\mathbf{r} - \mathbf{r}'; t - t') V(r') \psi(\mathbf{r}', t'), \tag{90}$$

where as in Eq. (75) we have identified $\lim_{t_1 \to -\infty} \phi(\mathbf{r}, t; t_1)$ with $\psi_i(\mathbf{r}, t)$. Consider next forming from this equation the operator $\psi^\alpha(t)$ according to the definition (61)

$$\psi^\alpha(t) = \psi_i^\alpha - i \int_{-\infty}^{t} dt' \int d^3r' \chi_\alpha^*(\mathbf{r}', t') V(r') \psi(\mathbf{r}', t'). \tag{91}$$

Now we form the matrix element of this equation between eigenstates $\Phi_1$ and $\Phi_2$ of $H$. We use the relation

$$\begin{aligned}\langle \Phi_1 | \psi(\mathbf{r}', t') | \Phi_2 \rangle &= \langle \Phi_1 | e^{iHt'} \psi(\mathbf{r}', 0) e^{-iHt'} | \Phi_2 \rangle\\ &= e^{i(E_1 - E_2)t'} \langle \Phi_1 | \psi(\mathbf{r}', 0) | \Phi_2 \rangle\end{aligned} \tag{92}$$

and consider then the limit $t \to -\infty$ in Eq. (91). Since the integrand in Eq. (91) is not a "wild function" of $t'$, the integral approaches zero as $t \to -\infty$. Then we have

$$\lim_{t \to -\infty} \langle \Phi_1 | \psi^\alpha(t) | \Phi_2 \rangle = \langle \Phi_1 | \psi_i^\alpha | \Phi_2 \rangle, \tag{93}$$

which is just the first of Eqs. (66). Finally, we note that if we take the limits $t_2 \to +\infty$, $t_1 \to -\infty$ in Eq. (88) and identify $\lim_{t_2 \to +\infty} \phi(\mathbf{r}, t; t_2)$ with $\psi_0(\mathbf{r}, t)$, we have

$$\psi_0(\mathbf{r}, t) = \psi_i(\mathbf{r}, t) - i \int_{-\infty}^{\infty} dt' \int d^3r' G_0(\mathbf{r} - \mathbf{r}'; t - t') V(r') \psi(\mathbf{r}', t'), \quad (94)$$

which shows the manner in which $\psi_o$ "knows" about $\psi_i$ and about the interaction.

**The *S*-Matrix**

We now define the matrix elements between the "in" and "out" states (72); the state $n_{\alpha 1} \cdots n_{\alpha \nu}$ is abbreviated by $(\alpha)$, and we write

$$\langle \Phi_{\text{out}}^{(\alpha)} \mid \Phi_{\text{in}}^{(\beta)} \rangle \equiv S_{(\alpha)(\beta)} = \langle \Phi_{\text{in}}^{(\alpha)} | \, S \, | \Phi_{\text{in}}^{(\beta)} \rangle = \langle \Phi_{\text{out}}^{(\alpha)} | \, S \, | \Phi_{\text{out}}^{(\beta)} \rangle. \quad (95)$$

Assuming the completeness of the states $\Phi_{\text{in}}^{(\alpha)}$ and $\Phi_{\text{out}}^{(\beta)}$, we have a unitary operator $S$ defined by

$$|\Phi_{\text{in}}^{(\alpha)} \rangle = S \, |\Phi_{\text{out}}^{(\alpha)} \rangle, \quad (96)$$

or

$$S = \sum_{(\alpha)} |\Phi_{\text{in}}^{(\alpha)} \rangle \langle \Phi_{\text{out}}^{(\alpha)}|. \quad (97)$$

It follows from these equations that

$$\psi_0{}^{\alpha} = S^{\dagger} \psi_i{}^{\alpha} S. \quad (98)$$

There is a slight complication associated with the fact that there may be bound states in the presence of $V$. We must add these explicitly to the $\psi_i{}^{\alpha}$ or $\psi_o{}^{\alpha}$ in order to have a complete set of operators. The presence of bound states does not really cause any complication in our previous manipulations, however, since we have been concerned solely with matrix elements of $S$ between continuum states, which are orthogonal to the discrete bound states.

Let us see now that our formalism yields familiar results. If our prescription for calculating $S$-matrix elements is correct, the one describing the scattering from momentum $\mathbf{k}$ to $\mathbf{k}'$ should be given by $\langle \mathbf{k}'_{\text{out}} \mid \mathbf{k}_{\text{in}} \rangle$. We choose for the function $\chi_{\mathbf{k}}$ a wave packet "aimed" at the potential and having a mean momentum $\mathbf{k}$ and energy $k^2/2m$; we shall ultimately allow it to approach $e^{i\mathbf{k}\cdot\mathbf{r} - ik^2t/2m}$. (We continue to take the quantization volume to be unity.) Similarly, $\chi_{\mathbf{k}'}$ represents a post collision packet with mean momentum $\mathbf{k}'$ and energy $k'^2/2m$. To create the state $|\mathbf{k}_{\text{in}}\rangle$ we use the operator $\psi_i{}^{\mathbf{k}}$ which is, as we have noted, independent of time. It is extremely convenient to choose the time to be $-\infty$ in order to facilitate the use of the asymptotic condition. We write

$$S_{\mathbf{k}'\mathbf{k}} = \langle \mathbf{k}'_{\text{out}}|\, \mathbf{k}_{\text{in}}\rangle$$
$$= \lim_{y_0 \to -\infty} \int d^3y \langle \mathbf{k}'_{\text{out}}|\, \psi_i^\dagger(\mathbf{y}, y_0)\, |0\rangle \chi_{\mathbf{k}}(\mathbf{y}, y_0) \tag{99}$$
$$= \lim_{y_0 \to -\infty} \int d^3y \langle \mathbf{k}'_{\text{out}}|\, \psi^\dagger(\mathbf{y}, y_0)\, |0\rangle \chi_{\mathbf{k}}(\mathbf{y}, y_0).$$

In passing from the second line to the third we have used the asymptotic condition (66) to replace $\psi_i^\dagger$ by $\psi^\dagger$. It is useful here to use a four-dimensional notation $(\mathbf{y}, y_0)$ in place of $\mathbf{r}, t$; the volume element $d^3y\, dy_0$ will be written $d^4y$.

We now rewrite Eq. (99) by expressing it as an integral over all space at time $y_0 = +\infty$ together with an integral over all space-time:

$$S_{\mathbf{k}'\mathbf{k}} = \lim_{y_0 \to +\infty} \int d^3y \langle \mathbf{k}'_{\text{out}}|\, \psi^\dagger(\mathbf{y}, y_0)\, |0\rangle \chi_{\mathbf{k}}(\mathbf{y}, y_0)$$
$$- \int d^4y \frac{\partial}{\partial y_0} \{\langle \mathbf{k}'_{\text{out}}|\, \psi^\dagger(\mathbf{y}, y_0)\, |0\rangle \chi_{\mathbf{k}}(\mathbf{y}, y_0)\}. \tag{100}$$

Since the packet $\chi_{\mathbf{k}}(\mathbf{y}, y_0)$ depends on time, for large times, like $|y_0|^{-3/2}$, there is no worry about the convergence of the time integral.[16] In the first term, since $y_0 \to +\infty$, we may replace $\psi^\dagger(\mathbf{y}, y_0)$ by $\psi_o^\dagger(\mathbf{y}, y_0)$. Then, for this we have

$$\int d^3y \langle \mathbf{k}'_{\text{out}}|\, \psi_o^\dagger(\mathbf{y}, y_0)\, |0\rangle \chi_{\mathbf{k}}(\mathbf{y}, y_0) = \langle \mathbf{k}'_{\text{out}}|\, \psi_o^{\mathbf{k}\dagger}|0\rangle$$
$$= \delta_{\mathbf{k}'\mathbf{k}}. \tag{101}$$

The second term is transformed as follows:

$$S_{\mathbf{k}'\mathbf{k}} = \delta_{\mathbf{k}'\mathbf{k}} + i \int d^4y \langle \mathbf{k}'_{\text{out}}|\, \psi^\dagger(\mathbf{y}, y_0)\, |0\rangle$$
$$\times \left[ i\frac{\partial}{\partial y_0} - \left(-\frac{\nabla^2}{2m}\right) + \left(-\frac{\nabla^2}{2m}\right)\right] \chi_{\mathbf{k}}(\mathbf{y}, y_0)$$
$$+ i \int d^4y \chi_{\mathbf{k}}(\mathbf{y}, y_0) i \frac{\partial}{\partial y_0} \langle \mathbf{k}'_{\text{out}}|\, \psi^\dagger(\mathbf{y}, y_0)\, |0\rangle$$
$$= \delta_{\mathbf{k}'\mathbf{k}} + i \int d^4y \left\{ \left[ i\frac{\partial}{\partial y_0} + \left(-\frac{\nabla^2}{2m}\right)\right] \langle \mathbf{k}'_{\text{out}}|\psi^\dagger(\mathbf{y}, y_0)\, |0\rangle \right\} \chi_{\mathbf{k}}(\mathbf{y}, y_0). \tag{102}$$

In the first line here we have added and subtracted a term in an obvious

[16] It is evident that the packet is quite essential at this point. [See Eq. (3.21).]

way, and to obtain the final form we have used the equation of motion for $\chi$ together with an integration by parts.[17] Using the equation of motion for $\psi^\dagger$ [which is the adjoint of Eq. (46)], we obtain

$$S_{\mathbf{k}'\mathbf{k}} = \delta_{\mathbf{k}'\mathbf{k}} - i \int d^4y \langle \mathbf{k}'_{\text{out}} | \psi^\dagger(\mathbf{y}, y_0) | 0 \rangle V(\mathbf{y}) \chi_{\mathbf{k}}(\mathbf{y}, y_0). \tag{103}$$

In the limit that the packet $\chi_{\mathbf{k}}$ becomes sharp, and using the fact that $\psi^\dagger(\mathbf{y}, y_0) = e^{iHy_0} \psi^\dagger(\mathbf{y}, 0) e^{-iHy_0}$, we obtain finally

$$S_{\mathbf{k}'\mathbf{k}} = \delta_{\mathbf{k}'\mathbf{k}} - 2\pi i \delta(E_{k'} - E_k) \int d^3y \langle \mathbf{k}'_{\text{out}} | \psi^\dagger(\mathbf{y}, 0) | 0 \rangle V(\mathbf{y}) e^{i\mathbf{k}\cdot\mathbf{y}}. \tag{104}$$

This indeed coincides with the first of Eqs. (32), provided we can show that

$$\langle \mathbf{k}'_{\text{out}} | \psi^\dagger(\mathbf{y}, 0) | 0 \rangle = [\psi_{\mathbf{k}'}^{-}(\mathbf{y})]^*. \tag{105}$$

The passage from Eq. (99) to Eq. (104) is an example of what is called in quantum field theory the application of a *reduction formula*. We refer to displaying the dependence of $S_{\mathbf{k}'\mathbf{k}}$ on $\chi_{\mathbf{k}}$ as "contracting on $\mathbf{k}$." In the same way, we now consider $\langle \mathbf{k}'_{\text{out}} | \psi^\dagger(\mathbf{y}, 0) | 0 \rangle$ and contract on $\mathbf{k}'$:

$$\langle k'_{\text{out}} | \psi^\dagger(\mathbf{y}, 0) | 0 \rangle = \lim_{x_0 \to +\infty} \int d^3x \chi_{\mathbf{k}'}^*(\mathbf{x}, x_0) \langle 0 | \psi_0(\mathbf{x}, x_0) \psi^\dagger(\mathbf{y}, 0) | 0 \rangle. \tag{106}$$

Since the time $x_0$ is surely greater than zero, there is no harm in replacing the operator product in (106) by $[\psi(\mathbf{x}, x_0)\psi^\dagger(\mathbf{y}, 0)]_+$, where the subscript $+$ designates the Dyson time-ordering operation. This implies that if one has two time-dependent operators, $A(t)$, $B(t')$,

$$\begin{aligned} [A(t)B(t')]_+ &= A(t)B(t'), \qquad t > t' \\ &= B(t')A(t), \qquad t' > t. \end{aligned} \tag{107}$$

If $t = t'$, we use one-half the sum of the two orderings. The general rule for $[A(t_1)B(t_2)\cdots]_+$ is to arrange the operators in such a way that the earliest time stands to the right, then the next earliest, and so on.

Now our formalism up to this point has been correct for bosons. We may make it correct for fermions also (although in this simple example of noninteracting particles it makes no difference). We must write, for fermions instead of just $[\psi(\mathbf{x}, x_0)\psi^\dagger(\mathbf{y}, 0)]_+$, $\epsilon(x_0)[\psi(\mathbf{x}, x_0)\psi^\dagger(\mathbf{y}, 0)]_+$, where $\epsilon(x_0)$ is as usual the sign function.[18] Although $\epsilon(x_0)$ is certainly positive in our case and it would appear to make no difference whether we supply it, it is very important that it be there in our subsequent manipulations if

[17] Again, the wave packet is essential here.

[18] This is defined in connection with Eq. (14).

we were to allow for interactions between the particles. For fermions, at equal times we then have the instruction $\frac{1}{2}[A(t)B(t) - B(t)A(t)]$; if it were not for the appearance of the commutator of the two operators, there would occur in certain later formulas some additional $\delta$-functions. These arise because the fundamental commutation relation (35) for fermions becomes replaced by the anticommutation relation

$$\{\psi(\mathbf{x}, x_0), \psi^\dagger(\mathbf{y}, y_0)\} = \delta(\mathbf{x} - \mathbf{y}), \qquad \text{for } x_0 = y_0. \tag{108}$$

The two possibilities, simple time-ordering for bosons and $\epsilon$ times the ordering for fermions, are summarized in the symbol $T[A(t)B(t')]$ invented by Wick.[19]

We write now for Eq. (106) the expression

$$\langle \mathbf{k}'_{\text{out}}|\,\psi^\dagger(\mathbf{y}, 0)\,|0\rangle = \lim_{x_0 \to \infty} \int d^3x \chi_{\mathbf{k}'}^*(\mathbf{x}, x_0)\langle 0|\, T[\psi(\mathbf{x}, x_0)\psi^\dagger(\mathbf{y}, 0)]\,|0\rangle, \tag{109}$$

and in a way quite similar to that used in connection with Eq. (99) replace the integral at $x_0 = +\infty$ by one at $x_0 = -\infty$, together with an integration over all space-time. Thus we find

$$\begin{aligned}\langle \mathbf{k}'_{\text{out}}|\psi^\dagger(\mathbf{y}, 0)\,|0\rangle &= i\int d^4x\left(-i\frac{\partial}{\partial x_0}\right)\{\chi_{\mathbf{k}'}^*(\mathbf{x}, x_0)\langle 0|\, T[\psi(\mathbf{x}, x_0)\psi^\dagger(\mathbf{y}, 0)]\,|0\rangle\} \\ &\quad + \lim_{x_0 \to -\infty} \int d^3x \chi_{\mathbf{k}'}^*(\mathbf{x}, x_0)\langle 0|\, T[\psi(\mathbf{x}, x_0)\psi^\dagger(\mathbf{y}, 0)]\,|0\rangle.\end{aligned} \tag{110}$$

The last term vanishes as may be seen in the following way: Since the time $x_0$ is certainly earlier than 0, we may write $T[\psi(\mathbf{x}, x_0)\psi^\dagger(\mathbf{y}, 0)]$ as $\pm\psi^\dagger(\mathbf{y}, 0)\psi(\mathbf{x}, x_0)$, where the plus sign holds for bosons, the minus sign for fermions; next, we replace $\psi(\mathbf{x}, x_0)$ by the "in" operator $\psi_i(\mathbf{x}, x_0)$ according to the asymptotic condition. Acting on the vacuum state, however, this yields zero. In our very simple problem, it so happens that $\psi(\mathbf{x}, x_0)$ operating on the vacuum also yields zero. This is not the case in more complicated theories, where the fact that it is really $\psi_i$ that enters is important. Using the equation of motion for $\chi_{\mathbf{k}'}^*$ and entirely legitimate spatial integrations by parts [see Eq. (102)], we obtain finally

$$\begin{aligned}\langle \mathbf{k}'_{\text{out}}|\psi^\dagger(\mathbf{y}, 0)\,|0\rangle &= -i\int d^4x \chi_{\mathbf{k}'}^*(\mathbf{x}, x_0)\left[i\frac{\partial}{\partial x_0} - \left(-\frac{\nabla_x^2}{2m}\right)\right] \\ &\quad \times \langle 0|\, T[\psi(\mathbf{x}, x_0)\psi^\dagger(\mathbf{y}, 0)]\,|0\rangle.\end{aligned} \tag{111}$$

[19] We have introduced this operation in Eq. (2.100).

The indicated differential operations on the matrix element in Eq. (111) can be readily carried out. We note that

$$\begin{aligned} T[\psi(\mathbf{x}, x_0)\psi^\dagger(\mathbf{y}, 0)] &= \tfrac{1}{2}\{\psi(\mathbf{x}, x_0), \psi^\dagger(\mathbf{y}, 0)\} \\ &\quad + \tfrac{1}{2}\epsilon(x_0)[\psi(\mathbf{x}, x_0), \psi^\dagger(\mathbf{y}, 0)] \qquad \text{(bosons)}, \\ &= \tfrac{1}{2}\epsilon(x_0)\{\psi^\dagger(\mathbf{x}, x_0), \psi^\dagger(\mathbf{y}, 0)\} \\ &\quad + \tfrac{1}{2}[\psi(\mathbf{x}, x_0), \psi^\dagger(\mathbf{y}, 0)] \qquad \text{(fermions)}. \end{aligned} \tag{112}$$

We see then, using $(\partial/\partial x_0)\epsilon(x_0) = 2\delta(x_0)$, that

$$\begin{aligned} \frac{\partial}{\partial x_0} T[\psi(\mathbf{x}, x_0)\psi^\dagger(\mathbf{y}, 0)] &= T\left[\frac{\partial\psi}{\partial x_0}(\mathbf{x}, x_0)\psi^\dagger(\mathbf{y}, 0)\right] \\ &\quad + \delta(x_0)\begin{Bmatrix} [\psi(\mathbf{x}, x_0), \psi^\dagger(\mathbf{y}, 0)] \\ \{\psi(\mathbf{x}, x_0), \psi^\dagger(\mathbf{y}, 0)\} \end{Bmatrix} \quad \begin{matrix} \text{(bosons)} \\ \text{(fermions)}. \end{matrix} \end{aligned} \tag{113}$$

Since the equal time commutator (anticommutator) for bosons (fermions) is just $\delta(\mathbf{x} - \mathbf{y})$, we have for both cases the result

$$\frac{\partial}{\partial x_0} T[\psi(\mathbf{x}, x_0)\psi^\dagger(\mathbf{y}, 0)] = T\left[\frac{\partial\psi}{\partial x_0}(\mathbf{x}, x_0)\psi^\dagger(\mathbf{y}, 0)\right] + \delta(x_0)\delta(\mathbf{x} - \mathbf{y}). \tag{114}$$

The spatial differentiations in (111) pass right through the time-ordering symbol and we have, using the equation of motion for $\psi(\mathbf{x}, x_0)$, the result

$$\begin{aligned} \langle \mathbf{k}'_{\text{out}} | \psi^\dagger(\mathbf{y}, 0) | 0 \rangle = \chi_{\mathbf{k}'}^*(\mathbf{y}, 0) - i \int d^4x \chi_{\mathbf{k}'}^*(\mathbf{x}, x_0) V(x) \\ \times \langle 0 | T[\psi(\mathbf{x}, x_0)\psi^\dagger(\mathbf{y}, 0)] | 0 \rangle. \end{aligned} \tag{115}$$

As we have noted before, $\psi(\mathbf{x}, x_0)$ annihilates the vacuum; we need retain only that part of the time integral in (115) for which $x_0 > 0$ and write $T[\psi(\mathbf{x}, x_0)\psi^\dagger(\mathbf{y}, 0)]$ as $\theta(x_0)\psi(\mathbf{x}, x_0)\psi^\dagger(\mathbf{y}, 0)$.[20] We note that the function

$$K(\mathbf{x}x_0; \mathbf{y}y_0) \equiv \theta(x_0 - y_0)\langle 0 | \psi(\mathbf{x}, x_0)\psi^\dagger(\mathbf{y}, 0) | 0 \rangle \tag{116}$$

is the solution for the Green's function in the Schrödinger theory which vanishes for $x_0 < y_0$, that is,

$$\left[i\frac{\partial}{\partial x_0} - \left(-\frac{\nabla_x^2}{2m}\right) - V(x)\right] K(\mathbf{x}x_0; \mathbf{y}y_0) = i\delta(x_0 - y_0)\delta(\mathbf{x} - \mathbf{y}) \tag{117}$$

and $K(\mathbf{x}x_0; \mathbf{y}y_0) = 0$ for $x_0 < y_0$. We may evidently rewrite Eq. (115) in terms of $K$ as

$$\langle \mathbf{k}'_{\text{out}} | \psi^\dagger(\mathbf{y}, 0) | 0 \rangle = \chi_{\mathbf{k}'}^*(\mathbf{y}, 0) - i \int d^4x \chi_{\mathbf{k}'}^*(\mathbf{x}, x_0) V(x) K(\mathbf{x}x_0; \mathbf{y}0). \tag{118}$$

[20] The theta-function is defined by Eq. (14).

In order to expose Eq. (115) as an old friend, we write

$$\langle 0|\,\psi(\mathbf{x}, x_0)\psi^\dagger(\mathbf{y}, 0)\,|0\rangle \qquad \text{as} \qquad \sum_\alpha \langle 0|\,\psi(\mathbf{x}, x_0)\,|\alpha\rangle\langle\alpha|\,\psi^\dagger(\mathbf{y}, 0)\,|0\rangle,$$

where the states $|\alpha\rangle$ are a complete set of one-particle states which may be taken as either in or out states. Bound states, in case they exist, are also included here. The dependence on $x_0$ may be exhibited explicitly by using $e^{iHx_0}\psi(\mathbf{x}, 0)e^{-iHx_0} = \psi(\mathbf{x}, x_0)$ and writing

$$\langle 0|\,\psi(\mathbf{x}, x_0)\,|\alpha\rangle = e^{-iE_\alpha x_0}\langle 0|\,\psi(\mathbf{x}, 0)\,|\alpha\rangle,$$

where $E_\alpha$ is the energy of the state $|\alpha\rangle$. We call the *wave function* $\langle 0|\,\psi(\mathbf{x}, 0)\,|\alpha\rangle$ simply $\phi_\alpha(\mathbf{x})$ and obtain for $\langle \mathbf{k}'_{\text{out}}|\,\psi^\dagger(\mathbf{y}, 0)\,|0\rangle$ the equation

$$\begin{aligned}\langle \mathbf{k}'_{\text{out}}|\,\psi^\dagger(\mathbf{y}, 0)\,|0\rangle &= \chi_{\mathbf{k}'}{}^*(\mathbf{y}, 0) - i\sum_\alpha \int_0^\infty dx_0 e^{i(E_{k'}-E_\alpha)x_0} \\ &\quad\times \int d^3x\,\chi_{\mathbf{k}'}{}^*(\mathbf{x}, 0)V(x)\phi_\alpha(\mathbf{x})\phi_\alpha{}^*(\mathbf{y}) \\ &= \chi_{\mathbf{k}'}{}^*(\mathbf{y}, 0) + \lim_{\eta\to 0}\sum_\alpha \frac{(\chi_{\mathbf{k}'}, V\phi_\alpha)\phi_\alpha{}^*(\mathbf{y})}{E_{k'} + i\eta - E_\alpha}, \end{aligned} \tag{119}$$

where we have used the formal device of writing

$$\begin{aligned} -i\int_0^\infty dx_0 e^{i(E_{k'}-E_\alpha)x_0} &= \lim_{\eta\to 0(+)} (-i)\int_0^\infty dx_0 e^{i(E_{k'}-E_\alpha+i\eta)x_0} \\ &= \lim_{\eta\to 0(+)} \frac{1}{E_{k'} - E_\alpha + i\eta}. \end{aligned} \tag{120}$$

Finally, we take the complex conjugate of Eq. (119) (and as usual omit the instruction $\lim_{\eta\to 0(+)}$) to obtain

$$\begin{aligned}\langle 0|\,\psi(\mathbf{y}, 0)\,|\mathbf{k}'_{\text{out}}\rangle &= \chi_{\mathbf{k}'}(\mathbf{y}, 0) + \sum_\alpha \frac{\phi_\alpha(\mathbf{y})(\phi_\alpha, V\chi_{\mathbf{k}'})}{E_{k'} - E_\alpha - i\eta} \\ &= \chi_{\mathbf{k}'}(\mathbf{y}, 0) + \frac{1}{E_{k'} - i\eta - H}\sum_\alpha \phi_\alpha(\mathbf{y})(\phi_\alpha, V\chi_{\mathbf{k}'}) \\ &= \left(1 + \frac{1}{E_{k'} - i\eta - H}V\right)\chi_{\mathbf{k}'}. \end{aligned} \tag{121}$$

This is precisely the formula which in Chapter 5 was referred to as $\psi_{\mathbf{k}'}{}^-$, so Eq. (105) is established.

We may now, from Eq. (104), write our $S$-matrix element in the standard form

$$S_{\mathbf{k}'\mathbf{k}} = \delta_{\mathbf{k}'\mathbf{k}} - 2\pi i\,\delta(E_{k'} - E_k)T_{\mathbf{k}'\mathbf{k}} \tag{122}$$

with

$$T_{\mathbf{k}'\mathbf{k}} \equiv \int d^3y \langle \mathbf{k}'_{\text{out}} | \psi^\dagger(\mathbf{y}, 0) | 0 \rangle V(y) \chi_{\mathbf{k}}(\mathbf{y}, 0)$$
$$= (\psi_{\mathbf{k}'}^-, V\chi_{\mathbf{k}})$$
$$= (\chi_{\mathbf{k}'}, V\chi_{\mathbf{k}}) + \sum_\alpha \frac{(\chi_{\mathbf{k}'}, V\phi_\alpha)(\phi_\alpha, V\chi_{\mathbf{k}})}{E_{k'} - E_\alpha + i\eta}. \tag{123}$$

This well-known formula plays a central role in our subsequent work. If it were not that we wanted to introduce the important concepts of the in and out fields in a familiar setting, we could have simply begun with it and eliminated the material presented thus far in this chapter.

Before describing the relation of Eq. (123) to dispersion theory, we note another form of our reduction formula for the $S$-matrix which appears frequently in the literature. According to Eq. (102),

$$S_{\mathbf{k}'\mathbf{k}} = \delta_{\mathbf{k}'\mathbf{k}} - i\int d^4y \langle \mathbf{k}'_{\text{out}} | \psi^\dagger(\mathbf{y}, y_0) | 0 \rangle \left[ -i\frac{\overleftarrow{\partial}}{\partial y_0} - \left( -\frac{\overleftarrow{\nabla}_y^2}{2m} \right) \right] \chi_{\mathbf{k}}(\mathbf{y}, y_0) \tag{124}$$

where the arrows indicate that the differentiations are applied to $\langle \mathbf{k}'_{\text{out}} | \psi^\dagger(\mathbf{y}, y_0) | 0 \rangle$ and *not* to $\chi_{\mathbf{k}}(\mathbf{y}, y_0)$. We do not carry out the differentiations as we did before, but proceed with the contraction of $\langle \mathbf{k}'_{\text{out}}|$ in precisely the same manner. We find

$$S_{\mathbf{k}'\mathbf{k}} = \delta_{\mathbf{k}'\mathbf{k}} + iF_{\mathbf{k}'\mathbf{k}}$$

$$F_{k'k} = i\int d^4x \int d^4y \chi_{\mathbf{k}'}^*(\mathbf{x}, x_0) \left[ i\frac{\partial}{\partial x_0} - \left( \frac{\nabla_x^2}{2m} \right) \right] \langle 0 | T[\psi(\mathbf{x}, x_0)\psi^\dagger(\mathbf{y}, y_0)] | 0 \rangle$$
$$\times \left[ -i\frac{\overleftarrow{\partial}}{\partial x_0} - \left( -\frac{\overleftarrow{\nabla}_x^2}{2m} \right) \right] \chi_{\mathbf{k}}(\mathbf{y}, y_0). \tag{125}$$

The quantity $F_{\mathbf{k}'\mathbf{k}}$ is called the "Feynman amplitude." It is simply $-2\pi\delta(E_{\mathbf{k}'} - E_{\mathbf{k}})T_{\mathbf{k}'\mathbf{k}}$ and has a quite simple interpretation. The matrix element $\langle 0 | T[\psi(\mathbf{x}, x_0)]\psi^\dagger(\mathbf{y}, y_0) | 0 \rangle$ is just the Green's function $K(x, y)$, Eq. (110) (we abbreviate $\mathbf{x}, x_0$ by $x$, $\mathbf{y}, y_0$ by $y$), which describes the propagation of a particle from space-time point $(\mathbf{y}, y_0)$ to $(\mathbf{x}, x_0)$. To lowest order in the potential we have

$$K(x; y) \simeq K_0(x; y) - i\int d^4z K_0(x; z)V(z)K_0(z; y), \tag{126}$$

where $K_0$ is the free-particle ($V = 0$) Green's function; the differential operators "knock off" the $K_0$'s which describe the propagation from $y$ to the point of first interaction and from the point of interaction to $x$.

Substitution of (126) into (125) leads directly to

$$F_{k'k} \approx -\int d^4z\chi_{\mathbf{k}'}{}^*(z)V(z)\chi_{\mathbf{k}}(z)$$
$$= -2\pi\,\delta(E_{k'} - E_k)\int d^3z\chi_{\mathbf{k}'}{}^*(\mathbf{z}, 0)V(z)\chi_{\mathbf{k}}(\mathbf{z}, 0) \qquad (127)$$

in our problem, where $V$ is independent of time. The higher order terms in the expansion of $K$ in powers of $V$ represent successive scatterings by the potential. The differential operators in (125) simply chop off the "legs" before the first interaction and after the last one.

## 10.4 DISPERSION RELATIONS IN NONRELATIVISTIC THEORY

We turn now to the main topic of this chapter: dispersion relations within the framework of the nonrelativistic theory. We shall confine our attention to the study of the elastic scattering of spinless particles by a spherically symmetric, local potential.[21] Then the scattering amplitude may be taken to depend on the energy and the scattering angle. (We have until now been discussing the scattering of particles by an external potential; there is no difficulty, however, in describing the scattering of two particles interacting by a central potential. We need only replace the wave number $\mathbf{k}$ by the wave number $\boldsymbol{\varkappa}$ characterizing the relative motion and the particle mass $m$ by the reduced mass $M_r$. We shall continue to use $\mathbf{k}$ and $m$, however, in this chapter). Alternatively, we may use as variables the energy and the momentum transfer, the latter being defined by $\Delta^2 \equiv (\mathbf{k} - \mathbf{k}')^2 = 2k^2(1 - \cos\theta)$, where $\theta$ is the scattering angle. The momentum transfer variable will play a very important role in our subsequent work.

### Dispersion Relations for Forward Scattering

We begin by considering forward scattering, where $\Delta^2 = 0$ and the scattering amplitude is a function of only the energy $E = k^2/2m$. We write in place of $T_{\mathbf{kk}}$, $T(E)$, and rewrite the fundamental equation (123) in the form

$$T(E) = \int d^3x V(x) + \int d^3x \int d^3y e^{-i\mathbf{k}\cdot\mathbf{x}} V(x) G(\mathbf{x}, \mathbf{y}, E) V(y) e^{i\mathbf{k}\cdot\mathbf{y}}, \qquad (128)$$

where the Green's function, $G(\mathbf{x}, \mathbf{y}, E)$ is given by

$$G(\mathbf{x}, \mathbf{y}, E) = \sum_\alpha \frac{\phi_\alpha(\mathbf{x})\phi_\alpha{}^*(\mathbf{y})}{E + i\eta - E_\alpha}. \qquad (129)$$

[21] We shall discuss more general interactions later.

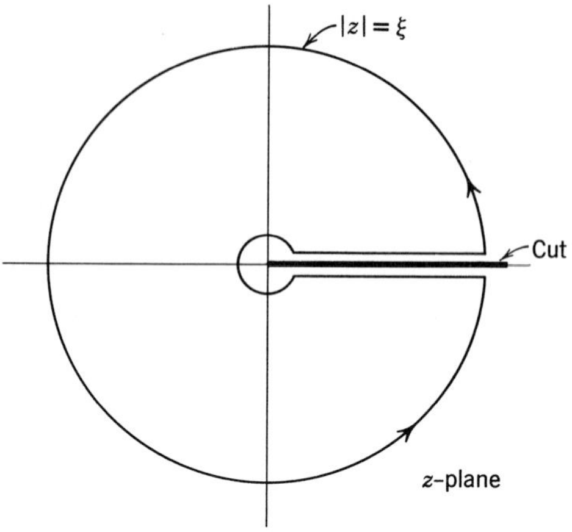


Fig. 10.1 The contour of integration for Eq. (131).

Here we have written $\chi_{\mathbf{k}} = e^{i\mathbf{k}\cdot\mathbf{x}}$, with the quantization volume again taken as unity. It is convenient to rewrite Eq. (128) as

$$\tilde{T}(E) \equiv T(E) - \int d^3x V(x) = \int d^3x \int d^3y \tilde{T}(\mathbf{x}, \mathbf{y}, E), \tag{130}$$

$$\tilde{T}(\mathbf{x}, \mathbf{y}, E) \equiv e^{-i\mathbf{k}\cdot\mathbf{x}} V(x) G(\mathbf{x}, \mathbf{y}, E) V(y) e^{i\mathbf{k}\cdot\mathbf{y}}. \tag{131}$$

We now regard $G$ as a function of the complex variable $z$, which we shall ultimately set equal to $E + i\eta$, with $\eta \to 0$. It is evidently analytic in the $z$-plane except for discrete poles at the positions of bound states (at $z = E_B < 0$) and a cut extending along the positive real axis from zero to infinity, corresponding to the positive energy continuum for scattering states. The wave number $k$, given by $\sqrt{2mE}$, becomes a complex variable also as $z$ roams over this cut plane and has *always* a positive imaginary part, since the entire cut $z$-plane is mapped into the upper-half $k$-plane by the relation $k = \sqrt{2mz}$, with $0 \leq \operatorname{Arg} z < 2\pi$.

Consider now the Cauchy integral

$$I_c(z) = \frac{1}{2\pi i} \int d\zeta \frac{\tilde{T}(\mathbf{x}, \mathbf{y}, \zeta)}{\zeta - z} \tag{132}$$

extended over the contour shown in Fig. 10.1. We obtain from this

$$\tilde{T}(\mathbf{x}, \mathbf{y}, z) = \sum_B \frac{\Gamma_B(\mathbf{x}, y)}{z - E_B} + \frac{1}{2\pi i} \int_0^\xi d\zeta \, \frac{\tilde{T}(\mathbf{x}, \mathbf{y}, \zeta_+) - \tilde{T}(\mathbf{x}, \mathbf{y}, \zeta_-)}{\zeta - z}$$
$$+ \frac{1}{2\pi i} \int_c d\zeta \frac{\tilde{T}(\mathbf{x}, \mathbf{y}, \zeta)}{\zeta - z}, \tag{133}$$

where $c$ is the large circle of radius $\xi$ shown in Fig. 10.1 and we have written $\zeta_{\pm}$ for the respective values of $\zeta$ just above and below the cut. Finally, we have

$$\Gamma_B(\mathbf{x}, \mathbf{y}) = [e^{-i\mathbf{k}\cdot\mathbf{x}}V(x)\phi_B(\mathbf{x})\phi_B^*(\mathbf{y})V(y)e^{i\mathbf{k}\cdot\mathbf{y}}]_{k\to k_B}, \tag{134}$$

$$k_B^{\ 2} = 2mE_B < 0. \tag{135}$$

In Eq. (132) we have discarded the contribution from a small circle about the edge of the cut and have assumed that there is a gap in the energy spectrum between the bound states and $\zeta = 0$. The latter condition is generally correct except for the Coulomb case, which is uninteresting. It is easy to see that $x^2\tilde{T}(\mathbf{x}, \mathbf{y}, \zeta')y^2$ [which is what is ultimately involved in computing $T(E)$] is finite near $\zeta' \sim 0$, provided only that $V$ is less singular than $x^{-3}$ for small $x$.

We wish now to relate $\tilde{T}(\mathbf{x}, \mathbf{y}, \zeta_+) - \tilde{T}(\mathbf{x}, \mathbf{y}, \zeta_-)$ to the imaginary part of $\tilde{T}(\mathbf{x}, \mathbf{y}, \zeta_+)$. Since $k(\zeta_-) = -k(\zeta_+)$, we need only show that

$$G(\mathbf{x}, \mathbf{y}, \zeta_-) = G^*(\mathbf{x}, \mathbf{y}, \zeta_+). \tag{136}$$

As far as the bound-state contributions to $G$ are concerned this is trivial, because the bound-state wave functions may be taken to be real. For the continuum states, we may use the condition of time reversal invariance[22] to write [see Eq. (4.232*a*)]

$$\phi_{\mathbf{k}}^- = \mu_k(\phi_{-\mathbf{k}}^{\ +})^*, \tag{137}$$

where $\mu_k$ is of modulus unity. We have then for the *continuum contribution* to $G$, called $G_c$,

$$\begin{aligned} G_c(\mathbf{x}, \mathbf{y}, \zeta_+) &= \sum_{\mathbf{k}} \frac{\phi_{\mathbf{k}}^-(\mathbf{x})\phi_{\mathbf{k}}^{-*}(\mathbf{y})}{\zeta_+ - E_k} = \sum_{\mathbf{k}} \frac{\phi_{-\mathbf{k}}^{+*}(\mathbf{x})\phi_{-\mathbf{k}}^{\ +}(\mathbf{y})}{\zeta_+ - E_k} \\ &= \left[\sum_{k} \frac{\phi_{\mathbf{k}}^{+}(\mathbf{x})\phi_{\mathbf{k}}^{+*}(\mathbf{y})}{\zeta_- - E_k}\right]^* = G_c^*(\mathbf{x}, \mathbf{y}, \zeta_-). \end{aligned} \tag{138}$$

In defining $G_c(\mathbf{x}, \mathbf{y}, \zeta_+)$, we have taken the liberty of using the wave functions $\phi_{\mathbf{k}}^-(x) = \langle 0|\,\psi(\mathbf{x})\,|\mathbf{k}_{\text{out}}\rangle$, since any complete set will do, and have used the time reversal relation (137) to express $\phi_{\mathbf{k}}^-$ in terms of $\phi_{\mathbf{k}}^+(\mathbf{x}) = \langle 0|\,\psi(\mathbf{x})\,|\mathbf{k}_{\text{in}}\rangle$. Since we sum over all $\mathbf{k}$, the direction is irrelevant and we may replace $\mathbf{k}$ by $-\mathbf{k}$. The final identity follows with the observation that we might have defined $G_c$ in terms of the $\phi_{\mathbf{k}}^+$ in the first place. We have then the result

$$\begin{aligned} \tilde{T}(\mathbf{x}, \mathbf{y}, \zeta_+) - \tilde{T}(\mathbf{x}, \mathbf{y}, \zeta_-) &= 2i \operatorname{Im} \tilde{T}(\mathbf{x}, \mathbf{y}, \zeta_+) \\ &= 2i \operatorname{Im} T(\mathbf{x}, \mathbf{y}, \zeta_+), \end{aligned} \tag{139}$$

since $\int d^3x V$ is real.

[22] It is assumed in this chapter that time reversal invariance obtains.

The next step in our derivation consists of integrating Eq. (132) for $\tilde{T}$ over $\mathbf{x}$ and $\mathbf{y}$ in order to form $\tilde{T}(E)$. We shall show that the integration over the bound-state terms converges and ultimately that the big circle makes no contribution. As for the integral along the real axis, our original integral representation is well defined and there is no difficulty in interchanging the orders of integration over $\zeta$ and over $\mathbf{x}$ and $\mathbf{y}$. The question of the convergence of the $\zeta$ integral, which remains when we allow $\xi \to \infty$, is easily resolved when we recall from Chapter 6 that for large energies the scattering amplitude approaches the first Born approximation, which is real. Thus Im $T(\zeta) \to 0$ as $\zeta \to \infty$ along the real axis. We have obtained then, so far,

$$\tilde{T}(z) = \sum_B \frac{\Gamma_B}{z - E_B} + \frac{1}{\pi}\int_0^\xi d\zeta \frac{\operatorname{Im} T(\zeta_+)}{\zeta - z} + R(\xi), \tag{140}$$

where

$$\begin{aligned} \Gamma_B &= \left| \int d^3x e^{-i\mathbf{k}\cdot\mathbf{x}} V(x)\phi_B(\mathbf{x}) \right|^2_{k \to k_B}, \\ R(\xi) &= \frac{1}{2\pi i}\int d^3x \int d^3y \int_c d\zeta \frac{\tilde{T}(x, y, \zeta)}{\zeta - z}. \end{aligned} \tag{141}$$

We consider first the evaluation of $\Gamma_B$. For an arbitrary real value of $|\mathbf{k}| = k$, the integral [see Eq. (133)] is

$$\begin{aligned} I &= \int d^3x e^{-i\mathbf{k}\cdot\mathbf{x}} V \phi_B(\mathbf{x}) \\ &= \int d^3x e^{-i\mathbf{k}\cdot\mathbf{x}}\left(E_B + \frac{\nabla^2}{2m}\right)\phi_B(\mathbf{x}) = \left(E_B - \frac{k^2}{2m}\right)\int d^3x e^{-i\mathbf{k}\cdot\mathbf{x}}\phi_B(\mathbf{x}), \end{aligned} \tag{142}$$

where we have used the Schrödinger equation to eliminate $V(x)$. In the computation of $\Gamma_B$ we require $I$ only in the limit $k \to k_B$, where $k_B{}^2 = 2mE_B$. Evidently it is only the asymptotic part of the wave function $\phi_B(\mathbf{x})$, the part that "knows" about $E_B$, that can contribute to $I$ in this limit. The exact form of the residue depends on the angular momentum of the bound state; in order to illustrate the method of computation, we assume an $S$-state. We then write, with $k_B = i\alpha_B = i(2m\,|E_B|)^{1/2}$,

$$\phi_B(x) = N\left[\frac{e^{-\alpha_B x}}{x} - g(x)\right], \tag{143}$$

where $x = |x|$ and $N$ is a normalization constant. This division of the wave function is meaningful provided

$$\lim_{x \to \infty} x e^{\alpha_B x} g(x) \to 0. \tag{144}$$

Substituting Eq. (143) into $I$, we find

$$I = \left(E_B - \frac{k^2}{2m}\right)\frac{4\pi N}{k^2 + {\alpha_B}^2} - N\left(E_B - \frac{k^2}{2m}\right)\int d^3x e^{-i\mathbf{k}\cdot\mathbf{x}} g(x). \quad (145)$$

Since by hypothesis $g \to 0$ more rapidly than $e^{-\alpha_B x}$, the integral over $g$ in (145) certainly exists as $k \to i\alpha_B$ from a real value. Therefore, in the limit,

$$I = -\frac{2\pi N}{m}. \quad (146)$$

and the residue $\Gamma_B$ is [see Eq. (141)]

$$\Gamma_B = \frac{4\pi^2 N^2}{m^2}. \quad (147)$$

We may express $N$ in terms of the effective range $r_e$ as defined in Section 6.6, using the bound-state wave function instead of that at zero energy. Thus we write [using Eq. (6.323)]

$$\tfrac{1}{2}r_e = \int_0^\infty dx\left[e^{-2\alpha_B x} - \frac{(x\phi_B)^2}{N^2}\right]. \quad (148)$$

Since $N$ is chosen so that

$$4\pi\int_0^\infty x^2\, dx \phi_B(x) = 1, \quad (149)$$

we have

$$N^2 = \frac{\alpha_B}{2\pi}\frac{1}{1 - \alpha_B r_e}, \quad (150)$$

and thus

$$\Gamma_B = \frac{2\pi\alpha_B}{m^2(1 - \alpha_B r_e)}. \quad (151)$$

We turn now to the final problem, which is the evaluation of $R(\xi)$. There are a variety of ways of showing that as $\xi \to \infty$, $R(\xi) \to 0$. The problem with which we are confronted is to show that the integral over $\mathbf{x}$ and $\mathbf{y}$ of $\tilde{T}(\mathbf{x}, \mathbf{y}, \zeta)$ exists and approaches zero sufficiently rapidly for the positive imaginary $k$'s which occur when $\zeta$ is on the large circle. Evidently, the potentially dangerous factors $e^{-i\mathbf{k}\cdot\mathbf{x}}$ and $e^{i\mathbf{k}\cdot\mathbf{y}}$ must be controlled by $G(\mathbf{x}, \mathbf{y}, \zeta)$. The most satisfactory demonstration is based on showing that for large $\zeta$'s $G$ approaches the value it would have if there were no interaction. This is (essentially) equivalent to the proof that the Born series converges for sufficiently large energies. (The fact that the wave number $k$ has a positive imaginary part only helps matters, as is almost self-evident.) The rather involved argument was outlined in Section 6.7 and is given in detail by Klein and Zemach [16]. We shall simply accept the result that as $E \to \infty$ ($k \to \infty$ with a positive imaginary part),

$$G(\mathbf{x}, \mathbf{y}, E) \to G_0(\mathbf{x}, \mathbf{y}, E) = -\frac{m}{2\pi}\frac{e^{ik|\mathbf{x}-\mathbf{y}|}}{|\mathbf{x}-\mathbf{y}|}, \tag{152}$$

as given by Eq. (6.364*b*). Then

$$\tilde{T}(E) = \int d^3x \int d^3y \tilde{T}(\mathbf{x}, \mathbf{y}, E) \to \int d^3x \int d^3y e^{i\mathbf{k}(\mathbf{y}-\mathbf{x})} V(x) \frac{e^{ik|\mathbf{x}-\mathbf{y}|}}{|\mathbf{x}-\mathbf{y}|} V(y)\left[-\frac{m}{2\pi}\right] \tag{153}$$

on the large circle. Now $k\,|\mathbf{x}-\mathbf{y}| \geq |\mathbf{k}\cdot(\mathbf{y}-\mathbf{x})|$, and since the equality holds only for that angle for which $\mathbf{k}$ is parallel to $(\mathbf{y}-\mathbf{x})$, we may conclude that as $k \to \infty$ in its upper half plane, $\tilde{T}(\mathrm{E}) \to 0$. Thus we see that in the limit as $\xi \to \infty$, $R(\xi) \to 0$. We are left finally with the dispersion relation for the physical, forward scattering $T$-matrix, obtained by letting $z \to E + i\eta$ in Eq. (140).

$$T(E) = \int d^3x V(x) + \sum_B \frac{\Gamma_B}{E - E_B} + \frac{1}{\pi}\int_0^\infty dE' \frac{\operatorname{Im} T(E')}{E' - E - i\eta}. \tag{154}$$

The precise conditions under which the limiting behavior (152) obtains have been given by Klein and Zemach [16]. They define $I_v(r)$ by

$$I_v(r) \equiv \int d^3r' \frac{|V(r')|}{|\mathbf{r}-\mathbf{r}'|} \tag{155}$$

and then show that the validity of Eq. (152) requires that

$$\begin{aligned} &I_v(r) < \infty \qquad \text{for all } r,\\ &I_v(r) \qquad \text{is continuous in } r,\\ &I_v(r) \cong 0\left(\frac{1}{r}\right) \qquad \text{as } r \to \infty. \end{aligned} \tag{156}$$

These are essentially the conditions that conventional scattering theory, without pathologies, shall hold. We postpone a discussion of Eq. (154) until after we have studied the much more difficult problem of nonforward scattering.

**Dispersion Relations for Nonforward Scattering**

The first problem which must be decided is that of the appropriate variables to be used to describe the $T$-matrix. The first thought might be to use the energy (or wave number) and the scattering angle. This, however, does not prove convenient. Instead we use the energy and the momentum transfer $\Delta$, where $\Delta = 2k \sin \theta/2$, and shall hold $\Delta$ constant to obtain a dispersion relation in the energy. Evidently, we face a new problem here because for a fixed value of $\Delta^2$ we must require that for a physical scattering event the wave number $k$ satisfies $k^2 \geq \Delta^2/4$, since the maximum momentum transfer in a collision is $2k$. It will be necessary for us to consider the scattering amplitude for *all* $k$, however, and we meet here for the first time an unphysical scattering amplitude (which we will hope to be able to define by a process of analytic continuation).

To begin, we express the $T$-matrix in terms of our chosen variables, $E$ and $\Delta^2$. From Eq. (123) we have

$$T(E, \Delta^2) = \int d^3x e^{i\boldsymbol{\Delta}\cdot\mathbf{x}} V(x) + \int d^3x \int d^3y e^{i\boldsymbol{\Delta}\cdot(\mathbf{x}+\mathbf{y})/2} V(x) G(\mathbf{x}, \mathbf{y}, E) V(y) e^{i\mathbf{q}\cdot(\mathbf{y}-\mathbf{x})}, \tag{157}$$

where

$$\begin{aligned} \boldsymbol{\Delta} &= \mathbf{k} - \mathbf{k}', \\ \mathbf{q} &= \frac{\mathbf{k} + \mathbf{k}'}{2}, \\ q^2 &= k^2 - \frac{\Delta^2}{4} = 2mE - \frac{\Delta^2}{4}. \end{aligned} \tag{158}$$

We have used the rotational invariance of $V$ here in order to assert that $T$ depends only on $\Delta^2$ and not on the direction of $\Delta$; it has "nothing about which to orient itself." Furthermore, we have taken $k^2 = k'^2$ in Eq. (158). Because of rotational invariance, the vector $\mathbf{q}$ can appear only as $q^2$, so there is no branch point associated with $|\mathbf{q}| = (2mE - \Delta^2/4)^{1/2}$.

It is evident that the representation (157) cannot be used in its present form for arbitrarily large $\Delta^2$ at fixed $k$, because $q^2$ becomes negative and there is an exponential increase associated with the term $e^{i\mathbf{q}\cdot(\mathbf{x}-\mathbf{y})}$. Of course for physical processes, where $\Delta \leq 2k$ this does not occur. In order to cope with this problem we shall use a technique which is more powerful than is required at present.[23] The method is to write $q^2 = 2mE - \zeta$, where we assume $\zeta < 0$. Then $T$ becomes a function of three variables $E$, $\Delta^2$, and $\zeta$ and is defined for all $\Delta^2$ and $\zeta < 0$ by Eq. (157). Our final task will be to see if we can ultimately effect an analytic continuation in $\zeta$ from negative values up to $\zeta = +\Delta^2/4$. Of course, with negative $\zeta$ there is no question of existence of the representation (157), so we may follow very much the same formal procedure that was used in deriving the forward dispersion relation. The equation which is the analog of Eq. (140) is

$$\begin{aligned} \tilde{T}(z, \Delta^2, \zeta) &\equiv T(z, \Delta^2, \zeta) - \int d^3x e^{i\boldsymbol{\Delta}\cdot\mathbf{x}} V(x) \\ &= \sum_B \frac{\Gamma_B(\Delta^2, \zeta)}{E - E_B} + \frac{1}{\pi} \int_0^{\xi} dz' \frac{\operatorname{Im} \tilde{T}(z', \Delta^2, \zeta)}{z' - z} + R(\xi), \end{aligned} \tag{159}$$

[23] It is, however, the same method that is used in relativistic dispersion theory.

where

$$\Gamma_B(\Delta^2, \zeta) = \left[\int d^3x e^{-i(\mathbf{q}-\Delta/2)\cdot\mathbf{x}} V(x)\phi_B(\mathbf{x})\right]\left[\int d^3y e^{i(\mathbf{q}+\Delta/2)\cdot\mathbf{y}} V(y)\phi_B^*(y)\right],$$

$$R(\xi) = \frac{1}{2\pi i}\int d^3x \int d^3y \int_c dz' \, \frac{\tilde{T}(\mathbf{x}, \mathbf{y}, z', \Delta^2, \zeta)}{z' - z}. \tag{160}$$

The same considerations which previously led to the vanishing of $R(\xi)$ in the limit of $\xi \to \infty$ apply here. That is, we use the fact that the Green's function approaches the free-particle value, and the resulting $T$-matrix element may therefore be studied explicitly.

The final step in carrying out the derivation is to effect the requisite analytic continuation in $\zeta$ from negative to positive values, in particular, to $\zeta = +\Delta^2/4$. We note that for $\operatorname{Re} z > 0$ [which is what is involved in the integral in Eq. (159)]

$$\operatorname{Im} \tilde{T}(z_+, \Delta^2, \zeta) = -\pi \sum_{\mathbf{k}} \delta(z - E_{\mathbf{k}})\left[\int d^3x e^{-i(\mathbf{q}-\Delta/2)\cdot\mathbf{x}} V(x)\phi_{\mathbf{k}}(\mathbf{x})\right] \times \left[\int d^3y \phi_{\mathbf{k}}^*(\mathbf{y}) V(y) e^{i(\mathbf{q}+\Delta\cdot 2)\cdot\mathbf{y}}\right], \tag{161}$$

where here $q^2 = 2mz - \zeta$. Now the integrals in (161) may be used to define an analytic continuation in $\zeta$ into the complex $\zeta$-plane provided

$$|\operatorname{Im} \sqrt{2mz - \zeta}| < \mu, \tag{162}$$

where $\mu$ is the exponential range of the potential in the sense that

$$\int d^3x e^{\mu' x} V(x) < \infty \qquad \text{for } \mu' < \mu. \tag{163}$$

In order to carry out our continuation we need only an infinitesimal strip in the $\zeta$-plane in the neighborhood of the real axis. For $\operatorname{Im} \zeta \approx 0$, the requirement (162) becomes

$$\operatorname{Re} \zeta - 2mz < \mu^2. \tag{164}$$

Thus, since we require the continuation in $\zeta$ for all $z$, $0 \leq z \leq \infty$, we see that we can reach the physical value $\operatorname{Re} \zeta = \Delta^2/4$, provided

$$\frac{\Delta^2}{4} < \mu^2. \tag{165}$$

The bound-state terms are quite innocuous because the presence of the bound-state wave function provides additional exponential damping. What we have shown then is that in the equation

$$\tilde{T}(z, \Delta^2, \zeta) = \sum_B \frac{\Gamma_B(\Delta^2, \zeta)}{E - E_B} + \frac{1}{\pi}\int_0^\infty dz' \, \frac{\operatorname{Im} \tilde{T}(z'_+, \Delta^2, \zeta)}{z' - z}, \tag{166}$$

the right-hand side may be continued in $\zeta$ to the physical value $\zeta = \Delta^2/4$ when the condition (165) is satisfied. [Of course, when $z = E + i\eta$ and $E > \Delta^2/8m$, there is no difficulty in continuing the expression (161) to $\zeta = \Delta^2/4$.] Since our representation gives us $\tilde{T}$ as an analytic function of $z$ as long as we stay off the obvious cuts and poles, and both sides of (166) may be continued in $\zeta$ over a common region in $z$, we may write for $z \to E + i\eta$

$$T(E, \Delta^2) = \int d^3x e^{i\Delta\cdot\mathbf{x}} V(x) + \sum_B \frac{\Gamma_B(\Delta^2)}{E - E_B} + \frac{1}{\pi}\int_0^\infty dE' \frac{\operatorname{Im} T(E', \Delta^2)}{E' - E - i\eta}. \tag{167}$$

It is convenient to rewrite this in terms of the scattering amplitude $f(E, \Delta^2)$, which with our current normalization of the plane wave solutions is[24]

$$f(E, \Delta^2) = -\frac{m}{2\pi} T(E, \Delta^2). \tag{168}$$

The final form of the dispersion relation which has been shown to be valid for $\Delta < 2\mu$ [Eq. (165)] is

$$\begin{aligned} f(E, \Delta^2) = &-\frac{m}{2\pi}\int d^3x e^{i\Delta\cdot\mathbf{x}} V(x) \\ &-\frac{m}{2\pi}\sum_B \frac{\Gamma_B(\Delta^2)}{E - E_B} + \frac{1}{\pi}\int_0^\infty dE' \frac{\operatorname{Im} f(E', \Delta^2)}{E' - E - i\eta}. \end{aligned} \tag{169}$$

**The Unitarity Condition**

We have temporarily exhausted what can be said concerning the analytic structure of the scattering amplitude. [Shortly we shall discuss analyticity as a function of momentum transfer.] We have said nothing as yet about the unitarity condition or about how the analytic continuation from the physical region $2mE > \Delta^2/4$ is to be carried out. The unitarity condition takes the following form in terms of the scattering amplitude $f$, which we (for the moment) write as $f(\mathbf{k}', \mathbf{k})$ instead of $f(E, \Delta^2)$:

$$\operatorname{Im} f(\mathbf{k}', \mathbf{k}) = \frac{k}{4\pi}\int d\Omega'' f^*(\mathbf{k}'', \mathbf{k}') f(\mathbf{k}'', \mathbf{k}), \tag{170}$$

where $|\mathbf{k}| = |\mathbf{k}'| = |\mathbf{k}''| = k$.[25] Now we would like to express this relation

[24] This is seen to agree with our conventional relation (6.70*b*) $f = -(2\pi)^2/k\ \rho_\epsilon$"$T$" $= -(2\pi)^2 m$"$T$", if it is recalled that "$T$" used elsewhere in the book equals the present one $\times(2\pi)^{-3}$.

[25] This is just Eq. (5.192) written in terms of $f$. We have also used the fact that $f(\mathbf{k}'', \mathbf{k}') = f(\mathbf{k}', \mathbf{k}'')$. This is evident from the fact that $f$ is a function of $k'^2$, $k''^2$ and $\mathbf{k}' \cdot \mathbf{k}''$ only when there is no spin dependence in the scattering. Note that when $\mathbf{k}' = \mathbf{k}$, Eq. (170) reduces to the optical theorem.

in terms of the variables which we are now using, $E$ and $\Delta^2$. Since the energy is fixed in Eq. (170), we may use the relation $\cos\theta = 1 - \Delta^2/2k^2$ to express $f$ in terms of $E$ and $\mu = \cos\theta$. Writing $f(\mathbf{k}', \mathbf{k}) \equiv g(E, \mu)$, Eq. (170) may be rewritten as

$$\begin{aligned}\operatorname{Im} g(E,\mu) &= \frac{k}{4\pi}\int_{-1}^{+1} d\mu_1 \int_{-1}^{+1} d\mu_2 \int_0^{2\pi} d\phi'' \int_{-1}^{+1} d\mu'' g^*(E,\mu_2) g(E,\mu_1)\delta(\mu_1 - \mu'') \\ &\quad \times \delta(\mu_2 - \mu\mu'' - \sqrt{1-\mu^2}\sqrt{1-\mu''^2}\cos\phi'') \\ &= \frac{k}{2\pi}\int_{-1}^{+1} d\mu_1 \int_{-1}^{+1} d\mu_2 \\ &\quad \times \frac{g^*(E,\mu_2)g(E,\mu_1)\theta(1-\mu_1^2-\mu_2^2-\mu^2+2\mu\mu_1\mu_2)}{\sqrt{1-\mu_1^2-\mu_2^2-\mu^2+2\mu\mu_1\mu_2}}, \qquad (171)\end{aligned}$$

where the limits of integration are purely formal, the actual limits being defined by the step-function in the numerator.[26]

Since Eq. (171) was deduced from the physical unitarity principle, it holds, of course, only in the physical region where $|\mu| \leq 1$. Formally, we find that the square root in (171) becomes imaginary if $|\mu| > 1$. We may also express (171) in terms of momentum transfer and energy variables with the result that

$$\operatorname{Im} f(E,\Delta^2) = \frac{1}{4\pi k}\int_0^{4k^2} d\Delta_1^2 \int_0^{4k^2} d\Delta_2^2 \frac{f^*(E,\Delta_2^2)f(E,\Delta_1^2)\theta(K)}{\sqrt{K(\Delta,\Delta_1,\Delta_2)}}, \qquad (172)$$

where $K$ is the symmetric function of its arguments given by

$$\begin{aligned}K(\Delta,\Delta_1,\Delta_2) &= 2(\Delta^2\Delta_1^2 + \Delta^2\Delta_2^2 + \Delta_1^2\Delta_2^2) \\ &\quad - \Delta_1^4 - \Delta_2^4 - \Delta^4 - \Delta^2\Delta_1^2\Delta_2^2/k^2. \qquad (173)\end{aligned}$$

We are free to use the expression given by Eq. (172) in our dispersion integral (169) once the variable of integration exceeds $\Delta^2/8m$. We are left, however, with the very embarrassing question of handling the region $0 < E < \Delta^2/8m$. The most obvious thought is to try to use the partial wave expansion [Eq. (6.70*a*)]

$$f(E,\Delta^2) = \sum_{l=0}^{\infty}(2l+1)e^{i\delta_l(k)}\frac{\sin\delta_l(k)}{k}P_l\left(1 - \frac{\Delta^2}{2k^2}\right). \qquad (174)$$

Although this series converges (in any problems of physical interest) for $\mu = 1 - \Delta^2/2k^2$ in the physical region, we have no assurance that it will do so as $k^2 \to 0$, $\mu \to -\infty$.

Incidentally, some insight as to why the kernels in Eqs. (171) and (172) are so complicated is obtained by noting the fact that (174), as is trivially

[26] The $\theta$-function in Eq. (171) is defined by Eq. (13).

checked, satisfies unitarity (and not only the optical theorem) identically. Using this in (171) leads to the remarkable formula

$$\frac{1}{2\pi}\frac{\theta(1-\mu^2-\mu_1^{\,2}-\mu_2^{\,2}+2\mu\mu_1\mu_2)}{\sqrt{1-\mu^2-\mu_1^{\,2}-\mu_2^{\,2}+2\mu\mu_1\mu_2}}=\sum_{l=0}^{\infty}\frac{2l+1}{4}P_l(\mu)P_l(\mu_1)P_l(\mu_2),$$

$$\frac{1}{\pi}\int_{-1}^{+1}d\mu\,\frac{P_l(\mu)\theta(1-\mu^2-\mu_1^{\,2}-\mu_2^{\,2}+2\mu\mu_1\mu_2)}{\sqrt{1-\mu^2-\mu_1^{\,2}-\mu_2^{\,2}+2\mu\mu_1\mu_2}}=P_l(\mu_1)P_l(\mu_2). \tag{175}$$

Thus the kernel in (171) has the task of expressing the product $P_l(\mu_1)P_l(\mu_2)$ as an integral over $P_l(\mu)$ for all $l$ even though it does not itself depend on $l$!

The question of whether the series (174) can be used to continue analytically the amplitude into the unphysical region will be answered in the affirmative. The tool for doing this is a mathematical theorem which states that a function of $\mu = \cos\theta$ which is analytic inside an ellipse in the complex $\mu$ plane with foci at $\pm 1$ may be expressed as a convergent series of Legendre polynomials throughout this region.[27] We write

$$\operatorname{Im} T(E,\theta)=\int d^3x\int d^3y e^{-i\mathbf{k}'\cdot\mathbf{x}}V(x)\operatorname{Im}[G(\mathbf{x},\mathbf{y},E)]V(y)e^{i\mathbf{k}\cdot\mathbf{y}}. \tag{176}$$

[See Eq. (130).] In order to make explicit the dependence on $\theta$, we choose a special coordinate system first used in a similar connection by Lehmann [18]:

$$\begin{aligned}\mathbf{k}&=k(1,0,0),\\ \mathbf{k}'&=k(\cos\theta,\sin\theta,0),\\ \mathbf{x}&=x(\sin\beta_1\cos\alpha_1,\sin\beta_1\sin\alpha_1,\cos\beta_1),\\ \mathbf{y}&=y(\sin\beta_2\cos\alpha_2,\sin\beta_2\sin\alpha_2,\cos\beta_2).\end{aligned} \tag{177}$$

Here $\theta$ is evidently the scattering angle and $\alpha_i, \beta_i (i = 1, 2)$ are conventional polar angles; the polar axis is perpendicular to $\mathbf{k}, \mathbf{k}'$. Now $G(\mathbf{x}, \mathbf{y}, E)$ is a function of $x^2$, $y^2$, $\mathbf{x}\cdot\mathbf{y}$, and $E$ and thus depends on the angles only through $\mathbf{x}\cdot\mathbf{y} = xy[\cos\beta_1\cos\beta_2 + \sin\beta_1\sin\beta_2\cos(\alpha_1-\alpha_2)] \equiv xy\cos\theta_{12}$. Thus, for the integrations implied in Eq. (176), if we introduce $\chi = \alpha_1 - \alpha_2$ and $\alpha_1$ in place of $\alpha_1$ and $\alpha_2$, the entire $\alpha_1$ dependence will be explicit in the exponentials, as is the $\theta$ dependence. We have then

$$\begin{aligned}\operatorname{Im} T(E,\theta)=&\int_0^\infty x^2\,dx\int_0^\infty y^2\,dy\int_0^\pi\sin\beta_1\,d\beta_1\int_0^\pi\sin\beta_2\,d\beta_2\int_0^{2\pi}d\chi\int_0^{2\pi}d\alpha_1\\ &\times e^{-ikx\sin\beta_1\cos(\theta-\alpha_1)}V(x)[\operatorname{Im} G(x,y,\theta_{12},E)]V(y)\\ &\times e^{iky\sin\beta_2\cos(\chi-\alpha_1)}.\end{aligned} \tag{178}$$

[27] See, for example, Chapter 15 of Ref. [17].

The integration over $\alpha_1$ may be easily carried out by writing the integrand in the form $e^{i\rho\cos(\alpha_1+\psi)}$and using the relation

$$J_0(\rho) = \int_0^{2\pi} \frac{d\alpha_1}{2\pi}\, e^{i\rho\cos(\alpha_1+\psi)}, \tag{179}$$

where $J_0$ is the usual Bessel function. We find then

$$\begin{aligned} \operatorname{Im} T(E, \theta) = 2\pi \int_0^\infty x^2\,dx \int_0^\infty y^2\,dy \int_0^\pi \sin\beta_1\,d\beta_1 \int_0^\pi \sin\beta_2\,d\beta_2 \int_0^{2\pi} d\chi \\ \times J_0[k\sqrt{2xy}\sqrt{\sin\beta_1 \sin\beta_2}\sqrt{\zeta - \cos(\theta - \chi)}]V(x) \\ \times [\operatorname{Im} G(x, y, \theta_{12}, E)]V(y), \end{aligned} \tag{180}$$

where

$$\zeta = \frac{y^2 \sin^2\beta_2 + x^2 \sin^2\beta_1}{2xy \sin\beta_1 \sin\beta_2}. \tag{181}$$

Now the minimum value of $\zeta$ is unity, which occurs when $\beta_1 = \beta_2 = \pi/2$ and $x = y$. We would like to allow $\cos\theta$ to take on complex values, so that $\cos(\theta - \chi)$ can become greater than unity. If it does so, $J_0$ increases exponentially for large $x$ with a maximum growth rate when $\zeta$ takes on its minimum value, $\zeta = 1$. If the integral is to exist, the exponential fall-off of the integrand due to the potential, $e^{-2\mu x}$ (since $x = y$ when $\zeta = 1$) must be sufficient to control the growth from $J_0$. Thus we must require that

$$\sqrt{2k}\,|\operatorname{Re}\sqrt{\cos(\theta - \chi) - 1}| < 2\mu. \tag{182}$$

Now if we write $\cos\theta = \cosh(\xi + i\eta)$, which is quite general, we note that $\cos(\theta - \chi) = \cosh[\xi + i(\eta + \chi)] \equiv \cosh(\xi + i\psi)$, where (since the range of $\chi$ is $0 < \chi < 2\pi$) we may regard $\psi$ as making an excursion over the interval $0 < \psi < 2\pi$. Simple substitution of $\cosh(\xi + i\psi)$ and evaluation of (182) leads to

$$(\cos\psi + 1)\sinh^2\frac{\xi}{2} < \frac{2\mu^2}{k^2}, \tag{183}$$

o that the greatest burden on $\xi$ is imposed by the choice $\cos\psi = 1$ or

$$\cosh\xi < 1 + \frac{2\mu^2}{k^2}. \tag{184}$$

The transformation $Z = \cos\theta = \cosh(\xi + i\eta)$ with fixed $\xi$ describes an ellipse in the $\cos\theta$ plane, since (here $Z_r$ and $Z_i$ are, respectively, the real and imaginary parts of $Z$)

$$\begin{aligned} Z_r &= \cosh\xi \cos\eta, \\ Z_i &= \sinh\xi \sin\eta, \\ \frac{Z_r^2}{\cos^2\xi} &+ \frac{Z_i^2}{\sinh^2\xi} = 1, \end{aligned} \tag{185}$$

which is an ellipse with foci at $\pm 1$, semimajor axis $\cosh \xi$, semiminor axis, $\sinh \xi$.

Since we have shown that Im $T(E, \theta)$ is analytic [it may be explicitly computed by our representation (178)] inside an ellipse with semimajor axis $1 + 2\mu^2/k^2$, semiminor axis $(2\mu/k)(1 + \mu^2/k^2)^{1/2}$, we may use the theorem on Legendre polynomials to effect the desired analytic continuation for (real) values of $\cos \theta$ bounded by

$$|\cos \theta| < 1 + \frac{2\mu^2}{k^2}. \tag{186}$$

In terms of $\Delta^2 = 2k^2(1 - \cos \theta)$, this is equivalent to

$$\Delta^2 < 4\mu^2 + 4k^2. \tag{187}$$

Since we require $k$ values for the dispersion integral over the range $0 < k^2 < \infty$, the Legendre series is valid for $\Delta^2 < 4\mu^2$, which is precisely the range of $\Delta^2$ for which the dispersion relation has been established.[28]

As we shall see shortly, the dispersion relations are, in fact, valid for all values of momentum transfer for a wide class of potentials. They have already been shown by our present derivation to hold if the potential decreases faster than any exponential. However, for potentials, with only exponential tails of range $\mu^{-1}$, the Legendre polynomial continuation will, in fact, fail for $\Delta^2 > 4\mu^2$.

The question arises as to whether the nonforward dispersion relations together with the unitarity condition provide a substitute for the Schrödinger equation. Aside from the problem of computing the bound-state energies (to which we return later) the limitation on momentum transfer poses a serious obstacle, for if we try to project partial wave amplitudes by writing

$$f_l(E) = \int_0^{4k^2} \frac{d\,\Delta^2}{4k^2} P_l\left(1 - \frac{\Delta^2}{2k^2}\right) f(E, \Delta^2), \tag{188}$$

we may, of course, use the dispersion relation only if $\Delta^2 < 4\mu^2$, which evidently means $k^2 < \mu^2$. In the dispersion integrals, however, there appear the $f_l(E)$ for $E$; and hence we do not really have a set of equations. We shall not stop to write down these relations, but merely remark that when they might be useful (for potentials of finite range, say) they yield formulas for the $f_l(E)$ expressed as integrals over all $f_{l'}(E)$ with $l' \geq l$. Thus there is no dispersion relation involving a single angular momentum. This is the price we must pay for having formulas where the size of the interaction region does not appear explicitly.

[28] Note incidentally that the fact that we were considering the imaginary part of $T$ has played a role only insofar as the first Born term did not have to be considered. It evidently is analytic in $\Delta^2$ only for $|\Delta^2| < \mu^2$.

There are two final remarks worth making. The first is that our derivation has shown that in the limit as $E \to \infty, f(E, \Delta^2)$ approaches the Born approximation $f_B(\Delta^2)$, given simply by

$$f_B(\Delta^2) = -\frac{m}{2\pi}\int d^3x e^{i\mathbf{\Delta}\cdot\mathbf{x}} V(x), \tag{189}$$

and thus there are no "subtractions" allowed; the value at infinity is fixed. This is a well-known result and was proven rigorously in Section 6.5, p. 309. The second is that if the potential contains both an ordinary *and* a space exchange part, the derivation fails. The point here is that $T(\mathbf{x}, \mathbf{y}, E)$ defined by Eq. (130) contains a term $V_{\text{exchange}}\ G(\mathbf{x}, \mathbf{y}, E)\ V_{\text{ordinary}}$ which no longer behaves nicely for $E$ large and complex. It is possible to derive a modified dispersion relation using techniques which we are about to develop, but no direct derivation has as yet been given.

We turn now to a discussion of the so-called Mandelstam (or double) dispersion relations which give the analytic dependence not only on the energy but also on the momentum transfer.

## 10.5 TWO-DIMENSIONAL REPRESENTATION

As we have seen in Section 10.4 the single variable dispersion relations are unsatisfactory as a dynamical tool. It is clear that if they could be established for all momentum transfers and if, in addition, the analytic dependence on momentum transfer were known, they would be far more useful. In 1958, in connection with the problem of pion-nucleon scattering, Mandelstam [13] conjectured a representation that expressed the analytic dependence of the scattering amplitude on energy and momentum transfer.

This Mandelstam, or double dispersion, representation has been vigorously studied and applied to many processes of physical interest in the past few years. The problem of deriving the representation within the framework of quantum field theory, even in the perturbation theory approximation (to all orders, of course), has not yet been solved. We shall not be able to go into these problems in this book. In spite of the theoretical difficulties, the double dispersion representation is extremely attractive and incorporates almost all the accumulated field theoretic intuition that has been developed in the past thirty years. We are therefore strongly inclined to accept the representation (with possibly certain modifications which for some problems can be justified on physical grounds) regardless of the connection with formal field theory.

We shall limit ourselves to a treatment of the double dispersion representation within the framework of nonrelativistic potential theory. Although the relativistic theory differs in important technical details (and is very much more complicated), many of the fundamental aspects of the theory may be illustrated quite clearly with the nonrelativistic theory. Furthermore, for whatever significance it may have, this representation may be derived in a relatively straightforward fashion from standard nonrelativistic quantum theory.

**The Second Born Approximation**

It is perhaps worth noting that our inability to give a direct proof that the dispersion relation in energy holds for arbitrary momentum transfer stems (at least in part) from the fact that we made very crude estimates of $\tilde{T}(\mathbf{x}, \mathbf{y}, E)$ and left several integrations undone. The apparently worrisome exponentials become quite harmless under certain circumstances. For example, consider the integral $I$ defined by

$$I = \int d\Omega_{\hat{\mathbf{x}}} e^{i(\mathbf{a}+\mathbf{b})\cdot\mathbf{x}}, \tag{190}$$

where $\mathbf{a}\cdot\mathbf{b} = 0$. The value is easily seen to be

$$I = 4\pi \frac{\sin |x|\sqrt{a^2 + b^2}}{|x|\sqrt{a^2 + b^2}}. \tag{191}$$

Now suppose $\mathbf{b} = i\boldsymbol{\beta}$, but still $\boldsymbol{\beta}\cdot\mathbf{a} = 0$. Casual inspection of Eq. (190) would lead one to believe that $I \to \infty$ as $|\mathbf{x}| \to \infty$ in this case. However, we see from (191) that as long as $a^2 > \beta^2$, there is an oscillatory dependence on $|\mathbf{x}|$ and, in fact, $I \to 0$ as $|\mathbf{x}| \to \infty$. The fact that something like this was going on and that the dispersion relations might be valid for arbitrary momentum transfers came from studying the second Born approximation for the Yukawa potential. We shall not give the details of the calculation,[29] but merely quote the result. Taking the potential to be

$$V = \frac{V_0 e^{-\mu r}}{r}, \tag{192}$$

we find for the second Born approximation to the scattering amplitude, $f_2$,[30]

$$f_2 = m^2 V_0^2 \int_{-1}^{+1} dy \frac{1}{\lambda(y)[\mu^2 - 2ik\lambda(y)]}, \tag{193}$$

where

$$\lambda(y) = \sqrt{\mu^2 + (\Delta^2/4)(1 - y^2)}. \tag{194}$$

[29] These may be found in Ref. [1], pp. 55–57.

[30] This calculation is also given in Morse and Feshbach [19]. Their method does not illustrate the point of Eqs. (190) and (191), however.

By inspection we see that this function is analytic in the upper-half $k$-plane and goes to zero nicely at infinity. Furthermore, this analyticity obtains for all $\Delta^2 > -4\mu^2$. We may then write

$$f_2(k^2, \Delta^2) = \int_0^\infty \frac{dk'^2}{\pi} \frac{w(k'^2, \Delta^2)}{k'^2 - k^2 - i\eta}, \tag{195}$$

where

$$\begin{aligned} w(k^2, \Delta^2) &= \operatorname{Im} f_2(k^2, \Delta^2) \\ &= 2m^2V_0^2 k \int_{-1}^{+1} dy \frac{1}{\mu^4 + 4k^2[\mu^2 + (\Delta^2/4)(1 - y^2)]} \\ &= \frac{2m^2V_0^2}{k} \int_0^\infty \frac{dt'}{t' + \Delta^2} \int_{-1}^{+1} \frac{1}{1 - y^2} \delta\left[t' - \frac{\mu^4 + 4k^2\mu^2}{k^2(1 - y^2)}\right] dy \\ &= 2m^2V_0^2 \int_0^\infty \frac{dt'}{t' + \Delta^2} \frac{1}{\sqrt{t'}} \frac{\theta(k^2t' - \mu^4 - 4k^2\mu^2)}{\sqrt{k^2t' - \mu^4 - 4k^2\mu^2}}. \end{aligned} \tag{196}$$

We now substitute $w(k^2, \Delta^2)$ into Eq. (195). It is convenient to make the following changes in notation (to conform with what appears in much of the field theoretical literature): $k^2 = s$, $\Delta^2 = -t[= 2s(1 - \cos\theta)]$ and $v_0 = 2mV_0$. We find

$$f_2(s, t) = \int_0^\infty \frac{ds'}{\pi} \int_0^\infty \frac{dt'}{\pi} \frac{\rho_2(s', t')}{(s' - s - i\eta)(t' - t)}, \tag{197}$$

where

$$\rho_2(s, t) = \frac{v_0^2}{2} \frac{\pi}{\sqrt{t}} \frac{\theta(st - \mu^4 - 4s\mu^2)}{\sqrt{st - \mu^4 - 4s\mu^2}}. \tag{198}$$

The region of the $s$, $t$-plane where $\rho_2$ fails to vanish is set by the $\theta$-function, $t > 4\mu^2 + \mu^4/s$, and is shown in Fig. 10.2. We note that $t \geq 4\mu^2$, attaining the lower limit only for infinite energy.

The representation (197) shows that $f_2$ may be extended to a function of two complex variables, $s$, $t$, which is analytic in the product of two cut planes. The $s$-plane is cut from 0 to $+\infty$, whereas the cut in the $t$-plane extends from $4\mu^2$ to $\infty$. The branch point at $t = 4\mu^2$ limits the Legendre polynomial expansion, as discussed in the previous section. The physical scattering amplitude is obtained from the extended function by taking $t$ negative and real and allowing $s$ to approach the positive real axis from above, as usual.[31]

The $s'$ integration in (197) may be carried out explicitly to give a dispersion relation in $t$, for fixed real $s$, which is the analogue of our

[31] That is, we replace $s$ by $s + i\eta$, with $s, \eta > 0$ and take the limit $\eta \to 0(+)$.

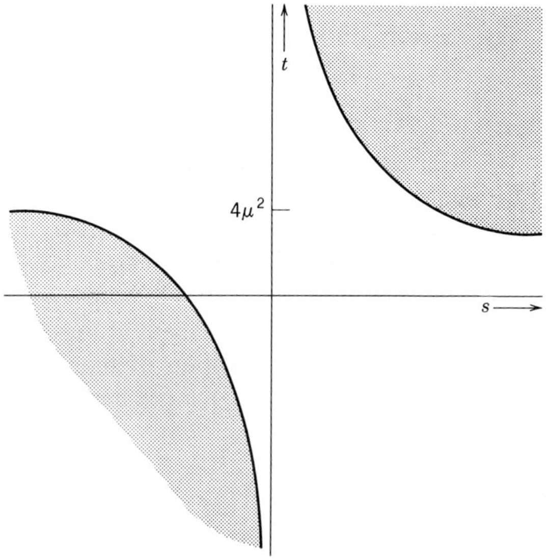


Fig. 10.2 Region of $s$, $t$ plane for which $\rho_2$ is nonvanishing.

earlier dispersion relation obtained for fixed real $t$. It takes the form

$$\begin{aligned} f_2(s, t) &= \pi v_0^2 \int_{4\mu^2}^{\infty} \frac{dt'}{\pi} \frac{1}{\sqrt{t'(t'-t)}} \frac{1}{\sqrt{\mu^4 - (t'-4\mu^2)(s+i\eta)}} \\ &= \pi v_0^2 \int_{4\mu^2}^{4\mu^2+\mu^4/s} \frac{dt'}{\pi} \frac{1}{\sqrt{t'(t'-t)}} \frac{1}{\sqrt{\mu^4 - (t'-4\mu^2)s}} \\ &\quad + i\pi v_0^2 \int_{4\mu^2+\mu^4/s}^{\infty} \frac{dt'}{\pi} \frac{1}{\sqrt{t'(t'-t)}} \frac{1}{\sqrt{(t'-4\mu^2)s - \mu^4}}. \end{aligned} \tag{199}$$

**The Double Dispersion Representation for Scattering by a Sum of Yukawa Potentials**

We turn now to the problem of proving that not only the second Born approximation but also the full amplitude can be written in the double dispersion representation analogous to Eq. (197). More precisely, we wish to establish for a pure Yukawa potential (and for the moment, no bound states) the representation

$$f(s, t) = -\frac{v_0}{\mu^2 - t} + \int_0^{\infty} \frac{ds'}{\pi} \int_0^{\infty} \frac{dt'}{\pi} \frac{\rho(s', t')}{(s' - s - i\eta)(t' - t)}, \tag{200}$$

where the first term is simply the first Born amplitude and $\rho(s, t)$ is a weight function which must be determined. Instead of considering a pure Yukawa potential, there is no difficulty in treating an arbitrary superposition of such

$$rV(r) = \int_{\mu_0}^{\infty} d\mu\sigma(\mu)e^{-\mu r}, \tag{201}$$

where we imagine a finite lower limit $\mu_0$. [Infinite range potentials (that is, $\mu_0 = 0$) are uninteresting.] It is convenient now to use units in which $2m = 1$. (This is effectively what we did before when we wrote $2mV_0 = v_0$ and, of course, now $s$ is just the energy.)

It will be useful to rederive by another method the analyticity of the scattering amplitude, regarded as a function of the scattering angle $\theta$. The standard integral equation for the $T$-matrix, namely,

$$T = V + V\frac{1}{E + i\eta - K}T, \tag{202}$$

[see (5.177)] becomes, when written in terms of the scattering amplitude,

$$f(\mathbf{k}', \mathbf{k}) = f_B(\mathbf{k}', \mathbf{k}) - \int \frac{d^3q}{(2\pi)^3}\, \tilde{V}(\mathbf{k}' - \mathbf{q})\, \frac{1}{q^2 - k^2 - i\eta} f(\mathbf{q}, \mathbf{k}). \tag{203}$$

Here

$$f_B(\mathbf{k}', \mathbf{k}) \equiv -\frac{1}{4\pi}\, \tilde{V}(\mathbf{k}' - \mathbf{k}) \equiv -\frac{1}{4\pi} \int d^3x e^{-i\mathbf{k}'\cdot\mathbf{x}} V(x) e^{i\mathbf{k}\cdot\mathbf{x}} \tag{204}$$

is just the first Born approximation to the scattering amplitude[32] and $\tilde{V}(\mathbf{k})$ is the Fourier transform of $V(r)$. The analogue of Eq. (5.67), which we repeat here,

$$T = V + V\frac{1}{E - K - V + i\eta}V, \tag{205}$$

is

$$f(\mathbf{k}', \mathbf{k}) = f_B(\mathbf{k}', \mathbf{k}) + \int d^3q_1 \int d^3q_2\, \frac{\tilde{V}(\mathbf{k}' - \mathbf{q}_1)}{4\pi}\, G(\mathbf{q}_1, \mathbf{q}_2, k^2)\, \frac{\tilde{V}(\mathbf{q}_2 - \mathbf{k})}{4\pi}, \tag{206}$$

where

$$G(\mathbf{q}_1, \mathbf{q}_2, k^2) = -4\pi \int \frac{d^3x_1\, d^3x_2}{(2\pi)^6}\, e^{-i\mathbf{q}_1\cdot\mathbf{x}_1} \left\langle \mathbf{x}_1 \middle| \frac{1}{k^2 + \nabla^2 + i\eta - V} \middle| \mathbf{x}_2 \right\rangle \times e^{i\mathbf{q}_2\cdot\mathbf{x}_2}. \tag{207}$$

[32] The numerical factor in Eq. (204) is seen to be consistent with Eq. (168) for units with $2m = 1$.

In terms of the more general potential (201), we have for $\bar{f} \equiv f - f_B$ the formal expression

$$\bar{f} = \int_{\mu_0}^{\infty} d\mu_1 \sigma(\mu_1) \int_{\mu_0}^{\infty} d\mu_2 \sigma(\mu_2) \int d^3q_1 \int d^3q_2 \times \frac{G(\mathbf{q}_1, \mathbf{q}_2, k^2)}{[\mu_1^2 + (\mathbf{k}' - \mathbf{q}_1)^2][\mu_2^2 + (\mathbf{q}_2 - \mathbf{k})^2]}. \tag{208}$$

To study the analyticity in $\cos\theta$ we again introduce Lehmann's coordinate system, as in connection with Eq. (176), writing

$$\begin{aligned} \mathbf{k} &= k(1, 0, 0), \\ \mathbf{k}' &= k(\cos\theta, \sin\theta, 0), \\ \mathbf{q}_i &= q_i(\sin\beta_i \cos\alpha_i, \sin\beta_i \sin\alpha_i, \cos\beta_i), \qquad i = 1, 2. \end{aligned} \tag{209}$$

Now replace the variables $\alpha_i$, $\beta_i$ by new ones, defined as

$$\begin{aligned} \alpha_1 &= \alpha_1, \\ \alpha_2 &= \chi - \alpha_1, \\ \lambda_i &= \frac{\mu_i^2 + k^2 + q_i^2}{2kq_i \sin\beta_i}. \end{aligned} \tag{210}$$

We note that for fixed $k$ and $\mu_i$, $\lambda_i$ takes on its minimum value when $q_i = (\mu_i^2 + k^2)^{1/2}$; since the smallest value of $\mu_i$ is $\mu_0$, the $\lambda_i$ have an absolute minimum $\lambda_0$, where

$$\lambda_0 = \left(1 + \frac{\mu_0^2}{k^2}\right)^{1/2} \geq 1. \tag{211}$$

In terms of these new variables we may replace (208) by

$$\bar{f}(k^2, \cos\theta) = \int_{\lambda_0}^{\infty} d\lambda_1 \int_{\lambda_0}^{\infty} d\lambda_2 \int_0^{2\pi} d\chi \int_0^{2\pi} d\alpha_1 \times \frac{w(\lambda_1, \lambda_2, k^2, \chi)}{[\lambda_1 - \cos(\theta - \alpha_1)][\lambda_2 - \cos(\chi - \alpha_1)]}, \tag{212}$$

where the weight function $w$ is related to integrals over $G$, $\sigma_i$, and so on, which need not be written explicitly. The fact that $G$ depends only on $\mathbf{q}_1 \cdot \mathbf{q}_2$ (as to angles) is crucial because we want the dependence on $\alpha_1$ to be explicit.

The integration over $\alpha_1$ may be carried out explicitly. (This is done by contour integration and is usually referred to as "elementary but tedious."[33])

[33] It is!

We find

$$\bar{f}(k^2, \cos\theta) = 2\pi \int_{\lambda_0}^{\infty} d\lambda_1 \int_{\lambda_0}^{\infty} d\lambda_2 \int_0^{2\pi} d\chi w(\lambda_1, \lambda_2, k^2, \chi)$$

$$\times \frac{(\lambda_1/\sqrt{\lambda_1^2 - 1}) + (\lambda_2/\sqrt{\lambda_2^2 - 1})}{\lambda_1\lambda_2 + \sqrt{\lambda_1^2 - 1}\sqrt{\lambda_2^2 - 1} - \cos(\theta - \chi)}. \tag{213}$$

Finally, let

$$y = \lambda_1\lambda_2 + \sqrt{\lambda_1^2 - 1}\sqrt{\lambda_2^2 - 1},$$

$$dy = \sqrt{\lambda_2^2 - 1}\left(\frac{\lambda_1}{\sqrt{\lambda_1^2 - 1}} + \frac{\lambda_2}{\sqrt{\lambda_2^2 - 1}}\right) d\lambda_1. \tag{214}$$

Imagining the integral over $\lambda_2$ to be performed, we have then

$$\bar{f}(k^2, \cos\theta) = \int_{y_0}^{\infty} dy \int_0^{2\pi} d\chi \frac{\bar{w}(y, k^2, \chi)}{y - \cos(\theta - \chi)}, \tag{215}$$

where $\bar{w}$ is the new weight function obtained after the $\lambda_2$ integration, and $y_0$ is the minimum value of $y_0 = 2\lambda_0^2 - 1 = 1 + 2\mu_0^2/k^2$.

The denominator in Eq. (215) is essentially the same quantity discussed in connection with Eq. (180). For variety we study it in a slightly different fashion. If the equation obtained by setting the denominator of (215) equal to zero is solved for $z = \cos\theta$, we obtain $z = y\cos\chi \pm i\sin\chi\sqrt{y^2 - 1}$. Thus as $\chi$ goes over its range, $z$ traces out an ellipse in the $z$-plane, with semimajor axis $y$, semiminor axis $(y^2 - 1)^{1/2}$. Provided $z$ stays inside such an ellipse with $y < y_0$, the denominator of (215) never vanishes. Thus the integral (215) defines $\bar{f}(k^2, \cos\theta)$ as a function analytic within an ellipse of semimajor axis $1 + 2\mu_0^2/k^2$, semiminor axis $(2\mu_0/k)(1 + \mu_0^2/k^2)^{1/2}$, just as before.

(It is important to note that if we knew $\bar{w}$ explicitly and could carry out the $\chi$ integration we might very well get a larger domain of analyticity. Our experience with the second Born approximation is an example of such a possibility.)

In order to see how to enlarge the domain of analyticity, we briefly reconsider the second Born approximation

$$f_2 = \int \frac{d^3q}{(2\pi)^3} \tilde{V}(\mathbf{k}' - \mathbf{q}) \frac{1}{q^2 - k^2 - i\eta} \frac{\tilde{V}(\mathbf{q} - \mathbf{k})}{4\pi}$$

$$= \frac{1}{4k^2} \int_0^{\infty} \frac{dq}{(2\pi)^3} \int_{\mu_0}^{\infty} d\mu_1 \sigma(\mu_1) \int_{\mu_0}^{\infty} d\mu_2 \sigma(\mu_2) \frac{I(\lambda_1, \lambda_2, \cos\theta)}{q^2 - k^2 - i\eta}, \tag{216}$$

where

$$\lambda_1 = \frac{\mu_1{}^2 + k^2 + q^2}{2qk}, \qquad \lambda_2 = \frac{\mu_2{}^2 + k^2 + q^2}{2qk}, \qquad \lambda_i \geq 1, \quad \text{all } k, q, \tag{217}$$

and

$$I(\lambda_1, \lambda_2, \cos\theta) = \int d\Omega_{\hat{\mathbf{q}}} \frac{1}{(\lambda_1 - \hat{\mathbf{k}}' \cdot \hat{\mathbf{q}})(\lambda_2 - \hat{\mathbf{k}} \cdot \hat{\mathbf{q}})}. \tag{218}$$

We shall show that $I(\lambda_1, \lambda_2, \cos\theta)$ defines a function of $\cos\theta$ analytic in the complex $\cos\theta$ plane, cut from $\lambda_1\lambda_2 + \sqrt{(\lambda_1{}^2 - 1)(\lambda_2{}^2 - 1)}$ along the real axis. To show this, we need merely evaluate the integral. It is simpler, however, not to do so but to get an integral representation for $I$. If we use the same coordinate system introduced in connection with Eq. (208) and carry out the azimuthal integration, we obtain

$$I(\lambda_1, \lambda_2, \cos\theta) = 4\pi \int_0^{\pi/2} \frac{d\beta}{\sin\beta} \frac{(\lambda'_1/\sqrt{\lambda'_1{}^2 - 1}) + (\lambda'_2/\sqrt{\lambda'_2{}^2 - 1})}{\lambda'_1\lambda'_2 + \sqrt{(\lambda'_1{}^2 - 1)(\lambda'_2{}^2 - 1)} - \cos\theta}, \tag{219}$$

where $\lambda'_i = \lambda_i/\sin\beta$. Making a simple transformation we have

$$I(\lambda_1, \lambda_2, \cos\theta) = 4\pi \int_{\lambda_{12}}^{\infty} d\zeta \frac{1}{\zeta - \cos\theta} \frac{1}{\kappa(\lambda_1, \lambda_2, \zeta)}, \tag{220}$$

where

$$\kappa(\lambda_1, \lambda_2, \zeta) = \lambda_1{}^2 + \lambda_2{}^2 + \zeta^2 - 2\lambda_1\lambda_2\zeta - 1, \qquad \lambda_{12} = \lambda_1\lambda_2 + \sqrt{(\lambda_1{}^2 - 1)(\lambda_2{}^2 - 1)}. \tag{221}$$

We see from (220) that $I$ has the advertised property. For completeness we quote the explicit form of $I$ which may be found from standard tables

$$I(\lambda_1, \lambda_2, \cos\theta) = \frac{2\pi}{\kappa(\lambda_1, \lambda_2, \cos\theta)} \ln \frac{\lambda_1\lambda_2 - \cos\theta + \kappa(\lambda_1, \lambda_2, \cos\theta)}{\lambda_1\lambda_2 - \cos\theta - \kappa(\lambda_1, \lambda_2, \cos\theta)}. \tag{222}$$

With this machinery in hand we look now at the third Born approximation

$$\begin{aligned} f_3 &= -\int \frac{d^3q_1}{(2\pi)^3} \int \frac{d^3q_2}{(2\pi)^3} \tilde{V}(\mathbf{k}' - \mathbf{q}_1) \frac{1}{q_1{}^2 - k^2 - i\eta} \tilde{V}(\mathbf{q}_1 - \mathbf{q}_2) \\ &\quad \times \frac{1}{q_2{}^2 - k^2 - i\eta} \frac{\tilde{V}(\mathbf{q}_2 - \mathbf{k})}{4\pi} \\ &= -\frac{1}{8k^2} \int_0^{\infty} \frac{dq_1}{(2\pi)^3} \int_0^{\infty} \frac{dq_2}{(2\pi)^3} \int_{\mu_0}^{\infty} d\mu_1\sigma(\mu_1) \int_{\mu_0}^{\infty} d\mu_2\sigma(\mu_2) \int_{\mu_0}^{\infty} d\mu_3\sigma(\mu_3) \\ &\quad \times \frac{J(\lambda_1, \lambda_2, \lambda_3, \cos\theta)}{(q_1{}^2 - k^2 - i\eta)(q_2{}^2 - k^2 - i\eta)}, \end{aligned} \tag{223}$$

where

$$\lambda_1 = \frac{\mu_1^{\ 2} + k^2 + q_1^{\ 2}}{2kq_1},$$

$$\lambda_2 = \frac{\mu_2^{\ 2} + q_1^{\ 2} + q_2^{\ 2}}{2q_1q_1},$$

$$\lambda_3 = \frac{\mu_3^{\ 2} + k^2 + q_2^{\ 2}}{2kq_2}, \tag{224}$$

and

$$J(\lambda_1, \lambda_2, \lambda_3, \cos\theta) \equiv \int d\Omega_{\hat{\mathbf{q}}_1} \int d\Omega_{\hat{\mathbf{q}}_2} \frac{1}{\lambda_1 - \hat{\mathbf{k}}' \cdot \hat{\mathbf{q}}_1} \frac{1}{\lambda_2 - \hat{\mathbf{q}}_1 \cdot \hat{\mathbf{q}}_2} \frac{1}{\lambda_3 - \hat{\mathbf{q}}_2 \cdot \hat{\mathbf{k}}}. \tag{225}$$

If we carry out the integration over $d\Omega_{q_2}$ using (220) we are led to

$$J(\lambda_1, \lambda_2, \lambda_3, \cos\theta) = 4\pi \int d\Omega_{\hat{\mathbf{q}}_1} \int_{\lambda_{23}}^{\infty} d\zeta \frac{1}{\lambda_1 - \hat{\mathbf{k}} \cdot \hat{\mathbf{q}}_1} \frac{1}{\zeta - \hat{\mathbf{q}}_1 \cdot \hat{\mathbf{k}}} \frac{1}{\kappa(\lambda_2, \lambda_3, \zeta)}, \tag{226}$$

with $\lambda_{23} = \lambda_2\lambda_3 + \sqrt{(\lambda_2^2 - 1)(\lambda_3^2 - 1)}$. Next we integrate over $d\Omega_{\hat{\mathbf{q}}_1}$, again using (220), to obtain

$$\begin{aligned} J(\lambda_1, \lambda_2, \lambda_3, \cos\theta) &= (4\pi)^2 \int_{\lambda_{23}}^{\infty} d\zeta \frac{1}{\kappa(\lambda_2, \lambda_3, \zeta)} \int_{\lambda_1\zeta}^{\infty} d\zeta' \frac{1}{\zeta' - \hat{\mathbf{k}}' \cdot \hat{\mathbf{k}}} \\ &\quad \times \frac{1}{\kappa(\lambda_1, \zeta, \zeta')} \\ &= (4\pi)^2 \int_{\lambda}^{\infty} d\zeta' \frac{1}{\zeta' - \cos\theta} \int_{\lambda_{23}}^{\lambda_1\zeta' - \sqrt{(\lambda_1^2 - 1)(\zeta'^2 - 1)}} \\ &\quad \times \frac{d\zeta}{\kappa(\lambda_2, \lambda_3, \zeta)\kappa(\lambda_1, \zeta, \zeta')}, \end{aligned} \tag{227}$$

where the second line has been found by interchanging the order of integration and setting $\lambda = \cosh(\xi_1 + \xi_2 + \xi_3)$, where $\cosh \xi_i = \lambda_i$. Finally, we must determine the minimum value of $\lambda$ as the $q_i$, $\mu_i$ go over their ranges. It is reasonable to expect that this comes near the symmetry point at $\lambda_1 = \lambda_2 = \lambda_3 = (1 + \mu^2/k^2)^{1/2}$, which yields the result $\lambda = (1 + \mu_0^2/k^2)^{1/2}(1 + 4\mu_0^2/k^2)$ for the absolute minimum. The important observation for our purposes is that $\lambda \geq 1 + 9\mu_0^2/2k^2$ and that it has moved out significantly from the second Born cut, which started at $1 + 2\mu_0^2/k^2$.

Evidently, with each iteration the cut will move out further and the $n$th approximation yields a cut starting at $1 + n^2\mu_0^2/2k^2$ and extending to $+\infty$.

If the Born series converged under all circumstances we could conclude, that $\tilde{f}$ was analytic in $\cos\theta$ except for a cut along the real axis extending from $1 + 2\mu_0^2/k^2$ to $+\infty$. We can, fortunately, come to the same conclusion without any such drastic assumption. For suppose we imagine iterating Eq. (203) $n$-times and then writing the exact remainder, analogous to Eq. (207). Calling the $n$th Born term $B_n$, we would find

$$f(\mathbf{k}', \mathbf{k}) = B_n(\mathbf{k}', \mathbf{k}) + \int d^3q_1 \int d^3q_2 B_n(\mathbf{k}', \mathbf{q}_1) G(\mathbf{q}_1, \mathbf{q}_2, \mathbf{k}^2) \frac{\tilde{V}(\mathbf{q}_2 - \mathbf{k})}{4\pi}. \tag{228}$$

Now we know by the preceding argument that $B_n$ is an analytic function of $\cos\theta$ except for the cut along the real axis from $1 + \mu_0^2/2k^2$ (including here the first Born term) to $+\infty$. By using the same method employed in the discussion of Eq. (208) we can prove that the integral in (228) is analytic in a gigantic ellipse with semimajor axis equal to $1 + (n+1)^2\mu_0^2/2k^2$. As $n \to \infty$ we conclude that $f(\mathbf{k}', \mathbf{k})$ is analytic everywhere in the cut $\cos\theta$-plane, including the remainder term which is an entire function of $\cos\theta$. Unfortunately, this argument gives no explicit information about the precise behavior of $f$ as $\cos\theta \to \infty$.

The original derivation of the analyticity of $f$ as a function of $\cos\theta$ was based on an analysis of the Friedholm series solution of the integral equation (203) and is mathematically somewhat more precise than the argument given.

Armed with this information concerning the analyticity in $\cos\theta$, we may proceed to show that the single variable dispersion relations previously derived for $-t > 4\mu^2$ hold for all values of $-t$. The cut in $\tilde{f} = f - f_B$ in $t$ (corresponding to that in $\cos\theta$ from $1 + 2\mu_0^2/k^2$ to $\infty$) becomes, since $t = -2k^2(1 - \cos\theta)$, one from $t = 4\mu_0^2$ to $+\infty$. The dispersion relation in energy becomes, in this notation,

$$f(s, t) = f_B(t) - \frac{1}{4\pi}\sum_B \frac{\Gamma_B(t)}{s - s_B} + \int_0^\infty \frac{ds'}{\pi}\frac{\operatorname{Im} f(s', t)}{s' - s - i\eta}. \tag{229}$$

Since we have shown that $\tilde{f}$ is analytic in the cut $t$-plane and $f_B$ is real, it is clear that $\operatorname{Im} f$ is also analytic. In addition, the $\Gamma_B(t)$ are polynomials in $t$, the highest power corresponding to the angular momentum state bound with the largest $l$. Thus the last two terms in Eq. (229) may be continued analytically to all values of $t$ not on the cut. That the function obtained is equal to the true $\tilde{f} = f - f_B$ [that is, still continues to satisfy (229)] is clear from the following argument: The two sides coincide for $0 < (-t) < 4\mu^2$ and are both analytic functions of $t$; from the fundamental theorem of analytic continuation it follows that they are equal everywhere. We conclude therefore that the dispersion relation in energy holds for all values of momentum transfer except those on the cut $4\mu^2 < t < \infty$, and certainly for all physical values, corresponding to $-t > 0$. Since we already knew this to be so for finite range potentials or ones that decrease faster than any exponential, we may safely conclude (since the Yukawa potential is the most dangerous potential with an exponential fall-off) that this result is generally true.

The final step in establishing a definite double dispersion representation for $f(s, t)$ is simply to write a representation for $\text{Im} f(s', t)$ in (229) that expresses the demonstrated analyticity in $t$. There is unfortunately a fundamental difficulty here. We do not know the behavior of $\text{Im} f(s', t)$ as $t \to \infty$. If we assumed that $\text{Im} f(s', t) \to 0$ as $t \to \infty$, we would be led to write

$$\text{Im} f(s', t) = \int_0^\infty \frac{dt'}{\pi} \frac{\rho(s', t')}{t' - t}. \tag{230}$$

We shall see shortly that this is definitely wrong if there are bound states, and it may very well be incorrect even if there are none. If $f(s', t)$ does not approach zero for large $t$, we may make a subtraction and write

$$\text{Im} f(s, t) = t \int_0^\infty \frac{dt'}{\pi} \frac{\rho(s', t')}{t'(t' - t)} + g(s), \tag{231}$$

where the subtraction was made at $t = 0$ for convenience only. [We have written the limits in (230) and (231) as zero; actually, as we have seen for the second Born approximation, Eq. (198), the $t'$ limit in general depends on $s'$, but is *never* less than $4\mu_0^2$.] Regge [21] has shown by an elegant but rather involved argument that $f(s, t) \to t^{\alpha(s)}$, where $\alpha(s)$ is complex and depends on the strength of the potential. It has also been shown by Blankenbecler and Goldberger [22] that $\alpha(s) \to -1$ as $s \to \infty$, but there is no reason to believe in general that $\text{Re}\, \alpha(s) < 1$ for all $s$.[34] For ease of writing we use Eq. (230) and by substitution into Eq. (229) obtain the final form for the double dispersion representation, valid for a superposition of Yukawa potentials

$$f(s, t) = f_B(t) - \frac{1}{4\pi} \sum_B \frac{\Gamma_B(t)}{s - s_B} + \int_0^\infty \frac{ds'}{\pi} \int_0^\infty \frac{dt'}{\pi} \frac{\rho(s', t')}{(s' - s - i\eta)(t' - t)}. \tag{232}$$

We would like to have a way of computing the weight function $\rho(s, t)$

[34] To pursue this matter further would lead us too far afield. For more detail, the reader is referred to Refs. [21] and [22].

in order to have a closed dynamical scheme. The only general principle which has not yet been employed is unitarity. We shall find that this indeed does the trick.

**The Role of the Unitarity Condition**

We consider first the case of no bound states and no subtractions. The unitarity condition is

$$\operatorname{Im} f(s, t) = \frac{f(s + i\eta, t) - f(s - i\eta, t)}{2i}$$

$$= \frac{\sqrt{s}}{4\pi} \int d\Omega_{\hat{\mathbf{k}}''} f^*[s, -(\mathbf{k}' - \mathbf{k}'')^2] f[s, -(\mathbf{k}'' - \mathbf{k})^2], \tag{233}$$

where as before $t = -(\mathbf{k}' - \mathbf{k})^2 = -2s(1 - \cos\theta)$, $k^2 = k'^2 = s$. Of course, as it stands this relation holds only in the physical region $0 < -t \le 4s$; both sides will be analytically continued with the aid of our representation (232). When we substitute (232) into (233), we encounter in every term [recall that $f_B(t) = -\int_{\mu_0}^{\infty} d\mu\sigma(\mu)(\mu - t)^{-1}$] an integral of the form

$$I = 4s^2 \int d\Omega_{\hat{\mathbf{k}}''} \frac{1}{[t_1 + (\mathbf{k}' - \mathbf{k}'')^2][t_2 + (\mathbf{k}'' - \mathbf{k})^2]} \tag{234}$$

(the factor of $4s^2$ is put in for convenience). Introducing $\lambda_i = 1 + t_i/2s$, we find

$$I(\lambda_1, \lambda_2. \hat{\mathbf{k}}' \cdot \hat{\mathbf{k}}) = \int d\Omega_{\hat{\mathbf{k}}''} \frac{1}{(\lambda_1 - \hat{\mathbf{k}}' \cdot \hat{\mathbf{k}}'')(\lambda_2 - \hat{\mathbf{k}}'' \cdot \hat{\mathbf{k}})}, \tag{235}$$

which is just the quantity defined in (218). It is most convenient for us to write it in terms of $t$ (using $\hat{\mathbf{k}}' \cdot \hat{\mathbf{k}} = \cos\theta = 1 + t/2s$) as

$$I = 4\pi \int_{t_0}^{\infty} dt' \frac{1}{t' - t} \frac{1}{\kappa(\lambda_1, \lambda_2, 1 + t'/2s)}, \tag{236}$$

with $t_0 = 2s(\lambda_1\lambda_2 - 1 + \sqrt{(\lambda_1{}^2 - 1)(\lambda_2{}^2 - 1)})$. With the aid of (236) all the terms on the right-hand side of (233) can be extended to all values of $t$. Similarly,

$$\operatorname{Im} f(s, t) = \int \frac{dt'}{\pi} \frac{\rho(s', t')}{t' - t} \tag{237}$$

may be used for the left-hand side of (233). The weight function $\rho(s, t)$ is then obtained by taking the discontinuity across the $t$ cut. Specifically, we use $\operatorname{Im} f(s, t + i\eta) - \operatorname{Im} f(s, t - i\eta) = 2i\rho(s, t)$. We find, finally,

$$\begin{aligned}\rho(s, t) = &\int_{\mu_0}^{\infty} d\mu_1 \sigma(\mu_1) \int_{\mu_0}^{\infty} d\mu_2 \sigma(\mu_2)\, K(s, t; \mu_1^2, \mu_2^2) \\ &- 2 \int_0^{\infty} \frac{ds_1}{\pi} \int_0^{\infty} \frac{dt_1}{\pi} \frac{\rho(s_1, t_1)}{s_1 - s} \int_{\mu_0}^{\infty} d\mu_2 \sigma(\mu_2)\, K(s, t; t_1, \mu_2^2) \\ &+ \int_0^{\infty} \frac{ds_1}{\pi} \int_0^{\infty} \frac{dt_1}{\pi} \int_0^{\infty} \frac{ds_2}{\pi} \int_0^{\infty} \frac{dt_2}{\pi} \\ &\times \frac{\rho(s_1, t_1)\rho(s_2, t_2)}{(s_1 - s - i\eta)(s_2 - s + i\eta)} K(s, t; t_1, t_2),\end{aligned} \tag{238}$$

where

$$K(s, t; t_1, t_2) = \frac{\pi}{2} \frac{\theta[t^{1/2} - t_1^{1/2}(1 + t_2/4s)^{1/2} - t_2^{1/2}(1 + t_1/4s)^{1/2}]}{\{s[t - (t_1^{1/2} + t_2^{1/2})^2][t - (t_1^{1/2} - t_2^{1/2})^2] - t\, t_1 t_2\}^{1/2}}. \tag{239}$$

[The denominator of $K$ is actually symmetric in the three $t$'s; it may be written as $\{s[t^2 + t_1^2 + t_2^2 - 2(t\, t_1 + t\, t_2 + t_1 t_2)] - t\, t_1 t_2\}^{1/2}$.] In the second line of (238) the $s_1$ integration is to be taken in the principal value sense.

The $K$ function has the property that it vanishes unless $t$, $s$ are sufficiently large for given $t_1$, $t_2$. Since the smallest values of $t_1$, $t_2$ that are possible occur in the first line of (238), the second Born term, we see that the exact $\rho(s, t)$ is bounded by

$$t^{1/2} > 2\mu_0 \left(1 + \frac{\mu_0^2}{4s}\right)^{1/2},$$

or

$$t > 4\mu_0^2 + \frac{\mu_0^4}{s}, \tag{240}$$

which agrees precisely with our direct second Born calculation, Eq. (199). The second and third terms in (238) evidently contribute only a region well inside the limiting curve. The second term sets in when $K(s, t; 4\mu_0^2, \mu_0^2) \neq 0$, and this describes the curve asymptotic to $t = 4\mu_0^4/s$ (small $s$) and $9\mu_0^2$ (large $s$). The third term in (238) does not contribute until $K(s, t; 4\mu_0^2, 4\mu_0^2) \neq 0$, which defines a curve asymptotic to $16\mu_0^4/s$ (small $s$) and $16\mu_0^2$ (large $s$). Clearly then, there is a finite region where $\rho \equiv \rho_2$, the latter computed from the first term in (238). [The exact region is that between the curves $t^{1/2} = 2\mu_0(1 + \mu_0^2/s)^{1/2}$ and $t^{1/2} = \mu_0(1 + 2\mu_0^2/s)^{1/2} + 2\mu_0(1 + \mu_0^2/s)^{1/2}$.]

Having once gotten started by evaluating $\rho_2$ we can evidently construct $\rho$ in a still larger region by substituting $\rho_2$ into the second term of (238) and integrating over the region where $\rho \equiv \rho_2$. This generates a finite region where the third term does not contribute. In terms of a strength parameter $V_0$ for the potential we have the following situation. At the first step we determine $\rho$ exactly in a region of the $s$-$t$-plane and have a term $\sim V_0^2$; at the second step, we have terms proportional to $V_0^2$ and $V_0^3$; at the third step, terms proportional to $V_0^2$, $V_0^3$, and $V_0^4$, and so on. We are gradually carving out the entire $s$-$t$-plane, expressing $\rho$ as a sequence of polynomials in $V_0$. If these $\rho$'s are substituted back into (232) and the integrations carried out over the regions where $\rho$ is known exactly, we will obtain $f(s, t)$ as the limit of a sequence of polynomials in $V_0$. If our "no-subtraction" philosophy is correct, the sequence will converge.

It is not hard to see that even if subtractions were required, the iterative construction of $\rho$ as described continues to hold. For any finite $(s, t)$ the integrations involved in obtaining $\rho$ extend only over a finite region because of the theta-function in $K(s, t; t_1, t_2)$. It is in the ultimate integrations over $\rho$ to form $f$ that we may encounter divergences. Of course, if we make a subtraction, as in Eq. (231), there is a new function introduced and this must be determined.

**The Case That Subtractions Are Required**

We turn now to the question of what modifications must be made in our procedure if subtractions are necessary. It is obvious that if there is an $s$-wave bound state, for example, that we cannot have an unsubtracted dispersion relation for $\operatorname{Im} f(s', t)$. This follows immediately from the unitarity condition. In the simplest situation we might write

$$\operatorname{Im} f(s', t) = \int_0^\infty \frac{dt'}{\pi} \frac{\rho(s', t')}{t' - t} + g_1(s'), \tag{241}$$

which could be termed a "subtraction at infinity," as contrasted with Eq. (231) which might be termed an "honest" subtraction. The only way in which the function $g_1$ in (241), or $g$ in (231), can be obtained is to use explicitly the unitarity condition for angular momentum zero. This will lead us to an integral equation for its determination, as we now show.

For definiteness we use Eq. (231) and, assuming an $s$-wave bound state only, we take for the representation the form

$$f(s, t) = -\frac{\tilde{V}(t)}{4\pi} - \frac{1}{4\pi} \frac{\Gamma_B}{s - s_B} + t \int_0^\infty \frac{ds'}{\pi} \int_0^\infty \frac{dt'}{\pi} \frac{\rho(s', t')}{(s' - s - i\eta)\, t'(t' - t)}$$
$$+ \int_0^\infty \frac{ds'}{\pi} \frac{g(s')}{s' - s - i\eta}. \tag{242}$$

The partial wave expansion is the standard one [Eq. (174)], to project out a given $f_l = e^{i\delta_l} \sin \delta_l / \sqrt{s}$, we write $t = -2s(1 - z)$, $z = \cos \theta$, multiply by $1/2P_l(z)$ and integrate over $z$ from $-1$ to $+1$. In these terms we encounter the following integral:

$$J_l = \frac{1}{2}\int_{-1}^{+1} dz \frac{P_l(z)}{t' + 2s(1 - z)} = \frac{1}{2s} Q_l\left(1 + \frac{t'}{2s}\right). \tag{243}$$

The $Q_l$'s are Legendre functions of the second kind; the only properties of these well-known functions that are needed for our purposes can be read from the integral representation given here. With this notation we obtain

$$f_0(s) = -\int_{\mu_0}^{\infty} d\mu\sigma(\mu)\frac{1}{2s}Q_0\left(1 + \frac{\mu^2}{2s}\right) - \frac{\Gamma_B/4\pi}{s - s_B} + \int_0^{\infty}\frac{ds'}{\pi}\frac{g(s')}{s' - s - i\eta}$$
$$+ \int_0^{\infty}\frac{ds'}{\pi}\int_0^{\infty}\frac{dt'}{\pi}\frac{\rho(s', t')}{s' - s - i\eta}\frac{1}{2s}\left[Q_0\left(1 + \frac{t'}{2s}\right) - \frac{2s}{t'}\right], \tag{244}$$
$$f_l(s) = -\int_{\mu_0}^{\infty} d\mu\sigma(\mu)\frac{1}{2s}Q_l\left(1 + \frac{\mu^2}{2s}\right) + \int_0^{\infty}\frac{ds'}{\pi}\int_0^{\infty}\frac{dt'}{\pi}\frac{\rho(s', t')}{s' - s - i\eta}$$
$$\times \frac{1}{2s}Q_l\left(1 + \frac{t'}{2s}\right), \qquad l \geq 1.$$

We see that in all equations except that for $l = 0$, the $f_l$ are obtained by quadratures involving the presumably known function $\rho(s, t)$. For $l = 0$, however, we have the unknown $g(s)$ appearing. But aside from a known contribution involving $\rho$, $g(s)$ is essentially equal to $\operatorname{Im} f_0(s)$. In fact, for $s > 0$

$$\operatorname{Im} f_0(s) = g(s) + \int \frac{dt'}{\pi}\rho(s, t')\frac{1}{2s}\left[Q_0\left(1 + \frac{t'}{2s}\right) - \frac{2s}{t'}\right]; \tag{245}$$

thus, given $f_0$ we can obtain $g$ in principle. The remaining relation that must be used is the unitarity condition for $f_0$, namely, $\operatorname{Im} f_0 = \sqrt{s}\,|f_0|^2$.

In order to solve for $f_0$ it is convenient to rewrite the representation (244) for $f_0$ as an integral equation. This may be done in several ways. One is to read from (244) the analytic properties of the $f_l$ (we discuss the general $l$ case for completeness) and then reexpress this information as a dispersion relation. In doing so we must make some statement concerning the behavior of the $f_l$ as $s \to \infty$. This is a somewhat subtle point which we shall not discuss in detail. We know that along the real $s$-axis, $f_l(s) \to 0$ at least as fast as $1/\sqrt{s}$, as $s \to \infty$. Suppose that there were some other behavior in another direction in the $s$-plane. If we were to write $\bar{s} \equiv 1/s$, this would mean the infinite value at $\bar{s} = 0$ would depend on the direction of approach to the $\bar{s}$-origin. This would imply an essential singularity at $\infty$, which we know not to be the case. We assume therefore that $f_l \to 0$ as $s \to \infty$ in any direction.

It is evident from (244) that the $f_l$ have a positive cut extending from 0 to $+\infty$ along the real axis. There is also a cut along the negative $s$-axis coming from the $Q_l$'s. From the definition (243) we find

$$\frac{1}{2i}\left\{Q_l\left(1+\frac{t'}{2(s+i\eta)}\right)-Q_l\left(1+\frac{t'}{2(s-i\eta)}\right)\right\}=\frac{\pi}{2}P_l\left(1+\frac{t'}{2s}\right)\theta\left(-s-\frac{t'}{4}\right). \tag{246}$$

Thus, from the first Born term in (244) there is a cut from $s = -\mu_0^2/4$ to $-\infty$, and from the double integral a cut from $-\mu_0^2$ to $-\infty$. Armed with this information we write a Cauchy integral for $(f_l - f_{lB})/s^l$ (where $f_{lB}$ is the first Born term) over a contour shown in Fig. 10.3; the factor of $s^l$ will be discussed shortly. We obtain

$$f_l(s) = f_{lB}(s) - \frac{1}{4\pi}\frac{\Gamma_B\delta_{l0}}{s-s_B} + s^l\int_0^\infty \frac{ds'}{\pi}\frac{\operatorname{Im} f_l(s')}{s'^l(s'-s-i\eta)} + s^l\int_{-\infty}^{-\mu_0^2}\frac{ds'}{\pi}\frac{\operatorname{Im}\,[f_l(s')-f_{lB}(s')]}{s'^l(s'-s)}. \tag{247}$$

The discontinuity across the negative axis may be expressed in terms of $\rho(s, t)$ using (244) and (246) we find

$$\operatorname{Im}\,[f_l(s) - f_{lB}(s)] = \int_0^\infty \frac{ds'}{\pi}\int_0^\infty \frac{dt'}{\pi}\frac{\pi}{4s}\frac{\rho(s',t')}{s'-s}P_l\left(1+\frac{t'}{2s}\right)\theta\left(-s-\frac{t'}{4}\right). \tag{248}$$

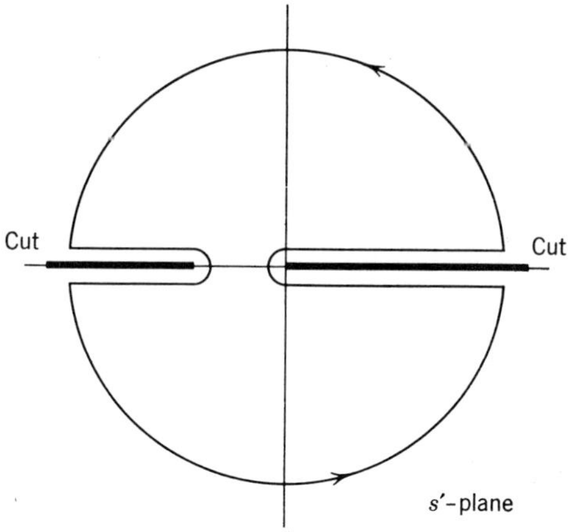


Fig. 10.3 Contour of integration leading to Eq. (247).

The factor of $s^l$ inserted introduces no new singularities, since $f_l \sim s^l$ near $s = 0$ and thus has the effect of guaranteeing the proper behavior at $s = 0$ even if we made some approximation in evaluating the left-hand cut. The representation (247) becomes an integral equation when we add the unitarity condition $\operatorname{Im} f_l = \sqrt{s}\,|f_l(s)|^2$. We shall return shortly to describe a method for solving such equations.

Before doing this, we indicate a more direct method of obtaining Eq. (247). In order to show that there is no trouble with the subtraction term that appears in the equation for $f_0$ [the first of Eqs. (244)], we discuss this equation. We rewrite it as (regarding $s$ as a general *complex* variable)

$$f_0(s) = f_{0B}(s) - \frac{\Gamma_B/4\pi}{s' - s} + \int_0^\infty \frac{ds'}{\pi} \int_0^\infty \frac{dt'}{\pi} \frac{\rho(s', t')}{s' - s} \frac{1}{2s'}$$
$$\times \left[Q_0\left(1 + \frac{t'}{2s'}\right) - \frac{2s'}{t'}\right] + \int_0^\infty \frac{ds'}{\pi} \frac{g(s')}{s' - s} + \int_0^\infty \frac{ds'}{\pi} \int_0^\infty \frac{dt'}{\pi}$$
$$\times \frac{\rho(s', t')}{s' - s} \left[\frac{1}{2s} Q_0\left(1 + \frac{t'}{2s}\right) - \frac{1}{2s'} \times Q_0\left(1 + \frac{t'}{2s'}\right)\right], \tag{249}$$

where we have added and subtracted a term in an obvious way. It is evident that, by construction, the last term has no discontinuity across the positive $s$-axis; it has only the negative discontinuity associated with $Q_0(1 + t'/2s)$. The discontinuity across the positive $s$-axis comes only from the third and fourth terms and reference to Eq. (245) shows that this discontinuity is simply $\operatorname{Im} f_0(s')$. Thus we have

$$f_0(s) = f_{0B} - \frac{\Gamma_B/4\pi}{s - s_B} + \int_0^\infty \frac{ds'}{\pi} \frac{\operatorname{Im} f_0(s')}{s' - s}$$
$$+ \int_0^\infty \frac{ds'}{\pi} \int_0^\infty \frac{dt'}{\pi} \frac{\rho(s', t')}{s' - s} \left[\frac{1}{2s} Q_0\left(1 + \frac{t'}{2s}\right) - \frac{1}{2s'} Q_0\left(1 + \frac{t'}{2s'}\right)\right]. \tag{250}$$

It is straightforward to show that the last line is just

$$\int_{-\infty}^{-\mu_0^2} \frac{ds'}{\pi} \frac{\operatorname{Im}\,[f_0(s') - f_{0B}(s')]}{s' - s}, \tag{251}$$

where $\operatorname{Im}\,[f_0(s') - f_{0B}(s)]$ is given by Eq. (248), and therefore Eq. (250) is the same as Eq. (247) for $l = 0$.

A very direct procedure which also verifies explicitly the statement made in connection with Eq. (251) is as follows. (Since it has been seen that the bound-state pole and subtraction terms are harmless, to save writing we assume $l \neq 0$.)

$$f_l(s) = \frac{1}{2}\int_{-1}^{+1} dz P_l(z) f(s, z) = f_{lB}(s) + \int_{-1}^{+1} \frac{dz}{2} P_l(z) \int_0^\infty \frac{ds'}{\pi} \int_{4\mu_0^2}^\infty \frac{dt'}{\pi}$$
$$\times \frac{\rho(s', t')}{s' - s} \frac{1}{t' + 2s'(1 - z)} + \int_{-1}^{+1} \frac{dz}{2} P_l(z) \int_{4\mu_0^2}^\infty \frac{dt'}{\pi} \int_0^\infty \frac{ds''}{\pi}$$
$$\times \frac{\rho(s'', t')}{s'' - s} \left[\frac{1}{t' + 2s(1 - z)} - \frac{1}{t' + 2s''(1 - z)}\right] \equiv f_{lB} + I_1 + I_2. \tag{252}$$

In writing this equation, we have again added and subtracted a term in an obvious way and have taken the liberty of carrying out the $s$, $t$ integrations in any order. To simplify a later step, we have written $s''$ in place of $s'$ in the last line. (It will also be helpful to remember the absolute limits on $t'$, as we have indicated.) If we imagine carrying out the $t'$ integration in $I_1$, it is clear that this term may be written as

$$I_1 = \int_{-1}^{+1} \frac{dz}{2} P_l(z) \int_0^\infty \frac{ds'}{\pi} \frac{\operatorname{Im} f(s', z)}{s' - s} = \int_0^\infty \frac{ds'}{\pi} \frac{\operatorname{Im} f_l(s')}{s' - s}, \tag{253}$$

since with $s'$ in the physical region we may write $\operatorname{Im} f(s', z) = \sum_{l'} (2l' + 1)$ $\operatorname{Im} f_{l'}(s') P_{l'}(z)$. In $I_2$, the last line of (252), we put things over a common denominator and write

$$I_2 = \int_{-1}^{+1} \frac{dz}{2} P_l(z) \int_{4\mu_0^2}^\infty \frac{dt'}{\pi} \int_0^\infty \frac{ds''}{\pi} \frac{\rho(s'', t')}{[t'/2(1 - z) + s][t' + 2s''(1 - z)]}$$
$$= -\int_{-1}^{+1} \frac{dz}{2} P_l(z) \int_{-\infty}^{-2\mu_0^2/1-z} \frac{ds'}{\pi} \frac{1}{s' - s} \int_0^\infty \frac{ds''}{\pi} \frac{\rho[s'', -2s'(1 - z)]}{s'' - s'}$$
$$= \int_{-\infty}^{-\mu_0^2} \frac{ds'}{\pi} \frac{1}{s' - s} \int_{-1}^{1+2\mu_0^2/s'} \frac{dz}{2} P_l(z) \int_0^\infty \frac{ds''}{\pi} \frac{\rho[s'', -2s'(1 - z)]}{s'' - s'}$$
$$= \int_{-\infty}^{-\mu_0^2} \frac{ds'}{\pi} \frac{1}{s' - s} \int_{4\mu_0^2}^{-4s'} \frac{dt''}{4s'} P_l\left(1 + \frac{t''}{2s'}\right) \int_0^\infty \frac{ds''}{\pi} \frac{\rho(s'', t'')}{s'' - s'}. \tag{254}$$

The second line follows by making the substitution $t' = -2s'(1 - z)$, the third by interchanging the order of integration, and the fourth by the variable change $z = 1 + t''/2s'$. The coefficient of $1/(s' - s)$ in the fourth line is just Eq. (248) with $s$ there replaced by $s'$, $s'$ by $s''$, $t'$ by $t''$.[35]

Now as we have noted, given the weight function $\rho(s, t)$, the partial

[35] The factors of $s^l/s'^l$ which appear in Eq. (247) are easily obtained on noticing that $f_l(s) \sim s^l$ for small $s$ and on making subtractions at $s = 0$.

wave amplitudes may all be obtained immediately except that corresponding to $l = 0$, and to find this we must solve an integral equation. The appropriate equation is (247) with $l = 0$, together with (248) and the unitarity condition. This is a mapping problem in complex variable theory. We seek a function $f_0$ which satisfies $\text{Im}\, f_0 = \sqrt{s}\,|f_0|^2$ for $s$ real, $> 0$; it has a given left-hand cut running from $-\infty$ to $-\mu_0{}^2/4$ and is otherwise analytic except perhaps for a discrete pole at $s = s_0 < 0$. To solve this we use a method first given by Chew and Mandelstam [23]. We write

$$f_0(s) = \frac{N(s)}{D(s)}, \tag{255}$$

where $N(s)$ has special properties: $N(s)$ is to be real for $s$ real and positive and is to contain the left-hand cut of $f_0$. $D$ is to be analytic in the whole $s$-plane except for a cut along $s$ real and $s \geq 0$. We shall demand that $D(s) \to 1$ as $s \to \infty$ and that $N \to 0$ as $s \to \infty$ in order that $f_0 \to 0$ as $s \to \infty$. That such a decomposition of $f_0$ is possible may be seen by observing that the function

$$N' = f_0(s) \exp\left[-\int_0^\infty \frac{ds'}{\pi}\,\frac{\delta_0(s')}{s' - s}\right] \tag{256}$$

has no right-hand cut, since the phase of $f_0$ is just given by $\delta_0$, the $s$-wave phase shift. Thus $N'$ has only a left-hand cut, and $f_0$ may be written in the form (255) with the appropriate phase insured. It is perhaps worth noting that (255) is exactly what we obtained in Chapter 6 in our discussion of the Jost function. In fact, $D$ is the same quantity introduced in Eq. (6.281). We recall also that the bound states correspond to the zeros of $D$.

The stated analyticity of $N$ and $D$ is now explicitly expressed in terms of an integral representation. (These forms follow instantly from Cauchy's theorem.)

$$\begin{aligned} D(s) &= 1 + \int_0^\infty \frac{ds'}{\pi}\,\frac{\text{Im}\, D(s')}{s' - s}, \\ N(s) &= \int_{-\infty}^{-\mu_0{}^2/4} \frac{ds'}{\pi}\,\frac{\text{Im}\, N(s')}{s' - s}. \end{aligned} \tag{257}$$

In these equations we regard $s$ as complex; the physical values are the boundary values as $s$ approaches the real axis from above. Now for $s > 0$ and real, since $N(s)$ is then real,

$$\begin{aligned} \text{Im}\, D(s) &= N(s)\, \text{Im}\, \frac{1}{f_0} \\ &= -\,N(s)\,\frac{\text{Im}\, f_0}{|f_0|^2} = -\,\sqrt{s}\, N(s), \qquad s > 0, \end{aligned} \tag{258}$$

where we have used $\operatorname{Im} f_0 = \sqrt{s}\,|f_0|^2$. To obtain $\operatorname{Im} N(s)$ we use the fact that $D$ is real for $s < 0$ and our expression (248) for $\operatorname{Im} f_0$

$$\operatorname{Im} N(s) = D(s)\frac{\pi}{4s}\left[\int_{\mu_0}^{\infty} d\mu\sigma(\mu)\,\theta\left(-s-\frac{\mu^2}{4}\right) + \int_0^{\infty}\frac{ds'}{\pi}\int_0^{\infty}\frac{dt'}{\pi}\frac{\rho(s',t')}{s'-s} \times \theta\left(-s-\frac{t'}{4}\right)\right] \equiv \pi D(s)\alpha(s). \tag{259}$$

We have then

$$D(s) = 1 - \int_0^{\infty}\frac{ds'}{\pi}\frac{\sqrt{s'}\,N(s')}{s'-s},$$
$$N(s) = \int_{-\infty}^{-\mu_0^2/4} ds'\,\frac{\alpha(s')D(s')}{s'-s}. \tag{260}$$

It is traditional to eliminate $N$ and get a single equation for $D$. If this is done, we find

$$D(-s) = 1 + \int_{\mu_0^2/4}^{\infty} ds'\,\frac{\alpha(-s')D(-s')}{\sqrt{s'}+\sqrt{s}}, \tag{261}$$

or for real positive $s$,

$$D(s) = 1 + \int_{\mu_0^2/4}^{\infty} ds'\,\frac{\alpha(-s')D(-s')}{\sqrt{s'}-i\sqrt{s}}. \tag{262}$$

By the substitution $s' = 1/x^2$, Eq. (261) may be transformed into a nonsingular integral equation of the Fredholm type, which is readily solved numerically.

If we flaunt tradition and eliminate $D$ instead of $N$, we find

$$N(s) = L(s) + \int_0^{\infty}\frac{ds'}{\pi}\sqrt{s'}\,\frac{L(s')-L(s)}{s'-s}N(s'), \tag{263}$$

where $L(s)$ is just the part of $f_0$ that has the left-hand cut, say the last term in Eq. (250). It appears in the derivation as

$$L(s) = \int_{-\infty}^{-\mu_0^2} ds'\,\frac{\alpha(s')}{s'-s}. \tag{264}$$

Equation (263) is also nonsingular and it is a matter of technical convenience whether one uses (261) or (268).

### The "Impact Parameter" Representation

Before leaving the subject of dispersion theory for nonrelativistic potential scattering, we mention briefly some recent work by Blankenbecler and Goldberger [22], which is closely related to our discussion of the eikonal approximation in Chapter 6. Let us assume that the dispersion relation in energy holds for all momentum transfers. (We know that it does do so for potentials which decrease faster than any exponential and also for those which are superpositions of Yukawa potentials.) For simplicity we assume no bound states in the following discussion, although they may be included without difficulty. We then imagine expanding $f(s, t)$ as a Fourier-Bessel transform

$$f(s, t) = \int_0^\infty b\, db H(b, s) J_0(b\sqrt{-t}). \tag{265}$$

From the single variable dispersion relation (229), writing $-t = \Delta^2$, multiplying by $\Delta J_0(b\Delta)$ and integrating over $\Delta$ from 0 to $\infty$ we obtain[36,37]

$$H(b, s) = -\int_0^\infty dz V(\sqrt{b^2 + z^2}) + \int_0^\infty \frac{ds'}{\pi} \frac{\operatorname{Im} H(b, s')}{s' - s - i\eta}. \tag{266}$$

It is important to note that, for fixed $b$, we have a dispersion relation which has only the physical cut. This is to be contrasted with those for fixed angular momentum [Eq. (247)].

In order to work with this new relation we must use the unitarity condition to determine Im $H$ in terms of $H$. The result is quite complicated.

$$\operatorname{Im} H(b\; s) = \int_0^\infty x\, dx \int_0^\infty y\, dy\, H^*(x, s) H(s, y)\, \mathscr{G}(b; x, y; s), \tag{267}$$

where

$$\mathscr{G}(b; x; y; s) = \frac{4s}{\pi} \int_0^\pi d\phi \int_0^\pi d\theta \sin\theta J_0\left(2kb \sin\frac{\theta}{2}\right) \frac{J_1[2kB(\theta, \phi)]}{B(\theta, \phi)}, \tag{268}$$

$$B(\theta, \phi) = x^2 + y^2 - 2xy \cos\frac{\theta}{2} \cos\phi. \tag{269}$$

For high energies, however, $\mathscr{G}$ becomes quite simple.

$$\mathscr{G} \cong \frac{1}{2kxy}\, \delta(x - y)\, \delta(x - b) + \mathscr{O}\left(\frac{1}{s}\right), \tag{270}$$

[36] The first term arises as follows:

$$-\int_0^\infty \Delta\, d\Delta J_0(b\Delta) \frac{1}{4\pi} \int d^3x e^{i\Delta\cdot \mathbf{x}} V(x) = -\int_0^\infty r\, dr V(r) \int_0^\infty d\Delta J_0(b\Delta) \sin \Delta r$$

$$= -\int_0^\infty \frac{dr r\, V(r)}{(r^2 - b^2)^{1/2}} \theta(r^2 - b^2) = -\int_0^\infty dz V(\sqrt{b^2 + z^2}).$$

[37] We note also the orthogonality relation $\displaystyle\int_0^\infty b\, db J_0(bx) J_0(by) = \frac{\delta(x - y)}{x}$.

and the unitarity condition becomes simply

$$\operatorname{Im} H(b, s) = \frac{1}{2\sqrt{s}} H^*(b, s)H(b, s). \tag{271}$$

The latter relation is very similar to $\operatorname{Im} f_l = \sqrt{s}\,|f_l|^2$; we shall see shortly that there is indeed an intimate relation between these.

If we use the high energy unitarity relation (271) for *all* energies, Eq. (266) becomes an exceedingly simple integral equation

$$H(b, s) = -\int_0^\infty dz V(\sqrt{b^2 + z^2}) + \int_0^\infty \frac{ds'}{\pi} \frac{|H(b, s')|^2}{2\sqrt{s'}\,(s' - s - i\eta)}. \tag{272}$$

To solve this we remark that

$$L(b, s) \equiv \frac{-\int_0^\infty dz V(\sqrt{b^2 + z^2})}{H(b, s)}, \tag{273}$$

regarded as a function of a complex variable $s$, is a function analytic in the entire cut $s$-plane which approaches unity as $s \to \infty$, assuming that $H(b, s)$ has no zeros. Any such function can be written as

$$L(b, s) = 1 + \int_0^\infty \frac{ds'}{\pi} \frac{\rho(b, s')}{s' - s}. \tag{274}$$

Now

$$\rho(b, s) = \operatorname{Im} L(b, s) = \frac{\int_0^\infty dz V(\sqrt{b^2 + z^2})}{|H(b, s)|^2} \operatorname{Im} H(b, s)$$

$$= \frac{\int_0^\infty dz V(\sqrt{b^2 + z^2})}{2\sqrt{s}}, \tag{275}$$

where we have used (271) to evaluate $\operatorname{Im} H(b, s)$. Substituting into (274), we find that

$$L(b, s) = 1 + \frac{\int_0^\infty dz V(\sqrt{b^2 + z^2})}{2\sqrt{-s}}, \tag{276}$$

which yields then finally

$$H(b, s) = \frac{-\int_0^\infty dz V(\sqrt{b^2 + z^2})}{1 + \dfrac{i}{2\sqrt{s}} \int_0^\infty dz V(\sqrt{b^2 + z^2})}. \tag{277}$$

This function has the proper analyticity and satisfies our high energy unitarity condition.

The scattering amplitude in this approximation becomes

$$f(s, t) = -\int_0^\infty b\, db J_0(b\sqrt{-t}) \frac{\displaystyle\int_0^\infty dz V(\sqrt{b^2 + z^2})}{1 + \dfrac{i}{2\sqrt{s}} \displaystyle\int_0^\infty dz V(\sqrt{b^2 + z^2})}. \qquad (278)$$

This may be written in essentially the form given for Eq. (6.511),

$$f(s, b) = -i\sqrt{s} \int_0^\infty b\, db J_0(b\sqrt{-t})[e^{i\chi(b,s)} - 1] \qquad (279)$$

with

$$\chi = -2 \tan^{-1} \frac{1}{2\sqrt{s}} \int_0^\infty dz V(\sqrt{b^2 + z^2}). \qquad (280)$$

The result of Section 6.7, p. 337 corresponds to replacing $\chi$ by the first term of the arctangent expansion.

The eikonal approximation discussed there was based on using an approximate scattering wave function (in our present units, $2m = 1$) of the form

$$\psi \cong \exp\left[ikz - \frac{i}{2\sqrt{s}} \int_{-\infty}^{z} dz' V(\sqrt{z'^2 + b^2})\right]. \qquad (281)$$

The current approximation may be expressed in terms of the wave function

$$\psi \cong \frac{e^{ikz}}{\left[1 + \dfrac{i}{4\sqrt{s}} \displaystyle\int_{-\infty}^{z} dz' V(\sqrt{z'^2 + b^2})\right]^2} \qquad (282)$$

These two evidently agree for sufficiently large $\sqrt{s}$, but they are quite different, in general, in their analytic dependence on $s$.

The approximate relation between $b$ and the classical impact parameter may be established by considering the partial wave amplitudes $f_l(s)$.

$$\begin{aligned} f_l(s) &= \int_{-1}^{+1} \frac{dz}{2} P_l(z) f[s, 2s(1 - z)] \\ &= \int_0^\infty \frac{db}{\sqrt{s}} H(b, s) J_{2l+1}(4\sqrt{s}\, b). \end{aligned} \qquad (283)$$

This is a rigorous expression. Consider now qualitatively the situation which obtains for large $l$. As we have often noted, $P_l(z) \cong J_0[(l + \frac{1}{2})\,\theta] \cong J_0(l\theta)$. Furthermore, for our representation

$$f[s, 2s(1 - \cos\theta] \cong \int_0^\infty b\, db H(b, s) J_0(b\sqrt{s}\,\theta),$$

since for large energies only small angles are important. Then

$$f_l(s) \cong \tfrac{1}{2}\int_0^\infty \theta d\,\theta J_0(l\theta) \int_0^\infty b\, db H(b, s) J_0(b\sqrt{s}\,\theta) \cong \left| \frac{1}{2s} H(b, s) \right|_{b=l/\sqrt{s}} . \tag{284}$$

Thus we see that $b$ has just the usual classical meaning of impact parameter in the limit of high $l$ and high energies. Note also that with this approximate relation between $H$ and $f_l$, Eq. (271) becomes exactly the partial wave unitarity condition.

It is clear that we can base a systematic construction scheme on the exact unitarity relation and thereby have a method in principle of discussing by dispersion theory any kind of potential, rather than just the superposition of Yukawa potentials as considered in connection with the Mandelstam representation. We note in passing that if $H(b, s)$ can be represented by

$$H(b, s) = \int_0^\infty dx \frac{\sigma(x^s)}{x + b^2}, \tag{285}$$

our representation (265) becomes

$$f(s, t) = \int_0^\infty dt' \frac{g(s, t')}{t' - t}, \tag{286}$$

where

$$g(s, t') = \tfrac{1}{2}\int_0^\infty dx \sigma(x^s) J_0(\sqrt{xt'}). \tag{287}$$

Thus we recover the old representation in terms of $t$.

We now bring to a close this section on nonrelativistic dispersion theory. There are many topics which could have been included, but with this introduction the literature should be quite readable. We mention one intriguing and, as far as we know, unsolved problem, which could be of importance in quantum field theory: Can the double dispersion representation be derived from the dispersion relation for fixed momentum transfer together with unitarity?

## 10.6 DISPERSION RELATIONS IN RELATIVISTIC FIELD THEORY

### Formal Field Theoretic Considerations

The methods of dispersion theory are used almost exclusively at the present time in attempts at carrying out quantitative calculations of processes involving elementary particles. To a very large extent these techniques have become a branch of theoretical physics independent of the quantum field theory on which they were originally based. By this we mean that we can give a set of reasonably logical computational rules by which physical reaction amplitudes may be obtained without explicit reference to the field concept. On the other hand, the experience and intuition which are required for engaging in this sport seem difficult to acquire without spending some time with field theory.[38] There is a distinct possibility that formal field theory as we now know it is meaningless. Nevertheless we shall find it possible to draw certain conclusions from this formalism that appear to be based on those aspects of the theory that we would imagine might well survive even very drastic modifications of the details.

We shall confine our attention to the simplest topics in relativistic dispersion theory. For example, we shall not study the problem of the derivation of dispersion relations, except for forward scattering. The existing derivations for nonforward scattering are extremely complicated and are largely uninteresting and uninstructive.[39] Because of space the double dispersion relations, which are presently the most widely used in practice, will not be treated at all. [It is perhaps worth mentioning that this representation has not yet been derived even within the framework of perturbation theory from field theoretical principles.]

We turn now to some preliminaries that will ultimately lead us to a formula for $S$-matrix elements from which dispersion relations may be inferred. The most important and difficult problem encountered is that of learning how to characterize physical states in quantum field theory. To resolve this we shall draw heavily on the intuition gained from our study of the nonrelativistic theory. It is important to recognize, however, that the problem here is much more subtle. Consider, for example, two nucleons which are interacting by way of a meson field. If the two are far apart, it is experimentally clear that we can associate a definite mass with each one, but the meson field which mediates the interaction is still present, interacting with the individual nucleons. In this sense the "potential" does not vanish when the particles are far apart. The mass is different from the mathematical (and essentially uninteresting) entity which exists when the coupling between mesons and nucleons is turned off. We must learn how to describe such complex situations.

The first people to make a serious effort at a correct physical particle

[38] We are unable to describe a systematic development of quantum field theory here. Such a presentation may, for example, be found in Refs. [24].

[39] Someone has said, "Derivations in dispersion theory are like a man's teats; they are neither useful nor ornamental."

description in field theory were Haag and Wightman [25]; their work was cast in its most useful form by Lehmann, Symanzik, and Zimmermann [26] in a classic paper. The first attempts at what we shall refer to as the *asymptotic condition* (to be explained shortly) were made by Yang and Feldman and by Källen [27]. There are some difficult technical points which still remain somewhat unsettled, but the basic ideas of this approach are generally accepted. The most careful treatment has been given by Greenberg and Wightman [28]. The problem of bound states has been largely solved by the work of Haag, Nishijima, and Zimmerman [29]. It is evident that for such states we can scarcely talk about "switching off" interactions, as was customary in many older formulations of the theory.

We shall restrict our attention to theories in which there are no bound states. We begin by considering a single, neutral, spinless meson field interacting with itself in an unspecified way. The extension of the formalism to charged bosons and to fermions will be made later. It is intuitively clear that in the remote past or in the remote future particles will be widely separated from each other and must behave essentially like free particles. We may then set up a basis system in Hilbert space labeling the states just as we would if there were no interaction. It is in terms of such states that we give meaning to the usual characterization of scattering experiments. From our experience with the nonrelativistic theory we would expect that if we start in the remote past with two well-localized wave packets describing particles "aimed at each other," they will collide and the packets overlap. Later, however, the two (or when new particles are produced, more than two) packets separate, and as the probability density of each packet decreases like $(1/t)^3$ [40] there is negligible probability that a subsequent collision can occur. Thus the particles again act as if free.

Before describing the construction of these *asymptotic states*, let us review briefly the discussion given in Section 3.1 of wave packet spreading, since this is of importance for our study. We saw in Section 1.4 that spinless, relativistic particles are described by the positive frequency solutions of the Klein-Gordon equation. A general wave packet for such a particle then has the form[41]

$$f(x) = \int d^4p\theta(p_0)\,\delta(p^2 + \mu^2)g(p)e^{ip\cdot x}\,, \tag{288}$$

where $\mu$ is the mass of the particle. The step function selects only that root of the $\delta$-function for which $p_0 = +\sqrt{\mu^2 + \mathbf{p}^2}$. We examine this in

[40] This result was obtained with Eq. (3.21).

[41] In the remainder of this chapter we shall set $c = 1$, where $c$ is the velocity of light, as well as continuing to take $\hbar = 1$. We shall also continue to write a four-vector $x$ as $(\mathbf{x}, ix_0)$ and the scalar product of two four-vectors $A$ and $B$ as $A \cdot B = \mathbf{A} \cdot \mathbf{B} - A_0B_0$.

the case where $|\mathbf{x}|$ and $x_0 \to \infty$. It is clear that the integral will be very small except in the neighborhood where the phase is stationary; that is, where $\nabla_p(\mathbf{p} \cdot \mathbf{x} - \sqrt{\mu^2 + \mathbf{p}^2}x_0) = 0$. This condition yields $\mathbf{v} = \mathbf{p}/p_0 = \mathbf{x}/x_0$. Expanding the exponent in a Taylor series about this point, and taking $g$ out of the integral, we find that for large $x$

$$f(x) \simeq \tfrac{1}{2}(2\pi)^{3/2} \frac{\sqrt{\mu}}{(ix_0)^{3/2}} e^{-i\mu(1-v^2)^{1/2}x_0} g\left(\frac{\mu\mathbf{v}}{\sqrt{1-v^2}}, \frac{\mu}{\sqrt{1-v^2}}\right). \tag{289}$$

Two such packets which have $g$'s that do not overlap will then diverge from each other. We note also the decrease in amplitude as $x_0^{-3/2}$. Except for Coulomb forces, the interaction energy will then go to zero as $x_0 \to \infty$. The purpose of this argument is to make reasonable the mathematical procedure now to be developed.

In nonrelativistic theory we used the concept of "what the wave function would have been had there been no interaction." We shall also use this idea in field theory. We shall characterize an incoming state (abbreviated as in) by saying that the particles were in such a state in the remote past that at some finite time $t$ they would have some given wave function. Similarly, an outgoing (out) state is one which would have had to have propagated into the future from a specified state at time zero.

To describe these initial and final configurations of our neutral spin-zero field we make use of the solutions of the Klein-Gordon equation. We consider a complete set of positive frequency solutions of this equation, call them $f_\alpha(x)$, where

$$(\mu^2 - \Box) f_\alpha(x) = 0 \tag{290}$$

and $\Box \equiv \nabla^2 - \partial^2/\partial x_0^2$. We normalize these functions according to the condition

$$i\int d^3x \left(f_\alpha^* \frac{\partial f_\beta}{\partial x_0} - \frac{\partial f_\alpha^*}{\partial x_0} f_\beta\right) \equiv i\int d^3x f_\alpha^* \frac{\overleftrightarrow{\partial}}{\partial x_0} f_\beta = \delta_{\alpha\beta}, \tag{291}$$

where the integration is carried out over any surface $x_0 =$ constant. To see that this normalization is preserved in time, we take the time derivative and use the Klein-Gordon equation

$$\begin{aligned} \int d^3x \frac{\partial}{\partial x_0}\left(f_\alpha^* \frac{\overleftrightarrow{\partial}}{\partial x_0} f_\beta\right) &= \int d^3x \left(f_\alpha^* \frac{\partial^2}{\partial x_0^2} f_\beta - \frac{\partial^2 f_\alpha^*}{\partial x_0^2} f_\beta\right) \\ &= \int d^3x [f^*_\alpha \nabla^2 f_\beta - (\nabla^2 f^*_\alpha) f_\beta] \\ &= \int d^3x \nabla \cdot [f_\alpha^* \nabla f_\beta - (\nabla f_\alpha^*) f_\beta] = 0, \end{aligned} \tag{292}$$

where the last step follows from the fact that the $f$'s are packets, confined in space. In practice we frequently allow such packets to approach plane waves $f_{\mathbf{k}}(x)$ given by

$$f_{\mathbf{k}}(x) = \frac{1}{\sqrt{2k_0\mathscr{V}}}\, e^{ik\cdot x}, \tag{293}$$

where $k_0 = +(\mu^2 + \mathbf{k}^2)^{1/2}$ and $\mathscr{V}$ is the quantization volume. The completeness relation for the $f_\alpha$'s is conveniently evaluated using the box eigenfunctions

$$\sum_\alpha f_\alpha(x) f_\alpha^*(y) = \frac{1}{\mathscr{V}} \sum_{\mathbf{k}} \frac{e^{ik\cdot(x-y)}}{2k_0} \to \int \frac{d^3k}{(2\pi)^3} \frac{e^{ik\cdot(x-y)}}{2k_0}$$
$$= \int \frac{d^4k}{(2\pi)^3}\, \theta(k_0)\, \delta(k^2 + \mu^2) e^{ik\cdot(x-y)} \equiv \Delta^+(x - y). \tag{294}$$

Properties of the singular function $\Delta^+(x - y)$ are discussed in many books.[42] The completeness relation together with the orthogonality condition yields

$$f_\alpha(x) = i \int d^3x'\, \Delta^+(x - x') \frac{\overleftrightarrow{\partial}}{\partial x_0'} f_\alpha(x'), \tag{295}$$

where the time $x_0'$ is arbitrary.

We next introduce a set of operators which when acting on the vacuum, or no particle state, creates states with particles having the single-particle wave functions $f_\alpha(x)$. Since we shall ultimately consider two such sets of operators and states, we designate the present ones as in states and label them with the letter $i$. Thus, if we take $|0\rangle$ as the vacuum state, we define a set of operators $a_\alpha^\dagger$ with the following properties:

$$\begin{aligned} a_\alpha^\dagger\, |0\rangle &= |\alpha; i\rangle, \\ a_\alpha^\dagger\, |\beta; i\rangle &= a_\alpha^\dagger a_\beta^\dagger\, |0\rangle = |\alpha, \beta; i\rangle, \qquad \alpha \neq \beta, \\ a_\alpha^\dagger\, |\beta, \gamma; i\rangle &= |\alpha, \beta, \gamma, i\rangle, \qquad \alpha \neq \beta \neq \gamma. \end{aligned} \tag{296}$$

If a state $\alpha$ is occupied $n_\alpha$ times, we have instead

$$a_\alpha^\dagger\, |(\alpha)^{n_\alpha}, \beta, \ldots; i\rangle = \sqrt{n_\alpha + 1}\, |(\alpha)^{n_\alpha+1}, \beta, \ldots; i\rangle. \tag{297}$$

The operator Hermitean conjugate to $a_\alpha^\dagger$ is $a_\alpha$; we have then the fundamental commutation relation

$$[a_\alpha, a_\beta^\dagger] = \delta_{\alpha\beta}, \tag{298}$$

all other commutators vanishing. From this it follows that the state $|\alpha, \beta, \gamma, \ldots; i\rangle$ is symmetric in the labels $\alpha, \beta, \gamma, \ldots$. It is clear that since the one-particle state is orthogonal to the vacuum, $\langle 0|\, a_\alpha^\dagger\, |0\rangle = 0$,

[42] See Refs. [24] or [30], for example.

that $a_\alpha |0\rangle = 0$, and furthermore that

$$a_\alpha |\beta; i\rangle = \delta_{\alpha\beta},$$

$$a_\alpha |\beta, \gamma; i\rangle = \delta_{\alpha\beta} |\gamma; i\rangle + \delta_{\alpha\gamma} |\beta; i\rangle, \tag{299}$$

but

$$a_\alpha |(\alpha)^{n_\alpha}, \beta, \ldots; i\rangle = \sqrt{n_\alpha}\, |(\alpha)^{n_\alpha - 1}, \beta, \ldots; i\rangle.$$

A word of caution is in order concerning use of the completeness relation for these states. We have, for example, $\sum_{\alpha,\beta} a_\alpha^\dagger a_\beta^\dagger |0\rangle \langle 0| a_\alpha a_\beta = 2$, as is easily verified by considering the expansion of an arbitrary two-particle state $|2\rangle = \sum_{\alpha,\beta} \phi_{\alpha\beta} a_\alpha^\dagger a_\beta^\dagger |0\rangle$, where $\phi_{\alpha\beta}$ is symmetric.

In terms of these operators we may define an "in" field operator $\phi_i(x)$ by

$$\phi_i(x) = \sum_\alpha [a_\alpha f_\alpha(x) + a_\alpha^\dagger f_\alpha^*(x)]. \tag{300}$$

Evidently, this object, by construction, satisfies the Klein-Gordon equation. The operator equation, $(\mu^2 - \Box)\phi_i(x) = 0$, is the analogue of the Heisenberg equation of motion for the nonrelativistic operators $\psi_i, \psi_o$, Eq. (58). It is easy to verify from the commutation rules for the $a_\alpha$ that

$$[\phi_i(x), \phi_i(y)] = i\Delta(x - y), \tag{301}$$

where

$$\begin{aligned}\Delta(x) &= -i[\Delta^+(x) - \Delta^+(-x)] \\ &= -\frac{i}{(2\pi)^3} \int d^4p\,\epsilon(p_0)\,\delta(p^2 + \mu^2) e^{ip\cdot x}.\end{aligned} \tag{302}$$

The only property of the function $\Delta$ needed is that it vanishes for spacelike $x$, $x^2 > 0$.[43,44] Using the orthogonality relation for the $f$'s we may express the $a$'s in terms of $\phi_i$. In fact, we could have begun with the basic commutation relation for the $\phi_i$ and worked backwards to obtain the properties of the $a$'s. We find

$$\begin{aligned}a_\alpha &= i\int d^3x f_\alpha^*(x) \frac{\overleftrightarrow{\partial}}{\partial x_0} \phi_i(x), \\ a_\alpha^\dagger &= i\int d^3x \phi_i(x) \frac{\overleftrightarrow{\partial}}{\partial x_0} f_\alpha(x).\end{aligned} \tag{303}$$

[43] To see this, we remark that $\Delta$ is invariant under any Lorentz transformation that does not reverse the sense of time. If therefore it vanishes on the plane $x_0 = 0$, it vanishes everywhere outside the light cone $x^2 = 0$. The vanishing of $\Delta(\mathbf{x}, 0)$ follows immediately from the integral representation, for when $x_0 = 0$, the integrand is an odd function of $p_0$. [Note also that $\Delta(x) = -\Delta(-x)$.]

[44] The sign function $\epsilon(p_0)$ is defined in connection with Eq. (14).

The time $x_0$ is arbitrary since both the $f$'s and $\phi_i$ are solutions of the Klein-Gordon equation. Evidently,

$$\langle 0|\, \phi_i(x)\, |\alpha, i\rangle = f_\alpha(x). \tag{304}$$

In a similar fashion, we can introduce a set of out operators, call them $b_\alpha$, $b_\alpha^\dagger$, which satisfy exactly the same commutation rules with each other as do the $a$'s with themselves. To summarize these properties we write

$$\phi_0(x) = \sum_\alpha [b_\alpha f_\alpha(x) + b_\alpha^\dagger f_\alpha\, f_\alpha{}^*(x)], \tag{305}$$

$$[\phi_0(x), \phi_0(y)] = i\Delta(x - y),$$
$$[b_\alpha, b_\beta^\dagger] = \delta_{\alpha\beta},\ [b_\alpha, b_\beta] = [b_\alpha^\dagger, b_\beta^\dagger] = 0,$$
$$b_\alpha = i\int d^3x f_\alpha{}^* \frac{\overleftrightarrow{\partial}}{\partial x_0} \phi_0(x),$$
$$b_\alpha^\dagger = i\int d^3x \phi_0(x) \frac{\overleftrightarrow{\partial}}{\partial x_0} f_\alpha(x),$$
$$|\alpha; \text{o}\rangle = b_\alpha^\dagger\, |\text{o}\rangle,$$
$$|\alpha, \beta; \text{o}\rangle = b_\alpha^\dagger b_\beta^\dagger\, |\text{o}\rangle, \qquad \alpha \neq \beta,$$
$$\sqrt{n_\alpha + 1}\, |(\alpha)^{n_\alpha+1}, \beta, \ldots; \text{o}\rangle = b_\alpha^\dagger\, |(\alpha)^{n_\alpha}, \beta, \ldots; \text{o}\rangle. \tag{306}$$

Since the $a_\alpha$, $b_\alpha$ satisfy the same commutation relations, we assume the existence of a unitary operator $S$ which has the property

$$S^{-1}a_\alpha S = b_\alpha, \tag{307}$$
$$S^{-1}a_\alpha^\dagger S = b_\alpha^\dagger.$$

This means that the general states

$$|\{\alpha\}; i\rangle \equiv |(\alpha_1)^{n_{\alpha_1}}(\alpha_2)^{n_{\alpha_2}}, \ldots; i\rangle = \frac{1}{\sqrt{n_{\alpha_1}!\, n_{\alpha_2}! \cdots}} (a_{\alpha_1}^\dagger)^{n_{\alpha_1}}(a_{\alpha_2}^\dagger)^{n_{\alpha_2}} \cdots |0\rangle \tag{308}$$

$$|\{\alpha\}; \text{o}\rangle \equiv |(\alpha_1)^{n_{\alpha_1}}(\alpha_a)^{n_{\alpha_2}}, \ldots; \text{o}\rangle = \frac{1}{\sqrt{n_{\alpha_1}!\, n_{\alpha_2}! \cdots}} (b_{\alpha_1}^\dagger)^{n_{\alpha_1}}(b_{\alpha_2}^\dagger)^{n_{\alpha_2}} \cdots |0\rangle$$

are related by

$$|\{\alpha\}; i\rangle = S\, |\{\alpha\}; \text{o}\rangle, \tag{309}$$

provided we adopt the phase convention $|0\rangle = S\,|0\rangle$. From the preceding relation we have

$$\langle\{\alpha\}; \text{o} \mid \{\beta\}; i\rangle = \langle\{\alpha\}; \text{o}|\, S\, |\{\beta\}; \text{o}\rangle = \langle\{\alpha\}; i|\, S\, |\{\beta\}; i\rangle$$
$$\equiv S_{\{\alpha\},\{\beta\}}. \tag{310}$$

Finally, since we assume the set of states to be complete,

$$S = \sum_{\{\alpha\}} |\{\alpha\}; i\rangle \langle\{\alpha\}; o|. \tag{311}$$

In order to give this scheme physical content we must have an explicit way of relating the in and out operators to each other. The only known way of doing this is by the introduction of a local field operator $\phi(x)$, which interpolates between these fields. The term local field operator here means only that $[\phi(x), \phi(y)] = 0$ if $(x - y)^2 > 0$. At a later time we shall discuss the meaning of this condition, but at present we shall say merely that it is concerned with the fact that measurements at two points do not influence each other if the points cannot be connected by a light signal. Our object is to show that in some sense the operator $\phi(x)$ may be taken to coincide in the remote past with what we have called $\phi_i(x)$ and in the remote future with $\phi_0(x)$. This is far from a trivial concept, and the existence of such a field operator $\phi(x)$ is not anywhere nearly so firmly established as that for the corresponding nonrelativistic problem discussed earlier in the chapter.

We would like to identify $\phi_i(x)$ with that field which would have propagated to $x$ from the remote past from some specified field. As a formal preliminary, we consider the integration of the equation $(\mu^2 - \Box)\,\phi_i(x) = 0$ in terms of a retarded Green's function of the Klein-Gordon operator, defined by

$$\begin{aligned} (\mu^2 - \Box)\Delta_r(x) &= \delta(x), \\ \Delta_r(x) &= 0 \quad \text{for } x_0 < 0, \\ &= 0 \quad \text{for } x^2 > 0. \end{aligned} \tag{312}$$

It is easy to write down an explicit solution to this equation. It is simply $\Delta_r(x) = -\theta(x_c)\Delta(x)$ or, in terms of a Fourier integral,

$$\Delta_r(x) = \int \frac{d^4p}{(2\pi)^4} \frac{e^{ip\cdot x}}{p^2 + \mu^2 - i\eta\epsilon(p_0)}. \tag{313}$$

In a similar fashion, we may introduce an advanced Green's function, $\Delta_a(x)$, which satisfies the same equation but is nonzero only if $x_0 < 0$. It is given by $\Delta_a(x) = \theta(-x_0)\Delta(x)$, so $\Delta_r(x) = \Delta_a(-x)$. Its Fourier representation is the same as that for $\Delta_r$ with the opposite sign of $\eta$. To derive a formal expression for $\phi_i(x)$ we write

$$\begin{aligned} (\mu^2 - \Box')\phi_i(x') &= 0, \\ (\mu^2 - \Box')\Delta_a(x' - x) &= \delta(x' - x), \end{aligned} \tag{314}$$

then multiply the first equation by $\Delta_a$ and the second by $\phi_i$, subtract, and integrate over all space between the planes $x'_0 = x_{20}$, $x_0' = x_{10}$, where $x_{20} > x_0$. We find

$$\theta(x_0 - x_{10})\phi_i(x) = \left(\int_{x_{20}} d^3x' - \int_{x_{10}} d^3x'\right)\phi_i(x')\frac{\overleftrightarrow{\partial}}{\partial x'_0}\Delta_a(x' - x). \quad (315)$$

Since $\Delta_a(y) = 0$ unless $y_0 < 0$, the surface integral at $x_{20}$ gives no contribution; then, when $x_0 > x_{10}$ we get a nonzero result. We write this as

$$\theta(x_0 - x'_0)\phi_i(x) = -\int d^3x'\phi_i(x')\frac{\overleftrightarrow{\partial}}{\partial x'_0}\Delta_a(x' - x). \quad (316)$$

In a similar fashion, using $\Delta_r$ where we had previously used $\Delta_a$, we find

$$\theta(x'_0 - x_0)\phi_i(x) = \int d^3x'\phi_i(x')\frac{\overleftrightarrow{\partial}}{\partial x'_0}\Delta_r(x' - x). \quad (317)$$

It is in terms of this first representation for $\phi_i(x)$ that we interpret $\phi_i$ as being a field configuration that develops from the past from some initial configuration. Actually, we shall always refer to $\phi_i$ as being that which results when $x'_0 \to -\infty$. We will need the second formula for $\phi_i$ shortly. If we were faced with trying to integrate the equation $(\mu^2 - \Box)\phi(x) = j(x)$, we would obtain by the preceding method

$$\phi(x) = -\int_{x_{10}=x'_0} d^3x'\phi(x')\frac{\overleftrightarrow{\partial}}{\partial x'_0}\Delta_r(x - x') + \int_{x_{10}}^{\infty} d^4x'\Delta_r(x - x')j(x'). \quad (318)$$

The first term is a solution of the homogeneous equation. If we allow $x_{10} \to -\infty$, it may evidently be called $\phi_i(x)$. We then obtain the formal result

$$\phi(x) = \phi_i(x) + \int d^4x'\,\Delta_r(x - x')(\mu^2 - \Box')\phi(x'). \quad (319)$$

We see that in a loose sense $\phi(x) \to \phi_i(x)$ as $x_0 \to -\infty$. This is *not* true as an operator relation nor do we need it.

We define for the field $\phi$ a set of operators analogous to the $a_\alpha$'s introduced in connection with $\phi_i$.

$$\phi_\alpha(x_0) \equiv i\int d^3x f_\alpha^*(x)\frac{\overleftrightarrow{\partial}}{\partial x_0}\phi(x),$$

$$\phi_\alpha^\dagger(x_0) \equiv i\int d^3x\phi(x)\frac{\overleftrightarrow{\partial}}{\partial x_0}f_\alpha(x). \quad (320)$$

Since the operator $\phi$ does not satisfy the equation $(\mu^2 - \Box)\phi(x) = 0$, it is a function of time. We now assume that the matrix elements of these operators between normalizable state vectors $|A\rangle$, $|B\rangle$ satisfy the relations

$$\begin{aligned} \lim_{x_0 \to -\infty} \langle A| \, \phi_\alpha(x_0) \, |B\rangle &= \langle A| \, a_\alpha \, |B\rangle, \\ \lim_{x_0 \to +\infty} \langle A| \, \phi_\alpha(x_0) \, |B\rangle &= \langle A| \, b_\alpha \, |B\rangle. \end{aligned} \tag{321}$$

Thus in this sense, and *in this sense only*, we say that $\phi(x)$ in the remote past coincides with $\phi_i(x)$ and in the remote future with $\phi_0(x)$. In mathematical language this is known as weak convergence, to be contrasted with strong convergence which would be the demand that

$$\lim_{x_0 \to -\infty} \langle A| \, (\phi_\alpha(x_0) - a_\alpha{}^\dagger)(\phi_\alpha(x_0) - a_\alpha) \, |A\rangle = 0 \tag{322}$$

for any (normalizable) $|A\rangle$. (The latter relation is almost surely wrong.)

To get some feeling for what is involved in this asymptotic condition, we substitute the formal relation (319) between $\phi(x)$ and $\phi_i(x)$ into Eq. (320) to obtain

$$\langle A| \, \phi_\alpha(x_0) \, |B\rangle = \langle A| \, a_\alpha \, |B\rangle + i \int d^3x f_\alpha{}^*(x) \frac{\overleftrightarrow{\partial}}{\partial x_0} \times \int d^4x' \, \Delta_r(x - x')(\mu^2 - \Box') \, \langle A| \, \phi(x') \, |B\rangle. \tag{323}$$

Assuming that the matrix element of $(\mu^2 - \Box')\phi(x')$ is sufficiently well behaved that we may interchange the orders of integration and on using Eq. (317) with $\phi_i$ replaced by $f_\alpha$, this becomes

$$\langle A| \, \phi_\alpha(x_0) \, |B\rangle = \langle A| \, a_\alpha \, |B\rangle + i \int d^4x' \theta(x_0 - x'_0) f_\alpha{}^*(x')(\mu^2 - \Box') \, \langle A| \, \phi(x') \, |B\rangle. \tag{324}$$

Since $f_\alpha{}^*(x')$ is a *packet* state and the matrix element under the integral sign must not be too crazy in the limit as $x_0 \to -\infty$, the integral vanishes. The precise justification of all these steps under what appear to be reasonable assumptions has been given in an unpublished manuscript by W. Zimmermann. In our description of the nonrelativistic theory, $(\mu^2 - \Box)\phi$ became somewhat similar to the potential times the field. Since we deal with short-range potentials, the properties of the integrals in Eqs. (323) and (324) are much easier to understand. In the usual Lagrangian formulation of field theory, $\phi$ satisfies a "field equation" of the form $(\mu^2 - \Box)\phi = j(x)$, where $j(x)$ contains the sources of the meson field (usually something like a bilinear form of field operators for nucleons or possibly a polynomial in $\phi$ itself). These objects are highly singular and not spatially confined, so the field theoretic problem is a very subtle one.

There is one further formal step which must be noted. Even though we do not specify any detailed equations of motion, we do assume the existence of an energy momentum four-vector operator $P_\mu$ that is the generator of infinitesimal translations. It is then true that

$$i\frac{\partial\phi(x)}{\partial x_\mu} = [P_\mu, \phi(x)], \tag{325}$$

or

$$\phi(x + a) = e^{-iP\cdot a}\phi(x)e^{iP\cdot a}. \tag{326}$$

We assume also that all the states of our Hilbert space are eigenvectors of $-P^2$ with nonnegative eigenvalues. It is then also true, just as in the nonrelativistic case, that

$$i\frac{\partial\phi_{i,0}(x)}{\partial x_\mu} = [P_\mu, \phi_{i,0}(x)], \tag{327}$$

the demonstration being similar to that given there. Finally, we note that in dealing with a one-particle state, there is no need to distinguish between in and out states. They can differ at most by a phase factor which is fixed by the asymptotic condition to be unity. To see this we remark that

$$(\mu^2 - \Box)\langle 0|\, \phi(x)\, |\alpha; i, \text{o}\rangle = 0. \tag{328}$$

This follows from the fact that we may write

$$\begin{aligned}(\mu^2 - \Box)\, \langle 0|\, e^{-iP\cdot x}\phi(0)e^{iP\cdot x}|\alpha; i, \text{o}\rangle &= (\mu^2 - \Box)e^{ip_\alpha\cdot x}\, \langle 0|\, \phi(0)\, |\alpha; i, \text{o}\rangle \\ &= (\mu^2 + p_\alpha^{\,2})\, \langle 0|\, \phi(0)\, |\alpha; i, \text{o}\rangle \\ &= 0,\end{aligned} \tag{329}$$

since for a one-particle state the four momentum $p_\alpha$ must satisfy the equation $p_\alpha^{\,2} + \mu^2 = 0$. From Eq. (319) and the corresponding one involving $\phi_0$ (and the advanced Green's function $\Delta_a$), we find

$$\langle 0|\, \phi(x)\, |\alpha; i\rangle - \langle 0|\, \phi(x)\, |\alpha; \text{o}\rangle = f_\alpha(x). \tag{330}$$

Physically this is to be expected—a one-particle state with the proper mass is steady; nothing happens to it. From now on we drop any distinction between one-particle states.

Our next task is to learn how to use the asymptotic condition to express matrix elements of physical interest in terms of the field operator $\phi$. The reason for doing so is to have a mechanism for imposing the locality condition $[\phi(x), \phi(y)] = 0$, $(x - y)^2 > 0$, which we shall refer to also as the *commutator condition*, or *causality condition*. As a typical example of what is called a reduction formula, we consider the $S$-matrix element describing a two-particle scattering process $\alpha + \beta \to \alpha' + \beta'$, where $\alpha$, $\beta$,

$\alpha'$, $\beta'$ represent wave packet states, which at an appropriate place will be allowed to become sharp momentum states.

Our procedure is step-by-step the same as that in the nonrelativistic treatment. The matrix element of interest is

$$\langle \alpha', \beta'; o \mid \alpha, \beta; i\rangle = \langle \alpha', \beta'; o| \, S \, |\alpha, \beta; o\rangle = \langle \alpha', \beta'; i| \, S \, |\alpha, \beta; i\rangle, \quad (331)$$

where $|\alpha, \beta; i\rangle = a_\alpha^\dagger a_\beta^\dagger |0\rangle$, $|\alpha', \beta'; o\rangle = b_{\alpha'}^\dagger b_{\beta'}^\dagger |0\rangle$. We write [see Eq. (303)]

$$\begin{aligned}\langle \alpha', \beta'; o \mid \alpha, \beta; i\rangle &= \langle \alpha', \beta'; o| \, a_\alpha^\dagger \, |\beta\rangle \\ &= i\int d^3y \, \langle \alpha', \beta'; o| \, \phi_i(y) \, |\beta\rangle \frac{\overleftrightarrow{\partial}}{\partial y_0} f_\alpha(y). \quad (332)\end{aligned}$$

The time $y_0$ is arbitrary, but it is clear that if we want to utilize the asymptotic condition, it is advantageous to take it in the remote past. If we do so, $\phi_i(y)$ may be replaced by $\phi(y)$. Following the nonrelativistic pattern, we then replace the surface integral at $y_0 = -\infty$ by a surface integral at $+\infty$ together with an integral over all space-time.

$$\begin{aligned}\langle \alpha', \beta'; o \mid \alpha, \beta; i\rangle = \lim_{y_0 \to +\infty} i\int d^3y \, \langle \alpha', \beta'; o| \, \phi(y) \, |\beta\rangle \frac{\overleftrightarrow{\partial}}{\partial y_0} f_\alpha(y) \\ -i\int d^4y \frac{\partial}{\partial y_0}\left[\langle \alpha', \beta'; o| \, \phi(y) \, |\beta\rangle \frac{\overleftrightarrow{\partial}}{\partial y_0} f_\alpha(y)\right]. \quad (333)\end{aligned}$$

In the first term, again using the asymptotic condition, $\phi(y)$ may be replaced by $\phi_0(y)$ and this term reduces to

$$\begin{aligned}\langle \alpha', \beta'; o| \, b_\alpha^\dagger \, |\beta\rangle = \langle \alpha', \beta'; o \mid \alpha, \beta; o\rangle &= \delta_{\alpha'\alpha}\delta_{\beta'\beta} + \delta_{\alpha'\beta}\delta_{\beta'\alpha} \\ &\equiv \delta_{\alpha',\beta';\alpha\beta}. \quad (334)\end{aligned}$$

If there were no interaction between particles this would be the value of the $S$-matrix element. The second, and evidently the interesting, term is rearranged as follows: The structure is

$$\begin{aligned}\int d^4y \frac{\partial}{\partial y_0}\left[G \frac{\overleftrightarrow{\partial}}{\partial y_0} f_\alpha(y)\right] &= \int d^4y \left[G \frac{\partial^2 f_\alpha}{\partial y_0^2} - \frac{\partial^2 G}{\partial y_0^2} f_\alpha\right] \\ &= \int d^4y \left[G(\nabla^2 - \mu^2) f_\alpha - \frac{\partial^2 G}{\partial y_0^2} f_\alpha\right] \\ &= -\int d^4y [(\mu^2 - \Box)G] f_\alpha(y). \quad (335)\end{aligned}$$

In going from the first line to the second we have used the fact that $f_\alpha$

satisfies the Klein-Gordon equation

$$\frac{\partial^2 f_\alpha}{\partial {y_0}^2} = (\nabla^2 - \mu^2) f_\alpha.$$

The last line follows by two spatial integrations by parts. The latter step is justified because $f_\alpha(y)$ is a *wave packet*. We have then

$$\langle \alpha', \beta'; 0|\, \alpha, \beta; i\rangle = \delta_{\alpha',\beta';\alpha\beta} + i \int d^4y\, [(\mu^2 - \Box_y)\langle \alpha', \beta'; 0|\, \phi(y)\, |\beta\rangle]\, f_\alpha(y). \tag{336}$$

The next step is to "contract on $\alpha'$." We write

$$\begin{aligned} G(y) &= \langle \alpha', \beta'; o|\, \phi(y)\, |\beta\rangle = \langle \beta'|\, b_{\alpha'}\phi(y)\, |\beta\rangle \\ &= i \int d^3x\, f_{\alpha'}{}^*(x)\, \frac{\overleftrightarrow{\partial}}{\partial x_0}\, \langle \beta'|\, \phi_0(x)\phi(y)\, |\beta\rangle \\ &= \lim_{x_0 \to +\infty} \int d^3x f_{\alpha'}{}^*(x)\, \frac{\overleftrightarrow{\partial}}{\partial x_0}\, \langle \beta'|\, \phi(x)\phi(y)\, |\beta\rangle \\ &= \lim_{x_0 \to +\infty} \int d^3x f_{\alpha'}{}^*(x)\, \frac{\overleftrightarrow{\partial}}{\partial x_0}\, \langle \beta'|[\phi(x)\phi(y)]_+\, |\beta\rangle \end{aligned} \tag{337}$$

where in the last line we have taken the liberty of writing the time-ordering symbol since $x_0$ is always greater than $y_0$. Here again

$$\begin{aligned} [\phi(x)\phi(y)]_+ &\equiv \theta(x_0 - y_0)\phi(x)\phi(y) + \theta(y_0 - x_0)\phi(y)\phi(x) \\ &= \tfrac{1}{2}\{\phi(x), \phi(y)\} + \tfrac{1}{2}\epsilon(x_0 - y_0)[\phi(x), \phi(y)] \\ &= \theta(x_0 - y_0)[\phi(x), \phi(y)] + \phi(y)\phi(x). \end{aligned} \tag{338}$$

We manipulate $G(y)$ just as before, writing it as a surface integral at $x_0 = -\infty$ plus an integral over all space-time.

$$\begin{aligned} G(y) = i \int d^4x\, \frac{\partial}{\partial x_0} \left[ f_{\alpha'}{}^*(x)\, \frac{\overleftrightarrow{\partial}}{\partial x_0}\, \langle \beta'|[\phi(x)\phi(y)]_+\, |\beta\rangle \right] \\ + \lim_{x_0 \to -\infty} i \int d^3x f_{\alpha'}{}^*(x) \frac{\overleftrightarrow{\partial}}{\partial x_0}\, \langle \beta'|\, \phi(y)\phi(x)\, |\beta\rangle. \end{aligned} \tag{339}$$

From the asymptotic condition, the last term is simply $\langle \beta'|\, \phi(y)a_\alpha|\beta\rangle = \langle \beta'|\, \phi(y)\, |0\rangle\delta_{\alpha'\beta} = f_{\beta'}{}^*(y)\delta_{\alpha'\beta}$. The volume integral is handled just as before, using the fact that $f_{\alpha'}{}^*$ satisfies the Klein-Gordon equation and is a wave packet. We obtain

$$G(y) = f_{\beta'}{}^*(y)\delta_{\alpha'\beta} + i \int d^4x f_{\alpha'}{}^*(x)(\mu^2 - \Box_x)\langle \beta'|\, [\phi(x)\phi(y)]_+\, |\beta\rangle. \tag{340}$$

Substituting this into our formula for the $S$-matrix element [Eq. (336)] and noting that $(\mu^2 - \Box_y)$ annihilates the first term in $G(y)$, we obtain the final formula

$$\langle \alpha', \beta'; 0 \mid \alpha, \beta; i \rangle = \delta_{\alpha',\beta';\alpha\beta} + (i)^2 \int d^4x \int d^4y f_{\alpha'}{}^*(x) \times \{(\mu^2 - \Box_x)(\mu^2 - \Box_y) \langle \beta' | \, [\phi(x)\phi(y)]_+ \, |\beta\rangle\} f_\alpha(y). \quad (341)$$

At this point, since all partial integrations are completed, we may allow the $f$'s to become plane-wave states.

It is clear now how to contract further if so desired or how to reduce any general matrix element in the same fashion.

There is an alternative form for the reduction formula which is much more useful in dispersion theory. We proceed just as before to Eq. (337), but then deviate as follows:

$$\begin{aligned} G(y) &= \langle \beta' | \, b_{\alpha'}\phi(y) \, |\beta\rangle = \langle \beta' | \, [b_{\alpha'}, \phi(y)] \, |\beta\rangle + \langle \beta' | \, \phi(y) b_{\alpha'} \, |\beta\rangle \\ &= \langle \beta' | \, [b_{\alpha'}, \phi(y)] \, |\beta\rangle + f_{\beta'}{}^*(y)\delta_{\alpha'\beta} \qquad (342) \\ &= \lim_{x_0 \to +\infty} \int d^3x f_{\alpha'}{}^*(x) \frac{\overleftrightarrow{\partial}}{\partial x_0} \langle \beta' | \, \theta(x_0 - y_0)[\phi(x), \phi(y)] \, |\beta\rangle + f_{\beta'}{}^*(y)\delta_{\alpha'\beta}. \end{aligned}$$

We have used the asymptotic condition and taken the liberty of inserting $\theta(x_0 - y_0)$, which is $+1$ under this circumstance. At this point we manipulate just as before, but now the surface integral encountered vanishes by virtue of the theta-function. We find, ultimately,

$$\langle \alpha', \beta'; 0 \mid \alpha, \beta; i \rangle = \delta_{\alpha',\beta';\alpha\beta} + (i)^2 \int d^4x \int d^4y f_{\alpha'}{}^*(x) \{(\mu^2 - \Box_x) \times (\mu^2 - \Box_y) \langle \beta' | \, \theta(x_0 - y_0)[\phi(x), \phi(y)] \, |\beta\rangle f_\alpha(y)\}. \quad (343)$$

This form for the $S$-matrix element derives its usefulness from the fact that the commutator of two field operators appears explicitly and this quantity is assumed to vanish for spacelike separation of $x$, $y$. Thus in the limit of sharp states for $f_{\alpha'}{}^*$, $f_{\alpha'}$ the $S$-matrix element takes the form of a Fourier transform of a *retarded* [because of the appearance of $\theta(x_0 - y_0)$] commutator which vanishes except in the forward light cone of $x - y$. As we shall see, this feature plays a crucial role in the derivation of dispersion relations. We will explain later why the first form of the $S$-matrix element, which is clearly numerically equal to the second one, is unsuitable in dispersion theory.

**An Example of a Dispersion Relation in Field Theory**

As our first example of a dispersion relation in field theory we consider the scattering of one of our neutral mesons by an external potential, or alternatively, pair annihilation in such a potential. If we assume that the coupling is described by a term in a Lagrangian of the form $j(x)\phi_e(x)$, where $j(x)$ is an abbreviation for $(\mu^2 - \Box)\phi(x)$, the $S$-matrix element for scattering to first order in $\phi_e$ is proportional to $\langle q'|\, j(x)\, |q\rangle = \langle q'|\, j(0)\, |q\rangle e^{i(q-q')x}$. We use the meson momenta to label the states, although they must be properly considered to be wave packets. The phase factor is irrelevant, so we drop it. We then choose to study

$$I = \sqrt{2q_0(2q'_0)}\langle q'|\, j(0)\, |q\rangle,$$

where $q_0(q'_0)$ is the mean energy of the initial (final) meson wave packet. The factor $(4q_0q'_0)^{1/2}$ is included so that in the limit of sharp states $I$ is an invariant. To see this, we use the standard reduction procedure as given previously to obtain

$$I = i^2(4q_0q'_0)^{1/2} \int d^4x \int d^4y f_{q'}{}^*(x)\{(\mu^2 - \Box_x)(\mu^2 - \Box_y) \times \langle 0|\, [\phi(x)j(0)\phi(y)]_+\, |0\rangle\} f_q(y). \quad (344)$$

In the limit of sharp states

$$f_p(x) = \frac{1}{(2p_0)^{1/2}} e^{ip\cdot x}, \quad (345)$$

if we take unit quantization volume. The unsightly square roots are removed, and what remains will be invariant, provided $\phi$ and $j$ commute for spacelike separations. Since $I$ is invariant, it must be a function of $(q - q')^2$, $q^2$ and $q'^2$, but since $q^2 = q'^2 = -\mu^2$ we may take it to be a function of $(q - q')^2$ only.

For the pair annihilation process we would be led to consider instead the quantity $J$ defined as

$$J = (4q_{10}q_{20})^{1/2}\langle 0|\, j(0)\, |q_1, q_2; i\rangle. \quad (346)$$

Here we must, of course, specify the character of the initial two-particle state as an in state. (It was unnecessary to do this for $I$ because only one-particle states appeared.) Using the reduction formula on $I$ and $J$, we find (taking sharp states) that

$$I = i(2q_0)^{1/2} \int d^4x e^{-iq'\cdot x}(\mu^2 - \Box)\langle 0|\, [\phi(x)j(0)]_+\, |q\rangle,$$

$$J = i(2q_{20})^{1/2} \int d^4x\{(\mu^2 - \Box)\langle 0|\, [j(0)\phi(x)]_+\, |q_{20}\rangle\} e^{iq_1\cdot x}. \quad (347)$$

Now the order of factors is irrelevant in the time-ordered product, so that if $q_2 = q$ and $q_1 = -q'$ the expressions coincide. In a sense to be explained, we need only compute $I$ or $J$.

The problem involved here is as follows: Let us regard $I$ as a function of the squared momentum transfer $\xi = (q' - q)^2$. Since $q^2 = q'^2 = -\mu^2$, this is necessarily a spacelike vector; that is, $\xi \geq 0$. The substitution $q \to q_2$, $q' \to -q_1$ carries $\xi$ into $(q_1 + q_2)^2$, which is timelike since $q_{10}$, $q_{20} > 0$, and $I$ is not defined for such vectors. Thus we encounter a problem of analytic continuation which must be resolved before our prescription for obtaining $J$ from $I$ (or vice versa) has any meaning. In elementary field theory, where we might compute these quantities using perturbation theory, such connections are well-known and go under the name of *the substitution law*. Here we have an explicit formula and the analytic continuation is made by hand. Evidently, if we know $I$ for an arbitrary four vector $q'$ (that is, not restricted to the mass shell $q'^2 = -\mu^2$), no very involved argument is required.

It is most convenient to discuss $J$ in the form in which a retarded commutator appears, rather than the time-ordered product representation already given. The steps leading to the desired form are given.

$$
\begin{aligned}
J &= (4q_{10}q_{20})^{1/2}\langle 0|\, j(0)a_{q_1}{}^{\dagger}\,|q_2\rangle = (4q_{10}q_{20})^{1/2}\langle 0|\,[j(0)\; a_{q_1}{}^{\dagger}]\,|q_2\rangle \\
&= \lim_{x_0 \to -\infty} i(4q_{10}q_{20})^{1/2} \int d^3x \langle 0|\,[j(0), \phi(x)]\theta(-x_0)\,|q_2\rangle \frac{\overleftrightarrow{\partial}}{\partial x_0} f_{q1}(x) \\
&= i(2q_{20})^{1/2} \int d^4x \{(\mu^2 - \Box)\langle 0|\,[j(0), \phi(x)]\theta(-x_0)\,|q_2\rangle\} e^{iq_1 \cdot x}. \qquad (348)
\end{aligned}
$$

The great advantage of this representation is in the appearance of the retarded commutator, for it suggests (according to our heuristic arguments in the first part of this chapter) the possibility of a dispersion relation in the timelike vector $q_1$. The one in which the time-ordered product appears gives no such hint. Although the two forms coincide in the physical region, when extended into the complex plane they are quite different.

Next we carry out the indicated differentiations in $J$. We find[45]

$$
\begin{aligned}
(\mu^2 - \Box)[j(0), \phi(x)]\theta(-x_0) &= [j(0), j(x)]\theta(-x_0) - [j(0), \phi'(x)]\delta(x_0) \\
&\quad - \frac{\partial}{\partial x_0}\{[j(0), \phi(x)]\delta(x_0)\}. \qquad (349)
\end{aligned}
$$

When the last term is substituted into $J$ and an integration by parts performed, we obtain a surface term which vanishes because of $\delta(x_0)$, whereas the integrated term has an integrand $-iq_{10}[j(0), \phi(x)]\delta(x_0)\, e^{iq_1 \cdot x}$. In most Lagrangian theories such equal time commutators are zero since $j$ does not usually contain time derivatives of fields. In general, the second term is expected to be nonvanishing. In fact, the equal time commutator

[45] We are again using the notation $\phi'(x) \equiv \partial\phi/\partial x_0$.

would most likely be proportional to $\delta(\mathbf{x})$, so the entire term would involve $\delta(x) = \delta(\mathbf{x})\delta(x_0)$. In $J$, such a term would contribute a function of $q_2$ alone. The most general thing we might encounter from the last two terms of Eq. (349) would be a finite number of derivatives of $\delta$-functions, which when substituted into $J$ would lead to polynomials in $(q_1 \cdot q_2)$ times functions of $q_2$ alone. We may write then quite generally,

$$J = i(2q_{20})^{1/2} \int d^4x \langle 0| [j(0), j(x)]\theta(-x_0) |q_2\rangle e^{iq_1 \cdot x} + R, \tag{350}$$

where

$$R = -i(2q_{20})^{1/2} \int d^4x \langle 0| [j(0), \phi'(x)]\delta(x_0) + \frac{\partial}{\partial x_0} \{[j(0), \phi(x)]\delta(x_0)\}|q_2\rangle e^{iq_1 \cdot x}.$$

To study $J$ further, let us choose a special reference frame, the one in which $\mathbf{q}_2 = 0$. In this rest-frame of $q_2$, the scalar variable $q_1 \cdot q_2 = -\mu q_{10}$. Although it is convenient to exhibit ultimately $J$ as a function of $(q_1 + q_2)^2$, we temporarily regard it as a function of $q_{10}$. In our chosen frame, $\langle 0| [j(0), j(x)]\theta(-x_0) |q_2\rangle$ may be taken to be a function of $x^2$ and $x_0$ alone and the integration over the angles of $\mathbf{x}$ carried out explicitly, with the result (here $|\mathbf{x}| \equiv r$, and $|q_1| = (q_{10}{}^2 - \mu^2)^{1/2}$)

$$J(q_{10}) = i(2\mu)^{1/2} \int_0^\infty 4\pi r^2 \, dr \frac{\sin |\mathbf{q}_1|r}{|\mathbf{q}_1|r} \times \int_{-\infty}^{\infty} dx_0 \langle 0| [j(0), j(x)]\theta(-x_0) |q_2\rangle \, e^{-iq_{10}x_0} + R. \tag{351}$$

Note that the limits on $x_0$ are formal since $\theta(-x_0) = 0$ unless $x_0 < 0$ and since the commutator vanishes unless $x_0{}^2 \geq r^2$; the actual limits are therefore $-\infty$ and $-r$. Consider now the function $J(q_{10}, r)$, defined as

$$J(q_{10}, r) \equiv i(2\mu)^{1/2} 4\pi r^2 \frac{\sin |\mathbf{q}_1|r}{|\mathbf{q}_1|r} \times \int_{-\infty}^{\infty} dx_0 \langle 0| [j(0), j(x)]\theta(-x_0) |q_2\rangle e^{-iq_{10}x_0}. \tag{352}$$

We note that

$$\int_0^\infty dr J(q_{10}, r) = J(q_{10}) - R. \tag{353}$$

We wish to study $J(q_{10}, r)$ as a function of $q_{10}$, unrestricted by the physical requirement that $q_{10} > +\mu$, and ultimately will extend $J(q_{10})$ to complex values of $q_{10}$. It is easy to verify by contour integration that if $x_0 > r$, [see Eq. (14)]

$$\epsilon(-x_0)\frac{\sin|\mathbf{q}_1|r}{|\mathbf{q}_1|}e^{-iq_{10}x_0} = \frac{P}{i\pi}\int_{-\infty}^{\infty} dq'_{10}\frac{\sin|\mathbf{q}'_1|r}{|\mathbf{q}'_1|}\frac{e^{-iq'_{10}x_0}}{q'_{10}-q_{10}}. \qquad (354)$$

Let us then write $J(q_{10}, r) \equiv D(q_{10}, r) + iA(q_{10}, r)$, where

$$D(q_{10}, r) = \frac{i}{2}(2\mu)^{1/2}4\pi r^2\frac{\sin|\mathbf{q}_1|r}{|\mathbf{q}_1|r}\int_{-\infty}^{\infty} dx_0\langle 0|\,[j(0), j(x)]\epsilon(-x_0)\,|q_2\rangle e^{-iq_{10}x_0},$$

$$(355)$$

$$A(q_{10}, r) = \frac{1}{2}(2\mu)^{1/2}4\pi r^2\frac{\sin|\mathbf{q}_1|r}{|\mathbf{q}_1|r}\int_{-\infty}^{\infty} dx_0\langle 0|\,[j(0), j(x)]\,|q_2\rangle e^{-iq_{10}x_0}.$$

[This separation is just that used in connection with Eqs. (15).] Using the identity (354) in $D(q_{10}, r)$ and interchanging the order of integration [which will be legitimate provided $\langle 0|\,[j(0), j(x)]\,|q_2\rangle$ has no $\delta$-function singularities], we find that

$$D(q_{10}, r) = \frac{P}{\pi}\int_{-\infty}^{\infty} dq'_{10}\frac{A(q'_{10}, r)}{q'_{10}-q_{10}},$$

or

$$J(q_{10}, r) = \frac{1}{\pi}\int_{-\infty}^{\infty} dq'_{10}\frac{A(q'_{10}, r)}{q'_{10}-q_{10}-i\eta}. \qquad (356)$$

The final step in the derivation of a dispersion relation for $J(q_{10})$ itself consists of integrating over $r$. There is no difficulty in doing this for $J(q_{10}, r)$ when $q_{10} > \mu$, since the original representation is well behaved in this case. In the dispersion relation for $J(q_{10}, r)$, however, we encounter $A(q'_{10}, r)$ where $q'_{10} < \mu$ and the integral over $r$ obviously does not exist. To see that this trouble is only apparent and not real we write

$$\langle 0|\,[j(0), j(x)]\,|q_2\rangle = \sum_n [\langle 0|\,j(0)\,|n\rangle\langle n|\,j(\mathbf{x}, 0)\,|q_2\rangle e^{-i(\mu-p_{n0})x_0}$$
$$-\ \langle 0|\,j(\mathbf{x}, 0)\,|n\rangle\langle n|\,j(0)\,|q_2\rangle e^{-ip_{n0}x_0}]. \qquad (357)$$

Here we have introduced a complete set of states $|n\rangle$ in order to set $1 = \Sigma\,|n\rangle\langle n|$ and have used the time translation operator to exhibit the time dependence explicitly. In this expression $p_{n0}$ is the energy of the state $|n\rangle$. Carrying out the time integration involved in $A(q_{10}, r)$, we find

$$A(q_{10}, r) = \pi(2\mu)^{1/2}4\pi r^2\frac{\sin|\mathbf{q}_1|r}{|\mathbf{q}_1|r}\sum_n\{\langle 0|\,j(0)\,|n\rangle\langle n|\,j(\mathbf{x}, 0)\,|q_2\rangle$$
$$\times\,\delta(p_{n0}-q_{10}-\mu) - \langle 0|\,j(x, 0)\,|n\rangle\langle n|\,j(0)\,|q_2\rangle\delta(p_{n0}+q_{10})\}. \qquad (358)$$

Now the least massive state $|n\rangle$ encountered is that corresponding to one meson. But $\langle 0|\, j(x)\, |1 \text{ meson}\rangle = (\mu^2 - \Box)\langle 0|\, \phi(x)\, |1 \text{ meson}\rangle = 0$, since $\langle 0|\, \phi(x)\, |1 \text{ meson}\rangle$ is a solution of the Klein-Gordon equation. Thus we must begin with a state $|n\rangle$ which has at least two mesons in it and hence a minimum energy $p_{n0} = 2\mu$. We see then that the first $\delta$-function vanishes unless $q_{10} > \mu$, whereas the second vanishes unless $q_{10} < -2\mu$, and in the potentially dangerous region $|q_{10}| < \mu$ (where $|\mathbf{q}_1|$ is imaginary) $A(q_{10}, r)$ is identically zero. The integral over $r$ may now be evaluated with impunity, with the result that

$$J(q_{10}) = \frac{1}{\pi}\left(\int_{-\infty}^{-2\mu} + \int_{\mu}^{\infty}\right) dq'_{10}\, \frac{A(q'_{10})}{q'_{10} - q_{10} - i\eta} + R. \tag{359}$$

Here, we have set[46] $A(q'_{10}) = \int_0^\infty dr\, A(q'_{10}, r)$.

The region $-\infty < q'_{10} < -2\mu$ gives no contribution to $J(q_{10})$. This follows from the fact that when the $r$ integration is carried out, we obtain in that range of $q'_{10}$, $\mathbf{p}_n = -\mathbf{q}'_1$, where $\mathbf{p}_n$ is the momentum of the state $|n\rangle$. This, together with the energy $\delta$-function, leads to $p_n^2 = \mathbf{p}_n^2 - p_{n0}^2 = \mathbf{q}'^2_1 - q_{10}^2 = -\mu^2$. However, the only states $|n\rangle$ for which this is true are one-meson states, and then we again encounter $\langle 0|\, j\, |1 \text{ meson}\rangle = 0$. The $r$ integration for the range $\mu < q'_{10} < \infty$ leads to $\mathbf{p}_n = \mathbf{q}'_1$, which together with the energy $\delta$-function tells us that if $m_n$ is the mass of state $|n\rangle$, the true limit of the remaining term is $q'_{10} \geq -\mu + m_n^2/2\mu$. Here, since the minimum value of $m_n$ is $2\mu$, we find $q'_{10} \geq \mu$, as already written. [This need not have happened; if for some reason (perhaps a selection rule) we had to take $m_n = 3\mu$, the tentative limit would have been $2\mu$ and the actual one would be $-\mu + 9\mu/2$.]

Our dispersion relation for $J$ may be expressed finally in terms of the invariant variable $(q_1 + q_2)^2 \equiv \xi$, which in the rest-frame of $q_2$ is $\xi = -2\mu^2 - 2\mu q_{10}$. Thus

$$J(\xi) = \frac{1}{\pi}\int_{\mu}^{\infty} dq'_{10}(2\mu)\, \frac{A(-2\mu^2 - 2\mu q'_{10})}{2\mu q'_{10} + 2\mu^2 + \xi - i\eta} + R, \tag{360}$$

or with $\xi' = 2\mu q'_0 + 2\mu^2$,

$$J(\xi) = \frac{1}{\pi}\int_{(2\mu)^2}^{\infty} d\xi'\, \frac{A(-\xi')}{\xi' + \xi - i\eta} + R. \tag{361}$$

The variable $\xi'$ may be interpreted as the $(\text{mass})^2$ of the intermediate states which the external potential may create and which can lead to a two-meson state.

Before continuing with this example, let us review what has been assumed

[46] This device was first used by Oehme [31].

either explicitly or tacitly in the derivation. Aside from epsilonics, the only serious assumption was that $\langle 0|\,[j(0), j(x)]\,|q_2\rangle$ contained no $\delta$-function singularities, such as $\delta(x^2)$. If this is correct, then $J$ approaches $R$ for large $\xi$. In most common instances $R$ itself will be a constant. There are various other alternatives possible. The original expression (351) for $J - R$ might approach a constant $K$, for example, as $\xi \to \infty$, and then our dispersion representation would have $K + R$ appearing in place of $R$. It is sometimes possible to identify $K + R$ with what old-fashioned field theorists call an *unrenormalized* coupling constant. It is sometimes even conjectured that $K + R = 0$, in which case we can compute $J(-\mu^2)$, which is conventionally called the *renormalized* coupling constant. A less daring assumption is that the commutator is so singular that the representation we have written does not exist and we must make one or more subtractions. Calling $J(-\mu^2) \equiv g$, we might then write (assuming that $R$ is a constant)

$$J(\xi) = g - \frac{\xi + \mu^2}{\pi} \int_{(2\mu)^2}^{\infty} d\xi' \, \frac{A(-\xi')}{(\xi' - \mu^2)(\xi' + \xi - i\eta)}, \tag{362}$$

which by construction satisfies the condition that $J(-\mu^2) = g$.

We have succeeded in deriving a dispersion relation for $J(\xi)$ and by this relation we may extend $J$ as a function of the complex variable $\xi$, analytic in the entire $\xi$-plane with a cut from $-\infty$ to $-4\mu^2$ along the real axis. The physically interesting value is obtained as the boundary value of this function from *below* the cut. It is clear that we have also succeeded in obtaining the function $I$ of Eq. (347), since our representation is valid for *all* $\xi$.

In order to illustrate how dispersion methods may be used to perform dynamical calculations, we pursue this example somewhat further. It is useful for this purpose to show that with $J(\xi) = D(\xi) + iA(\xi)$, $D$ and $A$ are real quantities. To see this we go again to the rest-frame of $q_2$ and write

$$D^*(q_{10}) = -i(2\mu)^{1/2} \int d^4x \left\{ (\mu^2 - \Box_x) \langle 0|\,[j(0), \phi(x)]\, \frac{\epsilon(-x_0)}{2}\, | q_2\rangle^* \right\} e^{-iq_1 \cdot x}. \tag{363}$$

(In this way of writing we include the terms previously labeled as $R$.) Using the fact that under the time reversal operation $T$, $T\,|0\rangle = |0\rangle$, we have

$$\langle 0|\,[j(0), \phi(x)]\,|q_2\rangle^* = \langle 0|\, T[j(0), \phi(x)]T^{-1}T\,|q_2\rangle. \tag{364}$$

Now for a neutral scalar field, we may choose phases so that

$$\begin{aligned} T\phi(x)T^{-1} &= \phi(\mathbf{x}, -x_0), \\ Tj(x)T^{-1} &= j(\mathbf{x}, -x_0) \end{aligned} \tag{365}$$

[since $j(x) = (\mu^2 - \Box)\phi(x)$]. It is consistent to take, for a particle at rest, $T\,|q_2\rangle = |q_2\rangle$. Hence

$$\langle 0|\,[j(0),\, \phi(x)]\,|q_2\rangle^* = \langle 0|\,[j(0),\, \phi(\mathbf{x},\, -x_0)]\,|q_2\rangle. \tag{366}$$

Substituting this into $D^*(q_{10})$ and changing $x \to -x$, we find

$$D^*(q_{10}) = i(2\mu)^{1/2} \int d^4x \left\{ (\mu^2 - \Box_x)\, \langle 0|\,[j(0),\, \phi(-\mathbf{x},\, x_0)]\, \frac{\epsilon(-x_0)}{2}\,|q_2\rangle \right\} e^{iq_1\cdot x}. \tag{367}$$

Finally, using the parity operator $\mathscr{P}$, with the property $\mathscr{P}^{-1}\phi(-\mathbf{x}, x_0)\mathscr{P} = \phi(\mathbf{x}, x_0) = \phi(x)$, and the fact that for a particle at rest $\mathscr{P}\,|q_2\rangle = |q_2\rangle$, we conclude $D^*(q_{10}) = D(q_{10})$. In an identical fashion we can prove that $A^*(q_{10}) = A(q_{10})$. Incidentally, in the standard terminology, $D$ is called the *dispersive* part of $J$ and $A$ the *absorptive* part.

A dynamical calculation evidently requires a computation of the absorptive amplitude $A$. To do this we introduce a complete set of states and, now that we know it is legitimate, fearlessly integrate over space and time immediately. We obtain, in an arbitrary Lorentz frame,

$$A = \pi(2q_{20})^{1/2} \sum_n \langle 0|\, j\,|n\rangle\, \langle n|\, j\,|q_2\rangle\, \delta(p_{n0} - q_{10} - q_{20})\, \delta_{\mathbf{p}_n,\mathbf{q}_1+\mathbf{q}_2}. \tag{368}$$

The states $|n\rangle$ may be conveniently chosen as in or out states, and $A$ will be real *provided* we sum over a complete set of states. If we wish every term in the sum to be manifestly real, we may take half the sum over *both* in and out states. [Note that since $j$ is a scalar under rotations, $\langle 0|\, j\,|n\rangle$ is nonzero only if $|n\rangle$ has angular momentum zero, that is, only $s$-states appear in the sum.] If the total energy of the meson pair, $q_{10} + q_{20}$, is less than $3\mu$ only the two-meson state makes a contribution to the sum [remember that the one-meson state drops out because $\langle 0|\, j\,|1 \text{ meson}\rangle = 0$]. Writing $|n\rangle = |p_1, p_2; i\rangle$, we encounter $\langle 0|\, j\,|p_1, p_2; i\rangle\langle p_1, p_2, i|\, j\,|q_2\rangle$. The first factor is just $J/(4q_{10}q_{20})^{1/2}$, whereas the second one is proportional to the complex conjugate of the $J$-matrix element for the process $p_1 + p_2 \to q_1 + q_2$. Since we are concerned only with $s$-waves, this is proportional to $e^{[(i\delta)\sin\delta]^*} = e^{-i\delta}\sin\delta$, at least in the region $4\mu^2 < -(q_1 + q_2)^2 < 9\mu^2$, where the phase shift $\delta$ is real. Since $A$ itself must be real, and in this energy region only the two meson state contributes, we conclude that the phase of $\langle 0|\, j\,|p_1, p_2; i\rangle$ is just $\delta$.[47] We shall see that there is a fairly natural generalization of this relation when $\delta$ is complex.

Suppose that we restrict our attention to two-particle intermediate states and actually evaluate $A$. With our quantization in a box of unit

[47] This same conclusion was obtained in Chapter 9, p. 533.

volume this becomes

$$A = \frac{\pi}{2}(2q_{20})^{1/2} \int \frac{d^3p_1}{(2\pi)^3}\, d^3p_2$$
$$\times \langle 0|\, j\, |p_1, p_2; i\rangle \langle p_1, p_2; i|\, j\, |q_2\rangle\, \delta(p_1 + p_2 - q_1 - q_2). \quad (369)$$

The apparently mysterious factor of $\frac{1}{2}$ arises from the completeness relation for two-particle states discussed earlier. [See the remark following Eq. (299).] The first matrix element in $A$ is simply related to $J$.

$$\langle 0|\, j\, |p_1, p_2; i\rangle = \frac{J[(p_1 + p_2)^2]}{(4p_{10}p_{20})^{1/2}}. \quad (370)$$

We next express $\langle p_1, p_2; i|\, j\, |q_2\rangle$ in terms of a scattering amplitude. From our fundamental reduction formula (336) we note that

$$\langle q_1, q_2; \text{o} \mid p_1, p_2; i\rangle = \langle q_1, q_2;\ \text{o}|\ p_1, p_2; \text{o}\rangle + \frac{i}{(2q_{10})^{1/2}} \int d^4x e^{-iq_1\cdot x}$$
$$\times \langle q_2|\, j(x)\, |p_1, p_2; i\rangle. \quad (371)$$

Using translational invariance to carry out the integration, this becomes

$$\langle q_1, q_2; \text{o} \mid p_1, p_2; i\rangle = \langle q_1, q_2; \text{o} \mid p_1, p_2; \text{o}\rangle$$
$$+ \frac{2\pi i\, \delta(q_{10} + q_{20} - p_{10} - p_{20})}{(2q_{10})^{1/2}}\, \delta_{\mathbf{q}_1+\mathbf{q}_2,\mathbf{p}_1+\mathbf{p}_2} \langle q_2|\, j(0)\, |p_1, p_2; i\rangle. \quad (372)$$

From the general theory of Chapter 8, the total transition probability per unit time is seen to be[48]

$$\dot{P} = \frac{1}{2} \sum_{q_1,q_2} 2\pi \left| \frac{1}{(2q_{10})^{1/2}} \langle q_2|\, j(0)\, |p_1, p_2; i\rangle \right|^2$$
$$\times\ \delta(q_{10} + q_{20} - p_{10} - p_{20})\, \delta_{\mathbf{q}_1+\mathbf{q}_2,\mathbf{p}_1+\mathbf{p}_2}. \quad (373)$$

[The origin of the factor of $\frac{1}{2}$ is the same as for Eq. (369).] This leads to the total scattering cross section, conveniently evaluated in the center-of-mass system,

$$\sigma = \int d\Omega\, \frac{p_{10}}{4} \left(\frac{1}{4\pi}\right)^2 |\langle q_2|\, j\, |p_1, p_2; i\rangle|^2. \quad (374)$$

Now, as we have remarked, in the evaluation of $A$ we have to deal with only the $s$-wave amplitude; in terms of the $s$-wave phase shift $\delta$, the cross section for this case of identical particles is

$$\sigma = \int \frac{d\Omega}{2} \left| \frac{2e^{i\delta} \sin \delta}{|\mathbf{p}_1|} \right|^2. \quad (375)$$

[48] See Eq. (8.126$a$).

From this we infer that

$$\langle q_2|\, j\, |p_1, p_2; i\rangle = 8\pi \frac{\sqrt{2}e^{i\delta}\sin\delta}{|\mathbf{p}_1|\,\sqrt{p_{10}}}. \tag{376}$$

Putting the pieces together, we obtain

$$A(\xi) = J(\xi)(e^{i\delta}\sin\delta)^*\theta(-\xi - 4\mu^2), \tag{377}$$

where $\delta$ is to be regarded as a function of the wave number $(-\xi/4 - \mu^2)^{1/2}$, with $\xi = (q_1 + q_2)^2$. If we remember our admonition to take one-half the sum of in and out intermediate states, we will finally obtain

$$A(\xi) = \mathrm{Re}\,[J(\xi)(e^{i\delta}\sin\delta)^*]\theta(-\xi - 4\mu^2), \tag{378}$$

since $\langle 0|\, j\, |p_1, p_2; o\rangle = \langle 0|\, j\, |p_1, p_2; i\rangle^*$. This relation follows from time reversal invariance and the fact that $T\,|\mathbf{p}_1, \mathbf{p}_2; 0\rangle = |-\mathbf{p}_1, -\mathbf{p}_2; i\rangle$, together with the observation that $J$ is a function of $(p_1 + p_2)^2$.

In this approximation of taking only two-particle intermediate states, and writing $J = D + iA$ as before, we have

$$\begin{aligned} A(\xi) &= \frac{\mathrm{Re}\,(e^{i\delta}\sin\delta)}{1 - \mathrm{Im}\,e^{i\delta}\sin\delta}\, D(\xi)\theta(-\xi - 4\mu^2), \\ &= \tan\phi(-\xi)\, D(\xi)\theta(-\xi - 4\mu^2), \end{aligned} \tag{379}$$

where

$$\tan\phi(-\xi) \equiv \frac{\mathrm{Re}\,(e^{i\delta}\sin\delta)}{1 - \mathrm{Im}\,e^{i\delta}\sin\delta}. \tag{380}$$

Evidently, if $\delta$ is real, as it will be $4\mu^2 < -\xi < 9\mu^2$, $\phi(-\xi) = \delta$ (to within a multiple of $\pi$) the angle $\phi$ provides the desired generalization of the final state theorem for complex $\delta$. Inserting $A$ into the dispersion relation (362), we find

$$J(\xi) = g - \frac{\xi + \mu^2}{\pi}\int_{(2\mu)^2}^{\infty} d\xi' \frac{\tan\phi(\xi')\, D(-\xi')}{(\xi' - \mu^2)(\xi' + \xi - i\eta)}. \tag{381}$$

This is a mapping problem in complex variable theory: $J(\xi)$ may be extended into the complex $\xi$-plane and is analytic everywhere except for a cut from $-\infty$ to $-4\mu^2$. Just below the cut we have $A(\xi) = \tan\phi(-\xi)D(\xi)$; we seek a solution for which $\tan\phi(\xi)D(-\xi)/\xi \to 0$, for large $\xi$, faster than $1/\ln\xi$. A solution which satisfies all these conditions, except perhaps for the one at infinity (which requires some knowledge of $\phi$), is

$$J(\xi) = g\exp\left[-\frac{\xi + \mu^2}{\tau}\int_{4\mu^2}^{\infty} d\xi' \frac{\phi(\xi')}{(\xi' - \mu^2)(\xi' + \xi - i\eta)}\right]. \tag{382}$$

More generally, $g$ could be replaced by a polynomial $P(\xi)$ with the property $P(-\mu^2) = g$. The preceding solution, however, is the only one which coincides with the iterative solution of the integral equation. In Appendix $G$ we discuss the problem of solving integral equations of this type which are frequently encountered in dispersion theory.

As a final remark concerning this solution, we suppose it were true that

$$\phi(\xi) = \frac{1}{2i} \ln \frac{\mathscr{D}^*(\xi)}{\mathscr{D}(\xi)}, \tag{383}$$

where $\mathscr{D}(\xi)$ is analytic in the $\xi$-plane cut from $-\infty$ to $-4\mu^2$. Furthermore, we assume that $\mathscr{D}$ has no zeros. This would be the case if $\phi(-\xi) = \delta$, with $\delta$ real, and with the scattering amplitude of the form

$$\frac{e^{i\delta} \sin \delta}{|\mathbf{p}|} = \frac{\mathscr{N}(-\xi)}{\mathscr{D}(-\xi)} \tag{384}$$

[as discussed in connection with the nonrelativistic Mandelstam representation, Eq. (255)]. We would then find

$$J(\xi) = g \frac{\mathscr{D}(\mu^2)}{\mathscr{D}(-\xi)}. \tag{385}$$

Our final solution for $J$, either in this form or as given in Eq. (382), is the dispersion theoretic version of the final state interaction theorem as has been discussed in Chapter 9 and is essentially the same result.

Precisely the same techniques employed in this example have been used successfully in studies of the electromagnetic structure of nucleons,[49] the decay of charged[50] and neutral[51] mesons, the influence of strong interactions in $\beta$-decay,[52] and the calculation of binding energies by dispersion theory.[53]

**Dispersion Relations for Forward Scattering**

We take up now the problem of deriving dispersion relations for two-particle scattering processes. We restrict the bulk of our attention to forward scattering and merely comment on (but do not treat in detail) the very much more complicated case of nonforward scattering. In order to avoid what are euphemistically referred to as "unessential complications," we treat first the scattering of neutral spin zero particles of mass $\mu$. We then take up somewhat more briefly the much more interesting case of pion-nucleon scattering. It is in connection with the latter that dispersion theory has scored its most clear-cut triumph, where it has played a significant role in the unraveling of a very complex experimental situation.

[49] See Ref. [32].
[50] See Ref. [33].
[51] See Ref. [34].
[52] See Ref. [35].
[53] See Ref. [36].

The $S$-matrix element for the process $p + k \to p' + k'$ has been given in Eq. (372), which we repeat here for convenient reference.

$$\langle p', k'; o \mid p, k; i\rangle = \langle p', k'; o \mid p, k; o\rangle + \frac{2\pi i\, \delta(p'_0 + k'_0 - p_0 - k_0)}{(2k_0)^{1/2}}\, \delta_{\mathbf{p}'+\mathbf{k}',\mathbf{p}+\mathbf{k}}\, \langle p', k'; o|\, j(0)\, |p\rangle. \quad (386)$$

It is convenient to define also what is sometimes called the *causal amplitude* $M(p', k'; p, k)$ by rewriting (386) in the form

$$\langle p', k'; o \mid p, k; i\rangle = \langle p', k'; o \mid p, k; o\rangle + 2\pi i\, \delta(p'_0 + k'_0 - p_0 - k_0)\, \delta_{\mathbf{p}'+\mathbf{k}',\mathbf{p}+\mathbf{k}}\, \frac{M(p', k'; p, k)}{(16 p'_0 k'_0 p_0 k_0)^{1/2}}, \quad (387)$$

where

$$M(p', k'; p, k) = (8p'_0 k'_0 p_0)^{1/2}\, \langle p', k'; o|\, j(0)\, |p\rangle = i(2p'_0 p_0)^{1/2} \int d^4x e^{-ik'\cdot x}(\mu^2 - \Box_x)\, \langle p'|\, [\phi(x), j(0)]\theta(x_0)\, |p\rangle. \quad (388)$$

We have used the reduction formula just as in Eq. (342) to exhibit the dependence of $M$ on $k'$ explicitly. Furthermore, we have gone to the limit of plane-wave states. Carrying out the differentiations as we did in connection with (348), we find

$$M = i(4p'_0 p_0)^{1/2} \int d^4x e^{-ik'\cdot x}\, \langle p'|\, [j(x), j(0)]\theta(x_0)\, |p\rangle + R,$$

$$R = i(4p'_0 p_0)^{1/2} \int d^4x e^{-ik'\cdot x}\, \langle p'|\, [\phi'(x), j(0)]\, \delta(x_0) + \frac{\partial}{\partial x_0}\{[\phi(x), j(0)]\, \delta(x_0)\}\, |p\rangle. \quad (389)$$

For simplicity we assume that $R$, which might, in general, be a polynomial in the components of $k'$, is independent of $k'$. (This is not an important point since such polynomials may always be taken care of by subtractions in the dispersion relation to be obtained.)

We are interested in forward scattering, where then $k' = k$, $p' = p$. This must be considered as the limit of nonforward scattering. To avoid the incorrect inclusion of vacuum fluctuation effects, we must subtract from Eq. (389) for forward scattering the same expression with the states

$|p\rangle$ replaced by the vacuum state $|0\rangle$.[54] For brevity we shall not make the subtraction explicitly, however, since it in no way affects our formal arguments.

It is instructive to compare the structure of $M$ to that which obtains if we were to use that form of the reduction formula leading to $[j(x)j(0)]_+$ in place of the retarded commutator in Eq.(389). If we call this new amplitude $F$ and introduce into it and into $M$ a complete set of states as in Eq. (357), we obtain (on carrying out the space-time integrations) for forward scattering,[55]

$$M = (4p_0{}^2)^{1/2} \sum_n |\langle p| j |n\rangle|^2 \left( \frac{\delta_{\mathbf{p}_n, \mathbf{p}+\mathbf{k}}}{p_{n0} - p_0 - k_0 - i\eta} + \frac{\delta_{\mathbf{p}_n, \mathbf{p}-\mathbf{k}}}{p_{n0} - p_0 + k_0 + i\eta} \right) + R$$

$$F = (4p_0{}^2)^{1/2} \sum_n |\langle p| j |n\rangle|^2 \left( \frac{\delta_{\mathbf{p}_n, \mathbf{p}+\mathbf{k}}}{p_{n0} - p_0 - k_0 - i\eta} + \frac{\delta_{\mathbf{p}_n, \mathbf{p}-\mathbf{k}}}{p_{n0} - p_0 + k_0 - i\eta} \right) + R. \tag{390}$$

These formula are identical except for the sign of $\eta$ in the second terms of the sums. [The $\eta$'s incidentally appear in the course of carrying out the time integrations, where to secure convergence we insert an exponential damping factor. We, of course, imagine that the limit $\eta \rightarrow 0(+)$ is to be understood.] As long as $k_0$ is physical, that is, $k_0 > +\mu$, the second denominator can never vanish: For it to vanish we must have $p_{n0}{}^2 = (p_0 - k_0)^2$, which is equivalent to the equation $(p - k)^2 + m_n{}^2 = 0$ in which $m_n$ is the mass of the state $|n\rangle$; but $(p - k)^2$ is a spacelike vector so $(p - k)^2 + m_n{}^2 > 0$. Consequently, whether there is an additive term $\pm i\eta$ is irrelevant. If, however, we allow $k_0$ to be negative or complex, there is a profound difference between $M$ and $F$. We see, for example, that both denominators in $M$ have their zeros in the lower half $k_0$-plane, whereas in $F$ the denominator of the first term vanishes in the lower half plane while that of the second vanishes in the upper half plane. Thus there is some hope that we may prove $M$ to be analytic in the upper half $k_0$-plane; there is no possibility, on the other hand, of having $F$ analytic in either half plane.

In order to prove that $M$ indeed satisfies a dispersion relation, we study it in the rest-frame of $\mathbf{p}$, in which case it may be regarded as a function of

[54] Actually, if the separation of $M$ into two terms is not made, this precaution is unnecessary, since

$$i \int d^4x e^{-ik\cdot x}(\mu^2 - \Box)\langle 0 | [\phi(x), j(0)]\theta(x_0) | 0 \rangle \equiv 0.$$

The separation is, however, rather convenient and, of course, if any approximations are made the vanishing of the preceding term may not be assured.

[55] The quantity $R$ in Eqs. (390) is just that defined in Eqs. (389).

the energy $k_0$ alone. The magnitude of $\mathbf{k}$ is, of course, given by $(k_0{}^2 - \mu^2)^{1/2}$. The true scalar variable on which $M$ depends in an arbitrary frame may be taken to be $-p \cdot k/\mu$, which reduces to $k_0$ in the rest-frame of $\mathbf{p}$. We note, incidentally, that $M$ is a scalar quantity, the factors of $(4p_0p'_0)^{1/2}$ introduced in the original definition being just what are required for this. [See the corresponding argument in connection with Eq. (344).] We now separate $M$ into dispersive and absorptive parts by writing $\theta(x_0) = [1 + \epsilon(x_0)]/2$, the "one" leading to the absorptive part, whereas $\epsilon(x_0)$ leads to the dispersive part. Then Eq. (388) and the first of Eqs. (389) may be written as

$$M(k_0) = D(k_0) + R + iA(k_0), \tag{391}$$

where

$$\begin{aligned} D(k_0) + R &= \frac{i}{2}(4\mu^2)^{1/2}\int d^4x e^{-ik\cdot x}(\mu^2 - \Box_x)\langle p|\,[\phi(x), j(0)]\epsilon(x_0)\,|p\rangle, \\ D(k_0) &= \frac{i}{2}(4\mu^2)^{1/2}\int d^4x e^{-ik\cdot x}\langle p|\,[j(x), j(0)]\epsilon(x_0)\,|p\rangle, \\ A(k_0) &= \frac{1}{2}(4\mu^2)^{1/2}\int d^4x e^{-ik\cdot x}\langle p|\,[j(x), j(0)]\,|p\rangle. \end{aligned} \tag{392}$$

It is quite straightforward to show that $D$, $R$, and $A$ are all real.

There is an important symmetry property that $M$ satisfies; namely, $M(-k_0) = M^*(k_0)$. The demonstration is facilitated by noting that $M$ is independent of the sign of $\mathbf{k}$. This physically obvious fact can be verified explicitly by noting that in the rest system of $\mathbf{p}$, the matrix element $(\mu^2 - \Box_x)\langle p|\,[\phi(x), j(0)]\theta(x_0)\,|p\rangle$ is a function of $x^2$ and $x_0$ alone; thus the integration over the angles of $\mathbf{x}$ in (389) may be carried out and a dependence only on $|\mathbf{k}|$ remains. If we now simply form $M^*(-k_0)$ and simultaneously change the sign of $\mathbf{k}$, the result is immediate. This relation is the analogue of the so-called *crossing symmetry*, which holds for the amplitude defined in terms of time-ordered operators. Quite generally, for non-forward scattering, we have

$$\begin{aligned} 2\pi\delta(p'_0 + k'_0 - p_0 - k_0)\,\delta_{\mathbf{p}'+\mathbf{k}',\mathbf{p}+\mathbf{k}}F(p', k'; p, k) \\ = i(4p'_0p_0)^{1/2}\int d^4x\int d^4y e^{-ik'\cdot x} \\ \times \{(\mu^2 - \Box_x)(\mu^2 - \Box_y)\langle p'|\,[\phi(x)\phi(y)]_+\,|p\rangle\}e^{ik\cdot y}. \end{aligned} \tag{393}$$

We see from this that, as an arithmetic identity,

$$F(p', k'; p, k) = F(p', -k; p, -k'). \tag{394}$$

This relation (first given by Gell-Mann and Goldberger [37]) as well as the one just proven for $M(k_0)$ [the corresponding generalization to non-forward scattering for $M$ is $M(p', k'; p, k) = M^*(p', -k; p, -k')$] must be carefully distinguished from symmetry operations relating physical processes. They are formal consequences of our definitions of these quantities and in that sense depend critically on the existence of a local field $\phi(x)$ which interpolates between $\phi_i(x)$ and $\phi_0(x)$. They cannot be given an abstract meaning in terms of operations on physical states, since it is difficult to produce particles in the laboratory with negative energy. Furthermore, we have no assurance that they can be realized in practice; we must be able to prove that it is possible to continue $M(k_0)$, for example, analytically from the physical domain of its original definition $k_0 > \mu$ to negative values of $k_0$. The same derivation that will lead to the dispersion relation will, however, prove that this continuation exists.

We next write, just as in Section 10.6, p. 638

$$M(k_0) - R = \int_0^\infty dr M(k_0, r),$$
$$M(k_0, r) = i(4\mu^2)^{1/2} 4\pi r^2 \frac{\sin |\mathbf{k}|\, r}{|\mathbf{k}|\, r} \int_{-\infty}^{\infty} dx_0 e^{ik_0x_0} \langle p|\, [j(x), j(0)]\theta(x_0)\, |p\rangle, \tag{395}$$

where $|\mathbf{k}| = (k_0^2 - \mu^2)^{1/2}$, and where we have carried out the integration over the angles of $\mathbf{x}$ in Eq. (389). Since the commutator fails to vanish only for $x^2 < 0$, $x_0 > 0$, we see that, just as in the corresponding discussion of Eq. (352), $M(k_0, r)$ may be continued into the upper half $k_0$-plane and a dispersion relation written for it. Following the argument leading to Eq. (356), again assuming that the commutator is not pathological, we find

$$D(k_0, r) = \frac{P}{\pi} \int_{-\infty}^{\infty} dk'_0 \frac{A(k'_0, r)}{k'_0 - k_0},$$
$$M(k_0, r) = \frac{1}{\pi} \int_{-\infty}^{\infty} dk'_0 \frac{A(k'_0, r)}{k'_0 - k_0 - i\eta}, \tag{396}$$

where $D(k_0, r)$ and $A(k_0, r)$ are defined in an obvious way from (392) by analogy with (395).

In order to get a physical dispersion relation we must integrate over $r$. There is no difficulty in doing this as long as $k_0$, $k'_0$ in (396) are greater in absolute value than $\mu$. The range $-\mu < k'_0 < \mu$ is superficially dangerous, and we can conclude at the moment only that

$$M(k_0) - R = \frac{1}{\pi} \int_0^\infty dr \int_{-\mu}^{\mu} dk'_0 \frac{A(k'_0, r)}{k'_0 - k_0 - i\eta} + \frac{1}{\pi} \left( \int_{-\infty}^{-\mu} + \int_{\mu}^{\infty} \right) dk'_0 \frac{A(k'_0)}{k'_0 - k_0 - i\eta}. \tag{397}$$

To see how deep the trouble is, let us express $A(k_0, r)$ in terms of a sum over states:

$$A(k_0, r) = \pi(4\mu^2)^{1/2} 4\pi r^2 \frac{\sin |\mathbf{k}|\, r}{|\mathbf{k}|\, r} \sum_n |\langle p|\, j\, |n\rangle|^2 \frac{\sin |\mathbf{p}_n|\, r}{|\mathbf{p}_n|\, r}$$

$$\times\, [\delta(k_0 + \mu - p_{n0}) - \delta(\mu - k_0 - p_{n0})]. \quad (398)$$

[We have used the fact that since $\langle p|\, [j(x), j(0)]\, |p\rangle$ depends only on $\mathbf{x}^2 = r^2$, it may be replaced by its angular average.] Here $|\mathbf{k}| = (k_0^2 - \mu^2)^{1/2}$ and $|\mathbf{p}_n| = (p_{n0}^2 - m_n^2)^{1/2}$, where $m_n$ is the mass of the state $|n\rangle$. The least massive state $|n\rangle$ is that of one meson, so the first $\delta$-function may contribute for $k_0 > 0$, the second for $k_0 < 0$. We are therefore in real trouble. It is interesting to note that if we assume that our particles have negative intrinsic parity, that is, they are pseudoscalar rather than scalar mesons, this difficulty disappears. The reason is that $\langle p|\, j\, |1 \text{ meson}\rangle = 0$ in this case. To see this, we remark that if $\mathbf{p} = 0$, and $j$ is a scalar under rotations, then the state $j\,|p\rangle$ has angular momentum zero and even parity. The one-meson state, which of necessity has angular momentum zero, has odd parity and the matrix element vanishes. Thus the least massive state that occurs is that involving two mesons. As a result, since $p_{n0} \geq 2\mu$, the dangerous region for $A(k'_0, r)$ is harmless and the final form of the dispersion relation is

$$M(k_0) = \frac{1}{\pi}\left(\int_{-\infty}^{-\mu} + \int_{\mu}^{\infty}\right) dk'_0 \frac{A(k'_0)}{k'_0 - k_0 - i\eta} + R. \quad (399)$$

From this representation it is clear that we can define $M(k_0) - R$ as a function of a complex variable $k_0$ which is analytic in the entire $k_0$-plane, except for cuts along the real axis $-\infty < k_0 < -\mu$ and $\mu < k_0 < \infty$.

It is worthwhile pointing out that the trouble we encounter with scalar mesons (for which there is an allowed one-particle intermediate state) is the rule rather than the exception. Suppose, for example, $|p\rangle$ described a one-nucleon state of mass $m > \mu$ and $j$ had the property of conserving nucleon number. Then the intermediate state $|n\rangle$ would have at least one nucleon. The $\delta$-functions in (398) would be replaced by $\delta(m + k_0 - p_{n0})$ and $\delta(m - k_0 - p_{n0})$, and if $m_n = m$, these arguments can vanish for $k_0$ in the range $-\mu < k_0 < \mu$. For the higher mass states, say one nucleon and one meson, there is no difficulty.

The fact that in our scalar meson problem only the one-particle state gives trouble suggests that a knowledge of the properties of $\langle p|\, j |1 \text{ meson}\rangle$ might enable us to circumvent the difficulty. We, can in fact, use the known analyticity of this matrix element to deform the $k'_0$ integration in the first term of (397) to a place where the $r$ integration may be carried out. This is the manner in which Symanzik first gave a rigorous proof of the pion-nucleon dispersion relations [38]. The method is, however, rather tedious.

We adopt instead a device used previously in the non-relativistic theory and first suggested by Bogoliubov [30]. We introduce into the original representation for $M - R$ in place of $|\mathbf{k}| = (k_0{}^2 - \mu^2)^{1/2}$, $|\mathbf{k}| = (k_0{}^2 - \zeta)^{1/2}$ and assume $\zeta < 0$. Since then there are no increasing exponentials, we can interchange orders of integration in a carefree fashion and deduce that

$$M(k_0, \zeta) - R = \frac{1}{\pi}\int_{-\infty}^{\infty} dk'_0 \frac{A(k'_0, \zeta)}{k'_0 - k_0 - i\eta}. \tag{400}$$

The parametric dependence on $\zeta$ is explicitly indicated. The absorptive part $A(k_0, \zeta)$ becomes, in the rest-frame of $\mathbf{p}$,

$$A(k_0, \zeta) = \pi(4\mu^2)^{1/2} \sum_n |\langle p|\, j\, |n\rangle|^2 \times [\delta(\mu + k_0 - p_{n0})\, \delta_{\mathbf{p}_n,\mathbf{k}} - \delta(\mu - k_0 - p_{n0})\, \delta_{\mathbf{p}_n,-\mathbf{k}}]. \tag{401}$$

Now, because we also have spatial momentum conservation as well as energy conservation, the first $\delta$-function is singular at $k_0 = (m_n{}^2 - \zeta - \mu^2)/2\mu$, whereas the second is singular at $k_0 = -(m_n{}^2 - \zeta - \mu^2)/2\mu$. There are discrete contributions coming from the one-particle state, for which $m_n = \mu$, and a continuum beginning with $k_0 = \pm(3\mu^2 - \zeta)/2\mu$ from the two and higher meson states. The matrix element $\langle p|\, j\, |1 \text{ meson}\rangle$ may be expressed in terms of the quantity $J$ of Section 10.6, p. 635. We note for the one-meson state

$$|\langle p|\, j\, |p_{n0}\rangle|^2\, \delta(\mu + k_0 - p_{n0})\, \delta_{\mathbf{p}_n,\mathbf{k}} = \frac{J^2(-\zeta)}{(4\mu p_{n0})}\frac{p_{n0}}{\mu}\, \delta\left(k_0 + \frac{\zeta}{2\mu}\right) \delta_{\mathbf{p}_n,\mathbf{k}},$$

$$|\langle p|\, j\, |p_{n0}\rangle|^2\, \delta(\mu - k_0 - p_{n0})\, \delta_{\mathbf{p}_n,-\mathbf{k}} = \frac{J^2(-\zeta)}{(4\mu p_{n0})}\frac{p_{n0}}{\mu}\, \delta\left(k_0 - \frac{\zeta}{2\mu}\right) \delta_{\mathbf{p}_n,-\mathbf{k}}, \tag{402}$$

since at these points $(p - p_n)^2 = -\zeta$. It is observed that we do not write $|J|^2$, since for $\zeta < 0$, $J(-\zeta)$ is *real*. The one-particle contribution to $A$ becomes

$$|A(k_0, \zeta)|_{1 \text{ meson}} = \frac{J^2(-\zeta)}{2\mu}\left[\delta\left(k_0 + \frac{\zeta}{2\mu}\right) - \delta\left(k_0 - \frac{\zeta}{2\mu}\right)\right]. \tag{403}$$

We have, finally, for the dispersion relation for $M(k_0, \zeta) - R$ the result

$$M(k_0, \zeta) - R = \frac{\zeta J^2(-\zeta)}{2\mu^2(k_0{}^2 - \zeta^2/4\mu^2)} + \frac{1}{\pi}\left(\int_{-\infty}^{-(3\mu^2-\zeta)/2\mu} + \int_{(3\mu^2-\zeta)/2\mu}^{\infty}\right) dk'_0 \frac{A(k'_0, \zeta)}{k'_0 - k_0 - i\eta}. \tag{404}$$

Now as long as $k_0 > \mu$, there is no difficulty in showing that $M(k_0, \zeta) - R$ is well defined from the original representation, even if $\zeta$ is increased from negative values to the physical one, $\zeta = \mu^2$. Similarly, in the integral in Eq. (404) there is no difficulty in allowing $\zeta \to \mu^2$.[56] Finally, $J(-\zeta)$ has been shown to be an analytic function of its argument all the way to $+4\mu^2$, so there is no difficulty in continuing it to $\zeta = \mu^2$. We conclude then, with $J(-\mu^2) \equiv g$, that

$$M(k_0) = \frac{g^2}{2} \frac{1}{k_0^2 - \mu^2/4} + \frac{1}{\pi}\left(\int_{-\infty}^{-\mu} + \int_{\mu}^{\infty}\right) dk'_0 \frac{A(k'_0)}{k'_0 - k_0 - i\eta} + R. \quad (405)$$

We see that $M(k_0)$ may be extended to a function analytic in the entire $k_0$-plane, except for poles at $k_0 = \pm\mu/2$ and for cuts along the real axis in the intervals $-\infty < k_0 < -\mu$ and $\mu < k_0 < +\infty$. The function thus extended satisfies $M(k_0) = M^*(-k_0)$, and thus $A(k_0) = -A(-k_0)$. The dispersion relation then takes the form

$$M(k_0) = \frac{g^2}{2} \frac{1}{k_0^2 - \mu^2/4} + \frac{2}{\pi} \int_0^{\infty} k'_0\, dk'_0 \frac{A(k'_0)}{k'^2_0 - k_0^2 - i\eta} + R. \quad (406)$$

The relation between $M$ and the forward scattering amplitude $f(k_0)$ in the laboratory system is

$$f(k_0) = \frac{M(k_0)}{8\pi\mu}, \quad (407)$$

so that in terms of this quantity our dispersion relation becomes [using the optical theorem, $\operatorname{Im} f = (|\mathbf{k}|/4\pi)\sigma$]

$$f(k_0) = \frac{g^2}{16\pi\mu^2} \frac{\mu}{k_0^2 - \mu^2/4} + \frac{1}{2\pi^2} \int_0^{\infty} \frac{k'_0\, dk'_0\, |\mathbf{k}'|\, \sigma(k'_0)}{k'^2_0 - k_0^2 - i\eta} + \frac{R}{8\pi\mu}. \quad (408)$$

(The constant $g$ has the dimensions of $\mu$.)

We have not paid attention to the question of subtractions, which may or may not have to be provided. Whereas our arguments about the passage of $\zeta$ from negative values up to $\zeta = \mu^2$ were slightly cavalier, they may easily be made rigorous; our deduction of Eq. (406) is really quite firm. We note in passing a rather neat way of writing Eq. (406), which removes the noncovariant appearance of the equation

$$M - R = -g^2\left[\frac{1}{(p + k)^2 + \mu^2} + \frac{1}{(p - k)^2 + \mu^2}\right] + \frac{1}{\pi} \int_{2\mu^2}^{\infty} dm^2 \rho(m^2)$$
$$\times \left[\frac{1}{(p + k)^2 + m^2 - i\eta} + \frac{1}{(p - k)^2 + m^2 + i\eta}\right], \quad (409)$$

[56] A little care is required for the integrals because of the appearance of $\zeta$ in the limits. We must make a change of variable and return to the original representation for $A$ to make sure that there is available analyticity in a strip along the real $\zeta$-axis through which we can pass from negative to positive values of $\zeta$. This is easily done.

where we have used $A(k_0) = -A(-k_0)$ and written

$$A\left(k_0 = \frac{m^2 - 2\mu^2}{2\mu}\right) \equiv \rho(m^2). \tag{410}$$

In this form of the dispersion relation, $m$ represents the mass of the various intermediate states which can be reached from the initial state. It should also be remembered, just as in Section 10.6, p. 640 that the constant $R$ does not necessarily represent the correct value of $M$ as $k_0 \to \infty$. It may be some entirely different constant, or perhaps we may even have to make a subtraction in order for any of these representations to have meaning.

When we go on to the physically more interesting case of pion-nucleon scattering, we find that this method of derivation fails if we regard the states $|p\rangle$ in the preceding derivation to be states of nucleons. The point is that we can no longer prove by elementary means that the matrix element analogous to $J$, namely $\langle 1 \text{ nucleon}| j | 1 \text{ nucleon}\rangle$ has the requisite analyticity. Strangely enough, this difficulty can be surmounted by regarding the states $|p\rangle$ as referring to the mesons and letting the $j$'s refer to nucleons. This is, in fact, what Symanzik did in his original proof. Actually, the demonstration can be carried out in the first form described, but it requires an apparatus too elaborate for us to develop here.

**Dispersion Relations for Pion-Nucleon Scattering**[57]

The most important application of dispersion relations for massive particles has been to the study of the scattering of pi-mesons (referred to frequently as pions) by nucleons. These relations have helped to simplify the analysis of experimental data and have provided an operational definition of the pion-nucleon coupling constant. In the extension to nonforward scattering they have also provided the basis for an approximate dynamical description of pion physics. This latter aspect is discussed only rather briefly because a much sounder formulation for a dynamical scheme is based on the double dispersion representation, which is beyond the scope of this book.

There are essentially no new principles involved in the pion-nucleon problem, as compared to the treatment given in the previous two sections. For this reason we are rather brief in our deduction of the dispersion relations.[58] The generalizations of our previous results that need be considered are to take account of the nucleon spin and isotopic spin and the isotopic spin of the mesons. We shall assume that isotopic spin is rigorously conserved in the following discussion.

[57] The original development of these relations is contained in Refs. [39].

[58] Details may be found in the discussion by Goldberger (1, Chapter IV).

In order to avoid repetition at a later stage we first discuss the general kinematical structure of the pion-nucleon system. Symbolically the process of interest may be written as $(p_1, \lambda_1) + (q_1, \beta) \rightarrow (p_2, \lambda_2) + (q_2, \alpha)$, where $p_1(p_2)$ is the four momentum of the initial (final) nucleon and $\lambda_1$, $\lambda_2$ are the corresponding spin and isotopic spin labels; $q_1(q_2)$ is the four momentum of the initial (final) pion and $\beta$, $\alpha$ label the isotopic spin. [The use of isotopic spin to label nucleons and mesons was described in Chapter 9, p. 521.] The nucleon mass is written as $m$, so that $p_1{}^2 = p_2{}^2 = -m^2$; the pion mass is then labeled as $\mu$, so $q_1{}^2 = q_2{}^2 = -\mu^2$.

Of the four momenta only three are independent, since $p_1 + q_1 = p_2 + q_2$. We choose as the independent momentum variables

$$P = \frac{p_1 + p_2}{2},$$
$$Q = \frac{q_1 + q_2}{2}, \tag{411}$$
$$K = q_1 - q_2 = p_2 - p_1.$$

These satisfy the following relations among themselves:

$$P^2 + \frac{K^2}{4} = -m^2,$$
$$Q^2 + \frac{K^2}{4} = -\mu^2, \tag{412}$$
$$P \cdot K = Q \cdot K = 0$$

There are thus, not surprisingly, only two scalar variables, which may be taken as $\nu \equiv -P \cdot Q/m$ and $K^2$. The variable $\nu$ is closely related to the total pion energy in the laboratory system, where the initial nucleon $p_1$ is at rest. $K^2$ is the square of the momentum transfer to the nucleon in the scattering. The total energy in the barycentric system, (where $\mathbf{p}_1 + \mathbf{q}_1 = 0$) $W$, is related to the variables (411) by

$$W^2 = -(p_1 + q_1)^2 = -(P + Q)^2,$$
$$= m^2 + \mu^2 + \frac{K^2}{2} + 2m\nu. \tag{413}$$

If we express $W^2$ in terms of the variables in the laboratory system, (that is, the system in which $\mathbf{p}_1 = 0$) designated by subscript $L$, we find

$$W^2 = m^2 + \mu^2 + 2mq_{0L}, \tag{414}$$

so

$$\nu = q_{0L} - \frac{K^2}{4m}. \tag{415}$$

For forward scattering, $K = 0$, $\nu = q_{0L}$. In terms of the pion momentum $q$ in the barycentric system and the scattering angle $\theta$ in that system we have the following:

$$K^2 = 2q^2(1 - \cos\theta),$$
$$q^2 = \frac{[W^2 - (m+\mu)^2][W^2 - (m-\mu)^2]}{4W^2}, \tag{416}$$
$$\nu = \frac{q^2 + E\omega - K^2/4}{m}.$$

Here the nucleon energy $E = (m^2 + q^2)^{1/2}$ and the pion energy $\omega = (\mu^2 + q^2)^{1/2}$ may be expressed in terms of $W$ as

$$E = \frac{W^2 + m^2 - \mu^2}{2W},$$
$$\omega = \frac{W^2 + \mu^2 - m^2}{2W}. \tag{417}$$

The general isotopic spin structure of the scattering amplitude is easily analyzed. We shall regard the amplitude $M$ as a $2 \times 2$ matrix in the two-component spinor space of the nucleons. The initial $\beta$ and final $\alpha$ pion isotopic labels take on the values 1, 2, 3, and appropriate combinations of these represent pions of definite charge. In place of the single real field operator $\phi$ used in the previous sections, we now have three real fields $\phi_1$, $\phi_2$, $\phi_3$; our convention shall be that the operator $(\phi_1 - i\phi_2)/\sqrt{2}$ annihilates positive mesons, $(\phi_1 + i\phi_2)/\sqrt{2}$ annihilates negative mesons, whereas $\phi_3$ annihilates neutral mesons. The scattering amplitude must be a tensor in the indices $\alpha$, $\beta$ and must be constructed from the operators $\tau_i (i = 1, 2, 3)$ in the nucleon isotopic spin space. Because such a second rank tensor can obviously be written as the sum of a symmetric and an antisymmetric part, we have

$$\begin{aligned} M_{\alpha\beta} &= M^{(+)}\tfrac{1}{2}\{\tau_\alpha, \tau_\beta\} + M^{(-)}\tfrac{1}{2}[\tau_\alpha, \tau_\beta] \\ &= M^{(+)}\delta_{\alpha\beta} + M^{(-)}\tfrac{1}{2}[\tau_\alpha, \tau_\beta], \end{aligned} \tag{418}$$

since $\{\tau_\alpha, \tau_\beta\} = 2\delta_{\alpha\beta}$.

Because we have assumed that isotopic spin is conserved, it is obviously useful to express $M_{\alpha\beta}$ in terms of amplitudes with given total isotopic spin. The nucleon has isotopic spin one-half, whereas the pion has isotopic spin one; the allowed values of the total isotopic spin are then $\frac{1}{2}$ and $\frac{3}{2}$. To make such a decomposition, we remark that $(\frac{1}{2})[\tau_\alpha, \tau_\beta] = i\epsilon_{\alpha\beta\delta}\tau_\gamma = -\boldsymbol{\tau}\cdot\mathbf{T}_{\alpha\beta}$, where $\mathbf{T}$ is the set of three spin-one rotation matrices given in Eq. (2-37):[59]

$$(T_j)_{\alpha\beta} = i\epsilon_{\alpha j\beta}, \qquad j = 1, 2, 3. \tag{419}$$

[59] This was called $\mathbf{S}_{ik}$ in Chapter 2, where it referred to rotations in ordinary space.

Next we form projection operators in the usual way: $\mathbf{I} = \boldsymbol{\tau}/2 + \mathbf{T}$ has eigenvalues $I^2 = \frac{1}{2}(\frac{1}{2} + 1), \frac{3}{2}(\frac{3}{2} + 1)$ so that $\boldsymbol{\tau} \cdot \mathbf{T} = 1$ for isotopic spin $\frac{3}{2}$, and $\boldsymbol{\tau} \cdot \mathbf{T} = -2$ for isotopic spin $\frac{1}{2}$. The the projection operators are evidently

$$P_{3/2} = \frac{2 + \boldsymbol{\tau} \cdot \mathbf{T}}{3}, \qquad P_{1/2} = \frac{1 - \boldsymbol{\tau} \cdot \mathbf{T}}{3}, \tag{420}$$

from which we deduce

$$\begin{aligned} \delta_{\alpha\beta} &= (P_{3/2} + P_{1/2})_{\alpha\beta}, \\ (\boldsymbol{\tau} \cdot \mathbf{T})_{\alpha\beta} &= (P_{3/2} - 2P_{1/2})_{\alpha\beta}. \end{aligned} \tag{421}$$

We now write $M$ as an operator in the combined space of nucleon and pion isotopic spin.

$$M = M^{(+)}(P_{3/2} + P_{1/2}) + M^{(-)}(2P_{1/2} - P_{3/2}). \tag{422}$$

Thus

$$\begin{aligned} M_{3/2} &= M^{(+)} - M^{(-)}, \\ M_{1/2} &= M^{(+)} + 2M^{(-)}, \end{aligned} \tag{423}$$

or

$$\begin{aligned} M^{(+)} &= \frac{2M_{3/2} + M_{1/2}}{3}, \\ M^{(-)} &= \frac{M_{1/2} - M_{3/2}}{3}. \end{aligned} \tag{424}$$

As a final exercise, we relate $M^{(\pm)}$ to the amplitudes for scattering pions of a given charge by nucleons of a given charge. For definiteness, let us suppose the nucleon to be a proton which has $I = \frac{1}{2}$, $I_3 = \frac{1}{2}$. A positive meson has $I_3 = +1$ so that the state $|\pi^+ p\rangle$ is pure $I = \frac{3}{2}$. Consequently, we have $M(\pi^+) = M_{3/2}$. The amplitude for the process $\pi^- + p \to \pi^- + p$ is computed as follows: since $|\pi^- p\rangle = (\phi_1 - i\phi_2)/\sqrt{2}\,|p\rangle$ ($(\phi_1 - i\phi_2)/\sqrt{2}$ is the creation operator for negative mesons), we have

$$\begin{aligned} M(\pi^-) = \langle \pi^- p|\, M\, |\pi^- p\rangle &= \frac{\langle p|\, [M_{11} + M_{22} - i(M_{12} - M_{21})]\, |p\rangle}{2} \\ &= \langle p|\, M^{(+)} + M^{(-)}\tau_3\, |p\rangle \\ &= M^{(+)} + M^{(-)} = \frac{M_{3/2} + 2M_{1/2}}{3}. \end{aligned} \tag{425}$$

Now $M(\pi^+) = M_{3/2} = M^{(+)} - M^{(-)}$, so

$$\begin{aligned} M^{(+)} &= M(\pi^-) + M(\pi^+), \\ M^{(-)} &= M(\pi^-) - M(\pi^+). \end{aligned} \tag{426}$$

In the same way we find, of course, that

$$M(\pi^0) = \langle p|\, M_{33}\,|p\rangle = M^{(+)} = M(\pi^-) + M(\pi^+),$$
$$M(\pi^0 n \leftarrow \pi^- p) = -\sqrt{2}M^{(-)} = \sqrt{2}(M_{3/2} - M_{1/2})/3. \tag{427}$$

The next problem to be discussed is that of determining the structure of the amplitude in spin space. Since we are concerned with an $S$-matrix element involving a single nucleon in the initial and final state, it is clear that the amplitude $M$ must be expressible as

$$M = \bar{u}_\rho(p_2)\mathcal{O}_{\rho\sigma}u_\sigma(p_1), \tag{428}$$

where $u(p_1)$, $\bar{u}(p_2) = u^*(p_2)\gamma_4$ are free-particle Dirac spinors of the form discussed in Section 2.2. Actually, here they are *eight* component quantities in the sense that they are two-component spinors in isotopic spin space, as has already been taken into account in the preceding paragraphs. We suppress the explicit nucleon spin labels. The matrix in Dirac spinor space $\mathcal{O}_{\rho\sigma}$ may be in principle a linear combination of the sixteen independent Dirac matrices with coefficients that are functions of the two scalar invariants $\nu$ and $K^2$. We shall show, in fact, that there are only two nonzero matrices which enter, by virtue of the fact that the spinors satisfy the equations $(i\gamma \cdot p_1 + m)u(p_1) = 0$ and $\bar{u}(p_2)(i\gamma \cdot p_2 + m) = 0$, together with the requirement that $M$ be a relativistic scalar quantity.

To see this we write, quite generally,

$$\mathcal{O} = g_1 + ig_\mu\gamma_\mu + g_{\mu\nu}\sigma_{\mu\nu} + ig_{5\mu}\gamma_\mu\gamma_5 + g_5\gamma_5, \tag{429}$$

where repeated indices are summed from 1 to 4 and the factors of $i$ are inserted for convenience. The four-dimensional space may be spanned by any four independent vectors, say, $P_\mu$, $Q_\mu$, $K_\mu$, and $L_\mu = \epsilon_{\mu\nu\lambda\sigma}P_\nu Q_\lambda K_\sigma$. (The last is a pseudovector, but this is irrelevant.) We then express $g_\mu$ in this basis system.

$$g_\mu = g^{(1)}P_\mu + g^{(2)}Q_\mu + g^{(3)}K_\mu + g^{(4)}L_\mu. \tag{430}$$

Next we observe that $i\gamma \cdot P$ between the free-particle spinors may be replaced by $-m$, so the $g^{(1)}$ term contributes something of the same structure as $g_1$ and may be dropped. Similarly, since $K = p_2 - p_1$, the $g^{(3)}$ term yields zero. Finally, the expression $\gamma \cdot L = \frac{1}{4}\{\gamma \cdot P, [\gamma \cdot Q, \gamma \cdot K]\}\gamma_5$ may be reduced to scalar multiples of $\gamma_5$ or $\gamma_5\gamma \cdot Q$, both of which appear later in either the $g_5$ or $g_{5\mu}$ terms. Thus, of all the $g_\mu$ terms only $g^{(2)}$ cannot be further reduced.

In a similar fashion, the tensor $g_{\mu\nu}$ may be expressed in terms of $P_\mu P_\nu$, $P_\mu K_\nu$, and so on, and leads to things which can be expressed as $a + b\gamma \cdot Q + c\gamma_5\gamma \cdot Q + d\gamma_5$. The first two are already counted, whereas the last are included again in $g_{5\mu}$ and $g_5$. Now since $M$ must be a scalar, and it is impossible to make a pseudoscalar out of the three vectors $P$, $Q$, $K$, the only thing that can enter into $g_{5\mu}$ is something proportional to $L_\mu$. However, $\gamma_5\gamma\cdot L$ can evidently be reduced to $\alpha + \beta\gamma\cdot Q$, which has already been counted. Finally, there is no pseudoscalar available to be $g_5$, hence it must vanish. The complete structure of the amplitude $M$ in both spin and isotopic spin space may now be given.

$$\begin{aligned} M_{\alpha\beta} &= \bar{u}(p_2)(A_{\alpha\beta} - i\gamma\cdot QB_{\alpha\beta})u(p_1), \\ A_{\alpha\beta} &= A^{(+)}\delta_{\alpha\beta} + A^{(-)}\tfrac{1}{2}[\tau_\alpha, \tau_\beta], \\ B_{\alpha\beta} &= B^{(+)}\delta_{\alpha\beta} + B^{(-)}\tfrac{1}{2}[\tau_\alpha, \tau_\beta], \end{aligned} \tag{431}$$

where the $A^{(\pm)}$ and $B^{(\pm)}$ are functions of the scalars $\nu$, $K^2$, or if one prefers, $W^2$, $K^2$.

The quantity that we have been calling $M_{\alpha\beta}$ will now be related to the physical scattering amplitude. Without specifying exactly what reduction formula we shall use, we know that the general form of the $S$-matrix element may be taken to be

$$\begin{aligned} S_{\alpha\beta} = \delta_{\mathbf{p}_2,\mathbf{p}_1}\,\delta_{\mathbf{q}_2,\mathbf{q}_1}\,\delta_{\alpha\beta} + 2\pi i\,\delta(p_{20} + q_{20} - p_{10} - q_{10})\,\delta_{\mathbf{p}_2+\mathbf{q}_2,\mathbf{p}_1+\mathbf{q}_1} \\ \times \left(\frac{m^2}{4p_{10}p_{20}q_{10}q_{20}}\right)^{1/2} M_{\alpha\beta}. \end{aligned} \tag{432}$$

Of course, the exact numerical factor in front of $M_{\alpha\beta}$ is not important, but the factor $(p_{10}p_{20}q_{10}q_{20})^{-1/2}$ is necessary if we want $M_{\alpha\beta}$ to be a Lorentz invariant. [This fact was proven quite generally in obtaining Eqs. (3.111) and (3.112) and has also emerged repeatedly in this chapter. The included factor $(m^2/4)^{1/2}$ is conventional.]

From this expression, the total transition probability per unit time is given by the expression

$$\dot{P} = (2\pi)^4 \int \frac{d^3p_2}{(2\pi)^3} \int \frac{d^3q_2}{(2\pi)^3} \frac{m^2}{4p_{10}p_{20}q_{10}q_{20}}\,\delta(p_2 + q_2 - p_1 - q_1)\,|M_{\alpha\beta}|^2. \tag{433}$$

Here we have replaced the Kronecker $\delta$-function on momentum transfer by a Dirac $\delta$-function and have inserted an integration over $p_2$ in order to write the answer quite symmetrically in the final particles. The total cross section is given by $v_{\text{rel}}\sigma = \dot{P}$, where the relative velocity $v_{\text{rel}}$ is chosen so as to make $\sigma$ an invariant. We have then,

$$\sigma = \left(\frac{m}{4\pi}\right)^2 \frac{1}{v_{\text{rel}}\,p_{10}q_{10}} \int \frac{d^3q_2}{q_{20}} \int \frac{d^3p_2}{p_{20}}\,\delta(p_2 + q_2 - p_1 - q_1)\,|M_{\alpha\beta}|^2. \tag{434}$$

In the system in which $\mathbf{p}_1 = 0$, $v_{\rm rel}$ is just the pion velocity, $|\mathbf{q}_1|/q_{10}$, so that in that system

$$v_{\rm rel}\, p_{10} q_{10} = \frac{|\mathbf{q}_1|}{q_{10}}\, m q_{10} = m\, |\mathbf{q}_1| = m\sqrt{{q_{10}}^2 - \mu^2}$$
$$= \sqrt{(p_1 \cdot q_1)^2 - m^2\mu^2}, \tag{435}$$

which is evidently an invariant. From the total cross section we infer the differential cross section for the final pion to be in the solid angle $d\Omega_{\hat{\mathbf{q}}_2}$.

$$\frac{d\sigma}{d\Omega_{\hat{\mathbf{q}}_2}} = |f|^2,$$

where

$$f = \frac{m}{4\pi}\sqrt{\frac{\rho}{v_{\rm rel}\cdot p_{10} q_{10}}}\, M, \tag{436}$$

$$\rho = \int \frac{|\mathbf{q}_2|^2\, d\,|\mathbf{q}_2|}{q_{20}} \int \frac{d^3 p_2}{p_{20}}\, \delta(p_2 + q_2 - p_1 - q_1). \tag{437}$$

In the center-of-mass system,

$$\rho = \frac{|\mathbf{q}_2|}{W},$$
$$v_{\rm rel} = \frac{W\,|\mathbf{q}_1|}{q_{10} p_{10}} = \frac{W\,|\mathbf{q}_2|}{q_{20} p_{20}}, \tag{438}$$
$$f = \frac{m}{4\pi W} M.$$

This is the desired relation between the barycentric scattering amplitude and the invariant amplitude $M$.

As a final bit of kinematics we must relate the scalars $A$ and $B$ to the quantities introduced in Section 7.2 to describe the scattering of spin zero particles by spin one-half targets. Since the description there was in terms of two-component Pauli spinors, we must evidently express our four-component spinors $\bar{u}(p_2)$, $u(p_1)$ in similar terms.[60]

$$u(p_1) = \frac{1}{[2m(E+m)]^{1/2}} \begin{pmatrix} E + m \\ -\boldsymbol{\sigma}\cdot\mathbf{q}_2 \end{pmatrix} \chi_1,$$
$$\bar{u}(p_2) = \frac{{\chi_2}^\dagger}{[2m(E+m)]^{1/2}} \begin{pmatrix} E + m \\ \boldsymbol{\sigma}\cdot\mathbf{q}_2 \end{pmatrix}, \tag{439}$$

[60] See, for example, Eqs. (2.69).

where $\mathbf{q}_1(\mathbf{q}_2)$ is the momentum of the initial (final) pion in the barycentric system, and $E = (m^2 + \mathbf{q}^2)^{1/2}$, $|\mathbf{q}_1| = |\mathbf{q}_2| = q$. The $\chi$'s are two component spinors. [Note also that in this system $Q_0 = \omega = (\mu^2 + q^2)^{1/2}$ and $W = E + \omega$.] Substituting these into the general form of $M$ [Eq. (431)] and using Eq. (438), we find that

$$f = \chi_2^\dagger\left\{\frac{E+m}{2W}\left[\frac{A + (W-m)B}{4\pi}\right] + \frac{E-m}{2W}\right.$$
$$\left.\times\left[\frac{-A + (W+m)B}{4\pi}\right]\boldsymbol{\sigma}\cdot\mathbf{q}_2\boldsymbol{\sigma}\cdot\mathbf{q}_1\right\}\chi_1. \tag{440}$$

In the forward direction, $\hat{\mathbf{q}}_1 = \hat{\mathbf{q}}_2$, so

$$f = \frac{m}{4\pi W}\chi_2^\dagger\left(A - \frac{p_1\cdot q_1}{m}B\right)\chi_1. \tag{441}$$

We have here suppressed all isotopic labels, such as the $\pm$, on the quantities $A$, $B$; it is most convenient to regard these formulas as applying to pure isotopic spin states ($I = \frac{1}{2}, \frac{3}{2}$) since we know how to compose everything required from these. Comparison of (440) with the form given in Eq. (7.87),

$$f = \chi_2^\dagger(f_1 + \boldsymbol{\sigma}\cdot\mathbf{q}_2\boldsymbol{\sigma}\cdot\mathbf{q}_1 f_2)\chi_1,$$

leads to

$$\begin{aligned} f_1 &= \frac{E+m}{2W}\left[\frac{A + (W-m)B}{4\pi}\right], \\ f_2 &= \frac{E-m}{2W}\left[\frac{-A + (W+m)B}{4\pi}\right]. \end{aligned} \tag{442}$$

We may also solve for $A$ and $B$ in terms of $f_1$ and $f_2$.

$$\begin{aligned} \frac{A}{4\pi} &= \frac{W+m}{E+m}f_1 - \frac{W-m}{E-m}f_2, \\ \frac{B}{4\pi} &= \frac{f_1}{E+m} + \frac{f_2}{E-m}. \end{aligned} \tag{443}$$

We recall for convenience the relation between the $f_i$ and the phase shifts [Eq. (7.90)].

$$\begin{aligned} f_1 &= \sum_{l=0}^{\infty} f_{l+}P'_{l+1}(\cos\theta) - \sum_{l=2}^{\infty} f_{l-}P'_{l-1}(\cos\theta), \\ f_2 &= \sum_{l=1}^{\infty}(f_{l-} - f_{l+})P'_l(\cos\theta), \\ f_{l\pm} &= e^{i\delta_{l\pm}}\frac{\sin\delta_{l\pm}}{q}. \end{aligned} \tag{444}$$

Finally, we remark that the forward scattering amplitude $f_L(0)$ in the laboratory system, where $\mathbf{p}_1 = 0$, is related to $f(0)$ by a factor $W/m$.

$$f_L(0) = \frac{W}{m} f(0) = \chi_2^\dagger\left(A - \frac{p_1 \cdot q_1}{m} B\right)\frac{\chi_1}{4\pi},$$

$$= \chi_2^\dagger(A + q_{0L}B)\frac{\chi_1}{4\pi}, \tag{445}$$

where $q_{0L}$ is the incident meson energy in the laboratory system. We shall return to these formulas after obtaining the dispersion relations for the scalar quantities $A$ and $B$.

We turn now to the problem of deriving dispersion relations for the $A$'s and $B$'s. As stated earlier, there are nothing but rather messy technical details which distinguish this problem from the ones treated in the previous sections. If, however, the derivation is to be given using only the elementary methods devised so far, we must find a reduction formula that exhibits the dependence of the $S$-matrix element on the nucleon momenta rather than on the pion momenta. The latter form, namely,

$$2\pi\delta(p_2 + q_2 - p_1 - q_1)M_{\alpha\beta}^{\lambda_2\lambda_1} = \left(\frac{p_{20}p_{10}}{m^2}\right)^{1/2}\int d^4x\int d^4y e^{-iq_2\cdot x}$$

$$\times \{(\mu^2 - \Box_x)(\mu^2 - \Box_y)\langle p_2, \lambda_2|\,[\phi_\alpha(x), \phi_\beta(y)]\theta(x_0 - y_0)\,|p_1, \lambda_1\rangle\}e^{iq_1\cdot y} \tag{446}$$

(where the three-momentum part of the $\delta$-function should be interpreted as a Kronecker $\delta$, and the nucleon spin and isotopic labels have been written explicitly) is singularly unsuited to the disentangling of the amplitudes $A$ and $B$ in spin and isotopic spin space.

To obtain the desired expression, we must introduce in and out nucleon field operators and apply the asymptotic condition to them. To save writing, we shall use plane-wave states rather than wave packets. Leaving aside an isotopic spin label, which is trivially appended, we would write for the nucleon in field, $\psi_i$,

$$\psi_i = \sum_{\mathbf{p},\lambda=1,2}\left(\frac{m}{p_0}\right)^{1/2}[A_{\mathbf{p}\lambda}u_{\lambda(p)}e^{ip\cdot x} + A_{\mathbf{p},-\lambda}u_{-\lambda(p)}e^{-ip\cdot x}] \tag{447}$$

with

$$A_{\mathbf{p}\lambda} = \left(\frac{m}{p_0}\right)^{1/2}\int d^3x\bar{u}_\lambda(p)\gamma_4\psi_i(x)e^{-i\epsilon(\lambda)p\cdot x},$$

$$A_{\mathbf{p}\lambda}^\dagger = \left(\frac{m}{p_0}\right)^{1/2}\int d^3x\bar{\psi}_i(x)\gamma_4u_\lambda(p)e^{i\epsilon(\lambda)p\cdot x}, \tag{448}$$

where $\bar{\psi} = \psi^* \gamma_4$. The spinors $u_\lambda(p)$ are just those given in Eq. (2.69), normalized so that $\bar{u}_\lambda(p) u_\lambda(p) = \epsilon(\lambda)$, $\bar{u}_\lambda(p)\gamma_4 u_\lambda(p) = p_0/m$. The (anti) commutation relation satisfied by the $A$'s is

$$\{A_{\mathbf{p}\lambda}, A^\dagger_{\mathbf{p}',\lambda'}\} = \delta_{\mathbf{p},\mathbf{p}'}\delta_{\lambda,\lambda'}; \tag{449}$$

all other pairs anticommute. As before, we have set the quantization volume equal to unity; otherwise a factor of $\mathscr{V}$ multiplies the $p_0$ in (447) and (448). There is an analogous set of formulas for the out field $\psi_0$ in terms of operators called $B_{\mathbf{p},\lambda}$, say. The asymptotic condition on the field operator $\psi$ which interpolates between $\psi_i$ and $\psi_o$ takes precisely the same form as for the scalar field, Eq. (321). That is, if we define

$$\begin{aligned} \psi_{p,\lambda}(x_0) &= \left(\frac{m}{p_0}\right)^{1/2} \int d^3x \bar{u}_\lambda \gamma_4 \psi(x) e^{-i\epsilon(\lambda)p\cdot x}, \\ \psi^\dagger_{p,\lambda}(x_0) &= \left(\frac{m}{p_0}\right)^{1/2} \int d^3x \bar{\psi}(x)\gamma_4 u_\lambda(p) e^{i\epsilon(\lambda)p\cdot x}, \end{aligned} \tag{450}$$

then

$$\begin{aligned} \lim_{x_0\to-\infty} \langle A| \,\psi_{\mathbf{p},\lambda}(x_0)\, |B\rangle &= A_{\mathbf{p},\lambda}, \\ \lim_{x_0\to+\infty} \langle A| \,\psi_{\mathbf{p},\lambda}(x_0)\, |B\rangle &= B_{\mathbf{p},\lambda}, \end{aligned} \tag{451}$$

and similarly for $\psi^\dagger_{\mathbf{p},\lambda}(x_0)$.

Physical states are described by allowing the in or out creation operators $A^\dagger_{\mathbf{p},\lambda}$, $B^\dagger_{\mathbf{p},\lambda}$ to act on the vacuum. Since these operators anticommute among themselves, we must be very careful about algebraic signs. For the same reason, as is well-known, we cannot here have multiple occupation of states (this is just the Pauli principle). We shall be concerned ultimately only with states that have one pion and one nucleon in them, so none of these annoyances will be present for us. For example, an in state of a pion in state $\alpha$ ($\alpha$ stands for both momentum and isotopic spin) and a nucleon in $(p, \lambda)$ would be written as

$$|(p, \lambda), \alpha; i\rangle = A^\dagger_{\mathbf{p},\lambda} a_\alpha^\dagger \,|0\rangle = A^\dagger_{\mathbf{p},\lambda}\,|\alpha\rangle. \tag{452}$$

Finally, we remark that the anticommutation relation among the $\psi_{i,o}$ and $\bar{\psi}_{i,o}$'s is

$$\{\psi^{(\alpha)}_{i,o}(x), \bar{\psi}^{(\beta)}_{i,o}(y)\} = i\left(-\gamma_\mu \frac{\partial}{\partial x_\mu} + m\right)_{\alpha\beta}\Delta(x-y), \tag{453}$$

where we have written explicitly the Dirac spinor indices in the only available place, and $\Delta$ is just the function given by Eq. (302), with $\mu$ replaced by $m$. We note that this anticommutator vanishes outside the light cone of $x - y$ and easily verify that

$$\{\psi_{i,o}(x), \bar{\psi}_{i,o}(y)\}\delta(x_0 - y_0) = \gamma_4\delta(x - y). \tag{454}$$

The fundamental locality or causality property for the interpolating field $\psi(x)$ is

$$[\psi(x), \bar{\psi}(y)] = 0, \qquad (x - y)^2 > 0. \tag{455}$$

The derivation of the desired reduction formula follows precisely the lines developed for the scalar field. We simply contract the nucleons instead of the pions from states like those in Eq. (452). We give only the final answer.

$$\begin{aligned} 2\pi\delta(p_2 + q_2 - p_1 - q_1)M_{\alpha\beta}^{\lambda_2\lambda_1} &= i(4q_{20}q_{10})^{1/2} \int d^4x \int d^4y [\bar{u}_{\lambda_2}(p_2)]_\rho e^{-ip_2\cdot x} \\ &\times \{[D_x]_{\rho\rho'}[\bar{D}_y]_{\sigma\sigma'}\langle q_2, \beta|\, T[\psi_{\rho'}(x)\bar{\psi}_{\sigma'}(y)]\,|q_1, \alpha\rangle\}[u_{\lambda_1}(p_1)]_\sigma e^{ip_1\cdot y} \quad (456) \\ &= i(4q_{20}q_{10})^{1/2} \int d^4x \int d^4y [\bar{u}_{\lambda_2}(p_2)]_\rho e^{-ip_2\cdot x}\{[D_x]_{\rho\rho'}[\bar{D}_y]_{\sigma\sigma'} \\ &\times \langle q_2, \beta|\, \{\psi_{\rho'}(x), \bar{\psi}_{\sigma'}(y)\}\theta(x_0 - y_0)\,|q_1, \alpha\rangle\}[u_{\lambda_1}(p_1)]_\sigma e^{ip_1\cdot y}, \quad (457) \end{aligned}$$

where

$$\begin{aligned} D_x &= \gamma_\mu \frac{\partial}{\partial x_\mu} = m, \\ \bar{D}_y &= -\gamma_\mu \frac{\partial}{\partial y_\mu} + m, \end{aligned} \tag{458}$$

and

$$T[\psi_{\rho'}(x)\bar{\psi}_{\sigma'}(y)] = \epsilon(x_0 - y_0)[\psi_{\rho'}(x)\bar{\psi}_{\sigma'}(y)]_+ = -T[\bar{\psi}_{\sigma'}(y)\psi_{\rho'}(x)]. \tag{459}$$

The form (457) is the time-ordered operator version of the reduction formula and, as explained in connection with the nonrelativistic theory, *must* contain the $T$-symbol rather than the simple time-ordering operation $(\cdots)_+$. Our matrix element otherwise would not be an invariant. From now on we shall usually suppress the Dirac spinor indices and also nucleon spin-isotopic spin labels when there is no ambiguity.

In order to deduce the crossing relations for these amplitudes, we use the invariance of the theory under the combined operations of charge conjugation and rotation by $\pi$ about the 2 axis in isotopic spin space. This so-called $G$-invariance is generated by the unitary operator

$$G = Ce^{i\pi I_2}, \tag{460}$$

where $C$ is the charge conjugation matrix introduced in Eq. (1.119). Under this operation, the nucleon and meson field operators transform according to

$$\begin{aligned} G^{-1}\phi_\alpha G &\equiv \phi'_\alpha = -\phi_\alpha, \\ G^{-1}\psi G &\equiv \psi' = iC\tau_2\bar{\psi}, \\ G^{-1}\bar{\psi}G &\equiv \bar{\psi}' = iC^{-1}\tau_2\psi. \end{aligned} \tag{461}$$

Here $\tau_2$ is the matrix

$$\tau_2 = \begin{pmatrix} 0 & -i \\ i & 0 \end{pmatrix}, \tag{462}$$

which operates on the isotopic spin part of $\psi$.

We are interested in the crossing relations analogous to Eq. (394) for the scalar amplitudes $A$, $B$ introduced earlier in this section. To do this, we split off the spinors $\bar{u}_{\lambda_2}(p_2)$, $u_{\lambda_1}(p_1)$ according to the definitions

$$\begin{aligned} M_{\alpha\beta}^{\lambda_2\lambda_1} &= [\bar{u}_{\lambda_2}(p_2)]_\rho \mathscr{F}_{\rho\sigma}(p_2, p_1; q_2\alpha, q_1\beta)[u_{\lambda_1}(p_1)]_\sigma \\ &= [\bar{u}_{\lambda_2}(p_2)]_\rho \mathscr{M}_{\rho\sigma}(p_2, p_1; q_2\alpha, q_1\beta)[u_{\lambda_1}(p_1)]_\sigma. \end{aligned} \tag{463}$$

The indices $\rho$, $\sigma$ are Dirac spinor indices and there are implied nucleon isotopic spinor indices on $\mathscr{F}_{\rho\sigma}$, $\mathscr{M}_{\rho\sigma}$; $\mathscr{F}$ refers to the time-ordered form (456), $\mathscr{M}$ to the retarded form (457). We may identify $\mathscr{F}$ and $\mathscr{M}$ with $A_{\alpha\beta} - i\gamma \cdot QB_{\alpha\beta}$, Eq. (431). Actually, as defined, $\mathscr{F}$ and $\mathscr{M}$ may contain terms which vanish when inserted between $\bar{u}(p_2)$, $u(p_1)$. With the aid of the $G$-symmetry operation, which implies, for example, that

$$\begin{aligned} \langle q_2\alpha| T[\psi(x)\bar{\psi}(y)] |q_1\beta\rangle &= \langle q_2\alpha| G^{-1}T[\psi(x)\bar{\psi}(y)]G |q_1\beta\rangle \\ &= \langle q_2\alpha| T[[iC\tau_2\bar{\psi}(x)][iC^{-1}\tau_2\psi(y)]] |q_1\beta\rangle, \end{aligned} \tag{464}$$

it is an exercise in patience with indices to deduce the relations

$$\begin{aligned} \mathscr{F}(p_2, p_1; q_2\beta, q_1\alpha) &= C\tau_2\tilde{\mathscr{F}}(-p_1, -p_2; q_2\alpha, q_1\beta)C^{-1}\tau_2, \\ \mathscr{M}(p_2, p_1; q_2\beta, q_1\alpha) &= C\tau_2\tilde{\gamma}_4\mathscr{M}^*(-p_2, -p_1; q_1\beta, q_2\alpha)\tilde{\gamma}_4C^{-1}\tau_2. \end{aligned} \tag{465}$$

We regard these objects as matrices in combined Dirac spinor and isotopic spinor space.

Whereas $\mathscr{F}$ and $\mathscr{M}$ are equal for physical scattering processes, we must distinguish between the $A$ and $B$ belonging to each of these when contemplating the extension of these amplitudes under operations of the variety expressed by Eq. (465). We may regard the $A^{(\pm)}$, $B^{(\pm)}$ as functions of the scalars $\nu$ and $K^2$. Using the relations

$$\begin{aligned} C\tilde{\gamma}_\mu C^{-1} &= -\gamma_\mu \\ \tau_2\tau_\alpha\tau_2 &= -\tau_\alpha, \end{aligned} \tag{466}$$

we find from $\mathscr{F}$,

$$\begin{aligned} (A^{(+)}(\nu, K^2), B^{(-)}(\nu, K^2)) &= (A^{(+)}(-\nu, K^2), B^{(-)}(-\nu, K^2)), \\ (A^{(-)}(\nu, K^2), B^{(+)}(\nu, K^2)) &= -(A^{(-)}(-\nu, K^2), B^{(+)}(-\nu, K^2)), \end{aligned} \tag{467}$$

and from $\mathscr{M}$,

$$\begin{aligned} (A^{(+)}(\nu, K^2), B^{(-)}(\nu, K^2)) &= (A^{(+)*}(-\nu, K^2), B^{(-)*}(-\nu, K^2)), \\ (A^{(-)}(\nu, K^2), B^{(+)}(\nu, K^2)) &= -(A^{(-)*}(-\nu, K^2), B^{(+)*}(-\nu, K^2)). \end{aligned} \tag{468}$$

Finally, if we break $\mathscr{M}$ into dispersive and absorptive parts in the now familiar fashion by writing $\theta(x_0 - y_0) = [1 + \epsilon(x_0 - y_0)]/2$, and $\mathscr{M} = \mathscr{D} + i\mathscr{A}$, we find

$$\begin{aligned} \gamma_4\mathscr{D}^*(p_2, p_1; q_2\alpha, q_1\beta)\gamma_4 &= \mathscr{D}(p_1, p_2; q_1\beta, q_2\alpha), \\ \gamma_4\mathscr{A}^*(p_2, p_1; q_2\alpha, q_1\beta)\gamma_4 &= \mathscr{A}(p_1, p_2; q_1\beta, q_2\alpha). \end{aligned} \tag{469}$$

Since $A^{(\pm)}$, $B^{(\pm)}$ are functions of $\nu$, $K^2$ which are unchanged by the indicated transformation of momenta, we find easily that the dispersive and absorptive parts of $A^{(\pm)}$ and $B^{(\pm)}$ are real.

We are now in position to discuss the dispersion relations satisfied by $A^{(\pm)}$, $B^{(\pm)}$ and for this purpose we use exclusively the quantity $\mathscr{M}$. We confine our attention for the present to forward scattering, corresponding to $K^2 = 0$, in which case $\nu$ becomes the energy of the pion in the system in which the nucleon $p_1$ is at rest—or $\mu/m$ times the energy of the nucleon in a system where the pion is at rest. For forward scattering there can be no change in spin orientation and we have

$$\begin{aligned} M_{\alpha\beta}{}^{\lambda\lambda} &= \bar{u}_\lambda(p)[A_{\alpha\beta} - i\gamma\cdot qB_{\alpha\beta}]u_\lambda(p) \\ &= A_{\alpha\beta} - \frac{p\cdot q}{m}B_{\alpha\beta} \\ &= A_{\alpha\beta} + q_{0L}B_{\alpha\beta}, \qquad \text{in the nucleon rest-frame,} \\ &= A_{\alpha\beta} + \frac{\mu}{m}p_{0L}B_{\alpha\beta}, \qquad \text{in the meson rest-frame} \end{aligned} \tag{470}$$

It is useful to deduce separate dispersion relations for the $A$'s and $B$'s even for forward scattering rather than to concentrate on just the particular combination (470), as we shall see.

To be completely cartesian we should write down explicit representations for $A^{(\pm)}$ and $B^{(\pm)}$ in terms of $\mathscr{M}$. This can be easily done. The reason for this is that we have no assurance *a priori* that our particular choice of amplitudes is free of what might be called kinematical singularities. They might, after all, have been chosen differently, for example, by taking $\gamma_5\gamma\cdot L$ instead of retaining the term $i\gamma\cdot Q$ as a fundamental form. When we solve for $A^{(\pm)}$ and $B^{(\pm)}$ in terms of $\mathscr{M}$, we discover by construction if this is the case. What we can and will show is that $\mathscr{M}$ itself is appropriately extendable to an analytic function of $\nu$ (or actually in the case that $K^2 = 0$, of $p_{0L}$) and, if no new singularities are introduced, $A^{(\pm)}$ and $B^{(\pm)}$ will share this property. We just sketch the way this is done, inserting the previously suppressed isotopic spinor labels (call them $\sigma$, $\sigma'$).

$$\begin{aligned} A^{(+)}\delta_{\alpha\beta}\delta_{\sigma'\sigma} + A^{(-)}\tfrac{1}{2}[\tau_\alpha, \tau_\beta]_{\sigma'\sigma} - i\gamma\cdot Q[B^{(+)}\delta_{\alpha\beta}\delta_{\sigma'\sigma} + B^{(-)}\tfrac{1}{2}[\tau_\alpha, \tau_\beta]_{\sigma'\sigma}] \\ = \mathscr{M}^{\sigma'\sigma}(p_2, p_1; q_2\alpha, q_1\beta). \end{aligned} \tag{471}$$

Now recall that we are interested only in the projection onto the spinors $\bar{u}(p_2)$, $u(p_1)$, so we multiply from the left by $(-i\gamma \cdot p_2 + m)$ and from the right by $(-i\gamma \cdot p_1 + m)$ and (471) becomes a proper matrix equation. We may easily separate out the parts involving $A^+$ and $B^+$ by setting $\alpha = \beta$, $\sigma' = \sigma$, and summing over these labels. If we take first simply the trace and then multiply by $-i\gamma \cdot Q$ and take the trace, we get two equations for $A^+$ and $B^+$ in terms of $\mathscr{M}$ and components of the momenta. Such components, after the trace is evaluated, must be expressible in terms of the fundamental scalars of our problem and cannot change the analytic structure of $\mathscr{M}$. The kinematical singularities appear from the determinant of the coefficients of $A^+$ and $B^+$ in the previously mentioned equations. In the case under discussion this reduces to $K^2(P^4Q^2 + m^4\nu^2)$. The factor of $K^2$ looks dangerous in the forward direction, but we know physically that the amplitudes $f_1$, $f_2$ [Eq. (442)] are regular at $K^2 = 0$, so such a singularity cannot occur. (It must cancel from the numerator determinant when we solve for $A^{(+)}$, $B^{(+)}$.) Similarly, in the forward direction, $m^4\nu^2 + P^4Q^2 = m^4(\nu^2 - \mu^2)$ would indicate a singularity at $\nu = q_{0L} = \mu$, which is also seen from physical considerations to be absent. We conclude therefore that our $A$'s and $B$'s were cleverly chosen and have no kinematic singularities.

Once one has gone through this rather tortuous preamble, he need not ever do it again.[61] Kinematic singularities will ultimately make themselves felt and all that really has to be done in practice is to establish the analyticity of $\mathscr{M}$, the task to which we now turn.

Using translational invariance we can remove the energy momentum transfer $\delta$-function from (457). In a coordinate system in which the meson is at rest, where $q = \{0, 0, 0, i\mu\}$, we have

$$\mathscr{M} = i(4\mu^2)^{1/2} \int d^4x e^{-ip\cdot x} D_x \langle q\alpha|\, \{\psi(x), \bar{f}(0)\}\theta(x_0)\, |q\beta\rangle, \qquad (472)$$

where $p_1 = p_2 = p$ is the nucleon four momentum in the system $\mathbf{q} = 0$ and we have written

$$\left(-\tilde{\gamma}_\mu \frac{\partial}{\partial y_\mu} + m\right)\bar{\psi}(y) \equiv \bar{f}(y).$$

It is unnecessary for us to specify $\bar{f}$ explicitly. The only properties of the operators $\psi$, $\bar{\psi}$ (or $\bar{f}$) of which we make use is that $\psi(\bar{\psi})$ acting on a state decreases (increases) the nucleon number by unity. (The nucleon number is defined as the number of nucleons minus the number of antinucleons. For reasons which are not understood in terms of any underlying symmetry principle, nucleon number seems to be a rigorously conserved quantity in

[61] Unless he writes a book.

all strong and weak interactions.) If we were to carry out the indicated differentiation in Eq. (472), because of the appearance of $\theta(x_0)$ we would encounter an equal time anticommutator, $\{\psi(x), \bar{f}(0)\}\delta(x_0)$. In the standard Lagrangian pseudoscalar meson theory this anticommutator gives a vanishing contribution to $\mathscr{M}$. This is a circumstance entirely without significance insofar as telling us whether or not any subtractions are required in the final dispersion relations. For example, in the Compton scattering amplitude a similar thing happens and we can prove that a subtraction *is* required [see Eq. (10)]. For ease of writing we shall drop the equal time anticommutator and work with the quantity

$$\mathscr{M}(p_0) = i(4\mu^2)^{1/2} \int d^4x e^{ip_0x_0 - i(p_0{}^2 - m^2)^{1/2}\hat{\mathbf{n}}\cdot\mathbf{x}} \langle q\alpha|\, [f(x), \bar{f}(0)]\theta(x_0)\,|q\beta\rangle, \tag{473}$$

where $D_x\psi(x) = f(x)$, and $\hat{\mathbf{n}}$ is a unit vector in the (irrelevant) direction of $\mathbf{p}$.

Now because of the causality condition (455) and the factor of $\theta(x_0)$, the integrand in (473) fails to vanish only if $x_0 > 0$ and inside the future light cone. All our experience would lead us to hope that $\mathscr{M}(p_0)$ could be extended to a function analytic in the upper half $p_0$-plane. There is, however, the unpleasant factor of $e^{-i(p_0{}^2 - m^2)^{1/2}\hat{\mathbf{n}}\cdot\mathbf{x}}$ which is potentially dangerous and which must be studied carefully. The absorptive part of $\mathscr{M}(p_0)$ can be written as

$$\mathscr{A}(p_0) = \pi(4\mu^2)^{1/2} \sum_n \int d^3x e^{-i\mathbf{p}\cdot\mathbf{x}} \{\langle q\alpha|\, f(\mathbf{x})\,|n\rangle\langle n|\, \bar{f}(0)\,|q\beta\rangle\delta(p_n - p_0 - \mu) + \langle q\alpha|\, \bar{f}(0)\,|n\rangle\langle n|\, f(\mathbf{x})\,|q\beta\rangle\delta(p_n + p_0 - \mu)\}. \tag{474}$$

Because of the conservation of nucleon number, the least massive state in the first term is that of a single nucleon, whereas in the second term it is a single antinucleon. The first term can therefore contribute for $p_0 > m - \mu$ and the second for $p_0 < -m + \mu$. For $p_0$ in the unphysical range, $-m < p_0 < m$, the spatial integral is not defined. For the more massive states, $m_n \geq m + \mu$, there is no trouble just as in the example of the previous section.

We now use the same trick as before, namely, of writing $\mathbf{p}^2 = p_0{}^2 - m^2$ as $\mathbf{p}^2 = p_0{}^2 - \zeta$, where $\zeta$ is first assumed negative. With this change everything is well behaved and, following our standard procedure, we may write immediately (aside from any worries concerning subtractions)

$$\mathscr{M}(p_0, \zeta) = \frac{1}{\pi} \int_{-\infty}^{\infty} dp'_0 \frac{\mathscr{A}(p'_0, \zeta)}{p'_0 - p_0 - i\eta}. \tag{475}$$

On carrying out the spatial integral in Eq. (474) we find a continuum contribution from states with $m_n \geq m + \mu$ extending from $-\infty < p_0 < -p_{0T}, p_{0T} < p_0 < +\infty$, where $p_{0T} = (m^2 - \zeta + 2m\mu)/2\mu$ (which reduces to $m$ as $\zeta \to m^2$). The discrete one nucleon and one antinucleon states make $\delta$-function contributions at $p_0 = \pm p_{0B}$, where $p_{0B} = (m^2 - \zeta - \mu^2)/2\mu \to -\mu/2$ as $\zeta \to m^2$. We have then

$$\left.\begin{aligned} \delta(p_{n0} - p_0 - \mu) &= \frac{p_{n0}}{\mu}\,\delta(p_0 - p_{0B}) \\ \delta(p_{n0} + p_0 - \mu) &= \frac{p_{n0}}{\mu}\,\delta(p_0 + p_{0B}) \end{aligned}\right\} p_{n0} = \mu + p_{0B}, \tag{476}$$

and the dispersion relation may be written as

$$\mathscr{M}(p_0, \zeta) = \frac{C_1(\zeta)}{p_{0B} - p_0} + \frac{C_2(\zeta)}{-p_{0B} - p_0} + \left(\int_{-\infty}^{-p_{0T}} + \int_{p_{0T}}^{\infty}\right) dp'_0 \frac{\mathscr{A}(p'_0, \zeta)}{p'_0 - p_0 - i\eta}. \tag{477}$$

The $C_i(\zeta)$ are defined by

$$C_1(\zeta) = (4\mu^2)^{1/2} \sum_{\text{spin}} \frac{p_{n0}}{\mu} \langle q\alpha| f\,|p_{n0}, \mathbf{p}\rangle\langle p_{n0}, \mathbf{p}|\,\bar{f}\,|q\beta\rangle,$$

$$C_2(\zeta) = (4\mu^2)^{1/2} \sum_{\text{spin}} \frac{p_{n0}}{\mu} \langle q\alpha| \bar{f}\,|p_{n0}, -\mathbf{p}\rangle\langle p_{n0}, -\mathbf{p}|\,f\,|q\beta\rangle, \tag{478}$$

where $p_{n0} = \mu + p_{0B}$ and $\mathbf{p}^2 = p_{0B}{}^2 - \zeta$. Here $|p_{n0}, \mathbf{p}\rangle$ represents the state of a single nucleon with the indicated energy and momentum, whereas $|p_{n0}, -\mathbf{p}\rangle$ is the state of a single antinucleon.

It is possible to discuss the analyticity of the relevant matrix elements in Eq. (478) by the methods used in Section 10.6, p. 634. We consider quite generally the spinor $F_\sigma$, defined as

$$F_\sigma \equiv \left(\frac{2q_0 p_0}{m}\right)^{1/2} \langle q\alpha| f_\sigma\,|p'\rangle, \tag{479}$$

where $|p'\rangle$ is a one nucleon state. From Lorentz and isotopic spin invariance it follows that the general structure of $F_\sigma$ is

$$F_\sigma = \{[i\gamma\cdot(p' - q) + m]F^{(1)}[(p' - q)^2] + F^{(2)}[(p' - q)^2]\}_{\sigma\sigma'}\,(-\,i\gamma_5\tau_\alpha)_{\sigma'\sigma''}[u(p')]_{\sigma''}, \tag{480}$$

where the appearance of the $\gamma_5$ here implies for the first time that the pion is a pseudoscalar quantity and the factor of $(-i)$ is written for convenience.

We recall that we are ultimately interested in the projection of $C_1$ on spinors $\bar{u}(p) \cdots u(p)$; by momentum conservation we will have $p' - q = p$, so the term involving $F^{(1)}$ will drop out. To follow the procedure of Section 10.6, p. 634 we "contract" the nucleon state $|p'\gamma\rangle$ and work in a system where $\mathbf{q} = 0$ to deduce a dispersion relation for $F^{(2)}$, which takes the form

$$F^{(2)}[(p' - q)^2] = -\frac{1}{\pi}\int_{(m+\mu)^2}^{\infty} d\xi \frac{\operatorname{Im} F^{(2)}(-\xi)}{\xi + (p' - q)^2 + i\eta}. \tag{481}$$

This may evidently be extended to a function of a complex variable $(p' - q)^2$ which is analytic except for a cut along the interval $-\infty < (p' - q)^2 < -(m + \mu)^2$ on the real axis. In our application we are concerned with $(p' - q)^2 = p^2 = -\zeta$ and we want $\zeta$ to vary from negative values of $\zeta = m^2$. There is no difficulty in making this continuation and $F^{(2)}(-\zeta)$ is real on the entire interval.

To complete the calculation of $C_1$, say, we use the fact that

$$[\bar{u}(p)\langle q\beta|\, f\, |p'\rangle]^* = \langle p'|\, \bar{f}\, |q\beta\rangle u(p)$$

and that we are really interested in $\bar{u}(p)C_1(\zeta)u(p)$.
We have then

$$\bar{u}(p)C_1(\zeta)u(p) = \frac{m}{\mu}\sum_{\text{spin}} \bar{u}(p)(-i\gamma_5\tau_\alpha)u(p')\bar{u}(p')(-i\gamma_5\tau_\beta)u(p)F^{(2)}(p^2)F^{(2)*}(p^2), \tag{482}$$

where $p' = p + q$ and the sum is over the spins of the intermediate states. Since $p^2 = -\zeta$ and the $F^{(2)}$'s are real, we find after a straightforward calculation that

$$\bar{u}(p)C_1u(p) = -\frac{[F^{(2)}(-\zeta)]^2}{2\mu}\bar{u}(p)i\gamma \cdot q\tau_\alpha\tau_\beta u(p), \tag{483}$$

and in a similar fashion that

$$\bar{u}(p)C_2u(p) = -\frac{[F^{(2)}(-\zeta)]^2}{2\mu}\bar{u}(p)i\gamma \cdot q\tau_\beta\tau_\alpha u(p). \tag{484}$$

We remark that if we were considering nonforward scattering we would replace $i\gamma \cdot q$ by $i\gamma \cdot Q$. The presence of the factor of $i\gamma \cdot q$ shows that in our general decomposition it is only the $B^{(\pm)}$ terms that get a contribution from the one-particle states and that the $A^{(\pm)}$ are void in the unphysical region. The final isotopic splitting is accomplished by writing $\tau_\alpha\tau_\beta = \delta_{\alpha\beta} + \frac{1}{2}(\tau_\alpha, \tau_\beta)$ and $\tau_\beta\tau_\alpha = \delta_{\alpha\beta} - \frac{1}{2}(\tau_\alpha, \tau_\beta)$.

Since the actual projection out of the $A^{(\pm)}$, $B^{(\pm)}$ does not, as we have shown, introduce any further singularities, we may simply accept the analyticity shown for $\mathscr{M}$ as being that for $A^{(\pm)}$ and $B^{(\pm)}$ and write

$$A^{(\pm)}(p_0, \zeta) = \frac{1}{\pi}\left(\int_{-\infty}^{-p_{0T}} + \int_{p_{0T}}^{\infty}\right) dp'_0 \, \frac{\operatorname{Im} A^{(\pm)}(p'_0, \zeta)}{p'_0 - p_0 - i\eta}, \tag{485}$$

$$\begin{aligned} B^{(\pm)}(p_0, \zeta) = {} & \frac{[F^{(2)}(-\zeta)]^2}{2\mu}\left(\frac{1}{p_{0B} - p_0} \mp \frac{1}{p_{0B} + p_0}\right) \\ & + \frac{1}{\pi}\left(\int_{-\infty}^{-p_{0T}} + \int_{p_{0T}}^{\infty}\right) dp'_0 \, \frac{\operatorname{Im} B^{(\pm)}(p'_0, \zeta)}{p'_0 - p_0 - i\eta}. \end{aligned}$$

Choosing $p_0 > m$, the analytic continuation of the left-hand sides of these equations to $\zeta = m^2$ is straightforward, since this may be done for the terms under the integrals. The function $F^{(2)}$ has been shown to be suitably analytic, the quantity $F^{(2)}(-m^2)$ being called the "coupling constant" $g$ and representing the amplitude for the unphysical transition nucleon + pion $\rightarrow$ nucleon. Using the even-odd properties of $A^{(\pm)}$ and $B^{(\pm)}$ [see Eq. (468)] and observing that in the limit $\zeta \rightarrow m^2$, $p_{0T} \rightarrow m$ and $p_{0B} \rightarrow -\mu/2$, we obtain the relations

$$\begin{aligned} A^{(\pm)}(p_0) = {} & \frac{1}{\pi}\int_m^{\infty} dp'_0 \operatorname{Im} A^{(\pm)}(p'_0)\left(\frac{1}{p'_0 - p_0 - i\eta} + \frac{1}{p'_0 + p_0 + i\eta}\right), \\ B^{(\pm)}(p_0) = {} & \frac{g^2}{2\mu}\left(\frac{1}{p_{0B} - p_0} \mp \frac{1}{p_{0B} + p_0}\right) \\ & + \frac{1}{\pi}\int_m^{\infty} dp'_0 \operatorname{Im} B^{(\pm)}(p'_0)\left(\frac{1}{p'_0 - p_0 - i\eta} \mp \frac{1}{p'_0 - p_0 + i\eta}\right). \end{aligned} \tag{486}$$

It is more conventional to express these dispersion relations in terms of the energy of the meson in the system in which the nucleon is at rest. Since the $A$'s and $B$'s are invariants, this amounts only to a change of variable, namely, $p_0 = (m/\mu)q_0$. (This follows from evaluating the invariant $p \cdot q$ in the two coordinate frames.) Without bothering to define new symbols for $A[(m/\mu)q_0]$, and so on, we write simply

$$\begin{aligned} A^{(\pm)}(q_0) = {} & \frac{1}{\pi}\int_\mu^{\infty} dq'_0 \operatorname{Im} A^{(\pm)}(q'_0)\left(\frac{1}{q'_0 - q_0 - i\eta} \pm \frac{1}{q'_0 + q_0 + i\eta}\right), \\ B^{(\pm)}(q_0) = {} & \frac{g^2}{2m}\left(\frac{1}{q_{0B} - q_0} \mp \frac{1}{q_{0B} + q_0}\right) \\ & + \frac{1}{\pi}\int_\mu^{\infty} dq'_0 \operatorname{Im} B^{(\pm)}(q'_0)\left(\frac{1}{q'_0 - q_0 - i\eta} \mp \frac{1}{q'_0 + q_0 + i\eta}\right), \end{aligned} \tag{487}$$

where $q_{0B} = -\mu^2/2m$.

The dispersion relations for forward pion-nucleon scattering are usually written in terms of the particular combination of $A$'s and $B$'s which make up the forward scattering amplitude in the laboratory system, that is, $(A + q_0B)/4\pi = f_L(0)$. To avoid subscripts and confusion with a well-established and conventional notation we call this quantity $F(q_0)$. We find then from Eq. (487) that

$$F^{(+)}(q_0) = \frac{f^2}{m}\frac{\mu^2}{q_{0B}^{\;2} - q_0^{\;2}} + \frac{2}{\pi}\int_\mu^\infty dq'_0 q'_0 \frac{\operatorname{Im} F^{(-)}(q'_0)}{q'^{\;2}_0 - q_0^{\;2} - i\eta}$$
$$- \frac{2}{\pi}\int_\mu^\infty dq_0 \operatorname{Im} B^{(+)}(q_0) - \frac{4m}{\mu^2} f^2, \tag{488}$$
$$F^{(-)}(q_0) = \frac{-2f^2}{\mu}\frac{\mu q_0}{q_{0B}^{\;2} - q_0^{\;2}} + \frac{2q_0}{\pi}\int_\mu^\infty dq'_0 \frac{\operatorname{Im} F^{(-)}(q'_0)}{q'^{\;2}_0 - q_0^{\;2} - i\eta},$$

where $f^2 = (\mu g/2m)^2/4\pi$. The fact that a constant appears in the first of Eqs. (488) (that is, the last two terms) is entirely without significance in that we could, for example, have added a constant in the $A^{(+)}$ dispersion relation which would have precisely cancelled this new one. We shall return to this point shortly.

Before going on to discuss applications of Eqs. (487) and (488), we summarize what went into the derivation. We have used the formalism of quantum field theory only in a rather general fashion. Aside from the generally accepted ideas, such as Lorentz invariance and the existence of a complete set of positive energy eigenstates of the energy momentum four-vector, our principal tool was the reduction formula for the $S$-matrix, which followed from the asymptotic condition. Another critical step was the causality, or local commutativity condition $\{\psi(x), \bar{\psi}(y)\} = 0$ for $(x - y)^2 > 0$. The physical significance of this condition is very far from clear. Actually, for the forward scattering dispersion relations we have required this condition for only special matrix elements and any experimental verification of our results certainly does not establish the condition for all matrix elements. It has not been necessary to specify any precise form of interaction between mesons and nucleons; only the specification of the mass-spectrum of the stationary states has played any role. Thus we conclude that the correctness of the dispersion relations, if established, is not a strong argument in favor of field theory since it explores so little of it. On the other hand, if we were to encounter a contradiction with experience, it would be shattering to much of present theoretical physics.

In order to compare Eqs. (488) with experiment we must determine all the constants appearing in these. We begin by making the most daring conjecture, namely, that the last two constant terms in the first of (488) cancel each other. In terms of the scattering amplitudes for pions of definite charge [see Eq. (426)] and with use of the optical theorem, $\sigma = 4\pi \operatorname{Im} F/q$, $q = (q_0^2 - \mu^2)^{1/2}$, we have

$$F^{(+)}(q_0) = \tfrac{1}{2}[F_-(q_0) + F_+(q_0)]$$
$$= \frac{f^2}{m}\frac{\mu^2}{q_{0B}^2 - q_0^2} + \frac{1}{4\pi^2}\int_\mu^\infty dq'_0 q'_0 q' \frac{\sigma_-(q'_0) + \sigma_+(q'_0)}{q'^2_0 - q_0^2 - i\eta},$$
$$F^{(-)}(q_0) = \tfrac{1}{2}[F_-(q_0) - F_+(q_0)] \qquad (489)$$
$$= -\frac{2f^2}{\mu}\frac{\mu q_0}{q_{0B}^2 - q_0^2} + \frac{q_0}{4\pi^2}\int_\mu^\infty dq'_0 q' \frac{\sigma_-(q'_0) - \sigma_+(q'_0)}{q'^2_0 - q_0^2 - i\eta},$$

where $F_- = F(\pi^- + p \to \pi^- + p)$, $F_+ = F(\pi^+ + p \to \pi^+ + p)$.

It is known experimentally that $\sigma_+ \simeq \sigma_- =$ constant at very high energies. In fact, $\sigma_+ \simeq \sigma_- \approx 30$ mb [62] from about 2 Bev to 20 Bev. In addition, there is no indication from cosmic ray interactions that $\sigma_+$ and $\sigma_-$ go to zero at even higher energies. As a result, the first of Eqs. (489) is certainly meaningless since the integral appears to diverge. Even if $\sigma_+$, $\sigma_- \to 0$ at some absurdly high energy, the equation would be quite useless to us as written. We are therefore forced in practice to make a subtraction, which we do at $q_0 = \mu$ $(q = 0)$ (that is, at zero kinetic energy).

$$F^{(+)}(q_0) = F^{(+)}(\mu) + \frac{f^2}{m}\frac{q^2\mu^2}{(\mu^2 - q_{0B}^2)(q_0^2 - q_{0B}^2)}$$
$$+ \frac{q^2}{4\pi^2}\int_\mu^\infty \frac{dq'_0 q'_0}{q'}\frac{\sigma_-(q'_0) + \sigma_+(q'_0)}{q'^2_0 - q_0^2 - i\eta}. \qquad (490)$$

It is sometimes argued that the appearance of the constant $F^{(+)}(\mu)$ is an explicit reflection of the fact that in conventional meson theory, in addition to the coupling constant $f$, there is another constant which is related to meson-meson scattering. The second of Eqs. (489) does not lead to any obvious trouble, and we shall see shortly that it appears to be consistent with experiment as it stands.

At the moment, there are two constants in the theory, namely $F^{(+)}(\mu)$ and $f^2$. We may go directly to experiment for $F^{(+)}(\mu)$, although the theoretical interpretation is somewhat troublesome because of electromagnetic effects which become important at low energies and which have been omitted in our theory. (Isotopic spin conservation is violated by electromagnetic interactions.) The determination of $f^2$ may be carried out

[62] One mb is the equivalent of $10^{-27}$ cm$^2$.

quite accurately using experiment and Eqs. (489). To do this, we rewrite the second of Eqs. (489) using the trivial identity,

$$\frac{1}{{q'_0}^2 - {q_0}^2} = \frac{1}{{q'_0}^2} + \frac{{q_0}^2}{{q'_0}^2({q'_0}^2 - {q_0}^2)} . \tag{491}$$

Taking the real part of this equation, we find that

$$\left[\frac{\text{Re}\, F^{(-)}(q_0)}{q_0} - \frac{{q_0}^2}{4\pi^2}\int_\mu^\infty dq'_0 \frac{q'}{{q'_0}^2} \frac{\sigma_-(q'_0) - \sigma_+(q'_0)}{{q'_0}^2 - {q_0}^2}\right]({q_0}^2 - {q_{0B}}^2)$$
$$= 2f^2 + \frac{{q_0}^2 - {q_{0B}}^2}{4\pi^2}\int_\mu^\infty dq'_0 \frac{q'}{{q'_0}^2}[\sigma_-(q'_0) - \sigma_+(q'_0)]. \tag{492}$$

This remarkable equation states that the reciprocal of the complicated quantity in brackets is a linear function of $q_0^2$. From a measurement of the total cross sections and the phase shift analysis of the experimental data, the left-hand side of the equation may be evaluated and $2f^2$ found by an extrapolation of the straight line to the point $q_0^2 = q_{0B}^2$. We may express Re $F^{(-)}$ in terms of phase shifts according to Eq. (445).

$$\text{Re}\, F^{(-)} = \frac{W}{m}\,\text{Re}\sum_{l=0}[(l+1)f_{l+}(q_c) + lf_{l-}(q_c)], \tag{493}$$

where $W^2 = m^2 + \mu^2 + 2mq_0$ and $q_c = (m/W)q_{1\text{ab}}$, and the appropriate isotopic combination is to be taken. Ideally we would try to use a direct measurement of the forward amplitude rather than a phase shift analysis, but this is experimentally difficult. The result of these considerations, first made by Haber-Schaim [40] is that $f^2 = 0.08$. This method is a very good one because the integral in the brackets in Eq. (492) is not sensitive to the very high-energy behavior of the cross sections.

The reason for the belief that the dispersion relation for $F^{(-)}(q_0)$ requires no subtraction is that if it is evaluated at $q_0 = \mu$, using experimental data for the cross sections and $f^2 = 0.08$, the value obtained agrees well with the measured value of $F^{(-)}(\mu)$. It is perhaps worth remarking that the very small value of $F^{(+)}(\mu)$, which is about $(-0.002/\mu)[(m + \mu)/\mu]$ as compared to $F^{(-)}(\mu) \approx (0.27/\mu)[(m + \mu)/\mu]$, is one of the great mysteries of pion physics. Such a very strong isotopic spin dependence at low energies was unexpected. The crossing relation, Eq. (465), may be shown to lead to the conclusion that at *zero* total energy, $F^{(-)}(0)$ should, in fact, be zero and that $F_{I=1/2}(0) = F_{I=3/2}(0)$.

Historically, one of the first applications of the forward dispersion relations[63] was to the unraveling of the energy dependence of the phase

[63] Reference [41].

shifts for pion-nucleon scattering. By insisting that any set obtained should satisfy the dispersion relations, one is able to discard four of six possible solutions which, within the experimental errors, fit the angular distribution data equally well. One is left with two sets: the Fermi set which is characterized by a large phase shift in the state $I = \frac{3}{2}$, $J = \frac{3}{2}$ and the Yang set which has a large one in $I = \frac{3}{2}$, $J = \frac{1}{2}$.

To distinguish between these two it is necessary to work with a different combination of phase shifts than appear in the forward scattering amplitude. This would be provided by a measurement of the polarization of recoil protons. We can do this "measurement" theoretically as follows: The forward amplitude is given in the center-of-mass system by $f_1 + f_2$, [See Eq. (442)] evaluated in the forward direction. We can, however, using the dispersion relations for $A$ and $B$, write a dispersion relation for $f_2$ alone. This was done by Davidon and Goldberger [42] with the result that the Fermi assignment is unambiguously selected.

It is impossible to attempt any sort of construction of a dynamical theory on the basis of the forward scattering dispersion relations. It turns out to be possible, just as in potential scattering, to derive analogous equations for fixed momentum transfer. The first attempts in this direction were made by Toll and Wheeler [10] and by Bethe and Rohrlich [43], who did not pursue the matter in detail. Dispersion relations for derivatives of the scattering amplitude at zero angle were discussed in the original paper of Gell-Mann, Goldberger, and Thirring [11]. Nonforward dispersion relations, as we now know them, were first presented in the fall of 1955 by Goldberger, Nambu and Oehme [37], by Gell-Mann and Polkinghorne [44], by Symanzik [45], by Salam [46], and by Capps and Takeda [47]. All these "derivations" were identical and were very similar to the heuristic derivation first given for forward scattering by Goldberger. The only modification required was the use of the Breit coordinate system, which is the center-of-mass system for the initial and final target particle. Holding the magnitude of this target momentum (which is one-half the momentum transfer in the scattering) fixed, we "prove" that the amplitude is an analytic function of projectile energy in this system. Since the results were correct, it was ultimately possible to give a rigorous proof. Somewhat after the important applications were made [48], Bogoliubov and co-workers [49], in an unbelievable *tour de force*, succeeded in giving a derivation from field theory. Considerably simpler derivations have been given since that time by Bremmerman, Oehme, and Taylor [50] and by Lehmann [51]. We shall content ourselves with a very brief summary of the method used by these people.

It is clear that there is a profound difference between the forward and nonforward scattering problems. In forward scattering we have seen that the unphysical region is void, except for the contribution of a one-particle state which gives rise to poles in the amplitude. We saw in the nonrelativistic theory that when the momentum transfer is finite, there is a continuum contribution corresponding to wave numbers less than half the fixed momentum transfer. One can easily handle (as we did) a discrete state, but a continuum is much more difficult.

Just to indicate some of the elements involved in nonforward dispersion relations, let us pretend that we wish to give a derivation of these. We would begin with Eqs. (457) and (463), written a little more symmetrically as

$$\mathscr{M}(p_1, p_2; q_2\alpha, q_1\beta) = i(4q_{10}q_{20})^{1/2} \int d^4x e^{-i[(p_1+p_2)/2]x} \left\langle q_2\alpha \middle| R\psi\left(\frac{x}{2}\right)\bar{\psi}\left(\frac{y}{2}\right) \middle| q_1\beta \right\rangle, \quad (494)$$

where

$$R[\psi(x)\bar{\psi}(y)] \equiv \{\psi(x), \bar{\psi}(y)\}\theta(x_0 - y_0). \quad (495)$$

Aside from certain degenerate terms arising from equal time anticommutators, $R[\psi(x/2)\bar{\psi}(-x/2)]$ vanishes, except for $x_0 > 0$ and $x^2 < 0$. This suggests looking for a dispersion relation in a variable related to $(p_1 + p_2)_0$. The most natural and symmetric timelike vector to use in defining such a variable (analogous to the use of the target rest system in forward scattering) is $q_1 + q_2$. In order to have all the relevant functional dependence of $\mathscr{M}$ explicit, we would like to hold the states $|q_2\alpha\rangle$ and $|q_1\beta\rangle$ fixed. It is natural then to choose the system so that $\mathbf{q}_1 + \mathbf{q}_2 = 0$ and to hold $\mathbf{q}_1 = -\mathbf{q}_2$ fixed. We write

$$\begin{aligned}
&\mathbf{q}_1 = \frac{\mathbf{\Delta}}{2}, \qquad \mathbf{q}_2 = -\frac{\mathbf{\Delta}}{2}, \\
&\mathbf{p}_1 = \mathbf{P} - \frac{\mathbf{\Delta}}{2}, \qquad \mathbf{p}_2 = \mathbf{P} + \frac{\mathbf{\Delta}}{2}, \\
&P \equiv \frac{p_1 + p_2}{2} = \left[\mathbf{P}, i\,\frac{(p_1 + p_2)_0}{2}\right], \\
&K \equiv q_1 - q_2 = (\mathbf{\Delta}, i0), \\
&Q \equiv \frac{q_1 + q_2}{2} = (0, i\sqrt{\mu^2 + \Delta^2/4}).
\end{aligned} \quad (496)$$

We note that $|\boldsymbol{P}| = (P_0{}^2 - m^2 - \Delta^2/4)^{1/2}$, so that for a physical scattering we must have $P_0 > (m^2 + \Delta^2/4)^{1/2}$ for given $\Delta$. The scalar variable that reduces to $P_0$ in our special reference frame is

$$\omega = \frac{-P \cdot Q}{(-Q^2)^{1/2}};$$

it is related to the scalar $\nu$ by the relation $\omega = m\nu/(-Q^2)^{1/2}$. Since we shall hold $\Delta$ fixed, the ratio of $\omega$ to $\nu$ is a constant. All the previously established crossing relations are true when expressed in terms of $\omega$.

We see from the representation (494) that we have potential trouble when the spatial exponential involving $\boldsymbol{P}$ can "blow up." This can happen for $P_0 < (m^2 + \Delta^2/4)$. If we look at the absorption amplitude, before carrying out the spatial integration, we find now in addition to the one-particle states encountered with forward scattering, a contribution from two and higher particle states. The one-particle states cause no trouble and may be handled as before. If $\Delta^2$ is slightly less than $12\mu^2$, only two-particle states contribute to the unphysical region and it is possible that special techniques could be devised to handle these. For larger $\Delta^2$, the situation looks hopeless by this method. If we simply close our eyes to these grave difficulties and formally carry out the spatial integrations, the correct answer emerges. We pretend to replace, in the dangerous exponential, $m^2$ by $\zeta$ and assume that $\zeta < -\Delta^2/4$; we obtain

$$\mathscr{M}(\omega, \Delta^2, \zeta) = \int_{-\infty}^{\infty} \frac{d\omega'}{\pi} \frac{\operatorname{Im} \mathscr{M}(\omega', \Delta^2, \zeta)}{\omega' - \omega - i\eta}.$$

Evaluating everything formally and setting $\zeta = m^2$ when it is all over leads finally to the nonforward dispersion relations, which we write in the form

$$A^{(\pm)}(w^2, \Delta^2) = \frac{1}{\pi} \int_{(m+\mu)^2}^{\infty} dw'^2 \operatorname{Im} A^{(\pm)}(w'^2, \Delta^2) \times \left( \frac{1}{w'^2 - w^2 - i\eta} \pm \frac{1}{w'^2 - 2m^2 - 2\mu^2 - \Delta^2 + i\eta} \right),$$

$$B^{(\pm)}(w^2, \Delta^2) = g^2 \left( \frac{1}{m^2 - w^2} \mp \frac{1}{m^2 - 2m^2 - 2\mu^2 - \Delta^2 + w^2} \right) + \frac{1}{\pi} \int_{(m+\mu)^2}^{\infty} dw'^2 \operatorname{Im} B^{(\pm)}(w'^2, \Delta^2) \times \left( \frac{1}{w'^2 - w^2 - i\eta} \mp \frac{1}{w'^2 - 2m^2 - 2\mu^2 - \Delta^2 + w^2 + i\eta} \right).$$

In writing these equations we have used the total center-of-mass energy squared, $w^2$, rather than $\omega$ to express the result. The coupling constant $g$ is exactly what appeared in the forward case. The imaginary parts of $A^{(\pm)}$, $B^{(\pm)}$ are to be obtained by appropriately disentangling the expression

$$\mathscr{A} = \tfrac{1}{2}(4q_{10}q_{20})^{1/2} \int d^4x e^{-i[(p_1+k)/2]x} \left\langle q_2\alpha \middle| f\left(\frac{x}{2}\right) \bar{f}\left(-\frac{x}{2}\right) \middle| q_1\beta \right\rangle,$$

where as usual we write $f(x) = D_x\psi(x)$, $\bar{f}(y) = \bar{D}_y\bar{\psi}(y)$.

This expression for the absorptive part of $\mathcal{M}$ is nothing more than the statement of unitarity and thus is valid only in the physical region. The absorptive parts are to be found in the unphysical region by analytic continuation. This has been shown by Lehmann [51] to be possible by means of a Legendre polynomial expansion so long as $\Delta^2$ does not exceed a certain value $\Delta^2_{\text{max}}$, which is precisely the maximum value for which the rigorous derivation of dispersion relations has been given.

$$\Delta^2_{\text{max}} = \frac{32}{3}\mu^2\frac{2m+\mu}{2m-\mu} \approx 12.4\mu^2.$$

There does not seem to be any physical basis for this limitation. It is widely believed that the relations themselves are true for arbitrary momentum transfers, although the Legendre polynomial continuation method has a definite limit.

There have been, unfortunately, no real experimental checks on the nonforward dispersion relations. Such tests are much more meaningful than for forward scattering, since we essentially require the causality condition, $[\psi(x), \bar{\psi}(y)] = 0$, $(x - y)^2 > 0$, as an operator relation (that is, for arbitrary matrix elements) in the rigorous derivation.

We now conclude this chapter on dispersion theory. A summary of what dispersion relations have been proved and by what methods may be found in a review by Goldberger [52]. We shall not discuss the attempts to use the nonforward relations as a basis for a dynamical theory because they are largely superseded by the double dispersion approach.[64]

## REFERENCES


1. M. L. Goldberger, Princeton University, 1959–1960, unpublished; *Dispersion Relations and Elementary Particles* (John Wiley and Sons, New York, 1961).
2. R. Kronig, *Physica* **12,** 543 (1946).
3. R. Kronig, *J. Amer. Optical Soc.* **12,** 547 (1926). H. A. Kramers, Atti del Congress Internationale de Fisici Como (1927).
4. E. Sellmeir, *Pogg. Ann.* **143,** 271 (1871).
5. See, for example, E. Fermi, *Lectures on Nuclear Physics* (University of Chicago Press, Chicago, 1950).
6. M. Gell-Mann, M. L. Goldberger, and W. Thirring, *Phys. Rev.* **95,** 1612 (1954).
7. W. Schutzer and J. Tiomno, *Phys. Rev.* **83,** 249 (1951).
8. N. G. van Kampen, *Phys. Rev.* **91,** 1267 (1953).
9. N. G. van Kampen, *Phys. Rev.* **89,** 1072 (1953).
10. E. P. Wigner, *Amer. J. Phys.* **23,** 371 (1955).
11. J. S. Toll, Princeton Thesis (1952). F. Rohrlich and R. L. Gluckstern, *Phys. Rev.* **86,** 1 (1952).
12. M. L. Goldberger, *Phys. Rev.* **97,** 508 (1955).


[64] These early efforts are described in Ref. [48].

13. S. Mandelstam, *Phys. Rev.* **112,** 1344 (1958); *ibid,* **115**, 1741, 1759 (1959).
14. N. Khuri, *Phys. Rev.* **107,** 1148 (1957).
15. G. Becker and E. Leibfried, *Phys. Rev.* **69,** 34 (1946).
16. A. Klein and C. Zemach, *Nuovo Cimento* **10,** 1078 (1958).
17. E. T. Whittaker and G. N. Watson, *Modern Analysis* (Cambridge University Press, 1943).
18. H. Lehmann, *Nuovo Cimento* **10,** 579 (1958).
19. P. M. Morse and H. Feshbach, *Methods of Theoretical Physics* (McGraw-Hill Book Co., New York, 1953).
20. R. Blankenbecler, M. L. Goldberger, N. Khuri and S. Treiman, *Ann. Phys.* **10,** 62 (1960).
21. T. Regge, *Nuovo Cimento* **14,** 951 (1959).
22. R. Blankenbecler and M. L. Goldberger, International Conference on Weak and Strong Interactions, La Jolla, California, June 1960, unpublished, and *Phys. Rev.* **126,** 766 (1962).
23. G. Chew and S. Mandelstam, *Phys. Rev.* **119,** 467 (1960).
24. E. Henley and W. Thirring, *Elementary Field Theory* (McGraw-Hill Book Co., New York, 1962) or F. Mandl, *Introduction to Quantum Field Theory* (Interscience Publishers Ltd., London, 1959).
25. R. Haag, Dan. Vid. Selsk. *Mat. Fys. Med.* **29,** No. 12 (1955). A. Wightman, 1953, unpublished; see also A. Wightman and S. Schweber, *Phys. Rev.* **98,** 812 (1955).
26. H. Lehmann, K. Symanzik, and W. Zimmerman, *Nuovo Cimento* **1,** 205 (1955)
27. C. N. Yang and D. Feldman, *Phys. Rev.* **79,** 972 (1952). G. Källen, *Arkiv Fysik* **2,** 371 (1950).
28. O. W. Greenberg and A. S. Wightman, unpublished.
29. R. Haag, *Suppl. Nuovo Cimento* **14,** 131 (1959). K. Nishijima, *Phys. Rev.* **122,** 298 (1961). W. Zimmermann, *Nuovo Cimento* **16,** 690 (1960).
30. N. N. Bogoliubov and D. V. Shirkov, *Introduction to the Theory of Quantized Fields* (Interscience Publishers, New York, 1959).
31. R. Oehme, *Nuovo Cimento* **4,** 1316 (1956).
32. P. Federbush, M. L. Goldberger, S. Treiman, *Phys. Rev.* **112,** 642 (1958). W. Frazer and J. Fulco, *Phys. Rev.* **117,** 1609 (1960).
33. M. L. Goldberger and S. Treiman, *Phys. Rev.* **110,** 1178 (1958).
34. M. L. Goldberger and S. Treiman, *Nuovo Cimento* **9,** 451 (1958).
35. M. L. Goldberger and S. Treiman, *Phys. Rev.* **111,** 354 (1958).
36. R. Blankenbecler and L. Cook, *Phys. Rev* **119,** 1745 (1960).
37. M. L. Goldberger, Proceedings of the Rochester Conference (1956).
38. K. Symanzik, *Phys. Rev.* **105,** 743 (1957).
39. These relations were first given explicitly by M. L. Goldberger, *Phys. Rev.* **99,** 979 (1955); the derivation presented there was a heuristic one and it remained for Symanzik [36] to provide a rigorous one. See also M. L. Goldberger, H. Miyazawa, and R. Oehme, *Phys. Rev.* **99,** 981 (1955), R. Karplus and M. Ruderman, *Phys. Rev.* **98,** 771 (1955).
40. U. Haber-Schaim, *Phys. Rev.* **104,** 1113 (1956).
41. H Anderson, W. Davidon, and U. Kruse, *Phys. Rev.* **100,** 339 (1955).
42. W. Davidon and M. L. Goldberger, *Phys. Rev.* **104,** 119 (1956).
43. H. A. Bethe and F. Rohrlich, *Phys. Rev.* **86,** 7 (1952).
44. M. Gell-Mann and J. Polkinghorn, unpublished.
45. K. Symanzik, unpublished.
46. A. Salam, *Nuovo Cimento* **3,** 424 (1956).

47. R. Capps and G. Takeda, *Phys. Rev.* **103,** 1877 (1956).
48. G. F. Chew, M. L. Goldberger, F. Low, and Y. Nambu, *Phys. Rev.* **106,** 1337, 1345 (1957).
49. N. N. Bogoliubov, B. V. Medvedev, M. K. Polivanov, *Fortschritte Physik* **6,** 169 (1958). This work was first reported at the Seattle Conference in September 1956.
50. H. Bremmermann, R. Oehme, and J. G. Taylor, *Phys. Rev.* **109,** 2178 (1958).
51. H. Lehmann, *Suppl. Nuovo Cimento* **14,** (1959).
52. M. L. Goldberger, *Proceedings of the International Conference on High Energy Physics*, Geneva, 1958.

# 11. SCATTERING BY SYSTEMS OF BOUND PARTICLES

In this chapter we shall describe techniques for studying scattering by a system of bound particles. Such interactions are among those most commonly encountered in physics—they include, for example, the entire fields of chemical reactions and nuclear reactions and also the study of microscopic mechanisms in irreversible thermodynamics. Unfortunately, however, no generally satisfactory methods are available for treating the scattering of composite systems, since such phenomena usually have the full complexity of the many-body problem in dynamics. On the other hand, for several limiting situations approximation methods are available. For a much wider class of interactions there are elaborate heuristic, or semi-heuristic, techniques. It will evidently not be possible for us to describe such varied phenomena in a uniform way. Instead we shall choose to study only certain of the methods which illustrate the theoretical development given in earlier chapters of this book and which have a considerable range of applications.

Perhaps the simplest conditions for studying scattering by a composite system are those in which the binding energy of a particle in the composite system (scatterer) is very small compared to the kinetic energy of the incident, elementary (scattered) particle. Then it is usually possible to make a quantitative theoretical study of the scattering, including an estimate of the accuracy of the approximations used.

We shall refer to this combination of conditions as the "weak binding" condition. Because of its relative simplicity and wide applicability, we shall describe scattering under conditions of "weak binding" in some detail. Classic illustrations include the study of crystalline structure and short-range order in liquids by x-ray or neutron scattering. Related techniques are available for investigating atomic and molecular structure by x-ray or electron scattering. Properties of atomic nuclei have also been studied by a variety of such methods using fast electrons, nucleons, and pi-mesons.

We shall begin our discussion of scattering by "weakly bound" systems in Section 11.1, where the scatterer will be assumed to consist of a single particle bound in a potential well. The relation to scattering by the same particle when it is not bound is studied. The asymptotic wave function is then obtained to exhibit the recoil kinematics of the scatterer. The discussion is extended in Section 11.2 to the description of scattering by a "small" system of weakly bound particles. Particular emphasis is placed on the effect of the structure of the bound system on the scattering cross section. In Sections 11.3, 11.4, and 11.5 the phenomena of multiple scattering and coherence are examined.

The "pick-up" and "stripping" reactions are described in Section 11.6. These illustrate our general discussion given in Chapter 4 of rearrangement collisions and of wave function symmetrization.

Finally, in Section 11.7 we review some special theoretical methods used in atomic scattering reactions. Here the rather detailed available knowledge of atomic structure is of course very helpful. Indeed, most methods used to study scattering by atoms exploit in some way the central field approximation. An exception to this, however, is the successful application of variational techniques to electron-hydrogen scattering.

As we have remarked, it will not be possible to describe here many of the (often quite specialized) techniques developed in nuclear, atomic, and molecular physics. In particular, we shall not discuss the elegant methods employed in analyzing nuclear reactions.[1] Also, we shall be unable to review the interesting, but largely undeveloped, theory of "near adiabatic" reactions involving atoms and molecules.[2]

## 11.1 SCATTERING FROM A SINGLE WEAKLY BOUND PARTICLE

In this section we consider the scattering of one elementary particle by another which is "weakly bound" in a fixed potential. By "weakly bound" we mean to imply that the binding energy of the bound scatterer is very small compared to the kinetic energy of the incident particle. Scattering in the limit of "quasi-classical" binding will also be discussed.

### General Description of the Scattering

The bound scatterer, hereafter called "target particle," is assumed to have a mass $M$ and spin $S_t$ and to be bound in a fixed potential $U(\mathbf{z})$, centered at the point $\mathbf{z} = 0$.[3] The target particle will be treated as nonrelativistic, so it has a kinetic energy operator

$$K_t = -\frac{1}{2M}\nabla_z^2. \tag{1}$$

[1] See, for example, Ref. [1].

[2] See, for example, Ref. [2, p. 446].

[3] We may, if we prefer, consider $U$ to result from the interaction with a very massive particle.

Its Hamiltonian is therefore

$$h = K_t + U. \tag{2}$$

In the initial bound state the wave function is $g_0(\mathbf{z}, s_t)$, where $s_t$ is the spin variable of the target particle (more generally, we interpret $s_t$ as representing all internal degrees of freedom). This satisfies the Schrödinger equation

$$hg_0 = W_0 g_0, \tag{3a}$$

where $W_0$ is the binding energy of the target particle.

After scattering has occurred, the state of the scatterer will in general be changed. To take account of the possibility that the target particle may recoil to an unbound state, we write $g_\gamma^\pm(\mathbf{z}, s_t)$ for the wave function following scattering. This satisfies the Schrödinger equation

$$hg_\gamma^\pm = W_\gamma g_\gamma^\pm, \tag{3b}$$

corresponding to eigenvariables $\gamma$ and energy $W_\gamma$.

The scattered (or "incident") particle has a mass $m$ and spin $S_i$. Its free states are of the form

$$\chi_{i;\mathbf{p},\nu} = (2\pi)^{-3/2}\, e^{i\mathbf{p}\cdot\mathbf{x}}\, u_{S_i,\nu}(s_i). \tag{4}$$

Here $s_i$ represents the spin (and any other) degrees of freedom and $\nu$ the eigenvalue of the azimuthal component of the spin $\mathbf{S}_i$. These states satisfy the Schrödinger equation

$$K_i \chi_{i;\mathbf{p},\nu} = \epsilon_p \chi_{i;\mathbf{p},\nu}, \tag{5}$$

where $K_i$ is the kinetic energy operator for the "incident" particle and

$$\epsilon_p = \frac{p^2}{2m} \quad \text{for} \quad \text{nonrelativistic energies,} \tag{6a}$$

$$= c(m^2c^2 + p^2)^{1/2} \quad \text{for} \quad \text{relativistic energies.} \tag{6b}$$

The initial state of the system prior to scattering is taken as

$$\chi_a = (2\pi)^{-3/2}\, e^{i\mathbf{p}\cdot\mathbf{x}}\, u_{S_i,\nu} g_0(\mathbf{z}, s_t). \tag{7}$$

A possible final state following the scattering is then

$$\chi_b^\pm = (2\pi)^{-3/2}\, e^{i\mathbf{k}\cdot\mathbf{x}}\, u_{S_i,\nu'} g_\gamma^\pm(\mathbf{z}, s_t). \tag{8}$$

The wave functions which describe the system before and after the scattering, Eqs. (7) and (8), satisfy the equation

$$K\chi_b^\pm = (\epsilon_k + W_\gamma)\chi_b^\pm, \tag{9}$$

with

$$K \equiv K_i + h = K_i + K_t + U. \tag{10a}$$

The interaction between the incident and target particles will be written as $V$. The Hamiltonian for the entire system is therefore

$$\begin{aligned} H &= K_i + h + V \\ &= K + V \qquad (10b) \\ &= K_i + K_t + V + U. \end{aligned}$$

The scattering process which we are describing provides an illustration of "scattering by two potentials," as was developed in Section 5.4. The states (8) represent a special case of those of Eq. (5.102). According to Eq. (5.110), the steady-state wave function for the complete scattering process is

$$\psi_a{}^+ = \chi_a{}^+ + \frac{1}{E_a + i\eta - K}\mathscr{T}_b\chi_a, \qquad (11)$$

where

$$E_a = \epsilon_p + W_0 \qquad (12)$$

and [see Eq. (5.111)]

$$\mathscr{T}_b = V + V\frac{1}{E_a + i\eta - H}V. \qquad (13)$$

The element of the scattering matrix for the transition from $a$ to $b$ is given by Eq. (5.111) as

$$\langle b|\mathscr{T}_b|a\rangle \equiv \langle \mathbf{k}, \nu', \gamma|\mathscr{T}_b|\mathbf{p}, \nu, 0\rangle = (\chi_b{}^-, \mathscr{T}_b\chi_a), \qquad (14)$$

since the first term in Eq. (5.111) vanishes for our application.[4]

The differential scattering cross section may be obtained from Eq. (3.137); it is

$$d\sigma = \frac{(2\pi)^4}{v_{\text{rel}}}\sum_{\nu',\gamma}\int d^3k\,\delta(\epsilon_k + W_\gamma - \epsilon_p - W_0)\,|\langle b|\mathscr{T}_b|a\rangle|^2. \qquad (15a)$$

Here $v_{\text{rel}} = d\epsilon_p/dp$ is the initial velocity of the incident particle $i$. The sum over $\nu'$ and $\gamma$ may be arbitrarily restricted here.[5] For a discrete, bound

[4] It should be noted that we are using the symbol $\chi_b{}^\pm$, and so on, for the states denoted by $\phi_\beta{}^\pm$ in Section 5.4. This change in notation is made in conformity with our conventional use of $\chi$ for an initial or final state.

The term $(\phi_\beta{}^-, U\chi_a)$ in Eq. (5.111) contains no operator to change the state of the incident particle. Thus, except in the singular case that $\mathbf{p} = \mathbf{k}$, $\nu' = \nu$, this term vanishes.

[5] The appearance of $\mathscr{T}_b$ rather than a momentum shell submatrix $T$ in Eqs. (15) may seem surprising. The reason for this is that with a fixed binding potential we do not have momentum conservation. Had we given the source of $U(\mathbf{z})$ a very large mass, our $\mathscr{T}_b$ would in fact be defined only on the momentum shell of the entire system and thus a $T$-quantity.

final state $\gamma$, we have

$$\frac{d\sigma(\gamma)}{d\Omega} = \frac{(2\pi)^4}{v_{\text{rel}}} \int k^2\, dk \delta(\epsilon_k + W_\gamma - \epsilon_p - W_0)\, |\langle \mathbf{k}, \nu', \gamma | \mathscr{T}_b | \mathbf{p}, \nu, 0\rangle|^2, \tag{15b}$$

where $d\Omega$ is an element of solid angle into which $\hat{\mathbf{k}}$ is directed. If $\gamma \neq 0$ this represents *inelastic scattering* from a state 0 to a state $\gamma$ of the bound particle. For $\gamma = 0$, the bound particle does not change its state, and the scattering is said to be *elastic*:

$$\frac{d\sigma_{\text{el}}}{d\Omega} = \frac{(2\pi)^4}{v_{\text{rel}}} \int k^2\, dk \delta(\epsilon_k - \epsilon_p)\, |\langle \mathbf{k}, \nu, 0 | \mathscr{T}_b | \mathbf{p}, \nu, 0\rangle|^2. \tag{15c}$$

For final states $\gamma$ in the continuum, the sum over $\gamma$ in Eq. (15*a*) is to be interpreted as an integral over some prescribed interval.

Our discussion to this point has been quite general. As stated in the introduction to this section, we wish to describe the scattering in the limit that the target particle is "weakly bound." To define this limit, we first note that the average kinetic and potential energies of the target particle $t$ in the initial bound state are

$$\begin{aligned} K_{\text{av}} &= (g_0, K_t g_0), \\ U_{\text{av}} &= (g_0, U g_0), \end{aligned} \tag{16}$$

respectively. The "weak binding" condition is defined by the inequalities

$$\frac{K_{\text{av}}}{\epsilon_p} \ll 1 \quad \text{and} \quad \frac{U_{\text{av}}}{\epsilon_p} \ll 1, \tag{17a}$$

where $\epsilon_p$ is given by Eq. (6*a*) if the incident particle $i$ is nonrelativistic and by Eq. (6*b*) if it is relativistic. Under most conditions, $K_{\text{av}}$, $U_{\text{av}}$, and $W_0$ are of the same order of magnitude and the conditions (17*a*) may be replaced by

$$\frac{|W_0|}{\epsilon_p} \ll 1. \tag{17b}$$

**The Impulse Approximation**

In the weak binding limit, as defined by Eqs. (17), the scattering matrix $\mathscr{T}_b$ can ordinarily be replaced by $\mathscr{T}$, the scattering matrix for a *free* target particle. This is called the "impulse approximation"; it implies that the potential $U$ be neglected when the scattering matrix (13) is evaluated.

The impulse approximation is very plausible and in many instances can be accepted as self-evident. For example, in studying the scattering of 1000-Mev pi-mesons by carbon nuclei bound in the crystalline lattice of a thin graphite target, we would not hesitate to neglect the binding effects on the carbon. The chemical binding (that is, $W_0$) is of the order of 1 ev, which is only one part in $10^9$ of the pion kinetic energy. The carbon nuclei in the target are separated by a distance about $10^5$ times greater than the range of the pion-nucleus interaction, which implies that a pion interaction can be localized at a *specific* nucleus. In this case, we do indeed feel confident that we can replace the scattering operator $\mathscr{T}_b$ for a chemically bound carbon nucleus by $\mathscr{T}$, the scattering operator for a free nucleus.

At this point we might be further tempted to relate the scattering operator for pion-carbon scattering to the free nucleon-pion scattering matrices. That is, we may very well ask if it is possible to describe pion-carbon scattering as a sequence of pion-nucleon scatterings, for each of which the free nucleon-pion scattering matrix is used. The 1000-Mev pion has an energy which is now two orders of magnitude larger than the nucleon binding energy. On the other hand, the mean spacing of nucleons in a nucleus is not large compared to the range of the pion-nucleon interaction. It is therefore not a priori clear that the free nucleon scattering matrix can be used to describe a given pion-nucleon encounter, since while they are interacting with each other either the pion or the nucleon may be interacting with other nucleons.

If we decide that we may indeed use free nucleon-pion scattering matrices here, we may be tempted to extend the impulse approximation even further. That is, we may ask if it is possible to describe pion-nucleon scattering in terms of the scattering of the pion with particles in the self-field of the nucleon—or, perhaps, in terms of nucleon scattering with particles in the self-field of the pion.

In addressing ourselves to these problems we are at the outset severely limited by the assumption that the scattering can be described by a conventional Schrödinger equation with *known* interactions, as in Eqs. (10). For the case of chemical binding this limitation does not appear to be troublesome. For the case of nuclear binding, however, we face difficulties, since nuclear binding forces are at best understood only in a very restricted way. In discussing the scattering of energetic particles by atomic nuclei we are therefore forced to rely in part on some plausibility arguments which require experimental justification.

The first explicit study of an impulse approximation seems to have been made by Fermi [3][6] in discussing the effect of molecular binding on neutron scattering by protons in hydrogen molecules. The term "impulse approximation" was introduced by Chew [4] in connection with an application to nucleon-deutron scattering. Careful studies of this approximation were subsequently made by Chew and Wick and by Ashkin and Wick

[6] Fermi also suggested to G. F. Chew and M. L. Goldberger in 1947 the possibility of extending his ideas to high-energy processes.

[5, 6]. Further advances, especially in the formal treatment, were made by Chew and Goldberger [7], who also estimated the contribution from multiple scattering.[7]

The scattering matrix corresponding to that of Eq. (13), but for a *free* target particle is

$$\mathscr{T} = V + V \frac{1}{E_a^{\,0} + i\eta - K_i - K_t - V} V. \tag{18}$$

Here the energy

$$E_a^{\,0} \equiv \epsilon_p + \epsilon_{t,P} \tag{19}$$

corresponds to a free target particle with momentum $P$ and energy $\epsilon_{t,P} = P^2/2M$. To assess the accuracy of the impulse approximation, we must find a quantitative relation between $\mathscr{T}_b$ and $\mathscr{T}$.

To do this, we first write

$$E_a = E_a^{\,0} + R_a, \tag{20}$$

where $R_a$ is an appropriate level shift, as described in Section 8.1, relating the energies $\epsilon_{t,P}$ and $W_0$. Using the familiar relation [see Eq. (5.114)]

$$\frac{1}{E_a + i\eta - H} = \frac{1}{E_a + i\eta - K_i - K_t - V} + \frac{1}{E_a + i\eta - K_i - K_t - V} U \frac{1}{E_a + i\eta - H},$$

we obtain by a straightforward algebraic calculation[8]

$$\mathscr{T}_b = \left(V + V \frac{1}{E_a + i\eta - K_i - K_t - V} V\right) \times \left(1 + \frac{1}{E_a + i\eta - K_i - K_t} U \frac{1}{d} \mathscr{T}_b\right), \tag{21}$$

[7] We are here considering the impulse approximation as distinct from that which neglects multiple scattering. This is in contrast to the use of the term "impulse approximation" in Refs. [3–7].

[8] Note the identities useful in verifying Eq. (21):

$$\frac{1}{E_a + i\eta - H} V = \frac{1}{E_a + i\eta - K_i - K_t - U}\left(V + V \frac{1}{E_a + i\eta - H} V\right)$$
$$= \frac{1}{E_a + i\eta - K_i - K_t - U} \mathscr{T}_b$$

$$V \frac{1}{E_a + i\eta - K_i - K_t - V} = \left(V + V \frac{1}{E_a + i\eta - K_i - K_t - V} V\right) \times \frac{1}{E_a + i\eta - K_i - K_t}.$$

where [see Eq. (10$a$)]

$$d \equiv E_a + i\eta - K. \tag{22}$$

The relation between the energy $E_a$ and the free-particle energy $E_a^0$, Eq. (20), may now be used to express the Green's functions in our expression for $\mathscr{T}_b$ [Eq. (21)] in terms of the free-particle "propagator" $d_0^{-1}$, where

$$d_0 \equiv E_a^0 + i\eta - K_i - K_t. \tag{23}$$

The right-hand side of Eq. (21) may be successively substituted for $\mathscr{T}_b$ to give the expansion

$$\begin{aligned} \mathscr{T}_b &= \mathscr{T} + \mathscr{T}\frac{1}{d_0}(R - R_a)\frac{1}{d_0}\mathscr{T} + \mathscr{O}\left[\mathscr{T}\frac{(R - R_a)^3}{d_0^3}\mathscr{T}\right] + \cdots \\ &\cong \mathscr{T} + \mathscr{T}\frac{1}{d_0}\left[R, \frac{1}{d_0}\mathscr{T}\right] + \mathscr{O}\left[\mathscr{T}\frac{(R - R_a)^3}{d_0^3}\mathscr{T}\right] + \cdots, \end{aligned} \tag{24}$$

where

$$R = U + U\frac{1}{d}U. \tag{25}$$

Equation (24) relates the two scattering matrices $\mathscr{T}_b$ and $\mathscr{T}$ in terms of the level shift matrix $R$. Since we anticipate that $R$ and $R_a$ will be comparable in magnitude, we may interpret (we shall discuss separately the case that $R$ and $R_a$ tend to cancel each other)

$$\Delta \equiv R_a\mathscr{T}\frac{1}{d_0^2}\mathscr{T} \tag{26}$$

as providing an estimate for the difference $\mathscr{T}_b - \mathscr{T}$. When $\Delta$ is negligibly small, Eq. (24) may be written as

$$\mathscr{T}_b \cong \mathscr{T}, \tag{27}$$

and $\mathscr{T}_b$ may be replaced by $\mathscr{T}$ in Eqs. (15). Equation (27) represents the precise statement of the impulse approximation. Because of the first of the weak binding conditions (17$a$), the precise value of $P$ to be used in Eq. (19) need not be known, since $P = \mathscr{O}(\sqrt{2MK_{\text{av}}})$. It is usually convenient in the weak binding limit, however, to choose this variable to agree with the corresponding $P$ appearing in Eq. (28).

To estimate the importance of $\Delta$, we write a matrix element of $\mathscr{T}$ as

$$\begin{aligned} &\langle \mathbf{k}, \nu', \mathbf{Q}, \nu'_t | \mathscr{T} | \mathbf{p}, \nu, \mathbf{P}, \nu_t \rangle \\ &\qquad = \delta(\mathbf{k} + \mathbf{Q} - \mathbf{p} - \mathbf{P})(\mathbf{k}, \nu', \mathbf{Q}, \nu'_t | T | \mathbf{p}, \nu, \mathbf{P}, \nu_t), \end{aligned} \tag{28}$$

where $T$ is the scattering matrix on the momentum shell, as usual [see Eq. (3.83)]. In Eq. (28) we have written $\mathbf{Q}$ for the final momentum of particle

$t$ and $\nu_t$ and $\nu'_t$ for the initial and final orientation of its spin. Then,[9]

$$\langle \mathbf{k}, \mathbf{Q}| \Delta |\mathbf{p}, \mathbf{P}\rangle = \delta(\mathbf{k} + \mathbf{Q} - \mathbf{p} - \mathbf{P}) R_a \int (\mathbf{k}, \mathbf{Q}|\, T \,|\mathbf{p}', \mathbf{P}') \times \frac{d^3p'\, d^3P' \delta(\mathbf{p}' + \mathbf{P}' - \mathbf{p} - \mathbf{P}) \;\; (\mathbf{p}', \mathbf{P}'|\, T\,|\mathbf{p}, \mathbf{P})}{(\epsilon_p + \epsilon_{t,P} + i\eta - \epsilon_{p'} - \epsilon_{t,P'})^2} . \quad (29)$$

We shall content ourselves with an order of magnitude evaluation of (29). Then, for very large $\epsilon_p$ we may assume that the principal contribution to the integration comes from domains near the singularity of the denominator, so the two matrices $T$ are restricted to the energy shell. Then, if the angular dependence of the $T$'s is ignored, we find

$$\Delta \approx \mathscr{T}\left[(2\pi)^2 R_a \frac{d}{d\epsilon_p}\left(\frac{p^2}{d\epsilon_p/dp} T\right)\right].$$

In the case that particle $i$ is relativistic, we see that

$$\begin{aligned} \frac{d}{d\epsilon_p}\left(\frac{p^2}{d\epsilon_p/dp} T\right) &= \frac{\epsilon_p}{c^2} pT\left(\frac{1}{\epsilon_p} + \frac{d}{d\epsilon_p} \ln T\right) \\ &= \mathscr{O}\left[\frac{\epsilon_p}{c^2} pT\left(\frac{1}{\epsilon_p} + Q\right)\right], \end{aligned}$$

where $Q$ is the *time delay* (8.285).[10] Now the scattering amplitude is $f \approx -(2\pi)^2(\epsilon_p/c^2)T$ and $t_C \equiv \hbar/\epsilon_p + Q$ is the time duration of the collision, so

$$\Delta \approx T\left(\frac{f}{\lambda\!\!\!^{-}} \frac{R_a t_C}{\hbar}\right)$$

where $\lambda\!\!\!^{-} = \hbar/p$. Therefore, Eq. (24) becomes

$$\mathscr{T}_b = \mathscr{T}\left[1 + \mathscr{O}\left(\left|\frac{f}{\lambda\!\!\!^{-}} \frac{R_a t_C}{\hbar}\right|\right)\right]. \quad (30a)$$

Ordinarily, in the absence of resonant scattering [see Eq. (8.339)] and in the weak binding limit, $t_C = \mathscr{O}(\hbar/\epsilon_p)$, so Eq. (30*a*) may be written as

$$\mathscr{T}_b = \mathscr{T}\left[1 + \mathscr{O}\left(\left|\frac{f}{\lambda\!\!\!^{-}} \frac{R_a}{\epsilon_p}\right|\right)\right]. \quad (30b)$$

[9] To simplify the notation, we shall omit spin labels while evaluating $\Delta$.

[10] Our identification of $(d/d\epsilon_p) \ln T$ with $Q$ here is recognized to be only an approximation. Strictly speaking, it is just the real part of $(d/d\epsilon_p) \ln T$ which is equal to the time delay. We expect the real and imaginary parts of this quantity to be of comparable magnitude, however.

For nonrelativistic particles $i$, it is easily seen that we must take $\epsilon_p = p^2/2m$ in $t_C$ and in Eq. (30*b*).

Equations (30) are fundamental to the theory of scattering by bound particles since they permit an evaluation of the error associated with the use of the impulse approximation (27).

Two assumptions have been essential for the derivation of Eq. (30). First, we have assumed that the operator $R$ [Eq. (25)] is of the same order of magnitude off the energy shell as is $R_a$ (which corresponds to a particular diagonal matrix element of $R$). This is quite reasonable unless the potential has singular properties which might lead to a very sensitive dependence of $R$ on energy. Ordinarily, we may assume that $R$ and $R_a$ are comparable in magnitude (for the relevant momenta) and of the order of

$$U_{\text{av}} = (g_0, Ug_0).$$

It has also been necessary to assume for the derivation of (30) that the interaction $V$ is not so singular that a very strong momentum dependence in $T$ introduces a new parameter in the value of (29). It is evident that these assumptions, although plausible, cannot at present be rigorously justified for the poorly understood *strong interactions* among particles. Considered as a semiheuristic relation, however, Eq. (30) has been generally useful and successful in studying such phenomena as pion and nucleon interactions with atomic nuclei.

It is perhaps well to mention some specific applications of Eqs. (30). First, we note the important conclusion that the impulse approximation is always valid when the first Born approximation may be used. This follows trivially, in fact, from Eqs. (13) and (18), since then $\mathscr{T}_b = V = \mathscr{T}$. An example is provided by the scattering of 1-kev x-rays by the atoms in a crystal. The scattering amplitude $f$ is of the order of $e^2/mc^2 \cong 3 \times 10^{-13}$ cm (times the atomic number), $\lambda\!\!\!^- \cong 2 \times 10^{-8}$ cm, and $R_a/\epsilon_p \cong 10^{-3}$. The relative error in the impulse approximation is therefore of the order $10^{-7}$ to $10^{-8}$. We mentioned earlier the example of 1000-Mev pions scattered by carbon nuclei in a graphite target. In this case $f \cong 5 \times 10^{-13}$ cm, $\lambda\!\!\!^- \cong 2 \times 10^{-14}$ cm, and $R_a/\epsilon_p \cong 10^{-9}$. The impulse approximation is seen to be valid, then, to within 1 part in $10^8$. If we wish to study 1-Bev pion-carbon scattering in terms of pion-nucleon scattering, we conclude from Eq. (30*b*) that the impulse approximation is somewhat marginal. Choosing $R_a = 7$ Mev, we find that

$$\left|\frac{fR_a}{\lambda\!\!\!^-\epsilon_p}\right| \cong \frac{1}{20}.$$

Near the ($\cong$200-Mev) resonance in pion-nucleon scattering the impulse approximation for pion-nucleus scattering would seem to fail. According to Eq. (8.331), the time delay $Q = (\Gamma/2)^{-1}$, where $\Gamma \simeq 60$ Mev is the resonance width. The scattering amplitude is $f \simeq 4 \times 10^{-13}$ cm, while $\lambdabar \simeq 1.4 \times 10^{-13}$ cm. Therefore, the correction term in Eq. (30*a*) is

$$\left| \frac{f}{\lambdabar} \frac{2R_a}{\Gamma} \right| \simeq 1,$$

which is insufficient to imply the validity of the impulse approximation.

Under some circumstances our considerations may overestimate the error in the impulse approximation. This may be the case when a "Brueckner model" [8] may be used to describe the scattering system. In this model the excitations of the scatterer are written as $\bar{K}_t$, where

$$\bar{K}_t \,|\mathbf{P}\rangle \equiv [K_t + R(P)]\,|\mathbf{P}\rangle$$

and $R$ is diagonal in momentum space. If $T$ is defined using $\bar{K}_t$ as an "equivalent kinetic energy," the leading correction term to the impulse approximation is of the relative order

$$\left| \frac{f}{\lambdabar} (\epsilon_p/R_a)^2 \right|.$$

There is another class of interactions of considerable practical importance for which the impulse approximation is valid, even when the weak binding conditions (17) are not met. For these interactions the quantities $R$ and $R_a$ nearly cancel in Eq. (24). In this case we say that the binding is "quasi-classical," since the binding potential $U(z)$ remains nearly constant over the region in which the scattering event occurs. For example, if we could treat $U(z)$ as being rigorously constant during the scattering interaction, we could define $E_a{}^0 = E_a - U$. This would immediately give $\mathcal{T}_b = \mathcal{T}$ [Eq. (18)], quite independently of the conditions (17).

To estimate the accuracy of the impulse approximation for quasi-classical binding, let us consider a process for which the scattering interaction of the bound particle is physically independent of its binding interaction. As an example, we mention neutron scattering by the atomic nuclei in a crystal, for which the electronic interactions responsible for chemical binding are essentially decoupled from the nuclear interaction. Then $[R, \mathcal{T}] = 0$ and [see Eq. (24)]

$$\left[R, \frac{1}{d_0}\mathcal{T}\right] = \left[R, \frac{1}{d_0}\right]\mathcal{T} = \frac{1}{d_0}\left[K_t, R\right]\frac{1}{d_0}\mathcal{T}.$$

Now $[K_t, R] \simeq -(1/2M)[\nabla_z{}^2, U(z)] \simeq \mathcal{O}[(1/2M)U\mathcal{R}^{-2}]$, where $\mathcal{R}$ is the "range" of the potential $U$. This permits us to write Eq. (24) as

$$\mathcal{T}_b \simeq \mathcal{T} + \left(\frac{1}{2M\mathcal{R}^2} U_{\text{av}}\right)\left(\mathcal{T} \frac{1}{d_0{}^3} \mathcal{T}\right).$$

Following the argument which led to Eq. (30*a*), we find that

$$\mathscr{T}\frac{1}{d_0^3}\mathscr{T} \cong \mathscr{T}\left(\frac{f}{\lambda}t_C^2\right).$$

Therefore, Eq. (24) becomes finally for "quasi-classical" binding

$$\mathscr{T}_b = \mathscr{T}\left[1 + \mathscr{O}\left(\left|\frac{\dfrac{f}{\lambda}\dfrac{\hbar^2}{2M\mathscr{R}^2}U_{\text{av}}t_C^2}{\hbar^2}\right|\right)\right]. \tag{31}$$

To illustrate Eq. (31), let us return to the example of neutron scattering in a crystal. Taking $t_C = \hbar/\epsilon_p, f = 10^{-12}$ cm, $\mathscr{R} = 10^{-8}$ cm and $U_{\text{av}} = 1$ ev we may express Eq. (31) as

$$\mathscr{T}_b = \mathscr{T}\left[1 + \mathscr{O}\left(\frac{10^{-6}}{\epsilon_{\text{ev}}^{3/2}}\right)\right],$$

where $\epsilon_{\text{ev}}$ is the neutron energy in electron volts. In this case the impulse approximation is valid even when the weak binding conditions (17) are severely violated.

**The Closure Approximation**

Having established the validity of the impulse approximation for the interaction considered in Section 11.1, p. 686, we have made an important step toward the evaluation of the scattering cross sections (15). The significance of this follows from the observation that the evaluation of $\mathscr{T}_b$ [Eq. (13)] requires in general the solution of a three-body problem, while $\mathscr{T}$ [Eq. (18)] is obtained as the solution of a two-body problem. To obtain the scattering cross sections (15), we must yet construct the wave functions $g_\gamma^{\pm}$. This involves no difficulty in principle, but it may be tedious. Fortunately, in the weak binding limit, as defined by Eqs. (17), evaluation of the scattering cross section need not require the wave functions $g_\gamma^{\pm}$ explicitly (except possibly $g_0$).

The reason for this is that the cross section (15*b*) tends to be sharply "peaked" at those states $\gamma$ which correspond to the recoil of the target particle as if it had been initially free. This suggests that we sum Eq. (15*a*) over all final states $\gamma$ of the target, restricting only the direction $\hat{\mathbf{k}}$ into which the incident particle $i$ is scattered. If the experimental counters which detect the scattered particles $i$ have an energy resolution which is broader than the "peak" in $d\sigma(\gamma)$, we will not lose useful information in carrying out this sum.

To evaluate this cross section, we first replace $\mathscr{T}_b$ by $\mathscr{T}$, in accordance with Eq. (27). It is convenient then to write $\mathscr{T}$ in the form

$$\mathscr{T} = \langle \mathbf{k}, \nu' | \mathscr{T} | \mathbf{p}, \nu \rangle, \tag{32}$$

where we imply a specific matrix element in the states of particle $i$, but that $\mathscr{T}$ is a general operator on the coordinates of the target particle $t$. Then, on setting

$$g_0(\mathbf{z}, s_t) \equiv |0\rangle, \tag{33}$$

we obtain from Eq. (15$a$) the cross section

$$\frac{d\bar{\sigma}}{d\Omega} = \frac{(2\pi)^4}{v_{\text{rel}}} \int k^2\, dk\, \langle 0|\, \mathscr{T}^\dagger\, \delta(\epsilon_k - \epsilon_p + h - W_0) \mathscr{T}\, |0\rangle_{\text{av}}. \tag{34}$$

Here the subscript av is interpreted to imply an *average* over initial spin orientations of the target and incident particle and a *sum* over the final spin orientations of both particles.[11]

To simplify Eq. (34) we introduce a Fourier expansion of the state $g_0$:

$$|0\rangle = (2\pi)^{-3/2} \int d^3P\, e^{i\mathbf{P}\cdot\mathbf{z}}\, |0; \mathbf{P}\rangle, \tag{35}$$

where the coefficient $|0; \mathbf{P}\rangle$ depends on the momentum $\mathbf{P}$ and spin variables $s_t$. Now using Eqs. (28) and (35) we have

$$\begin{aligned} \mathscr{T}|0\rangle &= \int d^3Q\, d^3P\, \frac{e^{i\mathbf{Q}\cdot\mathbf{z}}}{(2\pi)^{3/2}}\, \delta(\mathbf{k} + \mathbf{Q} - \mathbf{p} - \mathbf{P})(\mathbf{k}, \nu', \mathbf{Q}|\, T(\mathbf{S}_t)\, |\mathbf{p}, \nu, \mathbf{P})\, |0; \mathbf{P}\rangle \\ &= \int d^3P\, e^{-i\boldsymbol{\rho}\cdot\mathbf{z}}(\mathbf{k}, \nu', \mathbf{P} - \boldsymbol{\rho}|\, T(\mathbf{S}_t)\, |\mathbf{p}, \nu, \mathbf{P})\, \frac{e^{i\mathbf{P}\cdot\mathbf{z}}}{(2\pi)^{3/2}}\, |0; \mathbf{P}\rangle \\ &= e^{-i\boldsymbol{\rho}\cdot\mathbf{z}} T\, |0\rangle, \end{aligned} \tag{36}$$

where

$$\boldsymbol{\rho} \equiv \mathbf{k} - \mathbf{p} \tag{37}$$

is the momentum transferred to $i$. Rather than write the spin orientation labels $\nu_t$, $\nu'_t$ in Eq. (36) we have indicated that $T$ is a function of the spin operator $\mathbf{S}_t$. The quantity $T$ in the last line of Eq. (36) is, in detail,

$$T = (\mathbf{k}, \nu', \mathbf{P} - \boldsymbol{\rho}|\, T(\mathbf{S}_t)\, |\mathbf{p}, \nu, \mathbf{P}),$$

where $\mathbf{P} = (1/i)\nabla_z$ is the momentum operator for particle $t$. Thus in the last line of Eq. (36) we are using the notation introduced in Eq. (32) whereby $T$ represents a specific matrix element in the coordinates of $i$, but an operator in the coordinates of $t$.

Introducing Eq. (36) into Eq. (34), we obtain

$$\frac{d\bar{\sigma}}{d\Omega} = \frac{(2\pi)^4}{v_{\text{rel}}} \int k^2\, dk\, \langle 0|\, T^\dagger e^{i\boldsymbol{\rho}\cdot\mathbf{z}}\, \delta(\epsilon_k - \epsilon_p + h - W_0) e^{-i\boldsymbol{\rho}\cdot\mathbf{z}} T\, |0\rangle_{\text{av}}. \tag{38}$$

[11] If we are interested in polarization of either of the particles, the spin sums may evidently be restricted in an appropriate fashion in Eq. (34).

Here $T^\dagger$ represents the quantity

$$T^\dagger = (\mathbf{p}, \nu, \mathbf{P}|\, T^\dagger(\mathrm{S}_t)\, |\mathbf{k}, \nu', \mathbf{P} - \boldsymbol{\rho}),$$

where again $\mathbf{P}$ is the momentum operator for particle $t$.

To continue with the evaluation of the cross section, we recall anticipating that the final target states will be "peaked" in Eq. (38). This suggests that we may set $h = \overline{W} + (h - \overline{W})$, where $\overline{W}$ is the energy at which "peaking" occurs and $h - \overline{W}$ is to be considered a small quantity.[12] Thus we formally expand the $\delta$-function in Eq. (38) in a Taylor series.

$$\delta(\epsilon_k - \epsilon_p + h - W_0) = \sum_{n=0}^{\infty} \frac{1}{n!}\, \delta^n(e)(h - \overline{W})^n, \tag{39}$$

where

$$e \equiv \epsilon_{\mathrm{k}} - \epsilon_p + \overline{W} - W_0. \tag{40}$$

We have yet, of course, to determine the mean "excitation energy" $\overline{W}$.

On inserting the expansion (39) into Eq. (38), we obtain[13]

$$\begin{aligned} \frac{d\bar{\sigma}}{d\Omega} &= \sum_{n=0}^{\infty} \frac{1}{n!} \frac{(2\pi)^4}{v_{\mathrm{rel}}} \int k^2\, dk\, \delta^n(e)\Gamma_n \\ &= \sum_{n=0}^{\infty} \frac{(-1)^n}{n!} \frac{d^n}{de^n}\left[\left(\frac{k^2}{de/dk}\right)\frac{(2\pi)^4}{v_{\mathrm{rel}}}\Gamma_n\right]\Bigg|_{e=0}, \end{aligned} \tag{41}$$

where

$$\Gamma_n \equiv \langle 0|\, T^\dagger e^{i\boldsymbol{\rho}\cdot\mathbf{x}}(h - \overline{W})^n\, e^{-i\boldsymbol{\rho}\cdot\mathbf{z}} T\, |0\rangle_{\mathrm{av}} \tag{42}$$

and $e$ is to be set equal to zero after the indicated differentiations are carried out.[14] We shall attempt to choose $\overline{W}$ in such a way as to make the higher-order terms in Eq. (41) as small as possible.

Since [see Eq. (2)]

$$h = -\frac{1}{2M}\nabla_z^2 + U(z),$$

it follows that[15]

$$(h - \overline{W})\, e^{-i\boldsymbol{\rho}\cdot\mathbf{z}} = e^{-i\boldsymbol{\rho}\cdot\mathbf{z}}\left(\frac{\rho^2}{2M} - L - \overline{W} + h\right),$$

[12] We are following a slight modification of a technique developed by Placzek and Wick [9, 10] for evaluating expressions such as (38).

[13] Equation (41) is a generalization of a result given by Fowler [11].

[14] The expression (41) probably represents an asymptotic series, although the convergence has not been investigated. The moments $\Gamma_n$ may not exist for $n$ greater than a prescribed value. In this case the series must be terminated with a finite remainder. There is no practical consequence of this, since only the leading terms in the series (41) are ordinarily used.

[15] For bound systems of several particles $U$ may contain space-exchange forces. In this case an additional term is obtained in $[h, e^{-i\boldsymbol{\rho}\cdot\mathbf{z}}]$. [This term is given in Eq. (A.4) of Ref. [12].]

where

$$L \equiv \frac{\boldsymbol{\rho} \cdot \nabla_z}{iM} = \frac{\boldsymbol{\rho} \cdot \mathbf{P}}{M}. \tag{43}$$

By induction it then follows that

$$(h - \overline{W})^n e^{-i\boldsymbol{\rho}\cdot\mathbf{z}} = e^{-i\boldsymbol{\rho}\cdot\mathbf{z}} \left(h - \overline{W} + \frac{\rho^2}{2M} - L\right)^n, \tag{44}$$

so

$$\Gamma_n = \langle 0| \, T^\dagger \left(h - \overline{W} + \frac{\rho^2}{2M} - L\right)^n T \, |0\rangle_{\text{av}}. \tag{45a}$$

Equations (41) and (45$a$) permit us in principle to evaluate the scattering cross section in terms of averages over the initial state $g_0$.

The expressions (45$a$) are appropriately discussed in either of two limits. In the first limit, the momentum transfer $\rho$ is large compared to $\sqrt{2MK_{\text{av}}}$ [Eq. (16)], the characteristic momentum of $t$ in the initial bound state. Then we may neglect $L$, as compared to $\rho^2/2M$ in Eq. (45$a$). If we also neglect the commutator $[h, T]$, we see that the choice

$$\overline{W} = W_0 + \frac{\rho^2}{2M} \tag{46}$$

leads to $\Gamma_n = 0$ for $n \geq 1$. In this case the only nonvanishing term in Eq. (41) is that corresponding to $n = 0$, which as we shall see, is equivalent to the scattering cross section for a free target particle.

Even when we do not wish to neglect $L$ and $[h, T]$ in $\Gamma_n$, the definition (46) for $\overline{W}$ is appropriate for giving $\Gamma_n$ ($n \geq 1$) when [see Eq. (16)]

$$\begin{aligned} \frac{\rho^2}{2M} &\gg (\langle 0| \, L^2 \, |0\rangle_{\text{av}})^{1/2} = \left(\frac{2}{3}\frac{\rho^2}{M} K_{\text{av}}\right)^{1/2}, \\ \frac{\rho^2}{2M} &\gg |U_{\text{av}}| \geq \frac{|\langle 0| \, T^\dagger \, [h, T] \, |0\rangle_{\text{av}}|}{|\langle 0| \, T^\dagger T \, |0\rangle_{\text{av}}|}. \end{aligned} \tag{47}$$

When the conditions (47) are met, the scattering is called *quasi-free*, since then the excitation energy $\overline{W} - W_0$ is just equal to the recoil energy $\rho^2/2M$ of a free target particle. (That the target particle does indeed characteristically recoil as a free particle is shown in Section 11.1, p. 706.) For quasi-free scattering, we may then write the expressions (45$a$) as

$$\Gamma_n = \langle 0| \, T^\dagger (h - W_0 - L)^n T \, |0\rangle_{\text{av}}, \tag{45b}$$

using Eq. (46). The other limit, corresponding to a small $\rho^2/2M$, will be discussed later.

Let us first investigate the leading term, for which $n = 0$, in Eq. (41). Recalling the description of the subscript av following Eq. (34) and setting

$\mathbf{P} = 0$ in $T$, we obtain

$$\Gamma_0 = \langle 0| T^\dagger T |0\rangle_{\text{av}}$$
$$= \frac{1}{(2S_i + 1)(2S_t + 1)} \sum_{\substack{\nu,\nu_t \\ \nu',\nu'_t}} |(\mathbf{k}, \nu', -\boldsymbol{\rho}, \nu'_t| T |\mathbf{p}, \nu, 0, \nu_t)|^2$$

in the notation of Eq. (28). Because of Eq. (46), the first term in Eq. (41) is therefore

$$\frac{d\bar{\sigma}}{d\Omega} = \frac{(2\pi)^4}{v_{\text{rel}}} \frac{1}{(2S_i + 1)(2S_t + 1)} \sum_{\substack{\nu,\nu_t \\ \nu',\nu'_t}} \int k^2\, dk$$
$$\times\, \delta\left(\epsilon_k + \frac{\rho^2}{2M} - \epsilon_p\right) |(\mathbf{k}, \nu', -\boldsymbol{\rho}, \nu'_t| T |\mathbf{p}, \nu, 0, \nu_t)|^2 \equiv \frac{d\bar{\sigma}_f}{d\Omega}, \quad (48)$$

the differential cross section for a *free* target particle. That is, with Eq. (48), we have obtained the same cross section that would have resulted had we neglected the binding interaction $U$ at the outset.

Had we not neglected the **P**-dependence of $T$ in evaluating $\Gamma_0$, it would have been necessary to average the expression (48) over the momentum spectrum of the state $g_0$, using the transform (35). This correction will be described more carefully later, although it is ordinarily negligible under conditions of weak binding.

To determine how closely the expression (48) approximates the exact cross section (41), we must study the expressions (45*b*) for $n \neq 0$. These expressions are nonvanishing, as observed earlier, because even powers of $L$ have nonzero expectation values and because $[h, T] \neq 0$, which follows from the **P**-dependence of $T$ and the fact that $[\mathbf{P}, U] \neq 0$. The physical origin of the $L$-terms is the unpredictable motion of the target particle during the scattering event, while that of the $[h, T]$ terms is the distortion of the wave function $g_0$ due to preferential scattering for certain ranges of **P**.

The structure of the series (41) may be deduced by inspection (barring mathematical catastrophies!). The differential operator $d^n/d\epsilon^n$ will be characterized by a factor $\epsilon_p^{-n}$. Since $[h, T] = \mathcal{O}(UT)$[16] and since

$$\langle 0| L^2 |0\rangle = \frac{1}{3} \frac{\rho^2}{M^2} \langle 0| P^2 |0\rangle$$
$$= \frac{2}{3} \frac{\rho^2}{M} K_{\text{av}}$$

[16] This is often a gross overestimate of the commutator, as may be seen from Eq. (50).

[see Eq. (16)], we anticipate that the $n$th-order term in Eq. (41) will be a linear combination of expressions not greater than

$$\frac{d\bar{\sigma}_f}{d\Omega}\left\{\frac{1}{{\epsilon_p}^n}\left[{U_{\text{av}}}^l\left(\frac{\rho^2}{M}K_{\text{av}}\right)^q\right]\right\},$$

where $l + 2q = n$. Because of the weak binding conditions (17), these become increasingly small for larger values of $n$.

It is convenient to have a somewhat more careful estimate for the leading correction terms in Eq. (41). For this purpose we write

$$\frac{d\bar{\sigma}}{d\Omega} = \frac{d\bar{\sigma}_f}{d\Omega}(1 + \eta_1 + \eta_2 + \cdots), \tag{49}$$

where $d\bar{\sigma}_f/d\Omega$ is given by Eq. (48) and the $\eta$'s represent corrections to this.

Let $\eta_1$ give the error resulting from the failure to include the **P**-dependence of $T$ in Eq. (48). We assume that $T$ depends upon the momentum transfer $\boldsymbol{\rho}$ and the initial relative momentum

$$\boldsymbol{\varkappa} = \left(\frac{\mathbf{p}c^2}{\epsilon_p} - \frac{\mathbf{P}}{M}\right)\frac{\epsilon_p M}{\epsilon_p + Mc^2}.$$

(For nonrelativistic particles $i$, $\epsilon_p$ should be replaced by $mc^2$ here.) Then a Taylor expansion of $T$ about $\mathbf{P} = 0$ yields, if $p/\epsilon_p \gg P/Mc^2$,[17] the approximate result[18]

$$\begin{aligned} T &\cong T_{P=0}\left(1 + \frac{m}{M}\frac{\mathbf{p}\cdot\mathbf{P}}{p^2} + \cdots\right), &&\text{nonrelativistic,} \\ &\cong T_{P=0}\left(1 + \frac{\epsilon_p}{Mc^2}\frac{\mathbf{p}\cdot\mathbf{P}}{p^2} + \cdots\right), &&\text{relativistic.} \end{aligned} \tag{50}$$

From this we see that

$$\begin{aligned} \eta_1 &\cong \frac{m}{M}\frac{K_{\text{av}}}{\epsilon_p}, &&\text{nonrelativistic,} \\ &\cong \frac{K_{\text{av}}}{Mc^2}, &&\text{relativistic.} \end{aligned} \tag{51}$$

The leading corrections beyond $\eta_1$ arise from the term $\Gamma_1$ [Eq. (45$b$)], which may be written as

$$\Gamma_1 = \langle 0|\, T^\dagger[h, T]\, |0\rangle_{\text{av}} - \langle 0|\, T^\dagger TL\, |0\rangle_{\text{av}}. \tag{52}$$

[17] Other cases may easily be worked out.

[18] We recall that "relativistic" refers to the energy of particle $i$ only.

Now, if neither $U$ nor $T$ contains spin flip terms,

$$\begin{aligned}\langle 0|\, T^\dagger[h, T]\,|0\rangle_{\text{av}} &= \langle 0|\, T^\dagger[U, T]\,|0\rangle_{\text{av}} \\ &\cong T^\dagger(\nabla_P{}^2 T)\langle 0|\, \nabla_z{}^2 U\,|0\rangle \\ &\cong T^\dagger(\nabla_P{}^2 T) 2MK_{\text{av}}U_{\text{av}}.\end{aligned}$$

Using the expansion (50) and recalling that $\Gamma_0 \cong T^\dagger$, we obtain

$$\begin{aligned}\eta_2 &\cong \frac{m}{M}\frac{U_{\text{av}}}{\epsilon_p}\frac{K_{\text{av}}}{\epsilon_p}, \qquad \text{nonrelativistic,} \\ &\cong \frac{U_{\text{av}}}{\epsilon_p}\frac{K_{\text{av}}}{Mc^2}, \qquad \text{relativistic.}\end{aligned} \tag{53}$$

On the other hand, if $U$ and $T$ have respective spin flip parts $U_{\text{sp}}$ and $T_{\text{sp}}$, we must set

$$[U, T] \cong U_{\text{sp}}T_{\text{sp}}.$$

In this case we have a correction

$$\eta_3 \cong \frac{|\langle 0|\, U_{\text{sp}}\,|0\rangle|}{\epsilon_p}\frac{|T_{\text{sp}}|}{|T|}, \tag{54}$$

where the appropriate expression (6) for $\epsilon_p$ is to be used.

The second term on the right in Eq. (52) gives a contribution of order

$$\langle 0|\, T^\dagger TL\,|0\rangle_{\text{av}} \cong \text{Re}\,(T^\dagger\nabla_P T)\cdot \boldsymbol{\rho}\,\frac{\langle 0|\, P^2\,|0\rangle_{\text{av}}}{M}$$

Using the series (50), we find a correction

$$\begin{aligned}\eta_4 &\cong \frac{m}{M}\frac{\rho}{p}\frac{K_{\text{av}}}{\epsilon_p}, \qquad \text{nonrelativistic,} \\ &\cong \frac{\rho}{p}\frac{K_{\text{av}}}{Mc^2}, \qquad \text{relativistic.}\end{aligned} \tag{55}$$

The correction terms (51), (53), (54), and (55) resulted from either spin or momentum dependence in the $T$-matrix.[19] In addition to these some corrections arise solely from the operator $L$ in Eqs. (45$b$). These have been studied by Placzek and Wick [9, 10]. Following these authors we shall now set $\mathbf{P} = 0$ in $T$ and $T^\dagger$ and also set $[h, T] = 0$ in Eqs. (45$b$). Then

$$\begin{aligned}\Gamma_n &= \frac{1}{(2S_i + 1)(2S_t + 1)} \sum_{\substack{\nu,\nu_t \\ \nu',\nu'_t}} |(\mathbf{k}, \nu', -\boldsymbol{\rho}, \nu'_t|\, T\,|\mathbf{p}, \nu, 0, \nu_t)|^2 S_n \\ &\equiv |T|^2_{\text{av}}\, S_n,\end{aligned} \tag{56$a$}$$

[19] We note that when the Born approximation for $T$ is valid, there is no $\mathbf{P}$-dependence in $T$. In this case $\eta_1 = \eta_2 = \eta_4 = 0$.

where

$$S_n = \langle 0|\, (h - W_0 - L)^n \,|0\rangle_{\text{av}}. \tag{56b}$$

Continuing to follow Placzek and Wick, we consider $|T|^2_{\text{av}}$ to be a numerical constant and write Eq. (41) as

$$\frac{d\bar{\sigma}}{d\Omega} = \frac{(2\pi)^4}{v_{\text{rel}}} |T|^2_{\text{av}} \left[ \sum_{n=0}^{\infty} \frac{(-1)^n}{n!} \frac{d^n}{de^n} \left( \frac{k^2}{de/dk} S_n \right) \bigg|_{e=0} \right]. \tag{57}$$

The $S_n$ are readily expressed in the form

$$\begin{aligned} S_0 &= 1, \\ S_1 &= 0, \\ S_2 &= \frac{2}{3} \frac{\rho^2}{M} K_{\text{av}}, \\ S_3 &= \frac{\rho^2}{2M} B_{\text{av}}, \\ S_4 &= \frac{4}{5} \frac{\rho^2}{M} \langle 0|\, K_t^2 \,|0\rangle_{\text{av}} + \frac{\rho^2}{M} C_{\text{av}}, \end{aligned} \tag{58a}$$

where

$$\begin{aligned} B_{\text{av}} &\equiv \frac{1}{3M} \langle 0|\, \nabla_z^2 U \,|0\rangle_{\text{av}}, \\ C_{\text{av}} &\equiv \frac{1}{3M} \langle 0|\, (\nabla_z U)^2 \,|0\rangle_{\text{av}}. \end{aligned} \tag{58b}$$

Since $\Gamma_1 = 0$ in the approximation from which Eq. (57) was obtained, we may simply add the new terms in (57) to those corrections already found. From $S_2$ and $S_3$ we obtain, respectively,

$$\begin{aligned} \eta_5 &\cong \frac{\rho^2}{M\epsilon_p} \frac{K_{\text{av}}}{\epsilon_p}, \\ \eta_6 &\cong \frac{\rho^2}{M\epsilon_p} \frac{B_{\text{av}}}{\epsilon_p^2} \cong \frac{\rho^2}{M\epsilon_p} \frac{K_{\text{av}}}{\epsilon_p} \frac{U_{\text{av}}}{\epsilon_p}, \end{aligned} \tag{59}$$

where the appropriate expression (6) for $\epsilon_p$ should be used.

A quantitative evaluation of the terms $\eta_1, \eta_2, \ldots$ in Eq. (49) would require the specific properties of a given scattering matrix as well as a given wave function $g_0$. This is ordinarily unnecessary, since the series (49) is usually used merely to estimate the error in the first term [Eq. (48)]. For this purpose our estimates of $\eta_1$ to $\eta_6$ are usually sufficient.[20] From the form

[20] The largest terms in the series (49) may, in fact, be evaluated accurately, if they are needed. The series (57) is relatively simple, the first few terms for the total cross section having been given by Placzek and Wick [9, 10]. Their result is quoted in Note [13].

of the $\eta$'s it is evident that the "quasi-free" approximation is valid when the weak binding conditions (17) and the conditions (47) are satisfied. [The conditions (47) actually need not be satisfied for the cross section (48) to approximate (41) adequately. The physical interpretation given to the discussion requires modification, however, when the inequalities (47) are not valid.]

As we shall see from applications given later in this chapter, the quantitative basis for the quasi-free approximation (48), as provided by the series (49), is of profound importance for studying scattering from "weakly bound" systems.

The series (41) may also be summed into another form than that given by (49). To obtain the leading term in the new series we neglect the commutators $[h, T]$ and $[h, L]$ in the expressions (45$b$). These then become

$$\Gamma_n = \langle 0|\, T^\dagger(-L)^n T\,|0\rangle_{\text{av}} = \langle 0|\, T^\dagger\left(-\frac{\boldsymbol{\rho}\cdot\mathbf{P}}{M}\right)^n T\,|0\rangle_{\text{av}}.$$

This permits us to write (41) as

$$\begin{aligned}\frac{d\bar{\sigma}}{d\Omega} &= \sum_{n=0}^{\infty}\frac{(2\pi)^4}{v_{\text{rel}}}\frac{1}{n!}\int k^2\,dk\,\langle 0|\,T^\dagger\,\delta^n(e)\left(-\frac{\boldsymbol{\rho}\cdot\mathbf{P}}{M}\right)^n T\,|0\rangle_{\text{av}}\\ &= \frac{(2\pi)^4}{v_{\text{rel}}}\int k^2\,dk\,\langle 0|\,T^\dagger\,\delta\left(\epsilon_k-\epsilon_p+\frac{\rho^2-2\boldsymbol{\rho}\cdot\mathbf{P}}{2M}\right)T\,|0\rangle_{\text{av}} \qquad (60a)\\ &= \frac{(2\pi)^4}{v_{\text{rel}}}\int d^3P\,\langle 0;\mathbf{P}|\,T^\dagger\int k^2\,dk\,\delta\left(\epsilon_k+\frac{Q^2}{2M}-\epsilon_p-\frac{P^2}{2M}\right)T\,|0;\mathbf{P}\rangle_{\text{av}}.\end{aligned}$$

Here we have taken $\mathbf{Q} = \mathbf{P} - \boldsymbol{\rho}$, the recoil momentum of particle $t$, and have used the Fourier expansion (35).

Now the scattering cross section for a free target particle of initial momentum $\mathbf{P}$ is

$$\frac{d\bar{\sigma}_f(\mathbf{P})}{d\Omega} = \frac{(2\pi)^4}{v(\mathbf{P})}\frac{1}{(2S_i+1)(2S_t+1)}\sum_{\substack{\nu,\nu_t\\ \nu',\nu'_t}}$$
$$\times\int k^2\,dk\,\delta\left(\epsilon_k+\frac{Q^2}{2M}-\epsilon_p-\frac{P^2}{2M}\right)|(\mathbf{k},\nu',\mathbf{Q},\nu'_t|\,T\,|\mathbf{p},\nu,\mathbf{P},\nu_t)|^2, \qquad (61)$$

where $v(\mathbf{P}) \equiv (\mathbf{p}/\epsilon_p)c^2 - \mathbf{P}/M$. This expression enables us to abbreviate (60$a$):

$$\frac{d\bar{\sigma}}{d\Omega} = \int d^3P\,\langle 0;\mathbf{P}\,|\,0;\mathbf{P}\rangle_{\text{av}}\left[\frac{v(\mathbf{P})}{v_{\text{rel}}}\right]\frac{d\bar{\sigma}_f(\mathbf{P})}{d\Omega} \equiv \left\langle\frac{d\bar{\sigma}_f}{d\Omega}\right\rangle_{\text{av}}. \qquad (60b)$$

The quantity $\langle 0;\mathbf{P}\,|\,0;\mathbf{P}\rangle_{\text{av}}$ is an appropriately spin-averaged probability distribution for the target particle momentum in the state $g_0$.

Equation (60*b*) is more or less obvious from intuitive considerations.[21] It describes the cross section for two free particles, averaged over the momentum spectrum of the particle which is actually bound. The factor of reciprocal relative velocity is corrected to the velocity relative to the force center.

In constructing an expansion analogous to that of Eq. (49) to correct Eq. (60*b*), we must omit those $\eta$'s which did not involve one of the commutators $[h, T]$ or $[h, L]$. Among those already evaluated, this includes $\eta_1$, $\eta_4$, and $\eta_5$. The series (41) may therefore be written in the form

$$\frac{d\bar{\sigma}}{d\Omega} = \left\langle \frac{d\bar{\sigma}_f}{d\Omega} \right\rangle_{\text{av}} (1 + \eta_2 + \eta_3 + \eta_6 + \cdots). \tag{62}$$

When using either (49) or (62) for the "quasi-free" approximation, it is of course necessary to include the correction (30) to the impulse approximation when estimating errors.

In discussing the quasi-free approximation at the beginning of this section we remarked that the scattering cross section (15*a*) will be sharply "peaked" near a given domain of the $\gamma$. It is evident that this peaking will occur at the energy $\overline{W}$, as given by Eq. (46). It remains to find the width of this peak.

The $l$th moment ($l = 1, 2, \ldots$) of the excitation energy is obtained from the relation

$$\langle \Delta W^l \rangle_{\text{av}} = \frac{\sum_{\gamma} (W_\gamma - \overline{W})^l [d\bar{\sigma}(\gamma)/d\Omega]}{d\bar{\sigma}/d\Omega}, \tag{63a}$$

where $d\bar{\sigma}(\gamma)/d\Omega$ is the appropriate cross section (15) for a final state $\gamma$ of the target and $d\bar{\sigma}/d\Omega$ is the expression (34). On following the steps that led to Eq. (38), we obtain

$$\frac{d\bar{\sigma}}{d\Omega} \langle \Delta W^l \rangle_{\text{av}} = \frac{(2\pi)^4}{v_{\text{rel}}} \int k^2 \, dk \tag{63b}$$
$$\langle 0 | \, T^\dagger \, e^{i\mathbf{p}\cdot\mathbf{z}} (h - \overline{W})^l \, \delta(\epsilon_k - \epsilon_p + h - W_0) \, e^{-i\mathbf{p}\cdot\mathbf{z}} T \, | 0 \rangle_{\text{av}}.$$

To evaluate this, we use the expressions (39) and (44) to find

$$e^{i\mathbf{p}\cdot\mathbf{z}} (h - \overline{W})^l \, \delta(\epsilon_k - \epsilon_p + h - W_0) \, e^{-i\mathbf{p}\cdot\mathbf{z}} = \sum_{n=0}^{\infty} \frac{1}{n!} \, \delta^n(e)(h - L - W_0)^{n+l}. \tag{64}$$

[21] Equation (60*b*) was proposed by Goldberger [14] in his theory of high-energy nuclear reactions.

For the case that $l = 1$, we verify from Eq. (63*b*) that in the approximation of Eq. (48)

$$\langle \Delta W \rangle_{\text{av}} = \langle 0 | (h - L - W_0) | 0 \rangle_{\text{av}} = 0,$$

as anticipated. For $l = 2$, we find in the same approximation that

$$\langle \Delta W^2 \rangle_{\text{av}} = \langle 0 | (h - L - W_0)^2 | 0 \rangle_{\text{av}} = \langle 0 | L^2 | 0 \rangle_{\text{av}} = \frac{2}{3} \frac{\rho^2}{M} K_{\text{av}}. \tag{65}$$

When the quasi-free condition (47) obtains, we verify that the ratio

$$\frac{\langle \Delta W^2 \rangle_{\text{av}}}{(\overline{W} - W_0)^2} \cong \frac{K_{\text{av}}}{(\rho^2/2M)} \ll 1,$$

as required for sharp "peaking" of the cross section (15*a*) near those target states which correspond to the recoil of a free particle.

It may be recalled that the quasi-free conditions are expected to be valid when $\rho \gg \sqrt{2MK_{\text{av}}}$ and when $(\rho \gg \sqrt{2MU_{\text{av}}})$. In the other limit that $\rho \ll \sqrt{2MK_{\text{av}}}$ the scattering is very different from that under quasi-free conditions, although the expressions (49) and (62) still provide accurate numerical estimates for the scattering cross section. Intuitively, we would expect that the state of the target would experience little change as a result of the scattering when $\rho \ll \sqrt{2MK_{\text{av}}}$, since then the momentum transfer $\rho$ is small compared to the initial momentum of the target particle.

When $\rho \ll \sqrt{2MK_{\text{av}}}$, the term $\rho^2/2M$ should be considered as *small* compared to $L$ in Eq. (45*a*). It is then useless and misleading to cancel the term $\rho^2/2M$ exactly, as was done by Eq. (46). Instead, it is better to take $\overline{W} = W_0$, so that [see Eq. (40)]

$$e = \epsilon_k - \epsilon_p.$$

The leading term in Eq. (41), corresponding to Eq. (48), is now

$$\begin{aligned} \frac{d\bar{\sigma}}{d\Omega} &= \frac{(2\pi)^4}{v_{\text{rel}}} \int k^2 \, dk \, \delta(\epsilon_k - \epsilon_p) \langle 0 | T^\dagger T | 0 \rangle_{\text{av}} \\ &= (2\pi)^4 \left( \frac{\epsilon_p}{c^2} \right)^2 \langle 0 | T^\dagger T | 0 \rangle_{\text{av}}, \qquad \text{relativistic}, \\ &= (2\pi)^4 m^2 \langle 0 | T^\dagger T | 0 \rangle_{\text{av}}, \qquad \text{nonrelativistic}. \end{aligned} \tag{66}$$

This cross section differs from that given by Eq. (48) only in the kinematical relations required by the energy conservation condition $\epsilon_k = \epsilon_p$. This means, to the extent that Eq. (66) is correct, that the scattered particles have exactly the same energy after scattering as they did before.

This result provides the theoretical basis for the *Mossbauer effect* [15]. Here $\gamma$-rays undergo resonance scattering (see Section 8.6) by atomic nuclei bound in a crystal lattice. For the effect to work $|\epsilon_k - \epsilon_p|$ must be less than the width $\Gamma$ of the resonance, this corresponding to an exceedingly small energy. For lattice binding, the energy $K_{\text{av}}$ is usually of the order of 1 ev, or less. The condition $\rho \ll \sqrt{2MK_{\text{av}}}$ then limits the $\gamma$-ray energy to a value significantly less than $(2Mc^2K_{\text{av}})^{1/2}$, which is usually of the order of a few hundred kev.

We have yet to quantitatively justify the kinematical relations implied by Eq. (66). To do this, we return to Eqs. (63) for the moments of the excitation energy of the target. Since the condition $\rho \ll \sqrt{2MK_{\text{av}}}$ implies that $\rho^2/2M$ is negligible, we may accept our earlier estimate (65) for $\langle \Delta W^2 \rangle_{\text{av}}$. We see then that

$$|\epsilon_k - \epsilon_p| \cong \sqrt{\langle \Delta W^2 \rangle_{\text{av}}} \cong \left(\frac{2}{3}\frac{\rho^2}{M} K_{\text{av}}\right)^{1/2} \ll K_{\text{av}}. \tag{67}$$

When, and only when, $[\frac{3}{2}(\rho^2/M)K_{\text{av}}]^{1/2}$ is negligible can we then use Eq. (66) for the scattering cross section.[22] (We, of course, also require that the impulse approximation be valid and that the $\eta$-corrections in either Eq. (49) or (62) be negligible.) When

$$\left(\frac{2}{3}\frac{\rho^2}{M} K_{\text{av}}\right)^{1/2} < |W_0|, \tag{68a}$$

the target particle will ordinarily be left in a bound state, since it receives less energy than $|W_0|$, its binding energy. On the other hand, when

$$\left(\frac{2}{3}\frac{\rho^2}{M} K_{\text{av}}\right)^{1/2} > |W_0|, \tag{68b}$$

we anticipate that the target particle will usually escape from its binding potential following the scattering. Condition (68*b*) is seen to be satisfied whenever the quasi-free conditions (47) are met.

Before closing this section it is well to observe that scattering particles bound in a physical target at a finite temperature may not all be in the same state $\gamma = 0$. In this case our $\langle 0| \cdots |0\rangle_{\text{av}}$ must be replaced by an ensemble average. The modification required is almost trivial, since then we need only take

$$\langle 0| \cdots |0\rangle_{\text{av}} = \frac{\sum_{\gamma(0)} e^{-\beta W_{\gamma(0)}} \langle \gamma(0)| \cdots |\gamma(0)\rangle_{\text{av}}}{\sum_{\gamma(0)} e^{-\beta W_{\gamma(0)}}} \tag{69}$$

where $\beta^{-1}$ is the temperature of the target and $|\gamma(0)\rangle \equiv g_{\gamma(0)}$, a state of energy $W_{\gamma(0)}$.

[22] For the Mossbauer effect we require that the $\gamma$-ray width $\Gamma$ be greater than $[\frac{2}{3}(\rho^2/M)K_{\text{av}}]^{1/2}$.

**The Wave Function for the Scattered Particles**

We have given a systematic procedure for obtaining the cross section (34) in the weak binding limit, where the impulse approximation is valid. To extend our understanding of the physical significance of the result, we shall now evaluate the asymptotic form of the wave function (11). This will be important for later applications in describing the coherence and noncoherence of scattered waves.

We first discuss the quasi-free scattering, assuming that the conditions (47) obtain. Since we anticipate that the struck target particle will recoil as if free, it is perhaps well to review briefly the classical kinematical relations for this collision: Let the target particle be at rest and at the point $\mathbf{z}^0$ prior to the collision. It is struck by a nonrelativistic[23] beam particle $i$, which has an initial momentum $\mathbf{p}$. Following the collision, particles $i$ and $t$ have the respective momenta $\mathbf{k}_0$ and $\mathbf{Q}_0$. The initial and final relative momenta of the two particles are, respectively,

$$\begin{aligned} \boldsymbol{\varkappa}_0 &= \frac{M}{m + M}\,\mathbf{p}, \\ \boldsymbol{\varkappa}_{f0} &= \frac{M\mathbf{k}_0 - m\mathbf{Q}_0}{m + M}\,. \end{aligned} \tag{70}$$

A time $\Delta t$ after the collision the beam particle will be at a point $\mathbf{x}$, having traveled a distance $\mathbf{R}$, where

$$\mathbf{R} \equiv \mathbf{x} - \mathbf{z}^0 = \frac{\mathbf{k}_0}{m}\,\Delta t. \tag{71a}$$

Similarly, the target particle will have recoiled to a point $\mathbf{z}$, given by the relation

$$\mathbf{z} - \mathbf{z}^0 = \frac{\mathbf{Q}_0}{M}\,\Delta t. \tag{71b}$$

The separation between the two particles will then be

$$\mathbf{r} \equiv \mathbf{x} - \mathbf{z}, \tag{71c}$$

in terms of which the time elapsed is

$$\Delta t = \frac{r}{\kappa_0/M_r} \tag{71d}$$

where $M_r = mM/(m + M)$ is the reduced mass. The center of mass

$$\mathbf{C} \equiv \frac{m\mathbf{x} + M\mathbf{z}}{m + M} \tag{72a}$$

is seen to lie on the line through $\mathbf{z}^0$ directed parallel to $\mathbf{p}$.

[23] For simplicity, we shall assume in this section that both particles have nonrelativistic energies.

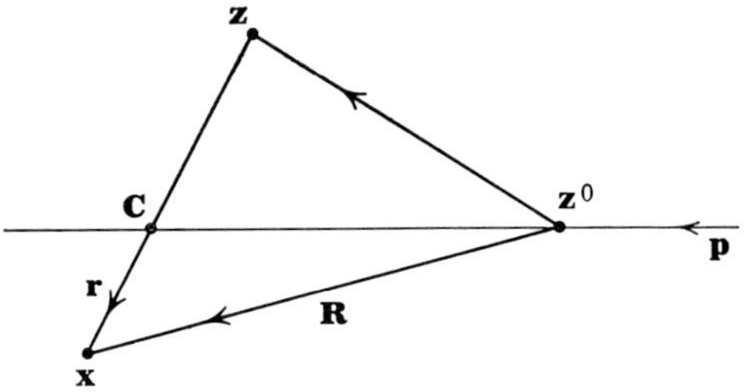


Fig. 11.1 Kinetic relations for collision with bound particle initially at $\mathbf{z}^0$.

The foregoing coordinate relations are illustrated in Fig. 11.1. We may evidently write

$$\begin{aligned} \mathbf{C} &= \mathbf{z} + \mathbf{r}\frac{m}{m+M} \\ &= \mathbf{x} - \mathbf{r}\frac{M}{m+M} \\ &= \mathbf{z}^0 + \left(\frac{\mathbf{p}}{m+M}\right)\frac{rM_r}{\kappa_0}. \end{aligned} \tag{72b}$$

Let us now turn to the evaluation of the asymptotic wave function (11), exhibiting explicitly the kinematical relations obtained above. This will be done in some detail, since the results are important for our later applications to scattering by systems of bound particles. In the interest of simplicity, we have agreed to limit our discussion to the case that both particles have nonrelativistic energies. The extension to scattering by relativistic particles or "quasi-particles" is straightforward.[24]

The wave function $\psi_a{}^+$, given by Eq. (11), becomes

$$\psi_a{}^+ = \chi_a + \frac{1}{E_a + i\eta - K}\mathscr{T}\chi_a, \tag{73}$$

if we make the impulse approximation [Eq. (27)]. Now, for a final spin orientation $\nu'$ of particle $i$, we may use Eq. (36) to obtain

$$(u_{S_i,\nu'}, \mathscr{T}\chi_a) = (2\pi)^{-3/2}\int d^3k\, e^{i\mathbf{k}\cdot\mathbf{x}}\, e^{-i\boldsymbol{\rho}\cdot\mathbf{z}} T g_0(\mathbf{z}, s_t),$$

where, as before, $(\mathbf{k}, \nu', \mathbf{P} - \boldsymbol{\rho}|\, T(\mathbf{S}_t)\, |\mathbf{p}, \nu, \mathbf{P})$ operates on the coordinates $(\mathbf{z}, s_t)$. The scattered wave in Eq. (73) is therefore

$$\begin{aligned} \psi_{\text{sc}}^{\nu'} &\equiv \left(u_{S_i,\nu'}, \frac{1}{E_a + i\eta - K}\mathscr{T}\chi_a\right) \\ &= (2\pi)^{-3/2}\int \frac{d^3k\, e^{i\mathbf{k}\cdot\mathbf{x}}}{\epsilon_p + i\eta - \epsilon_k - h + W_0}\, e^{-i\boldsymbol{\rho}\cdot\mathbf{z}}\, T\,|0\rangle. \end{aligned} \tag{74}$$

[24] It has been done explicitly in Ref. [16].

In the spirit of the expansion (39), we now set

$$(\epsilon_p - \epsilon_k + W_0 - h)^{-1} = \frac{1}{e}\left\{\sum_{n=0}^{\infty}\left[(h - \overline{W})\frac{1}{e}\right]^n\right\}, \tag{75}$$

where $e$ is defined by Eq. (40). In the last section we saw that when $\rho \gg \sqrt{2MK_{\text{av}}}$, if we take $\overline{W} = W_0 + \rho^2/2M$, the "fluctuations" $(h - \overline{W})^n$ are small. Then, using Eqs. (44), (46), and (75) in Eq. (74), we obtain

$$\psi_{\text{sc}}^{\nu'} = (2\pi)^{-3/2}\int d^3k\,\frac{e^{i\mathbf{k}\cdot\mathbf{x}}e^{-i\boldsymbol{\rho}\cdot\mathbf{z}}}{e}$$
$$\times\left\{1 + (h - L - W_0)\frac{1}{e} + \left[(h - L - W_0)\frac{1}{e}\right]^2 + \cdots\right\}T\,|0\rangle. \tag{76}$$

The terms involving powers of $h - L - W_0$ may be handled in a manner similar to that given for the expressions (45*b*). The leading term in Eq. (76) corresponds to the quasi-free approximation (48) and the kinematical relations (71) and (72). The higher-order terms in Eq. (76) are associated with a "spreading" in the classical kinematical relations (71) and (72). We shall not explicitly evaluate these higher-order terms here,[25] since the technique is similar to that used in Section 11.1, p. 692, and since their importance has already been determined.

Keeping only the leading term in Eq. (76) and introducing the Fourier expansion (35) for the bound-state wave function, we find

$$\psi_{\text{sc}} = (2\pi)^{-3}\int d^3k\,d^3P\,\frac{e^{i\mathbf{k}\cdot\mathbf{x}}\,e^{i\mathbf{Q}\cdot\mathbf{z}}}{\epsilon_p + P^2/2M + i\eta - \epsilon_k - Q^2/2M}$$
$$\times\sum_{\nu'}(\mathbf{k}, \nu', \mathbf{Q}|\,T(\mathbf{S}_t)\,|\mathbf{p}, \nu, \mathbf{P})\,|0; \mathbf{P}\rangle u_{S_{i,\nu'}}, \tag{77}$$

where

$$\mathbf{Q} \equiv \mathbf{P} - \boldsymbol{\rho} \tag{78}$$

is the recoil momentum of particle $t$.

To obtain the asymptotic value of $\psi_{\text{sc}}$, when $p\,|\mathbf{x} - \mathbf{z}| \gg 1$, it is convenient to introduce the relative momenta

$$\boldsymbol{\varkappa} = \frac{M\mathbf{p} - m\mathbf{P}}{m + M},$$
$$\boldsymbol{\varkappa}_f = \frac{M\mathbf{k} - m\mathbf{Q}}{m + M}, \tag{79a}$$

which reduce to those of Eq. (70) when $\mathbf{P} = 0$, and the total momentum

$$\mathbf{G} \equiv \mathbf{p} + \mathbf{P} = \mathbf{k} + \mathbf{Q}. \tag{79b}$$

[25] A discussion of these correction terms has been given in Ref. [16].

These relations permit us to rewrite Eq. (77) in the form

$$\psi_{sc} = (2\pi)^{-3} \sum_{v'} u_{S_i,v'} \int d^3\kappa_f\, d^3P \frac{e^{i\mathbf{G}\cdot\mathbf{C}} \quad e^{i\mathbf{k}_f\cdot\mathbf{r}}}{\dfrac{\kappa^2 - \kappa_f^{\,2}}{2M_r} + i\eta}$$

$$\times (\mathbf{k}, \nu', \mathbf{Q}|\, T\, |\mathbf{p}, \nu, \mathbf{P})\, |0; \mathbf{P}\rangle$$

$$= \lim_{r\to\infty} (2\pi)^{-3} \int d^3P\, e^{i\mathbf{G}\cdot\mathbf{C}} \quad \frac{e^{i\kappa r}}{r} f\, |0; P\rangle u_{S_i,\nu}(s_i). \tag{80}$$

Here we have introduced the scattering amplitude **f** [see Eq. (7.43)], having matrix elements

$$\langle \nu', \nu'_t|\, \mathbf{f}(\mathbf{k}, \mathbf{p}, \mathbf{S}_i, \mathbf{S}_t)\, |\nu, \nu_t\rangle$$
$$= -(2\pi)^2 M_r(\mathbf{k}, \nu', \mathbf{Q}, \nu'_t|\, T\, |\mathbf{p}, \nu, \mathbf{P}, \nu_t). \tag{81}$$

As before, $M_r$ represents the reduced mass of the two particles. Having evaluated the asymptotic intergal (80), we must, of course, henceforth consider **k** and **Q** as defined by

$$\mathbf{k} = \frac{m}{m+M}\mathbf{G} + \kappa\hat{\mathbf{r}},$$
$$\mathbf{Q} = \frac{M}{m+M}\mathbf{G} - \kappa\hat{\mathbf{r}}. \tag{82}$$

We shall next evaluate the **P**-integral in Eq. (80), treating $P$ as a "small" quantity, as is implied by the "weak binding" condition. Comparison of Eqs. (70) and (79) shows that

$$\boldsymbol{\varkappa} = \boldsymbol{\varkappa}_0 - \frac{M_r}{M}\mathbf{P},$$

so

$$\kappa \cong \kappa_0 - \frac{\mathbf{P}\cdot\mathbf{p}}{\kappa_0}\frac{M_r}{m+M}, \tag{83}$$

to within terms of $0(P^2)$, which we shall neglect.[26] Using these expressions and the definition (79*b*), we may write the argument of the exponentials in Eq. (80) as

$$\mathbf{G}\cdot\mathbf{C} + \kappa r = \mathbf{p}\cdot\mathbf{C} + \kappa_0 r + \mathbf{P}\cdot\left(\mathbf{C} - \frac{\mathbf{p}}{\kappa_0}\frac{M_r}{m+M}r\right)$$
$$= \mathbf{p}\cdot\mathbf{C} + \kappa_0 r + \mathbf{P}\cdot\mathbf{z}^0, \tag{84}$$

[26] The neglected terms would lead to "wave packet spreading" in the kinematical relations (71) and (72). The contribution from such terms is included in the $\eta$-corrections in Eq. (49). (See also the discussion in Ref. [16].)

using Eq. (72$b$).[27] This permits Eq. (80) to be evaluated in the form

$$\psi_{sc} = (2\pi)^{-3} \int d^3P \frac{e^{i\kappa_0 r}}{r} e^{i\mathbf{p}\cdot\mathbf{C}} \quad \mathbf{f}\, e^{i\mathbf{p}\cdot\mathbf{z}^0} \quad |0; \mathbf{P}\rangle u_{S_i,\nu}$$
$$= (2\pi)^{-3/2} \frac{e^{i\kappa_0 r}}{r} e^{i\mathbf{p}\cdot\mathbf{C}} \;\mathbf{f}\, u_{S_i,\nu} g_0(\mathbf{z}^0, s_t). \tag{85}$$

The appearance of $\mathbf{z}^0$ rather than $\mathbf{z}$ as the coordinate appearing in $g_0$ has an obvious physical interpretation. It is evidently the "initial coordinate" $\mathbf{z}^0$, and not the "recoil coordinate" $\mathbf{z}$, which determines the probability amplitude for scattering from the *initial state* $g_0$. This will be seen to have important implications when we study the interference of waves scattered by several bound target particles.

To put Eq. (85) into a more useful form, let us first define

$$\boldsymbol{\varkappa}_{f0} \equiv \kappa_0 \hat{\mathbf{r}}. \tag{86a}$$

Then

$$\boldsymbol{\varkappa}_{f0} = \frac{M\mathbf{k}_0 - m\mathbf{Q}_0}{m + M} \tag{86b}$$

defines $\mathbf{k}_0$, with

$$\mathbf{Q}_0 \equiv \mathbf{p} - \mathbf{k}_0 \equiv -\boldsymbol{\rho}. \tag{86c}$$

Therefore,

$$\kappa_0 r + \mathbf{p}\cdot\mathbf{C} = \boldsymbol{\varkappa}_{f0}\cdot\mathbf{r} + \mathbf{p}\cdot(\mathbf{C} - \mathbf{z}^0) + \mathbf{p}\cdot\mathbf{z}^0$$
$$= \mathbf{k}_0\cdot\left(\frac{M}{m+M}\mathbf{r} + \mathbf{C} - \mathbf{z}^0\right) + \mathbf{Q}_0\cdot\left(\mathbf{p}\frac{r}{\kappa_0}\frac{M_r}{m+M}\right.$$
$$\left. - \mathbf{r}\frac{m}{m+M}\right) + \mathbf{p}\cdot\mathbf{z}^0 = \mathbf{k}_0\cdot\mathbf{R} + \mathbf{Q}_0\cdot(\mathbf{z} - \mathbf{z}^0) + \mathbf{p}\cdot\mathbf{z}^0, \tag{87}$$

where Eqs. (71$a$) and (72$b$) have been used. Finally, we may use Eqs. (71$a$) and (71$b$) to write

$$\frac{R}{r} = \frac{v_i}{v_{\rm rel}}, \tag{88}$$

where $v_{\rm rel} = \kappa_0/M_r$ and $v_i = k_0/m$. Thus Eq. (85) is equivalent to

$$\psi_{sc} = (2\pi)^{-3/2} \frac{e^{ik_0 R}}{R} e^{i\mathbf{Q}_0\cdot(\mathbf{z}-\mathbf{z}^0)}\, e^{i\mathbf{p}\cdot\mathbf{z}^0}$$
$$\times \left(\frac{v_i}{v_{\rm rel}}\mathbf{f}\right) u_{S_i,\nu}(s_i) g_0(\mathbf{z}^0 s_t). \tag{89}$$

Equation (89) represents our final expression for $\psi_{sc}$. The phases of the exponentials have a direct physical significance. That of the incident wave

[27] We now consider (72$b$) to be the defining equation for $\mathbf{z}^0$.

at the initial position $\mathbf{z}^0$ of the particle $t$ is $\mathbf{p} \cdot \mathbf{z}^0$. In traveling to the point $\mathbf{x}$, a distance $\mathbf{R}$ from $\mathbf{z}^0$, it contributes an additional phase $k_0 R$. The target particle, in recoiling from $\mathbf{z}^0$ to $\mathbf{z}$, contributes a phase $\mathbf{Q}_0 \cdot (\mathbf{z} - \mathbf{z}^0)$ to $\psi_{sc}$.

The differential scattering cross section (48) may be obtained directly from Eq. (89) by finding the flux $j_{out}$ of particles $i$. For simplicity, we shall ignore spin degrees of freedom and set $\mathbf{P} = 0$ in $\mathbf{f}$. Then, according to Eq. (1.103$c$) [compare also Eq. (6.75$b$)],

$$\begin{aligned} \mathbf{j}_{out} &= \frac{k_0}{m} \hat{\mathbf{R}} \int \psi_{sc}{}^* \psi_{sc} \, d^3 z \\ &= \frac{k_0}{m} \frac{\hat{\mathbf{R}}}{(2\pi)^3} \left(\frac{v_i}{v_{rel}}\right)^2 \frac{|\mathbf{f}|^2}{R^2} \int g_0{}^*(\mathbf{z}^0) g_0(\mathbf{z}^0) \, d^3 z. \end{aligned} \tag{90}$$

Now an elementary calculation of the Jacobian of the transformation from $\mathbf{z}$ to $\mathbf{z}^0$ gives

$$d^3 z = \left(\frac{m}{M_r}\right)^3 \left(1 + \frac{\hat{\mathbf{r}} \cdot \mathbf{p}}{\kappa_0} \frac{M_r}{M}\right)^{-1} d^3 z^0.$$

Then on setting $\int |g_0(\mathbf{z}^0)|^2 \, d^3 z^0 = 1$ we obtain

$$\frac{d\bar{\sigma}}{d\Omega} = \left[\frac{(2\pi)^3}{v_{rel}} R^2\right] j_{out} = |f|^2 \left[\left(\frac{m v_i}{M_r v_{rel}}\right)^3 \Big/ \left(1 + \frac{M_r}{M} \frac{\hat{\mathbf{r}} \cdot \mathbf{p}}{\kappa_0}\right)\right] = \frac{d\bar{\sigma}_f}{d\Omega}, \tag{91}$$

which is readily seen to be identical with Eq. (48). Since $|\mathbf{f}|^2$ is the differential scattering cross section in the barycentric frame of reference, the factor multiplying $|\mathbf{f}|^2$ in Eq. (91) is just the appropriate factor (6.12) for transforming the differential cross section from the barycentric to the laboratory coordinate system.

In the previous section we saw that the scattering is very different when the momentum transfer $\rho \ll \sqrt{2MK_{av}}$. When this condition [which we shall now consider to imply the condition (68$a$)] is met, we have seen that we should set $e = \epsilon_p + i\eta - \epsilon_k$ in Eq. (76) [compare Eq. (66)]. This means that Eq. (77) should be replaced by

$$\begin{aligned} \psi_{sc} &= (2\pi)^{-3} \int d^3 k \, d^3 P \frac{e^{i\mathbf{k}\cdot\mathbf{x}} \, e^{i\mathbf{Q}\cdot\mathbf{z}}}{\epsilon_p + i\eta - \epsilon_k} \sum_{\nu'} (\mathbf{k}, \nu', \mathbf{Q}| \, T(\mathbf{S}_t) \, |\mathbf{p}, \nu, \mathbf{P}) \, |0; \mathbf{P}\rangle u_{S_i, \nu'} \\ &= (2\pi)^{-3/2} \frac{e^{ip|\mathbf{x}-\mathbf{z}|}}{|\mathbf{x} - \mathbf{z}|} e^{i\mathbf{p}\cdot\mathbf{z}} \mathbf{f}^e g_0(\mathbf{z}, s_t) u_{S_i, \nu}, \end{aligned} \tag{92}$$

where

$$\langle \nu', \nu'_t | \, \mathbf{f}^e(\mathbf{k}, \mathbf{p}, \mathbf{S}_i, \mathbf{S}_t) \, |\nu, \nu_t\rangle = -(2\pi)^2 m(\mathbf{k}, \nu', \mathbf{Q}, \nu'_t \, |T| \, \mathbf{p}, \nu, \mathbf{P}, \nu_t). \tag{93}$$

Since in this case the target particle is not expected to recoil from its bound state, the coordinate $\mathbf{z}$ coincides with the initial coordinate $\mathbf{z}^0$. [Indeed, the expression (92) vanishes when $\mathbf{z}$ becomes large.]

Equations (89) and (92) are readily modified for the case that the incident particle has a relativistic energy.[28] It is only necessary to take $v_i = c^2k_0/\epsilon_{k_0}$ and $v_{\text{rel}} = c^2p/\epsilon_p$ and to redefine the scattering amplitudes as

$$\mathbf{f} = -(2\pi)^2 \frac{\epsilon_p M}{\epsilon_p + Mc^2} (\mathbf{k}, \nu', \mathbf{Q}, \nu'_t \,|T|\, \mathbf{p}, \nu, \mathbf{P}, \nu_t), \tag{94}$$

$$\mathbf{f}^e = -(2\pi)^2 \left(\frac{\epsilon_p}{c^2}\right) (\mathbf{k}, \nu', \mathbf{Q}, \nu'_t \,|T|\, \mathbf{p}, \nu, \mathbf{P}, \nu_t). \tag{95}$$

Appropriate relativistic kinematical relations are of course also required.

## 11.2 SCATTERING BY A SMALL SYSTEM OF BOUND SCATTERERS

In Section 11.1 we described scattering in the weak binding limit by a single bound particle. As we shall now show, the approximation methods developed there may easily be adapted to the description of scattering by a system of bound scatterers.

We again consider the incident (or beam) particle to have a rest-mass $m$ and spin $S_i$. Its initial state is described by the product of a plane wave and a spin function as before [Eqs. (4), (5)]. The target particle is now assumed, however, to be composite, consisting of $N$ bound, elementary particles[29] of masses $M_1, \ldots, M_N$ and spins $S_1, \ldots, S_N$. In this and the following three sections we shall discuss scattering of the incident particle by this bound system in the weak binding limit.

### Properties of the Scattering System

The physical properties of the $N$-particle scatterer may often be largely unknown. For example, in the case of an atomic nucleus, we may know only its size, density, binding energy, and spin. Fortunately, to calculate the scattering from this system in the weak binding limit, it is usually unnecessary to know more than the most gross characteristics of the scatterer. Indeed, the scattering is often studied specifically to determine properties of the scatterer.

In this section we shall assume that the incident particle $i$ is distinguishable from each of the $N$ scattering particles. The $N$ particles in the scatterer need not be all different, however. When they are all identical, we shall take $M_\alpha = M (\alpha = 1, 2, \ldots, N)$. The wave function for the scatterer is,

[28] See Ref. [16] for the argument.

[29] It should be recalled that we use the term "elementary particle" in the special sense defined in Section 4.2.

of course, assumed to be suitably symmetrized in the coordinates of its identical constituents. The total mass and spin of the scatterer are taken to be $M_t$ and $S_t$, respectively. The space and spin (or other internal) coordinates of particle $\alpha$ in the scatterer are $(\mathbf{z}_\alpha, s_\alpha)$. The initial wave function describing the internal state of the scatterer is written as $g_0(\mathbf{z}_1, s_1, \mathbf{z}_2, s_2, \ldots, \mathbf{z}_N, s_N)$. A general (possibly continuum) state of the scatterer is $g_\gamma^{\pm}(\mathbf{z}_1, s_1, \ldots, \mathbf{z}_N, s_N)$. When it is desired to specify the orientation $\nu_t$ of the scatterer spin, the scatterer states will be written as $\gamma(\nu_t)$, $0(\nu_t)$, and so on. The wave functions $g_\gamma^{\pm}$ are characterized by the condition

$$D^{-1}(\boldsymbol{\lambda})g_\gamma^{\pm} = g_\gamma^{\pm}, \tag{96}$$

where $D(\boldsymbol{\lambda})$ is the displacement operator (2-5).[30]

The initial-state eigenfunction for the scatterer, moving with a momentum $\mathbf{P}$ and having a spin orientation $\nu_t$, is then

$$\chi_{t;0(\nu_t),\mathbf{P}} = (2\pi)^{-3/2}\, e^{i\mathbf{P}\cdot\mathbf{C}} g_0(\mathbf{z}_1, s_1, \ldots, \mathbf{z}_N, s_N), \tag{97a}$$

where $\mathbf{C}$ is the scatterer center of mass. An arbitrary state of the scatterer, corresponding to internal state $\gamma$, momentum $\mathbf{Q}$, and spin orientation $\nu'_t$, is

$$\chi^{\pm}_{t;\gamma(\nu'_t),\mathbf{Q}} = (2\pi)^{-3/2}\, e^{i\mathbf{Q}\cdot\mathbf{C}} g_\gamma^{\pm}(\mathbf{z}_1, s_1, \ldots, \mathbf{z}_N, s_N). \tag{97b}$$

The Hamiltonian for the scattering system is of the form

$$h \equiv \sum_{\alpha=1}^{N} K_\alpha + U, \tag{98a}$$

where $K_\gamma$ is the kinetic energy operator for particle $\alpha$ and $U$ is the interaction energy of the $N$ scattering particles. We suppose the scatterers to have nonrelativistic energies,[31] so

$$K_\alpha = -\frac{1}{2M_\alpha}\nabla_{z_\alpha}^{2}. \tag{98b}$$

The wave functions (97$a$) and (97$b$) then satisfy the equations

$$h\chi_{t;0(\nu_t),\mathbf{P}} = W_0\chi_{t;0(\nu_t),\mathbf{P}}, \tag{99a}$$

$$h\chi^{\pm}_{t;\gamma(\nu'_t),\mathbf{Q}} = W_\gamma\chi^{\pm}_{t;\gamma(\nu'_t),\mathbf{Q}}, \tag{99b}$$

where $W_0 = w_0 + P^2/2M_t$ and $W_\gamma = w_\gamma + Q^2/2M_t$. The quantities $w_0$ and $w_\gamma$ represent internal energies of the composite system.

The initial state of the entire system, prior to scattering, is [see Eq. (41)]

$$\chi_a = \chi_{i;\mathbf{p},\nu}\chi_{t;0(\nu_t),\mathbf{P}}. \tag{100}$$

[30] Gartenhaus and Schwartz [17] discuss the construction and use of the internal wave functions $g_\gamma$ for a multiparticle system.

[31] There are no kinematical difficulties associated with the removal of this restriction.

Possible final states are then

$$\chi_b^{\pm} = \chi_{i;\mathbf{k},\nu'}\chi^{\pm}_{t;\gamma(\nu'_t),\mathbf{Q}}. \tag{101}$$

These satisfy the respective equations

$$K\chi_a = E_a\chi_a = (\epsilon_p + W_0)\chi_a, \tag{102a}$$

$$K\chi_b^{\pm} = E_b\chi_b^{\pm} = (\epsilon_k + W_\gamma)\chi_b^{\pm}, \tag{102b}$$

where [see Eqs. (5) and (6)]

$$K = K_i + h. \tag{103}$$

The Hamiltonian for the interacting system is, finally,

$$H = K + V, \tag{104a}$$

where

$$V = \sum_{\alpha=1}^{N} V_\alpha. \tag{104b}$$

Here $V_\alpha$ represents the interaction between the incident particle $i$ and the target particle $\alpha$.

The energy required to remove particle $\alpha$ from the scatterer and give it a very small momentum $\mathbf{P}_\alpha$ will be called $R_{\alpha,a}$, in analogy to the quantity $R_a$ in Eq. (20) Similarly, as in Eq. (16), we define the average kinetic and potential energies of the particle $\alpha$ by

$$K_{\alpha,\text{av}} \equiv (g_0, K_\alpha g_0),$$
$$U_{\alpha,\text{av}} \equiv R_{\alpha,a} - K_{\alpha,\text{av}}. \tag{105}$$

For the present interaction, the weak binding conditions (17$a$) will be reexpressed as

$$\frac{K_{\alpha,\text{av}}}{\epsilon_p} \ll 1, \qquad \left|\frac{U_{\alpha,\text{av}}}{\epsilon_p}\right| \ll 1, \tag{106a}$$

for $\alpha = 1, 2, \ldots, N$. [The appropriate expression (6) must be used for $\epsilon_p$ here.] When $K_{\alpha,\text{av}}$, $U_{\alpha,\text{av}}$ and $R_{\alpha,a}$ are of comparable magnitude, we may write the conditions (106$a$) simply as

$$\frac{|R_{\alpha,a}|}{\epsilon_p} \ll 1. \tag{106b}$$

It will be convenient now, and for later applications, to construct a wave packet of initial target states. This will be formed in such a way that the most probable position of the target center of mass $\mathbf{C}$ is the coordinate origin at $\mathbf{x} = 0$. Such a state is

$$|0\rangle \equiv \int d^3P\, a(P)\chi_{t;0(\nu_t),\mathbf{P}}(\mathbf{z}_1, s_1, \ldots, \mathbf{z}_N, s_N); \tag{107a}$$

it is normalized to unity:

$$\langle 0 \mid 0 \rangle = 1. \tag{107b}$$

For this wave packet the density of particles $\alpha$ at the point $\mathbf{x}$ is

$$P_\alpha(\mathbf{x})|_{\text{w.p.}} = \langle 0| \, \delta(\mathbf{z}_\alpha - \mathbf{x}) \, |0\rangle,$$

which because of the condition (107*b*) is normalized to unity:

$$\int P_\alpha(\mathbf{x}) \, d^3x = 1.$$

As the width of the wave packet state (107*a*) shrinks to a very small value, so the center of mass $\mathbf{C}$ is localized at the point $\mathbf{x} = 0$, the function $P_\alpha(\mathbf{x})$ approaches a limit, characterized by the intrinsic geometrical size of the scatterer. This is

$$P_\alpha(\mathbf{x}) = \lim_{\text{w.p. size}\to 0} \langle 0| \, \delta(\mathbf{z}_\alpha - \mathbf{x}) \, |0\rangle. \tag{108a}$$

If we represent an average over scatterer spin orientations by a subscript av, we may define from (108*a*)

$$\bar{P}_\alpha(\mathbf{x}) = \langle 0| \, \delta(\mathbf{z}_\alpha - \mathbf{x}) \, |0\rangle_{\text{av}}, \tag{108b}$$

where the limit of a "sharp" wave packet is understood.

Equations (108) may be seen to depend only on the internal wave function $g_0$. Let $\mathbf{C}$, the center of mass, be chosen as an independent coordinate, the remaining $N - 1$ being constructed from the vectors $(\mathbf{z}_\alpha - \mathbf{C})$. Then in the limit that the center-of-mass is precisely at $\mathbf{C} = 0$, we may evidently write Eq. (108*b*) (for example) as

$$\bar{P}_\alpha(\mathbf{x}) = \langle 0| \, \delta(\mathbf{z}_\alpha - \mathbf{C} - \mathbf{x}) \, |0\rangle_{\text{av}}. \tag{108c}$$

Using the explicit expression for $|0\rangle$, Eqs. (97*a*) and (107), this may be reduced to the form

$$\bar{P}_\alpha(\mathbf{x}) = \frac{1}{(2S_t + 1)} \sum_{\nu t} (g_{0(\nu_t)}, \, \delta(\mathbf{z}_\alpha - \mathbf{C} - \mathbf{x}) g_{0(\nu_t)}). \tag{108d}$$

The probability of finding a particle $\alpha$ at $\mathbf{x}$ and a particle $\beta$ at $\mathbf{x}'$ is (particles $\alpha$ and $\beta$ may be identical, but we require that $\alpha \neq \beta$)

$$P_{\alpha\beta}(\mathbf{x}, \mathbf{x}') = \langle 0| \, \delta(\mathbf{z}_\alpha - \mathbf{x}) \, \delta(\mathbf{z}_\beta - \mathbf{x}') \, |0\rangle, \tag{109a}$$

where the limit of a very narrow wave packet [as in Eq. (108*a*)] is again implied. According to Eq. (107*b*) this is normalized to

$$\int P_{\alpha\beta}(\mathbf{x}, \mathbf{x}') \, d^3x \, d^3x' = 1.$$

On averaging over spin orientations as in Eq. (108*b*), we obtain from (109*a*)

$$\bar{P}_{\alpha\beta}(\mathbf{x}, \mathbf{x}') = \langle 0|\, \delta(\mathbf{z}_\alpha - \mathbf{x})\, \delta(\mathbf{z}_\beta - \mathbf{x}')\, |0\rangle_{\text{av}}. \tag{109b}$$

The argument leading to Eq. (108*d*) permits us to rewrite this as

$$\begin{aligned}\bar{P}_{\alpha\beta}(\mathbf{x}, \mathbf{x}') &= \langle 0|\, \delta(\mathbf{z}_\alpha - \mathbf{C} - \mathbf{x})\, \delta(\mathbf{z}_\beta - \mathbf{C} - \mathbf{x}')\, |0\rangle_{\text{av}} \\ &= \frac{1}{2S_t + 1} \sum_{\nu_t} (g_{0(\nu_t)}, \delta(\mathbf{z}_\alpha - \mathbf{C} - \mathbf{x})\, \delta(\mathbf{z}_\beta - \mathbf{C} - \mathbf{x}') g_{0(\nu_t)}). \end{aligned} \tag{109c}$$

The *single-particle* distribution function $P^{(1)}(\mathbf{x})$ for the scatterer is defined, using Eq. (108*a*), as

$$P^{(1)}(\mathbf{x}) = \frac{1}{N} \langle 0|\sum_{\alpha=1}^{N} \delta(\mathbf{z}_\alpha - \mathbf{x})\, |0\rangle \equiv \frac{1}{N} \rho(\mathbf{x}), \tag{110}$$

where $\rho(\mathbf{x})$ is the particle density in the scatterer. Similarly, the *pair distribution function* $P^{(2)}(\mathbf{x}, \mathbf{x}')$ is[32]

$$P^{(2)}(\mathbf{x}, \mathbf{x}') = \frac{1}{N(N-1)} \langle 0| \sum_{\alpha \neq \beta = 1}^{N} \delta(\mathbf{z}_\alpha - \mathbf{x})\, \delta(\mathbf{z}_\beta - \mathbf{x}')\, |0\rangle, \tag{111}$$

where the sum extends over all $\alpha$ and $\beta$, excluding only terms corresponding to $\alpha = \beta$, and the limit of a narrow wave packet is again implied. On averaging Eqs. (110) and (111) over spin orientations, we obtain the distribution functions $\bar{P}^{(1)}(\mathbf{x})$ and $\bar{P}^{(2)}(\mathbf{x}, \mathbf{x}')$.

Other pair distribution functions than that given by Eq. (111) may be defined. For example, we may be interested in those pairs of particles which are in some substate $\lambda$. If $\mathcal{N}_\lambda$ is the number of such pairs in the target, and if $\Lambda_{\beta\alpha,\lambda}$ is a projection operator onto the substate $\lambda$ for particles $\alpha$ and $\beta$, the required distribution is (in the limit of a narrow wave packet)

$$\bar{P}_\lambda(\mathbf{x}, \mathbf{x}') = \frac{1}{2\mathcal{N}_\lambda} \langle 0| \sum_{\alpha \neq \beta = 1}^{N} \Lambda_{\beta\alpha,\lambda}\, \delta(\mathbf{z}_\alpha - \mathbf{x})\, \delta(\mathbf{z}_\beta - \mathbf{x}') \Lambda_{\beta\alpha,\lambda}\, |0\rangle_{\text{av}}. \tag{112a}$$

More generally, a conditional pair distribution function, associated with (say) a spin matrix $\Omega_{\beta\alpha}{}^{\lambda}$ may be defined (again for a very narrow wave packet)

$$\bar{P}_{\beta\alpha,\lambda}(\mathbf{x}, \mathbf{x}') \langle 0|\, \Omega_{\beta\alpha}{}^{\lambda}\, |0\rangle_{\text{av}} \equiv \langle 0|\, \delta(\mathbf{z}_\alpha - \mathbf{x}) \delta(\mathbf{z}_\beta - \mathbf{x}') \Omega_{\beta\alpha}{}^{\lambda}\, |0\rangle_{\text{av}}. \tag{112b}$$

The single-particle distribution (110) in general vanishes for $x \gg \mathcal{R}$, the "size of the scatterer." A simple example is provided by a uniform

[32] When all $N$ particles are equal, $P_\alpha(\mathbf{x}) = P^{(1)}(\mathbf{x})$ and $P_{\alpha\beta}(\mathbf{x}, \mathbf{x}') = P^{(2)}(\mathbf{x}, \mathbf{x}')$ for all $\alpha \neq \beta$.

spherical system having a "sharp" boundary. Then

$$P^{(1)}(\mathbf{x}) = \frac{1}{\mathscr{V}} \quad \text{for} \quad x < \mathscr{R}$$
$$= 0 \quad \text{for} \quad x > \mathscr{R}, \tag{113}$$

where $\mathscr{V} = (4\pi/3)\mathscr{R}^3$ is the volume of the scatterer.

The pair distribution function $P^{(2)}(\mathbf{x}, \mathbf{x}')$ depends upon the dynamical properties of the scatterer. It can consequently be evaluated by direct calculation only for relatively simple systems. One of the more interesting of these is the ideal Fermi gas, which we now describe.

Let there be $N$ identical spin one-half particles in a box of volume $\mathscr{V}$. Plane-wave states for these particles are of the form

$$\omega_l(1) \equiv \omega_l(\mathbf{z}_1, s_1) = \frac{1}{\sqrt{\mathscr{V}}} e^{i\mathbf{l}\cdot\mathbf{z}_1} u_{\frac{1}{2},\nu}(s_1), \tag{114}$$

and so on. In the lowest state of the gas, these states will all be occupied for $l < p_F$, the Fermi momentum, which is

$$\frac{q_F}{\hbar} = 2\pi\left(\frac{3N}{8\pi\mathscr{V}}\right)^{1/3}.$$

Now, according to Eq. (4-47*b*), the lowest state of the gas is

$$|0\rangle = g_0 = \frac{1}{\sqrt{N!}} \sum_Q \epsilon_Q Q \omega_{l_1}(\mathbf{z}_1, s_1) \cdots \omega_{l_N}(\mathbf{z}_N, s_N), \tag{115}$$

where $l_1, \ldots, l_N$ represent all states having a momentum less than $p_F$. The $Q$'s represent the $N!$ permutations of the particle coordinates.

We shall first calculate the distribution of those pairs of particles which are in space-symmetric and in space-antisymmetric states. The space exchange operator[33] for a pair of Fermi particles $(\alpha, \beta)$ is

$$Q_{\alpha\beta}{}^x = -\tfrac{1}{2}(1 + \boldsymbol{\sigma}_\alpha \cdot \boldsymbol{\sigma}_\beta), \tag{116a}$$

where the $\boldsymbol{\sigma}_\alpha$ are the Pauli spin matrices, as usual.

According to Eq. (112*a*) the required distributions are therefore

$$\bar{P}_\pm(\mathbf{x}, \mathbf{x}') = \frac{1}{2\mathscr{N}_\pm} \langle 0| \sum_{\alpha\neq\beta=1}^{N} \delta(\mathbf{z}_\alpha - \mathbf{x})\, \delta(\mathbf{z}_\beta - \mathbf{x}') \frac{(1 \pm Q_{\alpha\beta}{}^x)}{2} |0\rangle_{\text{av}}, \tag{117a}$$

[33] If $\chi(\mathbf{z}_\alpha, s_\alpha, \mathbf{z}_\beta, s_\beta)$ is an antisymmetric wave function for two spin one-half particles, then $Q_{\alpha\beta}{}^x\chi(\mathbf{z}_\alpha, s_\alpha, \mathbf{z}_\beta, s_\beta) = \chi(\mathbf{z}_\beta, s_\alpha, \mathbf{z}_\alpha, s_\beta)$, which is $\pm\chi(\mathbf{z}_\alpha, s_\alpha, \mathbf{z}_\beta, s_\beta)$ if $\chi$ is space symmetric or space antisymmetric, respectively.

where the subscripts plus and minus refer to space-symmetric and space-antisymmetric states, respectively. Using Eq. (115), we obtain

$$\bar{P}_{\pm}(\mathbf{x}, \mathbf{x}') = \frac{1}{4\mathscr{N}_{\pm}} \sum_{l \neq m < p_F} [(\omega_l(1)\omega_m(2), (1 \pm Q_{12}{}^x) \times \delta(\mathbf{x} - \mathbf{z}_1)\, \delta(\mathbf{x}' - \mathbf{z}_2)\omega_l(1)\omega_m(2)) - (\omega_l(1)\omega_m(2), (1 \pm Q_{12}{}^x)\, \delta(\mathbf{x} - \mathbf{z}_1)\, \delta(\mathbf{x}' - \mathbf{z}_2)\omega_l(2)\omega_m(1))]. \tag{117b}$$

For $N \gg 1$, we see that for random spin orientations the numbers of symmetric and antisymmetric pairs are, respectively, $\mathscr{N}_+ = \frac{1}{8}N^2$ and $\mathscr{N}_- = \frac{3}{8}N^2$. Assuming that $\mathbf{x}$ and $\mathbf{x}'$ both lie within the container, we evaluate Eq (117$b$) to give

$$\bar{P}_{\pm}(\mathbf{x}, \mathbf{x}') = \frac{1}{\mathscr{V}^2}(1 \pm G_F), \tag{118}$$

where (note that $G_F = 1$ for $\mathbf{x} = \mathbf{x}'$)

$$G_F = \left[\frac{2\mathscr{V}}{(2\pi)^3 N} \int_{l < p_F} d^3l\, e^{i\mathbf{l}\cdot(\mathbf{x}-\mathbf{x}')}\right]^2 = \frac{9\pi}{2} \frac{[J_{3/2}(p_F\,|\mathbf{x} - \mathbf{x}'|)]^2}{(p_F\,|\mathbf{x} - \mathbf{x}'|)^3} \tag{119}$$

and $J_{3/2}$ is the Bessel function of order $\frac{3}{2}$. The distribution function (111) is, on the other hand,

$$\bar{P}^{(2)}(\mathbf{x}, \mathbf{x}') = \tfrac{3}{4}\bar{P}_- + \tfrac{1}{4}\bar{P}_+ = \frac{1}{\mathscr{V}^2}(1 - \tfrac{1}{2}G_F). \tag{120}$$

The functions $\bar{P}_{\pm}$ are illustrated in Fig. 11.2.

A closely related system is provided by the degenerate Fermi model for the ground state of atomic nuclei In this case, we set $N = A$, the mass number, and assume (for simplicity) that $A/2 = Z$, the charge number. The space exchange operator for two nucleons $\alpha$ and $\beta$ is [here $\boldsymbol{\tau}_\alpha$ is the isotopic spin operator for nucleon $\alpha$, as defined by Eqs (9-43)]

$$Q_{\alpha\beta}{}^x = -\tfrac{1}{4}(1 + \boldsymbol{\sigma}_\alpha \cdot \boldsymbol{\sigma}_\beta)(1 + \boldsymbol{\tau}_\alpha \cdot \boldsymbol{\tau}_\beta). \tag{116b}$$

The respective numbers of space-symmetric and space-antisymmetric pairs are[34] $\mathscr{N}_+ = \frac{3}{16}A^2 + \frac{3}{4}A$ and $\mathscr{N}_- = \frac{5}{16}A^2 - \frac{5}{4}A$. In this case Eq. (117$b$) yields

$$\begin{aligned} \bar{P}_- &= \frac{1}{\mathscr{V}^2}(1 - G_F) \equiv \bar{P}_a, \\ \bar{P}_+ &= \frac{1}{\mathscr{V}^2}\left(1 + \frac{1 - 4/A}{1 + 4/A}\, G_F\right) \equiv \bar{P}_s, \end{aligned} \tag{121}$$

[34] See Ref. [1, p. 239], where these expressions for $\mathscr{N}_{\pm}$ are obtained.

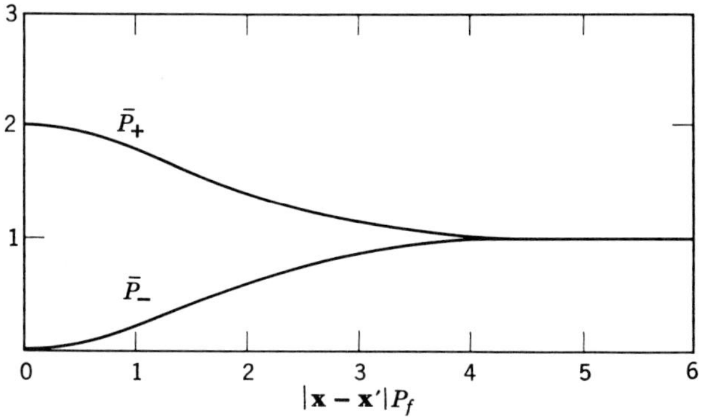


Fig. 11.2 The pair distributions $\bar{P}_\pm$ for a unit volume $\mathscr{V}$ as a function of $|\mathbf{x} - \mathbf{x}'|\ p_f$.

and

$$\bar{P}^{(2)} = \frac{1}{\frac{1}{2}A(A-1)}(\mathscr{N}_+\bar{P}_+ + \mathscr{N}_-\bar{P}_-) = \frac{1}{\mathscr{V}^2}\left(1 - \frac{A/4 - 1}{A - 1}\,G_F\right). \quad (122)$$

The function $G_F$, given by Eq. (119), is called a "pair correlation function." It expresses the influence of neighboring particles on each other in the pair distribution function. If the particles were uncorrelated, we would have

$$\bar{P}^{(2)}(\mathbf{x}, \mathbf{x}') = \bar{P}^{(1)}(\mathbf{x})\bar{P}^{(1)}(\mathbf{x}'). \quad (123)$$

Since the expression (119) vanishes for $|\mathbf{x} - \mathbf{x}'| \gg \hbar/p_F$, the particles in a Fermi gas are uncorrelated for separations large compared to $\hbar/p_F$.

For a general $N$-particle system, we may write Eqs. (111) and (112) as

$$\bar{P}^{(2)}(\mathbf{x}, \mathbf{x}') = \bar{P}^{(1)}(\mathbf{x})\bar{P}^{(1)}(\mathbf{x}')[1 + \bar{G}(\mathbf{x}, \mathbf{x}')] \quad (124a)$$

and

$$\bar{P}_\lambda(\mathbf{x}, \mathbf{x}') = \bar{P}^{(1)}(\mathbf{x})\bar{P}^{(1)}(\mathbf{x}')[1 + \bar{G}_\lambda(\mathbf{x}, \mathbf{x}')] \quad (124b)$$

where $\bar{G}$ and the $\bar{G}_\lambda$ describe correlations in particle positions. The normalization conditions on $\bar{P}^{(2)}$ and $\bar{P}_\lambda$ imply that

$$\begin{aligned} &\int \bar{P}^{(1)}(\mathbf{x})\bar{P}^{(1)}(\mathbf{x}')\bar{G}(\mathbf{x}, \mathbf{x}')\, d^3x\, d^3x' = 0, \\ &\int \bar{P}^{(1)}(\mathbf{x})\bar{P}^{(1)}(\mathbf{x}')\bar{G}_\lambda(\mathbf{x}, \mathbf{x}')\, d^3x\, d^3x' = 0. \end{aligned} \quad (125)$$

If the scattering medium is approximately "uniform" well inside its boundaries (as is the case for a liquid droplet, a crystal, or presumably a large atomic nucleus), considerations of translational invariance imply that

$$\begin{aligned} \bar{G}(\mathbf{x}, \mathbf{x}') &\cong \bar{G}(|\mathbf{x} - \mathbf{x}'|) \\ \bar{G}_\lambda(\mathbf{x}, \mathbf{x}') &\cong \bar{G}_\lambda(|\mathbf{x} - \mathbf{x}'|), \end{aligned} \tag{126}$$

except for $\mathbf{x}$ and $\mathbf{x}'$ near the boundaries of the scatterer.

We shall call the functions $G(|\mathbf{x} - \mathbf{x}'|)$, $G_\lambda(|\mathbf{x} - \mathbf{x}'|)$ the "pair correlation functions" for the medium, since they describe the deviation form "randomness" in particle positions. In general

$$G \to 0, \quad G_\lambda \to 0 \quad \text{for} \quad |\mathbf{x} - \mathbf{x}'| \gg \mathscr{R}_c, \tag{127}$$

where $\mathscr{R}_c$ is the "correlation range" in the medium. Equation (127) has meaning of course only if $\mathscr{R}_c \gg \mathscr{R}$, the size of the scatterer.

As was mentioned in the previous section and in Section 7.5, a scattering experiment involves many scattering events and should be described by a statistical ensemble. The ensemble of scatterers, ordinarily representing the physical target, is normally in thermal equilibrium with its surroundings and may be distributed over many states $\gamma$. Our considerations are easily extended to include a thermal average over scatterer states. We need only replace the state $|0\rangle$ given in Eq. (107$a$) by a state $|\gamma(0)\rangle$, defined with $g_0$ replaced by $g_{\gamma(0)}$. If we interpret the subscript av as including an ensemble average,[35] the particle distribution functions $\bar{P}^{(1)}$ and $\bar{P}^{(2)}$ are, for example,

$$\begin{aligned} \bar{\bar{P}}^{(1)}(\mathbf{x}) &= \frac{1}{\mathscr{Z}} \sum_{\gamma(0)} e^{-W_{\gamma(0)}/\theta} \left[ \frac{1}{N} \left\langle \gamma(0) \middle| \sum_{\alpha=1}^{N} \delta(\mathbf{z}_\alpha - \mathbf{x}) \middle| \gamma(0) \right\rangle \right] \\ \bar{\bar{P}}^{(2)}(\mathbf{x}, \mathbf{x}') &= \frac{1}{\mathscr{Z}} \sum_{\gamma(0)} e^{-W_{\gamma(0)}/\theta} \\ &\quad \times \left[ \frac{1}{N(N-1)} \left\langle \gamma(0) \middle| \sum_{\alpha \neq \beta = 1}^{N} \delta(\mathbf{z}_\alpha - \mathbf{x}) \delta(\mathbf{z}_\beta - \mathbf{x}') \middle| \gamma(0) \right\rangle \right] \end{aligned} \tag{128}$$

where $\theta$ is the temperature and

$$\mathscr{Z} \equiv \sum_{\gamma(0)} e^{-W_{\gamma(0)}/\theta}. \tag{129}$$

We may define Eqs. (112), (124), and (126) as thermal averages in the same manner. In this case the pair correlation functions and the correlation range $\mathscr{R}_c$ become functions of the temperature.

Some comment on the relevance of these considerations to the scattering problem is appropriate at this point. The mean particle density $\rho$ given in

[35] For a discussion of this, see Section 7.5, p. 402, and the references given there.

Eq. (110) permits us to define the scattering mean free path $\lambda$ in the scattering medium as

$$\lambda = \frac{1}{\rho \bar{\sigma}_f}, \tag{130}$$

where $\bar{\sigma}_f$ is the mean cross section for scattering on a free target particle. When

$$\lambda \gg \mathscr{R}, \tag{131}$$

the size of the scatterer, we expect the incident particle $i$ to scatter only once for each interaction with the medium. The condition (131) defines a "small scatterer" and limits the types of interaction to be considered in the remainder of this section.

In later sections of this chapter we shall not require the condition (131). We shall, however, usually assume that

$$\lambda \gg \frac{\hbar}{p}. \tag{132}$$

This implies that the energy of particle $i$ can be defined between scatterings, or that successive interactions take place "on the energy shell."

Scattering in the weak binding limit is much simplified if, in addition to the condition (132), we have also

$$\lambda \gg \mathscr{R}_c, \tag{133}$$

or a mean free path large compared to the correlation range. The inequality (133) implies that at a given interaction the scattering medium is not likely to be disturbed by previous scatterings. This means that in following a particle $i$ through a sequence of interactions in a given scattering system, we do not have to keep changing the properties of the system (such as introducing a new function $G$ after each interaction).

### The Scattering Cross Section

We are now in a position to evaluate the scattering cross section for a small scattering system [Eq. (131)] in the weak binding limit [Eqs. (106)]. In analogy to Eq. (74), we shall write

$$\psi_{\mathrm{sc},\alpha} = \frac{1}{E_a + i\eta - K}\mathscr{T}_\alpha \chi_a, \tag{134}$$

for the wave scattered from particle $\alpha$ ($\alpha = 1, 2, \ldots, N$). Here, again assuming the impulse approximation, we take $\mathscr{T}_\alpha$ as the scattering matrix for particle $i$ interacting with a free (that is, unbound) particle $\alpha$. It has the matrix elements [see Eq. (28)]

$$\langle \mathbf{k}, \nu', \mathbf{Q}_\alpha, \nu'_\alpha | \mathscr{T}_\alpha | \mathbf{p}, \nu, \mathbf{P}_\alpha, \nu_\alpha \rangle$$
$$= \delta(\mathbf{k} + \mathbf{Q}_\alpha - \mathbf{p} - \mathbf{P}_\alpha)(\mathbf{k}, \nu', \mathbf{Q}_\alpha, \nu'_\alpha | \mathscr{T}_\alpha | \mathbf{p}, \nu, \mathbf{P}_\alpha, \nu_\alpha), \tag{135}$$

which are obtained from a solution of the two-body problem. The error associated with the use of the impulse approximation is again[36] given by Eqs. (30) or (31), with $R_a$ now replaced by $R_{\alpha,a}$ and $f$ the scattering amplitude from particle $\alpha$.

For a sufficiently "small scatterer," characterized by the inequality (131), the rescattering of once-scattered waves may be neglected. This means that the complete wave scattered by the bound system is just the sum of the quantities (134):

$$\psi_{\text{sc}} = \sum_{\alpha=1}^{N} \psi_{\text{sc},\alpha} = \frac{1}{E_a + i\eta - K} \mathscr{T}_N \chi_a, \tag{136a}$$

where

$$\mathscr{T}_N = \sum_{\alpha=1}^{N} \mathscr{T}_\alpha \tag{136b}$$

is evidently the scattering matrix for the composite $N$-particle system.

Equations (136) follow from a simple assumption of additivity for the waves scattered by the separate target particles. A formal justification of these equations, in terms of the condition (131), will be given in Section 11.3.

For a final state $b$ [Eq. (102$b$)], we may quite generally write

$$\langle b | \mathscr{T}_N | a \rangle = (\chi_b^-, \mathscr{T}_N \chi_a)$$
$$= \delta(\mathbf{k} + \mathbf{Q} - \mathbf{p} - \mathbf{P})(\mathbf{k}, \nu', \mathbf{Q}, \gamma(\nu'_t) | T_N | \mathbf{p}, \nu, \mathbf{P}, 0(\nu_t)), \tag{137}$$

where $\mathbf{P}$ and $\mathbf{Q}$ are, respectively, the initial and final momenta of the $N$-particle system. Here $T_N$ is just the submatrix of $\mathscr{T}_N$ on the momentum shell, as usual.

The scattering cross section is most conveniently expressed in terms of the wave packet state (107). The width of this packet will now be so chosen, however, that[37]

$$\left\langle 0 \left| \frac{(\mathscr{P} - \bar{\mathbf{P}})^2}{2M_t} \right| 0 \right\rangle \ll |R_{\alpha,a}|, \tag{138}$$

where $\mathscr{P}$ is the total momentum operator for the system, $\bar{\mathbf{P}} \equiv \langle 0 | \boldsymbol{\rho} | 0 \rangle$, and $M_t$ is the mass of scatterer. We shall introduce a projection operator

[36] This will be shown in Section 11.3.

[37] The condition (138) is quite different from that imposed on the wave packet to define the particle distribution functions in the last section.

$\Lambda_\mu$ onto any desired set of final eigenstates $\chi_{t;\gamma(\nu'_t)\mathbf{Q}}$, as defined by Eqs. (102*b*). The scattering cross section, for these final states and the initial state $|0\rangle$, is then

$$\frac{d\sigma}{d\Omega} = \frac{(2\pi)^4}{v_{\text{rel}}} \int k^2\, dk \langle 0 | \mathscr{T}_N^{\dagger} \delta(\epsilon_k - \epsilon_p + h - W_0) \Lambda_\mu \mathscr{T}_N | 0 \rangle. \quad (139a)$$

Here $v_{\text{rel}}$ is the relative velocity of the beam particle and target system. The average cross section for an unpolarized initial ensemble of spins is

$$\frac{d\bar{\sigma}}{d\Omega} = \frac{(2\pi)^4}{v_{\text{rel}}} \int k^2\, dk \langle 0 | \mathscr{T}_N^{\dagger} \delta(\epsilon_k - \epsilon_p + h - W_0) \Lambda_\mu \mathscr{T}_N | 0 \rangle_{\text{av}}, \quad (139b)$$

where the subscript av again represents an average over the $(2S_i + 1) \times (2S_t + 1)$ initial spin orientations.

To show that the expressions (139) do, indeed, give the scattering cross section, we use Eqs (107) to write (139*a*) out in detail:

$$\begin{aligned} d\sigma = {} & \frac{(2\pi)^4}{v_{\text{rel}}} \sum_{\gamma,\nu'} \int d^3k\, d^3P\, d^3P'\, d^3Q a(\mathbf{P}) a^*(\mathbf{P}') \\ & \times \delta(\mathbf{k} + \mathbf{Q} - \mathbf{p} - \mathbf{P})\, \delta(\mathbf{k} + \mathbf{Q} - \mathbf{p} - \mathbf{P}')\, \delta(\epsilon_k - \epsilon_p + W_\gamma - W_0) \\ & \times (\mathbf{k}, \nu', \mathbf{Q}, \gamma \,|\, \mathscr{T}_N \,|\, \mathbf{p}, \nu, \mathbf{P}', 0)^* (\mathbf{k}, \nu', \mathbf{Q}, \gamma \,|\, \mathscr{T}_N \,|\, \mathbf{p}, \nu, \mathbf{P}, 0) \quad (140) \\ = {} & \frac{(2\pi)^4}{v_{\text{rel}}} \sum_{\gamma\, \nu'} \int d^3k\, \delta(\epsilon_k - \epsilon_p + W_\gamma - W_0) \\ & \times |(\mathbf{k}, \nu', \mathbf{P} - \boldsymbol{\rho}, \gamma|\, T_N\, |\mathbf{p}, \nu, \mathbf{P}, 0)|^2, \end{aligned}$$

which is our "standard form" for a cross section Here

$$\boldsymbol{\rho} \equiv \mathbf{k} - \mathbf{p}, \quad (141)$$

as before. In obtaining the final version of Eq. (140), we have used the condition (138). The sum over $\gamma$ and $\nu'$ in (140) extends over just those states onto which $\Lambda_\mu$ projects.

Let us now evaluate the cross section (139*b*) using the "small scatterer" approximation, Eq. (136*b*). We have, then,

$$\frac{d\bar{\sigma}}{d\Omega} = \frac{(2\pi)^4}{v_{\text{rel}}} \int k^2\, dk \sum_{\alpha,\beta=1}^{N} \langle 0 | \mathscr{T}_\beta^{\dagger}\, \delta(\epsilon_k - \epsilon_p + h - W_0) \Lambda_\mu \mathscr{T}_\alpha | 0 \rangle_{\text{av}}. \quad (142)$$

On introducing the partial Fourier expansion [which is our present version of Eq. (35)]

$$|0\rangle = (2\pi)^{-3/2} \int d^3P_\alpha\, e^{i\mathbf{P}_\alpha \cdot \mathbf{z}_\alpha} |0; \mathbf{P}_\alpha\rangle, \quad (143)$$

we may follow the arguments which led to Eq. (36) to obtain

$$\mathscr{T}_\alpha |0\rangle = e^{-i\boldsymbol{\rho}\cdot\mathbf{z}_\alpha} T_\alpha |0\rangle. \tag{144}$$

Here $T_\alpha$ represents the scattering matrix (35),

$$T_\alpha = (\mathbf{k}, \nu', \mathbf{P} - \boldsymbol{\rho} \,|T_\alpha(\mathbf{S}_\alpha)|\, \mathbf{p}, \nu, \mathbf{P}_\alpha), \tag{145}$$

considered as an operator on the variables $(\mathbf{z}_\alpha, s_\alpha)$.[38] This permits us to rewrite Eq. (142) as

$$\frac{d\bar{\sigma}}{d\Omega} = \frac{(2\pi)^4}{v_{\text{rel}}} \sum_{\alpha,\beta=1}^{N} \int k^2\, dk \langle 0|\, T_\beta^\dagger\, e^{i\boldsymbol{\rho}\cdot\mathbf{z}_\beta} \times \delta(\epsilon_k - \epsilon_p + h - W_0)\Lambda_\mu\, e^{-i\boldsymbol{\rho}\cdot\mathbf{z}_\alpha}\, T_\alpha|0\rangle_{\text{av}}. \tag{146a}$$

The corresponding expression for an unpolarized initial state is obtained by merely dropping the av instruction in Eq. (146). Equation (146*a*) is valid for a small scatterer whenever the impulse approximation may be used [Eqs. (30) or (31)].

For scattering from an ensemble of targets in thermal equilibrium, we have [compare Eqs. (128)]

$$\frac{d\bar{\bar{\sigma}}}{d\Omega} = \frac{1}{\mathscr{Z}} \sum_{\gamma(0)} e^{-W_{\gamma(0)}/\theta} \left[\frac{(2\pi)^4}{v_{\text{rel}}} \sum_{\alpha,\beta=1}^{N} \int k^2\, dk \langle \gamma(0)|\, T_\beta^\dagger\, e^{i\boldsymbol{\rho}\cdot\mathbf{z}_\beta} \times \delta(\epsilon_k - \epsilon_p + h - W_{\gamma(0)})\Lambda_\mu\, e^{-i\boldsymbol{\rho}\cdot\mathbf{z}_\alpha}\, T_\alpha\, |\gamma(0)\rangle\right], \tag{146b}$$

where $\mathscr{Z}$ is defined by Eq. (129).

### The Cross Section for Elastic Scattering

Elastic scattering is defined by the condition that the internal energy $w_0$ of the scatterer not be changed as a result of the interaction. A final state will then be labeled as $0(\nu'_t)$ to indicate that the spin orientation of the scatterer may change.

The cross section for elastic scattering may be obtained from Eq. (146*a*) by choosing[39]

$$\Lambda_\mu = \int d^3Q \sum_{\nu'_t} \chi_{t;\,0(\nu'_t),\mathbf{Q}}\, \chi_{t;\,0(\nu'_t),\mathbf{Q}}^\dagger \quad . \tag{147}$$

[38] As in Section 11.1, p. 691, we interpret $\mathbf{P}_\alpha$ as $(1/i)\nabla_{\mathbf{z}_\alpha}$ in Eq. (144).

[39] In this section we shall require that the small-scatterer approximation [Eq. (131)] and that the impulse approximation [Eqs. (30) or (31)] be valid, but we need not assume weak binding.

For a target initially at rest, Eq. (146$a$) then gives

$$\frac{d\bar{\sigma}_{\rm el}}{d\Omega} = \frac{(2\pi)^4}{v_{\rm rel}} \int k^2\,dk \int d^3Q\,\delta\left(\epsilon_k - \epsilon_p + \frac{Q^2}{2M_t}\right)\frac{1}{(2S_i+1)(2S_t+1)}$$
$$\times \sum_{\substack{\nu,\nu_t\\ \nu',\nu'_t}} \left|\sum_{\alpha=1}^{N} \int d^3P a(\mathbf{P})(\chi_{t;0(\nu'_t),\mathbf{Q}}, e^{-i\boldsymbol{\rho}\cdot\mathbf{z}_\alpha} T_\alpha \chi_{t;0(\nu_t),\mathbf{P}})\right|^2$$
$$= \frac{(2\pi)^4}{v_{\rm rel}} \frac{1}{(2S_i+1)(2S_t+1)} \sum_{\substack{\nu,\nu_t\\ \nu',\nu'_t}} \int k^2\,dk\,\delta\left(\epsilon_k - \epsilon_p + \frac{\rho^2}{2M_t}\right) \tag{148}$$
$$\times \left|\sum_{\alpha=1}^{N}(g_{0(\nu'_t)}, e^{-i\boldsymbol{\rho}\cdot(\mathbf{z}_\alpha-\mathbf{C})}\, T_\alpha g_{0(\nu_t)})\right|^2,$$

where $\boldsymbol{\rho}$ is $\mathbf{k} - \mathbf{p}$, as in Eq. (141). To obtain the final form of Eq. (148), we have used Eqs. (97) and (107), integrating over the center-of-mass coordinate $\mathbf{C}$.[40]

Since $T_\alpha$ in general depends upon the spin operator $\mathbf{S}_\alpha$, evaluation of the matrix element in Eq. (148) requires an explicit knowledge of the spin couplings in the wave function $g_0$. For many applications, however, there are adequate, simple approximations for evaluating this matrix element. We shall illustrate these with an example. Let us suppose that the $N$ scattering particles are identical and that $N \gg 1$. We shall invoke the weak binding assumption (106) to neglect the $\mathbf{P}_\alpha$-dependence of $T_\alpha$ [the resulting error has been estimated in Eq. (51)]. Finally, it is assumed that the scatterer is a "saturated spin system," in the sense that

$$S_t \ll NS_\alpha \tag{149}$$

(unless $S_\alpha = 0$). This means that each particle in the system has its spin very nearly averaged uniformly in all directions. The implication of this is that

$$(g_{0(\nu'_t)}, e^{-i\boldsymbol{\rho}\cdot(\mathbf{z}_\alpha-\mathbf{C})}\, T_\alpha g_{0(\nu_t)}) \cong \langle T_\alpha\rangle_\alpha (g_{0(\nu'_t)}, e^{-i\boldsymbol{\rho}\cdot(\mathbf{z}_\alpha-\mathbf{C})}\, g_{0(\nu_t)}), \tag{150}$$

where [see Eq. (135)]

$$\langle T_\alpha\rangle_\alpha \equiv \frac{1}{2S_t+1} \sum_{\nu_\alpha} (\mathbf{k}, \nu', -\boldsymbol{\rho}, \nu_\alpha|\, T_\alpha\, |\mathbf{p}, \nu, 0, \nu_\alpha). \tag{151}$$

Since we have assumed that $N \gg 1$ and that $S_t \ll NS_\alpha$, it follows that the contribution to $S_t$ from any given particle $\alpha$ is small. Thus,

$$[\mathbf{S}_t, e^{-i\boldsymbol{\rho}\cdot(\mathbf{z}_\alpha-\mathbf{C})}] \cong 0,$$

[40] We again consider the variable $\mathbf{z}_\alpha - \mathbf{C}$ as independent of $\mathbf{C}$.

from which it in turn follows that $\nu'_t = \nu_t$ in Eq. (150). Therefore, it may be concluded that [see Eq. (108*d*)]

$$(g_{0(\nu'_t)}, e^{-i\boldsymbol{\rho}\cdot(\mathbf{z}_\alpha - \mathbf{C})}T_\alpha g_{0(\nu_t)})$$
$$= \langle T_\alpha \rangle_\alpha \delta_{\nu'_t,\nu_t} \int d^3x\, e^{-i\boldsymbol{\rho}\cdot\mathbf{x}} (g_{0(\nu_t)}, \delta(\mathbf{z}_\alpha - \mathbf{C} - \mathbf{x}) g_{0(\nu_t)}) \quad (152)$$
$$\cong \langle T_\alpha \rangle_\alpha \delta_{\nu'_t,\nu_t} \int \bar{P}^{(1)}(\mathbf{x})\, e^{-i\boldsymbol{\rho}\cdot\mathbf{x}}\, d^3x.$$

In obtaining the final form here, we have argued from considerations of continuity that since $P^{(1)}(\mathbf{x}) = \bar{P}^{(1)}(\mathbf{x})$ for $S_t = 0$, $P^{(1)}(\mathbf{x}) \cong \bar{P}^{(1)}(\mathbf{x})$ for $S_t \ll NS$.

The use of Eq. (152) in Eq. (148) now leads to (when particle *i* is treated relativistically)

$$\frac{d\bar{\sigma}_{\text{el}}}{d\Omega} = \frac{(2\pi)^4}{v_{\text{rel}}} \left[\frac{k^2}{kc^2/\epsilon_k + \hat{\mathbf{k}}\cdot(\mathbf{k} - \mathbf{p})/M_t}\right] \frac{N^2}{2S_i + 1} \sum_{\nu,\nu'} |\langle T_\alpha \rangle_\alpha|^2 \, |c(\boldsymbol{\rho})|^2. \quad (153a)$$

Here the vector $\mathbf{k}$ is of course restricted by the condition of energy conservation

$$\epsilon_k + \frac{\rho^2}{2M_t} = \epsilon_p,$$

and the *scatterer form factor* $c(\boldsymbol{\rho})$ is defined as

$$c(\boldsymbol{\rho}) \equiv \int d^3x \bar{P}^{(1)}(\mathbf{x})\, e^{-i\boldsymbol{\rho}\cdot\mathbf{x}}. \quad (154)$$

The factor of $N^2$ in Eq. (153*a*) arises from the squared sum in Eq. (148), since the $\langle T_\alpha \rangle_\alpha$ are all identical.

For a *massive* scatterer, defined by the condition that

$$M_t \gg \frac{\epsilon_p}{c^2},$$

Eq. (153*a*) has the simpler form (now $v_{\text{rel}} = c^2 p/\epsilon_p$)

$$\frac{d\bar{\sigma}_{\text{el}}}{d\Omega} = \frac{(2\pi)^4}{(2S_i + 1)} N^2 \left(\frac{\epsilon_p}{c^2}\right)^2 \sum_{\nu,\nu'} |\langle T_\alpha \rangle_\alpha|^2 \, |c(\boldsymbol{\rho})|^2. \quad (153b)$$

The "saturated spin" condition leading to Eqs. (153) has a wide range of applicability, since it applies to most atoms, molecules, atomic nuclei, and macroscopic solids.

As a simple example, let us consider elastic scattering of a pion by an atomic nucleus, neglecting the isotopic spin dependence of the interactions. Then (see Section 7.2)

$$T_\alpha = a + \mathbf{b}\cdot\boldsymbol{\sigma}_\alpha, \quad (155)$$

where $a$ and $\mathbf{b}$ are functions of the energy and scattering angle. Using Eq. (153$b$), we obtain

$$\frac{d\bar{\sigma}_{\text{el}}}{d\Omega} = (2\pi)^4 \left(\frac{\epsilon_p}{c^2}\right)^2 N^2\, |a|^2\, |c(\rho)|^2. \tag{156}$$

The scattering matrix (155) is that appropriate for the laboratory frame of reference, since it corresponds to $\mathbf{P}_\alpha = 0$. The corresponding barycentric scattering matrix $(T_\alpha)_B$ is conveniently parametrized by Eq. (7.77), from which we obtain

$$\langle (T_\alpha)_B \rangle_\alpha = -\frac{\kappa}{(2\pi)^2 \rho_\epsilon} f, \tag{157}$$

where $f$ is defined by Eq. (7.78). Here $\kappa$ is the barycentric[41] momentum of the pion. According to the transformation equation (3.112) for the $T$-matrix, we have the relation

$$\langle T_\alpha \rangle_\alpha = \frac{\epsilon_\kappa}{\epsilon_p} \langle (T_\alpha)_B \rangle = -\frac{\kappa \epsilon_\kappa}{(2\pi)^2 \rho_\epsilon \epsilon_p} f, \tag{158}$$

where $\epsilon_\kappa = c(m^2c^2 + \kappa^2)^{1/2}$.[42] Inserting the expression (158) for $a$ in Eq. (156), and setting $\rho_\epsilon = \kappa\epsilon_\kappa(1 + \epsilon_\kappa/c^2M)^{-1}$, where $M$ is the nucleon mass, we find

$$\frac{d\bar{\sigma}_{\text{el}}}{d\Omega} = \frac{|f|^2 A^2}{(1 + \epsilon_\kappa/c^2M)^2} |c(\rho)|^2 \tag{159}$$

for the elastic pion-nucleus scattering cross section. Here we have set $N = A$, the nuclear mass number.

If we try representing the nucleon distribution in a nucleus by the function

$$\bar{P}^{(1)}(x) = \frac{2}{\pi^{3/4}\mathscr{R}^3} e^{-x^2/\mathscr{R}^2}, \tag{160}$$

then by Eq. (154)

$$c(\rho) = e^{-\rho^2\mathscr{R}^2/4}. \tag{161}$$

When $p\mathscr{R} \gg 1$, we see then that Eq. (159) will be sharply peaked at small scattering angles. In this case we may evaluate $f$ for zero scattering angle, and the entire angular dependence of $d\bar{\sigma}_{\text{el}}/d\Omega$ arises from the form factor

[41] The term "barycentric" here of course refers to that coordinate frame in which the sum of the pion and nucleon momenta is zero.

[42] The symbol $\epsilon_\kappa$ was used in Chapters 6 and 7 to represent the total barycentric energy. Here we use it, as indicated, for the energy of the pion in the pion-nucleon barycentric system.

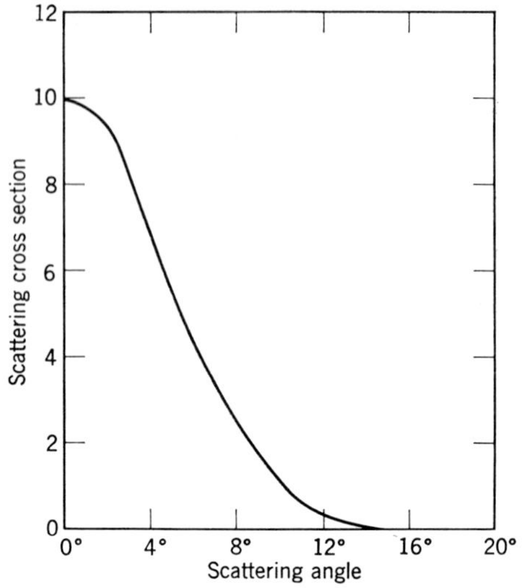


Fig. 11.3 The cross section for elastic scattering of 300-Mev nucleons by carbon nuclei (in arbitrary units), using the expressions (159) and (161).

$|c(\rho)|^2$. This is illustrated in Fig. 11.3, where we use Eqs. (159) and (161) to evaluate the scattering of 300-Mev nucleons by carbon nuclei.

The cross section (153*a*) differs in three essential respects from that for a free scattering particle $\alpha$. First, the form factor $|c(\rho)|^2$ of the bound system modifies the angular distribution. Second, the scattered waves from the $N$ particles interfere *coherently*, so the cross section is multiplied by $N^2$. Finally, the kinematics of the collision are those for a recoiling mass $M_t$, rather than $M$, the particle mass.

**The Quasi-free Scattering**

Let us next discuss the cross section (146*a*) in the limits of weak binding and quasi-free scattering. In the present context, the conditions (47) are reinterpreted as

$$\begin{aligned} \rho &\gg \sqrt{2M_\alpha K_{\alpha,\mathrm{av}}} \\ \rho &\gg \sqrt{2M_\alpha U_{\alpha,\mathrm{av}}} \end{aligned} \tag{162}$$

[see Eqs. (105)] for $\alpha = 1, 2, \ldots, N$. Then, the momentum $\boldsymbol{\rho}$ imparted to a scatterer $\alpha$ is large compared to its spread of momenta in the state $g_0$.

It was shown in Section 11.1, p. 700, that in the quasi-free limit the scattering cross section $d\bar{\sigma}(\gamma)$, for a final state $\gamma$, is sharply peaked about certain states $\gamma$ which correspond closely to the recoil of a free target particle. This suggests that we simply sum (146*a*) over all final states $\gamma$, which is accomplished by taking $\Lambda_\mu$ to be the unit operator [compare Eq. (38)]. This leads, for a target system initially at rest, to the expression[43]

$$\frac{d\bar{\sigma}}{d\Omega} = \frac{d\bar{\sigma}_d}{d\Omega} + \frac{d\bar{\sigma}_{nd}}{d\Omega}, \tag{163a}$$

where

$$\frac{d\bar{\sigma}_d}{d\Omega} = \frac{(2\pi)^4}{v_{\text{rel}}} \sum_{\alpha=1}^{N} \int k^2\, dk \times \langle 0|\, T_\alpha^\dagger\, e^{i\mathbf{p}\cdot\mathbf{z}_\alpha}\, \delta(\epsilon_k - \epsilon_p + h - W_0) e^{-i\mathbf{p}\cdot\mathbf{z}_\alpha} T_\alpha\, |0\rangle_{\text{av}} \tag{163b}$$

and

$$\frac{d\bar{\sigma}_{nd}}{d\Omega} = \frac{(2\pi)^4}{v_{\text{rel}}} \sum_{\alpha\neq\beta=1}^{N} \int k^2\, dk \times \langle 0|\, T_\beta^\dagger\, e^{i\mathbf{p}\cdot\mathbf{z}_\beta} \delta(\epsilon_k - \epsilon_p + h - W_0)\, e^{-i\mathbf{p}\cdot\mathbf{z}_\alpha} T_\alpha\, |0\rangle_{\text{av}}. \tag{163c}$$

[The notation in Eq. (163*c*) implies that the sum extends over those $N(N-1)$ terms for which $\alpha \neq \beta$.]

The cross section (163*c*) vanishes in the quasi-free limit. This can be seen most easily from Eq. (89). That is, $d\bar{\sigma}_{nd}/d\Omega$ results from the coherent interference of waves scattered from two different particles $\alpha$ and $\beta$ (coherence is implicit, since the same final states ($\mathbf{k}$, $\nu'$, $\gamma(\nu'_t)$) must result). The appearance of $\mathbf{z}^0$ in Eq. (89) implies that in the quasi-free limit the particle $\alpha$ recoils out of the scatterer in $|0\rangle$, while $\beta$ recoils out of the scatterer in $\langle 0|$. The *overlap* of $g_0^*$ and $g_0$ therefore vanishes. When the quasi-free conditions (162) are met, we may therefore take

$$\frac{d\bar{\sigma}}{d\Omega} = \frac{d\bar{\sigma}_d}{d\Omega}. \tag{164}$$

The evaluation given in Section 11.1, p. 692, may be adapted to Eq. (164) with no essential modification. In analogy to Eq. (39) we introduce into the $\alpha$th term in (163*b*) the expansion

$$\delta(\epsilon_k - \epsilon_p + h - W_0) = \sum_{n=0}^{\infty} \frac{1}{n!}\, \delta^n(e_\alpha)(h - \overline{W}_\alpha)^n, \tag{165}$$

where

$$\begin{aligned} e_\alpha &= \epsilon_k - \epsilon_p + \frac{\rho^2}{2M_\alpha}, \\ \overline{W}_\alpha &= W_0 + \frac{\rho^2}{2M_\alpha}. \end{aligned} \tag{166}$$

[43] The corresponding expressions for the ensemble average (146*b*) are obvious.

Therefore, as in Eq. (41),

$$\begin{aligned}\frac{d\bar{\sigma}_d}{d\Omega} &= \sum_{\alpha=1}^{N}\left[\sum_{n=0}^{\infty}\frac{1}{n!}\frac{(2\pi)^4}{v_{\text{rel}}}\int k^2\,dk\,\delta^n(e_\alpha)\Gamma_{\alpha;n}\right]\\ &= \sum_{\alpha=1}^{N}\left\{\sum_{n=0}^{\infty}\frac{(-1)^n}{n!}\frac{d^n}{de_\alpha{}^n}\left[\frac{k^2}{de_\alpha/dk}\frac{(2\pi)^4}{v_{\text{rel}}}\Gamma_{\alpha;n}\right]\right\}_{e_\alpha=0}.\end{aligned} \tag{167}$$

Here, in analogy to the expression (42), we have defined[44]

$$\begin{aligned}\Gamma_{\alpha;n} &\equiv \langle 0|\,T_\alpha{}^\dagger\,e^{i\boldsymbol{\rho}\cdot\mathbf{z}_\alpha}(h-\overline{W}_\alpha)^n\,e^{-i\boldsymbol{\rho}\cdot\mathbf{z}_\alpha}T_\alpha\,|0\rangle_{\text{av}}\\ &\equiv \langle 0|\,T_\alpha{}^\dagger(h-L_\alpha-W_0)^nT_\alpha\,|0\rangle_{\text{av}},\end{aligned} \tag{168}$$

with

$$L_\alpha = \frac{\boldsymbol{\rho}\cdot\nabla_{z\alpha}}{iM_\alpha} = \frac{\boldsymbol{\rho}\cdot\mathbf{P}_\alpha}{M_\alpha}. \tag{169}$$

If, as in deriving Eq. (48), we neglect the $\mathbf{P}_\alpha$-dependence of $T_\alpha$, the leading term (for which $n = 0$) in Eq. (167) gives[45]

$$\begin{aligned}\frac{d\bar{\sigma}_d}{d\Omega} &\cong \sum_{\alpha=1}^{N}\frac{(2\pi)^4}{v_{\text{rel}}}\int k^2\,dk\,\delta\left(\epsilon_k-\epsilon_p+\frac{\rho^2}{2M_\alpha}\right)\langle 0|\,T_\alpha{}^\dagger T_\alpha\,|0\rangle_{\text{av}}\\ &= \sum_{\alpha=1}^{N}\frac{d\bar{\sigma}_f(\alpha)}{d\Omega},\end{aligned} \tag{170}$$

where

$$\begin{aligned}\frac{d\bar{\sigma}_f(\alpha)}{d\Omega} &= \frac{(2\pi)^4}{v_{\text{rel}}}\int k^2\,dk\,\delta\left(\epsilon_k-\epsilon_p+\frac{\rho^2}{2M_\alpha}\right)\\ &\times\frac{1}{(2S_i+1)(2S_\alpha+1)}\sum_{\substack{\nu,\nu'\\ \nu_\alpha,\nu'_\alpha}}|(\mathbf{k},\nu',-\boldsymbol{\rho},\nu'_\alpha|\,T_\alpha\,|\mathbf{p},\nu,0,\nu_\alpha)|^2\end{aligned} \tag{171}$$

is the cross section for a free scatterer $\alpha$ initially at rest. As might have been anticipated, in the quasi-free limit, the scattering from the $N$-particle system is given as the sum of the free-particle cross sections of its $N$ constituent particles.

An expansion of the form (49) may now be obtained for $d\bar{\sigma}_d/d\Omega$. The derivation will not be repeated, since it differs in only trivial details from that given in Section 11.1, p. 695. The final expansion is of the form

$$\frac{d\bar{\sigma}_d}{d\Omega} = \sum_{\alpha=1}^{N}\frac{d\bar{\sigma}_f(\alpha)}{d\Omega}(1+\eta_{\alpha;1}+\eta_{\alpha;2}+\cdots), \tag{172}$$

[44] See footnote 15.

[45] Evaluation of $\langle 0|\,T_\alpha{}^\dagger T_\alpha\,|0\rangle_{\text{av}}$ is described in Note [18].

where the $\eta_{\alpha;i}$ differ from the $\eta_i$ of Section 11.1, p. 695, only in that $K_{av}$ and $U_{av}$ are replaced by $K_{\alpha,av}$ and $U_{\alpha,av}$ and so on.

Equation (170) is easily modified to treat scattering by an ensemble of polarized scattering systems. If $\rho_i$ is the appropriate initial density matrix (as described in Section 7.5), we need only set

$$\langle 0|\, T_\alpha{}^\dagger T_\alpha\, |\rangle_{av} = \mathrm{Tr}\,(\rho_i T_\alpha{}^\dagger T_\alpha) \tag{173a}$$

in Eq. (170). A final state of polarization is obtained from evaluation of the quantities

$$\langle 0|\, T_\alpha{}^\dagger \Omega^\lambda T_\alpha\, |0\rangle_{av} = \mathrm{Tr}\,(\rho_i T_\alpha{}^\dagger \Omega^\lambda T_\alpha), \tag{173b}$$

where the $\Omega^\lambda$ are appropriate spin matrices, as in Section 7.5, p. 410. Equation (170) is also easily modified to describe scattering by a thermodynamic ensemble, as prescribed by Eq. (146*b*).

If it is desirable to take account of the momenta of the scatterers in the state $g_0$, the cross section $d\bar{\sigma}_d/d\Omega$ may be expressed in the form (62):

$$\frac{d\bar{\sigma}_d}{d\Omega} = \sum_{\alpha=1}^{N} \left\langle \frac{d\bar{\sigma}_f(\alpha)}{d\Omega} \right\rangle_{av} (1 + \eta_{\alpha;2} + \eta_{\alpha;3} + \eta_{\alpha;6} + \cdots), \tag{174}$$

where

$$\left\langle \frac{d\bar{\sigma}_f(\alpha)}{d\Omega} \right\rangle_{av} \equiv \int d^3P\, \langle 0; \mathbf{P}_\alpha \mid 0; \mathbf{P}_\alpha \rangle_{av} \frac{1}{v_{rel}} \left| \frac{d\epsilon_k}{dk}\hat{\mathbf{k}} - \frac{\mathbf{P}_\alpha}{M_\alpha} \right| \frac{d\bar{\sigma}_f(\alpha, \mathbf{P}_\alpha)}{d\Omega}. \tag{175}$$

Here $|0; \mathbf{P}_\alpha\rangle$ is the Fourier coefficient (143) and $d\bar{\sigma}_f(\alpha; \mathbf{P}_\alpha)/d\Omega$ is the scattering cross section for a particle *i* striking a free particle $\alpha$ of initial momentum $\mathbf{P}_\alpha$.[46]

For the study of x-ray or neutron scattering by chemically bound systems, the lowest-order term in Eq. (172) or (174) is usually very accurate. The accuracy of these expressions for nuclear interactions is less well established. For incident particles of energies above 100 Mev, observed nuclear interactions have generally been consistent with Eqs. (172) and (174), however. This is illustrated in Fig. 11.4, where the scattering of 350-Mev protons by Li nuclei is compared with proton-proton scattering.[47] For this interaction the leading correction term in Eq. (174) appears to be $\eta_3$ [Eq. (54)]. Estimating $|T_{sp}|/|T| \simeq 1$ and $|\langle 0|\, U_{sp}\, |0\rangle| \simeq 40$ Mev, we anticipate that $\eta_3 \simeq 0.1$.

[46] The derivation of Eqs. (174) and (175) is so similar to that of Eqs. (60) and (62) that it need not be repeated.

[47] The observed cross section on Li are due to Chamberlain and Segré [19].

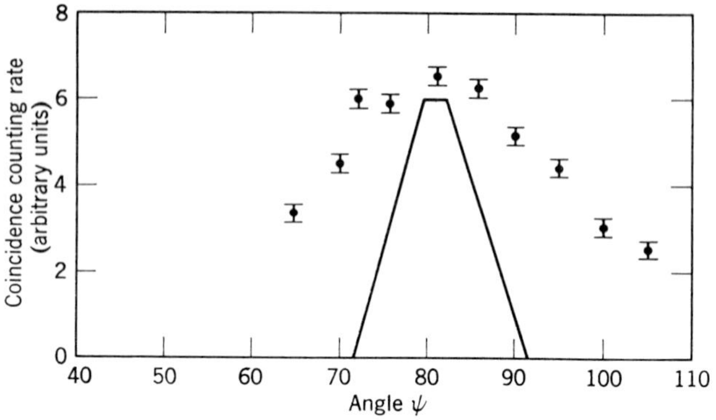


Fig. 11.4 Comparison of cross sections for scattering 340-Mev protons by protons and by lithium nuclei (taken from the first of Refs. [19]). Both the scattered and recoil protons are counted in coincidence. The angle $\psi$ is the angle between the two counters, which would be 90 degrees for nonrelativistic protons and counters of zero cross-sectional area. The solid curve corresponds to scattering by free protons, whereas the points represent the scattering from lithium nuclei.

As was noted in Section 11.1, p. 700, the accuracy of Eqs. (172) and (174) does not require that the quasi-free conditions (162) be met, but only that the weak binding conditions (106) be valid. The physical interpretation of the final state and Eq. (164) must in general be modified, however, when $\rho$ is not large compared to virtual momenta in the state $g_0$.

### The Quasi-elastic Scattering

We have just seen that in the weak binding and quasi-free [defined by the inequalitites (162)] limits the scattering cross section for a "small" system of bound particles is just the sum of the individual free-particle cross sections for the various target particles. In this case, the analysis given in Section 11.1 of the scattering from a single bound particle may be used, since the role of the other particles in a composite scatterer is just to provide the "binding potential" for the struck particle. Since this struck particle recoils as if free, it may be observed experimentally. Observation of a recoil particle of species $\alpha_0$ evidently restricts the sum in Eq. (170) to those particles of species $\alpha_0$.[48]

When the quasi-free conditions (162) do not obtain, the physical nature of the scattering interaction will ordinarily be profoundly changed. As we saw in Section 11.1, the target particles will not, in general, recoil from the scatterer. Indeed, the scattering may be elastic, corresponding to no

[48] Formally, this is equivalent to choosing the projection operator $\Lambda_\mu$ in Eqs. (139) so as to select only final states corresponding to the recoil of particles of a prescribed species.

excitation of the scatterer. When the scattering is inelastic, but a relatively small excitation energy is imparted to the scatterer, the process is said to be *quasi-elastic*. More precisely, quasi-elastic scattering is defined by the condition that the excitation energy of the scatterer $\langle(w_\gamma - w_0)\rangle$ be

$$\langle(w_\gamma - w_0)\rangle = \mathcal{O}(K_{\alpha,\text{av}}) = \mathcal{O}(U_{\alpha,\text{av}}). \tag{176}$$

The weak binding condition then implies that the incident particle lose only a small fraction of its energy $\epsilon_p$.

Equation (172) is generally valid even when the conditions (162) are not satisfied. Equation (164) will not ordinarily be correct, however. To obtain the cross section (163*a*) then we must evaluate the expression (163*c*). As a first step toward doing this, we expand the $\delta$-function in a manner analogous to that of Eqs. (39) and (165):

$$\delta(\epsilon_k - \epsilon_p + h - W_0) = \sum_{n=0}^{\infty} \frac{1}{n!}\, \delta^n(e_q)(h - W_q)^n, \tag{177}$$

where

$$e_q \equiv \epsilon_k - \epsilon_p + W_q - W_0 \tag{178}$$

and $W_q$ is to be determined. Equation (177) permits us to write (163*c*) as

$$\begin{aligned}\frac{d\bar{\sigma}_{nd}}{d\Omega} &= \sum_{n=0}^{\infty} \frac{1}{n!}\frac{(2\pi)^4}{v_{\text{rel}}} \int k^2\, dk\, \delta^n(e_q) Q_n \\ &= \sum_{n=0}^{\infty} \frac{(-1)^n}{n!} \frac{d^n}{de_q{}^n}\left[\frac{k^2}{de_q/dk}\frac{(2\pi)^4}{v_{\text{rel}}} Q_n\right]_{e_q=0}.\end{aligned} \tag{179}$$

Here we have introduced the abbreviation

$$Q_n \equiv \sum_{\alpha \neq \beta = 1}^{N} \langle 0|\, T_\beta{}^\dagger\, e^{i\boldsymbol{\rho}\cdot\mathbf{z}_\beta}(h - W_q)^n\, e^{-i\boldsymbol{\rho}\cdot\mathbf{z}_\alpha} T_\alpha\, |0\rangle_{\text{av}}. \tag{180}$$

Evaluation of the terms in Eq. (179) is similar to that of Eq. (167). First, an appropriate value for the energy $W_q$ must be found. Then the expressions (180) must be evaluated in a convenient form. As before, two classes of terms will be found to contribute to the $Q_n$ for $n \geq 1$. The first type results from the commutator $[h, T_\alpha]$ and the $\mathbf{P}_\alpha$-dependence of $T_\alpha$. These contributions do not differ from the corresponding ones found in Section 11.1, p. 695, so we may use Eqs. (51), (53), and (54) to estimate their importance. The second class of terms is obtained if $[h, T_\alpha]$ and the $\mathbf{P}_\alpha$-dependence of $T_\alpha$ are neglected.

To choose $W_q$ conveniently, we first observe that, as implied by Eq. (2.6),

$$e^{-i\boldsymbol{\rho}\cdot\mathbf{z}_\alpha} T_\alpha\, |0\rangle$$

is a state corresponding to a total momentum $-\boldsymbol{\rho}$ for the scatterer.[49]

[49] This is physically obvious, since $-\boldsymbol{\rho}$ is the momentum transferred to the scatterer, which was initially at rest.

Because of the factor $\langle 0|\, T_\beta^{\dagger}\, e^{i\boldsymbol{\rho}\cdot\mathbf{z}_\beta}$, this momentum is not associated with the single particle $\alpha$ in Eq. (180), however, as it was in the expressions (168). This suggests that we transform the state $e^{-i\boldsymbol{\rho}\cdot\mathbf{z}_\alpha}T_\alpha\,|0\rangle$ to rest with the operator $e^{i\boldsymbol{\rho}\cdot\mathbf{C}}$, where

$$\mathbf{C} = \frac{1}{M_t}\sum_{\lambda=1}^{N} M_\lambda \mathbf{z}_\lambda \tag{181}$$

is the scatterer center of mass.[50] It is easily seen that

$$e^{i\boldsymbol{\rho}\cdot\mathbf{C}}h\, e^{-i\boldsymbol{\rho}\cdot\mathbf{C}} = h + \frac{\rho^2}{2M_t} - L_N, \tag{182a}$$

where

$$L_N \equiv \frac{1}{M_t}\,\boldsymbol{\rho}\cdot\left(\sum_{\lambda=1}^{N}\mathbf{P}_\lambda\right). \tag{182b}$$

Using these equations, we can rewrite (180) as

$$Q_n = \sum_{\alpha\neq\beta=1}^{N}\left\langle 0|\, T_\beta^{\dagger}\, e^{i\boldsymbol{\rho}\cdot(\mathbf{z}_\beta-\mathbf{C})}\left(h + \frac{\rho^2}{2M_t} - W_q - L_N\right)^n e^{-i\boldsymbol{\rho}\cdot(\mathbf{z}_\alpha-\mathbf{C})}T_\alpha\,|0\right\rangle_{\text{av}}. \tag{183}$$

Now,

$$[L_N, h] = [L_N, T_\alpha] = [L_N, e^{-i\boldsymbol{\rho}\cdot(\mathbf{z}_\alpha-\mathbf{C})}] = 0$$

and $L_N\,|0\rangle = 0$. Therefore, Eq. (183) is equivalent to

$$Q_n = \sum_{\alpha\neq\beta=1}^{N}\left\langle 0|\, T_\beta^{\dagger}\, e^{i\boldsymbol{\rho}\cdot\mathbf{Z}_\beta}\left(h + \frac{\rho^2}{2M_t} - W_q\right)^n e^{-i\boldsymbol{\rho}\cdot\mathbf{Z}_\alpha}\, T_\alpha|0\right\rangle_{\text{av}}, \tag{184}$$

where

$$\mathbf{Z}_\alpha = \mathbf{z}_\alpha - \mathbf{C}, \qquad \alpha = 1, 2, \ldots, N. \tag{185}$$

It is evident from (184) that an appropriate choice for $W_q$ is

$$W_q = W_0 + \frac{\rho^2}{2M_t}, \tag{186}$$

so [see Eq. (178)]

$$\mathscr{e}_q = \epsilon_k - \epsilon_q + \frac{\rho^2}{2M_t}. \tag{187}$$

The kinematical relation implied by $\mathscr{e}_q = 0$ is seen [Eq. (148)] to be exact for elastic scattering and appropriate with weak binding for quasi-elastic scattering.[51] It implies that the excitation energy of the scatterer

[50] A much more thorough discussion of such transformations is given in Ref. [17].

[51] The modification if the scatterer recoils with relativistic energies is easily made. Using the kinetic energy operators $c(M_\alpha^2c^2 + P_\alpha^2)^{1/2}$ [Eq. (1.102$a$)] we find that $c(M_t^2c^2 + \rho^2)^{1/2} - M_tc^2$ replaces $\rho^2/2M_t$ in Eq. (187).

is just that which it receives as a result of its recoil. When the mass $M_t$ is very large compared with $\epsilon_p/c^2$, the condition $e_q = 0$ becomes equivalent to that used in Eq. (66).

It is convenient now to reexpress Eq. (179) in a modified form.[52] To do this, we write

$$\delta(e_q) = \frac{1}{2\pi}\int_{-\infty}^{+\infty} dt\ e^{ie_q t}$$

and so obtain

$$\frac{d\bar{\sigma}_{nd}}{d\Omega} = \frac{(2\pi)^3}{v_{\text{rel}}}\int k^2\,dk \int_{-\infty}^{+\infty} dt\ e^{ie_q t} \sum_{\alpha\neq\beta=1}^{N} \langle 0|\ T_\beta{}^\dagger\ e^{i\mathbf{p}\cdot\mathbf{Z}_\beta}\ e^{i(h-W_0)t}\, e^{-i\mathbf{p}\cdot\mathbf{Z}_\alpha} T_\alpha\ |0\rangle_{\text{av}}. \tag{188a}$$

Recalling the interpretation of $i(d/de_q)$ as a "time operator," we may interpret $t$ here as representing an "interaction time" for the collision.

Now, $e^{i(h-W_0)t}$ is a displacement operator in time, since

$$e^{i(h-W_0)t}\mathbf{Z}_\alpha\, e^{-i(h-W_0)t} = \mathbf{Z}_\alpha(t),$$

where $\mathbf{Z}_\alpha(t)$ is the Heisenberg operator which coincides with the Schrödinger operator $\mathbf{Z}_\alpha$ at $t = 0$. Similarly we may write

$$e^{i(h-W_0)t}T_\alpha\, e^{-i(h-W_0)t} = T_\alpha(t),$$

to indicate the time dependence of both $\mathbf{P}_\alpha(t)$ and $\mathbf{S}_\alpha(t)$, on which $T_\alpha(t)$ may depend. The relations permit us to rewrite (188*a*) in the compact form[53]

$$\frac{d\bar{\sigma}_{nd}}{d\Omega} = \frac{(2\pi)^3}{v_{\text{rel}}}\int k^2\,dk \int_{-\infty}^{+\infty} dt\ e^{ie_q t} \sum_{\alpha\neq\beta=1}^{N} \langle 0|\ T_\beta{}^\dagger\ e^{i\mathbf{p}\cdot\mathbf{Z}_\beta}\, e^{-i\mathbf{p}\cdot\mathbf{Z}_\alpha(t)} T_\alpha(t)\ |0\rangle_{\text{av}}. \tag{188b}$$

Equation (188*b*) depends upon the operators $T_\beta{}^\dagger$ and $\mathbf{Z}_\beta$ at a given time ($t = 0$) and upon the operators $T_\alpha$ and $\mathbf{Z}_\alpha$ at a relative time interval $t$. If the "motion" of particle $\alpha$ may be neglected during the time (this is ordinarily of order $\epsilon_p{}^{-1}$) of the collision, we may set $\mathbf{Z}_\alpha(t) = \mathbf{Z}_\alpha$ and $T_\alpha(t) = T_\alpha$. This leads to the first approximation to Eq. (188*b*).

$$\frac{d\bar{\sigma}^{(0)}_{nd}}{d\Omega} \equiv \frac{(2\pi)^4}{v_{\text{rel}}}\int k^2\,dk\ \delta\!\left(\epsilon_k + \frac{\rho^2}{2M_t} - \epsilon_p\right) \sum_{\alpha\neq\beta=1}^{N} \langle 0|\ T_\beta{}^\dagger\ e^{i\mathbf{p}\cdot(\mathbf{z}_\beta-\mathbf{z}_\alpha)} T_\alpha\ |0\rangle_{\text{av}}, \tag{189}$$

where it is implied that $\mathbf{P}_\alpha$ is neglected in the $T_\alpha$. The approximation (189) might of course have been obtained directly from the first term, corresponding to $n = 0$, in Eq. (179).

[52] We here follow the argument of Wick [10] and Fowler [11].

[53] It is evident that if the sum is extended over all $\alpha$ and $\beta$ in Eq. (188*b*) the complete cross section (163*a*) is obtained.

In the weak binding limit we expect the collision time to be very short and the expression (189) therefore to provide a good approximation for $d\bar{\sigma}_{nd}/d\Omega$. Taking account of the actual time dependence of $\mathbf{Z}_\alpha(t)$ and $T_\alpha(t)$, we are able to obtain an expansion of the form (49) for $d\bar{\sigma}_{nd}/d\Omega$ with the leading term given by Eq. (189).[54] The terms in this expansion may easily be identified with the corresponding terms in Eq. (49). For example, we may write to the first order in $t$

$$T_\alpha(t) \cong T_\alpha + \dot{T}_\alpha t \cong T_\alpha + \mathbf{P}_\alpha \cdot (\nabla_{\mathbf{P}_\alpha} \nabla_{\mathbf{P}_\alpha} T_\alpha)_{P_\alpha=0} \cdot \mathbf{P}_\alpha t + T_{\text{sp}}\dot{\mathbf{S}}_\alpha t.$$

Here $\dot{\mathbf{P}}_\alpha = i[h, \mathbf{P}_\alpha] \cong \mathbf{P}_\alpha U_{\alpha,\text{av}}$ and $\dot{\mathbf{S}}_\alpha = i[h, \mathbf{S}_\alpha] \cong U_{\text{sp}}$, where $U_{\text{sp}}$ is the spin flip part of $U_{\alpha,\text{av}}$ and $T_{\text{sp}}$ is the spin flip part of $T_\alpha$. On taking

$$t \equiv \frac{1}{i}\frac{d}{de_p} \cong \frac{1}{\epsilon_p}$$

and using Eq. (50), we obtain

$$T_\alpha(t) = (T_\alpha)_{P_\alpha=0}(1 + \eta_1 + \eta_2 + \eta_3),$$

where $\eta_1$, $\eta_2$, and $\eta_3$ are given by Eqs. (51), (53), and (54) respectively.[55] As before, the term $\eta_1$ arises solely from the momentum dependence of $T_\alpha$, while $\eta_2$ and $\eta_3$ result from the $\dot{\mathbf{P}}_\alpha$ and $\dot{\mathbf{S}}_\alpha$ terms given before. Similarly, to the first order in $t$

$$\mathbf{Z}_\alpha(t) \cong \mathbf{Z}_\alpha + \frac{\mathbf{P}_\alpha}{M_\alpha} t,$$

so

$$e^{-i\boldsymbol{\rho}\cdot\mathbf{Z}_\alpha(t)} \cong e^{-i(\boldsymbol{\rho}\cdot\mathbf{Z}_\alpha - L_\alpha t)},$$

where $L_\alpha$ is defined by Eq. (169).

These expansions permit us to write Eq. (188*b*) as

$$\frac{d\bar{\sigma}_{nd}}{d\Omega} \cong \frac{(2\pi)^3}{v_{\text{rel}}} \int k^2\, dk \int_{-\infty}^{+\infty} dt\, e^{i\epsilon_q t}$$
$$\times \sum_{\alpha \neq \beta = 1}^{N} \langle 0|\, T_\beta^\dagger\, e^{i[\boldsymbol{\rho}\cdot(\mathbf{z}_\beta - \mathbf{z}_\alpha) + L_\alpha t]} T_\alpha\, |0\rangle_{\text{av}} (1 + \eta_1 + \eta_2 + \eta_3). \quad (190)$$

On expanding the exponential, the linear term in $L_\alpha$ is seen to give us the correction $\eta_4$ [Eq. (55)]. The simplest of the second-order terms in $t$ is

$$\langle 0|\, (iL_\alpha t)^2\, |0\rangle_{\text{av}} = -\frac{2}{3}\frac{\rho^2}{M_\alpha} K_{\alpha,\text{av}} t^2,$$

[54] Van Hove [20] has observed that Eq. (188*b*) may be expressed as an integral over a time-dependent two-particle distribution function $\bar{P}^{(2)}(\mathbf{x}, \mathbf{x}', t)$.

[55] For convenience, we are omitting particle labels $\alpha$ from the $\eta$-corrections. The reader who so desires may insert such labels.

which gives us the term $\eta_5$ [Eq. (59)]. Proceeding in this manner, we obtain from (190) the series

$$\frac{d\bar{\sigma}_{nd}}{d\Omega} = \frac{d\bar{\sigma}_{nd}^{(0)}}{d\Omega}(1 + \eta_1 + \eta_2 + \cdots). \tag{191}$$

On comparing this with Eqs. (49) and (172), we see that the weak binding condition implies a well-defined approximation which is equivalent for both Eqs. (163*b*) and (163*c*). When $\eta_1, \eta_2, \ldots$ may be neglected, we are justified in replacing Eq. (191) by that which obtains in the weak binding limit:

$$\frac{d\bar{\sigma}_{nd}}{d\Omega} = \frac{d\bar{\sigma}_{nd}^{(0)}}{d\Omega}. \tag{192}$$

We shall presently discuss the cross section (189) for several special cases. To consider now a very simple example, however, let us suppose that the $N$ scattering particles are identical, each having a mass $M$, and that the scattering matrices $T_\alpha$ do not depend upon the spins $\mathbf{S}_\alpha$. Then $T_\alpha = T_\beta \equiv (\mathbf{k}, \nu'|\, T\, |\mathbf{p}, \nu)$, a function of $\mathbf{p}$, $\boldsymbol{\rho}$, and the spin $\mathbf{S}_i$. Then Eqs. (189) and (192) reduce to

$$\begin{aligned}\frac{d\bar{\sigma}_{nd}}{d\Omega} \cong \frac{d\bar{\sigma}_{nd}^{(0)}}{d\Omega} &= \frac{(2\pi)^4}{v_{\text{rel}}}\int k^2\, dk\, \delta\left(\epsilon_k - \epsilon_p + \frac{\rho^2}{2M_t}\right) \\ &\times \frac{1}{2S_i + 1}\sum_{\nu,\nu'}|T|^2 \sum_{\alpha\neq\beta=1}^{N}\langle 0|\, e^{i\boldsymbol{\rho}\cdot(\mathbf{z}_\beta - \mathbf{z}_\alpha)}|0\rangle_{\text{av}} \\ &= \frac{d\bar{\sigma}_f^{(0)}}{d\Omega}\sum_{\alpha\neq\beta=1}\langle 0|\, e^{i\boldsymbol{\rho}\cdot(\mathbf{z}_\beta - \mathbf{z}_\alpha)}|0\rangle_{\text{av}},\end{aligned} \tag{193}$$

where

$$\frac{d\bar{\sigma}_f^{(0)}}{d\Omega} \equiv \frac{(2\pi)^4}{v_{\text{rel}}}\frac{k^2}{kc^2/\epsilon_k + (\hat{\mathbf{k}}\cdot\boldsymbol{\rho})/M_t}\frac{1}{2S_i + 1}\sum_{\nu,\nu'}|T|^2. \tag{194}$$

In this expression we must of course relate $k$ to $p$ by the equation $\epsilon_k = \epsilon_p - \rho^2/2M_t$. [For a nonrelativistic incident particle, $\epsilon_k/c^2$ is to be replaced by $m$ in Eq. (194).]

Among the many applications of Eqs. (193) and (194), some of the most important are to the scattering of light quanta and x-rays by the atoms or molecules of a gas, liquid, or solid. It is often possible to neglect the scatterer spin interactions, in which case we write $T = (\mathbf{k}, \hat{e}'|\, T\, |\mathbf{p}, \hat{e})$ in terms of the photon polarization vectors $\hat{e}$ [see, for example, Eq. (8.170)].

The expression (193) may be expressed in terms of the pair distribution function (111), which is equivalent to (109*c*) since the target particles are assumed to be identical. Thus, using (109*c*),

$$F(\rho) \equiv \langle 0|\, e^{i\boldsymbol{\rho}\cdot(\mathbf{z}_\beta - \mathbf{C} - \mathbf{z}_\alpha + \mathbf{C})}|0\rangle_{\text{av}} = \int d^3x\, d^3x'\, \bar{P}^{(2)}(\mathbf{x}, \mathbf{x}')\, e^{i\boldsymbol{\rho}\cdot(\mathbf{x}-\mathbf{x}')} \tag{195}$$

If the expression (124$a$) for $\bar{P}^{(2)}$ is substituted into this equation, it may be rewritten in the form

$$F(\rho) = |c(\rho)|^2 + \frac{1}{\mathscr{V}} C(\rho). \tag{196}$$

Here $c(\rho)$ is defined by Eq. (154) and

$$\frac{1}{\mathscr{V}} C(\rho) \equiv \int \bar{P}^{(1)}(\mathbf{x})\bar{P}^{(1)}(\mathbf{x}')\bar{G}(\mathbf{x}, \mathbf{x}')\, e^{i\boldsymbol{\rho}\cdot(\mathbf{x}-\mathbf{x}')}\, d^3x\, d^3x', \tag{197}$$

where $\mathscr{V}$ is an appropriately defined volume of the scatterer. The expression (196) permits us to write (193) as

$$\frac{d\bar{\sigma}_{nd}}{d\Omega} = N(N-1)\left[|c(\rho)|^2 + \frac{1}{\mathscr{V}} C(\rho)\right] \frac{d\bar{\sigma}_f^{(0)}}{d\Omega}. \tag{198}$$

The normalization condition on the distribution functions imposes important restrictions on $F$, $c$ and $C$. For scattering in the forward direction $\mathbf{k} = \mathbf{p}$, so $\boldsymbol{\rho} = 0$, and we find [see Eqs. (125)]

$$F(0) = c(0) = 1, \qquad C(0) = 0. \tag{199}$$

The total cross section (163$a$) is obtained by adding Eqs. (170) and (198). Since the scatterers are identical, we have for all $\alpha$

$$\frac{d\bar{\sigma}_f(\alpha)}{d\Omega} \equiv \frac{d\bar{\sigma}_f}{d\Omega}.$$

When the incident energy is sufficiently high, we may use the fact that $d\bar{\sigma}_{nd}/d\Omega$ vanishes for $\rho \gg \sqrt{2MK_{\alpha,\mathrm{av}}}$ to set

$$\frac{d\bar{\sigma}_f^{(0)}}{d\Omega} = \frac{d\bar{\sigma}_f}{d\Omega}.$$

Then Eq. (163$a$) leads to

$$\frac{d\bar{\sigma}}{d\Omega} = \left\{N + N(N-1)\left[|c(\rho)|^2 + \frac{1}{\mathscr{V}} C(\rho)\right]\right\} \frac{d\bar{\sigma}_f}{d\Omega}, \tag{200}$$

which is valid for all $\boldsymbol{\rho}$ in the weak binding limit.

The cross section (200) resulted from a sum over all final states of the scatterer. It therefore contains the elastic scattering cross section (153$a$), which for the present example is

$$\frac{d\bar{\sigma}_{\mathrm{el}}}{d\Omega} = N^2\, |c(\rho)|^2 \frac{d\bar{\sigma}_f}{d\Omega}. \tag{201}$$

The cross section for inelastic scattering is therefore

$$\frac{d\bar{\sigma}_{\text{in}}}{d\Omega} = \frac{d\bar{\sigma}}{d\Omega} - \frac{d\bar{\sigma}_{\text{el}}}{d\Omega} = N\left[1 - |c(\rho)|^2 + \frac{N-1}{\mathscr{V}}\,C(\rho)\right]\frac{d\bar{\sigma}_f}{d\Omega}. \tag{202}$$

The conditions (199) imply that $d\bar{\sigma}_{in}/d\Omega = 0$ for $\rho = 0$, or for scattering in the forward direction.

For a scatterer which is approximately uniform in its interior and has a size $\mathscr{R} \gg \mathscr{R}_c$, the correlation range, we may use the approximation (126) for $\bar{G}$. This permits us to write Eq. (197) in the form

$$\frac{1}{\mathscr{V}}\,C(\rho) \cong \int d^3x[\bar{P}^{(1)}(x)]^2 \int d^3r G(r)\, e^{i\boldsymbol{\rho}\cdot\mathbf{r}}.$$

With the definition $\mathscr{V}^{-1} \equiv \int d^3x[\bar{P}^{(1)}(x)]^2$, this gives

$$C(\rho) = \int d^3r G(r)\, e^{i\boldsymbol{\rho}\cdot\mathbf{r}}. \tag{203}$$

It is expected that $C(\rho)$ will vanish for $\rho$ large compared to $\mathscr{R}_c^{-1}$ [which must be of the order of $\sqrt{2MK_{\alpha,\text{av}}}$, since the conditions (162) imply the vanishing of $d\bar{\sigma}_{nd}/d\Omega$].

From these considerations it is seen that the inelastic cross section falls to zero in the interval $0 < \rho < \mathscr{R}^{-1}$. For $\rho \gg \mathscr{R}_c^{-1}$, the expression (202) approaches $N(d\bar{\sigma}_f/d\Omega)$, in agreement with Eq. (170). This is illustrated in Fig. 11.5.

For a scatterer sufficiently large that $\mathscr{R} \gg \mathscr{R}_c$, we may rewrite Eq. (200) as

$$\frac{d\bar{\sigma}}{d\Omega} = N\left[1 + \frac{N-1}{\mathscr{V}}\,C(\rho)\right]\frac{d\bar{\sigma}_f}{d\Omega}, \qquad \rho \gg \mathscr{R}^{-1}. \tag{204}$$

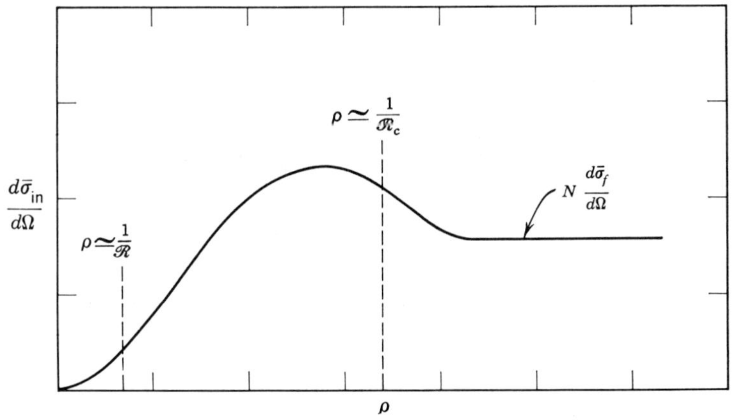


Fig. 11.5 The inelastic cross section (202) illustrated as a function of momentum transfer.

When $\rho < \mathscr{R}_c^{-1}$, we may estimate from Eq. (203) that $C(\rho) \sim (4\pi/3)\mathscr{R}_c^3$ times a numerical factor. Therefore,

$$\frac{N-1}{\mathscr{V}} C(\rho) \cong \text{order of the number of scatterers}$$

$$\text{within the volume } \frac{4\pi}{3} \mathscr{R}_c^{\,3},$$

to within a numerical factor.

The function $G(r)$ plays an important role in theories of imperfect gases, liquids, and amorphous solids. Elaborate experimental techniques[56] have been developed to determine $G(r)$ using Eq. (204). The measurement of both $d\bar{\sigma}/d\Omega$ and $d\bar{\sigma}_f/d\Omega$ permits $C(\rho)$ to be determined, from which $G(r)$ is found by a Fourier transform on Eq. (203).

To evaluate the cross section (189) in the general case of spin-dependent scattering, we observe that the quantity [see Eq. (145)]

$$\frac{1}{2S_i+1}\sum_{\nu,\nu'} (\mathbf{p}, \nu, 0\, |T_\beta^\dagger(\mathbf{S}_\beta)|\, \mathbf{k}, \nu', -\boldsymbol{\rho})$$

$$\times (\mathbf{k}, \nu', -\boldsymbol{\rho}\, |T_\alpha(S_\alpha)|\, \mathbf{p}, \nu, 0) \equiv \frac{1}{2S_i+1}\sum_{\nu,\nu'} T_\beta^\dagger T_\alpha$$

is a matrix in the space of the spins $\mathbf{S}_\alpha$ and $\mathbf{S}_\beta$.[57] It may therefore be expanded in terms of a complete set of orthogonal basis matrices $\Omega_{\beta\alpha}{}^{\lambda}$, as defined in connection with Eq. (7.272*b*):

$$\frac{1}{2S_i+1}\sum_{\nu,\nu'} T_\beta^\dagger T_\alpha = \sum_\lambda B_{\beta\alpha}{}^{\lambda}\Omega_{\beta\alpha}{}^{\lambda}. \tag{205}$$

Here the coefficients $B_{\beta\alpha}{}^{\lambda}$ are functions, say, of $\boldsymbol{\rho}$ and $\mathbf{p}$.

We may now write

$$\langle 0|\, e^{i\boldsymbol{\rho}\cdot(\mathbf{z}_\beta-\mathbf{z}_\alpha)}\, \Omega_{\beta\alpha}{}^{\lambda}\, |0\rangle_{\text{av}} \equiv F_{\beta\alpha}{}^{\lambda}(\rho)\langle 0|\, \Omega_{\beta\alpha}{}^{\lambda}\, |0\rangle_{\text{av}}. \tag{206}$$

Here

$$F_{\beta\alpha}{}^{\lambda}(\rho) = \int d^3x\, d^3x' \bar{P}_{\beta\alpha,\lambda}(\mathbf{x}, \mathbf{x}')\, e^{i\boldsymbol{\rho}\cdot(\mathbf{x}-\mathbf{x}')}, \tag{207}$$

where $\bar{P}_{\beta\alpha,\gamma}(\mathbf{x}, \mathbf{x}')$ is the conditional pair distribution function associated with the spin matrix $\Omega_{\beta\alpha}{}^{\lambda}$, as defined by Eq. (112*b*).

[56] See, for example, Chapter 9 in Ref. [21].

[57] When the scatterers have other internal degrees of freedom, this matrix must of course be generalized.

Using Eqs. (189), (192), (205), and (206), we have (for a relativistic incident particle)

$$\frac{d\bar{\sigma}_{nd}}{d\Omega} = \frac{(2\pi)^4}{v_{\text{rel}}} \frac{k^2}{kc^2/\epsilon_k + (\hat{\mathbf{k}} \cdot \boldsymbol{\rho})/M_t} \sum_{\alpha \neq \beta = 1}^{N} \sum_{\lambda} B_{\beta\alpha}{}^{\lambda} F_{\beta\alpha}{}^{\lambda} \langle 0 | \, \Omega_{\beta\alpha}{}^{\lambda} \, | 0 \rangle_{\text{av}}. \tag{208}$$

The total cross section (163*a*) is obtained on adding Eqs. (170) and (208). The accuracy of this cross section is well defined, under conditions of weak binding, by Eqs. (172) and (191).

The cross section (208) is much more complex than is that given by (170). The reason for this is, of course, that $d\bar{\sigma}_{nd}/d\Omega$ results from the interference of waves from different scattering particles and depends on the structure of the bound system. The usefulness of (208) results primarily from the fact that it provides a means for studying the structure of the scatterer.

When the target particles are identical, we have for all $\alpha$, $\beta$

$$B_{\beta\alpha}{}^{\lambda} \equiv B^{\lambda}, \qquad F_{\beta\alpha}{}^{\lambda} \equiv F^{\lambda} \tag{209}$$

and $\langle 0 | \, \Omega_{\beta\alpha}{}^{\lambda} \, | 0 \rangle_{\text{av}}$ independent of $\alpha$, $\beta$. Then (208) becomes simply

$$\frac{d\bar{\sigma}_{nd}}{d\Omega} = (2\pi)^4 \frac{[k(\epsilon_k/c^2)]N(N-1)}{kc^2/\epsilon_k + (\hat{\mathbf{k}} \cdot \boldsymbol{\rho})/M_t} \sum_{\lambda} B^{\lambda} F^{\lambda} \langle 0 | \, \Omega_{\beta\alpha}{}^{\lambda} \, | 0 \rangle_{\text{av}}, \tag{210}$$

where $\beta$ and $\alpha$ represent any two particles in the scatterer.

The moments (63*a*) of the excitation energy may be obtained as in Section 11.1, p. 699. For the *l*th moment, we have

$$\frac{d\bar{\sigma}_{nd}}{d\Omega} \langle \Delta W^l \rangle_q \equiv \frac{(2\pi)^4}{v_{\text{rel}}} \sum_{\alpha \neq \beta = 1}^{N} \int k^2 \, dk$$

$$\times \langle 0 | \, T_\beta{}^\dagger \, e^{i\boldsymbol{\rho} \cdot \mathbf{z}_\beta} \left( h - W_0 - \frac{\rho^2}{2M_t} \right)^l \delta(\epsilon_k - \epsilon_p + h - W_0) \, e^{-i\boldsymbol{\rho} \cdot \mathbf{z}_\alpha} T_\alpha \, | 0 \rangle_{\text{av}}. \tag{211}$$

For the second moment, we easily obtain[58] in the approximation of Eq. (192)

$$\langle \Delta W^2 \rangle_q \cong \left( \frac{2}{3} \frac{\rho^2}{M} K_{\alpha,\text{av}} \right)_{\alpha}, \tag{212}$$

where an average over the scatterers $\alpha$ is implied.

To summarize the results of this section, we recall that Eqs. (163*a*), (170), and (208) give the scattering cross section quite generally under conditions of weak binding. When the quasi-free conditions (162) are met, the expression (208) vanishes and the cross section is given by Eqs. (164) and (170).

[58] To obtain (212), we recall that $\rho \leq \sqrt{2M_\alpha K_{\alpha,\text{av}}}$ and follow the argument leading to Eq. (65). See, for example, Ref. [11].

Then a struck target particle, say $\alpha$, recoils as if free, the excitation energy of the target system being $\rho^2/2M_\alpha$, with mean-square spread [see Eq. (65)]

$$\langle \Delta W^2 \rangle \cong \frac{2}{3}\frac{\rho^2}{M_\alpha} K_{\alpha,\text{av}}.$$

When $\rho < (2M_\alpha K_{\alpha,\text{av}})^{1/2}$, for all $\alpha$, the mean-square excitation energy of the scatterer is given by Eq. (212). In this case the struck target particles are not expected to recoil from the scattering system. Since the expression (208) is then nonvanishing, the full expression (163*a*) must be used for the cross section.

**Quasi-classical Binding**

When the weak binding conditions (106) do not obtain, the techniques just described may, under some circumstances, still be useful for evaluating the cross section. Equations (30) and (31) indicate that the impulse approximation may very well be valid even when $R_a/\epsilon_p$ is not small. Only the impulse approximation and the small scatterer approximation (131) were required for the validity of Eqs. (148) and (163).

Reference to Eqs. (51), (53), (55), and (59) shows that $\eta_1$, $\eta_2$, $\eta_4$, $\eta_5$, and $\eta_6$ are small when the target particle masses $M_\alpha$ are sufficiently large. That this is a general conclusion is clear from Eq. (188*b*), since a massive particle (for moderate $K_{\alpha,\text{av}}$) is expected to have a small velocity and acceleration. The spin direction of a massive particle may change rapidly, however, as is clear from the expression (54) for $\eta_3$. From this we may conclude that our "weak binding approximation" will be valid, even in the absence of weak binding, if (1) the impulse approximation is valid, (2) the target particles are sufficiently massive that $\eta_1, \eta_2, \eta_4, \eta_5, \ldots$ are small, and (3) there are no spin flip interactions (either $U_{\alpha,\text{av}}$ or $T_\alpha$ has no spin flip dependence).

To investigate the case of quasi-classical binding [see the discussion leading to Eq. (31)], we write Eq. (163*a*) in the form of Eq. (188*b*):

$$\frac{d\bar{\sigma}}{d\Omega} = \frac{(2\pi)^3}{v_{\text{rel}}} \int k^2\, dk \int_{-\infty}^{+\infty} dt\, e^{i\epsilon_q t} \times \sum_{\alpha=1}^{N} \sum_{\beta=1}^{N} \langle 0|\, T_\beta^{\dagger}\, e^{i\mathbf{p}\cdot\mathbf{z}_\beta} e^{-i\mathbf{p}\cdot\mathbf{z}_\alpha(t)} T_\alpha(t)\, |0\rangle_{\text{av}}. \tag{213a}$$

Now $[K_\alpha, T_\alpha] = 0$, so the time dependence of $T_\alpha(t)$ arises from the commutator $[U_\alpha, T_\alpha]$, where $U_\alpha$ is the binding interaction for particle $\alpha$. If $T_\alpha$ and $U_\alpha$ do not both depend upon the spin $\mathbf{S}_\alpha$, then $[U_\alpha, T_\alpha] \cong (\partial T_\alpha/\partial P_\alpha)\, |\nabla_{z_\alpha} U_\alpha|$, which was assumed small under conditions of quasi-classical binding.

Under the same conditions, $\boldsymbol{\rho} \cdot \mathbf{Z}_\alpha(t) \cong \boldsymbol{\rho} \cdot \mathbf{Z}_\alpha + L_\alpha t$. Now (here $\mathbf{Q}_\alpha \equiv \mathbf{P}_\alpha - \boldsymbol{\rho}$)

$$\int_{-\infty}^{+\infty} dt\, e^{-i[\boldsymbol{\rho} \cdot \mathbf{z}_\alpha - (\epsilon_q - L_\alpha)t]} = 2\pi\, e^{-i\boldsymbol{\rho} \cdot \mathbf{z}_\alpha}\, \delta\left(\epsilon_k + \frac{Q_\alpha^{\,2}}{2M_\alpha} - \epsilon_p - \frac{P_\alpha^{\,2}}{2M_\alpha}\right),$$

so Eq. (213$a$) becomes

$$\frac{d\bar{\sigma}}{d\Omega} = \frac{(2\pi)^4}{v_{\text{rel}}} \int k^2\, dk \sum_{\alpha=1}^{N} \sum_{\beta=1}^{N}$$
$$\times \langle 0|\, T_\beta^{\dagger}\, e^{i\boldsymbol{\rho} \cdot (\mathbf{z}_\beta - \mathbf{z}_\alpha)} T_\alpha\, \delta\left(\epsilon_k + \frac{Q_\alpha^{\,2}}{2M_\alpha} - \epsilon_p - \frac{P_\alpha^{\,2}}{2M_\alpha}\right) |0\rangle_{\text{av}}, \tag{213b}$$

which is a generalization of Eqs. (60).

To obtain the cross section (213$b$) we have had to first assume that $U_\alpha$ and $T_\alpha$ do not both contain noncommuting spin flip parts. In addition, we have had to assume that the binding force $|\nabla_{z_\alpha} U_\alpha|$ on particle $\alpha$ is small. Terms neglected in (213$b$) are therefore of relative order

$$\eta_{c1} = \frac{\rho^2 (\nabla_{z_\alpha} U_\alpha)_{\text{av}}{}^2}{\hbar^2 M_\alpha^{\,2}}\, t_c^{\,4},$$

$$\eta_{c2} = \left| \langle 0|\, \frac{\partial^2 T_\alpha}{\partial P_\alpha^{\,2}} \nabla_{z_\alpha}{}^2 U_\alpha\, |0\rangle_{\text{av}} \right| \frac{t_c \hbar}{|T_\alpha|},$$

and so on, where $t_c$ is the collision time [see Eq. (30$a$)].

In the general case Eq. (213$a$) is not easy to evaluate. For some particularly simple scattering systems, however, other approximation methods may be available,[59] as will be illustrated in the next section.

The differential energy spectrum is obtained from Eq. (146$a$) by choosing $\Lambda_\mu$ to vanish outside a small momentum interval $\delta k$ for particle $i$. This gives, for example,

$$\frac{d\bar{\sigma}}{d\Omega\, dk} = \frac{(2\pi)^4 k^2}{v_{\text{rel}}} \sum_{\alpha,\beta=1}^{N} \langle 0|\, T_\beta^{\dagger}\, e^{i\boldsymbol{\rho} \cdot \mathbf{z}_\beta}\, \delta(\epsilon_k - \epsilon_p + h - W_0)\, e^{-i\boldsymbol{\rho} \cdot \mathbf{z}_\alpha}\, T_\alpha\, |0\rangle_{\text{av}}. \tag{213c}$$

Evaluation of this cross section is discussed in Ref. [11].

**Further Generalizations**

Van Hove has shown [20] that, as long as the impulse approximation is valid, the general expression (163) for the scattering cross section may be put into (an often) convenient form. We shall illustrate his argument for

[59] See, for example, Refs. [10] and [20] and the application to molecular scattering of Zemach and Glauber [22].

the special case that the scatter is very massive, so its recoil energy may be neglected, and there is no spin dependence of the scattering matrices $T_\alpha$.

What Van Hove shows is that an expansion [such as given in Eq. (165)] of the energy-conserving $\delta$-function is unnecessary if we introduce a generalized, time-dependent, two-particle correlation function. The scattering cross section may be expressed in terms either of this correlation function or of the *power spectrum* of the density fluctuations. Applications have included the scattering of neutrons and x-rays by solids and the scattering of light by a plasma. In the classical limit Van Hove's results have been very useful in studying radar backscatter from the ionosphere and radio-wave interactions with laboratory plasmas.

On setting $T_\alpha(t) \simeq T_\alpha = T_\beta$,[60] we may rewrite Eq. (213*a*) as

$$\frac{d\bar{\sigma}}{d\Omega\, d\epsilon_k} = \frac{d\sigma_f^{(0)}}{d\Omega} S(\boldsymbol{\rho}, \omega), \tag{214a}$$

where $d\sigma_f^{(0)}/d\Omega$ is given by Eq. (194), $\omega \equiv e_q = \epsilon_k - \epsilon_q$, and

$$S(\boldsymbol{\rho}, \omega) = \frac{1}{2\pi}\int d\tau\, e^{i\omega\tau} \sum_{\alpha,\beta=1}^{N} \langle 0|\, e^{i\boldsymbol{\rho}\cdot\mathbf{z}_\beta}\, e^{-i\boldsymbol{\rho}\cdot\mathbf{z}_\alpha(\tau)}\, |0\rangle_{\text{av}}$$
$$= \frac{1}{2\pi}\int d\tau\, e^{i\omega\tau} \int d^3x\, d^3x'\, e^{i\boldsymbol{\rho}\cdot(\mathbf{x}-\mathbf{x}')}$$
$$\times \langle 0| \left[\sum_{\beta=1}^{N} \delta(\mathbf{x} - \mathbf{Z}_\beta)\right]\left[\sum_{\alpha=1}^{N} \delta(\mathbf{x}' - \mathbf{Z}_\alpha(\tau))\right] |0\rangle_{\text{av}}.$$

In obtaining Eq. (214*a*) from (213*a*), we have restricted the $k$-integration to a narrow range $\delta\epsilon_k$ as was done in Eq. (213*c*).

Now,

$$n(\mathbf{x}, t) = \sum_{\beta=1}^{N} \delta(\mathbf{x} - \mathbf{Z}_\beta(t))$$

is the particle density operator at the position $\mathbf{x}$ and time $t$. Since the state $|0\rangle$ is stationary in time, we may rewrite $S(\boldsymbol{\rho}, \omega)$ as

$$S(\boldsymbol{\rho}, \omega) = \frac{1}{2\pi}\int d\tau\, e^{i\omega\tau} \int d^3x\, d^3x'\, e^{i\boldsymbol{\rho}\cdot(\mathbf{x}-\mathbf{x}')} \langle 0|\, n(\mathbf{x}, t) n(\mathbf{x}', t + \tau)\, |0\rangle_{\text{av}},$$

which is Van Hove's principal result.

On introducing the density function [Eq. (110)]

$$\rho(\mathbf{x}) = \langle 0|\, n(\mathbf{x}, t)\, |0\rangle_{\text{av}},$$

[60] A more general discussion is given in Ref. [20]. When it is not possible to neglect the difference between $T_\alpha(t)$ and $T_\alpha$ the discussion is somewhat more complicated.

which is independent of time, we have

$$\langle 0|\, n(\mathbf{x}, t)n(\mathbf{x}', t+\tau)\, |0\rangle_{\text{av}} = \rho(\mathbf{x})\rho(\mathbf{x}') + \langle 0|\, \delta n(\mathbf{x}, t)\, \delta n(\mathbf{x}', t+\tau)\, |0\rangle_{\text{av}},$$

where $\delta n \equiv n - \rho$ is the fluctuation in particle density in the scatterer. This permits rewriting Eq. (214$a$) in the form

$$\frac{d\bar{\sigma}}{d\Omega\, d\epsilon_k} = \frac{d\sigma_f^{(0)}}{d\Omega}\, [N^2\, \delta(\omega)\, |c(\rho)|^2 + \delta S(\boldsymbol{\rho}, \omega)], \tag{214b}$$

where

$$\delta S(\boldsymbol{\rho}, \omega) = \frac{1}{2\pi} \int d\tau\, e^{i\omega\tau} \int d^3x\, d^3x'\, e^{i\boldsymbol{\rho}\cdot(\mathbf{x}-\mathbf{x}')} \times \langle 0|\, \delta n(\mathbf{x}, t)\, \delta n(\mathbf{x}', t+\tau)\, |0\rangle_{\text{av}}. \tag{215a}$$

Comparison of the first term in brackets in Eq. (214$b$) with Eqs. (200), (201), and (202) shows that this gives the *elastic* scattering of the beam particle by the target system. [The factor $\delta(\omega)$ shows this directly.] The $\delta S(\boldsymbol{\rho}, \omega)$-term gives the inelastic scattering by the target.

The quantity $\delta S(\boldsymbol{\rho}, \omega)$ may be written in several forms. For a target which is uniform within its volume $\mathscr{V}$ we have for example

$$\delta S(\boldsymbol{\rho}, \omega) = \frac{\mathscr{V}}{2\pi} \int d\tau\, d^3r\, e^{i(\omega\tau - \mathbf{r}\cdot\boldsymbol{\rho})} \langle 0|\, \delta n(0, 0)\, \delta n(\mathbf{r}, \tau)\, |0\rangle_{\text{av}}.$$

A general expression for $\delta S$ may be obtained on supposing that $\delta n(\mathbf{x}, t)$ vanishes outside some very long time interval $T$. Then a Fourier transform may be introduced for $\delta n$.

$$\delta n(\mathbf{x}, t) \equiv (2\pi)^{-2} \int d^3l\, d\epsilon\, e^{-i(\mathbf{l}\cdot\mathbf{x} - \epsilon t)}\, \delta\tilde{n}(\mathbf{l}, \epsilon).$$

On inserting this into Eq. (215$a$) and on averaging over the long time $T$, we find that

$$\delta S(\boldsymbol{\rho}, \omega) = \frac{(2\pi)^3}{T} \langle 0|\, \delta\tilde{n}(\boldsymbol{\rho}, \omega)\, \delta\tilde{n}^*(\boldsymbol{\rho}, \omega)\, |0\rangle_{\text{av}}. \tag{215b}$$

This quantity, which is independent of $T$ for sufficiently large $T$, represents the *power spectrum* of the density fluctuations in the target.[61] An experimental measurement of the cross section (214$b$) permits us to determine this power spectrum.

[61] The notion of power spectrum is familiar from electric circuit theory. If $V(t)$ is a noise voltage across a unit resistance, we may take $V(t) = 0$ outside the interval $0 \le t \le T$ to write $V(t) = (2\pi)^{-1/2} \int d\omega\, \tilde{V}(\omega) e^{i\omega t}$. The mean power over the long interval $T$ is then $\int d\omega [|\tilde{V}(\omega)|^2/T]$; thus $|\tilde{V}(\omega)|^2/T$ is the power spectrum.

**Examples**

For scattering from macroscopic systems it is ordinarily necessary to carry out the statistical average implied by Eq. (146*b*). When the scattering may be described by Eq. (204) this is easily done, at least formally. Letting $G_{\gamma(0)}(r)$ be the pair correlation function for a pure state $\gamma(0)$, we define

$$G_{\text{th}}(r) \equiv \frac{1}{\mathscr{Z}} \sum_{\gamma(0)} e^{-W_{\gamma(0)}/\theta} G_{\gamma(0)}(r), \tag{216}$$

where $\mathscr{Z}$ is given by Eq. (129). Then the corresponding average of Eq. (203) is

$$C_{\text{th}}(\rho) \equiv \int d^3r G_{\text{th}}(r)\, e^{i\boldsymbol{\rho}\cdot\mathbf{r}} \tag{217}$$

and the thermal averaged cross section (204) is [for a macroscopic sample we may take $(N-1) \simeq N$]

$$\frac{d\bar{\bar{\sigma}}}{d\Omega} = N \frac{d\bar{\sigma}_f}{d\Omega}\left[1 + \frac{N}{\mathscr{V}} C_{\text{th}}(\rho)\right]. \tag{218}$$

The thermal pair correlation function $G_{\text{th}}(r)$ plays an important role in the statistical mechanics of nonideal gases, liquids, and amorphous solids. Consider, for example, a uniform medium for which the Hamiltonian has the form

$$h = \sum_{\alpha=1}^{N} \frac{P_\alpha^2}{2M} + \sum_{\alpha<\beta=1}^{N} v(\mathbf{z}_\alpha - \mathbf{z}_\beta),$$

where $v(\mathbf{z}_\alpha - \mathbf{z}_\beta)$ represents the interaction between a pair of particles. Then the energy of the system at a temperature $\theta$ is

$$\begin{aligned}\overline{W} &\equiv \frac{1}{\mathscr{Z}} \sum_{\gamma} e^{-W_\gamma/\theta} W_\gamma \\ &= \sum_{\alpha=1}^{N} K_{\alpha,\text{av}} + N(N-1) \int \bar{\bar{P}}^{(2)}(\mathbf{z}_\alpha, \mathbf{z}_\beta) v(\mathbf{z}_\alpha - \mathbf{z}_\beta)\, d^3z_\alpha\, d^3z_\beta.\end{aligned}$$

For classical systems, $K_{\alpha,\text{av}} = \frac{3}{2}\theta$, so $\overline{W}$ may be evaluated in terms of $G_{\text{th}}(|\mathbf{z}_\alpha - \mathbf{z}_\beta|)$. It is also possible to express the equation of state of the scatterer in terms of an integral over $G_{\text{th}}$.[62]

A further relation of importance to scattering theory is[63]

$$1 + \frac{N}{\mathscr{V}} \int G_{\text{th}}(r)\, d^3r = \frac{N}{\mathscr{V}}\, \theta \left(\mathscr{V} \frac{\partial^2 \mathscr{F}}{\partial \mathscr{V}^2}\right)^{-1}, \tag{219a}$$

[62] See, for example, Ref. [23].
[63] See, for example, Ref. [24].

where $\mathscr{F}$ is the free energy of the scatterer. Van Hove and McVoy [25] have observed from (219$a$) that as $\theta \to 0$, for those systems which have a finite sound speed at zero temperature, we obtain

$$1 + \frac{N}{\mathscr{V}} \int G_{\text{th}}(r)\, d^3r = 0. \tag{219b}$$

The implication of (219$b$) is that [see Eq. (218)]

$$\frac{d\bar{\bar{\sigma}}}{d\Omega} \cong 0 \tag{220}$$

when $\rho$ is significantly smaller than $\mathscr{R}_c^{-1}$. The result (220) is easily verified for the special example of a degenerate Fermi gas. In this case it follows from Eq. (120) that $G_{\text{th}} = -\frac{1}{2}G_F$, where $G_F$ is given by Eq. (119).

Scattering by a crystal provides another important example for which the scatterer is a thermodynamic system. For simplicity, we assume identical scattering particles and combine Eqs. (170) and (193) to give the thermal average (146$b$) as

$$\frac{d\bar{\bar{\sigma}}}{d\Omega} = \frac{d\bar{\sigma}_f}{d\Omega}\left[\frac{1}{\mathscr{Z}} \sum_{\gamma(0)} e^{-W_{\gamma(0)}/\theta} \langle\gamma(0)| \sum_{\alpha\neq\beta=1}^{N} e^{i\boldsymbol{\rho}\cdot(\mathbf{z}_\beta - \mathbf{z}_\alpha)}\, |\gamma(0)\rangle\right], \tag{221}$$

where again it is assumed[64] that

$$\frac{d\bar{\sigma}_f^{(0)}}{d\Omega} \cong \frac{d\bar{\sigma}_f^{(\alpha)}}{d\Omega} = \frac{d\bar{\sigma}_f}{d\Omega}.$$

The thermal energy in a crystal is associated with vibration of the atoms about their equilibrium positions. This suggests writing, for $\alpha = 1, 2, \ldots, N$,

$$\mathbf{z}_\alpha = \mathbf{y}_\alpha + \mathbf{q}_\alpha, \tag{222}$$

where the $\mathbf{y}_\alpha$ give the equilibrium positions of the atoms in the lattice and the $\mathbf{q}_\alpha$ are the displacements from equilibrium. Treating the $\mathbf{q}_\alpha$ as small and letting $\langle\cdots\rangle_{\text{av}}$ represent the thermal average in Eq. (221), we have

$$\begin{aligned}\langle e^{i\boldsymbol{\rho}\cdot(\mathbf{z}_\beta - \mathbf{z}_\alpha)}\rangle_{\text{av}} &= e^{i\boldsymbol{\rho}\cdot(\mathbf{y}_\beta - \mathbf{y}_\alpha)}\, \langle\{1 + i\boldsymbol{\rho}\cdot(\mathbf{q}_\beta - \mathbf{q}_\alpha) \\ &\qquad - \tfrac{1}{2}[\boldsymbol{\rho}\cdot(\mathbf{q}_\beta - \mathbf{q}_\alpha)]^2 + \cdots\}\rangle_{\text{av}} \\ &\cong e^{i\boldsymbol{\rho}\cdot(\mathbf{y}_\beta - \mathbf{y}_\alpha)}(1 - \tfrac{1}{3}\rho^2\langle q_\alpha^{\ 2}\rangle_{\text{av}} + \cdots).\end{aligned} \tag{223}$$

[64] This is essentially a weak binding approximation. [See the discussion preceding Eq. (200).]

Now[65]

$$2\mathscr{M} \equiv \tfrac{1}{3}\rho^2\langle q_\alpha{}^2\rangle_{\text{av}}. \tag{224a}$$

can be estimated by writing

$$\frac{M}{2}\langle \dot{q}_\alpha{}^2\rangle_{\text{av}} = \frac{M}{2}\,\omega^2\langle q_\alpha{}^2\rangle_{\text{av}} = \tfrac{3}{2}\theta,$$

assuming an energy $\frac{1}{2}\theta$ for each degree of freedom and a single vibration frequency $\omega$. Then

$$2\mathscr{M} = \frac{\theta}{M\omega^2}\,\rho^2, \tag{224b}$$

and Eq. (221) becomes

$$\frac{d\bar{\bar{\sigma}}}{d\Omega} \cong \frac{d\bar{\sigma}_f}{d\Omega}\,e^{-2\mathscr{M}}\left(\sum_{\alpha\neq\beta=1}^{N} e^{i\boldsymbol{\rho}\cdot(\mathbf{y}_\beta-\mathbf{y}_\alpha)}\right), \tag{225}$$

where the double sum runs over the lattice positions.

To illustrate scattering with spin interactions, we consider the case of a spin zero particle (such as a pi-meson or He atom) scattered by a system of $N$ identical spin one-half scatterers. The scattering matrices then have the form (155)

$$T_\alpha = a + \mathbf{b}\cdot\boldsymbol{\sigma}_\alpha.$$

For the present application, Eq. (205) becomes

$$T_\beta{}^\dagger T_\alpha = |a|^2 + b^*\cdot\boldsymbol{\sigma}_\beta\mathbf{b}\cdot\boldsymbol{\sigma}_\alpha + a^*\mathbf{b}\cdot\boldsymbol{\sigma}_\alpha + a\mathbf{b}^*\cdot\boldsymbol{\sigma}_\beta. \tag{226}$$

To evaluate the averages (206) we observe that because of the implied average over all spin orientations of the target

$$\langle 0|\,\mathbf{b}\cdot\boldsymbol{\sigma}_\alpha\,|0\rangle_{\text{av}} = 0 \tag{227a}$$

and

$$\langle 0|\,\mathbf{b}^*\cdot\boldsymbol{\sigma}_\beta\mathbf{b}\cdot\boldsymbol{\sigma}_\alpha\,|0\rangle_{\text{av}} = \tfrac{1}{3}\,|b|^2\,\langle 0|\,\boldsymbol{\sigma}_\alpha\cdot\boldsymbol{\sigma}_\beta\,|0\rangle_{\text{av}}. \tag{227b}$$

The unit matrix and $\boldsymbol{\sigma}_\alpha\cdot\boldsymbol{\sigma}_\beta$ may be expressed in terms of the singlet and triplet projection operators (7.168) for a pair $(\alpha, \beta)$ of particles.

$$\begin{aligned} 1 &= \Lambda_{\alpha\beta}^{(0)} + \Lambda_{\alpha\beta}^{(1)}, \\ \boldsymbol{\sigma}_\alpha\cdot\boldsymbol{\sigma}_\beta &= \Lambda_{\alpha\beta}^{(1)} - 3\Lambda_{\alpha\beta}^{(0)}. \end{aligned} \tag{228}$$

[65] The quantity $e^{-2\mathscr{M}}$ is called the "Debye-Waller" correction factor. An accurate evaluation of $\mathscr{M}$ would involve taking account of the crystal symmetry in Eq. (223) and the actual vibrational frequencies $\omega$. Methods for calculating $\mathscr{M}$ are discussed, for example, in Ref. [21].

These relations permit us to write Eq. (208) as

$$\frac{d\bar{\sigma}_{nd}}{d\Omega} = \frac{(2\pi)^4}{v_{\text{rel}}} \frac{k^2}{kc^2/\epsilon_k + (\hat{\mathbf{k}} \cdot \boldsymbol{\rho})/M_t} \sum_{\alpha \neq \beta = 1}^{N}$$
$$\times [(|a|^2 - |b|^2)\langle 0| \Lambda^{(0)}_{\alpha\beta} |0\rangle_{\text{av}} F^{(0)}(\rho)$$
$$+ (|a|^2 + \tfrac{1}{3}|b|^2)\langle 0| \Lambda^{(1)}_{\alpha\beta} |0\rangle_{\text{av}} F^{(1)}(\rho)], \tag{229}$$

where $F^{(0)}(\rho)$ and $F^{(1)}(\rho)$ represent the quantities (207) associated with the singlet and triplet projection operators, respectively.

Now [see Eqs. (7.168)]

$$\sum_{\alpha \neq \beta = 1}^{N} \boldsymbol{\sigma}_\alpha \cdot \boldsymbol{\sigma}_\beta = 4\mathbf{s}^2 - 3N, \tag{230}$$

where $\mathbf{s} = \frac{1}{2} \sum_{\alpha=1}^{N} \boldsymbol{\sigma}_\alpha$ is the sum of the spin operators of the bound particles.[66]

Using Eq. (230), we can write (229) in the more explicit form:

$$\frac{d\bar{\sigma}_{nd}}{d\Omega} = \frac{(2\pi)^4 k\epsilon_k}{kc^2/\epsilon_k + (\hat{\mathbf{k}} \cdot \boldsymbol{\rho})/M_t} \left[(|a|^2 - |b|^2)\frac{F^{(0)}(\rho)}{4}\right.$$
$$\times (N^2 + 2N - 4\langle 0| \mathbf{s}^2 |0\rangle_{\text{av}})$$
$$\left. + (|a|^2 + \tfrac{1}{3}|b|^2)\frac{F^{(1)}(\rho)}{4}(3N^2 - 6N + 4\langle 0| \mathbf{s}^2 |0\rangle_{\text{av}})\right]. \tag{231}$$

When $N \gg 1$ and for a saturated spin system [see the condition (149)] we may keep only the $N^2$ terms in Eq. (231). Then

$$\frac{d\bar{\sigma}_{nd}}{d\Omega} = \frac{(2\pi)^4 k\epsilon_k}{kc^2/\epsilon_k + (\hat{\mathbf{k}} \cdot \boldsymbol{\rho})/M_t} \frac{N^2}{4} [(|a|^2 - |b|^2)F^{(0)} + 3(|a|^2 + \tfrac{1}{3}|b|^2)F^{(1)}]. \tag{232}$$

This may be further simplified if we express the $F$'s in the form (196):

$$F^{(\nu)}(\rho) = |c(\rho)|^2 + \frac{1}{\mathscr{V}} C^{(\nu)}(\rho), \qquad \nu = 0, 1 \tag{233}$$

where the $C^{(\nu)}$ are defined [as in Eq. (203)] in terms of singlet and triplet pair correlation functions $G^{(0)}$ and $G^{(1)}$. Then, referring to Eq. (156), we find

$$\frac{d\bar{\sigma}_{nd}}{d\Omega} = \frac{d\bar{\sigma}_{\text{el}}}{d\Omega} + \frac{(2\pi)^4 k\epsilon_k}{kc^2/\epsilon_k + (\hat{\mathbf{k}} \cdot \boldsymbol{\rho})/M_t} \frac{N^2}{4\mathscr{V}}$$
$$\times [(|a|^2 - |b|^2)C^{(0)}(\rho) + 3(|a|^2 + \tfrac{1}{3}|b|^2)C^{(1)}(\rho)]. \tag{234}$$

Here the elastic cross section is given by Eq. (153*a*).

[66] When there is orbital angular momentum $\mathbf{s} \neq \mathbf{S}_t$, the target spin operator, we recall.

Finally, the cross section (170) is

$$\frac{d\bar{\sigma}_d}{d\Omega} = \frac{(2\pi)^4 k\epsilon_k}{kc^2/\epsilon_k + (\hat{\mathbf{k}} \cdot \boldsymbol{\rho})/M} N(|a|^2 + |b|^2) = N\frac{d\bar{\sigma}_f}{d\Omega}. \tag{235}$$

The complete cross section is now obtained using Eq. (163*a*).

When the scatterer is a degenerate Fermi gas, we see from Eq. (118) that $C^{(1)}(\rho) = -C^{(0)}(\rho)$. If we now, as previously, replace the recoil term in (234) by that in (235), the cross section (163*a*) becomes

$$\frac{d\bar{\sigma}}{d\Omega} - \frac{d\bar{\sigma}_{\text{el}}}{d\Omega} = N\frac{d\bar{\sigma}_f}{d\Omega}\left[1 - \frac{N}{2\mathscr{V}} C^{(0)}(\rho)\right], \tag{236}$$

which is equivalent to Eq. (204). The discussion following Eq. (220) is then applicable also to Eq. (236).[67]

The scattering of high-energy particles by atomic nuclei provides an important application of the theory developed in this section. To illustrate this we consider the scattering of pi-mesons by nuclei which are sufficiently small that the "small scatterer" approximation (131) is valid. Only scattering without charge exchange will be discussed (see Section 9.2, p. 520). Then the scattering matrix for nucleon $\alpha$ ($\alpha = 1, 2, \ldots, A \equiv N$) is of the general form

$$T_\alpha = a + \mathbf{b} \cdot \boldsymbol{\sigma}_\alpha + c\tau_{3\alpha} + \mathbf{d} \cdot \boldsymbol{\sigma}_\alpha \tau_{3\alpha}. \tag{237}$$

Here $\tau_{3\alpha}$ is the "third component" of the isotopic spin operator for nucleon $\alpha$ [see Eq. (9.43)]; it has the eigenvalue $+1$ for a proton and $-1$ for a neutron. The coefficients $a$, $\mathbf{b}$, $c$, and $\mathbf{d}$ are functions of $\mathbf{p}$ and $\boldsymbol{\rho}$, since in using Eq. (189) the $\mathbf{P}_\alpha$-dependence is neglected.

We shall treat the nucleus as a "saturated" spin and isotopic spin system [see Eq. (149)], so the resultant nuclear spin $\mathbf{s}$ and isotopic spin $I$ are considered to be small. Then the free nucleon cross section (171), averaged over equal numbers of neutrons and protons, is

$$\frac{d\bar{\sigma}_f}{d\Omega} = \frac{(2\pi)^4}{v_{\text{rel}}} \frac{k^2}{kc^2/\epsilon_k + (\hat{\mathbf{k}} \cdot \boldsymbol{\rho})/M}(|a|^2 + |b|^2 + |c|^2 + |d|^2), \tag{238}$$

where $M$ is the nucleon mass,[68] $v_{\text{rel}} = kc^2/\epsilon_k$, $\epsilon_k = c(m^2c^2 + k^2)^{1/2}$, and $m$ is the pion rest-mass. The cross section (170) is then

$$\frac{d\bar{\sigma}_d}{d\Omega} = A\frac{d\bar{\sigma}_f}{d\Omega}. \tag{239}$$

[67] Equation (236) of course reduces to Eq. (204), or (218) in the limit of low temperatures, when we set $b = 0$.

[68] It is convenient to neglect the neutron-proton mass difference.

Equations (189) and (192) give

$$\frac{d\bar{\sigma}_{nd}}{d\Omega} = (2\pi)^4 \left(\frac{\epsilon_p}{c^2}\right)^2 \sum_{\alpha \neq \beta = 1}^{A} \langle 0| \, e^{i\mathbf{p}\cdot(\mathbf{z}_\beta - \mathbf{z}_\alpha)} T_\beta^\dagger T_\alpha \, |0\rangle_{\text{av}}, \tag{240}$$

if the nuclear recoil energy $p^2/2AM$ is neglected.

To evaluate the expression (240) we shall use Eqs. (205), (206), and (207). Following an approximation of Lax and Feshbach [26], we shall assume that only two distribution functions $\bar{P}_\lambda(\mathbf{x}, \mathbf{x}')$ [Eq. (112*a*)] need be distinguished in atomic nuclei. These are $\bar{P}_s(\mathbf{z}_\alpha, \mathbf{z}_\beta)$ and $\bar{P}_a(\mathbf{z}_\alpha, \mathbf{z}_\beta)$, corresponding to the parts $|s\rangle$ and $|a\rangle$ of the state $|0\rangle$ which are, respectively, symmetric and antisymmetric in the variables $\mathbf{z}_\alpha$ and $\mathbf{z}_\beta$. These are

$$\begin{aligned} |s\rangle &\equiv \tfrac{1}{2}(1 + Q_{\alpha\beta}{}^x)\,|0\rangle, \\ |a\rangle &\equiv \tfrac{1}{2}(1 - Q_{\alpha\beta}{}^x)\,|0\rangle, \end{aligned} \tag{241}$$

where [see Eq. (116*b*)]

$$Q_{\alpha\beta}{}^x \equiv -\tfrac{1}{4}(1 + \boldsymbol{\sigma}_\alpha \cdot \boldsymbol{\sigma}_\beta)(1 + \boldsymbol{\tau}_\alpha \cdot \boldsymbol{\tau}_\beta) \tag{242}$$

is the operator which interchanges $\mathbf{z}_\alpha$ and $\mathbf{z}_\beta$ in a state such as $|0\rangle$. To verify that the definitions are correct, we use the equation $(Q_{\alpha\beta}{}^x)^2 = 1$ to show that $Q_{\alpha\beta}{}^x\,|s\rangle = |s\rangle$ and $Q_{\alpha\beta}{}^x\,|a\rangle = -\,|a\rangle$.

Equations (241) permit us to write

$$\begin{aligned} &\langle 0| \sum_{\alpha\neq\beta} e^{i\mathbf{p}\cdot(\mathbf{z}_\beta - \mathbf{z}_\alpha)} T_\beta^\dagger T_\alpha \,|0\rangle_{\text{av}} \\ &\quad = \langle s| \sum_{\alpha\neq\beta} e^{i\mathbf{p}\cdot(\mathbf{z}_\beta - \mathbf{z}_\alpha)} T_\beta^\dagger T_\alpha \,|s\rangle_{\text{av}} + \langle a| \sum_{\alpha\neq\beta} e^{i\mathbf{p}\cdot(\mathbf{z}_\beta - \mathbf{z}_\alpha)} T_\beta^\dagger T_\alpha \,|a\rangle_{\text{av}} \\ &\quad = F^{(s)}(p)\langle s| \sum_{\alpha\neq\beta} T_\beta^\dagger T_\alpha \,|s\rangle_{\text{av}} + F^{(a)}(p)\langle a| \sum_{\alpha\neq\beta} T_\beta^\dagger T_\alpha \,|a\rangle_{\text{av}}, \end{aligned} \tag{243}$$

where $F^{(s)}$ and $F^{(a)}$ are defined by Eq. (207), using $\bar{P}_s$ and $\bar{P}_a$, respectively. [We have here used the Lax-Feshbach assumption[70] that all $F_\lambda$ in Eq. (206) are equal for either states $|s\rangle$ or $|a\rangle$.]

Using Eqs. (241) again, we may put Eq. (243) in the form

$$\begin{aligned} \langle 0| \sum_{\alpha\neq\beta} e^{i\mathbf{p}\cdot(\mathbf{z}_\beta - \mathbf{z}_\alpha)} T_\beta^\dagger T_\alpha \,|0\rangle_{\text{av}} &= \tfrac{1}{2}(F^{(s)} + F^{(a)})\langle 0| \sum_{\alpha\neq\beta} T_\beta^\dagger T_\alpha \,|0\rangle_{\text{av}} \\ &\quad + \tfrac{1}{2}(F^{(s)} - F^{(a)})\langle 0| \sum_{\alpha\neq\beta} T_\beta^\dagger T_\alpha Q_{\alpha\beta}{}^x \,|0\rangle_{\text{av}}. \end{aligned} \tag{244}$$

Now the spin and isotopic spin operators $T_\beta^\dagger T_\alpha$ and $T_\beta^\dagger T_\alpha Q_{\alpha\beta}{}^x$ may be put into the form (205), using the expressions (237) and (242). The spin and isotopic spin expectation values (206) may be evaluated for a saturated system by a method due to Wigner [27].[69] The required relations are summarized in Table 11.1.

[69] A somewhat simpler, but less general, method is given by Blatt and Weisskopf [1, pp. 236–239]. When $A \gg 1$ the leading terms of order $A^2$ and of order $A$ may be obtained by elementary considerations.

[70] Reference [26].

**Table 11.1 Spin and Isotopic Spin Averages for Atomic Nuclei**

Terms of order unity (compared with $A$ and $A^2$) are neglected. We define the isotopic spin $I$ by $I(I+1) = \langle 0|\,(\sum_{\alpha=1}^{A} \frac{1}{2}\boldsymbol{\tau}_\alpha)^2\,|0\rangle$ and its three-component as $I_3 = \langle 0|\sum_{\alpha=1}^{A}\frac{1}{2}\tau_{3\alpha}\,|0\rangle = \frac{1}{2}(Z-N)$, where $Z$ is the number of protons and $N = A - Z$ is the number of neutrons in the nucleus. The operator $Q_{\alpha\beta}{}^{x}$ is defined by Eq. (242). The complete expressions are given in Ref. [12].

---

$$\langle 0|\sum_{\alpha\neq\beta=1}^{A}\boldsymbol{\sigma}_\alpha\cdot\boldsymbol{\sigma}_\beta\,|0\rangle_{\text{av}} \cong -3A$$

$$\langle 0|\sum_{\alpha\neq\beta}\boldsymbol{\tau}_\alpha\cdot\boldsymbol{\tau}_\beta\,|0\rangle_{\text{av}} = 4I(I+1) - 3A$$

$$\langle 0|\sum_{\alpha\neq\beta}\tau_{3\alpha}\tau_{3\beta}\,|0\rangle_{\text{av}} = 4I_3{}^2 - A$$

$$\langle 0|\sum_{\alpha\neq\beta}\boldsymbol{\sigma}_\alpha\cdot\boldsymbol{\sigma}_\beta\tau_{3\alpha}\tau_{3\beta}\,|0\rangle_{\text{av}} \cong -3A$$

$$\langle 0|\sum_{\alpha\neq\beta} Q_{\alpha\beta}{}^{x}\,|0\rangle_{\text{av}} \cong -\tfrac{1}{4}A^2 + 4A$$

$$\langle 0|\sum_{\alpha\neq\beta}\boldsymbol{\sigma}_\alpha\cdot\boldsymbol{\sigma}_\beta\boldsymbol{\tau}_\alpha\cdot\boldsymbol{\tau}_\beta\,|0\rangle_{\text{av}} \cong -9A - 4I(I+1)$$

$$\langle 0|\sum_{\alpha\neq\beta}\boldsymbol{\tau}_\alpha\cdot\boldsymbol{\tau}_\beta Q_{\alpha\beta}{}^{x}\,|0\rangle_{\text{av}} \cong -\tfrac{3}{4}A^2$$

$$\langle 0|\sum_{\alpha\neq\beta}\boldsymbol{\sigma}_\alpha\cdot\boldsymbol{\sigma}_\beta Q_{\alpha\beta}{}^{x}\,|0\rangle_{\text{av}} \cong -\tfrac{3}{4}A^2 - 4I(I+1)$$

$$\langle 0|\sum_{\alpha\neq\beta}\boldsymbol{\tau}_\alpha\cdot\boldsymbol{\tau}_\beta\tau_{3\alpha}\,|0\rangle_{\text{av}} = A^2 + A + 4I_3{}^2 - 4I(I+1)$$

$$\langle 0|\sum_{\alpha\neq\beta}\tau_{3\alpha}\tau_{3\beta}Q_{\alpha\beta}{}^{x}\,|0\rangle_{\text{av}} \cong -\tfrac{1}{4}A^2 - 2I_3{}^2$$

$$\langle 0|\sum_{\alpha\neq\beta}\boldsymbol{\sigma}_\alpha\cdot\boldsymbol{\sigma}_\beta\tau_{3\alpha}\tau_{3\beta}Q_{\alpha\beta}{}^{x}\,|0\rangle_{\text{av}} \cong -\tfrac{3}{4}A^2 + 4I(I+1) - 6I_3{}^2.$$

Using the relations in Table 11.1 and keeping only terms of order $A^2$ and $A$, we may evaluate (244) to write the expression (240) as

$$\frac{d\bar{\sigma}_{nd}}{d\Omega} = \tfrac{1}{2}(F^{(s)} + F^{(a)})(A^2\sigma_c - A\sigma_f) + \tfrac{1}{2}(F^{(s)} - F^{(a)})\left(4A\sigma_c - \frac{A^2}{4}\sigma_f\right). \qquad (245)$$

Here we have introduced the abbreviations

$$\sigma_c \equiv (2\pi)^4\left(\frac{\epsilon_p}{c^2}\right)^2|a|^2,$$
$$\sigma_f \equiv (2\pi)^4\left(\frac{\epsilon_p}{c^2}\right)^2(|a|^2 + |b|^2 + |c|^2 + |d|^2). \qquad (246)$$

Since $d\bar{\sigma}_{nd}/d\Omega$ is nonvanishing only for $\rho \lesssim \sqrt{2MK_{\alpha,\text{av}}}$, it is ordinarily possible to set $\sigma_f = d\sigma_f/d\Omega$ [Eq. (238)].

The elastic cross section (156), in the notation (246), is

$$\frac{d\bar{\sigma}_{\rm el}}{d\Omega} = |Ac(\rho)|^2 \, \sigma_c. \tag{247}$$

If $F^{(s)}$ and $F^{(a)}$ are written in the form (233), we may write Eq. (245) as

$$\frac{d\bar{\sigma}_{nd}}{d\Omega} = \frac{d\bar{\sigma}_{el}}{d\Omega} + \left[\frac{A^2}{2}(C^{(s)} + C^{(a)}) + 2A(C^{(s)} - C^{(a)})\right]\sigma_c$$
$$+ \left[\frac{A^2}{8}(C^{(a)} - C^{(s)}) - \frac{A}{2}(C^{(s)} + C^{(a)})\right]\sigma_f, \tag{248}$$

which shows that in the absence of correlations

$$\frac{d\bar{\sigma}_{nd}}{d\Omega} = \frac{d\bar{\sigma}_{\rm el}}{d\Omega}.$$

For a Fermi gas model of the nucleus, we see from Eqs. (121) that when $A \gg 1$

$$C^{(s)}(\rho) = -C^{(a)}(\rho) = \int d^3 r G_F(r) e^{i\boldsymbol{\rho}\cdot\mathbf{r}}.$$

Then

$$\frac{d\bar{\sigma}_{nd}}{d\Omega} = \frac{d\bar{\sigma}_{\rm el}}{d\Omega} - \frac{A^2}{4}\sigma_f C^{(s)}(\rho). \tag{249}$$

The cross section corresponding to that of Eq. (245) for nucleon-nucleus scattering has been given in Ref. [12].

There are many applications of Eq. (208). For example, the spin correlations in ferromagnetic and antiferromagnetic materials give a contribution to neutron scattering that has been useful in studying properties of such substances.[71]

## 11.3 MULTIPLE SCATTERING BY A SYSTEM OF BOUND PARTICLES

In this section we return to the same scattering interaction that was studied in Section 11.2, but we no longer assume that the scatterer is "small" in the sense that the mean free path is large compared to the size of the scatterer. In this case it is generally necessary to include a sequence of scatterings, or "multiple scattering," of the incident particle by the bound target particles. When, however, the impulse approximation is valid and when the conditions that the mean free path be large compared to both projectile wavelength and the target particle correlation range are met [the inequalities (132) and (133)], the modifications of the treatment given in

[71] Neutron diffraction is reviewed in Ref. [28].

Section 11.2 to account for multiple scattering are rather trivial; they amount essentially to bookkeeping on the sequence of interactions.

The phenomenon of multiple scattering is frequently encountered and is conventionally treated in a great variety of ways. Examples range from the scattering of neutrons in a moderator, described by classical transport theory, to the transmission of light through a lens, described by a refractive index. We shall now show how all of such phenomena may be exhibited in terms of a special solution of the Schrödinger equation.

#### Multiple Scattering Form for Solution of the Schrödinger Equation

As in Section 11.2, p. 708, we consider an incident particle of rest-mass $m$ and spin $S_i$ to interact with a scatterer consisting of $N$ bound particles of respective masses $M_1, \ldots, M_N$ and spins $S_1, \ldots, S_N$. The mass and spin of the scatterer are $M_t$ and $S_t$, as before; its wave functions are expressed by Eqs. (97) and its Hamiltonian $h$ by Eqs. (98). For the present, we continue to assume that the incident particle $i$ is distinguishable from each of the target particles.

The initial state prior to the scattering is $\chi_a$, the product of a plane wave for the projectile and the target wave function, as given in Eq. (100). Possible final states are of a similar form with in general a different momentum for the scattered particle and the target in some state of excitation. Then the Schrödinger wave function describing the entire event is obtained from Eq. (5.80) as

$$\psi_a^{+} = \chi_a + \frac{1}{d} V \psi_a^{+}, \tag{250}$$

where

$$d \equiv E_a + i\eta - K. \tag{251}$$

We recall that $V$ represents the sum of interactions $V_\alpha$ between the incident particle $i$ and target particle $\alpha$ and $K$ includes the kinetic energy of $i$ and the total target Hamiltonian [see Eqs. (102), (103), and (104)].

It is perhaps well to emphasize again the admonitions of Chapter 5 that care must be exercised in going consistently to the limit $\eta \to 0(+)$ in Eq. (250). The full complexity of the $(N + 1)$-body problem is of course concealed in this equation. Our choice of $d$ implies a representation in which particle $i$ is described by free (that is, plane-wave) states and the remaining $N$ particles by eigenstates of $h$, although some of these may of course recoil from the bound system as a result of the scattering. Since we have assumed for the present that $i$ is of a different species from any of the target particles, $\psi_a^{\dagger}$ is not to be symmetrized with respect to the coordinates $(\mathbf{x}, s_i)$ of particle $i$. We naturally assume, however, that the target wave functions $g_\gamma^{\pm}$ are appropriately symmetrized in the coordinates of identical bound particles.

Let us first define a scattering matrix $t_\alpha$ for scattering $i$ on a *bound particle* $\alpha$.

$$t_\alpha = V_\alpha + V_\alpha \frac{1}{E_a + i\eta - K - V_\alpha} V_\alpha. \tag{252a}$$

This may be written also in the form of an integral equation.

$$t_\alpha = V_\alpha + V_\alpha \frac{1}{d} t_\alpha. \tag{252b}$$

The scattering matrix $t_\alpha$ is seen to be a direct generalization of the single bound-particle scattering matrix $\mathscr{T}_b$ defined by Eq. (13). Just as in that case $t_\alpha$ differs from the scattering matrix $\mathscr{T}_\alpha$ for a *free target* particle $\alpha$ [Eq. (135)] in that it includes a modification for the influence of the binding of $\alpha$ on the scattering. We shall see presently that $t_\alpha$ and $\mathscr{T}_\alpha$ are related by the impulse approximation.

We now assert that a formally exact solution of Eq. (250) is provided by the set of coupled equations[72]

$$\psi_a^{+} = \chi_a + \sum_{\alpha=1}^{N} \frac{1}{d} t_\alpha \psi_\alpha, \tag{253a}$$

$$\psi_\alpha = \chi_a + \sum_{\beta(\neq\alpha)=1}^{N} \frac{1}{d} t_\beta \psi_\beta. \tag{253b}$$

The close similarity here to our treatment of small scatterers, Eqs. (136), is apparent. The exact scattered wave,

$$\psi_{\text{sc}} = \sum_{\alpha=1}^{N} \frac{1}{d} t_\alpha \psi_\alpha,$$

is the sum of waves scattered from all $N$ scatterers, as in (136). The wave incident on scatterer $\alpha$ is now not $\chi_a$, but an "effective wave" $\psi_\alpha$. The effective wave $\psi_\alpha$ is the sum of the incident wave $\chi_a$ plus the waves scattered from all scatterers other than $\alpha$.

The structure of Eqs. (253*b*) is seen more easily if we generate a multiple scattering series by successive substitution into the right-hand side of these equations.

$$\psi_\alpha = \chi_a + \sum_{\beta(\neq\alpha)=1}^{N} \frac{1}{d} t_\beta \chi_a + \sum_{\beta(\neq\alpha)=1}^{N} \frac{1}{d} t_\beta \frac{1}{d} \sum_{\lambda(\neq\beta)=1}^{N} t_\lambda \chi_a + \cdots. \tag{254}$$

[72] Equations having the structure of Eqs. (253) were proposed by Foldy [29] to describe the scattering of sound waves. Lax [30] proposed such equations on an empirical basis. A formal derivation was given by Watson [31]. Somewhat similar equations were derived by Eckstein [32].

It is evident that the incident particle undergoes a sequence of scatterings from one target particle to another, making any number of such scatterings before emerging from the target.[73]

To prove our assertion that the multiple scattering form for the wave function given by Eqs. (253) indeed provides a solution of the full Schrödinger equation, Eq. (250), we first substitute (253$a$) into the right-hand side of (250):

$$\psi_a{}^+ = \chi_a + \frac{1}{d} V\chi_a + \frac{1}{d} V \sum_{\alpha=1}^{N} \frac{1}{d} t_\alpha \psi_\alpha. \tag{255a}$$

Now, since $V = \sum_{\beta=1}^{N} V_\beta$, we have

$$V\frac{1}{d} t_\alpha = V_\alpha \frac{1}{d} t_\alpha + \sum_{\beta(\neq\alpha)=1}^{N} V_\beta \frac{1}{d} t_\alpha = t_\alpha - V_\alpha + \sum_{\beta(\neq\alpha)=1}^{N} V_\beta \frac{1}{d} t_\alpha, \tag{256}$$

using the definition of $t_\alpha$, Eq. (252$b$). Inserting (256) into (255$a$) and rearranging the sums, we find that

$$\begin{aligned}\psi_a{}^+ &= \chi_a + \frac{1}{d} V\chi_a + \sum_{\beta=1}^{N} \frac{1}{d} V_\beta \sum_{\alpha(\neq\beta)=1}^{N} \frac{1}{d} t_\alpha \psi_\alpha + \frac{1}{d} \sum_{\alpha=1}^{N} t_\alpha \psi_\alpha - \frac{1}{d} \sum_{\alpha=1}^{N} V_\alpha \psi_\alpha \\ &= \left(\chi_a + \frac{1}{d} \sum_{\alpha=1}^{N} t_\alpha \psi_\alpha\right) + \frac{1}{d} V\chi_a + \frac{1}{d} \sum_{\beta=1}^{N} V_\beta(\psi_\beta - \chi_a) - \frac{1}{d} \sum_{\alpha=1}^{N} V_\alpha \psi_\alpha = \psi_a{}^+,\end{aligned} \tag{255b}$$

which completes our proof. In obtaining the second form of this equation we have used Eqs. (253$b$). To obtain the third form, we have again used Eq. (253$a$). We emphasize that Eqs. (253) provide a formally exact solution of Eq. (250).

The differential scattering cross section may now be found, as in Section 11.2. The scattering matrix for the system, $\mathscr{T}_N$, of Eq. (136$b$) is generalized to the form

$$\begin{aligned}\langle b | \mathscr{T}_N | a \rangle &= (\chi_b{}^-, \mathscr{T}_N \chi_a) = \sum_{\alpha=1}^{N} (\chi_b{}^-, t_\alpha \psi_\alpha) \\ &= \delta(\mathbf{k} + \mathbf{Q} - \mathbf{p} - \mathbf{P})(\mathbf{k}, \nu', \mathbf{Q}, \gamma(\nu'_t)\, |T_N|\, \mathbf{p}, \nu, \mathbf{P}, 0(\nu_t)),\end{aligned} \tag{257}$$

where the notation of Eq. (137) is retained. The scattering cross section may be obtained using our standard rules as defined, for example, by Eqs. (139).

The multiple scattering equations (253), while appealing to the intuition, are, of course, difficult to solve in general. Most applications have consequently been made in the weak binding limit where all target particle

[73] An explicit demonstration of this using wave packets was given by Watson [33]. On replacing $\chi_a$ by an incident wave packet $\phi_a(t)$ [see Eq. (3.28$a$)] we see that the scattered wavelet (as obtained in Section 3.4) is rescattered, giving rise to secondary wavelets which in turn are rescattered, and so on.

energies are small compared to the projectile energy [Eqs. (106)], in which case they simplify considerably.

The first implication of the weak binding limit is that the bound scattering matrices $t_\alpha$ may be replaced by the free-particle scattering matrices $\mathscr{T}_\alpha$ [see Eq. (135)]:

$$\mathscr{T}_\alpha = V_\alpha + V_\alpha(\epsilon_p + \epsilon_{\alpha,P_\alpha} + i\eta - K_i - K_\alpha - V_\alpha)^{-1}V_\alpha, \tag{258}$$

where $\epsilon_{\alpha,P_\alpha}$ is the eigenvalue of $K_\alpha$ for a free state of momentum $\mathbf{P}_\alpha$. With an appropriate identification of symbols, we may adapt the discussion given in Section 11.1, p. 686, of the impulse approximation to our present problem.

To do this, we set the energy of the initial state $E_a = \epsilon_p + W_0$ [see Eq. (102$a$)] in Eq. (252$a$), writing

$$W_0 \equiv \epsilon_{\alpha,P_\alpha} + R_{\alpha,a} + \delta W_{\alpha,0}. \tag{259}$$

Here $R_{\alpha,a}$ is the appropriate level shift[74] for particle $\alpha$, as in Eqs. (20), it being supposed that $\epsilon_{\alpha,P_\alpha} \ll \epsilon_p$. Then

$$h - \delta W_{\alpha,0} \equiv K_\alpha + U_\alpha, \tag{260}$$

where $U_\alpha$ is an effective interaction energy of particle $\alpha$ with the remaining bound system. Equations (259) and (260) permit us to write, for the propagator $d$,

$$E_a + i\eta - K - V_\alpha$$
$$= (\epsilon_p + \epsilon_{\alpha,P_\alpha} + i\eta - K_i - K_\alpha - V_\alpha) + R_{\alpha,a} - U_\alpha. \tag{261}$$

On following the derivation given for a single scattering center [Eq. (24)] we are now led to the expansion

$$t_\alpha = \mathscr{T}_\alpha + \mathscr{T}_\alpha \frac{1}{d_{\alpha,0}}(R_\alpha - R_{\alpha,a})\frac{1}{d_{\alpha,0}}\mathscr{T}_\alpha + \dots, \tag{262}$$

where $d_{\alpha,0} = \epsilon_p + \epsilon_{\alpha,P_\alpha} + i\eta - K_i - K_\alpha$ and

$$R_\alpha = U_\alpha + U_\alpha \frac{1}{d} U_\alpha. \tag{263}$$

The discussion leading to Eq. (30$a$), that is an estimate of the correction to the approximation $t_\alpha \simeq \mathscr{T}_\alpha$, remains unchanged. We obtain

$$t_\alpha = \mathscr{T}_\alpha\left[1 + \mathscr{O}\left(\left|\frac{f_\alpha}{\bar{\lambda}}\frac{R_{\alpha,a}t_{\alpha,c}}{\hbar}\right|\right)\right], \tag{264}$$

[74] See Eq. (105).

where $f_\alpha$ is the scattering amplitude and $t_{\alpha,c}$ is the time duration for a collision with a free particle $\alpha$, as in Eqs. (30). In the limit of quasi-classical binding, that is, where $U_\alpha$ varies slowly over the scattering interaction region, the expression (31) again suffices to estimate the accuracy of the impulse approximation, with $f$, $M$, $U_{\text{av}}$, $t_c$, and $\mathscr{R}$ replaced by $f_\alpha$, $M_\alpha$, $U_{\alpha,\text{av}}$ [see Eq. (105)], $t_{\alpha,c}$, and $\mathscr{R}_\alpha$ (the range of the interaction between particles $i$ and $\alpha$).

When the impulse approximation is valid, we may set

$$\mathscr{t}_\alpha = \mathscr{T}_\alpha \tag{265}$$

in our formal multiple scattering equations (253). These then become

$$\psi_a^+ = \chi_a + \sum_{\alpha=1}^{N} \frac{1}{d} \mathscr{T}_\alpha \psi_\alpha, \tag{266a}$$

$$\psi_\alpha = \chi_a + \sum_{\beta(\neq\alpha)=1}^{N} \frac{1}{d} \mathscr{T}_\beta \psi_\beta. \tag{266b}$$

The simplification implied by the approximation (265) and leading to Eqs. (266) is a very significant one, since the *many-body* operators $\mathscr{t}_\alpha$ are replaced by the *two-particle* operators $\mathscr{T}_\alpha$.

For explicit evaluation of the scattering cross section it is equally important to be able to simplify the many-body propagators $d^{-1}$. The reciprocal of this propagator is just the quantity appearing in the $\delta$-function in the cross section evaluation carried out in Section 11.2, Eqs. (139), and may be handled in much the same manner.

### Multiple Scattering in the Quasi-free Limit

The assumption that the mean free path for scattering is large compared to the particle wavelength permits us to replace the propagators $d^{-1}$ in Eqs. (266) by their asymptotic, on the energy shell, values. This means that the scattering matrices $\mathscr{T}_\alpha$ are evaluated on the energy shell. In the weak binding limit, Eqs. (266) then reduce to coupled algebraic (rather than integral) equations.

We shall now illustrate these remarks, assuming that the weak binding conditions (106), condition (132), and the quasi-free conditions of momentum transfer large compared to target momenta (162) are satisfied. For simplicity, we shall continue to assume that the composite target is at rest before the scattering occurs. Generalization to a moving target is evidently straightforward.

Let us first discuss a typical scattered wave in Eqs. (266):

$$\psi_{\text{sc},\beta} \equiv \frac{1}{E_a + i\eta - K} \mathscr{T}_\beta \psi_\beta. \tag{267}$$

This may be evaluated, because of the assumption of projectile wavelength being small compared to the mean free path, in the same manner as was the scattered wave from a single target in Section 11.1 [Eqs. (74)]. On expanding $\psi_\beta$ in plane waves of particle $i$, we have

$$\psi_\beta = \int d^3k' \frac{e^{i\mathbf{k}'\cdot\mathbf{x}}}{(2\pi)^{3/2}} \psi_\beta(\mathbf{k}'), \tag{268}$$

where the Fourier coefficient $\psi_\beta(\mathbf{k}')$ is of course a function of the spin variable $s_i$ and the coordinates $(\mathbf{z}_1, s_1, \ldots, \mathbf{z}_N, s_N)$ of the bound particles. Then, according to Eq. (74),

$$\psi_{\mathrm{sc},\beta} = (2\pi)^{-3/2} \int d^3k' \int \frac{d^3k\, e^{i\mathbf{k}\cdot\mathbf{x}}}{\epsilon_p + i\eta - \epsilon_k - h + W_0}\, e^{-i\boldsymbol{\rho}\cdot\mathbf{z}_\beta} T_\beta \psi_\beta(\mathbf{k}'). \tag{269}$$

Here $T_\beta$ is the matrix $\langle\mathbf{k}|\, T_\beta\, |\mathbf{k}'\rangle$, which acts as an operator on the variables $(s_i, \mathbf{z}_\beta, s_\beta)$. It is defined in connection with Eq. (135). The momentum transfer $\boldsymbol{\rho}$ in Eq. (269) is now

$$\boldsymbol{\rho} \equiv \mathbf{k} - \mathbf{k}'. \tag{270}$$

The quasi-free conditions of large momentum transfer (162) may be used to simplify the propagator in Eq. (269). We may write

$$\frac{1}{\epsilon_p + i\eta - \epsilon_k - h + W_0} = -i \int_{-\infty}^{0} e^{-i(\epsilon_p + i\eta - \epsilon_k - h + W_0)t}\, dt,$$

which permits us to express Eq. (269) as

$$\psi_{\mathrm{sc},\beta} = (2\pi)^{-3/2}(-i) \int d^3k'\, d^3k \int_{-\infty}^{0} dt\, e^{i\mathbf{k}\cdot\mathbf{x}}\, e^{-i(\epsilon_p + i\eta - \epsilon_k + W_0 - W_{\mathrm{ex}})t}$$
$$\times\, e^{-i\boldsymbol{\rho}\cdot\mathbf{z}_\beta(t)} T_\beta(t)\, e^{i(h - W_{\mathrm{ex}})t} \psi_\beta(\mathbf{k}'). \tag{271}$$

Here $W_{\mathrm{ex}}$ is an effective excitation energy, to be determined according to our convenience, and $\mathbf{z}_\beta(t)$ and $T_\beta(t)$ are the Heisenberg operators

$$\mathbf{z}_\beta(t) = e^{i(h - W_{\mathrm{ex}})t} \mathbf{z}_\beta\, e^{-i(h - W_{\mathrm{ex}})t},$$

$$T_\beta(t) = e^{i(h - W_{\mathrm{ex}})t} T_\beta\, e^{-i(h - W_{\mathrm{ex}})t},$$

as in Eq. (188$b$).

To interpret Eq. (271), we recall from Eqs. (266) that $\psi_\beta(\mathbf{k}')$ represents the incident wave on particle $\beta$. It will consist of a sum of terms, each term corresponding to a particular sequence of scatterings by which particle $i$ may reach particle $\beta$. This includes direct arrival of the incident wave, arrival of once-scattered waves, twice-scattered waves, and so on. All possible such "paths" will be included in $\psi_\beta$. Because of the assumed short wavelength, condition (132), for each such path the integral over $\mathbf{k}'$ will be restricted to a narrow domain about that $\mathbf{k}'$ which corresponds to stationary phase along the given path.

Even for a given sequence of scatterings, $\psi_\beta(\mathbf{k}')$ will not be an eigenfunction of $h$. In the weak binding limit it is possible to choose $W_{\text{ex}}$ in such a manner that

$$e^{i(h-W_{\text{ex}})t}\psi_\beta(\mathbf{k}') \cong \psi_\beta(\mathbf{k}'),$$

however. We saw in Section 11.1, p. 693, that for a single quasi-free scattering on a nucleon, say $\gamma$, the scatterer receives a mean excitation energy $\rho_\gamma^2/2M_\gamma$, where $\rho_\gamma$ is the momentum transfer for the scattering. The mean square spread in the excitation energy is given by Eq. (65) as

$$K_{\gamma,\text{av}} \frac{\rho_\gamma^{\;2}}{2M_\gamma}.$$

For a quasi-elastic scattering the excitation energy was set equal to zero, with a mean square spread again given by Eq. (65). This suggests that we choose

$$W_{\text{ex}} - W_0 = \text{sum of excitation energies for each scattering along a given path.} \tag{272}$$

Then the linear term in $(h - W_{\text{ex}})t$ can be made to vanish in the expression for the cross section and we are left with the estimate

$$\begin{aligned} e^{i(h-W_{\text{ex}})t}\psi_\beta(\mathbf{k}') &= [1 + \mathcal{O}(|(h - W_{\text{ex}})t|)^2]\psi_\beta(\mathbf{k}') \\ &\cong \left[1 + \mathcal{O}\left(\sum_\gamma \frac{K_{\gamma,\text{av}}(\rho_\gamma^{\;2}/2M_\gamma)}{\epsilon_p^{\;2}}\right)\right]\psi_\beta(\mathbf{k}'), \end{aligned} \tag{273}$$

where the summation extends over the previous scatterings on a given path.[75] In writing the final expression here, we have taken $t \cong \epsilon_p^{\;-1}$.

The "time dependence" of $\mathbf{z}_\beta(t)$ and $T_\beta(t)$ leads to the correction terms found for Eq. (190). In the weak binding limit, we may therefore write Eq. (271) as

$$\begin{aligned} \psi_{\text{cs},\beta} = (2\pi)^{-3/2}\int \frac{d^3k'\, d^3k\, e^{i\mathbf{k}\cdot\mathbf{x}}\, e^{-i\mathbf{p}\cdot\mathbf{z}_\beta}}{\epsilon_p + i\eta - \epsilon_k - (W_{\text{ex}} - W_0) - \rho^2/2M_\beta} \\ \times T_\beta\psi_\beta(\mathbf{k}')(1 + \eta_1 + \eta_2 + \cdots), \end{aligned} \tag{274}$$

the series being of the form given in Eq. (191).[76] The expansion given here implies that the dependence of $T_\beta$ on the momentum $\mathbf{P}_\beta$ is to be neglected.

The result (274) is not unexpected. The assumption of short wavelength (132) implies that each of a sequence of scatterings is an isolated event, of which the single scatterings described in Section 11.2 are typical.

[75] To maintain a condensed notation we understand, without writing it explicitly, that for each path of the scattered particle a different $W_{\text{ex}}$ will in general be required.

[76] Strictly speaking, the errors from a sequence of scatterings are cumulative, but probably with random phases.

When the corrections $\eta_1, \eta_2, \ldots$ in Eq. (274) may be neglected, this expression may be evaluated just as was the wave function (76). It is necessary only to generalize the notation of Section 11.1, p. 706, by adding appropriate particle labels on the coordinates. For example, we rewrite Eq. (71*a*) as

$$\mathbf{R}_\beta = \mathbf{x} - \mathbf{z}_\beta{}^0, \tag{275a}$$

where $\mathbf{z}_\beta{}^0$ represents the "initial position" of particle $\beta$. The momentum variable $\mathbf{p}$ of Section 11.1, p. 702, is now replaced by $\mathbf{k}'$ and $\mathbf{k}_0$ is replaced by $\mathbf{k}_\beta$, with

$$\mathbf{k}_\beta \equiv k_\beta \hat{\mathbf{R}}_\beta. \tag{275b}$$

Similarly, the variable $\mathbf{Q}_0$ is replaced by

$$\mathbf{Q}_\beta = \mathbf{k}' - \mathbf{k}_\beta. \tag{275c}$$

Then, on transcribing Eq. (89) to our present notation, we obtain

$$\psi_{\text{sc},\beta} = (2\pi)^{-3/2} \int d^3k' \, \frac{e^{ik_\beta R_\beta}}{R_\beta} \, e^{i\mathbf{Q}_\beta \cdot (\mathbf{z}_\beta - \mathbf{z}_\beta{}^0)} \, e^{i\mathbf{k}' \cdot \mathbf{z}_\beta{}^0} \left( \frac{v_\beta}{v_{\beta,\text{inc}}} \, \mathbf{f}_\beta \right) \psi_\beta(\mathbf{k}'), \tag{276}$$

where in $\psi_\beta(\mathbf{k}')$ we must use the quantity $\mathbf{z}_\beta{}^0$ instead of $\mathbf{z}_\beta$ for the coordinate of particle $\beta$. Here $k_\beta$ is determined from the equation

$$\epsilon_{k\beta} = (\epsilon_p - W_{\text{ex}} + W_0) - \frac{Q_\beta{}^2}{2M_\beta},$$

$v_\beta = k_\beta/m$ (or $v_\beta = k_\beta c^2/\epsilon_{k\beta}$ for relativistic scattering), and $v_{\beta,\text{inc}}$ is the velocity of particle $i$ prior to scattering on particle $\beta$. The scattering amplitude $\mathbf{f}_\beta$ in Eq. (276) is [see Eq. (81)]

$$\mathbf{f}_\beta = \mathbf{f}_\beta(\hat{\mathbf{k}}_\beta, \hat{\mathbf{k}}', \mathbf{S}_i, \mathbf{S}_\beta), \tag{277}$$

appropriate for a free scatterer $\beta$.

To simplify Eq. (276) further, let us write[77]

$$\psi_\beta(\mathbf{k}') = (2\pi)^{-3/2} \sum_{\sigma(\neq\beta)=0}^{N} \psi_{\beta\sigma}(\mathbf{k}') \times g_0(\mathbf{z}_1, s_1, \ldots, \mathbf{z}_\beta{}^0, s_\beta, \ldots, \mathbf{z}_\sigma{}^0, s_\sigma, \ldots, \mathbf{z}_N, s_N) \tag{278}$$

where the $\psi_{\beta\sigma}(\mathbf{k}')$ are to be determined. According to our conclusions in Section 11.1, p. 706, we must use coordinates $\mathbf{z}_\alpha{}^0$ in $g_0$ for those target particles which for a given sequence of scatterings have been struck by the incident particle and which have therefore recoiled. Equation (278) is supplemented by the definition

$$\psi_{\beta 0}(\mathbf{k}') = \delta(\mathbf{k}' - \mathbf{p}) u_{S_i,\nu}(s_i), \tag{279}$$

which prescribes the boundary condition on the incident wave.

[77] The factor $(2\pi)^{-3/2} g_0$ in Eq. (278) is the complete wave function for a target at rest.

The interpretation of $\psi_{\beta\sigma}$ is that it represents a wave scattered from particle $\sigma$ to particle $\beta$. This implies that the $\mathbf{k}'$-integration in Eq. (278) will in effect be restricted to a narrow domain of $\hat{\mathbf{k}}'$ near

$$\hat{\mathbf{k}}' = \frac{\mathbf{z}_\beta{}^0 - \mathbf{z}_\sigma{}^0}{|\mathbf{z}_\beta{}^0 - \mathbf{z}_\sigma{}^0|} \equiv \hat{\mathbf{n}}_{\beta\sigma} \tag{280}$$

and at $|\mathbf{k}'| \cong |\mathbf{k}_{\beta\sigma}|$, as required by energy conservation.[78] This permits us to make the substitution in the integrand in Eq. (276),

$$\frac{v_\beta}{v_{\beta,\text{inc}}} \mathbf{f}_\beta \psi_\beta(\mathbf{k}') = \sum_{\sigma(\neq\beta)=0}^{N} \mathbf{f}_{\beta\sigma} \psi_{\beta\sigma}(\mathbf{k}') \frac{g_0}{(2\pi)^{3/2}} \tag{281}$$

where

$$\mathbf{f}_{\beta\sigma} \equiv \frac{v_\beta}{v_{\beta,\text{inc}}} \mathbf{f}_\beta(\hat{\mathbf{k}}_\beta, \hat{\mathbf{n}}_{\beta\sigma}, \mathbf{S}_i, \mathbf{S}_\beta) \tag{282}$$

is evaluated on the energy shell. We may therefore reexpress (276) as

$$\psi_{\text{sc},\beta} = (2\pi)^{-3} \int d^3k' \frac{e^{ik_\beta R_\beta}}{R_\beta} e^{i\mathbf{Q}_\beta \cdot (\mathbf{z}_\beta - \mathbf{z}_\beta{}^0)}$$
$$\times \sum_{\sigma(\neq\beta)=0}^{N} \mathbf{f}_{\beta\sigma}\, e^{i\mathbf{k}' \cdot \mathbf{z}_\beta{}^0} \psi_{\beta\sigma}(\mathbf{k}') g_0(\mathbf{z}_1, s_1, \ldots, \mathbf{z}_\beta{}^0, s_\beta, \ldots, \mathbf{z}_N, s_N). \tag{283}$$

The conclusion that the $\mathbf{k}'$-integral in Eq. (283) is restricted to a small domain near $\mathbf{k}' = k_{\beta\sigma}\hat{\mathbf{n}}_{\beta\sigma}$ [which follows from the condition (132)] permits us to evaluate the integral explicitly. Since $k_\beta R_\beta = \mathbf{k}_\beta \cdot \mathbf{R}_\beta$ and $\mathbf{R}_\beta = \mathbf{x} - \mathbf{z}_\beta{}^0$, we have to the first order in $\delta\mathbf{k}' \equiv \mathbf{k}' - \mathbf{k}_{\beta\sigma}\hat{\mathbf{n}}_{\beta\sigma}$,

$$\begin{aligned}\delta[k_\beta R_\beta &+ \mathbf{Q}_\beta \cdot (\mathbf{z}_\beta - \mathbf{z}_\beta{}^0) + \mathbf{k}' \cdot \mathbf{z}_\beta{}^0] \\ &= \delta\mathbf{k}_\beta \cdot R_\beta + \delta\mathbf{Q}_\beta \cdot (\mathbf{z}_\beta - \mathbf{z}_\beta{}^0) + \delta\mathbf{k}' \cdot \mathbf{z}_\beta{}^0 + (\mathbf{k}' - \mathbf{k}_\beta - \mathbf{Q}_\beta) \cdot \delta z_\beta{}^0.\end{aligned} \tag{284}$$

The last term here vanishes, since momentum conservation implies that $\mathbf{k}' = \mathbf{k}_\beta + \mathbf{Q}_\beta$. Reference to the kinematical discussion given in Section 11.1, Eqs. (71), shows that we may write the first two terms on the right side of (284) as

$$(\delta\mathbf{k}_\beta \cdot \mathbf{v}_\beta + \delta\mathbf{Q}_\beta \cdot \mathbf{V}_\beta)\, \Delta t, \tag{285a}$$

where $\mathbf{v}_\beta$ and $\mathbf{V}_\beta$ are the respective velocities, subsequent to the collision, of particles $i$ and $\beta$ and $\Delta t$ is the time lapse, as described by Eqs. (71). The expression (285*a*) vanishes, since it is equivalent to

$$\Delta t\, \delta\left(\epsilon_k + \frac{Q_\beta{}^2}{2M_\beta}\right), \tag{285b}$$

[78] The basis for this conclusion will become clear.

or $\Delta t$ times the variation of the total recoil energy, which is fixed by energy conservation ($\mathbf{k}'$ is a virtual momentum). The right-hand side of Eq. (284) is therefore just $\delta\mathbf{k}' \cdot \mathbf{z}_\beta{}^0$. This enables us to perform the $\mathbf{k}'$-integral in Eq. (283) to obtain

$$\psi_{\mathrm{sc},\beta} = \frac{e^{ik_\beta R_\beta}}{R_\beta} \sum_{\sigma(\neq\beta)=0}^{N} e^{i\boldsymbol{\rho}_{\beta\sigma}\cdot(\mathbf{z}_\beta - \mathbf{z}_\beta{}^0)} (2\pi)^{-3/2} \mathbf{f}_{\beta\sigma} \Psi_{\beta\sigma}(\mathbf{z}_\beta{}^0)$$
$$\times\, g_0(\mathbf{z}_1, s_1, \ldots, \mathbf{z}_\beta{}^0, s_\beta, \ldots, \mathbf{z}_\sigma{}^0, s_\sigma, \ldots, \mathbf{z}_N, s_N). \quad (286)$$

Here

$$\boldsymbol{\rho}_{\beta\sigma} \equiv \mathbf{k}_\beta - k_{\beta\sigma}\hat{\mathbf{n}}_{\beta\sigma} \quad (287)$$

and

$$\Psi_{\beta\sigma}(\mathbf{y}) \equiv (2\pi)^{-3/2} \int d^3k'\, e^{i\mathbf{k}'\cdot\mathbf{y}} \psi_{\beta\sigma}(\mathbf{k}'). \quad (288)$$

The variable $\mathbf{z}_\beta{}^0$ is now to be evaluated with $\mathbf{k}' = k_{\beta\sigma}\hat{\mathbf{n}}_{\beta\sigma}$. We emphasize that in $g_0$ we must use a coordinate $\mathbf{z}_\alpha{}^0$ in place of $\mathbf{z}_\alpha$ for each target particle $\alpha$ which has undergone a scattering with the incident particle in a particular sequence of scatterings (and for which the momentum transfer is large compared to $\sqrt{2M_\alpha K_{\alpha,\mathrm{av}}}$).

Returning to our approximate multiple scattering equations (266*b*), we use (286) to write $\psi_\alpha(\mathbf{x})$ at $\mathbf{x} = \mathbf{z}_\alpha{}^0$ as

$$\psi_\alpha(\mathbf{z}_\alpha{}^0) = \chi_a(\mathbf{z}_\alpha{}^0) + (2\pi)^{-3/2} \sum_{\beta(\neq\alpha)=1}^{N} \sum_{\sigma(\neq\beta)=0}^{N} E_{\alpha\beta,\sigma}\mathbf{f}(\alpha\beta, \beta\sigma)\Psi_{\beta\sigma}(\mathbf{z}_\beta{}^0) g_0, \quad (289)$$

where

$$E_{\alpha\beta,\sigma} \equiv \frac{e^{ik_\beta R_{\alpha\beta}}}{R_{\alpha\beta}} e^{-i\boldsymbol{\rho}_{\beta\sigma}\cdot(\mathbf{z}_\beta - \mathbf{z}_\beta{}^0)}, \quad (290)$$

$\mathbf{R}_{\alpha\beta} \equiv \mathbf{z}_\alpha{}^0 - \mathbf{z}_\beta{}^0$, $\boldsymbol{\rho}_{\beta 0} \equiv \mathbf{k}_\beta - \mathbf{p}$, and

$$\begin{aligned} \mathbf{f}(\alpha\beta, \beta\sigma) &\equiv \mathbf{f}_\beta(\hat{\mathbf{n}}_{\alpha\beta}, \hat{\mathbf{n}}_{\beta\sigma}, \mathbf{S}_i, \mathbf{S}_\beta), \\ \mathbf{f}(\alpha\beta, \beta 0) &\equiv \mathbf{f}_\beta(\hat{\mathbf{n}}_{\alpha\beta}, \hat{\mathbf{p}}, \mathbf{S}_i, \mathbf{S}_\beta). \end{aligned} \quad (291)$$

The appropriate variables $\mathbf{z}$ or $\mathbf{z}^0$ must of course be used in $g_0$.

Using Eq. (278), we may write

$$\psi_\alpha(\mathbf{z}_\alpha{}^0) = \chi_a(\mathbf{z}_\alpha{}^0) + (2\pi)^{-3/2} \sum_{\beta(\neq\alpha)=1}^{N} \Psi_{\alpha\beta}(\mathbf{z}_\alpha{}^0) g_0. \quad (292)$$

On comparing Eqs. (289) and (292), we obtain the set of coupled equations [$\chi_{i;\mathbf{p}\nu}$ is just the initial plane-wave state for $i$, Eq. (4)]

$$\Psi_{\alpha\beta}(\mathbf{z}_\alpha{}^0) = E_{\alpha\beta,0}\mathbf{f}(\alpha\beta, \beta 0)\chi_{i;\mathbf{p}\nu}(\mathbf{z}_\beta{}^0) + \sum_{\sigma(\neq\beta)=1}^{N} E_{\alpha\beta,\sigma}\mathbf{f}(\alpha\beta, \beta\sigma)\Psi_{\beta\sigma}(\mathbf{z}_\beta{}^0), \quad (293)$$

where $\alpha$, $\beta \neq 0$. From Eq. (279) we find that

$$\Psi_{\alpha 0}(\mathbf{z}_\alpha{}^0) = (2\pi)^{-3/2} e^{i\mathbf{p}\cdot\mathbf{z}_\alpha{}^0} u_{S_i,\nu}(s_i). \quad (294)$$

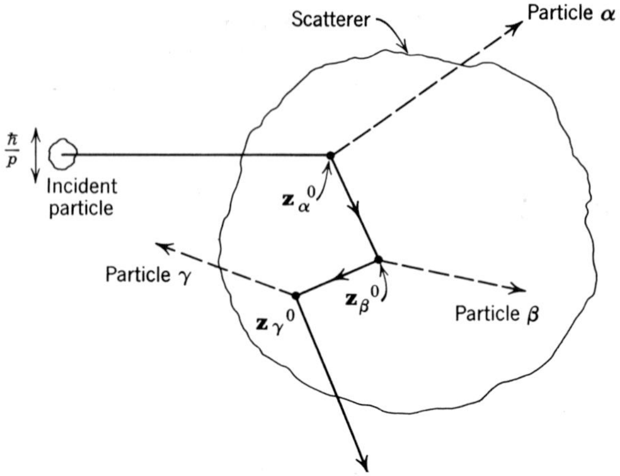


Fig. 11.6 Illustration of a typical orbit described by Eqs. (293) and (295).

This is a set of purely *algebraic equations* to determine the $\Psi_{\alpha\beta}(\mathbf{z}_\alpha^0)$. Once these have been found, we may use Eq. (266*a*) to give the asymptotic form of $\psi_a^+$.

$$\psi_a^+ = \chi_a + \sum_{\alpha=1}^{N} \frac{e^{ik_\alpha R_\alpha}}{R_\alpha} \sum_{\beta(\neq\alpha)=0}^{N} e^{-i\rho_{\alpha\beta}\cdot(\mathbf{z}_\alpha-\mathbf{z}_\alpha^0)} \mathbf{f}_{\alpha\beta}\Psi_{\alpha\beta}(\mathbf{z}_\alpha^0)\frac{g_0}{(2\pi)^{3/2}}$$

$$\cong \chi_a + \sum_{\alpha=1}^{N} \frac{e^{ik_\alpha x}}{x} e^{-ik_\alpha\hat{\mathbf{x}}\cdot\mathbf{z}_\alpha^0} \sum_{\beta(\neq\alpha)=0}^{N} e^{-i\rho_{\alpha\beta}\cdot(\mathbf{z}_\alpha-\mathbf{z}_\alpha^0)} \mathbf{f}_{\alpha\beta}\Psi_{\alpha\beta}(\mathbf{z}_\alpha^0)\frac{g_0}{(2\pi)^{3/2}}, \quad (295)$$

where [see Eq. (282)] $\mathbf{f}_{\alpha\beta} \equiv \mathbf{f}_\alpha(\hat{\mathbf{x}}, \hat{\mathbf{n}}_{\alpha\beta}, \mathbf{S}_i, \mathbf{S}_\alpha)$ and it has been assumed that the scatterer center-of-mass was localized at $\mathbf{x} = 0$ prior to the scattering. In Eqs. (295) we must take $\boldsymbol{\rho}_{\alpha\beta} = k_\alpha\hat{\mathbf{x}} - \mathbf{k}_{\alpha\beta}\hat{\mathbf{n}}_{\alpha\beta}$, a special form of Eq. (287).

We remark, finally, that our results are valid for relativistic energies of the incident particle if the scattering amplitude (94) is used and if relativistic kinematics are used to define $\mathbf{z}_\alpha^0$, and so on.[79]

Because the struck target particles recoil from the scatterer in the quasi-free limit, waves scattered from different particles in Eqs. (293) and (295) are strictly incoherent. This means that there is no interference between waves scattered by different particles, or along different "paths" in the scatterer. A special case of this general conclusion was obtained in Section 11.2, p. 725, where it was concluded that $d\sigma_{nd}$ vanished in the quasi-free limit [see Eq. (164)].

In principle it is possible to determine the orbit of the incident particle in passing through the scattering system (to within an accuracy of the order $\hbar/p$) by deducing from an observation of the recoil scatterers the successive points $\mathbf{z}_\alpha^0$ at which scattering has occurred. (This is illustrated in Fig. 11.6.)

[79] This is shown in Ref. [16].

Except for the dynamics of the actual binary collision processes, Eqs. (293) and (295) are very similar to a description of the interaction using classical mechanics. Indeed, in Section 11.5 we shall show how to obtain a classical transport equation from these wave equations.

The coefficient of the factors $(2\pi)^{-3/2}(e^{ik_\alpha x}/)x$ in Eq. (295) gives the scattering amplitude $\mathbf{f}^{(N)}$ for the interaction. Using Eqs. (293) and (294) we find that

$$\begin{aligned}\mathbf{f}^{(N)}u_{S_i,\nu}g_0 &= \sum_{\alpha=1}^{N} e^{-i(k_\alpha\hat{\mathbf{x}}-\mathbf{p})\cdot\mathbf{z}_\alpha}\mathbf{f}_\alpha(\hat{\mathbf{x}},\hat{\mathbf{p}},\mathbf{S}_i,\mathbf{S}_\alpha)u_{S_i,\nu}(s_i)\\ &\quad\times g_0(\mathbf{z}_1, s_1, \ldots, \mathbf{z}_\alpha{}^0, s_\alpha, \ldots, \mathbf{z}_N, s_N)\\ &\quad+\sum_{\alpha=1}^{N} e^{-ik_\alpha\hat{\mathbf{x}}\cdot\mathbf{z}_\alpha{}^0}\sum_{\beta(\neq\alpha)=1}^{N} e^{-i\hat{\mathbf{x}}_{\alpha\beta}\cdot(\mathbf{z}_\alpha-\mathbf{z}_{\alpha 0})}\\ &\quad\times \mathbf{f}_\alpha(\hat{\mathbf{x}},\hat{\mathbf{n}}_{\alpha\beta},\mathbf{S}_i,\mathbf{S}_\alpha)\frac{e^{ik_\beta R_{\alpha\beta}}}{R_{\alpha\beta}}e^{-i\mathbf{p}_{\beta 0}\cdot(\mathbf{z}_\beta-\mathbf{z}_\beta{}^0)}f(\alpha\beta,\beta 0)\,e^{i\mathbf{p}\cdot\mathbf{z}_\beta{}^0}\\ &\quad\times u_{S_i,\nu}(s_i)g_0(\mathbf{z}_1, s_1,\ldots,\mathbf{z}_\alpha{}^0, s_\alpha,\ldots,\mathbf{z}_\beta{}^0,\mathbf{s}_\beta,\ldots,\mathbf{z}_N, s_N)+\cdots.\end{aligned} \tag{296}$$

The differential cross section is given by the scalar product of (296) with its adjoint. To evaluate this, we recall that all cross terms vanish in the quasi-free limit and use the $\mathbf{z}_\alpha - \mathbf{z}_\alpha{}^0$ transformation derived for Eq. (90). Then,

$$\begin{aligned}\frac{d\bar{\sigma}}{d\Omega} &= (u_{S_i,\nu}g_0, f^{(N)\dagger}f^{(N)}u_{S_i,\nu}g_0)_{\text{av}}\\ &= \Bigg(u_{S_i,\nu}g_0, \Bigg[\sum_{\alpha=1}^{N}\mathbf{f}_\alpha{}^\dagger(\hat{\mathbf{x}},\hat{\mathbf{p}},\mathbf{S}_i,\mathbf{S}_\alpha)\mathbf{f}(\hat{\mathbf{x}},\hat{\mathbf{p}},\mathbf{S}_i,\mathbf{S}_\alpha)\\ &\quad+\sum_{\alpha=1}^{N}\sum_{\beta(\neq\alpha)=1}^{N}\mathbf{f}_\alpha{}^\dagger(\hat{\mathbf{x}},\hat{\mathbf{n}}_{\alpha\beta},\mathbf{S}_i,\mathbf{S}_\alpha)\mathbf{f}_\alpha(\hat{\mathbf{x}},\hat{\mathbf{n}}_{\alpha\beta},\mathbf{S}_i,\mathbf{S}_\alpha)\\ &\quad\times\frac{1}{R_{\alpha\beta}{}^2}\mathbf{f}_\beta{}^\dagger(\hat{\mathbf{n}}_{\alpha\beta},\hat{\mathbf{p}},\mathbf{S}_i,\mathbf{S}_\beta)\mathbf{f}_\beta(\hat{\mathbf{n}}_{\alpha\beta},\hat{\mathbf{p}},\mathbf{S}_i,\mathbf{S}_\beta)+\cdots\Bigg]u_{S_i,\nu}g_0\Bigg)_{\text{av}}\\ &=\sum_{\alpha=1}^{N}\frac{d\bar{\sigma}_f}{d\Omega}(\alpha;\hat{\mathbf{p}}\to\hat{\mathbf{x}})+\sum_{\alpha=1}^{N}\sum_{\beta(\neq\alpha)=1}^{N}\frac{d\bar{\sigma}_f}{d\Omega}(\alpha;\hat{\mathbf{n}}_{\alpha\beta}\to\hat{\mathbf{x}})\\ &\quad\times\frac{1}{R_{\alpha\beta}{}^2}\frac{d\bar{\sigma}_f}{d\Omega}(\beta;\hat{\mathbf{p}}\to\hat{\mathbf{n}}_{\alpha\beta})+\cdots,\end{aligned} \tag{297}$$

where the subscript av implies an average over initial spin directions and the differential cross sections

$$\frac{d\bar{\sigma}_f}{d\Omega}(\alpha;\hat{\mathbf{p}}\to\hat{\mathbf{x}})$$

are the free-particle cross sections (170) for scattering of $i$ from a direction $\hat{\mathbf{p}}$ to a direction $\hat{\mathbf{x}}$, and so on.[80]

Equation (297) might have been written down directly from more intuitive arguments. The first term is equivalent to the cross section given in Eqs. (164) and (170). The second term represents the contribution from two successive scatterings, an effect neglected in Section 11.2. To estimate the importance of this term, we neglect the angular and particle dependence, taking

$$\frac{d\bar{\sigma}_f(\alpha)}{d\Omega} = \frac{1}{4\pi}\,\bar{\sigma}_f,$$

where $\bar{\sigma}_f$ is the total cross section for an "average scatterer." We also suppose the scattering system to have a uniform particle density for $z_\alpha{}^0 < \mathscr{R}$, the size of the system. Then (297) gives

$$\begin{aligned}\bar{\sigma} &= N\bar{\sigma}_f\left[1 + \bar{\sigma}_f \int \frac{\rho(\mathbf{z}_\beta{}^0)\, d^3 z_\beta{}^0}{R_{\alpha\beta}{}^2} + \cdots\right] \\ &\cong N\bar{\sigma}_f(1 + \rho\bar{\sigma}_f\mathscr{R}) = N\bar{\sigma}_f\left(1 + \frac{\mathscr{R}}{\lambda}\right).\end{aligned} \tag{298}$$

The multiple scattering correction here is seen to be negligible when the mean free path is large compared to the size of the scatterer; this is not very surprising. We see, therefore, that the discussion of Section 11.2 is fully justified by our present more thorough study. When $\lambda \cong \mathscr{R}$ a few terms in the series (296) will often suffice to determine $d\bar{\sigma}/d\Omega$. If $\lambda \ll \mathscr{R}$, the transport equation derived in Section 11.5 may be used to obtain the scattering cross section.

#### Multiple Scattering in the Quasi-elastic Limit

When the momentum transfers $\boldsymbol{\rho}_{\beta\sigma}$ are less than the quantities $\sqrt{2M_\beta K_{\beta,\text{av}}}$ (and $\sqrt{2M_\beta U_{\beta,\text{av}}}$), the scattering tends to be elastic or quasi-elastic, as specified by the condition (176). In this case the excitation energy of the scattering system tends to be small compared to particle-binding energies, and waves scattered from different particles may interfere with each other.

In this section we shall discuss Eqs. (266), again assuming that the weak binding conditions (106) and the short wavelength condition (132) are satisfied, but that now the quasi-elastic limit is appropriate. Although Eqs. (293) and (295) may be modified almost trivially to describe elastic and quasi-elastic scattering, the physical significance of the change is profound.

[80] In obtaining (297), we have assumed for simplicity that following the first scattering particle $i$ is not polarized. If it is polarized, the second scattering must be expressed in the form of Eq. (7.261).

Once again, we must first evaluate a typical scattered wave, Eq. (276), a task that was accomplished in the present approximation already in Eq. (92). Consistent with the quasi-elastic condition, we must neglect the term $W_{ex} - W_0 + p^2/2M_\beta$ in the propagator in Eq. (274). This implies that the struck target particle does not recoil, so the coordinates $\mathbf{z}_\beta{}^0$ are equal to the coordinates $\mathbf{z}_\beta$, precisely as argued in connection with Eq. (92). In addition to replacing $\mathbf{z}_\beta{}^0$ by $\mathbf{z}_\beta$ in Eq. (296), we must then take $k_\beta = p$, $v_\beta = v_{\beta,\text{inc}}$ and replace $\mathbf{f}_\beta$ by $\mathbf{f}_\beta{}^e$, as defined by Eqs. (93) or (95).

Equation (292) now becomes

$$\psi_\alpha(\mathbf{z}_\alpha) = \chi_\alpha(\mathbf{z}_\alpha) + (2\pi)^{-3/2} \sum_{\beta(\neq\alpha)=1}^{N} \Psi_{\alpha\beta}(\mathbf{z}_\alpha) g_0(\mathbf{z}_1, s_1, \ldots, \mathbf{z}_N, s_N). \tag{299}$$

Similarly, the scattering amplitudes (291) are replaced by $\mathbf{f}^e(\alpha\beta, \beta\sigma)$, and so on, defined as in Eqs. (93) or (95). Then, in place of Eqs. (293) and (294), we have just (recalling that now $\mathbf{R}_{\alpha\beta} = \mathbf{z}_\alpha - \mathbf{z}_\beta$)[81]

$$\begin{aligned}\Psi_{\alpha\beta}(\mathbf{z}_\alpha) &= \frac{e^{ipR_{\alpha\beta}}}{R_{\alpha\beta}} \mathbf{f}^e(\alpha\beta, \beta 0) \chi_{i;\,\mathbf{p}\nu}(\mathbf{z}_\beta) \\ &\quad + \sum_{\sigma(\neq\beta)=1}^{N} \frac{e^{ipR_{\alpha\beta}}}{R_{\alpha\beta}} \mathbf{f}^e(\alpha\beta, \beta\sigma) \Psi_{\beta\sigma}(\mathbf{z}_\beta),\end{aligned} \tag{300}$$

for $\beta \neq 0$, and

$$\Psi_{\alpha 0}(\mathbf{z}_\alpha) = (2\pi)^{-3/2} e^{i\mathbf{p}\cdot\mathbf{z}_\alpha} u_{S_i,\nu}(s_i). \tag{301}$$

Equations (300) may be simplified if we set

$$\Psi_{\alpha\beta}(\mathbf{z}_\alpha) \equiv (2\pi)^{-3/2} \frac{e^{ipR_{\alpha\beta}}}{R_{\alpha\beta}} Q_{\alpha\beta} u_{S_i,\nu}(s_i) \tag{302}$$

for $\beta = 1, 2, \ldots, N$. Substitution into (300) then gives

$$Q_{\alpha\beta} = \mathbf{f}^e(\alpha\beta, \beta 0) e^{i\mathbf{p}\cdot\mathbf{z}_\beta} + \sum_{\sigma(\neq\beta)=1}^{N} \mathbf{f}^e(\alpha\beta, \beta\sigma) \frac{e^{ipR_{\beta\sigma}}}{R_{\beta\sigma}} Q_{\beta\sigma}. \tag{303}$$

This is a set of *coupled algebraic* equations to determine the $Q_{\beta\sigma}$.

The asymptotic wave function (295), evaluated at a point $\mathbf{x}$ far from the scatterer, is now of the form

$$\begin{aligned}\psi_a{}^+ = \chi_a + (2\pi)^{-3} \frac{e^{ipx}}{x} &\Bigg[ \sum_{\alpha=1}^{N} e^{-i(p\hat{\mathbf{x}}-\mathbf{p})\cdot\mathbf{z}_\alpha} \mathbf{f}_\alpha{}^e(\hat{\mathbf{x}}, \hat{\mathbf{p}}, \mathbf{S}_i, \mathbf{S}_\alpha) \\ &+ \sum_{\alpha=1}^{N} e^{-ip\hat{\mathbf{x}}\cdot\mathbf{z}_\alpha} \sum_{\beta(\neq\alpha)=1}^{N} \mathbf{f}_\alpha{}^e(\hat{\mathbf{x}}, \hat{\mathbf{n}}_{\alpha\beta}, \mathbf{S}_i, \mathbf{S}_\alpha) \frac{e^{ipR_{\alpha\beta}}}{R_{\alpha\beta}} Q_{\alpha\beta} \Bigg] \\ &\times u_{S_i,\nu}(s_i) g_0(\mathbf{z}_1, s_1, \ldots, \mathbf{z}_N, s_N).\end{aligned} \tag{304}$$

[81] We shall neglect the recoil of the target center of mass here. If this is not justified, we may instead consider Eqs. (299), (304), and the others as defined in the barycentric subspace of the $(N + 1)$-particle system (Section 4.6).

From this we see that the scattering amplitude $\mathbf{f}^{(N)}$ for elastic and quasi-elastic scattering is

$$\mathbf{f}^{(N)} = \sum_{\alpha=1}^{N} \Bigg[ \mathbf{f}_\alpha{}^e(\hat{\mathbf{k}}, \hat{\mathbf{p}}, \mathbf{S}_i, \mathbf{S}_\alpha)\, e^{-i(\mathbf{k}-\mathbf{p})\cdot \mathbf{z}_\alpha} + \sum_{\beta(\neq\alpha)=1}^{N} e^{-i\mathbf{k}\cdot\mathbf{z}_\alpha}\, \mathbf{f}_\alpha{}^e(\hat{\mathbf{k}}, \hat{\mathbf{n}}_{\alpha\beta}, \mathbf{S}_i, \mathbf{S}_\alpha)\, \frac{e^{ipR_{\alpha\beta}}}{R_{\alpha\beta}}\, Q_{\alpha\beta} \Bigg], \tag{305}$$

where $\mathbf{k} \equiv p\hat{\mathbf{x}}$ is the final momentum of particle $i$.

The differential cross section for an unpolarized initial state is (if we consider the target mass $M_t$ to be sufficiently large)

$$\frac{d\bar{\sigma}}{d\Omega} = \frac{1}{(2S_i + 1)(2S_t + 1)} \sum_{\nu,\nu_t} (u_{S_i,\nu} g_0, \mathbf{f}^{(N)\dagger}\mathbf{f}^{(N)} u_{S_i,\nu} g_0). \tag{306}$$

For a polarized initial state, we must suitably average the spin orientations as described in Section 7.5. To determine the final polarization following scattering, we need to evaluate

$$\langle\Omega^\lambda\rangle_f \frac{d\bar{\sigma}}{d\Omega} = \frac{1}{(2S_i + 1)(2S_t + 1)} \sum_{\nu,\nu_t} (u_{S_i,\nu} g_0, \mathbf{f}^{(N)\dagger}\Omega^\lambda \mathbf{f}^{(N)} u_{S_i,\nu} g_0), \tag{307}$$

where the $\Omega^\lambda$ are defined by Eqs. (7.272).

When multiple scattering may be neglected, so the $Q_{\alpha\beta}$ may be set equal to zero in Eq. (305), the cross section (306) agrees precisely with the corresponding scattering cross section obtained in Section 11.2 for small systems [see Eqs. (163*a*), (170), and (193)]. Several applications of the general equations will be described presently.

It is frequently the case that a multiple scattering interaction includes some quasi-free and some quasi-elastic interactions. Our results are easily extended to include such mixed situations. The excitation energy (272) was defined to include only the quasi-free excitations. Since Eqs. (294) and (300) have the same structure, it is only necessary, at each scattering, to use a $\mathbf{z}_\beta{}^0$ or a $\mathbf{z}_\beta$ and an $\mathbf{f}$ or an $\mathbf{f}^e$, depending upon whether the momentum transfer $\boldsymbol{\rho}_{\alpha\beta}$ satisfies the quasi-free or quasi-elastic condition. We shall return to this more general problem in Section 11.5.

### Quasi-elastic Scattering from Systems Containing a Small Number of Scatterers

When the number of scatterers $N$ is small and when the quasi-elastic conditions obtain, the coupled algebraic equations (303) may be solved explicitly. The first application of these equations (in a somewhat less general form) was made by Foldy [29] to the scattering of sound waves by two fixed scatterers. Subsequently, Brueckner [34] applied these equations to a study of the scattering of pi-mesons by deuterons. For Brueckner's application the spin and isotopic spin dependence of the scattering amplitudes **f** requires that the $Q_{\alpha\beta}$ be matrices in both spin and isotopic spin space. For application to scattering by systems such as molecules in a gas or crystals it is usually necessary to perform a thermal average of Eq. (306), as was done in Eq. (146*b*). For example, in studying x-ray or neutron scattering by a crystal, the thermal average leads to Debye-Waller factors, as in Eq. (225).

The simplest nontrivial illustration of Eqs. (303) is that considered by Foldy [29] for a system consisting of two scatterers 1 and 2. In this example the scattering amplitudes $\mathbf{f}_1^e \equiv f_1$ and $\mathbf{f}_2^e \equiv f_2$ are taken to be independent of spin and scattering angle. We may then write Eq. (305) as (on taking $\mathbf{R}_{12} \equiv \mathbf{R}$)

$$\begin{aligned}\mathbf{f}^{(2)} \equiv f^{(2)} &= \sum_{\alpha=1}^{2} f_\alpha\, e^{-\mathbf{k}\cdot\mathbf{z}_\alpha}\left(e^{i\mathbf{p}\cdot\mathbf{z}_\alpha} + \sum_{\beta(\neq\alpha)=1}^{2} \frac{e^{ipR}}{R} Q_{\alpha\beta}\right)\\ &= Q_{21}\, e^{-i\mathbf{k}\cdot\mathbf{z}_1} + Q_{12}\, e^{-i\mathbf{k}\cdot\mathbf{z}_2},\end{aligned} \tag{308}$$

using Eqs. (303). The two Eqs. (303) are now

$$\begin{aligned}Q_{12} &= f_2\, e^{i\mathbf{p}\cdot\mathbf{z}_2} + f_2 \frac{e^{ipR}}{R} Q_{21},\\ Q_{21} &= f_1\, e^{i\mathbf{p}\cdot\mathbf{z}_1} + f_1 \frac{e^{ipR}}{R} Q_{12}.\end{aligned} \tag{309}$$

From these we obtain

$$Q_{12} = f_2\, e^{i\mathbf{p}\cdot\mathbf{z}_2}\, \frac{1 + f_1 e^{i\mathbf{p}\cdot\mathbf{R}}\,(e^{ipR}/R)}{1 - f_1 f_2 (e^{2ipR}/R^2)}, \tag{310}$$

with a similar expression for $Q_{21}$. Inserting these expressions into Eq. (308) finally leads to

$$\begin{aligned}f^{(2)} = \Big[&(f_1\, e^{-i\boldsymbol{\rho}\cdot\mathbf{z}_1} + f_2\, e^{-i\boldsymbol{\rho}\cdot\mathbf{z}_2})\\ &+ f_1 f_2 \frac{e^{ipR}}{R}\,(e^{i(\mathbf{p}\cdot\mathbf{z}_1 - \mathbf{k}\cdot\mathbf{z}_2)} + e^{i(\mathbf{p}\cdot\mathbf{z}_2 - \mathbf{k}\cdot\mathbf{z}_1)})\Big]\left(1 - f_1 f_2 \frac{e^{2ipR}}{R^2}\right)^{-1},\end{aligned} \tag{311}$$

where $\boldsymbol{\rho} \equiv \mathbf{k} - \mathbf{p}$.

The differential scattering cross section is obtained using Eq. (306). For fixed positions of the scatterers, this is just

$$\frac{d\bar{\sigma}}{d\Omega} = |f^{(2)}|^2. \tag{312}$$

The total cross section may be obtained from the optical theorem [Eq. (6.74*d*)] as

$$\bar{\sigma} = \frac{4\pi}{p}\,\mathrm{Im}\, f^{(2)},$$

where $f^{(2)}$ is evaluated for $\mathbf{k} = \mathbf{p}$. In the limit that $pR$ becomes very small, $\bar{\sigma}$ approaches the limiting form,

$$\lim_{R\to 0} \bar{\sigma} = 8\pi R^2, \tag{313}$$

which is *independent* of $f_1$ and $f_2$. When $f_1$ and $f_2$ are sufficiently small that multiple scattering may be neglected, we recover the results of Section 11.2.

When the scattering amplitudes $\mathbf{f}^e(\alpha\beta, \beta\sigma)$ are isotropic, so

$$\mathbf{f}^e(\alpha\beta, \beta\sigma) = \mathbf{f}^e(\beta),$$

a function of $\beta$ only, the quantities $Q_{\alpha\beta}$ in Eqs. (303) are seen to be independent of the index $\alpha$. In this case this set of equations simplifies to the form

$$Q_\beta = \mathbf{f}^e(\beta)\, e^{i\mathbf{p}\cdot\mathbf{z}_\beta} + \sum_{\sigma(\neq\beta)=1}^{N} \mathbf{f}^e(\beta) \frac{e^{ipR_{\beta\sigma}}}{R_{\beta\sigma}} Q_\sigma, \tag{314}$$

where we have set

$$Q_{\alpha\beta} \equiv Q_\beta.$$

The algebraic equations (314) have been studied by Foldy [35] and by Weinberg [36] for several kinds of scattering arrays. In particular, Weinberg [36] has discussed the intensity and width of the "major lobes," defined by the condition that

$$e^{-i(p\hat{\mathbf{x}}-\mathbf{p})\cdot\mathbf{z}_\alpha}$$

be independent of $\alpha$.

### The Refractive Index of a Gas

We have seen that Eqs. (300) or (303) are appropriate to describe multiple scattering when the momentum transfers $\rho_{\alpha\beta}$ [see Eq. (287)] are much smaller than $\sqrt{2M_\beta K_{\beta,\mathrm{av}}}$, the "spread" of momentum in the state $g_0$. When the number of scatterers $N$ is not small, approximations are usually required to solve these equations.[82] A simple illustration, which we now describe, is provided by a random distribution of $N$ scatterers.

By a "random distribution of scatterers" we mean that the particles in the scattering medium are uncorrelated in position. In this case the $N$-particle distribution function $\bar{\bar{P}}^{(N)}(\mathbf{x}_1, \ldots, \mathbf{x}_N)$ has the form [here a thermal average is implied, as in Eq. (128)]

$$\begin{aligned}\bar{\bar{P}}^{(N)}(\mathbf{x}_1, \ldots, \mathbf{x}_N) &= \sum_{\gamma(0)} e^{-W_{\gamma(0)}/\theta} \langle\gamma(0)|\, \delta(\mathbf{x}_1 - \mathbf{z}_1)\cdots\delta(\mathbf{x}_N - \mathbf{z}_N)\, |\gamma(0)\rangle \\ &= \prod_{\alpha=1}^{N} \bar{\bar{P}}^{(1)}(\mathbf{x}_\alpha).\end{aligned} \tag{315}$$

Such a system is, in fact, an ideal gas.[83]

[82] We shall not discuss here scattering by uniform arrays. Some discussion of this is given, for example, in Ref. [36].

[83] It follows from Eq. (315) that $\bar{G}(\mathbf{x}, \mathbf{x}') = 0$, where $\bar{G}$ is the pair correlation function introduced in Eq. (124*a*).

To be more specific, we consider the scattering system to consist of a very large number $N$ of molecules of a single species. These are distributed uniformly throughout a box of "macroscopic" volume $\mathscr{V}$. The density of molecules in the box is then

$$\rho \equiv \frac{N}{\mathscr{V}}. \tag{316}$$

Finally, the walls of the box are assumed to be transparent to the incident particles $i$.

We shall see that for the system being studied, and in the quasi-elastic limit, the scattered waves interfere destructively in all but the forward direction. This leads to a "refractive index" for the propagation of waves through the scattering medium.

Application of the quasi-elastic condition in the case of a gas requires some care. Since the molecules are free to recoil, it might be thought that the quasi-free equations (293) should always be used. To resolve this question, we note that if the recoil distance $|\mathbf{z}_\beta - \mathbf{z}_\beta{}^0|$ of a struck molecule $\beta$ is much less than the dimensions in which it can be localized within the box (without disturbing the initial distribution) there is no need to distinguish between Eqs. (293) and (303). Now, the appropriate recoil distance is $(p_{\alpha\beta}/M)\Delta t$, where $\Delta t$ is the time between gas-kinetic collisions (that is, once a molecule collides with a second molecule in the gas its "memory" of the collision with particle $i$ is lost). Therefore, if $v_{\text{th}}$ is the thermal speed of the molecules and if $\lambda_g$ is the gas-kinetic mean free path, $\Delta t = \lambda_g/v_{\text{th}}$ and the recoil distance is

$$|\mathbf{z}_\beta - \mathbf{z}_\beta{}^0| \cong \frac{p_{\alpha\beta}}{M}\frac{\lambda_g}{v_{th}}.$$

The width of a wave packet which may be constructed for the molecules is $\Delta W \cong \hbar/Mv_{\text{th}}$. Now, if

$$\frac{|\mathbf{z}_\beta - \mathbf{z}_\beta{}^0|}{\Delta W} \cong \frac{p_{\alpha\beta}\lambda_g}{\hbar} \ll 1, \tag{317}$$

we may use Eqs. (303) to describe the scattering.

The condition (317) may be reexpressed in the form which we have used previously for quasi-elastic conditions. The effective "binding interaction" of a typical molecule $\beta$ in the gas is of the order of

$$\delta K_{\beta,\text{av}} = \frac{1}{2M}\left(\frac{\hbar}{\lambda_g}\right)^2,$$

so the condition $p_{\alpha\beta} \ll \sqrt{2M\,\delta K_{\beta,\text{av}}}$ is equivalent to

$$p_{\alpha\beta} \ll \frac{\hbar}{\lambda_g},$$

which is just the condition (317).

The condition (317) is always satisfied if the scattering angle is sufficiently small. Since the scatterings which contribute to the refractive index will be found to correspond to very nearly zero angles of deflection, we conclude that Eqs. (303) may be applied with great generality to studies of the refractive index.

When the momentum $p$ of particle $i$ is large compared to $\hbar/\lambda_g$, the small recoil condition (317) will be violated for large scattering angles. Then there will be both quasi-free (or incoherent) scattering at large angles and refraction of the incident wave along the path of propagation of the *coherent wave*.[84]

Returning to Eqs. (299) and (300), we define

$$\Psi_\alpha(\mathbf{z}_\alpha) = \chi_{i;\mathbf{p},\nu}(\mathbf{z}_\alpha) + \sum_{\beta(\neq\alpha)=1}^{N} \Psi_{\alpha\beta}(\mathbf{z}_\alpha). \tag{318}$$

We seek a solution corresponding to coherent propagation along the direction $\hat{\mathbf{p}}$ of the incident wave.[85] We shall assume that all waves scattered in directions other than $\hat{\mathbf{p}}$ interfere destructively; the self-consistency of this assumption will be verified during the course of our discussion. Then we may set [see Eqs. (291) for a definition of the amplitudes $\mathbf{f}^e(\alpha\beta, \beta\sigma)$, which describe elastic scattering from the incident direction $\hat{\mathbf{n}}_{\beta\sigma}$ to the final direction $\hat{\mathbf{n}}_{\alpha\beta}$]

$$\sum_{\sigma(\neq\beta)=0}^{N} \mathbf{f}^e(\alpha\beta, \beta\sigma)\Psi_{\beta\sigma}(\mathbf{z}_\beta) = \mathbf{f}^e(\alpha\beta, \beta 0)\Psi_\beta(\mathbf{z}_\beta) \tag{319}$$

in Eq. (300). This equation may now be summed over $\beta$ ($\neq\alpha$) and added to Eq. (301) to give

$$\Psi_\alpha(\mathbf{z}_\alpha) = \chi_{i;\mathbf{p},\nu}(\mathbf{z}_\alpha) + \sum_{\beta(\neq\alpha)=1}^{N} \frac{e^{ipR_{\alpha\beta}}}{R_{\alpha\beta}} \mathbf{f}^e(\alpha\beta, \beta 0)\Psi_\beta(\mathbf{z}_\beta). \tag{320}$$

To solve Eq. (320) we assume that $\Psi_\alpha(\mathbf{z}_\alpha)$ has the form

$$\Psi_\alpha(\mathbf{z}_\alpha) = (2\pi)^{-3/2} \exp\left[i\int^{\mathbf{z}_\alpha} \mathbf{l}(\mathbf{y})\cdot d\mathbf{y}\right] Q u_{S_i,\nu}, \tag{321}$$

where $\mathbf{l} = l\hat{\mathbf{p}}$ and $Q$ is a constant matrix. The choice of the exponential in Eq. (321) is suggested by the eikonal approximation [see Eq. (6.499$b$)], the

[84] The blueness of the sky, which results from the incoherent scattering of light, is an example of this phenomenon.

[85] For simplicity, we take $\hat{\mathbf{p}}$ normal to the boundary of the gas container so refraction at the boundary does not occur. This restriction will be removed in later discussions.

lower limit being fixed by the phase of the incident wave on entering the medium. The ansatz (321) enables us to write (320) as

$$Q = \exp\left[i\int^{\mathbf{z}_\alpha}(\mathbf{p}-\mathbf{l})\cdot d\mathbf{y}\right] + \sum_{\beta(\neq\alpha)=1}^{N} \frac{e^{ipR_{\alpha\beta}}}{R_{\alpha\beta}} \exp\left(-i\int_{\mathbf{z}_\beta}^{\mathbf{z}_\alpha}\mathbf{l}\cdot d\mathbf{y}\right)\mathbf{f}^e(\alpha\beta, \beta 0)Q. \tag{322}$$

We shall now replace the sum over $\beta$ here by an integral over $\mathbf{z}_\beta$. This may be done strictly only when evaluating the thermal average of the cross section (306). Because of our assumptions about the randomness of the scatterers, (315) and (316), an integral over the volume $\mathscr{V}$ must be performed for each coordinate $\mathbf{z}_\alpha$. Since $N \gg 1$ we may, however, carry out this integral for each amplitude $\mathbf{f}^{(N)}$ independently, the error being of relative order $N^{-1}$. Then taking

$$\sum_\beta = \rho\int d^3z_\beta,$$

we obtain from (322)

$$Q = \exp\left[i\int^{\mathbf{z}_\alpha}(\mathbf{p}-\mathbf{l})\cdot d\mathbf{y}\right] + \rho\int d^3z_\beta \frac{e^{ipR_{\alpha\beta}}}{R_{\alpha\beta}} \times \exp\left(-i\int_{\mathbf{z}_\beta}^{\mathbf{z}_\alpha}\mathbf{l}\cdot d\mathbf{y}\right)\mathbf{f}^e(\hat{\mathbf{n}}_{\alpha\beta}, \hat{\mathbf{p}}, \mathbf{S}_i)Q. \tag{323}$$

In obtaining this result we have used the fact that the thermal average will average out any dependence on the molecular spin $\mathbf{S}_\beta$, which permits us to replace $\mathbf{f}^e(\alpha\beta, \beta 0)$ by $\mathbf{f}^e(\hat{\mathbf{n}}_{\alpha\beta}, \hat{\mathbf{p}}, \mathbf{S}_i)$.

A ray path for the incident wave is illustrated in Fig. 11.7. To the right of the boundary we have $\mathbf{l} = \mathbf{p}$. Within the gas $\mathbf{l}$ is constant since the

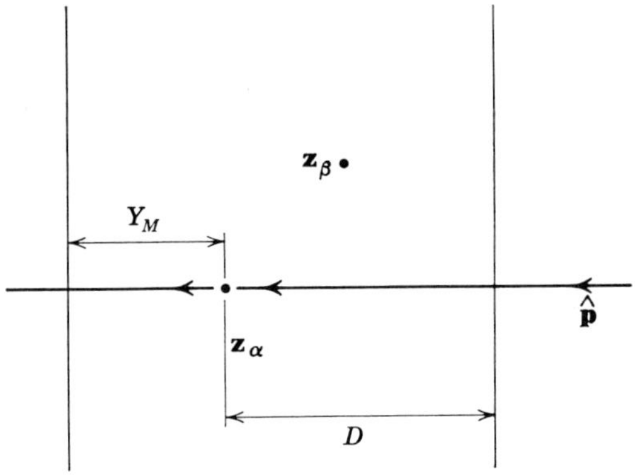


Fig. 11.7 Ray path for coherent scattering in a uniform medium.

medium has been assumed to be uniform. Therefore,

$$\int^{\mathbf{z}_\alpha} (\mathbf{p} - \mathbf{l}) \cdot d\mathbf{y} = (p - l)D, \tag{324}$$

where $D$ is the distance from $\mathbf{z}_\alpha$ to the boundary, measured in the direction $-\hat{\mathbf{p}}$. Similarly, we may write

$$\int_{\mathbf{z}_\beta}^{\mathbf{z}_\alpha} \mathbf{l} \cdot d\mathbf{y} = \mathbf{l} \cdot \mathbf{R}_{\alpha\beta}, \tag{325}$$

where $\mathbf{R}_{\alpha\beta} = \mathbf{z}_\alpha - \mathbf{z}_\beta$ as before.

The integrand in Eq. (323) is not correct for $R_{\alpha\beta} \lesssim 1/p$, since we are using the asymptotic form of the scattered wave, which requires that $pR_{\alpha\beta} \gg 1$.[86] Equation (323) will be correct then only if

$$\frac{\rho\,|f^e|}{p^2} \ll 1. \tag{326}$$

Assuming this to be the case, we may extend the integral in Eq. (323) over the entire volume of the scattering system.

On writing

$$\mathbf{Y} \equiv \mathbf{z}_\beta - \mathbf{z}_\alpha = -\mathbf{R}_{\alpha\beta},$$

the integral in Eq. (323) is

$$\int d^3Y \frac{e^{ipY}}{Y} e^{i\mathbf{l}\cdot\mathbf{Y}} \mathbf{f}^e(-\hat{\mathbf{Y}}, \hat{\mathbf{p}}, \mathbf{S}_i) \cong \frac{2\pi}{il} \left[ \frac{1}{i(p+l)} (e^{i(p+l)Y_M} - 1) \times \mathbf{f}^e(-\hat{\mathbf{p}}, \hat{\mathbf{p}}, \mathbf{S}_i) - \frac{1}{i(p-l)} (e^{i(p-l)D} - 1) \mathbf{f}^e(\hat{\mathbf{p}}, \hat{\mathbf{p}}, \mathbf{S}_i) \right]. \tag{327}$$

Here $Y_M$ is the distance from $\mathbf{z}_\alpha$ to the boundary of the medium in the direction $\hat{\mathbf{p}}$, as illustrated in Fig. 11.7. To obtain (327), we have assumed that $lY$ is a large number, consistent with the condition (326), and have used the asymptotic value for the integral over directions of $\hat{\mathbf{Y}}$.

The condition (326) will be seen [Eq. (332)] to imply that $l \cong p$, which in turn implies that the first term on the right-hand side of Eq. (327) may be neglected.[87] Furthermore, for a parity-conserving interaction, $\mathbf{f}^e(\hat{\mathbf{p}}, \hat{\mathbf{p}}, \mathbf{S}_i)$ is independent of $\mathbf{S}_i$. Thus we may set

$$\mathbf{f}^e(\hat{\mathbf{p}}, \hat{\mathbf{p}}, \mathbf{S}_i) \equiv f_+, \tag{328}$$

a complex number. Finally, it is customary to set

$$l = np, \tag{329}$$

[86] We shall take explicit account of corrections from small distances on p. 773.

[87] That is, $|p - l|^{-1} \gg |p + l|^{-1}$. We now see the basis for the assumption made in Eq. (320) that scattering only in the forward direction need be considered.

where $n$ is called the "refractive index." As we shall conclude presently, $n$ has a positive imaginary part.[88]

We may now insert the integral (327) into Eq. (323), taking account of the indicated approximations. Since $D$ depends on the variable $\mathbf{z}_\alpha$, to obtain a solution of Eq. (323) we must equate to zero the coefficient of $e^{i(p-l)D}$, which leads to the condition

$$1 - \frac{2\pi\rho f_+}{p^2 n(n-1)} Q = 0. \tag{330}$$

This means that the incident wave $\chi_{i;\mathbf{p},\nu}$ in Eq. (318) is *canceled* by *scattered waves*, a result known as the "extinction theorem" in optics. The remaining terms in Eq. (323) provide the equation

$$1 = \frac{2\pi\rho f_+}{p^2 n(n-1)}. \tag{331}$$

The solution of Eq. (331), taking account of the condition (326), is

$$n - 1 \cong \frac{2\pi\rho f_+}{p^2}. \tag{332}$$

This is a familiar expression; it may be derived by a variety of arguments. It is seen from the inequality (326) to be valid only for $|n-1| \ll 1$. On p. 775 and in Section 11.4 this restriction will be relaxed.

The expression (332) and the condition $|n-1| \ll 1$ permit us to solve Eq. (330) for $Q$ to obtain

$$Q \cong 1. \tag{333}$$

The wave function (321) therefore has the amplitude of the incident wave, except for attenuation resulting from the Im $n$. Finally, the condition that the scattering amplitude and projectile wavelength be small compared to the target particle separation, Eq. (326), implies that we need not distinguish between the wave functions $\psi_\alpha$ and the system wave function $\psi_a{}^+$ [see Eqs. (266)].

From the optical theorem [Eq. (6.74$d$)] we have $\bar{\sigma}_f = 4\pi/p \operatorname{Im}(f_+)$, where $\bar{\sigma}_f$ is the total cross section for scattering from a single molecule in the gas. This implies that the imaginary part of Eq. (332) is

$$\operatorname{Im}(n) = \frac{1}{2}\frac{\rho\bar{\sigma}_f}{p} = \frac{1}{2p\lambda}, \tag{334}$$

where $\lambda$ is the mean free path (130). From Eqs. (321) and (334) we see that the intensity of the coherent wave at a depth $D$ in the gas is

$$|\Psi'_\alpha(\mathbf{z}_\alpha)|^2 = (2\pi)^{-3}\, e^{-D/\lambda}. \tag{335}$$

[88] This implies that the neglected term $e^{i(p+l)Y_M}$ in Eq. (327) is exponentially small.

The coherent wave is therefore exponentially attenuated with a mean free path $\lambda$. The physical origin of this attenuation is the incoherent large angle scattering in the medium.

**The Refraction of Light**

We have just described the propagation of a particle through an ideal gas in the quasi-elastic limit. We shall now restrict our discussion to the propagation of light, but we shall generalize somewhat the model of the gas. The condition (326), which implies that the refractive index $n$ is nearly unity, will be relaxed. Instead of the assumption of complete randomness in the positions of the molecules, Eq. (315), we suppose that each of the $N$ identical molecules is an impenetrable sphere of diameter $d$. Thus no two molecules may have their centers a distance apart less than $d$. Aside from this restriction, the molecules will be assumed to be distributed at random in the volume $\mathscr{V}$. The walls of the container holding the $N$ molecules are supposed to be transparent to electromagnetic radiation and are not restricted in shape (except that the radius of curvature of the boundary at each point must be large compared to the wavelength of the radiation). As in the preceding section, we consider $N$ to be a large number and the mean molecular density $\rho$ to be a constant within the gas [see Eq. (316)].

The box containing the gas is placed in the path of a plane electromagnetic wave having wave number $\mathbf{p}/\hbar$. We assume that not only is the wavelength $h/p$ large compared to $d$, but that the energy $pc$ of a single photon is significantly less than the lowest electronic excitation energy of the molecules. It is evidently to be anticipated that the propagation of the radiation through the gas will be described by a refractive index.[89]

The wave function $\psi$ describing the interaction of the electromagnetic wave with the medium may be used to evaluate the expectation value of the operator $\mathscr{E}(\mathbf{x}) = 1/c[\partial \mathbf{A}(\mathbf{x})/\partial t] = -p\mathbf{A}(\mathbf{x})$, where $\mathbf{A}(\mathbf{x})$ is the operator associated with the vector potential of the radiation field [see Eq. (8.170)]. When the gas is in thermal equilibrium at a temperature $\theta$, this expectation value is [see, for example, Eq. (146$b$)]

$$\bar{\bar{\mathbf{E}}}(\mathbf{x}) = \frac{1}{\mathscr{Z}} \sum_{\gamma(0)} e^{-W_{\gamma(0)}/\theta}(\psi, \mathscr{E}(\mathbf{x})\psi), \tag{336}$$

where $\gamma(0)$ is a given "initial state" of the gas.

The wave function $\psi$ consists of a sum of terms describing a state with no photons, a state with one photon, a state with two photons, and so on.

[89] Our description may also be applied, with somewhat less accuracy, to the propagation of light in a liquid. For a detailed atomistic discussion of the refraction and dispersion of light see Ref. [37].

For a weak electromagnetic wave it suffices to consider only the zero and one photon states.[90] Thus we take

$$\psi = \psi_{\text{vac}} + \psi_a^+. \tag{337}$$

Here $\psi_{\text{vac}}$ describes just the gas in a state $\gamma(0)$ and no photons. The wave function $\psi_a^+$ describes the interaction of a single photon with the gas when it is in an initial state $\gamma(0)$. This wave function is determined by the multiple scattering equations (266), where $\chi_a$ is the single photon incident state.

Since the only nonvanishing matrix elements of $\mathscr{E}(\mathbf{x})$ increase or decrease the number of photons by unity, substitution of the expression (337) into Eq. (336) gives

$$\bar{\bar{\mathbf{E}}}(\mathbf{x}) = \frac{2}{\mathscr{L}} \sum_{\gamma(0)} e^{-W_{\gamma(0)}/\theta} \operatorname{Re}(\psi_{\text{vac}}, \mathscr{E}(\mathbf{x})\psi_a^+). \tag{338}$$

Equation (299) suggests that to solve the set of Eqs. (266*b*), we define a set of (complex) quantities

$$\bar{\bar{\mathbf{P}}}^{(1)}(\mathbf{y}_\alpha)\mathbf{E}_\alpha(\mathbf{y}_\alpha) \equiv \frac{2}{\mathscr{L}} \sum_{\gamma(0)} e^{-W_{\gamma(0)}/\theta}(\psi_{\text{vac}}, \delta(\mathbf{y}_\alpha - \mathbf{z}_\alpha)\mathscr{E}(\mathbf{y}_\alpha)\psi_a), \tag{339}$$

where $\bar{\bar{P}}^{(1)}(\mathbf{y}_\alpha)$ represents the probability of finding molecule $\alpha$ within a unit volume at the position $\mathbf{y}_\alpha$. The real part of $\mathbf{E}_\alpha(\mathbf{y}_\alpha)$ is the microscopic electric field at the position of molecule $\alpha$ resulting from interaction of the radiation with the other molecules. In using Eq. (339), we assume that the molecules are sufficiently massive that their recoil may be neglected.[91]

From the structure of the general multiple scattering equations (266*b*), we see that $\mathbf{E}_\alpha(\mathbf{y}_\alpha)$ will have the form

$$\mathbf{E}_\alpha(\mathbf{y}_\alpha) = \mathbf{E}_0(\mathbf{y}_\alpha) + \sum_{\beta(\neq\alpha)=1}^{N} \mathbf{E}_{\text{sc},\beta}(\mathbf{y}_\alpha), \tag{340}$$

where

$$\mathbf{E}_0(\mathbf{y}_\alpha) = \frac{2}{\mathscr{L}} \sum_{\gamma(0)} e^{-W_{\gamma(0)}/\theta}(\psi_{\text{vac}}, \mathscr{E}(\mathbf{y}_\alpha)\chi_a) \tag{341}$$

is the (complex) incident electric field vector at $\mathbf{y}_\alpha$ and $\mathbf{E}_{\text{sc},\beta}(\mathbf{y}_\alpha)$ is the electric field at $\mathbf{y}_\alpha$ due to scattering by molecule $\beta$.

Since we have assumed that the wavelength $h/p$ is large compared with the molecular diameter $d$, we may use the classical scattering formula[92] for $\mathbf{E}_{\text{sc},\beta}$. This is

$$\mathbf{E}_{\text{sc},\beta}(\mathbf{y}_\alpha) = \nabla_\alpha \times \left[\frac{\mathbf{P}_\beta \times \mathbf{R}_{\alpha\beta}}{R_{\alpha\beta}^2}\left(\frac{1}{R_{\alpha\beta}} - ip\right)e^{ipR_{\alpha\beta}}\right], \tag{342}$$

[90] Our results would not be modified by including larger numbers of photons as long as the number is small enough that nonlinear dependence on the field intensity may be neglected.

[91] This is the quasi-elastic approximation. The molecular coordinates $\mathbf{z}_\alpha$ then appear as parameters in the scattering equations, just as in Eqs. (300).

[92] See, for example, Panofsky and Phillips [38, p. 222].

where $\mathbf{P}_\beta$ is the dipole moment of molecule $\beta$ induced by the incident field $\mathbf{E}_\beta(\mathbf{y}_\beta)$, $\mathbf{R}_{\alpha\beta} \equiv \mathbf{y}_\alpha - \mathbf{y}_\beta$, and $\nabla_\alpha$ is the gradient operator acting on $\mathbf{y}_\alpha$. If the molecule has a polarizability $\bar{\bar{\alpha}}$, we have[93]

$$\mathbf{P}_\beta = \bar{\bar{\alpha}}\mathbf{E}_\beta(\mathbf{y}_\beta). \tag{343}$$

The sum over $\beta$ in Eq. (340) may be replaced by an integral, as in Eq. (323), to give

$$\mathbf{E}_\alpha(\mathbf{y}_\alpha) = \mathbf{E}_0(\mathbf{y}_\alpha) + \rho\int_{S_0} d^3y_\beta \mathbf{E}_{\mathrm{sc},\beta}(\mathbf{y}_\alpha). \tag{344}$$

Here the integral extends over the entire volume $\mathscr{V}$, except for the exclusion at $\mathbf{y}_\alpha$ of a sphere $S_0$ of radius $d$. It is because of the assumption that the molecules are impenetrable spheres of radius $d$ that the spherical volume $S_0$ must be excluded in (344). Equation (344) differs in two respects from the corresponding equation for the refractive index of a particle in a gas, Eq. (323). First, we are excluding the sphere $S_0$ from the domain of integration in Eq. (344). Second, we are now using the complete scattered wave (342) rather than just its asymptotic form, as in Eq. (323).

Equation (344), with the expressions (342) and (343), represents an integral equation for the field $\mathbf{E}_\alpha$. It is to be solved subject to the appropriate boundary conditions at the surface of $\mathscr{V}$. To do this we follow the conventional arguments of macroscopic electrodynamics, defining the *phenomenological field* $\mathbf{E}(\mathbf{x})$ as

$$\mathbf{E}(\mathbf{x}) = \mathbf{E}_0(\mathbf{x}) + \rho\int_n d^3y_\beta \mathbf{E}_{\mathrm{sc},\beta}(\mathbf{x}). \tag{345}$$

Here the integral extends over the volume $\mathscr{V}$, excluding now a *very small* needle-shaped region in which $\mathbf{x}$ lies, the axis of the needle being parallel to $\mathbf{E}(\mathbf{x})$.[94]

Now, using (345), we see that

$$\rho\int_{S_0} d^3y_\beta \mathbf{E}_{\mathrm{sc},\beta}(\mathbf{y}_\alpha) = \mathbf{E}(\mathbf{y}_\alpha) - \mathbf{E}_0(\mathbf{y}_\alpha) - \rho\int_{S_0-n} d^3y_\beta \mathbf{E}_{\mathrm{sc},\beta}(\mathbf{y}_\alpha), \tag{346}$$

where the integral on the right extends over the volume contained between the *very small* needle surface and the surface of the sphere of radius $d$. To evaluate this integral, we recall the assumption that $1/p \gg d$. This permits us to neglect $ip$ as compared with $1/R_{\alpha\beta}$ and to set $e^{ipR_{\alpha\beta}} = 1$ in the formula for $\mathbf{E}_{\mathrm{sc},\beta}(\mathbf{y}_\alpha)$, Eq. (342). We may also consider $\mathbf{P}_\beta = \bar{\bar{\alpha}}\mathbf{E}_\beta(\mathbf{y}_\beta)$ to be independent of $\mathbf{y}_\beta$ within the small sphere, replacing $\mathbf{P}_\beta$ by $\mathbf{P}_\alpha$. Then the expression (342) may be written, for small $R_{\alpha\beta}$, as

$$\mathbf{E}_{\mathrm{sc},\beta}(\mathbf{y}_\alpha) = \nabla_\beta \times \left(\frac{\mathbf{P}_\alpha \times \mathbf{R}_{\alpha\beta}}{R_{\alpha\beta}{}^3}\right) = \nabla_\beta\left(\frac{\mathbf{P}_\alpha \cdot \mathbf{R}_{\alpha\beta}}{R_{\alpha\beta}{}^3}\right).$$

[93] Our analysis can, of course, be carried through for a tensor polarizability.

[94] This provides the standard definition of $\mathbf{E}$. See Brown [39] for a more complete discussion.

The volume integral on the right side of Eq. (346) may therefore be written as a surface integral. This vanishes on the needle-shaped surface, so we obtain

$$\int_{S_0-n} \mathbf{E}_{\text{sc},\beta}(\mathbf{y}_\alpha)\, d^3y_\beta = \int_{S_0-n} d\mathbf{\Sigma}\left[\frac{\mathbf{R}_{\alpha\beta}\cdot\mathbf{P}_\alpha}{R_{\alpha\beta}{}^3}\right] = -\frac{4\pi}{3}\mathbf{P}_\alpha, \tag{347}$$

where $d\mathbf{\Sigma}$ is a surface element.

Now, the dipole moment per unit volume at $\mathbf{y}_\alpha$ is $\mathbf{P}(\mathbf{y}_\alpha) = \rho\mathbf{P}_\alpha = \rho\bar{\bar{\alpha}}\mathbf{E}_\alpha(\mathbf{y}_\alpha)$. Using the integral equation for $\mathbf{E}_\alpha(\mathbf{y}_\alpha)$, Eq. (344), and the evaluation of the integral over the fields emanating from scatterers other than $\alpha$, provided by Eqs. (346) and (347), we find that

$$\mathbf{P}(\mathbf{y}_\alpha) = \rho\bar{\bar{\alpha}}\mathbf{E}_0(\mathbf{y}_\alpha) + \rho\bar{\bar{\alpha}}\left[\mathbf{E}_\alpha(\mathbf{y}_\alpha) - \mathbf{E}_0(\mathbf{y}_\alpha) + \frac{4\pi}{3}\mathbf{P}(\mathbf{y}_\alpha)\right]. \tag{348}$$

The terms involving the incident field $\mathbf{E}_0$ are seen to cancel here (the extinction theorem again!).

To complete our solution, we set

$$\mathbf{P}(\mathbf{x}) = \frac{1}{4\pi}(n^2 - 1)\mathbf{E}(\mathbf{x}),$$

where $n^2$ is the dielectric constant, to obtain from Eq. (348)

$$\frac{n^2 - 1}{n^2 + 2} = \frac{4\pi}{3}\rho\bar{\bar{\alpha}}. \tag{349}$$

This is the celebrated Lorentz-Lorentz formula for the refractive index $n$ of light.[95]

In the limit of large wavelengths the forward scattering amplitude for light is

$$f_+ = p^2\bar{\bar{\alpha}}.$$

In this limit, and when $\rho\bar{\bar{\alpha}} \ll 1$, Eqs. (332) and (349) agree. To correct Eq. (349) for a real gas or liquid, we must take explicit account of molecular interactions.[96]

**Multiple Scattering of an Identical Particle**

We have assumed so far in this chapter that the incident particle $i$ is distinguishable from each of the particles in the bound target system. When this is not the case, and $i$ is dynamically equivalent to one or more of the target particles, we must symmetrize the wave function $\psi_a{}^+$ to give the

[95] See Ref. [37] for a review of other derivations of this expression.

[96] This is done, as a generalization of the methods described here, by Kaufman and Watson [40].

wave function $\psi^{+}_{\mathscr{S}(a)}$, as described in Chapter 4. This causes no difficulty in principle, but it often complicates the application of specific approximations used in scattering problems.

Such a complication appears to be encountered in writing $\psi^{+}_{\mathscr{S}(a)}$ in the multiple scattering form of Eqs. (253) or (266). Indeed, such equations have not been obtained in the general case. However, in the weak binding limit it is possible to express $\psi^{+}_{\mathscr{S}(a)}$ in the multiple scattering form (266). To do this, as we shall presently demonstrate, it is only necessary to use the appropriately symmetrized two-body scattering matrices $\mathscr{T}_\alpha$ whenever particles $i$ and $\alpha$ are identical.[97]

To restate this useful result, Eqs. (266) provide an expression for $\psi^{+}_{\mathscr{S}(a)}$ if only the two-body operators $\mathscr{T}_\alpha$ are properly symmetrized in the weak binding limit, and when the condition (132) of mean free path being large compared to projectile wavelength is satisfied. This conclusion is not surprising, since it is only at the time that a pair of particles are interacting that we anticipate having to symmetrize their wave function. To demonstrate the basis for these remarks, we shall assume (for simplicity) that the $N$ bound target particles and the incident beam particle $i$ represent a system of $(N + 1)$ identical particles. The incident particle is furthermore sufficiently energetic that the weak binding conditions (106) obtain. Examples of such interactions include the scattering of fast electrons by atoms, nucleons by atomic nuclei, and helium atoms by liquid helium.

From Eq. (4.125) we see that the complete symmetrized wave function has the form

$$\psi^{+}_{\mathscr{S}(a)} = C\mathscr{S}\psi_a{}^{+}, \tag{350a}$$

where $\psi_a{}^{+}$ is the unsymmetrized wave function with particle $i$ treated as distinguishable from the other particles, $C$ is a normalized constant, and $\mathscr{S}$ is the symmetrization operator. Using Eqs. (4.98), (4.99), and (4.131), we may write this as

$$\psi^{+}_{\mathscr{S}(a)} = \frac{1}{\sqrt{N+1}} \sum_{\alpha=0}^{N} \delta_\alpha \psi^{+}_{a(\alpha)}, \tag{350b}$$

where $\psi^{+}_{a(\alpha)}$ is the unsymmetrized wave function with particle $\alpha$ ($\alpha = 0, 1, 2, \ldots, N$) taken to be a distinguishable incident particle, and $\delta_\alpha$ takes on the value $+1$ for $\alpha = 0$ and $-1$ for $\alpha = 1, \ldots, N$ with Fermi-Dirac statistics, whereas $\delta_\alpha = 1$ ($\alpha = 0, 1, \ldots, N$) for Bose-Einstein statistics.

Now,

$$\psi^{+}_{a(\alpha)} = \chi_{a(\alpha)} + \frac{1}{E_a + i\eta - H} V_\alpha \chi_{a(\alpha)}, \tag{351}$$

[97] This was demonstrated by Takeda and Watson [41]. Their argument, which is restricted to the weak binding limit, seems to be the only one available. It is given later. See also the general discussion of Feshbach in the second of references [48].

where $\chi_{a(\alpha)}$ represents the incident state (100), with particle $\alpha$ as the incident particle, and $V_\alpha$ represents the interaction of particle $\alpha$ with the remaining $N$ particles. Equation (351) permits us to write (350*b*) in the form

$$\psi^+_{\mathscr{S}(a)} = \chi_{\mathscr{S}(a)} + \frac{1}{E_a + i\eta - H}\,|V\rangle. \tag{352}$$

Here $H$ is the complete Hamiltonian for the $(N + 1)$ particles,

$$|V\rangle \equiv \sum_{\alpha=0}^{N} \frac{\delta_\alpha}{\sqrt{N+1}}\, V_\alpha \chi_{a(\alpha)}, \tag{353}$$

and

$$\chi_{\mathscr{S}(a)} = \sum_{\alpha=0}^{N} \frac{\delta_\alpha}{\sqrt{N+1}}\, \chi_{a(\alpha)}, \tag{354}$$

as in Eq. (4.98).

Our objective is to express the symmetrized wave function (352) in multiple scattering form, always treating that particle with the label $\alpha = 0$ as the scattered particle. Since all $(N + 1)$ particles are, in fact, indistinguishable, we must agree on a convention for assigning the label $\alpha = 0$ following each scattering. Although other conventions might be used, we shall find it most convenient to call that particle with the highest energy, as it emerges from each scattering in the multiple scattering sequence, the particle 0.

We shall find that to obtain equations of the form (266) for $\psi^+_{\mathscr{S}(a)}$, we must omit certain kinds of interactions. These are called "target exchange scattering" and are illustrated by the following event. We suppose that a fast incident particle $\beta$ strikes a bound particle $\alpha$. Particles $\alpha$ and 0 recoil, leaving $\beta$ bound in place of 0. Because the particles are identical, this final state cannot be distinguished from that in which particles $\alpha$ and $\beta$ emerge from the scattering. The term "target exchange" is used, since particle $\beta$ has exchanged places with the target particle 0.

It is intuitively clear that target exchange scattering will tend to be small in the weak binding limit. Since particle 0 does not scatter with either $\beta$ or $\alpha$, it can recoil only because of the nonvanishing overlap of the bound-state wave function with that of the plane incident wave. At sufficiently high incident energies this overlap becomes arbitrarily small and the target exchange terms may be neglected.

To put the Schrödinger equation for $\psi^+_{\mathscr{S}(a)}$, Eq. (352), into a more useful form, we define

$$K^{(0)} \equiv H - V_0,$$

$$d(0) \equiv E_a + i\eta - K^{(0)}, \tag{355}$$

recalling that $V_0$ represents the interaction of particle 0 with the remaining particles. Now, we may use the relation

$$\frac{1}{E_a + i\eta - H} = \frac{1}{d(0)} + \frac{1}{d(0)} V_0 \frac{1}{E_a + i\eta - H}$$

to reexpress (352) as

$$\psi^+_{\mathscr{S}(a)} = \chi_{\mathscr{S}(a)} + \frac{1}{d(0)} [|V\rangle - V_0\chi_{\mathscr{S}(a)}] + \frac{1}{d(0)} V_0\psi^+_{\mathscr{S}(a)}. \tag{356}$$

This is an integral equation for the symmetrized wave function $\psi^+_{\mathscr{S}(a)}$ in which particle 0 is treated as the free particle between scatterings.

Let us first study the second term on the right-hand side of Eq. (356). To do this, we write

$$V_\alpha = \sum_{\beta(\neq\alpha)=0}^{N} V_{\alpha\beta},$$

where $V_{\alpha\beta}$ describes the interaction between particles $\alpha$ and $\beta$. This permits us to write, using Eqs. (353) and (354),

$$(|V\rangle - V_0\chi_{\mathscr{S}(a)}) = \sum_{\alpha=1}^{N} \frac{\delta_\alpha}{\sqrt{N+1}} \sum_{\beta(\neq\alpha)=1}^{N} (V_{\alpha\beta} - V_{0\beta})\chi_{a(\alpha)}. \tag{357}$$

The propagator $d(0)^{-1}$ in Eq. (356) implies that particle 0 recoils as a free particle following the interaction described by Eq. (357). We see therefore that this term is nonvanishing only because of target exchange terms. At sufficiently high energies this may then be neglected.[98]

To simplify the third term on the right side of Eq. (356), we introduce a complete set of coordinates $\xi_\alpha$ for particle $\alpha$ ($\alpha = 0, 1, 2, \ldots, N$) and write

$$V_0\psi^+_{\mathscr{S}(a)} = \sum_{\beta=1}^{N} \sum_{\xi'_0,\xi'_\beta} \langle\xi_0\xi_\beta|\, V_{0\beta}\, |\xi'_0\xi'_\beta\rangle \psi^+_{\mathscr{S}(a)}(\xi'_0\xi_1, \ldots, \xi'_\beta, \ldots, \xi_N). \tag{358}$$

Because $\psi^+_{\mathscr{S}(a)}$ is symmetrized in the coordinates $(\xi_0, \ldots, \xi_N)$, we may rewrite (358) as

$$V_0\psi^+_{\mathscr{S}(a)} = \bar{V}_0\psi^+_{\mathscr{S}(a)}, \tag{359}$$

where

$$\bar{V}_0 = \sum_{\beta=1}^{N} \tfrac{1}{2}[\langle\xi_0\xi_\beta|\, V_{0\beta}\, |\xi'_0\xi'_\beta\rangle + \delta_\beta \langle\xi_0\xi_\beta|\, V_{0\beta}\, |\xi'_\beta\xi'_0\rangle]. \tag{360}$$

Neglecting the expression (357), we may now write our approximate Schrödinger equation for $\psi^+_{\mathscr{S}(a)}$, Eq. (356), in the form

$$\psi^+_{\mathscr{S}(a)} = \chi_{\mathscr{S}(a)} + \frac{1}{d(0)} \bar{V}_0\psi^+_{\mathscr{S}(a)}, \tag{361}$$

[98] It should be recalled that particle 0 is, by definition, the particle with the highest energy following each scattering.

where the *symmetrized matrix elements* of the interaction $V_0$ are used at each scattering.

Finally, if we imagine that an order has been defined for the variables $\xi_0, \ldots, \xi_N$, so that sums over the $\xi$'s may be restricted to the domain $\xi_0 > \xi_1 > \xi_2 > \cdots > \xi_N$, we may drop the factor of $\frac{1}{2}$ in the definition (360) of $\bar{V}_0$. This ordering and the assumption that the high-energy incident particle is always chosen to be 0 permit us to replace $\chi_{\mathscr{S}(a)}$ by $\chi_{a(0)}$ in Eq. (361), the neglected terms corresponding to target exchange contributions. Equation (361) then becomes

$$\psi^+_{\mathscr{S}(a)} = \chi_{a(0)} + \frac{1}{d(0)} \bar{V}_0 \psi^+_{\mathscr{S}(a)}. \tag{362}$$

Except for the use of the symmetrized interaction $\bar{V}_0$, this equation is identical with Eq. (250). We may therefore apply *all results* of this chapter to scattering of identical particles if symmetrized two-body scattering matrices $\mathscr{T}_\alpha$ are used[99] and if the target exchange terms may be neglected.

To estimate the size of the neglected target exchange terms, we consider an initial state containing particles 0 and 1 bound and a fast incident particle 2 having momentum $\mathbf{p}$. In the final state, 2 is bound and 0 and 1 recoil with momenta $\mathbf{k}_0$ and $\mathbf{k}_1$.

Defining

$$\begin{aligned} \mathbf{r} &\equiv \mathbf{z}_0 - \mathbf{z}_1, \\ \bar{\mathbf{z}} &\equiv \tfrac{1}{2}(\mathbf{z}_0 + \mathbf{z}_1), \\ \mathbf{K} &\equiv \mathbf{k}_0 + \mathbf{k}_1, \\ \mathbf{k} &\equiv \tfrac{1}{2}(\mathbf{k}_0 - \mathbf{k}_1), \end{aligned}$$

the target exchange scattering matrix is, in first Born approximation,

$$T_x = (2\pi)^{-9/2} \int e^{-\mathbf{K}\cdot\bar{\mathbf{z}}}\, e^{-i\mathbf{k}\cdot\mathbf{r}} g_2^*(\mathbf{z}_2) V_{12}(r)\, e^{i\mathbf{p}\cdot\mathbf{z}_2} g_{01}(\bar{\mathbf{z}}, \mathbf{r})\, d^3\bar{\mathbf{z}}\, d^3r\, d^3z_2. \tag{363}$$

The corresponding direct process, where particle 0 strikes a bound particle 1 and both recoil, has the scattering matrix

$$T_d = (2\pi)^{-9/2} \int e^{-i\mathbf{K}\cdot\bar{\mathbf{z}}} e^{-i\mathbf{k}\cdot\mathbf{r}} V_{12}(r) g_1(\mathbf{z}_1)\, e^{i\mathbf{p}\cdot\mathbf{z}_0}\, d^3\bar{\mathbf{z}}\, d^3r. \tag{364}$$

If the bound particles are constrained to a volume $\mathscr{V} = (4\pi/3)\mathscr{R}^3$, and if characteristic momenta in the bound state are of order $\hbar/\mathscr{R}$, we deduce from Eqs. (363) and (364) that

$$\left|\frac{T_x}{T_d}\right| \simeq \frac{\hbar^3}{(p\mathscr{R})^3}. \tag{365a}$$

[99] We shall not use a separate symbol for a symmetrized two-particle matrix $\mathscr{T}_\alpha$. It will be understood that if particle $i$ and $\alpha$ are identical, the scattering matrix is appropriately symmetrized.

For nonrelativistic particles, we see from Eq. (106*a*) that

$$\left|\frac{T_x}{T_d}\right| \cong \left(\frac{K_{\alpha,\mathrm{av}}}{\epsilon_p}\right)^{3/2}. \tag{365b}$$

For relativistic particles, this ratio is

$$\left|\frac{T_x}{T_d}\right| \cong \left(\frac{\sqrt{2Mc^2K_{\alpha,\mathrm{av}}}}{\epsilon_p}\right)^{3}. \tag{365c}$$

In either case, the weak binding condition implies that the target exchange terms are small.[100]

## 11.4 USE OF A PSEUDOPOTENTIAL TO DESCRIBE ELASTIC SCATTERING

In this section we consider in some detail the *elastic scattering* of an incident particle $i$, having rest-mass $m$ and spin $S_i$, by the $N$-particle scatterer described in Section 11.2, p. 708. In contrast to the study made in Section 11.2, p. 717, we do not now assume that the mean free path in the scattering medium is necessarily large compared to the size of the scatterer.

We first introduce the notion of a "pseudopotential"[101] $\mathscr{V}_c$, from which the elastic scattering may be calculated. Such potentials often permit us to describe in a simple and intuitively appealing form the interaction between quite complex physical systems. One of the better known examples is provided by the introduction of a pseudopotential to represent the force between pairs of low-energy nucleons. Here the degrees of freedom associated with interacting fields are suppressed by expressing the net effect of the interaction in terms of a potential. Other examples include the use of a pseudopotential to describe the interaction of a particle $i$ with an extended system. In such cases the pseudopotential is often called an "optical model potential."

Throughout our work we have emphasized that the term "elementary particle" is to be applied as a matter of convenience, particularly when we are studying elastic scattering for which internal degrees of freedom are not excited. For the description of scattering by such particles as molecules, atoms, and atomic nuclei, whose properties are more or less well understood on the basis of a composite structure, it is somewhat uneconomical in postulates to insist on treating these as elementary particles. Instead, it

[100] An alternate formulation of scattering by identical particles is given in Ref. [42].

[101] The word "pseudopotential" has a special connotation in some of the low-energy nuclear physics literature. There it is used in conjunction with a mathematical description of an interaction such that the phase shift $\delta$ is given exactly by $k \cot \delta = a$, where $k$ is the wave number and $a$ the scattering length.

is customary to attempt to predict the scattering from such particles by considering them to be "composite." It is in this spirit, then, that we shall attempt to relate the properties of the pseudopotential $\mathscr{V}_c$ to the structure of the $N$-particle scattering system.

**General Properties of the Pseudopotential**

We again consider scattering from an initial state whose wave function is given by the product of a plane-wave state for the projectile and the ground-state wave function for the target, Eq. (100). Since we are now restricting our attention to elastic scattering, the final states will correspond to the internal energy of the target unchanged, but with the final spin orientation $\nu'_t$ in general different from $\nu_t$, the initial spin orientation [see Eq. (101)]. The scattering matrix $\mathscr{T}_c$ for elastic scattering is a submatrix of $\mathscr{T}_N$, the complete scattering matrix [see Eq. (257)]. Thus

$$\begin{aligned}\langle \mathbf{k}, \nu', \mathbf{Q}, \nu'_t | \mathscr{T}_c | \mathbf{p}, \nu, \mathbf{P}, \nu_t \rangle &\equiv \langle \mathbf{k}, \nu', \mathbf{Q}, 0(\nu'_t) | \mathscr{T}_N | \mathbf{p}, \nu, \mathbf{P}, 0(\nu_t) \rangle \\ &= \delta(\mathbf{k} + \mathbf{Q} - \mathbf{p} - \mathbf{P})(\mathbf{k}, \nu', \mathbf{Q}, \nu'_t | T_c | \mathbf{p}, \nu, \mathbf{P}, \nu_t), \quad (366)\end{aligned}$$

where $T_c$ is defined only on the momentum shell. Since the incident particle and target have respective spins $S_i$ and $S_t$, the matrix $T_c$ is a $(2S_i + 1)(2S_t + 1) \times (2S_i + 1)(2S_t + 1)$ matrix in spin space.[102] It may therefore be written in the form

$$T_c = T_c(\boldsymbol{\varkappa}_f, \boldsymbol{\varkappa}, \mathbf{S}_i, \mathbf{S}_t), \quad (367)$$

where $\boldsymbol{\varkappa}$ and $\boldsymbol{\varkappa}_f$ are the initial and final momenta in the barycentric coordinate system.[103] The differential cross section for elastic scattering is expressed as

$$\begin{aligned}\frac{d\bar{\sigma}_{\text{el}}}{d\Omega} &= \frac{(2\pi)^4}{v_{\text{rel}}} \int k^2 \, dk \, \delta\left(\epsilon_k + \frac{Q^2}{2M_t} - \epsilon_p - \frac{P^2}{2M_t}\right) \\ &\quad \times \frac{1}{(2S_i + 1)(2S_t + 1)} \sum_{\substack{\nu,\nu' \\ \nu_t,\nu'_t}} |T_c|^2,\end{aligned} \quad (368)$$

where $\mathbf{Q} = \mathbf{P} - \boldsymbol{\rho}$, $\boldsymbol{\rho} = \mathbf{k} - \mathbf{p}$ and the remaining notation is that of Eq. (148).

It is evident from Eqs. (366) and (367) that the elastic scattering is formally equivalent to a simple two-particle scattering involving only elementary particles. This is, of course, a consequence of the condition that internal degrees of freedom not be excited. It is natural, then, to seek

[102] We are assuming no other degeneracies than those associated with spin orientation.

[103] It should be recalled that we have assumed the target system to be sufficiently massive that the transformation from the laboratory frame to the barycentric frame is nonrelativistic even when particle $i$ has a relativistic energy.

a "potential" $\mathscr{V}_c$ from which $\mathscr{T}_c$ may be found. That is, we ask that $\mathscr{T}_c$ satisfy the equation

$$\mathscr{T}_c = \mathscr{V}_c + \mathscr{V}_c \frac{1}{e} \mathscr{T}_c, \tag{369a}$$

where $e^{-1}$ is just a two-body propagator.

$$e = \epsilon_P + \frac{P^2}{2M_t} + i\eta - \epsilon_{k'} - \frac{(Q')^2}{2M_t} \tag{370}$$

with $\mathbf{k}'$ and $\mathbf{Q}'$ virtual momenta. Alternatively, we may use the complete propagator $d$, defined by Eq. (251), to write Eq. (379*a*) as

$$\mathscr{T}_c = \mathscr{V}_c + \mathscr{V}_c \frac{1}{d} \mathscr{T}_c, \tag{369b}$$

since $d = e$ when operating on the wave functions $g_{0(\nu_t)}$. In writing Eq. (369*b*), we have generalized the definition (366) of $\mathscr{T}_c$ in a trivial way in that we consider the expression (366) to be multiplied by a projection operator onto the states $g_{0(\nu_t)}$.[104]

We shall return later to the problem of actually finding the appropriate pseudopotential $\mathscr{V}_c$. It is seen from Eq. (369*a*) that $\mathscr{V}_c$, like $\mathscr{T}_c$, is a $(2S_i + 1)(2S_t + 1) \times (2S_i + 1)(2S_t + 1)$ spin matrix, so it may be written as

$$\mathscr{V}_c = \langle \mathbf{k}_c, \nu', \mathbf{Q}', \nu'_t | \mathscr{V}_c | \mathbf{k}, \nu, \mathbf{Q}, \nu_t \rangle \equiv \delta(\mathbf{k}' + \mathbf{Q}' - \mathbf{k} - \mathbf{Q}) V_c. \tag{371}$$

The submatrix $V_c$ is defined only on the momentum shell. It may evidently be expressed in terms of the barycentric momenta $\boldsymbol{\varkappa}$ and $\boldsymbol{\varkappa}'$ and the spins $\mathbf{S}_i$ and $\mathbf{S}_t$, as in Eq. (367).

$$V_c = V_c(\boldsymbol{\varkappa}', \boldsymbol{\varkappa}, \mathbf{S}_i, \mathbf{S}_t). \tag{372}$$

The wave function $\psi^+_{c(a)}$ describing elastic scattering may be constructed from the complete wave function $\psi_a{}^+$, [see Eq. (250)] using the projection operator

$$\Lambda_0 \equiv \sum_{\nu'_t} g_{0(\nu'_t)} g_{0(\nu'_t)}{}^\dagger, \tag{373}$$

where the sum on $\nu'_t$ runs over all spin orientations of the target in its ground states $g_0$.[105] Then,

$$\psi^+_{c(a)} = \Lambda_0 \psi_a{}^+. \tag{374}$$

[104] Since the meaning is self-evident, we shall often use $\mathscr{T}_c$ in either of the forms implied by Eq. (369*a*) or by Eq. (369*b*) without explicitly mentioning the distinction.

[105] We shall continue to refer to $g_0$ as the ground state wave function, although it may in fact be taken as any initial state of the scatterer.

Now it follows from Eq. (4.9$a$) and from Eq. (366) that

$$\psi^{+}_{c(a)} = \left(1 + \frac{1}{d}\mathscr{T}_c\right)\chi_a. \tag{375$a$}$$

From Eqs. (369) and (375$a$) we see that $\psi^{+}_{c(a)}$ satisfies a Schrödinger equation with a potential $\mathscr{V}_c$.

$$\psi^{+}_{c(a)} = \chi_a + \frac{1}{d}\mathscr{V}_c\psi^{+}_{c(a)} = \chi_a + \frac{1}{e}\mathscr{V}_c\psi^{+}_{c(a)}. \tag{375$b$}$$

The many-body wave functions $g_0$ play no dynamical role in Eqs. (375) and may therefore be removed. To do this, we first rewrite Eq. (374) in the form

$$\psi^{+}_{c(a)} = \sum_{\nu'_t} \phi^{+}_{\mathbf{p},\nu,\mathbf{P},\nu_t} g_{0(\nu'_t)}, \tag{376}$$

where $\phi^{+}_{\mathbf{p},\nu,\mathbf{P},\nu_t}$ is a $(2S_i + 1)(2S_t + 1)$ column matrix in spin space.[106] From Eqs. (375$b$) and (376) we see that $\phi^{+}$ satisfies the two-body Schrödinger equation

$$\phi^{+}_{\mathbf{p},\nu,\mathbf{P},\nu_t} = \chi_{\mathbf{p},\nu,\mathbf{P},\nu_t} + \frac{1}{e}\mathscr{V}_c\phi^{+}_{\mathbf{p},\nu,\mathbf{P},\nu_t}. \tag{377}$$

Here [see Eq. (97$b$)]

$$\chi_{\mathbf{p},\nu,\mathbf{P},\nu_t} = (2\pi)^{-3} e^{i\mathbf{p}\cdot\mathbf{x}} e^{i\mathbf{P}\cdot\mathbf{C}}\, u_{\nu,\nu_t}, \tag{378}$$

where $u_{\nu,\nu_t}$ is a $(2S_i + 1)(2S_t + 1)$ column spin matrix, describing the spin orientations $(\nu, \nu_t)$.

For applications the Schrödinger equation for $\phi$, Eq. (377), is conveniently transformed into an equation in the barycentric subspace (see Section 4.6). We have

$$\langle \mathbf{k}', \nu', \mathbf{Q}', \nu'_t \mid \phi^{+}_{\mathbf{p}\,\nu,\mathbf{P},\nu_t}\rangle \equiv \delta(\mathbf{k}' + \mathbf{Q}' - \mathbf{p} - \mathbf{P})\langle \boldsymbol{\varkappa}', \nu', \nu'_t \mid \phi^{+}_{\boldsymbol{\varkappa},\nu,\nu_t}\rangle, \tag{379}$$

where $\phi^{+}_{\boldsymbol{\varkappa},\nu,\nu_t}$ satisfies the equation [see Eq. (371)]

$$\phi^{+}_{\boldsymbol{\varkappa},\nu\,\nu_t} = \chi_{\boldsymbol{\varkappa},\nu,\nu_t} + \frac{1}{e} V_c\phi^{+}_{\boldsymbol{\varkappa},\nu,\nu_t}. \tag{380}$$

Here ($\mathbf{r} \equiv \mathbf{x} - \mathbf{C}$)

$$\chi_{\boldsymbol{\varkappa},\nu,\nu_t} = (2\pi)^{-3/2} e^{i\boldsymbol{\varkappa}\cdot\mathbf{r}} u_{\nu,\nu_t}, \tag{381}$$

and now

$$e = \epsilon_K + \frac{\kappa^2}{2M_t} + i\eta - \epsilon_{K'} - \frac{(\kappa')^2}{2M_t} \tag{382}$$

in a barycentric momentum representation.

[106] The labels ($\mathbf{p}$, $\nu$, $\mathbf{P}$, $\nu_t$) characterize the initial state, of course.

For many applications we require the incoming wave functions $\psi_{c(a)}^{-}$ and $\phi^{-}$. These may be constructed from the outgoing wave functions using the time reversal operator $T$, according to Eq. (4.232*a*).[107]

$$\psi_{c(a)}^{-} = \mu_a{}^{*} T \psi_{c(-a)}^{+} = \left(1 + \frac{1}{d^{\dagger}} \mathscr{T}_c{}^{\dagger}\right) \chi_a \tag{383}$$

and

$$\phi_{\varkappa,\nu,\nu_t}^{-} = \mu_a{}^{*} T \phi_{-\varkappa,-\nu,-\nu_t}^{+}, \tag{384}$$

where $\mu_a$ is an appropriate phase factor.

We may very well ask at this point: Does a pseudopotential exist for any given elastic scattering process? Why introduce a pseudopotential? The first question may, in general, be answered in the affirmative, since Eq. (369*a*) may be considered as an integral equation to determine $\mathscr{V}_c$ once the scattering matrix $\mathscr{T}_c$ is known.[108] The answer to the second question depends on the scattering process being studied. An acceptable pseudopotential may have relatively simple properties, as in the example of nuclear forces mentioned previously, and it be used for convenience. The pseudopotential may have simple scaling laws for a class of scattering systems. This is, for example, the case for pseudopotentials used in the study of elastic scattering by atomic nuclei. When the scattering system is very nearly uniform in its properties over a distance that is large compared to a wavelength $\hbar/p$, the pseudopotential is ordinarily equivalent to a simple potential. To show this, we note that for such a system a wave packet for particle $i$ may be constructed in the neighborhood of a point $\mathbf{r}$. The interaction energy between the wave packet and the medium will then be some function $\mathscr{V}_c(\mathbf{r}, \mathbf{S}_i, \mathbf{S}_t)$ of the coordinate $\mathbf{r}$. This is equivalent, however, to the condition that the pseudopotential have the form

$$\langle \mathbf{r}', \nu', \nu'_t | \mathscr{V}_c | \mathbf{r}, \nu, \nu_t \rangle = \delta(\mathbf{r}' - \mathbf{r}) \mathscr{V}_c(\mathbf{r}, \mathbf{S}_i, \mathbf{S}_t), \tag{385}$$

or that it be *local*. Because of its great simplicity, most applications of pseudopotentials have been to systems for which the expression (385) is valid.

The eikonal approximation (see Section 6.7, p. 330) may often be used[109] when the expression (385) is applicable. The relative simplicity of the eikonal approximation has, in fact, provided much of the motivation for developing a systematic theory of pseudopotentials.

[107] It is easily verified from Eqs. (4.10) that the second of Eqs. (383) is correct. From Eq. (369*b*) we may obtain the relation $\mathscr{T}_c{}^{\dagger} = \mathscr{V}_c{}^{\dagger} + \mathscr{V}_c{}^{\dagger} \dfrac{1}{d^{\dagger}} \mathscr{T}_c{}^{\dagger}$.

[108] We are following here a discussion given by Francis and Watson [43].

[109] An extension of the eikonal approximation to spin-dependent interactions has been made by Greider [44] and by Weinberg [45].

The definition of the pseudopotential as the potential leading to $\mathcal{T}_c$ by no means guarantees that it will be Hermitian or that

$$\mathscr{V}_c^\dagger = \mathscr{V}_c.$$

Indeed, this will not generally be the case. The physical significance of this remark may be seen by recalling the *optical theorem* [see Eq. (5.39)], which for the present application reads [in the notation of Eq. (366)]

$$-2\,\text{Im}\,[(\mathbf{p}, \nu, \mathbf{P}, \nu_t|\; T_c\; |\mathbf{p}, \nu, \mathbf{P}, \nu_t)] = \frac{v_{\text{rel}}}{(2\pi)^3}\,\sigma_T. \tag{386}$$

Here $\sigma_T$ is the total cross section for scattering into all final channels, states of excitation, and spin orientations from the initial state [$\mathbf{p}$, $\nu$, $\mathbf{P}$, $0(\nu_t)$]. The total *elastic* scattering cross section $\sigma_{\text{el}}$ represents, in general, only a part of the contribution to $\sigma_T$, of course.

To study the relation between $\sigma_{\text{el}}$ and $\sigma_T$ we return to Eq. (369*a*), which we rewrite in the barycentric subspace as

$$T_c = V_c + V_c \frac{1}{e} T_c. \tag{387}$$

For convenience, we shall also rewrite Eq. (380) in the abbreviated notation

$$\phi_{\bar{a}}^+ = \chi_{\bar{a}} + \frac{1}{e} V_c \phi_{\bar{a}}^+, \tag{388a}$$

where

$$\phi_{\bar{a}}^+ \equiv \phi_{\varkappa,\nu,\nu_t}^+, \qquad \chi_{\bar{a}} \equiv \chi_{\varkappa,\nu,\nu_t}. \tag{388b}$$

Now, the total elastic cross section for a given initial state $\chi_{\bar{a}}$ is [compare Eq. (368)]

$$\sigma_{\text{el}} = \frac{(2\pi)^4}{v_{\text{rel}}}(\chi_{\bar{a}}, T_c^\dagger\, \delta(e) T_c \chi_{\bar{a}}). \tag{389}$$

According to Eq. (3.63), we may set

$$\delta(e) = \frac{i}{2\pi}\left(\frac{1}{e} - \frac{1}{e^\dagger}\right),$$

so

$$-2\pi i(\chi_{\bar{a}}, T_c^\dagger\, \delta(e) T_c \chi_{\bar{a}}) = \left(\chi_{\bar{a}}, T_c^\dagger\left(\frac{1}{e} - \frac{1}{e^\dagger}\right) T_c \chi_{\bar{a}}\right)$$
$$= (\chi_{\bar{a}}, T_c^\dagger \phi_{\bar{a}}^+) - (\phi_{\bar{a}}^+, T_c \chi_{\bar{a}}) + (\chi_{\bar{a}}, (T_c - T_c^\dagger)\chi_{\bar{a}}), \tag{390}$$

using Eq. (388) and the relation $T_c \chi_{\bar{a}} = \mathscr{V}_c \phi_{\bar{a}}^+$. From this result, the optical theorem [Eq. (386)] and the formula for the total elastic cross section [Eq. (389)] we obtain

$$\sigma_T = -2\frac{(2\pi)^3}{v_{\text{rel}}}\left\{\frac{1}{2i}(\chi_{\bar{a}}, (T_c - T_c^{\dagger})\chi_{\bar{a}})\right\} = \sigma_{\text{el}} - 2\frac{(2\pi)^3}{v_{\text{rel}}}\,\text{Im}\,(\phi_{\bar{a}}^+, \mathscr{V}_c\phi_{\bar{a}}^+). \tag{391}$$

Clearly, $\sigma_T = \sigma_{\text{el}} + \sigma_{ab}$, where $\sigma_{ab}$ is the total inelastic scattering cross section (including scattering to other channels as well as that leading to excitations of the target system). We see then from Eq. (391) that

$$\sigma_{\text{ab}} = -2\frac{(2\pi)^3}{v_{\text{rel}}}\,\text{Im}\,(\phi_{\bar{a}}^+, \mathscr{V}_c\phi_{\bar{a}}^+). \tag{392}$$

This would obviously vanish if $\mathscr{V}_c$ were Hermitian. We conclude therefore that if scattering can occur into other than the incident channel or if the scatterer can be left in an excited state, the pseudopotential $\mathscr{V}_c$ will *not* be Hermitian. Equation (392) is of considerable practical value since it permits us to find the total "absorption" cross section $\sigma_{\text{ab}}$ just from the elastic scattering.[110,111]

To illustrate an application of our formula for the absorption cross section, Eq. (392), we note that it is frequently the case that $\mathscr{V}_c$ is proportional to the single particle density $P^{(1)}(\mathbf{r})$, introduced in Section 11.2 [Eq. (110)]. Then we may write

$$\mathscr{V}_c = V_0\left(\frac{N}{\rho(0)}\right)P^{(1)}(\mathbf{r}), \tag{393}$$

where

$$V_0 \equiv V_r - iV_i \tag{394}$$

is a complex constant "well depth" ($V_r$ and $V_i$ are real constants and $V_i \geq 0$) and $\rho(0)$ is the particle density in the scatterer at its center-of-mass. The expressions (393) and (394) permit us to express (392) as

$$\sigma_{\text{ab}} = \frac{2(2\pi)^3}{v_{\text{rel}}}\left(\frac{N}{\rho(0)}\right)V_i\int d^3r P^{(1)}(\mathbf{r})\,|\phi_{\bar{a}}^+(\mathbf{r})|^2. \tag{395a}$$

When the scatterer has a uniform density for $r < \mathscr{R}$ and zero density for $r > \mathscr{R}$ [see Eq. (113)], we may simplify the expression (395*a*).

$$\sigma_{\text{ab}} = \frac{2(2\pi)^3}{v_{\text{rel}}}V_i\int_{r<\mathscr{R}} d^3r\,|\phi_{\bar{a}}^+(\mathbf{r})|^2. \tag{395b}$$

If the eikonal approximation [Eq. (6.499*a*)] may be used to evaluate

[110] Equation (391) was obtained first by Lax [46]. This derivation from the point of view of the many-body problem was given in Ref. [43].

[111] An elementary derivation and some applications of Eq. (391) to nuclear physics are reviewed in Ref. [47].

$\phi_{\bar{a}}{}^+$, we have for $r < \mathscr{R}$

$$\phi_{\bar{a}}{}^+(\mathbf{r}) \equiv \phi_{\boldsymbol{\kappa}}{}^+(\mathbf{r}) = (2\pi)^{-3/2} e^{i[\kappa(z-D_i)+n\kappa D_i]}, \tag{396}$$

where $\kappa$ is the relative momentum of the incident particle and target and [see Eq. (6.491)][112]

$$n\kappa \equiv (\kappa^2 - 2M_r V_0)^{1/2} \tag{397}$$

defines the "refractive index" $n$. In obtaining Eq. (396) from Eq. (6.499*a*) we have assumed that the incident relative momentum $\boldsymbol{\kappa}$ is parallel to the $z$-axis and have written $D_i(\mathbf{r})$ for the distance from $\mathbf{r}$ to the boundary of the scatterer in the direction $(-\boldsymbol{\kappa})$. This is illustrated in Fig. 11.8.

Use of the eikonal approximation in the form (396) implies that $\kappa^2 \gg 2M_r |V|$, so

$$n\kappa \cong \kappa - \frac{M_r V}{\kappa}. \tag{398}$$

Therefore Eq. (395*b*) may be written as

$$\sigma_{\text{ab}} = \frac{2V_i}{v_{\text{rel}}} \int_{r<\mathscr{R}} d^3r\, e^{-(2M_r/\kappa)V_i D_i}. \tag{399}$$

If we set $v_{\text{rel}} = \kappa/M_r$ and

$$\frac{1}{\lambda} = \frac{2M_r}{\kappa} V_i = \frac{2V_i}{v_{\text{rel}}}, \tag{400}$$

the mean free path, we see that Eq. (399) agrees precisely with Eq. (6.529).

In terms of the mean free path $\lambda$ and the representation of the single particle distribution given in Eq. (110), we may write Eq. (395*a*) in the sometimes convenient form

$$\sigma_{\text{ab}} = \frac{(2\pi)^3 \lambda}{\rho(0)} \sum_{\alpha=1}^{N} \langle 0 |\, |\phi_{\boldsymbol{\kappa}}{}^+(\mathbf{z}_\alpha)|^2 \,| 0 \rangle. \tag{401}$$

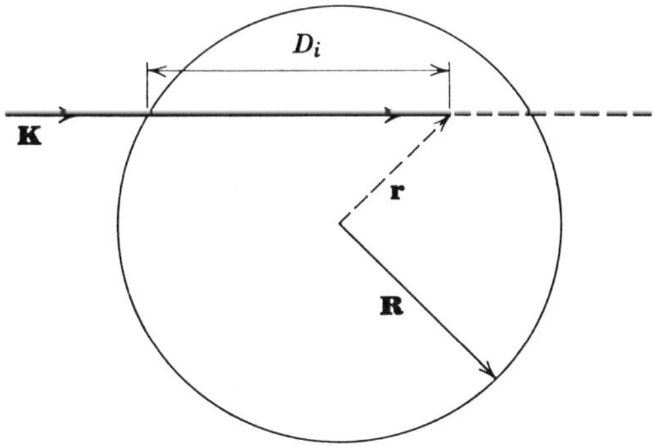


Fig. 11.8 Scattering described by wave function (396).

[112] If particle $i$ is nonrelativistic, $M_r$ is the reduced mass of $i$ and $t$. The appropriate relativistic form of $M_r$ is given in the equation following (6.533).

**Relation of Pseudopotential to Many-Body Problem**

In Section 11.4, p. 781 we have discussed the pseudopotential (371) in a purely formal manner. We turn now to the task of expressing this in terms of the fundamental interactions among the $(N + 1)$ interacting particles, as described by the full Schrödinger equation (250)

$$\psi_a^+ = \chi_a + \frac{1}{d} V \psi_a^+. \tag{402}$$

As we shall see, the pseudopotential $\mathscr{V}_c$ is closely related to the "level shift" operator and the "decay" operator, both called $R_a$ in Chapter 8 [see Eqs. (8.12) and (8.62)].

Let us first recall that the matrix $\mathscr{T}_c$ for elastic scattering [see Eq. (366)] may be expressed in terms of the solution to Eq. (402) as

$$\mathscr{T}_c = (\chi_{a'}, V\psi_a^+), \tag{403}$$

where [see Eqs. (97) and (100)]

$$\chi_{a'} = \chi_{i;\mathbf{k}\nu'}\chi_{t;0(\nu_t'),\mathbf{Q}} \tag{404}$$

is some final state corresponding to elastic scattering by the $N$-particle scatterer.

To put the expression (403) into the form of the two-body equation used in Section 11.4, p. 782, Eq. (369), we seek a solution of the Schrödinger equation (402) having the structure[113]

$$\psi_a^+ = F\psi_{c(a)}^+. \tag{405}$$

As we shall show shortly, $F$ is to be determined from the equation

$$F = 1 + \frac{1}{d}(1 - \Lambda_0)VF, \tag{406}$$

where $\Lambda_0$ is the projection operator onto the ground state of the target [see Eq. (373)], and the pseudopotential is given by

$$\mathscr{V}_c = (g_{0(\nu''_t)}, VFg_{0(\nu'_t)}). \tag{407a}$$

It should be emphasized again that $\mathscr{V}_c$ operates on the variables $\mathbf{S}_t$, $\mathbf{S}_i$ and the momenta of particles $i$ and $t$, but not on the internal coordinates of $t$. In matrix form we may write $\mathscr{V}_c$ as

$$\langle \mathbf{k}', \nu', \mathbf{Q}, \nu'_t | \mathscr{V}_c | \mathbf{p}, \nu, \mathbf{P}, \nu_t \rangle = (\chi_{a'}, VF\chi_a). \tag{407b}$$

When we wish to leave the representation in which $\mathscr{V}_c$ is given unspecified,

[113] The similarity of Eq. (405) to Eqs. (8.19*c*) and (8.61) is evident.

but indicate explicitly the expectation value for states $g_0$, we shall write Eq. (407*a*) simply as

$$\mathscr{V}_c = \langle , 0|\ VF\ |, 0\rangle. \tag{407c}$$

It follows from the definitions of $\psi_{c(a)}$ and $\mathscr{V}_c$ that

$$\Lambda_0 VF\psi^+_{c(a)} = \mathscr{V}_c\psi^+_{c(a)}. \tag{408}$$

To verify that with $F$ and $\mathscr{V}_c$ as given, Eq. (405) indeed does provide a solution to the Schrödinger equation, we substitute (405) into the right-hand side of (402).

$$\begin{aligned}\psi_a{}^+ = \chi_a + \frac{1}{d}VF\psi^+_{c(a)} &= \chi_a + \frac{1}{d}(1-\Lambda_0)VF\psi^+_{c(a)} + \frac{1}{d}\Lambda_0 VF\psi^+_{c(a)}\\ &= \chi_a + (F-1)\psi^+_{c(a)} + \frac{1}{d}\mathscr{V}_c\psi^+_{c(a)},\end{aligned} \tag{409}$$

where Eqs. (406) and (408) have been used. Now, using Eq. (375*b*), we see that (409) simplifies to

$$\psi_a{}^+ = F\psi^+_{c(a)} = \psi^+_a,$$

which completes our demonstration.

The expression (408) for $\mathscr{V}_c$ differs from that given for $R_a$ in Chapter 8 in that $\mathscr{V}_c$ is defined to be diagonal only in the internal states of the scatterer. It is not restricted, however, in its momentum or spin matrix elements.[114]

To illustrate the definition of $\mathscr{V}_c$ in terms of the actual potential $V$, let us consider a spin-independent interaction

$$V = \sum_{\alpha=1}^{N} V^{(1)}(\mathbf{x} - \mathbf{z}_\alpha)$$

for a bound system of $N$ equivalent particles. In a "first Born approximation" we may take $F = 1$ and write

$$\begin{aligned}\mathscr{V}_c(\mathbf{r}) &= \sum_{\alpha=1}^{N}\langle 0|\ V^{(1)}(\mathbf{r}-\mathbf{z}_\alpha)\ |0\rangle\\ &= \int d^3y\langle 0|\sum_\alpha \delta(\mathbf{y}-\mathbf{z}_\alpha)\ |0\rangle V^{(1)}(\mathbf{r}-\mathbf{y})\\ &= \int d^3y\rho(\mathbf{y})V^{(1)}(\mathbf{r}-\mathbf{y});\end{aligned} \tag{410}$$

[114] Derivations essentially equivalent to that given here for $\mathscr{V}_c$ were given in the Appendix to Ref. [43] and by Feshbach [48].

we have here introduced the particle density in the scatterer $\rho(\mathbf{y})$ just as in Section 11.2 [Eq. (110)] and have obtained an intuitively obvious first approximation to the pseudopotential. The corrections to this naive first step result from the distortion (sometimes referred to as polarization) of the scattering system by the incident particle. Note that to evaluate these we must return to the full equation for the operator $F$ [Eq. (406)].

When the incident particle is sufficiently massive, or when the $V^{(1)}$'s lead to only very small momentum transfers, $F$ may be treated in an *adiabatic approximation*. In this approximation we neglect the energy changes $\epsilon_p - K_i$ in the propagator $d$ in (406), writing this as

$$d \cong d_A \equiv W_0 - h, \tag{411}$$

where we recall that $h$ is the Hamiltonian for the target system [see Eqs. (98)]. The equation for determining $F$ becomes

$$F \cong 1 + \frac{1}{d_A}(1 - \Lambda_0)VF. \tag{412}$$

When $V$ is local, and thus a function of $(\mathbf{x} - \mathbf{C}) = \mathbf{r}$, we see from the expression (412) that $F$ commutes with $\mathbf{r}$. This means, in turn, that $\mathbf{r}$ commutes with $\mathscr{V}_c$. Therefore in the adiabatic approximation $\mathscr{V}_c$ is local,[115] as was the situation quite generally when the scattering system was uniform over the wavelength of the projectile.

It is frequently convenient to reexpress $\mathscr{V}_c$ in terms of two-body scattering operators $\ell_\alpha$, as in our treatment of the multiple scattering solution of the Schrödinger equation in Section 11.3. For this purpose we modify slightly the definitions of these operators previously given in Eqs. (252), defining

$$\ell_\alpha = V_\alpha + V_\alpha \frac{1 - \Lambda_0}{d} \ell_\alpha. \tag{413}$$

We shall see presently that in the weak binding limit the difference between the definitions (252) and (413) is negligible.

In analogy to the multiple scattering form of the wave function, Eqs. (253), we now write

$$\begin{aligned} F &= 1 + \frac{1}{d} \sum_{\alpha=1}^{N} (1 - \Lambda_0)\ell_\alpha F_\alpha, \\ F_\alpha &= 1 + \frac{1}{d} \sum_{\beta(\neq\alpha)=1}^{N} (1 - \Lambda_0)\ell_\beta F_\beta. \end{aligned} \tag{414}$$

To see that these equations provide a solution to the exact equation for $F$,

[115] Some applications of the adiabatic approximation are described in Ref. [49].

Eq. (406), we substitute the first of Eqs. (414) into the right-hand side of (406).

$$F = 1 + \frac{1}{d}(1 - \Lambda_0)V + \frac{1 - \Lambda_0}{d}\left[\sum_{\beta=1}^{N} V_\beta\right]\frac{1 - \Lambda_0}{d}\sum_{\alpha=1}^{N} t_\alpha F_\alpha. \tag{415}$$

Now, from the definition of $t_\alpha$,

$$\begin{aligned}\sum_{\beta=1}^{N} V_\beta \frac{1 - \Lambda_0}{d}\sum_{\alpha=1}^{N} t_\alpha F_\alpha &= \sum_{\alpha=1}^{N}(t_\alpha - V_\alpha)F_\alpha + \sum_{\beta=1}^{N} V_\beta \frac{1 - \Lambda_0}{d}\sum_{\alpha(\neq\beta)=1}^{N} t_\alpha F_\alpha \\ &= \sum_{\alpha=1}^{N}(t_\alpha - V_\alpha)F_\alpha + \sum_{\beta=1}^{N} V_\beta(F_\beta - 1) = \sum_{\alpha=1}^{N} t_\alpha F_\alpha - V. \end{aligned}\tag{416}$$

Substitution of this into Eq. (415) then gives

$$\begin{aligned}F &= 1 + \frac{1}{d}(1 - \Lambda_0)V + \frac{1}{d}(1 - \Lambda_0)\left[\sum_{\alpha=1}^{N} t_\alpha F_\alpha - V\right] \\ &= 1 + \frac{1}{d}\sum_{\alpha=1}^{N}(1 - \Lambda_0)t_\alpha F_\alpha,\end{aligned}$$

which shows the consistency of our expression for $F$ [Eqs. (414)].

The potential $\mathscr{V}_c$ may now be expressed in terms of the scattering matrices $t_\alpha$. Using the preceding equations for $F$ in terms of $t_\alpha$, we have

$$\begin{aligned}VF &= V + \left[\sum_{\beta=1}^{N} V_\beta\right]\left[\frac{1 - \Lambda_0}{d}\sum_{\alpha=1}^{N} t_\alpha F_\alpha\right] \\ &= V + \sum_{\alpha=1}^{N} t_\alpha F_\alpha - V = \sum_{\alpha=1}^{N} t_\alpha F_\alpha.\end{aligned}\tag{417}$$

Therefore the pseudopotential $\mathscr{V}_c$, defined by Eq. (407*c*) as an expectation value of $VF$, is equivalent to

$$\mathscr{V}_c = \langle , 0|\sum_{\alpha=1}^{N} t_\alpha F_\alpha |, 0\rangle. \tag{418}$$

Successive substitution of the equation (414) for $F_\alpha$ into this gives a multiple scattering series expansion for $\mathscr{V}_c$.

$$\mathscr{V}_c = \sum_{\alpha=1}^{N}\langle , 0|\, t_\alpha |, 0\rangle + \sum_{\alpha\neq\beta=1}^{N}\langle , 0|\, t_\alpha \frac{1 - \Lambda_0}{d} t_\beta |, 0\rangle + \cdots. \tag{419}$$

Except for the appearance of the projection operators $1 - \Lambda_0$, the structure of this equation is no different from that of the corresponding multiple scattering expansion of $\mathscr{T}_N$ given in Section 11.3, Eq. (257).

In the weak binding limit, where all the target particle energies are small compared to the projectile energy $\epsilon_p$, [see Eq. (106)], we may replace the

$\ell_\alpha$ by the two-body scattering operators $\mathcal{T}_\alpha$ [see Eq. (258)]. This is evidently an important simplification since it enables us to consider $\mathcal{V}_c$ in Eq. (418) as involving a sequence of two-body interactions, rather than at each step an $(N + 1)$-body interaction.

If we were to ignore the projection operator $1 - \Lambda_0$ in the present definition of the $\ell_\alpha$'s, (433), they would be identical with those defined previously by Eqs. (252). This suggests that the term $V_\alpha(-\Lambda_0/d)\ell_\alpha$ in Eq. (413) be treated as a small perturbation. To zeroth order in this quantity our earlier relation between $\ell_\alpha$ and the two-body operator $\mathcal{T}_\alpha$, Eq. (264), may be taken as providing a relation between the $\ell_\alpha$ used here and $\mathcal{T}_\alpha$. To first order we have therefore

$$\ell_\alpha = \mathcal{T}_\alpha\left[1 + \mathcal{O}\left(\left|\frac{f_\alpha}{\lambda}\frac{R_{\alpha,a}t_{\alpha,c}}{\hbar}\right|\right)\right] - V_\alpha\frac{\Lambda_0}{d}\mathcal{T}_\alpha + \cdots. \tag{420}$$

It remains to estimate the importance of the term $V_\alpha(\Lambda_0/d)\mathcal{T}_\alpha$.

It is expected that in the weak binding limit this term will, indeed, be relatively unimportant. The reason for this is that the high virtual energies of excitation involves states with little overlap of the ground states $g_0$. Reference to Eq. (419) suggests that it will be sufficient to estimate the importance of just the term

$$-\langle, 0|\, V_\alpha\frac{\Lambda_0}{d}\mathcal{T}_\alpha\,|, 0\rangle = -\langle, 0|\, V_\alpha\,|, 0\rangle\frac{1}{e}\,\langle, 0|\,\mathcal{T}_\alpha\,|, 0\rangle, \tag{421}$$

where the two-particle propagator $e$ is defined by Eq. (370). The remaining terms in the series (419) are, in general, associated with more complex excitations of the target and thus with less "overlap" of the states $g_0$. For an order-of-magnitude argument, we may replace $\langle, 0|\, V_\alpha\,|, 0\rangle$ by $\langle, 0|\,\mathcal{T}_\alpha\,|, 0\rangle$ in Eq. (421).[116] We shall also ignore spin interactions and the recoil energy of the target system.

Then in a momentum representation we may write the expression (421) as we did in our discussion of the elastic scattering from small systems [see Section 11.2, Eqs. (150), (152)].

$$\begin{aligned}\delta(\mathbf{k} + \mathbf{Q} - \mathbf{p} - \mathbf{P})T_\alpha^{\,2}\int\left[\int P^{(1)}(\mathbf{z}')\, e^{-i(\mathbf{k}-\mathbf{p}')\cdot\mathbf{z}'}\, d^3z'\right]\\ \times\frac{d^3p'}{\epsilon_p + i\eta - \epsilon_{p'}}\left[\int P^{(1)}(\mathbf{z})\, e^{-i(\mathbf{p}'-\mathbf{p})\cdot\mathbf{z}}\, d^3z\right]\\ \cong -[(2\pi)^2\epsilon_p T_\alpha]\mathcal{T}_\alpha\int\frac{d^3z\, d^3z'}{|\mathbf{z} - \mathbf{z}'|}\, e^{ip|\mathbf{z}-\mathbf{z}'|}\, e^{i(\mathbf{p}\cdot\mathbf{z}-\mathbf{k}\cdot\mathbf{z}')}P^{(1)}(\mathbf{z})P^{(1)}(\mathbf{z}').\end{aligned} \tag{422}$$

[116] It is always possible to resum the series (420) in such a manner that only $\mathcal{T}_\alpha$'s, and no $V_\alpha$'s, appear in the series. When $V_\alpha$ is singular, this must be done.

The asymptotic value of the $\mathbf{p}'$-integral has been used here, since the weak binding limit implies that $\hbar/p \ll \mathscr{R}$, the "size" of the target. The quantity $[-(2\pi)^2\epsilon_p T_\alpha]$ may be replaced by the scattering amplitude $f_\alpha$.

The expression (422) tends to be largest when $\mathbf{k} \cong \mathbf{p}$ or for forward scattering. Thus we may estimate this to be of the order of

$$\mathscr{T}_\alpha f_\alpha \int d^3z\, d^3z'\, P^{(1)}(\mathbf{z}) P^{(1)}(\mathbf{z}') \frac{e^{ipR}}{R} e^{i\mathbf{p}\cdot\mathbf{R}}$$
$$\cong \mathscr{T}_\alpha f_\alpha \left[\frac{2\pi i}{p} \int dR P^{(1)}(\mathbf{R})\right] \cong \mathscr{T}_\alpha f_\alpha \left(\frac{i}{p\mathscr{R}^2}\right), \qquad (423)$$

where $\mathbf{R}$ has been written for $(\mathbf{z} - \mathbf{z}')$ and $\mathscr{R}$ is the "size" of the scatterer. This result finally permits us to rewrite our estimate of $t_\alpha$, Eq. (420), in the more explicit form

$$t_\alpha = \mathscr{T}_\alpha \left[1 + \mathscr{O}\left(\left|\frac{f_\alpha}{\lambda} \frac{R_{\alpha,a} t_{\alpha,c}}{\hbar}\right|\right) + \mathscr{O}\left(\left|\frac{f_\alpha}{\lambda}\left(\frac{\hbar}{p\mathscr{R}}\right)^2\right|\right) + \cdots\right]. \qquad (424)$$

The first term here represents, of course, just the correction to the impulse approximation. The second term provides an estimate of the importance of the projection operator $\Lambda_0$ in the definition (423). This correction is usually not greater than the first correction. We conclude therefore that in the weak binding limit we may replace the $t_\alpha$ by the $\mathscr{T}_\alpha$ and replace our expressions for $F$ and $\mathscr{V}_c$, given in Eqs. (414) and (418), by the equations

$$F = 1 + \frac{1}{d}(1 - \Lambda_0) \sum_{\alpha=1}^{N} \mathscr{T}_\alpha F_\alpha,$$
$$F_\alpha = 1 + \frac{1}{d}(1 - \Lambda_0) \sum_{\beta(\neq\alpha)=1}^{N} \mathscr{T}_\beta F_\beta, \qquad (425)$$

and

$$\mathscr{V}_c = \langle, 0| \sum_{\alpha=1}^{N} \mathscr{T}_\alpha F_\alpha |, 0\rangle. \qquad (426)$$

These expressions are also ordinarily valid in the limit of quasi-classical binding where the binding potential $U$ in the target varies slowly over the scattering region [see Eq. (31)]. In this case we expect the states $g_\gamma$ to be densely spaced. Omitting one such level, as is implied by $\Lambda_0$ in Eq. (413), is not expected to have an appreciable effect on the $t_\alpha$.

**First Approximation to the Pseudopotential**

We shall now investigate the first, or "single scattering," approximation to $\mathscr{V}_c$, assuming that the weak binding limit obtains and that Eqs. (425)

and (426) may be used. In the single scattering approximation we set $F_\alpha = 1$, so Eq. (426) yields

$$\mathscr{V}_c = \sum_{\alpha=1}^{N} \langle, 0| \mathscr{T}_\alpha |, 0\rangle. \tag{427}$$

We shall study the validity of this approximation in Section 11.4, p. 801. To anticipate our conclusions there, however, we remark that both the conditions (132) and (133) are required to be met if the expression (427) is to provide an adequate representation for the pseudopotential. That the mean free path be much larger than target dimensions is *not* required may seem a little surprising in view of our previous estimate of multiple scattering corrections, which seem to be important if this is not true. The difference is due to the profound role played by the operator $1 - \Lambda_0$ in the weak binding equations for $F$ [Eqs. (425)]. The multiple scattering equations in the impulse approximation limit, described by Eqs. (266), could lead to any final state of the scatterer consistent with conservation laws; in the definition of $\mathscr{V}_c$, as expressed by Eq. (426), the scatterer is required to return to one of the initial states $g_0$. The operator $1 - \Lambda_0$ in Eqs. (425) forbids the system to return to any of these states, however, until after the final scattering has occurred. We see then that this last scattering must "undo" the excitations created by previous scatterings. This implies that all the scatterings described by a given term in (426) must occur "near each other," the term "near" implying a distance of the order of $\mathscr{R}_c$ over which correlations are important in the target.[117] The implication of this is that the size of the target is irrelevant for determining the validity of the single scattering approximation (427). The physical significance of this is that $\mathscr{V}_c$ depends only on *local conditions* near a given point in the scattering medium when $\mathscr{R} \gg \mathscr{R}_c$.

The expression (427) may be evaluated in the manner of the elastic scattering amplitude described in Section 11.2, p. 720. For simplicity we shall assume that the $N$ scatterers in the target are identical and that this is a "saturated spin system" in the sense that the target spin is very much less than the maximum possible value [see Eq. (149)]. If we exhibit the dependence of $\mathscr{T}_\alpha |0\rangle$ on the coordinate $\mathbf{z}_\alpha$, as in Eq. (144), we may write our single scattering approximation to $\mathscr{V}_c$ as

$$\begin{aligned}\langle \mathbf{q}'| \mathscr{V}_c |\mathbf{q}\rangle &= \sum_{\alpha=1}^{N} \langle \mathbf{q}', 0|\, e^{-i(\mathbf{q}'-\mathbf{q})\cdot(\mathbf{z}_\alpha - \mathbf{C})} T_\alpha \,|\mathbf{q}, 0\rangle \\ &= \sum_{\alpha=1}^{N} (g_{0(v'_t)}, e^{-i(\mathbf{q}'-\mathbf{q})\cdot(\mathbf{z}_\alpha-\mathbf{C})} T_\alpha g_{0(v_t)}).\end{aligned} \tag{428}$$

[117] Equations (425) suffer from the practical defect that they permit scatterings without change of state of the medium in excited states, corresponding to $\gamma \neq 0$. This can be easily corrected by reordering the terms in Eqs. (425) (see Section 11.5).

Here $(\mathbf{q}', \mathbf{q})$ represent two momentum states of the incident particle $i$, and $T_\alpha$ is the matrix element [see Eq. (145)]

$$\langle \mathbf{q}', \nu', \mathbf{q} - \mathbf{q}', \nu'_t | \, T_\alpha \, | \mathbf{q}, \nu, 0, \nu_t \rangle \equiv T_\alpha(\mathbf{q}', \mathbf{q}, \mathbf{S}_i, \mathbf{S}_\alpha), \tag{429}$$

in which the $\mathbf{P}_\alpha$-dependence is neglected.

The expression (428) is identical with that met in Section 11.2, p. 721 where we studied elastic scattering from small systems [Eq. (150)]. For the case that the size $\mathscr{R}$ of the medium is much greater than $d$, the range of the force between $i$ and a scatterer $\alpha$, and when $(\epsilon_p - mc^2) \gg |\langle \mathbf{p} | \mathscr{V}_c \, | \mathbf{p} \rangle|$,[118] the pseudopotential can be considerably simplified, however. These conditions imply that we may set $\mathbf{q}' \simeq \mathbf{q} \simeq p$ in $T_\alpha$, which then becomes

$$T_\alpha^{\,0} \equiv T_\alpha(\mathbf{p}, \mathbf{p}, \mathbf{S}_i, \mathbf{S}_\alpha), \tag{430}$$

using the notation of Eq. (429). The approximate pseudopotential (428) may now be transformed explicitly into a coordinate space representation.

$$\begin{aligned} \langle \mathbf{r}' | \, \mathscr{V}_c \, | \mathbf{r} \rangle &= (2\pi)^{-3} \int d^3q' \, d^3q \, e^{i\mathbf{q}' \cdot \mathbf{r}'} \, e^{-i\mathbf{q} \cdot \mathbf{r}} \langle \mathbf{q}' | \, \mathscr{V}_c \, | \mathbf{q} \rangle \\ &= (2\pi)^3 \sum_{\alpha=1}^{N} (g_0, \delta(\mathbf{r} - \mathbf{z}_\alpha + \mathbf{C}) \, \delta(\mathbf{r}' - \mathbf{z}_\alpha + \mathbf{C}) T_\alpha^{\,0} g_0) \\ &\equiv \delta(\mathbf{r}' - \mathbf{r}) \mathscr{V}_c(\mathbf{r}), \end{aligned} \tag{431a}$$

where

$$\mathscr{V}_c(\mathbf{r}) = (2\pi)^3 \sum_{\alpha=1}^{N} (g_0, \delta(\mathbf{r} - \mathbf{z}_\alpha + \mathbf{C}) T_\alpha^{\,0} g_0). \tag{431b}$$

Here $\mathscr{V}_c$ is equivalent to a simple potential (with possible spin dependence), as described in connection with Eq. (385).

The spin average implied by Eq. (431*b*) may be evaluated using the expressions given in Table 11.1.[119] For a spin-saturated system, we may remove $T_\alpha^{\,0}$ from the integrals in (431*b*), writing it as $\langle T_\alpha^{\,0} \rangle_\alpha$ as defined by Eq. (151). For a parity-conserving interaction[120]

$$\langle T_\alpha^{\,0} \rangle_\alpha = T^0,$$

which is independent of the spin $\mathbf{S}_i$. Then Eq. (431*b*) becomes simply

$$\mathscr{V}_c(\mathbf{r}) = (2\pi)^3 N T^0 (g_0, \delta(\mathbf{x} - \mathbf{z}_\alpha + \mathbf{C}) g_0) = (2\pi)^3 N T^0 \bar{P}^{(1)}(\mathbf{r}), \tag{432}$$

where we have introduced the spin-averaged single-particle distribution function, using Eq. (108*d*) and the condition of spin saturation.

[118] When $i$ is nonrelativistic, we, of course, write this condition simply as $\epsilon_p \gg |\langle \mathbf{p} | \mathscr{V}_c \, | \mathbf{p} \rangle|$, where $\epsilon_p$ is the kinetic energy of the incident particle $i$.

[119] If the scattering system is an atomic nucleus, $T_\alpha$ depends also on the isotopic spin operator $\boldsymbol{\tau}_\alpha$ of nucleon $\alpha$.

[120] When parity is not conserved, $T^0$ will, in general, have the form $T^0 = a + b\hat{\mathbf{p}} \cdot \mathbf{S}_i$.

Now, the density of scatterers in the medium is

$$\rho(\mathbf{r}) \equiv N\bar{P}^{(1)}(\mathbf{r}), \tag{433}$$

so $\mathscr{V}_c$ may be written as[121]

$$\mathscr{V}_c(\mathbf{r}) = -\frac{2\pi}{\epsilon_p/c^2} f_+ \rho(\mathbf{r}), \tag{434}$$

where $f_+ \equiv -(2\pi)^2(\epsilon_p/c^2)T^0$ [compare Eq. (328)] is the forward scattering amplitude for the interaction of $i$ with one of the particles $\alpha$.

It is often convenient to replace the scattering amplitude $f_+$, defined in the laboratory frame of reference, by the barycentric amplitude for forward scattering $f_B$, discussed at length in Chapters 6 and 7. Since the forward amplitude divided by the momentum is an invariant [see Eq. (6.23)], we have

$$\frac{1}{p} f_+ = \frac{1}{\kappa} f_B, \tag{435}$$

where $\kappa$ is the momentum of particle $i$ or $\alpha$ in the barycentric frame of the two particles $i$ and $\alpha$. This permits us to write Eq. (434) in the more familiar form

$$\mathscr{V}_c = V_0\left[\frac{\rho(\mathbf{r})}{\rho(0)}\right], \tag{436a}$$

where

$$V_0 = -\frac{2\pi c^2}{\epsilon_p}\rho(0)f_+ = -\frac{2\pi c^2}{\epsilon_p}\frac{p}{\kappa}\rho(0)f_B \tag{436b}$$

is the "well depth," as introduced in our discussion of the absorption cross section [Eq. (393)].[122]

The elastic scattering of particle $i$ is now obtained from our approximate one-particle relativistic equation, Eq. (6.533) (the nonrelativistic limit being trivial), with the potential given by

$$v(r) = \frac{2\epsilon_p}{c^2}\left(1 + \frac{\epsilon_p}{M_t c^2}\right)^{-1} \mathscr{V}_c(\mathbf{r}). \tag{437}$$

When the eikonal approximation is valid, we may use Eq. (6.494) to write Eq. (6.533) as

$$l^2(r) + v(r) = p^2 \tag{438}$$

and the wave function (380) becomes

$$\phi^+_{\varkappa,\nu,\nu_t} = (2\pi)^{-3/2} \exp\left[i\int^{\mathbf{r}} \mathbf{l}(\mathbf{y}) \cdot d\mathbf{y}\right] u_{S_i,\nu,\nu_t}. \tag{439}$$

[121] If $i$ has nonrelativistic energy, we may, of course, set $\epsilon_p/c^2 = m$ here.

[122] When $\rho(0)$ is not defined, we can define $\mathscr{V}_c$, but not $V_0$.

If the scattering medium is uniform for $r < \mathscr{R}$, so $v(r) = v(0)$ when $r < \mathscr{R}$, we may introduce the "refractive index" $n$, defined by Eq. (329), to rewrite Eq. (438) as

$$n^2 - 1 = -\frac{1}{p^2}\,v(0). \tag{440}$$

Because of the assumed condition $(\epsilon_p - mc^2) \gg |\mathscr{V}_c|$, this is equivalent to

$$n = 1 + \frac{2\pi\rho f_+}{p^2},$$

using the definition of $v(0)$ from Eq. (436*b*). This is precisely the same result as found in Section 11.3, p. 771, Eq. (332).

One of the more important applications of the theory developed here has been to the scattering of pi-mesons by atomic nuclei. In this case the Schrödinger equation (6.533) is

$$[\nabla^2 + \kappa^2 - v(r)]\phi_\kappa{}^+ = 0. \tag{441}$$

The pseudopotential $v(r)$ may be expressed in terms of its real and imaginary parts as

$$v(r) = v_R(r) - iv_I(r). \tag{442}$$

From the optical theorem [Eq. (6.74*d*)] we have $\operatorname{Im} f_B = (\kappa/4\pi)\bar{\sigma}_f$, where $\bar{\sigma}_f$ is the total pion-nucleon cross section averaged over the neutrons and protons in the nucleus. Now, $[\rho(0)\bar{\sigma}_f]^{-1} = \lambda$, the mean free path for scattering in the interior of the nucleus, so we may use Eqs. (436) to write

$$v_I(r) = \frac{p}{\lambda}\left[\frac{\rho(r)}{\rho(0)}\right]. \tag{443}$$

$v_R$ may be evaluated from the pion-nucleon phase shifts using the expressions derived in Chapter 7 for this process [Eqs. (7.78), (7.79), and (7.57*b*)]. More elegantly, we may use the pion-nucleon dispersion relations to obtain $v_R$. This has been done by several authors,[123] using the theory developed in Section 10.6, p. 671. The resulting well depth $V \equiv \operatorname{Re}(V_0)$ is shown in Fig. (11.9). The imaginary well depth $V_I = -\operatorname{Im}(V_0)$ is also shown in Fig. (11.9).[124]

Several corrections to Eqs. (436) have been made in studying the elastic scattering of pions by atomic nuclei. The first is a multiple scattering correction to Eq. (427). This will be described in Section 11.4, p. 801. Corrections to take account of the fact that $\mathbf{q}'$ is not exactly equal to $\mathbf{q}$ and

[123] See, for example, Sternheimer [50] or Frank, Gammel, and Watson [51].

[124] A correction for true absorption of pions has been included in $V_I$. The theoretical basis for this correction is given in Ref. [31].

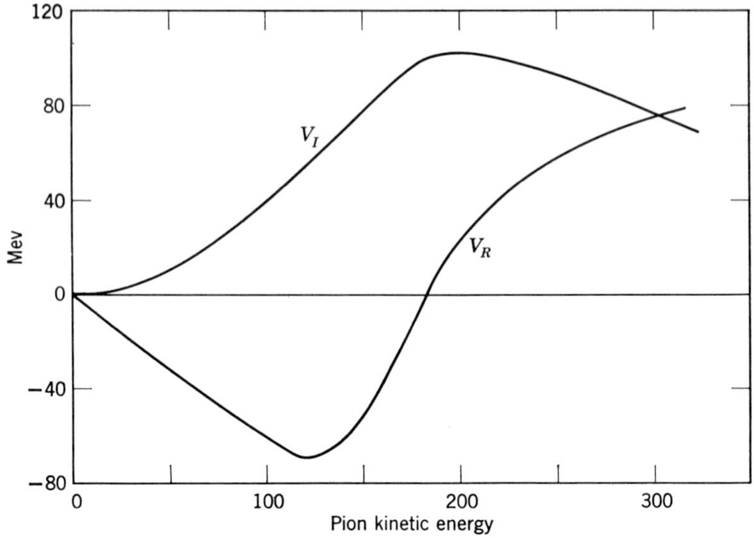


Fig. 11.9 The real part $V_R$ and the imaginary part $V_I$ of the potential well depth $V_0$ [Eq. (436)] for pion-nucleus scattering.

that neither is equal to **p** have been developed in a convenient form by Kisslinger [52] and by Rainwater and his collaborators.[125] The relation between the effective potential and the scattering amplitude as given by Eq. (436), modified by the preceding corrections, seems to be in quite satisfactory agreement with measured pion cross sections, even at somewhat lower energies than we might expect the impulse approximation to be valid.

### The Elastic Scattering of Nucleons by Atomic Nuclei

The use of a local pseudopotential or "optical model potential" for studying the elastic scattering of neutrons and protons by atomic nuclei has had considerable success for energies ranging from a few Mev to hundreds of Mev and for nuclei ranging from carbon to uranium. Phenomenological applications have included theories of Coulomb barrier penetration[126] and low-energy nuclear reactions.[127] The first attempt to relate the optical model potential to nucleon-nucleon forces seems to have been made by Fernbach, Serber, and Taylor [57], who studied the scattering of fast neutrons by atomic nuclei.

[125] See Ref. [53], where further references are given.
[126] Ostrofsky, Breit, and Johnson [54].
[127] See Refs. [55] and [56].

For neutron or proton energies in excess of 100 Mev we may anticipate that the weak binding approximation is reasonably well justified. In this case we might expect the single scattering approximation to $\mathscr{V}_c$, given by Eq. (427), to provide a suitable approximation for $\mathscr{V}_c$, having being recalled from Section 11.3, p. 779 that properly symmetrized scattering matrices $\mathscr{T}_\alpha$ are to be used. We shall generalize somewhat the approximation (430) for the scattering matrix to allow for nonforward scattering, however. Finally, our neglect of multiple scattering corrections in Eq. (427) must be justified.

According to Eq. (3.112), we can relate the scattering matrix $T_\alpha$ in Eq. (428) to the barycentric scattering matrix $T_B(\hat{\boldsymbol{\varkappa}}_f, \hat{\boldsymbol{\varkappa}}, \boldsymbol{\sigma}_i, \boldsymbol{\sigma}_\alpha, \boldsymbol{\tau}_i, \boldsymbol{\tau}_\alpha)$, which is a matrix in spin and isotopic spin space [in the notation of Eq. (7.191)]. To avoid the algebra of transforming spins, we shall restrict ourselves to nonrelativistic nucleon energies, however. Then

$$T_\alpha = T_B(\hat{\boldsymbol{\varkappa}}_f, \hat{\boldsymbol{\varkappa}}, \boldsymbol{\sigma}_i, \boldsymbol{\sigma}_\alpha, \boldsymbol{\tau}_i, \boldsymbol{\tau}_\alpha). \tag{444}$$

The nucleus will be treated as a spin and isotopic spin-saturated system containing $A$ nucleons, so that $\mathscr{V}_c$ in a momentum representation [Eq. (428)] may be written in the manner of Eq. (150), where the elastic scattering from a small spin-saturated system was discussed.

$$\langle \mathbf{q}' | \mathscr{V}_c | \mathbf{q} \rangle = \langle T_B \rangle_\alpha \sum_{\alpha=1}^{A} (g_0, e^{-i(\mathbf{q}'-\mathbf{q})\cdot(\mathbf{z}_\alpha - \mathbf{C})} g_0), \tag{445}$$

where the definition (151) of $\langle T_B \rangle_\alpha$ now includes an isotopic spin average.

The general structure of the nucleon-nucleon scattering matrix given by Eqs. (7.214), (7.215), and (7.220) permits us to set (here $M$ is the nucleon rest-mass)

$$\langle T_B \rangle_\alpha = -\left[(2\pi)^2 \frac{M}{2}\right]^{-1} \left(\frac{1}{4} \bar{A} - \frac{i\bar{C}}{\kappa^2 \sin\theta} \boldsymbol{\sigma}_i \cdot \boldsymbol{\varkappa}_f \times \boldsymbol{\varkappa}\right), \tag{446}$$

where $\bar{A}$ and $\bar{C}$ represent the quantities $A$ and $C$ in Eq. (7.215), appropriately symmetrized and averaged over the neutrons and protons in the nucleus.

In the spirit of the forward scattering approximation by which we were led to Eq. (430), we shall set $\boldsymbol{\varkappa}_f = \boldsymbol{\varkappa}$ in $\bar{A}$ and in $(\bar{C}/\sin\theta)$ and then set both equal to the incident barycentric momentum. In doing this we are neglecting terms of order $\theta^2$, where $\theta$ is the angle between $\boldsymbol{\varkappa}_f$ and $\boldsymbol{\varkappa}$. The term $\boldsymbol{\sigma}_i \cdot \boldsymbol{\varkappa}_f \times \boldsymbol{\varkappa}$ is of relative order $\theta$ and will be kept in (446), however. On transforming back to the laboratory frame of reference, we have

$$\frac{\boldsymbol{\sigma}_i \cdot \boldsymbol{\varkappa}_f \times \boldsymbol{\varkappa}}{\kappa^2} \cong 2 \frac{\boldsymbol{\sigma}_i \cdot \mathbf{q}' \times \mathbf{q}}{q^2} \cong 2 \frac{\boldsymbol{\sigma}_i \cdot \mathbf{q}' \times \mathbf{q}}{p^2}.$$

This permits us to write Eq. (445) as

$$\langle \mathbf{q}'|\mathscr{V}_c|\mathbf{q}\rangle = \left[V_c + \frac{V_S}{p^2}\, i\boldsymbol{\sigma}_i \cdot \mathbf{q}' \times \mathbf{q}\right]\frac{A}{(2\pi)^3\rho(0)}(g_0, e^{-i(\mathbf{q}'-\mathbf{q})\cdot(\mathbf{z}_\alpha - \mathbf{C})}g_0), \quad (447)$$

where

$$V_c = -\frac{\pi}{M}\bar{A}\rho(0),$$
$$V_S = \frac{8\pi}{M}\frac{\bar{C}}{\sin\theta}\rho(0), \quad (448)$$

and $\rho(0)$ is the nucleon density $\rho(r)$ at the center of the nucleus. Finally, the expression (447) may be transformed to a coordinate representation, as was done in Eq. (431). This leads to the potential[128]

$$\langle \mathbf{r}'|\mathscr{V}_c|\mathbf{r}\rangle = \delta(\mathbf{r}' - \mathbf{r})\mathscr{V}_c(\mathbf{r}), \quad (449a)$$

where

$$\mathscr{V}_c(\mathbf{r}) = V_c\frac{\rho(r)}{\rho(0)} + V_S\frac{1}{p^2\rho(0)}\frac{d\rho}{r\,dr}\,\boldsymbol{\sigma}_i \cdot \mathbf{L} \quad (449b)$$

and $\mathbf{L} = (\hbar/i)\mathbf{r} \times \nabla_r$ is the relative orbital angular momentum of the nucleon and nucleus.

The central and spin orbit well depths $V_c$ and $V_S$ are, in general, both complex. Thus $\mathscr{V}_c$ is characterized by four well depth parameters. Otherwise, the expression (469*b*) has the form of a simple potential between a spin zero and spin one-half particle with a central and a spin orbit interaction. This means, for example, that the elastic scatterings of neutrons and protons[129] may be evaluated using the theory developed for such problems in Chapter 7 [Eqs. (7.67)]. The scattering cross section is given in terms of phase shifts by Eqs. (7.80), (7.77), and (7.78). Because $\mathscr{V}_c$ is not Hermitean, the phase shifts $\delta_{l\pm}$ that appear in the scattering amplitude are complex. This implies that the "$S$-matrix" elements $S_{l\pm} = e^{2i\delta_{l\pm}}$ do not satisfy the condition $|S_{l\pm}| = 1$.[130]

The polarization of nucleons elastically scattered by nuclei has been observed experimentally,[131] as predicted by Eqs. (449). Detailed theoretical studies[132] have generally led to agreement with experimental observations.

[128] This is given in more detail in Ref. [58].

[129] Coulomb corrections may be required for proton scattering. These can be included by inserting the Coulomb interaction into Eqs. (7.67) and following the procedure given in Section 6.4, p. 263.

[130] The $S$-matrix obtained from the potential (449) is only a submatrix of the complete multichannel $S$-matrix for nucleon-nucleus scattering. The complete $S$-matrix is, of course, unitary.

[131] See Refs. [59], [60], and [61].

[132] See, for example, Refs. [62], [63], [64], and [58].

**The Second Approximation to the Pseudopotential**

The multiple scattering expansion for $\mathscr{V}_c$ may be found by successive substitutions of the right-hand side of the second of Eqs. (425) into Eq. (426).

$$\mathscr{V}_c = \sum_{\alpha=1}^{N} \langle, 0| \mathscr{T}_\alpha |, 0\rangle + \sum_{\alpha\neq\beta=1}^{N} \langle, 0| \mathscr{T}_\alpha \frac{1-\Lambda_0}{d} \mathscr{T}_\beta |, 0\rangle + \cdots. \tag{450}$$

The first term in this series was discussed in Sections 11.4, p. 794 and 11.4, p. 798. The second term,

$$\mathscr{V}_{c_2} \equiv \sum_{\alpha\neq\beta=1}^{N} \langle, 0| \mathscr{T}_\alpha \frac{1-\Lambda_0}{d} \mathscr{T}_\beta |, 0\rangle, \tag{451}$$

will now be investigated for that interaction discussed in Section 11.4, p. 795. In addition to the weak binding assumption and the conditions that the mean free path be large compared to both projectile wavelength $\hbar/p$ and target particle correlation range $\mathscr{R}_c$, we shall assume that $\hbar/p \ll \mathscr{R}_c$ and that $\mathscr{R} \gg \mathscr{R}_c$, which implies that $N \gg 1$. To simplify the discussion, we shall suppose there is only a single pair distribution function $P^{(2)}(\mathbf{z}_\alpha - \mathbf{C}, \mathbf{z}_\beta - \mathbf{C})$ [that is, all $P_\lambda$ in Eqs. (112) are supposed to be equal].[133]

The matrix element in Eq. (451) is similar to the "nondiagonal" terms encountered in the closure approximation for the scattering from small systems in Eqs. (163). That is, an excitation of the scattering medium due to scattering on particle $\beta$ must be removed by a scattering on particle $\alpha$. If particle $\beta$ were to recoil as in quasi-free scattering, the second scattering would be unable to return the system to a state $g_0$. Therefore the intermediate state following the first scattering must be "quasi-elastic" and we may neglect the target excitations in evaluating the energy denominator $d$ [see Eq. (187)].[134] In analogy to the closure approximation referred to, Eq. (163*c*), we then have [using the shorthand notation for the $T_\alpha$ matrices given in Eq. (429)]

$$\langle \mathbf{q}'| \mathscr{V}_{c_2} |\mathbf{q}\rangle = \sum_{\alpha\neq\beta=1}^{N} \int \frac{d^3k}{\epsilon_p + i\eta - \epsilon_k} (g_0, T_\alpha(\mathbf{q}', \mathbf{k}, \mathbf{S}_i, \mathbf{S}_\alpha)$$
$$\times\, e^{-i(\mathbf{q}'-\mathbf{k})\cdot(\mathbf{z}_\alpha-\mathbf{C})}(1-\Lambda_0)\, e^{-i(\mathbf{k}-\mathbf{q})\cdot(\mathbf{z}_\beta-\mathbf{C})}\, T_\beta(\mathbf{k}, \mathbf{q}, \mathbf{S}_i, \mathbf{S}_\beta) g_0). \tag{452}$$

[133] This restriction is easily generalized (see, for example, Ref. [65]). The case that all $P_\lambda$ are equal was described in Ref. [66]. Earlier discussions of the expression (451) were given by Lax [30], Francis and Watson [43], and by Glauber [67]. More general analyses have been published by Beg [68] and by Johnson and Watson [65].

[134] In the interest of simplicity, we shall take the mass $M_t$ sufficiently large that the recoil energy $p^2/2M_t$ may be neglected.

Since we have assumed that the scatterer is spin saturated and that all the $P_\lambda$ are equal, any dependence of $T_\alpha$ and $T_\beta$ on $\mathbf{S}_\alpha$ and $\mathbf{S}_\beta$ averages out in Eq. (452). As was discussed in Section 11.4, p. 795, we may replace $\mathbf{q}'$ and $\mathbf{q}$ by $\mathbf{p}$ in $T_\alpha$ and $T_\beta$. In addition, as will become clear, the dominant contribution to the integral over $\mathbf{k}$ in Eq. (452) comes from $\mathbf{k} \cong \mathbf{q}$, so we may set $\mathbf{k} = \mathbf{p}$ in $T_\alpha$ and $T_\beta$. This means that we may replace $T_\alpha$ and $T_\beta$ by the quantity $T^0$, which appears in Eq. (432), and remove these from the integrand, so

$$\langle \mathbf{q}' | \mathscr{V}_{c_2} | \mathbf{q} \rangle = (T^0)^2 \int \frac{d^3k}{\epsilon_p + i\eta - \epsilon_k} \times \sum_{\alpha \neq \beta = 1}^{N} (g_0, e^{-i(\mathbf{q}'-\mathbf{k}) \cdot (\mathbf{z}_\alpha - \mathbf{C})} (1 - \Lambda_0)\, e^{-i(\mathbf{k}-\mathbf{q}) \cdot (\mathbf{z}_\beta - \mathbf{C})} g_0). \tag{453}$$

In terms of the single-particle and two-particle distribution function described by Eqs. (108*d*), (109*c*), and (124*a*), we may write

$$\begin{aligned}(g_0, e^{-i(\mathbf{q}'-\mathbf{k}) \cdot (\mathbf{z}_\alpha - \mathbf{C})} (1 - \Lambda_0)\, e^{-i(\mathbf{k}-\mathbf{q}) \cdot (\mathbf{z}_\alpha - \mathbf{C})} g_0) &= \int d^3r\, d^3r' [\bar{P}^{(2)}(\mathbf{r}, \mathbf{r}') - \bar{P}^{(1)}(\mathbf{r})\bar{P}^{(1)}(\mathbf{r}')]\, e^{-i(\mathbf{q}' \cdot \mathbf{r}' - \mathbf{q} \cdot \mathbf{r})}\, e^{i\mathbf{k} \cdot (\mathbf{r}' - \mathbf{r})} \\ &= \int d^3r\, d^3r' P^{(1)}(\mathbf{r}) P^{(1)}(\mathbf{r}') G(|\mathbf{r} - \mathbf{r}'|)\, e^{-i(\mathbf{q}' \cdot \mathbf{r}' - \mathbf{q} \cdot \mathbf{r})}\, e^{i\mathbf{k} \cdot (\mathbf{r}' - \mathbf{r})}. \end{aligned} \tag{454}$$

Here we have used the approximation of writing the pair correlation function $G(\mathbf{r}, \mathbf{r}')$ as $G(|\mathbf{r} - \mathbf{r}'|)$, which is legitimate because of the assumption that $\mathscr{R} \gg \mathscr{R}_c$.

The assumption that $p\mathscr{R}_c \gg 1$ permits us to evaluate the $\mathbf{k}$-integral in Eq. (453) as

$$\int \frac{d^3k\, e^{i\mathbf{k} \cdot (\mathbf{r}' - \mathbf{r})}}{\epsilon_p + i\eta - \epsilon} \cong -(2\pi)^2 \epsilon_p \frac{e^{ip|\mathbf{r}' - \mathbf{r}|}}{|\mathbf{r}' - \mathbf{r}|}. \tag{455}$$

This result and the expression (454) permit us to write $\langle \mathbf{q}' | \mathscr{V}_{c_2} | \mathbf{q} \rangle$, Eq. (453), as

$$\langle \mathbf{q}' | \mathscr{V}_{c_2} | \mathbf{q} \rangle = -N(N-1)(T^0)^2[(2\pi)^2 \epsilon_p] \times \int d^3r\, d^3r' \bar{P}^{(1)}(\mathbf{r}) \bar{P}^{(1)}(\mathbf{r}') G(|\mathbf{r}' - \mathbf{r}|) \frac{e^{ip|\mathbf{r}' - \mathbf{r}|}}{|\mathbf{r}' - \mathbf{r}|}\, e^{-i(\mathbf{q}' \cdot \mathbf{r}' - \mathbf{q} \cdot \mathbf{r})}. \tag{456}$$

Because it may be assumed that $|q' - q| = \mathscr{O}(\hbar/\mathscr{R})$ and because (by hypothesis) $\mathscr{R} \gg \mathscr{R}_c$, we may set

$$\mathbf{q}' \cdot \mathbf{r}' - \mathbf{q} \cdot \mathbf{r} \cong \mathbf{p} \cdot \mathbf{R} + (\mathbf{q}' - \mathbf{q}) \cdot \mathbf{Y}, \tag{457}$$

where $\mathbf{R} \equiv \mathbf{r}' - \mathbf{r}$ and $\mathbf{Y} \equiv \frac{1}{2}(\mathbf{r}' + \mathbf{r})$. Furthermore, because $\mathscr{R} \gg \mathscr{R}_c$, we may take $\bar{P}^{(1)}(\mathbf{r}') = \bar{P}^{(1)}(\mathbf{r}) = \bar{P}^{(1)}(\mathbf{Y})$, so

$$\langle \mathbf{q}' | \mathscr{V}_{c_2} | \mathbf{q} \rangle \cong -(T^0)^2 (2\pi)^2 \epsilon_p \left[ \int e^{-i(\mathbf{q}'-\mathbf{q}) \cdot \mathbf{Y}} \rho^2(Y)\, d^3Y \right] \times \int d^3R G(R) \frac{e^{i(\mathbf{p} \cdot \mathbf{R} + pR)}}{R}, \tag{458}$$

where $\rho(\mathbf{Y}) = N\bar{P}^{(1)}(\mathbf{Y})$, as usual.

The condition $p\mathscr{R}_c \gg 1$ may be used again to obtain

$$\int \frac{d^3R}{R} G(R)\, e^{i(\mathbf{p} \cdot \mathbf{R} + pR)} \cong \frac{2\pi}{ip} \int dRG(R)\, e^{ipR}(e^{ipR} - e^{-ipR}) \cong -\frac{2\pi}{ip} \mathscr{R}_c, \tag{459}$$

where we now define $\mathscr{R}_c$ precisely as

$$\mathscr{R}_c \equiv \int dRG(R).$$

Finally, Eq. (458) may be transformed to coordinate space to give

$$\langle \mathbf{r}' | \mathscr{V}_{c_2} | \mathbf{r} \rangle = \delta(\mathbf{r}' - \mathbf{r}) \mathscr{V}_{c_2}(\mathbf{r}), \tag{460a}$$

with

$$\mathscr{V}_{c_2}(\mathbf{r}) = \left[ \frac{(2\pi)^6}{ip} \epsilon_p \mathscr{R}_c (T^0)^2 \right] \rho^2(\mathbf{r}). \tag{460b}$$

This may be combined with our first approximation, Eq. (434), to write $\mathscr{V}_c$ in the form

$$\mathscr{V}_c(\mathbf{r}) = \left( -\frac{2\pi c^2}{\epsilon_p} f_+ \right) \rho(\mathbf{r}) \left[ 1 + \frac{2\pi i \rho(\mathbf{r})}{p} \mathscr{R}_c f_+ + \cdots \right]. \tag{461}$$

The expression (461) may easily be generalized to the case that the pair correlation functions are spin dependent. The technique is similar to that used in connection with Eq. (208) and is not described here.[135] Evaluation of (461) when $p\mathscr{R}_c$ is not large compared to unity requires explicit knowledge of the momentum dependence of the $T_\alpha$.

To illustrate the importance of the multiple scattering correction in Eq. (461), we show as a function of energy the quantity

$$\Delta \equiv \text{Re} \left[ \frac{2\pi \rho(0)}{p} \mathscr{R}_c f_+ \right], \tag{462}$$

appropriate for pion-nucleus scattering, in Fig. 11.10.

[135] See Ref. [65] for a detailed discussion.

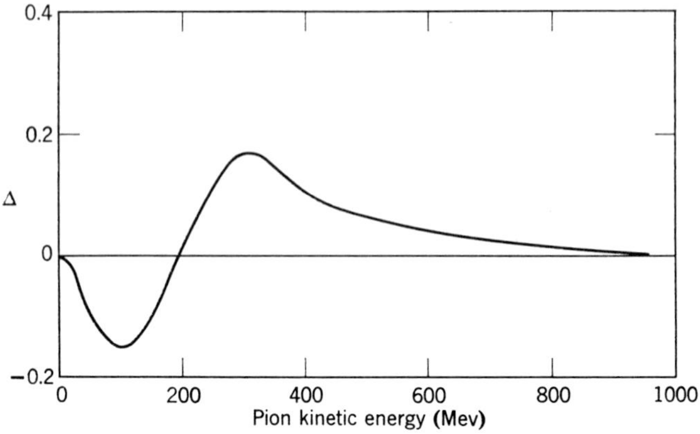


Fig. 11.10 The second-order correction (462) for pion-nucleus scattering as a function of pion kinetic energy.

The optical theorem enables us to estimate the importance of multiple scattering corrections in (461) from

$$\Delta' \equiv \mathrm{Im}\left[\frac{2\pi\rho(0)}{p}\mathscr{R}_c f_+\right] = \frac{\rho(0)}{2}\mathscr{R}_c\bar{\sigma}_f = \frac{\mathscr{R}_c}{2\lambda}, \tag{463}$$

where $\bar{\sigma}_f$ is the total cross section for scattering $i$ by one of the target particles and $\lambda = [\rho(0)\bar{\sigma}_f]^{-1}$ is the scattering mean free path for $i$ within the scatterer. According to our assumptions, $\lambda \gg \mathscr{R}_c$, hence $\Delta'$ is small.

Equation (461) provides, of course, a correction to the expression (332) for the refractive index of a gas. It is not correct, however, to apply this correction to the Lorentz-Lorentz formula (349), which was derived under the assumption that $p\mathscr{R}_c \ll 1$.[136]

## 11.5 COHERENCE AND INCOHERENCE IN MULTIPLE SCATTERING PHENOMENA

In Section 11.3 we gave separate discussions of the multiple scattering equations (266) for incoherent scattering and for coherent scattering. That is, the scattered waves described by Eqs. (293) are all mutually incoherent and also incoherent with the incident wave. On the other hand, the scattered waves described by Eqs. (300) interfere coherently with each other and with the incident wave unless coherence is destroyed by a change in internal state (such as spin flip) of one of the bound scatterers. It may be

[136] The pair correlation function plays an important role in the scattering of light by a gas near its critical point. See Ref. [37, p. 80], for a discussion of this subject. An exact series expansion in terms of correlation functions is given for the dielectric constant in Ref. [40].

easily seen that, in general, we must deal simultaneously with both coherence and incoherence. For example, the optical theorem [see Eq. (386)] states that if there is any scattering, the elastic scattering amplitude $T_c$ cannot vanish. Now, Eqs. (293) describe only inelastic scattering and (as they stand) are consequently inconsistent with the unitarity of the scattering matrix.[137] As another example, we recall that the refractive index obtained using Eqs. (300) in Section 11.3, p. 763 contained an imaginary term. This imaginary term represented the attenuation of the coherent wave due to incoherent scatterings. To describe explicitly these incoherent scatterings, Eqs. (293) would be needed, however.

As was remarked in Section 11.3, we must in general combine Eqs. (293) and (300) into a single set of multiple scattering equations. Since the structure of these equations is the same, this may often be done rather simply. Equations (425) and (426) represent a first step toward accomplishing this result in a precise and automatic fashion. By using the projection operators $\Lambda_0$ and $1 - \Lambda_0$ to control virtual states, scattering in the initial states $g_0$ is explicitly exhibited and determined by the pseudopotential $\mathscr{V}_c$. The unitarity condition on the $S$-matrix is automatically satisfied, as indicated by Eq. (391). Furthermore, because of the appearance of the operators $1 - \Lambda_0$, none of the scattered waves generated by $F$ can interfere with the incident wave.

When the scattering system is sufficiently large that the incident particle scatters many times in passing through it, it seems desirable to generalize the technique used in Eqs. (425). For example, let us suppose that the first inelastic scattering occurs on a bound particle $\alpha_1$, exciting the scatterer to a discrete state $\gamma_1$, and that the particle $i$ travels sufficiently far before scattering again that the scatterer may be observed to be in the state $\gamma_1$. Before the second scattering, but after the first scattering, the state $\gamma_1$ may be considered as an initial state, just as was 0 before the first scattering. That is, we may observe a particle $i$ about to interact with a system in the state $\gamma_1$. Thus a new unitarity condition and a new question of coherence must be studied. This suggests that we define a psuedopotential $\langle, \gamma_1 | \mathscr{V}_c |, \gamma_1\rangle$ for scattering in the state $\gamma_1$. Extending this argument, we may introduce a pseudopotential $\langle, \gamma | \mathscr{V}_c |, \gamma\rangle$ for elastic scattering in each state $\gamma$ that is excited by an inelastic scattering of particle $i$. Having done this, we shall have obtained a set of multiple scattering equations that account automatically for all coherent and incoherent scatterings.

Methods of doing this are discussed in this section. The scattering interaction is again just that described in Section 11.2, p. 708, that is, a particle $i$ having rest-mass $m$ and spin $S_i$ interacts with a composite system

[137] The reason for this is that in obtaining Eqs. (293) from Eqs. (266) we neglected all scatterings not satisfying the conditions (162).

of $N$ bound scatterers. Since the techniques to be developed are unnecessarily elaborate for application to small scatterers, we shall assume at the outset that

$$N \gg 1. \tag{464}$$

Otherwise, we shall assume the conditions described in the introduction to Section 11.3.

**The Generalized Pseudopotential and the Identification of Coherent Scatterings**

In pursuit of the objective just described, we shall show how to construct a pseudopotential for "elastic scattering" in excited states. The method will then be incorporated within the framework of the impulse approximation to express all interactions in terms of the two-body scattering operators $\mathscr{T}_\alpha$ ($\alpha = 1, 2, \ldots, N$).

To accomplish this, we again attempt to satisfy the Schrödinger equation (250) with a solution of the form (405)

$$\psi_a^{\ +} = F\psi_{c(a)}^+, \tag{465}$$

but we now write the equation for $F$ in a form differing from that given in Section 11.4. The pseudopotential $\mathscr{V}_c$ will be defined for scattering in any state $\gamma$ of the scattering system; that is, we shall generalize Eq. (371) by writing

$$\langle \mathbf{k}', \nu', \mathbf{Q}', \gamma', \nu'_t| \mathscr{V}_c |\mathbf{k}, \nu, \mathbf{Q}, \gamma, \nu_t\rangle$$
$$= \delta_{\gamma',\gamma}\langle \mathbf{k}', \nu', \mathbf{Q}', \gamma, \nu'_t| \mathscr{V}_c |\mathbf{k}, \nu, \mathbf{Q}, \gamma, \nu_t\rangle$$
$$= \delta_{\gamma',\gamma}\delta(\mathbf{k}' + \mathbf{Q}' - \mathbf{k} - \mathbf{Q})V_c, \tag{466a}$$

where

$$V_c = V_c[\boldsymbol{\varkappa}', \boldsymbol{\varkappa}, \mathbf{S}_i, \mathbf{S}_t(\gamma), \gamma]. \tag{466b}$$

Here $S_t(\gamma)$ is the spin operator for the target when in the state $\gamma$ and $\boldsymbol{\varkappa}$ and $\boldsymbol{\varkappa}'$ are, respectively, initial and final barycentric momenta, as in Eq. (372). The quantities $\mathscr{V}_c$ and $V_c$ are seen to be

$$(2S_i + 1)[2S_t(\gamma) + 1] \times (2S_i + 1)[2S_t(\gamma) + 1]$$

matrices in spin space.

For the states corresponding to $\gamma = 0$ the pseudopotential (466) will be taken as identical with that given by Eq. (371). Equations (375) are therefore unchanged. These are

$$\psi_{c(a)}^+ = \chi_a + \frac{1}{d}\mathscr{V}_c\psi_{c(a)}^+ = \chi_a + \frac{1}{e}\mathscr{V}_c\psi_{c(a)}^+, \tag{467}$$

where $d$ and $e$ are defined by Eqs. (251) and (370), respectively. We emphasize again that since $\mathscr{V}_c$ is diagonal in the states $\gamma$, it is only those states corresponding to $\gamma = 0$ in which scattering occurs in Eq. (467).

To obtain an integral equation for the operator $F$ in Eq. (465), we introduce a "counting operator" $P$ which will be defined in such a way as to restrict those virtual states $\gamma$ of the medium that can occur in a sequence of scatterings. The operator $P$ will not act on the spin orientation $\nu_t$ of the scattering system nor will it act on the dynamical variables of the scattered particle $i$. $P$ will be, in fact, just an "instruction symbol" that will tell us which intermediate states to keep in a sum over virtual states. The operator $1 - P$ will represent an "instruction symbol" to keep only those virtual states $\gamma$ which are to be omitted when $P$ appears. Thus

$$(1 - P) + P$$

is equivalent to the instruction that we sum over all states $\gamma$ when it appears between two operators in a matrix product.

With this understanding, and for the moment postponing a precise definition of $P$, we write

$$F = 1 + \frac{1}{\mathscr{d}} PVF \tag{468}$$

as the equation to determine $F$ in Eq. (465). Here

$$\mathscr{d} \equiv d - \mathscr{V}_c. \tag{469}$$

By successively substituting the expression (468) for $F$ into the right-hand side of this equation, we obtain the series

$$\begin{aligned} F &= 1 + \frac{1}{\mathscr{d}_1} PV + \frac{1}{\mathscr{d}_2} PV \frac{1}{\mathscr{d}_1} PV + \cdots \\ &= \sum_{n=0}^{\infty} F_n, \end{aligned} \tag{470a}$$

where

$$F_0 = 1, \qquad F_1 = \frac{1}{\mathscr{d}_1} PV, \tag{471a}$$

and

$$F_{n+1} = \frac{1}{\mathscr{d}_{n+1}} PVF_n, \qquad n \geq 1. \tag{471b}$$

Here we have introduced a subscript on each $\mathscr{d}$ to indicate the number of interactions that occur to the right of it. This is, in effect, a subscript on $\mathscr{V}_c$, since $\mathscr{d}_n = d - \mathscr{V}_{cn}$, etc.

We are now able to complete the definition of the counting operator $P$. This will be defined as an instruction to *prohibit the occurrence of any state $\gamma$ more than once* in a given term $F_n$ in the expansion (470*a*). Expressed somewhat differently, once a state $\gamma$ appears as a matrix label, $\langle, \gamma | V |, \gamma' \rangle$, in a series of interactions in a given $F_n$, we must omit $\gamma$ in all subsequent sums over states for interactions to the left of $\langle, \gamma| V |, \gamma'\rangle$. The operator $P$ places no other constraint on the sums over states in Eqs. (470$a$) or (471). To illustrate these instructions explicitly, we write out the first three terms in the expansion (470$a$). Using the notation introduced in connection with Eq. (407$c$), we do not explicitly indicate matrix labels or sums over the states of particle $i$ or the spin orientation $\nu_t$.

$$\begin{aligned}\langle, \gamma| F |, 0\rangle = \delta_{\gamma,0} &+ \frac{1}{\mathscr{d}_\gamma(0)} \langle, \gamma| V |, 0\rangle \\ &+ \sum_{\gamma'(\neq 0,\gamma)} \frac{1}{\mathscr{d}_\gamma(0, \gamma')} \langle, \gamma| V |, \gamma'\rangle \frac{1}{\mathscr{d}_{\gamma'}(0)} \langle, \gamma'| V |, 0\rangle \\ &+ \sum_{\gamma''(\neq 0,\gamma)} \sum_{\gamma'(\neq 0,\gamma,\gamma'')} \frac{1}{\mathscr{d}_\gamma(0, \gamma', \gamma'')} \langle, \gamma| V |, \gamma'\rangle \\ &\times \frac{1}{\mathscr{d}_{\gamma'}(0, \gamma'')} \langle, \gamma'| V |, \gamma''\rangle \frac{1}{\mathscr{d}_{\gamma''}(0)} \langle, \gamma''| V |, 0\rangle + \cdots.\end{aligned} \tag{470b}$$

We have here written the $\gamma$th matrix element of the diagonal matrix $\mathscr{d}$[138] as $\mathscr{d}_\gamma$. We have also anticipated that each propagator may depend on those states $\gamma$ to which previous scatterings have led, and we have therefore written $\mathscr{d}_\gamma(0, \gamma')$, and so on, in Eq. (470$b$).

We now assert that our definition of $P$ and Eq. (468) imply that

$$\langle, \gamma| (1 - P)VF |, 0\rangle = \langle, \gamma| \mathscr{V}_c |, \gamma\rangle\langle, \gamma| F |, 0\rangle, \tag{472a}$$

which in operator form reads

$$(1 - P)VF = \mathscr{V}_c F. \tag{472b}$$

We have indicated in Eqs. (470) and (471) that $\mathscr{V}_c$ depends on previous states $\gamma$ in each term of the perturbation expansion (470) for $F$. This means that we must interpret Eq. (472$b$) as reading, in more detail,

$$(1 - P)VF = \sum_{n=0}^{\infty} \mathscr{V}_{cn} F_n, \tag{472c}$$

where the $F_n$ are defined by Eqs. (471) and $\mathscr{V}_{cn}$ is the appropriate pseudopotential to act on $F_n$ [as in Eq. (470$b$)].

Let us now show that Eqs. (465), (468), and (472) provide a solution to the Schrödinger equation (250). On substituting Eq. (465) into the right-hand side of (250), we obtain

$$\psi_a^+ = \chi_a + \frac{1}{d} V\psi_a^+ = \chi_a + \frac{1}{d} VF\psi_{c(a)}^+. \tag{473}$$

[138] The $d$ are, of course, not generally diagonal in variables other than $\gamma$.

On using the relation

$$\frac{1}{\mathscr{d}_{n+1}} = \frac{1}{d} + \frac{1}{d}\mathscr{V}_{cn+1}\frac{1}{\mathscr{d}_{n+1}},$$

Eq. (470$a$), and the fact that $(1 - P) + P$ is a unit operator, it is seen that

$$\begin{aligned}\frac{1}{d}VF &= \frac{1}{d}PVF + \frac{1}{d}(1-P)VF\\ &= \sum_{n=0}^{\infty}\frac{1}{\mathscr{d}_{n+1}}PVF_n - \frac{1}{d}\sum_{n=0}^{\infty}\mathscr{V}_{cn+1}\frac{1}{\mathscr{d}_{n+1}}PVF_n + \frac{1}{d}(1-P)VF\\ &= F - 1 + \frac{1}{d}\left[(1-P)VF - \sum_{n=1}^{\infty}\mathscr{V}_{cn}F_n\right].\end{aligned} \tag{474}$$

Here we have used the recurrence relation (471$b$) for the $F_n$'s and Eq. (470$a$) to set $\sum_{n=1}^{\infty} F_n = F - 1$. Now, according to Eq. (472$c$), Eq. (474) is equivalent to

$$\frac{1}{d}VF = F - 1 + \frac{1}{d}\mathscr{V}_{c0}. \tag{475}$$

The pseudopotential $\mathscr{V}_{c0}$ is that which can act on $F_0 = 1$. This means that it is just $\langle, 0|\mathscr{V}_c|, 0\rangle$, or the pseudopotential appearing in Eq. (467). We may therefore substitute Eq. (475) into Eq. (473) and use Eq. (467) to obtain

$$\psi_a^{+} = \chi_a + (F-1)\psi_{c(a)}^{+} + \frac{1}{d}\mathscr{V}_c\psi_{c(a)}^{+} = F\psi_{c(a)}^{+}.$$

This completes our demonstration that Eqs. (465), (468), and (472) provide a solution to the Schrödinger equation (250).

It remains, however, for us to prove Eq. (472) and to define the complete pseudopotential $\mathscr{V}_c$. Returning to the form (472$a$) for this equation, we recall that the operator $1 - P$ instructs us to insert the virtual state $\gamma$ somewhere in the expansion (470) of the left-hand side of this equation. The rules for matrix multiplication imply that

$$\begin{aligned}\langle, \gamma|(1-P)VF|, 0\rangle &= \langle, \gamma|V|, \gamma\rangle\langle, \gamma|F|, 0\rangle\\ &\quad + \sum_{\gamma'(\neq\gamma)}\langle, \gamma|V|, \gamma'\rangle\langle, \gamma'|F|, 0\rangle.\end{aligned} \tag{476}$$

Because of the $P$-operator in Eq. (468), the state $\gamma$ does not appear as a virtual state in $\langle, \gamma|F|, 0\rangle$. The operator $1 - P$ requires, however, that $\gamma$ appears as a virtual state in $\langle, \gamma'|F|, 0\rangle$ in the second line of (476).

Therefore

$$\langle, \gamma'|\, F\,|, 0\rangle = \langle, \gamma'|\,, 0\rangle + \left\langle, \gamma' \left| \frac{1}{d} PV \right|, \gamma \right\rangle \langle, \gamma|\, F\,|, 0\rangle$$
$$+ \sum_{\gamma''(\neq \gamma, \gamma')} \left\langle, \gamma' \left| \frac{1}{d} PV \right|, \gamma'' \right\rangle \langle, \gamma''|\, F\,|, 0\rangle, \quad (477)$$

where $\langle, \gamma''|\, F\,|, 0\rangle$ must contain $\gamma$ as a virtual state. When the expression (477) is inserted into (476), the term $\langle, \gamma'\,|, 0\rangle$ vanishes since $1 - P$ requires that the state $\gamma$ be repeated. Thus

$$\langle, \gamma|\,(1 - P)VF\,|, 0\rangle = \left\langle, \gamma \left| \left(V + V\frac{P}{d}V\right) \right|, \gamma \right\rangle \langle, \gamma|\, F\,|, 0\rangle$$
$$+ \sum_{\gamma''(\neq \gamma)} \left\langle, \gamma|\,(1 - P)V\frac{P}{d}V\,|, \gamma'' \right\rangle \langle, \gamma''|\, F\,|, 0\rangle$$

Continuing in this manner, and using Eq. (477) at each step, we finally obtain

$$\langle, \gamma|\,(1 - P)VF\,|, 0\rangle = \langle, \gamma|\, VF\,|, \gamma\rangle\langle, \gamma|\, F\,|, 0\rangle. \quad (478)$$

Comparison of this expression with Eq. (472*a*) shows that

$$\langle, \gamma'|\,\mathscr{V}_c\,|, \gamma\rangle = \delta_{\gamma',\gamma}\langle, \gamma|\, VF\,|, \gamma\rangle, \quad (479)$$

which exhibits an explicit form for the pseudopotential (466).[139]

Equations (468) and (479) effect a complete separation of the coherent from the incoherent scattering. Since there is never a return to a previous state $\gamma$ in the expansion of Eq. (468), each scattering is *completely incoherent* with the incident wave and with each other scattering, all coherent interactions being taken into account by the propagators $\mathscr{d}^{-1}$. We observe, incidentally, that Eq. (468) is a rather straightforward generalization of Eq. (406).

Some care is evidently needed in evaluating the equation (468) for $F$, since the propagators $\mathscr{d}^{-1}$ [Eq. (469)] contain the pseudopotential $\mathscr{V}_c$, which in turn contains the operator $P$ and the propagators $\mathscr{d}^{-1}$ [see Eq. (479)]. Thus, in evaluating Eq. (468), we must omit virtual states $\gamma$ in $\mathscr{V}_c$ in the denominators when these states occur in the numerators, and so on.[140] It was for this reason that we placed labels on the $\mathscr{d}$'s in Eqs. (470) and (471). The procedure for evaluating the terms in the expansion (470*a*) is straightforward and may be done systematically. This may be illustrated with the simple example of a scatterer which has only two states $\gamma = 0$ and $\gamma = 1$.

[139] This expression, together with Eqs. (486) and (472), was first obtained in Ref. [33].

[140] We might express this by saying that one should "look only up into the numerators" and "not down into the denominators" in interpreting the instructions of the $P$-operator.

There are then four $F$-matrices,

$$F_{00}, F_{10}, F_{01}, \quad \text{and} \quad F_{11},$$

where $F_{10} = \langle, 1|\, F\, |, 0\rangle$, and so on. For an initial state $|, 0\rangle$, only $F_{00}$ and $F_{10}$ are, of course, required. Now, taking $K\,|, 1\rangle \equiv E_1\,|, 1\rangle$ in $\mathscr{d}$ [see Eq. (251)], Eq. (468) reads

$$F_{00} = 1,$$
$$F_{10} = \frac{1}{E_a + i\eta - E_1 - \langle, 1|\,\mathscr{V}_c\,|, 1\rangle}\,\langle, 1|\, V\,|, 0\rangle.$$

These expressions are exact, so it is seen that the series (470$a$) has in each case but one nonvanishing term. In the equation for $F_{10}$ we have

$$\langle, 1|\,\mathscr{V}_c\,|, 1\rangle = \langle, 1|\, V\,|, 1\rangle,$$

since the $P$-operator in $\langle, 1|\,\mathscr{V}_c\,|, 1\rangle$ forbids the reappearance of the state $\gamma = 0$ in the denominator. Equation (479) for $\langle, 0|\,\mathscr{V}_c\,|, 0\rangle$ now gives

$$\begin{aligned}\langle, 0|\,\mathscr{V}_c\,|, 0\rangle &= \langle, 0|\, VF\,|, 0\rangle \\ &= \langle, 0|\, V\,|, 0\rangle F_{10} + \langle, 0|\, V\,|, 0\rangle F_{00} \\ &= \langle, 0|\, V\,|, 0\rangle + \langle, 0|\, V\,|, 1\rangle \frac{1}{E_a + i\eta - E_1 - \langle, 1|\, V\,|, 1\rangle}\,\langle, 1|\, V\,|, 0\rangle \end{aligned} \tag{480}$$

which is exact.

When only a finite number of states $\gamma$ of the scatterer contribute significantly to the scattering, and thus need to be kept as virtual states, the series (470$a$) always terminates after a finite number of terms. In this case the pseudopotential and the propagators $\mathscr{d}^{-1}$ may be written out explicitly,[141] as in Eq. (480). When a large number of degrees of freedom of the composite system are excited by the scattering, this becomes impractical.

Fortunately, when complex excitations of the scatterer occur, we can ordinarily ignore this problem. That is, we can evaluate the pseudopotential $\mathscr{V}_c$ in the propagators $\mathscr{d}^{-1}$ in Eq. (468) just as if *no previous scatterings* had occurred. [This means that the subscripts $n$ on the $\mathscr{d}$'s may be omitted in Eqs. (470$a$) and (471) or that the dependence on previous scatterings in Eq. (470$b$) may be ignored.] This will generally be the case if the states $\gamma$ lie in the continuum, since the specific states to be omitted have zero measure in the integrals over virtual states. Again, when the condition (133) obtains, we can ignore previous states when evaluating the propagators. Here successive scatterings in the medium occur in uncorrelated

[141] The methods of "closely coupled states" and of "near adiabatic transitions" used in atomic and molecular interactions provide familiar examples of the use of only a finite number of states of the scatterer.

regions of the scattering system and therefore excite *different degrees of freedom* of the system. The condition (133) implies that the coupling between these degrees of freedom, which would be necessary to return the system to a previously occurring state $\gamma$, is missing. There are yet other conditions for which previous states may be ignored in the evaluation of $\mathscr{V}_c$ in the propagators. For example, the *statistical weight* in the sum-over-states of the (few) states excluded by $P$ may be negligible.

When a suitable approximation (as indicated earlier) is available to evaluate the pseudopotential in the $\mathscr{d}^{-1}$ in Eq. (468), this equation can provide a useful starting point for developing perturbation expansions for the scattering amplitude, because of the complete isolation of coherent from incoherent scattering in Eqs. (468) and (479). Since the waves scattered incoherently can never interfere with one another, we might anticipate that an expansion in a series of incoherent scatterings will be relatively simple. On the other hand, coherently scattered waves may reinforce one another and require consideration of very many scatterings.[142] Fortunately, the evaluation of $\mathscr{d}^{-1}$ [Eq. (469)] requires only the study of simple potential scattering and may be done exactly (if only by numerical methods)[143] once the pseudopotential has been obtained.

#### Introduction of Two-Body Scattering Matrices

For most applications when the weak binding limit obtains, it is desirable to reexpress Eqs. (468) and (479) in terms of the two-body scattering matrices $\mathscr{T}_\alpha$, using the impulse approximation. To do this, we first define a new set of scattering operators for bound particles.[144]

$$t_\alpha = V_\alpha + V_\alpha \frac{P_0}{\mathscr{d}} t_\alpha, \qquad \alpha = 1, 2, \ldots, N. \tag{481}$$

Here $\mathscr{d}$ is defined by Eq. (469) and $P_0$ is a counting operator yet to be defined. We shall assume that $P_0$ satisfies the condition (472), so Eq. (468) is valid with $P$ replaced by $P_0$. This equation is seen to be satisfied by that $F$ which is obtained from a solution of the equations

$$\begin{aligned} F &= 1 + \frac{P_0}{\mathscr{d}} \sum_{\alpha=1}^{N} t_\alpha F_\alpha, \\ F_\alpha &= 1 + \frac{P_0}{\mathscr{d}} \sum_{\beta(\neq\alpha)=1}^{N} t_\beta F_\beta. \end{aligned} \tag{482}$$

[142] For example, the motion of a particle in a potential in classical mechanics corresponds to the limit of very many "scatterings," each accompanied by a very small momentum transfer.

[143] Application of Eqs. (468) and (479) to atomic and molecular scattering has been considered in Ref. [49].

[144] It should be noted that we are using the same symbol $t_\alpha$ here that was used in Eqs. (252) and (413) to represent somewhat different scattering operators.

To show this,[145] we observe that the relation

$$V\frac{P_0}{d}\sum_{\alpha=1}^{N}t_\alpha F_\alpha = \sum_{\alpha=1}^{N}t_\alpha F_\alpha - V \tag{483}$$

follows from the definition (481), the argument being that given in Eq. (416). On substituting the first of Eqs. (482) into Eq. (468) and using the second of Eqs. (482) and Eq. (483), there results

$$\begin{aligned} F &= 1 + \frac{1}{d}P_0 V + \frac{1}{d}P_0 V\sum_{\alpha=1}^{N}t_\alpha F_\alpha \\ &= 1 + \sum_{\alpha=1}^{N}\frac{P_0}{d}t_\alpha F_\alpha, \end{aligned} \tag{484}$$

which completes our demonstration.

To satisfy the condition (472) we must give $P_0$ a suitable definition. It would evidently be very inconvenient to keep account of all virtual states that occur in a given scattering operator $t_\alpha$, as would be required if we chose $P_0$ to be the counting operator used in Section 11.5, p. 807. We therefore define $P_0$ to be that operator which prohibits repetition of those states $\gamma$ of the medium that occur as an initital or final state of any $t_\alpha$. Thus $P_0$ does not allow reoccurrence of the initial state $g_0$ or of any other state that has once appeared between two $t_\alpha$'s. The counting operator $P_0$ does not act on states of particle $i$ or restrict sums over the spin orientation $\nu_t$ of the target.

As was done in Eq. (470*b*), we illustrate the application of $P_0$ by exhibiting the first three terms in the expansion of the second of Eqs. (482). Following the notation of Eq. (470*b*), we obtain

$$\begin{aligned} \langle, \gamma|\, F_\alpha \,|, 0\rangle &= \delta_{\gamma,0} + \frac{1}{d_\gamma(0)}\sum_{\beta(\neq\alpha)=1}^{N}\langle, \gamma|\, t_\beta \,|, 0\rangle \\ &+ \sum_{\gamma'(\neq 0,\gamma)}\sum_{\beta(\neq\alpha)=1}^{N}\sum_{\sigma(\neq\beta)=1}^{N}\frac{1}{d_\gamma(0, \gamma')}\langle, \gamma|\, t_\beta \,|, \gamma'\rangle\frac{1}{d_{\gamma'}(0)}\langle, \gamma'|\, t_\sigma \,|, 0\rangle \\ &+ \sum_{\gamma''(\neq 0,\gamma)}\sum_{\gamma'(\neq 0,\gamma,\gamma'')}\sum_{\beta(\neq\alpha)}\sum_{\sigma(\neq\beta)}\sum_{\mu(\neq\sigma)}\frac{1}{d_\gamma(0, \gamma', \gamma'')} \\ &\times \langle, \gamma|\, t_\beta \,|, \gamma'\rangle\frac{1}{d_{\gamma'}(0, \gamma'')}\langle, \gamma'|\, t_\sigma \,|, \gamma''\rangle\frac{1}{d_{\gamma''}(0)}\langle, \gamma''|\, t_\sigma \,|, 0\rangle + \cdots. \end{aligned}$$

Equation (483) may be rewritten in the form

$$VF = \sum_{\alpha=1}^{N}t_\alpha F_\alpha, \tag{485}$$

[145] Equations (482) were obtained in Ref. [33].

which permits us to express the left-hand side of Eq. (472) [or Eq. (475)] as

$$\langle, \gamma|\,(1 - P_0)\sum_\alpha t_\alpha F_\alpha\,|, 0\rangle = \sum_\alpha \langle, \gamma|\, t_\alpha\,|, \gamma\rangle\langle, \gamma|\, F_\alpha\,|, 0\rangle + \sum_{\gamma'(\neq\gamma)} \langle, \gamma|\,(1 - P_0)\sum_\alpha t_\alpha|, \gamma'\rangle\langle, \gamma'|\, F_\alpha\,|, 0\rangle. \tag{486}$$

Since the operator $1 - P_0$ instructs us to locate a state $\gamma$ between a pair of $t$-matrices, we see that $\langle, \gamma'|\, F_\alpha\,|, 0\rangle$ must contain a virtual state $\gamma$ once, but only once. Because of the $P_0$-operator, $\langle, \gamma|\, F_\alpha\,|, 0\rangle$ will not contain a virtual state $\gamma$, except following the last scattering.[146]

Now

$$\sum_\alpha \langle, \gamma|\, t_\alpha\,|, \gamma\rangle\langle, \gamma|\, F_\alpha\,|, 0\rangle = \langle, \gamma|\sum_\alpha{}' t_\alpha|\,, \gamma\rangle\langle, \gamma|\, F\,|, 0\rangle, \tag{487}$$

where the prime on the summation symbol means that we must omit that index $\alpha$ that corresponds to the particle from which the last scattering in $F$ occurred. (For the unit matrix in $F$ the sum on $\alpha$ is unrestricted, of course.) We emphasize that Eq. (487) describes merely a somewhat trivial rearrangement of the sums over particles that appear in the first term on the right in Eq. (486).

We may write the second term on the right in Eq. (486) in the form

$$\begin{aligned}
&\sum_{\gamma'(\neq\gamma)}\sum_\alpha \langle, \gamma|\,(1 - P_0)t_\alpha\,|\,, \gamma'\rangle\langle, \gamma'|\, F_\alpha\,|, 0\rangle \\
&= \sum_{\gamma'(\neq\gamma)}\sum_{\alpha\neq\beta=1}^{N} \langle, \gamma|\,(1 - P_0)t_\alpha\,|, \gamma'\rangle\left\langle, \gamma'\left|\frac{P_0}{d}\, t_\beta F_\beta\right|, 0\right\rangle \\
&= \sum_{\gamma''(\neq\gamma',\gamma)}\sum_{\gamma'(\neq\gamma)} \langle, \gamma|\,(1 - P_0)t_\alpha|\,, \gamma'\rangle\left\langle, \gamma'\left|\frac{P_0}{d}\, t_\beta\right|, \gamma''\right\rangle \\
&\quad\times \langle, \gamma''|\, F_\beta,\,|\, 0\rangle + \sum_{\alpha\neq\beta=1}^{N}\left\langle, \gamma\left|\, t_\alpha\frac{P_0}{d}\, t_\beta\right|, \gamma\right\rangle\langle, \gamma|\, F_\beta\,|, 0\rangle.
\end{aligned} \tag{488}$$

Here we have been able to drop the unit matrix in $F_\alpha$ since this would give a term $\langle, \gamma|\,(1 - P_0)\, t_\alpha\,|, 0\rangle$ in Eq. (488), which vanishes because $1 - P_0$ requires another state $\gamma$. The argument given in connection with Eq. (487) permits us to reorder the sums over particles in the last term of Eq. (488) to write this as

$$\left\langle, \gamma\left|\sum_{\alpha\neq\beta}{}' t_\alpha\frac{P_0}{d}\, t_\beta\right|, \gamma\right\rangle\langle, \gamma|\, F\,|, 0\rangle. \tag{489}$$

Continuing in this manner, we obtain Eq. (472) with the pseudopotential defined as

$$\langle, \gamma|\,\mathscr{V}_c\,|, \gamma\rangle = \left\langle, \gamma\left|\sum_{\alpha=1}^{N}{}' t_\alpha F_\alpha\right|, \gamma\right\rangle. \tag{490a}$$

[146] We repeat that $P_0$ does not instruct us to prohibit reoccurrence of virtual states that have appeared within a $t$-matrix.

The prime here is interpreted, as in Eqs. (487) and (489), to imply that in $\mathscr{V}_c$ the first scattering from a single particle, say $\alpha_0$, is to be omitted if $\mathscr{V}_c$ follows immediately after a term $(P_0/\mathscr{d})\mathscr{t}_{\alpha_0}$ in $F$. We shall see that this rather cumbersome restriction may, in fact, be ignored.

To show this we recall the condition (464) that $N \gg 1$.[147] The effect of omitting scattering from a single particle is to modify $\mathscr{V}_c$ by an amount of relative order $1/N$. Since it has been assumed that $N \gg 1$, we may ignore this correction and thus the prime on the summation in Eq. (490*a*) to obtain simply

$$\langle, \gamma | \mathscr{V}_c |, \gamma \rangle = \left\langle, \gamma \left| \sum_{\alpha=1}^{N} \mathscr{t}_\alpha F_\alpha \right|, \gamma \right\rangle. \tag{490b}$$

That one additional particle will have a negligible effect on scattering from a system composed of many particles is not surprising.[148]

Let us next make use of the weak binding condition to simplify the scattering operators (481). This implies the validity of the impulse approximation, which permits us to neglect the binding of particle $\alpha$ in the medium and to write

$$\mathscr{d} \cong \epsilon_p + \epsilon_{\alpha, P_\alpha} + i\eta - K_i - K_\alpha - \mathscr{V}_c, \tag{491}$$

as in Eq. (262). The operator $P_0$ in Eqs. (481) plays a role not significantly different from that of $1 - \Lambda_0$ in Eqs. (413). The prohibition of states $\gamma$ resulting from scatterings at a distance greater than the correlation range $\mathscr{R}_c$ from particle $\alpha$ is of no consequence.[149] The effect on $\mathscr{t}_\alpha$ caused by $\boldsymbol{P_0}$ resulting from scatterings within a distance $\mathscr{R}_c$ from particle $\alpha$ is of relative order $(f/\mathscr{R})(\hbar/p\mathscr{R})$, as in Eq. (423). Unless the ratio of the scattering mean free path $\lambda$ to $\mathscr{R}_c$ is of this order, we may use Eq. (424) to write

$$\mathscr{t}_\alpha = t_\alpha \left[ 1 + \mathscr{O}\left( \left| \frac{f_\alpha}{\lambda} \frac{R_{\alpha,a} t_{\alpha,c}}{\hbar} \right| \right) + \mathscr{O}\left( \left| \frac{f_\alpha}{\lambda} \left( \frac{\hbar}{p\mathscr{R}} \right)^2 \right| \right) \right], \tag{492}$$

where

$$t_\alpha = V_\alpha + V_\alpha \frac{1}{\epsilon_p + \epsilon_{\alpha, P_\alpha} + i\eta - K_\alpha - (K_i + \mathscr{V}_c)} t_\alpha. \tag{493}$$

For many applications the potential $\mathscr{V}_c$ depends only on the gross properties of the scattering medium, so its dependence on the precise, instantaneous state $\gamma$ and on the previous scatterings[150] may be neglected

[147] We have not previously used this condition in this section.

[148] In special circumstances the omitted scatterer in Eq. (490*a*) might have an identifiable signature which would make (490*b*) invalid even when $N \gg 1$. In this case Eq. (490*a*) must be used.

[149] This argument was developed on p. 811.

[150] This has been discussed on p. 811. This point is again reviewed at the end of this section.

in Eq. (493). Thus $t_\alpha$ is a two-body scattering matrix that describes the scattering of particles $i$ and $\alpha$ while $i$ is moving in a prescribed potential $\mathscr{V}_c$.[151] When the scattering system is uniform, spin saturated, and large compared with the wavelength $\hbar/p$, the pseudopotential $\mathscr{V}_c$ leads only to forward scatterings within the medium. This means that $\mathbf{Q}' = \mathbf{Q}$ and $\boldsymbol{\varkappa}' = \boldsymbol{\varkappa}$ in Eqs. (466). Therefore

$$(K_i + \mathscr{V}_c)\,|\mathbf{q}, \gamma\rangle = [\epsilon_q + V_0(q)]\,|\mathbf{q}, \gamma\rangle \equiv \epsilon_q{}^*\,|\mathbf{q}, \gamma\rangle. \tag{494}$$

Here $|\mathbf{q}, \gamma\rangle$ describes a plane-wave state of momentum $\mathbf{q}$ for the incident particle and $V_0(q)$ is the value of the well depth $V_c$ [Eq. (466$b$)] when particle $i$ has a momentum $\mathbf{q}$. The incident particle thus behaves dynamically as if it were a quasi-particle of energy $\epsilon_q{}^* = \epsilon_q + V_0(q)$. This property is also maintained between scatterings, since the quantities $\mathscr{d}$ in Eqs. (482) contain the energy $K_i + \mathscr{V}_c$. In other words, the effect of $\mathscr{V}_c$ in a large uniform scattering system is to replace the kinetic energy $\epsilon_q$ of particle $i$ by an effective energy $\epsilon_q{}^*$.

It is often possible to write $V_0(q)$ in the approximate form

$$V_0(q) = a + bq^2,$$

where $a$ and $b$ are constants. Then if $\epsilon_q = (1/2m)q^2$, we have

$$\epsilon_q{}^* = \frac{1}{2m^*}q^2,$$

to within an additive constant, where

$$\frac{1}{m^*} = \frac{1}{m} + 2b.$$

The quantity $m^*$ is called an effective mass.

It was observed in Section 11.1, p. 689 that the weak binding restrictions on the impulse approximation may often be relaxed if we also assign a quasi-particle energy to the target particle $\alpha$. In this case $\epsilon_{\alpha,P_\alpha}$ in Eq. (491) is replaced by some function $\epsilon^*_{\alpha,P_\alpha}$ of the momentum $P_\alpha$. The two-body scattering matrices then describe the scattering of two quasi-particles.[152]

Extensive studies of scattering matrices using quasi-particle energies have been made by Brueckner and his collaborators[153] in their investigations of nuclear structure. The effective mass approximation is frequently

[151] Effective two-body scattering operators, as in Eq. (493), were introduced in Ref. [31].

[152] We saw in Chapter 3 and in Section 6.1 that the scattering of quasi-particles may be treated in much the same manner as that of ordinary particles.

[153] See Ref. [8], where further references may be found.

used in studying electron interactions in solids. Other applications of Eqs. (493) have been made in analyses of nuclear optical models.[154]

The interaction $\mathscr{V}_c$ may usually be neglected in Eq. (493) in the weak binding limit.[155] The condition for neglecting $\mathscr{V}_c$ in the two-body scattering matrices may be obtained on following the arguments leading to Eq. (264). We find

$$t_\alpha = \mathscr{T}_\alpha\left[1 + \mathscr{O}\left(\left|\frac{f_\alpha}{\lambda\!\!\!^{-}}\frac{V_c t_{\alpha,c}}{\hbar}\right|\right)\right], \tag{495}$$

where $\mathscr{T}_\alpha$ is the two-body operator (258). When the correction terms in Eqs. (492) and (495) are negligible, we may simply replace the $t_\alpha$'s by the $\mathscr{T}_\alpha$'s in Eqs. (482) and (490). Here the wave function (465) takes the form

$$\begin{aligned} \psi_a^{+} &= F\psi_{c(a)}^{+}, \\ F &= 1 + \frac{P_0}{\mathscr{d}}\sum_{\alpha=1}^{N}\mathscr{T}_\alpha F_\alpha, \\ F_\alpha &= 1 + \frac{P_0}{\mathscr{d}}\sum_{\beta(\neq\alpha)=1}^{N}\mathscr{T}_\beta F_\beta. \end{aligned} \tag{496}$$

The pseudopotential is now

$$\langle,\gamma'|\mathscr{V}_c|,\gamma\rangle = \delta_{\gamma',\gamma}\left\langle,\gamma\left|\sum_{\alpha=1}^{N}\mathscr{T}_\alpha F_\alpha\right|,\gamma\right\rangle. \tag{497}$$

If we define wave functions $\psi_\alpha \equiv F_\alpha\psi_{c(a)}^{+}$, we may reexpress Eqs. (496) as

$$\begin{aligned} \psi_a^{+} &= \psi_{c(a)}^{+} + \sum_{\alpha=1}^{N}\frac{P_0}{\mathscr{d}}\mathscr{T}_\alpha\psi_\alpha, \\ \psi_\alpha &= \psi_{c(a)}^{+} + \sum_{\beta(\neq\alpha)=1}^{N}\frac{P_0}{\mathscr{d}}\mathscr{T}_\beta\psi_\beta, \end{aligned} \tag{498}$$

which are similar in structure to Eqs. (266).[156]

When the particle $i$ is identical with some of the bound particles $\alpha$, $\psi_a^{+}$ must be symmetrized as described in Chapter 4. We saw in Section 11.3, p. 779 that in an approximation consistent with the weak binding condition this may be effected by merely using appropriately symmetrized two-body scattering operators $\mathscr{T}_\alpha$ in Eqs. (496), (497), and (498). A somewhat improved technique for Fermi-Dirac statistics was given in Ref. [42].

[154] See, for example, Ref. [51].

[155] The use of quasi-particles in the two-body scattering matrices is ordinarily made under conditions where the weak binding condition does not obtain.

[156] We have taken the liberty of using the symbol $\psi_\alpha$ to represent the wave functions defined by Eqs. (498), even though these are not equivalent to the $\psi_\alpha$ in Eqs. (266*b*).

More elaborate treatments of the pseudopotential for the scattering of a Fermi particle by a bound system of $N$ equivalent Fermi particles[157] require a detailed analysis of the $(N + 1)$ body problem. Such methods are at present cumbersome to use and are also incomplete.[158]

Equations (496) and (498) are evidently much simpler if we can consider the pseudopotential (497) to be a definite, fixed interaction, independent of the state $\gamma$ and the previous scatterings [see Eqs. (470)]. We argued at the close of Section 11.5, p. 811 that condition (133) is sufficient to permit neglecting the effect of previous scatterings on the $P_0$-operator when evaluating $\mathscr{V}_c$ in a given propagator $\mathscr{d}^{-1}$.[159] The condition (133) is also generally sufficient to enable us to set

$$\langle, \gamma | \mathscr{V}_c |, \gamma\rangle = \langle, 0 | \mathscr{V}_c |, 0\rangle \tag{499}$$

independent of $\gamma$. The reason for this is, as has been described before, that the condition (133) implies that each scattering tends to occur at a distance greater than $\mathscr{R}_c$ from the location of any previous scattering. Since $\mathscr{R}_c$ represents the distance over which the local state of the medium is communicated, disturbances resulting from previous scatterings should not affect any given scattering when the condition (133) is satisfied.

**The Single Inelastic Scattering Approximation**

In Section 11.2 we studied scattering from a "small scatterer," defined as being sufficiently small that rescattering of previously scattered waves may be neglected. This led to the expression given in Eqs. (136) for the scattered wave. The analogous approximation as obtained from Eqs. (498) is physically quite different. This is

$$\psi_a^{+} = \psi_{c(a)}^{+} + \frac{P_0}{\mathscr{d}} \sum_{\alpha=1}^{N} \mathscr{T}_\alpha \psi_{c(a)}^{+}. \tag{500}$$

Equation (500) contains, in contrast to Eqs. (136), a rigorous description of all scatterings that leave the system in the initial states $g_0$. It also provides an exact description of all interactions that involve just one inelastic scattering. That is, the factor $\psi_{c(a)}^{+}$ accounts for the elastic interaction of particle $i$ with the medium prior to the inelastic scattering, whereas the

[157] See Refs. [69], [70], and [71].

[158] For example, these methods begin with unperturbed single particle states that correspond to binding in a prescribed potential well. Gallilean invariance is therefore violated by the resulting solutions to the scattering problem.

[159] By this, we, of course, mean previous explicit scatterings in Eqs. (496) and (498). For those scatterings that occur within a *given* $\mathscr{V}_c$, we must take proper account of $P_0$ as was shown in Section 11.4.

propagator $\mathscr{d}^{-1}$ describes the subsequent elastic interaction of $i$ with the medium.

The elastic scattering given by the term $\psi_{c(a)}^{+}$ in Eq. (500) has been extensively discussed in Section 11.4 and is not considered here. To obtain the scattering matrix for inelastic scattering, we use Eqs. (3.78) and (470) to write

$$\frac{1}{\mathscr{d}} = \frac{1}{d} + \frac{1}{d}\mathscr{V}_c\frac{1}{\mathscr{d}},$$

where $d$ is defined by Eq. (251). This permits us to reexpress (500) as

$$\psi_a^{+} = \psi_{c(a)}^{+} + \frac{P_0}{\mathscr{d}}\left(1 + \mathscr{V}_c\frac{1}{\mathscr{d}}\right)\sum_{\alpha=1}^{N}\mathscr{T}_\alpha\psi_{c(a)}^{+}.$$

Then, for a final state $\chi_b^{-}$, as given by Eq. (101),[160] the scattering matrix is, in the single scattering approximation,

$$\begin{aligned}\langle b|\mathscr{T}_N|a\rangle &= \left(\chi_b^{-}, \left(1 + \mathscr{V}_c\frac{1}{\mathscr{d}}\right)\sum_{\alpha=1}^{N}\mathscr{T}_\alpha\psi_{c(a)}^{+}\right)\\ &= \left(\left(1 + \frac{1}{\mathscr{d}^\dagger}\mathscr{V}_c^\dagger\right)\chi_b^{-}, \sum_{\alpha=1}^{N}\mathscr{T}_\alpha\psi_{c(a)}^{+}\right). \qquad (501)\end{aligned}$$

Now [see Eq. (383)]

$$\psi_{c(b)}^{-} = \left(1 + \frac{1}{\mathscr{d}^\dagger}\mathscr{V}_c^\dagger\right)\chi_b^{-} = \chi_b^{-} + \frac{1}{d^\dagger}\mathscr{V}_c^\dagger\psi_{c(b)}^{-} \qquad (502)$$

is the incoming wave function for elastic scattering in the state $\gamma$. With this wave function we may put Eq. (501) into the concise form

$$\begin{aligned}\langle b|\mathscr{T}_N|a\rangle &\equiv \langle \mathbf{k}, \nu', \mathbf{Q}, \gamma(\nu'_t)|\mathscr{T}_N|\mathbf{p}, \nu, \mathbf{P}, 0(\nu_t)\rangle\\ &= \left(\psi_{c(b)}^{-}, \sum_{\alpha=1}^{N}\mathscr{T}_\alpha\psi_{c(a)}^{+}\right). \qquad (503)\end{aligned}$$

This is a rather obvious generalization of Eqs. (136*b*) and (137), the "plane waves" $\chi_a$ and $\chi_b^{-}$ now being replaced by the waves $\psi_{c(a)}^{+}$ and $\psi_{c(b)}^{-}$ coherently scattered by the pseudopotential $\mathscr{V}_c$.

Before describing applications of Eq. (503) it is well to return to a discussion of its physical significance. Since further *inelastic* scatterings are omitted in Eq. (500), we recognize that Eq. (503) is only an approximation to the exact scattering matrix $\mathscr{T}_N$. As remarked earlier, this approximation is quite different from that made in Eqs. (136). We saw in Section 11.3 that Eqs. (136) are valid only if condition (131) is met. Because of the $P_0$-operator in Eqs. (498), the validity of Eq. (503) does not

[160] The $P_0$-operator requires, of course, that $\gamma \neq 0$ in the state $\chi_b^{-}$.

necessarily require that $\lambda \gg \mathscr{R}$. We anticipate that this will be the case if observations on the systems are sufficient to determine that a single scattering has occurred at a point $\mathbf{z}_\alpha$ (or $\mathbf{z}_\alpha{}^0$). When the conditions (132) and (133) are satisfied, we may suppose that such observations can in principle be made for quasi-free scattering. For example, measurement of the recoil momenta $\hat{\mathbf{Q}}_\alpha$ and $\hat{\mathbf{k}}$ then permit us to reconstruct the location of the scattering. Measurement of the energy of the two recoil particles often gives good evidence that only a single quasi-free scattering has occurred.

The implication of this is that conditions (132) and (133) may be sufficient for the validity of the approximation (503) if appropriate observations are made on the system.[161] In practice, this often means detecting in coincidence the scattered and recoil scattering particles. It should by no means be concluded, however, that the conditions (132) and (133) always imply the validity of Eq. (503). For example, if the observations on the final state are so gross that the cross sections (163) are appropriate, Eqs. (500) and (503) will be valid, in general, only if $\lambda \gg \mathscr{R}$, the size of the medium.

Equations (500) and (503) were first obtained (with an approximate $\mathscr{V}_c$) in Ref. [31] and in the form given here in Ref. [33], where they were applied to a discussion of photomeson production in atomic nuclei. Extensive applications to the scattering of nucleons by light atomic nuclei have been made by Levinson and Banerjee,[162] who have found remarkable agreement with measured cross sections by using shell-model wave functions to represent the nuclear states $g_\gamma$. Further applications[163] to the scattering of high-energy protons by atomic nuclei, again using shell-model wave functions, agree well with experimental observations. Some illustrations are shown in Fig. 11.11.

We shall illustrate applications of Eq. (503) for a scatterer consisting of $N$ identical particles. Our use of Eq. (503) implies that the weak binding limit and conditions (132) and (133) are met. For simplicity, we shall also assume that the scatterer is sufficiently heavy that its recoil velocity may be neglected, that $\mathscr{V}_c$ is independent of spin, and that the scatterer is spherically symmetric. Finally, we shall consider the dimension $\mathscr{R}$ of the scatterer to be large enough and $|\mathscr{V}_c|$ small enough that the eikonal approximation may be used to evaluate $\psi_{c(a)}^{+}$ and $\psi_{c(b)}^{-}$ [see the conditions (6.488)].

[161] It may be convenient to construct "wave packets of excitation" from the states $\gamma$. In this case the operator $P_0$ may be so defined as to control virtual wave packet states.

[162] See Ref. [72]. The excellent agreement with measured cross sections is somewhat better than might have been anticipated from estimates of the validity of the impulse approximation.

[163] See Ref. [73].

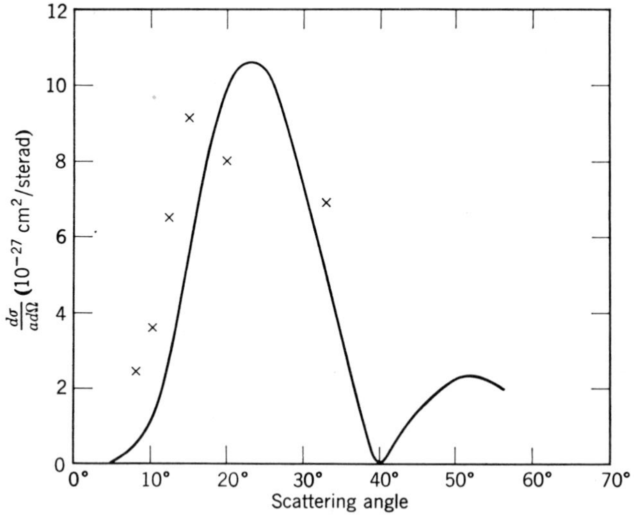


Fig. 11.11 Theoretical differential cross section for exciting the 4.43 Mev level of $C^{12}$ by 180 Mev protons.

We can then take [compare Eqs. (376) and (380)]

$$\begin{aligned} \psi^{+}_{c(a)} &= \phi_{\mathbf{p},v}(\mathbf{x})^{+}\chi_{t;0(v_t),\mathbf{P}}, \\ \psi^{-}_{c(b)} &= \phi_{\mathbf{k},v'}(\mathbf{x})^{-}\chi^{-}_{t;\gamma(v'_t),\mathbf{Q}}, \end{aligned} \tag{504}$$

where according to Eqs. (6.499$a$) and (6.500),

$$\begin{aligned} \phi^{+}_{p,v} &= (2\pi)^{-3/2}\, e^{i\mathbf{p}\cdot\mathbf{x}}\, e^{i\Delta\phi(\mathbf{x})} u_{S_i,v}, \\ \phi^{-}_{\mathbf{k},v'} &= (2\pi)^{-3/2}\, e^{i\mathbf{k}\cdot\mathbf{x}}\, e^{-i\Delta\phi^{-}(\mathbf{x})} u_{S_i,v}. \end{aligned} \tag{505}$$

Equation (6.498) implies that the path of integration defining $\Delta\phi$ runs from $-\infty$ to $\mathbf{x}$ along a straight line parallel to $\hat{\mathbf{p}}$. Similarly, we see from Eq. (6.501) that $\Delta\phi^{-}$ is defined by a line integral running from $\mathbf{x}$ to $+\infty$ along a straight line parallel to $\hat{\mathbf{k}}$.

The wave functions (505) and (504) permit us to give an explicit form for the scattering matrix (503) and to evaluate the scattering cross section in essentially the same manner as in Section 11.2. This becomes especially simple when the range $\mu^{-1}$ of the force between $i$ and a target particle $\alpha$ is sufficiently small that the phases $\Delta\phi(\mathbf{x})$ and $\Delta\phi^{-}(\mathbf{x})$ are effectively constant over an interval $|\delta\mathbf{x}| \simeq \mu^{-1}$. Then we may replace $\mathbf{x}$ by $\mathbf{z}_\alpha$ in $\Delta\phi(\mathbf{x})$ and $\Delta\phi^{-}(\mathbf{x})$ in Eq. (503). Using Eq. (144), we then obtain

$$\langle b|\mathscr{T}_N|a\rangle = \sum_{\alpha=1}^{N} (\chi_{t;\gamma(v'_t),\mathbf{Q}},\, e^{-i(\mathbf{k}-\mathbf{p})\cdot\mathbf{z}_\alpha}\, e^{i[\Delta\phi(\mathbf{z}_\alpha)+\Delta\phi^{-}(\mathbf{z}_\alpha)^*]}\, T_\alpha \chi_{t;0(v_t),\mathbf{p}}), \tag{506}$$

where $T_\alpha$ is the quantity (145). This differs from the corresponding matrix element which would have been obtained in Section 11.2 only in the appearance of the factor $e^{i[\Delta\phi(\mathbf{z}_\alpha)+\Delta\phi^-(\mathbf{z}_\alpha)]}$.

Let us first apply Eq. (506) to a reaction of the kind studied by Levinson and Banerjee [72]. The nuclear scattering system is assumed to be adequately described by single Hartree-Fock orbitals $g_{n_\alpha}(\mathbf{z}_\alpha, s_\alpha)$ $(\alpha = 1, 2, \ldots, N)$. Our assumption that the mass $M_t$ is so large that the recoil of the scatterer may be neglected permits us to imagine that the $g_{n_\alpha}$ describe particles bound to a fixed center in space (located at $\mathbf{z}_\alpha = 0$). The initial wave function [see Eq. (99*a*)] is then of the form

$$\chi_{t;0} = C\mathscr{S}g_{n_1}(\mathbf{z}_1, s_1)\cdots g_{n_N}(\mathbf{z}_N, s_N), \tag{507}$$

where $\mathscr{S}$ is the symmetrization operator and $C$ is a constant [see, for example, Eq. (4.52)]. The final state $\gamma$ is supposed to represent a discrete, excited state of the scatterer and is labeled by the quantum numbers $(n_1, \ldots, n_{\alpha-1}n'_\alpha n_{\alpha+1}, \ldots, n_N)$, corresponding to the single orbital transition $n_\alpha \to n'_\alpha$. Equation (507) and the corresponding final state wave function

$$\chi_{t,\gamma} = C\mathscr{S}g_{n_1}, \ldots, g_{n_{\alpha-1}}g_{n'_\alpha}g_{n_{\alpha+1}}, \ldots, g_{n_N},$$

permit us to simplify Eq. (506) to the form

$$\langle b|\mathscr{T}_N|a\rangle = (g_{n'_\alpha}(\mathbf{z}_\alpha, s_\alpha), e^{-i(\mathbf{k}-\mathbf{p})\cdot\mathbf{z}_\alpha} \times e^{i[\Delta\phi(\mathbf{z}_\alpha)+\Delta\phi^-(\mathbf{z}_\alpha)^*]}(\mathbf{k}, \nu', -\rho|\, T_\alpha(\mathbf{S}_\alpha)\,|\mathbf{p}, \nu, 0)g_{n_\alpha}(\mathbf{z}_\alpha, s_\alpha)). \tag{508}$$

The notation of Eq. (145) has been used here.

The differential scattering cross section for the transition $n_\alpha \to n'_\alpha$ is

$$\frac{d\sigma}{d\Omega} = \frac{(2\pi)^4}{v_{\text{rel}}}\int k^2\,dk\,\delta[\epsilon_k - \epsilon_p + \Delta W(n_\alpha \to n'_\alpha)]\,|\langle b|\mathscr{T}_N|a\rangle|^2, \tag{509}$$

where $\Delta W(n_\alpha \to n'_\alpha)$ is the excitation energy of the scatterer in the final state.

Evaluation of the integral in Eq. (508) is straightforward in principle, but may require numerical methods. A qualitative understanding of the effect of the pseudopotential $\mathscr{V}_c$ may usually be obtained from simple models, however. For example, let us suppose the scatterer to be uniform within a volume $(4\pi/3)\mathscr{R}^3$. Then from Eqs. (6.498) and (6.501) we have (when $M_r\,|V_0| \ll p^2$)

$$\begin{aligned} \Delta\phi(\mathbf{z}) &= -\frac{M_rV_0}{p}D_i, \\ \Delta\phi^-(\mathbf{z})^* &= -\frac{M_rV_0}{p}D_o, \end{aligned} \tag{510}$$

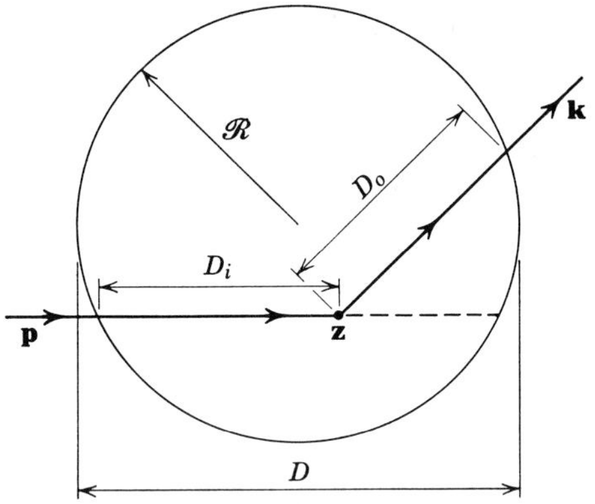


Fig. 11.12 Illustration of the distances $D_o$ and $D_i$ in Eqs. (510).

where $V_0 = V_c(\varkappa, \varkappa)$ is the well depth (466$b$) for forward scattering and $D_i$ and $D_o$ are the distances from the point $\mathbf{z}$ to the boundary of the scatterer in the directions $(-\hat{\mathbf{p}})$ and $\hat{\mathbf{k}}$, respectively. This is illustrated in Fig. 11.12.

From Eq. (400) we see that the imaginary part of $V_0$ causes exponential damping of the incident particle wave along its full path in the scatterer. If we neglect the spin dependence and set $\mathbf{p} \cong \mathbf{k}$ in Eq. (508), this becomes

$$\langle b| \mathscr{T}_N |a\rangle \cong T_\alpha^{\,0} \int d^3z\, e^{-i(M_r V_0/p)D}\, g_{n'_\alpha}(\mathbf{z})^* g_{n_\alpha}(\mathbf{z}), \tag{511}$$

where $T_\alpha^{\,0}$ is the function $T_\alpha$ for forward scattering [Eq. (430)] and $D$ is the total path length traveled through the scatterer, as illustrated in Fig. 11.12. The cross section (509) obtained from Eq. (511) is of the order of

$$\frac{d\sigma}{d\Omega} \cong \mathcal{O}\left\{\left[\left(\frac{\lambda}{\mathscr{R}}\right)^6 \frac{d\sigma_f}{d\Omega}\right] N\right\},$$

where $d\sigma_f/d\Omega$ is the cross section for forward scattering on a free target particle and the mean free path $\lambda \ll \mathscr{R}$.

In general, if the exponential attenuation as obtained from the expressions (510) dominates the exponential factor $e^{-i(\mathbf{k}-\mathbf{p})\cdot\mathbf{z}_\alpha}$ in Eq. (508), the cross section (509) will be very different from the corresponding cross section (146$a$). When the factor $e^{i(\Delta\phi+\Delta\phi^{-*})}$ in Eq. (508) does not differ significantly from unity, the cross section (509) will of course be equivalent to the corresponding cross section (146$a$).[164]

If we desire the cross section which is obtained from (506) by summing over all final states $\gamma$, we see from Eq. (508) that this is given by Eqs. (163) with $T_\alpha$ replaced by $e^{i[\Delta\phi(\mathbf{z}_\alpha)+\Delta\phi^-(\mathbf{z}_\alpha)^*]}T_\alpha$. For example, in the weak

[164] For a detailed numerical evaluation of the cross section (509) see Ref. [72].

binding limit Eq. (163*b*) becomes[165]

$$\frac{d\bar{\sigma}_d}{d\Omega} \cong \frac{(2\pi)^4}{v_{\text{rel}}} \sum_{\alpha=1}^{N} \int k^2\, dk\, \delta\left(\epsilon_k - \epsilon_p + \frac{\rho^2}{2M}\right)$$

$$\times \langle 0|\, T_\alpha^{\dagger}\, e^{-2\,\text{Im}\,[\Delta\phi(\mathbf{z}_\alpha)+\Delta\phi^-(\mathbf{z}_\alpha)^*]}\, T_\alpha\, |0\rangle_{\text{av}} \cong N \frac{d\bar{\sigma}_f}{d\Omega} \mathscr{A}, \quad (512)$$

where $d\bar{\sigma}_f/d\Omega$ is the free particle cross section (171) and

$$\mathscr{A} = \int d^3z \bar{P}^{(1)}(\mathbf{z})\, e^{-(1/\lambda)(D_0+D_i)}. \quad (513)$$

Here $1/\lambda = -\text{Im}\,(2M_r V_0/p)$, as in Eq. (400), and $\bar{P}^{(1)}$ is defined by Eq. (110).

Equation (512) differs from Eq. (170) only by the factor $\mathscr{A}$, which takes into account the attenuation of the incident particle wave along its trajectory through the scatterer. This should be compared with the factor

$$\int d^3z \bar{P}^{(1)}(\mathbf{z})\, e^{-(1/\lambda)D_i}, \quad (514)$$

which appears in Eq. (399). The physical distinction between these two factors is clear. The factor (514) represents the attenuation of the incident wave prior to the first inelastic scattering. Since Eq. (399) describes the *total* inelastic cross section, it is irrelevant for calculating $\sigma_{\text{ab}}$ if the particle *i* undergoes more than one inelastic scattering in the medium. In contrast to this, Eq. (512) gives the cross section for precisely *one* inelastic scattering. The factor $e^{-(1/\lambda)D_0}$ in the integrand of Eq. (513) then accounts for the probability that *i* travel to the boundary of the scatterer without further scatterings.

When the phase terms $\Delta\phi(\mathbf{z})$ and $\Delta\phi^-(\mathbf{z})$ vary by a negligible amount over a correlation distance $\mathscr{R}_c$, we may set $\Delta\phi(\mathbf{z}_\beta) = \Delta\phi(\mathbf{z}_\alpha)$ and $\Delta\phi^-(\mathbf{z}_\beta) = \Delta\phi^-(\mathbf{z}_\alpha)$ when evaluating $d\bar{\sigma}_{nd}/d\Omega$ [see Eq. (163*c*)]. In this case, we have

$$\frac{d\bar{\sigma}_{nd}}{d\Omega} \cong \frac{(2\pi)^4}{v_{\text{rel}}} \sum_{\alpha\neq\beta=1}^{N} \int k^2\, dk\, \delta(\epsilon_k - \epsilon_p)$$

$$\times \langle 0|\, T_\beta^{\dagger}\, e^{i\mathbf{p}\cdot\mathbf{z}_\beta}\, e^{-(1/\lambda)(D_0+D_i)} P_0\, e^{-i\mathbf{p}\cdot\mathbf{z}_\alpha}\, T_\alpha\, |0\rangle_{\text{av}}. \quad (515)$$

The effect of $P_0$ is to remove the elastic scattering from $d\bar{\sigma}_{nd}/d\Omega$.[166] Because $N$ has been assumed to be large, we may write Eq. (515) as

$$\frac{d\bar{\sigma}_{nd}}{d\Omega} \cong N^2 \frac{d\bar{\sigma}_f}{d\Omega} \int d^3z\, d^3z' \bar{P}^{(1)}(\mathbf{z}) \bar{P}^{(1)}(\mathbf{z}')$$

$$\times G(|\mathbf{z} - \mathbf{z}'|)\, e^{i\mathbf{p}\cdot(\mathbf{z}-\mathbf{z}')}\, e^{-(1/\lambda)[D_0(\mathbf{z})+D_i(\mathbf{z})]}, \quad (516)$$

[165] The $P_0$-operator may be neglected here.

[166] This has, of course, already been included in the $\psi^+_{c(a)}$ term in Eq. (500).

if the dependence of $T_\alpha$ on $\mathbf{S}_\alpha$ may be neglected, as in Eq. (198). Here we have neglected the difference between $d\bar{\sigma}_f^{(0)}/d\Omega$ [Eq. (194)] and $d\bar{\sigma}_f/d\Omega$, as in Eq. (200). If $\mathscr{R} \gg \mathscr{R}_c$, the exponential factor describing attenuation in Eq. (516) leads to just the factor (513). The sum of the cross sections (512) and (516) is then [in the notation of Eq. (202)]

$$\frac{d\bar{\sigma}_{\text{in}}}{d\Omega} = N\left[1 - |c(\rho)|^2 + \frac{N-1}{\mathscr{V}}\, C(\rho)\right]\frac{d\bar{\sigma}_f}{d\Omega}\mathscr{A}. \tag{517}$$

When $T_\alpha$ depends on the spin $\mathbf{S}_\alpha$, we may evaluate the cross section using the spin sums given in Section 11.2, p. 748. If $\mathscr{R}$ is not much larger than $\mathscr{R}_c$, edge effects give a somewhat different attenuation factor in Eq. (517).

As a final example of the use of Eq. (503), we consider the photoproduction of a pi-meson from a nucleus of mass number $A$. The interaction $V$ will now contain two types of terms: (1) operators $V_\alpha(\gamma)$ describing the interaction of a $\gamma$-ray with nucleon $\alpha$ to produce a pion; and (2) operators $V_\alpha(\pi)$ describing the interaction of a pion with nucleon $\alpha$, where in both cases $\alpha$ runs from 1 to $A$. Now, since the pion must be created in Eq. (503), we may take $\mathscr{T}_\alpha = \mathscr{T}_\alpha(\gamma)$ in Eq. (503), where $\mathscr{T}_\alpha(\gamma)$ is the matrix for producing the pion by a $\gamma$-ray striking nucleon $\alpha$.

Because the $\gamma$-ray interactions are weak, we may neglect $\mathscr{T}_\alpha(\gamma)$ in evaluating the wave functions $\psi^+_{c(a)}$ and $\psi^-_{c(b)}$. In particular, we may simply take $\psi^+_{c(a)} = \chi_a$. Therefore Eq. (503) becomes

$$\langle b|\,\mathscr{T}_N\,|a\rangle = \sum_{\alpha=1}^{A}(\psi^-_{c(b)}, \mathscr{T}_\alpha(\gamma)\chi_a). \tag{518}$$

We may interpret Eq. (518) as follows: A $\gamma$-ray enters the nucleus and strikes nucleon $\alpha$ to produce a pion. The meson then travels out of the nucleus under the influence of the pion-nuclear pseudopotential, as described by $\psi^-_{c(b)}$.

Photomeson production involves sufficiently violent excitations of the nucleus that we may reasonably neglect the nondiagonal terms when evaluating the cross section. If we then evaluate the cross section using Eq. (512), we obtain

$$\frac{d\bar{\sigma}(\gamma \to \pi)}{d\Omega} = A(2\pi)^3\,\frac{d\bar{\sigma}_f}{d\Omega}(\gamma \to \pi)\int d^3z\bar{P}^{(1)}(\mathbf{z})\,|\phi_{\mathbf{k}}^-(\mathbf{z})|^2, \tag{519}$$

where the attenuation factor has been written directly in terms of the final wave functions $\phi_k^-$, given by Eq. (505), and $d\bar{\sigma}_f/d\Omega(\gamma \to \pi)$ is the photopion cross section for a free nucleon, averaged over neutrons and protons. Now,

$$\int d^3z\bar{P}^{(1)}(\mathbf{z})\,|\phi_{\mathbf{k}}^-(\mathbf{z})|^2 = \int d^3z\bar{P}^{(1)}(\mathbf{z})\,|\phi_{\mathbf{k}}^+(z)|^2 = \frac{\lambda\sigma_{\text{ab}}\rho(0)}{A(2\pi)^3}, \tag{520}$$

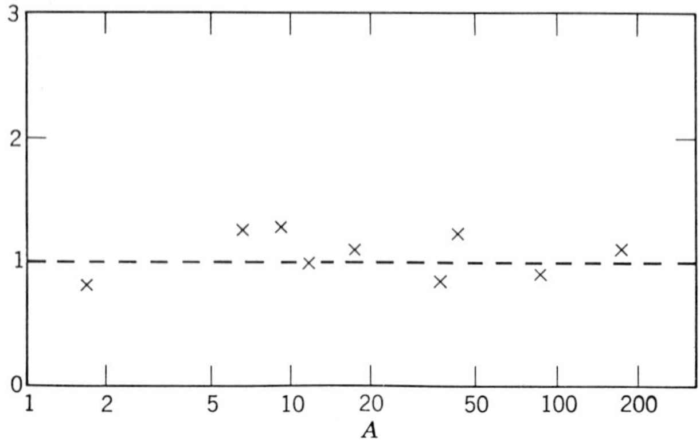


Fig. 11.13 The ratio $\frac{d\bar{\sigma}(\gamma \to \pi)d\Omega}{\bar{\sigma}_{\text{ab}}}$ shown as a function of the atomic mass number $A$ (multiplied by an arbitrary scale factor).

according to Eqs. (395*a*) and (400). This permits us to write Eq. (519) as

$$\frac{d\bar{\sigma}(\gamma \to \pi)}{d\Omega} = \frac{d\bar{\sigma}_f(\gamma \to \pi)}{d\Omega} [\lambda \sigma_{\text{ab}} \rho(0)]. \tag{521}$$

Here $\sigma_{\text{ab}}$ is the total inelastic cross section for pion scattering on the nucleus. Equation (521) therefore relates three experimental cross sections, each of which can be directly measured. In Fig. 11.13 we compare Eq. (521) with the observed relation.[167] The agreement is seen to be good.

**Transport of Particles in a Large Scattering System**

We observed in the previous section that Eq. (500), which describes the approximation where only one inelastic scattering is kept, is valid for small scatterers or when the observations on the system are such as to insure that only one inelastic scattering has occurred. When we are interested in the complete path of a particle $i$ in an extended scattering medium, the neglect of multiple scattering in Eq. (500) is ordinarily unjustified. In this case we must return to the general set of Eqs. (498).

Let us suppose that the scattering system is uniform within its interior and has a reasonably well-defined boundary. The linear dimensions of the scatterer are characterized by a length $\mathscr{R}$ and we suppose that $\mathscr{R} \gg \lambda$, the mean free path for scattering [as defined by Eq. (130)]. The scatterer will be considered to be at rest and sufficiently massive that its recoil may be neglected. Then the barycentric momentum $\boldsymbol{\varkappa}$, as in Eq. (466), is equal to the momentum $\mathbf{p}$ of particle $i$. We shall also assume that conditions (132) and (133) are both met. The weak binding condition and the quasi-free

[167] Equation (521) does not, in fact, require the eikonal approximation for its derivation. A derivation not using the eikonal approximation is given in Ref. [74].

conditions (162) are supposed valid for the inelastic scatterings described in Eqs. (498). Our assumption of a large scattering medium will be interpreted as implying the adequacy of the eikonal approximation for evaluating $\psi^{+}_{c(a)}$ and the propagators $\mathscr{d}^{-1}$ in Eqs. (498). Finally, we shall assume a spin-saturated scatterer [Eq. (149)], which means that the dependence of $\mathscr{V}_c$ on the spin of the scatterer may be neglected. This, and the validity of the eikonal approximation, in turn imply that the dependence of $\mathscr{V}_c$ on the spin $\mathbf{S}_i$ of particle $i$ may be neglected.[168]

These conditions permit us to reduce the coupled equations (498) to algebraic equations, as was done in Section 11.3, p. 759. Because of the appearance of the pseudopotential $\mathscr{V}_c$ in $\psi^{+}_{c(a)}$ and in the propagators $\mathscr{d}^{-1}$, the unitarity of the $S$-matrix is not violated by these equations. We shall see that the scattering may also be described by what is (essentially) a classical transport equation.

Equations (498) differ from the set of equations (266) in three respects. First, the propagators $\mathscr{d}^{-1}$ in Eqs. (498) contain the pseudopotential $\mathscr{V}_c$, whereas the $d^{-1}$ [Eq. (251)] in (266) do not. Because we have assumed that $\lambda \gg \mathscr{R}_c$ [condition (133)], we may take $\mathscr{V}_c$ to be a definite potential, independent of the state $\gamma$ and of the previous scatterings[169]. The eikonal approximation then permits us to express the $\mathscr{d}^{-1}$ in a very simple form in a coordinate representation. On introducing the refractive index $n$ from Eq. (397) and using Eq. (6.491), we may write the phase $\phi$ in Eq. (6.489) as

$$\phi = nk_\beta R_\beta,$$

where $k_\beta$ is the momentum of particle $i$ which would correspond to a refractive index of unity and $R_\beta$ is the distance of propagation. Thus, for example, the propagator $e^{ik_\beta R_\beta}/R_\beta$ in Eq. (286) is now replaced by $e^{ink_\beta R_\beta}/R_\beta$.

Equations (498) also differ from Eqs. (266) in that the counting operator $P_0$ appears in the former. The quasi-free condition (162) and the condition (133) that $\lambda \gg \mathscr{R}_c$ imply that $P_0$ will play a rather simple role in our considerations. We recall that $P_0$ instructs us to keep only inelastic scatterings in solving Eqs. (498) and to avoid returning to states $\gamma$ which have been encountered previously. Since the quasi-free condition corresponds to ejecting the struck particle from its bound state, the scattering is automatically *inelastic*. Condition (133) implies that a new scattering should occur out of the correlation range of previous scatterings, which means (as argued previously) that *different* degrees of freedom are excited. Thus the condition that each new state $\gamma$ be different from previous states $\gamma$ is also satisfied automatically.[170] Then, except for our bearing in mind

[168] This assumes that parity is conserved in the scattering, of course.

[169] The justification for this was described in Section 11.5, p. 818.

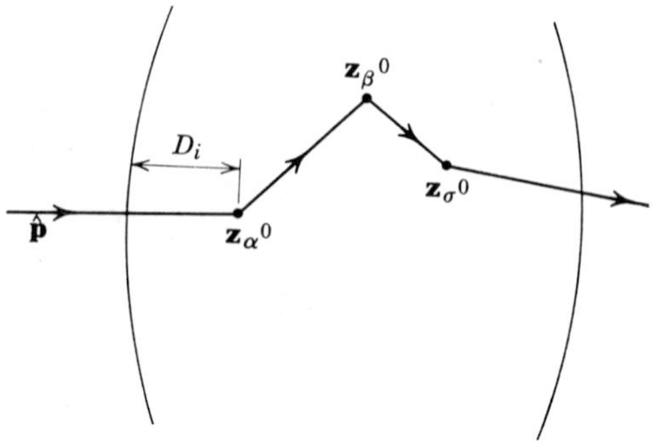


Fig. 11.14 A sequence of scatterings in an extended scatterer.

that no two scattered waves in Eqs. (498) can interfere coherently, we may simply ignore the $P_0$-operator in these equations.

Equations (498) and (266) finally differ in that in the former the coherent wave $\psi^+_{c(a)}$ replaces the incident wave $\chi_a$ in the latter. This suggests that Eq. (292) be replaced now by

$$\psi_\alpha(\mathbf{z}_\alpha{}^0) = \psi^+_{c(a)}(\mathbf{z}_\alpha{}^0) + \sum_{\beta(\neq\alpha)=1}^{N} \Psi_{\alpha\beta}(\mathbf{z}_\alpha{}^0)(2\pi)^{-3/2} g_0. \tag{522}$$

The assumed validity of the eikonal approximation permits us to use Eqs. (504) and (396) to write, since the scatterer is at rest,

$$\psi^+_{c(a)}(\mathbf{x}) = (2\pi)^{-3/2} \phi_{\mathbf{p},\nu}(\mathbf{x})^+ g_0 \tag{523a}$$

with

$$\phi^+_{\mathbf{p},\nu}(\mathbf{x}) = (2\pi)^{-3/2}\, e^{i[\mathbf{p}\cdot(\mathbf{x}-\mathbf{D}_i(\mathbf{x}))+n\mathbf{p}\cdot\mathbf{D}_i(\mathbf{x})]} u_{S_i,\nu}. \tag{523b}$$

Here $\mathbf{D}_i(\mathbf{x})$ is the distance directed along $\hat{\mathbf{p}}$ from the boundary of the medium to the point $\mathbf{x}$, as illustrated in Fig. 11.14. Equation (523*b*) is seen to be precisely equivalent to the first of Eqs. (505).

Our discussion has indicated that the derivation leading to Eqs. (293) for the $\Psi_{\alpha\beta}$ is unchanged except for the appearance of the refractive index $n$ in the propagators $E_{\alpha\beta,\sigma}$. Therefore we obtain without further calculation [see Eqs. (293)]

$$\Psi_{\alpha\beta}(\mathbf{z}_\alpha{}^0) = E_{\alpha\beta,0}\mathbf{f}(\alpha\beta,\beta 0)\phi^+_{\mathbf{p},\nu}(\mathbf{z}_\beta{}^0) + \sum_{\sigma(\neq\beta)=1}^{N} E_{\alpha\beta,\sigma}\mathbf{f}(\alpha\beta,\beta\sigma)\Psi_{\beta\sigma}(\mathbf{z}_\beta{}^0), \tag{524}$$

where now [see Eq. (290)]

$$E_{\alpha\beta,\sigma} = P_0 \frac{e^{ink_\beta R_{\alpha\beta}}}{R_{\alpha\beta}}\, e^{i\mathbf{p}_{\beta\sigma}\cdot(\mathbf{z}_\beta - \mathbf{z}_\beta{}^0)}. \tag{525}$$

[170] Condition (133) may be relaxed without great difficulty here. Then quasi-elastic scattering must also be considered. (See Ref. [16] for a complete discussion.)

As we have argued earlier, the operator $P_0$ may be ignored here. Equations (522) and (524) suffice to determine the wave functions $\psi_\alpha$ in Eqs. (498).

Equations (334) or (400) imply that

$$\operatorname{Im}(nk_\beta) = \frac{1}{2\lambda}.$$

where $\lambda$ is the mean free path for a quasi-free scattering in the medium. We therefore obtain from Eq. (525)

$$\Gamma_{\alpha\beta} \equiv |E_{\alpha\beta,\sigma}|^2 = \frac{e^{-R_{\alpha\beta}/\lambda}}{R_{\alpha\beta}^{\;2}}. \tag{526}$$

The quantity $\Gamma_{\alpha\beta}/4\pi$ represents the probability that a particle scattered isotropically at $\mathbf{z}_\beta{}^0$ will pass through a unit area at $\mathbf{z}_\alpha{}^0$ without having scattered enroute.

We emphasize that Eqs. (522) and (524) have a physical significance that was lacking in the corresponding Eqs. (292) and (293).[171] This can be seen from Eq. (526), which illustrates our "satisfying a new unitarity condition for each scattering," as was discussed in the introduction to this section.

To illustrate a rather simple application of Eqs. (524), we shall reexpress these in the form of a transport equation. To avoid a minor and irrelevant algebraic complication, we shall assume that particle $i$ has no spin and that the scattering amplitude $\mathbf{f}(\alpha\beta, \beta\sigma) = f(\alpha\beta, \beta\sigma)$ is spin independent. Then[172]

$$n_c(\mathbf{x}) = (2\pi)^3\, |\phi_{\mathbf{p}}{}^+(\mathbf{x})|^2 = e^{-D_i(\mathbf{x})/\lambda} \tag{527}$$

is the density of particles $i$ which have not undergone a quasi-free scattering at the point $\mathbf{x}$, normalized to a unit incident density. Similarly, we may define

$$n'(\mathbf{z}_\alpha{}^0, \hat{\mathbf{k}}) = (2\pi)^3 \sum_{\beta(\text{in }\delta\Omega_{\hat{\mathbf{k}}},\neq\alpha)=1}^{N} \left[\frac{1}{\delta\Omega_{\hat{\mathbf{k}}}}\, |\Psi_{\alpha\beta}(\mathbf{z}_\alpha{}^0)|^2\right]. \tag{528}$$

Here the sum over $\beta$ runs over all scatterers $\beta$ (except $\alpha$) lying in the small cone of solid angle $\delta\Omega_{\hat{\mathbf{k}}}$, with apex at $\mathbf{z}_\alpha{}^0$ and axis directed along the vector $-\hat{\mathbf{k}}$. (This is illustrated in Fig. 11.15.) Thus $n'(\mathbf{z}_\alpha{}^0, \hat{\mathbf{k}})$ represents the number of particles $i$ at $\mathbf{z}_\alpha{}^0$ per unit volume per unit solid angle with velocity vectors parallel to $\hat{\mathbf{k}}$ and which have undergone at least one quasi-free scattering.

[171] The exact scattering matrix for inelastic scattering is seen, by the arguments leading to Eqs. (501) and (503), to be

$$\langle b|\mathscr{T}_N|a\rangle = \left(\psi_{c(b)}^-, \sum_{\alpha=1}^{N} \mathscr{T}_\alpha \psi_\alpha\right).$$

[172] Since it is assumed that particle $i$ has no spin, we now omit the spin label $\nu$ on the wave function (523$b$).

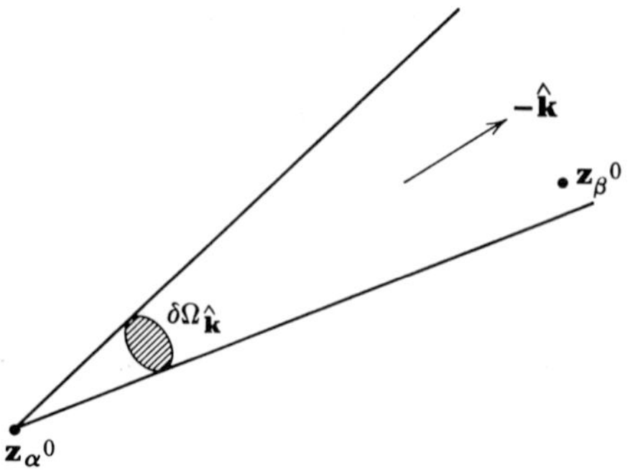


Fig. 11.15 Cone containing scatterers $\beta$ in sum in Eq. (528).

The total number of particles $i$ within the interval $d^3x\, d\Omega_{\hat{\mathbf{k}}}$ at a point $\mathbf{x}$ and having velocity parallel to $\hat{\mathbf{k}}$ is then $n(\mathbf{x}, \hat{\mathbf{k}})\, d^3x\, d\Omega_{\hat{\mathbf{k}}}$, where

$$n(\mathbf{x}, \hat{\mathbf{k}}) = n_c(\mathbf{x})\delta_{\hat{\mathbf{k}},\hat{\mathbf{p}}} + n'(\mathbf{x}, \hat{\mathbf{k}}). \tag{529}$$

[The $\delta$-function $\delta_{\hat{\mathbf{k}},\hat{\mathbf{p}}}$ is written in the notation of Eq. (7.4*b*).]

To obtain an integral equation for $n(\mathbf{x}, \hat{\mathbf{k}})$ we first insert the right-hand side of Eq. (524) for $\Psi'_{\alpha\beta}(\mathbf{z}_\alpha{}^0)$ in Eq. (528). We have argued that under the conditions stated for our problem, waves scattered from different scatterers do not interfere. This implies that all cross-terms in the squared sum vanish. Therefore,

$$n'(\mathbf{z}_\alpha{}^0, \hat{\mathbf{k}}) = \sum_\beta \frac{1}{\delta\Omega_{\hat{\mathbf{k}}}} \Gamma_{\alpha\beta}\left[|f(\alpha\beta, \beta 0)|^2\, n_c(\mathbf{z}_\beta{}^0) + \sum_{\sigma(\neq\beta)} |f(\alpha\beta, \beta\sigma)|^2\, |\Psi'_{\beta\sigma}(\mathbf{z}_\beta{}^0)|^2\right]. \tag{530}$$

The sum over $\beta$ here is restricted as described in connection with Eq. (528). This implies that the vector $(\mathbf{z}_\alpha{}^0 - \mathbf{z}_\beta{}^0)$ lies within the solid angle $\delta\Omega_{\hat{\mathbf{k}}}$ which is directed about $\hat{\mathbf{k}}$.

The second term in Eq. (530) may be expressed in terms of the density $n'$ if we first carry out the sum on $\sigma$ over a cone of solid angle $d\Omega_{\hat{\mathbf{k}}'}$ having its axis parallel to $(-\hat{\mathbf{k}}')$. Then

$$\sum_\sigma |f(\alpha\beta, \beta\sigma)|^2\, |\Psi'_{\beta\sigma}(\mathbf{z}_\beta{}^0)|^2 = \int d\Omega_{\hat{\mathbf{k}}'}\, |f(\alpha\beta, \beta\sigma)|^2\, n'(\mathbf{z}_\beta{}^0, \hat{\mathbf{k}}'), \tag{531}$$

using the definition (528) of $n'$. It is convenient to write this in terms of an effective "scattering cross section."[173]

$$\sigma_e(\hat{\mathbf{k}}, \hat{\mathbf{k}}') \equiv |f(\alpha\beta, \beta\sigma)|^2, \tag{532}$$

[173] It is seen from Eqs. (291) and (282) that the "cross section" (532) is the true scattering cross section multiplied by the square of the ratio of the final to initial relative velocities of the scattering particles.

recalling that $\mathbf{z}_\alpha{}^0 - \mathbf{z}_\beta{}^0$ and $\mathbf{z}_\beta{}^0 - \mathbf{z}_\sigma{}^0$ are parallel to $\hat{\mathbf{k}}$ and $\hat{\mathbf{k}}'$, respectively. Then, Eq. (531) is equivalent to the equation

$$\sum_\sigma |f(\alpha\beta, \beta\sigma)|^2 |\Psi_{\beta\sigma}(\mathbf{z}_\beta{}^0)|^2 = \int d\Omega_{\hat{\mathbf{k}}'} \sigma_e(\hat{\mathbf{k}}, \hat{\mathbf{k}}') n(\mathbf{z}_\beta{}^0, \hat{\mathbf{k}}') - \sigma_e(\hat{\mathbf{k}}, \hat{\mathbf{p}}) n_c(\mathbf{z}_\beta^0), \quad (533)$$

using Eq. (529).

Equations (529), (530), and (533) may now be combined to give

$$n(\mathbf{x}, \hat{\mathbf{k}}) = n_c(\mathbf{x})\delta_{\hat{\mathbf{k}},\hat{\mathbf{p}}} + \int R^2\, dR \int d\Omega_{\hat{\mathbf{k}}'} \frac{e^{-R/\lambda}}{R^2}\, \sigma_e(\hat{\mathbf{k}}, \hat{\mathbf{k}}') n(\mathbf{x}', \hat{\mathbf{k}}'). \quad (534)$$

Here we have written $R \equiv |\mathbf{x} - \mathbf{x}'|$ and have used Eq. (526) for $\Gamma_{\alpha\beta}$. The sum over $\beta$ in Eq. (530) has been replaced by an integral, using the relation

$$\frac{1}{d\Omega_{\hat{\mathbf{k}}}} \sum_\beta = \rho \int R^2\, dR,$$

where $\rho$ is the density of scatterers in the medium.[174] The integral over $R$ extends from the point $\mathbf{x}' = \mathbf{x}$ to the boundary of the scattering system along the direction of $-\hat{\mathbf{k}}$.

Equation (534) might, of course, have been derived directly from purely classical considerations, without the use of quantum mechanics (other than to obtain the cross section $\sigma_e$). The purpose of our discussion has been to illustrate the transition from quantum mechanical multiple scattering phenomena to classical transport theory.[175] To accomplish this it was first necessary to extract coherent scattering resulting from interference of elastically scattered waves using the $P_0$-operator. This led to the factor of $e^{-R/\lambda}$ in the integrand of Eq. (534) and to the attenuation of $n_c(\mathbf{x})$, as given by Eq. (527). The second step in the derivation followed from the observation that waves scattered from well-separated points in the medium do not interfere with each other. This permitted us to neglect the cross-terms in obtaining Eq. (530) from Eq. (528).

Because of the finite binding energy of the particles in the scatterer, there will be some quasi-elastic interference of scattered waves, as described in Section 11.2, p. 728. This will, in general, lead to corrections to the purely "classical" equation (534). These corrections were investigated in

174 Strictly speaking, we should have averaged Eq. (534) over the initial state $g_0$ of the scatterer. The assumption that the scatterer is uniform implies that $\rho$ is a constant.

175 A classical transport equation of the form (534) was obtained by Foldy [29]. This derivation was given in Ref. [16].

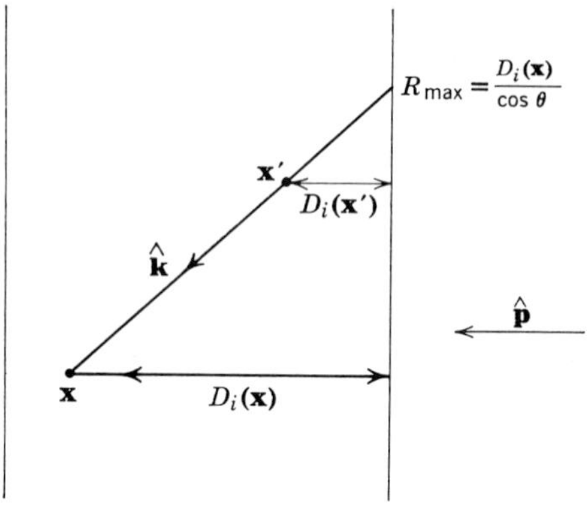


Fig. 11.16 An application of Eq. (534) to scattering by a slab of finite thickness.

Ref. [16], where it was shown that the cross section $\sigma_e$ in Eq. (534) should be modified by a factor[176]

$$\left[1 + \rho \int d^3y G(\mathbf{y})\, e^{ip(\hat{\mathbf{k}}-\hat{\mathbf{k}}')\cdot\mathbf{y}}\right].$$

Here $G(\mathbf{y})$ is the pair correlation function for the scattering system. This factor is seen to be just that appearing in Eq. (204).

As an elementary application of Eq. (534), we consider scattering by a system shaped as a flat slab. The incident momentum **p** is normal to the faces of the slab, as illustrated in Fig. 11.16. It is supposed that the cross section $\sigma_e$ is isotropic and that the scattering particles are sufficiently massive that the energy loss of particles $i$ on scattering may be neglected.

If we successively substitute the expression (534) for $n$ into the right-hand side, we obtain the series

$$n(\mathbf{x}, \hat{\mathbf{k}}) = e^{-(1/\lambda)D_i(\mathbf{x})}\, \delta_{\hat{\mathbf{k}},\hat{\mathbf{p}}} + \frac{1}{4\pi\lambda}\int_0^{R_{\max}} R^2\, dR \int d\Omega_{\hat{\mathbf{k}}'}\, \frac{e^{-(1/\lambda)D_i(\mathbf{x}')}}{R^2}\, \delta_{\hat{\mathbf{k}}',\hat{\mathbf{p}}} + \cdots$$
$$= e^{-(1/\lambda)D_i(\mathbf{x})}\delta_{\hat{\mathbf{k}},\hat{\mathbf{p}}} + \frac{1}{4\pi\lambda}\int_0^{R_{\max}} dR\, e^{-R/\lambda}\, e^{-D_i(\mathbf{x}')\lambda} + \cdots. \quad (535)$$

Here $R_{\max} = D_i(\mathbf{x})/\cos\theta$, where $\theta$ is the angle between **k** and **p**[177] and $\lambda^{-1} = 4\pi\rho\sigma_e$. On setting $D_i(\mathbf{x}') = D_i(\mathbf{x}) - R\cos\theta$, we may perform the $R$-integral in Eq. (535) to obtain

$$n(\mathbf{x}, \hat{\mathbf{k}}) = e^{-(1/\lambda)D_i(\mathbf{x})}\left[\delta_{\hat{\mathbf{k}},\hat{\mathbf{p}}} + \frac{(1 - e^{-(R_{\max}/\lambda)(1-\cos\theta)})}{1 - \cos\theta} + \cdots\right].$$

[176] If the energy lost by particle $i$ on scattering must be taken into account, this factor requires a further kinematical correction. (See Ref. [16], where the complete expressions are given.)

[177] When $\theta > \pi/2$, we must reexpress $R_{\max}$ in terms of the distance to the far face of the slab.

As might have been expected, the structure of the series (535) is very different from that in Eq. (298).

## 11.6 DEUTERON STRIPPING AND PICK-UP REACTIONS

We have so far in Chapter 11 discussed only simple scattering by composite systems, the entrance and exit channels remaining the same. In this section we shall illustrate the general theory described in Section 4.1 of multichannel reactions by studying the deuteron "stripping" and "pick-up" processes. In the stripping reaction a deuteron in collision with an atomic nucleus loses either its neutron [the $(d, p)$ process] or its proton [the $(d, n)$ process] to the nucleus, the other nucleon remaining free. The pick-up reactions are the inverses of the stripping reactions. In these either a proton captures a neutron [the $(p, d)$ process] or a neutron captures a proton [the $(n, d)$ process] from a target nucleus to form a deuteron.

It is customary to restrict the terms "stripping" and "pick-up" to refer not only to the two types of reaction just described but also to a specific mechanism by which these reactions occur. For example, in applying the term "stripping" to the $(d, p)$ process we imagine that the neutron is "suddenly" removed from the deuteron and "instantaneously" injected into a bound orbit in the nucleus. The proton continues as a free particle with the momentum which it had at the "instant the neutron was captured." The corresponding pick-up reaction may be described by application of time reversal to the stripping mechanism.

The impulse model just discussed for stripping and pick-up processes was first proposed by Chew and Goldberger [75] to account for experimental observations [76] of deuteron production by 90 Mev neutrons. A detailed theory of stripping reactions, based on what is essentially the same model, was given by Butler [77]. This theory has been quite successful in correlating measurements of the stripping cross section with the shell-model of nuclear structure.[178]

The description of the Chew-Goldberger model of stripping and pick-up processes will provide a good illustration of the general theory of reactions and of the effects of wave function symmetrization, as developed in Chapter 4. In addition, it will serve to illustrate in a new context some of the considerations of coherence and incoherence given in Section 11.5.

### General Description of the Stripping Reaction

To be specific, we consider the $(d, p)$ reaction, all energies being assumed to be nonrelativistic: A deuteron having momentum $\boldsymbol{\varkappa}$ in the barycentric frame of reference collides with an atomic nucleus containing $Z$ protons and $N$ neutrons. The neutron that was initially in the deuteron is captured

[178] A detailed review is given in Ref. [78].

into a bound state of the nucleus, leaving in the final state a free proton and a nucleus containing $Z$ protons and $(N + 1)$ neutrons.

It will prove convenient to treat neutrons and protons as distinguishable particles, separately antisymmetrizing the scattering amplitude in neutron and proton coordinates. We shall also describe the scattering in the barycentric subspace[179] and thus avoid $\delta$-functions associated with momentum conservation. Then the initial (unsymmetrized) wave function for the incident deuteron and nucleus is

$$\chi_a = (2\pi)^{-3/2}\, e^{i\boldsymbol{\kappa}\cdot\mathbf{r}} g_0(\xi) D(\mathbf{y}). \tag{536}$$

Here $\mathbf{r}$ is the vector extending from the center-of-mass of the initial nucleus to that of the deuteron and $\mathbf{y}$ is the relative coordinate of the neutron and proton in the deuteron. The quantities $g_0(\xi)$ and $D(\mathbf{y})$ represent, respectively, the wave functions for the nucleus and deuteron, whereas $\xi$ designates the set of coordinates for the $Z$ protons and $N$ neutrons in the nucleus. Letting $\mathbf{x}$ and $\mathbf{z}$ be the respective coordinate vectors from the nuclear center-of-mass to the proton and neutron in the deuteron, we have

$$\begin{aligned} \mathbf{r} &= \tfrac{1}{2}(\mathbf{x} + \mathbf{z}), \\ \mathbf{y} &= (\mathbf{x} - \mathbf{z}). \end{aligned} \tag{537}$$

Wave functions for the final state will be of the form

$$\chi'_b = (2\pi)^{-3/2}\, e^{i\boldsymbol{\kappa}_f\cdot\mathbf{x}} u_{1/2,\nu'}\, g'_0(\xi, \mathbf{z}, s). \tag{538}$$

Here $\boldsymbol{\kappa}_f$ is the momentum in the barycentric frame of the final free proton and $u_{1/2,\nu'}$ is its spin function, corresponding to a spin orientation $\nu'$. $g'_0(\xi, \mathbf{z}, s)$ is the wave function of the final nucleus, consisting of $Z$ protons and $(N + 1)$ neutrons, $(\mathbf{z}, s)$ being the coordinate and spin variable of the added neutron. As was implied in Chapters 3 and 4, we suppose that $g'_0$ is antisymmetrized in the coordinates of all its $(N + 1)$ neutrons.[180] The initial nucleus has been assumed to be sufficiently heavy that the shift in center-of-mass of the final nucleus resulting from the addition of the extra neutron may be neglected in the wave function (538).

As in Section 4.1, we shall label the initial and final channels as $c$ and $c'$, respectively. To obtain the scattering matrix we must exhibit the channel Hamiltonians $K_c$ and $K_{c'}$ and channel interactions $V_c, V_{c'}$. The Hamiltonian for the initial nucleus will be designated as $h$, so

$$hg_0(\xi) = W_0 g_0(\xi), \tag{539}$$

where $W_0$ is the nuclear ground-state energy. The kinetic energy operators for the neutron and proton in the deuteron are written as $K_n$ and $K_p$,

[179] The barycentric subspace is defined on p. 169.

[180] It is, of course, also antisymmetrized in the coordinates of its $Z$ protons.

respectively. Finally, we write $V_n$ for the interaction of the extra neutron with the initial nucleus, $V_p$ for the interaction of the extra proton with the initial nucleus, and $V_{pn}$ for the interaction between the neutron and proton in the deuteron. If we then define

$$K \equiv h + K_p + K_n, \tag{540}$$

the channel Hamiltonians are seen to be

$$\begin{aligned} K_c &= K + V_{pn}, \\ K_{c'} &= K + V_n. \end{aligned} \tag{541}$$

Thus, [see Eq. (4.69)]

$$\begin{aligned} (K + V_{pn})\chi_a &= E_a\chi_a, \\ (K + V_n)\chi'_b &= E_b\chi'_b, \end{aligned} \tag{542}$$

where

$$\begin{aligned} E_a &= W_0 + \epsilon_D + \frac{\kappa^2}{2M_a}, \\ E_b &= W'_0 + \frac{\kappa_f^2}{2M_b}. \end{aligned} \tag{543}$$

Here the binding energy of the deuteron is taken as $(-\epsilon_D)$ and the reduced masses in channels $c$ and $c'$ are taken as $M_a$ and $M_b$, where

$$\begin{aligned} M_a &= \frac{(2M)M_0}{2M + M_0}, \\ M_b &= \frac{M(M_0 + M)}{2M + M_0}, \end{aligned}$$

with $M$ the mass of the nucleon and $M_0$ that of the initial nucleus.[181] Finally, the binding energy of the final nucleus is written as $W'_0$ in the second of Eqs. (543). Energy conservation, of course, requires that

$$E_b = E_a \tag{544}$$

for asymptotic final states.

The channel interactions are seen to be

$$\begin{aligned} V_c &= V_p + V_n, \\ V_{c'} &= V_p + V_{pn}, \end{aligned} \tag{545}$$

and the complete Hamiltonian is

$$H = K + V_n + V_p + V_{pn}. \tag{546}$$

[181] We are neglecting the relativistic contributions to the mass due to the binding energy.

According to Eq. (4.8), the wave function describing the entire scattering event is

$$\psi_a^+ = \chi_a + \frac{1}{E_a + i\eta - H}(V_p + V_n)\chi_a. \tag{547}$$

The scattering matrix is obtained from Eq. (4.13) as

$$T_{ba} = (\chi'_b, (V_p + V_{pn})\psi_a^+). \tag{548}$$

The wave function (547) and scattering matrix (548) are not yet symmetrized as required for Fermi-Dirac statistics. This may be easily done, however, as was described in Section 4.4. It is apparent from Eq. (4.60*a*) that we may antisymmetrize the wave function and scattering matrix in neutron coordinates and in proton coordinates separately (we are not using the isotopic spin formalism). Let us first carry out symmetrization in the neutron coordinates.

To do this it is convenient to relabel the coordinates of the $(N + 1)$ neutrons as $(\mathbf{z}_0, s_0)$, $(\mathbf{z}_1, s_1)$, . . . , $(\mathbf{z}_N, s_N)$. As we have remarked, the nuclear wave functions $g_0$ and $g'_0$ are assumed to be appropriately symmetrized.

Now, when neutron $i$ $(i = 0, 1, 2, \ldots, N)$ is in the initial deutron and treated as distinguishable from the other neutrons, we may write the scattering matrix (548) as

$$T_{ba(i)} = (\chi'_b, (V_p + V_{pn_i})\psi^+_{a(i)}). \tag{549a}$$

In this expression we have written $V_{pn_i}$ for the interaction between the $i$th neutron and that proton which is in the deuteron. The wave function $\psi^+_{a(i)}$ is just the wave function (547), corresponding to having neutron $i$ in the incident deuteron. On letting $Q$ be one of the $(N + 1)!$ permutations of the neutron coordinates, we have

$$T_{ba[Q(i)]} \equiv (\chi'_b, Q(V_p + V_{pn_i})\psi^+_{a(i)}). \tag{549b}$$

Since the state $\chi'_b$ is antisymmetrized in all the neutron coordinates, we may rewrite Eq. (549*b*) in the form

$$\begin{aligned} T_{ba[Q(i)]} &= (Q^{-1}\chi'_b, (V_p + V_{pn_i})\psi^+_{a(i)}) \\ &= \delta_{Q^{-1}} T_{ba(i)} \\ &= \delta_Q T_{ba(i)}. \end{aligned} \tag{549c}$$

Here $\delta_Q$ is the quantity defined by Eq. (4.40) and we have used the condition that $\delta_{Q^{-1}} = \delta_Q$.

We shall write $T_{\mathscr{S}_n(ba)}$ for the scattering matrix which is symmetrized in the neutron coordinates. Because the $\chi'_b$ are already antisymmetrized in neutron coordinates, we may use Eq. (4.140*b*) to write

$$T_{\mathscr{S}_n(ba)} = \frac{C}{(N+1)!} \sum_Q \delta_Q T_{ba[Q(0)]}, \tag{550}$$

where the sum runs over all $(N + 1)!$ permutations $Q$ and $T_{ba[Q(0)]}$ is the scattering matrix (549*b*) with $i = 0$. From Eq. (4.93) we have $C = \sqrt{N+1}$. Then, using Eq. (549*c*) and the fact that $\delta_Q{}^2 = 1$, we obtain

$$T_{\mathscr{S}_n(ba)} = \sqrt{N+1}\, T_{ba(0)} \equiv \sqrt{N+1}\, T_{ba}. \tag{551}$$

Except for the numerical factor of $\sqrt{N+1}$ in Eq. (551), this is *identical* with the unsymmetrized scattering matrix (548). To see the significance of this factor, let us use the nuclear shell model for $g'_0$, assuming that the captured neutron is in a single orbital state $\omega_l(\mathbf{z}_0, s_0)$ about a "core" described by $g_0(\xi) \equiv g_0(\mathbf{z}_1, s_1, \ldots, \mathbf{z}_N, s_N)$. Then, according to Eq. (4.98),

$$g'_0 = \frac{1}{\sqrt{N+1}}\Big[\omega_l(\mathbf{z}_0, s_0) g_0(\mathbf{z}_1, s_1, \ldots, \mathbf{z}_N, s_N) - \sum_{j=1}^{N} \omega_l(\mathbf{z}_j, s_j)$$
$$\times\, g_0(\mathbf{z}_1, s_1, \ldots, \mathbf{z}_{j-1}, s_{j-1}, \mathbf{z}_0, s_0, \mathbf{z}_{j+1}, s_{j+1}, \ldots, \mathbf{z}_N, s_N)\Big]. \tag{552}$$

If the state $\omega_l$ is *orthogonal* to all the orbital states in $g_0$ and if the "nuclear core," described by $g_0$, does not become excited in $\psi^+_{a(0)}$, all terms but the first in Eq. (552) vanish on performing the integrals in $T_{ba(0)}$. Then the factor $(N+1)^{-1/2}$ in Eq. (552) is canceled by the factor $(N+1)^{1/2}$ in Eq. (551). The resulting scattering amplitude is in this case identical with that which would have been obtained if the extra neutron were consistently treated as distinguishable from the other neutrons.

We have yet to effect the antisymmetrization of the scattering matrix with respect to the proton coordinates. The result may be written down immediately using Eq. (4.164). The expression $\sqrt{N+1}\, T_{ba}$ represents the *direct* term, since that proton originally in the deuteron emerges as a free particle in the final state. By interchanging this proton with any proton in the nucleus, say the $j$th, we obtain a new final state $\chi'_{b(j)}$. The *exchange* scattering matrix is then

$$(T_{ba})_{\text{ex}} \equiv (\chi'_{b(j)}, (V_{p_j} + V_{p_jn})\psi_a{}^+), \tag{553}$$

where $V_{p_j} + V_{p_jn}$ represents the interaction of the $j$th proton with the remaining $(Z + N + 1)$ nucleons. We emphasize that the expression (553) is independent of the index $j$ for $j = 1, 2, \ldots, Z$. According to Eq. (4.164) the complete, symmetrized scattering matrix is then

$$T_{\mathscr{S}(ba)} = \sqrt{N+1}\,[T_{ba} - Z(T_{ba})_{\text{ex}}]. \tag{554}$$

Most studies of the deuteron pick-up and stripping processes have assumed that the exchange amplitude in Eq. (554) is negligible, since this term is not consistent with the simple model described in the introduction to this section. Because of this, we shall devote our attention principally to the term $T_{ba}$.

**Simplification of $T_{ba}$**

The expression (548) for $T_{ba}$ may be put into a more convenient form, following what is essentially the treatment of the two-potential scattering problem given in Section 5.4. To do this we first introduce the wave functions $\phi'_b{}^{\pm}$ which describe scattering of a free proton by the residual nuclear core:

$$\phi'_b{}^{\pm} = \chi'_b + \frac{1}{E_a \pm i\eta - (K + V_n)} V_p \phi'_b{}^{\pm} \tag{555}$$

where $\chi'_b$ is the "final" wave function (538).

Now, Eq. (555) may be used to eliminate $\chi'_b$ from the right-hand side of Eq. (548). This leads to the expression

$$\begin{aligned} T_{ba} &= (\phi'_b{}^-, (V_p + V_{pn})\psi_a{}^+) \\ &\quad - \left(\phi'_b{}^-, V_p \frac{1}{E_a + i\eta - (K + V_n)} (V_p + V_{pn})\psi_a{}^+\right) \\ &= (\phi'_b{}^-, V_{pn}\psi_a{}^+) \\ &\quad + \left(\phi'_b{}^-, V_p\left[1 - \frac{1}{E_a + i\eta - (K + V_n)}(V_p + V_{pn})\right]\psi_a{}^+\right). \end{aligned} \tag{556}$$

The last term here may be simplified if we use Eq. (547) to write

$$\begin{aligned} &\left[1 - \frac{1}{E_a + i\eta - (K + V_n)}(V_p + V_{pn})\right]\psi_a{}^+ \\ &\qquad = \left[1 + \frac{1}{E_a + i\eta - (K + V_n)}(V_n - V_{pn})\right]\chi_a. \end{aligned} \tag{557}$$

Thus Eq. (556) is equivalent to

$$\begin{aligned} T_{ba} &= (\phi'_b{}^-, V_{pn}\psi_a{}^+) + (\phi'_b{}^-, V_p\chi_a) \\ &\quad + \left(\phi'_b{}^-, V_p \frac{1}{E_a + i\eta - (K + V_n)}(V_n - V_{pn})\chi_a\right). \end{aligned} \tag{558}$$

Use of Eq. (555) in the last term here leads immediately to the result

$$\begin{aligned} T_{ba} &= (\phi'_b{}^-, V_{pn}\psi_a{}^+) + (\phi'_b{}^-, V_p\chi_a) \\ &\quad + (\phi'_b{}^-, (V_n - V_{pn})\chi_a) - (\chi'_b, (V_n - V_{pn})\chi_a). \end{aligned} \tag{559}$$

The last term on the right in Eq. (559) is seen to vanish. According to the first of Eqs. (542), $V_{pn}\chi_a = (E_a - K)\chi_a$. Therefore

$$(\chi'_b, (V_n - V_{pn})\chi_a) = (\chi'_b, (V_n + K - E_a)\chi_a) = 0,$$

using the second of Eqs. (542) and the condition that $E_a = E_b$. Equation (559) is then just

$$\begin{aligned} T_{ba} &= (\phi'_b{}^-, V_{pn}\psi_a{}^+) + (\phi'_b{}^-, (V_p + V_n - V_{pn})\chi_a) \\ &= (\phi'_b{}^-, V_{pn}\psi_a{}^+) + (\phi'_b{}^-, (K + V_p + V_n - E_a)\chi_a). \end{aligned} \quad (560)$$

Now, from Eq. (555) we see that

$$(K + V_p + V_n + i\eta - E_a)\phi'_b{}^- = -i\eta\chi'_b.$$

Since $(\phi'_b{}^-, \chi_a)$ and $(\chi'_b, \chi_a)$ are nonsingular, we conclude that the last term in Eq. (560) vanishes in the limit as $\eta \to 0(+)$. Therefore, in this limit,[182]

$$T_{ba} = (\phi'_b{}^-, V_{pn}\psi_a{}^+). \quad (561)$$

The expression (561) will provide the basis for our description of the stripping process. By eliminating an explicit appearance of the potential $V_p$ in Eq. (561), we have given this a form which suggests the "impulse" model described in the introduction to this section.[183]

**The Pick-Up Process**

The scattering matrix for the pick-up process may be obtained from Eq. (561) using the relation (4.235), expressing time reversal invariance. For the pick-up reaction we consider $\chi'_b$ to be the initial state and $\chi_a$ to be the final state [see Eqs. (536) and (538)]. Then, according to Eq. (4-235),

$$T_{ab} = \mu_a\mu'_b{}^* T_{-b-a} = \mu_a\mu'_b{}^*(T\phi'_{-b}{}^-, TV_{pn}\psi_{-a}{}^+)^*, \quad (562)$$

where $T$ is the time reversal operator and $\phi'_{-b}{}^-$ and $\chi_{-a}{}^+$ are the states obtained from $\phi'_b{}^-$ and $\psi_a{}^+$ by reversing all velocities and angular momenta.

Now, from the assumption of time reversal invariance, $TV_{pn} = V_{pn}T$ and [see Eqs. (4.232)]

$$T\phi'_{-b}{}^- = \mu'_b\phi'_b{}^+,$$
$$T\psi_{-a}{}^+ = \mu_a\psi_a{}^-.$$

Therefore

$$T_{ab} = (\phi'_b{}^+, V_{pn}\psi_a{}^-)^* = (\psi_a{}^-, V_{pn}\phi'_b{}^+). \quad (563)$$

[182] The expression (561) was given by Greider [79]. The derivation published in Ref. [79] contains an error, but the final expression [equivalent to Eq. (561)] is correct.

[183] Many discussions of the pick-up and stripping reactions have simply neglected the proton interaction $V_p$, in which case Eqs. (548) and (561) are trivially identical.

This is then the scattering matrix for the pick-up reactions, where a proton collides with a nucleus to capture a neutron.

**Optical Model of Stripping**

We have described the "stripping model" as follows: A deuteron passes near, or into, a nucleus and "suddenly" loses its neutron to the nucleus. The proton continues as a free particle. The assumed absence of any significant momentum transfer to the proton leads one to expect a characteristically forward-peaked angular distribution.

Now, the wave function $\psi_a^+$ in Eq. (561), of course, contains all possible transitions which can result from the deuteron scattering. It may be argued that those transitions which excite the nucleus, or break-up the deuteron, will lead to "compound nucleus formation" and a final state that is experimentally distinguishable from that characteristic of stripping. Only the terms in $\psi_a^+$ that correspond to *elastic* scattering of the deuteron by the nucleus are expected to avoid such "compound systems" and thus appreciable momentum transfers to the proton. We saw in Section 11.4 how to extract that part of a wave function $\psi_a^+$ which describes elastic scattering [see Eq. (374)], called there $\psi_{c(a)}^+$. This suggests that we replace $\psi_a^+$ in Eq. (561) by $\psi_{c(a)}^+$ which, in analogy to Eq. (375*b*), we now write as

$$\psi_{c(a)}^+ = \left[1 + \frac{1}{E_a + i\eta - (K + V_{pn} + \mathscr{V}_D)} \mathscr{V}_D\right]\chi_a. \tag{564}$$

Here $\mathscr{V}_D$ is the pseudopotential needed to describe elastic scattering of the deuteron by the initial nucleus, this representing a special case of the general pseudopotential introduced in Eqs. (369). We emphasize that $\mathscr{V}_D$ is *diagonal* in the internal states of both the nucleus and the deuteron. Thus Eq. (564) describes just the "potential" scattering of two "elementary particles."

By following similar arguments, we replace $\phi'^-_b$ in Eq. (561) by $\phi'^-_{c(b)}$, the wave function for elastic scattering of the final proton by the nuclear "core." Again, this is done in the spirit of omitting all terms in Eq. (561) which would correspond to large momentum transfers to the proton. The wave function $\phi'^-_{c(b)}$ may be expressed in terms of a pseudopotential $\mathscr{V}_p$, diagonal in nuclear states, as in Eq. (375*b*).

$$\phi'^-_{c(b)} = \left[1 + \frac{1}{E_a + i\eta - (W'_0 + K_p + \mathscr{V}_p^\dagger)} \mathscr{V}_p^\dagger\right]\chi'_b. \tag{565}$$

Here the kinetic energy $K_p$ of the proton and the final nuclear energy $W'_0$ are defined in connection with Eqs. (540) and (543). Equation (565) describes the potential scattering of two "particles."

On making the indicated substitutions, Eq. (561) is replaced by

$$T_{ba} = (\phi'^{-}_{c(b)}, V_{pn}\psi^{+}_{c(a)}). \tag{566}$$

The corresponding expression for the pick-up reaction, obtained from Eq. (563), is

$$T_{ab} = (\psi^{-}_{c(a)}, V_{pn}\phi'^{+}_{c(b)}). \tag{567}$$

The expressions (566) and (567) are very much simpler than are the corresponding expressions (561) and (563), since the complexity of the many-body problem does not appear in the evaluation of (566) or (567) (once a model has been chosen for the nuclear wave functions $g_0$ and $g'_0$ and for the pseudopotentials $\mathscr{V}_D$ and $\mathscr{V}_p$). Indeed, the wave functions (564) and (565) may always be explicitly evaluated, if only numerically. When the eikonal approximation may be used, these wave functions may be written out immediately in the form of Eqs. (504) and (505).

The expressions (566) and (567) are, of course, not exact. They have been obtained from the exact scattering matrices (561) and (563) by discarding inelastic scattering from the wave functions $\psi_a^+$ and $\phi'_b{}^-$. We have not argued that the neglected terms are necessarily small. Instead, we have suggested that they tend to lead to final states which can be experimentally distinguished from those reached by the stripping mechanism, as given by Eqs. (566) and (567).[184]

With the exception of the proton exchange terms [as included in Eq. (554)], Eqs. (566) and (567) include as special cases all the various models proposed for the stripping and pick-up reactions.[185]

### The Pick-Up Reaction at High Energies

Most discussions of the pick-up reactions have been based on a modification of Eq. (567) in which $\phi'^{+}_{c(b)}$ and $\psi^{-}_{c(a)}$ are replaced by the plane-wave states $\chi'_b$ and $\chi_a$; that is,[186]

$$T_{ab} = (\chi_a, V_{pn}\chi'_b). \tag{568}$$

The relation of Eq. (568) to Eq. (567) is similar to that of Eqs. (163) to Eq. (512), since in Eq. (567) the distortion of the incident and outgoing waves is taken into account. At sufficiently high energies that the eikonal

[184] A similar argument was given in Section 11.5 in discussing the scattering matrix (503).

[185] The argument by which Eqs. (566) and (567) are obtained from Eqs. (561) and (563) was first given by Francis and Watson [80], who obtained a slightly different form for the scattering matrices. The explicit Eqs. (566) and (567) were first given by Greider [79].

[186] Equation (568) is often called a "Born approximation," although this bears little relation to the usual Born approximation, as described in Section 6.7.

approximation may be used, Eq. (567) is not difficult to evaluate.[187] As might be anticipated from Eqs. (6.499) and (6.500), the result is usually not qualitatively different from that obtained using Eq. (568). When there is strong absorption of the incident and outgoing waves in traveling through the nucleus, there is a tendency, however, to favor short transnuclear trajectories and thus to enhance the number of deuterons appearing at large angles.[188]

We first discuss the pick-up reaction, assuming the expression (568) to be valid. Modifications arising from the use of Eq. (567) are then reviewed.

We shall neglect the $^3D_1$ admixture in the deuteron wave function, assuming that $D(\mathbf{y})$ is a pure $^3S_1$ state. This satisfies the Schrödinger equation

$$V_{pn}D(\mathbf{y}) = \left(\epsilon_D + \frac{1}{M}\nabla_y^2\right)D(\mathbf{y}). \tag{569}$$

Equations (536), (538), and (569) permit us to write Eq. (568) as

$$T_{ab} = (2\pi)^{-3}\int g_0^{\dagger}(\xi)\, e^{-i\boldsymbol{\varkappa}\cdot\mathbf{r}}\left[\left(\epsilon_D + \frac{1}{M}\nabla_y^2\right)D(\mathbf{y})\right]^{\dagger} \times e^{i\boldsymbol{\varkappa}_f\cdot\mathbf{x}}\, u_{\frac{1}{2},\nu'}g'_0(\xi, \mathbf{z}, s)\, d\xi\, d^3r\, d^3y, \tag{570}$$

where a matrix product is implied for the spin components of the wave functions.

The Fourier transform of $D(\mathbf{y})$ is

$$D(\mathbf{y}) = (2\pi)^{-\frac{3}{2}}\int d^3Q\, e^{i\mathbf{Q}\cdot\mathbf{y}}\,\bar{D}(Q)u_{1,\nu}, \tag{571}$$

where $u_{1,\nu}$ is the triplet spin function of the deuteron and $\bar{D}$ is the Fourier transform of the space dependent part of $D$. On introducing this expression into Eq. (570), we obtain

$$T_{ab} = -(2\pi)^{-\frac{3}{2}}\int d\xi\, d^3z u_{1,\nu}^{\dagger} g_0^{\dagger}(\xi)g'_0(\xi, \mathbf{z}, s) \times u_{\frac{1}{2},\nu'}\left[|\epsilon_D| + \frac{1}{M}\left(\boldsymbol{\varkappa}_f - \frac{\boldsymbol{\varkappa}}{2}\right)^2\right]\bar{D}^*\left(\left|\frac{\boldsymbol{\varkappa}}{2} - \boldsymbol{\varkappa}_f\right|\right)e^{i\mathbf{z}\cdot(\boldsymbol{\varkappa}_f-\boldsymbol{\varkappa})}. \tag{572}$$

To evaluate the integrals in this expression we assume that $g'_0$ has the "shell-model" form (552) and that the orbital state $\omega_l(\mathbf{z}, s)$ is orthogonal to the orbital states in $g_0$. Then, as noted in connection with Eq. (552), the

[187] This has been done by Greider [79].
[188] This is illustrated in Fig. 11.17.

exchange terms vanish and we find that

$$(2\pi)^{-3/2} \int d\xi\, d^3z\, e^{i\mathbf{z}\cdot(\boldsymbol{\varkappa}_f - \boldsymbol{\varkappa})}\, g_0^{\dagger}(\xi) g'_0(\xi, \mathbf{z}, s)$$

$$= \frac{(2\pi)^{-3/2}}{\sqrt{N+1}} \int d^3z\, e^{i\mathbf{z}\cdot(\boldsymbol{\varkappa}_f - \boldsymbol{\varkappa})}\, \omega_l(\mathbf{z}, s) \equiv \frac{1}{\sqrt{N+1}}\, \bar{\omega}_l(\boldsymbol{\varkappa}_f - \boldsymbol{\varkappa}, s), \quad (573)$$

where $\bar{\omega}_l$ is the Fourier transform of $\omega_l$. Thus,

$$T_{ab} = -\frac{1}{\sqrt{N+1}} \left[ |\epsilon_D| + \frac{1}{M}\left(\boldsymbol{\varkappa}_f - \frac{\boldsymbol{\varkappa}}{2}\right)^2 \right] \bar{D}^* \left( \left| \frac{\boldsymbol{\varkappa}}{2} - \boldsymbol{\varkappa}_f \right| \right)$$
$$\times\, (u_{1,\nu}, \bar{\omega}_l(\boldsymbol{\varkappa}_f - \boldsymbol{\varkappa}, s) u_{1/2,\nu'}). \quad (574)$$

The probability $\mathscr{P}(Q)$ of finding a momentum $Q$ per unit volume in momentum space in the deuteron is given by

$$\frac{3}{4}\, \mathscr{P}(Q) \equiv \frac{1}{4} \sum_{\nu,\nu',\nu_l} |(u_{1,\nu}, \bar{\omega}_l(\boldsymbol{\varkappa}_f - \boldsymbol{\varkappa}, s) u_{1/2,\nu'})|^2_{\text{av}}, \quad (575)$$

where the summation over $\nu_l$ represents a sum over the two spin orientations for the bound neutron and av implies an average over the orientations of the neutron orbital angular momentum. The factor of $\frac{3}{4}$ here gives just the probability of finding a triplet spin state in the initial random distribution of spins.

If we neglect the proton exchange terms in Eq. (554) and thus take $T_{\mathscr{S}(ab)} = \sqrt{N+1}\, T_{ab}$, the cross section for deuteron pick-up is

$$\left.\frac{d\bar{\sigma}}{d\Omega}\right|_{\text{pick-up}} = \frac{(2\pi)^4}{v_{\text{rel}}} \int \kappa^2\, d\kappa\, \delta\left( \frac{\kappa^2}{2M_a} + \epsilon_D + W_0 - W'_0 - \frac{\kappa_f^{\,2}}{2M_b} \right)$$
$$\times \left( \frac{1}{4} \sum_{\nu,\nu',\nu_l} |T_{\mathscr{S}(ab)}|^2 \right)_{\text{av}} = (2\pi)^4 \left( \frac{3M_b M_a \kappa}{4\kappa_f} \right) \left[ |\epsilon_D| + \frac{1}{M}\left( \frac{\boldsymbol{\varkappa}}{2} - \boldsymbol{\varkappa}_f \right)^2 \right]^2$$
$$\times \left| \bar{D}\left( \left| \frac{\boldsymbol{\varkappa}}{2} - \boldsymbol{\varkappa}_f \right| \right) \right|^2 \mathscr{P}(|\boldsymbol{\varkappa}_f - \boldsymbol{\varkappa}|), \quad (576)$$

since $v_{\text{rel}} = \kappa_f / M_b$ is the incident relative velocity of the proton and nucleus. The final deuteron momentum $\kappa$ is determined from the condition of energy conservation.

$$\frac{\kappa^2}{2M_a} = \frac{\kappa_f}{2M_b} + W'_0 - W_0 - \epsilon_D.$$

The expression (576), obtained originally by Chew and Goldberger [75] is remarkably simple. It depends on easily calculable kinematical factors and the deuteron and nuclear momentum distributions. In using Eq. (576)

it is customary to assume that $\bar{D}\left(\left|\frac{\varkappa}{2} - \varkappa_f\right|\right)$ is known. Then, comparison of Eq. (576) with experimental observations of deuteron pick-up permits the determination of $\mathscr{P}(|\varkappa - \varkappa_f|)$, the nuclear momentum distribution. The argument that $\bar{D}\left(\left|\frac{\varkappa}{2} - \varkappa_f\right|\right)$ is reasonably well known rests on the fact that models of nucleon-nucleon interactions are available which describe neutron-proton scattering quite well for energies ranging from zero to a few hundreds of Mev. Since the deuteron binding involves relatively low energies, it is to be expected that $|\bar{D}|^2$ is much better known than is the many-body distribution function $\mathscr{P}(|\varkappa_f - \varkappa|)$.

Frequently, the Hulthèn wave function is used for $D(\mathbf{y})$.[189] This is

$$D(\mathbf{y}) = \frac{A}{y}\,[e^{-\alpha y} - e^{-\beta y}]u_{1,\nu}, \tag{577}$$

where the normalization factor is[190]

$$A = \left[\frac{\alpha}{2\pi(1 - \alpha r_e)}\right]^{1/2}. \tag{578}$$

Here $\alpha = [M\,|\epsilon_D|]^{1/2}$ and $r_e$ is the effective range (see Section 6.6) for neutron-proton scattering. On normalizing Eq. (577) directly, we can evaluate $\beta$ in terms of $\alpha$ and $r_e$. This leads numerically to the result that

$$\beta \simeq 6\alpha. \tag{579}$$

Finally, on evaluating the Fourier transform of (577) and inserting the result into Eq. (579), we obtain the pick-up cross section

$$\left.\frac{d\bar{\sigma}}{d\Omega}\right|_{\text{pick up}} = \left(\frac{M_a M_b}{M^2}\right)\left(\frac{12\pi^2\alpha}{1 - \alpha r_e}\right)\left(\frac{\kappa}{\kappa_f}\right)\frac{(\beta^2 - \alpha^2)^2\mathscr{P}(|\varkappa_f - \varkappa|)}{\left[\beta^2 + \left(\frac{\varkappa}{2} - \varkappa_f\right)^2\right]^2}. \tag{580}$$

Equation (580) was obtained originally by Chew and Goldberger [75]. Because all quantities except $\mathscr{P}(|\varkappa_f - \varkappa|)$ appearing in it are known, it may be used to determine the nuclear momentum distribution from experimental observations of the pick-up reaction.

Since Eq. (580) was obtained from the approximate Eq. (568), it is important to study the more accurate Eq. (567). This has been done by Greider [79], who used the eikonal approximation for the wave functions $\psi_{c(a)}^{-}$ and $\phi'^{+}_{c(b)}$. At the high energies for which the eikonal approximation

[189] See Ref. [75] for a more complete discussion.

[190] We saw in Section 6.5 that the normalization of a bound-state wave function may be expressed in terms of the scattering parameters. Equation (578) follows from Eq. (6.255*b*).

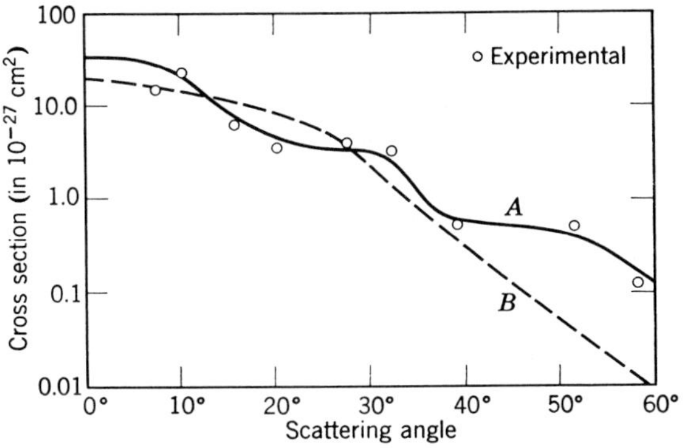


Fig. 11.17 Comparison of the deuteron pick-up cross sections as obtained from Eq. (567) (curve $A$), from Eq. (580) (curve $B$), and experiment for 95 Mev protons (taken from References [79] and [81]).

is valid, the principal effect of the distorted waves in Eq. (567) arises from absorption of the incident and outgoing waves. To compare his calculations with some experimental data (Selove [81]) from 95 Mev protons, Greider chose [see Eqs. (564) and (565)]

$$\mathscr{V}_p = (-25 - 15i)\text{ Mev}$$
$$\mathscr{V}_D = (-20 - 15i)\text{ Mev}$$

for the numerical values of the pseudopotentials inside the nucleus. In addition, he took

$$\mathscr{P}(q) = \frac{e^{-q^2/2ME_0}}{(2\pi ME_0)^{3/2}},$$

with $E_0 = 14$ Mev, for the nuclear momentum distribution. His results are compared with those from Eq. (580) (for the same momentum distribution) and with the experimental data in Fig. 11.17.

**The Butler Theory of Stripping**

Butler [77] has applied what is essentially the Chew-Goldberger model to stripping reactions for low energy (of the order of 10 Mev) deuterons. The use of plane waves for $\psi_{c(a)}^{+}$ and $\phi'^{-}_{c(b)}$ in Eq. (566) is not easily justified for such low energies. To provide a simple model for evaluating (566) Butler assumed that the coherent wave $\psi_{c(a)}^{+}$ is strongly absorbed once the neutron enters the nucleus, whereas $\psi_{c(a)}^{+}$ may be replaced by $\chi_a$ when the neutron is outside the nucleus.[191] Butler neglected the proton interaction $V_p$, which implies that $\phi'^{-}_{c(b)} = \chi'_b$ [see Eq. (565)].

[191] Butler stated his model somewhat differently. The relation to this discussion and to the pick-up reaction was demonstrated by Daitch and French [82] and by Francis and Watson [80].

The expression (566) may therefore be written in the form appropriate to Butler's model by taking

$$\begin{aligned} \psi^{+}_{c(a)} &= \chi_a \quad \text{for } z \geq \mathscr{R} \\ &= 0 \quad \text{for } z < \mathscr{R} \\ \phi'^{-}_{c(b)} &= \chi'_b, \end{aligned} \tag{581}$$

where $z$ is the neutron coordinate and $\mathscr{R}$ is the nuclear radius. This permits us to write

$$T_{ba} = (\chi'_b, V_{pn}\chi_a)|_{z\geq\mathscr{R}} = [(\chi_a, V_{pn}\chi'_b)|_{z\geq\mathscr{R}}]^*, \tag{582}$$

where the notation implies that the integral on $z$ extends only over the domain $z \geq \mathscr{R}$. Comparison of Eq. (582) with Eq. (568) suggests that we can obtain the scattering matrix for stripping from that evaluated in Section 11.6, p. 841 with only minor modifications.

The fundamental assumption of the Butler model [as expressed by the first of Eqs. (581)] that $\psi^{+}_{c(a)}$ vanish when the neutron is within the nucleus is certainly plausible. At energies of the order of 10 Mev nuclear interactions are very strong and are expected to lead to inelastic scattering of the deuteron and nucleus, which, as we saw in Section 11.4 implies absorption of the coherent wave. The neglect of the proton interaction is less easily justified, however. We may argue qualitatively that $\phi'^{-}_{c(b)}$ represents a "bundle of plane waves" having momentum spread $\hbar/\mathscr{R}$, the effect of which is to smooth the proton angular distribution over a range of angles of the order of $\hbar/\kappa_f\mathscr{R}$, which is ordinarily small.[192] The corresponding neglect of the proton interaction in $\psi^{+}_{c(a)}$ is also not unreasonable. Since the deuteron is bound to the neutron in $\psi^{+}_{c(a)}$ and since Eq. (581) implies that the neutron is outside the nucleus when $\psi^{+}_{c(a)}$ is nonvanishing, the proton cannot penetrate very far into the nucleus.[193] It would thus appear that the assumptions implied by Eqs. (581) are at least qualitatively justified.

To evaluate the expression (582), we again use the nuclear wave function (552). Then, in place of Eq. (574), we have

$$T_{ba} = -\left[|\epsilon_D| + \frac{1}{M}\left(\boldsymbol{\varkappa}_f - \frac{\boldsymbol{\varkappa}}{2}\right)^2\right] \frac{\bar{D}\left(\left|\boldsymbol{\varkappa}_f - \dfrac{\boldsymbol{\varkappa}}{2}\right|\right)}{\sqrt{N+1}} \times (2\pi)^{-3/2}\int_{z>\mathscr{R}} d^3z\, e^{i(\boldsymbol{\varkappa}-\boldsymbol{\varkappa}_f)\cdot\mathbf{z}}(u_{1/2,\nu'}\omega_l(\mathbf{z}, s), u_{1,\nu}), \tag{583}$$

[192] This argument was given in Ref. [80]. Calculations by Cheston [83] are consistent with this.

[193] This argument is rather weak when applied to small nuclei. Unfortunately, an adequate discussion of this point does not seem to be available. It is to a certain extent circumvented in Ref. [80], however.

which follows immediately on limiting the domain of integration in Eq. (573) to $z > \mathscr{R}$.

The neutron orbital state $\omega_l$ in Eq. (583) is written as[194]

$$\omega_l(\mathbf{z}, s) = R_l(z)\, Y_l^{\,m}(\hat{\mathbf{z}})u_{\frac{1}{2},\nu_n}, \tag{584}$$

where $R_l(z)$ is the radial wave function corresponding to an orbital angular momentum $l$. The quantity $u_{\frac{1}{2},\nu_n}$ represents the neutron spin function for orientation $\nu_n$. On introducing Eq. (584) into Eq. (583) and using Eq. (1.85$b$) for $e^{i(\boldsymbol{\varkappa}-\boldsymbol{\varkappa}_f)\cdot\mathbf{z}}$, we find

$$T_{ba} = -\frac{(-i)^l}{\sqrt{N+1}}\left[|\epsilon_D| + \frac{1}{M}\left(\boldsymbol{\varkappa}_f - \frac{\boldsymbol{\varkappa}}{2}\right)^2\right]\bar{D}\left(\left|\boldsymbol{\varkappa}_f - \frac{\boldsymbol{\varkappa}}{2}\right|\right)$$
$$\times\,(u_{\frac{1}{2},\nu'}u_{\frac{1}{2},\nu_n}, u_{1,\nu})I, \tag{585}$$

where [here $j_l$ is the spherical Bessel function, as defined by Eq. (6.125)]

$$I = \delta_{m,0}\sqrt{4\pi(2l+1)}\int_{\mathscr{R}}^{\infty} j_l(\rho z)R_l(z)z^2\,dz \tag{586}$$

and

$$\boldsymbol{\rho} \equiv \boldsymbol{\varkappa}_f - \boldsymbol{\varkappa}. \tag{587}$$

The integral (586) may be evaluated explicitly, since outside the nucleus (or, more precisely, outside the range of the nuclear binding forces) $R_l$ can be calculated. To do this, we write Eqs. (543) and (544) as[195]

$$E_a - W_0 = \epsilon_D + \frac{\kappa^2}{4M} = E_b - W_0$$
$$= \epsilon_N + \frac{\kappa_f^{\,2}}{2M}. \tag{588}$$

Here $-\epsilon_N$ is the binding energy of the neutron, which we shall write in the form

$$-\epsilon_N = \frac{t^2}{2M}. \tag{589}$$

Equations (588) and (589) permit us to write

$$\left[-\epsilon_D + \frac{1}{M}\left(\boldsymbol{\varkappa}_f - \frac{\boldsymbol{\varkappa}}{2}\right)^2\right] = \frac{1}{2M}(t^2 + \rho^2). \tag{590}$$

[194] Strictly speaking, we should couple the angular momentum of the neutron to the spin of the nuclear core to give a definite total spin for the nucleus. This is easily done and results only in a change in the numerical factors in the expression for the cross section.

[195] In writing these relations, we have assumed that the nuclear core is very heavy compared to the deuteron.

Now, the Schrödinger equations for $j_l$ and $R_l$ are

$$t^2 R_l = \frac{1}{z^2}\frac{d}{dz}\left(z^2 \frac{dR_l}{dz}\right) - \frac{l(l+1)}{z^2} R_l,$$
$$\rho^2 j_l = -\frac{1}{z^2}\frac{d}{dz}\left(z^2 \frac{dj_l}{dz}\right) + \frac{l(l+1)}{z^2} j_l. \tag{591}$$

Therefore

$$(\rho^2 + t^2) j_l R_l = \frac{1}{z^2}\left[j_l \frac{d}{dz}\left(z^2 \frac{dR_l}{dz}\right) - R_l \frac{d}{dz}\left(z^2 \frac{dj_l}{dz}\right)\right]$$
$$= \frac{1}{z^2}\frac{d}{dz}\left(j_l z^2 \frac{dR_l}{dz} - R_l z^2 \frac{dj_l}{dz}\right)$$

and [see Eq. (586)]

$$(\rho^2 + t^2) I = \delta_{m,0}\sqrt{4\pi(2l+1} \int_{\mathscr{R}}^{\infty} dz \frac{d}{dz}\left(j_l z^2 \frac{dR_l}{dz} - R_l z^2 \frac{dj_l}{dz}\right)$$
$$= -\delta_{m,0}\sqrt{4\pi(2l+1)}\mathscr{R}^2\left(j_l \frac{dR_l}{dz} - R_l \frac{dj_l}{dz}\right)\bigg|_{z=\mathscr{R}}. \tag{592}$$

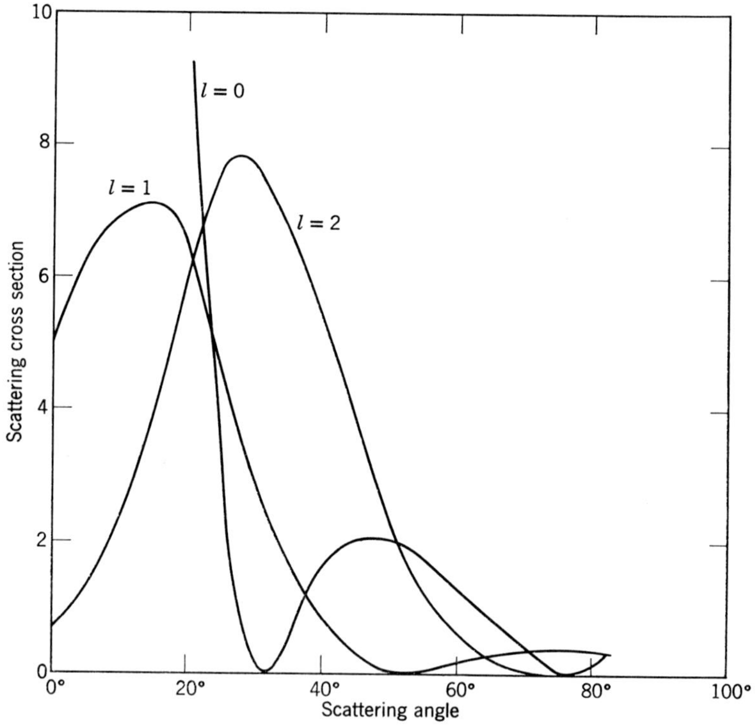


Fig. 11.18 Dependence of the deuteron stripping cross section (593) on the orbital angular momentum $l$ of the captured neutron (taken from Ref. [78], for $R = 5 \times 10^{-13}$ cm and an incident energy of 8.8 Mev).

If we combine Eqs. (585) and (592), we obtain for the stripping cross section

$$\left.\frac{d\bar{\sigma}}{d\Omega}\right|_{\text{stripping}} = \frac{(2\pi)^4}{v_{\text{rel}}}\frac{1}{3}\sum_{m,\nu',\nu_n,\nu}\int \kappa_f^2\,d\kappa_f\,\delta\left(\frac{\kappa_f^2}{2M}+\epsilon_N-\epsilon_D-\frac{\kappa^2}{4M}\right)|T_{ba}|^2$$

$$= C\frac{\kappa_f}{\kappa}\left|\bar{D}\left(\left|\boldsymbol{\varkappa}_f-\frac{\boldsymbol{\varkappa}}{2}\right|\right)\right|^2\left|j_l(\rho\mathscr{R})\frac{dR_l(\mathscr{R})}{d\mathscr{R}}-R_l(\mathscr{R})\frac{dj_l(\rho\mathscr{R})}{d\mathscr{R}}\right|^2, \tag{593}$$

where $C$ is a constant.[196] This is Butler's expression for the stripping cross section.

Equation (593) has proven to be very useful, and surprisingly successful,[197] in determining the angular momentum state of the captured neutron and thus in providing a comparison with predictions of the nuclear shell model. The striking dependence of the expression (593) on the orbital angular momentum $l$ of the captured neutron is illustrated in Fig. 11.18.

**The Proton Exchange Terms**

We have so far ignored the proton exchange term $ZT_{ba_{\text{ex}}}$ in Eq. (554). From Eq. (561), we obtain

$$T_{ba_{\text{ex}}} = (\phi'^{-}_{b(j)}, V_{p_j n}\psi_a^{+}), \tag{594}$$

where $j$ labels any one of the protons except that in the initial deuteron, $\phi'^{-}_{b(j)}$ describes the scattering of the $j$th proton by the residual nucleus, and $V_{p_j n}$ represents the interaction between the $j$th proton and that neutron which was in the initial deuteron.

Exchange terms such as (594) have been studied by French [84] and by Nakamura and Soga [85]. These authors have supposed that exchange can occur only with a single nuclear proton outside an inert core. Their results are based on Eq. (548) with $\psi_a^{+}$ replaced by $\chi_a$, rather than on Eq. (561), however.[198] We have not considered this a sufficiently plausible approximation to discuss here, and so would interpret the description of exchange in Refs. [84] and [85] with caution. If accepted as given, the numerical results in Refs. [85] do indicate a rather large exchange contribution to the stripping cross section. The available calculations are, however, too incomplete to permit any general conclusions as to the importance of the exchange terms.

[196] We have not attempted to retain the correct constant factors here, since we did not bother to couple appropriately the neutron angular momentum to that of the nuclear core.

[197] A thorough review of applications of Eq. (593) is given in Ref. [78].

[198] No numerical results are given in Ref. [84].

## 11.7 THE SCATTERING OF CHARGED PARTICLES BY ATOMS

Among the most important and extensive applications of scattering theory are those which attempt to describe atomic and molecular encounters. Because of the diversity of the phenomena no general mathematical techniques are available for such studies. Indeed, except for the simpler interactions, accurate computational methods do not exist.

We make no attempt in this section to give an exhaustive description of atomic reactions. Our discussion is restricted to the scattering of charged particles by neutral atoms; a few selected topics are included that illustrate the techniques developed earlier in this book.[199]

### The Use of a Pseudopotential for Elastic Scattering

Let us consider the elastic scattering of a nonrelativistic particle of mass $m$ and charge $Q$ by a neutral atom having $Z$ electrons. This is a special case of the interaction of a particle with a composite system, so the notation introduced in Section 11.2 can be applied here.

The Hamiltonian $h$ for the atom was described in connection with Eqs. (98). The complete Hamiltonian $H$ has the form (104),

$$H = K + V. \tag{595}$$

Here $V$ is the interaction between the scattered particle and the atom and

$$K = K_i + h, \tag{596}$$

where $K_i$ is the kinetic energy operator for the incident particle. For $V$ we shall neglect all but the Coulomb interactions between the charge $Q$ and the nucleus and the bound electrons of the atom. Thus

$$V = \frac{ZeQ}{r} + \sum_{\alpha=1}^{Z} V_\alpha, \tag{597$a$}$$

where

$$V_\alpha = V^{(1)}(|\mathbf{r} - \mathbf{z}_\alpha|) \equiv -\frac{eQ}{|\mathbf{r} - \mathbf{z}_\alpha|}. \tag{597$b$}$$

We have assumed here that the nucleus is at the origin of the coordinate system and sufficiently heavy that its recoil may be neglected. Consistent with our earlier notation, we have let $\mathbf{z}_\alpha$ ($\alpha = 1, 2, \ldots, Z$) represent the coordinate of the $\alpha$th electron and $\mathbf{r}$ that of the scattered particle.

[199] More complete discussions will be found in Refs. [86], [87], [88], and [89].

For convenient reference, we rewrite Eq. (98*a*) out explicitly in the form

$$h = \sum_{\beta=1}^{Z} K_\beta + U, \tag{598a}$$

$$U = \sum_{\alpha<\beta=1}^{Z} \frac{e^2}{|\mathbf{z}_\alpha - \mathbf{z}_\beta|} - \sum_\alpha \frac{Ze^2}{z_\alpha}, \tag{598b}$$

where $K_\beta$ is the kinetic energy operator of the $\beta$th electron.

The elastic scattering may evidently be described using the pseudopotential introduced in Section 11.4 [see Eqs. (407)]. In the lowest order approximation, obtained by taking $F = 1$, [Eq. (406)] this potential has a rather obvious form. Referring to Eqs. (410) and (597), we see that this is

$$\mathscr{V}_c(\mathbf{r}) \cong \mathscr{V}_{c1}(\mathbf{r}) \equiv -eQ \int \frac{\rho(\mathbf{y})\, d^3y}{|\mathbf{r} - \mathbf{y}|} + \frac{ZeQ}{r}, \tag{599}$$

where [see Eq. (110)]

$$\rho(\mathbf{y}) = \langle 0 | \sum_{\alpha=1}^{Z} \delta(\mathbf{y} - \mathbf{z}_\alpha) | 0 \rangle \tag{600}$$

is the electron density of the atom at the point $\mathbf{y}$.

The approximation (599) has been frequently used to describe scattering of charged particles by atoms. The electron density $\rho(\mathbf{y})$ is usually obtained from the Fermi-Thomas, the Hartree, or the Hartree-Fock approximation. This is illustrated in Fig. 11.19, where $\mathscr{V}_c(\mathbf{r})$ is shown for the electron-neon interaction, the electron density being obtained from Hartree-Fock wave functions.[200]

The potential (599) gives a qualitatively good description of the scattering at energies that are large compared to the first ionization potential of the atom. At lower energies it fails to account for the modification of the interaction resulting from distortion of the atom. At higher energies for which excitation and ionization predominate, the potential (599) fails to satisfy the condition (402), required by unitarity of the $S$-matrix. This suggests that we look next at the "second approximation" to $\mathscr{V}_c$, obtained by setting $F = 1$ in the right-hand side of Eq. (406). With the notation used in Eq. (407*c*), this second approximation has the form

$$\mathscr{V}_{c2} = \sum_{\gamma(\neq 0)} \langle , 0 | V | , \gamma \rangle [\epsilon_p + i\eta - K_i - (W_\gamma - W_0)]^{-1} \langle , \gamma | V | , 0 \rangle, \tag{601}$$

where $W_\gamma$ and $W_0$ are the atomic energy levels [see Eqs. (99)] of the atomic states $\gamma$ and 0.

The expression (601) is cumbersome to evaluate directly, although useful approximate methods,[201] including variational principles (Schwartz [93])

[200] These are taken from Ref. [90].

[201] See, for example, Refs. [91] and [92].

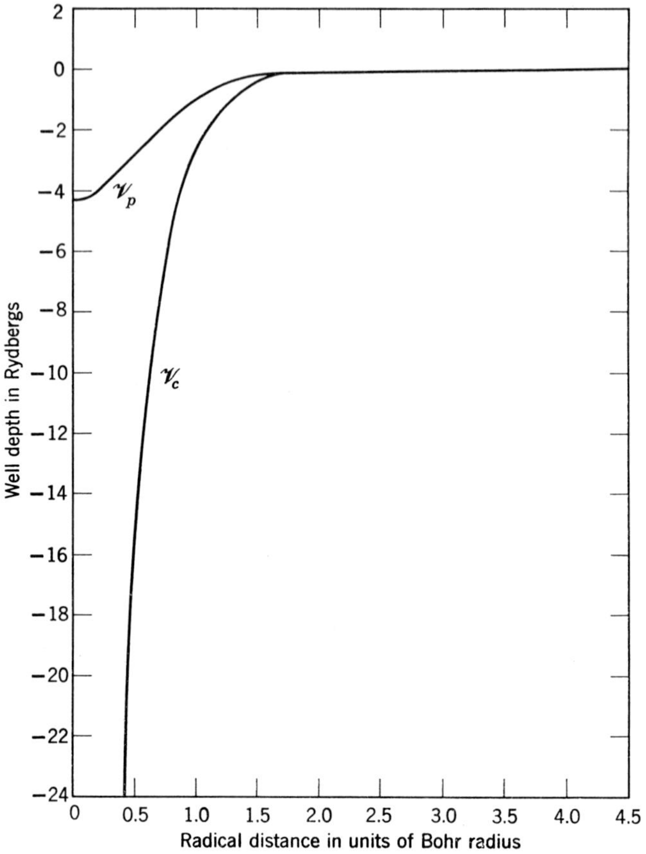


Fig. 11.19 Comparison of $V_c$ and $V_p$ in Eq. (607) for the electron-neon interaction.

are available. Consequently, most discussions of Eq. (601) have made the adiabatic approximation (411), neglecting the change $(\epsilon_p - K_i)$ in kinetic energy in the propagator $[\epsilon_p + i\eta - K_i - (W_\gamma - W_0)]^{-1}$. In addition, it is usually assumed that the coordinate $r$ is large compared to atomic dimensions, so the Coulomb interaction (597*b*) may be written as

$$V^{(1)}(|\mathbf{r} - \mathbf{z}_\alpha|) \cong -\frac{Qe}{r}\left(1 + \frac{\hat{\mathbf{r}} \cdot \mathbf{z}_\alpha}{r}\right).$$

Then the interaction (597*a*) simplifies to

$$V(r) = -\frac{Qe}{r^2}\left(\sum_{\alpha=1}^{Z} \hat{\mathbf{r}} \cdot \mathbf{z}_\alpha\right). \tag{602}$$

This, and the adiabatic approximation, permit writing Eq. (601) as

$$\mathscr{V}_{c2}(\mathbf{r}) \cong -\frac{\overline{\overline{\alpha}} Q^2}{2r^4}, \tag{603}$$

where

$$\overline{\overline{\alpha}} \equiv 2e^2 \sum_{\gamma(\neq 0)} \frac{\left|\left(g_\gamma, \hat{\mathbf{r}} \cdot \sum_{\alpha=1}^{Z} \mathbf{z}_\alpha\right)\right|^2}{W_\gamma - W_0} \tag{604}$$

is the atomic polarizability [which was introduced in Eq. (343)].

The atomic polarizability $\overline{\overline{\alpha}}$ of an atom may be directly measured, or it may be calculated (in some approximation) from the expression (604).[202] Methods for estimating atomic polarizabilities have been given by Mayer and Mayer [96] and by Buckingham, [97], who obtains the approximate expression

$$\overline{\overline{\alpha}} \cong \frac{4}{9a_0} \sum_{\alpha=1}^{Z} [(g_0, z_\alpha^2 g_0)]^2,$$

where $a_0$ is the Bohr radius.

The expression (603) is not appropriate for use in the Schrödinger equation because of the $r^{-4}$ singularity at $r = 0$.[203] This singularity is, of course, artificial, having arisen as a result of the expansion in powers of $(z_\alpha/r)$ which led to Eq. (602). Because of the complexity of the expression (601) (even in the adiabatic approximation) it has been customary to simply "cut-off" the singularity in an heuristic manner suggested by Buckingham [97].[204]

$$\mathscr{V}_{c2} \to \mathscr{V}_p \equiv -\frac{\overline{\overline{\alpha}} Q^2}{2(r^2 + s^2)^2}, \tag{605}$$

where $s$ is a length related to the size of the atom. Mittleman and Watson [49] have obtained

$$s^4 \cong \tfrac{1}{2} a_0 \overline{\overline{\alpha}} Z^{-1/3}, \tag{606}$$

where again $a_0$ is the Bohr radius.

Combining Eqs. (599) and (605), we obtain

$$\mathscr{V}_c(r) = \mathscr{V}_{c1}(r) + \mathscr{V}_p(r), \tag{607}$$

as the adiabatic approximation, valid to second order, for the pseudo-potential. The quantities $\mathscr{V}_c$ and $\mathscr{V}_p$ are illustrated in Fig. 11.19 for the electron-neon interaction.[205] The effect of these terms on electron-hydrogen

[202] See, for example, Refs. [91], [92], [94], and [95].

[203] The discussion given in Section 6.3 of singular potentials may be recalled.

[204] A natural cut-off may be obtained from the use of a variational principle, however (see Ref. [49]).

[205] For neon we have taken $\overline{\overline{\alpha}} = 2.6a_0^3$.

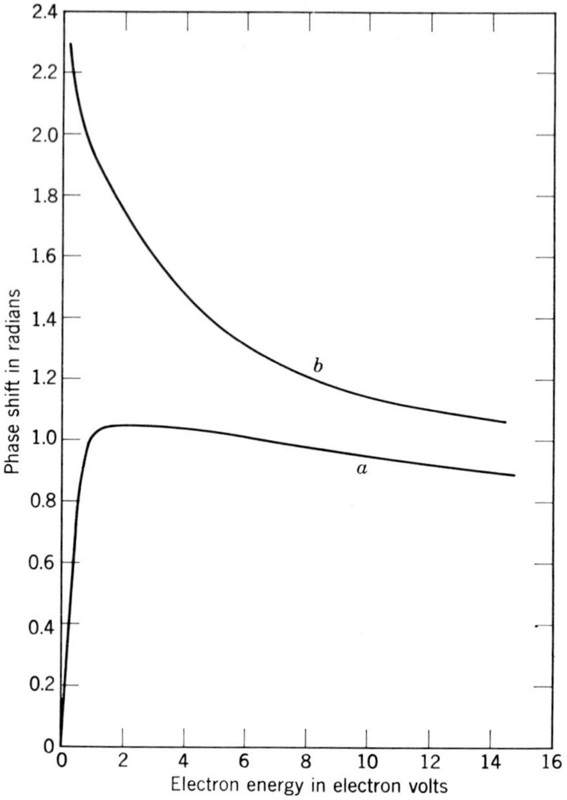


Fig. 11.20 Comparison of the $S$-wave phase shift for electron-hydrogen scattering neglecting $V_p$ (curve $a$) and keeping $V_p$ (curve $b$).

scattering is shown in Fig. 11.20,[206] where the $s$-wave phase shift is given as a function of the energy with and without the inclusion of $\mathscr{V}_p$.

We have yet to estimate the accuracy of the expression (607) for $\mathscr{V}_c$, which was obtained following several approximations: (1) the adiabatic approximation was made in evaluating $\mathscr{V}_{c2}$, (2) the (nonphysical) singularity in Eq. (603) was removed in a rather arbitrary manner, (3) the expansion for $\mathscr{V}_c$ was terminated by setting $F = 1$ in the right-hand side of Eq. (406), and (4) for electron scattering equivalence of the scattered and bound electrons has not been taken into account as required by the Pauli principle.

[206] This result is taken from the work of Malik and Trefftz [98]. For other applications of Eq. (607) see the review of Burke and Smith [89, 99], where electron scattering by oxygen is studied.

The accuracy of the adiabatic approximation has been discussed by Mittleman and Watson [49]. Following their argument, we write $\delta\mathscr{V}_{c2}$ for the difference between the exact value (601) and the corresponding adiabatic expression to obtain (here we write $m_e$ for the electron mass and $m$ for the mass of the incident particle)

$$\left|\frac{\delta\mathscr{V}_{c2}}{\mathscr{V}_c}\right| \cong \frac{m_e}{m} L(q, Z) + \sqrt{m_e\epsilon_p/m \text{ Ry}}\, M(q, Z), \tag{608}$$

where $L$ and $M$ are dimensionless functions of $q \equiv Z^{1/3}/0.9a_0 r$[207] and $Z$, Ry $= 13.57$ ev is the Rydberg, and $\epsilon_p$ is (as before) the kinetic energy of the scattered particle. The quantities $L$ and $M$ are shown in Figs. 11.21 as functions of $q$ for several values of $Z$.[208] It is seen that only for $Z < 10$ is the adiabatic approximation significantly in error[209] if the Fermi-Thomas estimate can be considered reliable. Although it is not apparent from Fig. 11.21 because of the limited range of $q$ shown, the expression (608) vanishes for $q \to \infty$. It is, of course, expected that for large enough impact parameters the adiabatic approximation should become exact.

The method used in Eq. (605) of removing the singularity of Eq. (603) at $r = 0$ is probably not seriously in error. Unfortunately, however, the scattering at very low energies is often rather sensitive to the actual potential strength. This is particularly so when there is a real (or virtual) negative ion of nearly zero binding energy.

A rough estimate of the error associated with setting $F = 1$ in the right-hand side of Eq. (406) was made by Mittleman and Watson [49]. The fractional error resulting from neglecting further terms in Eq. (607) was estimated to be of the order of $Z^{-4/3}$ near the center of the atom. At large distance ($r \to \infty$) Eq. (607) is exact.

When the scattered particle is an electron, and when it is necessary to include the coordinates of the scattered electron when antisymmetrizing the wave function, the analysis of the interaction becomes much more difficult. For the special case that Hartree-Fock wave functions are used for the atom, however, the scattering may be handled rather simply.[210] Here the pseudopotential is nondiagonal in coordinate space, having the form $(\mathbf{r}|\mathscr{V}_c|\mathbf{r}')$. To write it out, we first suppose that the atomic state $g_0$

[207] We again write the Bohr radius as $a_0$.

[208] The Fermi-Thomas model was used in Ref. [49] to evaluate $L$ and $M$. Thus for $Z = 1$ these quantities should not be taken as quantitatively accurate.

[209] Some care must be exercised in interpreting the significance of Eq. (608). For example, a large relative error may have little effect on the scattering for large $r$, where both $\mathscr{V}_c$ and $\delta\mathscr{V}_{c2}$ tend to be small.

[210] This has been shown in Ref. [42]. The method given here is applicable to the exact potential defined by Eqs. (406) and (407).

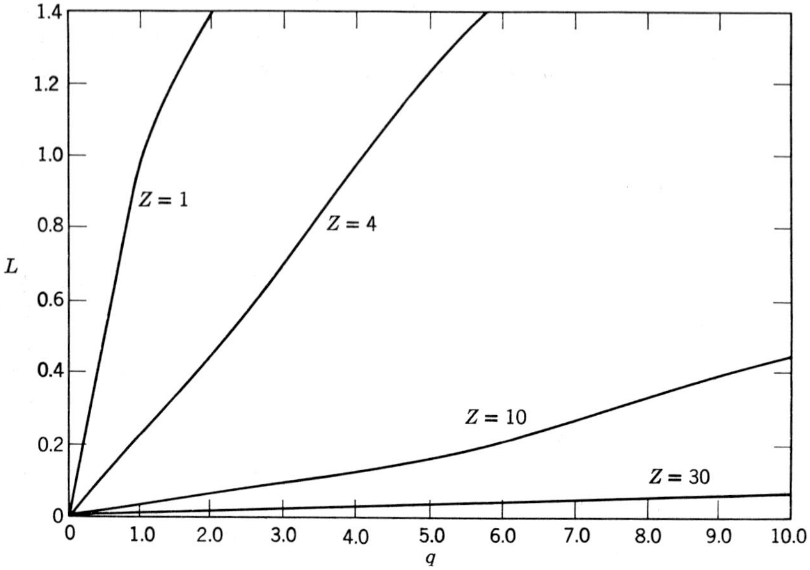


(*a*)

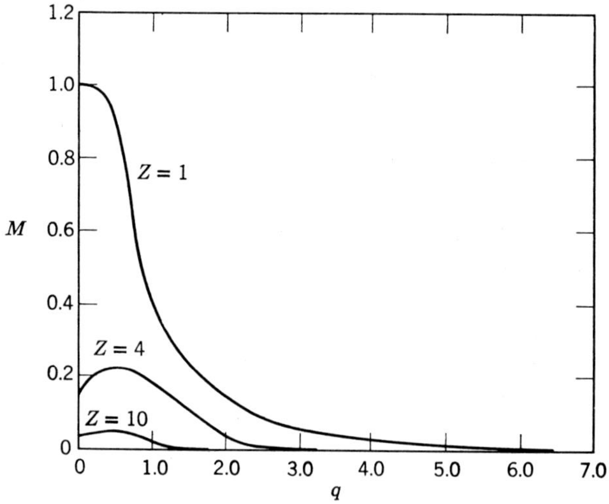


(*b*)

Fig. 11.21 The dimensionless functions $L$ (*a*) and $M$ (*b*) of Eq. (608).

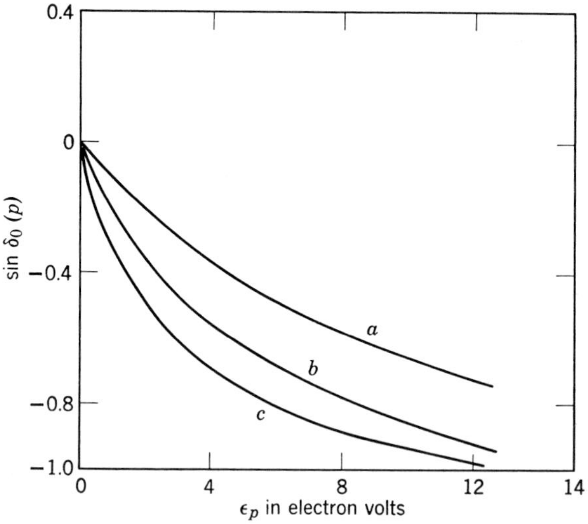


Fig. 11.22 The sine of the $S$-wave phase shift for electron-neon scattering. Curve $a$ corresponds to use of the complete interaction (609); curve $b$ results when $V_{\text{orth}}$ is dropped; curve $c$ corresponds to the case that neither $V_{\text{orth}}$ nor $V_{\text{ex}}$ is kept.

is an antisymmetrized product constructed from the Hartree-Fock orbitals $\omega_n(\mathbf{z}_\alpha)$ $(n, \alpha = 1, 2, \ldots, Z)$. Then

$$(\mathbf{r}|\mathscr{V}_c|\mathbf{r}') = \delta(\mathbf{r}' - \mathbf{r})[\mathscr{V}_{c1}(r) + \mathscr{V}_p(r)] + \mathscr{V}_{\text{ex}} + \mathscr{V}_{\text{orth}}, \tag{609}$$

where $\mathscr{V}_{c1}$ and $\mathscr{V}_p$ are given by Eqs. (599) and (605), respectively. The term $\mathscr{V}_{\text{ex}}$ is the usual Hartree-Fock exchange potential

$$(\mathbf{r}|\mathscr{V}_{\text{ex}}|\mathbf{r}') = -\sum_{n=1}^{Z} \frac{e^2}{|\mathbf{r} - \mathbf{r}'|} \omega_n(\mathbf{r})\omega_n^{\dagger}(\mathbf{r}'). \tag{610}$$

Similarly,[211]

$$(\mathbf{r}|\mathscr{V}_{\text{orth}}|\mathbf{r}') = -\sum_{n=1}^{Z} \omega_n(\mathbf{r})\omega_n^{\dagger}(\mathbf{r}')\mathscr{V}_p(\mathbf{r}'). \tag{611}$$

The Schrödinger equation to determine the scattered wave function is then, in the notation of Eq. (308),

$$\left(-\frac{1}{2m}\nabla_r^2 - \epsilon_p\right)\phi^+_{\mathbf{p},\nu,\nu_t}(\mathbf{r}) + \int d^3r'(\mathbf{r}|\mathscr{V}_c|\mathbf{r}')\phi^+_{\mathbf{p},\nu,\nu_t}(\mathbf{r}') = 0. \tag{612}$$

It is easily verified from Eq. (612) that the presence of $\mathscr{V}_{\text{orth}}$ in Eq. (609) insures us that $\phi^+_{\mathbf{p},\gamma,\gamma_t}$ is orthogonal to each of the Hartree-Fock orbitals $\omega_n$.

The effect of the various terms in Eq. (609) is illustrated in Fig. 11.22, where the $s$-wave phase shift for electron-neon scattering is shown as a

[211] A discussion of the physical significance of the interaction (609), as well as a "local" approximation to it, has been given in Ref. [100].

function of energy. The various curves indicate the importance of the several contributions in Eq. (609) to $\mathscr{V}_c$.[212]

**Effective Range Expansion**

We observed in Section 6.6 that the effective range expansion (6.326) has been quite useful in studying low-energy nucleon-nucleon scattering. Such an expansion can also be useful in describing atomic scatterings. Because of the long-range $r^{-4}$ dependence of the polarization potential as $r \to \infty$, however, an expansion such as that of Eq. (6.326) does not exist for charged particle scattering by neutral atoms.[213] A modified effective range expansion can be obtained, as has been shown by O'Malley, Spruch, and Rosenberg [102].

We shall briefly describe their argument. The wave function $\phi^+_{\mathbf{p},\nu,\nu_t}$ may be expressed as a sum of partial waves to obtain a set of radial equations from Eq. (611) [see Eqs. (6.110)]. For $s$-waves and for large $r$ the asymptotic form is

$$\left[\frac{d^2}{dr^2} + \frac{P^2}{r^4} + p^2\right]\bar{u} = 0, \tag{613}$$

a generalization of Eq. (6.311). In Eq. (613) we have set

$$P^2 \equiv me^2\bar{\bar{\alpha}}. \tag{614}$$

The substitution

$$r = \left(\frac{P}{p}\right)^{1/2} e^y$$

transforms Eq. (613) into the Mathieu equation

$$\left[\frac{d^2}{dy^2} - \frac{1}{4} + 2Pp\cosh(2y)\right]\bar{u} = 0, \tag{615}$$

which has solutions[214]

$$\bar{u}_\pm = M^{(1)}_{\pm\nu}(y). \tag{616}$$

From these functions a function analogous to that of Eqs. (6.312) [that is, having the asymptotic form of the solutions to Eq. (611)] may be constructed. With these functions the argument of Section 6.6, p. 289, may

[212] The curves shown in Fig. 11.22 are obtained by numerical solution of Eq. (612), using the wave functions of Ref. [90]. The computations are taken from Ref. [101].

[213] Some of the coefficients in the expansion are infinite (see Ref. [102]).

[214] See Ref. [103], Section 2.4. The relation between the index $\nu$ and the parameter $(Pp)$ is given in Eq. (2.2.44) of this reference.

be followed to give the expansion of O'Malley, Spruch, and Rosenberg.

$$p \cot \delta_0(p) = \frac{1}{a} + \frac{\pi P^2}{3a^2} p + \mathcal{O}\left[\frac{P^2 p^2}{a} \ln (Pp)\right], \tag{617}$$

where $a$ is the scattering length.

**Application of Variational Principles**

Several variational techniques were described in Section 6.7, p. 313, for use in connection with potential scattering. Each of these methods may be directly generalized for application to scattering by composite systems. The Kohn variational principle has, in particular, been extensively applied to atomic scattering phenomena. This includes establishing rigorous bounds on scattering lengths [104] and the direct evaluation of scattering phase shifts [105].

The most detailed and quantitative applications of variational methods have, for reasons of mathematical simplicity, been limited to electron-hydrogen and positron-hydrogen scattering. For more complicated interactions variational principles have been useful for semiquantitative studies and for estimating the accuracy of other calculational methods (such as that described in Section 11.7, p. 857).

We shall limit ourselves here to a very simple illustration of the Kohn variational method [described in Section 6.7, p. 313], considering only the elastic scattering of an electron by a neutral atom of charge number $Z$ and zero spin. More complex interactions involve complications in detail rather than of principle.

According to Eq. (4.98) the symmetrized wave function for the scattering (including inelastic scattering) is of the form

$$\psi^+_{\mathscr{S}(a)} = \frac{1}{\sqrt{Z+1}} \sum_{\alpha=1}^{Z} \delta_\alpha \psi^+_{a(\alpha)}, \tag{618}$$

where $\psi^+_{a(\alpha)}$ is the wave function for the scattering of electron $\alpha$ ($\alpha = 1, 2, \ldots, Z$) by the atom, this electron being treated as distinguishable from the others. The asymptotic form of $\psi^+_{\mathscr{S}(a)}$, as $z_\alpha \to \infty$, is

$$\psi^+_{\mathscr{S}(a)}\bigg|_{z_\alpha \to \infty} = \frac{(2\pi)^{-3/2}}{\sqrt{Z+1}} \delta_\alpha \sum_{\gamma, \nu'} \left(e^{ipz_\alpha}\delta_{\gamma,0} + \frac{e^{ik_\gamma z_\alpha}}{z_\alpha} f_{\gamma 0}\right) g_\gamma u_{1/2,\nu'}. \tag{619}$$

Here, as in Eq. (4.164),[215]

$$f_{\gamma 0} \equiv (f_d - Z f_{\text{ex}})_{\gamma 0}$$

[215] Following the notation of Eq. (4.164), we have again used subscripts $d$ and ex to indicate "direct" and "exchange" amplitudes, respectively.

is the symmetrized scattering amplitude for scattering to a final state $\gamma$ of the atom and $\nu'$ is a final spin orientation of the scattered electron.

For the elastic scattering from a spin-zero atom we keep only the term corresponding to $\gamma = 0$ in Eq. (619) and introduce a partial wave expansion

$$\psi^{+}_{c(\mathscr{S}(a))}\Big|_{z_\alpha \to \infty} = \delta_\alpha \sum_{l=0}^{\infty} \psi_l(\mathbf{z}_\alpha)[(i)^l \sqrt{(2l+1)/4\pi}(2mp^2)^{-1/2} e^{i\delta_l} \cos \delta_l] u_{1/2,\nu}. \tag{620}$$

In this equation $\delta_l$ is the phase shift for the $l$th partial wave, $u_{1/2\nu}$ is the wave function of the incident electron and

$$\psi_l(\mathbf{z}_\alpha) = \phi_l(\mathbf{z}_\alpha) g_0(\mathbf{z}_1, s_1, \ldots, \mathbf{z}_{\alpha-1}, s_{\alpha-1}, \mathbf{z}_0, s_0, \ldots, \mathbf{z}_Z, s_Z), \tag{621a}$$

$$\phi_l(\mathbf{r}) = \sqrt{4mp^2/\pi(Z+1)}\; Y_l^0(\hat{\mathbf{r}})[j_l(pr) - \tan \delta_l n_l(pr)]. \tag{621b}$$

Let us now consider two solutions of the same Schrödinger equation $\Psi_a$ and $\Phi_a$.

$$(H - E)\Psi_a = 0,$$
$$(H - E)\Phi_a = 0. \tag{622a}$$

If we consider $\Phi_a$ and $\Psi_a$ to be column matrices in spin space and $\tilde{\Phi}_a$ and $\tilde{\Psi}_a$ to be the corresponding transposed row matrices (in spin space only), then

$$(H - E)\tilde{\Phi}_a = 0,$$
$$(H - E)\tilde{\Psi}_a = 0, \tag{622b}$$

since $H$ does not depend upon spin variables.

In analogy to Eq. (6.407) we introduce the quantity

$$I \equiv \int \tilde{\Phi}_{a\,\mathrm{Tr}}(E - H)\Psi_{a\,\mathrm{Tr}}\, d\tau, \tag{623}$$

where

$$d\tau = \prod_{\alpha=0}^{Z} d^3 z_\alpha \tag{624}$$

and $\Psi_{a\,\mathrm{Tr}}$ and $\Phi_{a\,\mathrm{Tr}}$ are "trial wave functions" considered as approximations to $\Psi_a$ and $\Phi_a$. When the exact functions $\Psi_a$ and $\tilde{\Phi}_a$ are inserted in Eq. (623), $I$ vanishes.

Now, consider the variations

$$\Psi_{a\,\mathrm{Tr}} = \Psi_a + \delta\Psi_a,$$
$$\tilde{\Phi}_{a\,\mathrm{Tr}} = \tilde{\Phi}_a + \delta\tilde{\Phi}_a, \tag{625}$$

where the variations $\delta\Psi_a$ and $\delta\Phi_a$ are subject to the same boundary conditions at large and small $z_\alpha$ as are the exact wave functions [see Eqs. (6.409)].

Because of Eqs. (622), the first-order variation in $I$ is

$$\delta I = \int \tilde{\Phi}_a(E - H)\, \delta\Psi_a \, d\tau. \tag{626}$$

To simplify Eq. (626), we write the Hamiltonian (595) as $H = K_T + V_T$, where $V_T$ is the sum of all Coulomb interactions between the $(Z + 2)$ charged particles and

$$K_T = -\frac{1}{2m}\sum_{\alpha=0}^{Z} \nabla_\alpha^{\,2}. \tag{627}$$

Then, using Eq. (622*b*), we obtain

$$\begin{aligned} \delta I &= \int [(K_T\tilde{\Phi}_a)\, \delta\Psi_a - \tilde{\Phi}_a(K_T\, \delta\Psi_a)]\, d\tau \\ &= \sum_{\alpha=0}^{Z} \frac{1}{2m} \int d\tau \nabla_\alpha \cdot (\tilde{\Phi}_a \nabla_\alpha\, \delta\Psi_a - \delta\tilde{\Psi}_a \nabla_\alpha \Phi_a) \\ &= \frac{Z+1}{2m} \int \prod_{\alpha=1}^{Z} d^3 z_\alpha \int d\mathbf{\Sigma}_0 \cdot (\tilde{\Phi}_a \nabla_0\, \delta\Psi_a - \delta\tilde{\Psi}_a \nabla_0 \Phi_a). \end{aligned} \tag{628}$$

Here $d\mathbf{\Sigma}_0$ is an element of surface area on a very large sphere of constant $|\mathbf{z}_0|$. We have replaced the sum over $\alpha$ in Eq. (628) by a factor of $Z + 1$ and kept only the flux of electron 0, since all electrons contribute equally.

As an application of Eq. (628), we suppose the variation $\delta\Psi_a$ to occur only in the *elastic* scattering channel. Then, since only the asymptotic wave functions appear in the final expression (628), we may choose

$$\Psi_a = \Phi_a = \psi_l \tag{629}$$

[see Eqs. (621)] and let

$$\delta\Psi_a = \delta\phi_l g_0 + \phi_l\, \delta g_0. \tag{630}$$

On inserting the expressions (629) and (630) into Eq. (628) and on following the argument which led to Eq. (6.414), we obtain

$$\delta I = \frac{Z+1}{2m} \int d\mathbf{\Sigma}_0 \cdot (\phi_l \nabla_0\, \delta\phi_l - \delta\phi_l \nabla_0 \phi_l) = -\frac{2}{\pi}\, p\, \delta(\tan \delta_l). \tag{631}$$

The variational principle then reads

$$\delta\left(I + \frac{2}{\pi}\, p \tan \delta_l\right) = 0, \tag{632}$$

which is of the same form as Eq. (6.415).

When electron exchange can be neglected and when the interaction is given in the approximate form (607), it is seen that the variational principle (632) reduces to that for simple potential scattering, as given by

Eq. (6.415). The expression (632) is much more general, however, since its domain of applicability is not restricted to potential scattering.

Schwartz [105] has given an elaborate numerical discussion of electron-hydrogen scattering using the variational principle (632). He has chosen trial wave functions for $I$ [Eq. (623)] using up to fifty variable parameters and having asymptotic forms consistent with Eq. (613). In particular, he has obtained the respective singlet and triplet scattering lengths [see Eq. (617)]

$$\begin{aligned} a_s &= -(5.965 \pm 0.003)a_0, \\ a_t &= -(1.7686 \pm 0.0002)a_0, \end{aligned} \tag{633}$$

where $a_0$ is the Bohr radius.

## REFERENCES AND NOTES


1. J. M. Blatt and V. F. Weisskopf, *Theoretical Nuclear Physics*, John Wiley and Sons, New York, 1952.
2. H. S. Massey and E. H. S. Burhop, *Electronic and Ionic Impact Phenomena*, Clarendon Press, London, 1952.
3. E. Fermi, *Ric. Sci.* **VII-11,** 13 (1936).
4. G. F. Chew, *Phys. Rev.* **80,** 196 (1950).
5. G. F. Chew and G. C. Wick, *Phys. Rev.* **85,** 636 (1952).
6. J. Ashkin and G. C. Wick, *Phys. Rev.* **85,** 686 (1952).
7. G. F. Chew and M. L. Goldberger, *Phys. Rev.* **87,** 778 (1952).
8. K. A. Brueckner and W. Wada, *Phys. Rev.* **103,** 1008 (1956).
9. G. Placzek, *Phys. Rev.* **86,** 377 (1952).
10. G. C. Wick, *Phys. Rev.* **94,** 1228 (1954).
11. T. K. Fowler, *Phys. Rev.* **112,** 1325 (1958).
12. T. K. Fowler and K. M. Watson, *Nuclear Phys.* **13,** 549 (1959).
13. The expression (57) for the total cross section is

$$\bar{\sigma} = \int \frac{d\bar{\sigma}}{d\Omega}\, d\Omega = \bar{\sigma}_f \left[1 + \frac{1}{3}\frac{m}{M}\frac{K_{\mathrm{av}}}{\epsilon_p} - \frac{1}{32}\left(\frac{m}{M} + 1\right)^2 \frac{m}{M}\frac{C_{\mathrm{av}}}{\epsilon_p{}^3} + \cdots\right].$$

Here $\bar{\sigma}_f$ is the total free-particle cross section (48), while $K_{\mathrm{av}}$ and $C_{\mathrm{av}}$ are defined by Eqs. (16) and (58*b*). In deriving this result it was assumed that the incident particle had a nonrelativistic energy.

14. M. L. Goldberger, *Phys. Rev.* **74,** 1269 (1948).
15. See, for example, *The Mössbauer Effect*, ed. by H. Frauenfelder, W. A. Benjamin, New York, 1962.
16. K. M. Watson, *Phys. Rev.* **118,** 886 (1960).
17. S. Gartenhaus and C. Schwartz, *Phys. Rev.* **108,** 482 (1957).
18. Since the quantity $T_\alpha{}^\dagger T_\alpha$ is a matrix in the $(2S_\alpha + 1)$-dimensional spin space of particle $\alpha$, it may be expanded in terms of orthogonal basis matrices $\omega^\lambda(\alpha)$ of Section 7.5, p. 409. According to Eq. (7.270) this expansion has the form $T_\alpha{}^\dagger T_\alpha = \Sigma_\lambda b_\lambda \omega^\lambda(\alpha)$, where $\omega^1(\alpha) = 1$, the unit matrix. Now $\langle 0|\, \omega^\lambda(\alpha)\, |0\rangle_{\mathrm{av}} = 0$, unless $\lambda = 1$, since the $\omega^\lambda$ are irreducible tensors and the av implies an average over all coordinate orientations. Therefore

$$\langle 0|\, T_\alpha{}^\dagger T_\alpha\, |0\rangle_{\mathrm{av}} = \frac{1}{(2S_i + 1)(2S_\alpha + 1)} \operatorname{Tr}(T_\alpha{}^\dagger T_\alpha).$$

19. O. Chamberlain and E. Segré, *Phys. Rev.* **87,** 81 (1952). See also J. Cladis, W. Hess, and B. Moyer, *Phys. Rev.* **87,** 425 (1952).
20. L. Van Hove, *Phys. Rev.* **95,** 249 (1954).
21. R. W. James, *The Optical Principles of the Diffraction of X-Rays*, Bell and Sons, London, 1948.
22. A. C. Zemach and R. J. Glauber, *Phys. Rev.* **101,** 118 (1956); **101,** 129 (1956).
23. M. Born and H. S. Green, *A General Kinetic Theory of Liquids*, Cambridge University Press, New York, 1949.
24. G. Placzek, B. R. A. Nijboer, and L. Van Hove, *Phys. Rev.* **82,** 392 (1951).
25. L. Van Hove and K. W. McVoy, *Phys. Rev.* **125,** 1034 (1962).
26. M. Lax and H. Feshbach, *Phys. Rev.* **81,** 189 (1951).
27. E. P. Wigner, *Phys. Rev.* **51,** 947 (1937).
28. M. K. Wilkinson, E. O. Wollan, and W. C. Koehler, *Ann. Rev. Nuclear Sci.*, Annual Reviews, Palo Alto, California, 1961, p. 303.
29. L. L. Foldy, *Phys. Rev.* **67,** 107 (1945).
30. Melvin Lax, *Revs. Modern Phys.* **23,** 287 (1951).
31. K. M. Watson, *Phys. Rev.* **89,** 575 (1953).
32. Hans Ekstein, *Phys. Rev.* **87,** 31 (1952).
33. K. M. Watson, *Phys. Rev.* **105,** 1388 (1957).
34. K. A. Brueckner, *Phys. Rev.* **89,** 834 (1953).
35. L. L. Foldy, *Phys. Rev.* **67,** 107 (1954).
36. Steven Weinberg, unpublished.
37. L. Rosenfeld, *Theory of Electrons*, North-Holland Publishing Co., Amsterdam, 1951.
38. W. K. H. Panofsky and M. Phillips, *Classical Electricity and Magnetism* Addison-Wesley Publishing Co., Cambridge, Mass,, 1955.
39. W. Fuller Brown, Vol. 17, *Encyclopedia of Physics*, Springer-Verlag, Berlin 1956.
40. A. N. Kaufman and K. M. Watson, *Phys. Fluids* **4,** 931 (1961).
41. Gyo Takeda and K. M. Watson, *Phys. Rev.* **97,** 1336 (1955).
42. B. Lippmann, M. Mittleman, and K. Watson, *Phys. Rev.* **116,** 920 (1959).
43. N. C. Francis and K. M. Watson, *Phys. Rev.* **92,** 291 (1953).
44. K. Greider, *Nuclear Phys.* **14,** 498 (1959/60).
45. S. Weinberg, *Phys. Rev.* **126,** 1899 (1962).
46. M. Lax, *Phys. Rev.* **78,** 306 (1950).
47. N. C. Francis and K. M. Watson, *Am. J. Phys.* **21,** 659 (1953).
48. H. Feshbach, *Ann. Phys.* **5,** 357 (1958); **19,** 287 (1962).
49. M. H. Mittleman and K. M. Watson, *Phys. Rev.* **113,** 198 (1959).
50. R. Sternheimer, *Phys. Rev.* **101,** 384 (1956).
51. R. Frank, J. Gammel, and K. Watson, *Phys. Rev.* **101,** 891 (1956).
52. L. Kisslinger, *Phys. Rev.* **98,** 761 (1955).
53. R. Williams, W. Baker, and J. Rainwater, *Phys. Rev.* **104,** 1695 (1956).
54. M. Ostrofsky, G. Breit, and D. Johnson, *Phys. Rev.* **49,** 22 (1936).
55. H. Feshbach, R. Peaslee, and V. Weisskopf, *Phys. Rev.* **71,** 145 (1947).
56. H. A. Bethe and R. R. Wilson, *Phys. Rev.* **83,** 690 (1951).
57. S. Fernbach, R. Serber, and T. Taylor, *Phys. Rev.* **75,** 1352 (1949).
58. W. B. Riesenfeld and K. M. Watson, *Phys. Rev.* **102,** 1157 (1956).
59. J. Dickson, B. Rose, and D. Salter, *Proc. Roy. Soc.* (*London*) **68,** 361 (1955).
60. O. Chamberlain, E. Segré, R. Tripp, C. Wiegand, and T. Ypsilantis, *Phys. Rev.* **102,** 1659 (1956).
61. P. Hillman, A. Johanson, and Th. Tyrên, *Nuclear Phys.* **4,** 648 (1957).

62. H. A. Bethe, *Ann. Phys.* **3,** 190 (1958).
63. A. Kerman, H. McManus, and R. Thaler, *Ann. Phys.* **8,** 551 (1959).
64. F. Bjorklund, *Phys. Rev.* **108,** 795 (1957).
65. R. R. Johnston and K. M. Watson, *Nuclear Phys.* **28,** 583 (1961).
66. K. M. Watson, *Revs. Modern Phys.* **30,** 565 (1958).
67. R. Glauber, "High Energy Collision Theory," *Lectures in Theoretical Physics*, Vol. 1, ed. by W. E. Brittin and L. G. Dunham, Interscience Publishers, New York, 1959.
68. M. A. Baqi Beg, *Phys. Rev.* **120,** 1867 (1960).
69. L. M. Frantz and R. L. Mills, *Nuclear Phys.* **15,** 16 (1960).
70. J. S. Bell and E. J. Squires, *Phys. Rev. Letters* **3,** 96 (1959).
71. L. M. Frantz, R. L. Mills, R. G. Newton, and A. Sessler, *Phys. Rev. Letters* **1,** 340 (1958).
72. C. A. Levinson and M. K. Banerjee, *Ann. Phys.* **3,** 67 (1958).
73. H. Tyren and Th. A. J. Maris, *Nuclear Phys.* **3,** 35 (1957); H. Tyren, P. Hillman, and Th. A. J. Maris, *Nuclear Phys.* **7,** 10 (1958); Th. A. J. Maris, P. Hillman, and H. Tyren, *Nuclear Phys.* **7,** 1 (1958).
74. N. C. Francis and K. M. Watson, *Phys. Rev.* **89,** 328 (1953).
75. G. F. Chew and M. L. Goldberger, *Phys. Rev.* **77,** 470 (1950).
76. H. F. York, *Phys. Rev.* **75,** 1467A (1949); K. A. Brueckner and W. M. Powell, *Phys, Rev.* **75,** 1274 (1949).
77. S. T. Butler, *Proc. Roy. Soc.* (*London*) **208,** 559 (1951). See also Bhatia, Huang, Huby, and Newns, *Phil. Mag.* **43,** 485 (1952).
78. S. T. Butler, *Nuclear Stripping Reactions*, John Wiley and Sons, New York, 1957.
79. K. R. Greider, *Phys. Rev.* **114,** 786 (1959).
80. N. C. Francis and K. M. Watson, *Phys. Rev.* **93,** 313 (1954).
81. W. Selove, *Phys. Rev.* **101,** 231 (1956).
82. P. B. Daitch and J. B. French, *Phys. Rev.* **87,** 900 (1952).
83. W. B. Cheston, *Phys. Rev.* **96,** 1590 (1954).
84. A. P. French, *Phys. Rev.* **107,** 1655 (1957).
85. K. Nakamura and M. Soga, *Progr. Theoret. Phys.* **21,** 837 (1959).
86. D. R. Bates, *Atomic and Molecular Processes*, Academic Press, New York, 1962.
87. H. S. W. Massey, *Encyclopedia of Physics*, Springer-Verlag, Berlin · Göttinger · Heidelberg, 1956, p. 232.
88. H. S. W. Massey and E. H. S. Burhop, *Electronic and Ionic Impact Phenomena*, Clarendon Press, Oxford, 1956.
89. P. G. Burke and K. Smith, *Revs. Modern Phys.* **34,** 458 (1962).
90. Beatrice Worsley, *Can. J. Phys.* **36,** 289 (1958).
91. A. Dalgarro and J. T. Lewis, *Proc. Roy. Soc.* **233A,** 70 (1956); A. Dalgarro and A. L. Stewart, *Proc. Roy. Soc.* **238A,** 269 (1956).
92. R. Sternheimer, *Phys. Rev.* **96,** 951 (1954), *ibid.* **105,** 158 (1957) and previous references.
93. C. Schwartz, *Ann. Phys.* **2,** 156 (1959) and *ibid.* **2,** 170 (1959).
94. R. M. Sternheimer and H. M. Foley, *Phys. Rev.* **102,** 731 (1956) and previous references.
95. E. G. Wikner and T. P. Das, *Phys. Rev.* **109,** 360 (1958) and previous references.
96. J. Mayer and M. Mayer, *Phys. Rev.* **43,** 605 (1933).
97. R. A. Buckingham, *Proc. Roy. Soc.* **160A,** 94 (1937).
98. F. B. Malik and E. Trefftz, *Z. Astrophys.* **50,** 96 (1960).
99. P. Hammerling, W. Shine, and B. Kivel, *J. Appl. Phys.* **28,** 760 (1957).

100. M. H. Mittleman and K. M. Watson, *Ann. Phys.* **10,** 268 (1960).
101. M. H. Mittleman and K. M. Watson, unpublished.
102. T. F. O'Malley, L. Spruch, and L. Rosenberg, *J. Math. Phys.* **2,** 491 (1961).
103. J. Meixner and F. W. Schafke, *Die Grundlehren der Mathematischen Wissenschaften,* Springer-Verlag, Berlin, 1954, vol. 71.
104. L. Rosenberg, L. Spruch, and T. F. O'Malley, *Phys. Rev.* **119,** 164 (1960) and earlier references.
105. C. Schwartz, *Ann. Phys.* **16,** 36 (1961) and *Phys. Rev.* **124,** 1468 (1961).

# APPENDIX A TABLE OF WIGNER COEFFICIENTS

The Wigner coefficients, introduced in Eqs. (1.59), are given below for $j_2 = \frac{1}{2}$ and for $j_2 = 1$.

$$\langle j - \tfrac{1}{2}, \tfrac{1}{2}; m - \tfrac{1}{2}, \tfrac{1}{2} \mid j, m \rangle = \left(\frac{j + m}{2j}\right)^{1/2}$$

$$\langle j - \tfrac{1}{2}, \tfrac{1}{2}; m + \tfrac{1}{2}, -\tfrac{1}{2} \mid j, m \rangle = \left(\frac{j - m}{2j}\right)^{1/2}$$

$$\langle j + \tfrac{1}{2}, \tfrac{1}{2}; m - \tfrac{1}{2}, \tfrac{1}{2} \mid j, m \rangle = -\left[\frac{j - m + 1}{2(j + 1)}\right]^{1/2}$$

$$\langle j + \tfrac{1}{2}, \tfrac{1}{2}; m + \tfrac{1}{2}, -\tfrac{1}{2} \mid j, m \rangle = \left[\frac{j + m + 1}{2(j + 1)}\right]^{1/2}$$

$$\langle j - 1, 1; m - 1, 1 \mid j, m \rangle = \left[\frac{(j + m - 1)(j + m)}{2j(2j - 1)}\right]^{1/2}$$

$$\langle j - 1, 1; m, 0 \mid j, m \rangle = \left[\frac{(j - m)(j + m)}{j(2j - 1)}\right]^{1/2}$$

$$\langle j - 1, 1; m + 1, -1 \mid j, m \rangle = \left[\frac{(j - m - 1)(j - m)}{2j(2j - 1)}\right]^{1/2}$$

$$\langle j, 1; m - 1, 1 \mid j, m \rangle = -\left[\frac{(j + m)(j - m + 1)}{2j(j + 1)}\right]^{1/2}$$

$$\langle j, 1; m, 0 \mid j, m \rangle = \frac{m}{[j(j + 1)]^{1/2}}$$

$$\langle j, 1; m + 1, -1 \mid j, m \rangle = \left[\frac{(j - m)(j + m + 1)}{2j(j + 1)}\right]^{1/2}$$

$$\langle j + 1, 1; m - 1, 1 \mid j, m \rangle = \left[\frac{(j - m + 1)(j - m + 2)}{2(j + 1)(2j + 3)}\right]^{1/2}$$

$$\langle j + 1, 1; m, 0 \mid j, m \rangle = -\left[\frac{(j - m + 1)(j + m + 1)}{(j + 1)(2j + 3)}\right]^{1/2}$$

$$\langle j + 1, 1; m + 1, -1 \mid j, m \rangle = \left[\frac{(j + m + 1)(j + m + 2)}{2(j + 1)(2j + 3)}\right]^{1/2}$$

# APPENDIX B COLLISIONS INVOLVING THREE INCIDENT PARTICLES

Chemical reactions in gases often occur as a result of collisions involving three incident particles. The third particle frequently plays the role of a catalyst and is ordinarily essential for the energy-momentum balance of the reaction. Examples of particular importance include nonradiative electron attachment in the upper atmosphere, the third particle (a molecule or atom) being required for energy conservation.

Let us consider collisions of three incident particles, labeled as $B$, $T$, and $C$. We suppose the collisions to occur in a gas containing, respectively, $n_B$, $n_T$, and $n_C$ particles of each type per unit volume. Elementary considerations permit us to write the three-body reaction rate per unit volume in the form

$$R(a \to b) = n_B n_T v_{BT} \sigma_{BT} \left( n_C \frac{4\pi}{3} \mathscr{R}^3 \right), \tag{1}$$

where $v_{BT}$ is the relative velocity of particles $B$ and $T$, $\sigma_{BT}$ is a scattering cross section, and $\mathscr{R}$ is the range of the $BT$ interaction. The quantity $n_C[(4\pi/3)\mathscr{R}^3]$ is just the probability of finding a particle $C$ present when $B$ and $T$ scatter.[1]

The transition probability (3.113) is

$$\sum_b P_{ba} = \sum_b |(\chi'_b, SX_a)|^2, \tag{2}$$

where now wave packets are formed for each of the three incident particles.

$$X_a = \int d^3q_1\, d^3q_2\, d^3q_3 a_B(\mathbf{q}_1 - \mathbf{p}_1) a_T(\mathbf{q}_2 - \mathbf{p}_2) a_C(\mathbf{q}_3 - \mathbf{p}_3) \chi_a, \tag{3}$$

with $\chi_a$ an appropriate three-particle initial state. Equation (3.114) is therefore replaced by

$$\begin{aligned}(\chi'_b, SX_a) = -2\pi i \int d^3q_1\, d^3q_2\, d^3q_3 a_B(\mathbf{q}_1 - \mathbf{p}_1) a_T(\mathbf{q}_2 - \mathbf{p}_2) \\ \times\, a_C(\mathbf{q}_3 - \mathbf{p}_3)\, \delta(E_b - E_a)\, \delta(\mathbf{P}_b - \mathbf{P}_a) T_{ba}.\end{aligned} \tag{4}$$

[1] We assume here that the density $n_C$ is sufficiently low that $n_C\,[(4\pi/3)\mathscr{R}^3] \ll 1$.

To evaluate Eq. (4), we shall assume that the wave packet

$$G_C(\mathbf{r}) = \int d^3q_3 a_C(\mathbf{q}_3 - \mathbf{p}_3)e^{i(\mathbf{q}_3-\mathbf{p}_3)\cdot\mathbf{r}}$$

is so broad that it overlaps both other wave packets during the collision, which occurs at $\mathbf{r} = 0$. This implies that $a_C(\mathbf{q}_3 - \mathbf{p}_3)$ will be very sharply peaked at $\mathbf{q}_3 = \mathbf{p}_3$, which permits us to set $\mathbf{q}_3 = \mathbf{p}_3$ everywhere except in $a_C$ in the integrand of (4). The integral

$$\int d^3q_3 a_C(\mathbf{q}_3 - \mathbf{p}_3) = G_C(0) \tag{5}$$

may then be performed and Eq. (4) rewritten as

$$(\chi'_b, SX_a) = -2\pi i G_C(0) \int d^3q_1\, d^3q_2 a_B(\mathbf{q}_1 - \mathbf{p}_1) \times a_T(\mathbf{q}_2 - \mathbf{p}_2)\, \delta(E_b - E_a)\, \delta(\mathbf{P}_b - \mathbf{P}_a) T_{ba}. \tag{6}$$

Since $\mathbf{q}_3$ has been replaced by the constant momentum $\mathbf{p}_3$ in Eq. (6), this expression is formally equivalent to Eq. (3.114). On evaluating the expression (2) we obtain just the right-hand side of Eq. (3.134) multiplied by $|G_C(0)|^2$.

$$\sum_b P_{ba} = \sum_b [\delta(E_b - E_{\bar{a}})\, \delta(\boldsymbol{P}_b - \boldsymbol{P}_{\bar{a}})\, |T_{b\bar{a}}|^2]\, \frac{(2\pi)^4 F}{|\mathbf{v}_1 - \mathbf{v}_2|}\, |G_C(0)|^2. \tag{7}$$

Now $F$ is the flux of beam particles per beam particle and $(2\pi)^{-3}\, |G_C(0)|^2$ is the density of $C$ particles per $C$ particle [see Eq. (3.120) to verify normalization]. To reexpress Eq. (7) as a transition rate $R(a \rightarrow b)$ per unit volume per second, we scale $F$ up to $F = n_B\, |\mathbf{v}_1 - \mathbf{v}_2|$ and $(2\pi)^{-3}\, |G_C(0)|^2$ up to $n_C$ and then multiply by $n_T$. This gives

$$R(a \rightarrow b) = \left[\sum_b \delta(E_b - E_{\bar{a}})\, \delta(\mathbf{P}_b - \mathbf{P}_{\bar{a}})(2\pi)^7\, |T_{ba}|^2\right] n_B n_T n_C. \tag{8}$$

As a simple application, let us consider the recombination process[2]

$$e + e + A^+ \rightarrow e + A, \tag{9}$$

corresponding to the collision of two electrons with an ion $A^+$ to give a neutral atom $A$ and a single free electron. Since the atom is quite massive, we may consider it to be at rest. The two electrons in the initial state have momenta $\mathbf{p}_1$ and $\mathbf{p}_2$ and the final free electron has a momentum $\mathbf{k}$. If one

[2] Our example is stimulated by the work of R. G. Giovanelli, *Austral. J. Sci. Res.* **A1**, 275 (1948).

electron is captured into a state $\gamma$ of the atom, we may write Eq. (8) in the form

$$dR(\mathbf{p}_1, \mathbf{p}_2, + \rightarrow \mathbf{k}, \gamma) = n(+)\, dn(\mathbf{p}_1)\, dn(\mathbf{p}_2) \times \left[(2\pi)^7\, d^3k\, \delta(\epsilon_k - \epsilon_{p_1} - \epsilon_{p_2} - I_\gamma) \frac{1}{4g(+)} \sum_s |T_{ba}|^2\right]. \quad (10)$$

Here $n(+)$ is the number of ions per unit volume and

$$dn(p) = \frac{n_e}{(2\pi m\theta)^{3/2}}\, e^{-\epsilon_p/\theta}\, d^3p \quad (11)$$

is the number of electrons per unit volume in the momentum interval $d^3p$ when $n_e$ is the total number of free electrons per unit volume. We have written $\epsilon_k \equiv k^2/2m$, with $m$ the electron mass, and have called the temperature of the electrons $\theta$. Finally, in Eq. (10), $I_\gamma$ is the ionization energy of the state $\gamma$, $g(+)$ is the statistical weight of the initial ion state, and the sum on $s$ represents a sum over both initial and final spin states.

The inverse of the reaction (9) is the ionization process

$$e + A \rightarrow e + e + A^+. \quad (12)$$

The differential scattering cross section for this reaction is

$$d\sigma(\mathbf{k}, \gamma \rightarrow \mathbf{p}_1, \mathbf{p}_2, +) = \frac{(2\pi)^4}{v(k)}\, d^3p_1\, d^3p_2\, \delta(\epsilon_k - I_\gamma - \epsilon_{p_1} - \epsilon_{p_2}) \times \frac{1}{2g(\gamma)} \sum_s |T_{ab}|^2, \quad (13)$$

where $g(\gamma)$ is the statistical weight of the state $\gamma$ and $v(k) = k/m$ is the velocity of the initial electron.

Now, from the general reciprocity condition (4.235) we have

$$\sum_s |T_{ab}|^2 = \sum_s |T_{ba}|^2. \quad (14)$$

This permits us to write Eq. (10) as

$$dR(\mathbf{p}_1, \mathbf{p}_2, + \rightarrow \mathbf{k}, \gamma) = \frac{n(+) n_e^2}{(2\pi m\theta)^3}\, e^{-(\epsilon_{p_1}+\epsilon_{p_2})/\theta} \times \left[(2\pi)^3\, d^3k v(k) \frac{g(\gamma)}{2g(+)}\, d\sigma(\mathbf{k}, \gamma \rightarrow \mathbf{p}_1, \mathbf{p}_2, +)\right]. \quad (15)$$

The total ionization cross section is obtained from (13) by integrating over $\mathbf{p}_1$ and $\mathbf{p}_2$.

$$\sigma(\mathbf{k}\gamma \rightarrow +) = \int d\sigma(\mathbf{k}, \gamma \rightarrow \mathbf{p}_1, \mathbf{p}_2, +)$$

$$= \frac{(2\pi)^4}{v(k)} \int d^3p_1\, d^3p_2\, \delta(\epsilon_{p_1} + \epsilon_{p_2} - \epsilon_k + I_\gamma) \frac{1}{2g(\gamma)} \sum_s |T_{ab}|^2. \quad (16)$$

The total recombination rate to a state $\gamma$ per unit volume is then obtained by integrating Eq. (15) over $\mathbf{p}_1$, $\mathbf{p}_2$ and $\mathbf{k}$.

$$R(+ \rightarrow \gamma) = \frac{n(+)n_e^2 g(\gamma)}{2(m\theta)^2 g(+)} e^{I_\gamma/\theta} \int d^3k\, e^{-\epsilon_k/\theta} v(k)\sigma(k\gamma \rightarrow +). \quad (17)$$

The corresponding ionization rate is seen from Eq. (13) to be

$$R(\gamma \rightarrow +) = \frac{n(\gamma)n_e}{(2\pi m\theta)^{3/2}} \int d^3k\, e^{-\epsilon_k/\theta} v(k)\sigma(k\gamma \rightarrow +). \quad (18)$$

It is easily verified from Eqs. (17) and (18) that for statistical equilibrium

$$R(\gamma \rightarrow +) = R(+ \rightarrow \gamma),$$

implying no net rate of ionization or recombination.

# APPENDIX C COMPLETENESS AND ORTHOGONALITY OF CHANNEL STATES

We wish to show that the wave functions for different reaction channels are orthogonal when projected into the subspace of asymptotic scattering states. To do this, we prove Eq. (4.23$a$). In this subspace, the channel wave functions define a complete, orthonormal set. In addition, in this subspace, the $S$-matrix and scattering matrix $\mathcal{T}$ are linear operators (and may be represented by matrices).

Let us consider two different final state channels for which the eigenfunctions are $\chi'_b$ and $\chi''_d$, respectively.[1] Here $b$ and $d$ represent the appropriate labels for the eigenstates. The projection operators

$$\begin{aligned}\Lambda' &= \sum_b \chi'_b(\chi'_b)^\dagger,\\ \Lambda'' &= \sum_d \chi''_d(\chi''_d)^\dagger\end{aligned} \tag{1}$$

are two members of the set (4.21). The respective transition probabilities (4.22$a$) are

$$\begin{aligned}P_{ba} &= (\Psi(t), \Lambda'\Psi(t)),\\ P_{da} &= (\Psi(t), \Lambda''\Psi(t)),\end{aligned} \tag{2}$$

when $t \gg |t_c|$. We wish to show that

$$\lim_{t\to\infty} (\Psi(t), \Lambda'\Lambda''\Psi(t)) = 0, \tag{3}$$

which is just Eq. (4.23$a$).[2] Equation (3) is valid even when

$$(\chi'_b, \chi''_d) \neq 0,$$

[1] In accordance with our convention, the channel labels $c$ are not designated as such because we do not imply an enumeration of channels. Instead, we write $\chi_{c';b} = \chi'_b$, $\chi_{c'';d} = \chi''_d$.

[2] The "spreading" of the wave packets does not affect the validity of Eq. (3). We saw in Chapter 3 that this "spreading" may be chosen to be arbitrarily small for any finite, preassigned time interval $\Delta t$. We may formally neglect "spreading" of packets in Eq. (3) by employing a double limiting process.

or

$$\Lambda'\Lambda'' \neq 0. \tag{4}$$

To establish the validity of (3), consider

$$\begin{aligned} D &\equiv (\Psi(t), \Lambda''\Lambda'\Psi(t)) \\ &= \sum_b \sum_d (\chi''_d, \Psi(t))^*(\chi''_d, \chi'_b)(\chi'_b, \Psi(t)), \end{aligned} \tag{5}$$

where it is supposed that $t \gg |t_c|$. It is important here to recall from the definitions (4.21) that the summation on $b$ and $d$ covers only a restricted, finite range of states.

By Eqs. (3.93) and (3.95), we may write

$$(\chi'_b, \Psi(t)) = -2\pi i \int d^3q_1\, d^3q_2\, A(\mathbf{q}_1, \mathbf{q}_2)e^{-iE_a t} \times [\delta(\mathbf{P}_b - \mathbf{P}_a)\, \delta(E_b - E_a)T_{ba}] \tag{6}$$

and

$$(\chi''_d, \Psi(t)) = -2\pi i \int d^3q'_1,\, d^3q'_2\, A(\mathbf{q}'_1, \mathbf{q}'_2)e^{-iE_{a'}t} \times [\delta(\mathbf{P}_d - \mathbf{P}_{a'})\, \delta(E_d - E_{a'})T_{da'}]. \tag{7}$$

Here $E_b$ and $E_d$ are the respective eigenenergies, whereas $\mathbf{P}_b$ and $\mathbf{P}_d$ are the respective eigenmomenta for the states $\chi'_b$ and $\chi''_d$. The remaining notation represents an obvious extension of that of Chapter 3.

Equations (6) and (7) may be substituted into Eq. (5) to give

$$\begin{aligned} D = (2\pi)^2 \sum_b \sum_d \int d^3q_1\, d^3q_2\, d^3q'_1\, d^3q'_2\, a_B(\mathbf{q}_1 - \mathbf{p}_1)a_T(\mathbf{q}_2 - \mathbf{p}_2) \\ \times a_B^*(\mathbf{q}'_1 - \mathbf{p}_1)a_T^*(\mathbf{q}'_2 - \mathbf{p}_2)\, \delta(P_b - P_a)\, \delta(P_d - P_{a'}) \\ \times \delta(E_b - E_a)\, \delta(E_d - E_{a'})e^{i(E_{a'} - E_a)t} \\ \times T_{ba}T^*_{da'}(\chi''_d, \chi'_b). \end{aligned} \tag{8}$$

[A considerable similarity here to Eq. (3.122) may be noted.]

We now must consider the scalar product $(\chi''_d, \chi'_b)$. Let us suppose there to be $\nu$ particles in state $b$ and $\mu$ particles in state $d$. We may therefore write [Compare Eq. (3.185), for example]

$$\begin{aligned} \chi'_b &= \prod_{\alpha=1}^{\nu} (2\pi)^{-(3/2)\nu}\, e^{i\mathbf{l}_\alpha \cdot \mathbf{r}_\alpha}\, g_{s_\alpha}(\xi_\alpha), \\ \chi''_d &= \prod_{\beta=1}^{\mu} (2\pi)^{-(3/2)\mu}\, e^{i\mathbf{k}_\beta \cdot \mathbf{r}''_\beta} g''_{s_\beta}(\xi''_\beta). \end{aligned} \tag{9}$$

That is, we write $\chi'_b$ and $\chi''_d$ as a product of single-particle wave functions. The $\mathbf{l}_\alpha$ and $\mathbf{k}_\beta$ are the momenta of these particles, whereas $g_{s_\alpha}$ and $g''_{s_\beta}$ are their internal state wave functions.

When the $\nu$ particles in state $b$ and the $\mu$ particles in state $d$ are simple, or "elementary," they are described in orthogonal subspaces, so

$$(\chi'_b, \chi''_d) = 0$$

and our proof is trivial. When these particles are *composite*, so that $g_{s_\alpha}$ and $g''_{s\beta}$ represent *bound states* of elementary particles, the elementary particles will generally be arranged differently in the bound systems of states $b$ and $d$. Here, we anticipate that

$$(\chi'_b, \chi''_d) \neq 0.$$

Then it is clear that the scalar product $(\chi''_d, \chi'_b)$ will be a product of a number of factors, say $L$, corresponding to "clusters" of particles among which the elementary particles are rearranged. That is,

$$(\chi''_d, \chi'_b) = \prod_{j=1}^{L} (\chi''_{d,j}, \chi'_{b,j}), \tag{10}$$

where $\chi'_{b,j}$ and $\chi''_{d,j}$ are products of particle wave functions which are involved in the $j$th cluster in Eq. (10). Evidently, $1 \leq L \leq \nu$ and $1 \leq L \leq \mu$.

It is also evident that

$$(\chi''_{d,j}, \chi'_{b,j}) = \delta(\mathbf{P}_{d,j} - \mathbf{P}_{b,j})C_j, \tag{11}$$

where $\mathbf{P}_{d,j}$ and $\mathbf{P}_{b,j}$ are the respective total momenta of all particles in cluster $j$ for the respective states $d$ and $b$. $C_j$ is a nonsingular function of the variables labeling the eigenstates $\chi''_{d,j}$ and $\chi'_{b,j}$. Therefore Eq. (10) is of the form

$$(\chi''_d, \chi'_b) = \prod_{j=1}^{L} \delta(\mathbf{P}_{d,j} - \mathbf{P}_{b,j})C. \tag{12}$$

Again $C$ is nonsingular and of the form

$$C \approx [\tau]^{(\nu+\mu)/2-L}, \tag{13}$$

where $\tau$ is a "volume" of magnitude characterized by the "size" of the composite particles; for atomic nuclei $\tau \approx (10^{-13})^3$ cm$^3$, whereas for atoms or molecules $\tau \approx (10^{-8})^3$ cm$^3$. Equation (13) follows from the expression (12) by elementary dimensional considerations.

Let us next introduce into Eq. (8) the representations

$$\begin{aligned}
\delta(E_b - E_a) &= (2\pi)^{-1}\int d\lambda\, e^{i\lambda(E_a-E_b)},\\
\delta(E_d - E_{a'}) &= (2\pi)^{-1}\int d\lambda'\, e^{-i\lambda'(E_{a'}-E_d)},\\
\delta(\mathbf{P}_b - \mathbf{P}_a) &= (2\pi)^{-3}\int d^3y\, e^{-i\mathbf{y}\cdot(\mathbf{P}_a-\mathbf{P}_b)},\\
\delta(\mathbf{P}_d - \mathbf{P}_{a'}) &= (2\pi)^{-3}\int d^3y'\, e^{-i\mathbf{y}'\cdot(\mathbf{P}_{a'}-\mathbf{P}_d)}.
\end{aligned} \tag{14}$$

We define

$$\begin{aligned} \mathbf{q}_1 - \mathbf{p}_1 &= \boldsymbol{\rho}_1, \\ \mathbf{q}_2 - \mathbf{p}_2 &= \boldsymbol{\rho}_2, \\ \mathbf{q}'_1 - \mathbf{p}_1 &= \boldsymbol{\rho}'_1, \\ \mathbf{q}'_2 - \mathbf{p}_2 &= \boldsymbol{\rho}'_2, \end{aligned} \tag{15}$$

and write, valid for wave packets of small momentum spread,

$$\begin{aligned} E_a &= E_{\bar{a}} + \boldsymbol{\rho}_1 \cdot \mathbf{v}_1 + \boldsymbol{\rho}_2 \cdot \mathbf{v}_2, \\ E_{a'} &= E_{\bar{a}} + \boldsymbol{\rho}'_1 \cdot \mathbf{v}_1 + \boldsymbol{\rho}'_2 \cdot \mathbf{v}_2, \\ \mathbf{P}_a &= \mathbf{P}_{\bar{a}} + \boldsymbol{\rho}_1 + \boldsymbol{\rho}_2, \\ \mathbf{P}_{a'} &= \mathbf{P}_{\bar{a}} + \boldsymbol{\rho}'_1 + \boldsymbol{\rho}'_2. \end{aligned} \tag{16}$$

Here $\mathbf{P}_{\bar{a}} = \mathbf{p}_1 + \mathbf{p}_2$ and $E_{\bar{a}}$ is the value of $E_a$ when $\mathbf{q}_1 = \mathbf{p}_1$ and $\mathbf{q}_2 = \mathbf{p}_2$. [Our argument is similar to that employed to evaluate Eq. (3.127).]

On using Eqs. (14), (15), and (16) in Eq. (8), and setting $T_{ba} = T_{b\bar{a}}$, $T_{da} = T_{d\bar{a}}$, we obtain

$$\begin{aligned} D &= \sum_b \sum_d \int d^3\rho_1\, d^3\rho_2\, d^3\rho'_1\, d^3\rho'_2\, a_B(\boldsymbol{\rho}_1) a_T(\boldsymbol{\rho}_2) \\ &\quad \times a_B^*(\boldsymbol{\rho}'_1) a_T^*(\boldsymbol{\rho}'_2)(2\pi)^{-6} \int d\lambda\, d\lambda'\, d^3y\, d^3y' \\ &\quad \times \exp\{-i\lambda'[E_{\bar{a}} - E_d + \boldsymbol{\rho}'_1 \cdot \mathbf{v}_1 + \boldsymbol{\rho}'_2 \cdot \mathbf{v}_2] \\ &\qquad\qquad + i\lambda[E_{\bar{a}} - E_b + \boldsymbol{\rho}_1 \cdot \mathbf{v}_1 + \boldsymbol{\rho}_2 \cdot \mathbf{v}_2]\} \\ &\quad \times \exp\{-i\mathbf{y}' \cdot (\mathbf{P}_{\bar{a}} - \mathbf{P}_d) + i\mathbf{y} \cdot (\mathbf{P}_{\bar{a}} - \mathbf{P}_b) - i\mathbf{y}' \cdot (\boldsymbol{\rho}'_1 + \boldsymbol{\rho}'_2)\} \\ &\quad \times \exp\{i\mathbf{y} \cdot (\boldsymbol{\rho}_1 + \boldsymbol{\rho}_2) + i[(\boldsymbol{\rho}'_1 - \boldsymbol{\rho}_1) \cdot \mathbf{v}_1 + (\boldsymbol{\rho}'_2 - \boldsymbol{\rho}_2) \cdot \mathbf{v}_2]t\} \\ &\quad \times T_{b\bar{a}} T_{d\bar{a}}^*(\chi''_d, \chi'_b) \\ &= (2\pi)^{-6} \sum_b \sum_d \int d\lambda\, d\lambda'\, d^3y\, d^3y'\, G_B[\mathbf{y} - \mathbf{v}_1(t - \lambda)] G_T[\mathbf{y} - \mathbf{v}_2(t - \lambda)] \\ &\quad \times G_B^*[\mathbf{y}' - \mathbf{v}_1(t - \lambda')] G_T^*[\mathbf{y}' - \mathbf{v}_2(t - \lambda')] \\ &\quad \times \exp[-i\lambda'(E_{\bar{a}} - E_d) + i\lambda(E_{\bar{a}} - E_b) \\ &\qquad\qquad - i\mathbf{y}' \cdot (\mathbf{P}_{\bar{a}} - \mathbf{P}_d) + i\mathbf{y} \cdot (\mathbf{P}_{\bar{a}} - \mathbf{P}_b)] \\ &\quad \times \prod_{j=1}^{L} \delta(\mathbf{P}_{d,j} - \mathbf{P}_{b,j}) C T_{b\bar{a}} T_{d\bar{a}}^*. \end{aligned} \tag{17}$$

To continue, we write

$$\begin{aligned} \sum_b &= \sum_{s_b} \int d^3P_{b_1} \cdots d^3P_{b_L}\, d^3\kappa_{L+1} \cdots d^3\kappa_\nu, \\ \sum_d &= \sum_{s_d} \int d^3P_{d_1} \cdots d^3P_{d_L}\, d^3\rho'_{L+1} \cdots d^3\rho'_\mu, \end{aligned} \tag{18}$$

where the sums are over internal states of the particles and $\boldsymbol{\kappa}_{L+1} \cdots \boldsymbol{\rho}_\mu$ are certain momentum variables of relative motion for the particles. In analogy to Eqs. (3.212), we expand $E_b$ and $E_d$ in the neighborhood of $E_{\bar{a}}$. To the first order,

$$
\begin{aligned}
E_b &= E_{\bar{a}} + \sum_{j=L+1}^{\nu} \boldsymbol{\kappa}_j \cdot \mathbf{v}_{bj} + \sum_{j=1}^{L} (\mathbf{P}_{bj} - \bar{\mathbf{P}}_j) \cdot \mathbf{V}_j, \\
E_d &= E_{\bar{a}} + \sum_{j=L+1}^{\mu} \boldsymbol{\rho}'_j \cdot \mathbf{v}_{dj} + \sum_{j=1}^{L} (\mathbf{P}_{d,j} - \bar{\mathbf{P}}_j) \cdot \mathbf{V}_j
\end{aligned}
\tag{19}
$$

Here the $\bar{\mathbf{P}}_j$ are the mean values of $\mathbf{P}_{b,j}$ and $\mathbf{P}_{d,j}$, and the $\mathbf{V}_j$ are corresponding mean velocities for the clusters of Eq. (10). The expansion (19) implies that the $\boldsymbol{\kappa}_j$ and $\boldsymbol{\rho}'_j$ are defined as deviations from mean relative momenta. Similarly, we may write

$$
\begin{aligned}
\mathbf{P}_b &= \mathbf{P}_{\bar{a}} + \sum_{j=1}^{L} (\mathbf{P}_{b,j} - \bar{\mathbf{P}}_j), \\
\mathbf{P}_d &= \mathbf{P}_{\bar{a}} + \sum_{j=1}^{L} (\mathbf{P}_{d,j} - \bar{\mathbf{P}}_j).
\end{aligned}
\tag{20}
$$

Let us now specialize $D$, as given by Eq. (17), to that coordinate system for which $\mathbf{v}_2 = 0$. We may then substitute the expansions (19) and (20) into Eq. (17) to obtain

$$
\begin{aligned}
D = (2\pi)^{-6} &\sum_{s_b} \sum_{s_d} \int d\lambda\, d\lambda'\, d^3y\, d^3y'\, T_{b\bar{a}} T_{d\bar{a}}^* \\
&\times G_B[\mathbf{v}_1(t-\lambda)] G_B^*[\mathbf{v}_1(t-\lambda')] G_T(\mathbf{y}) G_T^*(\mathbf{y}') \\
&\times \int d^3\kappa_{L+1} \cdots d^3\kappa_\nu\, d^3\rho'_{L+1} \cdots d^3\rho'_\mu\, d^3P_{b,1} \cdots d^3P_{b,L} C \\
&\times \exp\left[-i(\mathbf{y}-\mathbf{y}') \cdot \sum_{j=1}^{L} (\mathbf{P}_{b,j} - \bar{\mathbf{P}}_j) + i(\lambda' - \lambda) \sum_{j=1}^{L} (\mathbf{P}_{b,j} - \bar{\mathbf{P}}_j) \cdot \mathbf{V}_j\right] \\
&\times \exp\left[i\lambda' \sum_{i=L+1}^{\mu} \boldsymbol{\rho}'_i \cdot \mathbf{v}_{d,i} - i\lambda \sum_{i=L+1}^{\nu} \boldsymbol{\kappa}_i \cdot \mathbf{v}_{b,i}\right].
\end{aligned}
\tag{21}
$$

We have introduced here Eqs. (18) for the sum over final states.

Now, $C$ is a nonsingular function of the $\boldsymbol{\kappa}_i$'s and $\boldsymbol{\rho}'_i$'s, so

$$
\begin{aligned}
\int d^3\rho'_i\, e^{i\lambda' \boldsymbol{\rho}'_i \cdot \mathbf{v}_{d,i}} &= \mathcal{O}\left(\frac{1}{(\lambda' v_{d,i})^3}\right), \\
\int d^3\kappa_i\, e^{-i\lambda \boldsymbol{\kappa}_i \cdot \mathbf{v}_{b,i}} &= \mathcal{O}\left(\frac{1}{(\lambda v_{b,i})^3}\right).
\end{aligned}
\tag{22}
$$

As $t \to \infty$, we may set $\lambda' = \lambda = t$ in Eqs. (22), because the arguments of $G_B$ and $G_B^*$ must be nonvanishing in Eq. (21). On setting $v_{d,i} \approx v_{b,i'} \approx \bar{v}$, an effective mean "velocity," Eq. (21) is seen to contain a factor

$$\left[\frac{1}{(\bar{v}t)^3}\right]^{\nu+\mu-2L} \tag{23}$$

We also set

$$\int d^3P_{b,1}\cdots d^3P_{b,L} = (\Delta P^3)^L, \tag{24}$$

where $\Delta P$ is determined by the (finite) range of the integral and by the exponentials in the integrand. Also,

$$\int d^3y\, d^3y'\, G_T(\mathbf{y})G_T^*(\mathbf{y}') \approx w_T{}^3$$

and

$$\int d\lambda\, d\lambda' G_B[\mathbf{v}_1(t-\lambda)]G_B^*[\mathbf{v}_1(t-\lambda')] \approx \frac{F}{v_1}\, w_B.$$

[See Eqs. (3.132) and (3.133).]

Therefore the order of magnitude of $D$ is

$$D \approx \frac{(\tau)^{(\nu+\mu)/2-L}}{[(\bar{v}t)^3]^{\nu+\mu-2L}}(\Delta P)^{3L}T_{b\bar{a}}T_{d\bar{a}}^*\left(w_T{}^3w_B\frac{F}{v_1}\right), \tag{25}$$

where the value (13) has been inserted for $C$. According to Eq. (3.134), the probabilities of Eq. (2) have the order of magnitude

$$\begin{aligned} P_{ba} &\approx (\Delta P)^{3\nu-4}\frac{F}{v_1}|T_{b\bar{a}}|^2, \\ P_{da} &\approx (\Delta P)^{3\mu-4}\frac{F}{v_1}|T_{d\bar{a}}|^2. \end{aligned} \tag{26}$$

Therefore

$$\frac{D}{\sqrt{P_{ba}P_{da}}} \approx \frac{(w_T{}^3w_B\,\Delta P^4)\left[\dfrac{\tau}{(\bar{v}t)^3}\right]^{(\nu+\mu)/2-L}}{[(\bar{v}t)^3\,\Delta P^3]^{(\nu+\mu)/2-L}}. \tag{27}$$

The case $L = (\nu + \mu)/2$ can occur only if $\mu = \nu$ and if no rearrangement of elementary particles among the composite particles occurs. In this trivial case $\chi'_b$ and $\chi''_d$ belong to the same orthonormal set of wave functions, and *thus* to the same channel. Our discussion is therefore irrelevant for the case that $L = \frac{1}{2}(\nu + \mu)$.

For relevant cases, $L < (\mu + \nu)/2$ and the ratio (27) approaches zero when $t \to \infty$. We can furthermore determine the rate that this quantity becomes small. The ratio $\tau/(\bar{v}t)^3$ is essentially the ratio of the volume occupied by a patricle to the volume within which it is contained. This is therefore a measure of the probability of overlap of the wave functions of two particles.

In Section 3.4 we discussed the use of wave packets for the final states as well as for the the initial states. This provides the most natural choice for the final states—and leads, as was shown, to the same result for the cross section as that obtained with eigenstates. We shall now show that the states $\chi'_b$ and $\chi''_d$ are effectively orthogonal when used in wave packets.

Let

$$\begin{aligned} \phi'_b &= \prod_{\alpha=1}^{\nu} [(2\pi)^{-(3/2)\nu} e^{i\mathbf{l}_\alpha \cdot \mathbf{r}_\alpha} g_{s_\alpha}(\xi_\alpha) G_\alpha(\mathbf{r}_\alpha - \mathbf{v}_\alpha t)], \\ \phi''_d &= \prod_{\beta=1}^{\mu} [(2\pi)^{-(3/2)\mu} e^{i\mathbf{k}_\beta \cdot \mathbf{r}''_\beta} g''_{s_\beta}(\xi''_\beta) G''_\beta(r''_\beta - v''_\beta t)]. \end{aligned} \tag{28}$$

These represent wave packets constructed from the $\chi'_b$ and $\chi''_d$ of Eq. (9). They are equivalent to the final state packets introduced by Eq. (3.195) and used in Section 3.4. When the wave packets do not overlap (to an arbitrary degree of precision), these are called "asymptotic wave packet states."

An "experimental volume" is defined by $\mathscr{V}_x = \mathscr{O}[(v_\alpha t)^3] \simeq \mathscr{O}[(v_\beta t)^3]$. The scalar produce $(\phi''_d, \phi'_b)$ will be nonvanishing only if $2[(\nu + \mu)/2 - L]$ wave packets overlap properly and if the "cluster momenta" satisfy the conservation restrictions described by Eq. (12). Then, using (12),

$$(\phi''_d, \phi'_b) = \left(\frac{\tau}{\mathscr{V}_p}\right)^{(\nu+\mu)/2-L}, \tag{29}$$

where $\mathscr{V}_p$ is a mean volume included within the wave packets described by $G_\alpha$ and $G''_\beta$. Now, the probability that the $[(\nu + \mu) - 2L]$ wave packets overlap, if $\mathscr{V}_p \ll \mathscr{V}_x$ and if the wave packets are distributed in random positions, is $(\mathscr{V}_p/\mathscr{V}_x)^{\nu+\mu-2L}$. The corresponding probability amplitude is $(\mathscr{V}_p/\mathscr{V}_x)^{(\nu+\mu)/2-L}$.

Thus, on averaging over a random choice of packets,

$$\begin{aligned} \overline{(\phi''_d, \phi'_b)} &= \left(\frac{\tau}{\mathscr{V}_p}\right)^{(\nu+\mu)/2-L} \left(\frac{\mathscr{V}_p}{\mathscr{V}_x}\right)^{(\nu+\mu)/2-L} \\ &= \left(\frac{\tau}{\mathscr{V}_x}\right)^{(\nu+\mu)/2-L} \end{aligned} \tag{30}$$

Therefore, in a macroscopic volume, the wave packets formed from different channel states may be treated as orthogonal. This is a very important conclusion, for it permits us to treat the asymptotic channel states as an orthonormal set. They are also complete, in the sense that any possible asymptotic state, resulting after interacting systems have separated, can be expressed in terms of them.

# APPENDIX D COMPUTATIONAL METHODS FOR SCATTERING AT LOW ENERGY

In this appendix we describe a systematic way of obtaining the expansions of wave functions in various parameters that have been alluded to in the section on effective range theory. The problem is to obtain a solution to the general Schrödinger equation

$$[D + U(x)]u(x) = -\epsilon g(x)u(x). \tag{1}$$

Here $D$ has the form

$$D = \frac{d^2}{dx^2} + f(x), \tag{2}$$

and $f(x)$ is no more singular than $x^{-2}$ near the origin; we assume $g(x)$ is no more singular than $x^{-1}$ as $x \to 0$; $\epsilon$ is a parameter of smallness. Our solution which must vanish at $x = 0$ is to be sought as a power series in $\epsilon$. The asymptotic form of our equation is

$$Dw(x) = -\epsilon g(x)w(x), \tag{3}$$

and in general this solution will be badly behaved at the origin (or at least not go to zero as a true eigenfunction), and we refer to it (in the standard notation) as an irregular solution. We demand $u(x) \to w(x)$ as $x \to \infty$. We need also the limiting forms of $u$ and $w$ as $\epsilon \to 0$. Thus $v_R$ is the regular solution of

$$[D + U(x)]v_R(x) = 0, \tag{4}$$

which has the asymptotic form $y_I(x)$, an irregular solution of

$$Dy_I(x) = 0. \tag{5}$$

We shall also have need for the irregular solution of (4) and the regular solution of (5). We call these $v_I(x)$ and $y_R(x)$ respectively, and the normalization of these is chosen in such a way that the Wronskian conditions

$$\begin{aligned} W_x(v_I, v_R) &= 1, \\ W_x(y_I, y_R) &= 1, \end{aligned} \tag{6}$$

obtain for all $x$. Since we have already said $v_R \to y_I$ asymptotically, we must have $v_I \to -y_R$. We may construct $v_I$ from $v_R$ as follows: We write the Wronskian condition as

$$\frac{d}{dx}\frac{v_I}{v_R} = -\frac{1}{v_R^{\,2}}, \tag{7}$$

and integrate it directly; we find

$$v_I = v_R\left[-\int^x dx' \frac{1}{v_R^{\,2}(x')} + \text{constant}\right], \tag{8}$$

and we may use the constant to insure that $v_I \to -y_R$. Next, we form the expression

$$W_{x_0}(y_I, w) - W_{x_0}(v_R, u) = W_{x_1}(y_I, w) - W_{x_1}(v_R, u) + \epsilon\int_{x_0}^{x_1} dx\, g(x)[y_I(x)w(x) - v_R(x)u(x)]. \tag{9}$$

In the limit as $x_0 \to 0$, $x_1 \to \infty$, we find then,

$$\lim_{x_0\to 0} W_{x_0}(y_I, w) = \lim_{x_0\to 0} \epsilon\int_{x_0}^{\infty} dx\, g(x)[y_I(x)w(x) - v_R(x)u(x)]. \tag{10}$$

The power series expansions of $u$ and $w$ may be obtained by replacing the differential equations by integral equations that incorporate the boundary conditions. These are obtained by the usual method of variation of parameters for solving inhomogeneous equations in terms of the solutions of the homogeneous equation. We find

$$u(x) = W_{x_1}(v_I, u)v_R(x) + \epsilon\left[v_I(x)\int_0^x dx'\, g(x')v_R(x')u(x') + v_R(x)\int_x^{x_1} dx'\, g(x')v_I(x')u(x')\right]. \tag{11}$$

Taking $x_1$ in the asymptotic region, we obtain

$$W_{x_1}(v_I, u) = W_{x_1}(-y_R, w) = -W_0(y_R, w) + \epsilon\int_0^{x_1} dx'\, g(x')y_R(x')w(x'), \tag{12}$$

and thus

$$u(x) = -W_0(y_R, w)v_R(x) + \epsilon\left\{v_I(x)\int_0^x dx'\, g(x')v_R(x')u(x') - v_R(x)\int_0^x dx'\, g(x')v_I(x')u(x') + v_R(x)\int_0^{x_1} dx'\, g(x')[y_R(x')w(x') + v_I(x')u(x')]\right\}. \tag{13}$$

We may now take $x_1 = \infty$, since the integral converges, and choose the normalization of $y_R$ and $w$ so that $W_0(y_R, w) = -1$. Our final formula for $u(x)$ is then

$$u(x) = v_R(x) + \epsilon\left\{v_I(x)\int_0^x dx'\, g(x')v_R(x')u(x') - v_R(x)\int_0^x dx'\, g(x')v_I(x')u(x') + v_R(x)\int_0^\infty dx'\, g(x')[y_R(x')w(x') + v_I(x')u(x')]\right\}. \quad (14)$$

The same general procedure applied to $w(x)$, taking into account the difference in boundary conditions at the origin, leads to

$$w(x) = y_I(x) + W_{x_0}(y_I, w)y_R(x) + \epsilon\left[y_I(x)\int_0^x dx'\, g(x')y_R(x')w(x') - y_R(x)\int_{x_0}^x dx'\, g(x')y_I(x')w(x')\right]. \quad (15)$$

Since $W_{x_0}(y_1, w)$ is proportional to $\epsilon$ according to Eq. (10), this equation reduces properly in the limit $\epsilon \to 0$ to $w(x) = y_1(x)$.

We shall not attempt to establish what conditions must be imposed on $U(x)$ and $g(x)$ in order that the iteration solution of (13) (which generates a series in $\epsilon$) converges. Actually (13), (15), and (10) must be evaluated simultaneously.

# APPENDIX E THE JACOB-WICK EXPANSION OF THE SCATTERING MATRIX

The structure of the general formulas developed in Chapter 7 to describe scattering processes and reactions involving particles of arbitrary spin is quite complicated. The complexity is to a certain extent of our own making. We have insisted on first coupling the initial spins and these in turn to the relative orbital angular momentum; in a similar way we handled the final reaction products. Furthermore, we used a fixed axis of spin quantization, which was natural for the initial state but unnatural for the final state. Although the derivation is quite straightforward, the result is ugly and awkward to work with because of the vector coupling coefficients that appear.

As we have seen in detail in the special case of the scattering of a spin one-half particle by a central field, the nonrelativistic appearance of the results is only superficial and that appropriately interpreted there is no restriction on their validity. In order to show this we made much use of the explicit plane-wave solutions of the Dirac equation, but, in fact, these were essentially geometrical aspects of the theory involving Lorentz transformations from the rest systems of the particles before and after the scattering.

In a brilliant paper Jacob and Wick[1] have shown that the vector coupling coefficients and the apparently nonrelativistic appearance may both be eliminated from the formulas. The trick is that the component of the angular momentum of a particle along its direction of motion is an invariant quantity. If, in addition, we do not couple the spin and orbital momenta explicitly, the vector coupling coefficients disappear. We introduce instead of the $S$-matrix elements between states of total channel spin, orbital angular momentum and total angular momentum, those between states labeled by total angular momenta and projections of the particle spins along the initial *and* final directions of the relative momenta. The final formula is no more complicated than that for spinless particles except that

[1] M. Jacob and G. C. Wick, Annals of Physics, **7**, 404 (1959).

in place of the simple Legendre polynomials these appear somewhat more complicated but easily computable functions of the scattering angle. The Jacob-Wick formulas are also very easy to use for massless particles and in connection with polarization phenomena.

The component of the spin along the direction of motion of a particle will be referred to as the helicity quantum number or simply as the helicity. Since the orbital angular momentum evidently has no component along the direction of motion, the helicity, call it $\lambda$, may also be regarded as the projection of the total angular momentum $J$ along the direction of motion. Since the component of angular momentum along the direction is invariant under rotations, we may classify states as usual according to specified values of the total angular momentum while still giving precise values to the helicities. For a general two-particle reaction $A + B \rightarrow C + D$ in the barycentric system we take the energy $E$, $J$, $M$ (that is, the eigenvalues of $J_z$) and the helicities $\lambda_a$, $\lambda_b$ to label the initial state and $E, J, M, \lambda_c, \lambda_d$ for the final state. The $S$-matrix element has the form

$$\langle E'J'M'\lambda_c\lambda_d| S |EJM\lambda_a\lambda_b\rangle = \delta(E' - E)\delta_{J'J}\, \delta_{M'M}\langle\lambda_c\lambda_d| S^J(E) |\lambda_a\lambda_b\rangle. \quad (1)$$

What we had done previously was to add the initial spins to form a resultant $S$, the final spins to a resultant $S'$, and introduce an element of $S^J(E)$ as $\langle L'S'| S^J(E) |LS\rangle$, where $L\,(L')$ is the initial (final) orbital angular momentum quantum number. As we remarked in Chapter 7, the introduction of the total spin is extremely awkward if we want to deal with polarization phenomena. In fact, we must essentially undo the addition of the spins.

It will not be necessary for us to use explicit wave equations for the description of free particles of mass $m$, spin $s$, although these exist. We need to know only certain facts about the plane-wave solutions corresponding to momentum $p$ and energy $E = +(m^2 + p^2)^{1/2}$. For each $\mathbf{p}, m \neq 0$, there are $2s + 1$ linear independent solutions which may be chosen to be states of definite helicity $\lambda$, which takes on values $\lambda = s,\ s - 1, \ldots, -s + 1, -s$. If $m = 0$, there are only two solutions, $\lambda = \pm s$. Next, under a rotation we obtain a state with a different direction of $\mathbf{p}$ but the same $\lambda$. Under space reflection, $\lambda$ changes sign. If a Lorentz transformation parallel to $\mathbf{p}$ is applied, we obtain a state with the same or opposite direction of $\mathbf{p}$ but different magnitude. If the transformed state has the same direction of $p$, $\lambda$ is unchanged.

We define a phase convention for the basic free-particle states by a series of steps. First, to describe a particle with momentum $\mathbf{p}'$ in a direction given by polar angles $\theta$, $\phi$ we apply the rotation operator to a state $\psi_{p\lambda}$ having momentum $p$ in the positive $z$-direction and helicity $\lambda$ and with $|\mathbf{p}| = |\mathbf{p}'| = p$. As we saw in Chapter 1 this is obtained by rotating $\psi_{p\lambda}$ through an angle $\theta$ about an axis parallel to $\mathbf{p} \times \mathbf{p}'$, the sense of the rotation being from $\mathbf{p}$ toward $\mathbf{p}'$ as a right-hand screw. Thus we would write

$$|p\theta\phi, \lambda\rangle = e^{-i\theta\mathbf{n}\cdot\mathbf{J}}\psi_{p\lambda}, \tag{2}$$

where we have used the rotation operator $R$ given by

$$R(\mathbf{n}, \theta) = e^{-i\theta\mathbf{n}\cdot\mathbf{J}} \tag{3}$$

discussed in Chapter 1, Eq. (1.52), and $\mathbf{n} = \mathbf{p} \times \mathbf{p}'/p^2 \sin\theta$.

We shall digress briefly to discuss some general properties of the rotation operator. It is convenient to relate $R$ to the rotation operator for rotation specified by three Euler angles $\alpha$, $\beta$, $\gamma$ defined as

$$R_{\alpha\beta\gamma} = e^{-i\alpha J_z}\, e^{-i\beta J_z}\, e^{-i\gamma J_z}. \tag{4}$$

We shall show that $R(\mathbf{n}, \theta) = R_{\phi,\theta,-\phi}$ although it is almost obvious. To see this, we remark that the unit vector $\mathbf{n}$ is given by $(-\sin\phi, \cos\phi, 0)$, so $\mathbf{n}\cdot\mathbf{J} = -J_x \sin\phi + J_y \cos\phi$. Consider then

$$\begin{aligned} e^{-i\phi J_z} J_y\, e^{i\phi J_z} &= J_y + (-i\phi)(J_z, J_y) + \frac{(-i\phi)^2}{2!}[J_z, (J_z, J_y)] + \cdots \\ &= J_y - \phi J_x - \frac{\phi^2}{2!} J_y + \frac{\phi^3}{3!} J_x + \cdots \\ &= -J_x \sin\phi + J_y \cos\phi. \end{aligned} \tag{5}$$

Evidently, then,

$$e^{-i\phi J_z}\, e^{-i\theta J_y}\, e^{i\phi J_z} = e^{-i\theta(-J_x \sin\phi + J_y \cos\phi)} = e^{-i\phi\mathbf{n}\cdot\mathbf{J}} \tag{6}$$

which completes the demonstration.

Consider now the action of $R_{\alpha\beta\gamma}$ on a state characterized by total angular momentum $J$, and $J_z = M$, call it $\mathscr{Y}_{JM}$. Acting on such a state $R$ evidently yields a linear combination of the magnetic substates $\mathscr{Y}_{JM'}$ so that

$$R_{\alpha\beta\gamma}\mathscr{Y}_{JM} = \sum_{M'} \mathscr{D}^J_{M'M}(\alpha\beta\gamma)\mathscr{Y}_{JM'}, \tag{7}$$

where the $\mathscr{D}^J_{M'M}(\alpha\beta\gamma)$ are a well-known set of matrices evidently defined by

$$\begin{aligned} \mathscr{D}^J_{M'M} &= (\mathscr{Y}_{JM'}, R_{\alpha\beta\gamma}\mathscr{Y}_{JM}) \\ &= e^{-i\alpha M'}\, d^J_{M'M}(\beta) e^{-i\gamma M}, \end{aligned} \tag{8}$$

with

$$d^J_{M'M}(\beta) = (\mathscr{Y}_{JM'}, e^{-i\beta J_y}\mathscr{Y}_{JM}). \tag{9}$$

We shall not give the explicit formula for the $d^J_{M'M}(\beta)$, but we merely note for the present a few symmetry properties.

$$\begin{aligned} d^J_{M'M}(\beta) &= d^J_{-M,-M'}(\beta) = (-1)^{M'-M}\, d^J_{MM'}(\beta), \\ d^J_{M'M}(\beta) &= (-1)^{J+M'}\, d^J_{M',-M}(\pi - \beta). \end{aligned} \tag{10}$$

For the special angle $\beta = \pi$ we have

$$d^J_{M'M}(\pi) = (-1)^{J-M}\,\delta_{M',-M}. \tag{11}$$

The derivation of these symmetry relations are taken up at the end of this Appendix together with some formulas for particular $d^J_{M'M}(\beta)$.

We are in position now to complete our specification of phases of wave functions. We have

$$\begin{aligned}|p\theta\phi;\lambda\rangle &= R_{\phi,\theta,-\phi}\psi_{p\lambda}\\ &= e^{i\lambda\phi}R_{\phi,\theta,0}\psi_{p\lambda},\end{aligned} \tag{12}$$

since by definition $\lambda$ is the component of angular momentum of $\psi_{p\lambda}$ along the $z$-axis. The state $\psi_{p\lambda}$ in turn may be generated by applying a Lorentz transformation in the $z$-direction to a fixed state $\psi_{p_0\lambda}$. For a particle with finite rest-mass this is conveniently chosen as that corresponding to zero momentum. Since the total angular momentum is then simply the spin $s$, the relative phases of the helicity states $\psi_{0\lambda}$ may be fixed by our standard convention.

$$(J_x \pm iJ_y)\psi_{0\lambda} = +\sqrt{(s \mp \lambda)(s \pm \lambda + 1)}\,\psi_{0\lambda\pm1}. \tag{13}$$

This method cannot be used, of course, for massless particles. There are, however, only two helicity values to be compared, namely $\pm\lambda$. Since a spatial reflection changes the sign of the helicity, we may use this fact to connect the two states. It is convenient to use the operator describing reflection in the $xz$-plane. This is obtained by following the parity operation $P$ ($\mathbf{r} \to -\mathbf{r}$) by a rotation of 180° about the $y$-axis. Thus ($x \to -x$, $y \to -y$, $z \to -z$) and then ($x \to -x$, $y \to y$, $z \to -z$) or ($x \to x$, $y \to -y$, $z \to z$). Call the combined operation $Y$.

$$Y = e^{-i\pi J_y}\mathscr{P}. \tag{14}$$

We remark then that

$$Y\psi_{p,s} = \eta\psi_{p,-s}, \tag{15}$$

where $\eta$ is a phase, the "parity factor," independent of the momentum $p$ (see Eq. 76). The latter statement follows from the fact that $Y$ commutes with a Lorentz transformation in the $z$-direction. For the plane-wave solutions of the Dirac equation for zero rest-mass used in this book, $\eta = -1$. Similarly, for photons we have as vector potential corresponding to $\lambda = \pm$, $A_{\pm1} = \mp(\mathbf{e}_x \pm i\mathbf{e}_y)e^{ipz}$ where the $\mathbf{e}$'s are unit vectors and again $\eta = -1$.

The inversion operator applied to a state describing a particle of nonzero rest-mass is easily found. Since the parity operator $\mathscr{P}$ commutes with $\mathbf{J}$, we may write

$$\mathscr{P}\psi_{0\lambda} = \eta\psi_{0\lambda} \tag{16}$$

with $\eta$ independent of $\lambda$. Also the rotation about the $y$-axis may be effected using Eq. (11).

$$e^{-i\pi J_y}\psi_{0\lambda} = (-1)^{s-\lambda}\psi_{0,-\lambda} \tag{17}$$

Finally, a Lorentz transformation up to momentum $p$ along $z$ yields

$$Y\psi_{p\lambda} = \eta(-1)^{s-\lambda}\psi_{p,-\lambda}. \tag{18}$$

Consider now two free particles $m_1$, $m_2$ with spins $s_1$, $s_2$. The wave function for this combined system may be written as direct product of wave functions obtained by our prescription, that is, as

$$|p_1\theta_1\phi_1, \lambda_1\rangle \times |p_2\theta_2\phi_2, \lambda_2\rangle.$$

We invariably work in the barycentric subspace where $\mathbf{p}_1 = -\mathbf{p}_2$; if we take $\mathbf{p}_1$ in the direction $\hat{\boldsymbol{\kappa}}$ characterized by polar angles $\theta$, $\phi$, we have $\theta_1 = \theta$, $\phi_1 = \phi$ and $\theta_2 = \pi - \theta$, $\phi_2 = \pi \pm \phi$. The two-particle state may then be formed by a rotation generated by the total angular momentum operator $J = J_1 + J_2$ acting on a suitably defined state corresponding to $p = (\theta = 0, \phi = 0)$. To define the latter we take for particle two (which in our standard state is moving along $-z$)

$$\phi_{p\lambda_2} = (-1)^{s_2-\lambda_2}\, e^{-i\pi J_y{}^{(2)}}\, \psi_{p\lambda_2}. \tag{19}$$

To understand this choice, consider the case of a massive particle and the limit $p \to 0$; then we see that $\phi_{0\lambda_2} = \psi_{0-\lambda_2}$ with no excess numerical factor. We define for the basic two-particle state in which $\mathbf{p}_1 = \boldsymbol{\kappa}$ is along the $+z$-axis

$$\psi_{\kappa\lambda_1\lambda_2} = \psi^{(1)}_{\kappa\lambda_1}\phi^{(2)}_{p\lambda_2}. \tag{20}$$

For $\boldsymbol{\kappa}$ in an arbitrary direction, then,

$$|\kappa\theta\phi, \lambda_1\lambda_2\rangle = R_{\phi,\theta,-\phi}\psi_{\kappa\lambda_1\lambda_2} = e^{i\lambda\phi}R_{\phi,\theta,0}\psi_{\kappa\lambda_1\lambda_2}, \tag{21}$$

where $\lambda = \lambda_1 - \lambda_2$ is the resultant angular momentum in the direction of relative motion. $R$ is, of course, $\exp[-i\theta\mathbf{n}\cdot(\mathbf{J}_1 + \mathbf{J}_2)]$ with $\mathbf{n}$ a unit vector in the direction $\hat{e}_z \times \boldsymbol{\kappa}$.

As usual, it is useful to construct states with total angular momentum $J$ and $J_z = M$. Since the helicities $\lambda_1$, $\lambda_2$ and magnitude of momentum $p$ are invariant under rotations, we may use all these quantities as quantum numbers and ask for the states $|p; JM, \lambda_1\lambda_2\rangle$. It is fairly obvious how to do this. We subject the standard state $\psi_{\kappa\lambda_1\lambda_2}$ to a rotation and then test the result with the matrix $\mathscr{D}^J_{MM'}$ to find out how much of the desired state is present.

We write this as

$$|\kappa; JM, \lambda_1\lambda_2\rangle = \frac{N_J}{2\pi}\int dR\mathscr{D}^{J*}_{MM'}(\alpha\beta\gamma)R_{\alpha\beta\gamma}\psi_{\kappa\lambda_1\lambda_2} \tag{22}$$

and we show in a moment that $M' = \lambda_1 - \lambda_2$. The region of integration over the Euler angles $\alpha\beta\gamma$ is the obvious one: $0 < \alpha < 2\pi$, $0 < \beta < \pi$,

$0 < \gamma < 2\pi$, so $dR = \sin\beta\, d\beta\, d\alpha\, d\gamma$ and $N_J$ is a normalization factor. To see that $M' = \lambda_1 - \lambda_2$ we remark that $R_{\alpha\beta\gamma}\psi_{\kappa\lambda_1\lambda_2} = e^{-i\gamma(\lambda_1-\lambda_2)}R_{\alpha\beta 0}\psi_{\kappa\lambda_1\lambda_2}$ and $\mathscr{D}^{J*}_{MM'}(\alpha\beta\gamma) = e^{iM\alpha}\, d^J_{MM'}e^{iM'\gamma}$. The integration over $\gamma$ becomes

$$\int_0^{2\pi} d\gamma\, e^{i\gamma(M'-\lambda_1+\lambda_2)} = 2\pi\, \delta_{M';\lambda_1-\lambda_2}. \tag{23}$$

Thus we have

$$|\kappa, JM, \lambda_1\lambda_2\rangle = N_J \int d\Omega\, \mathscr{D}^{J*}_{M\lambda}(\phi, \theta, -\phi)\, |\kappa\theta\phi, \lambda_1\lambda_2\rangle, \tag{24}$$

where we have used the definition, Eq. (21) for $|\kappa\theta\phi, \lambda_1\lambda_2\rangle$ and introduced $\lambda \equiv \lambda_1 - \lambda_2$.

The normalization of our states will be exactly the same as that employed in Chapter 7. We shall work exclusively in the barycentric coordinate system, so that, for example,

$$\begin{aligned}\langle \mathbf{p}'_1\mathbf{p}'_2\lambda'_1\lambda'_2 \mid \mathbf{p}_1\mathbf{p}_2\lambda_1\lambda_2\rangle &= \delta(\mathbf{P}'-0)\,\delta(\boldsymbol{\kappa}'-\boldsymbol{\kappa})\,\delta_{\lambda'_1\lambda_1}\,\delta_{\lambda'_2\lambda_2}\\ &= \delta(\mathbf{P}'-0)\frac{1}{\rho_\epsilon}\,\delta[\epsilon(\kappa')-\epsilon(\kappa)]\,\delta_{\hat{\boldsymbol{\kappa}}'\hat{\boldsymbol{\kappa}}}\,\delta_{\lambda'_1\lambda_1}\,\delta_{\lambda'_2\lambda_2},\end{aligned} \tag{25}$$

where $\kappa$, $\kappa'$ are the relative momenta, $\rho_\epsilon = \kappa^2/(d\epsilon/d\kappa)$ and $\delta_{\hat{\boldsymbol{\kappa}}'\hat{\boldsymbol{\kappa}}}$ is the $\delta$-function on the unit sphere. For any operator $\mathscr{O}$ which commutes with the four components of the total energy-momentum vector, we would write

$$\begin{aligned}\langle \mathbf{p}'_1\mathbf{p}'_2\lambda'_1\lambda'_2|\, \mathscr{O}\, |\mathbf{p}_1\mathbf{p}_2\lambda_1\lambda_2\rangle &= \delta(\mathbf{P}'-0)\frac{1}{\rho_\epsilon}\,\delta[\epsilon(\kappa')-\epsilon(\kappa)]\\ &\quad\times \langle \hat{\boldsymbol{\kappa}}'\lambda'_1\lambda'_2|\, \mathscr{O}(\kappa)\, |\hat{\boldsymbol{\kappa}}\lambda_1\lambda_2\rangle\end{aligned} \tag{26}$$

which implies that if $\mathscr{O} = 1$

$$\langle \hat{\boldsymbol{\kappa}}'\lambda'_1\lambda'_2 \mid \hat{\boldsymbol{\kappa}}\lambda_1\lambda_2\rangle = \delta_{\hat{\boldsymbol{\kappa}}'\hat{\boldsymbol{\kappa}}}\delta_{\lambda'_1\lambda_1}\delta_{\lambda'_2\lambda_2}. \tag{27}$$

The states $|\hat{\boldsymbol{\kappa}}\lambda_1\lambda_2\rangle$ may be identified with those previously designated by $|\kappa\theta\phi, \lambda_1\lambda_2\rangle$. We now determine the factors $N_J$ used in the definition of $|\kappa, JM, \lambda_1\lambda_2\rangle$ by the requirement that they satisfy the usual orthonormality condition

$$\begin{aligned}\delta_{J'J}\,\delta_{M'M}\,&\delta_{\lambda'_1\lambda_1}\,\delta_{\lambda'_2\lambda_2}\\ &= \langle J'M'\lambda'_1\lambda'_2 \mid JM\lambda_1\lambda_2\rangle\\ &= N^*_{J'}N_J \int d\Omega' \int d\Omega\, \mathscr{D}^{J'}_{M'\lambda'}(\phi', \theta', -\phi')\, \mathscr{D}^{J*}_{M\lambda}(\phi\theta - \phi)\langle \hat{\boldsymbol{\kappa}}'\lambda'_1\lambda'_2 \mid \hat{\boldsymbol{\kappa}}\lambda_1\lambda_2\rangle\\ &= N^*_{J'}N_J\,\delta_{\lambda'_1\lambda'_2}\,\delta_{\lambda'_2\lambda_2}\int d\Omega e^{-iM'\phi}\,e^{iM\phi}\,d^{J'}_{M'\lambda'}(\theta)\,d^J_{M\lambda'}(\theta) \qquad (28)\\ &= 2\pi N^*_{J'}N_J\,\delta_{\lambda'_1\lambda_1}\,\delta_{\lambda'_2\lambda_2}\,\delta_{M'M}\int_0^\pi \sin\theta\, d\theta\, d_{M\lambda}{}^{J'}(\theta)\, d_{M\lambda}{}^J(\theta)\\ &= \frac{4\pi}{2J+1}\,|N_J|^2\,\delta_{JJ'}\,\delta_{\lambda'_1\lambda_1}\,\delta_{\lambda'_2\lambda_2}\,\delta_{M'M} \qquad (29)\end{aligned}$$

where the standard normalization of the $d^J_{MM'}$'s has been used [see Eq. (90)]. Evidently, the choice $N_J = [(2J + 1)/4\pi]^{1/2}$ is indicated. The transformation matrix

$$\begin{aligned}\langle \hat{\boldsymbol{\varkappa}}, \lambda'_1\lambda'_2 \mid JM\lambda_1\lambda_2\rangle &= \langle \theta\phi, \lambda'_1\lambda'_2 \mid JM, \lambda_1\lambda_2\rangle \\ &= N_J\, \delta_{\lambda'_1\lambda_1}\, \delta_{\lambda'_2\lambda_2}\, \mathscr{D}^{J*}_{M\lambda}(\phi, \theta, -\phi) \qquad (30)\end{aligned}$$

satisfies the conditions of unitarity:

$$\int d\Omega \langle \hat{\boldsymbol{\varkappa}}, \lambda_1\lambda_2 \mid JM, \lambda_1\lambda_2\rangle\langle \hat{\boldsymbol{\varkappa}}, \lambda_1\lambda_2 \mid J'M', \lambda_1\lambda_2\rangle^* = \delta_{JJ'}\, \delta_{MM'}$$

$$\sum_{JM} \langle \hat{\boldsymbol{\varkappa}}, \lambda_1\lambda_2 \mid JM, \lambda_1\lambda_2\rangle\langle \hat{\boldsymbol{\varkappa}}', \lambda_1\lambda_2 \mid JM, \lambda_1\lambda_2\, \lambda\rangle^* = \delta_{\hat{\boldsymbol{\varkappa}}'\hat{\boldsymbol{\varkappa}}}. \qquad (31)$$

The last of these relations follows from the fact that

$$\sum_J \frac{2J+1}{2}\, d^J_{mm'}(\theta)\, d^J_{mm'}(\theta') = \delta(\cos\theta - \cos\theta') \qquad (32)$$

and that

$$d^J_{mm'} = 0 \quad \text{for } m \text{ or } m' > J.$$

Note then that the state $|\theta\phi, \lambda_1\lambda_2\rangle$ may be easily expressed in terms of the $|JM\lambda_1\lambda_2\rangle$.

$$\begin{aligned}|\theta\phi, \lambda_1\lambda_2\rangle &= \sum_{\substack{JM \\ \lambda'_1\lambda'_2}} |JM\lambda'_1\lambda'_2\rangle\langle JM\lambda'_1\lambda'_2 \mid \theta\phi, \lambda_1\lambda_2\rangle \\ &= \sum_{JM} |JM\lambda_1\lambda_2\rangle N_J\, \mathscr{D}_{M\lambda}{}^J(\phi, \theta, -\phi). \qquad (33)\end{aligned}$$

We wish now to express the $S$-matrix in terms of the helicity amplitudes. Consider a reaction of the form $A + B \rightarrow C + D$. In our standard notation $A + B$ is the channel $c$, $C + D$ is channel $c'$. We do not write the channel labels in explicitly; the momenta and helicity labels will give adequate warning about what channel we are concerned with. Now, however, we do not add the spins together but instead use the helicity labels $\lambda_A\lambda_B$, $\lambda_C\lambda_D$ to characterize the states. As in Chapter 7 we write in the barycentric system

$$\langle \mathbf{p}_C\mathbf{p}_D\lambda_C\lambda_D|\, S\, |\mathbf{p}_A\mathbf{p}_B\lambda_A\lambda_B\rangle = \delta(\mathbf{P}' - 0)\langle \boldsymbol{\varkappa}_f, \lambda_C\lambda_D|\, S\, |\boldsymbol{\varkappa}, \lambda_A\lambda_B\rangle, \quad (34)$$

where $\mathbf{P}' = \mathbf{p}_C + \mathbf{p}_D$ and $\boldsymbol{\varkappa}_f$, $\boldsymbol{\varkappa}$ are the final and initial relative momenta. We further define

$$\langle \boldsymbol{\varkappa}_f, \lambda_C\lambda_D|\, S\, |\boldsymbol{\varkappa}, \lambda_A\lambda_B\rangle = \frac{1}{\sqrt{\rho_f\rho}}\, \delta[\epsilon(\boldsymbol{\varkappa}_f) - \epsilon(\boldsymbol{\varkappa})]\langle \hat{\boldsymbol{\varkappa}}_f, \lambda_C\lambda_D|\, \mathbf{S}\, |\hat{\boldsymbol{\varkappa}}, \lambda_C\lambda_D\rangle \qquad (35)$$

where $\rho = \kappa^2/(d\epsilon/d\kappa)$, $\rho_f = \kappa_f^2/(d\epsilon/d\kappa_f)$. In terms of this reduced matrix, the unitarity relation takes the form

$$\sum_{\lambda''_1\lambda''_2} \int d\Omega'' \langle \hat{\boldsymbol{\kappa}}_f, \lambda_C\lambda_D | \mathbf{S}^\dagger | \hat{\boldsymbol{\kappa}}'', \lambda''_1\lambda''_2 \rangle \langle \hat{\boldsymbol{\kappa}}'', \lambda''_1\lambda''_2 | S | \hat{\boldsymbol{\kappa}}, \lambda_A\lambda_B \rangle = \delta_{AC}\,\delta_{BD}\,\delta_{\hat{\boldsymbol{\kappa}}_f\hat{\boldsymbol{\kappa}}}. \tag{36}$$

The $T$-matrix is introduced as usual.

$$\langle \mathbf{p}_C\mathbf{p}_D, \lambda_C\lambda_D | S | \mathbf{p}_A\mathbf{p}_B\lambda_A\lambda_B \rangle = \delta(\mathbf{P}' - 0)\{\delta_{AC}\,\delta_{BD}\,\delta(\boldsymbol{\kappa}_f - \boldsymbol{\kappa}) \\ -2\pi i\,\delta[\epsilon(\kappa_f) - \epsilon(\kappa)]\langle \hat{\boldsymbol{\kappa}}_f, \lambda_\epsilon\lambda_D | T | \hat{\boldsymbol{\kappa}}, \lambda_A\lambda_B \rangle\} \tag{37}$$

and

$$\langle \hat{\boldsymbol{\kappa}}_f, \lambda_C\lambda_D | S | \hat{\boldsymbol{\kappa}}, \lambda_A\lambda_B \rangle = \delta_{AC}\,\delta_{BD}\,\delta_{\hat{\boldsymbol{\kappa}}_f\hat{\boldsymbol{\kappa}}} - 2\pi i\sqrt{\rho_f\rho}\langle \hat{\boldsymbol{\kappa}}_f, \lambda_C\lambda_D | T | \hat{\boldsymbol{\kappa}}, \lambda_A\lambda_B \rangle. \tag{38}$$

The differential cross section becomes

$$\frac{d\sigma}{d\Omega_f} = \frac{(2\pi)^4}{\kappa^2}\rho_f\rho\, |\langle \hat{\boldsymbol{\kappa}}_f, \lambda_C\lambda_D | T | \hat{\boldsymbol{\kappa}}, \lambda_A\lambda_B \rangle|^2, \tag{39}$$

or if $A \neq C$, $B \neq D$,

$$\frac{d\sigma}{d\Omega} = \left(\frac{2\pi}{\kappa}\right)^2 |\langle \hat{\boldsymbol{\kappa}}_f, \lambda_C\lambda_D | \mathbf{S} | \hat{\boldsymbol{\kappa}}, \lambda_A\lambda_B \rangle|^2. \tag{40}$$

Next we remark that the transformation to the $JM$ representation may be effected with the aid of the transformation matrix, Eq. (30). We recall that

$$\langle JM'\lambda_C\lambda_D | S | JM\lambda_A\lambda_B \rangle = \delta_{M'M}\langle \lambda_C\lambda_D | S^J | \lambda_A\lambda_B \rangle, \tag{41}$$

and obtain

$$\langle \hat{\boldsymbol{\kappa}}_f, \lambda_C\lambda_D | S | \hat{\boldsymbol{\kappa}}, \lambda_A\lambda_B \rangle = \sum_{J,M} \langle \hat{\boldsymbol{\kappa}}_f, \lambda_C\lambda_D \mid JM\lambda_C\lambda_D \rangle \langle \lambda_C\lambda_D | S^J | \lambda_A\lambda_B \rangle \\ \times \langle JM\lambda_A\lambda_B \mid \hat{\boldsymbol{\kappa}}, \lambda_A\lambda_B \rangle. \tag{42}$$

The function

$$\sum_M \langle \hat{\boldsymbol{\kappa}}_f, \lambda_C\lambda_D \mid JM\,\lambda_C\lambda_D \rangle \langle JM\,\lambda_A\lambda_B \mid \hat{\boldsymbol{\kappa}}, \lambda_A\lambda_B \rangle \equiv \langle \hat{\boldsymbol{\kappa}}_f, \lambda_C\lambda_D | \mathscr{Y}_J | \hat{\boldsymbol{\kappa}}, \lambda_A\lambda_B \rangle$$

is the analogue here of (43)

$$\sum_{m'mM} Y_{l'}^{m'}(\hat{\boldsymbol{\kappa}}_f) Y_l^{m*}(\hat{\boldsymbol{\kappa}}) \langle l's'; m'\nu' \mid JM \rangle \langle JM \mid ls; m\nu \rangle \\ \equiv \langle \hat{\boldsymbol{\kappa}}_f, l's'\nu' | \mathscr{Y}_J | \hat{\boldsymbol{\kappa}}, ls\nu \rangle, \tag{44}$$

defined by (7-12$a$). Inserting the explicit formulas for the transformation coefficients, we have

$$\langle \hat{\boldsymbol{\kappa}}_f, \lambda_C\lambda_D | S | \hat{\boldsymbol{\kappa}}, \lambda_A\lambda_B \rangle = \sum_{J,M} \frac{2J+1}{4\pi} \mathscr{D}^{J*}_{M\mu}(\phi_f, \theta_f, -\phi_f) \\ \times \langle \lambda_C\lambda_D | S^J | \lambda_A\lambda_B \rangle \mathscr{D}_{M\lambda}{}^J(\phi, \theta, -\phi), \tag{45}$$

where $\mu = \lambda_C - \lambda_D$, $\lambda = \lambda_A - \lambda_B$. This is the generalization of the spin zero problem where we had

$$\langle \hat{\boldsymbol{\varkappa}}_f | \, S \, | \hat{\boldsymbol{\varkappa}} \rangle = \sum_{l,m} Y_l^m(\hat{\boldsymbol{\varkappa}}_f) S_l Y_l^{m*}(\hat{\boldsymbol{\varkappa}}), \tag{46}$$

and, in fact, reduces to precisely it when we recall that

$$Y_l^m(\theta, \phi) = \left(\frac{2l+1}{4\pi}\right)^{1/2} \mathscr{D}_{m0}{}^{l*}(\phi, \theta, -\phi) \tag{47}$$

[see Eq. (92)].

Our formulas become considerably simpler if we choose the initial direction $\hat{\boldsymbol{\varkappa}}$ to coincide with the $z$-axis, in which case

$$\mathscr{D}_{M\lambda}{}^J(\phi, 0, -\phi) = e^{i(\lambda - M)\phi} \, d_{M\lambda}{}^J(0) = \delta_{M\lambda}. \tag{48}$$

We obtain then

$$\langle \hat{\boldsymbol{\varkappa}}_f, \lambda_C\lambda_D | \, S \, | \hat{\boldsymbol{\varkappa}}, \lambda_A\lambda_B \rangle = \sum_J \frac{2J+1}{4\pi} \langle \lambda_C\lambda_D | \, S^J \, | \lambda_A\lambda_B \rangle \, d_{\lambda\mu}{}^J(\theta_f) e^{i(\lambda-\mu)\phi_f}. \tag{49}$$

This is a remarkably simple formula; the $d_{\lambda u}{}^J$ are a little complicated, but certainly not as bad to use as the ones involving products of vector coupling coefficients that appeared in Chapter 7.

We have an exactly analogous formula for $\langle \hat{\boldsymbol{\varkappa}}_f, \lambda_C\lambda_D | \, T \, | \hat{\boldsymbol{\varkappa}}, \lambda_A\lambda_B \rangle$. In terms of it we define a scattering amplitude, $f_{\lambda_C\lambda_D;\lambda_A\lambda_F}$, according to

$$f^{(\theta_f,\phi_f)}_{\lambda_C\lambda_D;\lambda_A\lambda_B} = \frac{-\pi\sqrt{\rho_f\rho}}{\kappa} \sum_J (2J+1) \langle \lambda_C\lambda_D | \, T^J \, | \lambda_A\lambda_B \rangle \, d_{\lambda\mu}{}^J(\theta_f) e^{i(\lambda-\mu)\phi_f}, \tag{50}$$

and

$$\frac{d\sigma}{d\Omega_f} = |f^{(\theta_f\phi_f)}_{\lambda_C\lambda_D;\lambda_A\lambda_B}|^2. \tag{51}$$

Using the fact that

$$\sum_{JM} \frac{2J+1}{4\pi} \mathscr{D}_{M\lambda}{}^{J*}(\phi_f, \theta_f, -\phi_f) \mathscr{D}_{M\lambda}{}^J(\phi, \theta, -\phi) = \delta_{\hat{\varkappa}_f\hat{\varkappa}}, \tag{52}$$

we find immediately that

$$\langle \lambda_C\lambda_D | \, S^J \, | \lambda_A\lambda_B \rangle = \delta_{AC} \, \delta_{BD} \, \delta_{\lambda_A\lambda_C} \, \delta_{\lambda_B\lambda_D} - 2\pi i\sqrt{\rho_f\rho} \langle \lambda_C\lambda_D | \, T^J \, | \lambda_A\lambda_B \rangle, \tag{53}$$

where the factor $\delta_{AC} \, \delta_{BD}$ has been written to emphasize that the first term is present only with elastic scattering.

The total cross section for unpolarized particles may be easily worked out if we use the following result for the product of two $d$'s:

$$d^J_{\lambda\mu}(\theta)\, d^{J'}_{\lambda'\mu'}(\theta) = \sum_{l=|J-J'|}^{J+J'} \langle l, \lambda - \lambda' \mid JJ'; \lambda, -\lambda' \rangle \langle l, \mu - \mu' \mid JJ'; \mu, -\mu' \rangle \times (-1)^{\lambda'-\mu'}\, d^l_{\lambda'-\lambda;\mu-\mu'}(\theta). \tag{54}$$

Then

$$\overline{\frac{d\sigma}{d\Omega_f}} = \frac{(\pi^2 \rho_f \rho/\kappa^2)}{(2S_A + 1)(2S_B + 1)} \sum_{\substack{\lambda_A\lambda_B \\ \lambda_C\lambda_D}} \sum_{JJ'} (2J + 1)(2J' + 1)(-1)^{\lambda-\mu} \times \langle \lambda_C\lambda_D |\, T^J \,| \lambda_A\lambda_B \rangle^* \langle \lambda_C\lambda_D |\, T^{J'} \,| \lambda_A\lambda_B \rangle \times \sum_l \langle l0 \mid JJ'; \lambda, -\lambda \rangle \langle l0 \mid JJ'; \mu, -\mu \rangle P_l(\cos\theta) \tag{55}$$

where we have used the fact that $d_{00}{}^l(\theta) = P_l(\cos\theta)$. We shall not stop here to discuss polarization phenomena; the reader is referred to the original paper of Jacob and Wick.

We turn now to the question of the limitations on the submatrices $S^J$, $T^J$ imposed by symmetry relations.

We consider first the time reversal invariance. Under the operation of time reversal the helicities of the particles are unchanged since both spin *and* momentum are changed. The direction of relative motion is, of course, reversed. Thus we write (with the polar angles of $\hat{\boldsymbol{\varkappa}}$, given by $\theta$, $\phi$)

$$T\,|\hat{\boldsymbol{\varkappa}}\lambda_1\lambda_2\rangle = |-\hat{\boldsymbol{\varkappa}}\lambda_1\lambda_2\rangle = |\pi - \theta, \pi + \phi, \lambda_1\lambda_2\rangle. \tag{56}$$

Let us see what this implies for the states $|JM\lambda_1\lambda_2\rangle$.

$$\begin{aligned} T\,|\hat{\boldsymbol{\varkappa}}\lambda_1\lambda_2\rangle &= T \sum_{JM} |JM\lambda_1\lambda_2\rangle N_J\, \mathscr{D}_{M\lambda}{}^J(\phi, \theta, -\phi) \\ &= \sum_{JM} T\,|JM\lambda_1\lambda_2\rangle N_J\, D_{M\lambda}{}^{J*}(\phi, \theta, -\phi) \\ &= \sum_{JM} T\,|J - M\lambda_1\lambda_2\rangle N_J\, D^J_{-M\lambda}{}^*(\phi, \theta, -\phi) \\ &= \sum_{JM} T\,|J - M\lambda_1\lambda_2\rangle N_J (-1)^{\lambda+M}\, D^J_{M-\lambda}(\phi, \theta, -\phi) \\ &= \sum_{JM} T\,|J - M\lambda_1\lambda_2\rangle N_J (-1)^{-J+M}\, D^J_{-M\lambda}(\pi + \phi, \pi - \theta, -\pi - \phi), \end{aligned} \tag{57}$$

where we have used the relations, easily checked,

$$\begin{aligned} D^J_{-M\lambda}(\phi, \theta, -\phi) &= (-1)^{\lambda+M}\, \mathscr{D}^{J*}_{M-\lambda}(\phi, \theta, -\phi), \\ D^J_{M-\lambda}(\phi, \theta, -\phi) &= (-1)^{-J-\lambda}\, D_{M\lambda}{}^J(\pi + \phi, \pi - \theta, -\pi - \phi). \end{aligned} \tag{58}$$

By hypothesis $T|\hat{\varkappa}\lambda_1\lambda_2\rangle$ is to be equal to

$$|-\hat{\varkappa}\lambda_1\lambda_2\rangle = |(\pi - \theta, \pi + \phi, \lambda_1\lambda_2\rangle = \sum_{JM} |JM\lambda_1\lambda_2\rangle \mathscr{D}_{M\lambda}{}^J(\pi + \phi, \pi - \theta, -\pi - \phi),$$

hence

$$T|JM\lambda_1\lambda_2\rangle = (-1)^{J+M}|JM\lambda_1\lambda_2\rangle, \tag{59}$$

which is a familiar result.

In the manner used by us many times we show that

$$\langle\hat{\varkappa}_f\lambda_C\lambda_D| S |\hat{\varkappa}\lambda_A\lambda_B\rangle = \langle-\hat{\varkappa}\lambda_A\lambda_B| S |-\hat{\varkappa}_f\lambda_C\lambda_D\rangle \tag{60}$$

since the relation between the $(\pm)$ states of scattering theory generated by the "plane waves" $|\hat{\varkappa}\lambda_1\lambda_2\rangle$ satisfy

$$T|\hat{\varkappa}\lambda_1\lambda_2\rangle^{\pm} = |-\hat{\varkappa}\lambda_1\lambda_2\rangle^{\mp}. \tag{61}$$

Now we see that

$$\begin{aligned}
&\langle\hat{\varkappa}_f\lambda_C\lambda_D| S |\hat{\varkappa}\lambda_A\lambda_B\rangle \\
&= \sum_{JM} \mathscr{D}^{J*}_{M\mu}(\phi_f, \theta_f, -\phi_f)\langle\lambda_C\lambda_D| S^J |\lambda_A\lambda_B\rangle \mathscr{D}_{M\lambda}{}^J(\phi, \theta, -\phi) \\
&= \sum_{JM} (-1)^{-\mu-M} \mathscr{D}^J_{M-\mu}(\phi_f, \theta_f, -\phi_f)\langle\lambda_C\lambda_D| S^J |\lambda_A\lambda_B\rangle \\
&\qquad\qquad \times (-1)^{\lambda+M} \mathscr{D}^{J*}_{M-\lambda}(\phi, \theta, -\phi) \\
&= \sum_{JM} \mathscr{D}^{J*}_{M\lambda}(\pi + \phi, \pi - \theta, -\pi - \phi)\langle\lambda_C\lambda_D| S^J |\lambda_A\lambda_B\rangle \\
&\qquad\qquad \times \mathscr{D}^J_{M\mu}(\pi + \phi_f, \pi - \theta_f, -\pi - \phi_f)
\end{aligned} \tag{62}$$

However, this is supposed to equal

$$\begin{aligned}
&\langle-\hat{\varkappa}\lambda_A\lambda_B| S |-\hat{\varkappa}_f\lambda_C\lambda_D\rangle \\
&\qquad = \sum_{JM} \mathscr{D}_{m\lambda}{}^*(\pi + \phi, \pi - \theta, -\pi - \phi)\langle\lambda_A\lambda_B| S^J |\lambda_C\lambda_D\rangle \\
&\qquad\qquad \times \mathscr{D}_{M\mu}{}^J(\pi + \phi_f, \pi - \theta_f, -\pi - \phi_f),
\end{aligned} \tag{63}$$

hence

$$\langle\lambda_C\lambda_D| S^J |\lambda_A\lambda_B\rangle = \langle\lambda_A\lambda_B| S^J |\lambda_C\lambda_D\rangle. \tag{64}$$

In a similar fashion, by considering reflections in the $x$-$z$ plane, assuming parity conservation, we find

$$\langle-\lambda_C - \lambda_D| S^J |-\lambda_A - \lambda_B\rangle = (-1)^{S_C+S_D-S_A-S_B} \frac{\eta_C\eta_D}{\eta_A\eta_B} \langle\lambda_C\lambda_D| S^J |\lambda_A\lambda_B\rangle, \tag{65}$$

where the $\eta$'s are the intrinsic parities of each particle.

For identical particles we must consider the effect of interchanging the two. The operator $P_{12}$ that does this has the effect on our basic two-particle state

$$P_{12}\psi_{\kappa\lambda_1\lambda_2} = (-1)^{2s-\lambda_1+\lambda_2}\, e^{i\pi J_y}\psi_{\kappa\lambda_2\lambda_1}. \tag{66}$$

Of course, $P_{12}$ commutes with the rotation operators so that we can easily study the effect of $P_{12}$ on $|JM\lambda_1\lambda_2\rangle$. We find

$$P_{12}\,|JM\lambda_1\lambda_2\rangle = (-1)^{J-2s}\,|JM\lambda_2\lambda_1\rangle. \tag{67}$$

The appropriately symmetrized states are

$$\{1 + (-1)^{2s}P_{12}\}\,|JM\lambda_1\lambda_2\rangle = |JM\lambda_1\lambda_2\rangle + (-1)^J\,|JM\lambda_2\lambda_1\rangle. \tag{68}$$

We note finally the relation between the customary states $|JMls\rangle$ formed by adding the spins $s_1$, $s_2$ to a resultant $S$ and total angular momentum $\mathbf{J} = \mathbf{l} + \mathbf{S}$. The transformation coefficients are

$$\langle JMls \mid JM\lambda_1\lambda_2\rangle = \left[\frac{2l+1}{2J+1}\right]^{1/2}\langle J\lambda \mid ls0\lambda\rangle\langle s\lambda \mid s_1s_2\lambda_1 - \lambda_2\rangle. \tag{69}$$

We return to the derivation of some of the properties of the $d^J_{M'M}$'s that we have used. First we note that they are, with our standard phase assignments, all real quantities. This follows from the fact that $J_y = (J_+ - J_-)/2i$ and that all the matrix elements of $J_+$, $J_-$ are real. Consider now, with $T$ the time reversal operator,

$$d^J_{M'M}(\beta) = (\mathscr{Y}_{JM'}, e^{-i\beta J_y}\mathscr{Y}_{JM}) = (T\mathscr{Y}_{JM'}, T e^{-i\beta J_y}\mathscr{Y}_{JM})^* \tag{70}$$

$$\begin{aligned}
&= (-1)^{M-M'}(\mathscr{Y}_{J-M'}, e^{-i\beta J_y}\mathscr{Y}_{J-M})^* \\
&= (-1)^{M-M'}(\mathscr{Y}_{J-M}, e^{i\beta J_y}\mathscr{Y}_{J-M'}) \\
&= (\mathscr{Y}_{J-M}, e^{-i\pi J_z}\, e^{i\beta J_y}\, e^{i\pi J_z}\mathscr{Y}_{J-M'}) \\
&= (\mathscr{Y}_{J-M}, e^{-i\beta J_y}\mathscr{Y}_{J-M'}) = d^J_{-M,-M'}(\beta).
\end{aligned} \tag{71}$$

All the steps should be self-evident. Next we use the reality of the $d^J$'s to assert that

$$\begin{aligned}
d^J_{M'M}(\beta) &= [d^J_{M'M}(\beta)]^* = (\mathscr{Y}_{JM'}, e^{-i\beta J_y}\mathscr{Y}_{JM})^* \\
&= (T\mathscr{Y}_{JM'}, T e^{-i\beta J_y}\mathscr{Y}_{JM}) \\
&= (-1)^{M-M'}(\mathscr{Y}_{J-M'}, e^{-i\beta J_y}\mathscr{Y}_{J-M}) \\
&= (-1)^{M-M'}\, d^J_{-M',-M}(\beta) = (-1)^{M-M'}\, d^J_{MM'}(\beta).
\end{aligned} \tag{72}$$

The value of $d^J_{M'M}(\pi)$ was quite important. It is clear physically that since $e^{-i\pi J_y}$ corresponds to a rotation about the $y$-axis by 180° that this operator acting on $\mathscr{Y}_{JM}$ must be proportional to $\mathscr{Y}_{J-M}$. Formally this is checked by noting that

$$e^{i\pi J_y}J_z\, e^{-i\pi J_y} = -J_z \tag{73}$$

so

$$e^{i\pi J_y}J_z\, e^{-i\pi J_y}\mathscr{Y}_{JM} = -J_z\mathscr{Y}_{JM} = -M\mathscr{Y}_{JM}, \tag{74}$$

or

$$J_z(e^{-i\pi J_y}\mathscr{Y}_{JM}) = -M(e^{-i\pi J_y}\mathscr{Y}_{JM}), \tag{75}$$

and thus

$$e^{-i\pi J_y}\mathscr{Y}_{JM} = \alpha(J, M)\mathscr{Y}_{J-M} \tag{76}$$

where $\alpha(J, M)$ is a real number of modulus unity.

Since

$$J_{\pm}e^{-i\pi J_y} = -e^{-i\pi J_y}J_{\mp}, \tag{77}$$

we deduce easily that

$$\alpha(J, M) = -\alpha(J, M - 1). \tag{78}$$

With our choice of phases, $\alpha(J, J) = +1$ and we may write

$$\alpha(J, M) = (-1)^{J-M}. \tag{79}$$

This may be easily checked for spin one-half where $e^{-i\pi J_y} = -i\sigma_y$. With this value of $\alpha$ we have then

$$d^J_{M'M}(\pi) = (-1)^{J-M}\, \delta_{M',-M}. \tag{80}$$

The final relation of this general type is

$$d^J_{M'M}(\beta) = (-1)^{J+M'}\, d^J_{M',-M}(\pi - \beta). \tag{81}$$

To derive it we write

$$\begin{aligned} d^J_{M',-M}(\pi - \beta) &= (\mathscr{Y}_{JM'}, e^{-i(\pi-\beta)J_y}\mathscr{Y}_{J-M}) \\ &= (e^{i\pi J_y}\mathscr{Y}_{JM'}, e^{i\beta J_y}\mathscr{Y}_{J-M}) \\ &= (-1)^{J+M'}(\mathscr{Y}_{J-M'}, e^{i\beta J_y}\mathscr{Y}_{J-M}) \\ &= (-1)^{J+M'}(\mathscr{Y}_{J-M}, e^{-i\beta J_y}\mathscr{Y}_{J-M'})^* \\ &= (-1)^{J+M'}[d^J_{-M,-M'}(\beta)]^* = (-1)^{J+M'}\, d^J_{-M,-M'}(\beta) \\ &= (-1)^{J+M'}\, d^J_{M'M}(\beta), \end{aligned} \tag{82}$$

where we have used the reality of the $d^J$'s and $d^J_{M'M} = d^J_{-M,-M'}$. Since $J' + M$ is integral, this is the desired relation.

The normalization of the $\mathscr{D}$'s is readily deduced. Consider a rotation $R$ through Euler angles $\alpha$, $\beta$, $\gamma$.

$$\begin{aligned} (R\mathscr{Y}_{J'M'}, R\mathscr{Y}_{JM}) = \delta_{M'M}\,\delta_{J'J} &= \sum_{N'N}(\mathscr{D}^{J'}_{N'M'}(\alpha\beta\gamma)\mathscr{Y}_{J'N'}, \mathscr{D}_{NM}{}^{J}(\alpha\beta\gamma)\mathscr{Y}_{JN}) \\ &= \delta_{J'J}\sum_N \mathscr{D}^{J^*}_{NM'}(\alpha\beta\gamma)\, \mathscr{D}^J_{NM}(\alpha\beta\gamma), \end{aligned} \tag{83}$$

or

$$\sum_N \mathscr{D}^{J^*}_{NM'}(\alpha\beta\gamma)\, \mathscr{D}^J_{NM}(\alpha\beta\gamma) = \delta_{M'M}. \tag{84}$$

Since this is true for any rotation $R$, choose it to be $-\gamma\ -\beta\ -\alpha$ and the obvious relation

$$\mathscr{D}^J_{\mu\nu}(-\gamma\ -\beta\ -\alpha) = \mathscr{D}^{J*}_{\nu\mu}(\alpha\beta\gamma) \tag{85}$$

to deduce

$$\sum_N \mathscr{D}^{J*}_{M'N}(\alpha\beta\gamma)\mathscr{D}^J_{MN}(\alpha\beta\gamma) = \delta_{M'M} \tag{86}$$

Now the $\mathscr{D}^J$'s form an $2J + 1$-dimensional irreducible representation of the rotation group and consequently satisfy the orthogonality relations shared by such group theoretical objects. We write this requirement

$$\int dR\, \mathscr{D}^{J'*}_{\mu'\nu'}(R)\, \mathscr{D}^J_{\mu\nu}(R) = \delta_{J'J}\, \delta_{\nu'\nu} g_{\mu'\mu}. \tag{87}$$

To determine $g_{\mu'\mu}$ we set $J' = J$, $\nu' = \nu$ and sum over $\nu$.

$$\int dR \sum_\nu \mathscr{D}^{J*}_{\mu'\nu}(R)\, \mathscr{D}^J_{\mu\nu}(R) = \delta_{\mu'\mu} \int dR = (2J + 1) g_{\mu'\mu}. \tag{88}$$

Since

$$\int dR = \int \sin\beta\, d\beta\, d\alpha\, d\gamma = 8\pi^2,$$

$$g_{\mu'\mu} = \frac{8\pi^2}{2J + 1}\, \delta_{\mu'\mu}. \tag{89}$$

This leads to the relation Eq. (29) used earlier.

$$\int_0^\pi \sin\theta\, d\theta\, d^J_{\mu\nu}(\theta)\, d^{J'}_{\mu\nu}(\theta) = \frac{2\delta_{JJ'}}{2J + 1}. \tag{90}$$

Some special values of the $d$'s are worth recording. We can write a general formula, but it is very involved. For $J = l$, an integer, we have

$$d_{m0}{}^l(\theta) = (-1)^m\, d_{0m}{}^l(\theta) = \left(\frac{4\pi}{2l + 1}\right)^{1/2} e^{-im\phi} Y_l^m(\theta, \phi) \tag{91}$$

or stated otherwise

$$Y_l^m(\phi, \theta) = \left(\frac{2l + 1}{4\pi}\right)^{1/2} [\mathscr{D}_{m0}{}^l(\phi\ \theta 0)]^*. \tag{92}$$

A few other relations are (again for $l$ integral)

$$\begin{aligned}
d_{00}{}^l(\theta) &= P_l(\cos\theta),\\
d_{10}{}^l(\theta) &= -[l(l + 1)]^{-1/2} \sin\theta P'_l(\cos\theta),\\
d_{20}{}^l(\theta) &= [(l - 1)l(l + 1)(l\\
&\qquad + 2)]^{-1/2}[2P'_{l-1}(\cos\theta) - l(l - 1)P_l(\cos\theta)],\\
d_{m1}{}^l(\theta) &= [l(l + 1)]^{-1/2}\left\{-m\left[\frac{(1 + \cos\theta)}{\sin\theta}\right] d_{m0}{}^l(\theta)\right.\\
&\qquad \left. - [(l - m)(l + m + 1)]^{1/2}\, d^l_{m+1,0}(\theta)\right\}.
\end{aligned} \tag{93}$$

For $j = l + \frac{1}{2}$, half integer, we have

$$d^j_{\frac{1}{2},\frac{1}{2}}(\theta) = (l+1)^{-1} \cos\frac{\theta}{2} [P'_{l+1}(\cos\theta) - P'_l(\cos\theta)],$$

$$d^j_{-\frac{1}{2},\frac{1}{2}}(\theta) = (l+1)^{-1} \sin\theta/2 \, [P'_{l+1}(\cos\theta) + P'_l(\cos\theta)]$$

$$d^j_{\frac{1}{2},\frac{3}{2}}(\theta) = (l+1)^{-1} \sin\theta/2 \times \left[\left(\frac{l}{l+2}\right)^{\frac{1}{2}} P'_{l+1}(\cos\theta) + \left(\frac{l+2}{l}\right)^{\frac{1}{2}} P'_l(\cos\theta)\right], \tag{94}$$

$$d^j_{-\frac{1}{2},\frac{3}{2}}(\theta) = (l+1)^{-1} \frac{\cos\theta}{2} \times \left[-\left(\frac{l}{l+2}\right)^{\frac{1}{2}} P'_{l+1}(\cos\theta) + \left(\frac{l+2}{l}\right)^{\frac{1}{2}} P'_l(\cos\theta)\right].$$

These tables are obtained from the following recursion relations:

$$2[(j+\mu)(j+\mu-1)]^{\frac{1}{2}} d_{\lambda\mu}{}^j(\theta) = [(j'+\lambda)(j+\lambda-1)]^{\frac{1}{2}} \times (1+\cos\theta)\, d^{j-1}_{\lambda-1,\mu-1}(\theta) + 2(j^2-\lambda^2)^{\frac{1}{2}} \sin\theta\, d^{j-1}_{\lambda,\mu-1}(\theta) + [(j-\lambda)(j-\lambda-1)]^{\frac{1}{2}}(1-\cos\theta)\, d^{j-1}_{\lambda+1,\mu-1}(\theta). \tag{95}$$

For integral $j = l$ we may start with $d_{m0}{}^l$ and build up. If $j$ is half integral, we begin with $d^j_{\lambda\frac{1}{2}}(\theta)$.

$$d^j_{\lambda\frac{1}{2}} = (j+\tfrac{1}{2})^{-\frac{1}{2}}\left[(j+\lambda)^{\frac{1}{2}} d^{j-\frac{1}{2}}_{\lambda-\frac{1}{2},0}(\theta) \cos\frac{\theta}{2} - (j-\lambda)^{\frac{1}{2}} d^{j-\frac{1}{2}}_{\lambda+\frac{1}{2},0}(\theta) \frac{\sin\theta}{2}\right]. \tag{96}$$

It is perhaps useful to discuss the derivation of our scattering formula in terms of wave functions. As we shall see, the argument is largely a repetition of what has already been done.

Suppose we have a function of space and spin $F_K(\mathbf{r}, \sigma)$ which, although it does not have a well-defined angular momentum, does have a specified projection $J_z = K$. A typical example is that of a plane wave in the $z$-direction times a spin function with projection $K$ along the $z$-axis. Then the projection of such a function which has a definite angular momentum $J$ and $z$-projection $M$ is given by

$$\psi_{MK}{}^J(\mathbf{r}, \sigma) = \frac{2J+1}{8\pi^2} \int dR \mathscr{D}_{MK}{}^{*J}(R) R_{\alpha\beta\gamma} F_K(\mathbf{r}, \sigma). \tag{97}$$

Except for a change in normalization, this is just Eq. (22). To verify that this function has the desired property, imagine writing

$$F_K(\mathbf{r}, \sigma) = \sum_J a_J \Phi_{JK}(\mathbf{r}, \sigma), \tag{98}$$

where the $\Phi_{JK}(\mathbf{r}, \sigma)$ are eigenfunctions of $J^2$, $J_z$; only the projection $K$ can enter because we assumed $J_zF_K = KF_K$. Now we know that

$$R_{\alpha\beta\gamma}\Phi_{JK} = \sum_L \Phi_{JL}\mathscr{D}_{LK}{}^J(R). \tag{99}$$

Using the orthogonality relation satisfied by the $D$'s namely

$$\frac{2J+1}{8\pi^2}\int dR\mathscr{D}_{MK}{}^{J*}(R)\, D_{LK}{}^{J'}(R) = \delta_{JJ'}\,\delta_{ML}, \tag{100}$$

we find immediately

$$\psi_{MK}{}^J(\mathbf{r}, \sigma) = a_J\Phi_{JM}(\mathbf{r}, \sigma), \tag{101}$$

independent of $K$. Evidently,

$$F_K(\mathbf{r}, \sigma) = \sum_{JM} \delta_{MK}\psi_{MK}{}^J(\mathbf{r}, \sigma). \tag{102}$$

Consider the case of a plane wave moving along $z$ times a spin function $u_\lambda$ with $z$-projection $\lambda$. As before, for a two-particle scattering in the center-of-mass system we have $\lambda = \lambda_1 - \lambda_2$. A rotation characterized by angle $\alpha$, $\beta$, 0 has the following effect on $e^{ikz}$:

$$R_{\alpha\beta 0}\, e^{ikz} = e^{ik\hat{\Omega}\cdot\mathbf{r}}, \tag{103}$$

where $\hat{\Omega}$ is the unit vector with polar angles $\beta$, $\alpha$. To verify this we write

$$\begin{aligned} R_{\alpha\beta 0}\, e^{ikr\cos\theta} &= R_{\alpha\beta 0}\sum_l (2l+1)i^l j'_l(kr)P_l(\cos\theta) \\ &= R_{\alpha\beta 0}\sum_l \sqrt{4\pi(2l+1)}\,i^l j_l(kr)Y_l{}^0(\hat{\mathbf{r}}) \\ &= \sum_{lm}\sqrt{4\pi(2l+1)}\,i^l j_l(kr)Y_l{}^m(\hat{\mathbf{r}})\,\mathscr{D}_{m0}{}^l(\alpha, \beta, 0) \\ &= 4\pi\sum_{lm} i^l j_l(kr)Y_l{}^m(\hat{\mathbf{r}})Y_l{}^{*m}(\hat{\Omega}) \\ &= \sum_l (2l+1)i^l j_l(kr)P_l(\hat{\Omega}\cdot\hat{\mathbf{r}}) = e^{ik\hat{\Omega}\cdot\mathbf{r}}. \end{aligned} \tag{104}$$

In deriving this result we have used the fact that

$$\mathscr{D}_{m0}{}^l(\alpha, \beta, 0) = \sqrt{4\pi/(2l+1)}\,Y_l{}^{*m}(\beta, \alpha) = \sqrt{4\pi/(2l+1)}\,Y_l{}^m(\hat{\Omega}) \tag{105}$$

and the addition theorem for spherical harmonics

$$\frac{2l+1}{4\pi}\,P_l(\hat{\Omega}\cdot\hat{\mathbf{r}}) = \sum_m Y_l{}^m(\hat{\mathbf{r}})Y_l{}^{*m}(\hat{\Omega}). \tag{106}$$

We would like to use this result to discuss the asymptotic form of the $\psi_{M\lambda}{}^J(\mathbf{r}, \sigma)$ which results from the angular momentum decomposition of the plane wave. We write

$$\psi_{M\nu}{}^{\lambda J} = \frac{2J+1}{8\pi^2}\int dR\,\mathscr{D}_{M\nu}{}^{J^*}(R)R_{\alpha\beta\gamma}(e^{ikz}u_\lambda)$$

$$= \delta_{\nu\lambda}\psi_M{}^{\lambda J} \tag{107}$$

since only $\nu = \lambda$ contributes as is seen by carrying out the integral over $\gamma$. Now suppose $\mathbf{r}$ is very large in some definite direction. Then,

$$\psi_M{}^{\lambda J} = \frac{2J+1}{8\pi^2}\int dR\,\mathscr{D}_{M\lambda}{}^{J*}(R)\,e^{ik\hat{\boldsymbol{\Omega}}\cdot\mathbf{r}}(R_{\alpha\beta\gamma}u_\lambda); \tag{108}$$

we integrate over the Euler angles by choosing $\mathbf{r}$ as the polar axis. Only those directions $\hat{\boldsymbol{\Omega}}$ parallel or antiparallel to $\mathbf{r}$ contribute for large $r$. This is a relation we have used repeatedly (see Eq. 3.173) and we find

$$\psi_n{}^{\lambda J} \to \frac{2J+1}{2ikr}[\mathscr{D}_{n\lambda}{}^{J*}(\phi,\theta,-\phi)(R_{\phi,\,-\phi}u_\lambda)\,e^{ikr}$$

$$-\mathscr{D}_{n\lambda}{}^{J*}(\pi+\phi,\pi-\theta,-\pi-\phi)(R_{\pi+\phi,\pi-\theta,-\pi-\phi}u_\lambda)\,e^{-ikr}], \tag{109}$$

where $\theta,\phi$ and the polar angles of the asymptotic direction of $\mathbf{r}$.

In the presence of interaction the ingoing and outgoing parts of the waves have different amplitudes, the coefficient of the outgoing wave being simply the $S$-matrix. This matrix element is, of course, independent of $M$, but it depends on the helicities $\lambda$. If $B_{M\mu}{}^J$ is the outgoing amplitude and $A_{M\lambda}{}^J$ is the incoming amplitude, we have quite generally

$$B_{M\mu}{}^J = \sum \langle\mu|\,\mathbf{S}^J\,|\lambda\rangle A_{M\nu}{}^J. \tag{110}$$

Since the plane wave is given by

$$e^{ikz}u_\lambda = \sum_J \psi_\lambda{}^{\lambda J}, \tag{111}$$

we see that

$$A_{M\nu}{}^J = \delta_{\nu\lambda}\,\delta_{M\lambda}. \tag{112}$$

The asymptotic form of the total wave function which is

$$\psi = \sum_{JM\mu}\psi_M{}^{\mu J} \tag{113}$$

becomes

$$\psi \to e^{ikz}u_\lambda + \frac{e^{ikr}}{r}\sum_{J,\mu}\frac{2J+1}{2ik}\{\langle\mu|\,S^J\,|\lambda\rangle - \delta_{\mu\lambda}\},\,\mathscr{D}_{\lambda\mu}{}^{J*}(\varphi,\theta,-\varphi)$$

$$\times\,(R_{\varphi,\theta,-\varphi}\,u_\mu). \tag{114}$$

Selecting a particular final helicity state, we find for the scattering amplitude $f_{\mu;\lambda}(\theta,\phi)$ the same expression given earlier (for simple scattering).

$$f_{\mu;\lambda}(\theta,\phi) = \sum_J\frac{2J+1}{2ik}\{\langle\mu|\,\mathbf{S}^J\,|\lambda\rangle - \delta_{\mu\lambda}\}\,\mathscr{D}_{\lambda\mu}{}^{J*}(\phi,\theta,-\phi). \tag{115}$$

# APPENDIX F EVALUATION OF THE AMPLITUDE FOR NUCLEON-NUCLEON SCATTERING IN TERMS OF PHASE SHIFTS

The expressions (7.218) may be obtained in a straightforward manner using the vector addition coefficients. They may also be found, as we shall now describe, using operator techniques such as those developed in Section 7.2.

The scattering matrix for singlet states is easily obtained from Eq. (7.27). We have

$$T_{s,s} = \langle \hat{\varkappa}_f, 0, 0 \,|T|\, \hat{\varkappa}, 0, 0\rangle = \sum_{J,m} Y_J{}^m(\hat{\varkappa}_f) T^J_{J0,J0} Y_J{}^{m*}(\hat{\varkappa})$$

$$= \sum_J \frac{(2J+1)}{4\pi} T^J_{J0,J0} P_J(\hat{\varkappa}_f \cdot \hat{\varkappa}). \tag{1}$$

For the triplet states, we use Eq. (7.27) to write

$$\begin{aligned}
\langle \hat{\varkappa}_f, 1, \nu' \,|T|\, \hat{\varkappa}, 1, \nu\rangle
&= \sum_{J,M,m',m} \{Y_J{}^{m'}(\hat{\varkappa}_f)\langle J, 1; m', \nu' \mid J, M\rangle T^J_{J1,J1} \\
&\qquad \times \langle J, 1; m, \nu \mid J, M\rangle Y_J{}^{m*}(\hat{\varkappa}) \\
&\quad + Y^{m'}_{J+1}(\hat{\varkappa}_f)\langle J+1, 1; m', \nu' \mid J, M\rangle T^J_{J+1\,1,J+1\,1} \\
&\qquad \times \langle J+1, 1; m, \nu \mid J, M\rangle Y^{m*}_{J+1}(\hat{\varkappa}) \\
&\quad + Y^{m'}_{J-1}(\hat{\varkappa}_f)\langle J-1, 1; m', \nu' \mid J, M\rangle T^J_{J-1\,1,J-1\,1} \\
&\qquad \times \langle J-1, 1; m, \nu \mid J, M\rangle Y^{m*}_{J-1}(\hat{\varkappa}) \\
&\quad + Y^{m'}_{J+1}(\hat{\varkappa}_f)\langle J+1, 1; m', \nu' \mid J, M\rangle T^J_{J+1,1,J-1,1} \\
&\qquad \times \langle J-1, 1; m, \nu \mid J, M\rangle Y^{m*}_{J-1}(\hat{\varkappa}) \\
&\quad + Y^{m'}_{J-1}(\hat{\varkappa}_f)\langle J-1, 1; m', \nu' \mid J, M\rangle T^J_{J-1,1,J+1,1} \\
&\qquad \times \langle J+1, 1; m, \nu \mid J, M\rangle Y^{m*}_{J+1}(\hat{\varkappa})\}.
\end{aligned} \tag{2}$$

The term on line one of the right here corresponds to $l = J$. This may be written in terms of the projection operator onto the state corresponding to $l = J$;

$$\Lambda_{J,J} = -\frac{\Sigma^2 + \Sigma - J(J+1)}{J(J+1)}, \tag{3}$$

where $\Sigma \equiv \mathbf{S}\cdot\mathbf{L} = \frac{1}{2}(\boldsymbol{\sigma}_1 + \boldsymbol{\sigma}_2)\cdot\mathbf{L}$. Now [see Eq. (7.71)]

$$\sum_M \langle J, 1; m', \nu' \mid J, M\rangle\langle J, 1; m, \nu \mid J, M\rangle = \langle J1m'\nu'|\, \Lambda_{J,J}\, |J1m\nu\rangle$$
$$= \left(u_{1,\nu'}, \int Y_J^{m'}(\hat{\mathbf{q}})^* \Lambda_{J,J} Y_J^{m}(\hat{\mathbf{q}})\, d\Omega_{\hat{q}} u_{1,\nu}\right). \tag{4}$$

The term on line one of the right side of Eq. (2) is therefore

$$\text{“line one”} = \left(u_{1,\nu'}, \sum_{J'Jm'm} \int Y_{J'}^{m'}(\hat{\boldsymbol{\varkappa}}_f) Y_{J'}^{m'*}(\hat{\mathbf{q}})\right.$$
$$\left. \times\, \Lambda_{J,J} Y_J^{m}(\hat{\mathbf{q}}) Y_J^{m*}(\hat{\boldsymbol{\varkappa}})\, d\Omega_{\hat{q}} u_{1,\nu}\right) T^J_{J1,J1}$$
$$= \frac{1}{4\pi}\Big(u_{1,\nu'}, \sum_J (2J+1) T^J_{J1,J1} \Lambda_{J,J} P_J(\hat{\boldsymbol{\varkappa}}\cdot\hat{\boldsymbol{\varkappa}}_f) u_{1,\nu}\Big), \tag{5}$$

where the $\mathbf{L}$ in $\Lambda_{J,J}$ is $i\boldsymbol{\varkappa} \times \nabla_{\hat{\varkappa}}$ and thus $\Lambda_{J,J}$ acts only on the spin variables and on the unit vector $\hat{\boldsymbol{\varkappa}}$. The final simplification of Eq. (5) is effected on writing

$$\Sigma = i\mathbf{S}\cdot\hat{\mathbf{n}} \sin\theta \frac{\partial}{\partial \cos\theta}. \tag{6}$$

Here, as in Chapter 7, $\hat{\mathbf{n}} \sin\theta = \hat{\boldsymbol{\varkappa}} \times \hat{\boldsymbol{\varkappa}}_f$ and $\cos\theta = \hat{\boldsymbol{\varkappa}}\cdot\hat{\boldsymbol{\varkappa}}_f$. Substitution of Eq. (6) into (3) and the result into Eq. (5) permits us to evaluate (5), just as was done in connection with Eq. (7.73).

The second and third lines of the right side of Eq. (2) are evaluated in the same manner, using the projection operators

$$\Lambda_{J,l} = \frac{\Sigma^2 - J\Sigma - J - 1}{(J+1)(2J+1)} \qquad \text{for } l = J+1$$
$$= \frac{\Sigma^2 + (J+1)\Sigma + J}{J(2J-1)} \qquad \text{for } l = J-1. \tag{7}$$

The last two terms in Eq. (2) are more difficult to simplify. To handle these we need an operator, analogous to that of Eq. (7.105), which will change the orbital angular momentum by two units. The expression (7.168) suggests that we begin by evaluating the matrix elements of the operator

$$s \equiv \boldsymbol{\sigma}_1\cdot\hat{\boldsymbol{\varkappa}}\boldsymbol{\sigma}_2\cdot\hat{\boldsymbol{\varkappa}}. \tag{8}$$

It is evident that $s^\dagger = s$ and that $s^2 = 1$.

Now, the wave functions

$$\Phi_J{}^M(l) \equiv i^l \mathscr{Y}^M_{Jl1} \tag{9}$$

[see Eq. (7.165)] are eigenfunctions of the operators $\mathbf{J}^2$ and $J_z$ and they satisfy the relations

$$T\Phi_J{}^M(l) = i^{J+M}\Phi_J{}^{-M}(l), \tag{10}$$

where $T$ is the time reversal operator. Letting $s$ operate on the $\Phi_J{}^M(l)$'s leads to equations of the form

$$\begin{aligned} s\Phi_J{}^M(J-1) &= a\Phi_J{}^M(J-1) + b\Phi_J{}^M(J+1), \\ s\Phi_J{}^M(J+1) &= b'\Phi_J{}^M(J-1) + a'\Phi_J{}^M(J+1), \end{aligned} \tag{11}$$

where the constants $a$, $b$, $a'$, $b'$ do not depend on $M$. [This follows since $s$ commutes with the angular momentum operator $\mathbf{J}$ and thus with the raising and lowering operators $J^{\pm} = J_x \pm iJ_y$ (see Eqs. (1.63)).] Because $s$ is Hermitean, it follows that $b' = b^*$. Application of the time reversal operator to Eqs. (11), with use of Eq. (10), shows that $a$, $b$, $a'$ and $b' = b$ are all real.

The coefficient $a$ may be evaluated from the relation

$$\begin{aligned} a &= (\Phi_J{}^M(J-1), s\Phi_J{}^M(J-1)) \\ &= \frac{1}{2J+1}\sum_M (\Phi_J{}^M(J-1), s\Phi_J{}^M(J-1)) \\ &= \frac{1}{2J+1}\,\mathrm{Tr}\,\{s\Lambda_{J,J-1}\}. \end{aligned} \tag{12}$$

To evaluate the trace here, we write

$$s = 2(\mathbf{S}\cdot\hat{\boldsymbol{\varkappa}})^2 - 1 \tag{13}$$

and use the general relations valid for any angular momentum operator $\mathbf{J}$($\alpha$, $\beta$, $\gamma$, and $\delta$ are vector indices running from one to three)

$$\begin{aligned} \mathrm{Tr}\,[J_\alpha] &= 0, \\ \mathrm{Tr}\,[J_\alpha J_\beta] &= \delta_{\alpha,\beta}\frac{J(J+1)(2J+1)}{3}, \\ \mathrm{Tr}\,[J_\alpha J_\beta J_\gamma] &= \epsilon_{\alpha\beta\gamma}\frac{J(J+1)(2J+1)}{6}, \\ \mathrm{Tr}\,[J_\alpha J_\beta J_\gamma J_\delta] &= \frac{J(J+1)(2J+1)}{15}\{[J(J+1)+\tfrac{1}{2}] \\ &\quad\times(\delta_{\alpha,\beta}\,\delta_{\gamma,\delta} + \delta_{\alpha,\delta}\,\delta_{\beta,\gamma}) + (J-1)(J+2)\,\delta_{\alpha,\gamma}\,\delta_{\beta,\delta}\}. \end{aligned} \tag{14}$$

[The quantity $\epsilon_{\alpha\beta\gamma}$ has been defined in connection with Eq. (2.35).] The structure of Eqs. (14) is fairly obvious; the numerical coefficients may be obtained by contracting indices and using the commutation relations for angular momenta.[1]

Evaluation of the trace in Eq. (12) is now straightforward. We use the expressions (7) and (13) and recall that we are working in the $\hat{\boldsymbol{\varkappa}}$ coordinate system, so that $i\hat{\boldsymbol{\varkappa}} \cdot \mathbf{L} = \hat{\boldsymbol{\varkappa}} \cdot \boldsymbol{\varkappa} \times \nabla_{\varkappa} = 0$. Then, taking a partial trace over spin states and identifying the $J_\alpha$ in Eqs. (14) with the components of $\mathbf{S}$, we obtain

$$\text{Tr (spins)}\, [(\mathbf{S} \cdot \hat{\boldsymbol{\varkappa}})^2(\mathbf{S} \cdot \mathbf{L})^2] = L^2 = l(l + 1) = J(J - 1), \tag{15}$$

and so on. From such relations, we are able to evaluate Eq. (12) to find

$$a = \frac{1}{2J + 1}. \tag{16}$$

Now, since $s^2 = 1$, we have $a^2 + b^2 = 1$, or

$$b = \pm 2\, \frac{[J(J + 1)]^{1/2}}{2J + 1}. \tag{17}$$

Our phase convention (9) requires that we use the *negative* sign here.[2] In a manner similar to that by which Eq. (16) was obtained, we are led to

$$a' = -\frac{1}{2J + 1}. \tag{18}$$

Having found $a$, $b$, $a'$ and $b'$, we can rewrite Eqs. (11) as

$$\Phi_J{}^M(J + 1) = -\frac{2J + 1}{2\sqrt{J(J + 1)}}\left(\boldsymbol{\sigma}_1 \cdot \hat{\boldsymbol{\varkappa}}\boldsymbol{\sigma}_2 \cdot \hat{\boldsymbol{\varkappa}} - \frac{1}{2J + 1}\right)\Phi_J{}^M(J - 1),$$

$$\Phi_J{}^M(J - 1) = -\frac{2J + 1}{2\sqrt{J(J + 1)}}\left(\boldsymbol{\sigma}_1 \cdot \hat{\boldsymbol{\varkappa}}\boldsymbol{\sigma}_2 \cdot \hat{\boldsymbol{\varkappa}} + \frac{1}{2J + 1}\right)\Phi_J{}^M(J + 1). \tag{19}$$

[The matrix elements of the tensor operator $S_{12}$, which were required in Section 7.4, may be trivially found using Eqs. (19).]

The remaining task to obtain Eqs. (7.218) is straightforward, but tedious. The projection operators (7) may be multiplied by those in Eqs. (19) to give the operators needed for the last two lines in Eq. (2).

[1] The general expression for the trace of the product of an arbitrary number of components of **J** does not seem to be known.

[2] This differs from the choice usually made in the literature. Our choice here is consistent with that given in Section 7.4 for the matrix elements of the tensor interaction $S_{12}$.

# APPENDIX G.1 NONRELATIVISTIC DISPERSION RELATIONS FOR COMBINATIONS OF ORDINARY AND EXCHANGE POTENTIALS

We turn to the derivation of the double dispersion representation and single variable dispersion relations for both ordinary and space exchange potentials. We write

$$V(\mathbf{x}) = V_1(x) + V_2(x)P_{\mathbf{x}} \equiv V_e(x) + V_o(x), \tag{1}$$

where $P_{\mathbf{x}}$ is the exchange operator defined by $\langle \mathbf{x}'| P_{\mathbf{x}} |\mathbf{x}''\rangle = \delta(\mathbf{x}' + \mathbf{x}'')$. We call $V_e = V_1$ the even potential and $V_2P_{\mathbf{x}} = V_o$ the odd potential.

It is easy to see that the Green's function $G(\mathbf{x}, \mathbf{y}, E)$ used in Chapter 10 satisfies the equation

$$G(\mathbf{x}, \mathbf{y}, E) = g(\mathbf{x}, \mathbf{y}, E) + \int d^3z\, g(\mathbf{x}, \mathbf{z}, E)[V_e(z) + V_0(z)]G(\mathbf{z}, \mathbf{y}, E), \tag{2}$$

where

$$g(\mathbf{x}, \mathbf{y}, E) = -\frac{1}{4\pi}\frac{e^{ik\,|\mathbf{x}-\mathbf{y}|}}{|\mathbf{x} - \mathbf{y}|},$$

$$k = \sqrt{E} \quad \text{in our units.} \tag{3}$$

Equation (3) is just the coordinate space representation of the operator relation (5.19).

$$\frac{1}{E + i\eta - H} = \frac{1}{E + i\eta - K} + \frac{1}{E + i\eta - K}V\frac{1}{E + i\eta - H}. \tag{4}$$

In an obvious operator notation, we rewrite (2) as $G = g + g(V_e + V_o)G$ and define an operator $R$ by

$$R = (V_e + V_o)G. \tag{5}$$

Multiplying Eq. (2) by $V_e + V_o$, we find that $R$ satisfies the equation

$$R = (L_e + L_o)(1 + R), \tag{6}$$

where

$$L_e = V_e g, \qquad L_o = V_o g. \tag{7}$$

Quite explicitly,

$$\langle \mathbf{x}|\, L_e\, |\mathbf{x}'\rangle = V_1(\mathbf{x})g(\mathbf{x}, \mathbf{x}', E),$$
$$\langle \mathbf{x}|\, L_o\, |\mathbf{x}'\rangle = V_2(\mathbf{x})g(-\mathbf{x}, \mathbf{x}', E). \tag{8}$$

Next write $R = R_e + R_o$, where $R_e$ and $R_o$ satisfy

$$R_e = L_e + L_eR_e + L_oR_o,$$
$$R_o = L_o + L_eR_o + L_oR_e. \tag{9}$$

A formal solution of these coupled equations is easily obtained in terms of "even" and "odd" operators $\mathscr{L}_e$, $\mathscr{L}_o$ defined by

$$\mathscr{L}_e = L_e + L_o \frac{1}{1 - L_e} L_o,$$

$$\mathscr{L}_0 = L_o + L_o \frac{1}{1 - L_e} L_e. \tag{10}$$

We find easily

$$R_e = \mathscr{L}_e + \mathscr{L}_e \frac{1}{1 - \mathscr{L}_e} \mathscr{L}_e,$$

$$R_o = \mathscr{L}_o + \mathscr{L}_o \frac{1}{1 - \mathscr{L}_e} \mathscr{L}_o. \tag{11}$$

The structure of these equations is easy to understand. If every term were expanded into powers of $V_e$, $V_o$, we would find that $R_e$ contains $V_o$ an even number of times, whereas $V_o$ contains it as an odd number. All the elements in (11) share the analyticity enjoyed by $g$, which is analyticity in the upper-half $k$-plane or in the whole $E$-plane cut in our usual fashion. Actually, we must project out the bound states before this statement is precise. It is also not hard to show that the Born series solutions converge in the whole $E$-plane and that for sufficiently large $E$

$$R_e \to V_1(\mathbf{x}) \frac{e^{ik\,|\mathbf{x}-\mathbf{y}|}}{|\mathbf{x} - \mathbf{y}|}, \qquad R_o = V_2(\mathbf{x}) \frac{e^{ik\,|\mathbf{x}+\mathbf{y}|}}{|\mathbf{x} + \mathbf{y}|}. \tag{12}$$

These conclusions may be put on a firm basis by using the Fredholm solutions to Eqs. (9).

The scattering amplitude becomes

$$f(\mathbf{k}', \mathbf{k}) = -\frac{\tilde{V}_e(\mathbf{k}', \mathbf{k})}{4\pi} - \frac{\tilde{V}_o(\mathbf{k}', \mathbf{k})}{4\pi}$$
$$+ \int d^3x \int d^3y\, e^{-i\mathbf{k}'\cdot\mathbf{x}} R(\mathbf{x}, \mathbf{y}, E)[V_e(y) + V_o(y)]e^{i\mathbf{k}\cdot\mathbf{y}} \tag{13}$$

and by analogy with Eq. (9) we write in abbreviated form $f = f_e + f_o$ with

$$f_e = -\frac{\tilde{V}_e(\mathbf{k}', \mathbf{k})}{4\pi} + \langle \mathbf{k}'|\, R_e V_e + R_o V_o \,|\mathbf{k}\rangle,$$

$$f_o = -\frac{\tilde{V}_o(\mathbf{k}', \mathbf{k})}{4\pi} + \langle \mathbf{k}'|\, R_o V_e + R_e V_o \,|\mathbf{k}\rangle. \tag{14}$$

It is clear that because of Eq. (12) we can easily derive dispersion relations in the energy just as we did in Chapter 10. For $f_e$, we hold fixed $-t = +(\mathbf{k} - \mathbf{k}')^2 \leq 4\mu_0^2$, whereas for $f_o$, we fix $-\bar{t} = (\mathbf{k} + \mathbf{k}')^2 \leq 4\mu_0^2$. Assuming no bound states we may write

$$f_e(s, t) = -\frac{\tilde{V}_e(t)}{4\pi} + \int_0^\infty \frac{ds'}{\pi} \frac{\operatorname{Im} f_e(s', t)}{s' - s - i\eta},$$

$$f_o(s, t) = -\frac{\tilde{V}_o(\bar{t})}{4\pi} + \int_0^\infty \frac{ds'}{\pi} \frac{\operatorname{Im} f_o(s', \bar{t})}{s' - s - i\eta}. \tag{15}$$

There is no question of subtraction here again because of Eqs. (12).

We turn next to the question of the analyticity in $t$, $\bar{t}$, respectively, of $f_e$ and $f_o$. Substitution of the integral equations for $R_e$ and $R_o$, Eqs. (9) into Eq. (14), leads to

$$f_e(\mathbf{k}', \mathbf{k}) = -\frac{\tilde{V}_e(\mathbf{k}', \mathbf{k})}{4\pi} + \int d^3p[L_e(\mathbf{k}', \mathbf{p}) f_e(\mathbf{p}, \mathbf{k}) + \bar{L}_o(\mathbf{k}', \mathbf{p}) f_o(\mathbf{p}, \mathbf{k})],$$

$$f_o(\mathbf{k}', \mathbf{k}) = -\frac{\tilde{V}_o(\mathbf{k}', \mathbf{k})}{4\pi} + \int d^3p[\bar{L}_e(\mathbf{k}', \mathbf{p}) f_o(\mathbf{p}, \mathbf{k}) + \bar{L}_o(\mathbf{k}'\mathbf{p}) f_e(\mathbf{p}, \mathbf{k})], \tag{16}$$

where

$$\bar{L}_e(\mathbf{k}'\mathbf{p}) = \frac{1}{(2\pi)^3} \frac{\tilde{V}_e(\mathbf{k}', \mathbf{p})}{p^2 - k^2 - i\eta}, \quad \text{and so on.} \tag{17}$$

Assuming that $V_e$ and $V_o$ are again superpositions of Yukawa potentials,

$$\frac{\tilde{V}_e}{4\pi}(\mathbf{p}_1, \mathbf{p}_2) = \int_{\mu_0}^\infty d\mu \frac{\sigma_e(\mu)}{\mu^2 + (\mathbf{p}_1 - \mathbf{p}_2)^2},$$

$$\frac{\tilde{V}_0}{4\pi}(\mathbf{p}_1, \mathbf{p}_2) = \int_{\mu_0}^\infty d\mu \frac{\sigma_0(\mu)}{\mu^2 + (\mathbf{p}_1 + \mathbf{p}_2)^2}. \tag{18}$$

We may again show as in Chapter 10 that

$$f_e(s, \cos\theta) + \frac{\tilde{V}_e(\mathbf{k}', \mathbf{k})}{4\pi} = \int_{\zeta_0}^\infty d\zeta \frac{\phi_e(\zeta, E)}{\zeta - \cos\theta},$$

$$f_o(s, \cos\theta) + \frac{\tilde{V}_o(\mathbf{k}', \mathbf{k})}{4\pi} = \int_{\zeta_0}^\infty d\zeta \frac{\phi_o(\zeta, E)}{\zeta + \cos\theta}; \tag{19}$$

# APPENDIX G.2 SOLUTION OF A CLASS OF INTEGRAL EQUATIONS

In Chapter 10 we encountered an integral equation for which the solution was simply stated. In this Appendix we shall derive this and also treat some slightly more general equations.

We begin with

$$J(\xi) = g - \frac{\xi + \mu^2}{\pi}\int_{4\mu^2}^{\infty} d\xi' \frac{\text{Re}\,[J(-\xi')f_o^*(-\xi')]}{(\xi' - \mu^2)(\xi' + \xi - i\eta)}, \tag{1}$$

where

$$f_o = e^{i\delta}\sin\delta, \qquad \delta = \delta\left[\left(-\frac{\xi}{4} - \mu^2\right)^{1/2}\right], \tag{2}$$

and $\delta$ need not be real. If we take the imaginary part of (2), we find

$$\text{Im}\,J(\xi) = \tan\phi(-\xi)\,\text{Re}\,J(\xi)\theta\left(-\xi - \frac{\mu^2}{4}\right), \tag{3}$$

where

$$\tan\phi(-\xi) = \frac{\text{Re}\,e^{i\delta}\sin\delta}{1 - \text{Im}\,e^{i\delta}\sin\delta}. \tag{4}$$

Thus the phase on the cut is $\phi(-\xi)$ to within a multiple of $\pi$. Clearly, (1) may be written with the aid of (3) as

$$J(\xi) = g - \frac{\xi + \mu^2}{\pi}\int_{4\mu^2}^{\infty} d\xi' \frac{\tan\phi(\xi')\,\text{Re}\,J(-\xi')}{(\xi' - \mu^2)(\xi' + \xi - i\eta)}. \tag{5}$$

Now define a function of a complex variable, $\xi$, $F(\xi)$ by

$$F(\xi) = -\frac{(\xi + \mu^2)}{2\pi i}\int_{4\mu^2}^{\infty} d\xi' \frac{\tan\phi(\xi')\,\text{Re}\,J(-\xi')}{(\xi' - \mu^2)(\xi' + \xi)}. \tag{6}$$

Then with $F_{\pm}(\xi) \equiv F(\xi \pm i\eta)$ for real $\xi$, we have from (5)

$$J(\xi) = g + 2iF_{-}(\xi), \tag{7}$$

and

$$F_{-}(\xi) - F_{+}(\xi) = -\tan\phi(-\xi)\,\text{Re}\,J(\xi)\theta(-\xi - 4\mu^2). \tag{8}$$

Note also

$$\operatorname{Re} J(\xi) = g + i[F_-(\xi) + F_+(\xi)]. \tag{9}$$

With the understanding that $\tan \phi(-\xi) = 0$ unless $\xi < -4\mu^2$, we drop the theta-function and write

$$F_-(\xi) - F_+(\xi) = \tan \phi(-\xi)\{g + i[F_-(\xi) + F_+(\xi)]\}, \tag{10}$$

or

$$F_+(\xi)\, e^{2i\phi(-\xi)} - F_-(\xi) = -g\, e^{i\phi(-\xi)} \sin \phi(-\xi). \tag{11}$$

Now write

$$F(\xi) = e^{-\Delta(\xi)}\chi(\xi), \tag{12}$$

where

$$\Delta(\xi) = \frac{\xi + \mu^2}{\pi}\int_{4\mu^2}^{\infty} d\xi' \,\frac{\phi(\xi')}{(\xi' - \mu^2)(\xi' + \xi)},$$

$$\Delta_{\pm}(\xi) = \pm i\phi(-\xi) + p(\xi),$$

$$p(\xi) = \frac{\xi + \mu^2}{\pi}\, P\!\int_{4\mu^2}^{\infty} d\xi' \,\frac{\phi(\xi')}{(\xi' - \mu^2)(\xi' + \xi)}. \tag{13}$$

Substituting (12) into (11), we obtain

$$\begin{aligned} \chi_+(\xi) - \chi_-(\xi) &= -g\, e^{p(\xi)} \sin \phi(-\xi) \\ &= -\frac{g}{2i}\,[e^{\Delta_+(\xi)} - e^{\Delta_-(\xi)}]. \end{aligned} \tag{14}$$

Now since $F(-\mu^2) = 0$ and $\Delta(-\mu^2) = 0$, we may satisfy (14) with the choice

$$\chi(\xi) = -\frac{g}{2i}\,[e^{\Delta(\xi)} - 1]. \tag{15}$$

Clearly, in place of the one in (15) we could have any entire function of $\xi$ which had the property of reducing to unity at $\xi = -\mu^2$. The ultimate decision on this point depends on the behavior of $F$ at infinity, which in turn depends on $\phi(\xi)$. We obtain for $J(\xi)$,

$$\begin{aligned} J(\xi) &= g + 2iF_- = g\, e^{-\Delta_-(\xi)} \\ &= g \exp\left[\frac{-(\xi + \mu^2)}{\pi}\int_{4\mu^2}^{\infty} d\xi' \,\frac{\phi(\xi')}{(\xi' - \mu^2)(\xi' + \xi - i\eta)}\right]. \end{aligned} \tag{16}$$

There are two other equations of this general variety frequently encountered. Actually, the second is a special case of the first which is conveniently solved by another technique. We consider first

$$f(x) = g(x) + \frac{1}{\pi}\int_{x_0}^{\infty} dx' \,\frac{e^{-i\delta(x')} \sin \delta(x') f(x')}{x' - x - i\eta}, \tag{17}$$

where $\delta(x)$ is a real function. In the special case that $g(x)$ has no cut along the positive $x$-axis from $x_0$ to $+\infty$, the phase of $f(x)$ along the cut is evidently equal to $\delta(x)$. We define a function of a complex variable, $z$, $F(z)$ by

$$F(z) = \frac{1}{2\pi i}\int_{x_0}^{\infty} dx' \, \frac{e^{-i\delta(x')} \sin \delta(x') F(x')}{x' - z} \tag{18}$$

and calling $F_{\pm}(x) = F(z = x \pm i\eta)$, we have

$$\begin{aligned} F_+(x) - F_-(x) &= e^{-i\delta(x)} \sin \delta(x) f(x) \theta(x - x_0) \\ &= e^{-i\delta(x)} \sin \delta(x)[g(x) + 2iF_+(x)]\theta(x - x_0), \end{aligned} \tag{19}$$

or

$$F_+(x)\, e^{-2i\delta(x)} - F_-(x) = g(x)\, e^{-i\delta(x)} \sin \delta(x) \theta(x - x_0). \tag{20}$$

Writing

$$\begin{aligned} F(z) &= e^{\Delta(z)} \chi(z), \\ \Delta(z) &= \frac{1}{\pi}\int_{x_0}^{\infty} dx' \, \frac{\delta(x')}{x' - z}, \\ \Delta_{\pm}(x) &= \pm i\, \delta(x) + p(x), \\ p(x) &= \frac{P}{\pi}\int_{x_0}^{\infty} dx' \, \frac{\delta(x')}{x' - x}, \end{aligned} \tag{21}$$

we find

$$\chi_+(x) - \chi_-(x) = -\frac{g(x)}{2i}[e^{-\Delta_+(x)} - e^{-\Delta_-(x)}]\theta(x - x_0). \tag{21}$$

Now $F(z)$ by construction is analytic in the complex $z$-plane except for a cut from $x_0$ to $+\infty$ along the real axis. Also $\Delta(z)$ has the same characteristics, and hence so does $\chi$. Aside from the question of behavior at infinity the solution to the mapping problem posed by (21) is trivially given by Cauchy's theorem, namely,

$$\chi(z) = -\frac{1}{2i}\cdot\frac{1}{2\pi i}\int_{x_0}^{\infty} dx' g(x') \frac{e^{-\Delta_+(x')} - e^{-\Delta_-(x')}}{x' - z}, \tag{22}$$

where in principle we could add any entire function of $z$ to this. The latter question again depends on behavior at infinity which must be specified from the outside. Ignoring any additive function, we have

$$\begin{aligned} f(x) &= g(x) + 2iF_+(x) = g(x) - \frac{e^{\Delta_+(x)}}{2\pi i}\int_{x_0}^{\infty} dx' g(x') \frac{e^{-\Delta_+(x')} - e^{-\Delta_-(x')}}{x' - x - i\eta} \\ &= g(x) + \frac{e^{\Delta_+(x)}}{\pi}\int_{x_0}^{\infty} dx' \, \frac{g(x')\, e^{-p(x')} \sin \delta(x')}{x' - x - i\eta} \\ &= e^{i\delta}\left[g(x) \cos \delta + \frac{e^{p(x)}}{\pi} P\int_{x_0}^{\infty} dx' \, \frac{g(x')\, e^{-p(x')} \sin \delta(x')}{x' - x}\right]. \end{aligned} \tag{23}$$

It is easy to show that if $g(x) = \text{constant} = g$, this may be reduced to

$$f(x) = ge^{\Delta_+(x)} = g \exp\left[\frac{1}{\pi}\int_{x_0}^{\infty} dx' \frac{\delta(x')}{x' - x - i\eta}\right], \tag{24}$$

which if $\delta(x) \to 0$ as $x \to \infty$ and $x \to 0$ is the solution which approaches $g$ as $x \to \infty$.

The final equation to be discussed is

$$f(x) = g(x) + \frac{1}{\pi}\int_{x_0}^{\infty} dx' \frac{\operatorname{Re}\{[e^{i\delta(x')}\sin\delta(x')]^* f(x')\}}{x' - x - i\eta}, \tag{25}$$

where $g(z)$ is an analytic function of $z$ except for a cut from $-\infty < x < -x_1$ along the negative real axis. If $\delta(x)$ is real, this is a special case of the previous problem. It is useful, however, to solve it over again in another way.

We consider the imaginary part of $f$ approached from opposite sides of the real axis.

$$\begin{aligned}
\operatorname{Im} f_+(x) &= \operatorname{Im} g(x)\theta(-x - x_1) \\
&\quad + [\operatorname{Re}(e^{i\delta}\sin\delta)\operatorname{Re} f + \operatorname{Im}(e^{i\delta}\sin\delta)\operatorname{Im} f_+]\theta(x - x_0), \\
\operatorname{Im} f_-(x) &= -\operatorname{Im} g(x)\theta(-x - x_1) \\
&\quad - [\operatorname{Re}(e^{i\delta}\sin\delta)\operatorname{Re} f - \operatorname{Im}(e^{i\delta}\sin\delta)\operatorname{Im} f_+]\theta(x - x_0).
\end{aligned} \tag{26}$$

From these equations we find

$$\begin{aligned}
\operatorname{Im} f_+(x) &= \operatorname{Im} g(x)\theta(-x - x_1) + \tan\phi(x)\operatorname{Re} f(x)\theta(x - x_0), \\
\operatorname{Im} f_-(x) &= -\operatorname{Im} g(x)\theta(-x - x_1) - \tan\phi(x)\operatorname{Re} f(x)\theta(x - x_0)
\end{aligned} \tag{27}$$

with

$$\tan\phi(x) = \frac{\operatorname{Re}(e^{i\delta}\sin\delta)}{1 - \operatorname{Im}(e^{i\delta}\sin\delta)}. \tag{28}$$

From these we find

$$\begin{aligned}
f_+ &= i\operatorname{Im} g(x)\theta(-x - x_1) + [1 + i\theta(x - x_0)\tan\phi(x)]\operatorname{Re} f(x), \\
f_- &= -i\operatorname{Im} g(x)\theta(-x - x_0) + [1 - i\theta(x - x_0)\tan\phi(x)]\operatorname{Re} f(x).
\end{aligned} \tag{29}$$

For ease of writing let us agree that $\tan\phi(x) = 0$ unless $x > x_0$.

Now suppose we had the homogeneous system

$$\begin{aligned}
\gamma_+ &= [1 + i\tan\phi(x)]\operatorname{Re}\gamma(x), \\
\gamma_- &= [1 - i\tan\phi(x)]\operatorname{Re}\gamma(x).
\end{aligned} \tag{30}$$

where we are given that $\gamma(z)$ is analytic except for a cut along the positive real axis $x_0 < x < \infty$. Then the jump condition

$$\ln\gamma_+ - \ln\gamma_- = \ln\frac{1 + \tan\phi}{1 - \tan\phi} = 2i\phi \tag{31}$$

is clearly satisfied by

$$\gamma(z) = P(z)\exp\left[\frac{1}{\pi}\int_{x_0}^{\infty} dx' \, \frac{\phi(x')}{x' - z}\right] \equiv P(z)\, e^{\Delta(z)}, \tag{32}$$

where $P(z)$ is an entire function of $z$.

In terms of $\gamma(z)$, the jump conditions on $f$, Eq. (29) may be written as

$$\frac{f_+}{\gamma_+} - \frac{f_-}{\gamma_-} = i \operatorname{Im} g(x)\left(\frac{1}{\gamma_+} + \frac{1}{\gamma_-}\right)\theta(-x - x_1). \tag{33}$$

Aside from any possible zeros of $\gamma$ we see that $f/\gamma$ is an analytic function of $z$ except for the indicated cut along the negative real axis with discontinuity across there being given by (31). By Cauchy's theorem then, to within an additive entire function of $z$,

$$\frac{f}{\gamma} = \frac{1}{2\pi}\int_{-\infty}^{-x_1} dx' \, \frac{\operatorname{Im} g(x')}{x' - x}\left(\frac{1}{\gamma_+} + \frac{1}{\gamma_-}\right) + \sum_i \frac{f(z_i)}{\prod_{j \neq i}(z_i - z_j)} \frac{e^{-\Delta(z_i)}}{z_i - z}, \tag{34}$$

where we have written $P(z) = \prod_i (z - z_i)$. In the interval $-\infty < x' < -x_1$, $\gamma_+ = \gamma_-$, so we have finally

$$f(x) = \frac{P(x)\, e^{\Delta_+(x)}}{\pi}\int_{-\infty}^{-x_1} dx' \, \frac{\operatorname{Im} g(x')}{x' - x} \frac{e^{-\Delta(x')}}{P(x')} + \sum_i \frac{f(x_i)}{\prod_{j \neq i}(z_i - z_j)} e^{-\Delta(z_i) + \Delta_+(x)} \prod_{k \neq i}(x - z_k). \tag{35}$$

In the special case of real $\delta$ representable in the form

$$\delta(x) = \frac{1}{2i}\ln\frac{\mathscr{D}^*(x)}{\mathscr{D}(x)}, \tag{36}$$

where $\mathscr{D}(z)$ in analytic except for a cut along the real $x$-axis from $x_0$ to $+\infty$,

$$e^{\Delta(x)} = \frac{\mathscr{D}(\infty)}{\mathscr{D}(x)}. \tag{37}$$

Then

$$f(x) = \frac{P(x)}{\pi \mathscr{D}(x)}\int_{-\infty}^{-x_1} dx' \, \frac{\operatorname{Im} g(x')\mathscr{D}(x')}{(x' - x)P(x')} - \sum_i f(z_i)\frac{\mathscr{D}(z_i)}{\mathscr{D}(x)} \frac{\prod_{k \neq i}(x - z_k)}{\prod_{j \neq i}(z_i - z_j)}. \tag{38}$$

A rigorous discussion of these types of equations may be found in the book of Mushkhelishvili.[1]

[1] N. I. Mushkhelishvili, *Singular Integral Equations* (P. Noordhoff Ltd., Groningen, The Netherlands, 1953).

# INDEX